Current Topics in Neurotoxicity

Volume 11

Series Editor
Richard Kostrzewa
Johnson City USA

Trevor Archer
Gothenburg Sweden

Current Topics in Neurotoxicity consists of volumes that cover the latest developments in neurotoxicity studies including neurodegeneration, neuroprotection, and neuroregeneration. The series has a translational focus and addresses the causes of neurotoxicity associated with diseases such as Alzheimers, Parkinsons Disease, autism, HIV associated dementia etc.

More information about this series at http://www.springer.com/series/8791

Dora Reglodi • Andrea Tamas
Editors

Pituitary Adenylate Cyclase Activating Polypeptide — PACAP

Springer

Editors
Dora Reglodi
Department of Anatomy
MTA-PTE "Lendulet" PACAP Research Team
University of Pecs, Pecs, Hungary

Andrea Tamas
Department of Anatomy
MTA-PTE "Lendulet" PACAP Research Team
University of Pecs, Pecs, Hungary

ISSN 2363-9563 ISSN 2363-9571 (electronic)
Current Topics in Neurotoxicity
ISBN 978-3-319-35133-9 ISBN 978-3-319-35135-3 (eBook)
DOI 10.1007/978-3-319-35135-3

Library of Congress Control Number: 2016948571

Printed on acid-free paper

This Springer imprint is published by Springer Nature
The registered company is Springer International Publishing AG Switzerland

Preface

Pituitary adenylate cyclase activating polypeptide (PACAP) was discovered in 1989 in the US-Japan Biomedical Research Laboratories of Tulane University, led by Akira Arimura. Atsuro Miyata and Akira Arimura isolated two peptides from ovine hypothalami, with 38 and 27 amino acid residues that could stimulate adenylate cyclase in pituitary cells, resulting in the name of the peptide (Miyata et al. Biochem Biophys Res Commun 1989). In the last 25 years since its discovery, research has proved that PACAP has a very widespread occurrence. Its receptors, the specific PAC1 receptor and the receptors also binding VIP (VPAC1 and 2 receptors), also have widespread distribution in various tissues. The distribution, the splice variants of the receptors, and the diverse signaling explain the miscellaneous actions of the neuropeptide. PACAP belongs to the vasoactive intestinal peptide/secretin/glucagon peptide family and is implicated in a large array of physiological and also pathological processes. Several review articles published in the last 10 years provide excellent sources for reviews on occurrence, as well as the different effects and therapeutic relations of PACAP (the most comprehensive review so far is seen in Vaudry et al. Pharmacol Rev 2009). The present book is a collection of reviews from leading PACAP research investigators from all around the world.

PACAP is one of the most well-conserved peptides throughout evolution, indicating important roles in basic biological processes. The first section of this volume contains reviews on evolutionary aspects of the peptide—from molecular evolution to different roles in annelids and molluscs. Soon after its discovery, it became evident that PACAP plays a role in neuronal cell proliferation, differentiation, and migration. Thanks to the vast amount of data in this area, PACAP is now considered a trophic factor. Reviews on these developmental aspects of the peptide will comprise the next section. The following section is focused on the theme of receptors and signaling, with overviews of novel directions in peptide transport and pharmacological properties. The sections following these contain reviews on the physiological functions in various organs and systems, starting with effects in the endocrine system, thermoregulation, circadian rhythm, and cardiac excitability. Afterwards, there are discussions about actions in the gastrointestinal and urinary tracts, normal and pathological bone and cartilage development, and roles in the reproductive system. PACAP has been shown also to play a role in various physiological barriers, especially the transport of PACAP through the blood brain barrier have been well studied. PACAP does not only affect neuronal function directly, but numerous glial effects have also been shown, as described in two further reviews. A separate chapter will deal with functions of PACAP in sensory systems, such as the olfactory, auditory, and visual system, the last focusing on the protective effects of PACAP in the retina. One of the most intensively studied aspects of PACAP is its strong neuroprotective effect, so the next section contains three reviews dedicated to these protective effects in both the central and peripheral nervous system. Very active investigation is currently underway regarding the role of PACAP in pain conditions, especially in migraines. A section is dedicated to chapters providing an overview of PACAP in pain. As PACAP partially shares receptors with vasoactive intestinal peptide, a neuropeptide closely related to PACAP and the inflammatory functions of which had long been known, the effects of PACAP in the immune system were described relatively early after the peptide's discovery. A vast amount of data have been collected since the first descriptions related to the immune functions of PACAP, summarized in three reviews of the next section. Another very intensively researched area is the role of PACAP in psychological processes and disorders, from depression to anxiety. A section consisting of six reviews provides an overview of these studies. Finally, there are three chapters that describe the clinical aspect of PACAP in relation to cancer, human milk and human blood, and the role of PACAP as a potential biomarker.

In essence, this volume is a detailed yet general reference resource, touching upon the different aspects of this interesting neuropeptide that has diverse effects and increasing therapeutic consideration.

Pécs, Hungary

Dóra Reglődi
Andrea Tamas

Acknowledgements

We thank the following researchers for their work in reviewing the chapter manuscripts.

University of Pecs, Department of Anatomy, Pecs, Hungary: Tamas Atlasz, Eszter Fabian, Jozsef Farkas, Daniel Balazs Fulop, Balazs Gaszner, Tamas Gaszner, Gabor Horvath, Gabriella Horvath, Ivan Ivic, Adel Jungling, Peter Kiss, Laszlo Akos Kovacs, Eszter Laszlo, Andrea Lubics, Andras D. Nagy, Dora Reglodi, Andrea Tamas, Alexandra Vaczy, Dora Werling

University of Pecs, Institute for Translational Medicine, Pecs, Hungary: Andras Garami

University of Pecs, Department of Physiology, Pecs, Hungary: Zsuzsanna Nagy, Istvan Abraham

University of Pecs, Department of Pharmacology and Pharmacotherapeutics, Pecs, Hungary: Balint Botz, Zsuzsanna Helyes, Gabor Petho

University of Pecs, Department of Medical Biology, Pecs, Hungary: Hajnalka Abraham

Department of Experimental Zoology, Hungarian Academy of Sciences, Tihany, Hungary: Zsolt Pirger

University of Debrecen, Department of Anatomy, Histology and Embryology, Debrecen, Hungary: Tamas Juhasz, Roza Zakany

Biological Research Centre, Hungarian Academy of Sciences, Szeged, Hungary: Imola Wilhelm

Neuroscience Research Group, University of Szeged, Hungarian Academy of Sciences, Szeged, Hungary: Bernadett Tuka

Department of Endocrine Neurobiology, Institute of Experimental Medicine, Hungarian Academy of Sciences, Budapest, Hungary: Peter Egri

Department of Human Morphology and Developmental Biology, Faculty of Medicine, Semmelweis University, Budapest, Hungary: Katalin Koves

Department of Pharmacology and Pharmacotherapy, Semmelweis University, Budapest, Hungary: Tibor Zelles

This work is dedicated to the 650th anniversary of the foundation of the University of Pecs, Hungary.

Contents

Part I
Evolutionary Aspects of PACAP

Chapter 1
Molecular Evolution of Pituitary Adenylyl Cyclase-Activating Polypeptide Subfamily and Cognate Receptor Subfamily

Jason S.W. On and Billy K.C. Chow

Abstract The evolution of the secretin/pituitary adenylyl cyclase-activating polypeptide (PACAP) family of peptides in relation to three rounds of genome duplication events occurred during vertebrate emergency has been an area of intense research focus in the past 10 years. This is possible mostly due to the advance in DNA sequencing technology; as a result, genomic DNA sequence data of representative species of evolutionarily important lineage that cover both vertebrates and invertebrates are released. By bioinformatics, data mining, molecular cloning, phylogenetic studies, and synteny analysis, we are now beginning to comprehend the evolution of various peptide ligands and their receptors from invertebrates to vertebrates. In summary, these pieces of information support the establishment of current vertebrate PACAP and PACAP receptor subfamilies via two whole-genome duplications. To date, the most ancient forms of PACAP/glucagon identified are from cephalochordate and urochordate prior to the 2R. The confirmation of bfPACAP/glucagon receptor-ligand pair indicates the origin of PACAP/glucagon peptides and receptor before the cephalochordate–vertebrate split. By gene and genome duplications, the ancestral PACAP/glucagon and receptor evolved to become PACAP and glucagon ligand and receptor subfamilies as observed in tetrapods today.

Keywords PACAP • Glucagon • Evolution • 2R

J.S.W. On
School of Biological Sciences, The University of Hong Kong, Pokfulam Road, Hong Kong, China

B.K.C. Chow (✉)
School of Biological Sciences, The University of Hong Kong, Pokfulam Road, Hong Kong, China
e-mail: bkcc@hku.hk

D. Reglodi, A. Tamas (eds.), *Pituitary Adenylate Cyclase Activating Polypeptide — PACAP*, Current Topics in Neurotoxicity 11,
DOI 10.1007/978-3-319-35135-3_1

Introduction

The peptide hormone pituitary adenylyl cyclase-activating polypeptide (PACAP) was coined initially by its effect in stimulating cAMP synthesis in anterior pituitary cells [1, 2]. After its discovery, this potent polypeptide has been shown to be pleiotropic with wide distribution in the central nervous system (CNS) and many peripheral organs. In addition to its action in the CNS, PACAP possesses physiological activities also in digestion, cardiovascular function, respiration, immune reaction, and cell cycle control [3, 4]. There are two functional isoforms of PACAP, PACAP-38 [2] and PACAP-27 [5], which are the result of progressive limited proteolysis of the same precursor structure and they share the same 27-amino acid N-terminal bioactive core [5]. PACAP-38 is the predominant form in both brain and peripheral tissues in mammals [6–10]. Since the first isolation of PACAP-38 from ovine hypothalamic tissues, much attention was paid to the amino acid sequence similarity between PACAP and vasoactive intestinal peptide (VIP); human PACAP-27 and VIP (28 amino acids) share 68 % amino acid sequence identity [2]. Their remarkable resemblance in structure has highlighted the plausibility of a close evolutionary relationship between PACAP and VIP and thus has prompted comparative analyses of these and related peptides in vertebrates. The peptide superfamily includes PACAP, PACAP-related peptide (PRP), VIP, VIP-related peptide (VRP) or peptide histidine methionine (PHM)/peptide histidine isoleucine (PHI), secretin (SCT), growth hormone-releasing hormone (GHRH), glucagon (GLUC), glucagon-like peptides (GLP1 and GLP2), gastric inhibitory polypeptide (GIP), parathyroid hormone (PTH), PTH2, PTH-related peptide (PTHrP), calcitonin (CALC), calcitonin gene-related peptides (CGRP-I and CGRP-II), adrenomedullin (ADM), ADM2, islet amyloid polypeptide (IAPP), corticotropin-releasing hormone (CRH), urocortin (UCN), UCN2, and UCN3 [11]. Based on structures, some of these peptides are further grouped into the secretin/PACAP subfamily (PACAP, VIP, PRP, VRP (PHM/PHI), GHRH, and secretin) and the glucagon subfamily (glucagon, GLP1, GLP2, GIP, and glucagon-related peptide (GCRP)) [11–13]. The emergence of these two subfamilies is now widely accepted as a result of gene and exon duplication events [3, 11, 14, 15]. In the past decade, the evolutionary hypothesis of the PACAP subfamily and their receptors has been proposed and immensely revised. This review aims to discuss the most recent discoveries regarding the evolution of PACAP subfamily peptides and their receptors.

Structural Change of PACAP in the Course of Evolution

PACAP, like other peptides in the same subfamily, possesses a similar structure with an N-terminal random coil region followed by an α-helical structure [16–19]. Recently, another N-terminal motif, helix N-capping motif, was additionally proposed to be the key element for receptor activation [20]. To fulfil the mission of

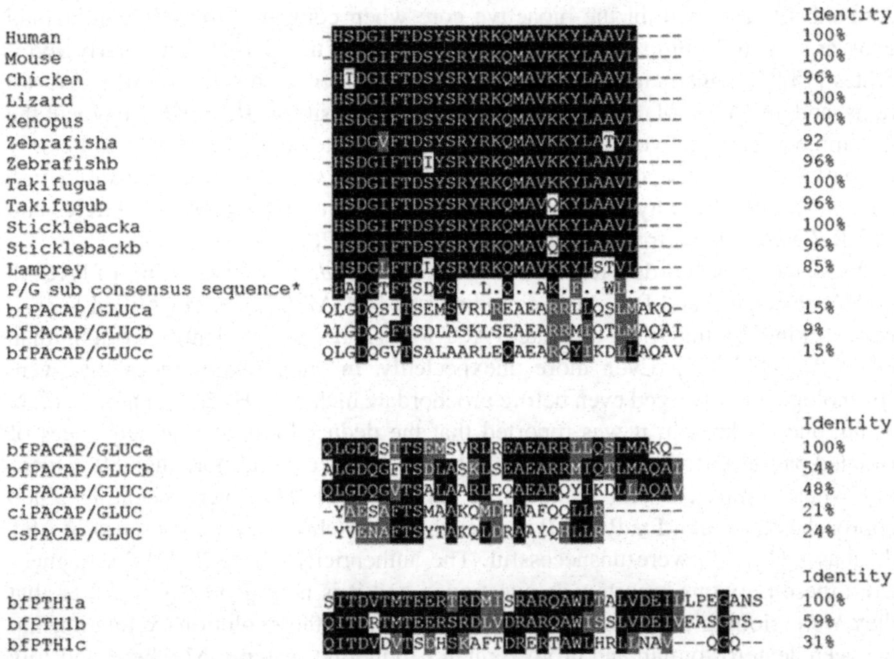

Fig. 1.1 (**a**) Amino acid comparison of human PACAP mature peptide (NM_001117) with orthologs from mouse (NM_009625), chicken (NM_001001291), lizard (XM_003219659), xenopus (NM_001102955), zebrafish (NM_152885 and NM_214715), takifugu (NM_001113191 and NM_001113192), stickleback (ENSGACG00000004163 and ENSGACG00000017084), and lamprey [15] as well as the (*asterisk*) predicted origin of PACAP and glucagon subfamilies [25] and PACAP/GLUCs from amphioxus (*Branchiostoma floridae*) [12, 29]. (**b**) Comparison of bfPACAP/GLUCa with other mature peptides from the same precursor and PACAP/GLUC from urochordates [29] (*Ciona intestinalis* (ci) *and Ciona savignyi* (cs)). (**c**) Comparison of bfPTH1 mature peptides [12, 29]

receptor activation, secretin/PACAP peptides adopt a "two-domain model" that their C-terminus initially interacts with the N-terminal extracellular domain of cognate receptors and then interaction between N-termini of peptides and transmembrane portion of receptors results in intracellular signaling [21–24]. Based on the observation of common structural features among evolutionarily distant secretin and related peptides with low sequence similarity, it was hypothesized that, instead of amino acid sequences, the three-dimensional structures were conserved during evolution of this peptide subfamily [11].

Sharing 100 % amino acid sequence identity between mammalian orthologs, PACAP is one of the most conserved members in the subfamily, and this high level of sequence conservation of PACAP mature peptide is also observed throughout vertebrates from different classes (Fig. 1.1) [13, 25]. PACAP mature peptides from amphibian, avian, teleost, and lamprey share at least 85 % identity with human PACAP-27. Japanese lamprey and sea lamprey PACAP have at most four amino acid substitutions within the bioactive core when compared to their mammalian counterparts, indicating the biological importance of this peptide since early vertebrates [15]. In agnathan, due to both relatively higher amino acid sequence variations and the lack of a typical peptide cleavage site (KRR/GKR) beyond the 27-amino acid bioactive core of PACAP, it is suggested that PACAP-27 is the only form of PACAP, while the extended form of PACAP with 38 amino acids may first emerge in chondrichthyes as PACAP processing two cleavage sites, "GKR" and "GRR," was isolated from the brain of stingray [15, 26].

In an ancestral chordate tunicate (*Chelyosoma productum*), two forms of PACAP cDNA, PACAP-1 and PACAP-2 encoding just PACAP-27, were identified [27]. It was striking to find that tunicate PACAP-1 shares 96 % identity with human PACAP-27 [13, 27]. Even more unexpectedly, in other invertebrates that were chronologically diverged even before urochordate including Hydra, planarian, crab, squid, and cockroach, it was reported that the deduced amino acid sequences of isolated partial mRNA (sequence deposited in NCBI) exhibit more than 89 % identity when compared with mammalian PACAP [28]. However, recent in silico approaches to search for PACAP in available invertebrate online genomes or EST databases [11, 15] were unsuccessful. The authenticity of the PACAP sequences from these previous reports was questioned and it is now generally believed that they were cloning artifacts. Recently, to investigate the evolutionary relationship between deuterostomian and protostomian peptidergic systems, Mirabeau and Joly [29] predicted large numbers of hormone precursor sequences from a wide spectrum of bilateral species by the use of hidden Markov model-based program. In cephalochordate, *Branchiostoma floridae* (bf), and urochordates, *Ciona intestinalis* (ci) and *Ciona savignyi* (cs), putative PACAP/glucagon-like peptides (bfPACAP/GLUCa, bfPACAP/GLUCb, bfPACAP/GLUCc, ciPACAP/GLUC, and csPACAP/GLUC), and putative PTH-like peptides (bfPTH1a, bfPTH1b, bfPTH1c, bfPTH2, and ciPTH) were identified. Between PACAP/GLUCs from the two lineages, only low levels of sequence similarity can be observed. Even though tunicates are phylogenetically closer to vertebrates [30, 31], amphioxi are thought to be a better representative of ancestral chordate due to the relatively lower degree of modifications of amphioxus genome in evolution [32, 33]. When comparing the newly identified bfPACAP/GLUCs with vertebrate orthologs, only restricted amino acid conservation was found, and this may explain our earlier failure in searching for PACAP-like sequences in invertebrate genomes. Nevertheless, in bfPACAP/GLUCs, a few conserved amino acid residues among members of PACAP/glucagon superfamily can still be recognized. Aligned with the predicted PACAP and glucagon peptides (HADGTFTSDYS..L.Q..AK.F..WL.) that was artificially built by Cardoso et al. [25], bfPACAP/GLUCs possesses some conserved residues (D^3, T^7, S^8, L^{14}, A^{19}, and L^{26}) in the first 27 amino acids (Fig. 1.1). To further functionally clarify the identity of these PACAP/GLUC peptides from amphioxus, we showed bfPACAP/GLUCs and bfPTHs to be bioactive and responsive to two distinctive endogenous receptors. We named these GPCRs bfPACAP/GLUCs receptor (bfPACAP/GLUCR) and bfPTH-like receptor (bfPTHR-like), respectively [12]. These two receptors are phylogenetically classified with other PTH receptors [11], and are now thought to be homologs of vertebrate PACAP/GLUC and PTH receptors [12]. The identification of the bfPACAP/GLUC and PTH ligands and receptors in cephalochordate not only helps to conclusively affirm the existence of PACAP/GLUC and PTH peptides before vertebrate, but also provides evidence for the presence of an ancestral form preceding the emergence of PACAP and glucagon subfamilies in vertebrates.

Speculated Evolutionary Trajectory of PACAP Subfamily

It is generally believed that the evolution of the PACAP subfamily from a common ancestral gene or exon took place in the beginning of vertebrate evolution and was facilitated by two rounds of whole-genome duplication (1R and 2R/2R hypothesis) and other gene duplication events. Moreover, the third round of genome duplication (3R) only in teleost lineage led to additional copies of genes, such as *PACAPa* and *PACAPb* in zebrafish [34–36]. The similarity in gene and synteny organization between members of PACAP subfamily is one of the facts that associate the molecular evolution of this subfamily with exon and gene duplication events. In human, the *PACAP* (PRPR-PACAP) and *VIP* (VRP-VIP) genes encode two mature peptides with high similarity, while *GHRH* and *SCT* genes each encodes only a single mature peptide together with an inactive C-terminal peptide [13]. Characterization of *PACAP* gene orthologs in several vertebrates reveals the same gene organization (Fig. 1.2). The open reading frame (ORF) of the gene contains four exons in which PACAP is encoded within exon 4 while PRP, which is nonfunctional in mammals due to the loss of its receptor PRP-R, is encoded by exon 3 and 4 [28]. Likewise, in tetrapods, the ORF of *VIP* gene contains five exons; it also encodes VIP in exon 4 after peptide PHM in humans or PHI in other vertebrates [28]. The physiological function of PHM/PHI still remains largely unclear, and a specific receptor for this peptide was first isolated from goldfish [37]. It was found that regarding sequence similarity, PACAP is more similar to VIP, and PRP actually shows higher similarity to PHM. This similarity in gene structure and amino acid sequences between the two genes were explained with an evolutionary scenario that primordial exon duplication occurred before the duplication of the ancestral gene [38].

Among the peptides from PACAP subfamily, except for VIP, evolution of PACAP was also described closely with GHRH due to the stimulatory effect of PACAP and PRP on pituitary growth hormone release in fish and birds [39–44]. Since in human PRP shows higher similarity to GHRH than PACAP, in the past without the identification of authentic GHRH in nonmammalian vertebrates, it had been believed that PRPs in nonmammalian vertebrates are functional homologs to mammalian GHRH. It was proposed later in mammalian evolution that separation of GHRH and PACAP genes occurred together with a transition of physiological niche that GHRH replaced PRP as the sole GH-releasing factor [45]. However, not until the identification of real GHRH in goldfish, zebrafish, African clawed frog, and chicken, as well as their cognate receptors in goldfish, zebrafish, and chicken in 2007 [46, 47], it was shown that GHRH actually is present as an individual gene in most nonmammalian

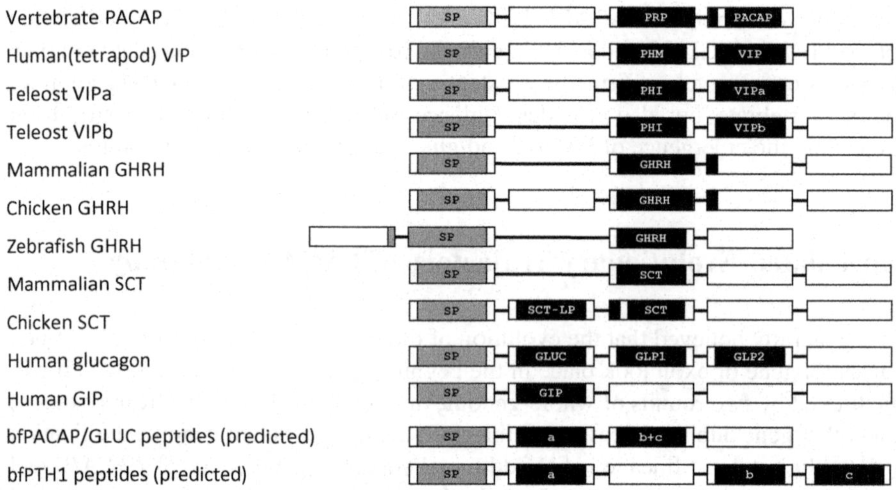

Fig. 1.2 Exon organization of *PACAP* and related genes from different vertebrates and *Branchiostoma floridae*. The open reading frame of *PACAP* and related genes are shown in the figure. Each *box* represents an exon and the *lines* between the *boxes* are the introns. Signal peptide and bioactive peptides in the genes are shown in *grey* and *black*, respectively. The structure of *bfPACAP/GLUCs* and *bfPTH1* genes is predicted from genome of *Branchiostoma florida* (http://genome.jgi-psf.org/Brafl1/Brafl1.home.html). The bfPACAP/GLUCb and bfPACAP/GLUCc were found in the same exon; hence, the actual gene organizations of *bfPACAP/GLUC* and *bfPTH1* remain to be confirmed by molecular cloning. The figure is not drawn to scale

vertebrates and PRP exists as part of PACAP gene throughout the vertebrate evolution [48]. It was also found that nonmammalian PRPs in fact cannot stimulate GH release as indicated by earlier studies [40, 47, 49, 50]. More interestingly, a GPCR highly specific for PRP was characterized in goldfish [51], zebrafish [52], fugu [35], and chicken [53] and recent studies have shown the potential of PRP as a reproductive modulator in fish [54].

As for the *SCT* gene, in chicken an extra exon between exon 3 and exon 5, encoding signal peptide and secretin, respectively, was found to contain a novel SCT-like peptide (SCT-LP) sharing 56 % identity with chicken SCT [55]. The presence of the extra exon fits into the hypothesis of PACAP subfamily evolution, though it misses in mammals. Considering the modest sequence identity between chicken PHI and chicken SCT-LP, it was suggested that chicken *SCT* gene could be a result of *VIP* gene duplication [55]. Nevertheless, because of the lack of SCT-LP in frog *SCT* gene and the unsuccessful search of *SCT* gene in lizard and ray-finned fishes probably due to incomplete genome assembly and/or gene losses, the question of whether SCT-LP was originated from VIP gene duplication or avian-specific exon duplication remains unclear [11, 55, 56]. Recently, putative *SCT* genes were identified in elephant shark and coelacanth genomes, and the coelacanth *SCT* gene was shown to share synteny with *SCT* genes from xenopus, chicken, and human [57]. Answer to whether there is presence of SCT-LP in these two putative *SCT* genes may help to elucidate the evolution of *SCT* gene.

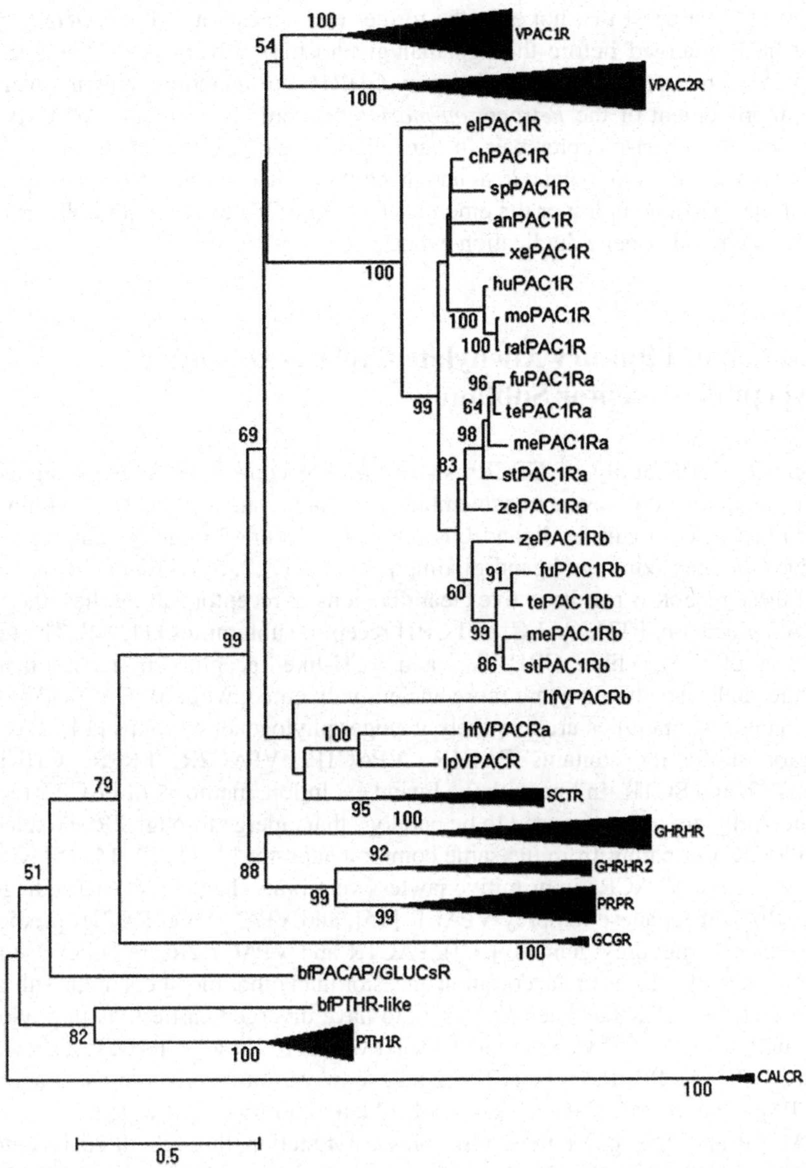

Fig. 1.3 A proposed evolutionary scheme of PACAP subfamily. The question mark indicates the unknown evolutionary situation before amphioxus and uncertain evolutionary path of secretin

With continuous effort on comparative study, the possible evolutionary courses of SCT, PACAP, VIP, and GHRH are much better understood now (Fig. 1.3). Recent genomic analysis concludes that lamprey has diverged after the 2R [58]. Molecular cloning of *PACAP* and *VIP* genes from primitive jawless vertebrates, hagfish, and Japanese lamprey [15] and the discovery of a predicted GHRH-like sequence in sea lamprey [11] suggest that not only the former two genes but also the *GHRH* gene could have emerged before the agnathan/gnathostome divergence. However, in-depth characterization of the sea lamprey GHRH-like including synteny analysis after improvement of the *Petromyzon marinus* genome is necessary to verify the existence of GHRH-receptor pair in agnathans. Regarding the origin of the *SCT* gene, the missing of *SCT* gene in agnathan could be due to either a lineage-specific loss of the fourth homolog or the emergence of the *SCT* gene only after divergence of agnatha by other gene duplication event.

Evolution of Pituitary Adenylate Cyclase-Activating Polypeptide Receptor Subfamily

Secretin/PACAP family of GPCRs shares several conserved common signature motifs including the seven transmembrane domains, an α-β-β/α fold within the N-terminal ectodomain for ligand binding, and disulfide-bridge-forming cysteine residues for stabilizing the ligand-binding pocket [21, 22, 59]. Phylogenetic analysis of these receptors results in five clear divisions or receptor subfamilies: they are PACAP, glucagon, PTH, CALC, and CRH receptor subfamilies [11, 12]. The identification of CALC-like, CRH-like, and PTH-like receptors in most bilateral genomes indicates that they are more anciently diverged, while PACAP and glucagon receptor subfamilies are relatively younger phylogenetic groups [14]. PACAP receptor subfamily contains PAC1R, VPAC1R, VPAC2R, PRPR, GHRHR, GHRHR2, and SCTR. In line with the ligand evolution, members of PACAP receptor subfamily are also suggested to be paralogs that emerge through 2R and teleost-specific 3R originating from the same common ancestor [3, 11, 12, 14, 15]. Given the presence of VPACRs in primitive jawless vertebrate (hagfish VPACRa, hagfish VPACRb, and Japanese lamprey VPACR [15]; and VPACR and PAC1R predicted from Japanese lamprey genome [57]), PAC1R and VPAC1/2R are believed to be phylogenetically closer to the common ancestor than other members in the subfamily. In fact, VPAC1R has been suggested to have diverged earliest, as in phylogenetic analyses VPAC1R was shown to cluster with PAC1R and VPAC2R [60], while SCTR, GHRHR, PRPR, and GHRHR2 that show various levels of cross-reactivity with PACAP in different species are evolved later chronologically [15].

PACAP and VIP exert their physiological function through three receptors, VPAC1R, VPAC2R, and PAC1R [61]. PAC1R binds its specific ligands, PACAP-27 and PACAP-38, with higher affinity than VIP and this preference of ligand binding appears to be consistent in most vertebrates [28, 61, 62]. On the other hand, ligand specificities of VPAC1R and VPAC2R vary in different vertebrate species. In tetrapods, the two receptors show similar binding affinities to PACAP and VIP [61, 63, 64]. However, VPAC1R from goldfish and dogfish [65, 66] and VPAC2R from goldfish and zebrafish [37, 52] bind VIP and PHI, respectively, suggesting that the capability of these receptors binding to PACAP may have emerged only after the divergence of tetrapod [3]. Because of the basal phylogenetic position of VPAC1R,

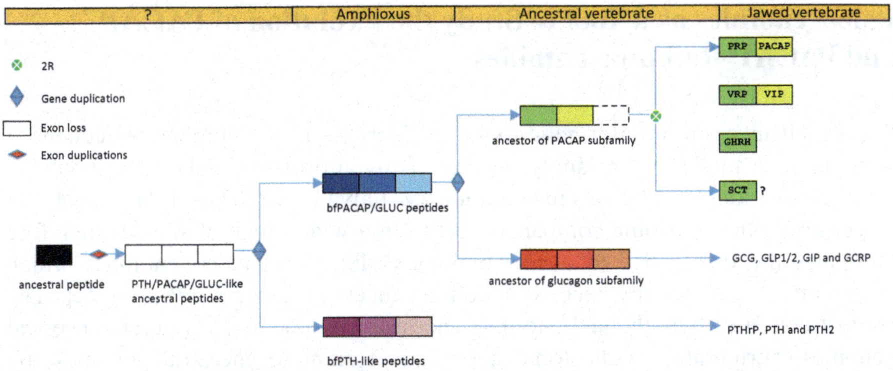

Fig. 1.4 Phylogenetic tree of PACAP receptor subfamily with bfPACAP/GLUCR and bfPTHR-like. The sequences used in the maximum likelihood tree are from our previous study [12] and the analysis was performed by using Mega 6.06. The tree was bootstrapped 500 times to test the reliability of branching. Only the bootstrap value >50 % is represented. Species abbreviation: anole lizard (an), *Branchiostoma floridae* (bf), chicken (ch), coelacanth (co), elephant shark (el), fugu (fu), hagfish (hg), human (hu), Japanese lamprey (lp), medaka (me), mouse (mo), stickleback (st), tetraodon (te), xenopus (xe), and zebrafish (ze)

it was once proposed that the binding affinity of ancestral VPACRs may have a higher preference for VIP [3]. However, this hypothesis is not consistent with data from the hagfish VPACRs [15]. Of the two hagfish VPACRs, hagfish VPACRa is coupled to both cAMP and PLC pathways whereas activation of hagfish VPACRb triggers only the PLC pathway. In vitro cAMP and PLC assays of hagfish VPACRa-transfected CHO-K1 cells show that although mammalian PACAP and VIP are equally potent, sea lamprey PACAP is more potent than VIP in stimulating both pathways. Therefore, in spite of the considerable similarity of agnathan VPACR in terms of amino acid sequence to vertebrate VPAC1R, its pharmacological properties are in fact similar to vertebrate PAC1R. To verify the ligand preference of the ancestral VPACR, characterization of the newly predicted PAC1R and VPACR in Japanese lamprey is needed.

To investigate the origin of the PACAP receptor subfamily, we have recently characterized two receptors, bfPACAP/GLUCR and bfPTHR-like, as well as their ligands from a cephalochordate amphioxus [12]. The bfPACAP/GLUCR was found to be specific to bfPACAP/GLUCs but not PACAPs and VIPs from teleost and sea lamprey. Likewise, bfPACAP/GLUCs exhibit no agonistic effect on human VPAC2R, PAC1R, and glucagon receptor in cAMP activation, indicating the structural differences between bfPACAP/GLUCs and vertebrate PACAP and VIP, as well as changes in the ligand-binding pocket of the receptors. Another interesting finding from our study was that sea lamprey PACAP, zebrafish PACAP, and goldfish VIP are all weak agonists of bfPTHR-like. Considering the close phylogenetic position of bfPACAP/GLUCR and bfPTHR-like (Fig. 1.4), it is possible that PACAP and glucagon receptor subfamilies may have diverged from the PTH receptor subfamily. Such possibilities have also been described in an evolutionary model that was built recently by comparing nematode and arthropod genomes with chicken and human genomes [14].

Paleogenomics as a Tool to Study the Evolution of PACAP and PACAP Receptor Families

Recent advances in bioinformatic tools and increase in the number of sequenced genomes have made synteny analysis significantly informative and nearly inevitable for clarifying identity of genes in evolutionary study [11, 12, 67–69]. The capability to perform whole-genome comparison between a wide range of species therefore facilitates the emergence of a new biological discipline, paleogenomics, which investigates evolution by reconstruction of ancestral genome [30, 70–72]. The reconstruction of hypothetical chordate ancestral genome and vertebrate ancestral genomes (vertebrate, gnathostome, teleost, and amniote ancestral genomes) by comparing vertebrate genomes with genomes of pre-2R species such as sea urchin, tunicate, and amphioxus allows the reconstruction of evolutionary rearrangements of gene families revealed at chromosomal level [30, 71]. This scientific method was recently utilized to trace the evolutionary history of relaxin hormone family and their receptors [67]. Hwang et al. were then inspired to perform a similar study for secretin peptide superfamily and their cognate receptors [11]. Through matching the evolutionarily conserved vertebrate chromosomal fragments where secretin gene family is located on the reconstructed post-2R gnathostome ancestral chromosomes (GAC) with 40 chromosomes, it was found that PACAP, VIP, and GHRH separately reside on three different ohnologous GAC, B0, B2, and B1, respectively. These three chromosomes are derived from chromosome B of chromosomes A–J in pre-2R vertebrate ancestral chromosomes (VAC). This analysis suggests that PACAP, VIP, and GHRH are ohnologous genes sharing the same origin. However, the fourth gene *SCT* gene is predicted to be located on a different GAC, D1. It is possible that the *SCT* gene was translocated from one of ohnologous B chromosomes to D1 after 2R. As an alternative explanation, all secretin superfamily genes are located on pre-2R VAC B, D, and E, and since most of the genes are distributed to post-2R GAC Ds (D0: UCN3, ADM2, IAPP, and PTHrP; D1: SCT, UCN2, ADM, CALCa (encode CALC and CGRP1), CALCb (encode CGRP2), and PTH; D3: PTH2), it was suggested that the ancestor of secretin gene superfamily may diverge via local gene duplications within VAC D and these genes were subsequently translocated to the other two VACs. As amphioxi and tunicates possess only one *PACAP/GLUC* gene which later on evolved to PACAP and glucagon subfamilies [12, 29] that are separately located on GAC Es and Bs, respectively, the translocation event in the hypothesis could have occurred after divergence of urochordate and prior to the 2R. Synteny analysis of *bfPACAP/GLUC* gene will be extremely useful to confirm the hypothesis, but unfortunately, due to incomplete amphioxi genome information, it is difficult to convincingly identify syntenic relationship of this gene with other species. To verify the above evolutionary scenario, high-resolution invertebrate genome sequences are needed.

Regarding the evolution of class B1 GPCRs, syntenic gene environment of bfPACAP/GLUCR and bfPTHR-like in amphioxus genomes (*Branchiostoma belcheri* and *Branchiostoma floridae*) [12] agree with the hypothesis by Hwang et al. [11]. It was proposed that vertebrate secretin receptor family may originate from VAC E as reported in Nakatani et al. [71] followed by gene duplications leading to the emergence of pre-2R lineages of different subfamilies. An interesting hypothesis by Hwang et al. [11] was that 2R may only generate SCTR, GHRHR2, and an ancestral VPAC receptor with the loss of the fourth receptor on GAC E3, whereas PAC1R, VPAC1R, VPAC2R, GHRHR, and PRPR on GAC E1 emerge later via local tandem gene duplication of the ancestral VPAC receptor. In the genome of spotted gar that has bypassed the teleost-specific 3R [73], the compact arrangement of GHRHR, PAC1R, PRPR, VPAC1R, and VPAC2R genes on the same linkage group (LG9) and the presence of SCTR and GHRHR2 on two other conserved loci (LG12 and LG15) fit well into the above hypothesis [12]. This new hypothesis revised the previous idea that 2R is the main driving force for the establishment of vertebrate PACAP receptor subfamily. In the view of the presence of only three VPAC receptors in agnathans [57] and six receptors belonging to PACAP subfamily in elephant shark [12, 57], if the hypothetic local gene duplication did happen, it likely occurred after divergence of agnathans and before emergence of chondrichthyes.

Conclusion

After years of studies, recently, significant progress has been made in understanding the evolution of PACAP subfamily and their cognate receptors. To date, the most ancient forms of PACAP/GLUC identified are from cephalochordate and urochordate prior to the 2R. The confirmation of bfPACAP/GLUCR-ligand pair indicates the origin of PACAP/GLUC peptides and receptor before the cephalochordate–vertebrate split. By gene and genome duplications, the ancestral PACAP/GLUC and receptor evolved to become PACAP and glucagon ligand and receptor subfamilies as observed in tetrapods today.

References

1. Arimura A. PACAP: the road to discovery. Peptides. 2007;28:1617–9.
2. Miyata A, Arimura A, Dahl RR, Minamino N, Uehara A, Jiang L, et al. Isolation of a novel 38 residue-hypothalamic polypeptide which stimulates adenylate cyclase in pituitary cells. Biochem Biophys Res Commun. 1989;164:567–74.
3. Vaudry D, Falluel-Morel A, Bourgault S, Basille M, Burel D, Wurtz O, et al. Pituitary adenylate cyclase-activating polypeptide and its receptors: 20 years after the discovery. Pharmacol Rev. 2009;61:283–357.
4. Vaudry D, Gonzalez BJ, Basille M, Yon L, Fournier A, Vaudry H. Pituitary adenylate cyclase-activating polypeptide and its receptors: from structure to functions. Pharmacol Rev. 2000;52:269–324.
5. Miyata A, Jiang L, Dahl RD, Kitada C, Kubo K, Fujino M, et al. Isolation of a neuropeptide corresponding to the N-terminal 27 residues of the pituitary adenylate cyclase activating

polypeptide with 38 residues (PACAP38). Biochem Biophys Res Commun. 1990;170:643–8. 0006-291X(90)92140-U[pii].

6. Piggins HD, Stamp JA, Burns J, Rusak B, Semba K. Distribution of pituitary adenylate cyclase activating polypeptide (PACAP) immunoreactivity in the hypothalamus and extended amygdala of the rat. J Comp Neurol. 1996;376:278–94. 10.1002/(SICI)1096-9861(19961209)376:2<278::AID-CNE9>3.0.CO;2-0 [pii].
7. Hannibal J, Mikkelsen JD, Clausen H, Holst JJ, Wulff BS, Fahrenkrug J. Gene expression of pituitary adenylate cyclase activating polypeptide (PACAP) in the rat hypothalamus. Regul Pept. 1995;55:133–48. 016701159400099 J [pii].
8. Masuo Y, Suzuki N, Matsumoto H, Tokito F, Matsumoto Y, Tsuda M, et al. Regional distribution of pituitary adenylate cyclase activating polypeptide (PACAP) in the rat central nervous system as determined by sandwich-enzyme immunoassay. Brain Res. 1993;602:57–63. 0006-8993(93)90241-E [pii].
9. Ghatei MA, Takahashi K, Suzuki Y, Gardiner J, Jones PM, Bloom SR. Distribution, molecular characterization of pituitary adenylate cyclase-activating polypeptide and its precursor encoding messenger RNA in human and rat tissues. J Endocrinol. 1993;136:159–66.
10. Arimura A, Somogyvari-Vigh A, Miyata A, Mizuno K, Coy DH, Kitada C. Tissue distribution of PACAP as determined by RIA: highly abundant in the rat brain and testes. Endocrinology. 1991;129:2787–9.
11. Hwang JI, Moon MJ, Park S, Kim DK, Cho EB, Ha N, et al. Expansion of secretin-like G protein-coupled receptors and their peptide ligands via local duplications before and after two rounds of whole-genome duplication. Mol Biol Evol. 2013;30:1119–30.
12. On JS, Duan C, Chow BK, Lee LT. Functional pairing of class B1 ligand-GPCR in cephalochordate provides evidence of the origin of PTH and PACAP/glucagon receptor family. Mol Biol Evol. 2015;32:2048–59.
13. Sherwood NM, Krueckl SL, McRory JE. The origin and function of the pituitary adenylate cyclase-activating polypeptide (PACAP)/glucagon superfamily. Endocr Rev. 2000;21:619–70.
14. Cardoso JC, Felix RC, Power DM. Nematode and arthropod genomes provide new insights into the evolution of class 2 B1 GPCRs. PLoS One. 2014;9(3):e92220.
15. Ng SY, Chow BK, Kasamatsu J, Kasahara M, Lee LT, Agnathan VIP. PACAP and their receptors: ancestral origins of today's highly diversified forms. PLoS One. 2012;7:e44691.
16. Underwood CR, Garibay P, Knudsen LB, Hastrup S, Peters GH, Rudolph R, et al. Crystal structure of glucagon-like peptide-1 in complex with the extracellular domain of the glucagon-like peptide-1 receptor. J Biol Chem. 2010;285:723–30.
17. Sun C, Song D, Davis-Taber RA, Barrett LW, Scott VE, Richardson PL, et al. Solution structure and mutational analysis of pituitary adenylate cyclase-activating polypeptide binding to the extracellular domain of PAC1-RS. Proc Natl Acad Sci U S A. 2007;104:7875–80.
18. Marx UC, Adermann K, Bayer P, Forssmann WG, Rosch P. Solution structures of human parathyroid hormone fragments hPTH(1–34) and hPTH(1–39) and bovine parathyroid hormone fragment bPTH(1–37). Biochem Biophys Res Commun. 2000;267:213–20.
19. Jin L, Briggs SL, Chandrasekhar S, Chirgadze NY, Clawson DK, Schevitz RW, et al. Crystal structure of human parathyroid hormone 1-34 at 0.9-A resolution. J Biol Chem. 2000;275:27238–44.
20. Neumann JM, Couvineau A, Murail S, Lacapere JJ, Jamin N, Laburthe M. Class-B GPCR activation: is ligand helix-capping the key? Trends Biochem Sci. 2008;33:314–9.
21. Pal K, Melcher K, Xu HE. Structure and mechanism for recognition of peptide hormones by Class B G-protein-coupled receptors. Acta Pharmacol Sin. 2012;33:300–11.
22. Moon MJ, Kim HY, Park S, Kim DK, Cho EB, Park CR, et al. Evolutionarily conserved residues at glucagon-like peptide-1 (GLP-1) receptor core confer ligand-induced receptor activation. J Biol Chem. 2012;287:3873–84.
23. Moon MJ, Kim HY, Kim SG, Park J, Choi DS, Hwang JI, et al. Tyr1 and Ile7 of glucose-dependent insulinotropic polypeptide (GIP) confer differential ligand selectivity toward GIP and glucagon-like peptide-1 receptors. Mol Cells. 2010;30:149–54.

24. Pioszak AA, Xu HE. Molecular recognition of parathyroid hormone by its G protein-coupled receptor. Proc Natl Acad Sci U S A. 2008;105:5034–9.
25. Cardoso JC, Vieira FA, Gomes AS, Power DM. The serendipitous origin of chordate secretin peptide family members. BMC Evol Biol. 2010;10:135.
26. Matsuda K, Yoshida T, Nagano Y, Kashimoto K, Yatohgo T, Shimomura H, et al. Purification and primary structure of pituitary adenylate cyclase activating polypeptide (PACAP) from the brain of an elasmobranch, stingray, Dasyatis akajei. Peptides. 1998;19:1489–95. S0196-9781(98)00091-6 [pii].
27. McRory J, Sherwood NM. Two protochordate genes encode pituitary adenylate cyclase-activating polypeptide and related family members. Endocrinology. 1997;138:2380–90.
28. Cardoso JC, Vieira FA, Gomes AS, Power DM. PACAP, VIP and their receptors in the metazoa: insights about the origin and evolution of the ligand-receptor pair. Peptides. 2007;28:1902–19. S0196-9781(07)00272-0 [pii].
29. Mirabeau O, Joly JS. Molecular evolution of peptidergic signaling systems in bilaterians. Proc Natl Acad Sci U S A. 2013;110:E2028–37.
30. Putnam NH, Butts T, Ferrier DE, Furlong RF, Hellsten U, Kawashima T, et al. The amphioxus genome and the evolution of the chordate karyotype. Nature. 2008;453:1064–71.
31. Delsuc F, Brinkmann H, Chourrout D, Philippe H. Tunicates and not cephalochordates are the closest living relatives of vertebrates. Nature. 2006;439:965–8.
32. Holland LZ. Evolution of new characters after whole genome duplications: insights from amphioxus. Semin Cell Dev Biol. 2013;24:101–9.
33. Tsagkogeorga G, Cahais V, Galtier N. The population genomics of a fast evolver: high levels of diversity, functional constraint, and molecular adaptation in the tunicate Ciona intestinalis. Genome Biol Evol. 2012;4:740–9. evs054 [pii].
34. Meyer A, Van de Peer Y. From 2R to 3R: evidence for a fish-specific genome duplication (FSGD). Bioessays. 2005;27:937–45.
35. Wang Y, Wong AO, Ge W. Cloning, regulation of messenger ribonucleic acid expression, and function of a new isoform of pituitary adenylate cyclase-activating polypeptide in the zebrafish ovary. Endocrinology. 2003;144:4799–810.
36. Fradinger EA, Sherwood NM. Characterization of the gene encoding both growth hormone-releasing hormone (GRF) and pituitary adenylate cyclase-activating polypeptide (PACAP) in the zebrafish. Mol Cell Endocrinol. 2000;165:211–9. S0303-7207(00)00251-3 [pii].
37. Tse DL, Pang RT, Wong AO, Chan SM, Vaudry H, Chow BK. Identification of a potential receptor for both peptide histidine isoleucine and peptide histidine valine. Endocrinology. 2002;143:1327–36.
38. Ohkubo S, Kimura C, Ogi K, Okazaki K, Hosoya M, Onda H, et al. Primary structure and characterization of the precursor to human pituitary adenylate cyclase activating polypeptide. DNA Cell Biol. 1992;11:21–30.
39. Toogood AA, Harvey S, Thorner MO, Gaylinn BD. Cloning of the chicken pituitary receptor for growth hormone-releasing hormone. Endocrinology. 2006;147:1838–46.
40. Parker DB, Power ME, Swanson P, Rivier J, Sherwood NM. Exon skipping in the gene encoding pituitary adenylate cyclase-activating polypeptide in salmon alters the expression of two hormones that stimulate growth hormone release. Endocrinology. 1997;138:414–23.
41. Vaughan JM, Rivier J, Spiess J, Peng C, Chang JP, Peter RE, et al. Isolation and characterization of hypothalamic growth-hormone releasing factor from common carp, *Cyprinus carpio*. Neuroendocrinology. 1992;56:539–49.
42. Harvey S, Scanes CG. Comparative stimulation of growth hormone secretion in anaesthetized chickens by human pancreatic growth hormone-releasing factor (hpGRF) and thyrotrophin-releasing hormone (TRH). Neuroendocrinology. 1984;39:314–20.
43. Leung FC, Taylor JE. In vivo and in vitro stimulation of growth hormone release in chickens by synthetic human pancreatic growth hormone releasing factor (hpGRFs). Endocrinology. 1983;113:1913–5.
44. Harvey S. GHRH: a growth hormone-releasing factor in birds? Neural regulation in the vertebrate endocrine system. New York: Kluwer; 1999.

45. Montero M, Yon L, Kikuyama S, Dufour S, Vaudry H. Molecular evolution of the growth hormone-releasing hormone/pituitary adenylate cyclase-activating polypeptide gene family. Functional implication in the regulation of growth hormone secretion. J Mol Endocrinol. 2000;25:157–68. JME00932 [pii].
46. Wang Y, Li J, Wang CY, Kwok AH, Leung FC. Identification of the endogenous ligands for chicken growth hormone-releasing hormone (GHRH) receptor: evidence for a separate gene encoding GHRH in submammalian vertebrates. Endocrinology. 2007;148:2405–16.
47. Lee LT, Siu FK, Tam JK, Lau IT, Wong AO, Lin MC, et al. Discovery of growth hormone-releasing hormones and receptors in nonmammalian vertebrates. Proc Natl Acad Sci U S A. 2007;104:2133–8.
48. Tam JK, Lee LT, Chow BK. PACAP-related peptide (PRP)—molecular evolution and potential functions. Peptides. 2007;28:1920–9.
49. Small BC, Nonneman D. Sequence and expression of a cDNA encoding both pituitary adenylate cyclase activating polypeptide and growth hormone-releasing hormone-like peptide in channel catfish (Ictalurus punctatus). Gen Comp Endocrinol. 2001;122:354–63.
50. Rousseau K, Le Belle N, Pichavant K, Marchelidon J, Chow BK, Boeuf G, et al. Pituitary growth hormone secretion in the turbot, a phylogenetically recent teleost, is regulated by a species-specific pattern of neuropeptides. Neuroendocrinology. 2001;74:375–85. nen74375 [pii].
51. Chan KW, Yu KL, Rivier J, Chow BK. Identification and characterization of a receptor from goldfish specific for a teleost growth hormone-releasing hormone-like peptide. Neuroendocrinology. 1998;68:44–56. nen68044 [pii].
52. Wu S, Roch GJ, Cervini LA, Rivier JE, Sherwood NM. Newly-identified receptors for peptide histidine-isoleucine and GHRH-like peptide in zebrafish help to elucidate the mammalian secretin superfamily. J Mol Endocrinol. 2008;41:343–66.
53. Wang Y, Li J, Wang CY, Kwok AY, Zhang X, Leung FC. Characterization of the receptors for chicken GHRH and GHRH-related peptides: identification of a novel receptor for GHRH and the receptor for GHRH-LP (PRP). Domest Anim Endocrinol. 2010;38:13–31.
54. Tam JK, Lee LT, Cheng CH, Chow BK. Discovery of a new reproductive hormone in teleosts: pituitary adenylate cyclase-activating polypeptide-related peptide (PRP). Gen Comp Endocrinol. 2011;173:405–10.
55. Wang Y, Huang G, Li J, Meng F, He X, Leung FC. Characterization of chicken secretin (SCT) and secretin receptor (SCTR) genes: a novel secretin-like peptide (SCT-LP) and secretin encoded in a single gene. Mol Cell Endocrinol. 2012;348:270–80.
56. Tam JK, Lau KW, Lee LT, Chu JY, Ng KM, Fournier A, et al. Origin of secretin receptor precedes the advent of tetrapoda: evidence on the separated origins of secretin and orexin. PLoS One. 2011;6:e19384.
57. Cardoso JC, Felix RC, Trindade M, Power DM. Fish genomes provide novel insights into the evolution of vertebrate secretin receptors and their ligand. Gen Comp Endocrinol. 2014;209:82–92.
58. Smith JJ, Kuraku S, Holt C, Sauka-Spengler T, Jiang N, Campbell MS, et al. Sequencing of the sea lamprey (Petromyzon marinus) genome provides insights into vertebrate evolution. Nat Genet. 2013;45:415–21, 421e1–2.
59. Grace CR, Perrin MH, DiGruccio MR, Miller CL, Rivier JE, Vale WW, et al. NMR structure and peptide hormone binding site of the first extracellular domain of a type B1 G protein-coupled receptor. Proc Natl Acad Sci U S A. 2004;101:12836–41.
60. Cardoso JC, Clark MS, Viera FA, Bridge PD, Gilles A, Power DM. The secretin G-protein-coupled receptor family: teleost receptors. J Mol Endocrinol. 2005;34:753–65.
61. Harmar AJ, Arimura A, Gozes I, Journot L, Laburthe M, Pisegna JR, et al. International Union of Pharmacology. XVIII. Nomenclature of receptors for vasoactive intestinal peptide and pituitary adenylate cyclase-activating polypeptide. Pharmacol Rev. 1998;50:265–70.
62. Wong AO, Leung MY, Shea WL, Tse LY, Chang JP, Chow BK. Hypophysiotropic action of pituitary adenylate cyclase-activating polypeptide (PACAP) in the goldfish: immunohistochemical demonstration of PACAP in the pituitary, PACAP stimulation of growth hormone release from pituitary cells, and molecular cloning of pituitary type I PACAP receptor. Endocrinology. 1998;139:3465–79.

63. Hoo RL, Alexandre D, Chan SM, Anouar Y, Pang RT, Vaudry H, et al. Structural and functional identification of the pituitary adenylate cyclase-activating polypeptide receptor VPAC2 from the frog Rana tigrina rugulosa. J Mol Endocrinol. 2001;27:229–38. JME01023 [pii].
64. Alexandre D, Anouar Y, Jegou S, Fournier A, Vaudry H. Molecular cloning, mRNA distribution and pharmacological characterization of a VIP/PACAP receptor in the frog Rana ridibunda. Ann N Y Acad Sci. 2000;921:300–3.
65. Bewley MS, Pena JT, Plesch FN, Decker SE, Weber GJ, Forrest Jr JN. Shark rectal gland vasoactive intestinal peptide receptor: cloning, functional expression, and regulation of CFTR chloride channels. Am J Physiol Regul Integr Comp Physiol. 2006;291:R1157–64.
66. Chow BK, Yuen TT, Chan KW. Molecular evolution of vertebrate VIP receptors and functional characterization of a VIP receptor from goldfish Carassius auratus. Gen Comp Endocrinol. 1997;105:176–85.
67. Yegorov S, Good S. Using paleogenomics to study the evolution of gene families: origin and duplication history of the relaxin family hormones and their receptors. PLoS One. 2012;7:e32923.
68. Kim DK, Cho EB, Moon MJ, Park S, Hwang JI, Kah O, et al. Revisiting the evolution of gonadotropin-releasing hormones and their receptors in vertebrates: secrets hidden in genomes. Gen Comp Endocrinol. 2011;170:68–78.
69. Abi-Rached L, Gilles A, Shiina T, Pontarotti P, Inoko H. Evidence of en bloc duplication in vertebrate genomes. Nat Genet. 2002;31:100–5.
70. Muffato M, Roest Crollius H. Paleogenomics in vertebrates, or the recovery of lost genomes from the mist of time. Bioessays. 2008;30:122–34.
71. Nakatani Y, Takeda H, Kohara Y, Morishita S. Reconstruction of the vertebrate ancestral genome reveals dynamic genome reorganization in early vertebrates. Genome Res. 2007;17:1254–65.
72. Birnbaum D, Coulier F, Pebusque MJ, Pontarotti P. "Paleogenomics": looking in the past to the future. J Exp Zool. 2000;288:21–2.
73. Amores A, Catchen J, Ferrara A, Fontenot Q, Postlethwait JH. Genome evolution and meiotic maps by massively parallel DNA sequencing: spotted gar, an outgroup for the teleost genome duplication. Genetics. 2011;188:799–808.

Chapter 2
Occurrence, Distribution, and Physiological Function of Pituitary Adenylyl Cyclase-Activating Polypeptide in Invertebrate Species

Zsolt Pirger, Nora Krajcs, and Tibor Kiss

Abstract The occurrence and distribution of pituitary adenylate cyclase-activating polypeptide (PACAP) are summarized in invertebrate species with special attention to annelids, mollusks, and arthropods in this review. Furthermore, the role of PACAP is highlighted in physiological and behavioral processes of oligochaete (*Lumbricus*), gastropods (*Helix*, *Lymnaea*), insect (*Drosophila*), as well as malacostraca (*Litopenaeus*). Since its discovery PACAP has become increasingly recognized for its important and diversified roles in the central and peripheral nervous system and in several peripheral organs of a variety of vertebrate and invertebrate species. Twenty-six years after its discovery, PACAP is now one of the most extensively studied neuropeptides both in invertebrate and vertebrate species. This review surveys the importance of PACAP or PACAP-like peptide(s) in invertebrates. The relevance of studies on lower vertebrates and invertebrates, which do not have a pituitary gland like higher vertebrate, is to contribute to the unraveling of fundamental effects of PACAP or PACAP-like peptide(s) and to provide a comparative view.

Keywords Invertebrate PACAP-like molecules • Morphology and physiology • Oligochaetes • Gastropods • Insects

---

Z. Pirger (✉) • N. Krajcs
MTA-ÖK BLI NAP_B Adaptive Neuroethological Research Group, Department of Experimental Zoology, Balaton Limnological Institute, MTA Centre for Ecological Research, Tihany 8237, Hungary
e-mail: pirger.zsolt@okologia.mta.hu

T. Kiss
Chemical Ecology and Neurobiology, Department of Experimental Zoology, Balaton Limnological Institute, MTA Centre for Ecological Research, Klebelsberg K. u. 3., Tihany 8237, Hungary

D. Reglodi, A. Tamas (eds.), *Pituitary Adenylate Cyclase Activating Polypeptide — PACAP*, Current Topics in Neurotoxicity 11,
DOI 10.1007/978-3-319-35135-3_2

Introduction

Pituitary adenylate cyclase-activating polypeptide (PACAP) was isolated first 26 years ago from ovine hypothalamic extract on the basis of stimulating cAMP formation in anterior pituitary cells [1, 2]. Thereafter, it was also discovered that PACAP has wide distribution not only in the hypothalamic nuclei, but also in the whole central (CNS) and peripheral nervous system (PNS) as well as in several peripheral organs. These include endocrine glands, gonads, respiratory and urogenital tracts, digestive system, skin, and muscles suggesting a broader function of PACAP than only the stimulation of pituitary gland [3–9]. PACAP has two biologically active isoforms, PACAP-38 [1] and PACAP-27 [2], which share the same N-terminal amino acids (AA) and are post-translational modifications of a single precursor, the preproPACAP [2, 10, 11]. In vertebrates the PACAP-38 form is the predominant isoform both in brain and peripheral tissues [4, 12, 13]. In contrast, in several invertebrate species the concentration of PACAP-27 is significantly higher, suggesting its pronounced role [4, 12, 14–18]. The discovery of PACAP was soon followed by identification of its receptors. Two types of receptors were characterized according to their relative affinities for PACAP: PAC1-R as well as VPAC1 and VPAC2 [4, 19]. PAC1-R is specific for PACAP and VPAC1, and VPAC2 receptors are activated by both PACAP and VIP molecules. PACAP receptors are members of G-protein-coupled receptor family and are unique in the sense that their complex genes are able to generate receptor splice variants, which have been reported for all three receptor types [20]. The possibility of genes being processed differently and thereby generating receptor splice variants could lead to alterations in pharmacology and signal transduction mechanisms [21]. Expression and distribution of PACAP and its receptors in the CNS and PNS of mammals have been described in detail [4]. Broad distribution of PACAP and its receptors is also observed in nonmammalian vertebrate species such as avians [22–24], reptiles [25, 26], amphibians [27–30], and fish [31–34]. Widespread occurrence and distribution of PACAP and its receptors suggest that the polypeptide exerts pleiotropic physiological functions [3, 4]. PACAP and its receptors are involved in numerous physiological functions, for example, as regulators of metabolism in the nervous, endocrine, cardiovascular, muscular, and the immune system. The physiological effects of PACAP in vertebrates are discussed by several excellent reviews [4, 35, 36]. Soon after the discovery of PACAP in vertebrates, the presence of PACAP-like peptide(s) is described in the fruit fly (*Drosophila melanogaster*) body wall neuromuscular junction and its possible role in neuronal plasticity as well as in the memory storage and retrieval [37, 38]. Thereafter, a number of studies have been published showing the presence and physiological role of PACAP or PACAP-like peptide(s) in several invertebrate species. In this review we summarize data obtained so far on the occurrence, distribution, and physiology of PACAP-like peptide(s) and its receptors in different invertebrates, such as oligochaetes, mollusks, insects, as well as malacostraca.

Expression and Localization of PACAP

Oligochaetes

The distribution of PACAP-like immunoreactivity (PACAP-IR) was studied in the CNS and PNS of three oligochaete (Annelida) worms with immunocytochemistry (IHC) [14]. Using PACAP-27 (n° 88121-5 antibody; [39, 40]) or PACAP-38 antibodies (anti-PACAP-38, Peninsula, CA, USA, [14, 39]) immunopositive cells and fibers were observed in cerebral, medial, and lateral parts of the subesophageal and ventral cord ganglia. In the peripheral nervous system, PACAP-IR was found in the enteric nervous system, epidermal sensory cells, and clitellum (reproductive organ). The distribution pattern of PACAP-IR was similar in all three species (*Lumbricus terrestris*, *Eisenia fetida*, *Lumbricus polyphemus*), suggesting a common distributional pattern of PACAP-like peptide(s) in oligochaete.

The levels of PACAP-27 (n° 88123-3 antibody) and PACAP-38 (n° 88111-3 antibody, [12]) isoforms were also measured in the nervous, intestinal, excretory, and reproductive systems of adult *Lumbricus* by RIA [15]. PACAP-27 and -38 isoforms were detected at significant amount in all of the examined tissues; however, their distribution was very heterogeneous. For example, the amount of PACAP-27 and PACAP-38 ranged between 0.31–17.12 and 0.02–1.51 ng/mg total protein, respectively. The level of PACAP-27-like immunoreactivity was approximately ten times higher than that of PACAP-38. In comparison, the highest PACAP-38 level in human is 4.7 ng/mg total protein in the bed nucleus of the stria terminalis [41]. These results suggest that, in contrast to vertebrates, the dominant isoform of the peptide is PACAP-27 in Oligochaeta.

PACAP immunopositive compounds were also observed in early stages (E1) of the embryonic development of the earthworm *Eisenia* using RIA (n° 88111-3 antibody), dot blot, and IHC (n° 9211-4 antibody, [42]) methods [17]. During embryonic development, the level of PACAP-like compounds decreased in cocoon fluids, while PACAP-IR cell bodies appeared in the developing body wall, prostomium, pharyngeal wall, and CNS. Furthermore, it was found that the clitellum of sexually mature worms contained significantly higher levels of PACAP-IR than other regions of the same animals or the clitellar region of a non-reproducing animal [17].

The presence and distribution of PAC1-R were described in the ventral nerve cord (VNC) of the adult *Eisenia* using IHC and commercial anti-PAC1-R [43]. Based on light and electron microscopic observations, the exact anatomical positions of labeled structures were established suggesting that PACAP mediates the activity of some interneurons, a few small motoneurons, and certain sensory fibers. High number of PAC1 receptors was found in both pre- and postsynaptic membranes in addition to extrasynaptic sites suggesting that PACAP acts as a neurotransmitter and neuromodulator in the earthworm nervous system. In early embryonic stages the first appearance of PAC1-R-like immunoreactivity was revealed by WB and Far WB methods as early as the E2 developmental stage. Immunolabeled CNS neurons were seen in the supraesophageal ganglion and distally in the subesophageal and ventral nerve cord ganglia. Ultrastructurally, PAC1-Rs were located mainly on plasma membranes and intracellular membranes, especially on cisternae of the endoplasmic reticulum [43, 44]. The authors conclude that PACAP-like compounds may affect the differentiation of germinal layers (at least that of the ectoderm) and of some neurons and they act as signaling molecules during earthworm embryonic development. Further details on the occurrence and functions of PACAP in Annelida can be found in another chapter of this book (Chap. 3).

Gastropods

IHC, RIA, WB, and mass spectrometric (MS) analysis revealed the presence of both the 27 and 38 AA isoforms of PACAP-like IR elements in CNS, PNS, and peripheral organs of simple garden snail, *Helix pomatia* [9, 16] and pond snail, *Lymnaea stagnalis* [45]. PACAP-containing neurons were present in each ganglion of the CNS but their distribution pattern was not homogenous. The majority of PACAP-IR neurons were observed in the areas where different peptide-containing neurons are located [46]. PACAP-like IR was observed in non-neuronal cells of the salivary gland, perineurium of the cerebral ganglion located around the blood vessel lacunae, and the wall of anterior aorta suggesting that non-neuronal PACAPs enter the circulatory system and thus may have humoral functions [8, 16]. Using IHC, it was shown that PACAP was absent in the muscle itself, but present in nerve fibers innervating the columellar, foot, heart, and tentacle flexor muscles [9, 16].

The concentration of PACAP-27-like polypeptide was significantly higher than that of PACAP-38, in contrast mammals, where PACAP-38 is the dominant isoform. However, the data obtained in gastropods correspond well those obtained in an oligochaete species (*Lumbricus*, *Eisenia*). Interestingly, both IHC and RIA studies revealed different expression levels of PACAP in active and non-active (hibernated or aestivated states) suggesting a dependence on behavioral state of the animal.

In a WB study a 14 kDa protein band was detected by PACAP-27 (n° 92112-4 antibody) and PACAP-38 antibodies (n° 88111-3). In addition, the anti-PACAP-38 reacted with a protein at 4.5 kDa. It is speculated that data obtained by both antibodies at 14 kDa represents an extended PACAP-like molluscan peptide or the precursor form of PACAP. In human prostate and prostate cancer cells a 14.6 kDa product was described, which is likely a product of the prePACAP protein (19.9 kDa), partially processed by convertases [47]. The assumption that extended PACAP-like molecules may exist is not unique. For example, in lower vertebrates, such as the stingray and catfish, 44- and 64-AA-long PACAP molecules were observed [42, 48]. Using the MS/MS fragment ion calculator the molecular weight based on sequence can be predicted accordingly, and the average mass of protonated quasimolecular ion ($[M^{+}H]^{+}$) of stingray and catfish PACAP would be *m/z* 5338.25 (5.3 kDa) and *m/z* 7856.25 (7.8 kDa), respectively. Based on the MALDI TOF/TOF measurement similar sequences of PACAP-27 and PACAP-38 can be identified from hemolymph and CNS samples of the snail with a molecular weight of 3147.1 and 4535.2, respectively. In addition, fragments of a PACAP-like molecule were found in *Helix* CNS homogenate with an identical AA sequence to mammalian PACAP-27 and -38 at positions 1–10 and 20–27 [16]. The AA sequence at 27–38 differs by only one AA (an iso-leucine to valin substitution) according to the mass calculation. Mass spectra of tryptic digest obtained by MALDI-TOF MS from *Lymnaea* CNS homogenate revealed complete sequence similarity of fragments between 1 and 32 AAs compared to mammalian PACAP-38 [45]. The average mass of $[M^{+}H]^{+}$ of synthetic mammalian PACAP-38 is *m/z* 4535.47 while in the pond snail, squid, planarian, and hydra the hypothetical average $[M^{+}H]^{+}$ of the PACAP-38-like molecule is *m/z* 4656.37. The reason for this difference could be the discrepancy in three AAs between synthetic mammalian PACAP-38 and isolated invertebrate PACAP-38-like molecule [49].

The PAC1-like receptor was also identified in the snail by IHC and biochemical methods [16, 45]. Similarly to vertebrate, PACAP activates a G-protein-coupled receptor and acts through the AC-cAMP pathway [45, 50]. In *Lymnaea* cerebral ganglia, the synthetic PACAP-38 incubation increased cAMP level by 82 %. In addition, both maxadilan, a specific PAC1-R agonist [51], and VIP, an agonist of VPAC1 and VPAC2 receptors also binding PACAP [19], increased cAMP synthesis by 47 % and 79 %, respectively. About 50 % of the cAMP-stimulating effect of PACAP-38 could be blocked by co-application of PACAP6-38 or maxadilan antagonist (M65). The biochemical results confirmed that PACAP-like peptides could increase cAMP levels through PACAP receptors in CNS or PNS. PAC1-like receptor expressing neuronal elements were observed in the CNS and a number of peripheral organs such as columellar muscle, heart, tentacles, and epithelial glandular cells. Far-WB experiments revealed three binding sites in snail brain homogenate. Two of these corresponded well to the VPAC1 (~45 kDa) and PAC1 (~60 kDa) receptors of vertebrates [16].

Insects

Using a vertebrate PACAP-38 antiserum (RIN-8920, Peninsula, CA, USA) PACAP-38-like IR was found in a subset of larval CNS neurons of *Drosophila* [37]. Furthermore, PACAP38-like IR was also found in nerve terminals innervating almost all muscle fibers in wild-type and NF1 mutant larvae [37, 52]. The staining appeared to be concentrated in varicosities where synaptic vesicles are localized. A comparison of the patterns of IR with previous anti-horseradish peroxidase staining, which reveals all nerve terminals arborized on muscle fibers, suggests that PACAP38-like IR is restricted to large-sized-type varicosities typical for neuropeptides.

In contrast to oligochaetes and gastropods, only the PACAP-38 isoform was present in the CNS of Insects [37]. In WB experiments an IR band was observed at 5.4 kDa that compared well with calculated mass of 4.5 kDa of mammalian PACAP-38. In addition a 19 kDa band was detected which may represent a possible PACAP precursor polypeptide in *Drosophila*. It is concluded that antibodies raised against mammalian PACAP38 identify an insect polypeptide with similar size [37].

In *Drosophila* a neuropeptide gene was identified that has some identity to PACAP [38]. This gene, named amnesiac, encodes a signal peptide followed by several possible peptides depending on the cleavage sites. One of the peptides deduced from the gene had 10 % identity with human PACAP-38 or 18 % with PACAP-27. This identity is too low to claim that amnesiac is homologous to PACAP in tunicates or vertebrates. However, the authors showed that an inserted space in PACAP after both amino acids 23 and 27 would increase the identity to 21 % for PACAP-38 and 30 % for PACAP-27. If amino acid similarity is used for the calculation, the relationship is higher [35].

Malacostraca

In some protostomes, such as the planarian (*Dugesia japonica*), the american cockroach (*Periplaneta americana*), and the bigfin reef squid (*Sepioteuthis lessoniana*) partial mRNAs corresponding to the highly conserved PACAP coding exon have been deposited in public databases [53]. Based on this public sequence information, Lugo and his coworkers [54] proposed a degenerative primer pair (F-LvPACAP and R-LvPACAP) and isolated for the first time the cDNA encoding the mature PACAP molecule from neural eyestalk tissue of a crustacean species, the white shrimp (*Litopenaeus vannamei*) by RT-PCR. Its high degree of sequence conservation is corroborated, when compared with sequences reported from tunicates (*Chelyosoma productum*) to mammalian vertebrates.

Functions of PACAP

The eukaryote *Tetrahymena thermophila* is a free-living ciliate protozoon widely used as an animal model in biological and biomedical research and exhibits a behavioral avoidance to PACAP-38. For example, the antagonists PACAP6-27 and 6-38, which inhibit PACAP receptors, serve as agonists for *Tetrahymena* [55]. The possibility cannot be excluded that PACAP is able to exert its action by directly activating the AC-cAMP pathways penetrating the cell membrane.

The high structural conservation and interphyletic distribution of a PACAP-like peptide and its receptor molecules suggest that this peptide is involved in the regulation of several basic physiological functions in invertebrates similar to those observed in vertebrates.

Effect in Regeneration

PACAP is involved in an array of physiological functions; thus, the role of the peptide is thought to be essential for cell survival. This is supported by the observation that the mortality of PACAP or PACAP receptor knockout mice is much higher than their wild-type mates [4, 19, 35, 56, 57]. Studies in PACAP knockout animals provide further evidence for the involvement of endogenous PACAP in regeneration processes. Upregulation of PACAP following nervous injuries has been shown in vertebrates by numerous previous studies [58]. It has been shown by RIA and IHC methods that the concentration of PACAP-like compounds increases in regenerating CNS and peripheral tissues of the earthworm *Eisenia* following injury indicating the possible role of PACAP in the regeneration [40]. Significant increase in the concentration of PACAP-like compounds was also observed in coelomocytes of regenerating earthworm [59]. Electron microscopic immunocytochemistry showed that PAC1 receptors are located on coelomocytes (mainly on amebocytes and on some granulocytes). Authors hypothesize a link between PACAP and coelomocytes, suggesting that PACAP modulates the function of amebocytes and certain granulocytes that play a role in regenerating earthworms. The data show that PACAP-like peptide(s) accumulate in the regenerating tissues of the earthworm suggesting trophic functions of these compounds in invertebrate tissues similarly to those observed on vertebrates.

Anti-apoptotic Effect

The anti-apoptotic effect of PACAP on vertebrate neuronal and non-neuronal cells is well documented [6, 7, 60, 61]. The anti-apoptotic effect of PACAP is mainly mediated by PAC1 receptor. The results imply that the anti-apoptotic effect of PACAP may be one of the basic functions of the peptide through evolution; both the peptide structure and this function have been conserved. PACAP has anti-apoptotic effect in the salivary gland cells of the snail [8]. In several gastropod species saliva or mucus release is performed by the holocrine release mechanism leading to cell destruction [62]. It has been suggested that cell death is indeed the physiological method of saliva release which takes place through a form of programmed cell death that is regulated by transmitters. It has been observed that stimulation of the salivary nerve or external application of dopamine elicits a change of mitochondrial membrane potential, and translocation of cytochrome-c from mitochondria to the cytoplasm is typical for the intrinsic mitochondrial pathway of programmed cell death. It has been observed that PACAP significantly attenuates the dopamine- and colchicine-induced apoptosis [8, 63].

Effect on Ion Channels

In snail (*Helix*) neurons expressing PAC1-like receptors synthetic PACAP-27 and -38 elicited membrane potential changes (both hyper- and depolarization) leading to significant changes in action potential frequency. PACAP6-38, as a receptor antagonist, powerfully antagonized the membrane effect of PACAP [16]. These results may suggest that PACAP is able to modulate the ion channels responsible for membrane and action potential generation. PACAP-like peptide has been found to modulate ionic conductance at the neuromuscular junction [37, 64]: in *Drosophila* larval muscles synthetic PACAP-38 enhanced L-type Ca^{2+}-current via AC-cAMP-PKA pathway [64]. Focal application of vertebrate PACAP-38 to the neuromuscular junction of *Drosophila* larval muscle triggered two temporally distinct responses: an immediate depolarization and a large enhancement of K-current. The enhancement occurred 12–14 min after the early depolarization. The effect of external PACAP-38 could be mimicked by high-frequency stimulation of motor nerve suggesting that PACAP or PACAP-like peptide is co-released and is functionally present in nerve terminals [37]. In the tentacle flexor muscles of the land snail, *Helix*, a potentiating effect of synthetic PACAP-27 was observed on cholinergic neuromuscular transmission. PACAP-27 presynaptically enhanced the release of acetylcholine by activating the AC-cAMP-PKA pathway. Postsynaptically, PACAP-27 enhanced muscle contractility by PKC-mediated signaling pathway resulting in an increased Ca^{2+} release from intracellular stores. These findings suggest that regulation of Ca^{2+} release may contribute to the stimulatory effect of PACAP [9].

Role of PACAP in Learning and Memory

PACAP activates molecular cascades leading to the execution of many physiological processes, including learning and memory [50, 65]. Feany and Quinn [38], cloning the memory gene in *Drosophila* responsible for the amnesiac mutation, observed that one of the "amnesiac" potential neuropeptides (AMN) had homology to the gene that encodes mammalian PACAP. The amn gene encodes a homolog of vertebrate PACAP, the AMN, strongly expressed in dorsal paired medial (DPM) neurons which is required for stable memory [66]. Furthermore, DMP activity is needed for middle-term memory, so suggesting that the PACAP-like AMN peptide release from the DPM neurons contributes to memory persistence [11, 66, 67].

The *Lymnaea* homologue of PACAP was found necessary for the acquisition and consolidation of long-term memory in the snail. We showed that systemic application of synthetic PACAP-27 or -38 accelerated the formation of transcription-dependent memory during single-trial reward chemical or multiple aversive tactile conditioning in *Lymnaea*. Using the antagonist PACAP6-38 it was also shown that the memory-accelerating effect of PACAP depended on G-protein-coupled PAC1-like receptors [50].

The age-related decline in memory performance could be reversed by administration of PACAP. Exogenous PACAP-38 boosted memory formation in aged *Lymnaea*, where endogenous PACAP-38 levels were significantly lower than in young snail based on WB experiments. In aged *Lymnaea*, there was a significant deficit of both intermediate- and long-term memory formation after one-trial reward conditioning. The deficit in both of these types of memory, however, was rescued by the application of the synthetic PACAP-38 peptide before training. Due to the evolutionarily conserved nature of these polypeptides and their established role in memory and synaptic plasticity, there was a very high probability that they could also act as "memory-rejuvenating" agents [68].

The role of PACAP and/or PACAP-like peptide(s) in acquisition and memory consolidation and recall in invertebrate model animals is discussed in detail in another chapter (Chap. 4).

Summary

The primary structure of PACAP has proved to be remarkably conserved during evolution not only in higher and lower vertebrates but also in invertebrates. In Table 2.1 different sequences of invertebrate PACAP molecules are aligned with human PACAP using ClustalW2—multiple sequence alignment. Detailed analysis revealed a high homology (>89 %) of inferred amino acid sequences: 35 AAs are conserved at the N-terminus and 3 AAs are variable at the C-terminus. Unfortunately, there is currently no definite sequence information about invertebrate PACAP or PACAP-like molecule(s) [3, 4, 69, 70]. Therefore, in IHC experiments different types of vertebrate antibodies are used which raises questions about the authenticity of the reported data even though antibodies with different epitopes produce the same effect. In WB experiments applying the same vertebrate antibodies the masses of immunopositive WB bands differ from those of expected. On the contrary, in most of the physiological experiments powerful and clear effect of the externally applied PACAP is observed suggesting the presence of specific receptor able to recognize synthetic PACAP. Although partial cDNA encoding PACAP-like peptide(s) in protostomes has been reported, no cDNA or gene encoding a PACAP-like peptide has been identified so far in species with fully/partially sequenced genomes [19, 53]. However, a highly conserved partial sequence corresponding to the exon encoding the mature PACAP peptide has been isolated in *Hydra magnipapillata*, in the tunicate, *Halocynthia roretzi* and in several protostomes such as

Table 2.1 Sequence comparison between in silico AA sequence of invertebrate PACAP peptides and human PACAP

		Identity (%)
Q8IU39_DUGJA	HSDGIFTDSYSRYRKQMAVKKYLAAVLGKRYRQRYRNK	92
Q75W94_HALRO	HSDGIFTDSYSRYRKQMAVKKYLAAVLGKRYRQRYRNE	89
Q8IU38_HYDMA	HSDGIFTDSYSRYRKQMAVKKYLAAVLGKRYRQRYRNK	92
Q75W88_EUCA	HSDGIFTDSYSRYREQMAVKKYLAAVLGKRYRQRYRNK	89
Q8IU37_SEPLE	HSDGIFTDSYSRYRKQMAVKKYLAAVLGKRYRQRYRNK	92
Q8IU36_PERAM	HSDGIFTDSYSRYRKQMAVKKYLAAVLGKRYRQRYRSK	89
PACAP_HUMAN	HSDGIFTDSYSRYRKQMAVKKYLAAVLGKRYKQRVKNK	100
	**************:****************:** :.:	

DUGJA—*Dugesia japonica* (AB083649), HALRO—*Halocynthia roretzi* (AB121759), HYDMA—*Hydra magnipapillata* (AB083650), EUCA—*Eriocheir japonica* (AB121765), SEPLE—*Sepioteuthis lessoniana* (AB083651), PERAM—*Periplaneta americana* (AB083652)

*—amino acid identity; :—replaceable amino acid

planarian (*Dugesia japonica*), crab (*Eriocheir japonica*), squid (*Sepioteuthis lessoniana*), and cockroach (*Periplaneta americana*).

In conclusion, PACAP or PACAP-like peptide(s) are present in invertebrates but the existence of a PACAP gene or peptide homologue remains to be convincingly demonstrated. However, the reported physiological effects of PACAP further confirm the presence of PACAP(-like) signaling pathways in invertebrates.

Acknowledgments P.Z's original research has been funded by the Hungarian Scientific Research Fund (PD-109099, Hungary), the Biology and Biotechnology Research Council (BB/KO18515/1, UK; PI: Michael O'Shea), the "Momentum" research project (LP2011/007, Hungary; PI: Dora Reglodi), and National Brain Project (KTIA_NAP_13-2-2014-0006, Hungary). K.T.'s original research was funded by the Hungarian Scientific Research Fund (OTKA, Hungary) and by the European Social Fund in the framework of TÁMOP-4.2.2.A-/11/1 KONV-2012-0024.

References

1. Miyata A, Arimura A, Dahl RR, Minamino N, Uehara A, Jiang L, et al. Isolation of a novel 38 residue-hypothalamic polypeptide which stimulates adenylate cyclase in pituitary cells. Biochem Biophys Res Commun. 1989;164:567–74.
2. Miyata A, Jiang L, Dahl RD, Kitada C, Kubo K, Fujino M, et al. Isolation of a neuropeptide corresponding to the N-terminal 27 residues of the pituitary adenylate cyclase activating polypeptide with 38 residues (PACAP38). Biochem Biophys Res Commun. 1990;170:643–8.
3. Vaudry D, Gonzalez BJ, Basille M, Yon L, Fournier A, Vaudry H. Pituitary adenylate cyclase-activating polypeptide and its receptors: from structure to functions. Pharmacol Rev. 2000;52:269–324.
4. Vaudry D, Falluel-Morel A, Bourgault S, Basille M, Burel D, Wurtz O, et al. Pituitary adenylate cyclase-activating polypeptide and its receptors: 20 years after the discovery. Pharmacol Rev. 2009;61:283–357.
5. Zhou CJ, Shioda S, Yada T, Inagaki N, Pleasure SJ, Kikuyama S. PACAP and its receptors exert pleiotropic effects in the nervous system by activating multiple signaling pathways. Curr Protein Pept Sci. 2002;3:423–39.
6. Somogyvari-Vigh A, Reglodi D. Pituitary adenylate cyclase activating polypeptide: a potential neuroprotective peptide. Curr Pharm Des. 2004;10:2861–89.
7. Shioda S, Ohtaki H, Nakamachi T, Dohi K, Watanabe J, Nakajo S, et al. Pleiotropic functions of PACAP in the CNS: neuroprotection and neurodevelopment. Ann N Y Acad Sci. 2006;1070:550–60.
8. Pirger Z, Racz B, Kiss T. Dopamine-induced programmed cell death is associated with cytochrome-c release and caspase-3 activation in snail salivary gland cells. Biol Cell. 2009;101:105–16.
9. Krajcs N, Hernádi L, Pirger Z, Reglődi D, Tóth G, Kiss T. PACAP modulates acetylcholine-elicited contractions at nicotinic neuromuscular contacts of the land snail. J Mol Neurosci. 2015;57:492–500.
10. Arimura A. Perspectives on pituitary adenylate cyclase activating polypeptide (PACAP) in the neuroendocrine, endocrine, and nervous systems. Jpn J Physiol. 1998;48:301–31.
11. Hashimoto H, Shintani N, Baba A. Higher brain functions of PACAP and a homologous *Drosophila* memory gene amnesiac: insights from knockouts and mutants. Biochem Biophys Res Commun. 2002;297:427–31.
12. Arimura A, Somogyvari-Vigh A, Miyata A, Mizuno K, Coy DH, Kitada C. Tissue distribution of PACAP as determined by RIA: highly abundant in the rat brain and testes. Endocrinology. 1991;129:2787–9.

13. Masuo Y, Matsumoto Y, Tokito F, Tsuda M, Fujino M. Effects of vasoactive ontestinal polypeptide (VIP) and pituitary adenylate cyclase activating polypeptide (PACAP) on the spontaneous release of acetylcholine from the rat hippocampus by brain microdialysis. Brain Res. 1993;611:207–15.
14. Reglodi D, Lengvari I, Szelier M, Vigh S, Arimura A. Distribution of PACAP-like immunoreactivity in the nervous system of oligochaeta. Peptides. 2000;21:183–8.
15. Somogyvari-Vigh A, Reglodi D, Li M, Lengvari I, Vigh S, Arimura A. Tissue distribution of PACAP27 and -38 in oligochaeta: PACAP27 is the predominant form in the nervous system of *Lumbricus polyphemus*. Peptides. 2000;21:1185–91.
16. Hernadi L, Pirger Z, Kiss T, Nemeth J, Mark L, Kiss P, et al. The presence and distribution of pituitary adenylate cyclase activating polypeptide and its receptor in the snail *Helix pomatia*. Neuroscience. 2008;155:387–402.
17. Boros A, Reglodi D, Herbert Z, Kiszler G, Nemeth J, Lubics A, et al. Changes in the expression of PACAP-like compounds during the embryonic development of the earthworm *Eisenia fetida*. J Mol Neurosci. 2008;36:157–65.
18. Pirger Z, Lubics A, Reglodi D, Laszlo Z, Mark L, Kiss T. Mass spectrometric analysis of activity-dependent changes of neuropeptide profile in the snail, *Helix pomatia*. Neuropeptides. 2010;44:475–83.
19. Cardoso JC, Vieira FA, Gomes AS, Power DM. PACAP, VIP and their receptors in the metazoa: insights about the origin and evolution of the ligand-receptor pair. Peptides. 2007;28: 1902–19.
20. Dickson L, Fynlayson K. VPAC and PAC receptors: from ligands to function. Pharmacol Ther. 2009;121:294–316.
21. Kilpatrick GJ, Dautzenberg FM, Martin GR, Eglen RM. 7TM receptors: the splicing on the cake. Trends Pharmacol Sci. 1999;20:294–301.
22. Mirabella N, Lamanna C, Assisi L, Botte V, Cecio A. The relationships of nicotinamide adenine dinucleotide phosphate-d to nitric oxide synthase, vasoactive intestinal polypeptide, galanin and pituitary adenylate activating polypeptide in pigeon gut neurons. Neurosci Lett. 2000;293:147–51.
23. Jozsa R, Hollosy T, Tamas A, Toth G, Lengvari I, Reglodi D. Pituitary adenylate cyclase activating polypeptide plays a role in olfactory memory formation in chicken. Peptides. 2005;26:2344–50.
24. Nowak JZ, Zawilska JB. PACAP in avians: origin, occurrence, and receptors -pharmacological and functional considerations. Curr Pharm Des. 2003;9:467–81.
25. Reglodi D, Somogyvari-Vigh A, Vigh J, Li M, Lengvari I, Arimura A. Pituitary adenylate cyclase activating polypeptide is highly abundant in the nervous system of anoxia-tolerant turtle, *Pseudemys scripta elegans*. Peptides. 2000;22:873–8.
26. Valiante S, Prisco M, Capaldo A, Zambrano I, De Falco M, Andreuccetti P, et al. Molecular characterization and gene expression of the pituitary adenylate cyclase-activating polypeptide (PACAP) in the lizard brain. Brain Res. 2007;1127:66–75.
27. Yon L, Feuilloley M, Chartrel N, Arimura A, Conlon JM, Fournier A, et al. Immunohistochemical distribution and biological activity of pituitary adenylate cyclase-activating polypeptide (PACAP) in the central nervous system of the frog *Rana ridibunda*. J Comp Neurol. 1992;324:485–9.
28. Hu ZT, Lelievre V, Chao A, Zhou XR, Waschek JA. Characterization and messenger ribonucleic acid distribution of a cloned pituitary adenylate cyclase-activating polypeptide type I receptor in the frog *Xenopus laevis* brain. Endocrinology. 2000;141:657–65.
29. Alexandre D, Vaudry H, Jegou S, Anouar Y. Structure and distribution of the mRNAs encoding pituitary adenylate cyclase-activating polypeptide and growth hormone-releasing hormone-like peptide in the frog, *Rana ridibunda*. J Comp Neurol. 2000;421:234–46.
30. Ciarlo M, Bruzzone F, Angelini C, Alexandre D, Anouar Y, Vallarino M, et al. Expression of PACAP receptors in the frog brain during development. Ann N Y Acad Sci. 2006;1070: 201–4.

31. Montero M, Yon L, Rousseau K, Arimura A, Fournier A, Dufour S, et al. Distribution, characterization, and growth hormone-releasing activity of pituitary adenylate cyclase-activating polypeptide in the European eel, *Anguilla anguilla*. Endocrinology. 1998;139:4300–10.
32. Adams BA, Lescheid DW, Vickers ED, Crim LW, Sherwood NM. Pituitary adenylate cyclase-activating polypeptide and growth hormone-releasing hormone-like peptide in sturgeon, whitefish, grayling, flounder and halibut: cDNA sequence, exon skipping and evolution. Regul Pept. 2002;109:27–37.
33. Jakab B, Reglodi D, Jozsa R, Hollosy T, Tamas A, Lubics A, et al. Distribution of PACAP-38 in the central nervous system of various species determined by a novel radioimmunoassay. J Biochem Biophys Methods. 2004;61:189–98.
34. Matsuda K, Nagano Y, Uchiyama M, Onoue S, Takahashi A, Kawauchi H, et al. Pituitary adenylate cyclase-activating polypeptide (PACAP)-like immunoreactivity in the brain of a teleost, *Uranoscopus japonicus*: immunohistochemical relationship between PACAP and adenohypophysial hormones. Regul Pept. 2005;126:129–36.
35. Sherwood NM, Krueckl SL, McRory JE. The origin and function of the pituitary adenylate cyclase-activating polypeptide (PACAP)/glucagon superfamily. Endocr Rev. 2000;21:619–70.
36. Reglodi D, Renaud J, Tamas A, Tizabi Y, Socias B, Del-Bel E, et al. Novel tactics for neuroprotection in Parkinson's disease: role of antibiotics, polyphenols and neuropeptides. Prog Neurobiol. 2015.
37. Zhong Y, Pena LA. A novel synaptic transmission mediated by a PACAP-like neuropeptide in *Drosophila*. Neuron. 1995;14:527–36.
38. Feany MB, Quinn WG. A neuropeptide gene defined by the *Drosophila* memory mutant amnesiac. Science. 1995;268:869–73.
39. Molnar L, Pollak E, Boros A, Reglodi D, Tamas A, Lengvari I, et al. Comparative anatomy of PACAP-immunoreactive structures in the ventral nerve cord ganglia of lumbricid oligochaetes. Ann N Y Acad Sci. 2006;1070:427–30.
40. Varhalmi E, Somogyi I, Kiszler G, Nemeth J, Reglodi D, Lubics A, et al. Expression of PACAP-like compounds during the caudal regeneration of the earthworm *Eisenia fetida*. J Mol Neurosci. 2008;36:166–74.
41. Palkovits M, Somogyvari-Vigh A, Arimura A. Concentrations of pituitary adenylate cyclase activating polypeptide (PACAP) in human brain nuclei. Brain Res. 1995;699:116–20.
42. Matsuda K, Yoshida T, Nagano Y, Kashimoto K, Yatohgo T, Shimomura H, et al. Purification and primary structure of pituitary adenylate cyclase activating polypeptide (PACAP) from the brain of an elasmobranch, stingray, *Dasyatis akajei*. Peptides. 1998;19:1489–95.
43. Molnar L, Pollak E, Boros A, Shioda S, Nakajo S, Tamas A, et al. PAC1 receptor localization in a model nervous system: light and electron microscopic immunocytochemistry on the earthworm ventral nerve cord ganglia. Regul Pept. 2008;145:96–104.
44. Boros Á, Somogyi I, Engelmann P, Lubics A, Reglodi D, Pollák E, et al. Pituitary adenylate cyclase-activating polypeptide type 1 (PAC1) receptor is expressed during embryonic development of the earthworm. Cell Tissue Res. 2010;339:649–53.
45. Pirger Z, Laszlo Z, Hiripi L, Hernadi L, Toth G, Lubics A, et al. Pituitary adenylate cyclase activating polypeptide (PACAP) and its receptors are present and biochemically active in the central nervous system of the pond snail *Lymnaea stagnalis*. J Mol Neurosci. 2010;42:464–71.
46. Hernadi L, Elekes K. Neurons with different immunoreactivity form clusters in the CNS of *Helix pomatia*. Acta Biol Hung. 1995;46:271–80.
47. Garcia-Fernandez MO, Bodega G, Solano RM, Ruiz-Villaespesa A, Sanchez-Chapado M, Carmena MJ, et al. Expression and distribution of pituitary adenylate cyclase-activating peptide in human prostate and prostate cancer tissues. Regul Pept. 2002;110:9–15.
48. McRory JE, Parker DB, Ngamvongchon S, Sherwood NM. Sequence and expression of cDNA for pituitary adenylate cyclase activating polypeptide (PACAP) and growth hormone-releasing hormone (GHRH)-like peptide in catfish. Mol Cell Endocrinol. 1995;108:169–77.
49. Kiss T, Pirger Z. Multifunctional role of PACAP-like peptides in molluscs. Protein Pept Lett. 2013;20:628–35.

50. Pirger Z, László Z, Kemenes I, Tóth G, Reglődi D, Kemenes G. A homolog of the vertebrate pituitary adenylate cyclase-activating polypeptide is both necessary and instructive for the rapid formation of associative memory in an invertebrate. J Neurosci. 2010;30:13766–73.
51. Lerner EA, Iuga AO, Reddy VB. Maxadilan, a PAC1 receptor agonist from sand flies. Peptides. 2007;28:1651–4.
52. Guo HF, The I, Hannan F, Bernards A, Zhong Y. Requirement of *Drosophila* NF1 for activation of adenylyl cyclase by PACAP38-like neuropeptides. Science. 1997;276:795–8.
53. Cardoso JC, Vieira FA, Gomes AS, Power DM. The serendipitous origin of chordate secretin peptide family members. BMC Evol Biol. 2010;10:135.
54. Lugo JM, Carpio Y, Morales R, Rodriguez-Ramos T, Ramos L, Estrada MP. First report of the pituitary adenylate cyclase activating polypeptide (PACAP) in crustaceans: conservation fits functions as growth promoting factor and immunomodulator in the white shrimp *Litopenaeus vannamei*. Fish Shellfish Immun. 2013;35:1788–96.
55. Lucas J, Riddle M, Barttholomew J, Thomas B, Forni J, Nickerson LE, et al. PACAP-38 signaling in *Tetrahymena thermophila* involves NO and cGMP. Acta Protozool. 2004;43:15–20.
56. Hoyle CHV. Neuropeptide families: evolutionary perspectives. Regul Pept. 1998;73:1–33.
57. Reglodi D, Kiss P, Szabadfi K, Atlasz T, Gabriel R, Horvath G, et al. PACAP is an endogenous protective factor-insights from PACAP-deficient mice. J Mol Neurosci. 2012;48:482–92.
58. Waschek JA. Multiple actions of pituitary adenylyl cyclase activating peptide in nervous system development and regeneration. Dev Neurosci. 2002;24:14–23.
59. Somogyi I, Boros A, Engelmann P, Varhalmi E, Nemeth J, Lubics A, et al. Pituitary adenylate cyclase-activating polypeptide-like compounds could modulate the activity of coelomocytes in the earthworm. Ann N Y Acad Sci. 2009;1163:521–3.
60. Rácz B, Gasz B, Borsiczky B, Gallyas F, Tamás A, Jozsa R, et al. Protective effects of pituitary adenylate cyclase activating polypeptide in endothelial cells against oxidative stress-induced apoptosis. Gen Comp Endocrinol. 2007;153:115–23.
61. Bourgault S, Vaudry D, Dejda A, Doan ND, Vaudry H, Fournier A. Pituitary adenylate cyclase-activating polypeptide: focus on structure-activity relationships of a neuroprotective peptide. Curr Med Chem. 2009;16:4462–80.
62. Pirger Z, Elekes K, Kiss T. Functional morphology of the salivary gland of the snail, *Helix pomatia*: a histochemical and immunocytochemical study. Acta Biol Hung. 2004;55:221–32.
63. Kiss T. Apoptosis and its functional significance in molluscs. Apoptosis. 2010;15:313–21.
64. Bhattacharya A, Lakhman SS, Singh S. Modulation of L-type calcium channels in *Drosophila* via a pituitary adenylyl cyclase-activating polypeptide (PACAP)-mediated pathway. J Biol Chem. 2004;279:37291–7.
65. Roberto M, Brunelli M. PACAP-38 enhances excitatory synaptic transmission in the rat hippocampal CA1 region. Learn Mem. 2000;7:303–11.
66. Keene AC, Stratmann M, Keller A, Perrat PN, Vosshall LB, Waddell S. Diverse odor-coonditioned memories require uniquely timed dorsa paired medial neuron output. Neuron. 2004;44:521–33.
67. Liu WJ, Guo F, Lu BK, Guo AK. Amnesiac regulates sleep onset and maintenance in *Drosophila melanogaster*. Biochem Biophys Res Commun. 2008;372:798–803.
68. Pirger Z, Naskar S, Laszlo Z, Kemenes G, Reglodi D, Kemenes I. Reversal of age-related learning deficiency by the vertebrate PACAP and IGF-1 in a novel invertebrate model of aging: the pond snail (*Lymnaea stagnalis*). J Gerontol A Biol. 2014;69:1331–8.
69. Gourlet P, Woussen-Colle MC, Robberecht P, de Neef P, Cauvin A, Vandermeers-Piret MC, et al. Structural requirements for the binding of the pituitary adenylate-cyclase-activating peptide to receptors and adenylate-cyclase activation in pancreatic and neuronal membranes. Eur J Biochem. 1991;195:535–41.
70. Vandermeers A, Vandeborre S, Hou X, de Neef P, Robberecht P, Vandermeers-Piret M, et al. Antagonistic properties are shifted back to agonistic properties by further N-terminal shortening of pituitary adenylate-cyclase-activating peptides in human neuroblastoma NB-OK-1 cell membranes. Eur J Biochem. 1992;208:815–9.

Chapter 3
PACAP-Like Compounds of Earthworms: Identification and Putative Functions from Embryonic Development to Brain Regeneration

Laszlo Molnar, Peter Engelmann, Anita Steib, and Edit Pollak

Abstract Presence and pattern of both PACAP38 and PACAP27 have been shown by immunocytochemistry, radioimmunoassay and Western blotting in various earthworm tissues. However, the identification of peptide coding mRNA of PACAP has not yet been carried out, so the exact experimental evidences of PACAP expression in this invertebrate group are missing. In spite of this, we have concluded that PACAP-like molecules (having recognizable epitopes for antibodies raised against PACAP) are expressed in earthworms. There are some strong supports of this hypothesis, namely (1) PACAP stimulates the cAMP synthesis of earthworm tissues in in vitro experiments, (2) occurrence of the specific PACAP receptor (PAC1) has been shown by electron microscopic immunocytochemistry in the central nervous system and (3) isolated PAC1 immunoreactive compound binds the synthetic PACAP molecules. Significant overexpression of PACAP-like molecules have been found in the body wall and injured central nervous system (CNS) of worms during the posterior segment regeneration. Accumulation of the PACAP-like molecules in the suboesophageal ganglion and the cauterized circumpharyngeal connectives has been shown in brain extirpated specimens, suggesting that these compounds are absolutely necessary for nervous tissue regeneration. Inhibitory effect of PACAP antagonists on the nervous tissue regeneration supports the putative morphogenetic role of PACAP-like peptides. Expressions of both PACAP-like compounds and PAC1 receptor have been characteristic signs of the increased activity of the specific immune cells (coelomocytes) during the regeneration and embryonic development. Similar effects and influences have been described in certain invertebrate and vertebrate species as well. Marked similarity between the invertebrate (cnidarian,

L. Molnar (✉) • A. Steib • E. Pollak
Department of Comparative Anatomy and Developmental Biology,
University of Pecs, Pecs, Hungary
e-mail: molnar@gamma.ttk.pte.hu; drmolnarl@gmail.com

P. Engelmann
Department of Immunology and Biotechnology, University of Pecs, Pecs, Hungary

D. Reglodi, A. Tamas (eds.), *Pituitary Adenylate Cyclase Activating Polypeptide — PACAP*, Current Topics in Neurotoxicity 11,
DOI 10.1007/978-3-319-35135-3_3

gastropod) chordate (tunicate) and vertebrate, PACAP has been identified by several research groups. These findings and our results strongly suggest that PACAP belongs to those highly conserved peptides that developed during the phylogeny of the metazoans.

Keywords Earthworm central nervous system • Regeneration blastema • Epimorphosis • Neoblasts • Coelomocyte • Radioimmunoassay • Western blot • Far Western blot • Morphogenesis • PAC1 receptor

Introduction

PACAP is thought to be a highly conserved peptide characterized by nearly the same sequence identity from Cnidarians to Mammals. Although most of the PACAP or PACAP-like peptides found in invertebrates were identified by mainly immunocytochemical methods there are some transcriptomic evidence of the occurrence of PACAP-like compounds also in lower invertebrates. A partial mRNA (114 bp) corresponding to the highly conserved PACAP coding exon has been isolated from a Cnidarian (*Hydra magnipapillata*) and deposited in public database [1].

The forms and multiple functions (e.g. signalling molecule of embryogenesis, neuroprotective and trophic effect, anti-apoptotic effect, etc.) of PACAP27 and 38 were overviewed by other chapters of this book so in this review we will focus on the expression and functions of PACAP-like peptides of earthworm that are model animals of both evolutionary neurobiology [2] and regeneration experiments [3].

Expression of PACAP-Like Peptides and PAC1-Receptor in Various Tissues of Adult Earthworms

In the central nervous system and peripheral tissues of adult earthworms, by means of immunohistochemistry, radioimmunoassay (RIA) and bioassay, the expression of both PACAP-38 and -27 like compounds was revealed [4, 5] suggesting that the latter form was the dominant [6]. The pattern of immunoreactive structures of the ventral nerve cord was identified by the investigation of whole mount preparations and it was found that PACAP forms were expressed in distinct neuron sets of earthworms [5].

Earthworms have no distinct endocrine organs, like pituitary or any endocrine gland of vertebrates, or corpus allatum and prothoracic gland of insects. Instead, their central nervous system contains neurosecretory cells that are not uniformly distributed among ganglia [7]. However, immunostaining was found not only in neurosecretory cells but also putative inter- and motoneurons (Fig. 3.1) [5]. We showed the occurrence of PACAP-like compounds in the body wall and in the albumen of

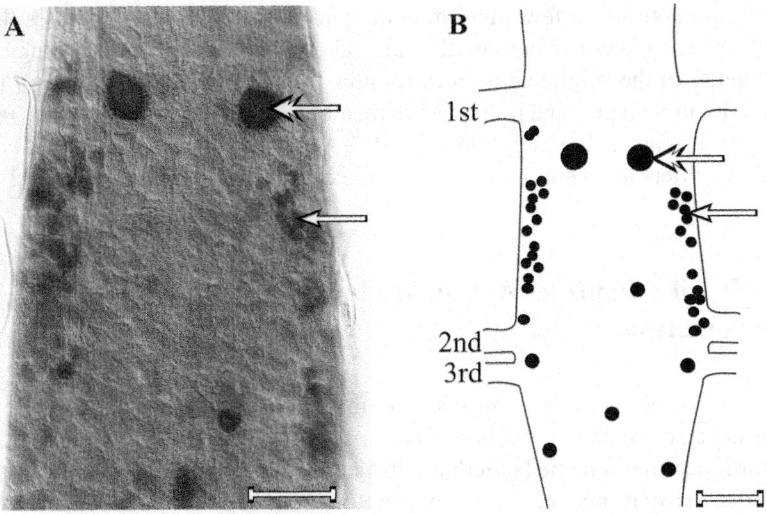

Fig. 3.1 Whole mount immunocytochemistry (**a**) revealed the distribution pattern of PACAP27 immunoreactive neurons in the ventral nerve cord ganglia (**b**) of *E. fetida*. *Arrows* indicate small interneurons, *double arrows* label giant motoneurons. Scale bars: **a** 50 μm, **b** 100 μm

deposited cocoons (eggs surrounded by nourishing albumen and protective proteinaceous sleeve), and in the developing embryos (see later) of the earthworm *E. fetida* [8]. PACAP-like compounds were continuously synthesized in all body wall tissues. Characteristic differences have been found between PACAP-like peptide content of various body parts of immature and sexually matured specimens. In immature worms, the level of PACAP-like peptides in the clitellar and postclitellar regions was detectable with no significant differences between these two body parts. Tissue extracts of clitellum isolated from sexually mature specimens, in which hypertrophy of the gland cells in clitellum appeared, showed extremely strong immune reaction. Extracts from the postclitellar segments gave significantly weaker immunoreactivity. These results indicate that the level of PACAP-like compounds significantly increased in the clitellum, while no characteristic changes were observed in the postclitellar segments of sexually immature worms [8].

According to our recent finding, the blood of earthworms contains high concentration of PACAP-like peptide (50 ± 9 fmol/mg) suggesting that certain neurosecretory cells of the earthworm CNS synthesize most of the PACAP-like compounds and they are transported to various tissues by the closed circulatory system so these compounds act as both transmitters and neurohormones.

PACAP can exert its effects through two types of G-protein-coupled receptors. VPAC1 and VPAC2 bind vasoactive intestinal peptide (VIP) and PACAP with similar affinities while PAC1 binds PACAP with a much higher affinity than VIP [9–11].

The presence and pattern of PAC1 receptor were identified by means of pre- and post-embedding immunocytochemistry in the ventral nerve cord ganglia. Both light and electron microscopic observations suggested that PACAP mediates the activity of some interneurons, a few small motoneurons and certain sensory fibres that are located in ventrolateral, ventromedial and intermediomedial sensory longitudinal axon bundles of the ventral nerve cord ganglia. Occurrence of PAC1 receptor immunoreactivity in both pre- and postsynaptic membranes of the central synapses in addition to extrasynaptic sites suggested that PACAP-like peptides of earthworms act as neurotransmitters and neuromodulators in the earthworm nervous system [12].

PACAP-Like Peptides Act as Mediators in Earthworm's Embryogenesis

Earthworm embryos develop in cocoons in which protective and nutritive compounds are also located [13]. Cocoon is a special product of the modified body part of mature specimens, named clitellum, that is rich in protein-secreting glands and its secretory activity is mediated by neurosecretory cells of the brain [7, 13]. The occurrence of PACAP-like peptides and the decrease of their concentrations during the embryonic development were shown in the cocoon fluid by radioimmunoassay [8].

Immunohistochemical observations revealed that PACAP-like immunoreactivity already appeared at E1 developmental stage in the developing body wall epithelium and ganglia of the central nervous system. During development, immunoreactivity became visible around the forming mouth, followed by the ventral pharyngeal wall. The developing stomatogastric ganglia also contained PACAP-like immunoreactive elements. Immunolabelled structures with morphological heterogeneity and random distribution in the body wall ectoderm were seen in extremely high number in E2–E3 embryos (Fig. 3.2). In embryos before hatching (E4), PACAP-like immunoreactive structures appeared at the entrance of the mouth cavity and in the prostomium. The central nervous system showed PACAP-like immunoreactivity just before hatching. Labelled cells were observed primarily in the supraoesophageal (so-called brain) and suboesophageal ganglia. Later, a few cells of the ventral nerve cord were also stained implying a rostrocaudal gradient during development.

The existence of a protein capable of synthetic PACAP38 binding in the earthworm in vitro and having the same molecular weight as that in Western blots for PAC1R was shown by Far Western blot. Expression of the PAC1 receptor was detected by Western blot as early as in E2 developmental stage embryos in which the extension of the ventral nerve cord ganglia and the differentiation of peripheral tissues are propagated. At this stage of the development immunopositive segmented body wall consisting of epithelium, muscles, metanephridia, developing alimentary canal, blood vessels and several coelomocytes in the body cavity could also be identified [14].

The first PAC1-labeled neural structures appeared in the brain at E1 developmental stage. These cells projected their relatively dense fibres, via the circumpharyngeal connectives, to the developing suboesophageal ganglion where they enter the dorsolateral fibre tracts of the ventral nerve cord. At this time there were no visible signs of PAC1 receptor immunoreactive cell bodies in the suboesophageal ganglion or

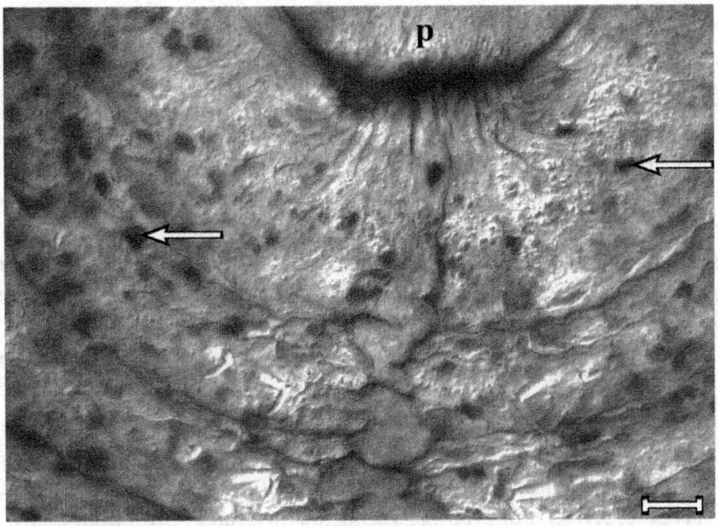

Fig. 3.2 Pattern of PACAP27 immunoreactive primary sensory cells (*arrows*) in the body wall of developing embryo at E3 developmental stage. Immunostaining was carried out by the same chemicals and methods applied by Molnar et al. [5]. *P* prostomium. Scale bar: 10 μm

any of the developing ventral nerve cord ganglia. A pair of PAC1-immunoreactive cell bodies was discernible only in the suboesophageal ganglion at later E2 developmental stage. Afterwards parallel with the segment and ganglion formation a pair of labelled cells was observed in each ganglion. While the number and pattern of labelled perikarya were identical in all ventral nerve cord ganglia, from the suboesophageal to the newly formed ganglia, the number of stained neural processes increased conspicuously during gangliogenesis.

The most extensive pattern of PAC1 receptor labelled neural structures was found in the ventral nerve cord of E4 developmental stage embryos, characterized by worm-like body form, that are ready for hatching. However, there were certain differences between the PAC1 receptor labelled patterns of ventral nerve cord in mature and newly hatched earthworms [12] proposing postembryonic developmental changes of the PAC1 receptor system. The anatomical position of stained fibres was apparently not changed across the ventral nerve cord. All neural processes were ipsilateral and they did not leave the central nervous system through the segmental nerves. This suggests that labelled neurons belong to the earthworm's interneuron system. The perikarya were weak while the neural processes strongly stained, we propose that latter structures possess significantly more receptor molecules than the former ones.

Ultrastructurally, PAC1 receptors were located both on plasma membranes and intracellular membranes of certain neural processes in the central neuropil. The ultrastructural distribution of the PAC1 receptor immunoreactivity was fairly similar to those found in the mature ventral nerve cord [12] suggesting that there were no conspicuous postembryonic changes in the PAC1 receptor system.

Effects of PACAP on Postembryonic Development from Segment Regeneration to Brain Restoration

Several experiments have been focused on the anterior and posterior segment regeneration in earthworms [7]. The gained information mainly concentrated on the histological and/or immunocytochemical observations of the intact and regenerating body parts.

There are detailed experimental evidences that all new tissues forming restored segments are derived from the regeneration blastema (epimorphosis). In this structure dedifferentiating tissue elements (epithelial, mesenchymal, etc.), free-floating coelomocytes (immune cells) and totipotent stem cells so-called neoblasts can be located (for review see [7]).

It is generally accepted that the formation of the regeneration blastema and later on the segment restoration is maintained by the central nervous system (the theory is called the nerve dependence of regeneration). Until now no direct experimental evidence has emerged to support this hypothesis (originally promulgated by Thomas Hunt Morgan [15]) since the possible role of elaborated tissue hormones, growth factors, etc. cannot be excluded.

Applying a set of experiments on regeneration process of earthworms our research group has focused on not only macroscopical and histological investigations but also the expression of specific compounds (for example on the known vertebrate neurotrophic factor-analogue PACAP) in the regenerating segments and the central nervous system.

By means of radioimmunoassay, Varhalmi et al. [16] revealed significant changes of PACAP-like immunoreactivity during the posterior segment regeneration. Both PACAP38 and PACAP27 could be detected in distinct tissues of earthworms like free-floating coelomocytes, ventral nerve cord ganglia, body wall and alimentary canal (midgut) tissues, and in the regeneration blastema. Marked differences in the expression of PACAP-like compounds isolated from various tissues were revealed. Expression of PACAP27-like compound in coelomocytes isolated from the last five intact segments increased significantly, while concentration of PACAP38-like compound did not change. In contrast, concentrations of both PACAP27- and PACAP38-like compounds in coelomocytes isolated from the rest of the body showed an approximately fourfold increase compared to controls. Following injury, concentrations of PACAP-like compounds markedly increased both in the regenerating body wall and in the alimentary canal, where the regeneration blastema is found. In contrast, no significant changes were shown in the regenerating ventral nerve cord ganglia suggesting that in comparison with body wall and alimentary canal there is no chemical gradient in the VNC ganglia during the posterior regeneration.

The regeneration blastema attached to the severed ganglion was interwoven with fine blood capillaries that formed a loose random network within the blastema (Fig. 3.3). Along with the capillaries, a high number of neoblasts, labelled for PACAP antibodies, scattered over the whole regenerating segment. On the 14th postoperative day of regeneration a distinct regeneration cone, containing extremely

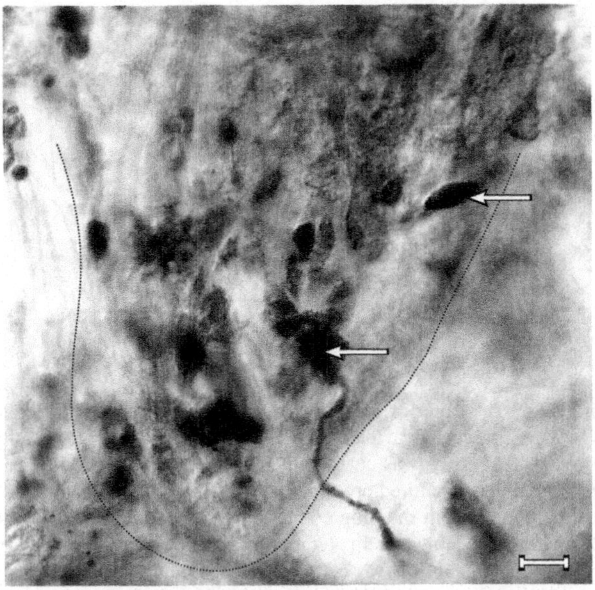

Fig. 3.3 Whole mount preparation of a regeneration cone in a severed ventral nerve cord ganglion with accumulated neoblasts (*arrows*) stained for PACAP38. Scale bar: 10 μm

high number of stained cells, appeared on the severed ganglion and attached to the peripheral regeneration blastema without sharp border. Most cells were round or oval in shape but elongated cells bearing processes were also found. Both round undifferentiated neoblasts and several elongated ones with identifiable processes were identified in this connective tissue. A few cells had large soma and an easily identifiable axon-like process that attached to other cells or fibres [16].

Increase in the concentration of PACAP-like compounds in the regenerating body wall and alimentary canal (consisting of regeneration blastema and the neighbouring original tissues) indicated that these compounds could influence the regeneration of earthworms. Since numerous immunolabelled neoblasts were found in the regenerating segments, and especially in the regeneration blastema, it seems that PACAP-like substances could have stimulating effects both on cell cycle and differentiation of neoblasts and other blastemal cells to various tissues from epithelial to nervous tissue.

A characteristic anteroposterior gradient was found in the scattered mass of coelomocytes in the regenerating earthworm's body cavity suggesting that coelomocytes could transport (perhaps also synthesize) PACAP-like compounds to the regenerating segments [17]. Trophic and transporting functions of coelomocytes during regeneration have been proposed earlier [18, 19]. A recent study has also suggested that coelomocytes play a role in hormone secretion, transport and hormone elimination in *E. fetida* [20]. PACAP-like compounds might act not only as mediators of regeneration but they could also modify the activity of certain coelomocytes characterized by phagocytic activity and might be potent anti-inflammatory peptides in earthworms.

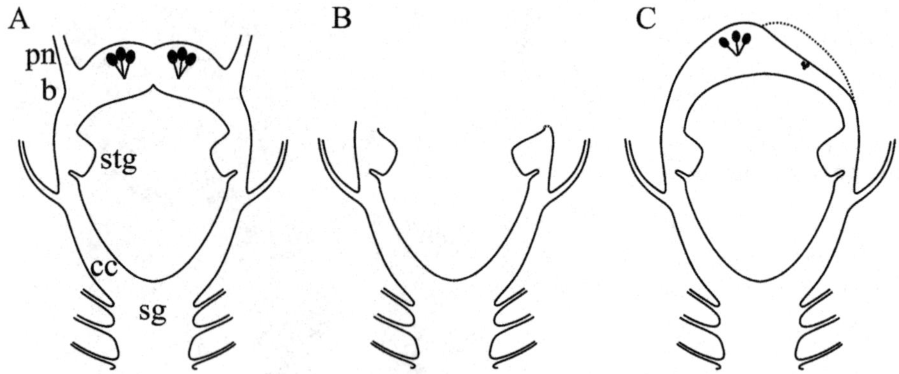

Fig. 3.4 The anterior part of the central nervous system of *E. fetida* with the landmark of GABAergic neurons in the brain (**a**). The effect of brain extirpation method on the brain regeneration (**b**, **c**). Left cc was cut with a sharp blade, the right one was cauterized. As a result of high temperature a denatured protein plug formed on the cc that temporarily blocked outflow of chemical compounds but the plug could be destroyed by phagocytes later so chemicals could outflow from the CNS. However, the temporary block resulted in distinctive changes in brain regeneration. The left hemiganglion was better developed than the right one (**c**) The line drawing shows the real (*solid line*) and expected (*dotted line*) size and form of regenerated structures. *Abbreviations*: *Pn* prostomial nerve, *b* brain, *stg* stomatogastric ganglion, *cc* circumpharyngeal connective, *sg* suboesophageal ganglion

Experimental data proved that epimorphic regeneration should really be determined by certain chemical factors, e.g. PACAP-like peptides that induce cell proliferation and differentiation during the tissue and organ remodelling. To clear whether the nervous system mediates only the reorganization of peripheral tissues (epithelial, connective and muscles tissues) or it can have an influence on own reorganization, we developed a brain regeneration model.

The regeneration of the cerebral ganglion (so-called brain) was investigated in the earthworm *E. fetida* applying two absolutely different extirpation methods, respectively (Fig. 3.4). In the first group of animals both of the circumpharyngeal connectives (ccs) were transected with sharp steel blade, in the second group the left cc was transected and the right one was cauterized by glowing tungsten wire on the level of the second segmental nerves. During regeneration, the concentration gradient of PACAP-like peptides along the central nervous system and the pattern of the characteristic GABAergic landmark structures were investigated.

Following brain extirpation a marked increase of PACAP-like peptide synthesis was determined by radioimmunoassay in the ventral nerve cord ganglia. However, a decreasing gradient of PACAP from the suboesophageal ganglion to cross-cut circumpharyngeal connectives was found in regenerating earthworms. While in the cauterized circumpharyngeal connectives the PACAP concentration was significantly higher than in the suboesophageal ganglion suggesting that neural processes transport PACAP-like peptides (and probably other neuropeptides, transmitters, etc.) to the site of regeneration. The elaboration of PACAP-like peptides and other neuroactive substances (transmitters, neurohormones, growth factors) from the central nervous system via the circumpharyngeal connectives could mediate the migration and attachment of earthworm stem cells (neoblasts) to the cut surface of circumpharyngeal connectives and mediate their differentiation to neuronal, glial, muscular and connective tissue cells resulting in the formation of a new brain and its capsule. This hypothesis is supported by the results of pharmacological experiments, namely injection of PACAP antagonist to the site of regeneration strongly inhibited differentiation of neural somata and growing of processes so the structure of the regenerated cerebral ganglion was significantly less organized than the original brain [21].

By means of immunocytochemistry, a high number of PAC1-receptor expressing cells were found in the regenerating blastema close to the cross-cut surface of the circumpharyngeal connectives. Most of them proved to be stem cells (neoblasts) of earthworms [21].

A recent finding that immune system of earthworms, in which PACAP-like peptides has an influence on the activity of various immune cells, can modify the effectiveness of regeneration, suggests that PACAP-like and other peptides have strong regulatory effect on cell differentiation and tissue development during the regeneration [22]. The importance of the PAC1 receptor expression in earthworm immunocytes was summarized recently [23].

Acknowledgement We gratefully acknowledge the support of the János Bolyai Research Scholarship of the Hungarian Academy of Sciences.

The present scientific contribution is dedicated to the 650th anniversary of the foundation of the University of Pecs, Hungary.

References

1. Cardoso JCR, Vieira FA, Gomes AS, Power DM. The serendipitous origin of chordate secretin peptide family members. BMC Evol Biol. 2010;10:135.
2. Herbert ZS, Zougman A, Pollak E, Boros A, Kapan N, Molnar L. Identification of novel neuropeptides in the ventral nerve cord ganglia and their targets in an annelid worm, *Eisenia fetida*. J Comp Neurol. 2009;514:415–32.
3. Bely AE. Early events in annelid regeneration: a cellular perspective. Integr Comp Biol. 2014;54:688–99.
4. Reglodi D, Szelier M, Lengvari I, Vigh S, Arimura A. PACAP38-like immunoreactivity in the nervous system of Oligochaeta. Peptides. 2000;21:183–8.
5. Molnar L, Pollak E, Boros A, Reglodi D, Tamas A, Lengvari I, et al. Comparative anatomy of PACAP-immunoreactive structures in the ventral nerve cord ganglia of lumbricid oligochaetes. Ann N Y Acad Sci. 2006;1070:427–30.
6. Somogyvari-Vigh A, Reglodi D, Li M, Lengvari I, Vigh S, Arimura A. Tissue distribution of PACAP27 and 38 in the Oligochaeta: PACAP27 is the dominant form in the nervous system of *Lumbricus polyphemus*. Peptides. 2000;21:1185–91.
7. Jamieson BGM. The ultrastructure of the oligochaeta. London: Academic; 1981. p. 245–67.
8. Boros A, Herbert ZS, Kiszler G, Nemeth J, Reglodi D, Lubics A, et al. Changes in the expression of PACAP-like compounds during the embryonic development of the earthworm *Eisenia fetida*. J Mol Neurosci. 2008;36:157–65.
9. Arimura A. Perspectives on pituitary adenylate cyclase activating polypeptide (PACAP) in the neuroendocrine, endocrine, and nervous systems. Jpn J Physiol. 1998;48:301–31.

10. Vaudry D, Gonzalez BJ, Basille M, Yon L, Fournier A, Vaudry H. Pituitary adenylate cyclase activating polypeptide and its receptors: from structure to functions. Pharmacol Rev. 2000;52:269–324.
11. Zhou CJ, Shioda S, Yada T, Inagaki N, Pleasure SJ, Kikuyama S. PACAP and its receptors exert pleiotropic effects in the nervous system by activating multiple signaling pathways. Curr Protein Pept Sci. 2002;3:423–39.
12. Molnar L, Pollak E, Boros A, Shioda S, Nakajo S, Tamas A, et al. PAC1 receptor localization in a model nervous system: light and electron microscopic immunocytochemistry on the earthworm ventral nerve cord ganglia. Regul Pept. 2008;145:96–104.
13. Devries J. The determination of the mesodermic metamerism specificity in the embryo of earthworm *Eisenia foetida* Sav. (effect of bromodeoxyuridine). Arch Biol (Bruxelles). 1983;94:23–40.
14. Boros A, Somogyi I, Engelmann P, Lubics A, Reglodi D, Pollak E, et al. Pituitary adenylate cyclase-activating polypeptide type 1 (PAC1) receptor is expressed during the embryonic development of the earthworm. Cell Tissue Res. 2010;339:649–53.
15. Morgan TH. Experimental studies of the internal factors of regeneration in the earthworm. Dev Genes Evol. 1902;14:562–91.
16. Varhalmi E, Somogyi I, Kiszler G, Nemeth J, Reglodi D, Lubics A, et al. Expression of PACAP-like compounds during the caudal regeneration of the earthworm *Eisenia fetida*. J Mol Neurosci. 2008;36:166–74.
17. Somogyi I, Boros A, Engelmann P, Varhalmi E, Nemeth J, Lubics A, et al. Pituitary adenylate cyclase activating polypeptide (PACAP)-like compounds could modulate the activity of coelomocytes in earthworm. Ann N Y Acad Sci. 2009;1163:521–3.
18. Liebmann E. New light on regeneration of *Eisenia fetida* (Sav.). J Morphol. 1943;73: 583–610.
19. Liebmann E. Experimental study on the role of the trophocytes in regeneration of *Eisenia foetida* (Sav.). Growth. 1945;9:43–54.
20. Wilhelm M, Koza A, Engelmann P, Nemeth P, Csoknya M. Evidence for the presence of thyroid-stimulating hormone, thyroglobulin and their receptors in *Eisenia fetida*: a multilevel hormonal interface between the nervous system and the peripheral tissues. Cell Tissue Res. 2006;324:535–46.
21. Steib A, Horvath B, Gunszt D, Somogyi I, Engelmann P, Nemeth J, et al. PACAP-dependent neural regeneration in the earthworm *Eisenia fetida*. J Mol Neurosci. 2013;51:230.
22. Molnar L, Pollak E, Skopek Z, Gutt E, Kruk J, Morgan AJ, et al. Immune system participates in brain regeneration and restoration of reproduction in the earthworm *Dendrobaena veneta*. Dev Comp Immunol. 2015;52:269–79.
23. Molnar L, Pollak E, Somogyi I, Engelmann P. On the existence of possible pituitary adenylate cyclase-activating polypeptide adenylate type 1 receptor in earthworms. Invertebrate Surviv J. 2015;12:173–5.

Chapter 4
PACAP and Learning in Invertebrates

Ildiko Kemenes and Gyorgy Kemenes

Abstract As one of the greatest achievements of medicine in the twentieth century, there has been a dramatic increase in life expectancy and this tendency continues even more rapidly in the twenty-first century. However, this positive achievement also has some negative consequences. The number of patients with clinical memory deficits, including those associated with Alzheimer's and Parkinson's diseases, is also on the rise and there is an urgent need for novel drugs and treatment methods. Neuropeptides are widely distributed in regions of the mammalian brain involved in learning and memory and represent targets of memory research. PACAP is one of the peptides that has been found to be vital in memory formation both in vertebrates and invertebrates and provides a promising avenue for research. The use of genetically and physiologically tractable invertebrate model systems will likely lead to the elucidation of fundamental cellular and molecular processes induced by PACAP, due to the numerical simplicity of the brain structures involved. The fruit fly (*Drosophila melanogaster*) and the pond snail (*Lymnaea stagnalis*) are two prominent invertebrate model systems that so far have provided valuable insights into the role of PACAP in ageing and memory-related processes, therefore this chapter concentrates on these species.

Keywords PACAP • Adenylate cyclase • Invertebrates • *Lymnaea* • *Drosophila* • *Amnesiac* • Classical conditioning • Learning • Memory • Ageing

Introduction

Since its discovery in 1989 [1] PACAP has been implicated in many physiological processes and has been demonstrated in several regions of the mammalian brain as well as other organs [2, 3]. However, much less is known about the expression and function of this peptide in invertebrates. Invertebrate homologues of the mammalian

I. Kemenes (✉) • G. Kemenes
Sussex Neuroscience, School of Life Sciences, University of Sussex, Brighton BN1 9QG, UK
e-mail: I.Kemenes@sussex.ac.uk

D. Reglodi, A. Tamas (eds.), *Pituitary Adenylate Cyclase Activating Polypeptide — PACAP*, Current Topics in Neurotoxicity 11,
DOI 10.1007/978-3-319-35135-3_4

PACAP peptide were first sequenced in the protochordates (tunicate/sea squirts) and it showed 96% identity with the bovine, human, and salmon PACAP(1–27) form [4]. The existence of PACAP peptide in the fruit fly (*Drosophila melanogaster*) [5] and in the fall armyworm (*Spodoptera frugiperda*) [6] was also recognised early after its discovery, followed by its identification in a number of other invertebrate species [7–11]. The fact that the homologue of the peptide is present in invertebrates as well as vertebrates indicates the evolutionarily highly conserved nature of PACAP. Molecular biological analyses indicate that there is little difference (only three or four amino acids) between the amino acid sequences of PACAP in vertebrate and invertebrate species (reviewed in [12]). Furthermore, PACAP activates similar molecular cascades leading to the execution of many physiological processes, including learning [10, 13]. Two invertebrate model systems, the fruit fly *Drosophila melanogaster* and the pond snail *Lymnaea stagnalis* have provided valuable insights into the evolutionarily conserved role of PACAP in memory formation and retention.

Drosophila melanogaster

Drosophila has been a favoured choice in the study of learning and memory due to the ease of training large numbers of animals combined with the possibility of genetic manipulation. A decade before the discovery of PACAP, Quinn et al. [14] identified a *Drosophila* memory mutant called "amnesiac", which learns normally but forgets rapidly. Experiments with the mutant showed that acquisition was not affected but memory retrieval was impaired. In 1995, Feany and Quinn [11] revisited their earlier experiments and investigated the idea that neuropeptides might be involved in memory storage and retrieval. By cloning the memory gene responsible for the amnesiac mutation they revealed a novel neuropeptide (AMN) that proved to be homologous to the mammalian PACAP. This was a very important step on the way towards elucidating the functional role of a peptide derived from the phenotype. In mammals there are a large number of well-characterised peptides and their receptors have been described but their physiological or behavioural role has not been described.

The identification of the gene mutation in a PACAP-like neuropeptide and its functional role in memory storage in *Drosophila* provided a new approach to dissect the functional role of specific neuropeptides in the brain. However, the identification of other mutations proved to be difficult at that time due to lack of efficient high-throughput behavioural testing methods and because the chemically mutated genes were not tagged, making cloning tedious. These methodological problems halted the progress of *Drosophila* research for several years.

In a more recent study, Waddell et al. [15] generated a specific antibody to detect the expression of the *amn* gene in the *Drosophila* brain. Surprisingly, they found that unlike all other proteins involved in olfactory learning that are preferentially expressed in the mushroom bodies (BM), AMN was strongly expressed in two large symmetrical neurons (DPM cells) outside these brain structures. These neurons however can strongly influence the mushroom bodies through their extensive projections. Genetically induced blockade of AMN neurosecretion from DPM cells shortened memory induced by shock-avoidance learning, similarly to the original *amn* mutation. Furthermore, memory in *amn* mutants can be restored by AMN expression in DPM neurons [15, 16]. Memory after appetitive conditioning, by pairing an odour with sugar, showed similar dependency on DPM neuron activity [17].

The original function of AMN is to act directly on the parts of the mushroom bodies where the expression of a gene (RUT), related to cyclase activity is abundant. Direct secretion of AMN peptide(s) into this region of the MB strengthens and potentiates behavioural associations that have been made by persistent activation of adenylyl cyclase. Experiments showed that DMP activity is needed for middle-term memory (60 min after training), suggesting that the PACAP-like AMN peptide release from the DPM neurons contributes to memory persistence.

Using the uas-shibirie^{ts1} transgene as a temperature sensitive blocker of DPM neurone function, Dubnau et al. [18] showed that DPM output is not required for the acquisition and storage of memory during the first 30 min after the training. Surprisingly, it was only retrieval that was affected indicating that the association of odour and shock and the storage of the olfactory memory both occur 'upstream' of the mushroom body. These experiments further support the notion that AMN is important in memory maintenance but not necessary during acquisition and early stages of memory formation. Surprisingly, blocking DPM output during another type of odour conditioning, pairing benzaldehyde (BA) with electric shock, resulted in blockade of BA memory indicating the requirement of DPM output during acquisition. This suggested that DPM neurons and presumably AMN peptides are differentially involved in odour memory. However, BA that has a bitter almond smell is a unique olfactory stimulant sensed by not just the olfactory system but several other sensory organs. This implies that *Drosophila* can use multiple neural pathways, similarly to the vertebrate trigeminal system to respond to several chemicals. This leads to the suggestion that DPM output is uniquely involved in memory acquisition of odours that require multisensory integration [18].

Waddell et al. [15] also showed that restoration of the *amn* gene expression in mutant DPM cells is also sufficient to re-establish wild-type memory throughout development but not when it is applied in the adult stage [19]. This suggests that a developmental defect in amnesiac mutants also affects the function of the DMP neuron and contributes to the expression of the mutant phenotype. These findings also indicate that PACAP is possibly required for normal brain function both during development and in adulthood.

Tamura et al. [16] further investigated this possibility and compared memory consolidation in flies of different ages. When old (more than 20 days) wild-type and *amn*-mutant animals were compared 3 h after a single training cycle of Pavlovian odour avoidance conditioning their memory retention was similarly impaired and was significantly lower than that of young (1-day-old) wild-type flies. This shows that memory retention in aged wild-type flies is similarly affected as those of the memory mutants, suggesting a link between age-dependent memory impairment and *amn*-dependent memory.

Lymnaea stagnalis

Observations concerning the role of PKA in memory consolidation in *Lymnaea* [20, 21] indicated that similar to other systems [22, 23], activation of adenylate cyclase (AC) is a key step in the formation of long-term memory (LTM). However, there was no information available on the molecules involved in the learning-induced activation of AC. A 2010 study then demonstrated both the presence and biochemical activity of a protein homologous to the vertebrate pituitary adenylate cyclase-activating polypeptide (PACAP) and the existence of its receptors in the *Lymnaea* nervous system [10]. Another study in 2010 showed that application of a PACAP receptor antagonist around the time of single-trial food-reward classical conditioning with a amyl-acetate as conditioned stimulus (CS) and sucrose as unconditioned stimulus (US) blocked associative LTM [24]. This latter finding suggested that in this 'strong' food-reward conditioning paradigm the activation of AC by PACAP was necessary for LTM to form. Interestingly, in a 'weak' multi-trial food-reward conditioning paradigm, lip touch paired with sucrose [25], memory formation was also dependent on PACAP [24]. Significantly, systemic application of PACAP at the beginning of multi-trial tactile conditioning accelerated the formation of transcription-dependent memory [24]. In PACAP-treated animals, robust LTM formed after just 3 trials, while control animals required >6 trials to form LTM. This memory-boosting effect of exogenously applied PACAP was blocked by the PACAP receptor antagonist PACAP6-38. These findings therefore provided the first evidence to show that in the same nervous system PACAP is both necessary and instructive for fast and robust memory formation after reward classical conditioning.

Pre-training application of the PACAP receptor antagonist resulted in a complete abolition of memory after both single-trial chemical and multi-trial tactile conditioning but not in a loss of the unconditioned feeding response. Based on this finding it is tempting to speculate that PACAP is released in response to the chemical and tactile conditioned stimuli, whereas similar to what was found in *Aplysia*, the effect of the unconditioned stimulus on AC may be mediated by different peptide or non-peptide transmitters, such as SCPs or 5-HT [26–28]. A likely scenario is that in *Lymnaea*, the PACAP-mediated effect of the chemical or tactile CS and the non-PACAP mediated effect of sucrose US converge on AC and this convergence provides the molecular basis for coincidence detection, a fundamental requirement for associative learning. *Lymnaea* is known to differentiate learning with amyl-acetate [29] from learning with touch [30] at the neuronal level within the same network (the feeding circuitry in this case), but there is no evidence for a similar differentiation at the molecular level within the same neuron. Thus, the same molecules (e.g. PACAP) can fulfil the same role (e.g. activation of AC) in different neurons leading to learning-induced changes in different pathways (e.g. activated by touch versus activated by amyl-acetate).

More recently, it also has been shown that the age-related decline in memory performance observed previously in *Lymnaea* [31, 32] can be reversed by systemic administration of PACAP [33]. Exogenous PACAP38 boosts memory formation in aged *Lymnaea*, where endogenous PACAP38 levels are low in the brain [33]. Treatment with insulin-like growth factor-1 (IGF-1), which in vertebrates was shown to transactivate PACAP type I (PAC1) receptors [34] also boosts memory formation in aged pond snails [33]. Due to the evolutionarily conserved nature of these polypeptides and their established role in memory and synaptic plasticity there is a very high probability that they could also act as 'memory rejuvenating' agents in humans.

Conclusions

Learning occurs in all animal species, from invertebrates to humans, as a result of associative or non-associative conditioning. In general, learning involves the activation of neural pathways responsible for the transmission of sensory information to sensory- to motor-neuron synapses and through modification of these synapses or the intrinsic properties of neurons results in altered neuronal and behavioural responses to the original sensory stimulus. In the case of classical conditioning, the originally neutral conditioned stimulus (CS) will elicit a response after it has been associated with the unconditioned stimulus (US). However, the mechanisms of this association were not well understood. The findings on the role of PACAP in learning and memory in *Drosophila* and *Lymnaea* now allow us to propose that it plays a key role in the initial steps of information processing during learning. As indicated in the proposed model in Fig. 4.1. PACAP can activate

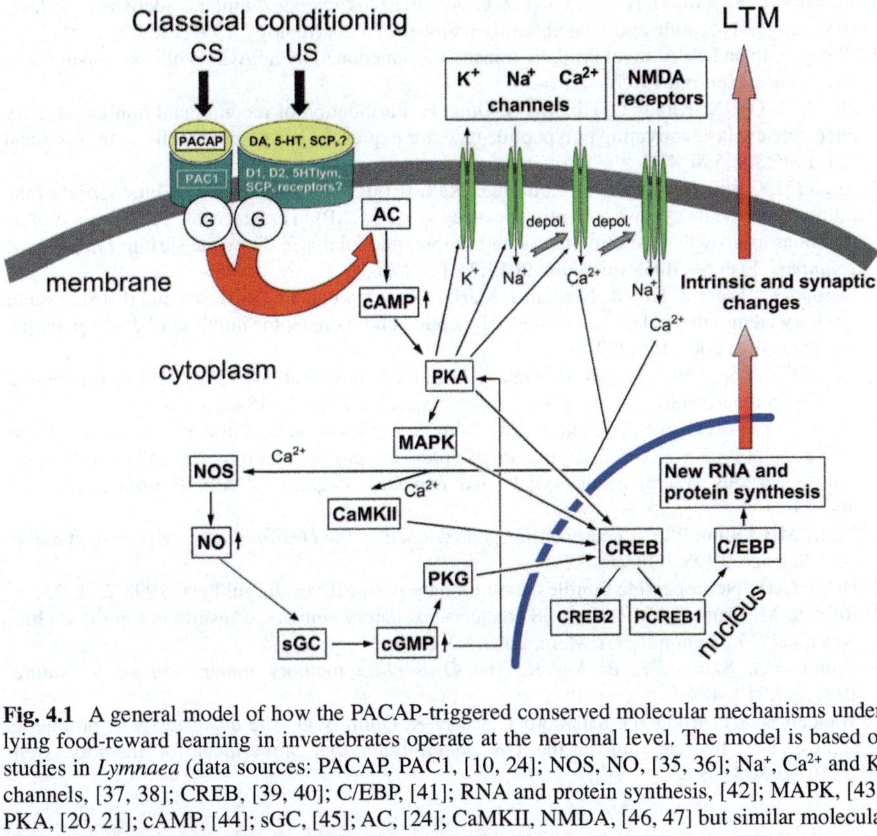

Fig. 4.1 A general model of how the PACAP-triggered conserved molecular mechanisms underlying food-reward learning in invertebrates operate at the neuronal level. The model is based on studies in *Lymnaea* (data sources: PACAP, PAC1, [10, 24]; NOS, NO, [35, 36]; Na^+, Ca^{2+} and K^+ channels, [37, 38]; CREB, [39, 40]; C/EBP, [41]; RNA and protein synthesis, [42]; MAPK, [43]; PKA, [20, 21]; cAMP, [44]; sGC, [45]; AC, [24]; CaMKII, NMDA, [46, 47] but similar molecular cascades underlie food-reward LTM in *Drosophila* [48, 49]

AC via the CS pathway and when this happens coincidentally with the activation of the US pathway also acting on AC, the induced molecular cascades can result in modification of neuronal activity that underlies the formation of long-term memory.

Acknowledgments I.K.'s original research has been funded by the Biology and Biotechnology Research Council (BBSRC, UK). G.K.'s original research was funded by the Medical Research Council (MRC, UK).

References

1. Miyata A, Arimura A, Dahl RR, Minamino N, Uehara A, Jiang L, et al. Isolation of a novel 38 residue-hypothalamic polypeptide which stimulates adenylate cyclase in pituitary cells. Biochem Biophys Res Commun. 1989;16(164):567–74.
2. Basille M, Vaudry D, Coulouarn Y, Jegou S, Lihrmann I, Fournier A, et al. Comparative distribution of pituitary adenylate cyclase-activating polypeptide (PACAP) binding sites and PACAP receptor mRNAs in the rat brain during development. J Comp Neurol. 2000;425:495–509.
3. Vaudry D, Falluel-Morel A, Bourgault S, Basille M, Burel D, Wurtz O, et al. Pituitary adenylate cyclase-activating polypeptide and its receptors: 20 years after the discovery. Pharmacol Rev. 2009;61:283–357.
4. McRory J, Sherwood NM. Two protochordate genes encode pituitary adenylate cyclase-activating polypeptide and related family members. Endocrinology. 1997;138:2380–90.
5. Zhong Y, Pena LA. A novel synaptic transmission mediated by a PACAP-like neuropeptide in *Drosophila*. Neuron. 1995;14:527–36.
6. Ohtaki T, Ogi K, Kitada C, Hinuma S, Onda H. Purification of recombinant human pituitary adenylate cyclase-activating polypeptide receptor expressed in Sf9 insect cells. Ann N Y Acad Sci. 1996;805:590–4.
7. Lugo JM, Carpio Y, Morales R, Rodriguez-Ramos T, Ramos L, Estrada MP. First report of the pituitary adenylate cyclase activating polypeptide (PACAP) in crustaceans: conservation of its functions as growth promoting factor and immunomodulator in the white shrimp *Litopenaeus vannamei*. Fish Shellfish Immunol. 2013;35:1788–96.
8. Hernádi L, Pirger Z, Kiss T, Németh J, Márk L, Kiss P, et al. The presence and distribution of pituitary adenylate cyclase activating polypeptide and its receptor in the snail *Helix pomatia*. Neuroscience. 2008;155:387–402.
9. Reglődi D, Lengvári I, Szelier M, Vigh S, Arimura A. Distribution of PACAP-like immunoreactivity in the nervous system of oligochaeta. Peptides. 2000;21:183–8.
10. Pirger Z, László Z, Hiripi L, Hernádi L, Tóth G, Lubics A, et al. Pituitary adenylate cyclase activating polypeptide (PACAP) and its receptors are present and biochemically active in the central nervous system of the pond snail *Lymnaea stagnalis*. J Mol Neurosci. 2010;42: 464–71.
11. Feany MB, Quinn WG. A neuropeptide gene defined by the *Drosophila* memory mutant *amnesiac*. Science. 1995;268:869–73.
12. Hoyle CHV. Neuropeptide families: evolutionary perspectives. Regul Pept. 1998;73:1–33.
13. Roberto M, Brunelli M. PACAP-38 enhances excitatory synaptic transmission in the rat hippocampal CA1 region. Learn Mem. 2000;7:303–11.
14. Quinn WG, Sziber PP, Booker R. The *Drosophila* memory mutant *amnesiac*. Nature. 1979;277:212–4.
15. Waddell S, Armstrong JD, Kitamoto T, Kaiser K, Quinn WG. The *amnesiac* gene product is expressed in two neurons in the *Drosophila* brain that are critical for memory. Cell. 2000;103:805–13.

16. Tamura T, Chiang AS, Ito N, Liu HP, Horiuchi J, Tully T, et al. Aging specifically impairs *amnesiac*-dependent memory in *Drosophila*. Neuron. 2003;40:1003–11.
17. Keene AC, Krashes MJ, Leung B, Bernard JA, Waddell S. *Drosophila* dorsal paired medial neurons provide a general mechanism for memory consolidation. Curr Biol. 2006;16:1524–30.
18. Dubnau J, Grady L, Kitamoto T, Tully T. Disruption of neurotransmission in *Drosophila* mushroom body blocks retrieval but not acquisition of memory. Nature. 2001;411:476–80.
19. DeZazzo J, Xia S, Christensen J, Velinzon K, Tully T. Developmental expression of an *amn*(+) transgene rescues the mutant memory defect of amnesiac adults. J Neurosci. 1999;19:8740–6.
20. Michel M, Kemenes I, Müller U, Kemenes G. Different phases of long-term memory require distinct temporal patterns of PKA activity after single-trial classical conditioning. Learn Mem. 2008;15(9):694–702.
21. Kemenes G, Kemenes I, Michel M, Papp A, Müller U. Phase-dependent molecular requirements for memory reconsolidation: differential roles for protein synthesis and protein kinase A activity. J Neurosci. 2006;26:6298–302.
22. Kandel ER. The molecular biology of memory storage: a dialogue between genes and synapses. Science. 2001;294:1030–8.
23. Margulies C, Tully T, Dubnau J. Deconstructing memory in *Drosophila*. Curr Biol. 2005;15:R700–13.
24. Pirger Z, László Z, Kemenes I, Tóth G, Reglődi D, Kemenes G. A homolog of the vertebrate pituitary adenylate cyclase-activating polypeptide is both necessary and instructive for the rapid formation of associative memory in an invertebrate. J Neurosci. 2010;30:13766–73.
25. Kemenes G, Benjamin PR. Appetitive learning in snails shows characteristics of conditioning in vertebrates. Brain Res. 1989;489:163–6.
26. Trudeau LE, Castellucci VF. Contribution of polysynaptic pathways in the mediation and plasticity of *Aplysia* gill and siphon withdrawal reflex: evidence for differential modulation. J Neurosci. 1992;12:3838–48.
27. Nargeot R, Baxter DA, Patterson GW, Byrne JH. Dopaminergic synapses mediate neuronal changes in an analogue of operant conditioning. J Neurophysiol. 1999;81:1983–7.
28. Brembs B, Lorenzetti FD, Reyes FD, Baxter DA, Byrne JH. Operant reward learning in *Aplysia*: neuronal correlates and mechanisms. Science. 2002;296:1706–9.
29. Kemenes I, Straub VA, Nikitin ES, Staras K, O'Shea M, Kemenes G, et al. Role of delayed nonsynaptic neuronal plasticity in long-term associative memory. Curr Biol. 2006;16:1269–79.
30. Jones NG, Kemenes I, Kemenes G, Benjamin PR. A persistent cellular change in a single modulatory neuron contributes to associative long-term memory. Curr Biol. 2003;13:1064–9.
31. Hermann PM, Lee A, Hulliger S, Minvielle M, Ma B, Wildering WC. Impairment of long-term associative memory in aging snails (*Lymnaea stagnalis*). Behav Neurosci. 2007;121:1400–14.
32. Audesirk TE, Alexander Jr JE, Audesirk GJ, Moyer CM. Rapid, nonaversive conditioning in a freshwater gastropod. I. Effects of age and motivation. Behav Neural Biol. 1982;36:379–90.
33. Pirger Z, Naskar S, László Z, Kemenes G, Reglődi D, Kemenes I. Reversal of age-related learning deficiency by the vertebrate PACAP and IGF-1 in a novel invertebrate model of aging: the pond snail (*Lymnaea stagnalis*). J Gerontol A Biol Sci Med Sci. 2014;69:1331–8.
34. Delcourt N, Thouvenot E, Chanrion B, Galeotti N, Jouin P, Bockaert J, et al. PACAP type I receptor transactivation is essential for IGF-1 receptor signalling and antiapoptotic activity in neurons. EMBO J. 2007;26:1542–51.
35. Kemenes I, Kemenes G, Andrew RJ, Benjamin PR, O'Shea M. Critical time-window for NO-cGMP-dependent long-term memory formation after one-trial appetitive conditioning. J Neurosci. 2002;22:1414–25.
36. Korneev SA, Straub V, Kemenes I, Korneeva EI, Ott SR, Benjamin PR, et al. Timed and targeted differential regulation of nitric oxide synthase (NOS) and anti-NOS genes by reward conditioning leading to long-term memory formation. J Neurosci. 2005;25:1188–92.
37. Staras K, Győri J, Kemenes G. Voltage-gated ionic currents in an identified modulatory cell type controlling molluscan feeding. Eur J Neurosci. 2002;15:109–19.
38. Nikitin ES, Vavoulis DV, Kemenes I, Marra V, Pirger Z, Michel M, et al. Persistent sodium current is a nonsynaptic substrate for long-term associative memory. Curr Biol. 2008;18: 1221–6.

39. Ribeiro MJ, Serfőző Z, Papp A, Kemenes I, O'Shea M, Yin JC, et al. Cyclic AMP response element-binding (CREB)-like proteins in a molluscan brain: cellular localization and learning-induced phosphorylation. Eur J Neurosci. 2003;18:1223–34.
40. Sadamoto H, Sato H, Kobayashi S, Murakami J, Aonuma H, Ando H, et al. CREB in the pond snail *Lymnaea stagnalis*: cloning, gene expression, and function in identifiable neurons of the central nervous system. J Neurobiol. 2004;58:455–66.
41. Hatakeyama D, Fujito Y, Sakakibara M, Ito E. Expression and distribution of transcription factor CCAAT/enhancer-binding protein in the central nervous system of *Lymnaea stagnalis*. Cell Tissue Res. 2004;318:631–41.
42. Fulton D, Kemenes I, Andrew RJ, Benjamin PR. A single time-window for protein synthesis-dependent long-term memory formation after one-trial appetitive conditioning. Eur J Neurosci. 2005;21:1347–58.
43. Ribeiro MJ, Schofield MG, Kemenes I, O'Shea M, Kemenes G, Benjamin PR. Activation of MAPK is necessary for long-term memory consolidation following food-reward conditioning. Learn Mem. 2005;12:538–45.
44. Nikitin ES, Kiss T, Staras K, O'Shea M, Benjamin PR, Kemenes G. Persistent sodium current is a target for cAMP-induced neuronal plasticity in a state-setting modulatory interneuron. J Neurophysiol. 2006;95:453–63.
45. Ribeiro M, Straub VA, Schofield M, Picot J, Benjamin PR, O'Shea M, et al. Characterization of NO-sensitive guanylyl cyclase: expression in an identified interneuron involved in NO-cGMP-dependent memory formation. Eur J Neurosci. 2008;28:1157–65.
46. Wan H, Mackay B, Iqbal H, Naskar S, Kemenes G. Delayed intrinsic activation of an NMDA-independent CaM-kinase II in a critical time window is necessary for late consolidation of an associative memory. J Neurosci. 2010;30:56–63.
47. Naskar S, Wan H, Kemenes G. pT305-CaMKII stabilizes a learning-induced increase in AMPA receptors for ongoing memory consolidation after classical conditioning. Nat Commun. 2014;5:3967.
48. Bhattacharya A, Lakhman SS, Singh S. Modulation of L-type calcium channels in *Drosophila* via a pituitary adenylyl cyclase-activating polypeptide (PACAP)-mediated pathway. J Biol Chem. 2004;279:37291–7.
49. Zhou CJ, Shioda S, Yada T, Inagaki N, Pleasure SJ, Kikuyama S. PACAP and its receptors exert pleiotropic effects in the nervous system by activating multiple signaling pathways. Curr Protein Pept Sci. 2002;3:423–39.

Part II
Effects of PACAP in Neuronal Development

Chapter 5
PACAP and Neural Progenitor Cells

Dan Lindholm, Johanna Mäkelä, and Laura Korhonen

Abstract PACAP is a neuropeptide with a multitude of functions on different cell types and organs including brain tissue. PACAP is relatively highly expressed in embryonic brain indicating a role in neuronal development. Particularly PACAP is expressed in the subventricular zone of lateral ventricles and in the dentate gyrus of hippocampus harboring the neural stem/progenitor cells (NPCs) that give rise to new neurons and glial cells in the developing brain. PACAP is known to stimulate the PACAP-1 receptors (PAC1R) that are expressed by the progenitor and other cells in the brain. Here we will shortly discuss the current view about the role of PACAP in NPCs and how PACAP affects cell proliferation, differentiation, specification, migration and survival of these cells as have been studied mainly in rodent brain. Available data suggests that PACAP acts in concert with other peptides, growth factors and other signaling molecules in governing the behavior of the NPCs both during development and in the adult brain. A better understanding about the action of PACAP in stem cells and its interactions with other factors will be helpful for the potential use of PACAP in treatment of different brain disorders and for understanding of neural repair mechanisms.

Keywords PACAP • PACAP receptor • cAMP • Protein kinase C • Progenitor cells • Cell proliferation • Cell differentiation • Neurogenesis • Brain development

D. Lindholm (✉) • J. Mäkelä
Medicum, Department of Biochemistry and Developmental Biology, Faculty of Medicine, University of Helsinki, Haartmaninkatu 8, 00290 Helsinki, Finland

Minerva Medical Research Institute, Biomedicum-2 Helsinki, Tukholmankatu 8, 00290 Helsinki, Finland
e-mail: dan.lindholm@helsinki.fi

L. Korhonen
Medicum, Department of Biochemistry and Developmental Biology, Faculty of Medicine, University of Helsinki, Haartmaninkatu 8, 00290 Helsinki, Finland

Minerva Medical Research Institute, Biomedicum-2 Helsinki, Tukholmankatu 8, 00290 Helsinki, Finland

Clinicum, Division of Child Psychiatry, Helsinki University Central Hospital, Helsinki, Finland

D. Reglodi, A. Tamas (eds.), *Pituitary Adenylate Cyclase Activating Polypeptide — PACAP*, Current Topics in Neurotoxicity 11,
DOI 10.1007/978-3-319-35135-3_5

Abbreviations

DCX	Doublecortin
DREAM	Downstream regulatory element antagonist modulator
ES	Embryonic stem
GFP	Green fluorescent protein
hMSCs	Human mesenchymal stem cells
iPS	Induced pluripotent stem
NPCs	Neural stem/progenitor cells
OLPs	Oligodendrocyte progenitor cells
PAC1R	PACAP receptor-1
PACAP	Pituitary adenylate cyclase-activating polypeptide
PKA	Protein kinase A
PKC	Protein kinase C
Ptc1	Patched receptor
RSM	Rostral migratory stream
Shh	Sonic hedgehog
VIP	Vasoactive intestinal peptide (VIP)

PACAP and Neural Progenitor Cells

Neural stem/progenitor cells (NPCs) are important constituents of the developing and mature brain contributing to neurogenesis, cell specification, neuronal plasticity, as well as glial cell functions and brain repair and regeneration in disease [1–3]. In the brain, NPCs are present in distinct neurogenic regions in the ventral neuroepithelium and in the dentate gyrus of the hippocampus [1]. Neurogenesis is particularly active during embryonic development but the two neurogenic regions are also present in the adult brain [2, 4]. The NPCs are part of a local microenvironment that forms a self-supporting niche in the brain that regulates stem cell proliferation and maintenance by the action of extrinsic and intrinsic factors [4, 5]. It is known that a number of trophic and other factor influence NPCs with some factors preferentially inducing neurogenesis whereas others affect mainly glial cell formation [2, 4, 5]. These factors have distinct signaling pathway with a complex interplay in the regulation of NPCs that is only partly understood. The elucidation of these pathways and their potential roles in different brain diseases represents major objectives of current studies on NPCs. PACAP acts via the PAC1-receptor (PAC1R) and may activate both cAMP/protein kinase A (PKA) and phospholipase C/protein kinase C (PKC) pathways that are major signaling pathways in NPCs and in neurons. Accumulated data also show that PACAP and PAC1R are expressed in the neurogenic regions indicating that PACAP plays a role in NPCs either alone or in conjunction with other factors. In the following, we will review the present literature on the action of PACAP in NPCs and in stem cells. Although many aspects of PACAP in stem cells biology have been covered, more experiments are clearly needed, using also gene-deleted mice and animal models, to define the underlying mechanisms by which PACAP and PAC1R influence NPCs and during development and adulthood and in different brain diseases.

PACAP and PACR1 Are Expressed in Developing Brain and in Neurogenic Regions

After its initial discovery in the hypothalamus [6] it became clear that PACAP is also present in other parts of the developing and mature nervous system (reviewed in [7]). In particular, in situ hybridization and other experiments revealed that PACAP is expressed in several regions of the adult brain [8] as well as in developing rat brain including the neurogenic regions, the ventral neuroepithelium, and the dentate gyrus of the hippocampus [9]. In addition, the high-affinity receptor PAC1R that is specific for PACAP was also expressed in adult [10] as well as in the developing rodent brain and with a distribution that partly overlaps that of PACAP [11]. The results of these studies localizing PACAP expression in brain tissue have recently been largely confirmed by the GENSAT (Gene Expression Nervous System Atlas) project at the Rockefeller University (see www.gensat.org). In this project, BAC-PACAP-EGFP mice were analyzed for expression of green fluorescent protein (GFP) under the endogenous PACAP promoter in BAC transgenic mice. Data showed that in these mice the neuroepithelium, parts of the hippocampus, as well as several other brain regions contain GFP positive cells indicative of PACAP expression. Particularly, there was a robust expression of PACAP in the neurogenic regions at early stages with the labeling of NPCs (www.gensat.org). Moreover, PACAP is abundant and show a more widespread expression in developing brain compared with vasoactive intestinal peptide (VIP). BAC-GFP mice for PAC1R are currently not available from GENSAT, but ample evidence suggests that different isoforms of PACR1 are expressed by NPCs and by other stem cells. Immunostaining experiments further revealed that PACR1 is expressed in the rostral migratory stream (RSM) that contains progenitor cells *en route* to the olfactory bulb. More specifically, both nestin and glial fibrillary acidic protein positive cells as well as differentiated doublecortin (DCX) and NeuN-positive neural cells were PACR1 positive in the RSM [12].

PACAP Influences NPCs and Neurogenesis

In the adult mouse, administration of PACAP stimulated the proliferation of NPCs in the subventricular zone of lateral ventricles and in the dentate gyrus of hippocampus shown by BrdU labeling [13]. Similar results were reported by Ohta et al. after intracerebroventricular infusions of PACAP into the adult forebrain [14]. In neurosphere cultures, the addition of PACAP or maxadilan, an agonist of the PAC1R, stimulated adult mouse NPCs proliferation through PKC-dependent pathway. Furthermore, PACAP acted synergistically with epidermal growth factor that is used as a mitogen for the NPCs. In addition, PACAP-stimulated NPCs had the capacity to generate both neuron or glial cells indicating that there was no clear bias to a particular cell fate after treatment with PACAP [13].

Hippocampal neurogenesis is known to be influenced by a variety of intrinsic and extrinsic factors. Rearing of mice in an enriched environment showed an increase in neurogenesis of both wild-type and PACAP gene-deleted mice with no obvious change in the basal rate of proliferation between the groups [15]. However, the survival of newly divided cells in the dentate gyrus of hippocampus was decreased in the PACAP knockout mice compared with controls [15]. This shows that endogenous PACAP is involved in the effects of enriched environment on neurogenesis in the hippocampus probably acting together with other factors. Collectively these results show that PACAP is able to stimulate proliferation, neurogenesis, as well as cell survival of NPCs in the adult brain that is of importance for brain repair and neuronal functions.

Recent data indicate that PACAP can influence NPCs by affecting cell migration in the brain. As discussed earlier, PAC1Rs are present in DCX positive cells [12] representing immature neuroblast migrating in the RSM. It was recently shown that PACAP induce phosphorylation of DCX via cAMP regulating microtubule and actin dynamics leading to increased neuroblast migration [16]. PACAP also takes part in neuronal laminar formation in the developing cortex through the activation of cAMP-PKA signaling [17]. In the postnatal cerebellum (see later), PACAP influences the migration of developing granule cells [18].

Effects of PACAP on Gliogenesis

PACAP has in several experiments been shown to selectively induce astrocyte differentiation of cultured cortical NPCs [19–21]. Elevation of cAMP was found to be essential for the PACAP-mediated stimulation of astrogenesis [20, 21]. In contrast, it was reported that particularly the β isoform of PKC, not directly cAMP, is crucially involved in differentiation of embryonic NSCs into astrocytes [22]. The reason behind these seemingly different results regarding the signaling cascades elicited by PACAP is not immediately obvious but could be due related to changes in cell composition or culture conditions used for the NPCs. Another aspect is that PACAP can increase calcium in cells both via a PKA and a PKC signaling pathway. PACAP acts also in concert with other signals and factors to regulate NPCs including ciliary neurotrophic factor and the bone morphogenetic proteins that are known to stimulate astrogenesis [4, 5]. It is not fully understood how PACAP-PAC1R interacts with the other factors and signaling pathways in the NPCs to either inhibit or promote astrogenesis and cell proliferation. A simplified scheme of some of these interactions for PACAP is given in Fig. 5.1 with regard to control of cell proliferation in NPCs (Fig. 5.1).

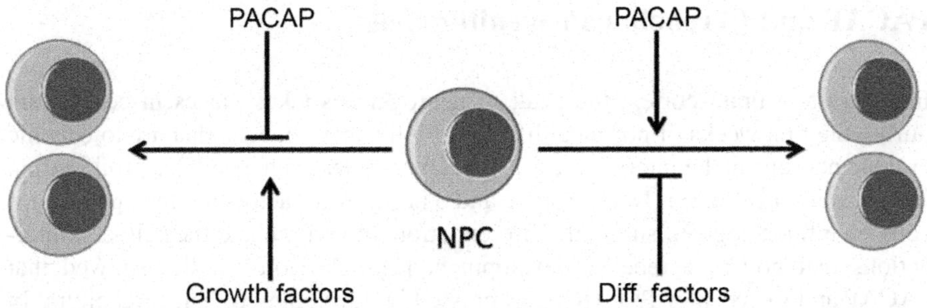

Fig. 5.1 Role of PACAP in regulation of NPC proliferation. A general theme for the action of PACAP in neurogenesis is that the neuropeptide acts together with other factors (growth factors, trophic factors, peptides, transmitters) to regulate cell proliferation and differentiation of NPCs. PACAP may inhibit the action of growth factors to inhibit (*left*) or alone stimulate cell proliferation (*right*). For cell differentiation similar schemes can be envisaged. More examples for these are discussed in the text. PACAP acts via the PAC1R that can exist in different isoforms in the NPCs and then signal either via the cAMP/PKA or PKA pathway in the progenitor cells. The precise signaling cascades for PACAP and the crosstalk with other factors in the NPCs are not fully understood

With regard to potential mechanisms underlying gliogenesis, it has been shown that PACAP can activate the small GTPases Rap1 and Ras causing increased calcium and to stimulation of the transcription factor, downstream regulatory element antagonist modulator (DREAM) in the NPCs [23]. DREAM in turn can bind to the promoter of the *GFAP* gene increasing its expression in differentiating astrocytes [24]. Regulation of cell differentiation is a complex process and it is likely that additional factors and signaling molecules are involved in determining the final outcome of PACAP signaling in the NPCs and in gliogenesis.

Role of PAC1R Isoforms in NPCs

Cloning of the PAC1R showed that the receptor is expressed as different isoforms arising from differential splicing [25]. The nature of the PAC1R isoform expressed by NPCs plays a crucial role in determining the effects of PACAP on cell proliferation and differentiation. Thus, sympathetic neuroblasts from the peripheral nervous system express mainly the hop subtype of PAC1R and respond to PACAP with an increase in cell proliferation subsequent to activation of PKC. In contrast, cultured cortical progenitors express mainly the short PAC1R and show a decrease in cell proliferation after PACAP that is due to enhanced cAMP/PKA signaling [26]. The antiproliferative effect of PACAP in the cortical progenitor cells was further related to increases in the cell cycle regulator, p57Kip2 [27]. As shown in cerebellar granule neurons, PACAP also increases the protein Lot1 in progenitor cells that can regulate the cell cycle (see later). It would be interesting to know whether there is a switch in the expression of the different PAC1R isoforms in vivo and how this may be regulated by various factors and under different conditions.

PACAP and Cerebellar Progenitor Cells

In contrast to brain cortex, the peak of neurogenesis takes places in cerebellum during the first weeks of postnatal life. During this period, cells that are to become granule neurons in the internal granule cell layer undergo an extensive proliferation in the external cell layer. Neurotrophic and other growth factors regulate progenitor cells by influencing cell survival, differentiation, migration, and the cell–cell interactions that govern cerebellar development [28]. Previous studies showed that PACAP and its receptor PAC1R are expressed in developing rodent cerebellum. In particular, in situ hybridization showed that PACAP was present in the developing Purkinje cells, whereas PAC1R was mainly expressed by the adjacent cells [29]. PACR1 is also expressed in the external cell layer harboring the progenitor cells during the peak of neurogenesis [29]. In adult, the expression of PACR1 and PACAP is decreased in the cerebellum [9–11].

Studies performed with cultured granule cells from cerebellum have further demonstrated that PACAP interacts with other signals to increase or inhibit cell proliferation. Sonic Hedgehog (Shh) is an important growth factor in the cerebellum and mutations in its signaling pathway are associated with occurrence of tumors such as medulloblastomas in children. Addition of PACAP to cerebellar granule progenitor cells was shown to reduce the Shh-induced cell proliferation in a cAMP-dependent manner [30–32]. This shows that there is a functional interaction between signals from PACAP-PAC1R and from the Shh receptor, Patched (ptc1) in the regulation of granule cell proliferation in developing cerebellum. Interestingly, in the absence of Shh, PACAP was found to modestly increase granule cell proliferation in a signaling cascade involving PKC [30]. This situation is similar to that observed with PACAP and cortical NPCs and indicates that it is the combined effects of different factors that ultimately determine the outcome of PACAP signaling in NPCs with regard to cell proliferation (Fig. 5.1). In view of its antiproliferative effect, PACAP may play a role in suppressing tumor formation in cerebellum by controlling Shh signaling. In an elegant study using single copy gene deleted *PACAP* mice mated with *ptc1* mutant mice, it was shown that the incidence of medulloblastoma tumors increased significantly compared with the corresponding control [33].

In addition to cell proliferation, PACAP can promote the survival of cultured granule progenitor cells [34]. More recent data have shown that PACAP protects cerebellar progenitor cells against apoptosis induced by the sphingolipid-derivate, ceramide partly by inhibiting Bax and caspase-3 activation [35]. Regarding potential mechanisms for increased cell survival, it has been shown that PACAP increases c-fos expression [34] as well as intracellular calcium in the cerebellar granule cells via activation of N-type calcium channels and by mobilization of cell calcium stores [36]. In addition, the tumor suppressor gene, *Lot1* was found to be important for the antiproliferative and cell survival promoting activities of PACAP in the cerebellar progenitor cells [37].

Taken together, these results support a role for PACAP and PAC1R in cerebellar development with major effects observed on cerebellar progenitor cells [38]. More credence for this view comes from studies of PACAP gene-deleted mice showing a reduction in the thickness of the external and internal granule cell layer at postnatal ages P4 and P7, respectively [39]. The activity of caspase-3 was also increased in the *PACAP* gene-deleted mice, revealing an enhancement of the naturally occurring cell death in the postnatal cerebellum. Subsequent studies have shown that PACAP counteracts the effect of the death-factor Fas ligand on caspase-3 and granule neuron cell death during cerebellar morphogenesis [40].

Effects of PACAP on Oligodendrocyte and Retinal Progenitor Cells

PACAP has been studied in the context of cultured oligodendrocyte progenitor cells (OLPs) that are known to express PAC1Rs. Furthermore, PAC1R is present in the optic chiasm that is a region rich in OLPs [41]. Addition of PACAP to OLP cultures had a dual effect increasing cell proliferation and decreasing myelination induced by differentiated oligodendrocytes [41]. Subsequent studies showed that PACAP acts in conjunction with other factors in influencing OLPs by enhancing the stimulatory effect of FGF2 on cell proliferation but inhibiting that of PDGF [42]. PACAP also antagonized the effect of Shh in OLPs, suggesting a complex crosstalk between PACAP and growth factors in the regulation of OLPs. So far few data on human OLP cells are available with regard to the expression or the effects of PACAP. In view of its role in OLPs, PACAP could play a role in brain diseases affecting myelin and the oligodendrocytes as observed, for example, in multiple sclerosis.

PACAP and PC1R are expressed in developing retina and PACAP exhibited a neuroprotective effect on retinal cells. It was further shown that PACAP regulates the proliferation of retinal progenitor cells through activation of the cAMP/PKA pathway [43]. The possible roles played by PACAP in retina disorders remain to be studied further.

Effects of PACAP on ES Cells and Human iPS Cells

In addition to NPCs, PACAP has been shown to influence other types of stem and progenitor cells. Using mouse embryonic stem (ES) cells cultured as embryoid bodies, it was shown that PACAP induced neuronal differentiation from cell aggregates after 1 week of treatment [44]. Similar effects were also observed with VIP and mouse ES expressing both PAC1 and VPAC2 receptors. A more close analysis revealed that PACR1 was upregulated by differentiation of ES cells into neurons and decreased after glial differentiation [45]. On the functional side it was shown that stimulations with PACAP or with VIP enhanced calcium currents in the ES cells partly by inducing particular calcium channel proteins [46]. PACAP also counteracted the Shh-mediated differentiation of mouse ES cells into motor neurons in culture, although this was not observed in spinal cord in vivo [47].

Human induced pluripotent stem (iPS) cells show great promises for modern medicine and for cell therapies in different diseases [48, 49]. Recent studies have shown the expression of PAC1R in human iPS cells. Stimulation with maxadilan activating the PAC1R protected the human iPS cells against cell death induced by ultraviolet light [50]. Moreover, the pluripotent state and the karyotype of the iPS cells were not influenced by maxadilan, indicating a specific effect of PACR1 activation on cell survival. The precise signaling and physiological roles of PACAP in human iPS cells warrant to be studied further.

Conclusion and Therapeutic Use of PACAP and Stem Cells

As shown earlier PACAP can influence NPCs and other stem/progenitor cell types in a variety of ways including effects on progenitor cell proliferation, differentiation, migration, and survival. Some of these actions of PACAP can be clearly beneficial for treatment of brain disease and for neuronal recovery in different clinical settings. Recent data have shown that administration of PACAP indirectly can influence brain repair by affecting homing of bone marrow-derived cells into the ischemic brain that can reduce effects of injury and promote neuronal recovery [51]. Neuroinflammation is a characteristic of several brain diseases and PACAP has a strong anti-inflammatory effect partly by influencing microglia cells. In this respect it was shown that PACAP can counteract the deleterious effects of cytokines on NPCs by interfering with proinflammatory cytokine interferon-γ [52]. Similarly PACAP was shown to protect NPCs against lipotoxicity by enhancing cell viability through activation of PAC1R [53]. In stroke, transplantation of PACAP-producing stem cells into mouse brain 3 days after injury produced functional recovery as shown by behavioral tests [54]. Furthermore, the beneficial effect of PACAP was associated with reduced brain inflammation and a change in microglia cells toward the neuroprotective M2 phenotype [54]. PACAP also enhanced repair mechanisms in spinal cord trauma in combination with the transplantation of human mesenchymal stem cells (hMSCs) [55]. Along the same line it was recently shown that transplanted hMSCs suppress inflammation after spinal cord injury in a crosstalk involving PACAP [56].

An increased understanding about the roles of PACAP and PAC1Rs in NPCs and other stem cells may pave the way for better therapies using stem cells and PACAP in different brain diseases.

Acknowledgement We thank Linnéa Lindholm for help with the editing of the manuscript. Supported by Parkinson Foundation, Finska Läkaresällskapet, Magnus Ehrnrooth Foundation, Liv och Hälsa Foundation, Minerva Foundation, Sigrid Juselius Foundation, and the Academy of Finland.

References

1. Decimo I, Bifari F, Krampera M, Fumagalli G. Neural stem cell niches in health and diseases. Curr Pharm Des. 2012;18:1755–83.
2. Taupin P, Gage FH. Adult neurogenesis and neural stem cells of the central nervous system in mammals. J Neurosci Res. 2002;69:745–9.
3. Temple S. The development of neural stem cells. Nature. 2001;414:112–7.
4. Kriegstein A, Alvarez-Buylla A. The glial nature of embryonic and adult neural stem cells. Annu Rev Neurosci. 2009;32:149–84.
5. Koivuniemi R, Mäkelä J, Hokkanen ME, Bruelle C, Ho TH, Ola R, et al. Hepatocyte growth factor activator inhibitor-1 is induced by bone morphogenetic proteins and regulates proliferation and cell fate of neural progenitor cells. PLoS One. 2013;8:e56117.
6. Miyata A, Arimura A, Dahl RR, Minamino N, Uehara A, Jiang L, et al. Isolation of a novel 38 residue-hypothalamic polypeptide which stimulates adenylate cyclase in pituitary cells. Biochem Biophys Res Commun. 1989;164:567–74.
7. Vaudry D, Falluel-Morel A, Bourgault S, Basille M, Burel D, Wurtz O, et al. Pituitary adenylate cyclase-activating polypeptide and its receptors: 20 years after the discovery. Pharmacol Rev. 2009;61:283–357.
8. Mikkelsen JD, Hannibal J, Larsen PJ, Fahrenkrug J. Pituitary adenylate cyclase activating peptide (PACAP) mRNA in the rat neocortex. Neurosci Lett. 1994;171:121–4.
9. Skoglösa Y, Takei N, Lindholm D. Distribution of pituitary adenylate cyclase activating polypeptide mRNA in the developing rat brain. Brain Res Mol Brain Res. 1999;65:1–13.
10. Hashimoto H, Nogi H, Mori K, Ohishi H, Shigemoto R, Yamamoto K, et al. Distribution of the mRNA for a pituitary adenylate cyclase-activating polypeptide receptor in the rat brain: an in situ hybridization study. J Comp Neurol. 1996;371:567–77.
11. Lindholm D, Skoglösa Y, Takei N. Developmental regulation of pituitary adenylate cyclase activating polypeptide (PACAP) and its receptor 1 in rat brain: function of PACAP as a neurotrophic factor. Ann N Y Acad Sci. 1998;865:189–96.
12. Matsuno R, Ohtaki H, Nakamachi T, Watanabe J, Yofu S, Hayashi D, et al. Distribution and localization of pituitary adenylate cyclase-activating polypeptide-specific receptor (PAC1R) in the rostral migratory stream of the infant mouse brain. Regul Pept. 2008;145:80–7.
13. Mercer A, Rönnholm H, Holmberg J, Lundh H, Heidrich J, Zachrisson O, et al. PACAP promotes neural stem cell proliferation in adult mouse brain. J Neurosci Res. 2004;76:205–15.
14. Ohta S, Gregg C, Weiss S. Pituitary adenylate cyclase-activating polypeptide regulates forebrain neural stem cells and neurogenesis in vitro and in vivo. J Neurosci Res. 2006;84:1177–86.
15. Ago Y, Yoneyama M, Ishihama T, Kataoka S, Kawada K, Tanaka T, et al. Role of endogenous pituitary adenylate cyclase-activating polypeptide in adult hippocampal neurogenesis. Neuroscience. 2011;172:554–61.
16. Toriyama M, Mizuno N, Fukami T, Iguchi T, Toriyama M, Tago K, et al. Phosphorylation of doublecortin by protein kinase A orchestrates microtubule and actin dynamics to promote neuronal progenitor cell migration. J Biol Chem. 2012;287:12691–702.
17. Ohtsuka M, Fukumitsu H, Furukawa S. PACAP decides neuronal laminar fate via PKA signaling in the developing cerebral cortex. Biochem Biophys Res Commun. 2008;369:1144–9.
18. Cameron DB, Raoult E, Galas L, Jiang Y, Lee K, Hu T, et al. Role of PACAP in controlling granule cell migration. Cerebellum. 2009;8:433–40.
19. Nishimoto M, Furuta A, Aoki S, Kudo Y, Miyakawa H, Wada K. PACAP/PAC1 autocrine system promotes proliferation and astrogenesis in neural progenitor cells. Glia. 2007;55:317–27.
20. Ohno F, Watanabe J, Sekihara H, Hirabayashi T, Arata S, Kikuyama S, et al. Pituitary adenylate cyclase-activating polypeptide promotes differentiation of mouse neural stem cells into astrocytes. Regul Pept. 2005;126:115–22.

21. Vallejo I, Vallejo M. Pituitary adenylate cyclase-activating polypeptide induces astrocyte differentiation of precursor cells from developing cerebral cortex. Mol Cell Neurosci. 2002;21:671–83.
22. Watanabe J, Ohba M, Ohno F, Kikuyama S, Nakamura M, Nakaya K, et al. Pituitary adenylate cyclase-activating polypeptide-induced differentiation of embryonic neural stem cells into astrocytes is mediated via the beta isoform of protein kinase C. J Neurosci Res. 2006;84:1645–55.
23. Vallejo M. PACAP signaling to DREAM: a cAMP-dependent pathway that regulates cortical astrogliogenesis. Mol Neurobiol. 2009;39:90–100.
24. Cebolla B, Fernández-Pérez A, Perea G, Araque A, Vallejo M. DREAM mediates cAMP-dependent, Ca^{2+}-induced stimulation of GFAP gene expression and regulates cortical astrogliogenesis. J Neurosci. 2008;28:6703–13.
25. Spengler D, Waeber C, Pantaloni C, Holsboer F, Bockaert J, Seeburg PH, et al. Differential signal transduction by five splice variants of the PACAP receptor. Nature. 1993;365:170–5.
26. Lu N, Zhou R, DiCicco-Bloom E. Opposing mitogenic regulation by PACAP in sympathetic and cerebral cortical precursors correlates with differential expression of PACAP receptor (PAC1-R) isoforms. J Neurosci Res. 1998;53:651–62.
27. Carey RG, Li B, DiCicco-Bloom E. Pituitary adenylate cyclase activating polypeptide antimitogenic signaling in cerebral cortical progenitors is regulated by p57Kip2-dependent CDK2 activity. J Neurosci. 2002;22:1583–91.
28. Lindholm D, Hamnér S, Zirrgiebel U. Neurotrophins and cerebellar development. Perspect Dev Neurobiol. 1997;5:83–94.
29. Skoglösa Y, Patrone C, Lindholm D. Pituitary adenylate cyclase activating polypepetide is expressed by developing rat Purkinje cells and decreases the number of cerebellar gamma-amino butyric acid positive neurons in culture. Neurosci Lett. 1999;265:207–10.
30. Nicot A, Lelièvre V, Tam J, Waschek JA, DiCicco-Bloom E. Pituitary adenylate cyclase-activating polypeptide and sonic hedgehog interact to control cerebellar granule precursor cell proliferation. J Neurosci. 2002;22:9244–54.
31. Niewiadomski P, Zhujiang A, Youssef M, Waschek JA. Interaction of PACAP with Sonic hedgehog reveals complex regulation of the hedgehog pathway by PKA. Cell Signal. 2013;25:2222–30.
32. Waschek JA, Dicicco-Bloom E, Nicot A, Lelievre V. Hedgehog signaling: new targets for GPCRs coupled to cAMP and protein kinase A. Ann N Y Acad Sci. 2006;1070:120–8.
33. Lelievre V, Seksenyan A, Nobuta H, Yong WH, Chhith S, Niewiadomski P, et al. Disruption of the PACAP gene promotes medulloblastoma in ptc1 mutant mice. Dev Biol. 2008;313:359–70.
34. Vaudry D, Gonzalez BJ, Basille M, Anouar Y, Fournier A, Vaudry H. Pituitary adenylate cyclase-activating polypeptide stimulates both c-fos gene expression and cell survival in rat cerebellar granule neurons through activation of the protein kinase A pathway. Neuroscience. 1998;84:801–12.
35. Zhokhov SS, Desfeux A, Aubert N, Falluel-Morel A, Fournier A, Laudenbach V, et al. Bax siRNA promotes survival of cultured and allografted granule cell precursors through blockade of caspase-3 cleavage. Cell Death Differ. 2008;15:1042–53.
36. Basille-Dugay M, Vaudry H, Fournier A, Gonzalez B, Vaudry D. Activation of PAC1 receptors in rat cerebellar granule cells stimulates both calcium mobilization from intracellular stores and calcium influx through N-type calcium hannels. Front Endocrinol (Lausanne). 2013;4:56.
37. Fila T, Trazzi S, Crochemore C, Bartesaghi R, Ciani E. Lot1 is a key element of the pituitary adenylate cyclase-activating polypeptide (PACAP)/cyclic AMP pathway that negatively regulates neuronal precursor proliferation. J Biol Chem. 2009;284:15325–38.
38. Falluel-Morel A, Chafai M, Vaudry D, Basille M, Cazillis M, Aubert N, et al. The neuropeptide pituitary adenylate cyclase-activating polypeptide exerts anti-apoptotic and differentiating effects during neurogenesis: focus on cerebellar granule neurones and embryonic stem cells. J Neuroendocrinol. 2007;19:321–7.

39. Allais A, Burel D, Isaac ER, Gray SL, Basille M, Ravni A, et al. Altered cerebellar development in mice lacking pituitary adenylate cyclase-activating polypeptide. Eur J Neurosci. 2007;25:2604–18.
40. Allais A, Burel D, Roy V, Arthaud S, Galas L, Isaac ER, et al. Balanced effect of PACAP and FasL on granule cell death during cerebellar development: a morphological, functional and behavioural characterization. J Neurochem. 2010;113:329–40.
41. Lee M, Lelievre V, Zhao P, Torres M, Rodriguez W, Byun JY, et al. Pituitary adenylyl cyclase-activating polypeptide stimulates DNA synthesis but delays maturation of oligodendrocyte progenitors. J Neurosci. 2001;21:3849–59.
42. Lelievre V, Ghiani CA, Seksenyan A, Gressens P, de Vellis J, Waschek JA. Growth factor-dependent actions of PACAP on oligodendrocyte progenitor proliferation. Regul Pept. 2006;137:58–66.
43. Njaine B, Martins RA, Santiago MF, Linden R, Silveira MS. Pituitary adenylyl cyclase-activating polypeptide controls the proliferation of retinal progenitor cells through downregulation of cyclin D1. Eur J Neurosci. 2010;32:311–21.
44. Cazillis M, Gonzalez BJ, Billardon C, Lombet A, Fraichard A, Samarut J, et al. VIP and PACAP induce selective neuronal differentiation of mouse embryonic stem cells. Eur J Neurosci. 2004;19:798–808.
45. Hirose M, Hashimoto H, Shintani N, Nakanishi M, Arakawa N, Iga J, et al. Differential expression of mRNAs for PACAP and its receptors during neural differentiation of embryonic stem cells. Regul Pept. 2005;126:109–13.
46. Chafai M, Louiset E, Basille M, Cazillis M, Vaudry D, Rostène W, et al. PACAP and VIP promote initiation of electrophysiological activity in differentiating embryonic stem cells. Ann N Y Acad Sci. 2006;1070:185–9.
47. Hirose M, Niewiadomski P, Tse G, Chi GC, Dong H, Lee A, et al. Pituitary adenylyl cyclase-activating peptide counteracts hedgehog-dependent motor neuron production in mouse embryonic stem cell cultures. J Neurosci Res. 2011;89:1363–74.
48. Cova L, Lindholm D. Stem cell therapy: how to do it right. Front Cell Dev Biol. 2014;2:66.
49. Inoue H, Nagata N, Kurokawa H, Yamanaka S. iPS cells: a game changer for future medicine. EMBO J. 2014;33:409–17.
50. Zhao Z, Yu R, Yang J, Liu X, Tan M, Li H, et al. Maxadilan prevents apoptosis in iPS cells and shows no effects on the pluripotent state or karyotype. PLoS One. 2012;7:e33953.
51. Lin CH, Chiu L, Lee HT, Chiang CW, Liu SP, Hsu YH, et al. PACAP38/PAC1 signaling induces bone marrow-derived cells homing to ischemic brain. Stem Cells. 2015;33:1153–72.
52. Mäkelä J, Koivuniemi R, Korhonen L, Lindholm D. Interferon-gamma produced by microglia and the neuropeptide PACAP have opposite effects on the viability of neural progenitor cells. PLoS One. 2010;5:e11091.
53. Mansouri S, Ortsäter H, Pintor Gallego O, Darsalia V, Sjöholm A, Patrone C. Pituitary adenylate cyclase-activating polypeptide counteracts the impaired adult neural stem cell viability induced by palmitate. J Neurosci Res. 2012;90:759–68.
54. Brifault C, Gras M, Liot D, May V, Vaudry D, Wurtz O. Delayed pituitary adenylate cyclase-activating polypeptide delivery after brain stroke improves functional recovery by inducing m2 microglia/macrophage polarization. Stroke. 2015;46:520–8.
55. Fang KM, Chen JK, Hung SC, Chen MC, Wu YT, Wu TJ, et al. Effects of combinatorial treatment with pituitary adenylate cyclase activating peptide and human mesenchymal stem cells on spinal cord tissue repair. PLoS One. 2010;5:e15299.
56. Tsumuraya T, Ohtaki H, Song D, Sato A, Watanabe J, Hiraizumi Y, et al. Human mesenchymal stem/stromal cells suppress spinal inflammation in mice with contribution of pituitary adenylate cyclase-activating polypeptide (PACAP). J Neuroinflammation. 2015;12:35.

Chapter 6
PACAP and Neural Development

Jun Watanabe, Tamotsu Seki, and Seiji Shioda

Abstract Brain development is a complex process, controlled in part by locally secreted factors that regulate proliferation, differentiation, migration, survival, and maturation. Pituitary adenylate cyclase-activating polypeptide (PACAP) is a neuropeptide that exerts a wide range of effects on different cell types in the brain as early as fetal stage. PACAP and its receptors are expressed in germinative neuroepithelia, suggesting that PACAP may be involved in neurogenesis. PACAP has recently been shown to regulate cell fate in various developmental contexts, in a manner dependent on dose, region, signaling, and receptor subtype. Interestingly, germ cells and embryonic stem (ES) cells also express PACAP receptors, and PACAP plays a crucial role in their development. This chapter reviews current knowledge on several aspects of PACAP in neural development, including adult neurogenesis and developmental neural diseases.

Keywords Pituitary adenylate cyclase-activating polypeptide (PACAP) • PAC1 receptor • Neural development • Neurogenesis • Neural disease

Abbreviations

CNS	Central nervous system
DCX	Doublecortin
DISC1	Disrupted-in-schizophrenia 1

J. Watanabe (✉)
Center for Biotechnology, Showa University,
1-5-8 Hatanodai, Shinagawa-ku, Tokyo 142-8555, Japan

Department of Anatomy, Showa University School of Medicine, Tokyo, Japan
e-mail: watanabe@med.showa-u.ac.jp

T. Seki
Department of Anatomy, Showa University School of Medicine, Tokyo, Japan

S. Shioda
Hoshi University School of Pharmacy and Pharmaceutical Sciences, Tokyo, Japan

D. Reglodi, A. Tamas (eds.), *Pituitary Adenylate Cyclase Activating Polypeptide — PACAP*, Current Topics in Neurotoxicity 11,
DOI 10.1007/978-3-319-35135-3_6

EB	Embryoid body
ES cell	Embryonic stem cell
FF	Follicular fluid
GFAP	Glial fibrillary acidic protein
IHC	Immunohistochemistry
iPS cell	Induced pluripotent stem cell
ISH	In situ hybridization
NB	Northern blot
NeuN	Neuronal nuclei
OB	Olfactory bulb
PAC1R-LI	PAC1R-like immunoreactivity
PACAP	Pituitary adenylate cyclase-activating polypeptide
PACAP-LI	PACAP-like immunoreactivity
PDGF	Platelet-derived growth factor
PGC	Primordial germ cell
PKA	Protein kinase A
PKC	Protein kinase C
PTSD	Posttraumatic stress disorder
RE	Rostral extension
RMS	Rostral migratory stream
RT-PCR	Reverse transcription-polymerase chain reaction
Shh	Sonic hedgehog
SVZ	Subventricular zone
VIP	Vasoactive intestinal peptide
WB	Western blot

Introduction

Pituitary adenylate cyclase-activating polypeptide (PACAP) is a neuropeptide, found in a wide variety of tissues, including the central nervous system (CNS) and peripheral tissues [1, 2]. Physiologically, PACAP has been reported to act as a hormone, a neuromodulator, a neurotransmitter, and a trophic factor, and has been implicated in a variety of developmental activities [3–5]. PACAP and its receptors, PAC1R, VPAC1R, and VPAC2R, are expressed in germ cells and embryonic stem (ES) cells and are distributed throughout the embryonic and postnatal brain during development (Tables 6.1, 6.2, and 6.3). PACAP has been shown to regulate neural precursor cell proliferation, differentiation, migration, maturation, and survival during neural development [6–10]. Despite current knowledge about PACAP, its overall role during neural development is not yet fully understood. This chapter describes the expression of PACAP and its receptors during neural development, especially on the role of PACAP as a regulatory factor.

Table 6.1 The expression of PACAP and its receptors in germ cells

Cell type	Spices	Origin	Ligand/ receptor	Method	Function	Reference
Primordial germ cell	Mouse	Dorsal mesentery and urogenital ridges of E10.5 and E11.5 mouse embryos	PACAP, PAC1R	IHC (PACAP), RT-PCR (PAC1-R)	PACAP stimulated in vitro proliferation of mouse primordial germ cells	[11]
Spermatogonia	Rat	Rat testes	PACAP	IHC, ISH	–	[13]
	Human	Normal human testes	PACAP	IHC	–	[14]
Primary spermatocytes	Rat	Rat testes	PACAP	IHC, ISH	–	[13]
Spermatids	Rat	Rat testes	PACAP	IHC, ISH	–	[13]
	Rat	Rat testes	PACAP, PAC1R	IHC (PACAP, PAC1R), WB (PAC1R)	–	[15]
	Human	Normal human testes	PACAP	IHC	–	[14]
Sperm	Mouse	Sperm from mouse epididymis	PACAP	IHC	PACAP enhanced fertilization and sperm penetration	[27]
Mature oocyte	Rat	Rat ovaries	PAC1R	RT-PCR	–	[22]

Abbreviations: *IHC* immunohistochemistry, *ISH* in situ hybridization, *RT-PCR* reverse transcription-polymerase chain reaction, *WB* western blot

Germ Cells

PACAP and PAC1R are detectable in the earliest stages of germ cell development in mice, as early as primordial germ cells (PGCs), the embryonic precursors of gametes of adult animals [11]. The gene encoding PAC1R is expressed by both PGCs and gonadal somatic cells. Furthermore, PACAP has been observed in gonadal ridges, mostly on germ cell surfaces, in E11.5–12.5 mouse embryos and could stimulate PGC proliferation in vitro. These findings suggested that PACAP may be an important regulator of PGC proliferation and functions in embryonic gonads.

During spermatogenesis, PACAP can act as a local regulator of testicular germ cell development and function [12]. PACAP-like immunoreactivity (PACAP-LI) has been observed in the spermatogonia and spermatids of rodents [13] and humans [14], and mRNAs encoding all three PACAP receptors have been detected in human

Table 6.2 The expression of PACAP and its receptors in ES cells and iPS cells

Cell type	Spices	Origin	Ligand/receptor	Method	Function	References
ES cell, EBs	Mouse	Mouse blastocysts	ES cells: PAC1R, VPAC1R (week), VPAC2R EBs: PAC1R	RT-PCR	PACAP and VIP induce the differentiation of ES cells into a neuronal phenotype	[29]
ES cell	Mouse	Mouse blastocysts	PACAP, VIP, PAC1R, VPAC1R, VPAC2R	RT-PCR	PACAP blocks Shh-dependent motor neuron generation	[30, 31]
ES cell	Mouse	Embryonic fibroblasts	PAC1R, VPAC1R, VPAC2R	RT-PCR	PACAP and VIP promote the generation of calcium currents in neuronal differentiating cells	[32]
iPS cell	Human	Umbilical cord matrix and amniotic membrane mesenchymal cells	PAC1R	RT-PCR, WB	Maxadilan prevents ultraviolet C-induced apoptosis	[33]

Abbreviations: *RT-PCR* reverse transcription-polymerase chain reaction, *WB* western blot

spermatogonia and spermatids [14]. Interestingly, PAC1-R-like immunoreactivity (PAC1R-LI) and PACAP-LI in rats were localized to the cytoplasm of round spermatids, which aggregated in the acrosome [15]. This finding suggested that endogenous PACAP may directly interact with the cytosolic PAC1-R-like protein without the ligand being released into the extracellular space. In contrast, VPAC2R was strongly expressed within the seminiferous tubuli, whereas the other two receptors were not found in germ cells [16]. Furthermore, VPAC1R was expressed in connective tissues, whereas PAC1R was not. Sperm head size was found to be smaller in PACAP-deficient mice than in wild-type mice, both in the longitudinal and transverse diameters, and PACAP treatment was shown to enhance human sperm motility [17]. These studies indicated that PACAP could influence the development and functioning of spermatozoa.

PACAP and its receptors are also expressed in rodent and human ovaries [18–20]. The role of PACAP in female fertility and reproduction has been thoroughly reviewed [21]. Assays of rat ovarian cells showed that both PACAP mRNA and PACAP-LI were present in the majority of granulosa and cumulus cells from large preovulatory follicles, in the majority of the cells comprising the interstitial glandular tissue and in solitary theca cells of growing and mature follicles [18]. PACAP-LI was also found in nerve fibers innervating the ovary. PAC1R and VPAC2R were

Table 6.3 The distribution of PACAP and its receptors during embryonic neural development in rodents

Developmental stage	Spices	Ligand/ receptor	Expressed region	Method	Reference
Primitive streak	E9 rat embryo	PAC1R	Neural fold, neural ectoderm, neuroepithelium in the primitive streak region and intraembryonic mesoderm	RT-PCR, IHC, ISH	[35]
	E8.5 mouse embryo	PAC1R(–)	Not detected	ISH	[36]
Neural tube closure	E9.5 mouse embryo	PACAP, PAC1R	Floor and roof plates of the neural tube and rhombencephalon, terminating abruptly at the boundary with the mesencephalon	RT-PCR	[38]
	E9.5 mouse embryo	PACAP, PAC1R		RT-PCR, ISH	[36]
	E10 rat embryo	PAC1R		RT-PCR	[39]
	E11 rat embryo	PAC1R		RT-PCR, IHC, ISH	[35]
Form of primary brain vesicle	E10.5 and E11.5 mouse embryo	PACAP, PAC1R	PACAP: dorsal root and trigeminal ganglia, rhombencephalon and cerebellum PAC1R: neural tube, rhombencephalon, dorsal root and trigeminal ganglia and sympathetic chain	RT-PCR	[37]
	E10.5 mouse embryo	PACAP, PAC1R		ISH, NB	[41]
	E12 rat embryo	PACAP, PAC1R		ISH	[40]
Form of Secondary brain vesicle	E14 and E16 rat embryo	PACAP, PAC1-R	PACAP: throughout the neuraxis PAC1R: neuroepithelium throughout the neuraxis including olfactory bulb, neuroepithelium, and external germinal layer of the cerebellum VPAC1R: similar to PAC1R VPAC2R: diencephalic nuclei	ISH	[40]
	E16 rat embryo	PAC1-R		ISH, IHC, RT-PCR	[35]
	E16/17 rat embryo	PACAP		ISH	[42]
	E14 and E17 rat embryo	PAC1R, VPAC1R, VPAC2R		ISH, RT-PCR	[39]
Late stage of embryonic neural development	E18 and E20 rat embryo	PACAP, PAC1R	PACAP: neocortex, pituitary, discrete PAC1R: ventricular zone throughout the nervous system; thalamic and brainstem nuclei and spinal cord VPAC1R: similar to PAC1R VPAC2R: diencephalic nuclei	ISH	[40]
	E20 rat embryo	PAC1R		ISH, IHC, RT-PCR	[35]
	E20 rat embryo	PACAP		ISH	[42]
	E21 rat embryo	PAC1R, VPAC1R, VPAC2R		ISH, RT-PCR	[39]

Abbreviations: *IHC* immunohistochemistry, *ISH* in situ hybridization, *RT-PCR* reverse transcription-polymerase chain reaction, *NB* northern blot

found to be expressed in granulosa cells, whereas VPAC1R and VPAC2R were expressed in theca cells [22]. Fully developed oocytes express PAC1R, and the addition of nanomolar concentrations of PACAP induced calcium release. However, PAC1R was not detected in the Met-I and -II phases of oocytes matured in vivo [22]. PACAP was reported to accelerate meiotic maturation in follicle- and cumulus-enclosed oocytes, while inhibiting meiotic maturation in denuded oocytes [23, 24]. PACAP was observed in human follicular fluid (FF) obtained from women undergoing hyperstimulation treatment [25, 26]. Interestingly, an inverse correlation was observed between the concentration of PACAP in FF and the number of recruited and retrieved oocytes. Low concentrations of PACAP in FF may correlate with the retrieval of a markedly higher number of ova, predicting a higher probability of ovarian hyperstimulation. Taken together, these findings indicate that PACAP is an important factor in developing oocytes.

Recent studies showed that PACAP also promotes fertilization. For example, PACAP-RI was detected in the anterior acrosomes of epididymal sperm fixed under mild conditions [27], and PAC1R mRNA and PAC1R-LI were observed in postovulatory cumulus cells. Pretreatment of cumulus–oocyte complexes with PACAP significantly enhanced the fertilization rate at low sperm concentrations. PACAP also enhanced sperm penetration through the oocyte investment, cumulus layer, and zona pellucida. Another study showed that PACAP could reverse the hypoxanthine-induced inhibition of oocyte meiotic maturation in cumulus cell–oocyte complexes and could efficiently promote male pronuclear formation after fertilization [28]. These findings suggest that PACAP may play a significant role in germ cell development and fertilization.

Embryonic Stem (ES) Cells, Embryoid Bodies (EBs), and Induced Pluripotent Stem (iPS) Cells

Mouse ES cells, which are derived from the inner mass of the blastocyst, are pluripotent cells that have the capacity for multilineage differentiation. Understanding ES cell differentiation can provide new perspectives on the cellular and molecular mechanisms of early development. PAC1R, VPAC1R, and VPAC2R are functionally expressed in mouse ES cells and EB-derived cells, and PACAP and vasoactive intestinal peptide (VIP) can induce the differentiation of ES cells into a neuronal phenotype [29]. PAC1R mRNA expression was further upregulated after differentiation of ES cells into a neuronal lineage, whereas the levels of PAC1R and PACAP mRNA were markedly reduced after glial differentiation [30]. Furthermore, PACAP was shown to completely block Sonic hedgehog (Shh)-dependent motor neuron generation from ES cell cultures and to reduce the expression of Gli-1 mRNA, a target of Shh [31]. Both PACAP and VIP were shown to modify the shape of undifferentiated ES cells, forming bipolar cells that express neuronal markers [32]. Electrophysiological recording showed that VIP and PACAP enhanced transient calcium current and that VIP generated a sustained calcium current. These findings demonstrate that PACAP and VIP induce morphological and functional differentiation of ES cells into a neuronal phenotype.

Induced pluripotent stem (iPS) cells were originally generated by reprogramming murine fibroblasts through the retrovirus-mediated transfection of four transcription factors, Oct4, SOX2, c-Myc, and Klf-4. iPS cells are similar to ES cells in morphology, proliferative abilities, gene expression, and differentiation abilities. PAC1R protein and mRNA were observed in human iPS cells [33]. In contrast to ES cells, maxadilan, a PAC1R specific agonist, effectively protected iPS cells against ultraviolet C-induced apoptotic cell death while not affecting the pluripotent state or karyotype. Further investigations are needed to understand the effect and molecular mechanism of PACAP in iPS cells.

Distribution of PACAP and Its Receptors During Neural Development in Rodents

As shown in our previous review [34], the PACAP/PAC1R system is expressed widely throughout the nervous system during development, and the activities of PACAP signaling are complex. For example, this peptide is involved in precursor cell proliferation, differentiation, and survival. Expression of the PAC1R gene was detected from the primitive streak stage in rat embryos, as early as embryonic day (E) 9.5 [35]. PAC1R mRNA was highly expressed in the neural fold, neural ectoderm, and the neuroepithelium in the primitive streak region, as well as in intraembryonic mesoderm adjacent to the headfold. In contrast, PAC1-R mRNA was not detected at this stage in E8.5 mouse embryos [36]. Both PACAP and its receptor genes were expressed during the stage of neural tube closure, as early as E9.5 in mouse and E10–11 in rat embryos [35–39]. In E9.5 mouse embryos, PAC1-R mRNA was strongly expressed in the floor and roof plates of the neural tube and the rhombencephalon, terminating abruptly at the boundary with the mesencephalon [36]. In E11 rat embryos, PAC1R mRNA was detected in neuroepithelium along the neural tube [35]. At the stage of primary brain vesicle formation, in E10.5 mouse and E12–13 rat embryos, PAC1R mRNA was most abundant in the neural tube and rhombencephalon and was also present in the dorsal root and trigeminal ganglia and in the sympathetic chain [35, 37, 40]. The distribution of PACAP mRNA overlapped in part with receptor expression, but PACAP mRNA was more extensively distributed in the rhombencephalon and in the developing hypothalamus [37, 40]. At this stage, PACAP down-regulated the expression of the *shh*- and protein kinase A (PKA)-dependent target gene *gli*-1 in cultured neuroepithelial cells, concomitant with a decrease in DNA synthesis [41]. These results suggest that PACAP may act in the neural tube during patterning to control cell proliferation and gene expression. PAC1-R mRNA was detected throughout the neuraxis in E14 rat brains, with enriched expression in the mesencephalon and rhombencephalon, with lower expression in the telencephalon [40]. The distribution of radiolabeled-PACAP binding sites in developing rat brains was found to overlap sites of PAC1R expression

site [39]. Although the distribution of PAC1-R in rat brains did not differ greatly from E14 to E16, its expression was elevated in the developing cerebral cortex and spinal cord [35, 40]. By E18, most PAC1-R mRNA was localized to the ventricular zone throughout the nervous system [40]. Expression was observed in the developing olfactory bulb and cerebellar primordium, areas of abundant cellular proliferation during this stage of development. PAC1-R mRNA expression was observed within the neocortical and mesencephalic ventricular zones of E20–E21 rat brains, with high levels in the olfactory bulb, hippocampus, and cerebellum [35, 39, 40]. In contrast to the expression of PAC1-R, enhanced PACAP mRNA expression in E16.5 mouse embryos (E18 rat embryos) was observed in the pituitary gland, in discrete thalamic and brainstem nuclei, and in the spinal cord [40, 42]. By birth, PACAP mRNA expression increased in the hippocampus, striatum, hypothalamus, and pontine gray nucleus [40, 42].

After birth, PAC1-R mRNA expression was predominantly limited to the subventricular zone (SVZ), olfactory bulb (OB), hippocampus, and cerebellum [35, 39, 40]. After birth, PAC1-R mRNA expression in the ventricular zone declined and could not be detected after postnatal day (P) 7. Its expression in the SVZ, the rostral migratory stream (RMS), and the OB layers peaked from P4 to P14. PAC1-R mRNA was expressed in all hippocampal subfields and the dentate gyrus during the first postnatal week. By P14, PAC1-R mRNA expression was restricted to granule cells of the dentate gyrus. In the cerebellum, PAC1-R mRNA was localized to the cerebellar granule cell and Purkinje cell layers as development proceeds. Over the course of postnatal development, the expression of PACAP was particularly strong in the hypothalamus, anterior olfactory nucleus, and subiculum [40, 42]. In the cerebellum, the postnatal expression of PACAP mRNA is very week, in contrast to the strong expression of PAC1-R.

Neurogenesis in the Rostral Migratory Stream (RMS)

Continuous neurogenesis has been observed in two restricted regions of the adult mammalian brain, the subventricular zone (SVZ) of the lateral ventricle and the hippocampal dentate gyrus. New neurons generated in the SVZ migrate through the rostral migratory stream (RMS) and are integrated into the neuronal circuits of the OB throughout life [43]. Assessment of the RMS of P10 mice showed that PAC1R-LI was strongly expressed in nestin-positive cells (neural stem cells) in the apical SVZ, and in Neuronal Nuclei (NeuN)-positive cells (mature neurons) in the OB [44]. However, the intensity of PAC1-R-LI in doublecortin (DCX)- and βIII-tubulin (Tuj1)-positive cells (immature neural progenitors) was weaker than that for other markers in the rostral extension (RE). These distributions of PAC1-R suggest that PACAP affects proliferating, but not migrating, cells in the RMS. In addition, DCX-positive cells showed strong PAC1R-LI expression in the granular layer of the OB. These cells also coexpressed NeuN. As the granular layer of the OB is thought to derive from the differentiation of migrated neuroblasts into mature neurons, this finding suggests that cells exhibiting PAC1-R-, DCX-, and NeuN-LI are in the process of differentiating from neuroblasts into mature neurons. The addition of PACAP to olfactory cell cultures was found to increase the number of neurons and to stimulate neuronal cell proliferation and survival [45]. Moreover, neutralization of PACAP in primary olfactory cultures resulted in neuron-specific losses. These results indicated the importance of PACAP in cell proliferation in the olfactory system. Cultured neural stem cells isolated from the lateral ventricle walls of adult mice were found to express PAC1R and to proliferate in vitro in response to physiologic concentrations of two PAC1 agonists, PACAP and Maxadilan, but not VIP, indicating that PAC1 mediates neural stem cell proliferation [46]. Taken together, these findings show that PACAP may participate in the stage-specific expression of PAC1-R during neurodevelopment, including both the proliferation and differentiation stages.

Neurogenesis in the Hippocampus

Assessments of the developing hippocampus showed that, at each postnatal stage, PAC1R mRNA was highly expressed in the pyramidal cell layer of the CA1–CA4 fields of the hippocampus and in the granule cell layers of the dentate gyrus, and was moderately expressed in other layers of the developing hippocampus [35]. A mature expression pattern was first observed at P0, with maximal expression at P60. PAC1R expression was also maintained in areas of neurogenesis in the adult CNS, whereas its ligand PACAP was expressed in the neighboring parenchyma [40], suggesting a role for PAC1 in adult neurogenesis. Intracerebroventricular infusion of PACAP increased cell proliferation in the ventricular zone of the lateral ventricle and in the dentate gyrus of the hippocampus [46, 47]. Interestingly, the proliferation of newly divided cells in the subgranular zone of the hippocampus and in the SVZ of lateral ventricles did not differ in wild-type and PACAP-deficient (PACAP$^{-/-}$) mice [48]. Rearing in enriched conditions enhanced the survival of newly divided wild-type cells, less so for cells in the granule cell layer of PACAP$^{-/-}$ mice. These findings suggest that endogenous PACAP is involved in the survival of cells generated in an enriched environment, but not basal cells, in the dentate gyrus of the adult hippocampus.

Cerebellum Development

During the development of the cerebellum, PACAP is produced by Purkinje cells [49] in the immature cerebellum and PAC1-R is expressed in the external and internal granule cell layers [39], strongly suggesting that PACAP may modulate the histogenesis of the cerebellar cortex. PACAP is also involved in the proliferation, survival, differentiation, neurite outgrowth, and motility of cerebellar granule cells

[5, 50–52], as well as markedly inhibiting the Shh-induced proliferation of rat and mouse cerebellar granule precursor cells [53]. PACAP blocks canonical Shh signaling through a mechanism involving the activation of PKA and inhibition of the translocation of the Shh-dependent transcription factor Gli2 into the primary cilium [54]. In the absence of Shh, however, PACAP and maxadilan modestly stimulated DNA synthesis, an effect reproduced by activating protein kinase C (PKC) [53]. These observations suggest that PAC1R acts as a sensor of environmental cues, coordinating diverse neurogenetic signals. Development of the CNS requires an equilibrium between cell survival and apoptosis. PACAP has been shown to prevent apoptosis induced by ceramides and FasL [55–57]. These data suggest that PACAP and apoptosis inducers interact during cerebellar development to control the apoptosis of granule cells and that they may affect some motor cerebellar functions. Cell migration and synaptogenesis represent pivotal processes in the maturation of the CNS. Exposure of granule neurones to PACAP inhibited cell displacement and strongly enhanced the number of cells bearing neurites [58]. Furthermore, endogenous PACAP was found to have short-term, cortical-layer-specific effects on granule cell migration in the early postnatal mouse cerebellum [59]. In vivo administration of PACAP induced a transient increase in the number of granule cells in the molecular layer and in the internal granule cell layer [60]. In contrast, PACAP did not affect the number of Purkinje cells. Administration of PACAP also induced a significant increase in the volume of the cerebellar cortex. These results indicate that PACAP increases the proliferation and/or inhibits the programmed death of granule cells, as well as stimulating neuronal migration from the external granule to the internal granule cell layer.

Effect of PACAP on Proliferation of Neural Progenitor Cells

As described earlier, PACAP has both pro- and antiproliferative effects on cerebellar granule precursor cells, as well as on corticogenesis [61]. PACAP acted as an antimitogenic signal, beginning from E13.5, both in culture and in vivo, and activated cAMP signaling through the short isoform of PAC1R. However, the number of BrdU-labeled proliferative cells in the E9.5 cortex was lower in PACAP$^{-/-}$ than in wild-type mice, suggesting that PACAP normally promotes proliferation at this stage of development. In early cultures of cortical precursors (E9.5 mice and E10.5 rats), PACAP induced an intracellular calcium flux and increased phospho-PKC levels, as well as it stimulated the production of G1 cyclin mRNA and protein, S-phase entry, and proliferation without affecting cell survival. Interestingly, the abundance of the hop receptor isoform was 24-fold greater than that of the short isoform at E10.5, whereas, at E14.5, expression of the short isoform was 15-fold greater and PACAP inhibited mitogenesis. These findings suggest that PACAP induces temporally specific effects on cortical proliferation via the developmentally regulated expression of specific receptor isoforms.

Effect of PACAP on Neuronal and Glial Differentiation of Neural Progenitor Cells

Inhibition of mitogenesis of neural precursor cells by PACAP may result in the generation of neuronal cells. PACAP has been shown to induce the neuronal differentiation and/or neurite outgrowth of ES cells [29, 30], cerebral neural progenitors [61, 62], cerebellar neural progenitors [58, 60], and hippocampal neural progenitors [63, 64]. PACAP increased the proliferation of adult neural stem cells, followed by neurogenesis [46–48]. In contrast, PACAP also enhanced the generation of astrocytes and oligodendrocytes. Neural progenitor cells derived from E14.5 mouse and rat telencephalons differentiated into astrocytes in response to PACAP [65–69]. Our previous study showed that the PACAP-generated signal in neural progenitor cells was mediated via the activation of phospholipase C, followed by calcium- and phospholipid-dependent PKC and resulting in the generation of astroglia [67, 68]. PACAP has also been reported to stimulate the production of cAMP, activating the small GTPases Rap1 and Ras and allowing extracellular calcium into the cell [70]. Calcium, in turn, stimulates the transcription factor downstream regulatory element antagonist modulator, which is bound to specific sites of the promoter of the gene encoding glial fibrillary acidic protein (GFAP), stimulating its expression during astrocyte differentiation. In oligodendrocyte progenitors, PACAP showed growth factor-dependent activity [71].

PACAP alone potently stimulates the proliferation of oligodendrocyte progenitors and enhances bFGF-induced DNA synthesis in these cells. In contrast, PACAP strongly antagonizes the mitogenic effects of platelet-derived growth factor (PDGF). As PDGF receptor expression is also regulated by cAMP/PKA pathways, PACAP may inhibit the mitogenic activities of PDGF via a direct effect of cAMP on proliferation and a reduction of the number of PDGF receptors at the cell surface. PACAP was also reported to stimulate proliferation and delay myelinogenesis in cultured postnatal rat oligodendrocyte progenitors [72]. Altogether, PACAP may play an important role as a modulator in neurogenesis through the developmentally regulated expression of specific receptor isoforms, as well as through developmentally regulated signaling and niches.

Developmental Abnormalities of PACAP-Deficient Mice

PACAP-deficient ($PACAP^{-/-}$) mice show various morphological, biochemical, and behavioral abnormalities [73–77], including a decreased fertility rate [78, 79] and a higher mortality rate, with the latter at least partially due to their temperature sensitivity [80], dysfunctions in lipid and carbohydrate metabolism [73], and respiratory abnormalities [81, 82]. $PACAP^{-/-}$ mice showed marked behavioral changes compared with wild type, including locomotor hyperactivity, explosive jumping behavior, increased exploratory behavior, and less anxiety and memory dysfunction [74, 83].

Studies investigating morphological alterations in the cerebellum of PACAP$^{-/-}$ mice found significant reductions in the thickness of the external granule cell layer at P4 and of the internal granule cell layer at P7 [84]. PACAP$^{-/-}$ mice are also reported to show abnormal axonal arborization in the dentate gyrus [85] and an earlier onset of myelination and dense myelinated fibers in the developing brain [86].

PACAP in Developmental Neurological Disorders

Recent studies suggest that PACAP signaling abnormalities may contribute to schizophrenia [87] and posttraumatic stress disorder (PTSD) [88] and may possibly contribute to attention-deficit hyperactivity disorder and autism [89, 90]. Associations between certain single nucleotide polymorphisms of the *PACAP* gene and schizophrenia have been suggested [87], and two PACAP-signaling pathways have been linked to schizophrenia [91]. One pathway regulates the association between disrupted-in-schizophrenia 1 (DISC1) and DISC1-binding zinc-finger protein via PACAP [92], and the other inhibits stathmin1 expression via PACAP [85]. Schizophrenia-like behavior in PACAP-deficient mice could be treated with a selective metabotropic glutamate 2/3 receptor agonist [93], a selective 5-HT$_7$ antagonist [94], a mixture of D$_2$ and serotonin 5-HT$_2$ antagonist, and a selective serotonin 5-HT$_2$ receptor antagonist [95]. PACAP signaling may therefore contribute to the pathogenesis of certain depressive conditions amenable to atypical antipsychotic drugs.

PTSD was shown to be associated with PACAP and PAC1R in females [88, 96]. Recent studies suggest that PACAP is a master regulator of central and peripheral stress responses [97–99]. A single nucleotide polymorphism in PAC1R was associated with increased reactivity of the amygdala and hippocampus to threat stimuli and decreased functional connectivity between the amygdala and hippocampus in PTSD patients [100]. Intriguingly, PACAP also acts on brain structures that mediate anxiety- and fear-related behaviors, where it may influence both hard-wired (genetically determined) stress responses and gene–environment interactions in stress-related psychopathology [99].

Conclusion

PACAP and its receptors are expressed in germinative neuroepithelia involved in neurogenesis. This neuropeptide takes part in the control of neurogenesis at different stages and locations and in different cell types during brain development. Furthermore, although the effects of PACAP on both proliferation and differentiation may be contradictory, these conflicting functions strongly suggest that the actual role of PACAP is highly influenced by receptor subtypes and other trophic factors or signal transduction molecules present. Future studies addressing the molecular basis and pathophysiological implications of PACAP-associated responses may contribute to the development of treatments for developmental neural disorders.

References

1. Arimura A, Somogyvári-Vigh A, Miyata A, Mizuno K, Coy DH, Kitada C. Tissue distribution of PACAP as determined by RIA: highly abundant in the rat brain and testes. Endocrinology. 1991;129:2787–9.
2. Arimura A, Shioda S. Pituitary adenylate cyclase activating polypeptide (PACAP) and its receptors: neuroendocrine and endocrine interaction. Front Neuroendocrinol. 1995;16: 53–88.
3. Arimura A. Perspectives on pituitary adenylate cyclase activating polypeptide (PACAP) in the neuroendocrine, endocrine, and nervous systems. Jpn J Physiol. 1998;48:301–31.
4. Sherwood NM, Krueckl SL, McRory JE. The origin and function of the pituitary adenylate cyclase-activating polypeptide (PACAP)/glucagon superfamily. Endocr Rev. 2000;21: 619–70.
5. Falluel-Morel A, Chafai M, Vaudry D, Basille M, Cazillis M, Aubert N, et al. The neuropeptide pituitary adenylate cyclase-activating polypeptide exerts anti-apoptotic and differentiating effects during neurogenesis: focus on cerebellar granule neurones and embryonic stem cells. J Neuroendocrinol. 2007;19:321–7.
6. Shen S, Gehlert DR, Collier DA. PACAP and PAC1 receptor in brain development and behavior. Neuropeptides. 2013;47:421–30.
7. Nakamachi T, Farkas J, Watanabe J, Ohtaki H, Dohi K, Arata S, et al. Role of PACAP in neural stem/progenitor cell and astrocyte-from neural development to neural repair. Curr Pharm Des. 2011;17:973–84.
8. Seaborn T, Masmoudi-Kouli O, Fournier A, Vaudry H, Vaudry D. Protective effects of pituitary adenylate cyclase-activating polypeptide (PACAP) against apoptosis. Curr Pharm Des. 2011;17:204–14.
9. Vaudry D, Gonzalez BJ, Basille M, Yon L, Fournier A, Vaudry H. Pituitary adenylate cyclase-activating polypeptide and its receptors: from structure to functions. Pharmacol Rev. 2000;52:269–324.
10. Vaudry D, Falluel-Morel A, Bourgault S, Basille M, Burel D, Wurtz O, et al. Pituitary adenylate cyclase-activating polypeptide and its receptors: 20 years after the discovery. Pharmacol Rev. 2009;61:283–357.
11. Pesce M, Canipari R, Ferri GL, Siracusa G, De Felici M. Pituitary adenylate cyclase-activating polypeptide (PACAP) stimulates adenylate cyclase and promotes proliferation of mouse primordial germ cells. Development. 1996;122:215–21.
12. Li M, Arimura A. Neuropeptides of the pituitary adenylate cyclase-activating polypeptide/vasoactive intestinal polypeptide/growth hormone-releasing hormone/secretin family in testis. Endocrine. 2003;20:201–14.
13. Shioda S, Legradi G, Leung WC, Nakajo S, Nakaya K, Arimura A. Localization of pituitary adenylate cyclase-activating polypeptide and its messenger ribonucleic acid in the rat testis by light and electron microscopic immunocytochemistry and in situ hybridization. Endocrinology. 1994;135:818–25.
14. Nakamura K, Nakamachi T, Endo K, Ito K, Machida T, Oka T, et al. Distribution of pituitary adenylate cyclase-activating polypeptide (PACAP) in the human testis and in testicular germ cell tumors. Andrologia. 2014;46:465–71.
15. Li M, Funahashi H, Mbikay M, Shioda S, Arimura A. Pituitary adenylate cyclase activating polypeptide-mediated intracrine signaling in the testicular germ cells. Endocrine. 2004;23:59–75.
16. Krempels K, Usdin TB, Harta G, Mezey E. PACAP acts through VIP type 2 receptors in the rat testis. Neuropeptides. 1995;29:315–20.
17. Brubel R, Kiss P, Vincze A, Varga A, Varnagy A, Bodis J, et al. Effects of pituitary adenylate cyclase activating polypeptide on human sperm motility. J Mol Neurosci. 2012;48:623–30.
18. Gräs S, Hannibal J, Georg B, Fahrenkrug J. Transient periovulatory expression of pituitary adenylate cyclase activating peptide in rat ovarian cells. Endocrinology. 1996;137:4779–85.

19. Scaldaferri L, Arora K, Lee SH, Catt KJ, Moretti C. Expression of PACAP and its type-I receptor isoforms in the rat ovary. Mol Cell Endocrinol. 1996;117:227–32.
20. Morelli MB, Barberi M, Gambardella A, Borini A, Cecconi S, Coticchio G, et al. Characterization, expression, and functional activity of pituitary adenylate cyclase-activating polypeptide and its receptors in human granulosa-luteal cells. J Clin Endocrinol Metab. 2008;93:4924–32.
21. Reglodi D, Tamas A, Koppan M, Szogyi D, Welke L. Role of PACAP in female fertility and reproduction at gonadal level—recent advances. Front Endocrinol (Lausanne). 2012;3:155.
22. Vaccari S, Latini S, Barberi M, Teti A, Stefanini M, Canipari R. Characterization and expression of different pituitary adenylate cyclase-activating polypeptide/vasoactive intestinal polypeptide receptors in rat ovarian follicles. J Endocrinol. 2006;191:287–99.
23. Apa R, Lanzone A, Mastrandrea M, Miceli F, de Feo D, Caruso A, et al. Control of human luteal steroidogenesis: role of growth hormone-releasing hormone, vasoactive intestinal peptide, and pituitary adenylate cyclase-activating peptide. Fertil Steril. 1997;68:1097–102.
24. Cecconi S, Rossi G, Barberi M, Scaldaferri L, Canipari R. Effect of pituitary adenylate cyclase-activating polypeptide and vasoactive intestinal polypeptide on mouse preantral follicle development in vitro. Endocrinology. 2004;145:2071–9.
25. Brubel R, Boronkai A, Reglodi D, Racz B, Nemeth J, Kiss P, et al. Changes in the expression of pituitary adenylate cyclase-activating polypeptide in the human placenta during pregnancy and its effects on the survival of JAR choriocarcinoma cells. J Mol Neurosci. 2010;42: 450–8.
26. Koppan M, Varnagy A, Reglodi D, Brubel R, Nemeth J, Tamas A, et al. Correlation between oocyte number and follicular fluid concentration of pituitary adenylate cyclase-activating polypeptide (PACAP) in women after superovulation treatment. J Mol Neurosci. 2012;48:617–22.
27. Tanii I, Aradate T, Matsuda K, Komiya A, Fuse H. PACAP-mediated sperm-cumulus cell interaction promotes fertilization. Reproduction. 2011;141:163–71.
28. Barberi M, Di Paolo V, Latini S, Guglielmo MC, Cecconi S, Canipari R. Expression and functional activity of PACAP and its receptors on cumulus cells: effects on oocyte maturation. Mol Cell Endocrinol. 2013;375:79–88.
29. Cazillis M, Gonzalez BJ, Billardon C, Lombet A, Fraichard A, Samarut J, et al. VIP and PACAP induce selective neuronal differentiation of mouse embryonic stem cells. Eur J Neurosci. 2004;19:798–808.
30. Hirose M, Hashimoto H, Shintani N, Nakanishi M, Arakawa N, Iga J, et al. Differential expression of mRNAs for PACAP and its receptors during neural differentiation of embryonic stem cells. Regul Pept. 2005;126:109–13.
31. Hirose M, Niewiadomski P, Tse G, Chi GC, Dong H, Lee A, et al. Pituitary adenylyl cyclase-activating peptide counteracts hedgehog-dependent motor neuron production in mouse embryonic stem cell cultures. J Neurosci Res. 2011;89:1363–74.
32. Chafai M, Basille M, Galas L, Rostene W, Gressens P, Vaudry H, et al. Pituitary adenylate cyclase-activating polypeptide and vasoactive intestinal polypeptide promote the genesis of calcium currents in differentiating mouse embryonic stem cells. Neuroscience. 2011;199: 103–15.
33. Zhao Z, Yu R, Yang J, Liu X, Tan M, Li H, et al. Maxadilan prevents apoptosis in iPS cells and shows no effects on the pluripotent state or karyotype. PLoS One. 2012;7:e33953.
34. Watanabe J, Nakamachi T, Matsuno R, Hayashi D, Nakamura M, Kikuyama S, et al. Localization, characterization and function of pituitary adenylate cyclase-activating polypeptide during brain development. Peptides. 2007;28:1713–9.
35. Zhou CJ, Shioda S, Shibanuma M, Nakajo S, Funahashi H, Nakai Y, et al. Pituitary adenylate cyclase-activating polypeptide receptors during development: expression in the rat embryo at primitive streak stage. Neuroscience. 1999;93:375–91.
36. Sheward WJ, Lutz EM, Harmar AJ. Expression of pituitary adenylate cyclase activating polypeptide receptors in the early mouse embryo as assessed by reverse transcription polymerase chain reaction and in situ hybridisation. Neurosci Lett. 1996;216:45–8.

37. Sheward WJ, Lutz EM, Copp AJ, Harmar AJ. Expression of PACAP, and PACAP type 1 (PAC1) receptor mRNA during development of the mouse embryo. Brain Res Dev Brain Res. 1998;109:245–53.
38. Shuto Y, Uchida D, Onda H, Arimura A. Ontogeny of pituitary adenylate cyclase activating polypeptide and its receptor mRNA in the mouse brain. Regul Pept. 1996;67:79–83.
39. Basille M, Vaudry D, Coulouarn Y, Jegou S, Lihrmann I, Fournier A, et al. Comparative distribution of pituitary adenylate cyclase-activating polypeptide (PACAP) binding sites and PACAP receptor mRNAs in the rat brain during development. J Comp Neurol. 2000;425: 495–509.
40. Jaworski DM, Proctor MD. Developmental regulation of pituitary adenylate cyclase-activating polypeptide and PAC(1) receptor mRNA expression in the rat central nervous system. Brain Res Dev Brain Res. 2000;120:27–39.
41. Waschek JA, Casillas RA, Nguyen TB, DiCicco-Bloom EM, Carpenter EM, Rodriguez WI. Neural tube expression of pituitary adenylate cyclase-activating peptide (PACAP) and receptor: potential role in patterning and neurogenesis. Proc Natl Acad Sci U S A. 1998;95:9602–7.
42. Skoglösa Y, Takei N, Lindholm D. Distribution of pituitary adenylate cyclase activating polypeptide mRNA in the developing rat brain. Mol Brain Res. 1999;65:1–13.
43. Sakamoto M, Kageyama R, Imayoshi I. The functional significance of newly born neurons integrated into olfactory bulb circuits. Front Neurosci. 2014;8:121.
44. Matsuno R, Ohtaki H, Nakamachi T, Watanabe J, Yofu S, Hayashi D, et al. Distribution and localization of pituitary adenylate cyclase-activating polypeptide-specific receptor (PAC1R) in the rostral migratory stream of the infant mouse brain. Regul Pept. 2008;145:80–7.
45. Hansel DE, May V, Eipper BA, Ronnett GV. Pituitary adenylyl cyclase-activating peptides and alpha-amidation in olfactory neurogenesis and neuronal survival in vitro. J Neurosci. 2001;21:4625–36.
46. Mercer A, Rönnholm H, Holmberg J, Lundh H, Heidrich J, Zachrisson O, et al. PACAP promotes neural stem cell proliferation in adult mouse brain. J Neurosci Res. 2004;76:205–15.
47. Ohta S, Gregg C, Weiss S. Pituitary adenylate cyclase-activating polypeptide regulates forebrain neural stem cells and neurogenesis in vitro and in vivo. J Neurosci Res. 2006;84:1177–86.
48. Ago Y, Yoneyama M, Ishihama T, Kataoka S, Kawada K, Tanaka T, et al. Role of endogenous pituitary adenylate cyclase-activating polypeptide in adult hippocampal neurogenesis. Neuroscience. 2011;172:554–61.
49. Gao WQ, Hatten ME. Immortalizing oncogenes subvert the establishment of granule cell identity in developing cerebellum. Development. 1994;120:1059–70.
50. Waschek JA. VIP and PACAP receptor-mediated actions on cell proliferation and survival. Ann N Y Acad Sci. 1996;805:290–300.
51. Gonzalez BJ, Basille M, Mei YA, Vaudry D, Fournier A, Cazin L, et al. Ontogeny of PACAP and PACAP receptors in the rat brain: role of PACAP in the cerebellum during development. Ann N Y Acad Sci. 1996;805:302–13.
52. Cameron DB, Raoult E, Galas L, Jiang Y, Lee K, Hu T, et al. Role of PACAP in controlling granule cell migration. Cerebellum. 2009;8:433–40.
53. Nicot A, Lelièvre V, Tam J, Waschek JA, DiCicco-Bloom E. Pituitary adenylate cyclase-activating polypeptide and sonic hedgehog interact to control cerebellar granule precursor cell proliferation. J Neurosci. 2002;22:9244–54.
54. Niewiadomski P, Zhujiang A, Youssef M, Waschek JA. Interaction of PACAP with Sonic hedgehog reveals complex regulation of the hedgehog pathway by PKA. Cell Signal. 2013;25:2222–30.
55. Allais A, Burel D, Roy V, Arthaud S, Galas L, Isaac ER, et al. Balanced effect of PACAP and FasL on granule cell death during cerebellar development: a morphological, functional and behavioural characterization. J Neurochem. 2010;113:329–40.
56. Falluel-Morel A, Aubert N, Vaudry D, Basille M, Fontaine M, Fournier A, et al. Opposite regulation of the mitochondrial apoptotic pathway by C2-ceramide and PACAP through a

MAP-kinase-dependent mechanism in cerebellar granule cells. J Neurochem. 2004;91:1231–43.
57. Vaudry D, Falluel-Morel A, Basille M, Pamantung TF, Fontaine M, Fournier A, et al. Pituitary adenylate cyclase-activating polypeptide prevents C2-ceramide-induced apoptosis of cerebellar granule cells. J Neurosci Res. 2003;72:303–16.
58. Falluel-Morel A, Vaudry D, Aubert N, Galas L, Benard M, Basille M, et al. Pituitary adenylate cyclase-activating polypeptide prevents the effects of ceramides on migration, neurite outgrowth, and cytoskeleton remodeling. Proc Natl Acad Sci U S A. 2005;102:2637–42.
59. Cameron DB, Galas L, Jiang Y, Raoult E, Vaudry D, Komuro H. Cerebellar cortical-layer-specific control of neuronal migration by pituitary adenylate cyclase-activating polypeptide. Neuroscience. 2007;146:697–712.
60. Vaudry D, Gonzalez BJ, Basille M, Fournier A, Vaudry H. Neurotrophic activity of pituitary adenylate cyclase-activating polypeptide on rat cerebellar cortex during development. Proc Natl Acad Sci U S A. 1999;96:9415–20.
61. Yan Y, Zhou X, Pan Z, Ma J, Waschek JA, DiCicco-Bloom E. Pro- and anti-mitogenic actions of pituitary adenylate cyclase-activating polypeptide in developing cerebral cortex: potential mediation by developmental switch of PAC1 receptor mRNA isoforms. J Neurosci. 2013;33:3865–78.
62. Dicicco-Bloom E, Lu N, Pintar JE, Zhang J. The PACAP ligand/receptor system regulates cerebral cortical neurogenesis. Ann N Y Acad Sci. 1998;865:274–89.
63. Ogata K, Shintani N, Hayata-Takano A, Kamo T, Higashi S, Seiriki K, et al. PACAP enhances axon outgrowth in cultured hippocampal neurons to a comparable extent as BDNF. PLoS One. 2015;10:e0120526.
64. Kambe Y, Miyata A. Role of mitochondrial activation in PACAP dependent neurite outgrowth. J Mol Neurosci. 2012;48:550–7.
65. Nishimoto M, Furuta A, Aoki S, Kudo Y, Miyakawa H, Wada K. PACAP/PAC1 autocrine system promotes proliferation and astrogenesis in neural progenitor cells. Glia. 2007;55:317–27.
66. Ohno F, Watanabe J, Sekihara H, Hirabayashi T, Arata S, Kikuyama S, et al. Pituitary adenylate cyclase-activating polypeptide promotes differentiation of mouse neural stem cells into astrocytes. Regul Pept. 2005;126:115–22.
67. Watanabe J, Ohno F, Shioda S, Kikuyama S, Nakaya K, Nakajo S. Involvement of protein kinase C in the PACAP-induced differentiation of neural stem cells into astrocytes. Ann N Y Acad Sci. 2006;1070:597–601.
68. Watanabe J, Ohba M, Ohno F, Kikuyama S, Nakamura M, Nakaya K, et al. Pituitary adenylate cyclase-activating polypeptide-induced differentiation of embryonic neural stem cells into astrocytes is mediated via the beta isoform of protein kinase C. J Neurosci Res. 2006;84:1645–55.
69. Vallejo I, Vallejo M. Pituitary adenylate cyclase-activating polypeptide induces astrocyte differentiation of precursor cells from developing cerebral cortex. Mol Cell Neurosci. 2002;21:671–83.
70. Vallejo M. PACAP signaling to DREAM: a cAMP-dependent pathway that regulates cortical astrogliogenesis. Mol Neurobiol. 2009;39:90–100.
71. Lelievre V, Ghiani CA, Seksenyan A, Gressens P, de Vellis J, Waschek JA. Growth factor-dependent actions of PACAP on oligodendrocyte progenitor proliferation. Regul Pept. 2006;137:58–66.
72. Lee M, Lelievre V, Zhao P, Torres M, Rodriguez W, Byun JY, et al. Pituitary adenylyl cyclase-activating polypeptide stimulates DNA synthesis but delays maturation of oligodendrocyte progenitors. J Neurosci. 2001;21:3849–59.
73. Gray SL, Cummings KJ, Jirik FR, Sherwood NM. Targeted disruption of the pituitary adenylate cyclase-activating polypeptide gene results in early postnatal death associated with dysfunction of lipid and carbohydrate metabolism. Mol Endocrinol. 2001;15:1739–47.
74. Hashimoto H, Shintani N, Tanaka K, Mori W, Hirose M, Matsuda T, et al. Altered psychomotor behaviors in mice lacking pituitary adenylate cyclase-activating polypeptide (PACAP). Proc Natl Acad Sci U S A. 2001;98:13355–60.

75. Colwell CS, Michel S, Itri J, Rodriguez W, Tam J, Lelièvre V, et al. Selective deficits in the circadian light response in mice lacking PACAP. Am J Physiol Regul Integr Comp Physiol. 2004;287:R1194–201.
76. Reglodi D, Kiss P, Szabadfi K, Atlasz T, Gabriel R, Horvath G, et al. PACAP is an endogenous protective factor-insights from PACAP-deficient mice. J Mol Neurosci. 2012;48: 482–92.
77. Shintani N, Tomimoto S, Hashimoto H, Kawaguchi C, Baba A. Functional roles of the neuropeptide PACAP in brain and pancreas. Life Sci. 2003;74:337–43.
78. Isaac ER, Sherwood NM. Pituitary adenylate cyclase-activating polypeptide (PACAP) is important for embryo implantation in mice. Mol Cell Endocrinol. 2008;280:13–9.
79. Shintani N, Mori W, Hashimoto H, Imai M, Tanaka K, Tomimoto S, et al. Defects in reproductive functions in PACAP-deficient female mice. Regul Pept. 2002;109:45–8.
80. Gray SL, Yamaguchi N, Vencová P, Sherwood NM. Temperature-sensitive phenotype in mice lacking pituitary adenylate cyclase-activating polypeptide. Endocrinology. 2002;143:3946–54.
81. Cummings KJ, Willie C, Wilson RJ. Pituitary adenylate cyclase-activating polypeptide maintains neonatal breathing but not metabolism during mild reductions in ambient temperature. Am J Physiol Regul Integr Comp Physiol. 2008;294:R956–65.
82. Arata S, Nakamachi T, Onimaru H, Hashimoto H, Shioda S. Impaired response to hypoxia in the respiratory center is a major cause of neonatal death of the PACAP-knockout mouse. Eur J Neurosci. 2013;37:407–16.
83. Takuma K, Maeda Y, Ago Y, Ishihama T, Takemoto K, Nakagawa A, et al. An enriched environment ameliorates memory impairments in PACAP-deficient mice. Behav Brain Res. 2014;272:269–78.
84. Allais A, Burel D, Isaac ER, Gray SL, Basille M, Ravni A, et al. Altered cerebellar development in mice lacking pituitary adenylate cyclase-activating polypeptide. Eur J Neurosci. 2007;25:2604–18.
85. Yamada K, Matsuzaki S, Hattori T, Kuwahara R, Taniguchi M, Hashimoto H, et al. Increased stathmin1 expression in the dentate gyrus of mice causes abnormal axonal arborizations. PLoS One. 2010;5:e8596.
86. Vincze A, Reglodi D, Helyes Z, Hashimoto H, Shintani N, Abrahám H. Role of endogenous pituitary adenylate cyclase activating polypeptide (PACAP) in myelination of the rodent brain: lessons from PACAP-deficient mice. Int J Dev Neurosci. 2011;29:923–35.
87. Hashimoto R, Hashimoto H, Shintani N, Chiba S, Hattori S, Okada T, et al. Pituitary adenylate cyclase-activating polypeptide is associated with schizophrenia. Mol Psychiatry. 2007;12:1026–32.
88. Ressler KJ, Mercer KB, Bradley B, Jovanovic T, Mahan A, Kerley K, et al. Post-traumatic stress disorder is associated with PACAP and the PAC1 receptor. Nature. 2011;470:492–7.
89. Masuo Y, Morita M, Oka S, Ishido M. Motor hyperactivity caused by a deficit in dopaminergic neurons and the effects of endocrine disruptors: a study inspired by the physiological roles of PACAP in the brain. Regul Pept. 2004;123:225–34.
90. Nijmeijer JS, Arias-Vásquez A, Rommelse NN, Altink ME, Anney RJ, Asherson P, et al. Identifying loci for the overlap between attention-deficit/hyperactivity disorder and autism spectrum disorder using a genome-wide QTL linkage approach. J Am Acad Child Adolesc Psychiatry. 2010;49:675–85.
91. Tohyama M, Miyata S, Hattori T, Shimizu S, Matsuzaki S. Molecular basis of major psychiatric diseases such as schizophrenia and depression. Anat Sci Int. 2015;90:137–43.
92. Hattori T, Baba K, Matsuzaki S, Honda A, Miyoshi K, Inoue K, et al. A novel DISC1-interacting partner DISC1-binding zinc-finger protein: implication in the modulation of DISC1-dependent neurite outgrowth. Mol Psychiatry. 2007;12:398–407.
93. Ago Y, Hiramatsu N, Ishihama T, Hazama K, Hayata-Takano A, Shibasaki Y, et al. The selective metabotropic glutamate 2/3 receptor agonist MGS0028 reverses psychomotor abnormalities and recognition memory deficits in mice lacking the pituitary adenylate cyclase-activating polypeptide. Behav Pharmacol. 2013;24:74–7.
94. Tajiri M, Hayata-Takano A, Seiriki K, Ogata K, Hazama K, Shintani N, et al. Serotonin 5-HT(7) receptor blockade reverses behavioral abnormalities in PACAP-deficient mice and

receptor activation promotes neurite extension in primary embryonic hippocampal neurons: therapeutic implications for psychiatric disorders. J Mol Neurosci. 2012;48:473–81.
95. Hashimoto H, Hashimoto R, Shintani N, Tanaka K, Yamamoto A, Hatanaka M, et al. Depression-like behavior in the forced swimming test in PACAP-deficient mice: amelioration by the atypical antipsychotic risperidone. J Neurochem. 2009;110:595–602.
96. Almli LM, Mercer KB, Kerley K, Feng H, Bradley B, Conneely KN, et al. ADCYAP1R1 genotype associates with post-traumatic stress symptoms in highly traumatized African-American females. Am J Med Genet B Neuropsychiatr Genet. 2013;162B:262–72.
97. Smith CB, Eiden LE. Is PACAP the major neurotransmitter for stress transduction at the adrenomedullary synapse? J Mol Neurosci. 2012;48:403–12.
98. Mustafa T. Pituitary adenylate cyclase-activating polypeptide (PACAP): a master regulator in central and peripheral stress responses. Adv Pharmacol. 2013;68:445–57.
99. Hashimoto H, Shintani N, Tanida M, Hayata A, Hashimoto R, Baba A. PACAP is implicated in the stress axes. Curr Pharm Des. 2011;17:985–9.
100. Stevens JS, Almli LM, Fani N, Gutman DA, Bradley B, Norrholm SD, et al. PACAP receptor gene polymorphism impacts fear responses in the amygdala and hippocampus. Proc Natl Acad Sci U S A. 2014;111:3158–63.

Chapter 7
PACAP Modulates Distinct Neuronal Components to Induce Cell-Specific Plasticity at Central and Autonomic Synapses

Eric R. Starr and Joseph F. Margiotta

Abstract Despite its widespread expression and diverse functional roles, PACAP is well suited to mediate and modulate synaptic transmission since it is often localized in presynaptic neuron terminals, with cognate receptors present on postsynaptic neurons. Here, we review select cases where PACAP signals via its associated receptors and intracellular pathways to induce acute or persistent changes in cellular components involved in synaptic transmission, thereby supporting its role as a neurotransmitter and/or neuromodulator. Two themes that emerge are that PACAP can rapidly induce synaptic plasticity by altering the function of presynaptic and postsynaptic components, or change general membrane excitability by altering the balance of voltage-dependent Ca^{2+} and K^{+} channel contributions to action potential firing. We discuss the need to determine how these two themes are related, to extend studies to endogenous PACAP, and to determine how the synaptic modifications induced by PACAP contribute to its postulated roles in widespread behaviors and stress disorders such as PTSD.

Keywords PACAP • PAC_1R • Synapse • Synaptic transmission • Synaptic plasticity • Neuromodulator

Abbreviations

AC	Adenylate cyclase
ACh	Acetylcholine
AChR	ACh receptor
AKAP	PKA anchoring protein

E.R. Starr • J.F. Margiotta (✉)
Department of Neurosciences, University of Toledo College of Medicine and Life Sciences, 3000 Transverse Drive, Toledo, OH 43614, USA
e-mail: Joseph.Margiotta@Utoledo.edu

D. Reglodi, A. Tamas (eds.), *Pituitary Adenylate Cyclase Activating Polypeptide — PACAP*, Current Topics in Neurotoxicity 11,
DOI 10.1007/978-3-319-35135-3_7

AMPAR	α-amino-3-hydroxy-5-methyl-4-isoxazolepropionic acid receptor
BLA	Basolateral amygdala subnucleus
BNST	Bed nucleus of the stria terminalis
CA	Catecholamine
CA3 or CA1	Cornu ammonis (Ammon's horn) hippocampus region 3 or 1
CAMKII	Ca^{2+}/Calmodulin kinase II
cAMP	Cyclic adenosine monophosphate
CdG	Cardiac ganglion
CeA	Central amygdala subnucleus
CG	Ciliary ganglion
CN	Cranial nerve
CREB	cAMP response element binding protein
Epac	Exchange protein factor directly activated by cyclic AMP
EPSC or EPSP	Excitatory postsynaptic current or potential
EW	Edinger–Westphal
GluA1	AMPAR subunit 1
GPCR	G-protein coupled receptor
G-protein	Guanine nucleotide binding protein
G_q	G-protein PLC-stimulatory α-subunit
G_s	G-protein AC-stimulatory α-subunit
HCN	Hyperpolarization-activated cyclic nucleotide-gated
IML	Spinal cord intermediolateral
IP	Inositol phosphate
IP3	IP triphosphate
lCeA	Lateral CeA
LTP or LTD	Long-term potentiation or depression
m	Quantal content
MAPK	Mitogen activated protein kinase
mEPSC or mEPSP	Miniature EPSC or EPSP
nAChR	Nicotinic AChR
NMDAR	*N*-methyl-D-aspartate receptor
NO	Nitric oxide
NOS1	Neuronal NO synthase
p	Probability of neurotransmitter release
PAC_1R	Type I PACAP receptor
PACAP	Pituitary adenylate cyclase activating polypeptide
pCREB	Phospho-CREB
PKA	Protein kinase A
PKC	Protein kinase C
PLC	Phospholipase-C
PP1/PP2A	Protein phosphatases 1 and 2A

PTSD	Post-traumatic stress disorder
PVN	Paraventricular nucleus
q	Quantal size
Ras-Raf-MEK-ERK-	MAPK pathway effectors
RGC	Retinal ganglion cell
SCG	Superior cervical ganglion
SCN	Suprachiasmatic nucleus
sEPSC or sEPSP	Spontaneous EPSC or EPSP
SMG	Submandibular ganglion
TTX	Tetrodotoxin
VDCC	Voltage dependent Ca^{2+} channel
VIP	Vasoactive intestinal peptide
$VPAC_1R$ $VPAC_2R$	Type II PACAP receptors
α3*-nAChR	Heteropentameric α3-nAChR subunit containing nAChR
α7-nAChR	Homopentameric α7-containing nAChR

Introduction

As indicated by work in this volume and elsewhere, PACAP is a widely expressed neuropeptide that displays a startling array of functions. While impressive in scope, such features challenge attempts to develop hypotheses about the mechanisms and significance of endogenous PACAP signaling in health and disease. The challenge is apparent in the nervous system where PACAP is expressed in central and autonomic neurons and has been implicated in neurotrophic, neuroinflammatory, anxiogenic, and neuromodulatory functions that utilize an array of context-dependent mechanisms for their cell-specific expression. At synapses in particular the very nature of PACAP as a neurotransmitter or neuromodulator is still debated, PACAP can alter membrane excitability as well as presynaptic and postsynaptic function, and such actions can be acute or persistent, and caused by rapidly activated intracellular signal pathways or long-term transcription-dependent mechanisms, respectively. Despite the relevance of synapses to behavior, and the recent highlighting of PACAP as a possible biomarker for post-traumatic stress disorder (PTSD), the precise nature of the impact of PACAP on synaptic transmission remains an emerging picture. The goal of this chapter is to frame the picture by focusing on selected case studies where PACAP has been shown to alter synaptic function and components underlying transmission, and to illuminate relevant cellular mechanisms. We anticipate that doing so will provide insights into the cell-specific role of endogenous PACAP in evoking synaptic modifications. It may also suggest possible clues about the relevance of such modifications to PACAP's emerging role in social, motor, feeding, and anxiogenic behaviors, and in stress disorders including PTSD. The results are summarized in Table 7.1 and a simplified rendition provided in Fig. 7.1.

Table 7.1 Summary of PACAP's synaptic localization and related functions

PACAP/receptor localization						PACAP effects					
Region Pre→Post Conv Pre Tm Post R	Pre PACAP (Y/N/?)	Pre PACAP release (Y/N/?)	Post GPCR	PACAP= NT? (Y/N/?)	Refs	PACAP source	Δ Synaptic/ membrane effector	Δ Synaptic/ membrane outcome	Signaling	PACAP =NM? (effect)	Refs
Hypothalamus RGC→SCN Glutamate AMPAR NMDAR	Y	≈Y	PAC_1R $VPAC_2R$	≈Y	[23–25]	Applied	↑q ↑TM release	$\uparrow A_{mEPSP(AMPAR)}$ $\uparrow F_{mEPSC(AMPAR)}$	PAC_1→AC→PKA	Y (+)	[24]
Hippocampus CA3→CA1 Glutamate AMPAR NMDAR	Y	?	PAC_1 R $VPAC_1R$ $VPAC_2R$	?	[35, 36, 119, 120]	Applied [Low] Applied [High]	? ↑GluA1-P_{S845} ↓GluA1-P_{T840} ? ↑GluA1-P_{S845} ↓GluA1-P_{T840}	$\uparrow I_{NMDAR}$ $\uparrow I_{AMPAR}$ $\downarrow I_{AMPAR}$	PAC_1→PLC→Pyk2→Src PAC_1→AC→PKA PAC_1/ $VPAC_2$→AC→PKA PAC_1→PP1/PP2A $VPAC_2$→AC→Epac→MAPK PAC_1/ $VPAC_2$→AC→PKA PAC_1→PP1/PP2A	Y (+) Y (−)	[29, 30, 37, 39–42]
Amygdala BLA→lCeA Glutamate AMPAR NMDAR	Y, DVN of CN X	?	PAC_1 R $VPAC_1R$ $VPAC_2R$	?	[47, 49, 55]	Applied	≠*p* ↑[AMPAR]	$\uparrow F_{EPSP(AMPAR)}$ $\uparrow A_{EPSP(AMPAR)}$	$VPAC_1$→AC→PKA→CaMKII	Y (+)	[47]
BNST GABA Glut GABAR AMPAR NMDAR AChR	Y	?	PAC_1 R $VPAC_1R$	?	[55–57, 61]	Applied	?	?	PAC_1 R→Anxiogenic response	?	[61]
Ciliary ganglion CN III→CG CG→CG ACh α3*- α7-nAChR	Y ?	Y ?	PAC_1 R PAC_1 R	Y ?	[10, 14, 19, 75]	Endogenous Applied (acute) Applied (persistent)	? α3*-nAChR α7-nAChR ↑m ≠*q* VDCC ↑m ↑*q* VDCC	? $\uparrow I_{\alpha 3^*/\alpha 7\text{-}nAChR}$ $\uparrow F_{sEPSC(\alpha 3^*R)}$ $\uparrow A_{sEPSC(\alpha 3^*R)}$ $\downarrow I_{Ca}$ & Depol ↓Excitability $\uparrow F_{sEPSC(\alpha 3^*R)}$ $\uparrow\uparrow A_{sEPSC(\alpha 3^*R)}$ $\uparrow I_{Ca}$	? PAC_1→AC→PKA PAC_1→AC→PKA/AKAP→ nAChR/Ca^{2+}→ NOS1→NO PAC_1→AC PAC_1→AC→PKA→mRNA ?	? Y(+) Y(+) Y (−) Y (++) ?	[14, 19, 20, 75, 118]

Cardiac ganglion CN X→CdG ACh α3*- α7-nAChR	Y	Y	PAC_1 R	Y	[5, 85–89, 121]	Applied	VDCC HCN nAChR	↑I_{Ca} Depolarization ↑Excitability ↑I_{nAChR}	PAC_1R→AC→PKA→MEK $PAC_1R/VPAC_2R$→G_o(?)	Y(+)	[5, 91, 92, 95]
Submandibular ganglion CN VII→SMG ACh α3*- α7-nAChR	?	?	PAC_1 R VPACR	?	[97–101]	Applied	VDCC nAChR	↓I_{Ca} Depolarization ↑Excitability ↑I_{nAChR}	$VDCC_{P,Q,N}$: PAC_1R→ $G_s\beta\gamma$ $VDCC_L$: VPACR→AC→PKA/PLC/PKC $PAC_1R/VPAC_2R$→G_o(?)	Y(+)	[95, 103, 122]
Sup. Cervical ganglion IML→SCG ACh α3*- α7-nAChR	Y	?	PAC_1 R	?	[106–109, 112, 113, 116]	Applied	?	Depolarization ↑g_{Na}, ↓g_K ↑CA release ↑Excitability?	PAC_1R→PLC→IP3→IP3R	Y(+)	[116]

Rows depict nervous system location and major columns PACAP localization and effects on excitability and synaptic function. See text for details. Table-specific abbreviations: Presynaptic (Pre); Postsynaptic (Post); Conventional (Conv); Neurotransmitter (NT), Receptor (R); References (Refs); Dorsal ventral nucleus (DVN); Change (Δ); Neuromodulator (NM); Increase (↑); Decrease (↓); $Current_{channel\ type}$ ($I_{channel\ type}$); Synaptic event frequency (F); Synaptic event amplitude (A); VDCC subtypes ($VDCC_{P,Q,N,L}$); Na^+ or K^+ conductance (g_{Na}, g_K)

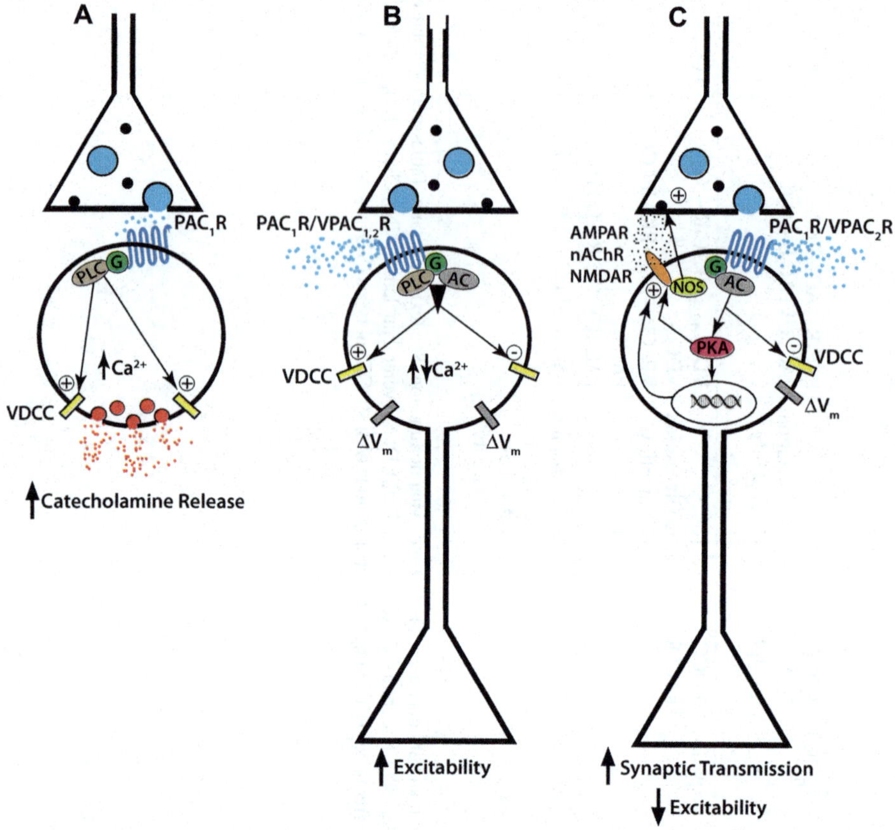

Fig. 7.1 Simplified summary of PACAP actions at synapses where PACAP is detected in presynaptic terminals and its receptor isoforms are found on postsynaptic cells. (**a**) In the adrenal gland, PACAP is an endogenous neurotransmitter that acts via chromaffin cell PAC_1Rs to stimulate CA release. (**b**) Notably in autonomic ganglion neurons, applied PACAP acts via PAC_1Rs and/or VPACRs to alter VDCC function that causes membrane depolarization and in most cases enhances excitability (see however **c**). (**c**) Notably in central and parasympathetic neurons, PACAP acts via PAC_1Rs and $VPAC_2Rs$ to induce synaptic plasticity in a time-dependent fashion. Acute effects are mediated largely via PAC_1Rs and AC and can be postsynaptic, featuring enhanced ionotropic receptor currents, and/or presynaptic featuring enhanced transmitter release, mediated as exemplified by CG neurons via nAChR dependent activation of NOS1. Persistent synaptic plasticity, requiring gene transcription and exemplified by CG neurons, features increases in both conventional presynaptic transmitter (ACh) release and postsynaptic unitary nAChR-mediated mEPSC amplitudes. In CG neurons acute PACAP application also *reduces* VDCC currents and *depresses* membrane excitability. Filled *dark circles* depict PACAP-, conventional transmitter- or CA-containing vesicles (*black*, *blue*, or *red*, respectively) with their respective cargos depicted in the same colors. ΔV_m refers to a change in resting potential. Positive and negative actions are indicated by encircled + and – symbols

Is PACAP a Neurotransmitter, a Neuromodulator or Both?

Put briefly, the answer depends on whether and how you look. The three minimal criteria for a neurotransmitter are its presence in and activity-dependent release from presynaptic nerve terminals (1, 2) and the expression of cognate receptors on adjacent postsynaptic cells (3) [1]. Neuromodulator criteria in some cases include, while in other cases exclude, those of neurotransmitters but the main distinctions are that neuromodulators can alter the presynaptic release or postsynaptic efficacy of conventional neurotransmitters, and can have effects that are spatially and temporally more widespread, impacting extrasynaptic receptors even those on non-adjacent cells, and evoking changes that can persist for minutes or, through transcription-dependent processes, even days [2–4]. The existence of dual transmission where peptides such as PACAP and conventional neurotransmitters such as acetylcholine (ACh) are released from the same nerve terminal under different stimulus conditions (reviewed by [4]) complicates a clear neurotransmitter/neuromodulator distinction. At all of the nerve–nerve synapses considered here, including those on hypothalamic, hippocampal, amygdala, and autonomic neurons, PACAP signals via high-affinity guanine nucleotide binding protein coupled receptors (GPCRs) on postsynaptic neurons (satisfying neurotransmitter criterion 3 above) and acts as a neuromodulator because it alters membrane properties or the release or efficacy conventional neurotransmitters. In many of the cases cited PACAP and conventional transmitters are also co-expressed in the same presynaptic terminal (satisfying criterion 1 above). For most, it is inferred or indirectly determined that PACAP is also co-released (criterion 2). In parasympathetic cardiac and ciliary ganglia, criterion 2 has been satisfied directly by demonstrating activity-dependent release of PACAP from ACh containing nerve terminals [5] qualifying it as a definitive neurotransmitter in these systems.

In the adrenal gland, PACAP satisfies the three criteria as a neurotransmitter since it is expressed in presynaptic splanchnic nerve terminals [6] and likely co-released with ACh onto postsynaptic chromaffin cells that express PACAP receptors and nicotinic ACh receptors (nAChRs). Splanchnic nerve stimulation as well as exogenous ACh and PACAP stimulate catecholamine (CA) secretion from chromaffin cells but CA secretion occurs even when nAChRs are desensitized or blocked suggesting ACh and PACAP operate independently as dual neurotransmitters [4, 7]. Acting as a neurotransmitter, PACAP is thought to stimulate its selective high-affinity receptor (PAC_1R) and associated phospholipase-C (PLC) signaling pathways (See Below), leading to voltage dependent Ca^{2+} channel (VDCC) activation and subsequent CA secretion [4] (Fig. 7.1a). The adrenal gland may be a unique case, however, since chromaffin cells unlike neurons are specialized for CA secretion, rather than synaptic communication with a postsynaptic target, and may be maximally increased by PACAP-induced Ca^{2+} mobilization, masking any contribution on secretion from accompanying modulatory effects on nAChR mediated transmission. Thus, with the possible exception of the specialized adrenal-chromaffin synapse where it apparently acts solely as a neurotransmitter, PACAP fulfills criteria as a neuromodulator *and* neurotransmitter at nerve–nerve synapses, with the caveat that for the latter case the release criterion has been verified in a limited number of cases.

General Remarks About PACAP Signaling

PACAP rapidly stimulates intracellular signal cascades, and over the long term alters gene transcription. PACAP is recognized by specific Class B secretin-like GPCRs that are classified according to their affinity for PACAP over its close relative, vasoactive intestinal peptide (VIP), and by their potencies for linking peptide binding to activation of canonical adenylate cyclase (AC) and PLC effectors within seconds [8–14]. Type I receptors (PAC_1Rs) have ≈1000-fold higher affinity for PACAP over VIP, while Type II receptors (VPACRs) have similar affinities for the two peptides. At PAC_1Rs, PACAP signals via G-protein α-subunit G_s to activate AC, increasing cyclic adenosine monophosphate (cAMP) production with ≈1000-fold greater potency than VIP; this difference is less dramatic at $VPAC_1Rs$ and the two peptides display similar potencies at $VPAC_2Rs$. PAC_1R signaling via G_q activates PLC causing inositol phosphate (IP) turnover, IP triphosphate (IP3) production and increasing intracellular Ca^{2+} levels with much greater potency than VIP while this ability is less apparent at $VPAC_1Rs$ and disappears at $VPAC_2Rs$. By virtue of activating PLC, PAC_1Rs and $VPAC_1Rs$ also couple via diacylglycerol to stimulate protein kinase C (PKC). AC activation and the resultant increase in cAMP levels stimulates protein kinase A (PKA) and associated mitogen activated protein kinase- (MAPK-; Ras-Raf-MEK-ERK-) signaling as well as exchange protein factor directly activated by cyclic AMP- (Epac-) mediated signaling [9]. Given the wide reach of these intracellular systems, it is not surprising that PACAP potently regulates synaptic components by virtue of its ability to rapidly activate AC, Epac, PKC, and/or PLC effectors as well as their downstream targets, but it does introduce a conundrum. Specifically, while PAC_1Rs and VPACRs can activate both AC and PLC, mechanisms whereby one effector's actions on downstream targets are specified or predominate remain largely unknown. For example, PACAP rapidly activates AC and PLC signaling in ciliary ganglion (CG) neurons, but while AC signaling resulting from PAC_1R activation normally predominates and is responsible for increasing the sensitivity of both heteropentameric $\alpha3\beta4\alpha5\pm\beta2$-containing and homopentameric α7-containing nAChRs (α3*- and α7-nAChRs, respectively), blocking AC to allow PAC_1R-induced PLC signaling to predominate results in rapid α7-nAChR inhibition [14].

Paralleling its ability to recruit intracellular signaling effectors and cause rapid changes in synaptic components, PACAP alters gene transcription as indicated by its ability to induce phosphorylation and nuclear translocation of the transcription factor cAMP response element binding protein (CREB→pCREB) [15–17]. Gene array studies in autonomic neuron cultures indicate PACAP increases the transcription of several genes including those encoding growth and differentiation factors, peptides, channels, cytoskeletal proteins, and transporters [17, 18]. In CG neuron cultures such increases are observed 48 h after even brief (15 min) exposure to PACAP [17] and correlate with long-term, transcription-dependent alterations in synaptic function that occur over the same time course and differ from those seen immediately following brief exposure [19, 20]. While these studies are intriguing, the requisite signal pathways, and alterations in synaptic gene transcription and translation relevant to PACAP inducing long-term synaptic plasticity have yet to be determined.

PACAP Signaling at Central Synapses

Hypothalamus

The hypothalamus receives abundant PACAP innervation. In one example photosensitive retinal ganglion cells (RGCs) utilizing melanopsin as photopigment [21] project to hypothalamic suprachiasmatic nucleus (SCN) neurons that function to synchronize hypothalamic neuron activity to the circadian daily cycle of light and dark [22]. PACAP is co-expressed with glutamate in the RGC projections [23] and postsynaptic SCN neurons express ionotropic glutamate receptors (*N*-methyl-D-aspartate and α-amino-3-hydroxy-5-methyl-4-isoxazolepropionic acid receptors; NMDARs and AMPARs), as well as PAC_1Rs and $VPAC_2Rs$ (reviewed by [24]). Evidence for PACAP release from RGC axon terminals is indirect [25] but exogenous PACAP does potently modulate glutamate-mediated transmission in the SCN [24]. Here, both the frequency and amplitude of AMPAR-mediated spontaneous miniature excitatory postsynaptic currents (mEPSCs) recorded from SCN neurons increased minutes after 10 nM PACAP application. The effects are consistent with PACAP enhancing presynaptic glutamate release and postsynaptic AMPAR and NMDAR sensitivity, with the latter mediated via PAC_1R activation of AC and PKA. These findings indicate that PACAP is a potent neuromodulator and a possible neurotransmitter at RGC→SCN hypothalamic synapses. Interestingly, PACAP phase shifts the glutamate-induced circadian rhythm in SCN neurons [26] and the circadian system is perturbed in mice deficient in PAC_1R or PACAP [27, 28] suggesting that such modulation may actively influence synchronization of SCN neurons to the light dark cycle in vivo.

Hippocampus

The hippocampus regulates cognitive functions including memory formation, storage and retrieval. Glutamatergic synapses connecting CA3 to CA1 hippocampal pyramidal neurons (CA3→CA1) are important models for synaptic plasticity and memory [1, 29–32]. CA3 neuron axons (Schaffer collaterals) produce NMDAR- and AMPAR-mediated EPSCs in pyramidal CA1 neurons. Depending on stimulus rate, Schaffer-collateral stimulation can produce homosynaptic long-term potentiation or depression (LTP or LTD) at the CA3→CA1 synapse. LTP can be induced by a 10–50 Hz tetanic stimulation of Schaffer collaterals for 5–15 min [33]. Stimulation of Schaffer collaterals at 1 Hz for 10 s to 15 min, however, is sufficient to induce LTD [34]. Physiologically, LTP is characterized by enhanced postsynaptic NMDAR and AMPAR sensitivity, increased AMPAR trafficking and expression in postsynaptic densities, and enhanced glutamate release from presynaptic CA3 terminals [32]. Conversely, LTD features reduced AMPAR sensitivity and expression in postsynaptic CA1 terminals and is associated with reduced quantal release of transmitter from presynaptic CA3 neuron terminals [32].

PACAP is synthesized in CA3 Schaffer collateral neurons [35], and CA1 pyramidal neurons express PAC_1Rs, $VPAC_1Rs$, and $VPAC_2Rs$ [36] but activity dependent PACAP release has not been demonstrated. Interestingly, exogenous PACAP induces synaptic potentiation or depression depending on the concentration applied. At low concentrations, (<10 nM) PACAP enhances CA3→CA1 synaptic transmission, characterized by a prolonged increase NMDAR and AMPAR sensitivity [29, 30, 37]. The PACAP-induced increase in NMDAR sensitivity involved a PAC_1R-induced PLC- and PKC-dependent activation of non-receptor tyrosine kinase, Pyk2, and subsequent recruitment and phosphorylation of NMDARs by Src kinase (Reviewed in [29, 38]). The PACAP induced increase in AMPAR-sensitivity is activated through a PAC_1R induced AC and PKA signal transducing cascade [30] as is increased phosphorylation at S845 on the GluA1 AMPAR subunit, while an accompanying decrease in phosphorylation of GluA1 T840 is mediated via protein phosphatases 1 and 2A (PP1/PP2A) [39]. It is still unknown if the PACAP-induced enhancements in AMPAR sensitivity or inverse phosphorylation of sites on GluA1 are correlated with alterations in AMPAR synthesis, trafficking or changes in channel properties. Contrasting with the low concentration effects, PACAP applied at ≥10 nM induces a sustained depression of CA3→CA1 synaptic transmission that is associated with an AC, Epac, p38 MAPK-mediated reduction in AMPAR ion currents [30, 40–42]. Surprisingly, the synaptic plasticity induced by high PACAP concentrations was independent of PAC_1R, PKA, PKC, and NMDAR activation but mimicked by activation of $VPAC_2Rs$ concluded using the selective $VPAC_2R$ agonist Bay 55-9837 [30, 42]. It is still unknown whether the reduction of AMPAR currents produced by high concentrations of PACAP is associated with changes in the surface expression or channel properties of AMPA receptors or whether there is an accompanying reduction in glutamate release from presynaptic Schaffer collateral axon terminals.

Taken together these studies demonstrate that PACAP signaling exerts potentiation or depression in a concentration-dependent fashion at CA3→CA1 synapses. While the outcomes of PACAP signaling appear inversely related to LTP and LTD, it will be important to determine how different stimulation rates producing LTP versus LTD correlate with actual PACAP levels released by Schaffer collateral terminals. In addition, while the PACAP-induced synaptic potentiation and depression effects are mediated by the activation of PAC_1Rs and $VPAC_2Rs$, respectively, the mechanisms whereby PACAP will potentiate or depress synaptic transmission remain unknown. Experiments with VIP provide a clue since VIP applied at low concentration (1 nM) potentiates CA3→CA1 synaptic transmission to a similar extent as low concentrations of PACAP. The effects of VIP were mediated by both $VPAC_1Rs$ and $VPAC_2Rs$ and by subsequent activation of PKA mediated signal cascades [43, 44]. These findings indicate that $VPAC_2R$ activation does not necessarily result in synaptic depression making it seem unlikely that the PACAP-induced potentiation or depression at CA3→CA1 synapses is mediated by the activation of distinct PACAP receptor subtypes. Instead, a possible explanation is that the mechanisms regulating the differential effects of PACAP are associated with intracellular cAMP-induced signal transducing pathways. Because of the role for PKA signaling in the PACAP-induced potentiation and the role for Epac in the PACAP-induced depression, it would be useful to examine possible connections between PACAP receptor surface localization, and intracellular cAMP dynamics in mediating the effects of PACAP on CA3→CA1 terminals. Lastly since both low and high PACAP concentrations produced similar AMPAR GluA1 S845 phosphorylation and T840 dephosphorylation, further study of possible concentration dependent actions on AMPAR channel properties or insertion as proposed outcomes [39] seem warranted. Such studies could clarify how PACAP signaling can dynamically alter synaptic transmission at CA3→CA1 synapses and help elucidate relevant mechanisms.

Amygdala

The amygdala is a subcortical nucleus associated with motivational and emotional behaviors and processing anxiogenic stimuli, notably those producing innate and learned fear [45, 46]. Neurons in the basolateral amygdala (BLA) subnucleus functionally integrate anxiogenic sensory stimuli, and project glutamatergic outputs to neurons in the central amygdala (CeA) subnucleus that modify anxiogenic and fear associated behaviors [46]. PACAP is expressed in inputs to the lateral CeA (lCeA) [47] likely arising from PACAP-containing neurons in the vagus nerve (CN X) brainstem dorsal complex [48, 49]. In addition CeA neurons express both PAC1Rs and VPACRs [9, 36]. Consistent with these observations, exogenous PACAP robustly alters activity at BLA→lCeA synapses [47]. Specifically, transient PACAP exposure (5 nM, 10 min) increased the number and amplitude of excitatory postsynaptic potentials (EPSPs) in lCeA neurons evoked by stimulating presynaptic BLA neurons for up to 1 h. Subsequent whole-cell recordings revealed that PACAP signaling enhanced AMPAR-mediated evoked EPSC amplitudes in lCeA neurons. Paired-pulse experiments revealed no significant differences between PACAP-treated and control neuronal responses, making it unlikely that PACAP alters the probability (p) of glutamate release from presynaptic BLA terminals, and supporting a postsynaptic mechanism. The effects of PACAP were dependent on a VPAC1R induced, cAMP/PKA and Ca^{2+}/Calmodulin kinase II (CAMKII) mediated signal transduction cascade. Interestingly, pretreatment with Pep1-TGL (200 μM), a synthetic peptide that inhibits GLU1R trafficking by disrupting the interaction of the C-terminus end of the GluA1 subunit with synapse associated protein SAP97, and critical for intracellular AMPAR trafficking and surface expression [50] blocked the effects of PACAP. Taken together these results suggest that PACAP induces plasticity at BLA→lCeA synapses by increasing AMPAR density at postsynaptic sites through cAMP/PKA and CAMKII dependent trafficking of AMPARs [47]. Further experiments will be required to verify whether PACAP increases the number of AMPARs expressed in postsynaptic densities. It would also be interesting to determine if higher PACAP concentrations induce synaptic depression in BLA→lCeA synapses as is observed at hippocampal CA3→CA1 synapses. Such studies might elucidate common signal transducing mechanisms underlying PACAP induced potentiation and depression at glutamatergic synapses throughout the central nervous system. Lastly, while the amygdala processes anxiogenic sensory cues linked to PTSD, and fear conditioning increases PAC1R mRNA in mouse amygdala [51, 52] a direct relationship between amygdaloid synaptic plasticity induced by PACAP and fear consolidation has yet to be established.

As part of the "extended amygdala" the bed nucleus of the stria terminalis (BNST) is a crucial integration center that regulates visceral homeostasis and behavioral responses to unpredictable and chronic stress [53, 54]. BNST axons project to several nuclei in the hypothalamus and the brainstem that mediate homeostasis and autonomic circuitry. Among hypothalamic targets is the paraventricular nucleus, a region believed responsible for activating the hypothalamic–pituitary axis during responses to stress [54]. BNST neurons receive functional GABAergic and glutamatergic synaptic inputs. Additionally, presynaptic terminals to BNST neurons express several neuropeptides that regulate synaptic transmission [53, 54]. While PACAP is present in presynaptic terminals localized to the lateral BNST [55, 56] it is not known if PACAP is co-released by presynaptic activity with conventional neurotransmitters. Autoradiographic experiments have demonstrated that BNST neurons express $VPAC_1Rs$ and not $VPAC_2Rs$ [57].

Several recent studies have supported a neuromodulatory role for PACAP signaling in the BNST. Of particular relevance to PTSD and stress-related disorders, rats exposed to chronic unpredictable stress developed concurrent increases in both PAC_1R and PACAP mRNA synthesis in the BNST. Subsequent experiments utilizing a single bilateral BNST infusion of PACAP identified that PACAP enhanced acoustic startle response [58] and caused significant weight loss in male and female rats [59]. Furthermore single infusions of PACAP to the dorsal lateral BNST elevated blood plasma corticosterone [60]. Interestingly, these anxiogenic PACAP effects were found to be mediated by PAC_1R activation [61] since infusions of the PAC_1R agonist maxadilan mimicked the effects while infusion of the PAC_1R antagonist PACAP(6-38) (40 μM) blocked them [61]. Taken together these findings identify that PACAP signaling in the BNST can exert an anxiogenic influence on behavior. Additionally, these studies highlight critical questions regarding whether the behavioral effects of PACAP on the BNST are associated with physiological alterations in the BNST. Specifically, it remains to be seen if PACAP exerts these anxiogenic effects by altering the excitable and/or synaptic properties of BNST neurons. Such studies could implicate a potential endogenous role for PACAP signaling in regulating BNST neuronal activity underlying behavioral responses to stress and illuminate the basis of stress disorders such as PTSD.

PACAP Signaling at Autonomic Synapses

Ciliary Ganglion

The parasympathetic avian CG is a useful model for understanding the development and function of fast chemical synapses in the nervous system (reviewed in [62, 63]). Presynaptic accessory oculomotor cranial nerve (CN) III (Edinger–Westphal, EW) nucleus inputs to postganglionic ciliary and choroid neuron populations within the CG release ACh, activating postsynaptic α3*- and perisynaptic α7-nAChR subtypes that underlie synaptic transmission through the ganglion [64–67]. At the giant, calyx-like presynaptic terminals of EW axons that innervate ciliary neurons, PACAP- and synapsin-I-immunolabeling co-localize, and KCl depolarization abolishes PACAP- without affecting synapsin-I-immunolabeling [19]. These results demonstrate that PACAP and ACh are present in the same presynaptic terminals within the CG, and that PACAP can be released by depolarizing stimuli. Neurotransmission from EW axons to ciliary neurons reliably follows stimuli delivered at frequencies up to 50 Hz [68]. Moreover 16–20 Hz stimulation is sufficient to trigger PACAP release from splanchnic nerve terminals and intercardiac nerve fibers [69] making it likely that PACAP is released from EW nerve terminals in vivo. Since CG neurons express abundant PAC_1Rs [10, 14] PACAP fulfills criteria as a neurotransmitter in this system, and is also well positioned to modulate fast nAChR mediated transmission at CG neuron synapses.

Our past and recent findings support a role for PACAP as a potent agent of both acute and persistent synaptic plasticity. PAC_1Rs on CG neurons couple to AC and PLC effectors and efficiently mobilize cAMP and induce IP turnover ($EC_{50} \approx 1$ nM for both) [10, 14]. Consistent with the ability of 8-Bromo-cAMP to increase the nAChR sensitivity of CG neurons [70] PACAP enhanced responses from both α3*- and α7-nAChRs in a PAC_1R-, AC-, and PKA-dependent fashion [10, 14]. As discussed above for NMDARs and AMPARs on hippocampal CA1 neurons, however, the downstream mechanisms and relevant phosphorylation targets responsible for the PACAP-induced enhanced sensitivity of α3*- and α7-nAChRs have yet to be determined. Since nAChR-mediated synapses form between CG neurons in culture, and resemble those seen on the neurons in vivo, particularly with regard to the differential expression of α3*- and α7-nAChRs at post- and peri/extra-synaptic sites, respectively [71, 72] we have used this accessible system to evaluate the impact of PACAP on nAChR-mediated synaptic transmission. Applying 100 nM PACAP for 5–15 min markedly enhanced synaptic activity in CG cultures, increasing the frequency and amplitude of spontaneous impulse-dependent nAChR-mediated EPSCs (sEPSCs) by 100–200 % and 40–50 %, respectively [19] and doing so without increasing neuronal excitability as with other autonomic neurons (See Below). The rapidly induced synaptic plasticity in CG neurons occurred by PAC_1R activation and was AC- and PKA-dependent, and PLC-independent. Despite the ability of similar PACAP treatments to enhance α3*- and α7-nAChR sensitivity, analysis of stimulus-evoked nAChR-mediated EPSCs revealed that PACAP enhanced synaptic function by increasing *pre*synaptic ACh release (quantal content, *m*) without affecting the unitary *post*synaptic response (quantal size, *q*) assessed from the amplitude of mEPSCs acquired in the presence of tetrodotoxin (TTX) [19]. The PACAP-induced presynaptic plasticity resulted from activation of Ca^{2+}/Calmodulin- and PKA-dependent neuronal nitric oxide synthase (NOS1) [73, 74] and subsequent increased production of NO [19] that was detected in CG neurons using the fluorescent NO indicator, *4-amino-5-methylamino-2′,7′-difluorofluorescein* [75, 76]. Applying the NO scavenger *2-4-carboxyphenyl-4,4,5,5- tetramethylimidazoline-1-oxyl-3-oxide* blocked the ability of PACAP to alter sEPSC frequency or amplitude, consistent with a mechanism where NO produced in postsynaptic neurons diffuses to presynaptic terminals thereby enhancing ACh release in a retrograde fashion [77–79]. Interestingly, co-immunoprecipitation experiments indicated that nAChRs and NOS1 associate in a molecular complex, and functional studies revealed that α7-nAChR activity was required for PACAP to induce PKA-dependent synaptic plasticity by a process localized to the cell membrane by A-kinase anchoring proteins (reviewed by [80]). These results support a mechanism where PACAP enhances presynaptic ACh release via retrograde NO action, and where NO is produced by upregulating Ca^{2+} permeable NOS1 associated nAChRs, with the largest contribution from α7-nAChRs present at perisynaptic and extrasynaptic membranes of postsynaptic neurons [75].

In accord with its neurotrophic actions, PACAP activates MAPK dependent pro-survival pathways [81] and as mentioned above, induces nuclear pCREB [17] indicating it can potently alter gene transcription, and suggesting it may have long-term transcription dependent effects on synaptic function and structure. Indeed, PACAP enhanced synaptic function in CG cultures >48 h after a transient 15 min exposure [19, 20]. The persistent plasticity was not simply a prolonged version of that seen immediately after PACAP exposure. While featuring similar increases in sEPSC frequency and quantal content, there was a more profound increase in sEPSC amplitude compared to similar measurements made immediately after PACAP exposure, and, consistent with a postsynaptic effect, a significant increase in postsynaptic quantal size. Moreover, unlike the acute PACAP-induced plasticity, the sustained plasticity was NOS1-independent and required gene transcription [20]. In a gene array study we detected a number of transcripts relevant to ACh processing that were upregulated 96 h after transient exposure to PACAP including the catalyzing enzyme, choline acetyltransferase (1.5-fold after 15 min and 2.2-fold after 24 h PACAP exposure), with nearly identical increases seen for the high-affinity choline transporter, responsible for rate-limiting active transport of choline into the nerve terminal [17]. Relevant to ACh storage, the vesicular acetylcholine transporter transcript increased

2.8-fold at 24 h. While the ability of PACAP to upregulate gene transcripts associated with ACh release, synthesis, and storage is consistent with its persistent positive presynaptic influence on nicotinic neurotransmission, the overall picture is still incomplete. Further studies are required to explore relevant cellular mechanisms, particularly to identify upregulated genes and, since nAChR transcripts were not altered by any of the PACAP treatments [17], to assess whether structural changes in synapse morphology or nAChR clustering may account for the postsynaptic changes associated with the sustained synaptic plasticity induced by PACAP.

Cardiac Ganglion

The parasympathetic cardiac ganglion (CdG) controls heart rate and blood flow to the cardiac muscle. CdG neurons are innervated by CN X axons and supply the sinoatrial node, the atrioventricular nodes, and the atrial and ventricular musculature [1, 82–84]. In guinea pig, PACAP is co-expressed with choline acetyltransferase in CN X input terminals, released in an activity dependent fashion, and activates PAC_1Rs on CdG neurons [5, 85–89]. Consistent with these neurotransmitter characteristics, focal PACAP application (50 μM, 500 ms) caused a slow depolarization (slow EPSP) in CdG neurons that was reduced by PACAP antagonist and that elicited bursts of action potentials, similar to those seen upon high frequency preganglionic stimulation. The results reveal an additional neuromodulatory action whereby the PACAP-induced depolarization reflects increased CdG neuron excitability [5]. Pharmacological studies indicate that both of these actions are mediated by a PAC_1R-induced AC, PKA, MEK signal transduction pathway [5, 90] requiring extracellular Ca^{2+} entry [91]. In accord with the dependence of increased excitability on AC activation, PACAP signaling induces a depolarizing shift in the activation of hyperpolarization-activated cyclic nucleotide-gated (HCN) channels and contributing to the development of the PACAP-induced depolarization [92]. Treatment with the cAMP analog, 8-br-cAMP or AC agonist, forskolin only partially mimicked the effects of PACAP, however, suggesting that other cellular mechanisms in addition to AC signaling contribute to the effects of PACAP on CdG neuron excitability [90]. Other studies have expanded upon these results showing that PACAP-induced increases in neuron excitability also depend on Ca^{2+} influx via T- and R-type voltage dependent Ca^{2+} channels (VDCC) that are linked to slow depolarization and repetitive firing [91, 93] as well as PAC_1R internalization [94]. While the increases in membrane excitability elicited by the PACAP-induced depolarization would be expected to affect nAChR mediated transmission in the CdG, this possibility has not been tested. Performing such studies is supported by the finding that PACAP increases nAChR currents in CdG and submandibular ganglion (SMG) neurons [95] and could elucidate a dual role for PACAP as a transmitter in mediating slow EPSCs and as a modulator regulating the excitability of these neurons and nAChR-mediated ganglionic transmission.

Submandibular Ganglion

The parasympathetic SMG increases salivary secretion from and blood flow to the submandibular gland [96]. The SMG receives preganglionic cholinergic fibers originating from the superior salivary nucleus of CN VII [97]. Several neuropeptides are expressed in these input terminals to the SMG [97] but neither the presence of PACAP in nor its release from these terminals has been demonstrated. Imaging and electrophysiological studies indicate that SMG neurons express nAChRs [98, 99] as well as PAC_1Rs and VPACRs, but it is not known which VPACR subtypes are expressed [100, 101].

PACAP application typically reduces, and in fewer demonstrated cases enhances, VDCC currents. In hamster SMG neuronal cultures [100] 100 nM PACAP inhibited P/Q-, N- and L-type VDCCs, with P/Q- and N-type inhibition mediated by PAC_1Rs-via G_s-protein $\beta\gamma$ subunits, and L-type inhibition dependent on a VPACR-mediated, AC/PKA- and PLC/PKC- dependent signal transduction. Taken together, these studies indicate that PACAP can potently alter VDCC activation in SMG neurons via distinct pathways. Similar to the CdG and the superior cervical ganglion (SCG; See below), focal ejection of 10 μM PACAP (10 at 50 ms and 1 Hz) to dissociated SMG neurons induces a depolarizing current sufficient to generate bursts of action potentials [101] indicative of increased neuronal excitability. Inhibition of VDCCs in SMG neurons could lead to a depolarization and increased excitability by depressing activation of Ca^{2+}-activated K^+ channels, particularly those responsible for the after-hyperpolarization [100] but this has yet to be demonstrated. It is not known if the PACAP-induced effects on VDCCs or excitability can alter the output of nAChR-mediated synapses on SMG neurons. Similar to the modulatory actions of PACAP in CG and CdG neurons, however, PACAP also enhances nAChR-mediated currents in SMG neurons. Following ectopic exposure to 10 μM PACAP nAChR mediated currents induced by focal perfusion of 100 μM ACh were enhanced, reflecting an increase in nAChR affinity [95]. Finally, whole animal experiments support a role for PACAP as a potent neuromodulator of SMG transmission in vivo. Specifically, PACAP injections increase salivary secretion from and blood flow to the submandibular gland [102], consistent with enhanced parasympathetic innervation of the SMG. Additionally, intra-arterial PACAP injections alone did not induce submandibular salivary secretion in ferret while combined ACh and PACAP injections did, and PACAP injections during "parasympathetic nerve evoked flow of saliva" also increased salivary secretion [103]. Taken together, these findings indicate that PACAP inhibits Ca^{2+} channels and increases excitability in SMG neurons. They further suggest that PACAP can alter nAChR-mediated synaptic activity in the SMG with concomitant regulation of submandibular salivary secretion in vivo.

Superior Cervical Ganglion

The SCG is a prototypical sympathetic paravertebral ganglion that innervates extensive targets in the head, neck and heart [104, 105]. Composed of multiple CAergic neuronal subpopulations SCG neurons express postsynaptic $\alpha3^*$-nAChRs [106] and α7-nAChRs [107] and principally receive preganglionic cholinergic inputs from neurons in the spinal cord intermediolateral (IML) cell column [108–111]. IML neurons synthesize and store PACAP in cholinergic preganglionic terminals within the SCG [112, 113]. While PACAP release from IML terminals has not been verified, SCG neurons do express PAC_1Rs [113–116].

PACAP has a significant neuromodulatory impact on SCG neuron excitability similar to that in the CdG. Focal PACAP perfusion (50 μM, 1 s) produced a prolonged depolarizing inward current in SCG neurons in culture that was insensitive to TTX indicative of activity-independence. Pharmacological studies indicated that the PACAP-induced inward current was mediated by a PAC_1R induced PLC-IP3-IP3R signal transduction pathway involving sodium and potassium dependent components [116]. In Na^+ deficient solution, the PACAP induced inward currents were reduced by 70 %, while increasing extracellular K^+ concentration reduced PACAP-induced currents by 35 %. When extracellular Ca^{2+} and Na^+ were reduced, the PACAP-induced reduction in inward current was not significantly different from the sodium deficient condition alone. These findings indicate that the PACAP-induced depolarization of SCG neurons is mediated by extracellular Na^+ entry and inhibition of a K^+ outward conductance. However, other possible options are cited such that the specific ion channels underlying these PACAP-induced effects remain unknown.

The ability of PACAP to elicit a sustained depolarization in SCG neurons suggests it can increase SCG neuron excitability and output. Indeed, when total medium CA metabolites (DOPAC and HVA) were measured in dissociated SCG neuronal cultures, ectopic exposure to concentrations of PACAP (≥10 nM, 48 h) resulted in a threefold increase in extracellular CA metabolite concentrations [117] suggesting that PACAP enhanced CA release from SCG neuronal cultures. Although more detailed studies remain to be conducted, these findings suggest that PACAP can induce changes in postganglionic SCG neuronal output, consistent with enhanced SCG neuronal excitability. Alternatively, PACAP could act as a secretagogue in the SCG as it does in the closely related sympathetic-like adrenal gland. Unfortunately, no experiments have been conducted to determine whether PACAP also alters nAChR-mediated synaptic properties of SCG neurons or ACh release from their cholinergic IML inputs. Such findings would complement those conducted in parasympathetic ganglia and might reveal multiple roles for PACAP in modulating excitability, secretion and synaptic transmission in sympathetic ganglia.

Conclusions and Future Directions

Despite its ubiquitous distribution and diverse functions, PACAP is well positioned to mediate and modulate synaptic transmission via its cognate GPCRs and their associated intracellular signaling effectors. The above summary indicates that in most cases PACAP satisfies two of the three criteria to qualify it as an endogenous neurotransmitter. We predict that future experiments will demonstrate its release from PACAP-containing axon terminals as has been shown for presynaptic terminals in the adrenal gland and CG. Our summary further reveals that applied PACAP is also a potent neuromodulator that triggers acute and persistent alterations of membrane components underlying neuronal synaptic transmission. While studies using knockout mice lacking PACAP support its role as a synaptic neurotransmitter and/or neuromodulator in vivo, a critical caveat to this conclusion is that the ability of *endogenous* PACAP to modulate synaptic properties has not been demonstrated directly. Electrophysiological experiments combining stimulation sufficient to release endogenous PACAP with postsynaptic recording and PACAP receptor antagonists would help address this concern.

Two themes that emerge from PACAP's actions are its ability to rapidly alter membrane excitability and specific synaptic components. The first theme has been demonstrated in autonomic neurons and appears entwined with the activation or inhibition of VDCCs. Here, the resulting depolarization or indirect action of Ca^{2+} influx on K^+ or other channels enhances excitability by increasing the capacity for repetitive firing (Fig. 7.1b). Interestingly the coupling of increased excitability to VDCCs and depolarization exemplifies the cell-specific nature of PACAP signaling since, while VDCC modulation is a common outcome, resultant increased excitability is seen in some autonomic neurons but not in CG neurons where PACAP induces a sustained $\approx$5 mV depolarization [19] and VDCC current inhibition but these effects are associated with *reduced* action potential firing [118] (Starr and Margiotta, unpublished). In accord with other PACAP-signaling outcomes, the cellular mechanisms integrating its actions on specific classes of ion channels to produce increased excitability in some autonomic neurons but not others remain to be determined, as does the possible extension of these effects to central neurons.

The second theme involves the ability of applied PACAP to rapidly induce synaptic plasticity in both the central and autonomic nervous system by altering the postsynaptic response to conventional neurotransmitters and/or the release of conventional neurotransmitters from presynaptic nerve terminals (Fig. 7.1c). A common outcome is the concentration dependent regulation of postsynaptic AMPAR and NMDAR responses in all central neurons examined and the upregulation of nAChR responses in autonomic CdG, SMG, and CG neurons, possibly driven by their differential phosphorylation and/or insertion into postsynaptic sites as has been demonstrated for hippocampal neurons. Less common is the ability of PACAP to increase the release of conventional neurotransmitter from presynaptic terminals as seen in the hypothalamus and CG neurons in culture. The latter case reveals that PACAP signaling via PKA integrates AKAP-targeted nAChR upregulation, and NOS1 activation to focally release NO and thereby enhance ACh release from juxtaposed presynaptic terminals in a retrograde fashion. It will be interesting to determine if this coordinated signaling also applies at other synapses. In addition, while it is reasonable to surmise that alterations in membrane excitability would influence synaptic transmission, studies have not been performed to test how these two themes resulting from PACAP signaling may interact. Lastly, our experiments with CG neurons indicate that even a brief, transient exposure to PACAP can induce persistent synaptic plasticity indicative of long-term behavioral consequences. Since PACAP signaling and stress- and fear-related disorders such as PTSD are interrelated, more detailed experiments expanding on those conducted in the amygdala are required to determine if its ability to induce persistent synaptic plasticity is the basis for its contribution. Obvious study candidates are other limbic system components such as the hippocampus and the BNST where PACAP is well positioned to modulate synaptic transmission and where it is believed to be involved in fear memory consolidation and responses to chronic stress.

Acknowledgements Supported by the National Science Foundation (Grant # 0951549) and by a University of Toledo Research Innovation Award.

References

1. Purves D, Augustine GJ, Fitzpatrick D, Hall W, LaMantia A-S, White LE. Neuroscience, vol. 5. Sunderland, MA: Sinauer Associates Incorporated; 2012. 1 p.
2. Iversen LL. Chemical messengers in the brain. Trends Neurosci. 1978;1:15–6.
3. Burrows M. The neurobiology of an insect brain. Oxford, UK: Oxford University Press; 1996.
4. Smith CB, Eiden LE. Is PACAP the major neurotransmitter for stress transduction at the adrenomedullary synapse? J Mol Neurosci. 2012;48:403–12.
5. Tompkins JD, Ardell JL, Hoover DB, Parsons RL. Neurally released pituitary adenylate cyclase-activating polypeptide enhances guinea pig intrinsic cardiac neurone excitability. J Physiol. 2007;582:87–93.
6. Hamelink C, Tjurmina O, Damadzic R, Young WS, Weihe E, Lee H-W, et al. Pituitary adenylate cyclase-activating polypeptide is a sympathoadrenal neurotransmitter involved in catecholamine regulation and glucohomeostasis. Proc Natl Acad Sci U S A. 2002;99:461–6.
7. Kuri BA, Chan S-A, Smith CB. PACAP regulates immediate catecholamine release from adrenal chromaffin cells in an activity-dependent manner through a protein kinase C-dependent pathway. J Neurochem. 2009;110:1214–25.
8. Spengler D, Waeber C, Pantaloni C, Holsboer F, Bockaert J, Seeburg PH, et al. Differential signal transduction by five splice variants of the PACAP receptor. Nature. 1993;365:170–5.
9. Vaudry D, Falluel-Morel A, Bourgault S, Basille M, Burel D, Wurtz O, et al. Pituitary adenylate cyclase-activating polypeptide and its receptors: 20 years after the discovery. Pharmacol Rev. 2009;61:283–357.
10. Margiotta JF, Pardi D. Pituitary adenylate cyclase-activating polypeptide type I receptors mediate cyclic AMP-dependent enhancement of neuronal acetylcholine sensitivity. Mol Pharmacol. 1995;48:63–71.
11. Hashimoto H, Ishihara T, Shigemoto R, Mori K, Nagata S. Molecular cloning and tissue distribution of a receptor for pituitary adenylate cyclase-activating polypeptide. Neuron. 1993;11:333–42.
12. Rawlings SR. At the cutting edge: PACAP, PACAP receptors and intracellular signaling. Mol Cell Endocrinol. 1994;101:C5–9.
13. Vaudry D, Gonzalez BJ, Basille M, Yon L, Fournier A, Vaudry H. Pituitary adenylate cyclase-activating polypeptide and its receptors: from structure to functions. Pharmacol Rev. 2000;52:269–324.
14. Pardi D, Margiotta JF. Pituitary adenylate cyclase-activating polypeptide activates a phospholipase C-dependent signal pathway in chick ciliary ganglion neurons that selectively inhibits alpha7-containing nicotinic receptors. J Neurosci. 1999;19:6327–37.
15. Deisseroth K, Bito H, Tsien RW. Signaling from synapse to nucleus: postsynaptic CREB phosphorylation during multiple forms of hippocampal synaptic plasticity. Neuron. 1996;16:89–101.

16. Zhou X, Nai Q, Chen M, Dittus JD, Howard MJ, Margiotta JF. Brain-derived neurotrophic factor and trkB signaling in parasympathetic neurons: relevance to regulating alpha7-containing nicotinic receptors and synaptic function. J Neurosci. 2004;24:4340–50.
17. Sumner AD, Margiotta JF. Pituitary adenylate cyclase-activating polypeptide (PACAP) alters parasympathetic neuron gene expression in a time-dependent fashion. J Mol Neurosci. 2008;36:141–56.
18. Braas KM, Schutz KC, Bond JP, Vizzard MA, Girard BM, May V. Microarray analyses of pituitary adenylate cyclase activating polypeptide (PACAP)-regulated gene targets in sympathetic neurons. Peptides. 2007;28:1856–70.
19. Pugh PC, Jayakar SS, Margiotta JF. PACAP/PAC1R signaling modulates acetylcholine release at neuronal nicotinic synapses. Mol Cell Neurosci. 2009;43:244–57.
20. Starr ER, Margiotta J. PACAP induces acute and sustained synaptic plasticity by distinct cellular mechanisms. Society for Neuroscience Abstract. Washington, DC: Society for Neuroscience Abstract Viewer; 2014. p. B.07.g.
21. Pickard GE, Sollars PJ. Intrinsically photosensitive retinal ganglion cells. Rev Physiol Biochem Pharmacol. 2012;162:59–90.
22. Schmidt TM, Chen S-K, Hattar S. Intrinsically photosensitive retinal ganglion cells: many subtypes, diverse functions. Trends Neurosci. 2011;34:572–80.
23. Hannibal J, Møller M, Ottersen OP, Fahrenkrug J. PACAP and glutamate are co-stored in the retinohypothalamic tract. J Comp Neurol. 2000;418:147–55.
24. Michel S, Itri J, Han JH, Gniotczynski K, Colwell CS. Regulation of glutamatergic signalling by PACAP in the mammalian suprachiasmatic nucleus. BMC Neurosci. 2006;7:15.
25. Hannibal J. Neurotransmitters of the retino-hypothalamic tract. Cell Tissue Res. 2002;309:73–88.
26. Harrington ME, Hoque S, Hall A, Golombek D, Biello S. Pituitary adenylate cyclase activating peptide phase shifts circadian rhythms in a manner similar to light. J Neurosci. 1999;19:6637–42.
27. Hannibal J, Jamen F, Nielsen HS, Journot L, Brabet P, Fahrenkrug J. Dissociation between light-induced phase shift of the circadian rhythm and clock gene expression in mice lacking the pituitary adenylate cyclase activating polypeptide type 1 receptor. J Neurosci. 2001;21:4883–90.
28. Kawaguchi C, Tanaka K, Isojima Y, Shintani N, Hashimoto H, Baba A, et al. Changes in light-induced phase shift of circadian rhythm in mice lacking PACAP. Biochem Biophys Res Commun. 2003;310:169–75.
29. Macdonald DS, Weerapura M, Beazely MA, Martin L, Czerwinski W, Roder JC, et al. Modulation of NMDA receptors by pituitary adenylate cyclase activating peptide in CA1 neurons requires G alpha q, protein kinase C, and activation of Src. J Neurosci. 2005;25:11374–84.
30. Costa L, Santangelo F, Li Volsi G, Ciranna L. Modulation of AMPA receptor-mediated ion current by pituitary adenylate cyclase-activating polypeptide (PACAP) in CA1 pyramidal neurons from rat hippocampus. Hippocampus. 2009;19:99–109.
31. Varela JA, Hirsch SJ, Chapman D, Leverich LS, Greene RW. D1/D5 modulation of synaptic NMDA receptor currents. J Neurosci. 2009;29:3109–19.
32. Nicholls J, Martin R, Fuchs P, Brown DA, Diamond ME, Weisblat D. From neuron to brain, vol. 5. Sunderland, MA: Sinauer Associates, Inc; 2012.
33. Andersen P, Sundberg SH, Sveen O, Wigström H. Specific long-lasting potentiation of synaptic transmission in hippocampal slices. Nature. 1977;266:736–7.
34. Mulkey RM, Malenka RC. Mechanisms underlying induction of homosynaptic long-term depression in area CA1 of the hippocampus. Neuron. 1992;9:967–75.
35. Skoglösa Y, Takei N, Lindholm D. Distribution of pituitary adenylate cyclase activating polypeptide mRNA in the developing rat brain. Mol Brain Res. 1999;65:1–13.
36. Joo KM, Chung YH, Kim MK, Nam RH, Lee BL, Lee KH, et al. Distribution of vasoactive intestinal peptide and pituitary adenylate cyclase-activating polypeptide receptors (VPAC1, VPAC2, and PAC1 receptor) in the rat brain. J Comp Neurol. 2004;476:388–413.

37. Macdonald JF, Jackson MF, Beazely MA. G protein-coupled receptors control NMDARs and metaplasticity in the hippocampus. Biochim Biophys Acta. 1768;2007:941–51.
38. Ali DW, Salter MW. NMDA receptor regulation by Src kinase signalling in excitatory synaptic transmission and plasticity. Curr Opin Neurobiol. 2001;11:336–42.
39. Toda AMA, Huganir RL. Regulation of AMPA receptor phosphorylation by the neuropeptide PACAP38. Proc Natl Acad Sci U S A. 2015;112:6712–7.
40. Roberto M, Scuri R, Brunelli M. Differential effects of PACAP-38 on synaptic responses in rat hippocampal CA1 region. Learn Mem. 2001;8:265–71.
41. Ster J, De Bock F, Bertaso F, Abitbol K, Daniel H, Bockaert J, et al. Epac mediates PACAP-dependent long-term depression in the hippocampus. J Physiol. 2009;587:101–13.
42. Kondo T, Tominaga T, Ichikawa M, Iijima T. Differential alteration of hippocampal synaptic strength induced by pituitary adenylate cyclase activating polypeptide-38 (PACAP-38). Neurosci Lett. 1997;221:189–92.
43. Cunha-Reis D, Ribeiro JA, Sebastião AM. VIP enhances synaptic transmission to hippocampal CA1 pyramidal cells through activation of both VPAC1 and VPAC2 receptors. Brain Res. 2005;1049:52–60.
44. Cunha-Reis D, Ribeiro JA, Sebastião AM. VPAC2 receptor activation mediates VIP enhancement of population spikes in the CA1 area of the hippocampus. Ann N Y Acad Sci. 2006;1070:210–4.
45. LeDoux J. The emotional brain, fear, and the amygdala. Cell Mol Neurobiol. 2003;23:727–38.
46. Adhikari A. Distributed circuits underlying anxiety. Front Behav Neurosci. 2014;8(112):1–6.
47. Cho J-H, Zushida K, Shumyatsky GP, Carlezon WA, Meloni EG, Bolshakov VY. Pituitary adenylate cyclase-activating polypeptide induces postsynaptically expressed potentiation in the intra-amygdala circuit. J Neurosci. 2012;32:14165–77.
48. Piggins HD, Stamp JA, Burns J, Rusak B, Semba K. Distribution of pituitary adenylate cyclase activating polypeptide (PACAP) immunoreactivity in the hypothalamus and extended amygdala of the rat. J Comp Neurol. 1996;376:278–94.
49. Das M, Vihlen CS, Legradi G. Hypothalamic and brainstem sources of pituitary adenylate cyclase-activating polypeptide nerve fibers innervating the hypothalamic paraventricular nucleus in the rat. J Comp Neurol. 2007;500:761–76.
50. Rumbaugh G, Sia G-M, Garner CC, Huganir RL. Synapse-associated protein-97 isoform-specific regulation of surface AMPA receptors and synaptic function in cultured neurons. J Neurosci. 2003;23:4567–76.
51. Mahan AL, Ressler KJ. Fear conditioning, synaptic plasticity and the amygdala: implications for posttraumatic stress disorder. Trends Neurosci. 2012;35:24–35.
52. Ressler KJ, Mercer KB, Bradley B, Jovanovic T, Mahan A, Kerley K, et al. Post-traumatic stress disorder is associated with PACAP and the PAC1 receptor. Nature. 2011;470:492–7.
53. Hammack SE, Roman CW, Lezak KR, Kocho-Shellenberg M, Grimmig B, Falls WA, et al. Roles for pituitary adenylate cyclase-activating peptide (PACAP) expression and signaling in the bed nucleus of the stria terminalis (BNST) in mediating the behavioral consequences of chronic stress. J Mol Neurosci. 2010;42:327–40.
54. Crestani CC, Alves FH, Gomes FV, Resstel LB, Correa FM, Herman JP. Mechanisms in the bed nucleus of the stria terminalis involved in control of autonomic and neuroendocrine functions: a review. Curr Neuropharmacol. 2013;11:141–59.
55. Kozicz T, Vigh S, Arimura A. Immunohistochemical evidence for PACAP and VIP interaction with met-enkephalin and CRF containing neurons in the bed nucleus of the stria terminalis. Ann N Y Acad Sci. 1998;865:523–8.
56. Kozicz T, Arimura A. Synaptic interaction between galanin immunoreactive neurons and axon terminals immunopositive for VIP and PACAP in the bed nucleus of the stria terminalis in the rat. Ann N Y Acad Sci. 2000;921:327–32.
57. Vertongen P, Schiffmann SN, Gourlet P, Robberecht P. Autoradiographic visualization of the receptor subclasses for vasoactive intestinal polypeptide (VIP) in rat brain. Peptides. 1997;18:1547–54.

58. Hammack SE, Cheung J, Rhodes KM, Schutz KC, Falls WA, Braas KM, et al. Chronic stress increases pituitary adenylate cyclase-activating peptide (PACAP) and brain-derived neurotrophic factor (BDNF) mRNA expression in the bed nucleus of the stria terminalis (BNST): Roles for PACAP in anxiety-like behavior. Psychoneuroendocrinology. 2009;34:833–43.
59. Kocho-Schellenberg M, Lezak KR, Harris OM, Roelke E, Gick N, Choi I, et al. PACAP in the BNST produces anorexia and weight loss in male and female rats. Neuropsychopharmacology. 2014;39:1614–23.
60. Lezak KR, Roman CW, Braas KM, Schutz KC, Falls WA, Schulkin J, et al. Regulation of bed nucleus of the stria terminalis PACAP expression by stress and corticosterone. J Mol Neurosci. 2014;54:477–84.
61. Roman CW, Lezak KR, Hartsock MJ, Falls WA, Braas KM, Howard AB, et al. PAC1 receptor antagonism in the bed nucleus of the stria terminalis (BNST) attenuates the endocrine and behavioral consequences of chronic stress. Psychoneuroendocrinology. 2014;47:151–65.
62. Stanley E. The calyx-type synapse of the chick ciliary ganglion as a model of fast cholinergic transmission. Can J Physiol Pharmcol. 1992;70:S73–7.
63. Dryer SE. Functional development of the parasympathetic neurons of the avian ciliary ganglion: a classic model system for the study of neuronal differentiation and development. Prog Neurobiol. 1994;43:281–322.
64. Martin AR, Pilar G. Quantal components of the synaptic potential in the ciliary ganglion of the chick. J Physiol. 1964;175:1–16.
65. Landmesser L, Pilar G. The onset and development of transmission in the chick ciliary ganglion. J Physiol. 1972;222:691–713.
66. Ullian EM, McIntosh JM, Sargent PB. Rapid synaptic transmission in the avian ciliary ganglion is mediated by two distinct classes of nicotinic receptors. J Neurosci. 1997;17:7210–9.
67. Sargent PB. Nicotinic receptors concentrated in the subsynaptic membrane do not contribute significantly to synaptic currents at an embryonic synapse in the chicken ciliary ganglion. J Neurosci. 2009;29:3749–59.
68. Dryer SE, Chiappinelli VA. Properties of choroid and ciliary neurons in the avian ciliary ganglion and evidence for substance P as a neurotransmitter. J Neurosci. 1985;5:2654–61.
69. Tornøe K, Hannibal J, Jensen TB, Georg B, Rickelt LF, Andreasen MB, et al. PACAP-(1-38) as neurotransmitter in the porcine adrenal glands. Am J Physiol Endocrinol Metab. 2000;279:E1413–25.
70. Margiotta JF, Berg DK, Dionne VE. Cyclic AMP regulates the proportion of functional acetylcholine receptors on chicken ciliary ganglion neurons. Proc Natl Acad Sci U S A. 1987;84:8155–9.
71. Margiotta JF, Berg DK. Functional synapses are established between ciliary ganglion neurones in dissociated cell culture. Nature. 1982;296:152–4.
72. Chen M, Pugh PC, Margiotta JF. Nicotinic synapses formed between chick ciliary ganglion neurons in culture resemble those present on the neurons in vivo. J Neurobiol. 2001;47:265–79.
73. Bredt DS, Ferris CD, Snyder SH. Nitric oxide synthase regulatory sites. Phosphorylation by cyclic AMP-dependent protein kinase, protein kinase C, and calcium/calmodulin protein kinase; identification of flavin and calmodulin binding sites. J Biol Chem. 1992;267:10976–81.
74. Hurt KJ, Sezen SF, Lagoda GF, Musicki B, Rameau GA, Snyder SH, et al. Cyclic AMP-dependent phosphorylation of neuronal nitric oxide synthase mediates penile erection. Proc Natl Acad Sci U S A. 2012;109:16624–9.
75. Jayakar SS, Pugh PC, Dale Z, Starr ER, Cole S, Margiotta JF. PACAP induces plasticity at autonomic synapses by nAChR-dependent NOS1 activation and AKAP-mediated PKA targeting. Mol Cell Neurosci. 2014;63:1–12.
76. Kojima H, Nakatsubo N, Kikuchi K, Kawahara S, Kirino Y, Nagoshi H, et al. Detection and imaging of nitric oxide with novel fluorescent indicators: diaminofluoresceins. Anal Chem. 1998;70:2446–53.

77. Bright DP, Brickley SG. Acting locally but sensing globally: impact of GABAergic synaptic plasticity on phasic and tonic inhibition in the thalamus. J Physiol. 2008;586:5091–9.
78. Etherington SJ. Postsynaptic production of nitric oxide implicated in long-term depression at the mature amphibian (Bufo marinus) neuromuscular junction. J Physiol. 2004;559:507–17.
79. Xue JG, Masuoka T, Gong XD, Chen KS, Yanagawa Y, Law SKA, et al. NMDA receptor activation enhances inhibitory GABAergic transmission onto hippocampal pyramidal neurons via presynaptic and postsynaptic mechanisms. J Neurophysiol. 2011;105:2897–906.
80. Wong W, Scott JD. AKAP signalling complexes: focal points in space and time. Nat Rev Mol Cell Biol. 2004;5:959–70.
81. Pugh PC, Margiotta JF. PACAP support of neuronal survival requires MAPK- and activity-generated signals. Mol Cell Neurosci. 2006;31:586–95.
82. Mitchell GAG. The innervation of the heart. Br Heart J. 1953;15:159–71.
83. Wallis D, Watson AH, Mo N. Cardiac neurones of autonomic ganglia. Microsc Res Tech. 1996;35:69–79.
84. Singh S, Johnson PI, Lee RE, Orfei E, Lonchyna VA, Sullivan HJ, et al. Topography of cardiac ganglia in the adult human heart. J Thorac Cardiovasc Surg. 1996;112:943–53.
85. Calupca MA, Vizzard MA, Parsons R. Origin of pituitary adenylate cyclase-activating polypeptide (PACAP)-immunoreactive fibers innervating guinea pig parasympathetic cardiac ganglia. J Comp Neurol. 2000;423:26–39.
86. Braas KM, May V, Harakall SA, Hardwick JC, Parsons RL. Pituitary adenylate cyclase-activating polypeptide expression and modulation of neuronal excitability in guinea pig cardiac ganglia. J Neurosci. 1998;18:9766–79.
87. Parsons RL, Locknar SA, Young BA, Hoard JL, Hoover DB. Presence and co-localization of vasoactive intestinal polypeptide with neuronal nitric oxide synthase in cells and nerve fibers within guinea pig intrinsic cardiac ganglia and cardiac tissue. Cell Tissue Res. 2006;323:197–209.
88. Dehaven WI, Cuevas J. Heterogeneity of pituitary adenylate cyclase-activating polypeptide and vasoactive intestinal polypeptide receptors in rat intrinsic cardiac neurons. Neurosci Lett. 2002;328:45–9.
89. Parsons R, Rossignol T, Calupca M, Hardwick J, Braas K. PACAP peptides modulate guinea pig cardiac neuron membrane excitability and neuropeptide expression. Ann N Y Acad Sci. 2000;921:202–10.
90. Tompkins JD, Parsons RL. Identification of intracellular signaling cascades mediating the PACAP-induced increase in guinea pig cardiac neuron excitability. J Mol Neurosci. 2008;36:292–8.
91. Hardwick JC, Tompkins JD, Locknar SA, Merriam LA, Young BA, Parsons RL. Calcium influx through channels other than voltage-dependent calcium channels is critical to the pituitary adenylate cyclase-activating polypeptide-induced increase in excitability in guinea pig cardiac neurons. Ann N Y Acad Sci. 2006;1070:317–21.
92. Merriam LA, Barstow KL, Parsons RL. Pituitary adenylate cyclase-activating polypeptide enhances the hyperpolarization-activated nonselective cationic conductance, Ih, in dissociated guinea pig intracardiac neurons. Regul Pept. 2004;123:123–33.
93. Tompkins JD, Merriam LA, Girard BM, May V, Parsons RL. Nickel suppresses the PACAP-induced increase in guinea pig cardiac neuron excitability. Am J Physiol Cell Physiol. 2015;308:C857–66.
94. Merriam LA, Baran CN, Girard BM, Hardwick JC, May V, Parsons RL. Pituitary adenylate cyclase 1 receptor internalization and endosomal signaling mediate the pituitary adenylate cyclase activating polypeptide-induced increase in guinea pig cardiac neuron excitability. J Neurosci. 2013;33(10):4614–22.
95. Liu DM, Cuevas J, Adams DJ. VIP and PACAP potentiation of nicotinic ACh-evoked currents in rat parasympathetic neurons is mediated by G-protein activation. Eur J Neurosci. 2000;12:2243–51.
96. Segal K, Lisnyansky I, Nageris B, Feinmesser R. Parasympathetic innervation of the salivary glands. Oper Tech Otolaryngol Head Neck Surg. 1996;7:333–8.

97. Endoh T. Modulation of voltage-dependent calcium channels by neurotransmitters and neuropeptides in parasympathetic submandibular ganglion neurons. Arch Oral Biol. 2004;49:539–57.
98. Shida T, Tokunaga A, Kondo E, Ueda Y, Ohno K, Saika T, et al. Expression of muscarinic and nicotinic receptor mRNA in the salivary gland of rats: a study by in situ hybridization histochemistry. Mol Brain Res. 1993;17:335–9.
99. McCann CM, Lichtman JW. In vivo imaging of presynaptic terminals and postsynaptic sites in the mouse submandibular ganglion. Dev Neurobiol. 2008;68:760–70.
100. Kamaishi H, Endoh T, Suzuki T. Multiple signal pathways coupling VIP and PACAP receptors to calcium channels in hamster submandibular ganglion neurons. Auton Neurosci. 2004;111:15–26.
101. Hayashi K, Endoh T, Shibukawa Y, Yamamoto T, Suzuki T. VIP and PACAP inhibit L-, N- and P/Q-type Ca2+ channels of parasympathetic neurons in a voltage independent manner. Bull Tokyo Dent Coll. 2002;43:31–9.
102. Mirfendereski S, Tobin G, Håkanson R, Ekström J. Pituitary adenylate cyclase activating peptide (PACAP) in salivary glands of the rat: origin, and secretory and vascular effects. Acta Physiol Scand. 1997;160:15–22.
103. Tobin G, Asztély A, Edwards AV, Ekström J, Håkanson R, Sundler F. Presence and effects of pituitary adenylate cyclase activating peptide in the submandibular gland of the ferret. Neuroscience. 1995;66:227–35.
104. Li C, Horn JP. Physiological classification of sympathetic neurons in the rat superior cervical ganglion. J Neurophysiol. 2006;95:187–95.
105. Lazarov NE, Reindl S, Gratzl M. Neurotransmitters and their receptors in the rat superior cervical ganglion. J Biomed Clin Res. 2009;2:19–25.
106. Mao D, Perry DC, Yasuda RP, Wolfe BB, Kellar KJ. The alpha4beta2alpha5 nicotinic cholinergic receptor in rat brain is resistant to up-regulation by nicotine in vivo. J Neurochem. 2008;104:446–56.
107. Skok MV, Voitenko LP, Voitenko SV, Lykhmus EY, Kalashnik EN, Litvin TI, et al. Alpha subunit composition of nicotinic acetylcholine receptors in the rat autonomic ganglia neurons as determined with subunit-specific anti-alpha(181-192) peptide antibodies. Neuroscience. 1999;93:1427–36.
108. Krishnaswamy A, Cooper E. An activity-dependent retrograde signal induces the expression of the high-affinity choline transporter in cholinergic neurons. Neuron. 2009;61:272–86.
109. Rando TA, Bowers CW, Zigmond RE. Localization of neurons in the rat spinal cord which project to the superior cervical ganglion. J Comp Neurol. 1981;196:73–83.
110. Reuss S, Johnson RF, Morin LP, Moore RY. Localization of spinal cord preganglionic neurons innervating the superior cervical ganglion in the golden hamster. Brain Res Bull. 1989;22:289–93.
111. Strack AM, Sawyer WB, Marubio LM, Loewy AD. Spinal origin of sympathetic preganglionic neurons in the rat. Brain Res. 1988;455:187–91.
112. Beaudet MM, Braas KM, May V. Pituitary adenylate cyclase activating polypeptide (PACAP) expression in sympathetic preganglionic projection neurons to the superior cervical ganglion. J Neurobiol. 1998;36:325–36.
113. May V, Beaudet M, Parsons R, Hardwick J, Gauthier E, Durda J, et al. Mechanisms of pituitary adenylate cyclase activating polypeptide (PACAP)-induced depolarization of symapthetic superior cervical ganglion (SCG) neurons. Ann NY Acad Sci. 1998;865:164–75.
114. Nogi H, Hashimoto H, Hagihara N, Shimada S, Yamamoto K, Matsuda T, et al. Distribution of mRNAs for pituitary adenylate cyclase-activating polypeptide (PACAP), PACAP receptor, vasoactive intestinal polypeptide (VIP), and VIP receptors in the rat superior cervical ganglion. Neurosci Lett. 1997;227:37–40.
115. Braas KM, May V. Pituitary adenylate cyclase-activating polypeptides directly stimulate sympathetic neuron neuropeptide Y release through PAC(1) receptor isoform activation of specific intracellular signaling pathways. J Biol Chem. 1999;274:27702–10.

116. Beaudet M, Parsons R, Braas K, May V. Mechanisms mediating pituitary adenylate cyclase activating polypeptide depolarization of rat sympathetic neurons. J Neurosci. 2000;20:7353–61.
117. May V, Braas KM. Pituitary adenylate cyclase-activating polypeptide (PACAP) regulation of sympathetic neuron neuropeptide Y and catecholamine expression. J Neurochem. 1995;65:978–87.
118. Pugh P, Sweet B, Margiotta J. PACAP's ups &downs: biphasic regulation of voltage-dependent calcium channels. San Diego, CA: Society for Neuroscience Annual Meeting; 2007. p. 1.
119. Hannibal J. Pituitary adenylate cyclase-activating peptide in the rat central nervous system: an immunohistochemical and in situ hybridization study. J Comp Neurol. 2002;453:389–417.
120. Shioda S, Ozawa H, Dohi K, Mizushima H, Matsumoto K, Nakajo S, et al. PACAP protects hippocampal neurons against apoptosis: Involvement of JNK/SAPK signaling pathway. Ann N Y Acad Sci. 1998;865:111–7.
121. Bibevski S, Zhou Y, McIntosh JM, Zigmond RE, Dunlap ME. Functional nicotinic acetylcholine receptors that mediate ganglionic transmission in cardiac parasympathetic neurons. J Neurosci. 2000;20:5076–82.
122. Segal RA, Greenberg ME. Intracellular signaling pathways activated by neurotrophic factors. Annu Rev Neurosci. 1996;19:463–89.

Part III
PACAP Receptors and Signalling

Chapter 8
The Pharmacophoric Determinants of PACAP

Alain Fournier, Steve Bourgault, and David Chatenet

Abstract The pituitary adenylate cyclase-activating polypeptide (PACAP) is a hormone that belongs to the secretin/glucagon/growth hormone-releasing hormone (GHRH)/vasoactive intestinal peptide (VIP) superfamily. It shows a widespread distribution and a large array of physiological actions. Accordingly, this peptide is linked to various pathological conditions and PACAP-derived drugs are promising compounds for the development of potent therapies. The effects of PACAP are mediated through three types of class B-G protein-coupled receptors identified as PAC1, VPAC1, and VPAC2. PACAP exhibits a larger affinity for PAC1 than VIP, whereas VPAC1 and VPAC2 are equally recognized by both ligands. The three receptors possess distinct pharmacophoric requirements and in particular, residues His^1, Asp^3, Phe^6, Thr^7, Tyr^{22}, and Leu^{23}, as well as structural motifs such as a highly stabilized α-helix and an N-terminal β-coil abutted to an N-capping-like arrangement, would be crucial for receptor selectivity. Moreover, PACAP is vulnerable to proteolysis and sites of enzymatic degradation were identified to develop metabolically stable compounds. The present review summarizes structure–activity relationships of PACAP and makes an overview of the conformational features of the molecule.

Keywords Pituitary adenylate cyclase-activating polypeptide • PACAP • Structure–activity relationship • SAR • Analogs • Fragments • Conformation • Receptors • GPCR

A. Fournier (✉)
Laboratoire d'études moléculaires et pharmacologiques des peptides,
Institut national de la recherche scientifique, INRS — Institut Armand-Frappier,
531 boul. des Prairies, Laval, QC, Canada, H7V 1B7
e-mail: alain.fournier@iaf.inrs.ca

S. Bourgault
Département de Chimie, Université du Québec à Montréal,
C.P. 8888, Succ. Centre-Ville, Montréal, QC, Canada, H3C 3P8

Quebec Network for Research on Protein Function, Engineering and Applications, PROTEO,
Québec, QC, Canada

D. Chatenet
Groupe de Recherche en Ingénierie des Peptides et en Pharmacothérapie,
Institut national de la recherche scientifique, INRS — Institut Armand-Frappier,
531 boul. des Prairies, Laval, QC, Canada, H7V 1B7

D. Reglodi, A. Tamas (eds.), *Pituitary Adenylate Cyclase Activating Polypeptide — PACAP*, Current Topics in Neurotoxicity 11,
DOI 10.1007/978-3-319-35135-3_8

Abbreviations

Aad	L-α-aminoadipic acid
Aca	Aminocaproic acid
BBB	Blood–brain barrier
Bip	L-biphenylalanine
Bpa	*p*-benzoyl-L-phenylalanine
CT	Calcitonin
Disc	L-1,3-dihydro-2*H*-isoindole carboxylic acid
GLP-1	Glucagon-like peptide-1
GPCR(s)	G protein-coupled receptor(s)
Hyp	Hydroxyproline
Iaa	4-Imidazole acetic acid
Iac	4-Imidazole acrylic acid
R-IBTM	2(*R*)-amino-3-oxohexahydroindolizino[8,7-*b*]indole-5(*R*)-carboxylic acid
S-IBTM	2(*S*)-amino-3-oxohexahydroindolizino[8,7-*b*]indole-5(*S*)-carboxylic acid
Ind	L-indoline-2-carboxylic acid
Nal	L-naphthylalanine
Nle	L-norleucine
NOESY	Nuclear Overhauser Effect spectroscopy
PACAP	Pituitary adenylate cyclase-activating polypeptide
PACAP27	27-amino acid isoform of PACAP
PACAP38	38-amino acid isoform of PACAP
PAC1	Pituitary adenylate cyclase-activating polypeptide type 1 receptor
Phe(*p*-I)	*Para*-iodo-L-phenylalanine
PTH	Parathyroid hormone
PTS-6	Peptide transport system-6
SAR	Structure–activity relationships
TFE	Trifluoroethanol
Tic	L-1,2,3,4-tetrahydroisoquinoline-3-carboxylic acid
Tiq	L-tetrahydroisoquinoline-1-carboxylic acid
VIP	Vasoactive intestinal peptide
VPAC1	VIP/PACAP type 1 receptor
VPAC2	VIP/PACAP type 2 receptor

Introduction

The pituitary adenylate cyclase-activating polypeptide (PACAP) exhibits multiple functions in various physiological systems. In this context, it is associated with several pathological conditions, and PACAP receptors appear as promising targets to treat various disorders such as neurodegenerative diseases [1–4], stroke [3, 5–8], cancers [9–12], and diabetes [13, 14]. The effects of PACAP are observed in the central nervous system, as well as in periphery, and they are mediated through three types of class B-G protein-coupled receptors identified as PAC1, VPAC1, and VPAC2. On the one hand, PAC1, which exhibits a greater affinity for PACAP than for VIP, is abundantly expressed in the central nervous system and is particularly associated with neuroprotective and neurotrophic effects. On the other hand, VPAC1 and VPAC2 are rather related to peripheral actions and are equally recognized by both ligands [3]. Thus, the three receptors possess distinct pharmacophoric requirements and accordingly, their peptide ligands contain precise molecular determinants, including key chemical functions at particular positions, and specific structural motifs. Noteworthy, it is postulated that PACAP activates its receptors following a two-step binding model in which the C-terminal segment of the ligand binds to the N-extracellular domain, which then positions the amino-terminal portion of the peptide hormone in close proximity to the transmembrane region of the receptor to initiate signaling [15]. This "two-domain" model involves a specific network of ligand–receptor intermolecular interactions that necessitates a well-defined spatial positioning of the N-terminal ligand side-chains [16–21]. This structural motif with a specific folded backbone conformation governs the pharmacological properties of the class B-GPCR agonists and antagonists, and must be considered to design selective and potent ligands. Moreover, PACAP is vulnerable to proteolysis and sites of enzymatic degradation must be identified to develop metabolically stable compounds.

Structure–Activity Relationship Studies Related to the PACAP Receptors

The initial characterization of PACAP showed that it is a 38-amino acid C-terminally α-amidated peptide (PACAP38) [22]:

```
    5         10        15        20        25        30        35
H-S-D-G-I-F-T-D-S-Y-S-R-Y-R-K-Q-M-A-V-K-K-Y-L-A-A-V-L-G-K-R-Y-K-Q-R-V-K-N-K
```

Furthermore, based on the existence of an internal cleavage amidation site (Gly^{28}–Lys^{29}–Arg^{30}), a truncated amidated isoform (PACAP27) was isolated and shown to be, like its precursor, a potent activator of adenylyl cyclase in cultured pituitary cells [23]. It has been observed that PACAP27 shares 68 % sequence conservation with vasoactive intestinal polypeptide (VIP), thereby identifying PACAP as a member of the secretin/glucagon/growth hormone-releasing hormone/vasoactive intestinal peptide superfamily. Comparison of the primary structures of the peptides belonging to this family [3, 16, 17, 24] shows that their N-terminal domain is largely conserved while the central and C-terminal segments significantly diverge. For instance, an aromatic residue (His or Tyr) is usually found at the N-terminal end, whereas Phe and Thr frequently occupy positions 6 and 7, respectively.

In addition, a negatively charged residue (Asp or Glu) is commonly located at position 3 and low molecular weight amino acids (Ala, Gly, or Ser) are observed at positions 2 and 4. Following the discovery of PACAP, two types of binding sites were characterized on the basis of their relative affinities for VIP and PACAP. Type I shows a high affinity for the 27- and 38-amino acid PACAP isoforms and a much lower affinity for VIP, while type II recognizes both PACAP and VIP with similar high affinity [3, 16–18, 20, 25, 26]. However, the latter binding site was subdivided into two classes depending on the relative affinity for secretin and helodermin. Subsequently, the three PACAP/VIP binding sites were cloned and named PAC1, VPAC1, and VPAC2 [27].

N- and C-Terminal PACAP Fragments

The typical first step used for the identification of key residues that are essential for activity and/or affinity consists in the truncation of the peptide from the amino- or carboxy-terminal end. When applied to PACAP, and as anticipated from the phylogenetic evolution of PACAP among vertebrates, the N-terminal domain appeared to be essential for the biological activity. As a matter of fact, removal of His^1 in both PACAP isoforms reduced the potency and affinity [28–31]. This is well illustrated with PACAP(2–27) that showed for PAC1 an affinity 80 times lower and an activity 100 times weaker than those of the parent molecule [17, 29]. This effect though was less pronounced with the 38-amino acid isoform since the affinity and activity of PACAP(2–38) were decreased by 13- and 30-fold, respectively, as compared to those of PACAP38. Additional deletions in the N-terminal domain of PACAP27 and PACAP38 led not only to further decreases of potency and affinity but also to the identification of fragments exhibiting antagonistic activity. Indeed, removal in PACAP27 of the two first residues (His^1 and Ser^2) caused a 2000-fold decrease of potency in a PAC1-cAMP production assay using AR4-2J rat pancreatic acinar cells [29, 30] and similarly, PACAP(3–27) and PACAP(3–38) were unable to stimulate adenylyl cyclase activity in rat brain hippocampal membrane [31]. Nonetheless, some residual affinity was measured, especially with PACAP(3–38) (100-fold decrease), thereby showing that the detrimental effect of the double deletion was less marked in PACAP38 than in PACAP27 [17]. As abovementioned, the further progressive deletions of N-terminal residues in both PACAP isoforms led to the identification of fragments displaying antagonistic activity [28, 31]. In particular, PACAP(6-38) binds significantly to PAC1, with an affinity only about ten times lower than that of PACAP38 [17]. Consequently, it is used frequently as a PAC1 receptor antagonist [28]. Nonetheless, although it is to a weaker extent than PACAP38, this fragment also binds to VPAC1 (≈300-fold) and VPAC2 (≈40-fold) [32]. Moreover, biological activities were described for this truncated peptide [33–35], thus suggesting that it behaves as a full or partial agonist in some paradigms. Additional sequential deletions in PACAP27, from Phe^6 to Tyr^{13}, gave fragments still able to bind, but very weakly, to human neuroblastoma NB-OK-1 cell membranes [36], indicating that the 14–27 segment of PACAP was sufficient for recognition by

PAC1. When applied to the 38-amino acid isoform, those truncations were much less detrimental towards affinity. Furthermore, some fragments retained the capacity to stimulate the formation of cAMP in the NB-OK-1 cell membranes [36]. Those results strongly suggest that the C-terminal 28–38 stretch facilitates the anchorage of PACAP to the PAC1 receptor [28, 31, 36], and that this affinity improvement causes residual activity [17]. Interestingly, a chimeric peptide formed by coupling the PACAP(28–38) sequence to the VIP molecule displayed for PAC1 an affinity 100-fold higher than VIP [37], which provided additional evidence that the C-terminal 28–38 region of PACAP38 strengthens the binding efficiency of this ligand. It is also worth to mention that in human plasma, ceruloplasmin binds to PACAP38 but not to PACAP27 [38], that the 38-residue PACAP isoform passes through the blood–brain barrier (BBB) by means of a transporter identified as PTS-6 while PACAP27 crosses the BBB by passive diffusion [39], and that PACAP38 is much less resistant to degradation in human plasma than the 27-amino acid counterpart [40]. It thus appears that the 28–38 tail of PACAP is important for blood transportation, BBB crossing, and degradation by plasma endopeptidases [3, 38–40].

When the truncation strategy was applied at the C-terminal end of PACAP27, it was observed that the fragments spanning from 1–26 to 1–23 could discriminate the three PACAP receptor subtypes, with the highest affinity for VPAC1 [41]. In addition, all truncated peptides were full agonists of the PAC1 and VPAC1 receptors, whereas the 1–25 and 1–26 fragments were partial agonists of VPAC2. Accordingly, Bourgault et al. [42] showed that PACAP(1–23) still displayed a good affinity for PAC1 expressed in transfected CHO cells and a very significant activity in calcium mobilization and PC12 cell differentiation assays, as compared to the native ligand. However, further deletions of amino acids beyond position 23 produced a drastic loss of affinity and potency. Thus, Leu^{23} could play a major role towards PAC1, as its exclusion to create the 1–22 fragment gave rise to a large decrease of binding (≈23-fold) and biological activity (≈19-fold), when using PACAP(1–23) as the reference compound. In fact, the hydrophobic residue leucine-23 is highly conserved within the structurally related VIP/secretin/glucagon/PACAP peptide superfamily, suggesting an essential role for this amino acid in the regulation of the biological functions of these peptides [42, 43]. Interestingly, supporting a foremost contribution of residue 23, [Lys^{23}]PACAP27 was devoid of affinity towards the PAC1 receptor [44], and [Ala^{23}]VIP showed poor affinities and activities in VPAC1 and VPAC2 paradigms [45].

Analogs of PACAP Containing Modifications in the N-Terminal Segment

The relative importance of some physicochemical properties of each residue in a peptide is frequently explored by achieving an L-alanine scanning, in which each amino acid of the primary structure is sequentially replaced with an Ala residue, and by performing a D-amino acid scanning, in which each L-residue is successively substituted with its D-counterpart. Considering the specific role of the N-terminal domain for the biological activity of PACAP, Ala- and D-scans were conducted on the His^1 to Thr^7 segment [46, 47]. This strategy revealed that the imidazole ring of His^1, the carboxylic function of Asp^3 and the phenyl group of Phe^6 are key chemical moieties for binding and activation. In contrast, replacement of Gly^4 with Ala had no significant effect, and Ala substitution of Ser^2 did not affect the binding properties of both PACAP isoforms towards PAC1and VPAC1 receptors, but slightly decreased the affinity and potency for VPAC2. For position 2, very similar results were reported by Robberecht et al. [29] and by Hou et al. [31]. Finally, it is worth to point out that the introduction in PACAP27 of an L-alanine at position 7 generated an analog virtually devoid of VPAC2 agonistic activity and acting as a potent VPAC1 and PAC1 agonist.

Some insights concerning the biologically active conformation of the N-terminal stretch of PACAP were obtained using the D-scan approach, which consists in serially reversing the configuration of each peptide residue to introduce local structural constraints [46, 47]. This methodology not only allows the assessment of the side-chain orientation but also provides information regarding the secondary structure that is important for the affinity and agonistic activity. The scanning revealed two distinct domains for the N-terminal segment of PACAP27. So, on the one hand, D-substitution of His^1, Ser^2, or Asp^3 did not affect much the binding properties and the biological activity of the peptide towards the three PACAP receptors, especially when PAC1 or VPAC1 were involved [46–48]. Worth to mention, in the context of stability improvement, [D-Ser^2]PACAP27 was extremely resistant against the proteolytic action of dipeptidyl peptidase IV (DPP IV), an enzyme breaking down PACAP by cleaving N-terminal dipeptides [40]. On the other hand, incorporation of a D-amino acid at position Phe^6 or Thr^7 was deeply deleterious. As observed before with PACAP fragments, PACAP38 analogs exhibited profiles not always identical to those of the PACAP27 equivalents, in particular when the D-substitution was applied at position 7. As a matter of fact, [D-Thr^7]PACAP38 behaved as a potent partial agonist with a potency only 3.5 times lower than that of the parent molecule (EC_{50} = 7.3 nM for the analog vs. 2.1 nM for PACAP38; E_{max} = 63 %), thus demonstrating again the particular contribution of the 28–38 portion. Further studies of the molecular structure of the His^1 to Thr^7 segment of PACAP were carried out by N-methylation of the polypeptide backbone, a scheme that introduces a restriction in the amide bond geometry and abolishes the hydrogen bonding ability. The N-methyl-scan revealed that the attachment of a methyl group on the nitrogen atom of the amide bond connecting His^1 to Ser^2, or Ser^2 to Asp^3, decreased by a little to a moderate extent the binding affinity and the activity of both PACAP isoforms [46, 47]. Moreover, N-methylation of PACAP at positions Ile^5, Phe^6, or Thr^7 yielded inactive compounds or extremely weak agonists, with very low binding affinities, suggesting that PACAP is sensitive to backbone restrictions in this region. Most interestingly, substitution of Gly^4 with sarcosine (Sar), an N-methylated glycine residue, inhibited the capacity of PACAP38 and PACAP27 to activate the PAC1 receptor, and decreased by 15- and 40-fold the affinity of the 38- and 27-amino acid isoforms, respectively [46]. Because substitution of Gly^4 with L-alanine was without any significant effect, these results suggest an important role for the amide proton of the Asp^3–Gly^4 bond, such as an involvement in hydrogen bonding. Furthermore, the analog [Sar^4]PACAP38, appeared as a potent PAC1 receptor antagonist with a pA_2 value calculated at 8.39 [46].

Additional residue modifications encompassing the N-terminal domain of PACAP were performed to further evaluate their importance for biological activity and affinity. For instance, a free N-terminal amino group is not essential for PAC1 activation since N-acetylation did not alter the affinity and potency of the native peptide [29, 40, 46]. Moreover, N-terminal acylations with larger moieties would be tolerated, as suggested by the data obtained with N-hexanoyl-derivatives of PACAP27 and PACAP38, which showed pharmacological properties similar or slightly weaker to those of the parent compounds [40]. Indeed, the binding affinity remained the same; whilst the activities in calcium mobilization and PC12 cell proliferation assays were decreased by 3.5-fold, thereby suggesting that PAC1 is sensitive to the steric hindrance and/or hydrophobicity introduced by an N-acylation with a 6-carbon aliphatic chain. Moreover, it was reported that a [Nle^{17}]PACAP27 analog (Nle: norleucine, an isostere of Met; the related Nle-analog is as potent as PACAP27 [49]) to which was coupled at the N-terminus the photoreactive probe, *p*-benzoyl-phenylalanine (Bpa), exhibited an EC_{50} of 135 nM and allowed the photolabeling of the PAC1 receptor [49]. Noteworthy, N-terminal acylation produced analogs exceptionally resistant towards DPP IV [40]. Substitution of His^1 with the aromatic residue Phe reduced but did not abolish the potency of PACAP27 to stimulate adenylyl cyclase activity in rat hippocampus and pancreatic AR4-2J cell membranes [29, 31], while the incorporation at this position of 4-imidazole acetic acid (Iaa) or 4-imidazole acrylic acid (Iac) had limited impact on affinity and biological activity, on either PACAP receptors [48]. These results, combined with the pharmacological data of [Ala^1]PACAP, indicate a participation of the imidazole group and a potential contribution of aromaticity. Thus, the implication of the distal (τ) and proximal (π) nitrogen atoms of the imidazole ring was explored and it appeared that π-N-methylation altered much more the affinity and activity of PACAP than τ-N-methylation [46]. Also, π-N-methylation reduced drastically the efficacy of PACAP27 and PACAP38 to activate PAC1, and the corresponding analogs behaved as weak partial agonists. Robberecht et al. [29] reported that a 2.4-fold rise in the affinity of PACAP27 towards the PAC1 receptor was obtained when the pH was raised from 6.0 to 7.5, and this phenomenon was not observed with [Phe^1]PACAP27 and PACAP(2–27). They postulated that the deprotonation state of the imidazole ring of His^1 was a major determinant for optimal binding of PACAP27. Chemically, the deprotonated condition is required to maintain the aromaticity of histidine. Following τ- or π-N-methylation of His^1, the aromatic attribute of this residue is preserved but tautomerization of the imidazole ring is inhibited and its pKa is increased. However, to explain the significant reduction of efficacy of PACAP, neither the tautomerization nor the increase of pKa would be suitable parameters to consider because it is expected that the τ- and π-N-methylated His-derivatives of PACAP would exhibit very similar properties. Thus, the methyl group attached to the π-nitrogen atom of the imidazole ring would not interfere much in the binding process of PACAP ligands but would cause sufficient hindrance to deeply disturb a key molecular interaction involving the His^1 side-chain with another residue within the PACAP molecule.

Substitutions at position Ser^2 with α-aminoisobutyric acid (Aib) partially improved the stability of PACAP towards DPP IV [40], and had no major effects on the potency of the 38- and 27-amino acid peptide ligands [40, 46–48]. Nonetheless, the binding affinity, the potency and the efficacy of PACAP were reduced when Ser^2 was replaced with residues bearing bulkier side-chains, such as Arg, Phe or substituted with their D-counterparts or Lys [29, 31, 44]. These observations indicated that the hydroxyl group in position 2 was not required for PAC1 receptor activation whereas an increase of the residue bulkiness, and possibly the incorporation of a positive charge were highly unfavorable. In contrast, introduction at position 2 of a structural constraint by substituting Ser with Pro or Hyp (hydroxyproline) did not influence much the affinity and activity of both PACAP isoforms for PAC1 and VPAC1, but decreased substantially the binding and virtually abolished the activity towards VPAC2 [46, 47]. Therefore, it seems that the structural restrictions caused by Pro or Hyp at position 2 induced an N-terminal conformation that might be closely related to the bioactive one. Interestingly, the inversion of chirality of proline-2 ([D-Pro^2]PACAP) yielded analogs of the isoforms behaving like antagonists. Accordingly, conformational analysis by NMR revealed that the introduction of Pro in position 2 of PACAP27 promoted an N-terminal turn structure [3, 46, 47], and that this molecular arrangement was even more strongly stabilized with the incorporation of D-Pro [46]. Also, as expected, it was determined that the spatial positioning of His^1 in [Pro^2]PACAP27 was very diverging to that observed in the [D-Pro^2]-analog.

The presence of a negative charge in position 3 and the distance of this charge from the peptide backbone are important features for the biological activity of both PACAP isoforms [29, 31, 44, 46–48]. Thus, it was shown that the PACAP27 or PACAP38 analogs obtained by substituting Asp^3 with L-alanine, L-asparagine, L-lysine, L-glutamic acid, or L-α-aminoadipic acid (Aad) displayed much weaker binding affinity and potency in PAC1, VPAC1, and/or VPAC2 assays. In contrast, the side-chain orientation was not a key element because [D-Asp^3]PACAP27 and [D-Asp^3]PACAP38 were potent analogs in any of the three receptor paradigms. However, the structural arrangement and/or the presence of the amide proton in the Ser^2–Asp^3 bond would be crucial constituents for activity because analogs containing modifications altering either of these two characteristics were poor agonists [46–48]. Indeed, compounds such as [Pro^3]-, [Sar^3]-, [Pip^3/Pipecolic acid]-, and [N-Me-Asp^3]PACAP exhibited weak or no biological activities. Nevertheless, depending on the receptor type and the nature of the alteration, the impact on the recognition process was highly variable, as demonstrated with the very significant binding measured with [N-Me-Asp^3]PACAP27 and [N-Me-Asp^3]PACAP38 [46, 47], and the large decrease of affinity of [Pro^3]PACAP27 and [Sar^3]PACAP27, especially towards the PAC1 receptor [48].

Further transformations within the Gly^4 to Thr^7 segment revealed additional information [44, 46, 47]. Thus, successive replacements of the amino acids of this stretch with Lys produced analogs virtually devoid of affinity towards all three types of PACAP receptor [44]. In addition, their sequential substitution with L-alanine or inversion of their chirality showed that residues Gly^4 and Ile^5 were rather tolerant towards those alterations, but that Phe^6 and Thr^7 were not. As a matter of fact, modifications at position 6 yielded PACAP27 and PACAP38 analogs with very weak affinity and activity, whereas the introduction of L-alanine at position 7 of PACAP27 gave rise to a potent selective VPAC1/PAC1 agonist. Also, the inversion of configuration of Thr^7 produced a PACAP27 derivative displaying poor affinity in the three PACAP receptor binding assays [46, 47]. The N-methyl-scan of the amide bonds of the 4–7 region demonstrated that this segment of PACAP is very sensitive to structural changes induced by N-methylation. This chemical modification increases the steric hindrance at the peptide backbone, eliminates hydrogen sharing at the substituted amide bond and allows the isomerization of the peptide bond (all *trans* to a mixture of *cis* and *trans*). Although the precise physicochemical effect(s) of the modification is/are not known, all four analogs obtained by successively N-methylating the peptide bonds of residues Gly^4 to Thr^7 were found inactive in the PAC1, VPAC1, and VPAC2 assays [46, 47]. Ramos-Álvarez et al. [48] also observed the deleterious effect of N-methylation in this portion of the molecule since they reported no biological activity for [Sar^4]PACAP27 in their paradigms for the three PACAP receptors.

As aforementioned, position Phe^6 was very sensitive to the Ala or Lys substitutions, and inversion of configuration. Hence, various additional replacements were explored to better define the role of this residue [46, 47]. Substitution of Phe^6 with the non-aromatic L-cyclohexylalanine (Cha) residue did not produce a large decrease of biological activity, as observed with L-alanine. Likewise, as demonstrated with [Tyr^6]PACAP27 and [Tyr^6]PACAP38, both isoforms were tolerant to the presence of a hydroxyl group at the *para* position of the aromatic ring of Phe. Furthermore, the incorporation at position 6 of the bulky aromatic residues L-biphenylalanine (Bip) or L-naphthylalanine (Nal) gave analogs at least as potent as the native peptide hormones in a PAC1 assay, but with reduced activity and affinity towards VPAC1 and VPAC2. Also, the photoprobe [Bpa^6, Nle^{17}]PACAP27, designed to label PAC1, displayed potent binding affinity and activity in PAC1 paradigms [49]. Overall, these results suggest that aromaticity is not essential at position 6 but that the presence of a hydrophobic side-chain, still bulky, maintains or even improves the binding affinity and the potency of PACAP towards the PAC1 receptor. Considering this finding and the observation that Phe^6 was very sensitive to structural changes, thus suggesting the participation of this residue to a particular molecular arrangement, Bourgault et al. [46] and Doan et al. [47] introduced at this position a few constrained residues containing aromatic/hydrophobic moieties, i.e., L-indoline-2-carboxylic acid (Ind), L-tetrahydroisoquinoline-1-carboxylic acid (Tiq), L-1,2,3,4-tetrahydroisoquinoline-3-carboxylic acid (Tic), and L-1,3-dihydro-2*H*-isoindole carboxylic acid (Disc). All these analogs containing static "side-chain," either of the PACAP38 or PACAP27 isoform, were essentially unable to recognize the PACAP receptors. Therefore, such a constraint could not be applied at position 6, where in fact it appeared that the freedom and/or the exact positioning of the hydrophobic side-chain would be a key feature for the ligand recognition process.

Analogs of PACAP Containing Modifications in the Central and C-Terminal Segment

Although they were utilized less frequently than N-terminal modifications, central and C-terminal chemical and structural alterations were also performed in PACAP. Indeed, SAR analyses including truncations studies [41, 42], as well as an Ala-scan applied to PACAP(15–31) [50], and a Lys-scan carried out on PACAP27 [44], revealed that residues Tyr^{10}, Arg^{14}, Val^{19}, Lys^{20}, Tyr^{22}, and Leu^{23} would play important roles regarding the chemical and/or structural properties of the central and C-terminal segments of PACAP, both forming a region that was shown by conformational studies [42, 46, 50–55] to exhibit an evident propensity to adopt an α-helix arrangement. This helical region would be an influential feature for the binding and activity of PACAP, and therefore modifications were introduced at key positions. Hence, incorporation of a CH_2–NH peptide bond surrogate between residues Lys^{21} and Tyr^{22}, an alteration destabilizing secondary helical structures, has decreased the biological activity of PACAP38, as well as its affinity for PAC1, thereby suggesting that the integrity of the α-helix is essential [40]. In addition, replacement of Val^{19} with the helix-breaking amino acid Gly was deleterious for the binding of PACAP(6-38) to the isolated extracellular N-terminal domain of the PAC1 receptor [54]. Surprisingly though, swap of Val^{19} with the potent helix promoter Ala diminished significantly the ability of PACAP(15–31) to bind to PAC1 [50]. As well, exchange in PACAP27 of residue Lys^{20}, an α-helix stabilizer, with Gly, an α-helix breaker, gave an analog that was still potent, with reduction of affinity and activity by only eightfold and fivefold ($IC_{50} = 4$ nM vs. 0.5 nM and $K_{act} = 5$ nM vs. 1 nM), respectively [29]. These two last data indicate that local alterations could be introduced in the helical segment without completely losing the pharmacological properties. Nonetheless, the simultaneous switch of Lys^{20} and Lys^{21} with Gly reduced the affinity by 200-fold and the activity by 80-fold for the PAC1 receptor [29], whereas substitution with Ala at either of these two positions of PACAP38, produced peptides that exhibited slightly better binding affinity and activity than the parent molecule [40]. Moreover, the introduction of Ala^{20} in PACAP38 derivatives designed to resist to enzyme degradation, did not affect significantly the peptide binding and biological characteristics [40]. Likewise, Yung et al. [44] as well as Sun et al. [54] explored the successive replacement of Lys^{20} and Lys^{21} with Glu, an anionic residue that is also an α-helix promoter better than Lys. They observed that the analog $[Glu^{21}]$PACAP27 displayed affinities ranging from 10 to 60 nM, which corresponded to reductions of 6- to 20-fold, depending of the receptor subtype [44], whereas $[Glu^{21}]$PACAP(6-38) was 2.4-fold less potent than PACAP(6–38) to bind to PAC1 [54]. Moreover, the analog $[Glu^{20}]$PACAP27 appeared as a very weak ligand (80-, -200, and 180-fold decrease towards PAC1, VPAC1, and VPAC2, respectively) [44], while $[Glu^{20}]$PACAP(6-38) exhibited a 18-fold reduction of binding to PAC1 [54]. Data from Kumar et al. [50] brought supplementary insights. Indeed, binding experiments carried out with a library that was built using an Ala-scan of the PACAP(15–31) fragment showed that $[Ala^{21}]$PACAP(15–31) was an excellent ligand for PAC1, whereas its $[Ala^{20}]$-equivalent was devoid of binding affinity.

Consequently, the Lys^{20} contribution in PACAP appeared to vary in function of the size of the peptide ligand and the physicochemical properties of the residue that occupies this position. It could be hypothesized that the Lys^{20} side-chain would be oriented towards the receptor surface and that the introduction of a negative charge at position 20 obtained with the substitution with Glu, would inhibit the interaction with the PAC1 receptor. Accordingly, Kumar et al. [50] proposed a salt bridge between Lys^{20} of PACAP and Glu^{104} of PAC1 and this ionic interaction would stabilize the ligand–receptor complex. Nevertheless, such an interaction would not be essential for affinity and activity, as demonstrated by the pharmacological effects obtained with [Gly^{20}]PACAP27 and [Ala^{20}]PACAP38 [29, 40].

Further residue exchanges demonstrated that the switch of residue Tyr^{22} to Lys produced a PACAP27 analog that was essentially recognized only by VPAC1 [44]. In addition, it was reported that [Ala^{22}]PACAP38 was mostly active at the PAC1 and VPAC1 sites, but not at VPAC2, with potencies being 11-fold and 2-fold lower than those of PACAP38 for the PAC1 and VPAC1 receptors, respectively [48]. In contrast, [Ala^{22}]PACAP(6–38) was described as a fragment that was not recognized by the PAC1 receptor [54]. Hence, this position seemed to favor VPAC1 selectivity, and this was supported by the results of Nicole et al. [45] who developed [Ala$^{12, 22, 28}$]VIP, a selective VPAC1 agonist. Studies on another tyrosine residue of PACAP demonstrated that the simultaneous replacement of Tyr^{10} and Ser^{11} with a type II (*S*-IBTM: 2(*S*)-amino-3-oxohexahydroindolizino[8,7-*b*]indole-5(*S*)-carboxylic acid) or II' (*R*-IBTM: 2(*R*)-amino-3-oxohexahydroindolizino[8,7-*b*]indole-5(*R*)-carboxylic acid) β-turn dipeptide mimetic almost inhibited the binding affinity of PACAP27 to PAC1 and effectively decreased it towards VPAC1, albeit the analog structure was only a little distorted compared to that of the parent molecule [51]. Also, Sun et al. [54] reported that the switch of Tyr^{10} for Ala decreased the affinity of PACAP(6–38) to PAC1 by only twofold, whereas Yung et al. [44] described [Lys^{10}]PACAP27 as a peptide for which bindings to the PAC1, VPAC1, and VPAC2 receptors were abolished. So far, it is difficult to precisely determine the role of Tyr^{10} in PACAP. Nevertheless, the data suggest that the Tyr^{10} side-chain participates in an intramolecular interaction or to close contacts within the PACAP receptor. Moreover, PAC1, and to a lesser extent VPAC1, would not tolerate structural distortions, even subtle, in the position-10 region of the peptide molecule.

On the one hand, successive substitution of the cationic residues Arg^{14} and Lys^{15} with Ala did not affect the potency of PACAP. Indeed, the analogs Ac-[Ala^{14}, Ala^{20}]PACAP38-propylamide and Ac-[Ala^{15}, Ala^{20}]PACAP38-propylamide were at least as potent as Ac-PACAP38-propylamide, a derivative that showed binding and activity properties equal or even slightly better than PACAP38, while being resistant to DPP IV action [40]. In addition, Sun et al. [54] reported that [Ala^{14}]PACAP(6–38) retained a high affinity for PAC1 while Kumar et al. [50] described [Ala^{15}]PACAP(15–31) as a ligand with excellent binding properties for this receptor. On the other hand, alike to what was observed with the substitution of Lys^{20} with Glu in PACAP27, the swap of Arg^{14} with Glu practically abolished the binding to all three PACAP receptors [44]. In contrast, the incorporation of Glu at position 15 led to a substantial loss of affinity of the related analog for VPAC1 (120-fold) but only to moderate decreases (fivefold) of binding towards PAC1 and VPAC2 [44]. Hence, Lys^{15} could be a target for improving the selectivity of PACAP analogs. Replacement of Arg^{30}, another cationic residue that would participate to the VPAC2 selectivity [44], showed that this position is rather tolerant to amino acid substitutions. For instance, it was observed that Ala could be introduced at position 30 in PACAP(6–38) and PACAP(15–31) without deleterious effects on PAC1 binding [50, 54]. Altogether, these results indicate that the helical segment of PACAP can tolerate modifications, as long as they do not involve a charge swapping at a cationic position assumed to participate in an interaction with the receptor. Nonetheless, it seems that specific domains involved in receptor recognition are present in the helical region of PACAP. This is supported by the results of Bourgault et al. [42] who showed that the incorporation of an aminocaproic acid (Aca) spacer, designed to replace six amino acid residues at a time in the helical core [for example, PACAP(1–8)-Aca-(15–27)], was inhibiting the activity and the binding affinity of the corresponding analogs, albeit the molecules maintained the propensity to adopt a partial α-helix structure in a membrane mimicking environment.

Molecular Structure of PACAP and Receptor Interaction

Structural analyses of PACAP, using biophysical and in silico methodologies such as nuclear magnetic resonance (NMR) and circular dichroism (CD) spectroscopies as well as molecular modeling and dynamics, revealed key conformational arrangements for this polypeptide. Thus, CD spectroscopy showed that PACAP is mostly disordered in aqueous solution [52, 55]. However, upon addition of structure-inducing co-solvents [for instance, methanol (MeOH) or trifluoroethanol (TFE)] or micelles mimicking the membrane environment, the molecule exhibits an evident propensity to adopt an α-helix secondary structure [42, 52, 55]. This was reinforced by NMR data of PACAP isoforms and fragments demonstrating that the peptide contains a helical core, which varies in length depending on the medium used for the analysis [46, 51–56]. For instance, an α-helix spanning from residues 9–26 or 7–27, with a discontinuity between Lys^{20} and Lys^{21} was reported when TFE was used as a co-solvent [51, 55]. In these studies, the results supported the presence of a helix that is preceded by an unstructured N-terminal segment. In an aqueous mixture containing 25 % MeOH, the disordered N-terminal domain of PACAP27 was followed by two distinct structural regions, i.e., a β-turn-like bend (residues 9–12) and an α-helix encompassing amino acids 12–24 [52], characterized by a segment (residues 15–20) where the conformation was looser. When the NMR analysis of PACAP27 was carried out in a biomimetic environment obtained with dodecylphosphocholine micelles, the α-helix extended from the C-terminus to residue Ile^5 and the N-terminal tetrapeptide was freely moving and disordered [46, 56]. Noteworthy, Inooka et al. [56] reported for fragment PACAP(1-21) when bound to PAC1 that the region 8–21 forms an α-helix similar to that observed with micelle-bound PACAP27. However, in contrast to what was observed with the 27-amino acid isoform in presence of micelles, residues 3–7 of PACAP(1-21) adopted a unique β-coil structure made by adjoining two β-turns, a type II' (Asp^3–Gly^4–Ile^5–Phe^6) and a type I (Gly^4–Ile^5–Phe^6–Thr^7), when the peptide was attached to the PAC1 receptor. This compact molecular arrangement would bring together residues Ile^5, Phe^6, and Tyr^{10} to form a hydrophobic cluster that could interact with a receptor hydrophobic patch or side chain.

NMR studies of PACAP38 revealed that the structure of its 1–27 segment overlays that of the 27-residue isoform, and that the 28–38 C-terminal extension exhibits a short helix (28–34) and is connected to the other part of the molecule through a flexible hinge [55]. Wray et al. [55] also observed an absence of NOE connectivities between the amino acids of the central segment of human PACAP38 and those of the C-terminal tail, thereby suggesting an extended conformation. Interestingly, an NMR analysis carried out with the grass carp PACAP38 isoform, which possesses 89 % sequence identity with the human ortholog, showed that the C-terminal 29–34 helical stretch is in the vicinity of the central α-helix core, thus leading to a bent conformation similar to a "fishhook" [57]. Finally, Sun et al. [54] reported that PACAP(6–38) adopts a helical structure when bound to the isolated N-terminal domain of PAC1, with a bend at the level of Ala^{18}. This molecular organization would favor hydrophobic and electrostatic contacts (for instance, Tyr^{10}, Arg^{14}, and Lys^{21}) with residues of the N-terminal domain of PAC1. Such interactions were also depicted by Kumar et al. [50] who postulated the participation of the hydrophobic residues Val^{19} and Val^{26} of PACAP, together with a salt bridge involving Lys^{20} of the ligand and Glu^{104} of the receptor. So far, conformational analyses demonstrated that the overall secondary structure of PACAP exhibits a high homology with that observed for other members of the secretin/glucagon/GHRH superfamily, which display a random architecture for the N-terminal segment and an extended helical arrangement covering a large span of the molecule [16, 17, 25, 58].

Even when performed in structure-inducing or membrane-mimicking environments, conformational studies suggested that the N-terminal segment of PACAP (His^1 to Gly^4) is disordered [46, 52, 55]. However, considering its essential role in biological activity, it most likely adopts a precise molecular arrangement upon binding to the receptor. Indeed, the Chou and Fasman's empirical method predicts a β-turn for the sequence His^1–Ser^2–Asp^3–Gly^4. Also, Bourgault et al. [46] showed that the incorporation of turn-promoting residues such as D-amino acids, or Pro, Hyp, or Aib at position 2, produced potent agonists. Moreover, the NMR-NOESY spectrum of the potent and constraint agonist [Pro^2]PACAP27 exhibited the typical connectivities of a β-turn involving residues His^1 to Gly^4. In addition, the group demonstrated that π-N-methylation of His^1 reduced drastically the efficacy of PACAP27 and PACAP38 to activate PAC1, and the corresponding analogs behaved as weak partial agonists. This nitrogen atom of a His imidazole ring is frequently involved in a hydrogen bonding with the backbone amide proton of the third residue of a β-turn when His is at position 1 [59]. In particular, this type of bonding is observed in Asx-turns that are characterized by stabilization of the β-turn structure by two hydrogen bonds involving amino acids 1 and 3, and 2 and 4, respectively. Such a molecular organization was hypothesized for PACAP [46, 47] and is illustrated in

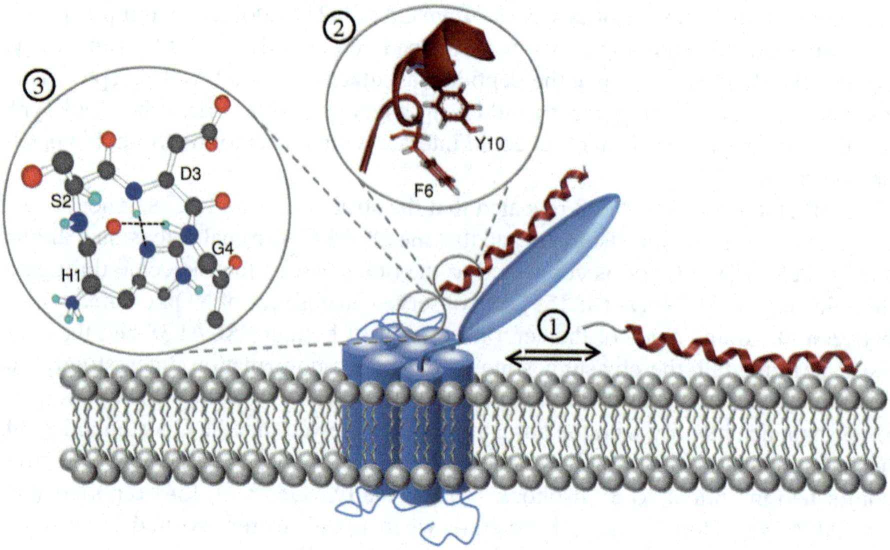

Fig. 8.1 Schematic representation of the "two-domain" model for receptor activation. (1) The helical segment of PACAP binds to the N-terminal extracellular domain of the receptor. (2) PACAP-receptor interaction favors the formation of a β-coil structure produced by adjoining a type II′ β-turn (residues Asp^3 to Phe^6) and a type I β-turn (residues Gly^4 to Thr^7). This coiled conformation, similar to a "helix N-capping" motif, creates a hydrophobic cluster. (3) As inferred from SAR studies and supported by structural analyses, the N-terminal tetrapeptide of PACAP (residues His^1 to Gly^4) would adopt an Asx-turn-like conformation stabilized by hydrogen bonds

Fig. 8.1 (enlargement 3). As a matter of fact, because of the N-terminal sequence homology between PACAP and current members of the secretin/glucagon/GHRH peptide hormone superfamily, it was suggested that an Asx-turn would be crucial for receptor activation, not only for PACAP but also for other family members [16].

Peptide ligands of class B-GPCRs would adopt a common helix N-capping motif that would constitute a fundamental element for receptor activation [21, 25]. This structure is frequently observed at the beginning of an α-helix, where it stabilizes the N-helical end through hydrogen bonds and hydrophobic interactions [60, 61]. In PACAP, this molecular arrangement (Fig. 8.1, enlargement 2) would bring closely the Ile^5, Phe^6, and Tyr^{10} side-chains to shape a hydrophobic cluster [56], and a hydrogen bond between the Thr^7 side-chain and the backbone amide proton of Tyr^{10} would strengthen the conformation [21]. Some results support the presence of an N-capping motif in PACAP, postulated to be a type IA because of its configuration. For instance, replacement of Phe^6 with hydrophobic residues such as Tyr, Cha, Bip, and Nal did not alter significantly the binding properties of PACAP analogs, whereas the introduction of Ala at position 6 reduced considerably the binding to all PACAP receptors [47]. However, as aforementioned, Sun et al. [54] reported that the switch of Tyr^{10} for Ala decreased the affinity of PACAP(6-38) to PAC1 by only twofold, whereas Yung et al. [44] described [Lys^{10}]PACAP27 as a ligand for which bindings to the PAC1, VPAC1, and VPAC2 receptors were abolished. So far, it seems that the presence at position 10 of the non-bulky, nonpolar residue alanine, which is also a much better helix-stabilizer than tyrosine, thereby reinforcing the α-helix N-boundary, could preserve the ability of PACAP to adopt an N-capping motif. Lysine is also a much better helix-former than Tyr. Nonetheless, this cationic residue would be too hydrophilic to promote a hydrophobic clustering. Noteworthy, it has been reported that Thr^{7} in VIP represents a specific key pharmacophore for VPAC2 selectivity [62, 63]. Furthermore, the analog [Ala^{7}]PACAP38 showed affinity and activity as potent as that of the native molecule in PAC1 assays, while the analog [Ala^{7}]PACAP27 bound and activated efficiently PAC1 and VPAC1, but exhibited very poor binding and almost no activation with VPAC2 [46, 47]. Hence, a key conformational feature for the PAC1 and VPAC1 receptors would be a hydrophobic cluster, for which structure consolidation through a hydrogen bond involving the hydroxyl group of Thr^{7} and the Tyr^{10} backbone amide proton would not be mandatory. Interestingly, using NMR, data for PACAP(1-21) bound to PAC1 were consistent with a hydrophobic clustering of amino acids 5, 6, and 10, while Thr^{7} did not show the structural parameters expected for a participation in an N-capping arrangement [21, 56]. Therefore, taken together, the [Ala^{7}]PACAP27 pharmacological results and the NMR spectroscopic data suggest that for PAC1 and VPAC1, an "N-capping-like" motif, characterized by the sole presence of a hydrophobic core, would be sufficient for binding and biological activity.

Biological and conformational studies led to the identification of three major structural domains in PACAP, i.e., a disordered/Asx-turn N-terminal tetrapeptide, an "N-capping-like" motif involving residues Ile^{5} to Tyr^{10}, and an α-helix spanning from position 7 or 9 up to the C-terminus. This peptide ligand topography is well adapted for the binding process described for the class B-GPCRs, which is known as the "two-domain" model [15–17, 20, 21]. As illustrated in Fig. 8.1, the central and C-terminal segments of the peptide bind to the large extracellular N-terminal domain of the receptor and after this initial docking, the N-terminal portion of the ligand interacts with the juxtamembrane region of the receptor to initiate activation and intracellular signaling [15]. Accordingly, a study by Bergwitz et al. [64] demonstrated that peptide hybrids of parathyroid hormone (PTH) and calcitonin (CT), which do not stimulate their wild-type class B receptors, could activate chimeric CT-PTH receptors. Moreover, C-terminal truncations of peptide ligands decrease the binding affinity towards class B-GPCRs, whereas deletions of N-terminal portions usually suppress their biological activity [29, 41, 42, 65, 66]. Finally, photoaffinity labeling studies carried out with various class B receptor ligands (PTH, CT, GLP-1, secretin, PACAP, VIP) generally supported the "two-domain" model [49, 67–85], although some particularities in the physical ligand–receptor interactions were observed, especially with PACAP and VIP, suggesting that the binding domains of the PAC1, VPAC1, and VPAC2 receptors could slightly differ from those of the other class B-GPCRs [16].

Conclusion

Biological and pharmacological studies have suggested that PACAP would be of therapeutic value for the treatment of several neurodegenerative and acute neurological disorders. Especially, its ability to prevent neuronal loss and its capacity to readily cross the BBB strongly support the potential of this peptide to be applied as a neuroprotective drug. Nonetheless, before considering its use, numerous physiological characterizations are essential and in particular, it is crucial to develop selective pharmacological tools targeting the PACAP receptors. So far, mainly because of their two-domain topography, class B-GPCRs remain difficult targets for drug development, and identification of small molecules that could mimic the biological actions of the peptide ligands was virtually unsuccessful. Nonetheless, a few nonpeptide antagonists were described, as reported for instance by Beebe et al. [86] who discovered hydrazide-based antagonists of the PAC1 receptor, and by Chu et al. [87] who identified substituted pentanamide derivatives that exhibited VPAC2 antagonism. On the one hand, according to Wu et al. [88], the hydrazide compounds would bind PAC1 at the boundary between the N-terminal extracellular domain and the receptor core formed by the 7-transmembrane helices (7TM) of the GPCR. On the other hand, the pentanamide-derived antagonists would interact with an allosteric site of the 7TM region and would act as noncompetitive inhibitors [87]. Given that PACAP agonists rather than antagonists are mostly needed for clinical applications, as described for some other class B receptor peptide ligands (CT, PTH, GLP-1) [15], it seems that PACAP-derived drugs would be most likely peptide compounds. Thus, the main challenge is to develop stable, potent and selective agonists towards the PACAP receptors. Moreover, if CNS actions are required, the molecules should be able to cross the BBB. The rational design of such peptide derivatives relies on SAR studies that provide information about the significant pharmacophoric elements for binding affinity and activation of the PAC1, VPAC1, and VPAC2 receptors. Thus far, many reports already described some selective agonists and antagonists for the PACAP/VIP receptors [3, 18, 20, 26], albeit some data, such as those about the selectivity of PACAP(6–38) or PG-99–465, an N-myristoylated VIP-derived analog, were subsequently revisited following further characterizations [87, 89]. The present review described numerous conformational and physicochemical features related to PACAP. Those key findings were obtained using various strategies, including single-point substitutions, conformational constraints and identification of enzymatic cleavage sites in PACAP. When applied to the design of pharmacological tools, agonists showing a clear preference for the PAC1/VPAC1 receptors, such as [Ala^7]PACAP27, [Ala^7]PACAP38, [Pro^2]PACAP27, and [Ala^{22}]PACAP38 were produced [46–48]. Moreover, a metabolically stable analog, Ac-[Phe(*p*-I)^6, Nle^{17}]PACAP27, was shown to protect in vivo, as potently as PACAP38, the dopaminergic neurons of the *substantia nigra* against cell death induced by a neurotoxic agent, but with a significantly decreased hypotensive side effect, compared to the native 38-amino acid isoform [90]. Interestingly, probably through a VPAC1 activation, this PACAP analog also reduced the production of pro-inflammatory cytokines, thereby indicating that a simultaneous action on PAC1 and VPAC1 could be a better approach to treat neurodegenerative and neurological disorders. So far, no complete PAC1 selectivity was still attained. Nevertheless, analogs displaying a predominant preference for this receptor were reported in the literature [47, 48]. For instance, [Hyp^2]PACAP27 exhibited pEC_{50} of 8.21 and 7.03 for PAC1 and VPAC1, respectively, while showing no activity in the VPAC2 assay (it binds though VPAC2 with a pIC_{50} of 6.89) [47]. Similarly, [Bip^6]PACAP27 displayed pEC_{50} of 9.23, 7.33, and 7.09 for PAC1, VPAC1, and VPAC2, respectively [47]. Also, the analog [Ala^{22}]PACAP38 showed pEC_{50} of 9.49, 8.97, and 6.55, and affinities corresponding to pIC_{50} of 8.43, 7.76, and >6 for the three PACAP receptors [48]. Therefore, current data are promising for developing selective PAC1 agonists. However, particular attention will have to be paid because the PAC1 receptor is expressed in a large set of splice variants that exhibit various affinities for the PACAP isoforms and different abilities to activate intracellular signaling cascades [3, 18, 48]. Hence, depending on the tissues and the PAC1 splice variants that are involved, the selectivity of PAC1 agonists might vary significantly.

Acknowledgements This work was supported by the Canadian Institutes of Health Research (AF; FRN102734), the Fonds de recherche du Québec-Nature et technologies (DC; 1888882), and the Natural Sciences and Engineering Research Council of Canada (SB; 1557119).

References

1. Reglodi D, Kiss P, Lubics A, Tamas A. Review on the protective effects of PACAP in models of neurodegenerative diseases in vitro and in vivo. Curr Pharm Des. 2011;17:962–72.
2. Reglodi D, Tamas A, Lubics A, Szalontay L, Lengvari I. Morphological and functional effects of PACAP in 6-hydroxydopamine-induced lesion of the substantia nigra in rats. Regul Pept. 2004;123:85–94.
3. Vaudry D, Falluel-Morel A, Bourgault S, Basille M, Burel D, Wurtz O, et al. Pituitary adenylate cyclase-activating polypeptide and its receptors: 20 years after the discovery. Pharmacol Rev. 2009;61:283–357.
4. Wang G, Pan J, Tan YY, Sun XK, Zhang YF, Zhou HY, et al. Neuroprotective effects of PACAP27 in mice model of Parkinson's disease involved in the modulation of K(ATP) subunits and D2 receptors in the striatum. Neuropeptides. 2008;42:267–76.
5. Chen Y, Samal B, Hamelink CR, Xiang CC, Chen Y, Chen M, et al. Neuroprotection by endogenous and exogenous PACAP following stroke. Regul Pept. 2006;137:4–19.
6. Dejda A, Seaborn T, Bourgault S, Touzani O, Fournier A, Vaudry H, et al. PACAP and a novel stable analog protect rat brain from ischemia: insight into the mechanisms of action. Peptides. 2011;32:1207–16.
7. Reglodi D, Somogyvari-Vigh A, Vigh S, Kozicz T, Arimura A. Delayed systemic administration of PACAP38 is neuroprotective in transient middle cerebral artery occlusion in the rat. Stroke. 2000;31:1411–7.
8. Reglodi D, Tamas A, Somogyvari-Vigh A, Szanto Z, Kertes E, Lenard L, et al. Effects of pretreatment with PACAP on the infarct size and functional outcome in rat permanent focal cerebral ischemia. Peptides. 2002;23:2227–34.
9. Hessenius C, Bader M, Meinhold H, Bohmig M, Faiss S, Reubi JC, et al. Vasoactive intestinal peptide receptor scintigraphy in patients with pancreatic adenocarcinomas or neuroendocrine tumours. Eur J Nucl Med. 2000;27:1684–93.

10. Leyton J, Gozes Y, Pisegna J, Coy D, Purdom S, Casibang M, et al. PACAP(6-38) is a PACAP receptor antagonist for breast cancer cells. Breast Cancer Res Treat. 1999;56(2):177–86.
11. Moretti C, Mammi C, Frajese GV, Mariani S, Gnessi L, Arizzi M, et al. PACAP and type I PACAP receptors in human prostate cancer tissue. Ann N Y Acad Sci. 2006;1070:440–9.
12. Reubi JC, Laderach U, Waser B, Gebbers JO, Robberecht P, Laissue JA. Vasoactive intestinal peptide/pituitary adenylate cyclase-activating peptide receptor subtypes in human tumors and their tissues of origin. Cancer Res. 2000;60:3105–12.
13. Banki E, Kovacs K, Nagy D, Juhasz T, Degrell P, Csanaky K, et al. Molecular mechanisms underlying the nephroprotective effects of PACAP in diabetes. J Mol Neurosci. 2014;54:300–9.
14. Sanlioglu AD, Karacay B, Balci MK, Griffith TS, Sanlioglu S. Therapeutic potential of VIP vs PACAP in diabetes. J Mol Endocrinol. 2012;49:R157–67.
15. Hoare SR. Mechanisms of peptide and nonpeptide ligand binding to class B G-protein-coupled receptors. Drug Discov Today. 2005;10:417–27.
16. Bourgault S, Chatenet D, Wurtz O, Doan ND, Leprince J, Vaudry H, et al. Strategies to convert PACAP from a hypophysiotropic neurohormone into a neuroprotective drug. Curr Pharm Des. 2011;17:1002–24.
17. Bourgault S, Vaudry D, Dejda A, Doan ND, Vaudry H, Fournier A. Pituitary adenylate cyclase-activating polypeptide: focus on structure-activity relationships of a neuroprotective peptide. Curr Med Chem. 2009;16:4462–80.
18. Harmar AJ, Fahrenkrug J, Gozes I, Laburthe M, May V, Pisegna JR, et al. Pharmacology and functions of receptors for vasoactive intestinal peptide and pituitary adenylate cyclase-activating polypeptide: IUPHAR review 1. Br J Pharmacol. 2012;166:4–17.
19. Laburthe M, Couvineau A. Molecular pharmacology and structure of VPAC receptors for VIP and PACAP. Regul Pept. 2002;108:165–73.
20. Laburthe M, Couvineau A, Tan V. Class II G protein-coupled receptors for VIP and PACAP: structure, models of activation and pharmacology. Peptides. 2007;28:1631–9.
21. Neumann JM, Couvineau A, Murail S, Lacapère JJ, Jamin N, Laburthe M. Class-B GPCR activation: is ligand helix-capping the key? Trends Biochem Sci. 2008;33:314–9.
22. Miyata A, Arimura A, Dahl RR, Minamino N, Uehara A, Jiang L, et al. Isolation of a novel 38 residue-hypothalamic polypeptide which stimulates adenylate cyclase in pituitary cells. Biochem Biophys Res Commun. 1989;164:567–74.
23. Miyata A, Jiang L, Dahl RD, Kitada C, Kubo K, Fujino M, et al. Isolation of a neuropeptide corresponding to the N-terminal 27 residues of the pituitary adenylate cyclase activating polypeptide with 38 residues (PACAP38). Biochem Biophys Res Commun. 1990;170:643–8.
24. Sherwood NM, Krueckl SL, McRory JE. The origin and function of the pituitary adenylate cyclase-activating polypeptide (PACAP)/glucagon superfamily. Endocr Rev. 2000;21:619–70.
25. Couvineau A, Laburthe M. VPAC receptors: structure, molecular pharmacology and interaction with accessory proteins. Br J Pharmacol. 2012;166:42–50.
26. Dickson L, Finlayson K. VPAC and PAC receptors: from ligands to function. Pharmacol Ther. 2009;121:294–316.
27. Harmar AJ, Arimura A, Gozes I, Journot L, Laburthe M, Pisegna JR, et al. International Union of Pharmacology. XVIII. Nomenclature of receptors for vasoactive intestinal peptide and pituitary adenylate cyclase-activating polypeptide. Pharmacol Rev. 1998;50:265–70.
28. Robberecht P, Gourlet P, De Neef P, Woussen-Colle MC, Vandermeers-Piret MC, Vandermeers A, et al. Structural requirements for the occupancy of pituitary adenylate-cyclase-activating-peptide (PACAP) receptors and adenylate cyclase activation in human neuroblastoma NB-OK-1 cell membranes. Discovery of PACAP(6-38) as a potent antagonist. Eur J Biochem. 1992;207:239–46.
29. Robberecht P, Gourlet P, De Neef P, Woussen-Colle MC, Vandermeers-Piret MC, Vandermeers A, et al. Receptor occupancy and adenylate cyclase activation in AR 4-2J rat pancreatic acinar cell membranes by analogs of pituitary adenylate cyclase-activating pep-

tides amino-terminally shortened or modified at position 1, 2, 3, 20, or 21. Mol Pharmacol. 1992;42:347–55.
30. Gourlet P, Woussen-Colle MC, Robberecht P, de Neef P, Cauvin A, Vandermeers-Piret MC, et al. Structural requirements for the binding of the pituitary adenylate-cyclase-activating peptide to receptors and adenylate-cyclase activation in pancreatic and neuronal membranes. Eur J Biochem. 1991;195:535–41.
31. Hou X, Vandermeers A, Gourlet P, Vandermeers-Piret MC, Robberecht P. Structural requirements for the occupancy of rat brain PACAP receptors and adenylate cyclase activation. Neuropharmacology. 1994;33:1189–95.
32. Gourlet P, Vandermeers A, Vandermeers-Piret MC, Rathe J, De Neef P, Robberecht P. Fragments of pituitary adenylate cyclase activating polypeptide discriminate between type I and II recombinant receptors. Eur J Pharmacol. 1995;287:7–11.
33. Ichinose M, Asai M, Imai K, Sawada M. Enhancement of phagocytosis in mouse macrophages by pituitary adenylate cyclase activating polypeptide (PACAP) and related peptides. Immunopharmacology. 1995;30:217–24.
34. Reglodi D, Borzsei R, Bagoly T, Boronkai A, Racz B, Tamas A, et al. Agonistic behavior of PACAP6-38 on sensory nerve terminals and cytotrophoblast cells. J Mol Neurosci. 2008;36:270–8.
35. Walker CS, Sundrum T, Hay DL. PACAP receptor pharmacology and agonist bias: analysis in primary neurons and glia from the trigeminal ganglia and transfected cells. Br J Pharmacol. 2014;171:1521–33.
36. Vandermeers A, Vandenborre S, Hou X, de Neef P, Robberecht P, Vandermeers-Piret MC, et al. Antagonistic properties are shifted back to agonistic properties by further N-terminal shortening of pituitary adenylate-cyclase-activating peptides in human neuroblastoma NB-OK-1 cell membranes. Eur J Biochem. 1992;208:815–9.
37. Gourlet P, Vandermeers A, Vandermeers-Piret MC, De Neef P, Robberecht P. Addition of the (28-38) peptide sequence of PACAP to the VIP sequence modifies peptide selectivity and efficacy. Int J Pept Protein Res. 1996;48:391–6.
38. Tams JW, Johnsen AH, Fahrenkrug J. Identification of pituitary adenylate cyclase-activating polypeptide1-38-binding factor in human plasma, as ceruloplasmin. Biochem J. 1999;341(Pt 2):271–6.
39. Banks WA, Kastin AJ, Komaki G, Arimura A. Passage of pituitary adenylate cyclase activating polypeptide1-27 and pituitary adenylate cyclase activating polypeptide1-38 across the blood-brain barrier. J Pharmacol Exp Ther. 1993;267:690–6.
40. Bourgault S, Vaudry D, Botia B, Couvineau A, Laburthe M, Vaudry H, et al. Novel stable PACAP analogs with potent activity towards the PAC1 receptor. Peptides. 2008;29:919–32.
41. Gourlet P, Vandermeers A, Vandermeers-Piret MC, Rathe J, De Neef P, Robberecht P. C-terminally shortened pituitary adenylate cyclase-activating peptides (PACAP) discriminate PACAP I, PACAP II-VIP1 and PACAP II-VIP2 recombinant receptors. Regul Pept. 1996;62:125–30.
42. Bourgault S, Vaudry D, Guilhaudis L, Raoult E, Couvineau A, Laburthe M, et al. Biological and structural analysis of truncated analogs of PACAP27. J Mol Neurosci. 2008;36:260–9.
43. Vaudry D, Gonzalez BJ, Basille M, Yon L, Fournier A, Vaudry H. Pituitary adenylate cyclase-activating polypeptide and its receptors: from structure to functions. Pharmacol Rev. 2000;52:269–324.
44. Yung SL, Dela Cruz F, Hamren S, Zhu J, Tsutsumi M, Bloom JW, et al. Generation of highly selective VPAC2 receptor agonists by high throughput mutagenesis of vasoactive intestinal peptide and pituitary adenylate cyclase-activating peptide. J Biol Chem. 2003;278:10273–81.
45. Nicole P, Lins L, Rouyer-Fessard C, Drouot C, Fulcrand P, Thomas A, et al. Identification of key residues for interaction of vasoactive intestinal peptide with human VPAC1 and VPAC2 receptors and development of a highly selective VPAC1 receptor agonist. Alanine scanning and molecular modeling of the peptide. J Biol Chem. 2000;275:24003–12.

46. Bourgault S, Vaudry D, Segalas-Milazzo I, Guilhaudis L, Couvineau A, Laburthe M, et al. Molecular and conformational determinants of pituitary adenylate cyclase-activating polypeptide (PACAP) for activation of the PAC1 receptor. J Med Chem. 2009;52:3308–16.
47. Doan ND, Bourgault S, Dejda A, Lètourneau M, Detheux M, Vaudry D, et al. Design and in vitro characterization of PAC1/VPAC1-selective agonists with potent neuroprotective effects. Biochem Pharmacol. 2011;81:552–61.
48. Ramos-Alvarez I, Mantey SA, Nakamura T, Nuche-Berenguer B, Moreno P, Moody TW, et al. A structure-function study of PACAP using conformationally restricted analogs: Identification of PAC1 receptor-selective PACAP agonists. Peptides. 2015;66:26–42.
49. Dejda A, Bourgault S, Doan ND, Lètourneau M, Couvineau A, Vaudry H, et al. Identification by photoaffinity labeling of the extracellular N-terminal domain of PAC1 receptor as the major binding site for PACAP. Biochimie. 2011;93:669–77.
50. Kumar S, Pioszak A, Zhang C, Swaminathan K, Xu HE. Crystal structure of the PAC1R extracellular domain unifies a consensus fold for hormone recognition by class B G-protein coupled receptors. PLoS One. 2011;6:e19682.
51. Gonzalez-Muniz R, Martin-Martinez M, Granata C, de Oliveira E, Santiveri CM, Gonzalez C, et al. Conformationally restricted PACAP27 analogues incorporating type II/II′ IBTM beta-turn mimetics. Synthesis, NMR structure determination, and binding affinity. Bioorg Med Chem. 2001;9:3173–83.
52. Inooka H, Endo S, Kitada C, Mizuta E, Fujino M. Pituitary adenylate cyclase activating polypeptide (PACAP) with 27 residues. Conformation determined by 1H NMR and CD spectroscopies and distance geometry in 25% methanol solution. Int J Pept Protein Res. 1992;40:456–64.
53. Komi N, Okawa K, Tateishi Y, Shirakawa M, Fujiwara T, Akutsu H. Structural analysis of pituitary adenylate cyclase-activating polypeptides bound to phospholipid membranes by magic angle spinning solid-state NMR. Biochim Biophys Acta. 1768;2007:3001–11.
54. Sun C, Song D, Davis-Taber RA, Barrett LW, Scott VE, Richardson PL, et al. Solution structure and mutational analysis of pituitary adenylate cyclase-activating polypeptide binding to the extracellular domain of PAC1-RS. Proc Natl Acad Sci U S A. 2007;104:7875–80.
55. Wray V, Kakoschke C, Nokihara K, Naruse S. Solution structure of pituitary adenylate cyclase activating polypeptide by nuclear magnetic resonance spectroscopy. Biochemistry. 1993;32:5832–41.
56. Inooka H, Ohtaki T, Kitahara O, Ikegami T, Endo S, Kitada C, et al. Conformation of a peptide ligand bound to its G-protein coupled receptor. Nat Struct Biol. 2001;8:161–5.
57. Sze KH, Zhou H, Yang Y, He M, Jiang Y, Wong AO. Pituitary adenylate cyclase-activating polypeptide (PACAP) as a growth hormone (GH)-releasing factor in grass carp: II. Solution structure of a brain-specific PACAP by nuclear magnetic resonance spectroscopy and functional studies on GH release and gene expression. Endocrinology. 2007;148:5042–59.
58. Wray V, Nokihara K, Naruse S. Solution structure comparison of the VIP/PACAP family of peptides by NMR spectroscopy. Ann N Y Acad Sci. 1998;865:37–44.
59. Hutchinson EG, Thornton JM. A revised set of potentials for beta-turn formation in proteins. Protein Sci. 1994;3:2207–16.
60. Aurora R, Rose GD. Helix capping. Protein Sci. 1998;7:21–38.
61. Dasgupta S, Bell JA. Design of helix ends. Amino acid preferences, hydrogen bonding and electrostatic interactions. Int J Pept Protein Res. 1993;41:499–511.
62. Igarashi H, Ito T, Pradhan TK, Mantey SA, Hou W, Coy DH, et al. Elucidation of the vasoactive intestinal peptide pharmacophore for VPAC(2) receptors in human and rat and comparison to the pharmacophore for VPAC(1) receptors. J Pharmacol Exp Ther. 2002;303:445–60.
63. Igarashi H, Ito T, Hou W, Mantey SA, Pradhan TK, Ulrich 2nd CD, et al. Elucidation of vasoactive intestinal peptide pharmacophore for VPAC(1) receptors in human, rat, and guinea pig. J Pharmacol Exp Ther. 2002;301:37–50.
64. Bergwitz C, Gardella TJ, Flannery MR, Potts Jr JT, Kronenberg HM, Goldring SR, et al. Full activation of chimeric receptors by hybrids between parathyroid hormone and calcitonin. Evidence for a common pattern of ligand-receptor interaction. J Biol Chem. 1996;271:26469–72.

65. Dennis T, Fournier A, Cadieux A, Pomerleau F, Jolicoeur FB, St Pierre S, et al. hCGRP8-37, a calcitonin gene-related peptide antagonist revealing calcitonin gene-related peptide receptor heterogeneity in brain and periphery. J Pharmacol Exp Ther. 1990;254:123–8.
66. Gefel D, Hendrick GK, Mojsov S, Habener J, Weir GC. Glucagon-like peptide-I analogs: effects on insulin secretion and adenosine 3′,5′-monophosphate formation. Endocrinology. 1990;126:2164–8.
67. Gensure RC, Shimizu N, Tsang J, Gardella TJ. Identification of a contact site for residue 19 of parathyroid hormone (PTH) and PTH-related protein analogs in transmembrane domain two of the type 1 PTH receptor. Mol Endocrinol. 2003;17:2647–58.
68. Dong M, Pinon DI, Cox RF, Miller LJ. Molecular approximation between a residue in the amino-terminal region of calcitonin and the third extracellular loop of the class B G protein-coupled calcitonin receptor. J Biol Chem. 2004;279:31177–82.
69. Dong M, Pinon DI, Cox RF, Miller LJ. Importance of the amino terminus in secretin family G protein-coupled receptors. Intrinsic photoaffinity labeling establishes initial docking constraints for the calcitonin receptor. J Biol Chem. 2004;279:1167–75.
70. Chen Q, Pinon DI, Miller LJ, Dong M. Molecular basis of glucagon-like peptide 1 docking to its intact receptor studied with carboxyl-terminal photolabile probes. J Biol Chem. 2009;284:34135–44.
71. Dong M, Lam PC, Pinon DI, Hosohata K, Orry A, Sexton PM, et al. Molecular basis of secretin docking to its intact receptor using multiple photolabile probes distributed throughout the pharmacophore. J Biol Chem. 2011;286:23888–99.
72. Dong M, Pinon DI, Miller LJ. Development of a biologically active secretin analogue incorporating a radioiodinatable photolabile p-(4-hydroxybenzoyl)phenylalanine in position 10. Regul Pept. 2002;109:181–7.
73. Dong M, Li Z, Pinon DI, Lybrand TP, Miller LJ. Spatial approximation between the amino terminus of a peptide agonist and the top of the sixth transmembrane segment of the secretin receptor. J Biol Chem. 2004;279:2894–903.
74. Dong M, Lam PC, Pinon DI, Sexton PM, Abagyan R, Miller LJ. Spatial approximation between secretin residue five and the third extracellular loop of its receptor provides new insight into the molecular basis of natural agonist binding. Mol Pharmacol. 2008;74:413–22.
75. Dong M, Lam PC, Gao F, Hosohata K, Pinon DI, Sexton PM, et al. Molecular approximations between residues 21 and 23 of secretin and its receptor: development of a model for peptide docking with the amino terminus of the secretin receptor. Mol Pharmacol. 2007;72:280–90.
76. Couvineau A, Tan YV, Ceraudo E, Laburthe M. Strategies for studying the ligand binding site of GPCRs: photoaffinity labeling of the VPAC1 receptor, a prototype of class B GPCRs. Methods Enzymol. 2013;520:219–37.
77. Ceraudo E, Hierso R, Tan YV, Murail S, Rouyer-Fessard C, Nicole P, et al. Spatial proximity between the VPAC1 receptor and the amino terminus of agonist and antagonist peptides reveals distinct sites of interaction. FASEB J. 2012;26:2060–71.
78. Couvineau A, Ceraudo E, Tan YV, Laburthe M. VPAC1 receptor binding site: contribution of photoaffinity labeling approach. Neuropeptides. 2010;44:127–32.
79. Ceraudo E, Tan YV, Nicole P, Couvineau A, Laburthe M. The N-terminal parts of VIP and antagonist PG97-269 physically interact with different regions of the human VPAC1 receptor. J Mol Neurosci. 2008;36:245–8.
80. Ceraudo E, Murail S, Tan YV, Lacapère JJ, Neumann JM, Couvineau A, et al. The vasoactive intestinal peptide (VIP) alpha-Helix up to C terminus interacts with the N-terminal ectodomain of the human VIP/Pituitary adenylate cyclase-activating peptide 1 receptor: photoaffinity, molecular modeling, and dynamics. Mol Endocrinol. 2008;22:147–55.
81. Tan YV, Couvineau A, Lacapère JJ, Laburthe M. Characterization of the new photoaffinity probe (Bz2-K24)-VIP. Ann N Y Acad Sci. 2006;1070:575–80.
82. Ceraudo E, Tan YV, Couvineau A, Lacapère JJ, Laburthe M. Spatial approximation between the C-terminus of VIP and the N-terminal ectodomain of the VPAC1 receptor. Ann N Y Acad Sci. 2006;1070:180–4.
83. Tan YV, Couvineau A, Laburthe M. Diffuse pharmacophoric domains of vasoactive intestinal peptide (VIP) and further insights into the interaction of VIP with the N-terminal ectodomain

of human VPAC1 receptor by photoaffinity labeling with [Bpa6]-VIP. J Biol Chem. 2004;279:38889–94.
84. Tan YV, Couvineau A, Van Rampelbergh J, Laburthe M. Photoaffinity labeling demonstrates physical contact between vasoactive intestinal peptide and the N-terminal ectodomain of the human VPAC1 receptor. J Biol Chem. 2003;278:36531–6.
85. Tan YV, Couvineau A, Murail S, Ceraudo E, Neumann JM, Lacapère JJ, et al. Peptide agonist docking in the N-terminal ectodomain of a class II G protein-coupled receptor, the VPAC1 receptor. Photoaffinity, NMR, and molecular modeling. J Biol Chem. 2006;281:12792–8.
86. Beebe X, Darczak D, Davis-Taber RA, Uchic ME, Scott VE, Jarvis MF, et al. Discovery and SAR of hydrazide antagonists of the pituitary adenylate cyclase-activating polypeptide (PACAP) receptor type 1 (PAC1-R). Bioorg Med Chem Lett. 2008;18:2162–6.
87. Chu A, Caldwell JS, Chen YA. Identification and characterization of a small molecule antagonist of human VPAC(2) receptor. Mol Pharmacol. 2010;77:95–101.
88. Wu L, Guang W, Chen X, Hong A. Homology modeling and molecular docking of human pituitary adenylate cyclase-activating polypeptide I receptor. Mol Med Rep. 2014;10:1691–6.
89. Dickson L, Aramori I, McCulloch J, Sharkey J, Finlayson K. A systematic comparison of intracellular cyclic AMP and calcium signalling highlights complexities in human VPAC/PAC receptor pharmacology. Neuropharmacology. 2006;51:1086–98.
90. Lamine A, Létourneau M, Doan ND, Maucotel J, Couvineau A, Vaudry H et al. Characterizations of a synthetic pituitary adenylate cyclase-activating polypeptide analog displaying potent neuroprotective activity and reduced in vivo cardiovascular side effects in a Parkinson's disease model.Neuropharmacology. 2016;108:440-450.

Chapter 9
PACAP-Derived Carriers: Mechanisms and Applications

David Chatenet, Alain Fournier, and Steve Bourgault

Abstract One of the main problems in therapeutic efficiency lies in the crossing of physiological barriers and cellular membranes. Therefore, significant efforts have been made to develop agents that cross these barriers and deliver therapeutic agents into intracellular compartments. In recent years, a large amount of data on the use of peptides as delivery agents has accumulated. Among the known cell-penetrating peptides (CPP), sequences derived from the native peptide hormone pituitary adenylate cyclase-activating polypeptide (PACAP) have recently proven to translocate different bioactive molecules across cellular membranes. PACAP, a hypophysiotropic neurohormone, participates in the regulation of multiple functions. The recent discovery of intracellular PACAP receptors in the brain and the testis as well as the physicochemical characteristics of PACAP, i.e., extended α-helix containing basic residues, prompted the evaluation of its cell-penetrating properties in a receptor-independent manner. In this review, we cover the current knowledge concerning the structural requirements, mechanistic assumptions, and metabolic features of these peptides as well as experiments demonstrating their unique carrier potential.

Keywords Pituitary adenylate cyclase-activating polypeptide • Cell-penetrating peptide • Glucosaminoglycan • Drug delivery • Metabolic stability • Cell transfection • Endocytosis • Non-covalent delivery system • Intracrine

D. Chatenet (✉)
Groupe de Recherche en Ingénierie des Peptides et en Pharmacothérapie,
Institut national de la recherche scientifique, INRS—Institut Armand-Frappier,
531 boul. des Prairies, Laval, QC, Canada, H7V 1B7
e-mail: david.chatenet@iaf.inrs.ca

A. Fournier
Laboratoire d'études moléculaires et pharmacologiques des peptides,
Institut national de la recherche scientifique, INRS—Institut Armand-Frappier,
531 boul. des Prairies, Laval, QC, Canada, H7V 1B7

S. Bourgault
Département de Chimie, Université du Québec à Montréal, Montréal, QC, Canada, H3C 3P8

Quebec Network for Research on Protein Function, Engineering and Applications, PROTEO, Québec, QC, Canada

D. Reglodi, A. Tamas (eds.), *Pituitary Adenylate Cyclase Activating Polypeptide — PACAP*, Current Topics in Neurotoxicity 11,
DOI 10.1007/978-3-319-35135-3_9

Abbreviations

BBB	Blood–brain barrier
BiFC	Bimolecular fluorescence complementation
cAMP	Cyclic adenosine monophosphate
CNS	Central nervous system
CPP	Cell-penetrating peptides
Disc	1,3-dihydro-*2H*-isoindole carboxylic acid
DPC	Dodecylphosphocholine
DPP IV	Dipeptidylpeptidase IV
FITC	Fluorescein isothiocyanate
GAGs	Glycosaminoglycans
GPCR	G protein-coupled receptors
hCT	Human calcitonin
LDH	Lactate dehydrogenase
MTT	3-(4,5-Dimethylthiazol-2-yl)-2,5-diphenyltetrazolium bromide
NEP	Neutral endopeptidase
PAC1	Pituitary adenylate cyclase-activating polypeptide type 1 receptor
PACAP	Pituitary adenylate cyclase-activating polypeptide
PACAP27	27-amino acid isoform of PACAP
PACAP38	38-amino acid isoform of PACAP
PTS-6	Peptide transport system-6
VIP	Vasoactive intestinal peptide
VPAC1	VIP/PACAP type 1 receptor
VPAC2	VIP/PACAP type 2 receptor

Introduction

Advances in genomics, biotechnology, and chemistry have led to an important increase in the identification and design of potent therapeutic and diagnostic molecules that interact with intracellular targets [1, 2]. Unfortunately, their inability to reach the cytoplasm due to their low cell membrane permeability precludes their usage into the clinic. Drug delivery systems are therefore representing a cornerstone in biomedical development. Several techniques have been developed to address the issue of efficient drug delivery, including physical methods that mainly consist of photo-, magnetic-, thermal-, ultrasound-, and electrical-based modulation systems [3], chemical approaches including liposomes [4] as well as pathways adopted from nature, e.g., receptor-mediated endocytosis [5] or viral delivery systems [6]. However, these approaches are largely if not exclusively limited to in vitro system due to their limited efficiencies or their severe toxic or immunogenic effects [7, 8].

Initially originating from transcription factor domain, cell-penetrating peptides (CPP) are now widely recognized as prominent drug delivery systems. The term CPP represents a collective denomination that comprises a growing number of short, generally polycationic peptides (10–30 amino acid residues in length) and/or amphipathic water-soluble peptides rich in basic amino acids, which are all able to translocate not only themselves but also macromolecular cargoes across biological membranes [7].

CPPs are classified according to their origin. For instance, the first CPPs were truncated versions of full-length proteins. Indeed, initial observation that homeodomains and homeodomain-containing proteins enter live cells through a receptor- or energy-independent mechanism has paved the way toward the identification of CPP with the discovery of penetratin, derived from the third α-helix of the Antennapedia homeodomain, VP22, derived from the transduction domain of the herpes simplex virus type 1 [9] and the trans-activating transcriptional activator (TAT), derived from the human immunodeficiency virus 1 (HIV-1) [10]. More recently, new cell-penetrating peptides derived from the N-terminal region of the X-protein of hepatitis B virus (Xentry) [11] and the flock house virus [12] have been described. Later, it was observed that the combination of different proteins could lead to chimeric peptides that can also have cell-penetrating properties [13]. For instance, transportan, a 27-amino acid peptide, contains 12 functional amino acids from the amino terminus of the neuropeptide galanin, connected via a lysine residue to mastoparan (a peptide derived from the venom of the *Vespula lewisii* wasp) [14]. Owing to the emergence of CPPs from natural or unnatural proteins, as well as from chimeric sequences, various research groups have developed, using predictive algorithms and rational design strategies, purely synthetic sequences that can translocate within cell cytoplasm [1]. Finally, the last class of CPPs is represented by polyarginine peptides, i.e., R9, or amphipathic peptides, i.e., model amphipathic peptide (MAP) [2]. The latter group can be further divided into primary amphipathic CPPs such as Pep-1 and pVEC, secondary amphipathic α-helical CPPs such as human calcitonin, β-sheet amphipathic CPPs like VT5, and proline-rich CPPs including Bac7 [15]. Interestingly, over the past decades, various endogenous ligands of G protein-coupled receptors have also been characterized as potential CPPs. Notably, dynorphins, which are endogenous ligands of the k-opioid receptors, are short and highly basic peptides that can translocate across plasma membranes of live neurons and non-neural cells [16]. Numerous studies have shown that human calcitonin (hCT), a 32-amino acid peptide involved in calcium homeostasis, and selected C-terminal sequences translocate in nasal epithelium therefore opening new vistas toward the development of hCT-derived CPPs [17]. In this review, we summarize and discuss the current knowledge of a new member of this class of CPPs derived from the human pituitary adenylate cyclase-activating polypeptide (PACAP). Furthermore, the mechanism of cellular uptake is described, and examples are also provided for the applications of these peptides as efficient cargo transporters.

Origins of PACAP-Derived Cell-Penetrating Peptides

The pituitary adenylate cyclase-activating polypeptide was originally isolated from ovine hypothalamic extracts on the basis of its ability to stimulate cyclic adenosine monophosphate (cAMP) production in cultured rat anterior pituitary cells [18]. This endogenous ligand exists in two isoforms of 38 (PACAP38) and 27 (PACAP27) amino acids; the latter originating from the internal cleavage of PACAP38 (Gly^{28}–Lys^{29}–Arg^{30}) [19]. These two biologically active peptides have been almost totally conserved during evolution, from fish to mammals [20]. The only exception, the positively charged C-terminal extension, i.e., PACAP(28-38), which is not essential for the biological activity of the peptide, is more variable [20]. However, the strong preservation of the N-terminal primary sequence (residues 1–27) of PACAP throughout evolution clearly supports an important role of this region for its biological activity. Both isoforms are widely distributed in the central nervous system (CNS) and in peripheral tissues [20] but differences between PACAP38 and PACAP27 in terms of abundance have been reported [20]. PACAP isoforms exert a large range of activities, including functions as a hypophysiotropic hormone, a neuromodulator, and a neurotrophic factor [21–23], and all these actions are mediated by the activation of three B family G protein-coupled receptors (GPCR) ubiquitously expressed in the CNS and in peripheral tissues [20]. The PACAP type 1 (PAC1) receptors exhibit a high affinity and specificity for PACAP whereas the VIP/PACAP type 1 (VPAC1) and type 2 (VPAC2) receptors have similar affinity for PACAP and the vasoactive intestinal peptide (VIP) [20]. Over the years, PACAP and its cognate receptors have been abundantly investigated for their potential role as apoptosis and inflammatory regulators. Particularly, PACAP, which can cross the blood–brain barrier (BBB), is well recognized for its neuroprotective action through PAC1 receptor activation and caspase-3 inhibition [20, 24, 25]. Worth to mention, over the past decade, PACAP derivatives have been studied for the treatment of brain and spinal cord traumas as well as neurodegenerative diseases, including Alzheimer's and Parkinson's disorders [20].

Until recently, GPCRs were only regarded as plasma membrane proteins modulating the activity of membrane-associated second messengers. However, the discovery of functional intracellular receptors within different cytoplasmic organelles, including the nucleus and the mitochondria, has almost become "a classic GPCR paradigm" [26]. These intracellular GPCRs are believed to be involved in the control of several processes including regulation of gene transcription, ion homeostasis, as well as cellular proliferation and remodeling [26]. Their mode of activation is still hypothetical but could involve ligands internalized from the extracellular space in a receptor-dependent or -independent manner, or synthesized within the cell. Known as "intracrine," these ligands relate to intracellular molecules binding to and activating intracellular receptors [27].

In recent years, various reports revealed the existence of functional nuclear PACAP receptors in testicular germ cells [28, 29], human breast cancer cells [30], human colonic adenocarcinoma cells [31], as well as in various tissue extracts [32]. Notably, higher levels of PACAP binding sites were observed in cytosolic and nuclear spermatid-enriched fractions while such binding sites were almost inexistent in membrane extracts [29]. Interestingly, it was later demonstrated, using bimolecular fluorescence complementation (BiFC) combined with fluorescence confocal microscopy image, that PAC1 receptors present at the nucleus were in fact PAC1 dimers [33]. Currently, it is believed that distinct functions of PAC1 dimers present on the membranes and the nucleus are the reasons why PACAP often displays "bell shape" actions and that a concentration threshold is necessary to activate those nuclear receptors [33]. At low concentration, PACAP triggers the activation of membrane receptors, while at high concentration, its effects are probably also mediated by the activation of nuclear PAC1. Altogether, these data strongly suggested the possibility that PACAP could act as an intracrine factor. While receptor-mediated endocytosis followed by endosomal escape could represent a possible way for PACAP to enter within the internal cell compartment, the helical amphipathic structure of PACAP as well as its high content in cationic amino acids (11 residues in PACAP38) motivated our groups to investigate the cell-penetrating properties of this peptide and some related fragments [32, 34, 35].

PACAP and [Arg^{17}]PACAP(11–38): The First PACAP-Related Cell-Penetrating Peptides

Based on the propensity of PACAP38 to interact with the cellular membrane [36], its amphipathic character, and the numerous basic residues dispersed throughout its C-terminal helical domain [37, 38], it was highly conceivable that PACAP38 could cross the plasma membrane in a receptor-independent manner, as observed with human calcitonin, a peptide belonging to the same family [39]. The data presented herein summarize the current knowledge regarding the mechanistic and structural understanding of the action of PACAP38, PACAP27, and related analogs as potent cell-penetrating peptides. Furthermore, approaches aimed at improving their metabolic stability are discussed, as well as the potential of such peptides to cross the blood–brain barrier. Finally, we present studies using PACAP38 and its inactive analog [Arg^{17}]PACAP(11–38) as delivery vectors for cargoes.

Studies investigating the cell-penetrating potential of any supposed CPP generally rely on fluorescence microscopy as well as flow cytometry with peptides to which a fluorescent probe is covalently attached. To avoid an artefactual redistribution of the membrane bound fluorescent peptides into the cytoplasm and nucleus caused by the fixation step [40], peptide uptake evaluation should be performed in living cells. Finally, to prevent the over-interpretation resulting from the insufficient discrimination between cell membrane-associated fluorescence and internalized fluorescence, proteolytic digestion or quenching steps should be included in the experimental protocol before CPP internalization is evaluated by flow cytometry and microscopy. Consequently, nowadays, all studies investigating the cellular uptake of PACAP or its derivatives take under consideration these different problems.

First internalization studies with both PACAP isoforms were conducted using CHO-K1 cells, a cell line that does not express PACAP receptors, as demonstrated by competitive binding assays [32]. Confocal microscopy experiments initially revealed the ability of both PACAP isoforms to cross the plasma membrane in a receptor-independent fashion. This effect was not associated with cell death or pore forming since no variation in MTT assay or LDH release was recorded. Subsequent experiments involving inactive analogs of PACAP38, i.e., [Disc^{6}]PACAP38 and [Arg^{17}]PACAP(11–38) [34, 37], confirmed the absence of participation of the receptors in the cellular uptake. Indeed, while these derivatives were unable to efficiently bind PACAP receptors, they were still able to massively enter within the cell. The fluorescent signal associated with each PACAP isoform as well as their related analogs was detected in the cytoplasm and the nucleus as a mixed distribution, punctuated (cytoplasm) and homogeneous, that could reflect the involvement of multiple uptake mechanisms [32, 34]. However, the inability of [Disc^{6}]PACAP27 or PACAP27 to modify significantly the uptake of PACAP27 demonstrated the non-saturable character of this uptake and therefore a PACAP receptor-independent endocytotic mechanism of action [32]. Interestingly, the uptake efficiency of both PACAP isoforms, which increases in a concentration-dependent manner, was generally 3 times higher for PACAP38 than for PACAP27, as assessed by flow cytometry [32]. Accordingly, using a physiological concentration of radiolabeled PACAP27 or PACAP38 (50 pM), it was observed that these neuropeptides quickly enter into cells via direct translocation, making these peptides available to interact with their intracellular endogenous receptors [32, 33]. Overall, these differences between PACAP27 and PACAP38 pointed toward a critical role of the C-terminal PACAP(28-38) segment for an efficient translocation of the peptide across the plasma membrane. Hence, previous studies have highlighted the propensity of this segment to interact with the phospholipid membrane in order to facilitate the interaction of the peptide with its cognate receptors [37]. Surprisingly, the segment PACAP(28-38), which possesses 6 basic residues and a helical secondary structure, showed no cell-penetrating properties [32]. Over the years, differences in CPP translocation between “leaky” and “non-leaky” cells were demonstrated. For instance, HeLa cells lack the ability to form tight junctions and therefore, they possess a highly permeable layer compared for instance to MDCK cells [41, 42]. While similar results, i.e., cellular uptake and distribution, were observed with PACAP27 and PACAP38 in HEK-293 and HeLa cells, uptake efficiency was different based on the cellular background [32, 34]. In conclusion, PACAP and its related analogs were shown to internalize in different cell lines in a concentration-, temperature-, and time-dependent manner following a receptor-independent, endocytotic uptake.

PACAP Structural Aspects and Internalization Properties

PACAP38 sequence (HSDGIFTDSYSRYRKQMAVKKYLAAVLGKRYKQRVKNK) includes an internal cleavage amidation site (Gly^{28}-Lys^{29}-Arg^{30}) leading to the formation of the N-terminal 27-residue sequence of PACAP38 that was named PACAP27 [20]. PACAP27 shows a high sequence identity with vasoactive intestinal polypeptide (VIP), thus identifying PACAP as a member of the secretin/glucagon/growth hormone-releasing hormone (GHRH) superfamily. Alignment of the amino acid sequences of the peptides from the secretin-glucagon-GHRH superfamily shows a strong homology at the N-terminal domain [37]. Analysis of PACAP38 primary structure reveals the high cationic nature of the peptide, which displays a net charge of +10 at physiological pH. Particularly, the 29-38 C-terminal segment includes 6 basic residues, conferring an unusually polycationic-rich domain for a natural polypeptide. Both PACAP isoforms do not show any specific conformation in aqueous solution whereas in structure-inducing environments, such as some water-miscible organic solvents or lipid micelles, these peptides exhibit a clear propensity to adopt an α-helical conformation. Indeed, conformational analyses have indicated that the secondary structure of PACAP27 is mainly characterized by an α-helix, for which the length varies in function of the milieu used for the spectroscopic analysis. The conformation of the 1–27 segment of PACAP38 superimposes that of PACAP27, and the C-terminal extension (28-38) exhibits a short helix connected to the 1-27 region by a flexible hinge [43]. The secondary structure of PACAP in a membrane-mimicking environment is similar to those described for other members of the secretin/glucagon/GHRH superfamily [24]. Overall, the high polycationic nature of PACAP as well as its environment-induced helical conformation constitute key elements to its unique cell-penetrating properties.

According to their physicochemical and structural properties, CPPs can be classified into three major classes: (1) cationic, (2) amphipathic and (3) hydrophobic [44]. Cationic CPPs are usually short peptides that are rich in arginine and lysine residues. Polyarginine CPPs tend to have a better uptake profile in comparison to polylysine peptides. The first CPP discovered more than 25 years ago was a cationic peptide derived from the HIV-1 protein TAT [45]. Amphipathic CPPs, which can be further divided into four classes (i.e., primary amphipathic α-helical, secondary amphipathic α-helical, β-sheet amphipathic and proline-rich amphipatic), are composed of hydrophobic and hydrophilic domains that are both necessary for peptide translocation.

Finally, hydrophobic CPPs are mainly composed of apolar amino acids and have a low net charge, usually less than 20 % of the peptide sequence [44]. Over the last two decades, numerous studies have been conducted in order to delineate the molecular basis of the cell-penetrating properties of CPPs according to their physicochemical and conformational properties, because CPPs usually reach the intracellular space by a combination of mechanisms. These processes can be divided into two main categories: endocytosis (active transport) and direct translocation across the plasma membrane (passive diffusion) [2]. Direct translocation of CPPs, which is still today the matter of controversial debates, involves the destabilization of the plasma membrane integrity by the peptide in a temperature- and energy-independent manner. This passive mechanism is usually ascribed to the uptake occurring at 4 °C and/or in absence of active transport, i.e., in ATP depletion conditions [2, 46]. Different models have been suggested to explain the passive translocation of CPPs across the lipid bilayer, including pore formation and adaptive translocation, as well as the barrel-stave and inverted micelle formations [2]. Besides, active transport of CPPs from the extracellular space to the cytoplasmic environment takes place via a common and naturally occurring process known as endocytosis. Studies have shown that CPPs usually use a combination of inter-connected endocytic routes such as micropinocytosis and phagocytosis, as well as clathrin-mediated and caveolae-mediated endocytosis [2]. Experimentally, the distinction between the different pathways of active uptake is generally performed using chemical inhibitors of endocytic routes, although these compounds are known for their somewhat low specificity and cytotoxicity.

The mechanism of PACAP cellular entry was initially investigated by evaluating whether uptake is temperature- and/or energy-dependent. By performing the cellular uptake assay at 4 °C or in the presence of sodium azide and 2-deoxy-D-glucose as ATP depletors, it was observed that both PACAP isoforms can reach the intracellular space, although the uptake was reduced by 80 % [32]. This indicates that PACAP cellular uptake is mainly mediated by endocytosis while direct translocation contributes to a lesser extent. The involvement of different endocytic pathways in PACAP uptake was evaluated by studying the extent of peptide internalization in the presence of classical inhibitors targeting specific endocytic mechanisms. It was shown that calveolae-mediated endocytosis does not contribute significantly whereas inhibition of clathrin-mediated endocytosis with chlorpromazine led to a 40 % decrease of PACAP uptake [32]. Moreover, pretreatment of cells with amiloride, a macropinocytosis inhibitor, caused a marked reduction in cellular uptake, indicating that the internalization of both PACAP isoforms is also dependent of the macropinocytosis pathway. Thus, as observed for other cationic CPPs, PACAP uptake does not rely on a unique cell entry mechanism but rather depends on process involving multiple pathways, including direct translocation, clathrin-mediated endocytosis, and micropinocytosis.

PACAP38 encompasses 11 basic residues, i.e., 4 arginine and 7 lysine residues, conferring a high polycationic nature to this peptide. Moreover, in a membrane-mimicking environment such as dodecylphosphocholine (DPC) micelles, this peptide readily adopts a helical structure spanning approximately from residues Thr-7 to the C-terminus [24]. Although PACAP38 displays a partial amphipathic character, helical wheel representation of this putative helical segment shows that the cationic residue side-chains are dispersed on both sides of the α-helix [35]. For cationic CPPs, it has been described that membrane-associated glycosaminoglycans (GAGs) contribute significantly to their cellular uptake [47]. GAGs, including heparan sulfate and chondroitin sulfate, are long and linear polysaccharides composed of repeating disaccharide units and some GAGs can contain up to 150 repeating disaccharide units. They are abundant on the outer leaflet of the plasma membrane of every cell type of metazoan organisms where they are O-linked to the protein core of proteoglycans. Owing to their high content in carboxyl and sulfate groups, GAGs are highly negatively charged biopolymers that constitute a major reservoir of polyanions surrounding cells. Accordingly, it was postulated that cationic CPPs could bind electrostatically to the sugar domains of proteoglycans allowing the initial adsorption to the plasma membrane; thus favoring and/or mediating the endocytosis of positively charged peptides. It has been proposed that the roles of GAGs in the endocytosis of CPPs could be ascribed to either (1) GAGs clustering upon peptide binding, (2) co-clustering of a receptor and GAGs upon CPP binding, and/or (3) GAGs mediating peptide adsorption to the plasma membrane and bringing it in the vicinity of a receptor that mediates endocytosis [47]. In this context, the roles of cell surface GAGs in the adsorption of PACAP38 to the plasma membrane and its subsequent uptake has been recently investigated. As reported for other cationic CPPs, it was observed by means of heparan and chondroitin sulfate deficient cells (CHO pgs-A-745) and heparinase treatment, that cell surface GAGs play a key role in the cellular uptake of PACAP38 [35]. As a matter of fact, CHO pgs-A-745 cells, which are deficient in xylosyltransferase [48], an enzyme that catalyzes the transfer of a D-xylosyl group to the side-chain of serine residues of proteoglycans, showed a two- to threefold lower uptake of PACAP38 compared to their unmuted CHO-K1 counterpart. It was also observed that plasma membrane adsorption of PACAP38 (4 °C, 15 min) was higher in wild type (WT) CHO-K1 cells than in CHO pgs-A-745 [35]. Interestingly, the binding of PACAP to heparin, employed as a model of the sulfated domains of heparan sulfate proteoglycans, induced a conformational transition of the peptide from a random coil to a α-helical secondary structure [35]. The contribution of this structural change in the cellular uptake of PACAP in the context of GAGs-mediated endocytosis was addressed by means of conformationally restricted PACAP derivatives. Thus, PACAP38 analogs in which two or three pairs of adjacent residues were substituted with their corresponding D-enantiomers were developed and their cellular uptake was measured in CHO-K1 and CHO pgs-A-745 cells. Double D-substitutions introduced into the putative helical region of PACAP resulted in local disturbance of the helical structure, without modifying other properties of the peptide such as hydrophobicity and net charge. In CHO-K1 cells, these helix-perturbed PACAP derivatives were significantly less internalized compared to PACAP38, with relative cellular uptakes varying between 10 and 20 % of that of PACAP38 [35]. In contrast, in GAGs-defective CHO cells, the uptake efficacies of the doubly D-substituted PACAP analogs were between 60 and 80 % of that of WT PACAP. Thus, the lower extent of uptake of PACAP D,D-derivatives in CHO-K1 cells, in contrast to pgs-A-745 cells, indicates that the random coil-to-α-helix conformational conversion is critical for GAGs-mediated endocytosis whereas GAGs-independent internalization is less affected by the random coil structure of PACAP.

Cargo Delivery

Yet a few studies have demonstrated that PACAP and its related analogs can mediate cargo delivery. First of all, the internalization studies carried out with fluorescein-labeled peptides showed that fluorophores are small cargoes that are easily internalized. Nonetheless, due to the presence of a high number of basic amino acids (net charge of +10), PACAP38 and [Arg^{17}]PACAP(11–38) provide the possibility of simple non-covalent complexation of a cargo, as reported for some CPPs such as calcitonin [17]. Thus, PACAP-based cellular delivery vectors would enable the use of covalent or non-covalent strategies for the delivery of a wide range of cargoes such as plasmid DNAs, small interfering RNAs, drugs, nanoparticulate pharmaceutical carriers, and anticancer medications.

Doan et al. initially demonstrated the potential as a carrier peptide of [Arg^{17}]PACAP(11–38), an inactive analog of PACAP38, by translocating fluorescein, a hydrophilic impermeable dye, and a non-permeable peptide aptamer, i.e., Id1/3-PA7 [49], known to induce cell cycle arrest. While the fluorescent probe was attached at the CPP N-terminus through an amide bond, the Id1/3-PA7 and [Arg^{17}]PACAP(11–38) peptides were fused using a disulfide bridge [34]. The use of a Cys–Cys bond was ensuring the release of the bioactive molecule within the reducing environment of the cytoplasm. Another example for the successful application of this PACAP-based CPP was the transfer of Alexa Fluor® 568-labeled streptavidin using biotinylated analogs of PACAP38 and [Arg^{17}]PACAP(11–38). Hence, these examples highlighted the possibility of using covalent or non-covalent strategies with this derivative for the efficient delivery of cargoes of variable sizes. Current methods of gene delivery, including viral and nonviral systems (lipoplexes, polyplexes) as well as physical techniques like electroporation, have some limitations that prevent their effective use. Since PACAP and its derivatives share some properties similar to poly/lipoplexes, like their cationic nature and high cellular association, it was postulated that they could serve as good DNA or oligonucleotide delivery agents. Also, because nuclear delivery is necessary for biological activity, the nuclear localization property of PACAP and its analogs could be exploited for this purpose. One study using [Arg^{17}]PACAP(11–38) to complex with DNA, and more particularly to the green fluorescent protein (GFP) plasmid, showed similar efficiency in transfection, as compared to the commonly used Lipofectamine™ method. Worth to mention, no cell death was observed with [Arg^{17}]PACAP(11–38) compared to Lipofectamine™ (almost 40 %) [34]. Gene regulation at the RNA level using siRNA, ribozymes or antisense oligonucleotide analogs, does not require nuclear uptake. Nonetheless, cellular uptake is necessary but it is limited by the negatively charged nature of the oligonucleotides. As such, non-covalent strategies often appear more suitable for their delivery [50]. Hence, using a fluorescently labeled ribozyme and [Arg^{17}]PACAP(11–38), Doan et al. also demonstrated a concentration-dependent increase of the ribozyme cellular uptake. Altogether, those investigations have highlighted the cell-penetrating properties of [Arg^{17}]PACAP(11–38) and revealed its potency and usability to successfully deliver, via covalent or non-covalent strategies, various cargoes including peptides (Id1/3-PA7), proteins (streptavidin), and polynucleotides (fluorescein-labeled ribozyme and GFP plasmid); thereby illustrating the cell penetration potency and versatility of PACAP-based CPPs.

Transport of PACAP Across Physiological Barriers

Among endogenous peptidic (neuro)hormones, PACAP is not only unique for its capacity to translocate plasma membrane, but also for its unusually high ability to cross physiological barriers, including the blood–testis barrier [51] and the BBB [52].

In particular, the BBB is a highly selective barrier composed of endothelial cells and astrocytes that separate the CNS from the blood stream. On the one hand, this barrier allows the passive diffusion of small lipidic molecules, salts and water as well as the selective transport of metabolically important biomolecules, such as amino acids and glucose. On the other hand, this barrier prevents the entry of various (macro)molecules and microorganisms into the brain. While the BBB plays vital physiological roles to preserve the homeostasis of the CNS and protects the brain from neurotoxins, it precludes the delivery of most therapeutic and diagnostic molecules in the CNS to treat or detect brain-related diseases. Thus, over the last decades, several molecules have been identified and/or developed as carriers in order to overcome the difficulty of delivering therapeutic agents to the CNS. Among possible molecular shuttles, numerous peptides have received a great deal of attention for their capacity to cross the BBB through various mechanisms. The majority of these peptides displaying a relevant BBB-penetrating capacity are transported from the blood circulation to the brain by means of transcytosis initiated by an interaction with specific receptor located on the endothelial cell surface [53]. It was reported that both PACAP isoforms readily cross the BBB in both directions, although they employ different mechanisms. PACAP38 enters the CNS by means of the specific peptide transporter system-6 (PTS-6) whereas PACAP27 crosses the BBB by passive diffusion [52]. Particularly, the influx of PACAP38 from the blood circulation into the CNS is very high. For instance, it is the highest among peptides of the secretin/glucagon/GHRH superfamily, and it is even higher than the influx reported for the small therapeutic molecule morphine [52, 54]. This relatively high CNS influx of PACAP38 is most likely related to the fact that this neuropeptide is the only member of the secretin/glucagon/GHRH superfamily to enter the brain by means of a specific saturable transport system [54]. Interestingly, PACAP38 crosses the BBB faster at the hippocampus and hypothalamus brain region and this characteristic can be related to the strong neuroprotective effect of PACAP on hippocampal neurons after i.v. injection [55, 56]. Similarly, it was reported that after a neurological insult, such as a stroke, the PACAP influx in the CNS is increased and this effect is not connected to a disruption of the BBB but rather to an increase activity of the PTS-6 transporter [57, 58]. Overall, these data indicate that PACAP38 derivatives, inactive towards specific GPCRs and with enhanced BBB transport, could be ultimately designed as potent molecular shuttles to vehicle therapeutics/diagnostics molecules into the brain. However, it will be important to delineate the molecular determinants allowing the interaction of PACAP38 with PTS-6 and to validate if the covalent linkage of a cargo to PACAP does not affect its capacity to cross the BBB.

PACAP Metabolic Stability and CPP Properties

In order to be used as an efficient in vivo carrier for the transport of (macro)molecules across membranes and physiological barriers, PACAP derivatives need to display appropriate pharmacokinetics parameters. Particularly, as a peptide, PACAP rapidly undergoes enzymatic degradation after its administration into the systemic circulation. Several enzymatic cleavage sites within PACAP38 and PACAP27 sequences have been described so far [37]. For instance, upon i.v. injection, it was reported in mice that PACAP38 is metabolized by dipeptidyl peptidase IV (DPP IV) [59], a ubiquitous amino-terminal dipeptidase that liberates dipeptides from the N-terminus of various regulatory peptides. Interestingly, the presence of the C-terminal segment hastened the breakdown of PACAP by DPP IV, even if this segment is distant from the main initial scissile bond [60]. Alternatively, the endopeptidase NEP, an active enzyme in the lung, degrades PACAP27 to generate C-terminally truncated fragments whereas this peptidase is inactive against the 38-residue isoform [61]. In isolated human plasma, PACAP38 was readily metabolized by endopeptidase(s) that cleave at the dibasic sites Arg^{14}–Lys^{15} and Lys^{20}–Lys^{21} [62]. Interestingly, PACAP27 was resistant towards this plasmatic enzyme(s), suggesting a key role for the 28–38 fragment in the molecular recognition. Similarly, the blood transporter ceruloplasmin is known to bind to the 38-residue isoform of PACAP, while it does not interact with the 27-residue native peptide [63]. In a study involving 12 healthy volunteers, the plasmatic half-life of PACAP38 was estimated to be 3.5 min [64] whereas a blood half-life of about 5 to 10 min was reported in patients afflicted with multiple myeloma and myeloma kidney [65]. In mice, a plasmatic half-life of less than 2 min was obtained for PACAP38 [59]. Overall, these studies indicate that upon its injection into the blood stream, PACAP is rapidly metabolized by a combination of peptidases. Thus, this reinforces the necessity to design PACAP derivatives with targeted chemical modifications that increase their enzymatic stability while maintaining and even enhancing the cell-penetrating properties and/or the capacity to cross physiological barriers.

Conclusion

An efficient delivery of many promising therapeutics remains a major problem and consequently it represents a limitation in their clinical applications. Due to their physicochemical properties, these compounds reach their site of action often just to a very minor degree. During the last years, cell-penetrating peptides have been developed and optimized as tools to shuttle charged or hydrophilic drugs across biomembranes. Carrier peptides derived from the native peptide hormone PACAP have just started to demonstrate their potential to transport several types of cargoes efficiently into a number of different cell lines. Additionally, all established PACAP-derived carrier peptides have been shown to be devoid of significant cytotoxicity [32, 34]. One of the reasons for this observation might be the proposed endocytotic internalization pathway that does not disrupt the cellular membrane. As one of the first members of the endogenous CPP family, PACAP and its inactive derivatives might help to avoid possible side effects such as immunogenicity and in vivo toxicity that currently limit the usage of CPPs as drug carriers in clinics. Moreover, PACAP and its receptors are broadly expressed in the CNS and in most peripheral organs [20] where they exert multiple activities. Especially, with its potent anti-apoptotic and anti-inflammatory properties, as well as its ability to cross the blood–brain barrier, PACAP is currently a promising candidate to safely reverse or slow down the course of disabling neurological illnesses [20]. Hence, the combination of native PACAP isoforms with other intracellular-targeting bioactive compounds acting synergistically or on complementary pathways might provide multifunctional drug candidates for the treatment of neurodegenerative diseases. This attracting avenue is currently under investigation in our groups.

Acknowledgements This work was supported by the Canadian Institutes of Health Research (AF; FRN102734), the Fonds de recherche du Québec-Nature et technologies (DC; 188882), and the Natural Sciences and Engineering Research Council of Canada (SB; 1557119).

References

1. Copolovici DM, Langel K, Eriste E, Langel U. Cell-penetrating peptides: design, synthesis, and applications. ACS Nano. 2014;8:1972–94.
2. Bechara C, Sagan S. Cell-penetrating peptides: 20 years later, where do we stand? FEBS Lett. 2013;587:1693–702.
3. Rodriguez-Devora JI, Ambure S, Shi ZD, Yuan Y, Sun W, Xui T. Physically facilitating drug-delivery systems. Ther Deliv. 2012;3:125–39.
4. Tiwari G, Tiwari R, Sriwastawa B, Bhati L, Pandey S, Pandey P, et al. Drug delivery systems: an updated review. Int J Pharm Invest. 2012;2(1):2–11.
5. Accardo A, Aloj L, Aurilio M, Morelli G, Tesauro D. Receptor binding peptides for target-selective delivery of nanoparticles encapsulated drugs. Int J Nanomed. 2014;9:1537–57.
6. Medina-Kauwe LK. Development of adenovirus capsid proteins for targeted therapeutic delivery. Ther Deliv. 2013;4:267–77.
7. Heitz F, Morris MC, Divita G. Twenty years of cell-penetrating peptides: from molecular mechanisms to therapeutics. Br J Pharmacol. 2009;157:195–206.
8. Juliano RL, Alam R, Dixit V, Kang HM. Cell-targeting and cell-penetrating peptides for delivery of therapeutic and imaging agents. Wiley Interdiscip Rev Nanomed Nanobiotechnol. 2009;1:324–35.
9. Elliott G, O'Hare P. Intercellular trafficking and protein delivery by a herpesvirus structural protein. Cell. 1997;88:223–33.
10. Lindgren M, Hallbrink M, Prochiantz A, Langel U. Cell-penetrating peptides. Trends Pharmacol Sci. 2000;21:99–103.
11. Montrose K, Yang Y, Sun X, Wiles S, Krissansen GW. Xentry, a new class of cell-penetrating peptide uniquely equipped for delivery of drugs. Sci Rep. 2013;3:1661.
12. Nakase I, Hirose H, Tanaka G, Tadokoro A, Kobayashi S, Takeuchi T, et al. Cell-surface accumulation of flock house virus-derived peptide leads to efficient internalization via macropinocytosis. Mol Ther. 2009;17:1868–76.
13. Soomets U, Lindgren M, Gallet X, Hallbrink M, Elmquist A, Balaspiri L, et al. Deletion analogues of transportan. Biochim Biophys Acta. 2000;1467:165–76.
14. Pooga M, Hallbrink M, Zorko M, Langel U. Cell penetration by transportan. FASEB J. 1998;12:67–77.
15. Farkhani SM, Valizadeh A, Karami H, Mohammadi S, Sohrabi N, Badrzadeh F. Cell penetrating peptides: efficient vectors for delivery of nanoparticles, nanocarriers, therapeutic and diagnostic molecules. Peptides. 2014;57:78–94.

16. Marinova Z, Vukojevic V, Surcheva S, Yakovleva T, Cebers G, Pasikova N, et al. Translocation of dynorphin neuropeptides across the plasma membrane. A putative mechanism of signal transmission. J Biol Chem. 2005;280:26360–70.
17. Schmidt MC, Rothen-Rutishauser B, Rist B, Beck-Sickinger A, Wunderli-Allenspach H, Rubas W, et al. Translocation of human calcitonin in respiratory nasal epithelium is associated with self-assembly in lipid membrane. Biochemistry. 1998;37:16582–90.
18. Miyata A, Arimura A, Dahl RR, Minamino N, Uehara A, Jiang L, et al. Isolation of a novel 38 residue-hypothalamic polypeptide which stimulates adenylate cyclase in pituitary cells. Biochem Biophys Res Commun. 1989;164:567–74.
19. Miyata A, Jiang L, Dahl RD, Kitada C, Kubo K, Fujino M, et al. Isolation of a neuropeptide corresponding to the N-terminal 27 residues of the pituitary adenylate cyclase activating polypeptide with 38 residues (PACAP38). Biochem Biophys Res Commun. 1990;170:643–8.
20. Vaudry D, Falluel-Morel A, Bourgault S, Basille M, Burel D, Wurtz O, et al. Pituitary adenylate cyclase-activating polypeptide and its receptors: 20 years after the discovery. Pharmacol Rev. 2009;61:283–357.
21. Arimura A, Somogyvari-Vigh A, Miyata A, Mizuno K, Coy DH, Kitada C. Tissue distribution of PACAP as determined by RIA: highly abundant in the rat brain and testes. Endocrinology. 1991;129:2787–9.
22. Koves K, Arimura A, Gorcs TG, Somogyvari-Vigh A. Comparative distribution of immunoreactive pituitary adenylate cyclase activating polypeptide and vasoactive intestinal polypeptide in rat forebrain. Neuroendocrinology. 1991;54:159–69.
23. Ghatei MA, Takahashi K, Suzuki Y, Gardiner J, Jones PM, Bloom SR. Distribution, molecular characterization of pituitary adenylate cyclase-activating polypeptide and its precursor encoding messenger RNA in human and rat tissues. J Endocrinol. 1993;136:159–66.
24. Bourgault S, Chatenet D, Wurtz O, Doan ND, Leprince J, Vaudry H, et al. Strategies to convert PACAP from a hypophysiotropic neurohormone into a neuroprotective drug. Curr Pharm Des. 2011;17:1002–24.
25. Seaborn T, Masmoudi-Kouli O, Fournier A, Vaudry H, Vaudry D. Protective effects of pituitary adenylate cyclase-activating polypeptide (PACAP) against apoptosis. Curr Pharm Des. 2011;17:204–14.
26. Boivin B, Vaniotis G, Allen BG, Hebert TE. G protein-coupled receptors in and on the cell nucleus: a new signaling paradigm? J Recept Signal Transduct Res. 2008;28:15–28.
27. Re RN, Cook JL. The intracrine hypothesis: an update. Regul Pept. 2006;133:1–9.
28. Tominaga A, Sugawara H, Futagawa T, Inoue K, Sasaki K, Minamino N, et al. Characterization of the testis-specific promoter region in the human pituitary adenylate cyclase-activating polypeptide (PACAP) gene. Genes Cells. 2010;15:595–606.
29. Li M, Funahashi H, Mbikay M, Shioda S, Arimura A. Pituitary adenylate cyclase activating polypeptide-mediated intracrine signaling in the testicular germ cells. Endocrine. 2004;23:59–75.
30. Valdehita A, Bajo AM, Fernandez-Martinez AB, Arenas MI, Vacas E, Valenzuela P, et al. Nuclear localization of vasoactive intestinal peptide (VIP) receptors in human breast cancer. Peptides. 2010;31:2035–45.
31. Omary MB, Kagnoff MF. Identification of nuclear receptors for VIP on a human colonic adenocarcinoma cell line. Science. 1987;238:1578–81.
32. Doan ND, Chatenet D, Letourneau M, Vaudry H, Vaudry D, Fournier A. Receptor-independent cellular uptake of pituitary adenylate cyclase-activating polypeptide. Biochim Biophys Acta. 1823;2012:940–9.
33. Yu R, Zhong J, Li M, Guo X, Zhang H, Chen J. PACAP induces the dimerization of PAC1 on the nucleus associated with the cAMP increase in the nucleus. Neurosci Lett. 2013;549:92–6.
34. Doan ND, Letourneau M, Vaudry D, Doucet N, Folch B, Vaudry H, et al. Design and characterization of novel cell-penetrating peptides from pituitary adenylate cyclase-activating polypeptide. J Control Release. 2012;163:256–65.

35. Tchoumi Neree A, Nguyen PT, Chatenet D, Fournier A, Bourgault S. Secondary conformational conversion is involved in glycosaminoglycans-mediated cellular uptake of the cationic cell-penetrating peptide PACAP. FEBS Lett. 2014;588:4590–6.
36. Umetsu Y, Tenno T, Goda N, Shirakawa M, Ikegami T, Hiroaki H. Structural difference of vasoactive intestinal peptide in two distinct membrane-mimicking environments. Biochim Biophys Acta. 1814;2011:724–30.
37. Bourgault S, Vaudry D, Dejda A, Doan ND, Vaudry H, Fournier A. Pituitary adenylate cyclase-activating polypeptide: focus on structure-activity relationships of a neuroprotective peptide. Curr Med Chem. 2009;16:4462–80.
38. Sze KH, Zhou H, Yang Y, He M, Jiang Y, Wong AO. Pituitary adenylate cyclase-activating polypeptide (PACAP) as a growth hormone (GH)-releasing factor in grass carp: II. Solution structure of a brain-specific PACAP by nuclear magnetic resonance spectroscopy and functional studies on GH release and gene expression. Endocrinology. 2007;148:5042–59.
39. Trehin R, Krauss U, Beck-Sickinger AG, Merkle HP, Nielsen HM. Cellular uptake but low permeation of human calcitonin-derived cell penetrating peptides and Tat(47-57) through well-differentiated epithelial models. Pharm Res. 2004;21:1248–56.
40. Drin G, Cottin S, Blanc E, Rees AR, Temsamani J. Studies on the internalization mechanism of cationic cell-penetrating peptides. J Biol Chem. 2003;278:31192–201.
41. Foerg C, Ziegler U, Fernandez-Carneado J, Giralt E, Merkle HP. Differentiation restricted endocytosis of cell penetrating peptides in MDCK cells corresponds with activities of Rho-GTPases. Pharm Res. 2007;24:628–42.
42. Zhang X, Wan L, Pooyan S, Su Y, Gardner CR, Leibowitz MJ, et al. Quantitative assessment of the cell penetrating properties of RI-Tat-9: evidence for a cell type-specific barrier at the plasma membrane of epithelial cells. Mol Pharm. 2004;1:145–55.
43. Wray V, Kakoschke C, Nokihara K, Naruse S. Solution structure of pituitary adenylate cyclase activating polypeptide by nuclear magnetic resonance spectroscopy. Biochemistry. 1993;32:5832–41.
44. Milletti F. Cell-penetrating peptides: classes, origin, and current landscape. Drug Discov Today. 2012;17:850–60.
45. Green M, Ishino M, Loewenstein PM. Mutational analysis of HIV-1 Tat minimal domain peptides: identification of trans-dominant mutants that suppress HIV-LTR-driven gene expression. Cell. 1989;58:215–23.
46. Jones AT, Sayers EJ. Cell entry of cell penetrating peptides: tales of tails wagging dogs. J Control Release. 2012;161:582–91.
47. Favretto ME, Wallbrecher R, Schmidt S, van de Putte R, Brock R. Glycosaminoglycans in the cellular uptake of drug delivery vectors—bystanders or active players? J Control Release. 2014;180:81–90.
48. Esko JD, Stewart TE, Taylor WH. Animal cell mutants defective in glycosaminoglycan biosynthesis. Proc Natl Acad Sci U S A. 1985;82:3197–201.
49. Mern DS, Hoppe-Seyler K, Hoppe-Seyler F, Hasskarl J, Burwinkel B. Targeting Id1 and Id3 by a specific peptide aptamer induces E-box promoter activity, cell cycle arrest, and apoptosis in breast cancer cells. Breast Cancer Res Treat. 2010;124:623–33.
50. Simeoni F, Morris MC, Heitz F, Divita G. Insight into the mechanism of the peptide-based gene delivery system MPG: implications for delivery of siRNA into mammalian cells. Nucleic Acids Res. 2003;31:2717–24.
51. Banks WA, Kastin AJ, Komaki G, Arimura A. Pituitary adenylate cyclase activating polypeptide (PACAP) can cross the vascular component of the blood-testis barrier in the mouse. J Androl. 1993;14:170–3.
52. Banks WA, Kastin AJ, Komaki G, Arimura A. Passage of pituitary adenylate cyclase activating polypeptide1-27 and pituitary adenylate cyclase activating polypeptide1-38 across the blood-brain barrier. J Pharmacol Exp Ther. 1993;267:690–6.
53. Spencer BJ, Verma IM. Targeted delivery of proteins across the blood-brain barrier. Proc Natl Acad Sci U S A. 2007;104:7594–9.
54. Dogrukol-Ak D, Tore F, Tuncel N. Passage of VIP/PACAP/secretin family across the blood-brain barrier: therapeutic effects. Curr Pharm Des. 2004;10:1325–40.

55. Nonaka N, Banks WA, Mizushima H, Shioda S, Morley JE. Regional differences in PACAP transport across the blood-brain barrier in mice: a possible influence of strain, amyloid beta protein, and age. Peptides. 2002;23:2197–202.
56. Uchida D, Arimura A, Somogyvari-Vigh A, Shioda S, Banks WA. Prevention of ischemia-induced death of hippocampal neurons by pituitary adenylate cyclase activating polypeptide. Brain Res. 1996;736:280–6.
57. Mizushima H, Banks WA, Dohi K, Shioda S, Matsumoto H, Matsumoto K. The effect of cardiac arrest on the permeability of the mouse blood-brain and blood-spinal cord barrier to pituitary adenylate cyclase activating polypeptide (PACAP). Peptides. 1999;20:1337–40.
58. Somogyvari-Vigh A, Pan W, Reglodi D, Kastin AJ, Arimura A. Effect of middle cerebral artery occlusion on the passage of pituitary adenylate cyclase activating polypeptide across the blood-brain barrier in the rat. Regul Pept. 2000;91:89–95.
59. Zhu L, Tamvakopoulos C, Xie D, Dragovic J, Shen X, Fenyk-Melody JE, et al. The role of dipeptidyl peptidase IV in the cleavage of glucagon family peptides: in vivo metabolism of pituitary adenylate cyclase activating polypeptide-(1-38). J Biol Chem. 2003;278:22418–23.
60. Lambeir AM, Durinx C, Proost P, Van Damme J, Scharpe S, De Meester I. Kinetic study of the processing by dipeptidyl-peptidase IV/CD26 of neuropeptides involved in pancreatic insulin secretion. FEBS Lett. 2001;507:327–30.
61. Gourlet P, Vandermeers A, Robberecht P, Deschodt-Lanckman M. Vasoactive intestinal peptide (VIP) and pituitary adenylate cyclase-activating peptide (PACAP-27, but not PACAP-38) degradation by the neutral endopeptidase EC 3.4.24.11. Biochem Pharmacol. 1997;54:509–15.
62. Bourgault S, Vaudry D, Botia B, Couvineau A, Laburthe M, Vaudry H, et al. Novel stable PACAP analogs with potent activity towards the PAC1 receptor. Peptides. 2008;29:919–32.
63. Tams JW, Johnsen AH, Fahrenkrug J. Identification of pituitary adenylate cyclase-activating polypeptide1-38-binding factor in human plasma, as ceruloplasmin. Biochem J. 1999;341:271–6.
64. Birk S, Sitarz JT, Petersen KA, Oturai PS, Kruuse C, Fahrenkrug J, et al. The effect of intravenous PACAP38 on cerebral hemodynamics in healthy volunteers. Regul Pept. 2007;140:185–91.
65. Li M, Maderdrut JL, Lertora JJ, Batuman V. Intravenous infusion of pituitary adenylate cyclase-activating polypeptide (PACAP) in a patient with multiple myeloma and myeloma kidney: a case study. Peptides. 2007;28:1891–5.

Chapter 10
Effects and Development of TAT-Tagged PACAP/VIP and Related Peptides

Rongjie Yu

Abstract This chapter introduces the construction of cell penetrating peptide TAT (YGRKKRRQRRR) tagged PACAP/VIP and related peptides, the extra activities endowed by TAT sequence, and the potential application of TAT-fused PACAP/VIP. The theoretical analysis indicated that TAT has a similar structure and similar binding site on PAC1 to PACAP(28–38), and TAT actually facilitating the binding and activation of VIP on PAC1. Furthermore, the addition of cell penetrating peptide TAT to PACAP/VIP and related peptides may help increase the activation efficiency of peptides on their intracellular and intranuclear receptors, which may bring extra activities to the peptides. So the TAT-tagged PACAP/VIP and related peptides deserve more detailed research and further development.

Keywords Cell penetrating peptide • TAT • PAC1 • Bio-barriers • Affinity

Construction of TAT-Tagged PACAP/VIP and Related Peptides

Cell penetrating peptide TAT sequence (YGRKKRRQRRR) rich in basic amino acid is the core domain (48–57 aa) of TAT protein from HIV 1 (human immunodeficiency virus type 1) [1, 2] belonging to the protein transduction domain family, which is widely used to help the linked macromolecules such as proteins, peptides and nucleotides traverse bio-membranes and biological barriers [3]. In 2012, in order to improve the traversing activity of PACAP, especially across blood–brain barrier (BBB) for the potential drug development of PACAP in central nervous system (CNS), TAT sequence was tagged at the C-terminus of PACAP to construct peptide PACAP-TAT [4, 5]. It was found that consistent with the previous design, TAT endowed PACAP with nearly 2.5-fold increased efficiency for traversing bio-barriers including BBB, BAB (blood–air barrier) and BTB (blood–testis

---

R. Yu (✉)
Biomedicine Institute of Jinan University, Guangzhou, Guangdong, China
e-mail: rongjie_yu1123@163.com

D. Reglodi, A. Tamas (eds.), *Pituitary Adenylate Cyclase Activating Polypeptide — PACAP*, Current Topics in Neurotoxicity 11,
DOI 10.1007/978-3-319-35135-3_10

barrier) after intraperitoneal (i.p.) injection. Although it has been showed that the location of TAT at the C- or N-terminal of the protein does not influence the transduction function [6], TAT was designed to be added at the C-terminal of PACAP to keep the intact domain PACAP(1–5) responsible for the activation of PAC1 (PACAP preferring receptor) [7].

Following the design principle of PACAP-TAT, VIP-TAT was designed and prepared with improved efficiency to cross the BBB after i.p. injection similarly to PACAP-TAT [8]. We also tried to fuse TAT sequence to N-terminus of the PAC1 specific agonist maxadilan (MAX) to construct the fusion protein PTD-maxadilan (PTD-MAX) [9]. It was shown that the traversing efficiency across the BBB of PTD-MAX increased not more than two-fold of MAX. TAT helps to promote the bio-barrier traversing ability of MAX slightly but not significantly and less effectively than that of PACAP or VIP. This may be due to the greater molecular weight of MAX (73 aa), two-fold of PACAP or VIP. Transferring larger molecules requires more TAT sequences. It has been shown that both terminal addition of TAT sequences to superoxide dismutase (SOD) increases traversing efficiency twofold of just one terminal fusion with TAT [6] and approximately 6–7 TAT sequences are essential to achieve the successful transferring of magnetic nanoparticles into cells [10]. So maybe fusion with TAT sequences at both terminals of MAX would help to improve its traversing activity across bio-membranes or bio-barriers more effectively.

Effects of TAT on the Activity of PACAP/VIP and Related Peptides

Originally, TAT sequence was used as a tool to improve the traversing abilities of PACAP and its related peptides. However, subsequent assays showed that the addition of TAT influenced the activities of PACAP and VIP on the receptor activation. PACAP-TAT (1 nM) stimulated the cell viability of PAC1-CHO (CHO cells with high expression of PAC1) more effectively than PACAP (1 nM) [4, 5] indicating that TAT increased the activation of peptides on PAC1. Similar phenomena were also found in VIP-TAT: TAT endowed VIP with extra activity of promoting proliferation of PAC1-CHO and Neuro2a cells, PAC1 antagonist PACAP (6–38) inhibited the anti-apoptotic effects of VIP-TAT significantly, and competition binding assay for PAC1 and cAMP assays further confirmed the affinity and activation bias of VIP-TAT for PAC1 [8]. All these results showed that TAT sequence promotes the activation of peptides on PAC1.

Some hypotheses were deduced for the working mechanism of TAT to promote the receptor activation:

1. The secondary structure analysis of TAT showed that TAT has an alpha-helix structure similar to PACAP (27–38) (Fig. 10.1). Moreover, our recent unpublished computer molecular docking (LibDock) results showed that TAT binds PAC1 at the binding site same as the site where PACAP(28–37) binds (Fig. 10.2a)

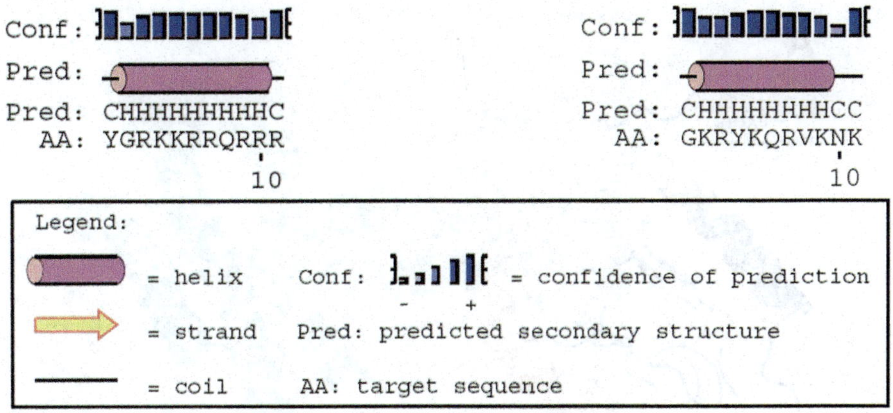

Fig. 10.1 The secondary structure analysis results by online PredictProtein software of TAT sequence (*left*) and PACAP(27-38) show that they have similar alpha-helix structures

by forming hydrogen bond interactions with nine residues in PAC1, such as ASP24, ASP111, ASP116, GLU117, GLU119, SER120, GLU121, GLN125, and GLU359 (Fig. 10.2b). And as we known, the presence of the PACAP (28–38) extension has been proven not only increase affinity for PAC1 at least 100-fold of VIP or VIP fragments, but it also stabilize the structure of N-terminally truncated VIP [11]. So TAT theoretically plays similar role to PACAP (28–38) with activity of facilitating PAC1 binding.

2. As cell penetrating peptide, TAT has been recently shown to penetrate into the cells via two pathways, one is the instantaneous direct temperature-independent membrane translocation process, and the other way is the longer-term energy-dependent process different from macropinocytosis [12], which is mediated mainly by caveolae [13, 14]. The formation of caveolae induced by PACAP/VIP is shown to be involved in the long-term activation of their G protein receptor coupled receptor (GPCR) and the effective downstream signal transduction [15–17]. Our recent unpublished data showed that VIP-TAT also had stronger proliferative activities on VPAC1-CHO and VPAC2-CHO cells than VIP, but the promoting effect of TAT on VIP to activate PAC1 was much stronger than that for VPAC1 and VPAC2 (Fig. 10.3). It needs more detailed research to elucidate whether the endocytosis of PAC1 induced by TAT is involved in both the ligand-dependent and the ligand-independent intrinsic activities of PAC1 [18].
3. The location of PACAP's receptor on the nuclear membrane or inside the nucleus has been observed and accepted. Particularly, PAC1 was found in the nuclei of testicular germ cells [19] and on the nuclear membrane of testis nuclei [20], while VPAC1 and VPAC2 were discovered in nuclei isolated from mouse splenic CD4 T cells and HEK293 stable transfectants [21], glioma cell lines [22] and breast cancer cells [23]. Furthermore, the nuclear translocation of PACAP/VIP receptors from the plasma membrane activated by the extracellular ligands and

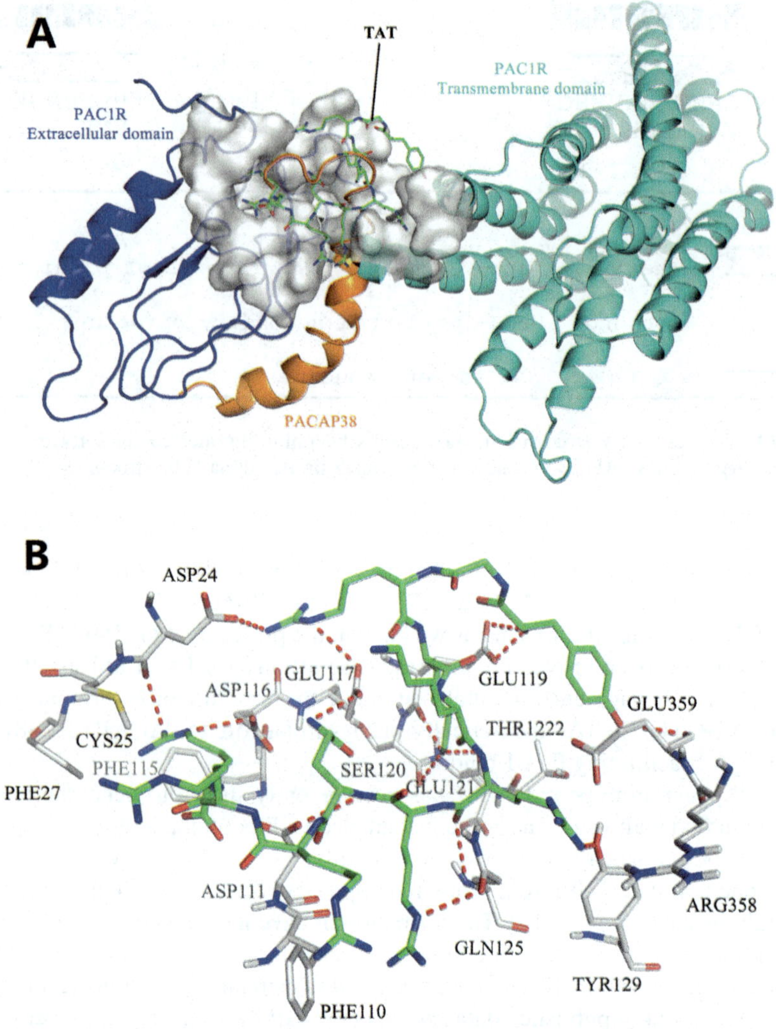

Fig. 10.2 The results of the computer molecular docking of TAT with PAC1. (**a**) The predicted binding site of TAT in PAC1, with PACAP38 superimposed according to resolved complex 3D structure of PACAP38 and extracellular domain of PAC1R (PDB code: 2JOD). PAC1 and PACAP38 are shown as cartoon style and TAT is shown as sticks. The binding site is represented by surface model. (**b**) The details of predicted binding mode of TAT and PAC1. The contact residues are shown and labeled by type and number. The *red dotted line* illustrates the hydrogen bond interaction

the further intranuclear translocation induced by intracellular ligands were not only hypothesized [21] but also observed [24]. TAT facilitates both the cell uptake and the nuclear uptake of the peptides [25, 26], thus may promote the activation of receptors on/in nucleus and induce corresponding signal pathways

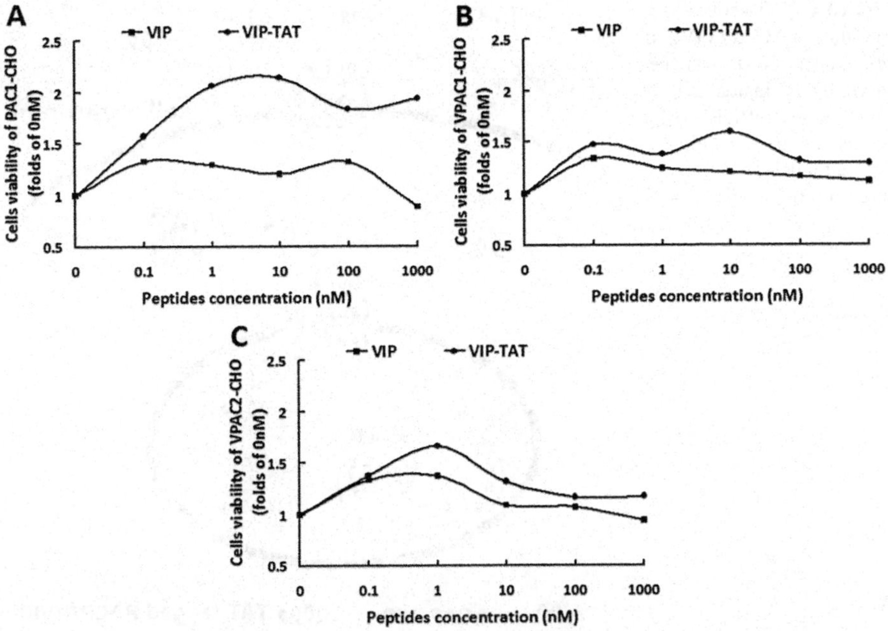

Fig. 10.3 The effects of VIP-TAT and VIP on the cell viabilities of PAC1-CHO, VPAC1-CHO and VPAC2-CHO show that TAT varies the activity of VIP and influences the activation of VIP on all three receptors, while TAT promotes the activation on PAC1 more significantly than it dose on VPAC1 and VPAC2

by increasing the intracellular or intranuclear ligands levels (as shown in Fig. 10.4). The extra activation of receptors inside the cells or on/in nucleus besides the activation of receptors on the cytomembrane may contribute to the different activities of TAT-tagged PACAP/VIP from that of intact PACAP/VIP.

Potential Application of TAT-Tagged PACAP/VIP and Related Peptides

Firstly, TAT sequence helps to amplify the drug application scope of PACAP/VIP and related peptides by endowing them with an increased traversing efficiency across bio-barriers. In detail, the enhanced traversing BBB ability and the activation bias on PAC1, which mediated the most effects of neuropeptide PACAP in brain as neurotransmitter, neuron-regulator, and neuron-protector, make TAT-tagged PACAP/VIP and related peptides display much more potential in drug development for the diseases in brain, such as neurodegenerative disease, stroke, nerve injury in CNS, amnesia, and so on. For example, the downregulation of PACAP was reported to be associated with the progress of Alzheimer's disease (AD) [27, 28], so the

Fig. 10.4 TAT sequence facilitates PACAP/VIP and related peptides to activate their GPCRs inside cells or on/in nucleus efficiently by increasing the intracellular and intranuclear concentration of ligands

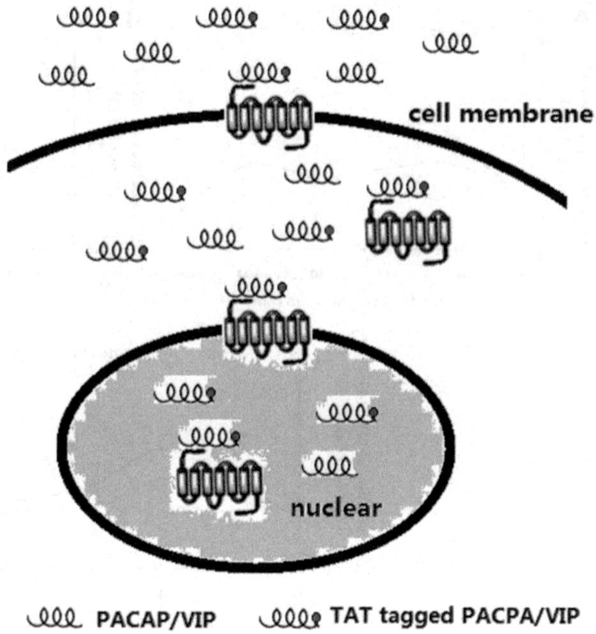

implement of PACAP or related peptides may help to ameliorate the AD. We have shown that the i.p. injection of VIP-TAT helps to protect mice against scopolamine-induced amnesia much more significantly than VIP, implying VIP-TAT may be a potential drug for prevention or treatment of AD [8]. The increased traversing activity of TAT-tagged PACAP/VIP and related peptides also facilitates the medication of peptides in the eye. PTD-MAX reached the retina effectively after intravitreous injection and inhibited retinal degeneration induced by MSG (monosodium glutamate), while MAX could not traverse to the retina effectively after intravitreous injection (Fig. 10.5) [29].

Secondly, TAT may aid to simplify the administration way of PACAP/VIP and related peptides. For example, application of VIP-TAT or PACAP-TAT in brain can be achieved by intranasal administration, the medication into the eye behind cornea maybe by eye drops and the local subcutaneous medication by plaster, because cell penetrating peptides have been proven to make percutaneous and intranasal macromolecular delivery practicable [30]. The application of TAT-tagged PACAP/VIP and related peptides with simpler route of administration deserve further development.

Finally, the construction of TAT-tagged PACAP/VIP and related peptides with enhanced cell uptake efficiency may offer novel tools for the study on intracellular or nuclear located PACAP/VIP receptors, which have been considered associated with new activity profiles of GPCRs. For example, TAT sequence may facilitate efficient imaging of receptors that localize intracellularly or on/in cell nucleus by promoting ligands into cells or nucleus.

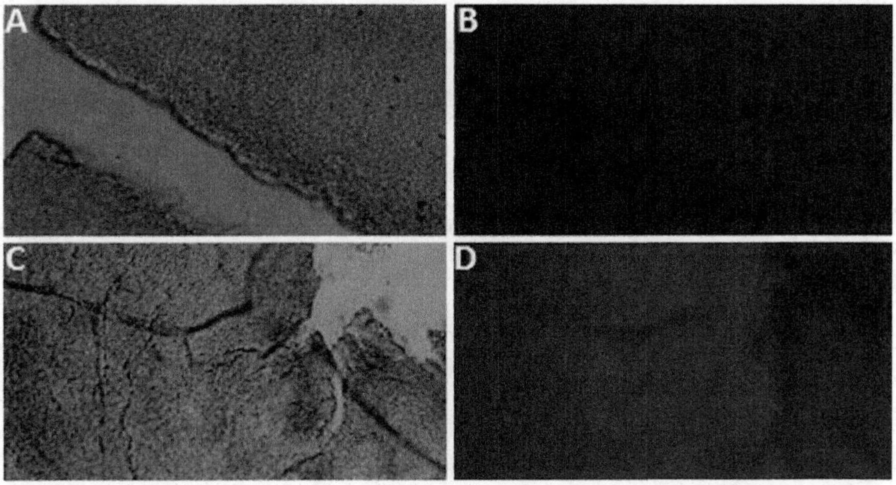

Fig. 10.5 Stretched preparations of retinas after intravitreous injection of FITC-labeled MAX (**a**, **b**) and PTD-MAX (**c**, **d**). When the extracted retinas after intravitreous injection were imaged under fluorescence inverted microscope, no significant green fluorescence signals presenting FITC signals were observed in MAX-FITC group (**b**), while significant green fluorescence signals (FITC signals) were captured in PTD-MAX-FITC group (**d**). The results indicated that PTD-MAX transpose to retina more significantly than MAX (reedited from Zhao [29])

Conclusion

The combination of TAT with PACAP/VIP and related peptides brings us novel peptides with extra bioactivities including increased traversing efficiency and bias for receptor PAC1. Further research and usage of TAT-tagged PACAP/VIP will contribute to more detailed profiles of PACAP/VIP and their GPCRs, and the application of TAT-tagged PACAP/VIP will offer novel potential drugs with more efficient and simpler route of administration.

References

1. Fawell S, Seery J, Daikh Y, Moore C, Chen LL, Pepinsky B, et al. Tat-mediated delivery of heterologous proteins into cells. Proc Natl Acad Sci U S A. 1994;91:664–8.
2. Vivès E, Brodin P, Lebleu B. A truncated HIV-1 Tat protein basic domain rapidly translocates through the plasma membrane and accumulates in the cell nucleus. J Biol Chem. 1997;272:16010–7.
3. Brooks H, Lebleu B, Vivès E. Tat peptide-mediated cellular delivery: back to basics. Adv Drug Deliv Rev. 2005;57:559–77.

4. Yu R, Guo X, Huang L, Zeng Z, Zhang H. The novel peptide PACAP-TAT with enhanced traversing ability attenuates the severe lung injury induced by repeated smoke inhalation. Peptides. 2012;38:142–9.
5. Yu R, Zeng Z, Guo X, Zhang H, Liu X, Ding Y, et al. The TAT peptide endows PACAP with an enhanced ability to traverse bio-barriers. Neurosci Lett. 2012;527:1–5.
6. Eum WS, Jang SH, Kim DW, Choi HS, Choi SH, Kim SY, et al. Enhanced transduction of Cu, Zn-superoxide dismutase with HIV-1 Tat protein transduction domains at both termini. Mol Cells. 2005;19:191–7. http://europepmc.org/abstract/med/15879701.
7. Yu R, Wang J, Li J, Wang Y, Zhang H, Chen J, et al. A novel cyclopeptide from the cyclization of PACAP(1-5) with potent activity towards PAC1 attenuates STZ-induced diabetes. Peptides. 2010;31(6):1062–7.
8. Yu R, Yang Y, Cui Z, Zheng L, Zeng Z, Zhang H. Novel peptide VIP-TAT with higher affinity for PAC1 inhibited scopolamine induced amnesia. Peptides. 2014;60:41–50.
9. Zeng L, Yu R, Xu M, Chen J, Wang J, Li J. Preparation and identification of recombinant PTD-maxadilan. Sheng Wu Gong Cheng Xue Bao. 2009;25:1739–45. http://europepmc.org/abstract/med/20222476.
10. Lewin M, Carlesso N, Tung CH, Tang XW, Cory D, Scadden DT, et al. Tat peptide-derivatized magnetic nanoparticles allow in vivo tracking and recovery of progenitor cells. Nat Biotechnol. 2000;18:410–4.
11. Gourlet P, Vandermeers A, Vandermeers-Piret MC, De Neef P, Robberecht P. Addition of the (28–38) peptide sequence of PACAP to the VIP sequence modifies peptide selectivity and efficacy. Int J Pept Protein Res. 1996;48:391–6.
12. Ben-Dov N, Korenstein R. The uptake of HIV Tat peptide proceeds via two pathways which differ from macropinocytosis. Biochim Biophys Acta. 1848;2015:869–77.
13. Wei L, Yang Q, Xiao L. Tempo-spatially resolved cellular dynamics of human immunodeficiency virus transacting activator of transcription (Tat) peptide-modified nanocargos in living cells. Nanoscale. 2014;6:10207–15.
14. Yamano S, Dai J, Hanatani S, Haku K, Yamanaka T, Ishioka M, et al. Long-term efficient gene delivery using polyethylenimine with modified Tat peptide. Biomaterials. 2014;35:1705–15.
15. Mahavadi S, Bhattacharya S, Kim J, Fayed S, Al-Shboul O, Grider JR, et al. Caveolae-dependent internalization and homologous desensitization of VIP/PACAP receptor, VPAC2, in gastrointestinal smooth muscle. Peptides. 2013;43:137–45.
16. Zhang W, Duan W, Cheung NS, Huang Z, Shao K, Li QT. Pituitary adenylate cyclase-activating polypeptide induces translocation of its G-protein-coupled receptor into caveolin-enriched membrane microdomains, leading to enhanced cyclic AMP generation and neurite outgrowth in PC12 cells. J Neurochem. 2007;103:1157–67.
17. Zhang W, Smith A, Liu JP, Cheung NS, Zhou S, Liu K, et al. GSK3beta modulates PACAP-induced neuritogenesis in PC12 cells by acting downstream of Rap1 in a caveolae-dependent manner. Cell Signal. 2009;21:237–45.
18. Yu R, Cui Z, Li M, Yang Y, Zhong J. Dimer-dependent intrinsic/basal activity of the class B G protein-coupled receptor PAC1 promotes cellular anti-apoptotic activity through Wnt/β--catenin pathways that are associated with dimer endocytosis. PLoS One. 2014;9:e113913.
19. Li M, Funahashi H, Mbikay M, Shioda S, Arimura A. Pituitary adenylate cyclase activating polypeptide-mediated intracrine signaling in the testicular germ cells. Endocrine. 2004;23:59–75.
20. Doan ND, Chatenet D, Létourneau M, Vaudry H, Vaudry D, Fournier A. Receptor-independent cellular uptake of pituitary adenylate cyclase-activating polypeptide. Biochim Biophys Acta. 1823;2012:940–9.
21. Goetzl EJ. Hypothesis: VPAC G protein-coupled receptors for vasoactive intestinal peptide constitute a dynamic system for signaling T cells from plasma membrane and nuclear membrane complexes. Regul Pept. 2006;137:75–8.
22. Barbarin A, Séité P, Godet J, Bensalma S, Muller JM, Chadéneau C. Atypical nuclear localization of VIP receptors in glioma cell lines and patients. Biochem Biophys Res Commun. 2014;454:524–30.

23. Valdehita A, Bajo AM, Fernández-Martínez AB, Arenas MI, Vacas E, Valenzuela P, et al. Nuclear localization of vasoactive intestinal peptide (VIP) receptors in human breast cancer. Peptides. 2010;31:2035–45.
24. Yu R, Zhong J, Li M, Guo X, Zhang H, Chen J. PACAP induces the dimerization of PAC1 on the nucleus associated with the cAMP increase in the nucleus. Neurosci Lett. 2013;549:92–6.
25. Potocky TB, Menon AK, Gellman SH. Cytoplasmic and nuclear delivery of a TAT-derived peptide and a beta-peptide after endocytic uptake into HeLa cells. J Biol Chem. 2003;278:50188–94. 10.1074/jbc.M308719200.
26. Smith CA, de la Fuente J, Pelaz B, Furlani EP, Mullin M, Berry CC. The effect of static magnetic fields and tat peptides on cellular and nuclear uptake of magnetic nanoparticles. Biomaterials. 2010;31:4392–400.
27. Han P, Caselli RJ, Baxter L, Serrano G, Yin J, Beach TG, et al. Association of pituitary adenylate cyclase-activating polypeptide with cognitive decline in mild cognitive impairment due to Alzheimer disease. JAMA Neurol. 2015;72:333–9.
28. Han P, Liang W, Baxter LC, Yin J, Tang Z, Beach TG, et al. Pituitary adenylate cyclase-activating polypeptide is reduced in Alzheimer disease. Neurology. 2014;82:1724–8.
29. Xiuli Zhao. Study on the protective effect of neuropeptide PACAP on nerve injury of eye. Master Thesis of Jinan University, 2011. http://cdmd.cnki.com.cn/Article/CDMD-10559-1011128321.htm.
30. Wang Z, Chen Y, Liu E, Gong J, Shin MC, Huang Y. CPP-mediated protein delivery in a non-covalent form: proof-of-concept for percutaneous and intranasal delivery. Protein Pept Lett. 2014;21:1129–36.

Part IV
Presence and Physiological Functions of PACAP in Diverse Systems

Chapter 11
Presence and Role of PACAP in Endocrine Glands of Mammals

Katalin Koves

Abstract PACAP is expressed and secreted by pituitary gonadotropes and folliculostellate cells, adrenal chromaffin cells, α- and β-cells of the Langerhans-islets, and some enteroendocrine cells. Low levels of PACAP produced by the abovementioned cells imply autocrine and paracrine actions. PACAP is also released into the portal circulation from some hypophysiotropic neurons and regulates pituitary functions. Because PACAP level is very low in the systemic circulation, other endocrine organs may not be under the influence of hypothalamic PACAP; however, they are innervated by PACAP immunoreactive sensory and the parasympathetic fibers. This means that PACAP is a neuromodulator in both the sensory and autonomic nervous systems of endocrine organs. The effect of PACAP is mediated through its specific receptors. The proliferating and differentiating effects of PACAP on endocrine tissues have also been demonstrated.

Keywords Pituitary • Adrenal • Endocrine pancreas • Thyroid • Parathyroid • Enteroendocrine cells • RIA • Immunohistochemistry

Pituitary

Soon after the isolation and characterization of PACAP, it became evident that this peptide in rats is released from the hypothalamus via the hypophyseal portal circulation into the sinusoids of the anterior pituitary gland [1]. PACAP stimulates growth hormone (GH), prolactin (PRL), adrenocorticotropic hormone (ACTH), and luteinizing hormone (LH) release from superfused rat pituitary cells at a very low dose (10^{-10} M GH, PRL, ACTH or 10^{-9} M LH) [2]. Radioimmunoassay (RIA) revealed that the level of PACAP in the anterior pituitary is very low; however, in the

K. Koves (✉)
Department of Anatomy, Histology and Embryology, Faculty of Medicine,
Semmelweis University, Tűzoltó u. 58, Budapest 1094, Hungary
e-mail: koves.katalin@med.semmelweis-univ.hu

D. Reglodi, A. Tamas (eds.), *Pituitary Adenylate Cyclase Activating Polypeptide — PACAP*, Current Topics in Neurotoxicity 11,
DOI 10.1007/978-3-319-35135-3_11

posterior pituitary it was similar to the extrahypothalamic levels [3]. PACAP in the posterior pituitary is derived from the hypothalamic magnocellular neurons. It was questionable whether PACAP in the anterior lobe is derived only from the hypophysiotropic hypothalamic neurons [4] or whether it is also produced locally. In the human pituitary, it was undetectable [5]. Oka and his coworkers [6] demonstrated PACAP mRNA in the normal human hypothalamus, but not in the normal human anterior pituitary. Another research group, Radleff-Schlimme and his coworkers [7] could demonstrate PACAP mRNA in the rat anterior pituitary and in an immortalized gonadotrope-derived cell line, alphaT3-1, by reverse transcript-polymerase chain reaction (RT-PCR). AlphaT3-1 cells release PACAP into the culturing medium and it has been measured by RIA. The controversial results may be explained by a difference in the technical details or the different physiological conditions of the animals from which the pituitary is derived. The different conditions may explain the negative result. In our laboratory, we found that PACAP is transiently expressed in the anterior pituitary of rats. PACAP containing cells could be found only on the day of proestrus of intact female rats using immunohistochemistry [8], and in situ hybridization [9, 10]. In this stage of the estrous cycle, PACAP immunoreactivity colocalized with LH and FSH immunoreactivities [8]. On the contrary, PACAP is expressed at a high level in the fetal pituitary and dramatically decreases at birth, as it was demonstrated by PCR technique [11].

With the use of cell immunoblot assay (CIBA) we have shown that PACAP is not only present but also released from the gonadotropes in static cell cultures [4, 12, 13]. An enriched population of folliculostellate (FS) cells was obtained using laser capture microdissection [14]. RT-PCR analysis revealed PACAP38 in this cell population. PACAP level has to be very low in FS cells because up to now immunohistochemistry did not confirm this result. From the abovementioned data it seems that the local source of PACAP in the anterior pituitary is the gonadotropes and FS cells [15]. CIBA also showed that the number of PACAP releasing cells depended on the gender, the stage of estrous cycle of females and the time of day when the animals were sacrificed and the pituitaries were removed for culturing [12]. The number of PACAP releasing cells was high in proestrous (37, 43, and 440 when the pituitaries were removed at 10, 16, and 20 h, respectively), very low in diestrous (1, 1, and 3) and in male rats (6, 5, and 11) in the whole culture (out of about 40,000 cells). These data suggested that the PACAP release in vivo had a circadian rhythm and the removed cells kept this ability in vitro. The gonadotropes always released LH in cell cultures but the responsiveness of LH cells to PACAP was different in the three groups. In the cultures removed in the morning, PACAP was stimulatory on LH release in females and inhibitory in males. In the afternoon, PACAP did not affect LH release. In the evening, PACAP was uneffective in proestrous and inhibitory in diestrous and male cultures. In vivo in proestrous rats, the LH releasing effect of gonadotropic hormone-releasing hormone (GnRH) and PACAP depends on the rate of proportion of the two peptides (see later). It seems possible that there is a high LH release (LH surge during the proestrous afternoon) when the level of PACAP is low and the level of GnRH is high. When the level of PACAP overrides the dropping level of GnRH, PACAP is also ineffective on gonadotropes and the LH surge ceases.

Tsujii and coworkers [16, 17] investigated the effect of GnRH and PACAP on gonadotropes in male perifused pituitary cell cultures. Intermittent administration of GnRH or PACAP (for 9 h with hourly pulses of 2.5 nM or 10 nM, respectively) initiated episodes of LH, FSH, and alpha-subunit secretion. PACAP was much less effective than GnRH. PACAP was a slightly more effective stimulator of LH release by pituitary cells from castrated than intact rats. This fact well correlates with our observation that the removed pituitary cells keep some information what was imprinted in vivo. Pulsatile PACAP stimulated alpha-subunit and LHbeta mRNA levels, but did not affect FSHbeta mRNA. By contrast, continuous PACAP increased alpha-subunit mRNA levels, but suppressed FSHbeta mRNA without affecting LHbeta mRNA. It was concluded that pulsatile PACAP is a relatively weak stimulator of gonadotropin secretion when administered alone, and regulates gonadotropin subunit gene expression quite differently than continuous PACAP.

Vigh and his coworkers [18] revealed that all cell types of the anterior pituitary have PACAP binding sites. Nearly all FS cells bound biot-PACAP38. A considerable number of GH and PRL cells and a lesser number of ACTH cells also bound biot-PACAP38, whereas only a few LH, FSH, or TSH cells bound biot-PACAP38. These results suggest that FS cells are a major target cell type for PACAP. RT-PCR found three variants of PAC1 and VPAC2 in lactotrope-enriched cell populations, whereas only PAC1-hop was detected in somatotropic-enriched cell populations [19]. PAC1-hopl and VPAC2 was demonstrated in GH3 cells [20]. Bresson-Bepoldin and his coworkers [21] employed a single cell RT-PCR technique and found that individual rat somatotropic cells expressed PAC1-hop or PAC1-short but not PAC1-hip. PACAP can release ACTH from corticotrope cell line (AtT-20) [22]; however, the receptor subtypes which mediate its effect are not identified to this day. Rawlings and his colleagues [23] demonstrated that the alpha T3-1 cells expressed mRNA for PAC1 and VPAC2, but not VPAC1. The predominant splice variant forms of PAC1 observed were short and hop variants, although the other forms were also seen at much lower levels. Stimulation of short and hop1 variants potently increased adenylyl cyclase (AC) and phospholipase-C (PLC). Hip variants stimulated only AC. Radleff-Schlimme and his coworkers [7] also showed PAC1 variants on alpha T3-1 cells. Another research group identified the region of the extracellular N-terminal domain of the PACAP receptor which is the major binding sites for PACAP [24]. Three peptide derivatives, containing a photoreactive p-benzoyl-phenylalanine residue were used. These photoreactive peptides linked to three fragments of the extracellular domain: the Ser (98)–Met (111) segment, the Ser (124)–Glu (125) dipeptide, and the Ser (141)–Met (172) fragment.

PACAP in the pituitary gland has a multifunctional role, as it is suggested by the presence of its receptors on all pituitary cell types. As it was demonstrated at the beginning of PACAP research, PACAP influenced the release of several tropic hormones from superfused rat pituitary cells [2]. Propato-Mussafiri and his coworkers [22] used AtT-20 rat and GH3 mouse clonal cell lines to investigate the effect of PACAP on ACTH, GH, and PRL release. PACAP38 and PACAP27 stimulated GH and PRL or ACTH secretion with a similar efficacy over the 2-h incubation period from GH3 and AtT-20 cells, respectively. Addition of PACAP to corticotropic hormone releasing hormone (CRH-41) resulted in an additive effect on ACTH secretion. No synergism was observed when PACAP was added together with thyrotropic hormone-releasing hormone (TRH) either on GH or PRL secretion. PACAP-stimulated ACTH release from AtT-20 cells was reduced by a somatostatin analog. This experiment does not provide evidence of whether hypothalamic or pituitary PACAP participates in vivo in the regulation of GH, PRL, or ACTH release. Another research group [25] compared the effect of PACAP in vitro in cell culture (where the release of GH and PRL was examined by reverse hemolytic plaque assay) and in vivo in medial basal hypothalamic-lesioned rats. In this animal model, the dopaminergic neurons are lesioned and as a consequence, PRL level is tonically high. In vitro, PRL release was inhibited and there was no effect on GH release, and in vivo the release of both hormones was stimulated. According to the authors, these experiments suggest that PACAP may stimulate the release of a paracrine, yet unknown factor, which in the intact pituitary overrides the direct inhibitory action of PACAP on the lactotropes. The same or another paracrine factor may also enhance GH release in vivo. In cell cultures, the paracrine factor is diluted by the medium. Therefore, the factor never reaches effective concentrations which are present within the pituitary tissue in vivo. A controversial effect of PACAP on GH release in cell culture was published by Góth and his coworkers [26]. They also used the reverse hemolytic plaque assay which allows determination of the number of secreting cells and the amount of hormone secreted by each cell. In their in vitro model PACAP stimulated GH secretion in a dose-dependent fashion. PACAP increased both the number of secreting cells and the mean amount of hormone secreted per cell. Velkeniers and his coworkers [27] found similar stimulation of GH synthesis and release in static cultures, but not in dynamic superfusion systems. PRL was not altered in either system. It was found by Yamauchi and his coworkers [28] that PACAP (6–38), a PACAP antagonist, blunted spontaneous pulsatile GH secretion in conscious male rats and PACAP exerted its effect through a serotoninergic mechanism.

PACAP is involved in the regulation of gonadotropin biosynthesis and secretion, both alone and in concerted action with GnRH [29]. PACAP in a high dose (10 venous bolus of 10 μg PACAP each of 40-min intervals) induced a very high LH release in 3-month-old Sprague-Dawley male rats [7]. The interaction of PACAP and GnRH was demonstrated in a gonadotropic cell line (Lβ-T2). Lariviere and his coworkers [30] observed that PACAP38 treatment increased the intracellular cAMP 120 times above basal level within 30 min while GnRH treatment reached a maximum of 2.7 times effect within 4 h. On the contrary, GnRH very potently enhanced the inositol phospholipid (IP3) turnover and PACAP had a very weak effect in this relation. PACAP itself is weakly stimulatory on LH secretion, but it strongly potentiates GnRH-induced LH secretion [31].

The intracellular mechanism by which PACAP exerts its effect on different pituitary cell types was also investigated in rat anterior pituitary primary culture by monitoring changes in cytosolic Ca^{2+} concentration ($[Ca^{2+}]i$) in single fura-2-loaded cells [32]. The cells were identified by immunocytochemistry at the end of the Ca^{2+} measurements. Stimulation of Ca^{2+} entry mediated by cAMP was the main mechanism in somatotropes. Ca^{2+} release from the intracellular Ca^{2+} stores mediated by phospholipase-C was the dominant mechanism in gonadotropes. Stimulation of Ca^{2+} entry not mediated by cAMP was the main mechanism in lactotropes. Corticotropes and thyrotropes exhibited weak responses to PACAP and this response was mediated by cAMP.

PACAP is also able to stimulate its own expression. PACAP in a high dose (10 venous bolus of 10 μg PACAP each of 40-min intervals) enhanced its expression level in the anterior pituitary about six times in 3-month-old Sprague-Dawley male rats [7].

Pineal Body

The pineal body in mammals functions as an endocrine tissue. The pinealocytes secrete melatonin in a circadian rhythm. The highest melatonin level occurs during the night time [33]. The pineal body also contains neurons in several species including humans [34], cotton rats [35] and monkeys [36], but not in albino rats. Recently it has been demonstrated that the pineal neurons in hamsters send information through a multisynaptic pathway to the retina [37]. The pineal organ receives sensory and autonomic fibers [38–40]. PACAP immunoreactive sensory fibers derive from the trigeminal ganglion, sympathetic fibers from the superior cervical ganglion and parasympathetic fibers from the sphenopalatine ganglion. PACAP immunoreactive cell bodies were not described up to now.

The presence of PACAP fibers implies the presence of PACAP receptors. Indeed, PACAP receptors were also found in the pineal organ [41]. Most potent was the VPAC1 receptor through which PACAP strongly enhanced melatonin release from cultured dissociated pinealocytes [42]. This stimulation was mediated via the activation of two enzymes serotonin-*N*-acetyltransferase [43] and hydroxyindole-O-methyltransferase [44]. The stimulatory effect of PACAP was accompanied by calcium influx through L-type channels [45].

Thyroid and Parathyroid Glands

In the thyroid gland three populations of PACAP fibers were observed. These may derive from the thyroid ganglion and are parasympathetic in nature. Sympathetic fibers associated with vessels contained NPY. A few fibers were sensory and also showed CGRP immunoreactivity. PAC1 and VPAC1 receptor mRNA occurred in follicular cells and blood vessels, whereas the expression of the VPAC2 receptor was also found, but it was very low [46]. The presence of PACAP receptors implies that PACAP influences the function of the thyroid gland as well. Indeed, it was demonstrated that PACAP increased cAMP production in porcine thyroid cells in vitro and PACAP also increased T4 production from mouse thyroid in vivo similar to TSH, but PACAP inhibited the binding of 125I-labeled TSH to thyroid cells.

These facts show that PACAP has a thyroid stimulating effect alone; however, it has an inhibitory action on the binding of TSH to thyroid receptors [47].

In the parathyroid gland, PACAP-immunoreactive fibers were found beneath the capsule, around blood vessels and close to glandular cells. These fibers were mainly CGRP immunoreactive and capsaicin sensitive sensory fibers. The parathyroid glandular cells expressed PAC1 and VPAC2 receptors [46].

Adrenal Gland

CRH is the main regulator of corticotropes, but its actions are supplemented by arginine-vasopressin (AVP). These two peptides control ACTH release in a concerted action and in turn, ACTH regulates the glucocorticoid synthesis and release (reviewed by [48, 49]). Catecholamine secretion from chromaffin cells of the adrenal medulla is also controlled by CRH. CRH fibers terminate in the locus ceruleus. These noradrenergic neurons send fibers to the intermediolateral cell column (IML) where cholinergic neurons provide the origin of the preganglionic splanchnic nerve fibers which innervate the adrenal medulla [50]. Half a decade after its discovery PACAP was demonstrated in rat, human [5] and pig adrenal gland [51] using RIA and northern blot analysis. The concentration of PACAP was very low compared to other nervous tissues. PACAP was also found in the adrenal gland of other species such as mouse, hamster and cow [52]. PACAP was also revealed in rat adrenal gland by PCR technique and S-EIA [53]. Immunohistochemistry revealed that PACAP immunoreactive fibers innervated adrenal chromaffin cells (often co-localized with choline acetyltransferase). Subcapsular fibers traversed the cortex innervating endocrine cells and blood vessels. These fibers contain CGRP or VIP. PACAP fibers were demonstrated in the splanchnic nerve fibers which are axons of the preganglionic neurons locating in the lateral horn of the Th9-12 spinal cord segments. These preganglionic neurons directly innervate the adrenal gland and co-store acetylcholine. Adrenal nerve cell bodies do not contain PACAP [51]; however, PACAP was shown in noradrenaline-containing chromaffin cells. These cells are the other source of PACAP in the adrenal gland [52]. This observation suggests that PACAP in the adrenal medulla, similarly to the pituitary gland, is also an autocrine regulator [54].

PACAP receptors were also demonstrated in the adrenal medulla. PAC_1 was exclusively found in the adrenal medulla, while VPAC1 and VPAC2 are expressed in both the adrenal cortex and the medulla. From the abovementioned data it is evident that PACAP is involved in the regulation of the hypothalamic–pituitary–adrenal axis at all the three levels. PACAP [55] and its receptor PAC1 [56] are expressed in the hypothalamic PVN. Stroth and his coworkers [57] and Hamelink and his coworkers [58] thoroughly investigated the effect of PACAP in the hypothalamic–pituitary–adrenal axis. PACAP appears to be an important regulator of CRH secretion from the PVN leading to ACTH secretion from the pituitary and in turn, secretion of corticosterone from the adrenal cortex. The abovementioned authors summarized the major effects of PACAP in the stress response of the organism.

(1) In the adrenal gland PACAP upregulates two enzymes involved in epinephrine biosynthesis: tyrosine hydroxylase and phenylethanolamine N-methyltransferase. (2) Stress hormone biosynthesis via upregulation of PACAP acts at the level of the hypothalamus to control induction of CRH mRNA (feedback mechanism). (3) PACAP control of stress hormone biosynthesis appears to be specialized for prolonged versus acute stress axis activation. (4) PACAP utilizes a novel ERK-mediated cAMP-dependent/PKA-dependent signaling pathway to activate a "stress response transcriptome" that includes other neuropeptides, neuroprotective factors, and prohormone convertases. (5) The modulation of stress response by PACAP in the brain may extend even beyond the hypothalamic–pituitary–adrenal and hypothalamic–sympathoadrenal axes.

Girard and his coworkers [59] investigated the effect of a series of neuropeptides on the induction of PACAP in cultured post-ganglionic sympathetic neurons. These results imply an action of PACAP to mediate stress responses specific to the sympathetic nervous system, although these have not yet been investigated in vivo despite the critical role of sympathetic activation in acute and chronic homeostatic and allostatic responses to a wide range of stressful stimuli [60, 61]. Hammack and colleagues [62, 63] observed that PACAP expression in the bed nucleus of the stria terminalis is greatly enhanced by prolonged stress, and postulated a potential anxiogenic role for PACAP in the limbic system. PACAP appears to be associated with the "fight-or-flight" response partly through its effect on adrenal production of cortisol and catecholamines.

Selenoprotein T is present in differentiating tissues. It is highly expressed in the adrenomedullary cells. Tanguy and his coworkers [64] demonstrated that PACAP regulates this protein. This also supports the view that PACAP has an essential role in the differentiation of tissues among them in the adrenal medulla as well.

Endocrine Pancreas

The pancreas is richly innervated by noradrenergic sympathetic nerves originating from the celiac and superior mesenteric ganglia [65]. The parasympathetic fibers derive from the vagus nerve and the postganglionic fibers from the intramural pancreatic ganglia [66]. Besides the autonomic innervation, the pancreas receives sensory fibers from the lower lumbar nerves [67]. In rats, PACAP immunoreactive nerve fibers were identified via immunohistochemistry in both the exocrine and endocrine pancreas [68, 69]. These fibers innervated vessels and ran among the exocrine acini and the cells of islets of Langerhans. There are differences among species in the density of PACAP innervation of the endocrine pancreas. The mouse pancreas is more abundantly innervated by PACAP nerves than the rat pancreas. PACAP fibers are parasympathetic and sensory [70, 71]. Yada and his coworkers [72] revealed PACAP immunoreactivity in the β-cells of the islet of Langerhans as well. With the use of an improved immunofluorescence method (enhancement of the signal by catalyzed reporter deposition technique) and electron microscopy

PACAP immunoreactivity was observed in all insulin and glucagon cells in human and rat islets, but not in somatostatin or pancreatic polypeptide (PP) cells [73]. PACAP immunoreactivity was associated with secretory granules. After the discovery of PACAP in the islets of Langerhans, their receptors were also demonstrated. In situ hybridization studies of pancreatic islets demonstrated the presence of both PAC1 and VPAC2 expressions in islets, but not VPAC1 [74]. Later, with the use of very sensitive RT-PCR methods PAC1, VPAC1, and VPAC2 expressions were confirmed in pancreatic β-cells. PACAP activates adenylyl cyclase in these cells [75]. It also causes calcium influx through L-type calcium channels [76].

The effect of PACAP on insulin and glucagon release was demonstrated in in vivo and in vitro models. In vivo PACAP38 injected *iv* in mice potentiated plasma glucagon response to the cholinergic agonist carbachol. This potentiation was reduced by a preceding combined α- and β-adrenergic blockade. PACAP also stimulated basal insulin levels [68]. It was also shown that administration of glucose inhibits PACAP-induced glucagon secretion, proving that the action is glucose sensitive [77]. The effect of PACAP on pancreatic functions was also investigated in dogs. An *iv* bolus injection of 6 pmol/kg PACAP27 or PACAP38 elicited a transient increase in plasma insulin, epinephrine, and norepinephrine concentrations, with a peak value at 2 min after injection. Injections of 60 and 600 pmol/kg caused greater increase in these hormone concentrations in a dose-dependent manner [78]. Filipsson and her coworkers [77] examined healthy human volunteers. They found that the highest tolerable infusion rate of PACAP was 3.5 pmol/kg/min. Higher doses cause intense flushing of the skin. The results showed that PACAP slightly increased serum insulin levels in overnight-fasted healthy volunteers and potentiated the insulin response to an intravenous glucose challenge. Finally, PACAP seems to have vasodilatory action in the pancreas, as demonstrated in vivo in rats. The vasodilator effect of PACAP was previously demonstrated in the cerebral blood flow as well [79]. Carlsson and his coworker [80] tested the effect of 5 and 10 nmol/kg PACAP on splanchnic blood flow in anesthetized rats. It was found that 5 nmol/kg induced a transient initial decrease in mean arterial blood pressure; however, 10 nmol/kg was able to induce an increase in the pancreatic blood flow studied by radiolabeled microsphere technique.

In vitro, in the presence of glucose PACAP27 in cultured rat islet cells stimulated insulin release in a biphasic manner: an initial sharp rise followed by a moderate elevation [72]. The stimulatory effect of PACAP was also shown in hamster clonal β-cell lines [81], in normal conscious dogs [78] and pig pancreas [82]. Depending on the above-mentioned experimental models, PACAP was effective in 10 fmol–1 nmol/l dose in stimulating glucose dependent insulin release.

Nakata and his coworkers later demonstrated [83] that intra-islet PACAP protects pancreatic β-cells against glucotoxicity and lipotoxicity. In cultured pancreas islets glucose increased the cytosolic Ca^{2+} concentration. In the islets derived from PACAP-null mice, the glucose-induced first phase increase in the cytosolic Ca^{2+} were severely impaired while they were preserved in islets of wild-type mice. Treatment with high glucose or palmitate (16-carbon saturated fatty acid) also impaired the glucose-induced insulin secretion in islets and increased mRNA expression of uncoupling protein 2 (UCP2) in islets of PACAP-null, but not of wild-type mice. Increased uncoupling protein-2 levels in β-cells are associated with impaired glucose-stimulated insulin secretion. These data indicate that islet-produced PACAP protects β-cells from deteriorating action of high glucose and palmitate at least partly by blocking the elevation of UCP2, suggesting an antidiabetic role for PACAP. When PACAP antibodies were added to freshly isolated islets, an inhibition of glucose (11.1 mmol/l)-stimulated insulin secretion was observed. This fact supports the view that PACAP is an autocrine and/or paracrine acting islet neuropeptide, necessary for optimal glucose induced insulin secretion [84]. It was also shown that in pig pancreatic cell cultures, PACAP improved the survival of β-cells. To this day PACAP is accepted as a potent insulinotropic hormone.

PAC1 deletion in mice resulted in a 50 % decrease in PACAP-mediated glucose-induced insulin secretion but basal insulin or plasma glucose levels were not affected [85]. This means, that the defective insulinotropic action of glucose was associated with marked glucose intolerance after both intravenous and gastric glucose administration. In addition, reduced glucagon response was reported in insulin-mediated experimentally induced hypoglycemia in PAC1-null mice [86]. Thus, PACAP-stimulated PAC1 receptor activation mediates glucagon response to insulin-induced hypoglycemia. Assessments of VIP receptor knockout mice revealed improved lean mass but decreased fat mass associated with a reduction in body weight in VPAC2-deficient mice [87]. Although glucose-induced insulin secretion was decreased, no alteration on glucose tolerance was reported in oral glucose tolerance tests performed in VPAC2-null mice. Moreover, the fact that glucose is cleared faster from the blood by insulin injections suggests that the VPAC2 knockout phenotype increased insulin sensitivity. These results demonstrate that VPAC2 receptors are also needed for optimal insulin release from pancreatic islets as well. VPAC1-null mutant mice have recently been generated. This mutation resulted in fetal, neonatal, and postnatal death due to growth retardation, intestinal obstructions, and hypoglycemia [88]. Intriguingly, VPAC1 knockout mice manifested lower baseline blood glucose levels compared with wild-type littermates and responded to oral glucose challenge with normal rise in blood glucose followed by rapid hypoglycemia and failure to restore baseline glucose levels. Insulin challenge resulted in profound hypoglycemia and an inadequate glucose homeostasis in VPAC1-null mutant animals. These results demonstrated that VPAC1 is required for both embryonic/neonatal development and proper function of endocrine pancreas.

Though PACAP has been primarily evaluated as an insulinotropic hormone, transgenic expression of PACAP revealed that PACAP also has a proliferative effect on islet cells. Transgenic mice overexpressing PACAP in pancreatic β-cells showed elevated insulin secretion after oral glucose administration, and there was no change in plasma glucose and glucagon levels. In addition, increased β-cell mass was found [89]. In the case of artificially induced apoptosis of pancreatic islets in transgenic mice, PACAP is extremely upregulated in the islets studied by DNA microarray analysis [90]. These facts suggest that PACAP has an essential role in proliferation and differentiation of islet cells and that in the future PACAP may be a therapeutic agent to promote islet survival [91].

Table 11.1 Occurrence of PACAP in endocrine organs

Organs	PACAP ir specific cell types	PACAP ir nerve fibers	References
Anterior lobe of PG	Gonadotropes, FS cells	Not observed	[8, 14]
Pineal body	No cells	Sensory and autonomic	[38]
Thyroid gland	No cells	Autonomic fibers	[46]
Parathyroid gland	No cells	Autonomic fibers	[46]
Adrenal medulla	Chromaffin cells	Sympathetic fibers	[51, 52]
Adrenal cortex	No cells	Sensory	[100]
Islets of Langerhans	α- and β-cells	Sensory and parasympathetic	[70–73]
Enteroendocrine cells	Cells in chicken provent. Parietal cells in human and cat stomach	Parasympathetic fibers	[92–94]

FS folliculostellate cells, *ir* immunoreactive, *PG* pituitary gland, *provent* proventriculus

PACAP in Enteroendocrine Cells

In the chicken proventriculus, many PACAP immunoreactive endocrine cells were observed besides the fibers in the smooth muscle layer. PACAP cells also exhibited serotonin and gastrin-releasing peptide immunoreactivity [92]. The role of PACAP in these cells was not investigated. In 18- and 20-week old fetuses, but not earlier, PACAP as well as secretin and gastrin appeared in the developing glands of the stomach. The presence of these neuropeptides in the developing glands suggests their role in the proliferation and differentiation of the epithelial derivates [93]. PACAP immunoreactive cells were also shown in adult human and cat gastric glands, but not in intact rats [94]. CIBA revealed that in human static gastric mucosa culture PACAP was released from some cells. Electron microscopic investigation clearly showed that PACAP immunoreactivity in the human gastric tissue was found in parietal cells. This finding is not surprising because it is well known that the parietal cells produce the intrinsic factor as well [95]. This clearly shows that the parietal cells are able to produce some peptides besides secreting acid. The role of PACAP in acid secretion was thoroughly investigated. PACAP exerts its effect on acid secretion via histamine producing enterochromaffin-like (ECL) and somatostatin producing D cells [96–99]. Autocrine regulation of acid secretion by PACAP is not excluded, but it was not investigated.

Conclusion

Table 11.1 summarizes the occurrence of PACAP in the endocrine glands.

PACAP has a role in the regulation of all endocrine functions.

1. In the pituitary gland PACAP exerts autocrine and paracrine actions on LH, FSH, GH, and ACTH synthesis and release (Table 11.2.). Hypothalamic-born PACAP

Table 11.2 Effect of PACAP38 on pituitary hormone release or mRNA level in in vivo and in vitro models

	In vivo	In vitro	References
LH	*iv* LH - f rat		[101][a]
	icv LH↓ f rat		[101][a]
	icv LH↓ f ewe		[102][b]
	iv LH↑ m rat #dose		[7]
	ia LH↑ m rat		[103]
	icv LH- m rat		[103]
		Balancing effect on LH f rat (depends on stage (primary rat cell-culture) of EC)	[13]
		Lβ-T2 cells LH↑	[30]
LHβ mRNA		P LHβ mRNA ↑	[16, 17]
		C LHβ mRNA (perifused rat system)	
α-subunit mRNA		P α-subunit mRNA↑	[16, 17]
		C α-subunit mRNA↑ (perifused rat system)	
FSHβ mRNA		P FSHβ mRNA -	[16, 17]
		C FSHβ mRNA↑ (perifused rat system)	
PRL	*iv* PRL↓ intact m rat		[25]
	iv PRL↑ h lesioned m rat	PRL ↓ R-HPA rat	[25]
		PRL - static rat culture	[27]
		PRL↑ GH3 mouse cells	[22]
GH	*iv* GH↑ intact m rat		[25]
		GH↑ R-HPA rat	[26]
		GH↑ GH3 mouse cells	[22]
ACTH		ACTH↑ AtT-20 rat cells	[22]
TSH	Not investigated		

f female, *m* male, *C* continuous administration, *P* pulsatile administration
↑=elevation of plasma level; ↓=depression of plasma level; -=not changed; #=extremely high
EC estrous cycle, *h* hypothalamus, *R-HPA* reverse hemolytic plaque assay
[a]Given before the critical period of proestrous stage
[b]Given into ovariectomized ewe

may influence pituitary functions directly by being released into the portal circulation or indirectly by acting on the releasing hormone producing cells. The effect of PACAP on LH secretion depends on the gender, the time of the day and in female on the stage of the estrous cycle.

Table 11.3 shortly summarizes the major effects of PACAP on endocrine organs other than the pituitary.

2. In the pineal body PACAP enhances melatonin synthesis via activation of two enzymes (hydroxyindole-O-methyl transferase and serotonin *N*-acetyltransferase) and melatonin release.
3. In the thyroid gland PACAP stimulates T4 production, but inhibits TSH binding to thyrocytes.

Table 11.3 Effect of PACAP on various hormone release

Organs	In vivo	In vitro	References
Pineal body	Melatonin ↑		[42]
Thyroid gland	T4 ↑		[47]
	TSH binding ↓		
Parathyroid gland	Not investigated		
Adrenal medulla		Epinephrine ↑	[57]
Adrenal cortex	Corticosteron ↑ (through HPA)		[56]
Islets of Langerhans	Glucagon ↑ mice		[67]
	Insulin ↑ mice		[67]
	Insulin ↑ dog		[78]
	Insulin ↑ human		[78]
		Insulin ↑ rat	[71, 104]
		Insulin ↑ hamster	[74]
		Insulin ↑ pig	[82]
Enteroendocrine cells	Not investigated		

HPA hypothalamic–pituitary–adrenal axis

4. In the adrenal medulla PACAP acts in an autocrine manner, upregulating two enzymes (tyrosine hydroxylase and phenylethanolamine *N*-methyltransferase). It also regulates CRH synthesis and release and in this way it is involved in the stress response.
5. In the endocrine pancreas PACAP in vivo stimulates basal glucagon and insulin release. However, PACAP is inhibitory on glucose induced glucagon and stimulatory on glucose induced insulin release. PACAP is an insulinotropic and antidiabetic hormone.
6. The role of PACAP in enteroendocrine cells is not clarified.

References

1. Dow RC, Bennie J, Fink G. Pituitary adenylate cyclase-activating peptide-38 (PACAP)-38 is released into hypophysial portal blood in the normal male and female rat. J Endocrinol. 1994;142:R1–4.
2. Miyata A, Arimura A, Dahl RR, Minamino N, Uehara A, Jiang L, et al. Isolation of a novel 38 residue-hypothalamic polypeptide which stimulates adenylate cyclase in pituitary cells. Biochem Biophys Res Commun. 1989;184:567–74.
3. Arimura A, Somogyvári-Vigh A, Miyata A, Mizuno K, Coy DH, Kitada C. Tissue distribution of PACAP as determined by RIA: highly abundant in the rat brain and testes. Endocrinology. 1991;129:2787–9.
4. Köves K, Kántor O, Lakatos A, Szabó E, Kirilly E, Heinzlmann A, et al. Advent and recent advances in research on the role of pituitary adenylate cyclase-activating polypeptide (PACAP) in the regulation of gonadotropic hormone secretion of female rats. J Mol Neurosci. 2014;54:494–511.

5. Ghatei MA, Takahashi K, Suzuki Y, Gardiner J, Jones PM, Bloom SR. Distribution, molecular characterization of pituitary adenylate cyclase-activating polypeptide and its precursor encoding messenger RNA in human and rat tissues. J Endocrinol. 1993;136:159–66.
6. Oka H, Jin L, Reubi JC, Qian X, Scheithauer BW, Fujii K, et al. Pituitary adenylate-cyclase-activating polypeptide (PACAP) binding sites and PACAP/vasoactive intestinal polypeptide receptor expression in human pituitary adenomas. Am J Pathol. 1998;153:1787–96.
7. Radleff-Schlimme A, Leonhardt S, Wuttke W, Jarry H. Evidence for PACAP to be an autocrine factor on gonadotrope cells. Ann NY Acad Sci. 1998;865:486–91.
8. Köves K, Kántor O, Scammel JG, Arimura A. PACAP colocalizes with luteinizing and follicle-stimulating hormone immunoreactivities in the anterior lobe of the pituitary gland. Peptides. 1998;19:1069–72.
9. Heinzlmann A, Kirilly E, Meltzer K, Szabó E, Baba A, Hashimoto H, et al. PACAP is transiently expressed in anterior pituitary gland of rats. In situ hybridization and cell immunoblot assay studies. Peptides. 2008;29:571–7.
10. Moore Jr JP, Burger LL, Dalkin AC, Winters SJ. Pituitary adenylate cyclase activating polypeptide messenger RNA in the paraventricular nucleus and anterior pituitary during the rat estrous cycle. Biol Reprod. 2005;73:491–9.
11. Winters SJ, Ghooray DT, Yang RQ, Holmes JB, O'Brien AR, Morgan J, et al. Dopamine-2 receptor activation suppresses PACAP expression in gonadotrophs. Endocrinology. 2014;155:2647–57.
12. Szabó E, Nemeskéri Á, Arimura A, Köves K. Effect of PACAP on LH release, studied by cell immunoblot assay, depends on the gender, on the time of day and in female rats on the day of estrous cycle. Regul Pept. 2004;123:139–45.
13. Szabó E, Nemeskéri A, Heinzlmann A, Suzuki N, Arimura A, Köves K. Cell immunoblot assay study demonstrating the release of PACAP from individual anterior pituitary cells of rats and the effect of PACAP on LH release. Regul Pept. 2002;109:75–81.
14. Jin L, Tsumanuma I, Ruebel KH, Bayliss JM, Lloyd RV. Analysis of homogenous population of anterior pituitary folliculostellate cells by laser capture microdissection and reverse transcription polymerase chain reaction. Endocrinology. 2001;142:1703–9.
15. Winters SJ, Moore Jr JP. PACAP, an autocrine/paracrine regulator of gonadotrophs. Biol Reprod. 2011;84:844–50.
16. Tsujii T, Winters SJ. Effects of pulsatile pituitary adenylate cyclase activating polypeptide (PACAP) on gonadotropin secretion and subunit mRNA levels in perifused rat pituitary cells. Life Sci. 1995;56:1103–11.
17. Tsujii T, Ishizaka K, Winters SJ. Effects of pituitary adenylate cyclase-activating polypeptide on gonadotropin secretion and subunit messenger ribonucleic acids in perifused rat pituitary cells. Endocrinology. 1994;135:826–33.
18. Vigh S, Arimura A, Gottschall PE, Kitada C, Somogyvári-Vigh A, Childs GV. Cytochemical characterization of anterior pituitary target cells for the neuropeptide, pituitary adenylate cyclase activating polypeptide (PACAP), using biotinylated ligands. Peptides. 1993;14:59–65.
19. Vertongen P, Velkeniers B, Hooghe-Peters E, Robberecht P. Differential alternative splicing of PACAP receptor in pituitary cell subpopulations. Mol Cell Endocrinol. 1995;113:131–5.
20. Coleman DT, Chen X, Sassaroli M, Bancroft C. Pituitary adenylate cyclase-activating polypeptide regulates prolactin promoter activity via a protein kinase A-mediated pathway that is independent of the transcriptional pathway employed by thyrotropin-releasing hormone. Endocrinology. 1996;137:1276–85.
21. Bresson-Bepoldin L, Jacquot MC, Schlegel W, Rawlings SR. Multiple splice variants of the pituitary adenylate cyclase-activating polypeptide type 1 receptor detected by RT-PCR in single rat pituitary cells. J Mol Endocrinol. 1998;21:109–20.
22. Propato-Mussafiri R, Kanse SM, Ghatei MA, Bloom SR. Pituitary adenylate cyclase-activating polypeptide releases 7B2, adrenocorticotrophin, growth hormone and prolactin from the mouse and rat clonal pituitary cell lines AtT-20 and GH3. J Endocrinol. 1992;132:107–13.

23. Rawlings SR, Piuz I, Schlegel W, Bockaert J, Journot L. Differential expression of pituitary adenylate cyclase-activating polypeptide/vasoactive intestinal polypeptide receptor subtypes in clonal pituitary somatotrophs and gonadotrophs. Endocrinology. 1995;136:2088–98.
24. Dejda A, Bourgault S, Doan ND, Létourneau M, Couvineau A, Vaudry H, et al. Identification by photoaffinity labeling of the extracellular N-terminal domain of PAC1 receptor as the major binding site for PACAP. Biochimie. 2011;93:669–77.
25. Jarry H, Leonhardt S, Schmidt WE, Creutzfeldt W, Wuttke W. Contrasting effects of pituitary adenylate cyclase activating polypeptide (PACAP) on in vivo and in vitro prolactin and growth hormone release in male rats. Life Sci. 1992;51:823–30.
26. Góth MI, Lyons CE, Canny BJ, Thorner MO. Pituitary adenylate cyclase activating polypeptide, growth hormone (GH)-releasing peptide and GH-releasing hormone stimulate GH release through distinct pituitary receptors. Endocrinology. 1992;130:939–44.
27. Velkeniers B, Zheng L, Kazemzadeh M, Robberecht P, Vanhaelst L, Hooghe-Peters EL. Effect of pituitary adenylate cyclase-activating polypeptide 38 on growth hormone and prolactin expression. J Endocrinol. 1994;143:1–11.
28. Yamauchi K, Murakami Y, Koshimura K, Nishiki M, Tanaka J, Kato Y. Involvement of pituitary adenylate cyclase-activating polypeptide (PACAP) in growth hormone secretion induced by serotoninergic mechanisms in the rat. Endocrinology. 1996;137:1693–7.
29. Grafer CM, Thomas R, Lambrakos L, Montoya I, White S, Halvorson LM. GnRH stimulates expression of PACAP in the pituitary gonadotropes via both the PKA and PKC signaling systems. Mol Endocrinol. 2009;23:1022–32.
30. Larivière S, Garrel G, Robin MT, Counis R, Cohen-Tannoudji J. Differential mechanisms for PACAP and GnRH cAMP induction contribute to cross-talk between both hormones in the gonadotrope LbetaT2 cell line. Ann N Y Acad Sci. 2006;1070:376–9.
31. Culler MD, Paschall CS. Pituitary adenylate cyclase-activating polypeptide (PACAP) potentiates the gonadotropin-releasing activity of luteinizing hormone-releasing hormone. Endocrinology. 1991;129:2260–2.
32. Alarcón P, García-Sancho J. Differential calcium responses to the pituitary adenylate cyclase-activating polypeptide (PACAP) in the five main cell types of rat anterior pituitary. Pflugers Arch. 2000;440:685–91.
33. Reiter RJ. Pineal melatonin: cell biology of its synthesis and of its physiological interactions. Endocr Rev. 1991;12:151–80.
34. Moore RY, Sibony P. Enkephalin-like immunoreactivity in neurons in the human pineal gland. Brain Res. 1988;457:395–8.
35. Matsushima S, Sakai Y, Hira Y, Oomori Y, Daikoku S. Immunohistochemical studies on sympathetic and non-sympathetic nerve fibers and neuronal cell bodies in the pineal gland of cotton rats. Sigmodon Hispidus Arch Histol Cytol. 1994;57:47–58.
36. Ichimura T, Arikuni T, Hashimoto PH. Fine-structural study of the pineal body of 3 monkey (Macaca fuscata) with special reference to synaptic formations. Cell Tissue Res. 1986;244:569–76.
37. Csáki Á, Vígh B, Boldogkői Z, Vereczki V, Szél Á, Köves K. Is a neuronal chain between the pineal body and the retina in rats and hamsters? Transneural tracing studies. Neurosci Lett. 2015;588:1–6.
38. Baeres FM, Møller M. Demonstration of PACAP-immunoreactive intrapineal nerve fibers in the golden hamster (Mesocricetus auratus) originating from the trigeminal ganglion. J Pineal Res. 2005;38:116–22.
39. Moller M, Baeres FM. The anatomy and innervation of the mammalian pineal gland. Cell Tissue Res. 2002;309:139–50.
40. Phansuwan-Pujito P, Møller M, Govitrapong P. Cholinergic innervation and function in the mammalian pineal gland. Microsc Res Tech. 1999;46:281–95.
41. Simonneaux V, Kienlen-Campard P, Loeffler JP, Basille M, Gonzalez BJ, Vaudry H, et al. Pharmacological, molecular and functional characterization of vasoactive intestinal polypeptide/pituitary adenylate cyclase-activating polypeptide receptors in the rat pineal gland. Neuroscience. 1998;85:887–96.

42. Simonneaux V, Ouichou A, Pévet P. Pituitary adenylate cyclase-activating polypeptide (PACAP) stimulates melatonin synthesis from rat pineal gland. Brain Res. 1993;603:148–52.
43. Yuwiler A, Brammer GL, Bennett BL. Interaction between adrenergic and peptide stimulation in the rat pineal: pituitary adenylate cyclase-activating peptide. J Neurochem. 1995;64: 2273–80.
44. Ribelayga C, Pévet P, Simonneaux V. Adrenergic and peptidergic regulations of hydroxyindole-O-methyltransferase activity in rat pineal gland. Brain Res. 1997;777:247–50.
45. Chik CL, Liu QY, Li B, Klein DC, Zylka M, Kim DS, et al. Alpha 1D L-type Ca(2+)-channel currents: inhibition by a beta-adrenergic agonist and pituitary adenylate cyclase-activating polypeptide (PACAP) in rat pinealocytes. J Neurochem. 1997;68:1078–87.
46. Fahrenkrug J, Hannibal J. Localisation of the neuropeptide PACAP and its receptors in the rat parathyroid and thyroid glands. Gen Comp Endocrinol. 2011;171:105–13.
47. Chen W, Inui T, Hachiya T, Ochi Y, Nakajima Y, Kajita Y. Stimulatory action of pituitary adenylate cyclase-activating polypeptide (PACAP) on thyroid gland. Biochem Biophys Res Commun. 1993;194:923–9.
48. Watts AG, Sanchez-Watts G. Physiological regulation of peptide messenger RNA colocalization in rat hypothalamic paraventricular medial parvicellular neurons. J Comp Neurol. 1995;352:501–14.
49. Watts AG. Glucocorticoid regulation of peptide genes in neuroendocrine CRH neurons: a complexity beyond negative feedback. Front Neuroendocrinol. 2005;26:109–30.
50. Eiden LE. Neuropeptide–catecholamine interactions in stress. Adv Pharmacol. 2013;68:399–404.Tornøe K, Hannibal J, Jensen TB, Georg B, Rickelt LF, Andreasen MB, et al. PACAP-(1-38) as neurotransmitter in the porcine adrenal glands. Am J Physiol Endocrinol Metab. 2000;279:E1413–25.
51. Tornøe K, Hannibal J, Jensen TB, Georg B, Rickelt LF, Andreasen MB, et al. PACAP-(1-38) as neurotransmitter in the porcine adrenal glands. Am J Physiol Endocrinol Metab. 2000;279:E1413–25.
52. Tabarin A, Chen D, Håkanson R, Sundler F. Pituitary adenylate cyclase-activating peptide in the adrenal gland of mammals: distribution, characterization and responses to drugs. Neuroendocrinology. 1994;59:113–9.
53. Kántor O, Heinzlmann A, Suzuki N, Vincze E, Kocsis K, Köves K. Distribution of PACAP and its mRNA in several nonneural tissues of rats demonstrated by sandwich enzyme immunoassay and RT-PCR technique. Regul Pept. 2002;109:103–5.
54. Conconi MT, Spinazzi R, Nussdorfer GG. Endogenous ligands of PACAP/VIP receptors in the autocrine–paracrine regulation of the adrenal gland. Int Rev Cytol. 2006;249:1–51.
55. Hannibal J. Pituitary adenylate cyclase-activating peptide in the rat central nervous system: an immunohistochemical and in situ hybridization study. J Comp Neurol. 2002;453:389–417.
56. Hashimoto H, Nogi H, Mori K, Ohishi H, Shigemoto R, Yamamoto K, et al. Distribution of mRNA for a pituitary adenylate cyclase-activating polypeptide receptor in the rat brain: an in situ hybridization study. J Comp Neurol. 1996;371:567–77.
57. Stroth N, Holighaus Y, Ait-Ali D, Eiden LE. PACAP: a master regulator of neuroendocrine stress circuits and the cellular stress response. Ann N Y Acad Sci. 2011;1220:49–59.
58. Hamelink C, Weihe E, Eiden LE. PACAP: an "emergency response" co-transmitter in the adrenal medulla. In: Vaudry H, Arimura A, editors. Pituitary adenylate cyclase-activating polypeptide. Norwell, MA: Kluwer-Academic Press; 2003. p. 227–50.
59. Girard BM, May V, Bora SH, Fina F, Braas KM. Regulation of neurotrophic peptide expression in sympathetic neurons: quantitative analysis using radioimmunoassay and real-time quantitative polymerase chain reaction. Regul Pept. 2002;109:89–101.
60. Kvetnanský R, Pacák K, Fukuhara K, Viskupic E, Hiremagalur B, Nankova B, et al. Sympathoadrenal system in stress. Interaction with the hypothalamic-pituitary-adrenocortical system. Ann N Y Acad Sci. 1995;771:131–58.
61. Kvetnansky R, Sabban EL, Palkovits M. Catecholaminergic systems in stress: structural and molecular genetic approaches. Physiol Rev. 2009;89:535–606.

62. Hammack SE, Cheung J, Rhodes KM, Schutz KC, Falls WA, Braas KM, et al. Chronic stress increases pituitary adenylate cyclase-activating peptide (PACAP) and brain-derived neurotrophic factor (BDNF) mRNA expression in the bed nucleus of the stria terminalis (BNST): roles for PACAP in anxiety-like behavior. Psychoneuroendocrinology. 2009;34:833–43.
63. Hammack SE, Roman CW, Lezak KR, Kocho-Shellenberg M, Grimmig B, Falls WA, et al. Roles for pituitary adenylate cyclase-activating peptide (PACAP) expression and signaling in the bed nucleus of the stria terminalis (BNST) in mediating the behavioral consequences of chronic stress. J Mol Neurosci. 2010;42:327–40.
64. Tanguy Y, Falluel-Morel A, Arthaud S, Boukhzar L, Manecka DL, Chagraoui A, et al. The PACAP-regulated gene selenoprotein T is highly induced in nervous, endocrine, and metabolic tissues during ontogenetic and regenerative processes. Endocrinology. 2011;152:4322–35.
65. Woods SC, Porte Jr D. Neural control of the endocrine pancreas. Physiol Rev. 1974;54:596–619.
66. Yi E, Smith TG, Love JA. Noradrenergic innervation of rabbit pancreatic ganglia. Auton Neurosci. 2005;117:87–96.
67. Sharkey KA, Williams RG, Dockray GJ. Sensory substance P innervation of the stomach and pancreas. Demonstration of capsaicin-sensitive sensory neurons in the rat by combined immunohistochemistry and retrograde tracing. Gastroenterology. 1984;87:914–21.
68. Fridolf T, Sundler F, Ahren B. Pituitary adenylate cyclase-activating polypeptide (PACAP): occurrence in rodent pancreas and effects on insulin and glucagon secretion in the mouse. Cell Tissue Res. 1992;269:275–9.
69. Köves K, Arimura A, Vigh S, Somogyvári-Vigh A, Miller J. Immunohistochemical localization of PACAP in the ovine digestive system. Peptides. 1993;14:449–55.
70. Filipsson K, Sundler F, Hannibal J, Ahren B. PACAP and PACAP receptors in insulin producing tissues: localization and effects. Regul Pept. 1998;74:167–75.
71. Filipsson K, Kvist-Reimer M, Ahren B. The neuropeptide pituitary adenylate cyclase-activating polypeptide and islet function. Diabetes. 2001;50:1959–69.
72. Yada T, Sakuradan M, Ihidall K, Nakata M, Muratall F, Arimura A, et al. Pituitary adenylate cyclase activating polypeptide is an extraordinarily potent intra-pancreatic regulator of insulin secretion from islet P-cells. J Biol Chem. 1994;269:1290–3.
73. Portela-Gomes GM, Lukinius A, Ljungberg O, Efendic S, Ahrén B, Abdel-Halim SM. PACAP is expressed in secretory granules of insulin and glucagon cells in human and rodent pancreas. Evidence for generation of cAMP compartments uncoupled from hormone release in diabetic islets. Regul Pept. 2003;113:31–9.
74. Inagaki N, Yoshida H, Mizuta M, Mizuno N, Fujii Y, Gonoi T, et al. Cloning and functional characterization of a third pituitary adenylate cyclase-activating polypeptide receptor subtype expressed in insulin-secreting cells. Proc Natl Acad Sci U S A. 1994;91:2679–83.
75. Borboni P, Porzio O, Pierucci D, Cicconi S, Magnaterra R, Federici M, et al. Molecular and functional characterization of pituitary adenylate cyclase-activating polypeptide (PACAP-38)/vasoactive intestinal polypeptide receptors in pancreatic beta-cells and effects of PACAP-38 on components of the insulin secretory system. Endocrinology. 1999;140:5530–7.
76. Yada T, Sakurada M, Nakata M, Yaekura K, Kikuchi M. PACAP as low as 10(-13) M raises cytosolic Ca2+ activity in pancreatic B-cells by augmenting Ca2+ influx through L-type Ca2+ channels to trigger insulin release. Adv Exp Med Biol. 1997;426:165–71.
77. Filipsson K, Tornoe K, Holst JJ, Ahrén B. Pituitary adenylate cyclase activating polypeptide stimulates insulin and glucagon secretion in humans. J Clin Endocrinol Metab. 1997;82:3093–8.
78. Kawai K, Yokota C, Ohashi S, Isobe K, Suzuki S, Nakai T, et al. Pituitary adenylate cyclase-activating polypeptide: effects on pancreatic-adrenal hormone secretion and glucose-lipid metabolism in normal conscious dogs. Metabolism. 1994;43:739–44.
79. Seki Y, Suzuki Y, Baskaya MK, Kano T, Saito K, Takayasu M, et al. The effects of pituitary adenylate cyclase-activating polypeptide on cerebral arteries and vertebral artery blood flow in anesthetized dogs. Eur J Pharmacol. 1995;275:259–66.
80. Carlsson PO, Ostenson CG, Efendic S, Langel U, Jansson L. Pituitary adenylate cyclase activating polypeptide (PACAP) redistributes the blood within the pancreas of anesthetized rats.

Regul Pept. 1996;63:123–8.Kulkarni RN, Smith DM, Ghatei MA, Bloom SR. A 67 kDa protein mediates pituitary adenylate cyclase-activating polypeptide and vasoactive intestinal peptide-stimulated insulin secretion in a hamster clonal beta-cell line. J Endocrinol. 1995;147:120-30.

81. Kulkarni RN, Smith DM, Ghatei MA, Bloom SR. A 67 kDa protein mediates pituitary adenylate cyclase-activating polypeptide and vasoactive intestinal peptide-stimulated insulin secretion in a hamster clonal beta-cell line. J Endocrinol. 1995;147:120–30.
82. Tornøe K, Hannibal J, Fahrenkrug J, Holst JJ. PACAP-(1-38) as neurotransmitter in pig pancreas: receptor activation revealed by the antagonist PACAP-(6-38). Am J Physiol. 1997;273:G436–46.
83. Nakata M, Shintani N, Hashimoto H, Baba A, Yada T. Intra-islet PACAP protects pancreatic β-cells against glucotoxicity and lipotoxicity. J Mol Neurosci. 2010;42:404–10.
84. Filipsson K, Sundler F, Ahrén B. PACAP is an islet neuropeptide which contributes to glucose-stimulated insulin secretion. Biochem Biophys Res Commun. 1999;256:664–7.
85. Jamen F, Persson K, Bertrand G, Rodriguez-Henche N, Puech R, Bockaert J, et al. PAC1 receptor-deficient mice display impaired insulinotropic response to glucose and reduced glucose tolerance. J Clin Invest. 2000;105:1307–15.
86. Persson K, Ahrén B. The neuropeptide PACAP contributes to the glucagon response to insulin-induced hypoglycaemia in mice. Acta Physiol Scand. 2002;175:25–8.
87. Asnicar MA, Köster A, Heiman ML, Tinsley F, Smith DP, Galbreath E, et al. Vasoactive intestinal polypeptide/pituitary adenylate cyclase-activating peptide receptor 2 deficiency in mice results in growth retardation and increased basal metabolic rate. Endocrinology. 2002;143:3994–4006.
88. Fabricius D, Karacay B, Shutt D, Leverich W, Schafer B, Takle E, et al. Characterization of intestinal and pancreatic dysfunction in VPAC1-null mutant mouse. Pancreas. 2011;40: 861–71.
89. Yamamoto K, Hashimoto H, Tomimoto S, Shintani N, Miyazaki J, Tashiro F, et al. Overexpression of PACAP in transgenic mouse pancreatic beta-cells enhances insulin secretion and ameliorates streptozotocin-induced diabetes. Diabetes. 2003;52:1155–62.
90. Han B, Wu J. DcR3 protects islet β cells from apoptosis through modulating *Adcyap1* and *Bank1* expression. J Immunol. 2009;183:8157–66.
91. Sanlioglu AD, Karacay B, Balci MK, Thomas S, Griffith TS, Sanlioglu S. Therapeutic potential of VIP vs PACAP in diabetes. J Mol Endocrinol. 2012;49:R157–67.
92. Sundler F, Ekblad E, Absood A, Håkanson R, Köves K, Arimura A. Pituitary adenylate cyclase activating peptide: a novel vasoactive intestinal peptide-like neuropeptide in the gut. Neuroscience. 1992;46:439–54.
93. Vincze E, Kántor O, Kausz M, Németh J, Arimura A, Gonda P, et al. Comparative study on the appearance of various bioactive peptides in foregut derivates during the ontogenesis. J Physiol Paris. 2001;95:99–103.
94. Vincze É, Kántor O, Kiss A, Gonda G, Gombás P, Kiss J, et al. Pituitary adenylate cyclase activating polypeptide (PACAP) is present in human and cat gastric glands. Peptides. 1999;20:937–41.
95. Smolka A, Donaldson Jr RM. Monoclonal antibodies to human intrinsic factor. Gastroenterology. 1990;98:607–14.
96. Oh DS, Lieu SN, Yamaguchi DJ, Tachiki K, Lambrecht N, Ohning GV, et al. PACAP regulation of secretion and proliferation of pure populations of gastric ECL cells. J Mol Neurosci. 2005;26:85–97.
97. Piqueras L, Taché Y, Martínez V. Peripheral PACAP inhibits gastric acid secretion through somatostatin release in mice. Br J Pharmacol. 2004;142:67–78.
98. Pisegna JR, Ohning GV, Athmann C, Zeng N, Walsh JH, Sachs G. Role of PACAP1 receptor in regulation of ECL cells and gastric acid secretion by pituitary adenylate cyclase activating peptide. Ann N Y Acad Sci. 2000;921:233–41.
99. Sandvik AK, Cui G, Bakke I, Munkvold B, Waldum HL. PACAP stimulates gastric acid secretion in the rat by inducing histamine release. Am J Physiol Gastrointest Liver Physiol. 2001;281:G997–1003.

100. Holgert H, Holmberg K, Hannibal J, Fahrenkrug J, Brimijoin S, Hartman BK, et al. PACAP in the adrenal gland relationship with choline acetyltransferase, enkephalin and chromaffin cells and effects of immunological sympathectomy. Neuroreport. 1996;8:297–301.
101. Köves K, Molnár J, Kántor O, Lakatos A, Görcs T, Somogyvári-Vigh A, et al. PACAP participates in the regulation of the hormonal events preceeding the ovulation. Acta Biol Hung. 1996;47:239–49.
102. Sawangjaroen K, Curlewis JD. Effects of pituitary adenylate cyclase-activating polypeptide (PACAP) and vasoactive intestinal polypeptide (VIP) on prolactin, luteinizing hormone and growth hormone secretion in the ewe. J Neuroendocrinol. 1994;6:549–55.
103. Osuga Y, Mitsuhashi N, Mizuno M. In vivo effect of pituitary adenylate cyclase activating polypeptide 38 (PACAP 38) on the secretion of luteinizing hormone (LH) in male rats. Endocrinol Jpn. 1992;39:153–6.
104. Kawai K, Ohse C, Watanabe Y, Suzuki S, Yamashita K, Ohashi S. Pituitary adenylate cyclase activating polypeptide stimulates insulin release from the isolated perfused rat pancreas. Life Sci. 1992;50:257–61.

Chapter 12
Distribution of PACAP in the Mammalian Nervous System

Katalin Koves

Abstract PACAP is widely distributed in the central and peripheral nervous systems of various mammalian species. To demonstrate PACAP or its mRNA several methods were used: immunohistochemistry, radioimmunoassay, sandwich-enzyme immunoassay, in situ hybridization, and reverse transcription polymerase chain reaction. High levels of PACAP measured by radioimmunoassay are due to the great number of PACAP containing cell bodies or dense fiber network. PACAP immunoreactive cell bodies were observed in several hypothalamic nuclei and in various extrahypothalamic structures, mainly in the limbic system. PACAP is also present in the retinohypothalamic tract and in the centrifugal visual system. In the periphery, PACAP was demonstrated in the sensory, motor, and autonomic nervous systems. Autonomic PACAP immunoreactive fibers innervate glands and smooth muscle layers of viscera and vessels. This chapter summarizes the most important data available in the literature up to now.

Keywords Immunohistochemistry • In situ hybridization • RIA • RT-PCR • S-EIA

Abbreviations

ARC	Arcuate nucleus
BBB	Blood–brain barrier
BDNF	Brain-derived neurotrophic factor/nerve growth factor
BNST	Bed nucleus of stria terminalis
CGRP	Calcitonin gene-related peptide
ELISA	Enzyme-linked immunosorbent assay
FG	Fluoro-Gold
HPLC	High-performance liquid chromatography

K. Koves (✉)
Department of Anatomy, Histology and Embryology, Faculty of Medicine, Semmelweis University, Tűzoltó u. 58, Budapest 1094, Hungary
e-mail: koves.katalin@med.semmelweis-univ.hu

D. Reglodi, A. Tamas (eds.), *Pituitary Adenylate Cyclase Activating Polypeptide — PACAP*, Current Topics in Neurotoxicity 11,
DOI 10.1007/978-3-319-35135-3_12

icv	Intracerebroventricular
IHC	Immunohistochemistry
IML	Intermediolateral cell column
ip	Intraperitoneal
ISH	In situ hybridization
LC	Locus ceruleus
ME	Median eminence
OVLT	Organon vasculosum laminae terminalis
PACAP21-38	Pituitary adenylate cyclase activating polypepide with 18 amino acid residues
PACAP27	Pituitary adenylate cyclase activating polypepide with 27 amino acid residues
PACAP38	Pituitary adenylate cyclase activating polypepide with 38 amino acid residues
PACAP6-16	Pituitary adenylate cyclase activating polypepide with 11 amino acid residues
PCR	Polymerase chain reaction
Pe	Hypothalamic periventricular nucleus
Pf	Hypothalamic perifornical region
PVN	Hypothalamic paraventricular nucleus
RHT	Retinohypothalamic tract
RIA	Radioimmunoassay
RT-PCR	Reverse transcription polymerase chain reaction
SCN	Hypothalamic suprachiasmatic nucleus
S-EIA	Sandwich enzyme immunoassay
SON	Hypothalamic supraoptic nucleus
SP	Substance P
VIP	Vasoactive intestinal polypeptide
WGA	Wheat germ agglutinin

Introduction

After the discovery of PACAP [1, 2] several polyclonal antibodies were raised in rabbits against PACAP in Arimura's laboratory. Because the amino acid sequence of PACAP is the same in all mammalian species studied (ovine and human: [3]; rat: [4]; murine: [5, 6]), the antibodies against ovine PACAP were suitable for staining in all mammals. The antibodies were characterized using enzyme linked immunosorbent assay (ELISA) [7] and radioimmunoassay (RIA) [8]. Antibody No. 88121-3 was the most potent for immunohistochemistry (IHC). This antibody was raised against PACAP27. It did not cross-react with other peptides present in the hypothalamus and with those showing similar amino acid sequence [7]. This antibody was widely used by the research community [9–16]. The most potent antibody for RIA was antibody No. 88111-3, raised against ovine PACAP38 and directed to the C-terminal portion of the peptide [8, 17, 18]. Other researchers raised polyclonal rabbit antibodies against PACAP21-38 and PACAP27 for measuring PACAP in human central nervous system and rat tissues [19]. Piggins and his coworkers [20] purchased PACAP antibody from Penninsula (code No. IHC8920 and IHC8922) and used it for IHC. Hannibal and his coworkers [21] raised a monoclonal mouse antibody against PACAP6-16 (code MabJHH1) which was also suitable for IHC. A few years after its discovery, the presence of PACAP was demonstrated by sandwich enzyme immunoassay (S-EIA) [22, 23], by polymerase chain reaction (PCR) [22, 24, 25] and in situ hybridization (ISH) [21, 24, 26, 27] as well.

PACAP in the Central Nervous System

Immunohistochemical and In Situ Hybridization Findings

PACAP in the Hypothalamus

Because PACAP was first isolated from ovine hypothalami by screening for pituitary adenylate cyclase stimulating activity, we decided to collect sheep brain from a slaughterhouse. IHC revealed that PACAP containing neurons are present in the hypothalamic magnocellular nuclei including the paraventricular nucleus (PVN) and the perifornical region (Pf) (Fig. 12.1a), the supraoptic nucleus (SON), the periventricular nucleus (Pe) (Fig. 12.1b), the anterior commissural nucleus, and many immunoreactive fibers were also detected in the median eminence (ME). In adrenalectomized rats, the fibers were seen in both layers of the ME (Fig. 12.1c) and posterior pituitary and in intact rats only in the internal zone and posterior pituitary (Fig. 12.1d, e). In intact cats, PACAP immunoreactivity was always present in both layers of the ME and also at the periphery of the posterior pituitary (Fig. 12.1f, g). A dense fiber network was also observed in the suprachiasmatic nucleus (SCN) of intact sheep (Fig. 12.1h) and not of intact, but of colchicine treated rats (Fig. 12.1i).

PACAP immunoreactivity in the hypothalamic-hypophyseal magnocellular system was also demonstrated in human, spider monkey [28] and cat brains [29] without any pretreatment. In intact rats the level of PACAP has to be relatively low because we could get good quality staining in rats in which the neuronal transport was inhibited by intracerebroventricular (*icv*) colchicine pretreatment or hypophysectomy [11]. Adrenalectomy also enhanced the intensity of immunostaining [30]. PACAP immunoreactive fibers in the external zone have to be the axons of hypophysiotropic neurons. In our experiment we proved this hypothesis [31] by injecting Fluoro-Gold (FG) tracer intraperitoneally (*ip*). This tracer is known to enter the blood stream and then the central nervous system where the blood–brain barrier (BBB) is missing; that is, through the ME, posterior pituitary, vascular organ of lamina terminalis (OVLT), and other circumventricular organs [32]. In control animals, FG spreading in a retrograde manner labeled cell bodies in several hypothalamic regions including Pe, arcuate nucleus (ARC), SON, and both parvocellular and magnocellular portions

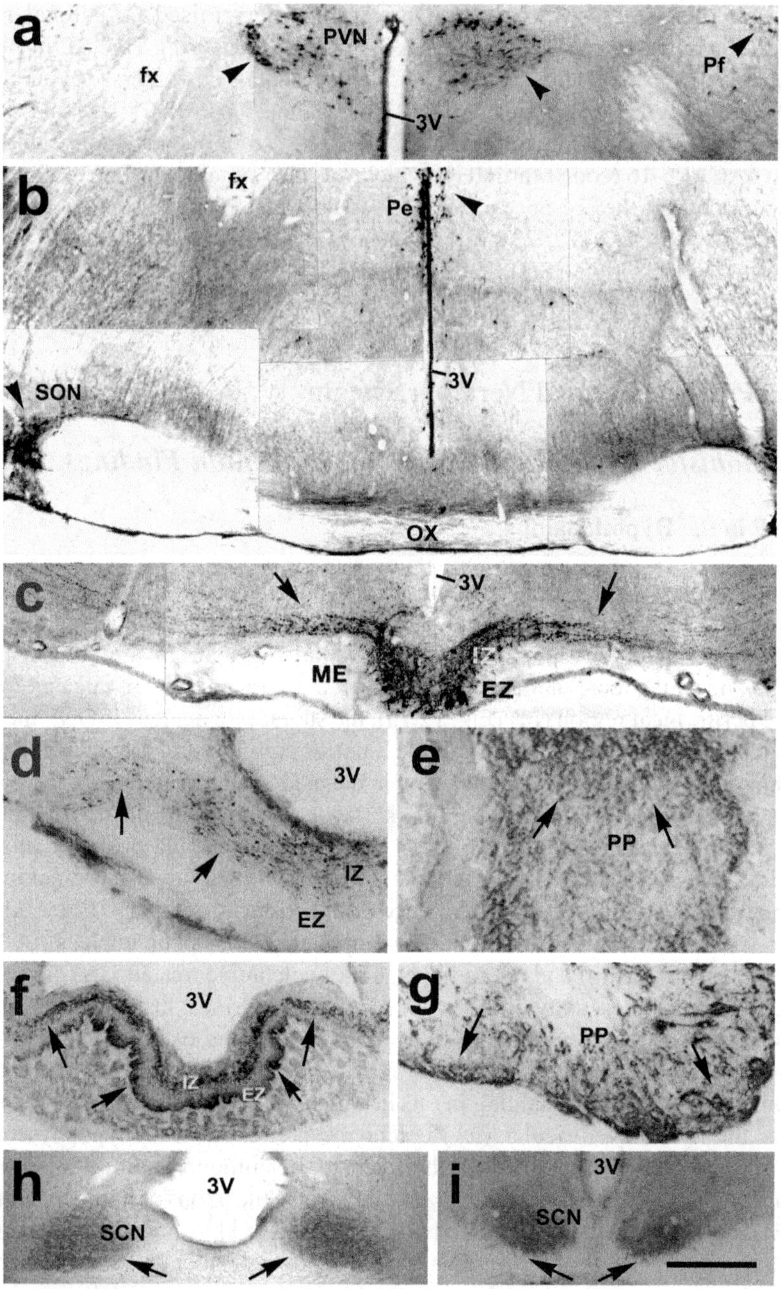

Fig. 12.1 Microphotographs demonstrating the distribution of PACAP immunoreactivity in frontal sections of the hypothalamus and median eminence of a Sprague-Dawley male rat 3 weeks after adrenalectomy (**a**–**c**), in the median eminence and posterior pituitary of an intact rat (**d**, **e**) and an intact cat (**f**, **g**) and in the suprachiasmatic nucleus of an intact sheep and colchicine treated rat (**h**, **i**).

of the PVN. In pituitary stalk-sectioned rats (a 2-mm-wide piece of razor blade was implanted in the internal lamina of the base of the skull at 5.0 mm in front of the ear bar), FG could reach the hypothalamus only through the capillary loops of the ME because the blade prevented the tracer from entering the hypothalamus through the posterior pituitary. In this latter model, the parvocellular nuclei and the ventral portion of the magnocellular part of PVN were labeled. The SON was completely empty. Double labeling revealed that in the Pe and the ventral magnocellular portion of the PVN, PACAP immunoreactive cells contained FG. These neurons are hypophysiotropic. It was also demonstrated that not only the hypothalamic PACAP regulates the pituitary functions, but PACAP produced by the anterior pituitary itself functions as an autocrine and paracrine factor [25].

PACAP in the Extrahypothalamic Regions

In colchicine treated rats PACAP immunoreactive neuronal cell bodies and fibers were also observed in many extrahypothalamic regions [11, 33]. In our laboratory we have observed a very dense fiber network in the central medial, central lateral, and paraventricular nuclei of the thalamus (Fig. 12.2a), the lateral habenula (Fig. 12.2b), the central amygdala (Fig. 12.2c), the lateral septum (Fig. 12.2d, e), the bed nucleus of stria terminalis (BNST) (Fig. 12.2f), and the dorsal endopiriform nucleus (Fig. 12.2g). Immunoreactive cells were observed in the lateral amygdala (Fig. 12.2c), hippocampus (Fig. 12.2h), dentate gyrus (Fig. 12.2i), cingulate and entorhinal cortices. In the lateral septum the PACAP immunoreactive fibers lined out neuronal cell bodies (Fig. 12.2e). With the use of semiquantitative ISH Mikkelsen and his colleagues [34] found PACAP mRNA in the cingulate and frontal cortices. The hybridization signal was observed in low concentrations in other cortical areas as well. The signal was detected over cell bodies in cortical layers II, III, and IV and occasionally over the cell bodies in the outer pyramidal layer.

Fig. 12.1 (continued) ABC technique with nickel intensification. (**a**) PACAP immunoreactive cell bodies are seen mainly in the magnocellular subdivision of the PVN. A few immunopositive cell bodies are also present in the parvocellular subdivision of PVN and in the lateral hypothalamus. (**b**) PACAP cell bodies in the Pe and SON. (**c**) Dense PACAP fibers in the median eminence. (**d, e**) PACAP fibers in the internal zone of the median eminence and the posterior pituitary of an intact rat. The axons are fine and beaded. (**f**) PACAP fibers in both layers of the median eminence of an intact cat and (**g**) PACAP fibers in the posterior pituitary of the same animal. (**h**) Dense fiber network in the SCN of an intact sheep and (**i**) of a colchicine treated rat. This nucleus is the termination of the retinohypothalamic tract. *Arrowheads* show immunoreactive cell bodies, *arrows* show fibers. *3V* third ventricle, *EZ* external zone, *fx* fornix, *IZ* internal zone, *ME* median eminence, *OX* optic chiasm, *Pe* periventricular nucleus, *Pf* perifornical area, *PP* posterior pituitary, *PVN* paraventricular nucleus, *SCN* suprachiasmatic nucleus, *SON* supraoptic nucleus. Scale: 800 μm in **a–c**; 300 μm in **d** and **e**; 900 μm in **f** and **g**; 500 μm in **h** and **i**

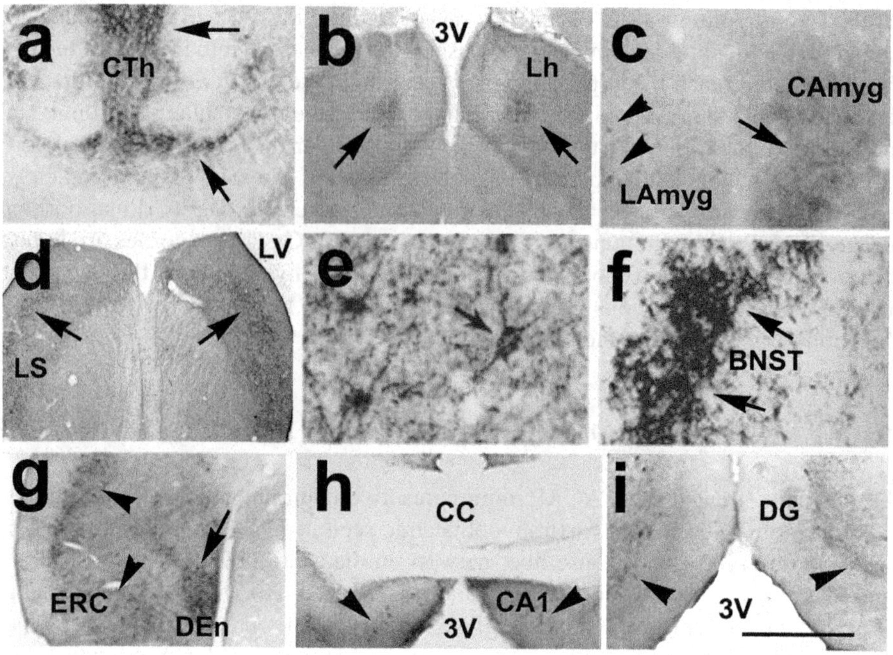

Fig. 12.2 Microphotographs demonstrating the distribution of PACAP immunoreactive cell bodies and fiber networks in the extrahypothalamic structures in frontal sections of the forebrain of a Sprague-Dawley male rat (**a**–**i**). ABC technique with nickel intensification. (**a**) Fiber network is seen in the central thalamus and (**b**) in the medial part of the lateral habenula. (**c**) In the central amygdala there is a dense fiber network and in the lateral amygdala scattered cell bodies are seen. (**d**) In the lateral septum many immunoreactive fibers are present. (**e**) High power detail of d. Immunoreactive fibers surround perikarya and dendrites of unlabeled cells. (**f**) Dense fiber network in the BNST. (**g**) In the entorhinal cortex many immunoreactive cells are seen and many fibers form a network deep in the endopiriform nucleus. (**h**, **i**) Immunoreactive cells in the hippocampal formation (CA1 region of the hippocampus and the dentate gyrus). *Arrowheads* show immunoreactive cell bodies, *arrows* show immunoreactive fibers. *3V* third ventricle, *BNST* bed nucleus of stria terminalis, *CAmyg* central amygdala, *CC* corpus callosum, *CTh* central thalamus, *DEn* endopiriform nucleus, *DG* dentate gyrus, *ERC* entorhinal cortex, *LAmyg* lateral amygdala, *Lh* lateral habenula, *LH* lateral hypothalamus, *LS* lateral septum, *LV* lateral ventricle. Scale: 2000 μm in **a** and **d**; 1500 μm in **b**; 1000 μm in **c**, **h**, and **i**; 100 μm in **e**–**g**

PACAP in the Retino-hypothalamic Tract (RHT) and the Centrifugal Visual System

In intact rats, we did not see PACAP immunoreactive cells in the retina; however, five days after the interruption of the optic nerve, PACAP immunoreactivity appeared in the retinal ganglion cells. To find evidence of whether the axons of these PACAP immunoreactive cells form a terminal-like fiber network in the SCN or not, a retrograde tracer was injected into the SCN. Two days later the optic nerve was transected. The tracer colocalized with PACAP immunoreactivity. This clearly shows that PACAP is present in the RHT [35]. Later, Hannibal and his coworkers [36] confirmed our results and elaborated on the role of PACAP in circadian rhythms. Furthermore, immunohistochemical studies revealed that PACAP-positive nerve fibers were present in the nerve fiber layer and positive cell bodies in the ganglion cell layer. Positive nerve fibers were also seen in the inner plexiform layer. PACAP-positive neuronal cell bodies were found in amacrine and horizontal cell layers and in the inner nuclear layer. No PACAP positivity was seen in the photoreceptors or retinal pigmented epithelium [13, 37].

With the use of monoclonal antibody, Hannibal's group demonstrated wider distribution of PACAP immunoreactivity [21, 27, 36, 38]. Their results showed that PACAP-containing nerve fibers were present in the visual system (the ventral part of the lateral geniculate nucleus and the nucleus of the optic tract) and in the termination of the retinohypothalamic tract (right and left sides of the SCN and dominantly the contralateral intergeniculate leaflet of the thalamus, the olivary pretectal nucleus, the ventrolateral preoptic nucleus, the anterior and lateral hypothalamic area, and the subparaventricular zone). Retinal projections, not previously described in rats, also contained PACAP (the lateral posterior nucleus of thalamus, the posterior limitans nucleus, the dorsal part of the anterior pretectal nucleus and the posterior and medial pretectal nuclei, and few in the superior colliculus) [39].

In our laboratory we demonstrated that PACAP is also present in a reverse connection between the central nervous system and the retina (centrifugal visual system) [40]. It was found that BDA injected into the vitreous body of the eye of rats was transported in ante- and retrograde manners. A considerable number of retrogradely labeled nerve cell bodies were seen in many structures such as the dentate gyrus (DG), CA1 and CA3 regions of the hippocampus, the SON and the PVN, the habenular complex, the indusium griseum, and the olfactory tubercle. The injection of tracer to one eye resulted in cell body labeling at both sides of the forebrain. Selected sections showing BDA labeling were stained for PACAP. Some BDA-labeled cells also showed PACAP immunoreactivity in the SON and the dentate gyrus. It was concluded that PACAP is present not only in the RHT, but in the centrifugal visual system as well.

PACAP in the Cerebellum

PACAP was also seen in the cerebellar Purkinje cells [41].

Figure 12.3 schematically illustrates the distribution of PACAP in frontal sections of the forebrain using a polyclonal rabbit antibody (a–d) and in situ hybridization (e–h).

PACAP in the Brainstem

PACAP immunoreactive cell bodies and fibers were also observed in brainstem structures in rats and cats using IHC [42, 43]. The following structures were PACAP immunoreactive: the dorsomedial and ventrolateral cell columns of the motor nuclei of cranial nerves, the primary somatosensory cells in the mesencephalic nucleus of the trigeminal nerve and the central axons of the branchial cranial nerves in the

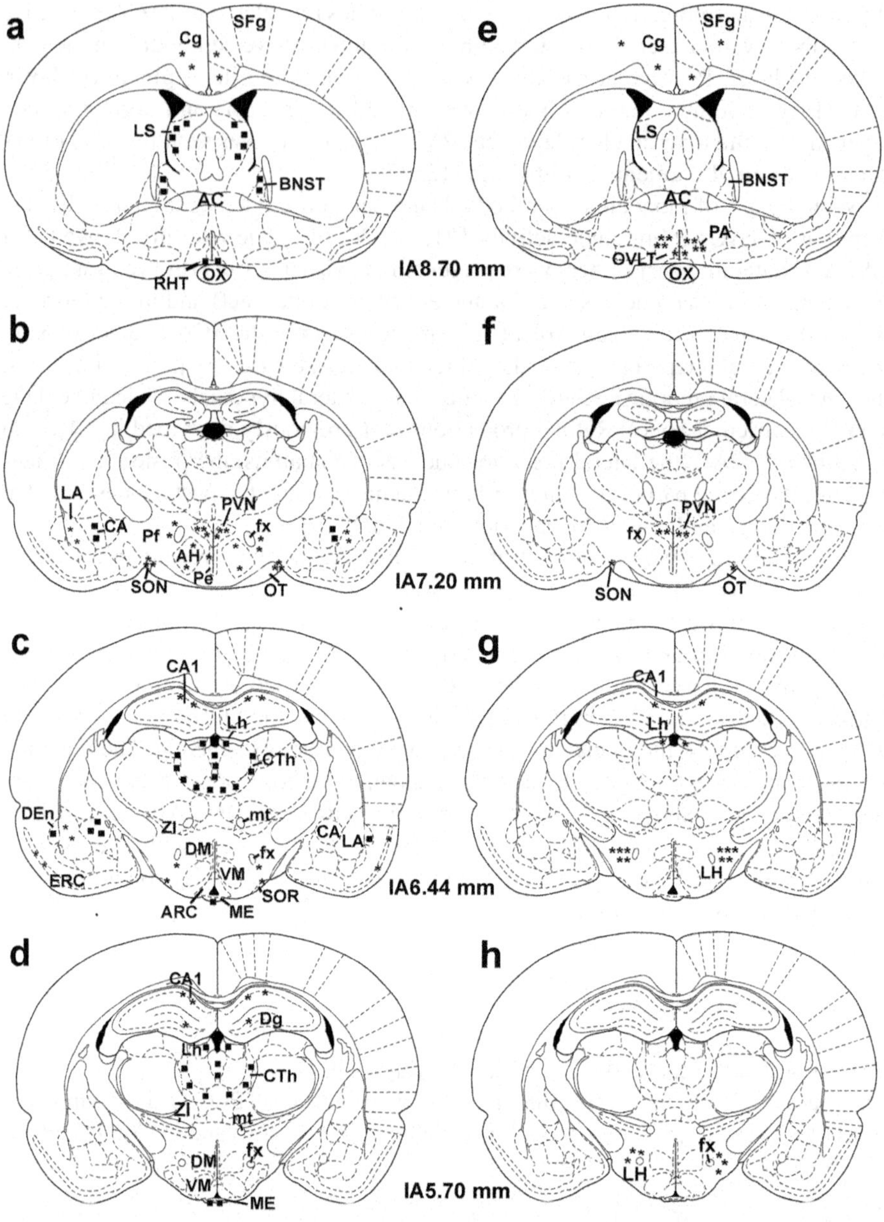

Fig. 12.3 Schematic illustration of the distribution of PACAP immunoreactive cells and fibers and PACAP mRNA in four representative frontal sections of the forebrain according to Paxinos and Watson's stereotaxic coordinates [109]. *Asterisks* show occurrence of PACAP immunoreactive cell bodies or PACAP mRNA and squares show fibers. **a–d** summarize the immunohistochemical data of Köves et al. [11], Kivipelto et al. [10] and Tamada et al. [15], **e–h** summarize the occurrence of PACAP mRNA obtained by Mikkelsen et al. [34], Hannibal [27], and Moore et al. [24] using in situ hybridization. *3V* third ventricle, *AC* anterior commissure, *AH* anterior hypothalamus, *ARC* arcuate

spinal trigeminal tract, visceral afferent fibers in the solitary tract and cell bodies in the dorsal motor nucleus of the vagus, second and third order sensory neurons of the cochlear and vestibular systems, scattered fibers in various regions of the brainstem and well-defined fiber bundles in the interpeduncular area, scattered cell bodies in the red nucleus, the substantia nigra, in some cell groups of the reticular formation and in the raphe nuclei, as well as in the pontine dorsolateral tegmentum and the locus ceruleus (LC). Figure 12.4 demonstrates PACAP immunoreactive cell bodies and fiber networks in some representative areas of the brainstem. Figure 12.5 schematically illustrates the distribution of PACAP in cross sections of the brainstem.

Results Obtained by RIA and S-EIA

A heterologous RIA method for PACAP38 and a homologous RIA method for a shorter form of PACAP27 revealed that in rat the highest concentration of radioimmunoassayable PACAP38 was found in the hypothalamus, but other brain regions also contained considerable amounts of PACAP38 (extrahypothalamic structures and brainstem). The PACAP38 concentration found in the posterior pituitary was comparable with that in the extrahypothalamic brain [8]. Ghatei and his coworkers [19], using an antibody to PACAP21–38, examined several extrahypothalamic regions of human brains. The highest level of PACAP immunoreactivity (similar to the hypothalamus) was found in the accumbens nucleus, septum, amygdala, striatum, and the globus pallidus. Moderate levels were found in many other regions (frontal cortex, cerebellum, hippocampus, brainstem, sacral spinal cord). In this study the authors could not detect PACAP in the pituitary gland and in several peripheral tissues such as pancreas, lung, heart, kidney, and spleen.

A Japanese research group measured PACAP level by S-EIA in 62 brain regions of rats [23]. Two antibodies were used in the assay: PA-6N (directed to the N terminal portion) and PA-2C (directed to the C terminal portion). In the forebrain, the highest level of PACAP was found in the hypothalamus (SON, PVN, Pe, ARC, and SCN), septum, BNST, and the amygdala. In the brainstem, a considerable amount was measured in the central gray matter, interfascicular, interpeduncular and dorsal raphe nuclei. During ontogenesis PACAP levels were quite low at postnatal day 2,

Fig. 12.3 (continued) nucleus, *BNST* bed nucleus of stria terminalis, *CA* central amygdala, *CA1* hippocampal region, *Cg* cingulate gyrus, *CTh* central thalamus, *Den* dorsal endopiriform nucleus, *Dg* dentate gyrus, *DM* dorsomedial nucleus, *ERC* entorhinal cortex, *fx* fornix, *IA* interaural line, *LA* lateral amygdala, *Lh* lateral habenula, *LH* lateral hypothalamus, *LS* lateral septum, *ME* median eminence, *mt* mammillothalamic tract, *OT* optic tract, *OVLT* organon vasculosum laminae terminalis, *OX* optic chiasm, *PA* preoptic area, *Pe* periventricular nucleus, *Pf* perifornical region, *PVN* paraventricular nucleus, *RHT* retinohypothalamic tract, *SFg* superior frontal gyrus, *SON* supraoptic nucleus, *SOR* retrochiasmatic portion of supraoptic nucleus, *VM* ventromedial nucleus, *ZI* zona incerta

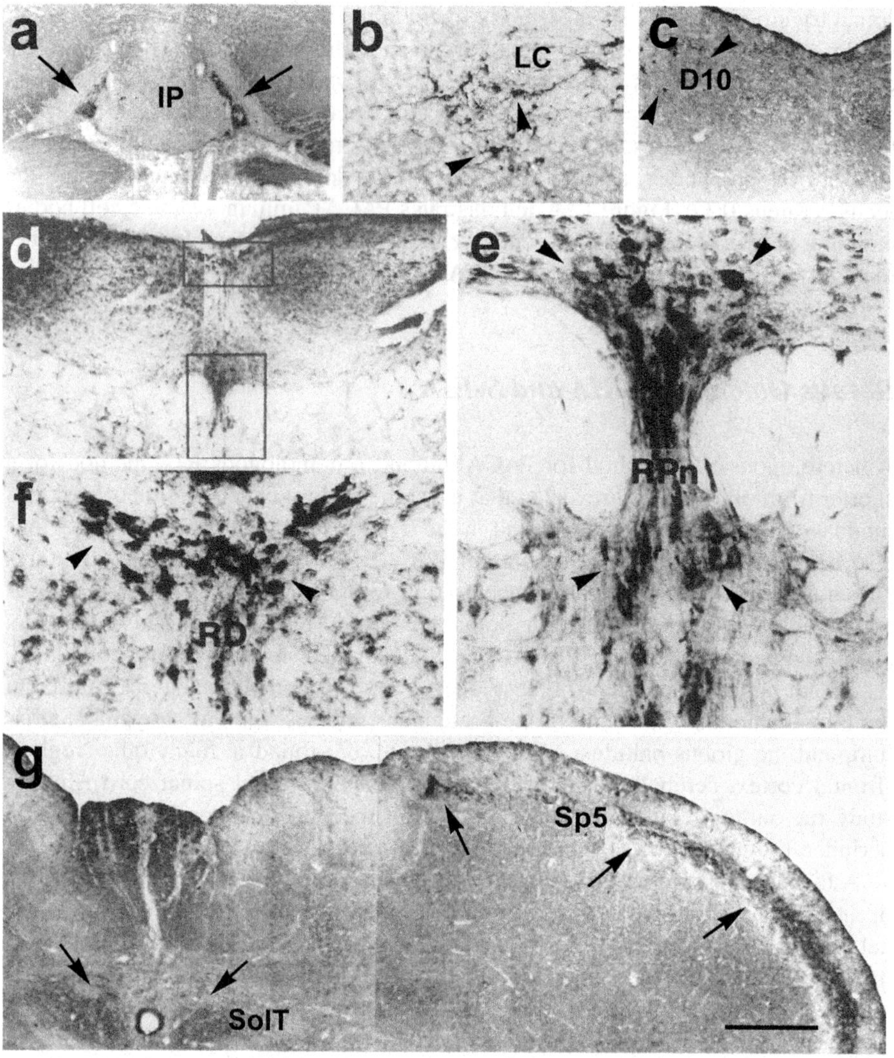

Fig. 12.4 Microphotographs demonstrating PACAP immunoreactive cell bodies and fibers in the cross sections of rat brainstem. (**a**) A dense fiber network is seen at both sides of the interpeduncular nucleus. (**b**) Immunoreactive cell bodies in the locus ceruleus and (**c**) in the dorsal nucleus of vagus. (**d**) In the pons two areas in the rectangles are seen with high magnification in e and f. (**e**) Immunoreactive cells in the raphe dorsalis and (**f**) in the raphe pontis. (**g**) Immunoreactive fibers in the spinal trigeminal tract formed by the central axons of the primary sensory neurons located in the trigeminal ganglion. *D10* dorsal motor nucleus of vagus, *IP* interpeduncular nucleus, *LC* locus ceruleus, *SolT* solitary tract, *SP5* spinal trigeminal tract, *RD* raphe dorsalis, *RPn* raphe pontis. Scale: 1000 μm in **a**, **d**, and **g**; 120 μm in **b** and **c**; 75 μm in **e** and **f**

peaked at 30–60 days, and then remained constant in most regions during adulthood [44]. Palkovits and his coworkers [45] mapped PACAP content in 79 regions of the adult human brain in samples taken by microdissection [46]. PACAP was measured by RIA. The highest concentrations were found in the hypothalamus (PVN, ME,

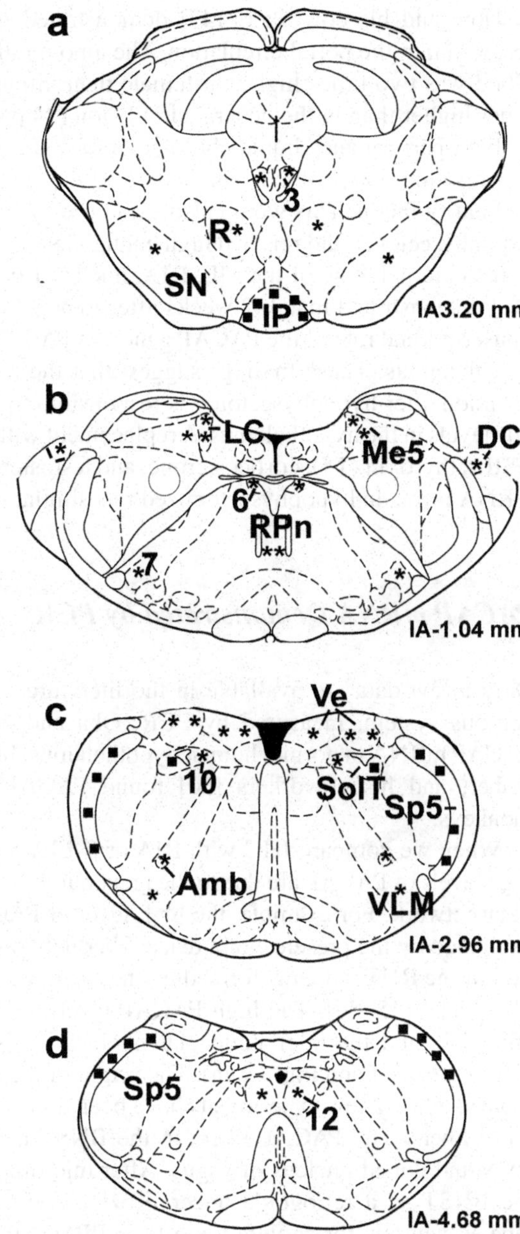

Fig. 12.5 Schematic illustration of the distribution of PACAP immunoreactive cells and fibers in the frontal sections of the brainstem according to Paxinos and Watson's stereotaxic coordinates [109]. *Asterisks* show occurrence of PACAP immunoreactive cell bodies and squares show fibers. (**a–d**) summarize the immunohistochemical data obtained by Kausz et al. [42] and Légrádi et al. [43]. Three motor nucleus of oculomotorius; 6 motor nucleus of abducens; 7 motor nucleus of facialis; 10 dorsal vagus nucleus; 12 motor nucleus of hypoglossus; *Amb* ambiguus nucleus, *DC* dorsal cochlear nucleus, *IP* interpeduncular nucleus, *LC* locus ceruleus, *Me5* mesencephalic nucleus of vagus, *R* red nucleus, *RPn* raphe pontis, *SN* substantia nigra, *SolT* solitary tract, *Sp5* spinal trigeminal tract, *Ve* vestibular nuclei, *VLM* ventrolateral medulla

pituitary stalk, Pe, and BNST) and in the vagal complex of the brainstem, although PACAP was also measurable in other regions such as the septum pellucidum, SON, ventromedial nucleus, periaqueductal and spinal gray matters, the motor nucleus of facial nerve, and in the spinal nucleus of the trigeminal nerve.

Dow and his coworkers [47] demonstrated for the first time, that PACAP is released into the portal circulation. The amount of PACAP38 in hypophysial portal blood was two times higher in female than in male rats (107 vs. 54 pM) and four times higher than in the general circulation (24 pM) measured by RIA. As assessed by high-performance liquid chromatography (HPLC), PACAP38 is the major form found in the portal blood. This indicated that PACAP38 might be a hypothalamic regulatory factor. This would also explain why PACAP accumulated in fibers after hypophysectomy and reached high enough levels to be detected by immunostaining in the external zone of the ME. RIA, nuclear run-on, and RNase protection assays [48] demonstrated that 1–2 weeks after hypophysectomy, the level of PACAP, the transcriptional rate of the PACAP gene and PACAP mRNA content decreased in the hypothalamus. These findings suggest that the reduced rate of PACAP gene transcription after hypophysectomy is responsible for the decrease of mRNA and peptide levels in the hypothalamus. A replacement with growth hormone (GH), prolactin (PRL), thyroxin (T4), corticosterone, and testosterone significantly restored PACAP mRNA levels in hypophysectomized rats similar to those in control animals.

PACAP mRNA Demonstrated by PCR

Only a few data are available in the literature concerning PACAP mRNA in the nervous system, measured by PCR. Oka and his coworkers [49] demonstrated PACAP mRNA in normal human hypothalamus, but not in normal anterior pituitary. Aubert and his coworkers [50] found PACAP in cerebellar Purkinje cells of monkeys.

When we compare IHC with RIA or S-EIA results, it becomes evident that the high levels of PACAP are due to the great number of PACAP cell bodies or the dense fiber network. For example, the high level of PACAP in the SCN (Fig. 12.1h, i) is explained by the presence of a dense fiber network [23]. This is the major termination of the RHT which is formed by the axons of the PACAP immunoreactive ganglion cells [35, 36]. The high PACAP level in the septum, the BNST, the central amygdala and the interpeduncular nucleus is also due to a dense fiber network (Fig. 12.2c, d, f and Fig. 12.4a). The origin of these PACAP fibers is not well documented. Only a few experiments have been done which directly certify the origin of these fibers. The PACAP fibers in the BNST are derived from the hypothalamic PVN and dorsal nucleus of vagus. After microinjecting cholera toxin B subunit in the BNST as a retrograde tracer, Kozicz and his coworkers [51] found double labeled neurons for cholera toxin B and PACAP in the PVN and the dorsal vagal complex. We can only speculate what the origin of PACAP fibers is in areas such as the septum, central amygdala, the interpeduncular nucleus, lateral habenula, and central thalamus. Our hypothesis is based on previous tract tracing studies. The lateral septum is very heavily innervated. These fibers may come from different sources [52]. After the transection of the fornix, many degenerated terminals were observed in the lateral septum. These fibers were connected to GABAergic neurons [53]. These hippocampo-septal fibers may contain PACAP since many PACAP immunoreactive cells were demonstrated in the hippocampus [11]. Iontophoretic application of horseradish peroxidase in the lateral septum revealed retrogradely labeled perikarya in the PVN, LC, and raphe nuclei [54]. All these regions contain PACAP immunoreactive cell bodies. The origin of PACAP fibers in the central amygdala may be the PACAP neurons located in the basolateral nucleus of the amygdala, since some efferents from this nucleus project to the central amygdala [55]. PACAP fibers present in the interpeduncular nucleus may originate from the dorsal tegmental nucleus (nucleus incertus in cat), the raphe nuclei and the LC because FG injected into the interpeduncular nucleus resulted in retrogradely labeled neurons in the abovementioned regions [56], where PACAP cell bodies were also described. Gruber and his colleagues [57] revealed that the origin of a dopaminergic fiber network in the lateral habenula (showing similar localization to the PACAP fiber network) is the ventral tegmental area of the mesencephalon where labeled cell bodies were seen. The study was done by a retrograde tracing approach. Gold-coupled wheat germ agglutinin (WGA) was injected in the medial portion of the lateral habenula. Labeling was also observed in the incertohypothalamic and Pe nuclei. The presence of PACAP cell bodies in the hypothalamic Pe is well documented. These cells may also send fibers to the lateral habenula. By utilization of anterograde transport of cholera toxin subunit B, direct retinal projections to midline and intralaminar thalamic nuclei were demonstrated in marmoset brains. As mentioned above, PACAP immunoreactive retinal ganglionic cells send fibers to many regions of the forebrain [58, 59].

PACAP in the Peripheral Nervous System

Immunohistochemical and In Situ Hybridization Findings

The presence of PACAP in the peripheral nervous system was also demonstrated by IHC [12, 14, 16, 60–68] and ISH [69].

PACAP in the Sensory Nervous System

In the sensory ganglia of the cranial nerves and in the spinal ganglia, small size pseudounipolar neurons were PACAP immunoreactive [60, 70–72]. The central axons of these sensory neurons enter the brainstem and the dorsal horn of the spinal cord where they form a dense network. Figure 12.6 shows PACAP immunoreactive cell bodies in cat trigeminal ganglion and immunoreactive fibers in the spinal trigeminal tract. Similarly, a dense fiber network was demonstrated in the dorsal horn of the spinal cord [71–73]. The mesencephalic nucleus of the trigeminal nerve also contained PACAP immunoreactive cell bodies as mentioned above [42, 74]. This nucleus is composed of pseudounipolar cells analog to primary sensory neurons. With the use of ISH, PACAP mRNA was also demonstrated in the primary sensory ganglia [63].

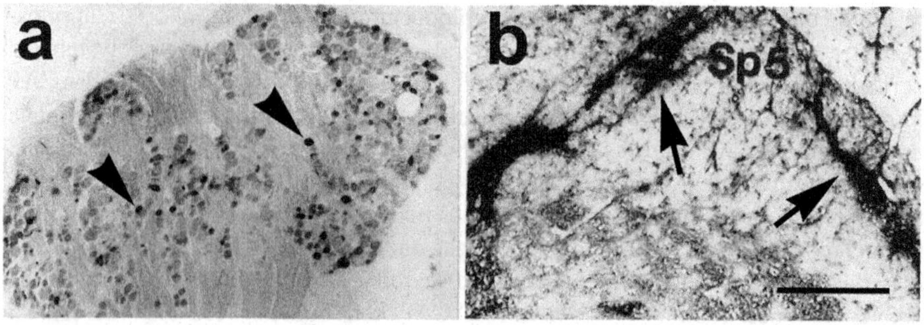

Fig. 12.6 Microphotographs showing PACAP immunoreactive cell bodies in the trigeminal ganglion (**a**) and immunoreactive fibers in the spinal trigeminal tract (Sp5) of a domestic cat (**b**). *Arrowheads* indicate cell bodies, *arrows* indicate fibers. Scale: 300 μm

After nerve injury mRNA level is upregulated and appears in the superficial layer of the dorsal horn and in some motor neurons in the ventral horn [75].

PACAP immunoreactive fibers were described in many regions of the head which are innervated by the trigeminal nerve. Elsås and his coworkers [9] described numerous PACAP-immunoreactive nerve fibers in the lacrimal gland and choroid which are partially sensory. They found a sparse supply of PACAP-containing nerve fibers in the iris, ciliary body, and conjunctiva as well. Baeres and Møller [76] described PACAP immunoreactive fibers in the pineal organ. These fibers originate from the trigeminal ganglion. Kausz and her coworkers [60] using a retrograde tracing technique and immunohistochemistry in cats and colchicine treated rats demonstrated that small and middle sized PACAP immunoreactive neurons located in the trigeminal ganglion innervate various regions of the forehead, eye, chin, mucosa of tongue, and muscles of mastication and whiskers. PACAP immunoreactive fibers were described in oral structures including the gingiva, tooth pulp [77], tongue [71] and palate [78]. It is well established that the nodose ganglion of the vagus provides sensory fibers for the respiratory, cardiac and digestive systems. In this ganglion, PACAP is present and colocalizes with CGRP, SP and VIP. This suggests that PACAP is involved in nociceptive vagal transduction [79].

PACAP-positive nerve fibers are present in the hairy skin of rats, forming a subepithelial plexus from which delicate varicose nerve fibers enter the overlying epithelium [66]. PACAP-immunoreactive sensory fibers were also observed in the human skin close to the dermal–epidermal border and around the hair follicles [80]. Capsaicin induced inflammation enhanced the level of PACAP (measured by RIA) in mouse plantar skin [81]. Papka and his coworkers [82] determined the expression pattern of PACAP in the cervix and its nerves during pregnancy and the influence of estrogen on this expression using IHC, RIA, and RT-PCR. PACAP immunoreactivity was detected in nerves of the cervix, lumbosacral (L6-S1) dorsal root ganglia and in the spinal cord of rats. PACAP immunoreactivity was highest on day 15 of pregnancy in the cervix and dorsal spinal cord, but then decreased over the last trimester of pregnancy. However, levels of PACAP mRNA increased in the L6-S1 dorsal root ganglia during late pregnancy compared to early pregnancy. Dorsal root ganglia of ovariectomized rats treated with estrogen showed increased PACAP mRNA synthesis in a dose-related manner, an effect partially blocked by the estrogen receptor antagonist.

PACAP was demonstrated not only in pseudounipolar sensory neurons, but also in the bipolar sensory neurons in the spiral and vestibular ganglia [83]. PACAP was found in the spiral bundle and beneath the inner and outer hair cells within the organ of Corti. Furthermore, evidence was obtained that PACAP is present in type I afferent axons leaving the spiral ganglion en route to the auditory nerve.

PACAP in the Autonomic Nervous System

PACAP is also present in the autonomic nervous system. PACAP mRNA in adult rat spinal cord is limited to the preganglionic sympathetic neurons of the intermediolateral cell column (IML) [12]. PACAP immunoreactive cell bodies were also observed in the para- and prevertebral sympathetic ganglia. Wojtkiewicz and his colleagues [84] injected fast blue in the parotid gland. Fast blue labeled neurons in the superior cervical sympathetic ganglion showed PACAP immunoreactivity as well. PACAP was shown in IML of the Th9-12 spinal cord segment. These preganglionic neurons directly innervate the adrenal gland [85]. PACAP immunoreactive nerve fibers innervating blood vessels were found in the cartilage canals [86] and around blood vessels and sweat glands in human skin [80]. These latter fibers may originate from the paravertebral sympathetic ganglia. Baeres and her coworkers [87] investigated the innervation of subarachnoidal vessels using FG tracer and immunohistochemical identification of the labeled neurons. It was found that the PACAP immunoreactive sensory fibers derive from the trigeminal ganglion and very scarce PACAP immunoreactive parasympathetic fibers from the otic and sphenopalatine ganglia.

Many neuronal cell bodies in craniosacral preganglionic parasympathetic nuclei and neurons in intramural parasympathetic ganglia also store PACAP [67]. Mulder and his coworkers [12] demonstrated a considerable number of PACAP immunoreactive cell bodies in the otic and sphenopalatine ganglia; it was later confirmed by Uddman and his coworkers [68]. Elsås and his coworkers [9] found a few PACAP positive cells in the ciliary ganglion. It was also shown that PACAP fibers innervate salivary glands [84, 88]. All the three parasympathetic ganglia providing the innervation of the salivary glands contain PACAP immunoreactive cell bodies [12, 67, 68, 88]. Milfendereski and his coworkers [89] measured PACAP38 and 27 level in the parotid gland using RIA. The concentration of the former was about twice that of the latter. Parasympathetic denervation reduced the total contents of PACAP38 and PACAP27 of parotid gland by 23 % and 44 %, respectively. These observations clearly show that PACAP is present in the parasympathetic fibers.

The cardiac ganglia are distributed in different regions of the atria of a number of mammalian species, surrounding the sinoatrial node, around the roots of the venae cavae and pulmonary veins, in the interatrial septum, in the proximity of the atrioventricular node and in the auricular appendages. These ganglia receive PACAP immunoreactive preganglionic fibers from the dorsal nucleus of vagus [90]. RT-PCR revealed that cardiac ganglia expressed proPACAP transcripts and that these ganglia

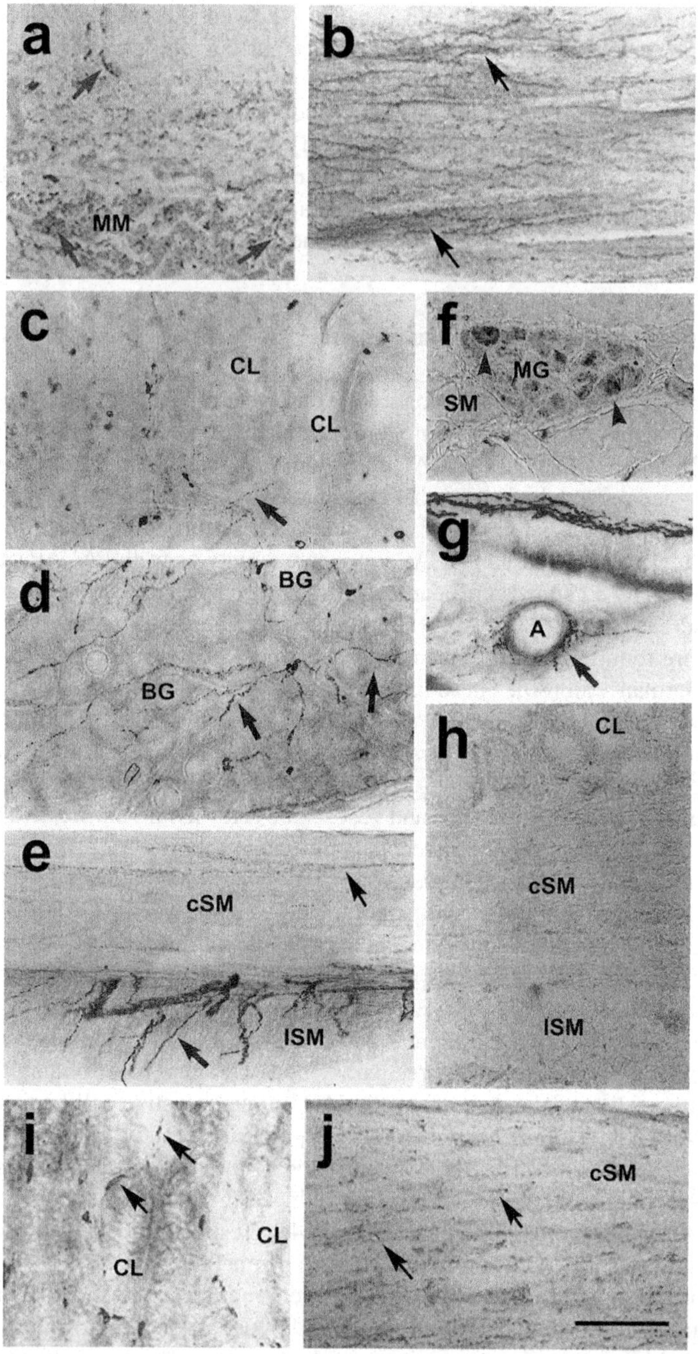

Fig. 12.7 PACAP immunoreactivity in ovine gastrointestinal tract. (**a**) Esophageal mucosa with muscularis mucosae. (**b**) Inner layer of external muscularis of stomach. (**c**) Duodenal mucosa. (**d**) Duodenal submucosa with the Brunner's glands. (**e**) Duodenal external muscularis. (**f**) Myenteric

have PACAP peptide biosynthetic capabilities. The in vivo sources of PACAP were localized immunocytochemically to neuronal fibers and a subpopulation of intrinsic postganglionic cardiac neurons [91].

During the last three decades it became evident that it is a general rule that PACAP immunoreactive nerve fibers innervate smooth muscle layers and glands in many organs. First PACAP fibers were demonstrated in the smooth muscle layers of the gastrointestinal tract [14, 62]. Later, PACAP immunoreactive nerve fibers were found in the respiratory tract [16], in the female [92] and male reproductive [93] and the urinary system [94] and in the smooth muscle layer of arteries [95]. In the abovementioned experiments the source of PACAP fibers innervating the smooth muscle layers was also investigated. It was previously demonstrated that capsaicin treatment of newborn rats resulted in the degeneration of primary sensory neurons involved in the mediation of pain. Capsaicin induced degeneration of unmyelinated primary afferent fibers, terminating in Rexed's laminae I and II of the spinal cord, left the parasympathetic fibers to further exist [96]. These fibers are the axons of intramural ganglia in the viscera [14] or the cranial ganglia in the head regions [67].

Sundler and his coworkers [14] mapped PACAP immunoreactivity in chicken, mouse, rat, hamster, guinea pig, ferret, cat, pig, sheep, and human stomach, small and large intestines. It was generally found that the number of PACAP immunoreactive nerve cell bodies in the myenteric ganglia is greater than in the submucosal ganglia, and the number of PACAP immunoreactive fibers is greater in the smooth muscle layers than in the mucosa. Portbury and his colleagues [65] determined the projections of PACAP-immunoreactive neurons in guinea pig small and large intestines. PACAP fibers were seen in the myenteric and submucous plexus, in the longitudinal and circular muscle layers and around blood vessels of the submucosa throughout the gut. In both, the small and large intestines, PACAP cell bodies were seen in the myenteric ganglia following colchicine treatment. Myotomy (cutting through the muscle layer) and myectomy (removal a piece of muscle layer) revealed that PACAP interneurons projected caudally in the ileum and colon. Myectomy resulted in a loss of PACAP fibers in the circular muscle below the surgery, whereas PACAP fibers remained in the submucosa and around blood vessels. Following extrinsic denervation of the ileum, the number of PACAP fibers in the submucosal ganglia and around blood vessels decreased. This suggests that a portion of PACAP fibers supplying the submucosal ganglia and blood vessels have an extrinsic source. These fibers may be parasympathetic from the vagus and sensory from the spinal nerves. Ny and coworkers [97] revealed PACAP innervation of the lower esophageal sphincter where PACAP has a relaxing effect.

Figure 12.7 shows representative photos from the ovine digestive system, demonstrating PACAP immunoreactivity in the esophagus, stomach, small and

Fig. 12.7 (continued) ganglion in the duodenum. (**g**) A PACAP innervated artery in the subperitoneal layer of the duodenum. (**h**) Specificity test. (**i**) mucosa of the colon. (**j**) external muscle layer of the colon. For specificity test PACAP antibody was removed with solid phase adsorption method [110]. *A* artery, *BG* Brunner's gland, *CL* Crypt of Lieberkühn, *cSM* circular smooth muscle layer, *MG* myenteric ganglion, *MM* muscularis mucosae, *lSM* longitudinal smooth muscle, *SM* smooth muscle. *Arrowheads* indicate PACAP immunoreactive cell bodies, *arrows* indicate immunoreactive fibers. Scale: 150 μm

large intestines. We found PACAP immunoreactive fibers in all layers, and PACAP immunoreactive neuronal cell bodies in the intramural ganglia of the submucosal and myenteric plexus.

A moderate number of PACAP immunoreactive nerve fibers were seen in the smooth muscle layer of bronchi, vessels and around seromucous glands of the respiratory tract [16, 61]. With the use of RIA, both PACAP27 and PACAP38 were found throughout the genital tract and the ovaries of human females. The predominant form was PACAP38 [98]. Immunocytochemistry revealed that PACAP was located in delicate varicose nerve fibers that were most abundant in the internal cervical os, where they mainly seemed to innervate blood vessels and smooth muscle cells. PACAP induced concentration-dependent smooth muscle relaxation [92]. Fahrenkrug and his coworkers [99] further investigated the human genital tract. They examined samples of cervix, uterine body, fallopian tube (isthmus, ampulla, infundibulum), vagina, and ovary. With the use of RIA the highest PACAP level was found in the ovary and considerable level was present in the uterus and the tube. Immunohistochemistry confirmed the previous observations that PACAP immunoreactive fibers are found in the smooth muscle layers and in the wall of vessels, not only in the wall of the uterus, but in the parametrium as well. In the ovary, immunoreactive fibers were associated with the wall of vessels in the hilum. Skakkebaek and his coworkers [66] described PACAP fibers surrounding the lactiferous duct of the nipple and blood vessels of the mammary gland. A few delicate varicose fibers were observed between secretory alveoli. Immunohistochemical analysis revealed that besides PACAP-immunoreactive fibers in the dermis close to the dermal–epidermal border, hair follicles, blood vessels and sweat glands are also innervated by PACAP fibers [80]. In the male genital organ, RIA revealed that the level of PACAP38 is much higher than that of PACAP27. With the use of IHC in the human penil erectile tissue (corpus cavernosum), gracile PACAP fibers were found in moderate numbers along bundles of smooth muscle cells, and dispersed in the adventitia; and along the borders of the adventitia and media in arteries of various sizes. In vitro PACAP induced relaxation of the smooth muscle bundles [93].

PACAP in Motor Neurons

PACAP immunoreactivity was shown in the motoneurons of the cranial nerves. In colchicine treated rats the dorsomedial (III, IV, VI, and XII cranial nerve motor nuclei) and ventrolateral (V, VII and ambiguus nucleus) cell columns exhibited PACAP immunoreactivity using IHC [42]. PACAP mRNA expression was demonstrated by ISH in adult rat motor and sensory neurons after L4–6 lumbar nerve injury. The incidence and level of PACAP mRNA expression were dramatically reduced in response to intrathecal anti-brain derived nerve growth factor (BDNF) administration, indicating that this factor is able to induce PACAP expression [100].

Results Obtained by RIA

PACAP levels measured by RIA in the mouse stomach, small and large intestines are much higher than in the rat, hamster, and guinea pig. PACAP level is higher in the smooth muscle layer than in the mucosa [14].

Results Obtained by PCR

Boeshore and his coworkers [101] used DNA microarray technology to study the changes in gene expression in superior cervical ganglia within 48 h of transecting the postganglionic trunks of the adult rat. The expression of more than 200 known genes changed in the ganglion, and most of these genes were not previously associated with a response to injury. In contrast, only 10 genes changed following transection of the preganglionic cervical sympathetic trunk. Real-time RT-PCR analysis verified the upregulation of a number of the axotomy-induced genes, including PACAP. PACAP mRNA was found in the otic and sphenopalatine ganglia using RT-PCR method [68]. With the use of RT-PCR, Pettersson and her coworkers [100] found an elevated PACAP level after L4–L6 spinal nerve injury in the related sensory and motor neurons.

PACAP in Fluids Studied by RIA

PACAP was also found in the blood plasma and milk [102], in kidney [103] and lung and colon cancer homogenates [104]. In these fluids and tissues the source of PACAP was not identified. It may originate from the innervating nerve fibers or it may be due to the release of PACAP from endocrine cells.

The data concerning the distribution of PACAP in mammals were surveyed in previous reviews and book chapters [30, 105–108]. During the last decade several new morphological data were added to the previous ones. It seems that some techniques became more sensitive and some new ones are introduced in this research work.

References

1. Miyata A, Arimura A, Dahl RR, Minamino N, Uehara A, Jiang L, et al. Isolation of a novel 38 residue-hypothalamic polypeptide which stimulates adenylate cyclase in pituitary cells. Biochem Biophys Res Commun. 1989;164:567–74.
2. Miyata A, Jiang L, Dahl RD, Kitada C, Kubo K, Fujino M, et al. Isolation of a neuropeptide corresponding to the N-terminal 27 residues of the pituitary adenylate cyclase activating polypeptide with 38 residues (PACAP38). Biochem Biophys Res Commun. 1990;170: 643–8.

3. Kimura C, Ohkubo S, Ogi K, Hosoya M, Ithoh Y, Onda H, et al. A novel peptide which stimulates adenylate cyclase: molecular cloning and characterization of the ovine and human cDNAs. Biochem Biophys Res Commun. 1990;166:81–9.
4. Ogi K, Kimura C, Onda H, Arimura A, Fujino M. Molecular cloning and characterization of cDNA for the precursor of rat pituitary adenylate cyclase-activating polypeptide (PACAP). Biochem Biophys Res Commun. 1990;173:1271–9.
5. Okazaki K, Itoh Y, Ogi K, Ohkubo S, Onda H. Characterization of murine PACAP mRNA. Peptides. 1995;16:1295–9.
6. Yamamoto K, Hashimoto H, Hagihara N, Nishino A, Fujita T, Matsuda T, et al. Cloning and characterization of the mouse pituitary adenylate cyclase-activating polypeptide (PACAP) gene. Gene. 1998;211:63–9.
7. Köves K, Arimura A, Vigh S, Somogyvári-Vigh A, Miller J. Immunohistochemical demonstration of a novel peptide, pituitary adenylate cyclase activating polypeptide, in the ovine hypothalamus. Endocrinology. 1990;127:264–71.
8. Arimura A, Somogyvári-Vigh A, Miyata A, Mizuno K, Coy DH, Kitada C. Tissue distribution of PACAP as determined by RIA: highly abundant in the rat brain and testes. Endocrinology. 1991;129:2787–9.
9. Elsås T, Uddman R, Sundler F. Pituitary adenylate cyclase-activating peptide-immunoreactive nerve fibers in the cat eye. Graefes Arch Clin Exp Ophthalmol. 1996;234:573–80.
10. Kivipelto L, Absood A, Arimura A, Sundler F, Håkanson R, Panula P. The distribution of pituitary adenylate cyclase-activating polypeptide-like immunoreactivity is distinct from helodermin and helospectin-like immunoreactivities in the rat brain. J Chem Neuroanat. 1992;5:85–94.
11. Köves K, Arimura A, Görcs TJ, Somogyvári-Vígh A. Comparative distribution of immunoreactive pituitary adenylate cyclase activating polypeptide and vasoactive intestinal polypeptide in rat forebrain. Neuroendocrinology. 1991;54:159–69.
12. Mulder H, Uddman R, Moller K, Elsås T, Ekblad E, Alumets J, et al. Pituitary adenylate cyclase activating polypeptide is expressed in autonomic neurons. Regul Pept. 1995;59:121–8.
13. Seki T, Shioda S, Izumi S, Arimura A, Koide R. Electron microscopic observation of pituitary adenylate cyclase-activating polypeptide (PACAP)-containing neurons in the rat retina. Peptides. 2000;21:109–13.
14. Sundler F, Ekblad EA, Absood R, Hakanson R, Köves K, Arimura A. Pituitary adenylate cyclase activating polypeptide: a novel vasoactive intestinal polypeptide-like neuropeptide in the gut. Neuroscience. 1992;46:439–54.
15. Tamada Y, Tanaka M, Ichitani Y, Okamura H, Yanaihara N, Ibata Y. Pituitary adenylate cyclase-activating polypeptide (PACAP)-like immunoreactive neuronal elements in rat hypothalamus and median eminence with special reference to morphological background of its effect on anterior pituitary—light and electron microscopic immunocytochemistry. Neurosci Lett. 1994;180:105–8.
16. Uddman R, Luts A, Arimura A, Sundler F. Pituitary adenylate cyclase-activating peptide (PACAP), a new vasoactive intestinal peptide (VIP)-like peptide in the respiratory tract. Cell Tissue Res. 1991;265:197–201.
17. Jakab B, Reglodi D, Józsa R, Hollósy T, Tamás A, Lubics A, et al. Distribution of PACAP-38 in the central nervous system of various species determined by a novel radioimmunoassay. J Biochem Biophys Methods. 2004;61:189–98.
18. Takahashi K, Totsune K, Murakami O, Satoh F, Sone M, Ohneda M, et al. Pituitary adenylate cyclase activating polypeptide (PACAP)-like immunoreactivity in human hypothalamus: co-localization with arginine vasopressin. Regul Pept. 1994;50:267–75.
19. Ghatei MA, Takahashi K, Suzuki Y, Gardiner J, Jones PM, Bloom SR. Distribution, molecular characterization of pituitary adenylate cyclase-activating polypeptide and its precursor encoding messenger RNA in human and rat tisues. J Endocrinol. 1993;136:159–66.
20. Piggins HD, Stamp JA, Burns J, Rusak B, Semba K. Distribution of pituitary adenylate cyclase activating polypeptide (PACAP) immunoreactivity in the hypothalamus and extended amygdala of the rat. J Comp Neurol. 1996;376:278–94.

21. Hannibal J, Mikkelsen JD, Fahrenkrug J, Larsen PJ. Pituitary adenylate cyclase-activating peptide gene expression in corticotropin-releasing factor-containing parvicellular neurons of the rat hypothalamic paraventricular nucleus is induced by colchicine, but not by adrenalectomy, acute osmotic, ether, or restraint stress. Endocrinology. 1995;136:116–24.
22. Kántor O, Heinzlmann A, Suzuki N, Vincze E, Kocsis K, Köves K. Distribution of PACAP and its mRNA in several nonneural tissues of rats demonstrated by sandwich enzyme immunoassay and RT-PCR technique. Regul Pept. 2002;109:103–5.
23. Masuo Y, Suzuki N, Matsumoto H, Tokito F, Matsumoto Y, Tsuda M, et al. Regional distribution of pituitary adenylate cyclase activating polypeptide (PACAP) in the rat central nervous system as determined by sandwich-enzyme immunoassay. Brain Res. 1993;602:57–63.
24. Moore Jr JP, Burger LL, Dalkin AC, Winters SJ. Pituitary adenylate cyclase activating polypeptide messenger RNA in the paraventricular nucleus and anterior pituitary during the rat estrous cycle. Biol Reprod. 2005;73:491–9.
25. Radleff-Schlimme A, Leonhardt S, Wuttke W, Jarry H. Evidence for PACAP to be an autocrine factor on gonadotrope cells. Ann N Y Acad Sci. 1998;865:486–91.
26. Hannibal J, Mikkelsen JD, Clausen H, Holst JJ, Wulff BS, Fahrenkrug J. Gene expression of pituitary adenylate cyclase activating polypeptide (PACAP) in the rat hypothalamus. Regul Pept. 1995;55:133–48.
27. Hannibal J. Pituitary adenylate cyclase-activating peptide in the rat central nervous system: an immunohistochemical and in situ hybridization study. J Comp Neurol. 2002;453:389–417.
28. Vigh S, Arimura A, Köves K, Somogyvári-Vigh A, Sitton J, Fermin CD. Immunohistochemical localization of the neuropeptide, pituitary adenylate cyclase activating polypeptide (PACAP), in human and primate hypothalamus. Peptides. 1991;12:313–8.
29. Vereczki V, Köves K, Tóth ZS, Baba A, Hashimoto H, Fógel K, et al. Pituitary adenylate cyclase-activating polypeptide does not colocalize with vasoactive intestinal polypeptide in the hypothalamic magnocellular nuclei and posterior pituitary of cats and rats. Endocrine. 2003;22:225–37.
30. Köves K, Kántor O, Heinzlmann A, Lakatos A, Szabó E, Kirilly E, et al. Advent and recent advance of the research on the role of pituitary adenylate cyclase activating polypeptide (PACAP) in the gonadotropic hormone secretion. J Mol Neurosci. 2014;54:494–511.
31. Köves K, Vereczki V, Kausz M, Kántor O, Molnár J, Nemeskéri A, et al. PACAP and VIP in the photoneuroendocrine system (PNES). Med Sci Monit. 2002;8:SR5–20.
32. Leong SK, Ling EA. Labelling neurons with fluorescent dyes administered via intravenous, subcutaneous or intraperitoneal route. J Neurosci Methods. 1990;32:15–23.
33. Köves K, Kántor O, Vereczki V, Kausz M, Nemeskéri A, Fógel K, et al. PACAP and VIP in the photoneuroendocrine system. From the retina to the pituitary gland. Ann N Y Acad Sci. 2000;921:321–6.
34. Mikkelsen JD, Hannibal J, Larsen PJ, Fahrenkrug J. Pituitary adenylate cyclase activating peptide (PACAP) mRNA in the rat neocortex. Neurosci Lett. 1994;171:121–4.
35. Köves K, Kausz M, Fogel K, Arimura A. Presence of PACAP in the retinohypothalamic pathway. Regul Pept. 1996;64:98.
36. Hannibal J, Ding JM, Chen D, Fahrenkrug J, Larsen PJ, Gilete MU, et al. Pituitary adenylate cyclase activating polypeptide (PACAP) in the retinohypothalamic tract. A daytime regulator of the biological clock. J Neurosci. 1997;17:2637–44.
37. Izumi S, Seki T, Shioda S, Zhou CJ, Arimura A, Koide R. Ultrastructural localization of PACAP immunoreactivity in the rat retina. Ann N Y Acad Sci. 2000;921:317–20.
38. Dürr K, Norsted E, Gömüç B, Suarez E, Hannibal J, Meister B. Presence of pituitary adenylate cyclase-activating polypeptide (PACAP) defines a subpopulation of hypothalamic POMC neurons. Brain Res. 2007;1186:203–11.
39. Hannibal J, Fahrenkrug J. Target areas innervated by PACAP-immunoreactive retinal ganglion cells. Cell Tissue Res. 2004;316:99–113.
40. Vereczki V, Köves K, Csáki A, Grósz K, Hoffman GE, Fiskum G. Distribution of hypothalamic, hippocampal and other limbic peptidergic neuronal cell bodies giving rise to retinopetal fibers: anterograde and retrograde tracing and neuropeptide immunohistochemical studies. Neuroscience. 2006;140:1089–100.

41. Nielsen HS, Hannibal J, Fahrenkrug J. Expression of pituitary adenylate cyclase activating polypeptide (PACAP) in the postnatal and adult rat cerebellar cortex. Neuroreport. 1998;9:2639–42.
42. Kausz M, Murai Z, Arimura A, Köves K. Distribution of pituitary adenylate cyclase activating polypeptide (PACAP) immunoreactive elements in the brain stem of rats studied by immunohistochemistry. Neurobiology. 1999;7:19–31.
43. Légrádi G, Shioda S, Arimura A. Pituitary adenylate cyclase-activating polypeptide-like immunoreactivity in autonomic regulatory areas of the rat medulla oblongata. Neurosci Lett. 1994;176:193–6.
44. Masuo Y, Tokito F, Matsumoto Y, Shimamoto N, Fujino M. Ontogeny of pituitary adenylate cyclase-activating polypeptide (PACAP) and its binding sites in the rat brain. Neurosci Lett. 1994;170:43–6.
45. Palkovits M, Somogyvári-Vigh A, Arimura A. Concentrations of pituitary adenylate cyclase activating polypeptide (PACAP) in human brain nuclei. Brain Res. 1995;699:116–20.
46. Palkovits M. Isolated removal of hypothalamic or other brain nuclei of the rat. Brain Res. 1973;59:449–50.
47. Dow RC, Bennie J, Fink G. Pituitary adenylate cyclase-activating peptide-38 (PACAP)-38 is released into hypophysial portal blood in the normal male and female rat. J Endocrinol. 1994;142:R1–4.
48. Shuto Y, Somogyvári-Vigh A, Onda H, Arimura A. Effect of hypophysectomy on pituitary adenylate cyclase activating polypeptide gene expression in the rat hypothalamus. Peptides. 1995;16:407–13.
49. Oka H, Jin L, Reubi JC, Qian X, Scheithauer BW, Fujii K, et al. Pituitary adenylate-cyclase-activating polypeptide (PACAP) binding sites and PACAP/vasoactive intestinal polypeptide receptor expression in human pituitary adenomas. Am J Pathol. 1998;153:1787–96.
50. Aubert N, Basille M, Falluel-Morel A, Vaudry D, Bucharles C, Jolivel V, et al. Molecular, cellular, and functional characterizations of pituitary adenylate cyclase-activating polypeptide and its receptors in the cerebellum of New and Old World monkeys. J Comp Neurol. 2007;504:427–39.
51. Kozicz T, Vigh S, Arimura A. The source of origin of PACAP- and VIP-immunoreactive fibers in the laterodorsal division of the bed nucleus of the stria terminalis in the rat. Brain Res. 1998;810:211–9.
52. Swanson LW, Cowan WM. The connections of the septal region in the rat. J Comp Neurol. 1979;186:621–55.
53. Leranth C, Frotscher M. Organization of the septal region in the rat brain: cholinergic-GABAergic interconnections and the termination of hippocampo-septal fibers. J Comp Neurol. 1989;289:304–14.
54. Staiger JF, Nürnberger F. Pattern of afferents to the lateral septum in the guinea pig. Cell Tissue Res. 1989;257:471–90.
55. Dong YL, Fukazawa Y, Wang W, Kamasawa N, Shigemoto R. Differential postsynaptic compartments in the laterocapsular division of the central nucleus of amygdala for afferents from the parabrachial nucleus and the basolateral nucleus in the rat. J Comp Neurol. 2010;518: 4771–91.
56. Hamill GS, Jacobowitz DM. A study of afferent projections to the rat interpeduncular nucleus. Brain Res Bull. 1984;13:527–39.
57. Gruber C, Kahl A, Lebenheim L, Kowski A, Dittgen A, Veh RW. Dopaminergic projections from the VTA substantially contribute to the mesohabenular pathway in the rat. Neurosci Lett. 2007;427:165–70.
58. Cavalcante JS, Costa MS, Santee UR, Britto LR. Retinal projections to the midline and intralaminar thalamic nuclei in the common marmoset (Callithrix jacchus). Brain Res. 2005;1043:42–7.
59. de Sousa TB, de Santana MA, Silva Ade M, Guzen FP, Oliveira FG, Cavalcante JC, et al. Mediodorsal thalamic nucleus receives a direct retinal input in marmoset monkey (Callithrix jacchus): a subunit B cholera toxin study. Ann Anat. 2013;195:32–8.

60. Kausz M, Arimura A, Köves K. Distribution and somatotopical localization of pituitary adenylate cyclase activating polypeptide (PACAP) in the trigeminal ganglion of cats and rats. Ann N Y Acad Sci. 1998;865:529–32.
61. Kinhult J, Andersson JA, Uddman R, StjaÈrne P, Cardell L-O. Pituitary adenylate cyclase-activating peptide 38 a potent endogenously produced dilator of human airways. Eur Respir J. 2000;15:243–7.
62. Köves K, Arimura A, Vigh S, Somogyvári-Vigh A, Miller J. Immunohistochemical localization of PACAP in the ovine digestive system. Peptides. 1993;14:449–55.
63. Mulder H, Uddman R, Moller K, Zhang YZ, Ekblad E, Alumets J, et al. Pituitary adenylate cyclase activating polypeptide expression in sensory neurons. Neuroscience. 1994;63:307–12.
64. Nagahama M, Tsuzuki M, Mochizuki T, Iguchi K, Kuwahara A. Light and electron microscopic studies of pituitary adenylate cyclase-activating peptide (PACAP)-immunoreactive neurons in the enteric nervous system of rat small and large intestine. Anat Embryol (Berl). 1998;198:341–52.
65. Portbury AL, McConalogue K, Furness JB, Young HM. Distribution of pituitary adenylyl cyclase activating peptide (PACAP) immunoreactivity in neurons of the guinea-pig digestive tract and their projections in the ileum and colon. Cell Tissue Res. 1995;279:385–92.
66. Skakkebaek M, Hannibal J, Fahrenkrug J. Pituitary adenylate cyclase activating polypeptide (PACAP) in the rat mammary gland. Cell Tissue Res. 1999;298:153–9.
67. Sundler F, Ekblad E, Hannibal J, Moller K, Zhang YZ, Mulder H, et al. Pituitary adenylate cyclase-activating peptide in sensory and autonomic ganglia: localization and regulation. Ann N Y Acad Sci. 1996;805:410–26.
68. Uddman R, Tajti J, Möller S, Sundler F, Edvinsson L. Neuronal messengers and peptide receptors in the human sphenopalatine and otic ganglia. Brain Res. 1999;826:193–9.
69. Pettersson LM, Heine T, Verge VM, Sundler F, Danielson N. PACAP mRNA is expressed in rat spinal cord neurons. J Comp Neurol. 2004; 471: 85–96.
70. Liu W, Møller M. Innervation of the rat pineal gland by PACAP-immunoreactive nerve fibers originating in the trigeminal ganglion: a degeneration study. Cell Tissue Res. 2000; 301:369–73.
71. Moller K, Zhang YZ, Hakanson R, Luts A, Sjölund B, Uddman R, et al. Pituitary adenylate cyclase activating peptide is a sensory neuropeptide: immunocytochemical and immunochemical evidence. Neuroscience. 1993;57:725–32.
72. Zhang YZ, Hannibal J, Zhao Q, Moller K, Danielsen N, Fahrenkrug J, et al. Pituitary adenylate cyclase activating peptide expression in the rat dorsal root ganglia: up-regulation after peripheral nerve injury. Neuroscience. 1996;74:1099–110.
73. Dun NJ, Miyazaki T, Tang H, Dun EC. Pituitary adenylate cyclase activating polypeptide immunoreactivity in the rat spinal cord and medulla: implication of sensory and autonomic functions. Neuroscience. 1996;73:677–86.
74. Larsen JO, Hannibal J, Knudsen SM, Fahrenkrug J. Expression of pituitary adenylate cyclase-activating polypeptide (PACAP) in the mesencephalic trigeminal nucleus of the rat after transsection of the masseteric nerve. Brain Res Mol Brain Res. 1997;46:109–17.
75. Pettersson LM, Dahlin LB, Danielsen N. Changes in expression of PACAP in rat sensory neurons in response to sciatic nerve compression. Eur J Neurosci. 2004;20:1838–48.
76. Baeres FM, Møller M. Demonstration of PACAP-immunoreactive intrapineal nerve fibers in the golden hamster (Mesocricetus auratus) originating from the trigeminal ganglion. J Pineal Res. 2005;38:116–22.
77. Ichikawa H, Sugimoto T. Pituitary adenylate cyclase-activating polypeptide-immunoreactive nerve fibers in rat and human tooth pulps. Brain Res. 2003;980:288–92.
78. Kato J, Uddman R, Sundler F, Kurisu K. Immunohistochemical study of the innervation of the boundary area of the hard and soft palates of the rat. Acta Anat (Basel). 1998;163:92–8.
79. Rytel L, Palus K, Całka J. Co-expression of PACAP with VIP, SP and CGRP in the porcine nodose ganglion sensory neurons. Anat Histol Embryol. 2015;44:86–91.

80. Steinhoff M, McGregor GP, Radleff-Schlimme A, Steinhoff A, Jarry H, Schmidt WE. Identification of pituitary adenylate cyclase activating polypeptide (PACAP) and PACAP type 1 receptor in human skin: expression of PACAP-38 is increased in patients with psoriasis. Regul Pept. 1999;80:49–55.
81. Helyes Z, Kun J, Dobrosi N, Sándor K, Németh J, Perkecz A, et al. Pituitary adenylate cyclase-activating polypeptide is upregulated in murine skin inflammation and mediates transient receptor potential vanilloid-1-induced neurogenic edema. J Invest Dermatol. 2015;135:2209–18.
82. Papka RE, Workley M, Usip S, Mowa CN, Fahrenkrug J. Expression of pituitary adenylate cyclase activating peptide in the uterine cervix, lumbosacral dorsal root ganglia and spinal cord of rats during pregnancy. Peptides. 2006;27:743–52.
83. Drescher MJ, Drescher DG, Khan KM, Hatfield JS, Ramakrishnan NA, Abu-Hamdan MD, et al. Pituitary adenylyl cyclase-activating polypeptide (PACAP) and its receptor (PAC1-R) are positioned to modulate afferent signaling in the cochlea. Neuroscience. 2006;142: 139–64.
84. Wojtkiewicz J, Juranek JK, Kowalski I, Bladowski M, Całka J, Majewski M. Immunohistochemical characterization of superior cervical ganglion neurons supplying porcine parotid salivary gland. Neurosci Lett. 2011;500:57–62.
85. Tornoe K, Hannibal J, Jensen TB, Georg B, Rickelt LF, Andreasen MB, et al. PACAP-(1–38) as neurotransmitter in the porcine adrenal glands. Am J Physiol Endocrinol Metab. 2000; 279:E1413–25.
86. Strange-Vognsen HH, Arnbjerg J, Hannibal J. Immunocytochemical demonstration of pituitary adenylate cyclase activating polypeptide (PACAP) in the porcine epiphyseal cartilage canals. Neuropeptides. 1997;31:137–41.
87. Baeres FM, Møller M. Origin of PACAP-immunoreactive nerve fibers innervating the subarachnoidal blood vessels of the rat brain. J Cereb Blood Flow Metab. 2004;24:628–35.
88. Tobin G, Asztély A, Edwards AV, Ekström J, Håkanson R, Sundler F. Presence and effects of pituitary adenylate cyclase activating peptide in the submandibular gland of the ferret. Neuroscience. 1995;66:227–35.
89. Mirfendereski S, Tobin G, Håkanson R, Ekström J. Pituitary adenylate cyclase activating peptide (PACAP) in salivary glands of the rat: origin, and secretory and vascular effects. Acta Physiol Scand. 1997;160:15–22.
90. Calupca MA, Vizzard MA, Parsons RL. Origin of pituitary adenylate cyclase-activating polypeptide (PACAP)-immunoreactive fibers innervating guinea pig parasympathetic cardiac ganglia. J Comp Neurol. 2000;423:26–39.
91. Braas KM, May V, Harakall SA, Hardwick JC, Parsons RL. Pituitary adenylate cyclase- activating polypeptide expression and modulation of neuronal excitability in guinea pig cardiac ganglia. J Neurosci. 1998;18:9766–79.
92. Steenstrup BR, Alm P, Hannibal J, Jørgensen JC, Palle C, Junge J, et al. Pituitary adenylate cyclase-activating polypeptide: occurrence and relaxant effect in female genital tract. Am J Physiol. 1995;269:E108–17.
93. Hedlund P, Alm P, Ekström P, Fahrenkrug J, Hannibal J, Hedlund H, et al. Pituitary adenylate cyclase-activating polypeptide, helospectin, and vasoactive intestinal polypeptide in human corpus cavernosum. Br J Phamacol. 1995;116:2258–66.
94. Fahrenkrug J, Hannibal J. PACAP in visceral afferent nerves supplying the rat digestive and urinary tracts. Ann N Y Acad Sci. 1998;865:542–6.
95. Uddman R, Goadsby PJ, Jansen I, Edvinsson L. PACAP, a VIP-like peptide: immunohistochemical localization and effect upon cat pial arteries and cerebral blood flow. J Cereb Blood Flow Metab. 1993;13:291–7.
96. Jancsó G. Selective degeneration of chemosensitive primary sensory neurones induced by capsaicin: glial changes. Cell Tissue Res. 1978;195:145–52.
97. Ny L, Larsson B, Alm P, Ekström P, Fahrenkrug J, Hannibal J, et al. Distribution and effects of pituitary adenylate cyclase activating peptide in cat and human lower oesophageal sphincter. Br J Pharmacol. 1995;116:2873–80.

98. Fahrenkrug J, Hannibal J. Pituitary adenylate cyclase activating polypeptide innervation of the rat female reproductive tract and the associated paracervical ganglia: effect of capsaicin. Neuroscience. 1996;73:1049–60.
99. Fahrenkrug J, Steenstrup BR, Hannibal J, Alm P, Ottesen B. Role of PACAP in the female Reproductive organs. Ann N Y Acad Sci. 1996;805:394–407.
100. Pettersson LM, Geremia NM, Ying Z, Verge VM. Injury-associated PACAP expression in rat sensory and motor neurons is induced by endogenous BDNF. PLoS One. 2014;9:e100730.
101. Boeshore KL, Schreiber RC, Vaccariello SA, Sachs HH, Salazar R, Lee J, et al. Novel changes in gene expression following axotomy of a sympathetic ganglion: a microarray analysis. J Neurobiol. 2004;59:216–35.
102. Borzsei R, Mark L, Tamas A, Bagoly T, Bay C, Csanaky K, et al. Presence of pituitary adenylate cyclase activating polypeptide-38 in human plasma and milk. Eur J Endocrinol. 2009;160:561–5.
103. Brubel R, Horvath G, Reglodi D, Lubics A, Tamas A, Kiss P, et al. Presence of pituitary adenylate cyclase activating polypeptide and its type I receptor in the rat kidney. Transplant Proc. 2011;43:1297–9.
104. Szanto Z, Sarszegi Z, Reglodi D, Nemeth J, Szabadfi K, Kiss P, et al. PACAP immunoreactivity in human malignant tumor samples and cardiac diseases. J Mol Neurosci. 2012;48: 667–73.
105. Arimura A, Shioda S. Pituitary adenylate cyclase activating polypeptide (PACAP) and its receptors: neuroendocrine and endocrine interaction. Front Neuroendocrinol. 1995;16:53–88.
106. Vaudry H, Arimura A, editors. Pituitary adenylate cyclase-activating polypeptide. New York: Kluver Academic Publishers; 2003.
107. Vaudry D, Gonzales BJ, Basille M, Yon L, Fournier A, Vaudry H. Pituitary adenylate cyclase-activating polypeptide and its receptors: from structure to functions. Pharmacol Rev. 2000;52:269–324.
108. Vaudry H, Laburthe M, editors. VIP, PACAP, and related peptides from gene to therapy. New York: Annals of the New York Academy of Sciences; 2006.
109. Paxinos G, Watson C. The rat brain in stereotaxic coordinates. San Diego: Academic; 2005.
110. Köves K, Arimura A. Solid-phase absorption method for removing undesired antibodiesfrom polyclonal antiserum. J Histochem Cytochem. 1989;37:33–37.

Chapter 13
Multiple Mechanisms Contribute to the PAC_1 Modulation of Parasympathetic Cardiac Neuron Excitability

Rodney L. Parsons, John D. Tompkins, Jean C. Hardwick, Laura A. Merriam, Beatrice M. Girard, and Victor May

Abstract Pituitary adenylate cyclase-activating polypeptide (PACAP) is a potent intercellular signaling molecule that regulates a variety of central and peripheral neuronal circuits important for behavior and physiological homeostasis. Although central neurons are not readily accessible for mechanistic studies, the ability for PACAP/PAC_1 receptor signaling to increase neuronal excitability in guinea pig parasympathetic cardiac ganglia provides a unique means to establish intracellular PACAP mechanisms in neuronal function. The guinea pig cardiac neurons predominantly express a very short null PAC_1 receptor isoform which is coupled to the adenylyl cyclase and MEK/ERK signaling cascades. PACAP/PAC_1 receptor activation of adenylyl cyclase and the resulting rise in intracellular cAMP enhances the nonselective cationic current I_h; treatment with I_h inhibitors diminishes the PACAP-induced increase in excitability. Thus, a shift in the voltage-dependence of I_h activation is one ionic mechanism contributing to the PACAP-induced increase in cardiac neuron excitability. Low concentrations of nickel also blunt the peptide-induced increase in excitability, suggesting that a PACAP enhanced calcium influx through T-type voltage-dependent calcium channels contributes to the modulation of excitability. Reducing ambient temperature and treatments with endocytosis inhibitors Pitstop2 or dynasore efficaciously block PACAP modulation of excitability suggesting PACAP/PAC_1 receptor internalization for endosomal MEK/ERK activation is requisite for the PACAP responses. In sum, the results presented in this review

R.L. Parsons (✉) • L.A. Merriam • B.M. Girard • V. May
Department of Neurological Sciences, University of Vermont College of Medicine, 89 Beaumont Avenue, C427 Given, Burlington, VT 05405, USA
e-mail: rodney.parsons@uvm.edu

J.D. Tompkins
Department of Medicine (Cardiology), David Geffen School of Medicine, University of California, Los Angeles, CA 90095, USA

J.C. Hardwick
Department of Biology, Ithaca College, Ithaca, NY 14850, USA

D. Reglodi, A. Tamas (eds.), *Pituitary Adenylate Cyclase Activating Polypeptide — PACAP*, Current Topics in Neurotoxicity 11,
DOI 10.1007/978-3-319-35135-3_13

demonstrate that the PACAP/PAC_1 receptor interactions can activate multiple intracellular signaling cascades to selectively modulate ionic conductances that gate neuronal excitability.

Keywords PACAP • PAC_1 receptor • Parasympathetic cardiac neurons • Neuronal excitability • Adenylyl cyclase • cAMP • Hyperpolarization-activated nonselective cationic currents • Hyperpolarization-induced rebound depolarization • T-type low voltage-dependent calcium currents • Receptor internalization • MEK kinase • Signaling endosome

Introduction

Pituitary adenylate cyclase-activating polypeptide (PACAP) peptides, members of the VIP/secretin/glucagon family of neuropeptides, are potent trophic and intercellular signaling molecules that are widely distributed within neural and endocrine tissues across species [1, 2]. PACAP effects have been shown to be critically important in central stress responses, peripheral sensory and autonomic function, and maintenance of multiple physiological systems critical for homeostasis [2–9]. The PACAP (*Adcyap1*) peptide amino acid sequence is highly conserved and the action of PACAP is tissue specific dependent on the activation of the different isoforms of the seven transmembrane G-protein-coupled PACAP-selective PAC_1 receptor (*Adcyap1r1*) and/or the $VPAC_1$- and $VPAC_2$-nonselective receptors [1, 2, 10–12].

PACAP has been identified within neural elements innervating both vascular tissues and the heart and PACAP peptides exert multiple and potent species-specific effects on cardiovascular regulation [2, 13]. PACAP is a potent vasodilator in most vascular beds [2]. Direct exogenous application of PACAP peptides on cardiac muscle can have negative inotropic or positive inotropic/chronotropic actions. In the dog and guinea pig, PACAP peptides can both increase the threshold for acetylcholine-induced atrial fibrillations as well as initiate spontaneous fibrillations [14–19]. PACAP peptides stimulate cAMP synthesis by cardiac myocytes and modulate L-type calcium channels in vascular smooth muscle [13, 20].

In addition to its direct effects on cardiac or smooth muscle cells, PACAP modulates cardiovascular function via potent excitatory effects on parasympathetic postganglionic neurons within the intrinsic cardiac nervous system (cardiac neurons); thus, they modulate cardiac output by regulating cardiac neuron activity [18, 19, 21, 22]. Studies of isolated canine and guinea pig hearts demonstrate that PACAP causes a biphasic change in heart rate, initially tachycardia followed by bradycardia [14–16, 18, 19, 21]. In contrast, PACAP initiates only tachycardia in rats and mice [18, 23]. The tachycardia common across species is due to a PACAP activation of VPAC receptors on atrial pacemaker tissues, an effect also produced by vasoactive intestinal polypeptide (VIP) [24]. The PACAP-induced bradycardia in guinea pig and canine hearts was blocked by atropine, an observation suggesting that the peptide-induced decrease in heart rate was due to the stimulation of cardiac neurons, leading to release of acetylcholine from parasympathetic postganglionic terminals and activation of atrial muscarinic receptors [14–16, 18, 19, 21]. Direct evidence for this hypothesis was derived from the observations of Braas et al. [22] that showed exogenous application of PACAP27 depolarized cardiac neurons leading in turn to multiple action potential generation. Note that in all of our studies on guinea pig cardiac neurons described in this chapter, PACAP27 was used and is referred to as PACAP throughout the text. In addition, it was determined that guinea pig cardiac neurons express PAC_1 selective receptors and that PACAP was co-localized with choline acetyltransferase in the cholinergic parasympathetic preganglionic terminals innervating virtually all guinea pig cardiac neurons [22, 25]. Subsequently, Tompkins et al. [26] demonstrated that both neurally released and exogenously applied PACAP could depolarize guinea pig cardiac neurons sufficiently to initiate a burst of action potentials.

PACAP Modulates Neuronal Excitability

In addition to initiating membrane depolarization, both endogenously released and exogenously applied PACAP significantly increases cardiac neuron excitability through PAC_1 receptor activation [19, 22, 26, 27]. The change in excitability is evident from the PACAP-induced increase in action potential generation in response to long suprathreshold, depolarizing steps [19, 22, 26, 27] (Fig. 13.1A_1–C_1). The increase in excitability can occur within minutes of bath application of nanomolar concentrations of PACAP (1–20 nM) in some cells whereas in other cells the shift in excitability occurs only after a few minute delay. In both cases, the effect of PACAP on excitability often can require 10–12 min to develop fully. Also, following a brief, local puffer application of PACAP, the increase in excitability can last for tens of minutes. Both the time course and long duration suggest possible involvement of intracellular signaling cascades. In the example shown in Fig. 13.1A_1–C_1, prior to PACAP, this neuron produced 1 action potential during a 1 s long 0.4 nA current pulse, whereas after 5 min of bath application of 20 nM PACAP, the same depolarizing current step elicited 12 action potentials. The magnitude of the shift in excitability is evident when excitability curves are constructed by plotting the number of action potentials generated by a series of long depolarizing pulses of increasing current intensities (Fig. 13.1d).

The PACAP enhanced excitability of guinea pig cardiac neurons is mediated solely through activation of PAC_1 receptors as maxadilan, a specific PAC_1 receptor agonist mimics the PACAP effect, but VIP does not [19, 22]. Figure 13.2 illustrates the change in excitability following local puffer application of PACAP (A, B) or maxadilan (C, D) whereas local application of VIP does not cause a shift in excitability (E, F). Tompkins et al. [26] also found that the PAC_1 receptor antagonist M65 blunted the PACAP-induced depolarization and increase in excitability providing further evidence that both actions of PACAP are mediated through the PAC_1 receptor.

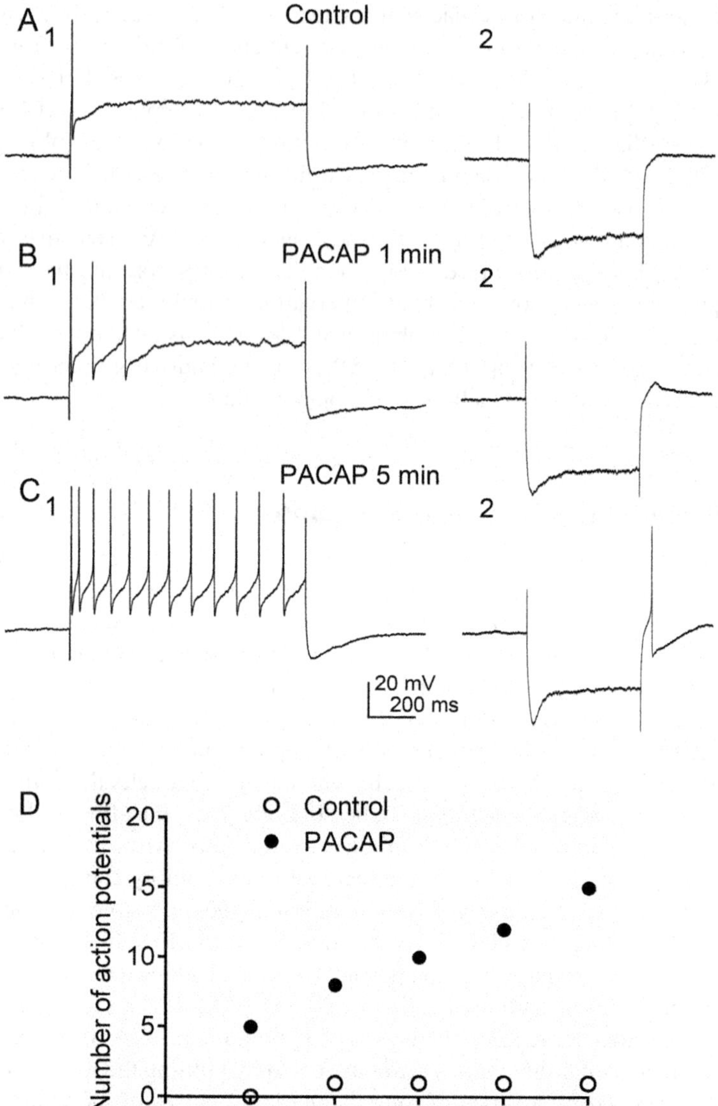

Fig. 13.1 PACAP enhances excitability, rectification and a hyperpolarization-induced rebound depolarization in guinea pig cardiac neurons. Panels **A_1–C_1** show the 20 nM PACAP-induced, time-dependent shift from phasic to multiple action potential generation. Prior to PACAP application, a 1 s, 0.4 nA depolarizing constant current pulse elicited 1 action potential. After 5 min in PACAP, the number of action potentials generated by this same depolarizing current pulse increased to 12. Panels **A_2–C_2** show that over this same time period the rectification in the hyperpolarization elicited by a 500 s constant current pulse increased progressively as well. Likewise, PACAP progressively enhanced the hyperpolarization-induced rebound depolarization. (**D**) An excitability curve showing the PACAP (at 5 min) enhancement of action potentials generated by 1 s depolarizing current steps of increasing intensity

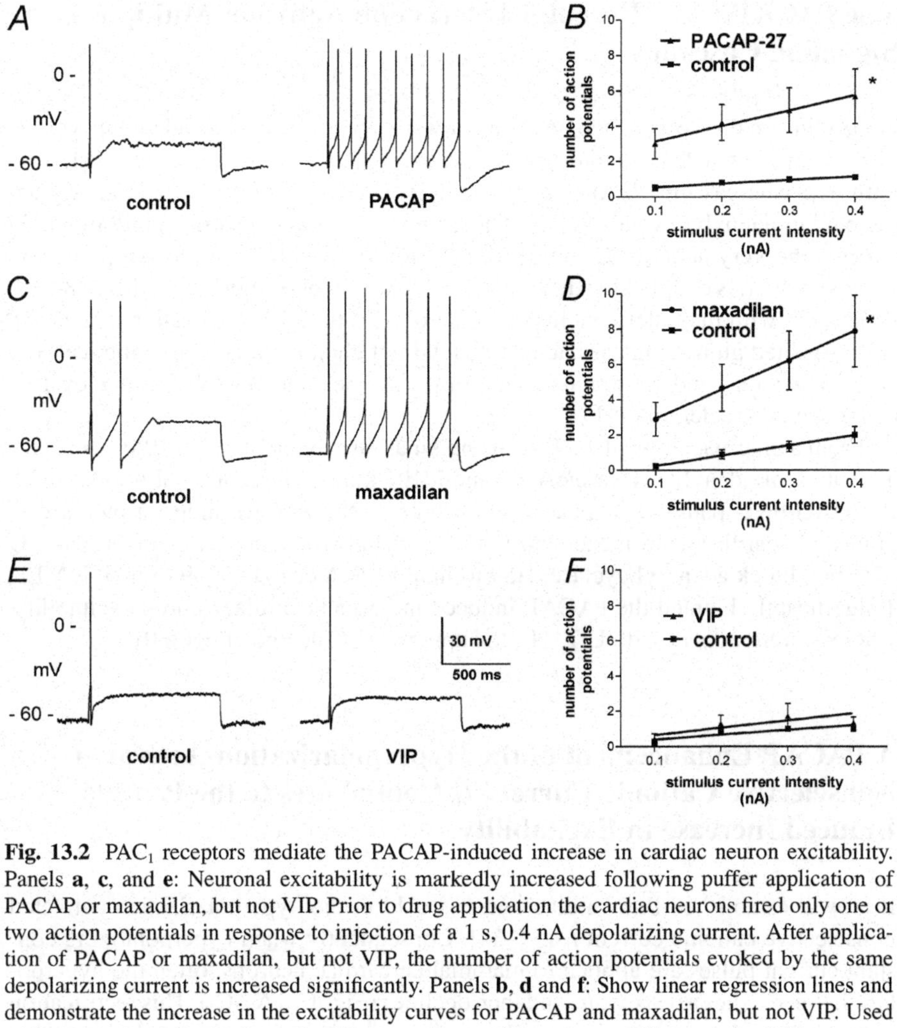

Fig. 13.2 PAC_1 receptors mediate the PACAP-induced increase in cardiac neuron excitability. Panels **a**, **c**, and **e**: Neuronal excitability is markedly increased following puffer application of PACAP or maxadilan, but not VIP. Prior to drug application the cardiac neurons fired only one or two action potentials in response to injection of a 1 s, 0.4 nA depolarizing current. After application of PACAP or maxadilan, but not VIP, the number of action potentials evoked by the same depolarizing current is increased significantly. Panels **b**, **d** and **f**: Show linear regression lines and demonstrate the increase in the excitability curves for PACAP and maxadilan, but not VIP. Used with permission from the Journal of Pharmacology and Experimental Therapeutics [19]

PACAP/PAC_1 receptor signaling occurs within peripheral and central sensory and autonomic circuits that not only regulate cardiac function, but also support the function of other organ systems that maintain homeostasis normally and that are involved in the development of the stress response [2–5, 7–9]. However, establishing PACAP neuronal signaling events within central nervous system circuits has been difficult because of the morphological complexity. Accordingly, these cardiac neurons have provided an excellent neuronal system to establish PACAP/PAC_1 receptor signaling cascades activated and ionic conductances modulated that contribute to the regulation of neuronal function.

The PACAP/PAC$_1$ Receptor Interaction Activates Multiple Signaling Cascades

Depending on the isoform, the PAC$_1$ selective receptor can activate adenylyl cyclase, phospholipase C (PLC), and/or the Ras/mitogen activated protein kinase (MAPK) pathway, whereas the VPAC$_1$- and VPAC$_2$-nonselective receptors commonly are coupled to adenylyl cyclase [1, 2]. The guinea pig cardiac neurons predominantly express the very short PAC$_1$ null receptor isoform (neither Hip nor Hop cassette inserts), which is coupled to adenylyl cyclase, but not phospholipase C [22, 28–30]. Consistent with the latter conclusion is the observation that application of PACAP to dissociated guinea pig cardiac neurons did not elicit a global rise in intracellular calcium whereas under the same conditions, caffeine elicited a significant elevation of intracellular calcium [29].

From a pharmacological analysis using kinase inhibitors, PACAP/PAC$_1$ receptor activation of adenylyl cyclase/cAMP and MEK kinase signaling, but not the PLC/IP3/DAG/PKC pathway, appear to contribute to the PACAP-induced increase in guinea pig cardiac neuron excitability [29, 30]. In particular, pretreatment with SQ 22536 to block adenylyl cyclase, H89 to inhibit PKA or PD 98059 to inhibit MEK 1 significantly blunted the PACAP-induced increase in cardiac neuron excitability, whereas pretreatment with U 73112 to suppress PLC had no effect [30].

A PACAP Enhancement of the Hyperpolarization-Activated Nonselective Cationic Current I_h Contributes to the Peptide-Induced Increase in Excitability

It is well established that cardiac neurons exhibit the hyperpolarization-activated nonselective cationic current I_h [31–34]. Consequently when hyperpolarizing constant current pulses are applied to mammalian cardiac neurons, often the hyperpolariziation reaches a peak value and then declines (Fig. 13.1A$_2$–C$_2$). This rectification or "sag" in the hyperpolarization is due to the activation of I_h, an inward current that opposes the hyperpolarization caused by the constant current pulse. There are four hyperpolarization cyclic nucleotide-gated channel (HCN) subunits that can form I_h channels [35]. However, it is not known whether all or just some of the subunits are expressed by cardiac neurons. Using semi-quantitative PCR, we determined that transcripts for all four HCN subunits were present in extracts from laser captured cardiac neurons (Fig. 13.3). Laser captured neurons from cardiac ganglia whole mounts was used in order to minimize inclusion of cardiac muscle, which also expresses HCN channels.

The PACAP/PAC$_1$ receptor interaction activates adenylyl cyclase leading to the generation of cAMP and the rise in cytosolic cAMP can enhance I_h [35]. Consequently, one mechanism that was postulated to contribute to the PACAP-induced increase in excitability was a cAMP-mediated enhancement of I_h. Evidence

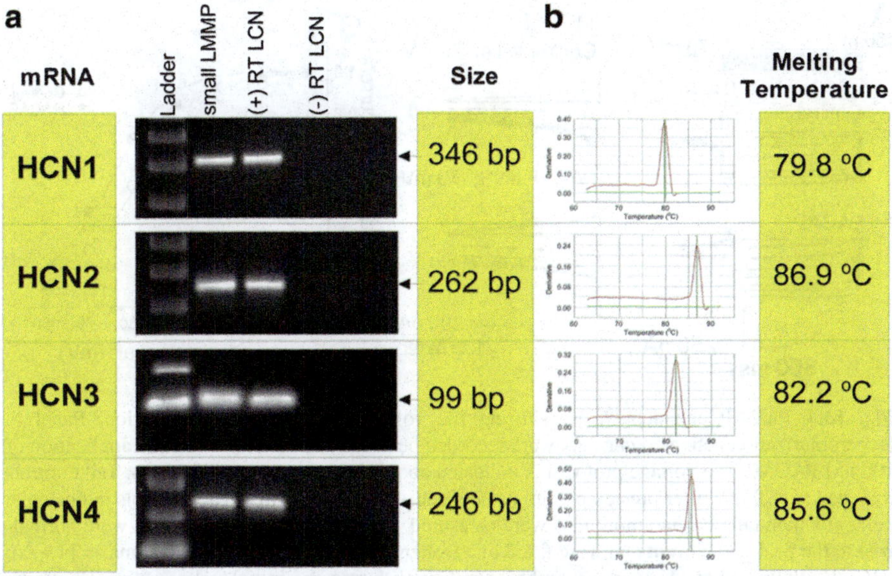

Fig. 13.3 Laser captured cardiac neurons express multiple HCN channel transcripts. (**a**) PCR products viewed on an ethidium bromide gel after amplification of cDNA from laser captured cardiac neurons (LCN). LMMP: extract of small intestine longitudinal muscle myenteric plexus as a positive control. (+) RT LCN: cDNA from laser captured neurons. (–) RT LCN: the mRNA from LCN neurons subjected to RT with omission of the enzyme therefore checking for genomic contamination of the LCN sample. NTC: no template control assessing the absence of primer dimers or other contaminations. (**b**) Melting point profiles for the different HCN isoforms. The amplified product for each gene was subjected to SYBR Green I melting point analysis by ramping the temperature of the reaction samples from 60 to 95 °C. A single DNA melting profile was observed under these conditions, demonstrating an amplification of a single unique product, free of primer dimmers or other anomalous products. The following primers were used to test for the presence of HCN subunit transcripts: HCN1 upper (5′-CATGAATTTTAGGACTGGGACT-3′), lower (5′-CTGCACTGGCGAGATCATAA-3′); HCN2 upper (5′-GATTCACCAAGATCCTCAGC-3′), lower (5′-GTAGAGCTCGCTCCACGAGT-3′); HCN3 upper (5′-CTGCGCATTGTTCGA TTCACCAAG-3′), lower (5′-TGTGAAAGATCTCCTCCCACTGGT-3′); HCN4 upper (5′-TGCGGGAGGCGATCATTAAC-3′), lower (5′-CCAAAATAGGAGCCATCAGC-3′)

for activation of I_h by PACAP is evident in the change in the profile of the hyperpolarizations noted in Fig. 13.1A$_2$–C$_2$. In this example, hyperpolarizing current steps (500 ms) were used to test for rectification in the current-induced hyperpolarization before and during exposure to 20 nM PACAP. Note that over time in PACAP, the rectification progressively increased along with the increase in action potential generation (Fig. 13.1A$_2$–C$_2$).

Direct evidence that PACAP enhanced I_h was obtained using hyperpolarizing voltage step protocols on whole cell clamped dissociated cardiac neurons [34]. As seen in Fig. 13.4, from analysis of tail currents, PACAP caused a positive shift in the voltage-dependence of I_h activation. This effect of PACAP was mimicked by the adenylyl cyclase activator forskolin and blocked by I_h inhibitors [34]. In addition,

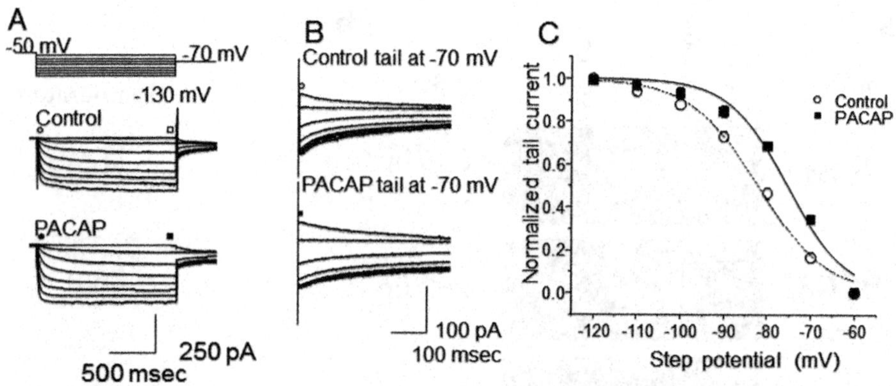

Fig. 13.4 PACAP enhances I_h by shifting the voltage dependence of activation. Panel **a**: Hyperpolarization activated currents were recorded before (control) and during the application of 100 nM PACAP. The voltage protocol is indicated above the current traces. Panel **b**: Tail currents recorded at −70 mV after prestep potentials between −60 and −130 mV. The symbols indicate the point at which tail current amplitudes were measured. (**c**) The resultant tail currents were normalized to the peak tail current and were fitted to a Boltzmann function of the form: $I/I\text{max} = \{1 + \exp[(V_m - V_{1/2})/k]\}^{-1}$ where V_m is the membrane potential during the initial voltage step, $V_{1/2}$ is the membrane potential at which I_h is half-activated, and k is the slope factor. The curves represent averaged results from 9 cells. PACAP significantly shifts the half activation voltage 6.6 mV. Data presented in (**c**) are mean ± SEM, but the error bars are smaller than the symbols in some cases. Used with permission from Regulatory Peptides [34]

the PACAP-induced increase in excitability in neurons within whole mount cardiac ganglia preparations was significantly suppressed by pretreatment with I_h blockers [36]. Results from intracellular recordings that demonstrate the suppression of the PACAP-induced increase in excitability by pretreatment with the I_h blocker cesium chloride are shown in Fig. 13.5).

Enhancement of a Nickel-Sensitive Current by PACAP Contributes to the Increase in Cardiac Neuron Excitability

The results described above provided strong evidence that one mechanism contributing to the PACAP-induced increase in cardiac neuron excitability is an adenylyl cyclase/cAMP-mediated enhancement of I_h [34, 36]. In direct measurements of I_h in dissociated neurons, application of the potent adenylyl cyclase activator forskolin produced a shift in the voltage-dependence of activation comparable to that produced by PACAP [34]. However, application of forskolin or exposure to 8-bromo-cAMP, the cell permeable analogue of cAMP, only partially recapitulated the effect of PACAP on excitability of neurons in the whole mount preparation [30]. For instance, with bath application of PACAP (10–20 nM), excitability is increased in 80–90 % of the cardiac neurons, whereas exposure to forskolin (5 μM) or treatment with 8-bromo-cAMP (1 mM) only enhances excitability in ~35 % of the guinea pig

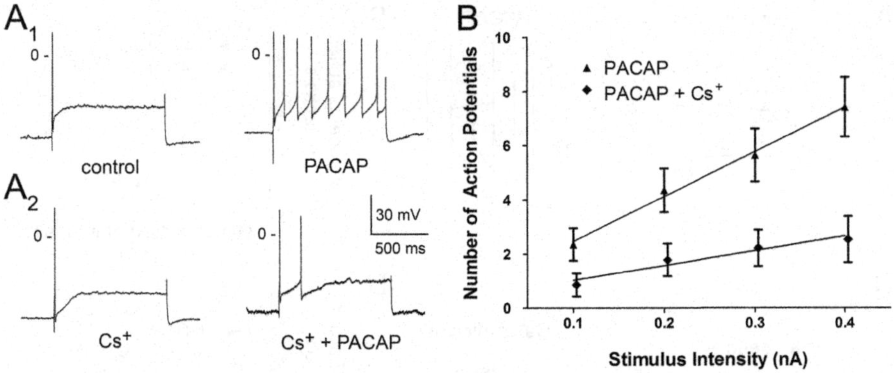

Fig. 13.5 Cesium (Cs^+) suppresses the PACAP-induced increase in excitability. **A_1** In a control cell, the response to a 1 s, 0.3 nA depolarizing current pulse was enhanced during exposure to 10 nM PACAP as evidenced by the increased number of action potentials elicited. **A_2** In a second cell bathed in a solution containing 2.5 mM Cs^+, only 2 action potentials were elicited by the depolarizing current pulse during exposure to 10 nM PACAP. (**B**) Excitability curves summarizing the effect of 10 nM PACAP on excitability for 14 cells kept in control solution and 14 cells bathed in a solution containing 2.5 mM Cs^+. Data are plotted as the mean number of action potentials generated by a 1 s depolarizing current pulse at multiple stimulus intensities with the lines fit with a linear regression. The slopes were significantly different (PACAP alone = 16.50 ± 0.92; PACAP in Cs^+-containing solution = 5.35 ± 0.92, $P = 0.001$). Used with permission from the American Journal of Physiology Integrative Comparative Physiology [36]

cardiac neurons in whole mount preparations [30]. These observations suggest that PACAP enhancement of other membrane conductances, in addition to I_h, also contributes to the peptide-induced modulation of excitability.

In 2006, Tompkins et al. [29] determined that the PACAP-induced increase in cardiac neuron excitability is dependent on calcium influx, but not calcium release from intracellular stores. For example, when cardiac ganglia whole mounts were maintained in a calcium-deficient solution, PACAP did not increase cardiac neuron excitability (Fig. 13.6c, d). In contrast, depletion of intracellular stores by pretreatment with caffeine and ryanodine did not eliminate the ability of PACAP to increase cardiac neuron excitability [29]. A voltage dependent calcium channel (VDCC) was suggested not to be the calcium influx pathway as PACAP decreased whole cell barium currents. Subsequently, it was suggested that the calcium influx might occur through receptor-activated nonselective cationic channels based on the observation that different putative nonselective cationic channel inhibitors could blunt the PACAP-induced increase in cardiac neuron excitability [37]. Unfortunately, many of these putative nonselective cationic channel inhibitors also can block voltage dependent calcium channels. Consequently, identification of the calcium influx pathway required further study. While most of the macroscopic calcium current in cardiac neurons is carried by N-type calcium channels, other VDCC types contribute to the total calcium current [38]. Thus, a possibility considered was that while decreasing N-type calcium currents, a PACAP potentiation of another smaller VDCC component may have been obscured. L-type calcium currents contribute a

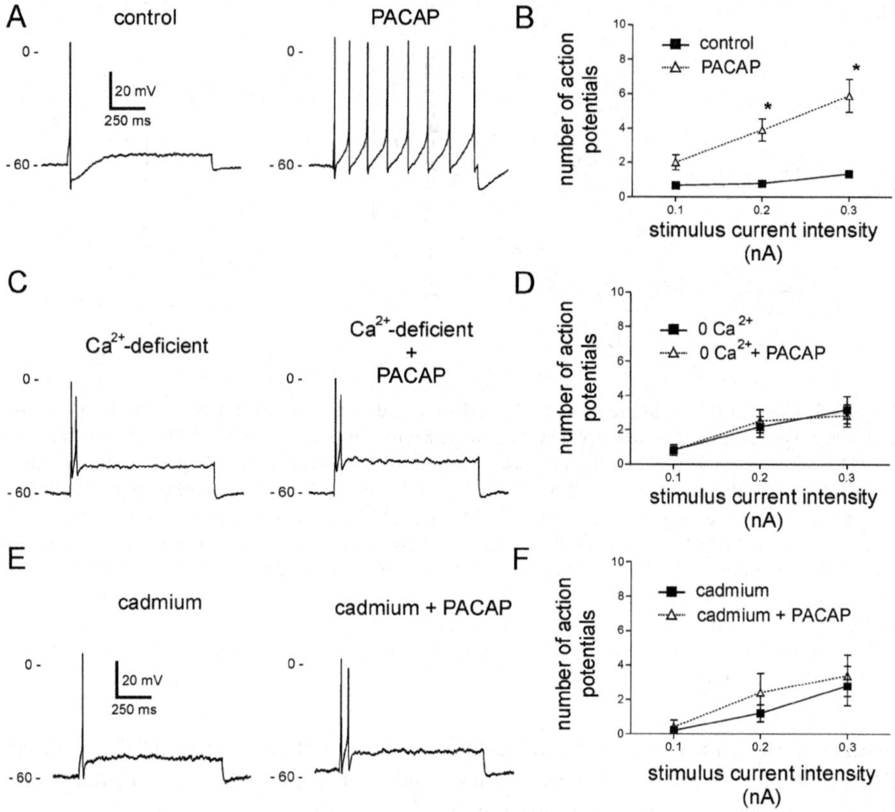

Fig. 13.6 The PACAP-induced increase in excitability was suppressed by substitution of Mg^{2+} for Ca^{2+} or addition of Cd^{2+} to the bath solution. (**a**) Membrane potential recording from a phasic intracardiac neuron in response to a long duration depolarizing current pulse (0.3 nA, 1 s) before (*left panel*) and after PACAP (*right panel*). The PACAP-induced increase in neuronal excitability is evident by the increased spike number in response to the depolarizing stimulus. (**b**) Excitability curve obtained prior to and following puffer application of PACAP under control conditions (*asterisk* indicates $P<0.05$, $n=9$ cells). (**c**) In a Ca^{2+}-deficient solution in which Mg^{2+} was substituted for Ca^{2+}, neuronal excitability was similar before (*left panel*; record illustrates response to a 1 s, 0.2 nA depolarizing step) and following PACAP application (*right panel*). (**d**) Averaged results for 6 cells. (**e**) Membrane potential recorded in response to a 1 s, 0.2 nA depolarizing current step in the same cell immediately before and after PACAP application in the presence of 200 μM Cd^{2+}. (**f**) The excitability curve summarized for data from 5 cells. Scale bar is the same for all records. Used with permission from the Journal of Neurophysiology [29]

small component to the total calcium current recorded in cardiac neurons [38] and PACAP can enhance L-type calcium channel currents in some cells [39, 40]. However, pretreatment with the L-type calcium channel blocker nifedipine failed to blunt the PACAP-induced modulation of excitability [29], eliminating L-type currents as a means of PACAP modulation of excitability.

Tompkins et al. [29] also demonstrated that pretreatment with 200 μM cadmium potently blocked the PACAP effect on excitability (Fig. 13.6e, f). At this concentration, cadmium effectively blocks all classes of VDCCs whereas low concentrations of nickel (Ni^{2+}) preferentially inhibits R- and T-type calcium channels [41–43]. R-type calcium currents are expressed in rat cardiac neurons, although there was no evidence of T-type calcium channel currents [38]. However, using QPCR analysis, Tompkins et al. [44] determined that transcripts for both T-type calcium channel and R-type channel isoforms were present in the guinea pig cardiac neurons.

Evidence from recent intracellular recordings also indicated that functional low voltage-activated T-type calcium channels were expressed in the guinea pig cardiac neurons [44]. As shown in Fig. 13.1A_2–C_2, following termination of the hyperpolarization initiated by a constant current pulse, a hyperpolarization-induced rebound depolarization was evident. The hyperpolarization-induced rebound depolarization, which is present in many guinea pig cardiac neurons, could potentially be produced by inward current continuing to flow through I_h channels immediately after the hyperpolarizing pulse is terminated. However, Tompkins et al. [44] noted that the hyperpolarization-induced rebound depolarization was present in cells pretreated with cesium chloride to block I_h (Fig. 13.7b). Consequently, some other conductance, in addition to I_h, must also contribute to the hyperpolarization-induced rebound depolarization in guinea pig cardiac neurons. A hyperpolarization-induced rebound depolarization is also a signature feature of T-type calcium channels [41, 45, 46]. Indeed, the rebound depolarization was sensitive to Ni^{2+} and mibefradil (Fig. 13.7c, d), two well-established blockers of T-type calcium channels [47, 48]; an observation consistent with the presence of these channels in guinea pig cardiac neurons. Following application of either Ni^{2+} or mibefradil, cells were hyperpolarized by 2–4 mV, which suggested that these blockers reduced an inward current flowing under resting conditions with the membrane potential ranging from −55 to −65 mV. This range of membrane potentials is similar to the window current for T-type calcium channels, providing further support for T-type calcium channel expression in the cardiac neurons.

Consequently, to test whether an inward current flowing through T-type calcium channels, or potentially also through R-type calcium channels (both T-type and R-type calcium channels can be blocked by low concentrations of Ni^{2+}), might contribute to the PACAP-induced increase in excitability, the ability of low concentrations of Ni^{2+} to blunt the PACAP effect was investigated. PACAP increased action potential generation produced by a suprathreshold depolarizing pulse (Fig. 13.8A1a, 2a). In contrast, after pretreatment with 50 μM Ni^{2+}, the 20 nM PACAP-induced increase in excitability was significantly blunted (Fig. 13.8B1a, 2a). In addition to increasing excitability, PACAP often potentiated the hyperpolarization-induced rebound depolarization and in many cells, this resulted in multiple action potentials being elicited by the rebound depolarization (Fig. 13.8A1b, 2b). Ni^{2+} also blunted the PACAP enhancement of the rebound depolarization (Fig. 13.8B1b, 2b). Averaged excitability curves that show the extent of the Ni^{2+}-induced suppression of the PACAP effect on excitability are presented in Fig. 13.8c.

Tompkins et al. [44] also showed in initial whole cell voltage clamp recordings on dissociated cardiac neurons that PACAP enhanced a low voltage-activated current. It was suggested that this low voltage-activated inward current is a calcium current, possibly through T-type calcium channels or perhaps also R-type calcium currents. However, a more extensive analysis of VDCCs in the guinea pig cardiac neurons is required to establish unequivocally the identification of the PACAP-enhanced low voltage-activated current.

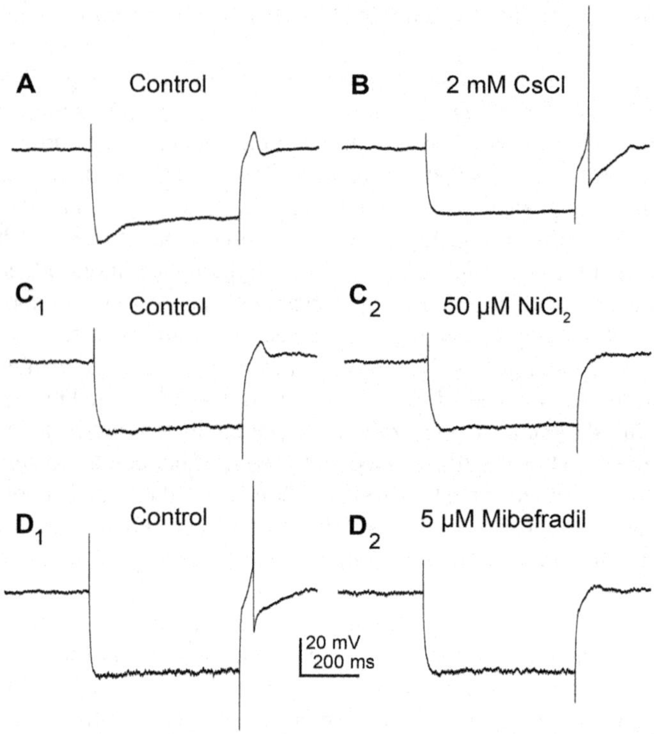

Fig. 13.7 Guinea pig cardiac neurons exhibit hyperpolarization-induced rebound depolarizations. (**a**) An example of the transient rebound depolarization recorded after termination of a 500 ms hyperpolarizing step. Rectification (sag) was evident in the hyperpolarization indicating the presence of I_h. (**b**) The hyperpolarization-induced rebound depolarization was also seen in preparations pretreated with 2 mM CsCl to block HCN channels. Note the lack of rectification in the hyperpolarizing trace. In this example the rebound depolarization was sufficient to elicit an action potential. (**c**) An example in which the rebound depolarization before nickel ($\mathbf{C_1}$) was blocked after addition of 50 μM nickel ($\mathbf{C_2}$). (**d**) An example in which the rebound depolarization before mibefradil ($\mathbf{D_1}$) was blocked after addition of 5 μM mibefradil ($\mathbf{D_2}$). Used with permission from the American Journal of Physiology Cell Physiology [44]

In sum, these observations suggest that a PACAP enhancement of a Ni^{2+}-sensitive current, possibly a T-type calcium current (or perhaps also R-type calcium currents) is a second current component that contributes to the PACAP modulation of cardiac neuron excitability.

Fig. 13.8 (continued) * From comparison of the solid circles and solid triangles, nickel significantly decreased the PACAP enhanced excitability at all current steps ($P < 0.05$). *Hyperpolarization-induced rebound depolarization.* (**a**) Prior to PACAP, a hyperpolarizing step to ~−100 mV evoked a rebound depolarization sufficiently large to evoke an action potential ($\mathbf{A_{1b}}$) whereas in 20 nM PACAP the rebound depolarization elicited a train of action potentials ($\mathbf{A_{2b}}$). Note also that in PACAP, the rectification in the hyperpolarization was noticeably enhanced. (**b**) Pretreatment with 50 μM nickel blunted the ability of PACAP to enhance the rebound depolarization ($\mathbf{B_{1b,2b}}$). Used with permission from the American Journal of Physiology Cell Physiology [44]

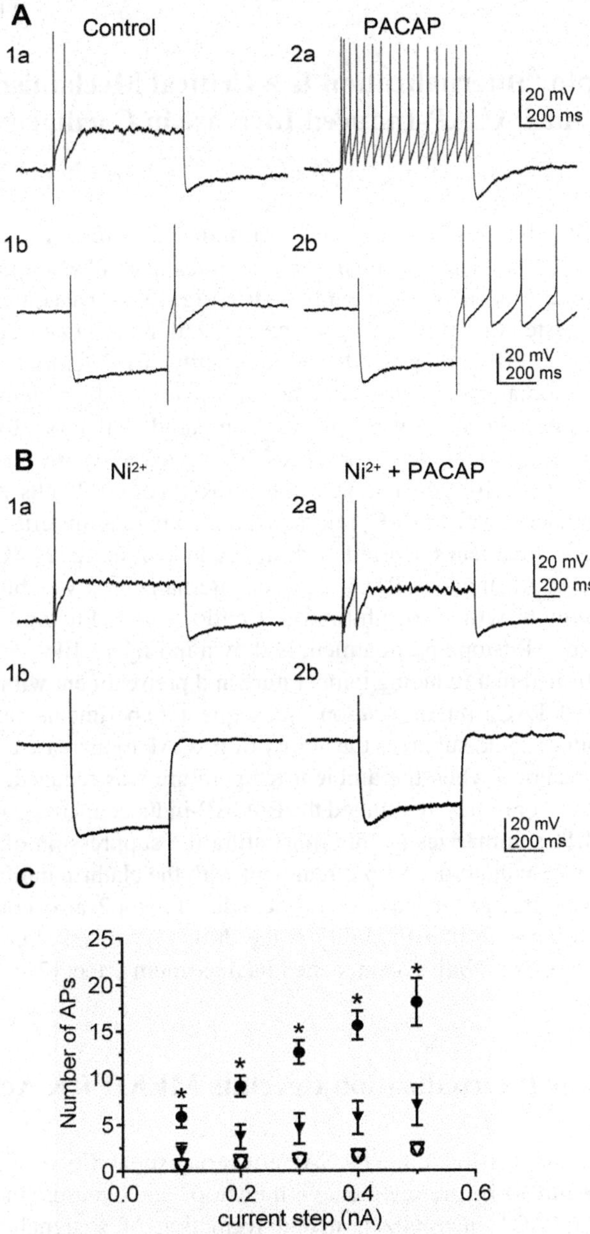

Fig. 13.8 Nickel suppresses the PACAP-induced increase in excitability and the PACAP-enhanced hyperpolarization-induced rebound depolarization. *Excitability*. (**a**) Prior to PACAP, a 1 s, 0.3 nA depolarizing step elicited two action potentials (**A$_{1a}$**), whereas during application of 20 nM PACAP, multiple action potentials were generated (**A$_{2a}$**). (**b**) Following pretreatment with 50 μM nickel, the ability of PACAP to increase excitability was suppressed (**B$_{1a,2a}$**). (**c**) Averaged excitability curves showing that 50 μM nickel depressed the PACAP-induced increase in excitability. *Open circles*: excitability curve for eight control cells before PACAP exposure. *Open inverted triangles*: excitability curve for seven nickel-pretreated cells before PACAP exposure. Note that the excitability curves for the naive cells and nickel-pretreated cells (both prior to PACAP) overlap. *Solid inverted triangles*: excitability curve for seven cells pretreated with 50 μM nickel and exposed to 20 nM PACAP. *Solid circles*: excitability curve for eight control cells during exposure to 20 nM PACAP.

PAC1 Receptor Internalization Is a Critical Mechanism Supporting the PACAP-Induced Increase in Cardiac Neuron Excitability

Merriam et al. [27] determined that the PACAP-induced increase in excitability was very temperature dependent; occurring consistently near physiological temperatures, but essentially absent at room temperature (Fig. 13.9). This rather unexpected observation suggested that PAC_1 receptor internalization might be a critical mechanism mediating the effect of PACAP on excitability. To determine if a PACAP-induced PAC_1 internalization was effected by lowing ambient temperature, the extent of PAC_1 internalization at 37 °C was compared to that occurring at 24 °C using HEK293 cells that stably express a GFP-tagged PAC_1 receptor [27]. It is apparent from Fig. 13.10b, c, that with the HEK-PAC1GFP cells maintained at 37 °C PACAP induced marked PAC_1 internalization, whereas this effect was blunted when the PACAP treatment occurred with the cells kept at 22–24 °C. It was also found that the PACAP-induced PAC_1 receptor internalization was blunted equally well by pretreatment with two inhibitors of endocytosis, Pitstop2 or dynasore. The suppression by Pitstop2 pretreatment is shown in Fig. 13.10d.

Having confirmed that reducing temperature and pretreatment with Pitstop2 and dynasore blunted PAC_1 internalization, subsequent experiments tested whether these two inhibitors could suppress the ability of PACAP to increase cardiac neuron excitability as occurred when the ambient temperature was reduced. Pretreatment with either Pitstop2 or dynasore blunted the PACAP-induced increase in excitability [27]. Figure 13.11 summarizes results illustrating the suppression of the PACAP-induced increase in excitability by pretreatment with the clathrin inhibitor Pitstop2. In additional experiments, it was determined that Pitstop2 also could reverse a PACAP-induced increase in excitability if the drug was added after the PACAP-induced increase in excitability had reached its maximum effect [27].

PAC_1 Receptor Internalization Recruits MEK/ERK Activation

Collectively, the suppression of the PACAP effect on excitability by either reducing ambient temperature or by pretreatment with Pitstop2 and dynasore indicated that a PACAP-induced PAC1 internalization and formation of a signaling endosome appeared to be a key mechanism mediating the PACAP modulation of excitability. Thus, it was considered important to identify what intracellular cascade(s) might be recruited following PAC_1 receptor internalization and formation of a signaling endosome. In earlier studies, Tompkins and Parsons [30] had reported that the MEK kinase inhibitor PD98059 significantly blunted the PACAP-induced increase in cardiac neuron excitability; a result implicating involvement of the MEK kinase intracellular cascade in the PACAP modulation of excitability. However, how the MEK kinase cascade was activated in the cardiac neurons was not established. To explore

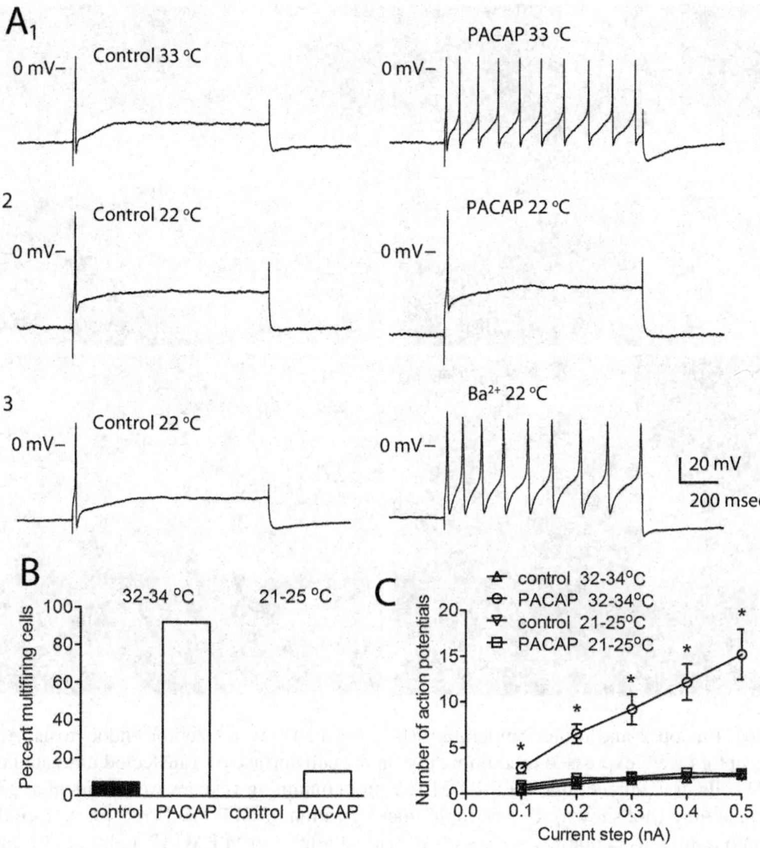

Fig. 13.9 The PACAP-induced increase in excitability was temperature-sensitive. (**a**) Recordings from different cells showed that 20 nM PACAP increased excitability when the bath solution was 33 °C (**A$_1$**), but not when the bath temperature was 22 °C (**A$_2$**). The recording in A$_3$ demonstrated that the addition of 1 mM $BaCl_2$ (Ba^{2+}) increased excitability at 22 °C. In all three recordings, the cells exhibited a phasic firing pattern prior to the addition of PACAP (**A$_1$**, **A$_2$**) or barium (**A$_3$**). The firing pattern shifted to multiple firing in **A$_1$** and **A$_3$**, but not in **A$_2$**. The amplitude of the 1 s depolarizing current pulse was 0.3 nA in each experiment. (**b**) The percentage of cells exhibiting multiple firing in 20 nM PACAP was significantly greater when the temperature was 32–34 °C ($n=13$ cells) than when the bath temperature was 21–25 °C ($n=23$ cells; Fisher's exact test, $P<0.0001$; Also, at 32–34 °C: control ($n=28$) vs. PACAP ($n=13$), significantly different, $P<0.0001$; at 21–25 °C: control ($n=10$) vs. PACAP ($n=23$), not significant, $P=0.5363$). (**c**) Averaged excitability curves generated in the cells maintained at either 33–34 °C or 21–25 °C prior to and during exposure to 20 nM PACAP. The number of action potentials generated at each current step was significantly greater (indicated by *asterisks*) at the warmer temperature in the presence of PACAP ($n=13$) when compared to control at warm temperature ($n=28$; unpaired *t*-test, $P=0.0011$ at 0.1 nA; $P=0.0002$ at 0.2 nA; $P=0.0004$ at 0.3 nA; $P=0.0002$ at 0.4 nA; $P=0.0005$ at 0.5 nA), and to PACAP at room temperature ($n=23$; unpaired *t*-test, $P=0.0029$ at 0.1 nA; $P=0.0003$ at 0.2 nA; $P=0.0005$ at 0.3 nA; $P=0.0003$ at 0.4 nA; $P=0.0005$ at 0.5 nA). The number of action potentials generated at each current step was not different when compared between control ($n=10$) and PACAP ($n=23$) at room temperature ($P=0.3502$ at 0.1 nA; $P=0.4040$ at 0.2 nA; $P=0.7132$ at 0.3 nA; $P=0.4495$ at 0.4 nA; $P=0.6642$ at 0.5 nA). Used with permission from the Journal of Neuroscience [27]

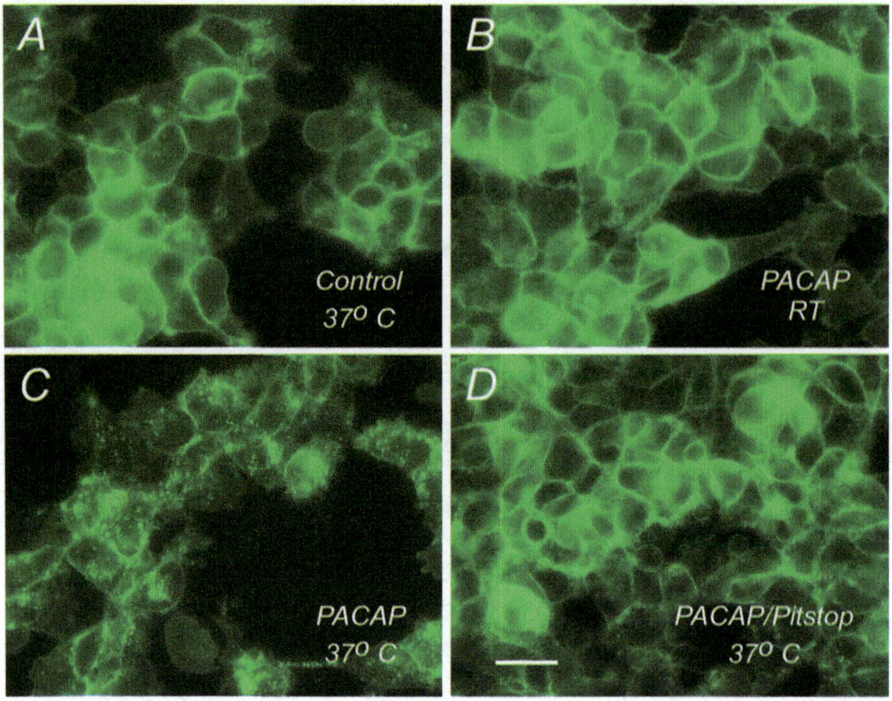

Fig. 13.10 Pitstop 2 and room temperature (RT) blocked PAC1 receptor endocytosis. *A*, GFP-PAC1 receptors were expressed predominantly on the cell surface of transfected untreated control HEK 293 cells; few intracellular GFP-PAC1 receptor containing vesicles were evident at 37 °C or room temperature (not shown). *C*, The rapid internalization of GFP-PAC1 receptors from the cell surface into numerous endocytic vesicles was evident after 25 nM PACAP addition (20 min). By contrast, maintaining the cultures at ambient room temperature (22 °C; panel **b**) or pretreatment of cells with 20 μM Pitstop 2 (10 min; panel **d**) followed by 25 nM PACAP addition (20 min) blocked GFP-PAC1 receptor endocytosis. In both instances, the GFP-PAC1 receptor fluorescence largely remained on the plasma membrane. Calibration bar in **d** equals 20 μm. Used with permission from the Journal of Neuroscience [27]

possible mechanisms, May et al. [49] used the HEK-PAC1GFP cells to show that cooling and pretreatment with Pistop2 suppressed the PACAP-induced increase in ERK activation. More recent, preliminary immunostaining studies on whole mount preparations indicate that PACAP also activates ERK in the cardiac neurons; an effect blunted both at room temperature and by pretreatment with Pitstop2 (Clason and Parsons, unpublished observations). Thus, PAC_1 internalization appears to be one mechanism used by PACAP to recruit the MEK kinase signaling cascade in guinea pig cardiac neurons. Studies now need to establish what ionic conductances are regulated by MEK kinase signaling that contribute to the PACAP-induced increase in excitability.

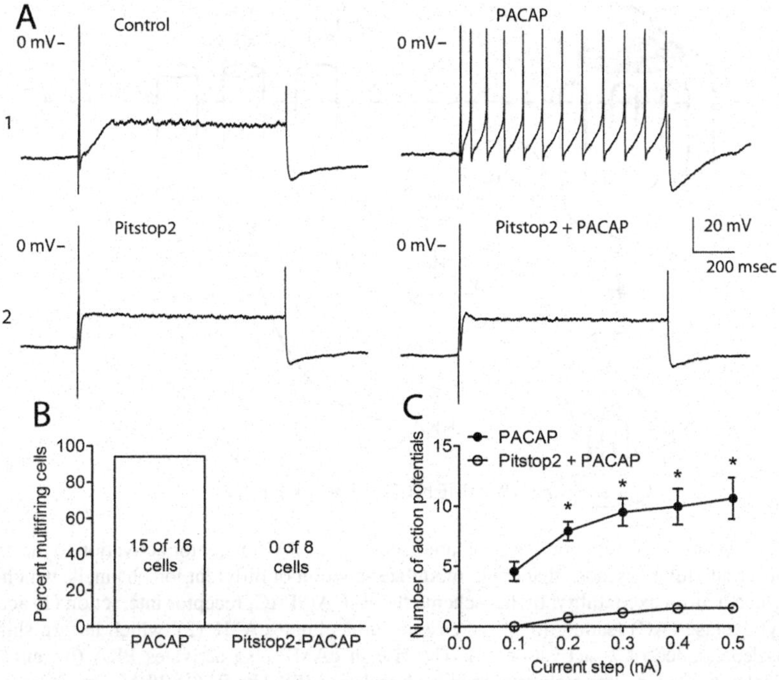

Fig. 13.11 Pretreatment with Pitstop 2 suppressed the PACAP-induced increase in excitability. (**a**) Recordings from two different cells. $\mathbf{A_1}$ illustrates the increase in excitability induced by 20 nM PACAP. $\mathbf{A_2}$ shows that pretreatment with 15 μM Pitstop 2 blocks the increase in excitability induced by 20 nM PACAP. For all recordings, a 1 s, 0.4 nA depolarizing current step was used to initiate action potential activity. (**b**) The percentage of cells exhibiting multiple firing when exposed to PACAP alone (n=16) was significantly greater than when exposed to PACAP after pretreatment with 15 μM Pitstop 2 (n=8; Fisher's exact test, P<0.0001). (**c**) Averaged excitability curves show that Pitstop 2 greatly suppressed the PACAP-induced increase in excitability. *Asterisks* indicate the number of action potentials generated at each current step was significantly greater in PACAP (n=16) than in PACAP and Pitstop 2 (n=8; unpaired t-test, P<0.0001 for steps 0.2–0.5 nA). Used with permission from the Journal of Neuroscience [27]

Summary: Multiple Mechanisms Contribute to the PACAP-Induced Increase in Cardiac Neuron Excitability

From studies on the guinea pig cardiac neurons, it is apparent that the PACAP/PAC$_1$ receptor interaction activates multiple intracellular signaling cascades that could regulate neuronal excitability through selective modulation of different membrane ion channels. A schematic that summarizes potential mechanisms contributing to the PACAP modulation of cardiac neuron excitability is presented in Fig. 13.12. To date our results indicate that PACAP activation of at least two ionic currents, the hyperpolarization-activated nonselective cationic current I_h and a Ni^{2+}-sensitive

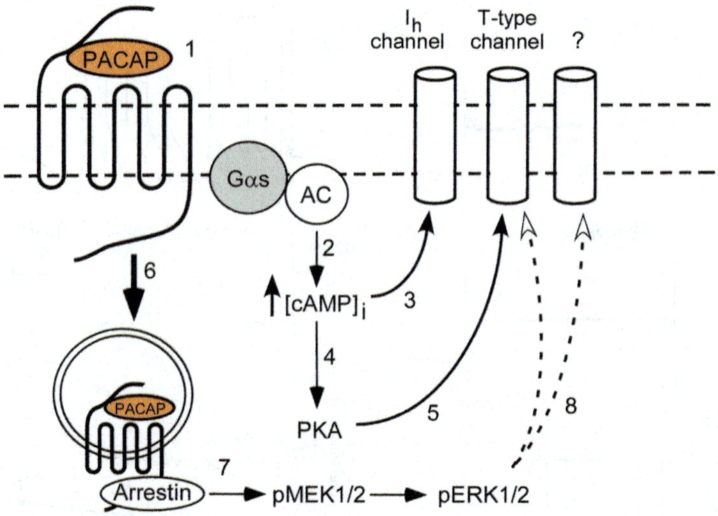

Fig. 13.12 A schematic showing the potential coupling of PAC_1 receptor activation to the recruitment of intracellular cascades that could modulate function of different ion channels, which regulate cardiac neuron excitability. In this scheme the PACAP/PAC_1 receptor interaction (1) activates adenylyl cyclase (AC) leading to an increase in intracellular cAMP (2), which in turn shifts the voltage-dependence of I_h activation (3). The rise in cAMP also activates PKA (4), enhancing T-type calcium currents through protein phosphorylation (5). The PACAP/PAC_1 receptor interaction also stimulates receptor internalization and formation of a signaling endosome (6), which recruits MEK kinase signaling (7). Activation of the MEK kinase signaling cascade (8) could modulate T-type calcium channel trafficking or alter the gating properties of an unidentified channel (?) through protein phosphorylation

low voltage-activated calcium current, contribute to the PACAP-induced increase in cardiac neuron excitability. We tentatively postulate that this low voltage-activated current is a T-type calcium current because of its established role in mediating pacemaker activity and repetitive firing in many neurons. Both of these currents can be modulated by the intracellular signaling cascades activated by PACAP/PAC_1 receptor interaction [44]. A rise in intracellular cAMP can directly enhance I_h and a subsequent activation of PKA can modulate T-type calcium channel function through phosphorylation. Also, it is very likely that additional ionic conductances, likely those modulated by the MEK kinase cascade following recruitment through PAC_1 receptor internalization, contribute to the PACAP-induced increase in excitability. Identification of other ionic conductances that potentially contribute to the PACAP modulation of neuronal excitability will be the focus of future studies.

Acknowledgements The authors thank Ms. Sandra Bossick for her assistance in preparation of this chapter for publication. Studies from the Parsons' laboratory described in this chapter were supported in part by NIH grants NS 23978, HL 65481, P20 RR16435, and P30 GM103498/P30 RR032135.

References

1. Arimura A. Perspectives on pituitary adenylate cyclase activating polypeptide (PACAP) in the neuroendocrine, endocrine, and nervous systems. Jpn J Physiol. 1998;48:301–31.
2. Vaudry D, Falluel-Morel A, Bourgault S, Basille M, Burel D, Wurtz O, et al. Pituitary adenylate cyclase-activating polypeptide and its receptors: 20 years after the discovery. Pharmacol Rev. 2009;61:283–357.
3. Cho J-H, Zushida K, Shumyatsky GP, Carlezon WA, Meloni EG, Bolshakov VY. Pituitary adenylate cyclase-activating polypeptide induces postsynaptically expressed potentiation in the intra-amygdala circuit. J Neurosci. 2012;32:14165–77.
4. Hammack SE, Chung J, Rhoades KM, Schutz KC, Falls WA, Braas KM, et al. Chronic stress increases pituitary adenylate cyclase-activating polypeptide (PACAP) and brain-derived neurotrophic factor (BDNF) mRNA expression in the bed nucleus of the stria terminalis(BNST); roles for PACAP in anxiety-like behavior. Psychoneuroendocrinology. 2009;34:833–43.
5. Hammack SE, May V. Pituitary adenylate cyclase activating polypeptide in stress-related disorders: data convergence from animal and human studies. Biol Psychiatry. 2014;S0006-3223:00962-7.
6. Hill J, Chan S-A, Kuri B, Smith C. Pituitary adenylate cyclase-activating peptide (PACAP) recruits low voltage-activated T-type calcium influx under acute sympathetic stimulation in mouse adrenal chromaffin cells. J Biol Chem. 2011;286:42459–69.
7. Legradi G, Das M, Giunta B, Hirani K, Mitchell EA, Diamond DM. Microinfusion of pituitary adenylate cyclase-activating polypeptide into the central nucleus of amygdale of the rat produces a shift from an active to passive mode of coping in the shock-probe fear/defensive burying test. Neural Plast. 2007;2007:79102.
8. Ressler KJ, Mercer KB, Bradley B, Javovanovic T, Mahan A, Kerley K, et al. Post-traumatic stress disorder is associated with PACAP and the PAC1 receptor. Nature. 2011;470:492–7.
9. Stroth N, Kuri BA, Mustafa T, Chan SA, Smith CB, Eiden LE. PACAP controls adrenomedullary catecholamine secretion and expression of catecholamine biosynthetic enzymes at high splanchnic nerve firing rates characteristic of stress transduction in male mice. Endocrinology. 2013;154:330–9.
10. Braas KM, May V. Pituitary adenylate cyclase-activating polypeptides directly stimulate sympathetic neuron neuropeptide Y release through PAC_1 receptor isoform activation of specific intracellular signaling pathways. J Biol Chem. 1999;274:27702–10.
11. Harmar T, Lutz E. Multiple receptors for PACAP and VIP. Trends Pharmacol Sci. 1994;15:97–9.
12. Spengler D, Waeber C, Pantaloni C, Holsboer F, Bockaert J, Seeburg PH, et al. Differential signal transduction by five splice variants of the PACAP receptor. Nature. 1993;365:170–5.
13. Ishizuka Y, Kashimoto K, Mochizuki T, Sato K, Ohshima K, Yanihara N. Cardiovascular and respiratory actions of pituitary adenylate cyclase activating polypeptides. Regul Pept. 1992;40:29–39.
14. Hirose M, Furukawa Y, Nagashima Y, Yamazaki K, Hoyano Y, Chiba S. Effects of PACAP-38 on the SA nodal pacemaker activity in autonomically decentralized hearts of anesthetized dogs. J Cardiovasc Pharmacol. 1997;29:216–21.
15. Hirose M, Furukawa Y, Nagashima Y, Lakhe M, Chiba S. Pituitary adenylate cyclase-activating polypeptide-27 causes a biphasic chronotropic effect and atrial fibrillation in autonomically decentralized, anesthetized dogs. J Pharmavol Exp Ther. 1997;283:478–87.
16. Hirose M, Furukawa Y, Lakhe M, Chiba SJ. Regional differences in cardiac effects of pituitary adenylate cyclase-activating polypeptide-27 in the isolated dog heart. Eur J Pharmacol. 1998;349:269–76.
17. Yonezawa T, Furukawa Y, Lakhe M, Nagashima Y, Hirose M, Chiba S. PACAP-38 activates parasympathetic nerves in isolated, blood-perfused dog atria. Eur J Pharmacol. 1996;315:289–96.

18. Chang Y, Lawson LJ, Hancock JC, Hoover DB. Pituitary adenylate cyclase-activating polypeptide localization and differential influence on isolated hearts from rats and guinea pigs. Regul Pept. 2005;129:139–46.
19. Hoover DB, Tompkins JD, Parsons RL. Differential activation of guinea pig intrinsic cardiac neurons by the PAC1 agonists maxadilan and pituitary adenylate cyclase-activating polypeptide 27 (PACAP27). J Pharmacol Exp Ther. 2009;331:197–203.
20. Chik CL, Li B, Ogiwara T, Ho AK, Karpinski E. PACAP modulates L-type Ca^{2+} channel currents in vascular smooth muscle cells: involvement of PKC and PKA. FASEB J. 1996;10:1310–7.
21. Seebeck J, Schmidt WE, Kilbinger H, Neumann J, Zimmermann N, Herzig S. PACAP induces bradycardia in guinea-pig heart by stimulation of atrial cholinergic neurons. Arch Pharmacol. 1996;354:424–30.
22. Braas KM, May V, Harakall SA, Hardwick JC, Parsons RL. Pituitary adenylate cyclase-activating polypeptide expression and modulation of neuronal excitability in guinea pig cardiac ganglia. J Neurosci. 1998;18:9766–79.
23. Hoover DB, Girard BM, Hoover JL, Parsons RL. PAC1 receptors mediate positive chronotrophic responses to PACAP-27 and VIP in isolated mouse atria. Eur J Pharmacol. 2013;713:25–30.
24. Hoover DB. Effects of guinea pig vasoactive intestinal peptide on the isolated perfused guinea pig heart. Peptides. 1989;10:343–7.
25. Calupca MA, Vizzard MA, Parsons RL. Origin of pituitary adenylate cyclase-activating polypeptide (PACAP)- immunoreactive fibers innervating guinea pig parasympathetic cardiac ganglia. J Comp Neurol. 2000;423:26–39.
26. Tompkins JD, Ardell JL, Hoover DB, Parsons RL. Neurally released pituitary adenylate cyclase-activating polypeptide enhances guinea pig intrinsic cardiac neurone excitability. J Physiol. 2007;582:87–93.
27. Merriam LA, Baran CN, Girard BM, Hardwick JC, May V, Parsons RL. Pituitary adenylate cyclase 1 receptor internalization and endosomal signaling mediate the pituitary adenylate cyclase activating polypeptide-induced increase in guinea pig cardiac neuron excitability. J Neurosci. 2013;33:4614–22.
28. Braas KM, Rossignol TM, Girard BM, May V, Parsons RL. Pituitary adenylate cyclase activating polypeptide (PACAP) decreases neuronal somatostatin immunoreactivity in cultured guinea-pig parasympathetic cardiac ganglia. Neuroscience. 2004;126:335–46.
29. Tompkins JD, Hardwick JC, Locknar SA, Merriam LA, Parsons RL. Ca^{2+} influx, but not Ca^{2+} release from internal stores, is required for the PACAP-induced increase in excitability in guinea pig intracardiac neurons. J Neurophysiol. 2006;95:2134–42.
30. Tompkins JD, Parsons RL. Identification of intracellular signaling cascades mediating the PACAP-induced increase in guinea pig cardiac neuron excitability. J Mol Neurosci. 2008;36:292–8.
31. Edwards FR, Hirst GDS, Klemm MF, Steele PA. Different types of ganglion cell in the cardiac plexus of guinea pigs. J Physiol. 1995;486:453–71.
32. Xi-Moy SX, Dun NJ. Potassium currents in adult rat intracardiac neurones. J Physiol. 1995;486:15–31.
33. Cuevas J, Harper AA, Trequattrini C, Adams DJ. Passive and active membrane properties of isolated rat intracardiac neurons: regulation by H- and M-currents. J Neurophysiol. 1997;78:1890–902.
34. Merriam LA, Barstow KL, Parsons RL. Pituitary adenylate cyclase-activating polypeptide enhances the hyperpolarization-activated nonselective cationic conductance, I_h, in dissociated guinea pig intracardiac neurons. Regul Pept. 2004;123:123–33.
35. Robinson RB, Sigelbaum SA. Hyperpolarization-activated cation currents: from molecules to physiological function. Annu Rev Physiol. 2003;65:453–80.
36. Tompkins JD, Lawrence YT, Parsons RL. Enhancement of I_h, but not inhibition of I_M, is a key mechanism underlying the PACAP-induced increase in excitability of guinea pig intrinsic cardiac neurons. Am J Physiol Regul Integr Comp Physiol. 2009;297:R52–9.

37. Merriam LA, Roman CW, Baran CN, Girard BM, May V, Parsons RL. Pretreatment with non-selective cationic channel inhibitors blunts the PACAP-induced increase in guinea pig cardiac neuron excitability. J Mol Neurosci. 2012;48:721–9.
38. Jeong S-W, Wurster RD. Calcium channel currents in acutely dissociated intracardiac neurons from adult rats. J Neurophysiol. 1997;77:1769–78.
39. Bhattacharya A, Lakhman SS, Singh S. Modulation of L-type calcium channels in Drosophila via a pituitary adenylyl cyclise activating polypeptide (PACAP) -mediated pathway. J Biol Chem. 2004;279:37291–7.
40. Dziema H, Obrietan K. PACAP potentiates L-type calcium channel conductance in suprachiasmatic nucleus neurons by activating the MAPK pathway. J Neurophysiol. 2002;88:1374–86.
41. Iftinca MC, Zamponi GW. Regulation of neuronal T-type calcium channels. Trends Pharmacol Sci. 2009;30:32–40.
42. Soong TW, Stea A, Hodson CD, Dubel SJ, Vincent SR, Snutch TP. Structure and functional expression of a member of the low voltage-activated calcium channel family. Science. 1993;260:1133–6.
43. Zhang J-F, Randall AD, Ellinor PT, Horne WA, Sather WA, Tanabe T, et al. Distinctive pharmacology and kinetics of cloned neuronal Ca^{2+} channels and their possible counterparts in mammalian CNS neurons. Neuropharmacology. 1993;32:1075–88.
44. Tompkins JD, Merriam LA, Girard BM, May V, Parsons RL. Nickel suppresses the PACAP-induced increase in guinea pig cardiac neuron excitability. Am J Physiol Cell Physiol. 2015:ajpcell.00403.2014.
45. Simms BA, Zamponi GW. Neuronal voltage-gated calcium channels: Structure, function and dysfunction. Neuron. 2014;82:24–45.
46. Talavera K, Nilius B. Biophysics and structure-function relationships of T-type Ca^{2+} channels. Cell Calcium. 2006;40:97–114.
47. Lacinova L. T-type calcium channel blockers-new and notable. Gen Physiol Biophys. 2011;30:403–9.
48. Lee J-H, Gomora JC, Cribbs LL, Perez-Reyes E. Nickel block of three cloned T-type calcium channels: low concentrations selectively block α1H. Biophys J. 1999;77:3034–42.
49. May V, Buttolph TR, Girard BM, Clason TA, Parsons RL. PACAP-induced ERK activation in HEK cells expressing PAC1 receptors involves both receptor internalization and PKC signaling. Am J Physiol Cell Physiol. 2014;306:C1068–79.

[illegible] channels [illegible] Neurosci [illegible]

[illegible]

[illegible] tract [illegible]

40. [illegible] MAPK [illegible]

[illegible] channels [illegible]

41. [illegible]

43. [illegible]

[illegible]

Chapter 14
PACAP in the Circadian Timing System: Learning from Knockout Models

Jens Hannibal

Abstract Glutamate and PACAP are neurotransmitters of the retinohypothalamic tract (RHT). This retinofugal pathway originates from melanopsin expressing intrinsically photosensitive retinal ganglion cells (ipRGCs) innervating areas in the brain which regulate the circadian timing system and areas involved in non-image forming photoperception (e.g., negative masking, pupillary light reflex). The major target is the suprachiasmatic nucleus (SCN), which generates circadian rhythms in behavior and physiology. The SCN is daily entrained to the light–dark cycle due to the release of glutamate and PACAP and activation of NMDA and the PACAP specific (PAC1) receptor on SCN neurons. The chapter briefly reviews in vitro *and* in vivo experiments and compares these data with physiology and gene expression data obtained from mice lacking PACAP, PAC1 receptor or glutamate signaling (VGLUT2-deficient mice). Together, the data provides evidence for PACAP being a neurotransmitter in the RHT acting together with glutamate regulating light entrainment, masking behavior, and the pupillary light reflex.

Keywords PACAP • Glutamate • VGLUT2 • Circadian rhythms • Entrainment • PAC1 receptor • Running wheels • RHT

Introduction

Temporal organization of behavior and physiology in order to prepare the individual for upcoming events is necessary to ensure optimal conditions for survival and reproduction. Circadian timing is a basic property of nearly all cells of the mammalian body and drives temporal organization of physiology and behavior with a cycle of approximately 24 h. In mammals, the main director controlling circadian rhythms is located in neurons of the hypothalamic suprachiasmatic nucleus (SCN) [1].

J. Hannibal (✉)
Department of Clinical Biochemistry, Faculty of Health Sciences, Bispebjerg Hospital, University of Copenhagen, Bispebjerg Bakke 23, Copenhagen, NV 2400, Denmark
e-mail: j.hannibal@dadlnet.dk

D. Reglodi, A. Tamas (eds.), *Pituitary Adenylate Cyclase Activating Polypeptide — PACAP*, Current Topics in Neurotoxicity 11,
DOI 10.1007/978-3-319-35135-3_14

These neurons considered as the master clock of the brain act in a synchronized harmony in which gene expression, metabolism, and firing rhythms of the neurons oscillate with a period length close to 24 h (circa diem = 1 day). The signals generated by the SCN neurons control the inter-SCN synchronization of the nearly 10,000 neurons of the SCN [2]. In additions, the SCN synchronizes other areas of the brain via neuronal projections and by hormonal signals and core body temperature conducting phase of peripheral cellular clocks regulating physiology of every organ of the body [3]. The molecular machinery of the biological clock has been known for more than a decade and consists of a group of clock genes, which by complex feedback interactions between the encoded clock proteins drive circadian rhythmicity within each neuron of the SCN [2, 3]. The endogenous period length of the clock deviates slightly from 24 h and neurons of the SCN are daily synchronized (entrained) to the astronomical day by light, which is the most important "Zeitgeber" for entrainment [4]. In mammals, light information for entrainment is transmitted from intrinsically photosensitive retinal ganglion cells (ipRGCs) containing the photoreceptor melanopsin rendering these cells photosensitive, and by signals from the classical photoreceptors, the rods and cones [5]. The ipRGCs project as the retinohypothalamic tract (RHT) to the brain including the SCN and provide light information for the entrainment process, negative masking, and the pupillary light reflex [6]. Two neurotransmitters, glutamate and PACAP, several subtypes of glutamate receptors and the PACAP specific receptor, the PAC1 receptor, convey the light signals to the SCN neurons [7]. In the present chapter, the role of PACAP as neurotransmitter in the circadian timing system and in regulation of other non-image forming functions (negative masking, pupillary light reflex) is reviewed with focus on the knowledge obtained from genetically modified mice lacking PACAP, PAC1 receptor, or glutamate signaling (VGLUT2 knockout mice).

PACAP in the Retinohypothalamic Tract (RHT)

Retinal projection reaching the SCN of ventral hypothalamus forebrain was identified in 1970 [8], and in 1972 this observation led to the discovery of the circadian brain clock of the SCN [9, 10]. By various tracer methods a distinct neuronal pathway was identified and a subpopulation of retinal ganglion cells was identified [11]. The pathway was intensely studied by electron microscopy, functional neuroanatomy, in vivo and in vitro studies and evidence was provided that the main neurotransmitter of the RHT was glutamate [12]. Glutamate application in vitro and in vivo mimics the effects of light on the circadian phase [13] and by inducing immediate early genes such as *Fos* [14]. Glutamate signaling in the SCN occurs by activation of several subtypes of glutamate receptors including both ionotrope and metabotropic receptors (reviewed in ref. [15]), which seem to target the light inducible clock genes *Per1* and *Per2* [16].

Many central—and peripheral neuronal systems can be characterized by their content of neurotransmitters, and often the classical amino acid derived neurotransmitters are found together with peptide neurotransmitters. In many cases, the neuropeptide transmitters seem to be able to modulate the effects of the primary classical transmitter [17]. In 1997, we showed that PACAP was located in a subset of retinal ganglion cells (RGCs) of the rat retina, and that these cells project to the SCN and the intergeniculate leaflet (IGL) of the lateral geniculate nucleus. Furthermore, it was shown that application of PACAP in vitro was able to phase shift the circadian rhythm in SCN neurons [18]. The PACAP containing RGCs which co-stored glutamate [19, 20] were light responsive as evaluated by their expression of FOS, and it was suggested that the PACAP RGCs could be light responsive, independent of the classical photoreceptors [21]. Light-responsive RGCs could explain observations in mice and humans with retinal degeneration of the outer retina who were able to entrain to light, had intact pupil reflex and were able to light suppress nocturnal melatonin secretion [22, 23]. The light responsiveness was confirmed by showing that only PACAP containing RGCs co-expressed a new photoreceptor, melanopsin [24]. Melanopsin expressing RGCs constitute the RHT in mice. This was elegantly demonstrated in genetically modified mice carrying a tau-LacZ promotor construct inserted into melanopsin expressing cells and β-GAL staining, which identified axonal melanopsin projections to the SCN [25]. Furthermore, another elegant study showed that the melanopsin containing RGCs were intrinsically photosensitive (ipRGCs) [26]. Several studies in different species have confirmed that PACAP is found exclusively in the ipRGCs [27–30] and that these cells project to many light-responsive areas in the brain involved in non-image forming photoperception [30, 31].

Indirect evidence supports that PACAP is released from RHT nerve terminals in the SCN upon light stimulation. This has been done in vitro by applying PACAP and glutamate and glutamate and the PAC1 receptor antagonist PACAP6-38 on in vitro brain slice models containing the SCN. By using this model, PACAP in micromolar concentrations was shown to play a modulatory role on glutamate induced phase shift at night [32] while in nanomolar concentrations PACAP had a light like effect on clock phasing [33]. The light like effect of PACAP on the circadian phase was also demonstrated in vivo by intracerebroventricular (icv) injections of either PACAP, the PACAP antagonist PACAP6-38, or specific PACAP antibodies, which modulate the effect of light on circadian phase [27, 32, 34, 35]. The molecular mechanisms leading to light-induced phase shift after the release of glutamate and PACAP involve several signaling transduction pathways initiated by NMDA—and PAC1 receptor activation, Homer-1A, cAMP, Ca^{2+}, and CREB [7]. The targets seem to be the clock genes *Per1* and *Per2*. The expression level of both genes was studied in in vitro brain slice models and the modulator role of PACAP on glutamate induced *Per* gene expression, corresponding to the effect on phase shifts of the circadian rhythm was confirmed in these studies [16].

Circadian Rhythms, Negative Masking, and Pupillary Light Reflex (PLR) in Mice Lacking PACAP, PAC1 Receptor, or Glutamate Signaling

In vivo studies of behavior and gene expression in gene modified animals have been shown useful for the understanding of the role of PACAP and the PAC1 receptor in the circadian timing system and of PACAP as neurotransmitter in other non-image forming light regulated systems controlling negative masking behavior (ie. inhibition of night activity in nocturnal animals by light) and the pupillary light reflex. Circadian rhythms in mice are typically examined by keeping the mice in cages carrying a running wheel, and running wheel (RW) activity is used as a parameter for circadian activity rhythms. Results obtained from the various knockout mice (KO) models have been compared with data from electrophysiological studies on circadian phase shifts measured after application of PACAP and/or glutamate, and from studies in which PACAP, glutamate antagonists or PACAP antibodies have been injected icv. These in vitro and in vivo studies have indicated that PACAP can potentiate glutamate induced phase delay during early night while decrease glutamate induced phase advance at late night. Application of glutamate together with PACAP6-38, a PAC1 receptor antagonist, decreases glutamate induced phase delays at early night and potentiates glutamate induced phase advances at late night (reviewed in ref. [7]).

Two independent groups have generated mice lacking the PACAP gene, one group mice lacking the PAC1 receptor gene, and two independent groups mice lacking glutamate signaling by removing the gene encoding the vesicular glutamate transporter2 (VGLUT2).

PACAP-Deficient Mice

Circadian Regulation by Light

PACAP-deficient mice raised on a C57 background and backcrossed into an ICR background were used in circadian studies on PACAP-deficient mice [36–38]. In their first study, Kawaguchi et al. [38] showed that PACAP KO mice had significant reduced phase advance when exposed to light at late subjective night while no difference was found in light response with light stimulation at early night. In both situations, light intensity was 20 lux, which may explain the unexpected results (see below). In contrast to the RW experiments, light-induced FOS in the SCN was significantly reduced at early night while no difference was found at late night in PACAP KO mice compared to wild type mice. At both time points light of 20 lux induced FOS expression in SCN neurons of both wild type and KO mice compared to no light stimulation [38]. In a follow up study using the same mouse model, it was taken into consideration that light intensity may be of importance for PACAP signaling to the clock. Reexamination of light-induced phase delays at early night during light intensities between 10 and 100 lux showed that while no difference in phase delay occurs at 20 lux, the phase delay at 100 lux was significantly smaller in PACAP KO compared to wild type mice [37]. Similar findings were observed at the end of the night, where the phase advance was significantly reduced by 30 min illumination at 100 lux.

Negative Masking Behavior

Another significant change in behavior has been observed in PACAP KO mice. When kept in 12:12 LD cycles, wild type mice start RW activity when light is turned off while PACAP KO mice have a positive phase angle with activity onset advanced by 2–3 h starting night activity significantly before the subjective night. This abnormal phase onset in PACAP KO mice of the ICR strain of mice is most likely explained by reduced negative masking found in the PACAP KO mice [37].

The Pupillary Light Reflex

The pupillary light reflex is dependent of the melanopsin ipRGCs and input from the classical photoreceptors, the rods and cones [39]. The PLR is driven by neurons located in the olivary pretectal nucleus (OPN) which is densely innervated by retinal PACAP fibers in rats, mice, and monkeys [28, 30, 31]. However, examination of the PLR in PACAP KO mice did not reveal any role of PACAP in the PLR [37] (but see discussion below).

When investigating circadian behavior in another strain of PACAP-deficient mice on a C57BL/6 background [40] similar observations were found, the light-induced phase shifts was found to be compromised at both early and late subjective night [40]. Despite the reduced light sensitivity in PACAP KO mice they were able to re-entrain when the external LD cycle was shifted by either by 8 h advance or delay (jetlag experiments) [40]. The PACAP KO mice used by Colwell et al. had no deficit in masking response and PLR was not studied [40].

PACAP Receptor 1 (PAC1)-Deficient Mice

Circadian Regulation by Light

Like the PACAP-deficient mice, PAC1 KO mice entrain to the LD cycle and keep a stable predictable RW activity during constant darkness [41, 42]. This indicates that both genotypes have a stable biological clock also indicated by the stable expression of clock genes found in PACAP KO mice [37]. Although PAC1 KO mice can entrain to LD cycles similarly to wild type animals they show, like PACAP KO mice, a significantly reduced response to light stimulation at early subjective night [41]. When placed in T-cycles (25–27 h) at different light intensities (10–300 lux) limit of entrainment was demonstrated in the PAC1 KO mice [41]. The reduced light sensitivity in PAC1 KO mice resulted in a decreased ability to re-entrain to an 8 h phase shift of the external LD cycle (jetlag experiment), which was more pronounced at low light intensities [41]. As indicated by in vitro and in vivo studies, examination of molecular expression in PAC1 mice supports that light signaling mediated by the PACAP release is targeting the light sensitive clock genes, *Per1* and *Per2* [42].

Negative Masking Behavior

As the PACAP KO mice [37], PAC1 KO mice had an impaired negative masking being more affected at low light intensity [41]. Together, the studies of mice lacking either PACAP or the PAC1 receptor support the results obtained from in vitro and in vivo studies and show that PACAP plays an important role during light regulation of the SCN activity at early night and being most essential at low light intensities. At late night, PACAP signaling seems to be less important or more variable at least in studies of mice. In mice, light produces large phase delays of the circadian rhythm in the early night and small phase advances at late night. Other rodent species like hamsters respond with large phase advances at late night and small phase delays at early night [15]. This difference is a property of the clock and the endogenous period (tau). Mice have a fast running clock (tau shorter than 24 h) and therefore need to slow down the speed of the clock (phase delay) to keep entrained. Hamsters, on the contrary, have a slow running clock (tau longer than 24 h) and therefore need to speed up their clock (phase advance) to keep entrained to the LD cycle [4, 15]. It is possible that it is due to the less sensitivity of the clock in mice that the phase shift differences at late night in both PACAP KO—and PAC1 KO compared to wild type mice are too small to demonstrate any significance [4, 15].

The Pupillary Light Reflex

The OPN which control the PLR receives retinal PACAP input and expresses the PAC1 receptor [28]. In contrast to the observation in PACAP KO mice which show normal PLR, the PLR in PAC1 KO was significantly attenuated compared to wild type mice and the difference became more pronounced by increasing light intensities [28] suggesting that PAC1 receptor signaling is involved in the PLR. The discrepancy in PLR found between PACAP KO—and PAC1 KO mice could be due to the different experiment paradigms used. It is important to note that although melanopsin in ipRGCs are necessary for an intact PLR, ipRGCs receive input from the classical photoreceptors [39]. PACAP KO mice were exposed to blue light ((λ)

460–490 nm) which selectively stimulate the melanopsin photoreceptor [6]. PAC1 KO mice were stimulated by white light which also activates rods and cones. It is not known which subtypes of melanopsin cells that control the PLR [43], but these ipRGCs may be dependent of input signals from the deep retina (rods and cones) and being more sensitive to white light. It is likely that the melanopsin/PACAP ipRGCs involved in the regulation of the PLR are different from the ipRGCs that regulate light entrainment and masking. The PAC1 KO PLR phenotype was more pronounced when exposed to light at higher intensity [28] while the PAC1 KO phenotype of entrainment and negative masking was displayed primary at low light intensity [41].

Vesicular Glutamate Transporter2 (VGLUT2)-Deficient Mice

The VGLUT2 transporter is a member of a family of vesicular glutamate transporters that pack glutamate into synaptic vesicles [44]. The different transporters show distinct expression within the CNS and retina [45]. VGLUT2 is expressed in ipRGCs [46] and are co-stored with PACAP in melanopsin containing ipRGCs [19]. Mating gene modified mice in which Cre recombinase is expressed exclusively in ipRGCs with mice carrying two loxP elements flanking exon 2 of *Vglut2,* result in mice with cell specific loss of VGLUT2 in ipRGCs which were examined for circadian behavior [47, 48, 49].

Circadian Regulation by Light

Circadian behavior in VGLUT2-deficient mice was investigated using RW at different light intensities, a factor which may explain some discrepancies in behavior found between the two stains of mice [47, 49]. Purrier et. al. performed their study at 900 lux (approx. 3.15 W/m^2) while Gompf et al. used two different light intensities of 130 lux (0.35 W/m^2) and 2000 lux (6.53 W/m^2), respectively. Most of the VGLUT2 KO mice entrained to the LD cycle at 900 lux although a subpopulation of the KO mice showed a more irregular circadian locomotor activity displaying RW activity also during the subjective day [47]. All VGLUT2 KO mice showed normal free-running activity when exposed to constant darkness indicating an intact circadian clock [47]. The altered sensitivity to light stimulation displayed during LD cycles also led to a decreased ability to re-entrain after an 8 h shift in the external LD cycle (jetlag experiment) which caused VGLUT2 KO mice to use significantly more LD cycles for re-entrainment [47]. VGLUT2 KO mice also fail to entrain to a skeleton photoperiod ((SPP) consisting of 1 h light in the morning and 1 h in the evening separated by 10 h of darkness simulating dawn and dusk) suggesting that the VGLUT2 KO mice had a disturbed settings for dusk and dawn signals [47].

The other strain of VGLUT2 KO mice [49] demonstrated a more severe phenotype. When placed in 12:12 LD conditions at 130 lux most of the VGLUT2 KO mice failed to entrain to the LD cycle and exhibited free-running periods as normally seen only in constant darkness [49]. When exposed to light at early night wild type mice phase delayed the circadian rhythm and induction of FOS was found in the SCN while the VGLUT2 KO mice showed no shift in the circadian phase and no induction of FOS in the SCN, an indication for a strongly compromised light sensitivity in these mice [49]. The decreased light sensitivity in these animals was substantiated by placing them in constant light (LL). During LL, wild type mice alter their free running endogenous rhythm (tau) which becomes longer than 25 h. The tau of VGLUT2 KO mice remained unchanged as during constant darkness [49]. However, although seemingly relatively less sensitive to light, a subpopulation of the VGLUT2 mice was able to re-entrain to a new LD cycle (8 h phase shift, jetlag experiment) indicating a remaining light sensitivity most likely due to PACAP signaling in these mice [49]. When exposing the population of VGLUT2 KO mice not responding to 130 lux light to constant light at 2000 lux, the tau was increased thus indicating light sensitivity remains also with a higher threshold in this subpopulation of VGLUT2 KO mice.

Negative Masking Behavior

In both strains of VGLUT2 KO mice negative masking was impaired despite the light intensities used [47, 49]. This was examined by placing the animals in an ultrashort LD cycle (Gompf; 2:2 LD, Purrier; 3.5:3.5 LD). Mice with normal masking behavior display activity only during the dark period, while disturbed masking as seen in the VGLUT2 KO mice, is displayed by activity in both the light and dark phase [47, 49].

The Pupillary Light Reflex

The PLR was also examined in VGLUT2 KO mice by Purrier et al. and since this strain of VGLUT2 KO mice displayed impaired light sensitivity when entrained to LD cycle and negative masking behavior, it would be expected that they also had impaired PLR at both high and low light intensity. Indeed, a significantly attenuated PLR was found in the VGLUT2 KO mice [47].

Summary and Conclusions

By using gene-modified mice lacking PACAP signaling (i.e., PACAP or PAC1 KO mice) and glutamate signaling (VGLUT2 KO mice), behavior, physiological, and gene expression studies extend and confirm previous in vitro and in vivo studies showing that PACAP is a neurotransmitter in the RHT, which mediates non-image forming light information to the circadian timing system, regulates negative masking and the pupillary light reflex. PACAP is a co-transmitter in the RHT acting together with glutamate, the primary neurotransmitter of the RHT. This notion was supported by the recent studies of VGLUT2 KO mice. These mice have a severe reduced light sensitivity regarding non-image forming photoperception (photoentrainment, negative masking, PLR) and the most affected animals were unable to entrain to LD cycles. These observations are in agreement with in vitro observations that PACAP acts as a co-neurotransmitter modulating light signals mediated by glutamate signaling. Future studies are needed to clarify under which stimulus conditions PACAP is released from RHT nerve terminals. Low light intensity seems to be one condition in which PACAP signaling is of importance. However, not only light intensity but also wavelength (color) seem to be of major importance in regulation of circadian timing [50].

Acknowledgements The Danish Biotechnology Center for Cellular Communication supported this work.

References

1. Reppert SM, Weaver DR. Coordination of circadian timing in mammals. Nature. 2002;418:935–41.
2. Mohawk JA, Takahashi JS. Cell autonomy and synchrony of suprachiasmatic nucleus circadian oscillators. Trends Neurosci. 2011;34:349–58.
3. Mohawk JA, Green CB, Takahashi JS. Central and peripheral circadian clocks in mammals. Annu Rev Neurosci. 2012;35:445–62.
4. Golombek DA, Rosenstein RE. Physiology of circadian entrainment. Physiol Rev. 2010;90:1063–102.
5. Sand A, Schmidt TM, Kofuji P. Diverse types of ganglion cell photoreceptors in the mammalian retina. Prog Retin Eye Res. 2012;31:287–302.
6. Do MT, Yau KW. Intrinsically photosensitive retinal ganglion cells. Physiol Rev. 2010;90:1547–81.
7. Hannibal J. Roles of PACAP-containing retinal ganglion cells in circadian timing. Int Rev Cytol. 2006;251:1–39.
8. Sousa-Pinto A, Castro-Correia J. Light microscopic observations on the possible retinohypothalamic projection in the rat. Exp Brain Res. 1970;11:515–27.
9. Hendrickson AE, Wagoner N, Cowan WM. An autoradiographic and electron microscopic study of retino-hypothalamic connections. Z Zellforsch Mikrosk Anat. 1972;135:1–26.
10. Moore RY, Lenn NJ. A retinohypothalamic projection in the rat. J Comp Neurol. 1972;146:1–14.
11. Pickard GE. Morphological characteristics of retinal ganglion cells projecting to the suprachiasmatic nucleus: a horseradish peroxidase study. Brain Res. 1980;183:458–65.
12. Ebling FJ. The role of glutamate in the photic regulation of the suprachiasmatic nucleus. Prog Neurobiol. 1996;50:109–32.
13. Ding JM, Chen D, Weber ET, Faiman LE, Rea MA, Gillette MU. Resetting the biological clock: mediation of nocturnal circadian shifts by glutamate and NO. Science. 1994;266:1713–7.
14. Kornhauser JM, Nelson DE, Mayo KE, Takahashi JS. Photic and circadian regulation of c-fos gene expression in the hamster suprachiasmatic nucleus. Neuron. 1990;5:127–34.

15. Hannibal J. Neurotransmitters of the retino-hypothalamic tract. Cell Tissue Res. 2002; 309:73–88.
16. Nielsen HS, Hannibal J, Knudsen SM, Fahrenkrug J. Pituitary adenylate cyclase-activating polypeptide induces period1 and period2 gene expression in the rat suprachiasmatic nucleus during late night. Neuroscience. 2001;103:433–41.
17. Nassel DR. Neuropeptide signaling near and far: how localized and timed is the action of neuropeptides in brain circuits? Invert Neurosci. 2009;9:57–75.
18. Hannibal J, Ding JM, Chen D, Fahrenkrug J, Larsen PJ, Gillette MU, et al. Pituitary adenylate cyclase activating peptide (PACAP) in the retinohypothalamic tract. A daytime regulator of the biological clock. J Neurosci. 1997;17:2637–44.
19. Engelund A, Fahrenkrug J, Harrison A, Hannibal J. Vesicular glutamate transporter 2 (VGLUT2) is co-stored with PACAP in projections from the rat melanopsin-containing retinal ganglion cells. Cell Tissue Res. 2010;340:243–55.
20. Hannibal J, Moller M, Ottersen OP, Fahrenkrug J. PACAP and glutamate are co-stored in the retinohypothalamic tract. J Comp Neurol. 2000;418:147–55.
21. Hannibal J, Vrang N, Card JP, Fahrenkrug J. Light dependent induction of c-Fos during subjective day and night in PACAP containing retinal ganglion cells of the retino-hypothalmic tract. J Biol Rhythms. 2001;16:457–70.
22. Freedman MS, Lucas RJ, Soni B, von Schantz M, Muñoz M, David-Gray Z, et al. Regulation of mammalian circadian behavior by non-rod, non-cone, ocular photoreceptors. Science. 1999;284:502–4.
23. Lucas RJ, Freedman MS, Munoz M, Garcia-Fernandez JM, Foster RG. Regulation of the mammalian pineal by non-rod, non-cone, ocular photoreceptors. Science. 1999;284:505–7.
24. Hannibal J, Hindersson P, Knudsen SM, Georg B, Fahrenkrug J. The photopigment melanopsin is exclusively present in PACAP containing retinal ganglion cells of the retinohypothalamic tract. J Neurosci. 2002;22(RC191):1–7.
25. Hattar S, Liao HW, Takao M, Berson DM, Yau KW. Melanopsin-containing retinal ganglion cells: Architecture, projections, and intrinsic photosensitivity. Science. 2002;295:1065–70.
26. Berson DM, Dunn FA, Takao M. Phototransduction by retinal ganglion cells that set the circadian clock. Science. 2002;295:1070–3.
27. Bergström AL, Hannibal J, Hindersson P, Fahrenkrug J. Light-induced phase shift in the Syrian hamster (*Mesocricetus auratus*)is attenuated by the PACAP receptor antagonist PACAP6-38 or PACAP immunoneutralization. Eur J Neurosci. 2003;9:2552–62.
28. Engelund A, Fahrenkrug J, Harrison A, Luuk H, Hannibal J. Altered pupillary light reflex in PACAP receptor 1-deficient mice. Brain Res. 2012;1453:17–25.
29. Hannibal J, Hindersson P, Ostergaard J, Georg B, Heegaard S, Larsen PJ, et al. Melanopsin is expressed in PACAP-containing retinal ganglion cells of the human retinohypothalamic tract. Invest Ophthalmol Vis Sci. 2004;45:4202–9.
30. Hannibal J, Kankipati L, Strang CE, Peterson BB, Dacey D, Gamlin PD. Central projections of intrinsically photosensitive retinal ganglion cells in the macaque monkey. J Comp Neurol. 2014;522:2231–48.
31. Hannibal J, Fahrenkrug J. Target areas innervated by PACAP immunoreactive retinal ganglion cells. Cell Tissue Res. 2004;316:99–113.
32. Chen D, Buchanan GF, Ding JM, Hannibal J, Gillette MU. PACAP: a pivotal modulator of glutamatergic regulation of the suprachiasmatic circadian clock. Proc Natl Acad Sci U S A. 1999;96:13409–14.
33. Harrington ME, Hoque S, Hall A, Golombek D, Biello S. Pituitary adenylate cyclase activating peptide phase shifts circadian rhythms in a manner similar to light. J Neurosci. 1999;19:6637–42.
34. Minami Y, Furuno K, Akiyama M, Moriya T, Shibata S. Pituitary adenylate cyclase-activating polypeptide produces a phase shift associated with induction of mPer expression in the mouse suprachiasmatic nucleus. Neuroscience. 2002;113:37–45.
35. Piggins HD, Marchant EG, Goguen D, Rusak B. Phase-shifting effects of pituitary adenylate cyclase activating polypeptide on hamster wheel-running rhythms. Neurosci Lett. 2001;305:25–8.

36. Hashimoto H, Shintani N, Tanaka K, Mori W, Hirose M, Matsuda T, et al. Altered psychomotor behaviors in mice lacking pituitary adenylate cyclase-activating polypeptide (PACAP). Proc Natl Acad Sci U S A. 2001;98:13355–60.
37. Kawaguchi C, Isojima Y, Shintani N, Hatanaka M, Guo X, Okumura N, et al. PACAP-deficient mice exhibit light parameter-dependent abnormalities on nonvisual photoreception and early activity onset. PLoS One. 2010;5:e9286.
38. Kawaguchi C, Tanaka K, Isojima Y, Shintani N, Hashimoto H, Baba A, et al. Changes in light-induced phase shift of circadian rhythm in mice lacking PACAP. Biochem Biophys Res Commun. 2003;310:169–75.
39. Guler AD, Ecker JL, Lall GS, Haq S, Altimus CM, Liao HW, et al. Melanopsin cells are the principal conduits for rod-cone input to non-image-forming vision. Nature. 2008;453:102–5.
40. Colwell CS, Michel S, Itri J, Rodriguez W, Tam J, Lelievre V, et al. Selective deficits in the circadian light response in mice lacking PACAP. Am J Physiol Regul Integr Comp Physiol. 2004;287:R1194–201.
41. Hannibal J, Brabet P, Fahrenkrug J. Mice lacking the PACAP type I receptor have impaired photic entrainment and negative masking. Am J Physiol Regul Integr Comp Physiol. 2008;295:R2050–8.
42. Hannibal J, Brabet P, Jamen F, Nielsen HS, Journot L, Fahrenkrug J. Dissociation between light induced phase shift of the circadian rhythm and clock gene expression in mice lacking the PACAP type 1 receptor (PAC1). J Neurosci. 2001;21:4883–90.
43. Schmidt TM, Chen SK, Hattar S. Intrinsically photosensitive retinal ganglion cells: many subtypes, diverse functions. Trends Neurosci. 2011;34:572–80.
44. Takamori S. VGLUTs: "exciting" times for glutamatergic research? Neurosci Res. 2006;55:343–51.
45. Fujiyama F, Hioki H, Tomioka R, Taki K, Tamamaki N, Nomura S, et al. Changes of immunocytochemical localization of vesicular glutamate transporters in the rat visual system after the retinofugal denervation. J Comp Neurol. 2003;465:234–49.
46. Johnson J, Fremeau Jr RT, Duncan JL, Renteria RC, Yang H, Hua Z, et al. Vesicular glutamate transporter 1 is required for photoreceptor synaptic signaling but not for intrinsic visual functions. J Neurosci. 2007;27:7245–55.
47. Purrier N, Engeland WC, Kofuji P. Mice deficient of glutamatergic signaling from intrinsically photosensitive retinal ganglion cells exhibit abnormal circadian photoentrainment. PLoS One. 2014;9:e111449.
48. Tong Q, Ye C, McCrimmon RJ, Dhillon H, Choi B, Kramer MD, et al. Synaptic glutamate release by ventromedial hypothalamic neurons is part of the neurocircuitry that prevents hypoglycemia. Cell Metab. 2007;5:383–93.
49. Gompf HS, Fuller PM, Hattar S, Saper CB, Lu J. Impaired circadian photosensitivity in mice lacking glutamate transmission from retinal melanopsin cells. J Biol Rhythms. 2015;30:35–41.
50. Walmsley L, Hanna L, Mouland J, Martial F, West A, Smedley AR, et al. Colour as a signal for entraining the mammalian circadian clock. PLoS Biol. 2015;13:e1002127.

Chapter 15
The Role of PACAP in the Regulation of Body Temperature

Andras Garami, Eszter Pakai, Zoltan Rumbus, and Margit Solymar

Abstract Studies aiming at the investigation of the thermal effects of the pituitary adenylate cyclase-activating polypeptide (PACAP) revealed that PACAP plays an important role in the regulation of body temperature. We review literature data on the effects of pharmacological modulation of PACAP signaling on deep body temperature as well as on the influence of PACAP-signaling deficiency on thermoregulation in animals and humans. We describe the contribution of behavioral and autonomic thermoeffectors to the hyperthermic effect of PACAP and the thermoregulatory phenotype of mice genetically lacking the peptide. We propose that behavioral (hyperactivity, wet-dog shakes) and autonomic (non-shivering thermogenesis and cutaneous vasoconstriction) cold-defense responses are recruited in PACAP-induced hyperthermia. The absence of PACAP results in hypometabolism and as a compensatory mechanism in increased locomotor activity. We hypothesize that the thermal effects of PACAP are evoked through modulation of the cold-activated pathway in the preoptic area of the hypothalamus. Hyperthermia in response to exogenous PACAP administration develops through activation of γ-aminobutyric acid-ergic neurons located in the median preoptic nucleus, while the hypometabolism in PACAP deficiency is caused by the absent suppression of tonically activated γ-aminobutyric acid-ergic neurons in the medial preoptic area, which leads to enhanced inhibition of non-shivering thermogenesis. The contribution of other central nervous system regions to the thermoregulatory effects of PACAP is also discussed.

Keywords Body temperature • Thermoregulation • Hyperthermia • Homeostasis • Thermoeffectors • Thermogenesis • Metabolic rate • Heat loss • Skin vasomotor tone • Locomotor activity • *Pacap* knockout • PAC1 receptor • Brown adipose tissue • Hypometabolism

---

A. Garami (✉) • E. Pakai • Z. Rumbus • M. Solymar
Institute for Translational Medicine, Medical School, University of Pecs,
12 Szigeti Str., Pecs 7624, Hungary
e-mail: andras.garami@aok.pte.hu

D. Reglodi, A. Tamas (eds.), *Pituitary Adenylate Cyclase Activating Polypeptide — PACAP*, Current Topics in Neurotoxicity 11,
DOI 10.1007/978-3-319-35135-3_15

Abbreviations

BAT	Brown adipose tissue
CNS	Central nervous system
GABA	γ-aminobutyric acid
i.c.v.	Intracerebroventricular(ly)
i.v.	Intravenous(ly)
KO	Knockout
LPB	Lateral parabrachial nucleus
MnPO	Median preoptic nucleus
MPO	Medial preoptic area
PAC1R	PAC1 receptor
PACAP	Pituitary adenylate cyclase-activating polypeptide
Pacap$^{-/-}$, *Pacap*$^{+/+}$	Mice with the *Pacap* gene homozygously missing or present, respectively
POA	Preoptic area
T_b	Body temperature
TRP	Transient receptor potential (channel)
TRPV1	TRP vanilloid-1
VIP	Vasoactive intestinal polypeptide
VMN	Ventromedial nuclei
VO_2	Oxygen consumption
VPAC1R, VPAC2R	VIP/PACAP receptor 1 and 2, respectively

Introduction

Temperature is life! Endothermic organisms like humans maintain a relatively stable deep body temperature (T_b) under even extreme environmental conditions with the help of behavioral and autonomic thermoeffectors [1]. The activity level of most thermoeffectors can be influenced by a variety of neuropeptides located both within and outside of the central nervous system (CNS) [2]. The effects of a number of these substances have been studied on thermoregulation [3], which can be also regarded as part of complex energy balance [4]. Underlying the complexity of energy balance regulation, how the activity of the effector(s) is affected by the different neuropeptides shows a great variability. As part of their thermoregulatory actions, some substances such as endothelin-1 activate mainly a single mechanism like constriction in cutaneous vessels to increase T_b [5], others recruit more effectors with synergistic effects, e.g., hypermetabolism and skin vasonconstriction in the hyperthermia induced by alarin [6], yet others induce the activation of disparate mechanisms like the increased heat loss with simultaneously elevated thermogenesis in response to α-melanocyte-stimulating hormone administration [7]. Further complicating the picture, some substances (e.g., leptin) can be important mediators in disease states such as systemic inflammation, in addition to their key role in energy balance homeostasis [8].

Since the identification of the two isoforms of the pituitary adenylate cyclase-activating polypeptide (PACAP) over 2 decades ago [9, 10], its involvement in the regulation of energy balance has been supported with multiple lines of evidence ranging from tissue distribution [11, 12], through thermoregulatory characterization in animal experiments (see Sects. 15.3.1.1 and 15.3.2.1), to thermal effects in human subjects (see Sects. 15.3.1.2 and 15.3.2.2). In addition to the maintenance of bodily homeostasis, the contribution of PACAP to systemic inflammation, which can be accompanied by hypothermia or fever or both in rats [13] and mice [14], has been also established [15, 16]. Although two major forms of the peptide, namely PACAP38 and PACAP27, have been identified (for a review, see ref. [17]), in most tissues PACAP38 is the predominant form [17, 18] and a variety of physiological effects of PACAP38 and PACAP27 are similar [19–21], therefore and for simplicity, PACAP in this chapter refers to PACAP38, unless specified otherwise.

This chapter focuses on the role of PACAP in the regulation of deep T_b under physiological conditions. First, we provide a short overview of the modern concepts in thermoregulation, and then we describe in detail what is known about the thermoregulatory role of PACAP from animal experiments and human studies. For that, we first review the data obtained from the modulation of PACAP signaling with the administration of agonists and antagonists. Second, we analyze how genetic modulation of PACAP signaling affects the maintenance of deep T_b. Last, we summarize the most important findings in discussion and propose a concept of how PACAP signaling contributes to thermoregulation.

Overview of the Architecture of the Thermoregulation System

The regulation of complex energy balance consists of a short-term and a long-term regulatory circle, which are in close interaction with each other through the common effector, metabolic rate [4]. As part of the short-term regulatory circle, deep T_b is maintained with the use of autonomic (thermogenesis and heat loss) and behavioral thermoeffectors (e.g., cold/warmth seeking).

According to the modern theory of thermoregulation, T_b is maintained with the help of relatively independent thermoeffector loops [22]. Each loop consists of a thermosensor, an afferent and efferent neural pathway, and a thermoeffector [23]. Thermosensors can be located in the core or on the periphery of the body. Central thermosensors are mostly warmth-sensitive, thus participate typically in warmth-defense responses and contribute more to the activation of autonomic effectors, while peripheral thermosensors are mainly cold-sensitive activating cold-defense mechanisms and trigger rather behavioral responses [1]. Accordingly, the activation of each thermoeffector loop is achieved by a combination of superficial and deep T_bs, which determines the actual stimulation pattern of central and peripheral thermosensors [24–26].

In a thermal response, the activation of behavioral effectors is preferred over that of autonomic ones, because behavioral effectors have less impact on the energy and water resources of the body than autonomic ones [23]. Thermoregulatory behaviors, which range from simple (e.g., warmth seeking) to complex mechanisms such as turning on a heating unit, are activated mainly from peripheral thermosensors, i.e., before any change in deep T_b occurs, therefore they can be used efficiently to maintain deep T_b by escaping a forthcoming thermal stimulus. Importantly, gross locomotor activity can also contribute to the regulation of T_b [27–29], thus it can be regarded as a behavioral thermoeffector in small rodents [30–32]. Autonomic thermoeffectors become activated mostly from central thermosensors, i.e., by changes of deep T_b. As part of cold defense, thermogenesis increases, while due to cutaneous vasoconstriction heat loss decreases [1]. Non-shivering thermogenesis in brown adipose tissue (BAT) is the prominent source of heat in infants, small rodents, and to a lesser extent in adult humans, while shivering in skeletal muscle is the major form of heat production in adult humans [33]. In warmth defense, heat dissipating mechanisms are activated involving skin vasodilation and elevated evaporative heat loss (e.g., sweating, panting) [1, 34].

The afferent and efferent pathways involved in the regulation of T_b have been recently reviewed elsewhere [1, 23, 35–37], their detailed description is beyond the scope of this chapter. In short, thermoeffector pathways begin with either cold- or warmth-sensitive primary afferent pseudounipolar neurons. Their cell bodies are located in the dorsal-root ganglia and they transmit signals from their thermosensory endings to secondary monopolar neurons in lamina-I of the spinal dorsal horn. Lamina-I neurons carry signals to neurons in the lateral parabrachial nucleus (LPB), which is situated in the rostral portion of the pons. The glutaminergic neurons of the LPB, which receive signals originating from cold-sensitive primary neurons, excite γ-aminobutyric acid (GABA)-ergic neurons in the median preoptic nucleus (MnPO) of the preoptic hypothalamic area (POA). These neurons then inhibit warmth-sensitive GABA-ergic neurons in the medial preoptic area (MPO), which is also located in the POA (Fig. 15.1). The glutaminergic neurons of the LPB, which receive signals originating from the warmth-sensitive primary neurons, excite glutaminergic neurons in the MnPO. These neurons then excite, rather than inhibit, the warmth-sensitive GABA-ergic neurons in the MPO (Fig. 15.1).

The warmth-sensitive neurons of the MPO tonically suppress BAT thermogenesis and skin vasoconstriction [38], which are cold-defense thermoeffectors. A different population of MPO neurons inhibits each thermoeffector. The BAT-specific population sends tonic inhibitory signals to neurons in the dorsomedial hypothalamus, whereas the skin-vasoconstriction-specific population inhibits neurons in the rostral raphe pallidus nucleus. Neurons of the dorsomedial hypothalamus and the rostral raphe pallidus nucleus both excite neurons in the intermediolateral column of the spinal cord, which then stimulate the sympathetic ganglia, controlling the thermoeffectors.

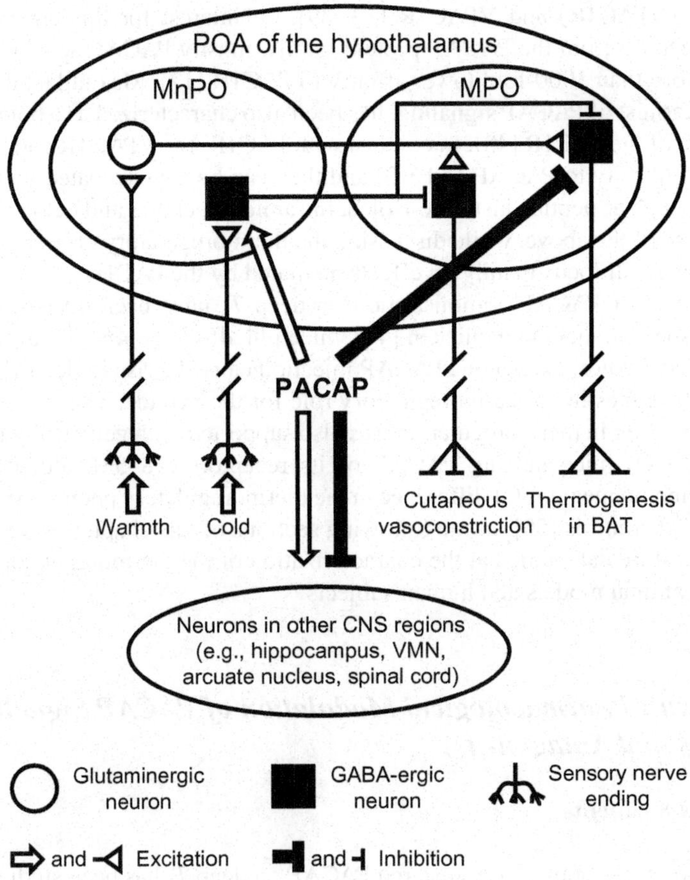

Fig. 15.1 Schematic representation of the proposed actions of the pituitary adenylate cyclase-activating polypeptide (PACAP) on the thermoregulation system. Exogenous PACAP administration induces the activation of the inhibitory γ-aminobutyric acid (GABA)-ergic neurons in the median preoptic nucleus (MnPO) of the preoptic area (POA) of the hypothalamus, which leads to cutaneous vasoconstriction and brown adipose tissue (BAT) thermogenesis and results in hyperthermia. Endogenous PACAP continuously suppresses the activation level of GABA-ergic neurons in the medial preoptic area (MPO), which inhibit BAT thermogenesis. When PACAP is absent, the result is hypometabolism. Neurons in other areas of the central nervous system (CNS), e.g., hippocampus, ventromedial nuclei (VMN), arcuate nucleus, and spinal cord, can also contribute to the thermoregulatory actions of PACAP. Schematic of neural pathways adapted and simplified by authors from Romanovsky et al. [35] with permission

PACAP in the Regulation of Normal Body Temperature

PACAP belongs to the secretin/glucagon/vasoactive intestinal polypeptide (VIP) family [17, 18]. Its biological effects are attributed mainly to three different G protein-coupled receptors: the PAC1 receptor (PAC1R), and the VIP/PACAP receptor 1 (VPAC1R) and VPAC2R [17, 39]. Of interest for this chapter is the PAC1R, because from the PACAP protein family, mostly PACAP acts on it, while VIP has more than 1000-fold lower affinity to PAC1R [17]. Maxadilan, a pharmacological agonist of PACAP signaling, has been also characterized as a potent selective agonist of the PAC1R [40]. In contrast with PAC1R, the VPAC1R and VPAC2R show equal affinity to PACAP and VIP and they can be also activated with further substances such as peptide histidine-isoleucine amide, secretin, and helodermin [41, 42]. Because of the above, while discussing the thermoregulatory effects of PACAP signaling, we will focus mainly on effects mediated by the PAC1R.

The effects of PACAP administration on deep T_b have been investigated in a variety of species, including chicken [43], mice [44, 45], rats [46–48], and humans [21]. In most studies, exogenous PACAP injection caused a rise in deep T_b [21, 43, 44, 46–48], suggesting a thermoregulatory role for the peptide. The importance of PACAP signaling in thermoregulation was also supported with genetically modified (knockout, KO) mice lacking PACAP or its receptors [48–53], although some authors found no meaningful difference in the thermoregulatory phenotype between KO mice and controls [50]. In the following sections of this chapter we review and discuss literature data aimed at the characterization of the thermoregulatory role of PACAP in animal models and human subjects.

Studies with Pharmacological Modulation of PACAP Signaling (Agonists and Antagonists)

Animal Experiments

The effect of exogenously administered PACAP on deep T_b has been studied in different animal models (chicks, mice, rats) and in most of the cases a hyperthermic effect was reported after intrahypothalamic [46, 54], intracerebroventricular (I.C.V.) [43, 44, 47, 48, 55], and intrathecal [56] injection of peptide, although the thermal effects of central versus peripheral PACAP administration have been directly compared only recently [48]. In that study, the authors showed that I.C.V. administration of PACAP (~7 nmol) to rats caused markedly more pronounced hyperthermia than intravenous (I.V.) infusion of the same dose of the peptide [48], which results clearly indicate a central rather than peripheral site of action. Such centrally mediated thermoregulatory effect of PACAP is also in accordance with several other studies showing that PACAP exerts its neurotrophic [57], anti-inflammatory [15], and neuroprotective effects [58] within the CNS.

How the activation of behavioral and autonomic thermoeffectors contributes to PACAP-induced hyperthermia has been investigated in different studies (see below). A behavioral mechanism, which has been commonly studied in the hyperthermic response to PACAP in small rodents is locomotor activity [45, 46, 54, 59]. In rats, an increased locomotor activity has been shown to take part in the hyperthermic response to the injection of PACAP into the lateral ventricle (1–2 nmol) [45] or into the ventromedial nuclei (VMN, 0.05 nmol) [46, 54]. Interestingly, no significant change in the locomotor activity of rats was recently reported by an independent group after microinfusion of PACAP (0.2 nmol) into the central nucleus of the amygdala [59], but the motor activity of the PACAP-treated rats was nearly 44 % higher (2973.6 ± 228.4) than that of the vehicle-treated rats (2070.8 ± 173.4) at 1 h after the infusion. It has to be also noted that microinfusion of PACAP (0.03 nmol) into the paraventricular nuclei of rats suppressed exploratory activity and elevated self-grooming behavior, suggesting complex behavioral effects of PACAP [60]. The differences in the applied administration routes and doses may also account for the discrepancy of the obtained results. In addition to changes in locomotor activity, an increased wet-dog shake behavior of rats in response to i.c.v. PACAP administration has been also reported [61]. The wet-dog shake behavior can be elicited in rats by cold and chemical agents such as the super-cooling substance, icilin [62]. It shows similarities with shivering in certain aspects [63] and recently it has been reported that icilin-induced wet-dog shakes are mediated by the transient receptor potential (TRP) melastatin-8 channel [64], which is a universal cold receptor in the thermoregulation system [65]. Although warmth seeking could be also assumed to play a part in a hyperthermic response, to our knowledge no studies have been performed in which the thermopreferendum of the animals was determined after PACAP administration.

The most studied thermoeffector in the development of PACAP-induced hyperthermia is non-shivering thermogenesis. Exogenous PACAP administration was reported to result in elevated oxygen consumption (VO_2) in chicks [43], rats [48], and mice [44]. A higher end-tidal CO_2 was also found in rats [56]. While VO_2 can reflect both shivering and non-shivering thermogenesis [33], the elevated BAT sympathetic nerve activity [66], increased uncoupling protein 1 mRNA expression in BAT [46], and decreased triglyceride stores in BAT [54], which were measured after PACAP injection indicate that in rats the elevation of non-shivering thermogenesis is responsible for the higher PACAP-induced VO_2. Several different sites within the CNS have been proposed to mediate the elevation of thermogenesis in response to PACAP. Inglott et al. [56, 67] suggested that PACAP acts through the excitation of sympathetic neurons in the spinal cord, because the increased BAT sympathetic nerve activity after intrathecal PACAP administration could be observed even after cervical transection of the spinal cord. Resch et al. [46, 54] concluded that the hyperthermic and hypermetabolic effect of PACAP is evoked from the VMN of the hypothalamus as microinfusion of the peptide into that region was effective, but when PACAP was infused into the closely located paraventricular nuclei it had no effect. As an approach to resolve the controversies, Banki et al. [48] proposed that PACAP acts on GABA-ergic neurons of the MnPO, which are situated in the common part of the cold-activated pathways for the two principal autonomic thermoeffectors [68]. These neurons receive inputs from peripheral sensory afferents through the LPB and connect to tonically activated GABA-ergic neurons in the MPO, which, in turn, are the first efferent neurons innervating the BAT and skin vessels [35]. When the activation level of GABA-ergic MPO neurons is reduced, for example because of PACAP-induced activation of the inhibitory GABA-ergic neurons in the

MnPO, BAT thermogenesis and cutaneous vasoconstriction are activated, resulting in hyperthermia. Spinal sympathetic neurons are more downstream part of the same pathway, which could explain why their activation could lead to the same hyperthermic effect as described by Inglott et al. [56]. In harmony with the findings of Resch et al. [54], besides the LPB, MnPO neurons also receive signals from other brain nuclei such as the VMN [69, 70], the activation of which could then also cause the same effect as activation of the MnPO neurons. While the distribution of PACAP and PAC1R (i.e., the most specific receptor of PACAP) in the CNS is in accordance with such proposed sites of action, for they are expressed in the MnPO, VMN, and spinal cord in addition to other brain areas [11, 12, 46], further studies are needed to specify which neuron(s) respond directly to PACAP and trigger hyperthermia.

In contrast with the effect of PACAP on thermogenesis, how the peptide influences cutaneous heat loss has remained more controversial. When the vasomotor response to PACAP was investigated, a vasodilator effect was found in a variety of vascular beds, including those of the brain, eye, heart, lungs, kidney, gastrointestinal tract, and the urogenital system [17, 18]. The studied vessels originated usually from the core of the body, thus from tissues which do not have a main function in heat loss of the body. In some studies increases in skin perfusion were also recorded, but only after local or systemic application of PACAP [71, 72], which administration routes are not necessarily appropriate to analyze the effects of PACAP on complex neural pathways, especially on their central elements. When PACAP was applied directly to vessels in vitro, it caused vasodilation in cerebral [73], meningeal [74], and tail arteries [75]. Suggesting a distinct, centrally mediated vasomotor effect, when PACAP was administered i.c.v. to rats, it caused an initial skin vasoconstriction (as indicated by a decrease of tail skin temperature), which was then followed by vasodilation [48]. Biphasic vasomotor responses to PACAP were also observed in dogs [76] and in cats [77–79]. In the feline hindquarters vascular bed, the vasoconstrictor part of the biphasic effect was blocked by α-adrenergic antagonist [77, 78], while the vasodilator part was abolished by a VIP receptor antagonist [79], which indicates that the two phases are mediated by distinct mechanisms. The finding that activation of VPAC1R and VPAC2R with VIP decreased mean arterial pressure, while the PAC1R agonist maxadilan caused a pressor response through tachycardia and sympathoexcitation [67], also supports distinctly mediated mechanisms. The role of catecholamines in PACAP-induced vasoconstriction is of high importance, because it can also suggest that through the activation of the sympathetic nervous system PACAP can induce cutaneous vasoconstriction, which, as an autonomic cold-defense effector, can contribute to the development of the hyperthermia induced by the peptide. In accordance with an action of PACAP on elements of the sympathetic nervous system, PACAP was shown to depolarize sympathetic neurons [80], to stimulate the release of catecholamines and neuropeptide Y from superior cervical ganglion neurons [81], and to exert a PAC1R-mediated excitatory effect on sympathetic preganglionic neurons, resulting in sympathetic vasoconstriction [67].

The blockade of PACAP signaling has been used mainly to specify which receptor (usually whether PAC1R or not) mediates the effect of PACAP (for reviews, see refs. [17, 18]). Thermophysiological studies aiming at the investigation of the thermal effects of PACAP antagonists alone are scarce in the literature. Diane et al. [53] mentioned a decreasing effect of PACAP antagonists on energy expenditure, but in the references cited by the authors [43, 56], no effect of PACAP6-38 alone was reported. In fact, intra-brain injection of PACAP6-38 alone did not influence deep T_b or metabolic rate in chicks [43], mice [44], and rats [54]. Although the negative results with the antagonist alone contradict a role for tonic activation of PACAP signaling in the regulation of T_b, further studies are warranted to investigate such mechanism, because the decreasing effect of a PACAP antagonist on energy expenditure could suggest that PAC1R expressing neurons are tonically activated and continuously stimulate heat production. When activation of the neurons is blocked, thermogenesis decreases. In harmony with our earlier proposal that PACAP causes hyperthermia primarily through the activation of GABA-ergic neurons in the MnPO [48], the blockade of tonically activated GABA-ergic neurons located in the same neural pathway in the MPO could also explain the lower metabolic rate in response to either pharmacological PACAP antagonism or genetic PACAP-signaling deficiency (see Sect. 15.3.2.1). Although, the tonically activated state of GABA-ergic MPO neurons has been established [38], it has not been investigated in a decisive experiment whether PACAP or its antagonists can influence the activation level of these neurons directly. Interestingly, in the hippocampus neurons tonically activated by endogenous PACAP have been discovered [82] and the hippocampus has been implied to be involved in controlling T_b [83], presumably through its neural connections with the hypothalamus [84, 85]. Whether the PACAP-activated hippocampal neurons play a definite role in thermoregulation remains subject for future studies. It has to be also noted that the interpretation of the results obtained with PACAP antagonists such as PACAP6-38 can be difficult as PACAP6-38 blocks both PAC1R and VPAC2R [86], furthermore it has been recently shown that PACAP6-38 can act as a potent agonist on sensory nerves [87].

Human Studies

In humans, an i.v. injection of PACAP27 resulted in a dose-dependent rise of rectal temperature by up to 2.7 °C within 30 min [21]. The increase of T_b was accompanied by erythema, edema, and hyperemia, predominantly in the skin of the face, neck, and upper trunk, however no change in heart rate or blood pressure was detected, indicating that the cutaneous vasodilation was a direct, local rather than a secondary systemic effect [21]. Similar findings of increased local skin blood flow without a change in systemic hemodynamics in response to i.v., intra-arterial, and intradermal administration of PACAP (either 27 or 38) and maxadilan were also observed in further studies in humans [88–91]. Supporting a direct action of PACAP on skin blood flow, PACAP and PAC1R are expressed on dermal nerve fibers in the proximity of blood vessels in human skin [92]. Since active cutaneous vasodilation is also used as an autonomic effector during heat exposure [34], one of the physiological functions of PACAP-induced direct skin vasodilation could be a contribution to warmth-defense responses. This hypothesis was tested in the studies by Kellog et al. [88, 93] with the intradermal application of PACAP6-38 during whole body heating in humans. Confirming a role for PACAP signaling in heat loss mechanisms, PACAP6-38 attenuated the heat-induced increase in cutaneous vascular conductance [88, 93], presumably through nitric oxide-mediated mechanisms [93]. PACAP expression has been also found in nerve fibers surrounding human sweat glands [92, 94], which are also effector organs in warmth defense, and direct PACAP application was shown to stimulate their function [95].

Taken together, data obtained from human studies suggest that PACAP administration leads to hyperthermia in humans, while through direct actions in the skin it causes activation of warmth defenses (skin vasodilation and sweating). It can be assumed that the direct effects are evoked by local release of PACAP from cutaneous nerve endings, for example in response to heat, and that they are independent from the systemic (thermoregulatory) effects of the peptide.

Genetic Studies on PACAP Signaling in Body Temperature Regulation

Animal Experiments

The most commonly used experimental model to study how the absence of PACAP signaling affects thermoregulation is the comparison of mice genetically lacking PACAP ($Pacap^{-/-}$) with their wild type littermates ($Pacap^{+/+}$). Though PAC1R KO mice also exist, their thermoregulatory phenotype has remained largely unknown, probably due to their poor survival rates [96]. One of the earliest thermophysiological findings in $Pacap^{-/-}$ mice was their impaired cold tolerance [49]. When the mice were housed at an ambient temperature of 21 °C instead of 24 °C, the survival rate of $Pacap^{-/-}$ pups drastically decreased compared to $Pacap^{+/+}$ mice [49]. The deep T_b of 7-d-old $Pacap^{-/-}$ mice pups dropped substantially more than that of controls, when exposed to an ambient temperature of 21 °C [49]. One of the mechanisms, which could be responsible for the higher lethality of the $Pacap^{-/-}$ mice is an impaired adaptive thermogenesis during cold exposure, which was supported by reduced levels of norepinephrine and dopamine in postganglionic nerve terminals innervating BAT [49] and by disruption of intermediary metabolism [97].

In later studies, T_b of adult $Pacap^{-/-}$ and $Pacap^{+/+}$ mice was also compared and led to seemingly contradictory results. In different experimental models, the deep T_b of $Pacap^{-/-}$ mice was unchanged [50], lower [48, 52] or higher [48] than that of their wild-type littermates. Based on the initial (i.e., before cold exposure) rectal temperature values reported by Adams et al. [51], the average deep T_b of the $Pacap^{-/-}$ mice used in that study was ~0.5 °C higher than that of the $Pacap^{+/+}$ mice, which difference is also statistically significant ($p<0.05$, Student's t test) based on our calculation. The discrepancy in the results may originate, at least in part, from the different experimental conditions (e.g., freely moving vs. restrained animals, cold vs. thermally neutral environment) applied by the authors. For example, Cummings et al. [50] measured rectal temperature in decapitated mice, which technique is inadequate for the detection of relatively small alterations in T_b, especially if those are caused, at least in part, by changes of locomotor activity. In freely moving mice, Hashimoto et al. [52] observed that the normally occurring rise of deep T_b was absent in *Pacap*$^{-/-}$ mice, while it was present in the controls. The reduced T_b of *Pacap*$^{-/-}$ mice during the night could be, however, associated with the relatively low housing temperature of 23 °C, resulting in a chronic, mild cold stress, which could also contribute to the lower T_b of *Pacap*$^{-/-}$ mice because of their impaired cold-defense responses as newborns [49] and adults [51]. When T_b of freely moving adult mice was measured under thermoneutral conditions, there was no difference between deep T_b of *Pacap*$^{-/-}$ and *Pacap*$^{+/+}$ mice during the night, while in the light phase of the day, the T_b of the *Pacap*$^{-/-}$ mice was higher than that of controls [48]. Surprisingly, when basal deep T_b was recorded in loosely restrained mice, when the influence of locomotor activity on thermoregulation is minimal, the T_b of *Pacap*$^{-/-}$ mice was lower compared to *Pacap*$^{+/+}$ controls during the light phase of the day [48]. The diverse findings on the alterations of deep T_b in *Pacap*$^{-/-}$ mice suggest that the deficiency of PACAP signaling results in a complex thermoregulatory phenotype.

Behaviorally, one of the common findings in studies with *Pacap*$^{-/-}$ mice is their increased locomotor activity [98, 99]. The hyperactive phenotype was also observed in PAC1R KO mice [100, 101]. Though the exact mechanistic basis and pathophysiological consequences are not fully clarified, the increased locomotor activity of the PACAP-deficient mice is usually associated with psychomotor disorders [96]. However, one of the consequences of hyperactivity is related to the thermoregulation system. In particular, it has been shown that changes in locomotor activity are reflected in parallel alterations of deep T_b in freely moving mice throughout the day and the effects of locomotion on T_b are higher during the light (inactive) phase than during the night (active) phase [29, 102]. In harmony with such correlation between locomotor activity and T_b, under thermoneutral conditions *Pacap*$^{-/-}$ mice were hyperactive compared to their wild-type littermates during the whole day and their T_b was higher than that of controls during the light phase, but not during the night [48]. From a thermoregulatory point of view, such elevated locomotor activity could be a primary change or a secondary compensatory mechanism for a primary alteration in an autonomic thermoeffector. For example, the changes observed in the gross locomotor activity of mice lacking the TRP vanilloid-1 (V1) channel are presumably primary [30, 103], because—despite their slightly lower resting metabolic rate—profound deficiencies in the two major autonomic effectors of TRPV1 KO mice could be excluded directly [30]. However, the *Pacap*$^{-/-}$ mice have severe deficiencies in a principal autonomic effector (thermogenesis), suggesting that their hyperactivity is rather a secondary, compensatory mechanism. Interestingly, when TRPV1 KO mice are young they are hyperactive [30], but become hypoactive and obese at an older age as compared to controls [103]. The close interaction between PACAP signaling and the TRPV1 channel [15, 104], the similar thermoregulatory phenotype (hyperactive and hypometabolic) of PACAP-deficient and TRPV1 KO mice at a young age, and the age-dependent change in the activity level of TRPV1 KO mice warrant for further thermophysiological studies in aged PACAP-deficient mice.

Among the autonomic thermoeffectors, non-shivering and in a lesser extent shivering thermogenesis was studied in adult *Pacap*$^{-/-}$ mice [48, 51, 53]. Similarly to the cold-sensitive phenotype of newborn *Pacap*$^{-/-}$ mice, in adulthood *Pacap*$^{-/-}$ mice cannot defend their deep T_b when exposed to severe (4 °C) cold acutely [51]. In a later study, the VO_2 of cold-exposed *Pacap*$^{+/+}$ mice increased significantly in response to cold, but that of the *Pacap*$^{-/-}$ did not [53], which supports the impaired adaptive thermogenesis in the absence of PACAP. The norepinephrine-induced increase of thermogenesis was also blunted in the *Pacap*$^{-/-}$ mice, probably due to lower β_3-adrenergic receptor gene expression in BAT in response to cold, suggesting an inability to upregulate β_3-adrenergic receptor mRNA in the absence of PACAP [53]. During chronic, mild cold exposure the expression of BAT uncoupling protein 1 mRNA was lower in the *Pacap*$^{-/-}$ mice than in controls, but their oxidative gene expression was normal [51], which results suggest impaired BAT heat production, but normal shivering thermogenesis in cold-exposed *Pacap*$^{-/-}$ mice. In conscious, restrained mice, the VO_2 and deep T_b of *Pacap*$^{-/-}$ mice was lower than that of controls in the study by Banki et al. [48] and since freely moving *Pacap*$^{-/-}$ mice had normal or even higher deep T_b than controls, the authors suggested that in the absence of PACAP the mice compensate for their lower basal metabolic rate with elevated locomotor activity [48]. Shortly after, an independent group found that the metabolic rate of conscious, unrestrained *Pacap*$^{-/-}$ mice was unaltered [53], which supports the role of locomotor activity for the compensation of reduced basal metabolism as proposed originally by Banki et al. [48]. Although under anesthesia the metabolic rate of *Pacap*$^{-/-}$ and *Pacap*$^{+/+}$ mice did not differ significantly [53], the interpretation of this negative finding is problematic, because the applied anesthesia (sodium pentobarbital) has been repeatedly shown to reduce thermoregulatory responses [105, 106], thus it could account for the diminishment of the inter-genotype difference in the anesthetized mice. The exact mechanism of the reduced metabolic rate in *Pacap*$^{-/-}$ mice needs to be further clarified, but an altered neural control of non-shivering thermogenesis can be suspected, as it was shown that the number of c-fos positive cells is significantly higher in the MPO of *Pacap*$^{-/-}$ mice than in that of controls [48] and inhibitory (GABA-ergic) neurons in the MPO tonically suppress BAT thermogenesis [38]. A higher activation level of the GABA-ergic MPO neurons (as indicated by increased c-fos positive cells) could result in an enhanced suppression of the metabolic rate in the absence of PACAP.

Human Studies

Studies investigating the association between variations of the *PACAP* gene and thermoregulatory mechanisms in humans are scarce in the literature. Gene polymorphism of the PACAP-signaling pathway has been found to be involved in pathological conditions such as sudden infant death syndrome [107, 108] and post-traumatic stress disorder [109–112], which can also have consequences on T_b regulation, but the direct impact of *PACAP* gene polymorphism on the thermoregulation system in humans has remained unknown.

Discussion

In this chapter we review literature data on pharmacological and genetic modulation of PACAP signaling obtained in animal models and in human subjects and discussed the relation of the findings to the functional architecture of the thermoregulation system. In summary, administration of exogenous PACAP leads to elevation of deep T_b in animals and humans (see Sect. 15.3.1). The hyperthermia is brought about by the activation of behavioral and autonomic cold-defense effectors, as PACAP administration results in the rise of BAT thermogenesis [46, 54, 66], increased peripheral (e.g., skin) vasoconstriction [48, 67, 77–79], stimulation of wet-dog shake behavior [61], and elevated locomotor activity [45, 46, 54]. While the PACAP-induced vasoconstriction was observed in several studies of independent groups [48, 67, 77–79], in a number of different experimental models regarding the vascular bed and the administration route a vasodilatory effect was found (for reviews, see refs. [17, 18]), which underlies the complex (local, vascular versus systemic, central) effects of PACAP. The thermoregulatory effect of PACAP is PAC1R dependent [43, 54] and the site of action is located within the CNS ([48, 54, 56], also see Sect. 15.3.1.1). Different regions within the CNS can play a role in the mediation of the hyperthermic response to PACAP such as the VMN [46], MnPO [48], and spinal cord [56]. In contrast with exogenous PACAP administration, when PACAP is absent, lower basal metabolic rate [48] and impaired adaptive thermogenesis were detected [51, 53]. Together with the increased number of c-fos positive cells in the MPO [48], these results suggest that when PACAP is present it suppresses the activation level of inhibitory (GABA-ergic) MPO neurons, thus contributes to the maintenance of normal metabolic rate. In the absence of PACAP, however, thermogenesis decreases due to the increased activation level of these inhibitory neurons. Although the PACAP-induced tonic activation of the GABA-ergic MPO neurons has not been investigated yet, endogenous PACAP has been shown to maintain the tonic activation of hippocampal neurons [82], which, in turn, are wired to thermosensitive neurons in the POA [85], thus could participate in the thermoregulatory actions of PACAP. Interestingly, deficiency of PACAP signaling also resulted in hyperactivity ([98, 99], also see Sect. 15.3.2.1), which can be assumed to serve as a compensatory thermoregulatory mechanism for reduced basal metabolic rate [48], although alternative explanations (e.g., psychomotor disorders) also exist [96]. The proposed influence of PACAP signaling on the thermoregulation system is summarized in Fig. 15.1. Exogenously administered PACAP activates GABA-ergic neurons in the MnPO, at which site the cold-activated thermoregulatory pathways of the two principal autonomic thermoeffectors dissect. Activation of GABA-ergic MnPO neurons leads to the blockade of the GABA-ergic neurons in the MPO, which tonically suppress both BAT thermogenesis and skin vasoconstriction. As a result, the suppression of thermogenesis and heat conservation will be reduced and hyperthermia develops. In addition, the PACAP signaling pathway can also have a tonic suppressing effect on the GABA-ergic MPO neurons innervating the BAT. In the absence of PACAP, the activity level of the GABA-ergic MPO neurons will increase, which leads to enhanced suppression of BAT thermogenesis and hypometabolism. Activation of other regions within the CNS (e.g., hippocampus, VMN, arcuate nucleus, spinal cord) by PACAP can also contribute to the thermoregulatory actions of the peptide either through neuroendocrinological connection with the POA or through alternative mechanisms. Examples of such mechanism include leptin- [44], melanocortin- [59, 113], cyclooxygenase- [47], catecholamine- [53], and thyroid hormone-mediated pathways [51], but their detailed discussion would be beyond the scope of this chapter.

Acknowledgements The authors' research summarized in this review has been supported in part by grants from the Hungarian Scientific Research Fund Grant PD105532 to A.G. and the Janos Bolyai Research Scholarship of the Hungarian Academy of Sciences to A.G. We thank Drs. Dora Reglodi and Andrea Tamas for thoroughly editing the manuscript and giving important feedback.

References

1. Romanovsky AA. Temperature regulation. In: Petersen OH, editor. Lecture notes: human physiology. Oxford: Wiley-Blackwell; 2007. p. 603–15.
2. Szekely M, Szelenyi Z. Regulation of energy balance by peptides: a review. Curr Protein Pept Sci. 2005;6:327–53.
3. Szekely M, Petervari E, Szelenyi Z. Orexigenic vs. anorexigenic peptides and feeding status in the modulation of fever and hypothermia. Front Biosci. 2004;9:2746–63.
4. Garami A, Szekely M. Body temperature: its regulation in framework of energy balance. Temperature. 2014;1:28–9.
5. Fabricio AS, Rae GA, D'Orleans-Juste P, Souza GE. Endothelin-1 as a central mediator of LPS-induced fever in rats. Brain Res. 2005;1066:92–100.
6. Miko A, Balla P, Tenk J, Balasko M, Soos S, Szekely M, et al. Thermoregulatory effect of alarin, a new member of the galanin peptide family. Temperature. 2014;1:51–6.
7. Balasko M, Garami A, Soos S, Koncsecsko-Gaspar M, Szekely M, Petervari E. Central alpha-MSH, energy balance, thermal balance, and antipyresis. J Therm Biol. 2010;35: 211–7.
8. Steiner AA, Romanovsky AA. Leptin: at the crossroads of energy balance and systemic inflammation. Prog Lipid Res. 2007;46:89–107.
9. Miyata A, Arimura A, Dahl RR, Minamino N, Uehara A, Jiang L, et al. Isolation of a novel 38 residue-hypothalamic polypeptide which stimulates adenylate cyclase in pituitary cells. Biochem Biophys Res Commun. 1989;164:567–74.
10. Miyata A, Jiang L, Dahl RD, Kitada C, Kubo K, Fujino M, et al. Isolation of a neuropeptide corresponding to the N-terminal 27 residues of the pituitary adenylate cyclase activating polypeptide with 38 residues (PACAP38). Biochem Biophys Res Commun. 1990;170:643–8.
11. Joo KM, Chung YH, Kim MK, Nam RH, Lee BL, Lee KH, et al. Distribution of vasoactive intestinal peptide and pituitary adenylate cyclase-activating polypeptide receptors (VPAC1, VPAC2, and PAC1 receptor) in the rat brain. J Comp Neurol. 2004;476:388–413.
12. Palkovits M, Somogyvari-Vigh A, Arimura A. Concentrations of pituitary adenylate cyclase activating polypeptide (PACAP) in human brain nuclei. Brain Res. 1995;699:116–20.
13. Pakai E, Garami A, Nucci TB, Ivanov AI, Romanovsky AA. Hyperbilirubinemia exaggerates endotoxin-induced hypothermia. Cell Cycle. 2015;14:1260–7.
14. Rudaya AY, Steiner AA, Robbins JR, Dragic AS, Romanovsky AA. Thermoregulatory responses to lipopolysaccharide in the mouse: dependence on the dose and ambient temperature. Am J Physiol Regul Integr Comp Physiol. 2005;289:R1244–52.

15. Helyes Z, Pozsgai G, Borzsei R, Nemeth J, Bagoly T, Mark L, et al. Inhibitory effect of PACAP-38 on acute neurogenic and non-neurogenic inflammatory processes in the rat. Peptides. 2007;28:1847–55.
16. Nemeth J, Reglodi D, Pozsgai G, Szabo A, Elekes K, Pinter E, et al. Effect of pituitary adenylate cyclase activating polypeptide-38 on sensory neuropeptide release and neurogenic inflammation in rats and mice. Neuroscience. 2006;143:223–30.
17. Vaudry D, Falluel-Morel A, Bourgault S, Basille M, Burel D, Wurtz O, et al. Pituitary adenylate cyclase-activating polypeptide and its receptors: 20 years after the discovery. Pharmacol Rev. 2009;61:283–357.
18. Vaudry D, Gonzalez BJ, Basille M, Yon L, Fournier A, Vaudry H. Pituitary adenylate cyclase-activating polypeptide and its receptors: from structure to functions. Pharmacol Rev. 2000;52:269–324.
19. Nowak JZ, Kuba K. Pituitary adenylate cyclase-activating polypeptide and vasoactive intestinal peptide-stimulated cyclic AMP synthesis in rat cerebral cortical slices: interaction with noradrenaline, adrenaline, and forskolin. J Mol Neurosci. 2002;18:47–52.
20. Lenti L, Domoki F, Kis D, Hegyi O, Toth GK, Busija DW, et al. Pituitary adenylate cyclase-activating polypeptide induces pial arteriolar vasodilation through cyclooxygenase-dependent and independent mechanisms in newborn pigs. Brain Res. 2007;1165:81–8.
21. Seeliger S, Buddenkotte J, Schmidt-Choudhury A, Rosignoli C, Shpacovitch V, von Arnim U, et al. Pituitary adenylate cyclase activating polypeptide: an important vascular regulator in human skin in vivo. Am J Pathol. 2010;177:2563–75.
22. Satinoff E. Neural organization and evolution of thermal regulation in mammals. Science. 1978;201:16–22.
23. Romanovsky AA. Thermoregulation: some concepts have changed. Functional architecture of the thermoregulatory system. Am J Physiol Regul Integr Comp Physiol. 2007;292: R37–46.
24. Jessen C. Independent clamps of peripheral and central temperatures and their effects on heat production in the goat. J Physiol. 1981;311:11–22.
25. Sakurada S, Shido O, Fujikake K, Nagasaka T. Relationship between body core and peripheral temperatures at the onset of thermoregulatory responses in rats. Jpn J Physiol. 1993;43:659–67.
26. Roberts WW. Differential thermosensor control of thermoregulatory grooming, locomotion, and relaxed postural extension. Ann NY Acad Sci. 1988;525:363–74.
27. Mount L, Willmott JV. The relation between spontaneous activity, metabolic rate and the 24 hour cycle in mice at different environmental temperatures. J Physiol. 1967;190:371–80.
28. Brown D, Livesey G, Dauncey M. Influence of mild cold on the components of 24 hour thermogenesis in rats. J Physiol. 1991;441:137–54.
29. Weinert D, Waterhouse J. Diurnally changing effects of locomotor activity on body temperature in laboratory mice. Physiol Behav. 1998;63:837–43.
30. Garami A, Pakai E, Oliveira DL, Steiner AA, Wanner SP, Almeida MC, et al. Thermoregulatory phenotype of the Trpv1 knockout mouse: thermoeffector dysbalance with hyperkinesis. J Neurosci. 2011;31:1721–33.
31. Kanizsai P, Garami A, Solymar M, Szolcsanyi J, Szelenyi Z. Energetics of fasting heterothermia in TRPV1-KO and wild type mice. Physiol Behav. 2009;96:149–54.
32. Szentirmai E, Kapas L, Sun Y, Smith RG, Krueger JM. Restricted feeding-induced sleep, activity, and body temperature changes in normal and preproghrelin-deficient mice. Am J Physiol Regul Integr Comp Physiol. 2010;298:R467–77.
33. Cannon B, Nedergaard J. Brown adipose tissue: function and physiological significance. Physiol Rev. 2004;84:277–359.
34. Szekely M, Carletto L, Garami A. The pathophysiology of heat exposure. Temperature. 2015;2:452.
35. Romanovsky AA, Almeida MC, Garami A, Steiner AA, Norman MH, Morrison SF, et al. The transient receptor potential vanilloid-1 channel in thermoregulation: a thermosensor it is not. Pharmacol Rev. 2009;61:228–61.

36. Morrison SF, Nakamura K. Central neural pathways for thermoregulation. Front Biosci. 2011;16:74.
37. Garami A, Shimansky YP, Pakai E, Oliveira DL, Gavva NR, Romanovsky AA. Contributions of different modes of TRPV1 activation to TRPV1 antagonist-induced hyperthermia. J Neurosci. 2010;30:1435–40.
38. Osaka T. Cold-induced thermogenesis mediated by GABA in the preoptic area of anesthetized rats. Am J Physiol Regul Integr Comp Physiol. 2004;287:R306–13.
39. Dickson L, Finlayson K. VPAC and PAC receptors: from ligands to function. Pharmacol Ther. 2009;121:294–316.
40. Moro O, Lerner EA. Maxadilan, the vasodilator from sand flies, is a specific pituitary adenylate cyclase activating peptide type I receptor agonist. J Biol Chem. 1997;272:966–70.
41. Christophe J, Chatelain P, Taton G, Delhaye M, Waelbroeck M, Robberecht P. Comparison of VIP-secretin receptors in rat and human lung. Peptides. 1981;2:253–8.
42. Robberecht P, Waelbroeck M, Camus JC, De Neef P, Coy D, Christophe J. Effects of HIS 1 modifications on the ability of vasoactive intestinal peptide to stimulate adenylate cyclase from rat and human tissues. Peptides. 1984;5:877–81.
43. Tachibana T, Oikawa D, Adachi N, Boswell T, Furuse M. Central administration of vasoactive intestinal peptide and pituitary adenylate cyclase-activating polypeptide differentially regulates energy metabolism in chicks. Comp Biochem Physiol A Mol Integr Physiol. 2007;147:156–64.
44. Hawke Z, Ivanov TR, Bechtold DA, Dhillon H, Lowell BB, Luckman SM. PACAP neurons in the hypothalamic ventromedial nucleus are targets of central leptin signaling. J Neurosci. 2009;29:14828–35.
45. Masuo Y, Noguchi J, Morita S, Matsumoto Y. Effects of intracerebroventricular administration of pituitary adenylate cyclase-activating polypeptide (PACAP) on the motor activity and reserpine-induced hypothermia in murines. Brain Res. 1995;700:219–26.
46. Resch JM, Boisvert JP, Hourigan AE, Mueller CR, Yi SS, Choi S. Stimulation of the hypothalamic ventromedial nuclei by pituitary adenylate cyclase-activating polypeptide induces hypophagia and thermogenesis. Am J Physiol Regul Integr Comp Physiol. 2011;301: R1625–34.
47. Pataki I, Adamik A, Jaszberenyi M, Macsai M, Telegdy G. Pituitary adenylate cyclase-activating polypeptide induces hyperthermia in the rat. Neuropharmacology. 2000;39:1303–8.
48. Banki E, Pakai E, Gaszner B, Zsiboras C, Czett A, Bhuddi PRP, et al. Characterization of the thermoregulatory response to pituitary adenylate cyclase-activating polypeptide in rodents. J Mol Neurosci. 2014;54:543–54.
49. Gray SL, Yamaguchi N, Vencova P, Sherwood NM. Temperature-sensitive phenotype in mice lacking pituitary adenylate cyclase-activating polypeptide. Endocrinology. 2002;143:3946–54.
50. Cummings KJ, Willie C, Wilson RJ. Pituitary adenylate cyclase-activating polypeptide maintains neonatal breathing but not metabolism during mild reductions in ambient temperature. Am J Physiol Regul Integr Comp Physiol. 2008;294:R956–65.
51. Adams BA, Gray SL, Isaac ER, Bianco AC, Vidal-Puig AJ, Sherwood NM. Feeding and metabolism in mice lacking pituitary adenylate cyclase-activating polypeptide. Endocrinology. 2008;149:1571–80.
52. Hashimoto H, Hashimoto R, Shintani N, Tanaka K, Yamamoto A, Hatanaka M, et al. Depression-like behavior in the forced swimming test in PACAP-deficient mice: amelioration by the atypical antipsychotic risperidone. J Neurochem. 2009;110:595–602.
53. Diane A, Nikolic N, Rudecki AP, King SM, Bowie DJ, Gray SL. PACAP is essential for the adaptive thermogenic response of brown adipose tissue to cold exposure. J Endocrinol. 2014;222:327–39.
54. Resch JM, Maunze B, Gerhardt AK, Magnuson SK, Phillips KA, Choi S. Intrahypothalamic pituitary adenylate cyclase-activating polypeptide regulates energy balance via site-specific actions on feeding and metabolism. Am J Physiol Endocrinol Metab. 2013;305:E1452–63.
55. Pataki I, Adamik A, Jaszberenyi M, Macsai M, Telegdy G. Involvement of transmitters in pituitary adenylate cyclase-activating polypeptide-induced hyperthermia. Regul Pept. 2003;115:187–93.

56. Inglott MA, Farnham MM, Pilowsky PM. Intrathecal PACAP-38 causes prolonged widespread sympathoexcitation via a spinally mediated mechanism and increases in basal metabolic rate in anesthetized rat. Am J Physiol Heart Circ Physiol. 2011;300:H2300–7.
57. Reglodi D, Kiss P, Lubics A, Tamas A. Review on the protective effects of PACAP in models of neurodegenerative diseases in vitro and in vivo. Curr Pharm Des. 2011;17:962–72.
58. Reglodi D, Somogyvari-Vigh A, Vigh S, Kozicz T, Arimura A. Delayed systemic administration of PACAP38 is neuroprotective in transient middle cerebral artery occlusion in the rat. Stroke. 2000;31:1411–7.
59. Iemolo A, Ferragud A, Cottone P, Sabino V. Pituitary adenylate cyclase-activating peptide in the central amygdala causes anorexia and body weight loss via the melanocortin and the TrkB systems. Neuropsychopharmacology. 2015;40:1846–55.
60. Norrholm SD, Das M, Legradi G. Behavioral effects of local microinfusion of pituitary adenylate cyclase activating polypeptide (PACAP) into the paraventricular nucleus of the hypothalamus (PVN). Regul Pept. 2005;128:33–41.
61. Agarwal A, Halvorson LM, Legradi G. Pituitary adenylate cyclase-activating polypeptide (PACAP) mimics neuroendocrine and behavioral manifestations of stress: evidence for PKA-mediated expression of the corticotropin-releasing hormone (CRH) gene. Mol Brain Res. 2005;138:45–57.
62. Tse S, Wei E. Inhibition of the shake response in rats by adenosine and 2-chloroadenosine. Psychopharmacology (Berl). 1986;90:322–6.
63. Wei E, Sigel S, Way EL. Regional sensitivity of the rat brain to the inhibitory effects of morphine on wet shake behavior. J Pharmacol Exp Ther. 1975;193:56–63.
64. Colburn RW, Lubin ML, Stone DJ, Wang Y, Lawrence D, D'Andrea MR, et al. Attenuated cold sensitivity in TRPM8 null mice. Neuron. 2007;54:379–86.
65. Almeida MC, Hew-Butler T, Soriano RN, Rao S, Wang W, Wang J, et al. Pharmacological blockade of the cold receptor TRPM8 attenuates autonomic and behavioral cold defenses and decreases deep body temperature. J Neurosci. 2012;32:2086–99.
66. Tanida M, Shintani N, Morita Y, Tsukiyama N, Hatanaka M, Hashimoto H, et al. Regulation of autonomic nerve activities by central pituitary adenylate cyclase-activating polypeptide. Regul Pept. 2010;161:73–80.
67. Inglott MA, Lerner EA, Pilowsky PM, Farnham MM. Activation of PAC1 and VPAC receptor subtypes elicits differential physiological responses from sympathetic preganglionic neurons in the anaesthetized rat. Br J Pharmacol. 2012;167:1089–98.
68. Nakamura K, Morrison SF. Central efferent pathways mediating skin cooling-evoked sympathetic thermogenesis in brown adipose tissue. Am J Physiol Regul Integr Comp Physiol. 2007;292:R127–36.
69. Imai-Matsumura K, Matsumura K, Tsai C-L, Nakayama T. Thermal responses of ventromedial hypothalamic neurons in vivo and in vitro. Brain Res. 1988;445:193–7.
70. Thornhill J, Jugnauth A, Halvorson I. Brown adipose tissue thermogenesis evoked by medial preoptic stimulation is mediated via the ventromedial hypothalamic nucleus. Can J Physiol Pharmacol. 1994;72:1042–8.
71. Gardiner S, Rakhit T, Kemp P, March J, Bennett T. Regional haemodynamic responses to pituitary adenylate cyclase-activating polypeptide and vasoactive intestinal polypeptide in conscious rats. Br J Pharmacol. 1994;111:589–97.
72. Banki E, Hajna Z, Kemeny A, Botz B, Nagy P, Bolcskei K, et al. The selective PAC1 receptor agonist maxadilan inhibits neurogenic vasodilation and edema formation in the mouse skin. Neuropharmacology. 2014;85:538–47.
73. Vamos Z, Ivic I, Cseplo P, Toth G, Tamas A, Reglodi D, et al. Pituitary adenylate cyclase-activating polypeptide (PACAP) induces relaxations of peripheral and cerebral arteries, which are differentially impaired by aging. J Mol Neurosci. 2014;54:535–42.
74. Syed AU, Koide M, Braas KM, May V, Wellman GC. Pituitary adenylate cyclase-activating polypeptide (PACAP) potently dilates middle meningeal arteries: implications for migraine. J Mol Neurosci. 2012;48:574–83.

75. Absood A, Chen D, Wang Z-Y, Hakanson R. Vascular effects of pituitary adenylate cyclase activating peptide: a comparison with vasoactive intestinal peptide. Regul Pept. 1992;40:323–9.
76. Ishizuka Y, Kashimoto K, Mochizuki T, Sato K, Ohshima K, Yanaihara N. Cardiovascular and respiratory actions of pituitary adenylate cyclase-activating polypeptides. Regul Pept. 1992;40:29–39.
77. Champion H, Santiago J, Garrison E, Cheng D, Coy D, Murphy W, et al. Analysis of cardiovascular responses to PACAP-27, PACAP-38, and vasoactive intestinal polypeptide. Ann N Y Acad Sci. 1996;805:429–41.
78. Santiago J, Kadowitz PJ. Analysis of responses to pituitary adenylate cyclase activating polypeptide-38 in the feline hindquarters vascular bed. Eur J Pharmacol. 1993;243:291–4.
79. Minkes R, McMahon T, Higuera T, Murphy W, Coy D, Kadowitz P. Analysis of systemic and pulmonary vascular responses to PACAP and VIP: role of adrenal catecholamines. Am J Physiol Heart Circ Physiol. 1992;263:H1659–69.
80. Beaudet MM, Parsons RL, Braas KM, May V. Mechanisms mediating pituitary adenylate cyclase-activating polypeptide depolarization of rat sympathetic neurons. J Neurosci. 2000;20:7353–61.
81. May V, Braas KM. Pituitary adenylate cyclase-activating polypeptide (PACAP) regulation of sympathetic neuron neuropeptide Y and catecholamine expression. J Neurochem. 1995;65:978–87.
82. Costa L, Santangelo F, Li Volsi G, Ciranna L. Modulation of AMPA receptor-mediated ion current by pituitary adenylate cyclase-activating polypeptide (PACAP) in CA1 pyramidal neurons from rat hippocampus. Hippocampus. 2009;19:99–109.
83. Horowitz J, Saleh M, Karem R. Correlation of hippocampal theta rhythm with changes in cutaneous temperature. Am J Physiol. 1974;227:635–42.
84. Hori T, Osaka T, Kiyohara T, Shibata M, Nakashima T. Hippocampal input to preoptic thermosensitive neurons in the rat. Neurosci Lett. 1982;32:155–8.
85. Boulant J, Demieville H. Responses of thermosensitive preoptic and septal neurons to hippocampal and brain stem stimulation. J Neurophysiol. 1977;40:1356–68.
86. Dickinson T, Fleetwood-Walker S, Mitchell R, Lutz E. Evidence for roles of vasoactive intestinal polypeptide (VIP) and pituitary adenylate cyclase activating polypeptide (PACAP) receptors in modulating the responses of rat dorsal horn neurons to sensory inputs. Neuropeptides. 1997;31:175–85.
87. Reglodi D, Borzsei R, Bagoly T, Boronkai A, Racz B, Tamas A, et al. Agonistic behavior of PACAP6-38 on sensory nerve terminals and cytotrophoblast cells. J Mol Neurosci. 2008;36:270–8.
88. Kellogg DL, Zhao JL, Wu Y, Johnson JM. VIP/PACAP receptor mediation of cutaneous active vasodilation during heat stress in humans. J Appl Physiol. 2010;109:95–100.
89. Warren JB, Cockcroft JR, Larkin SW, Kajekar R, Macrae A, Ghatei MA, et al. Pituitary adenylate cyclase activating polypeptide is a potent vasodilator in humans. J Cardiovasc Pharmacol. 1992;20:83–7.
90. Dorner GT, Wolzt M, Schmetterer L, Eichler H, Eichler H-G. Effect of pituitary adenylate cyclase activating polypeptide 1–27 on ocular, cerebral and skin blood flow in humans. Naunyn Schmiedebergs Arch Pharmacol. 1998;358:657–62.
91. Grevelink SA, Osborne J, Loscalzo J, Lerner EA. Vasorelaxant and second messenger effects of maxadilan. J Pharmacol Exp Ther. 1995;272:33–7.
92. Steinhoff M, McGregor GP, Radleff-Schlimme A, Steinhoff A, Jarry H, Schmidt WE. Identification of pituitary adenylate cyclase activating polypeptide (PACAP) and PACAP type 1 receptor in human skin: expression of PACAP-38 is increased in patients with psoriasis. Regul Pept. 1999;80:49–55.
93. Kellogg DL, Zhao JL, Wu Y, Johnson JM. Nitric oxide and receptors for VIP and PACAP in cutaneous active vasodilation during heat stress in humans. J Appl Physiol. 2012;113:1512–8.
94. Ødum L, Petersen L, Skov P, Ebskov L. Pituitary adenylate cyclase activating polypeptide (PACAP) is localized in human dermal neurons and causes histamine release from skin mast cells. Inflamm Res. 1998;47:488–92.

95. Rambotti M, Simonetti S, Spreca A. PACAP activated adenylate cyclase in human sweat glands. An ultracytochemical study. Eur J Histochem. 2002;46:223–8.
96. Hashimoto H, Shintani N, Baba A. New insights into the central PACAPergic system from the phenotypes in PACAP-and PACAP receptor-knockout mice. Ann N Y Acad Sci. 2006;1070:75–89.
97. Gray SL, Cummings KJ, Jirik FR, Sherwood NM. Targeted disruption of the pituitary adenylate cyclase-activating polypeptide gene results in early postnatal death associated with dysfunction of lipid and carbohydrate metabolism. Mol Endocrinol. 2001;15:1739–47.
98. Shibasaki Y, Hayata-Takano A, Hazama K, Nakazawa T, Shintani N, Kasai A, et al. Atomoxetine reverses locomotor hyperactivity, impaired novel object recognition, and prepulse inhibition impairment in mice lacking pituitary adenylate cyclase-activating polypeptide. Neuroscience. 2015;297:95–104.
99. Hammack SE, May V. Pituitary adenylate cyclase activating polypeptide in stress-related disorders: data convergence from animal and human studies. Biol Psychiatry. 2015;78: 167–177.
100. Otto C, Martin M, Wolfer DP, Lipp H-P, Maldonado R, Schutz G. Altered emotional behavior in PACAP-type-I-receptor-deficient mice. Mol Brain Res. 2001;92:78–84.
101. Nicot A, Otto T, Brabet P, DiCicco-Bloom EM. Altered social behavior in pituitary adenylate cyclase-activating polypeptide type I receptor-deficient mice. J Neurosci. 2004;24:8786–95.
102. Weinert D, Waterhouse J. Daily activity and body temperature rhythms do not change simultaneously with age in laboratory mice. Physiol Behav. 1999;66:605–12.
103. Wanner SP, Garami A, Romanovsky AA. Hyperactive when young, hypoactive and overweight when aged: connecting the dots in the story about locomotor activity, body mass, and aging in Trpv1 knockout mice. Aging. 2011;3:450.
104. Helyes Z, Kun J, Dobrosi N, Sandor K, Nemeth J, Perkecz A, et al. Pituitary adenylate-cyclase activating polypeptide is up-regulated in murine skin inflammation and mediates transient receptor potential vanilloid-1-induced neurogenic edema. J Invest Dermatol. 2015;135:2209–18.
105. Refinetti R, Carlisle H. Thermoregulation during pentobarbital and ketamine anesthesia in rats. J Physiol Paris. 1987;83:300–3.
106. Kiyatkin EA, Brown PL. Brain and body temperature homeostasis during sodium pentobarbital anesthesia with and without body warming in rats. Physiol Behav. 2005;84:563–70.
107. Wilson RJ, Cummings KJ. Pituitary adenylate cyclase-activating polypeptide is vital for neonatal survival and the neuronal control of breathing. Respir Physiol Neurobiol. 2008;164: 168–78.
108. Cummings KJ, Klotz C, Liu W-Q, Weese-Mayer DE, Marazita ML, Cooper ME, et al. Sudden infant death syndrome (SIDS) in African Americans: polymorphisms in the gene encoding the stress peptide pituitary adenylate cyclase-activating polypeptide (PACAP). Acta Paediatr. 2009;98:482–9.
109. Ressler KJ, Mercer KB, Bradley B, Jovanovic T, Mahan A, Kerley K, et al. Post-traumatic stress disorder is associated with PACAP and the PAC1 receptor. Nature. 2011;470:492–7.
110. Stevens JS, Almli LM, Fani N, Gutman DA, Bradley B, Norrholm SD, et al. PACAP receptor gene polymorphism impacts fear responses in the amygdala and hippocampus. Proc Natl Acad Sci U S A. 2014;111:3158–63.
111. Almli LM, Mercer KB, Kerley K, Feng H, Bradley B, Conneely KN, et al. ADCYAP1R1 genotype associates with post-traumatic stress symptoms in highly traumatized African-American females. Am J Med Genet B Neuropsychiatr Genet. 2013;162:262–72.
112. Wang L, Cao C, Wang R, Qing Y, Zhang J, Zhang XY. PAC1 receptor (ADCYAP1R1) genotype is associated with PTSD's emotional numbing symptoms in Chinese earthquake survivors. J Affect Disord. 2013;150:156–9.
113. Mounien L, Bizet P, Boutelet I, Gourcerol G, Fournier A, Vaudry H, et al. Pituitary adenylate cyclase-activating polypeptide directly modulates the activity of proopiomelanocortin neurons in the rat arcuate nucleus. Neuroscience. 2006;143:155–63.

Part V
Presence and Functions of PACAP in the Gastrointestinal and Urinary Tracts

Chapter 16
PACAP Regulation of Gastrointestinal Function and Obesity

John P. Vu, Jihane N. Benhammou, Deepinder Goyal, Leon Luong, Suwan Oh, Patrizia Germano, and Joseph R. Pisegna

Abstract Pituitary adenylate cyclase activating polypeptide (PACAP) is a 27- or 38-amino acid peptide that is widely distributed in both the peripheral and central nervous systems. PACAP has been found to be expressed within the enteric nervous system and gastric mucosa and has profound physiological effects in the gastrointestinal tract. We have previously shown that PACAP regulates gastric acid secretion by activating its high affinity PAC1 receptors expressed on gastric enterochromaffin-like cells (ECL). However, the peripheral mechanisms involved in PACAP regulation of appetite and feeding are unknown. Vasoactive intestinal peptide (VIP) is a 28-amino acid peptide abundantly expressed in the central nervous system as well as in the gastrointestinal tract, where it regulates different physiological functions. VIP inhibits gastric acid secretion via its VPAC1 receptors expressed on gastric D cells. VIP also regulates intestinal motility and VIP gene deletion results in the development of intestinal ileus. VIP is involved in the control of appetite/satiety, feeding behavior and in the secretion of some key regulatory metabolic hormones. VIP plays a very important role in the regulation of body weight and mass composition by significantly enhancing body weight and fat mass. Therefore, both PACAP and VIP neuropeptides could be crucial targets for the regulation of appetite/satiety, body phenotype and for the treatment of obesity.

Keywords Pituitary adenylate cyclase activating polypeptide (PACAP) • Vasoactive intestinal polypeptide (VIP) • Metabolic syndrome • Obesity • Gastrointestinal hormones

J.P. Vu, • J. N.Benhammou • D. Goyal • L. Luong
S. Oh • P. Germano • J.R. Pisegna (✉)
Division of Gastroenterology, Hepatology and Parenteral Nutrition, Department of Veterans Affairs, VA Greater Los Angeles Healthcare System, 11301 Wilshire Blvd (691/111C), Los Angeles, CA 90403, USA

Department of Medicine, Molecular, Cellular and Integrative Physiology, David Geffen School of Medicine at UCLA, Los Angeles, CA 90073, USA
e-mail: jpisegna@mednet.ucla.edu

D. Reglodi, A. Tamas (eds.), *Pituitary Adenylate Cyclase Activating Polypeptide — PACAP*, Current Topics in Neurotoxicity 11,
DOI 10.1007/978-3-319-35135-3_16

Obesity is a serious and growing public health issue, and its increasing prevalence is a major contributor to the development of metabolic syndrome and chronic disorders such as diabetes mellitus, hypertension, osteoarthritis, cancer, coronary artery and cerebrovascular diseases [1]. The clinical treatments of these conditions generate immense health care costs, posing one of the most serious threats to the global healthcare systems. The estimated direct and indirect costs for obesity treatment exceed $100 billion/year, and consumers spend about one-third of this amount for weight loss products and services [1, 2]. Therefore, there is a need for a better understanding of the pathogenesis of metabolic syndrome in order to promote more effective treatment options.

Peptide hormones produced by the gastrointestinal (GI) tract are considered to be the major peripheral regulators of appetite and satiety [3]. The GI tract acts as a nutrient sensor, releasing gastrointestinal hormones and neurotransmitters in response to luminal stimuli [4, 5]. These hormones are categorized as either anorexigenic, which increase satiety and suppress appetite, or orexigenic, which suppress satiety and increase appetite. Anorexigenic hormones include glucagon-like peptide-1 (GLP-1), peptide YY (PYY), cholecystokinin (CCK), and pancreatic polypeptide (PP) [5], and also the adipocytes-released leptin and the pancreatic β-cells-derived insulin. Orexigenic hormones include α-ghrelin, secreted by human gastric fundus P/D1 cells and pancreatic epsilon cells [6]. Gastrin is a gastric-released peptide produced in the G cell of the stomach in response to intraluminal amino acids. CCK increases secretion of pancreatic enzymes and bicarbonate, inhibits gastric acid secretion, reduces food intake whereas gastrin promotes gastric mucosal proliferation and stimulates gastric acid secretion. Gastrin may directly influence the release of ghrelin from the stomach [7]). The receptors for CCK and gastrin are structurally similar; G protein coupled and classified as CCK1R and CCK2R, respectively [8–11]. The clinical effectiveness of CCK as a potential weight loss target is limited by its relatively short-lasting effects on appetite. More recently, a specific CCK2R antagonist has been developed and is being tested in humans. The polypeptide-fold family of peptides includes NPY, pancreatic polypeptide (PP) and peptide YY (PYY). These peptides bind to receptors coupled to inhibitor G-proteins and, therefore, reduce intracellular Ca^{++}. Perhaps, the most extensively studied hormone is PYY, which is released following meal ingestion, independently from protein content [12]. PYY is released from the intestinal mucosa and is thought to act as an "ileal brake" to reduce the motility of the small intestine. Infused PYY results in suppression of gastrin-induced gastric acid secretion and in a delay in gastric emptying. PYY has been studied for its ability to suppress appetite. Intraperitoneal injections of PYY result in a reduction of eating behavior during the dark phase [13]. In clinical trials, infusion of PYY3-36 resulted in reduced calorie intake [14]. Pancreatic polypeptide is secreted from the pancreas as well as the distal ileum and released in response to food and vagal stimulation [15]. Secreted PP results in a reduction in gastric emptying and a delay in the release of meal-stimulated insulin release [16]. In patients with Prader–Willi Syndrome, infused PP resulted in a reduced appetite and food intake [17]. However, the effects of PP infused within the CNS vs. peripherally are unequivocal. GLP-1 is secreted within the intestine by the endocrine L-cells and its structure is highly conserved amongst species. GLP-1 is released in response to a meal and its release is proportional to the calories intake [18]. Like other members of the PACAP, VIP, and secretin family, GLP-1 peptide stimulates G protein coupling leading to cAMP release [19]. Administration of GLP-1 results in a dose-dependent reduction in appetite and consequent weight loss [20]. In one study, prandial subcutaneous injections of GLP-1 resulted in weight loss in obese subjects [21, 22]. Ghrelin has been a subject of intense interest. The major source of this peptide is the gastric X/A cells that release it to stimulate appetite [23]. The levels are mostly elevated during periods of fasting and are suppressed with meal ingestion. In patients with neuroendocrine tumors the levels of ghrelin appear increased and they may correlate with tumor bulk and with obesity [24].

Pituitary adenylate cyclase activating polypeptide (PACAP) is a gastrointestinal neuropeptide, which belongs to the family of secretin, glucagon, and GLP-1 hormones. This neuropeptide was identified in 1989 by Arimura and colleagues from ovine hypothalamus [25]. PACAP occurs naturally as 27- or 38-amino acid peptides named PACAP-27 and PACAP-38 respectively that have equal biological potencies. Most of the endogenous PACAP exists as PACAP-38. The primary sequence of PACAP is 68 % identical to its closest hormone relative, VIP as described below [26]. The high affinity type I PACAP receptor, named PAC1, was cloned, pharmacologically characterized and demonstrated to be a heptahelical, G protein coupled receptor of the Type 2 family, that is related molecularly and pharmacologically to receptors in the VIP, secretin, glucagon, and GLP-1 superfamily [27]. PAC1 receptor binds PACAP with preferential affinity compared to VIP and is therefore considered to be specific for PACAP hormones. We identified the gene for the human PAC1 to be localized on chromosome 7 and demonstrated that the gene contains an intron–exon splice variant which can be differentially expressed and coupled to a dual signaling pathway. The PAC1 receptor is unique in that it possesses a dual signal transduction coupling to either Gs or Gq/11 signal transduction pathways to stimulate cAMP and/or inositol phosphate and Ca^{++}. The pharmacology and functions of the VIP-PACAP receptor family have been well characterized [28].

PACAP and PAC1 are expressed in both the peripheral and central nervous systems as well as in peripheral tissues. In the GI tract immunoreactive PACAP and PAC1 are present in the esophagus, stomach, duodenum, small and large intestines, and liver [29, 30]. PACAP is expressed in both the endocrine and exocrine pancreas and in pancreatic beta islet cells [31, 32]. PACAP has trophic effect on the release of insulin in both humans and rodents [33–37]. This activity is regulated through the release of cAMP and through K_{ATP} channels resulting in the release of insulin containing granules in the beta cells. PACAP also appears to regulate beta cell transcription of insulin, GLUT-1, and hexokinase 1 [38]. Thus, PACAP plays a major role in the regulation of pancreatic beta cell function as evidenced by its stimulatory and trophic effects in isolated pancreatic islets [39, 40]. In whole animal studies, the PACAP partial antagonist, PACAP 6-38, or a genetic PAC1 deficient mouse model have a blunted glucose-induced insulin response. PACAP is also a potent stimulant for glucagon secretion. In the exocrine pancreas, PACAP was shown to be a potent stimulator of acinar release of amylase [41]. The use of the pancreatic AR42J pancreatic cancer cell line proved crucial to the cloning of PAC1 since this cell line express this receptor in high numbers. Whole animal studies have demonstrated that PACAP administration results in vasodilation, pancreatic blood flow, and the release of amylase, effects which can be blocked by the use of the partial antagonist, PACAP 6–38 [42, 43]. In pathological conditions, PACAP has been implicated in chronic pancreatitis-induced abdominal pain [44]. PACAP and VIP receptors are expressed on pancreatic adenocarcinoma cells as determined using in vivo scintigraphy, thus suggesting that targeting these receptors may prove clinically useful [45]. We have previously demonstrated that PAC1 are expressed on the pancreatic neuroendocrine tumor cell line, BON, and interestingly regulate the release of PACAP from these tumors in an autocrine-like fashion [46, 47].

The primary site for gastric acid secretion is the corpus of the stomach, whereas the site for regulating the hormonal control of gastric acid secretion resides in the gastric antrum. In general the regulation of gastric acid secretion can be divided into the cephalic phase in which vagal efferents stimulate acid secretion through the release of acetylcholine and PACAP. The second phase of gastric acid secretion is commonly referred to as the gastric phase which relies on intraluminal stimulation of gastrin to increase gastric acid secretion. The gastric mucosa contains paracrine and endocrine cells, namely the histamine-containing enterochromaffin-like (ECL) cells, the somatostatin containing D cells and the gastrin-containing G cells, that altogether regulate gastric acid secretion. ECL cells release histamine which acts at the H2 histamine receptors expressed on the parietal cells to stimulate gastric acid secretion. The G cell releases gastrin in response to intraluminal amino acids, that in turn stimulate the CCK2 receptor expressed on the ECL cell to release histamine. Vagal efferents release acetylcholine and stimulate gastric acid secretion by acting at specific muscarinic (M3) receptors expressed on the parietal cell. The cephalic phase of gastric acid secretion is also the likely source of PACAP, which acts specifically on the PAC1 receptors expressed on the gastric ECL cell [48]. In the stomach, PACAP is expressed in the myenteric neurons, whereas PAC1 is localized on the gastric enterochromaffin-like cell [29]. Gastric D cells are primarily responsible for the release of somatostatin which inhibits gastrin release. We have demonstrated that gastric D cells express VPAC1 receptors and a likely explanation is that PACAP stimulates gastric secretion acting on ECL cell and, later, through a less robust pathway inhibits gastric acid secretion by activating VPAC1 receptor on the gastric D cell. We have therefore hypothesized that prior to the ingestion of a meal there is neural release of acetylcholine and PACAP that result in a small increase of gastric acid secretion. This small amount of gastric acid secretion permits the dissimilation of protein and the release of amino acids that in turn activate the release from G cell of gastrin, which is a more potent secretagogue. Immediately after meal ingestion, the release of PACAP stimulates the VPAC1 receptor on the gastric D cell to release somatostatin and thereby reduce gastrin release and turn off gastric acid secretion. PAC1−/− mice were shown to develop gastric acid hypersecretion and could be a model for the study of clinical conditions such as Zollinger–Ellison Syndrome [49].

PACAP plays an important role in the regulation of food intake and thermogenesis by intracerebroventricular (ICV) injection. However, the peripheral mechanisms involved in PACAP regulation of appetite and feeding are unknown. To explore the role of PACAP and PAC1 in the regulation of appetite and food intake we performed an analysis of metabolic hormone release, food intake, and feeding behavior in WT and PAC1 −/− mice treated by IP injection of PACAP38 and PACAP27. PACAP38 and PACAP27 (100 nM, 1 and 10 μM in 200 μL) injected by IP in WT mice induced a dose-related decrease in cumulative food intake and significantly reduced bout duration, bout frequency, meal size, time spent in feeding, time spent in meals, total meal time, eating rate, and meal duration post injection compared to vehicle injected WT mice. Furthermore, PACAP38 IP injected into overnight fasted WT mice, significantly reduced the plasma levels of active-ghrelin compared to vehicle. In PAC1−/− mice, fasting levels of active-ghrelin, GLP-1, insulin, and leptin, as well as postprandial levels of α-ghrelin and insulin, were significantly altered compared to WT mice. Based on these studies we concluded that PACAP is an important regulator of appetite/satiety and energy homeostasis.

Vasoactive intestinal peptide (VIP) is a highly conserved 28-amino acid neuropeptide widely distributed in the CNS and GI tract neurons. VIP binds with equal high affinity to its G protein-coupled receptors VPAC1 and VPAC2 [50, 51]. Physiologically, VIP has a diverse role regulating gastrointestinal function. VIP is coupled to G proteins linked to the stimulation of cAMP and regulates both motility and secretion. VIP was originally identified from porcine intestine where it was isolated. Based on structural homology, VIP belongs to the glucagon/secretin family of peptides whose members include GLP-1 and GLP-2, glucagon, and gastric inhibitory peptide (GIP). Neuroendocrine tumors that secrete VIP (VIPomas) result in hypersecretion and severe diarrhea. These patients will typically develop hypokalemia and volume depletion. VIP receptors may also be overexpressed in some gastrointestinal tumors. The role of VIP in regulating appetite has been controversial. In animal models, ICV injections of VIP have been shown to decrease food intake in vertebrates, suggesting an anorexigenic role [52]. The recent development of a VIP−/− mouse has permitted the elucidation of the gastrointestinal role for VIP [53]. To evaluate the anatomy of the VIP−/− mice we performed autopsies in both knockout and WT mice. This resulted in an overall decreased length in the intestinal length. Morphometric analyses performed on the luminal organs of the gastrointestinal tract revealed an increase of the cross-sectional diameter of the histopathological regions of the gut. Furthermore, there was a significant increase in the thickness of the muscularis propria. To evaluate whether there is dysfunction of the goblet cells, we performed Alcian blue staining. In VIP deficient mice there was striking mucus accumulation inside of the goblet cells. These mice were noted to have delayed motility compared to the WT mice as determined using a fluorescent dextran method. Additionally, there was a 36 % decrease in the bolus transit in the VIP-deficient mice.

VIP−/− mice ($n=8$) fed a standard diet *ad libitum* for 22 weeks developed a significantly lower body weight compared to WT mice [54]. At 5 weeks of age, the body weight of VIP−/− was lower compared to WT mice. Furthermore, at each 5.5

week interval measurement, VIP–/– body weights were found to be significantly lower than WT mice: at 10 weeks of age. During of the course of the ensuing weeks the VIP–/– mice consistently showed a reduced tendency to accumulate body fat mass. Serum levels of α-ghrelin, GLP-1, gastrin, PYY, insulin, glucagon, leptin, and adiponectin in both VIP–/– and WT mice were measured following fasting and postprandial conditions. Following a 12–16 h fasting period, the fasting gastrin levels in the VIP–/– mice were increased in VIP–/– mice compared to the WT mice (150 pg/ml vs. 100 pg/ml) and the post prandial gastrin levels increased to over 300 pg/ml indicating an exaggerated meal-induced gastrin response, despite possessing an intact vagal stimulated gastric acid secretion response. VIP–/– mice have been shown previously to have developmental defects and deficits in social behavior [55–57].

The elucidation of the pathways regulating appetite is critical to our ability to treat obesity disorders. Previously, there have been intense investigations to understand the role of ghrelin, CCK, GLP-1, and PYY as potential targets to treat obesity and metabolic disorders. There is a relative paucity of studies to investigate the role of PACAP and VIP in the treatment of metabolic syndrome. Our results highlight a novel key role for both PACAP and VIP in regulating the release of key metabolic mediators of satiety and fat metabolism. VIP may be an important contributor to the appetite and obesity disorders. The development of genetically engineered PAC1 and VIP knockout mouse has allowed for an in-depth study on the role of the respective hormone and receptors in a murine model. Studies utilizing genome wide association analysis have correlated VIP genes to development of obesity [58]. In this study, the authors evaluated 500,000 SNPs from 1000 obese individuals to demonstrate an association. Another important GWAS study showed an association between polymorphisms in the PAC1, in a subgroup of patients with post-traumatic stress disorder [59]. Further, the role of the VPAC2 has not been fully investigated and may play an important role in mediating lipolysis [60]. These results provide convincing evidence that the use of pharmacological agents to inhibit the physiological effects of PACAP and VIP may prove to be an effective treatment of metabolic syndrome and obesity.

Acknowledgements This work received grant support from: Department of Veterans Affairs RR&D Merit Review (JRP and PMJ); Animal Core Services performed through CURE: Digestive Diseases Research Center supported by NIH grant P30DK41301; NIH NIDDK DK07180 T32 (JB).

References

1. Wang YC, McPherson K, Marsh T, Gortmaker SL, Brown M. Health and economic burden of the projected obesity trends in the USA and the UK. Lancet. 2011;378:815–25.
2. Azagury DE, Lautz DB. Obesity overview: epidemiology, health and financial impact, and guidelines for qualification for surgical therapy. Gastrointest Endosc Clin N Am. 2011;21:189–201.

3. Greenwood HC, Bloom SR, Murphy KG. Peptides and their potential role in the treatment of diabetes and obesity. Rev Diabet Stud. 2011;8:355–68.
4. Woods SC, Seeley RJ, Porte Jr D, Schwartz MW. Signals that regulate food intake and energy homeostasis. Science. 1998;280:1378–83.
5. Dockray G. Gut endocrine secretions and their relevance to satiety. Curr Opin Pharmacol. 2004;4:557–60.
6. Date Y, Kojima M, Hosoda H, Sawaguchi A, Mondal MS, Suganuma T, et al. Ghrelin, a novel growth hormone-releasing acylated peptide, is synthesized in a distinct endocrine cell type in the gastrointestinal tracts of rats and humans. Endocrinology. 2000;141:4255–61.
7. Fukumoto K, Nakahara K, Katayama T, Miyazatao M, Kangawa K, Murakami N. Synergistic action of gastrin and ghrelin on gastric acid secretion in rats. Biochem Biophys Res Commun. 2008;374:60–3.
8. Wank SA, Pisegna JR, de Weerth A. Molecular cloning of the rat cholecystokinin type B receptor; structure, and functional expression. Proc Natl Acad Sci U S A. 1992;89:8691–5.
9. Huppi K, Siwarski D, Pisegna JR, Wank SA. Chromosomal localization of the gastric and brain receptors for cholecystokinin (CCKAR and CCKBR) in human and mouse. Genomics. 1995;25:727–9.
10. De Weerth A, Pisegna JR, Wank SA. Guinea pig gallbladder and pancreas possess identical CCK-A receptor subtypes: receptor cloning and expression. Am J Physiol. 1993;265:G1116–21.
11. Pohl M, Poirot SS, Pisegna JR, Tarasova NI, Wank SA. The carboxy-terminus of the cholecystokinin B receptor (CCKBR), unlike that of the type A receptor (CCKAR), is essential for ligand-induced internalization. J Biol Chem. 1997;272:18179–84.
12. Adrian TE, Ferri GL, Bacarese-Hamilton AJ, Fuessl HS, Polak JM, Bloom SR. Human distribution and release of a putative new gut hormone, peptide YY. Gastroenterology. 1985;89:1070–7.
13. Batterham RL, Cowley MA, Small CJ, Herzog H, Cohen MA, Dakin CL, et al. Gut hormone PYY(3-36) physiologically inhibits food intake. Nature. 2002;418:650–4.
14. Batterham RL, Cohen MA, Ellis SM, Le Roux CW, Withers DJ, Frost GS, et al. Inhibition of food intake in obese subjects by peptide YY3-36. N Engl J Med. 2003;349:941–8.
15. Ekblad E, Sundler F. Distribution of pancreatic polypeptide and peptide YY. Peptides. 2002;23:251–61.
16. Schmidt PT, Naslund E, Gryback P, Jacobsson H, Holst JJ, Hilsted L, et al. A role for pancreatic polypeptide in the regulation of gastric emptying and short term metabolic control. J Clin Endocrinol Metab. 2005;90:5241–6.
17. Zipf WB, O'Dorisio TM, Cataland S, Sotos J. Blunted pancreatic polypeptide responses in children with obesity of Prader–Willi syndrome. J Clin Endocrinol Metab. 1981;52:1264–6.
18. Orskov C, Rabenhoj L, Wettergren A, Kofod H, Holst JJ. Tissue and plasma concentrations of amidated and glycine-extended glucagon-like peptide I in humans. Diabetes. 1994;43:535–9.
19. Gallwitz B, Witt M, Folsch UR, Creutzfeldt W, Schmidt WE. Binding specificity and signal transduction of receptors for glucagon-like peptide-1(7-36)amide and gastric inhibitory polypeptide on RINm5F insulinoma cells. J Mol Endocrinol. 1993;10:259–68.
20. Holst JJ. Glucagon-like peptide-1: physiology and therapeutic potential. Curr Opin Endocrinol Diab. 2005;12:56–62.
21. Gutzwiller JP, Drewe J, Goke B, Schmidt H, Rohrer B, Lareida J, et al. Glucagon-like peptide-1 promotes satiety and reduces food intake in patients with diabetes mellitus type 2. Am J Physiol. 1999;276:R1541–4.
22. Naslund E, King N, Mansten S, Adner N, Holst JJ, Gutniak M, et al. Prandial subcutaneous injections of glucagon-like peptide-1 cause weight loss in obese human subjects. Br J Nutr. 2004;91:439–46.
23. Kojima M, Hosoda H, Date Y, Nakazato M, Matsuo H, Kangawa K. Ghrelin is a growth-hormone-releasing acylated peptide from stomach. Nature. 1999;402:656–60.
24. Wang HS, Oh DS, Ohning GV, Pisegna JR. Elevated serum ghrelin exerts an orexigenic effect that may maintain body mass index in patients with metastatic neuroendocrine tumors. J Mol Neurosci. 2007;33:225–31.

25. Miyata A, Arimura A, Dahl RR, Minamino N, Uehara A, Jiang L, et al. Isolation of a novel 38 residue-hypothalamic polypeptide which stimulates adenylate cyclase in pituitary cells. Biochem Biophys Res Commun. 1989;164:567–74.
26. Vaudry D, Falluel-Morel A, Bourgault S, Basille M, Burel D, Wurtz O, et al. Pituitary adenylate cyclase-activating polypeptide and its receptors: 20 years after the discovery. Pharmacol Rev. 2009;61:283–357.
27. Pisegna JR, Wank SA. Cloning and characterization of the signal transduction of four splice variants of the human pituitary adenylate cyclase activating polypeptide receptor. Evidence for dual coupling to adenylate cyclase and phospholipase C. J Biol Chem. 1996;271:17267–74.
28. Harmar AJ, Fahrenkrug J, Gozes I, Laburthe M, May V, Pisegna JR, et al. Pharmacology and functions of receptors for vasoactive intestinal peptide and pituitary adenylate cyclase-activating polypeptide: IUPHAR review 1. Br J Pharmacol. 2012;166:4–17.
29. Miampamba M, Germano PM, Arli S, Wong HH, Scott D, Taché Y, et al. Expression of pituitary adenylate cyclase-activating polypeptide and PACAP type 1 receptor in the rat gastric and colonic myenteric neurons. Regul Pept. 2002;105:145–54.
30. Köves K, Arimura A, Vigh S, Somogyvári-Vigh A, Miller J. Immunohistochemical localization of PACAP in the ovine digestive system. Peptides. 1993;14:449–55.
31. Buscail L, Gourlet P, Cauvin A, De Neef P, Gossen D, Arimura A, et al. Presence of highly selective receptors for PACAP (pituitary adenylate cyclase activating peptide) in membranes from the rat pancreatic acinar cell line AR 4-2J. FEBS Lett. 1990;262:77–81.
32. Tornøe K, Hannibal J, Fahrenkrug J, Holst JJ. PACAP-(1–38). Am J Physiol. 1997;273:G436–46.
33. Fridolf T, Sundler F, Ahrén B. Pituitary adenylate cyclase activating polypeptide (PACAP): occurrence in rodent pancreas and effects on insulin and glucagon secretion in the mouse. Cell Tissue Res. 1992;269:275–9.
34. Filipsson K, Tornøe K, Holst J, Ahrén B. Pituitary adenylate cyclase-activating polypeptide stimulates insulin and glucagon secretion in humans. J Clin Endocrinol Metab. 1997;82:3093–8.
35. Persson-Sjögren S, Forsgren S, Lindström P. Vasoactive intestinal polypeptide and pituitary adenylate cyclase activating polypeptide: effects on insulin release in isolated mouse islets in relation to metabolic status and age. Neuropeptides. 2006;40:283–90.
36. Nakata M, Yada T. PACAP in the glucose and energy homeostasis: physiological role and therapeutic potential. Curr Pharm Des. 2007;13:1105–12.
37. Ahrén B. Role of pituitary adenylate cyclase-activating polypeptide in the pancreatic endocrine system. Ann N Y Acad Sci. 2008;1144:28–35.
38. Borboni P, Porzio O, Pierucci D, Cicconi S, Magnaterra R, Federici M, et al. Molecular and functional characterization of pituitary adenylate cyclase-activating polypeptide (PACAP-38)/vasoactive intestinal polypeptide receptors in pancreatic beta-cells and effects of PACAP-38 on components of the insulin secretory system. Endocrinology. 1999;140:5530–7.
39. Sakurai Y, Shintani N, Hayata A, Hashimoto H, Baba A. Trophic effects of PACAP on pancreatic islets: a mini-review. J Mol Neurosci. 2011;43:3–7.
40. Jamen F, Persson K, Bertrand G, Rodriguez-Henche N, Puech R, Bockaert J, et al. PAC1 receptor-deficient mice display impaired insulinotropic response to glucose and reduced glucose tolerance. J Clin Invest. 2000;105:1307–15.
41. Raufman JP, Malhotra R, Singh L. PACAP-38, a novel peptide from ovine hypothalamus, is a potent modulator of amylase release from dispersed acini from rat pancreas. Regul Pept. 1991;36:121–9.
42. Mungan Z, Ertan A, Hammer RA, Arimura A. Effect of pituitary adenylate cyclase activating polypeptide on rat pancreatic exocrine secretion. Peptides. 1991;12:559–62.
43. Alonso RM, Rodríguez AM, García LJ, López MA, Calvo JJ. Comparison between the effects of VIP and the novel peptide PACAP on the exocrine pancreatic secretion of the rat. Pancreas. 1994;9:123–8.
44. Michalski CW, Selvaggi F, Bartel M, Mitkus T, Gorbachevski A, Giese T, et al. Altered anti-inflammatory response of mononuclear cells to neuropeptide PACAP is associated with dereg-

ulation of NF-{kappa}B in chronic pancreatitis. Am J Physiol Gastrointest Liver Physiol. 2008;294:G50–7.
45. Reubi JC. In vitro evaluation of VIP/PACAP receptors in healthy and diseased human tissues. Clinical implications. Ann N Y Acad Sci. 2000;921:1–25.
46. Germano PM, Lieu SN, Xue J, Cooke HJ, Christofi FL, Lu Y, et al. PACAP induces signaling and stimulation of 5-hydroxytryptamine release and growth in neuroendocrine tumor cells. J Mol Neurosci. 2009;39:391–401.
47. Lieu SN, Oh DS, Pisegna JR, Germano PM. Neuroendocrine tumors express PAC1 receptors. Ann N Y Acad Sci. 2006;1070:399–404.
48. Zeng N, Kang T, Wong H, Walsh JH, Sachs GA, Pisegna JR. PACAP Type I receptor activation regulates ECL cells and gastric acid secretion. J Clin Invest. 1999;104:1383–91.
49. Lu Y, Germano P, Ohning GV, Vu JP, Pisegna JR. PAC1 Deficiency in a murine model induces gastric mucosa hypertrophy and higher basal gastric acid output. J Mol Neurosci. 2011;43:76–84.
50. Said SI, Mutt V. Polypeptide with broad biological activity: isolation from small intestine. Science. 1970;169:1217–8.
51. Vaudry D, Gonzalez BJ, Basille M, Yon L, Fournier A, Vaudry H. Pituitary adenylate cyclase-activating polypeptide and its receptors: from structure to functions. Pharmacol Rev. 2000;52:269–324.
52. Tachibana T, Saito S, Tomonaga S, Takagi T, Saito ES, Boswell T, et al. Intracerebroventricular injection of vasoactive intestinal peptide and pituitary adenylate cyclase-activating polypeptide inhibits feeding in chicks. Neurosci Lett. 2003;339:203–6.
53. Lelievre V, Favrais G, Abad C, Adle-Biassette H, Lu Y, Germano PM, et al. Gastrointestinal dysfunction in mice with a targeted mutation in the gene encoding vasoactive intestinal polypeptide: a model for the study of intestinal ileus and Hirschsprung's Disease. Peptides. 2007;28:1688–99.
54. Vu JP, Larauche M, Flores M, Luong L, Norris J, Oh S, et al. Regulation of appetite, body composition, and metabolic hormones by vasoactive intestinal polypeptide (VIP). J Mol Neurosci. 2015;56:377–87.
55. Lim MA, Stack CM, Cuasay K, Stone MM, McFarlane HG, Waschek JA, et al. Regardless of genotype, offspring of VIP-deficient female mice exhibit developmental delays and deficits in social behavior. Int J Dev Neurosci. 2008;26:423–34.
56. Girard BA, Lelievre V, Braas KM, Razinia T, Vizzard MA, Ioffe Y, et al. Noncompensation in peptide/receptor gene expression and distinct behavioral phenotypes in VIP- and PACAP-deficient mice. J Neurochem. 2006;99:499–513.
57. Hill JM, Mervis RF, Politi J, McCune SK, Gozes I, Fridkin M, et al. Blockade of VIP during neonatal development induces neuronal damage and increases VIP and VIP receptors in brain. Ann N Y Acad Sci. 1994;739:211–25.
58. Liu YJ, Guo YF, Zhang LS, Pei YF, Yu N, Yu P, et al. Biological pathway-based genome-wide association analysis identified the vasoactive intestinal peptide (VIP) pathway important for obesity. Obesity (Silver Spring). 2010;18:2339–46.
59. Ressler KJ, Mercer KB, Bradley B, Jovanovic T, Mahan A, Kerley K, et al. Post-traumatic stress disorder is associated with PACAP and the PAC1 receptor. Nature. 2011;470:492–7.
60. Akesson L, Ahrén B, Edgren G, Degerman E. VPAC2-R mediates the lipolytic effects of pituitary adenylate cyclase-activating polypeptide/vasoactive intestinal polypeptide in primary rat adipocytes. Endocrinology. 2005;146:744–50.

Chapter 17
Protective Intestinal Effects of Pituitary Adenylate Cyclase Activating Polypeptide

Gabriella Horvath, Anita Illes, Markus M. Heimesaat, Attila Bardosi, Sebastian Bardosi, Andrea Tamas, Balazs D. Fulop, Balazs Opper, Jozsef Nemeth, Andrea Ferencz, and Dora Reglodi

Abstract Pituitary adenylate cyclase activating polypeptide (PACAP) is an endogenous neuropeptide widely distributed throughout the body, including the gastrointestinal tract. Several effects have been described in human and animal intestines. Among others, PACAP influences secretion of intestinal glands, blood flow, and smooth muscle contraction. PACAP is a well-known cytoprotective peptide with strong anti-apoptotic, anti-inflammatory, and antioxidant effects. The present review gives an overview of the intestinal protective actions of this neuropeptide. Exogenous PACAP treatment was protective in a rat model of small bowel autotransplantation. Radioimmunoassay (RIA) analysis of the intestinal tissue

Andrea Ferencz and Dora Reglodi contributed equally

G. Horvath • A. Tamas • D.B. Fulop • B. Opper • D. Reglodi (✉)
Department of Anatomy, MTA-PTE "Lendulet" PACAP Research Group, University of Pecs, Pecs, Hungary
e-mail: dora.reglodi@aok.pte.hu

A. Illes
Department of Anatomy, MTA-PTE "Lendulet" PACAP Research Group, University of Pecs, Pecs, Hungary

Department of Internal Medicine, University of Pecs, Pecs, Hungary

M.M. Heimesaat
Department of Microbiology and Hygiene, Charite-University Medicine, Berlin, Germany

A. Bardosi • S. Bardosi
Jakob Henle Institute for Pathology and Neuropathology, Limbach Group, Heidelberg, Germany

J. Nemeth
Department of Pharmacology and Pharmacotherapy, University of Debrecen, Debrecen, Hungary

A. Ferencz
Department of Surgical Research and Techniques, Semmelweis University, Budapest, Hungary

D. Reglodi, A. Tamas (eds.), *Pituitary Adenylate Cyclase Activating Polypeptide — PACAP*, Current Topics in Neurotoxicity 11,
DOI 10.1007/978-3-319-35135-3_17

showed that endogenous PACAP levels gradually decreased with longer-lasting ischemic periods, prevented by PACAP addition. PACAP counteracted deleterious effects of ischemia on oxidative stress markers and cytokines. Another series of experiments investigated the role of endogenous PACAP in intestines in PACAP knockout (KO) mice. Warm ischemia–reperfusion injury and cold preservation models showed that the lack of PACAP caused a higher vulnerability against ischemic periods. Changes were more severe in PACAP KO mice at all examined time points. This finding was supported by increased levels of oxidative stress markers and decreased expression of antioxidant molecules. PACAP was proven to be protective not only in ischemic but also in inflammatory bowel diseases. A recent study showed that PACAP treatment prolonged survival of *Toxoplasma gondii* infected mice suffering from acute ileitis and was able to reduce the ileal expression of proinflammatory cytokines. We completed the present review with recent clinical results obtained in patients suffering from inflammatory bowel diseases. It was found that PACAP levels were altered depending on the activity, type of the disease, and antibiotic therapy, suggesting its probable role in inflammatory events of the intestine.

Keywords Ileitis • Intestinal ischemia • Crohn's disease • Colitis • PACAP

Introduction

Pituitary adenylate cyclase activating polypeptide (PACAP) is a widely recognized neuroprotective peptide. The neuroprotective actions of PACAP were soon recognized after its isolation from ovine hypothalami in 1989 [1, 2]. Subsequent studies showed that the occurrence of PACAP is widespread, it can be found in all organ systems, not only in the nervous system and so its general anti-apoptotic, antioxidant, and anti-inflammatory effects can be observed in numerous different organs and tissues besides the nervous system. The neuroprotective and general cytoprotective effects of PACAP have been reviewed in the recent years, and several aspects of this action can be found in other chapters of this book [2–5].

PACAP is also present in the gastrointestinal system where it exerts several actions. PACAP influences motility of the intestinal wall [6], an action phylogenetically well conserved [7], and it also regulates sphincter muscles [8–10]. PACAP acts on intestinal secretion, among others, it has been shown that PACAP stimulates secretion of growth factors, like brain-derived neurotrophic factor (BDNF) from intestinal smooth muscle cells [11]; furthermore, it inhibits pacemaker activity of interstitial cells of Cajal [12]. In the present review we briefly summarize protective actions of PACAP in the intestinal system, mainly in ischemic and inflammatory bowel disease models and give a brief summary of studies done in PACAP gene deficient mice. We also briefly review pathological findings on PACAP and/or receptor expression in both animal models and in human biopsies and we report on our recent findings in human samples from patients with inflammatory bowel disease.

Effects of Exogenous PACAP in Small Bowel Ischemia–Reperfusion Injury in Rats

Intestinal warm ischemia–reperfusion (I/R) is associated with high morbidity and mortality in some critical clinical settings, such as hemorrhagic shock, strangulation, obstruction, cardiovascular surgery, and severe trauma. Several factors are involved in the cascade of events resulting in intestinal death and critical condition of the patient following mesenteric infarction. These include production of reactive oxygen species leading to oxidative stress, inflammation, and finally tissue necrosis with septic signs [13, 14].

PACAP has been shown to protect against ischemic injuries in several tissues. The first studies showing the neuroprotective effects of PACAP investigated these effects in global and focal brain ischemia [15, 16]. Subsequently, numerous other studies confirmed that PACAP exerts strong protective effects in neuronal ischemic injuries [2, 17–20]. PACAP's efficacy in protecting tissues against ischemic lesions has become evident in the retina [21–23] and in peripheral tissues, such as liver [24], kidney [25–28], and heart [29]. Thus, it is not surprising that PACAP is also protective in intestinal I/R injury. In the following sections, we give a brief overview of our studies regarding the effects of exogenously administered PACAP in small intestinal ischemia, summarized in Fig. 17.1.

Our first studies showed that PACAP is protective in a rat model of small bowel autotransplantation, a cold ischemia injury model [30–32]. Small bowel was resected and grafts were preserved in University of Wisconsin solution, as commercially available, standard solution during transplantation, with or without additional PACAP. Grafts were autotransplanted after 1, 2, 3, or 6 h with a reperfusion lasting 3 h, a procedure called cold preservation injury. Radioimmunoassay (RIA) analysis of the intestinal tissue showed that endogenous PACAP levels gradually decreased with longer-lasting warm and cold ischemic periods, but tissue levels of both PACAP38 and PACAP27 remained higher when PACAP38 was added to the solution [30, 33]. Histological analysis showed that cold ischemia led to a destruction of the mucous, submucous, and muscularis layers, further deteriorated by the end of the reperfusion. PACAP protected the intestinal structure: a better preserved mucous layer and crypt morphology could be observed. These results clearly showed that PACAP was protective in intestinal cold I/R injury [30, 31]. They also demonstrated that adding PACAP to the preservation solution did not increase lipid peroxidation and prevented the reduction of endogenous scavenger capacity.

Studying the mechanism of action, we used biochemical assays to measure the oxidative stress markers and antioxidant enzyme levels, and molecular biological methods to study the expression of apoptotic signaling. Biochemical assays from the samples obtained after 3 h of reperfusion following 1, 2, 3, and 6 h of cold intestinal ischemia showed that malondialdehyde (MDA), an oxidative stress marker, gradually increased with the duration of the ischemia, while this increase was markedly less in the PACAP-treated groups. In contrast, levels of the endogenous scavenger molecule, reduced glutathione (GSH), showed a significant decrease with long-lasting I/R in the

Animal models	Wistar rat			CD1 PACAP$^{-/-}$ and PACAP$^{+/+}$ mouse			
Surgical protocol	**Warm I/R**	**Tx UW**	**Tx UW+PACAP-38**	**Warm I/R PACAP$^{+/+}$**	**Warm I/R PACAP$^{-/-}$**	**Tx (UW) PACAP$^{+/+}$**	**Tx (UW) PACAP$^{-/-}$**
	warm ischemia reperfusion	cold ischemia reperfusion	cold ischemia reperfusion	warm ischemia reperfusion	warm ischemia reperfusion	cold ischemia reperfusion	cold ischemia reperfusion
Small bowel tissue							
Oxidative stress							
MDA	↑↑↑↑*	↑↑↑*	↑*	↑↑↑*	↑↑↑↑*	↑↑*	↑↑↑*
GSH	↓↓↓*	↓↓*	↓*	↓*	↓↓*	↓*	↓*
SOD	↓↓↓↓*	↓↓↓*	↓↓*	↓↓↓*	↓↓↓↓*	↓↓*	↓↓↓*
Structural injury							
Qualitative Park's score (G0→G5)	G1→G5	G1→G4	G1→G3	G1→G5	G2→G5	G1→G3	G2→G4
Quantitative analysis							
Mucosa	↓↓↓*	↓↓*	↓*	↓↓↓*	↓↓↓↓*	↓↓*	↓↓↓*
Submucosa	↓↓↓*	↓↓*	↓*	↓↓↓*	↓↓↓↓*	↓↓*	↓↓↓*
Crypts	↓↓↓*	↓↓	↓	↓↓↓*	↓↓↓↓*	↓↓*	↓↓↓*
Muscle	↓	↓	↓	↓	↓	↓	↓
Endogenous PACAP-38	↓↓↓*	↓↓*	↑↑*				
Endogenous PACAP-27		↓↓*	↑↑↑*				
Anti-apoptotic effect							
pERK1/2		↓*	↑*				
pJNK		↑*	↓*				
p38 MAPK		↑*	↓*				
Citokines							
sICAM1		↑↑*	↓↓*				
L-Selectin		↑↑*	↓↓*				
TIMP-1		↑↑*	↓↓*				

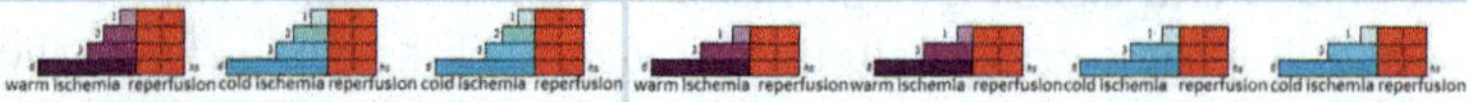

Fig. 17.1 Cytoprotective, anti-apoptotic, and anti-inflammatory effect of PACAP in intestinal warm ischemia–reperfusion (I/R) and transplantation (Tx)/cold ischemia animal models. Warm ischemia and cold storage in University of Wisconsin (UW) preservation solution lasted 1, 2, 3, or 6 h, while reperfusion took 3 h in each groups. According to Park's classification tissue injury was analyzed from grade zero (control) to the highest grade of injury (grade 5: G5). (*PACAP$^{+/+}$* PACAP wild type mice, *PACAP$^{-/-}$* PACAP knockout mice, *MDA* malondialdehyde, *GSH* reduced glutathione, *SOD* superoxide dismutase, *pERK1/2* phosphorylated-ERK1/2, *pJNK* phosphorylated-Jun-N-terminal kinase, *p38 MAPK* p38 Mitogen Activated Protein Kinase, *sICAM1* soluble Intercellular Adhesion Molecule-1, *TIMP-1* Tissue Inhibitor of MetalloProteinase-1)

control groups. In PACAP-treated groups, GSH levels were near baseline values and the level of this endogenous scavenger was higher in all PACAP-treated groups than in the ischemic control groups. A dramatic decrease was observed in activity of antioxidant enzyme: superoxide dismutase (SOD) in the control ischemic groups. In PACAP-treated groups only a slight decrease could be observed.

In another study, we examined changes of cytokine expression in the intestinal tissue after different ischemic periods in University of Wisconsin solution with or without additional PACAP using cytokine array and Luminex multiplex immunoassay [33]. It was found that the cold ischemia-induced changes in some cytokines could be ameliorated by PACAP. Namely, elevations in expression level of the sICAM (CD54) and L-selectin (CD62L/LECAM-1) regulated upon activation could be decreased by PACAP treatment, similarly to a tendency observed in renal ischemic injury [25]. These adhesion molecules are known to be involved in crosstalk between immune cells and endothelial cells, and their overexpression can lead to microvascular dysfunction and reperfusion damage. PACAP led to a slight reduction also in RANTES levels and a significant decrease of the strongly induced TIMP levels [33]. Studying the mitogen activated protein (MAP) kinase phosphorylation, we found that phosphorylated ERK1/2 levels gradually decreased with the ischemic time, while JNK1/2 and p38 MAP kinase phosphorylation increased. These changes were counteracted by addition of PACAP in the preservation solution [32].

Altogether, these studies showed that PACAP is protective in intestinal I/R injury partly through increasing the antioxidant status of the intestinal tissue, via anti-inflammatory actions and via anti-apoptotic signaling. This combination of anti-apoptotic, anti-oxidant, and anti-inflammatory actions of PACAP have been confirmed by several other studies. One of the best-known protective effects of PACAP is its anti-apoptotic effect [34], which has been extensively studied in the nervous system [35–37], sensory organs like inner ear [38] and retina [39] and has also been shown in peripheral organ lesions, such as in diabetic nephropathy [40], and in oxidative stress-induced cardiomyocyte injury [41, 42].

The antioxidant efficacy of PACAP has also been described in other studies. Miyamoto et al. [43] showed that PACAP suppressed cortical damage in mice with traumatic brain injury by enhancing antioxidant activity. In diabetic nephropathy, Banki et al. [40] showed that PACAP treatment increased glutathione levels. PACAP also stimulated glutathione formation, and blocked H_2O_2-evoked reactive oxygen species (ROS) accumulation and glutathione content reduction in oxidative stress-induced astroglial injury [44]. In kidney ischemia–reperfusion, similar changes were observed: PACAP treatment decreased MDA levels while increased GSH levels [45]. These studies suggest that PACAP has anti-oxidant effects via indirect action of mechanism, through stimulating endogenous antioxidant capacity. This hypothesis is also supported by the observation that direct scavenging action of PACAP can only be observed at very high (micromolar) concentrations of the peptide [46].

The anti-inflammatory actions of PACAP have been more widely investigated. It has been shown that the anti-inflammatory effects of the peptide are partially mediated via suppressing chemokine/cytokine production [47]. PACAP plays an important regulatory role in the immune system, a function reviewed several times by different groups [48, 49]. The role of PACAP in cytokine and chemokine expression has been investigated in the nervous system as well as in peripheral organs. Microglial cytokine and chemokine expression was shown to be altered by PACAP resulting in a neuroregeneration-favoring environment [50–52]. In the retina, PACAP increased anti-inflammatory cytokine expression [53]. Furthermore, PACAP led to an anti-inflammatory cytokine balance in kidney ischemia [25], in diabetic nephropathy model [54] and in retinal ischemia [55].

Small Intestinal Ischemic Injury in PACAP Knockout (KO) Mice

To answer the question of whether the intestinal protective action of PACAP can also be observed endogenously, we studied the results of intestinal I/R in PACAP KO mice, summarized in Fig. 17.1 [56–58]. No difference in the histological structure between young wild type and PACAP KO mice (heterozygous or homozygous) could be observed under normal circumstances (Fig. 17.2). However, in ischemic injuries, PACAP KO mice showed a higher degree of tissue injury. Cold preservation injury was studied by removing small bowel from wild type and PACAP KO mice and placing them in University of Wisconsin preservation solution for 1, 3 and 6 h. Small bowel biopsies were collected after the ischemic periods and after laparotomy for controls. We investigated the histological structure as well as the above-mentioned oxidative stress parameters. Most severe injury was observed after 6 h of cold storage, as expected. After 1 h of cold preservation, only minor clefting with the villus epithelium adjacent to the crypts left intact. Changes were more severe in PACAP KO mice at all examined time points. After 6 h, severely injured crypt appeared with denuded villi corresponding to most severe tissue injury (Grade 4), while in wild type mice this was less severe, Grade 3 injury. Quantitative histological analysis also confirmed this observation: mucosal thickness showed a more severe decrease with more prolonged ischemic times, and this decrease was more severe in PACAP KO mice. Similarly, depth of crypts and submucosal thickness were also more severely affected in PACAP-deficient mice. Measuring the parameters of oxidative stress revealed that the increase of MDA was higher, while the decrease in GSH and SOD was more pronounced in PACAP KO mice, indicating an elevated oxidative stress level in mice lacking endogenous PACAP. This corresponds well with the increased tissue injury revealed by the histological analysis [56].

We also performed warm ischemic injury in PACAP KO mice [57]. Warm ischemia was induced by occlusion of the superior mesenteric artery for 1, 3 or 6 h. Similarly to the above studies, histological and biochemical analyses were performed. Histology showed that PACAP KO mice suffered a more severe tissue injury at all time points. For example, 3 h of ischemia resulted in severely injured crypts and denuded villi in KO mice, while this histological picture was characteristic in the wild type group only after 6 h of ischemic time (Fig. 17.2). Quantitative histological analysis confirmed these observations: both mucosal and submucosal thickness as well as crypt depth were more severely decreased in mice lacking endogenous PACAP. Studying the oxidative stress markers, results similar to those observed in cold ischemic injury were obtained: levels of MDA were more significantly increased in the KO groups, while those of the antioxidant SOD and GSH were more severely decreased than in the wild type animals ([57], Fig. 17.1).

These results are in agreement with the general observation that PACAP KO mice are more vulnerable to tissue injuries than wild type mice [59]. This is also true for ischemic injuries. Ohtaki and coworkers showed that PACAP KO mice have increased infarct volume in a model of stroke [60] with increased edema formation [61]. Similar results were obtained in the retina: in bilateral carotid artery occlusion-induced retinal ischemia PACAP KO mice showed increased tissue damage [62]. In peripheral organ ischemia, it has been shown that PACAP KO mice react with increased tissue injury in ischemia–reperfusion-induced kidney lesion [27, 63]. Increased inflammatory response has also been demonstrated in endotoxin-induced lung injury [64] and in contact dermatitis [65]. In the background of this increased susceptibility, a decrease in the endogenous reserve capacity, to withstand the tissue

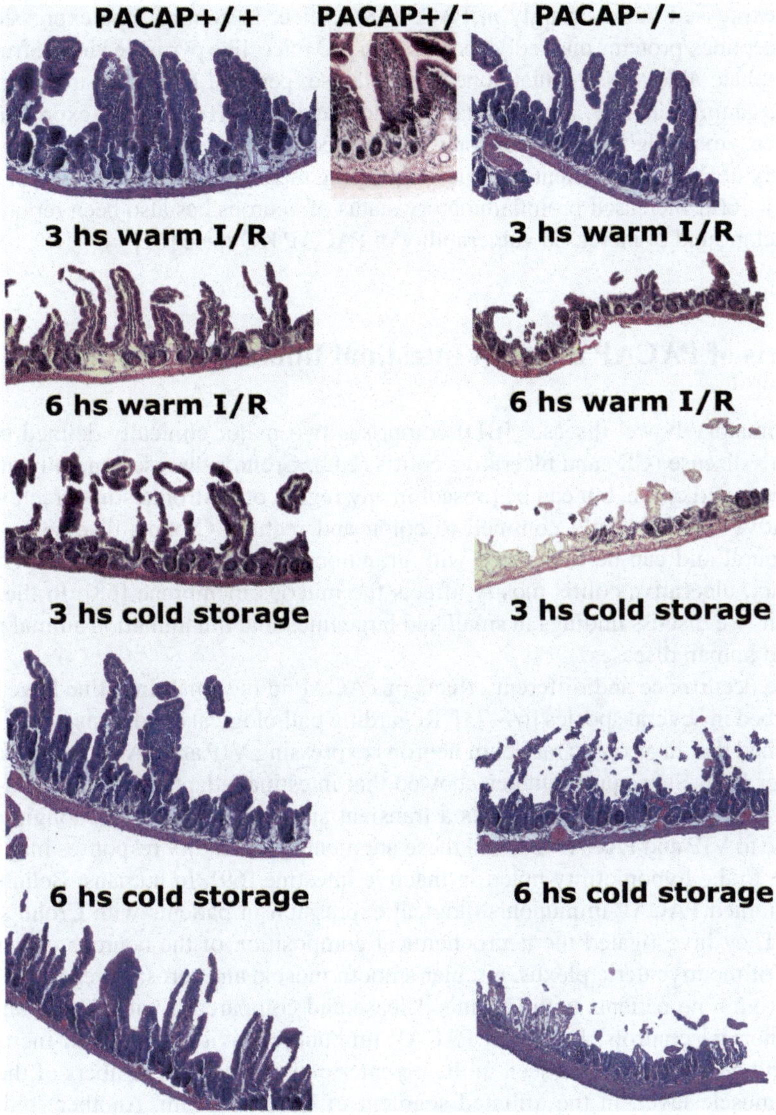

Fig. 17.2 Representative photomicrographs of small bowel sections from young adult PACAP$^{+/+}$, PACAP$^{+/-}$, and PACAP$^{-/-}$ mice. Tissue injury of small bowel after 3 and 6 h warm I/R and 3 h and 6 h cold preservation in UW in PACAP$^{+/+}$ and in PACAP$^{-/-}$ mice (Hematoxylin and eosin staining). *PACAP*$^{+/+}$: PACAP wild type mice, *PACAP*$^{-/-}$: PACAP knockout mice, *PACAP*$^{+/-}$: PACAP heterozygous mice

damaging effect of different harmful stimuli, is suggested. Some studies have already sought to elucidate detailed mechanisms. A proteomic analysis of PACAP KO mice showed that several altered proteins are involved in metabolic processes, energy homeostasis, and structural integrity [66]. ATP-synthase and tubulin beta-2A were expressed more strongly in PACAP KO mice. In contrast, the expression of more peptides/proteins markedly decreased in KO mice, like pyruvate kinase, fructose biphosphate aldolase-A, glutathione S-transferase, peptidyl propyl cis-trans isomerase-A, gamma enolase, and aspartate amino transferase. The altered expression of these enzymes might partially account for the decreased antioxidant and detoxifying capacity of PACAP-deficient mice accompanying the increased vulnerability of these animals [66]. Increased proinflammatory status of neurons has also been reported in the background of increased vulnerability of PACAP KO mice [67].

Effects of PACAP in Small Intestinal Inflammation

Inflammatory bowel disease (IBD) comprises two major clinically defined types: Crohn's disease (CD) and ulcerative colitis (UC). Crohn's disease generally affects the ileum and colon, but can be present in any region of gastrointestinal tract, while ulcerative colitis is only confined to colon and rectum. Crohn's disease is often transmural and can be associated with granulomas, strictures and fistulas. On the contrary, ulcerative colitis mostly affects the mucous membrane [68]. In the next sections we discuss findings in small and large intestinal inflammation animal models and human diseases.

The occurrence and different effects of PACAP in the small intestine have been described in several species [69–73]. Regarding pathological conditions, it has been described that in atrophic rat ileum neurons expressing VIP and PACAP decrease in number [69]. Subsequent studies showed that intestinal adaptation after bypassing the distal ileum of the rat includes a transient supersensitivity of the longitudinal muscle to VIP and PACAP-27, and these augmented relaxatory responses may contribute to the hypomotility noted in inactive intestine [69]. In humans, Belai et al. [74] studied PACAP immunohistological expression in patients with Crohn's disease. They investigated the neurochemical composition of the neurons and nerve fibers of the myenteric plexus, circular smooth muscle and serosa layer of the ileal wall from nine patients with Crohn's disease and compared it with seven samples from normal controls. Increased PACAP immunoreactivity along with increased VIP and NO synthase was seen in the myenteric plexus and nerve fibers of the circular muscle layer, in the afflicted segment of Crohn's ileum. Another study on human small intestine reported a decrease of PACAP immunoreactivity along with loss of interstitial cells of Cajal in diabetic gastroenteropathy [75].

In our recent study applying a *Toxoplasma gondii*-induced acute ileitis model, mice were treated with PACAP38 once daily via the intraperitoneal route in two different regimens; namely, between days 1–6 postinfection (as prophylaxis) or starting on day 4 (till day 6, as therapeutic treatment), when first histopathological changes are observed in the ileal mucosa [76]. The treatment prolonged the survival of the infected mice: while all placebo-treated control mice died by day 9.5, 80 % of mice with PACAP prophylaxis and 40 % with PACAP therapeutic treatment survived the acute phase. Twenty percent of the prophylaxis group survived till the end of experiment. This inflammation ameliorating (anti-inflammatory) effect of PACAP was further supported by less pronounced weight loss of treated mice, and by less distinct shrinkage of the small intestines. Histopathological examination of the ileal mucosa revealed that 7 days postinfection, placebo-treated control mice displayed severe necrosis of the mucous membrane, whereas moderate ileitis was found in the PACAP therapeutic group and only mild inflammation in the prophylaxis group. Given that PACAP did not exert direct effects on the intestinal microbiota composition, the beneficial effects are rather due to the anti-inflammatory, protective actions of the peptide as confirmed by studies of defined pro- and anti-inflammatory markers. The local increase in CD3+ T cells, for instance, and the local decrease in regulatory T cells was less marked in the PACAP-treated groups. In addition, lower ileal neutrophils, monocytes and macrophages were observed after PACAP treatment. Furthermore, ileal expression of proinflammatory cytokines including IL-23p19 (interleukin- 23p19), IL-22 (interleukin-22), IFN-gamma (interferon gamma), and MCP-1 (monocyte chemoattractant protein-1) was reduced in ileal ex vivo biopsies derived from PACAP-treated mice. The diminished pro-inflammatory response upon PACAP was further underlined by lower protein concentrations of IFN-gamma and NO in ex vivo biopsies of mesenteric lymph nodes taken from PACAP-treated mice. Our results from extraintestinal compartments revealed that PACAP prophylaxis also abrogated systemic inflammatory responses, as demonstrated by the decreased level of TNF-alpha (tumor necrosis factor alpha) and NO (nitrogen monoxide) in the spleen and of TNF-alpha, IFN-gamma, MCP-1, and IL-6 (interleukin-6) in the liver. Higher levels of anti-inflammatory cytokines such as IL-10 and IL-4 were measured in the serum/spleen and in the liver/mesenteric lymph node, respectively, of PACAP-treated animals. Furthermore, kidney and lung histology revealed less tissue damage after PACAP prophylaxis. Altogether, these results show that both PACAP therapeutic and prophylactic treatments can reduce the symptoms and pathological signs of ileitis in a murine model by a complex mechanistic network. The prophylactic treatment resulted to be more effective (in a time-of-treatment dependent manner), and could also reduce the extraintestinal sequelae of the *T. gondii* induced inflammation [76].

Effects of PACAP in Large Intestinal Inflammation

Similarly to the small intestine, occurrence of PACAP and its receptors as well as different functions of PACAP have been described in the large intestine of different species, including humans [77–82]. Regarding pathological changes, only a few data are available. Gonkowski and Calka [81] reported that immunoreactive PACAP changes in different large intestinal pathologies in the pig. They found that in the porcine descending colon chemically induced inflammation, nerve injury, and proliferative enteropathy (a naturally occurring inflammation of the porcine digestive tract) resulted in an increase in the number of PACAP-27-like immunoreactive (LI) nerve fibers in the circular muscle and mucosal layers. The number of PACAP-containing perikarya also increased in all cases in the myenteric and submucosal plexus, except for the outer submucosal plexus in case of axotomy-induced pathology. In another study, Wojtkiewicz et al. [83] described inflammation-induced changes in the neurochemical composition of the inferior mesenteric ganglia supplying the lower portion of the colon in pigs. Although PACAP was not investigated, they found that the number of VIP immunoreactive neurons was reduced in proliferative enteropathy. Changes in PACAP and receptor expression were also demonstrated in dextran sodium sulfate (DSS)-induced colitis of mice. A marked upregulation of PACAP (and of VIP) mRNA expression was shown in TRPA1 (transient receptor potential Ankyrin type 1) KO mice, while this was not observed in wild type controls [84]. Expression levels of VPAC2 and PAC1R mRNA were not altered, whereas that of VPAC1 was downregulated in TRPA1 KO mice 10 days following colitis induction [84]. An earlier study revealed that mice lacking the TRPV-1 (transient receptor potential cation channel subfamily V member 1) receptor exhibit a defect in PACAP expression. The authors argue that this might contribute to the development of a local pro-inflammatory environment characteristic for these mice. This might also be an additional factor for colitis-associated cancer development [85].

Regarding human colonic samples, in an earlier study we found significantly lower levels of PACAP38-LI and PACAP27-LI in tumoral and peripheral samples compared with normal healthy tissue in colon cancers, similarly to some urogenital tumors [86, 87]. This observation is in accordance with immunohistochemical analysis of human sigmoid colon and rectum tumors, where a less dense PACAP immunoreactive nerve fiber network was observed in the myenteric and submucosal plexi than in normal controls [88]. Another study found that in patients with symptomatic diverticular disease significantly higher PACAP levels can be detected along with increases in other neuropeptides within the mucosal plexus. Patients with symptomatic diverticular disease exhibit increased neuropeptides in mucosal biopsies, which may reflect resolved prior inflammation, as it parallels the changes seen in acute and chronic diverticulitis [89]. In contrast, another study found decreased density of PACAP containing nerve fibers in colon mucosal membrane of children with ulcerative colitis [90, 91]. Decreased immunoreactivity (measured both by immunohistochemistry and RIA) was also observed in the aganglionic segment of human colon with Hirschsprung's disease [92].

Measurement of PACAP Levels in Human Inflammatory Bowel Diseases

Our group performed a human study to determine PACAP levels in inflammatory conditions. We examined samples taken from patients suffering from ulcerative colitis ($n=30$) or Crohn's disease ($n=22$) and used samples obtained from patients having diarrhea without any morphological abnormality as controls ($n=8$), after obtaining patients' consent. In both groups of IBD 3 grades were distinguished: (1) chronic inflammation with lymphocyte and plasma cell infiltration without cryptitis or crypt abscess; (2) acute inflammation with cryptitis or crypt abscess and neutrophil

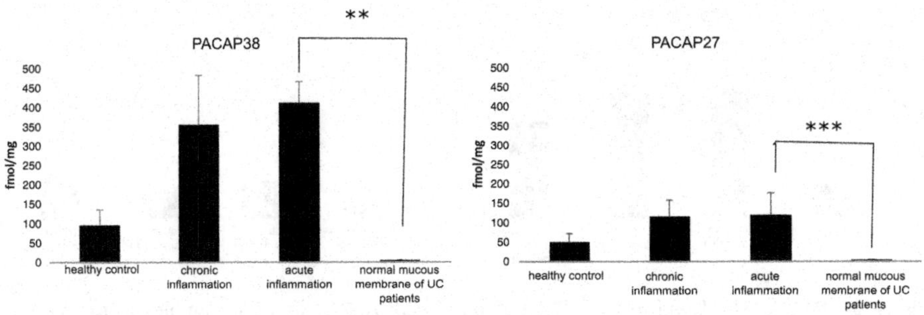

Fig. 17.3 PACAP38-like immunoreactivity (LI) and PACAP27-LI as measured by RIA in colon samples from patients suffering from ulcerative colitis. Data are expressed as mean fmol/mg tissue weight ± SEM. $^{**}p<0.01$; $^{***}p<0.001$

infiltration; (3) normal mucous membrane showing no signs of inflammation. For observing the effects of antibiotic therapy on PACAP level, colon samples of IBD patients treated with metronidazol or ciprofloxacin/metronidazol combination (CD: $n=5$; UC: $n=7$) were also examined. The procedure used was in accordance with protocols approved by the ethical committee (no. 2784,3117, University of Pecs; 8-28/92 009-10 I 8EKU, ETT TUKEB, Ministry of Health, Hungary). Tissue samples were weighed and homogenized in ice-cold distilled water. The homogenate was centrifuged (12,000 rpm, 4 °C, 30 min), and the supernatant was further processed for RIA analysis of PACAP38 and PACAP27 contents, as previously described [93, 94]. For these measurements antiserum PACAP38 "88111-3" (working dilution, 1:10,000) and PACAP27 "88123" (dilution: 1:45,000) were used.

These results show marked increase in both PACAPs' levels in ulcerative colitis (Fig. 17.3), but only a slight increase in PACAP38 could be detected in Crohn's disease (Fig. 17.4) compared to normal mucous membrane of these patients showing no inflammatory signs. We found that both PACAP38 and PACAP27 levels of normal mucous membrane samples of UC patients were much lower than in the inflamed samples, but there were no differences in case of CD patients. Measuring PACAP levels, significant elevation of both PACAPs could be detected in samples obtained from colon of UC patients showing acute inflammation. Studying the possible effect of antibiotics on PACAP levels in IBD patients revealed a significant decrease of both PACAP38 and PACAP27 level in UC patients receiving antibiotics treatment (Fig. 17.5). A moderate, but detectable decrease of PACAPs was seen in samples of CD patients.

Effects of Endogenous PACAP

The protective effects of endogenous PACAP have also been shown in mouse models of inflammatory colon diseases performed in PACAP KO mice [59]. Two groups, independently of each other and at the same time, reported similar results in DSS-induced colonic inflammation, a model of inflammatory bowel disease [95, 96].

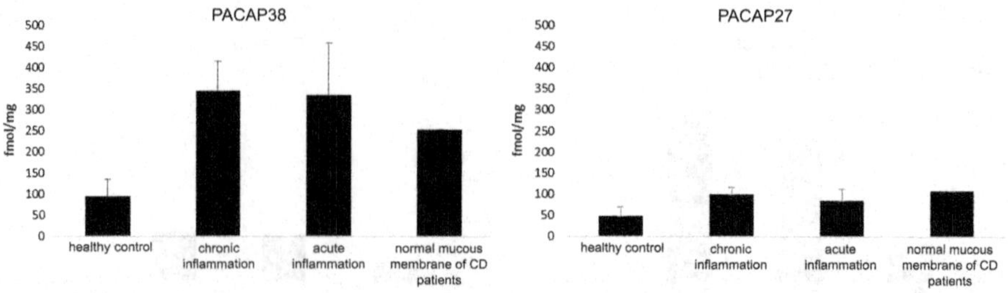

Fig. 17.4 PACAP38-LI and PACAP27-LI as measured by RIA in colon samples from patients suffering from Crohn's disease. Data are expressed as mean fmol/mg tissue weight ± SEM

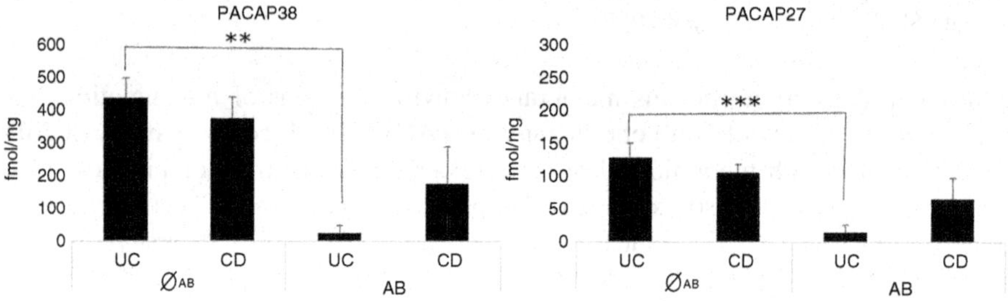

Fig. 17.5 PACAP38-LI and PACAP27-LI in IBD patients treated with antibiotics determined by RIA. Data are expressed as mean fmol/mg tissue weight ± SEM. $^{**}p<0.01$; $^{***}p<0.001$ AB: colon samples of patients receiving antibiotic therapy, ØAB: patients without antibiotic therapy

Azuma and coworkers [95], using a 7-day exposure (short-term) to DSS, found that PACAP KO mice exhibited higher mortality (increased by about 50 %) and increased weight loss, bleeding and diarrhea, symptoms that are used to determine the disease activity index. The large intestines were more shortened in PACAP KO mice, to one third of wild type mice. Similarly to our observations, the authors did not find morphological differences in the colonic sections in intact PACAP KO mice. However, histological scoring of the proximal and distal colon parts in DSS-induced colitis showed an increased number of infiltrating cells and crypt damage in PACAP-deficient mice. Furthermore, production of IL-1beta and IL-6 was upregulated by 50 % in the proximal colon of PACAP-deficient mice, and that of IFN-gamma, IL-1beta, IL-6, IL-12, and keratinocyte-derived chemokine expression was upregulated in the distal colon by 50 % [95]. Conversely, levels of anti-inflammatory cytokines such as IL-10 were only one third of that of wild type mice in the distal colon. Using a longer exposure to DSS (2 months), Nemetz and coworkers [96] described similar results. After induction of colitis by exposure to DSS, PACAP knockout animals displayed more severe clinical symptoms and higher histological inflammation scores (50–70 % higher), restricted to the distal colon. In addition, induction of IL-1beta and IL-6 mRNA expression levels were significantly higher (70 %) in PACAP-deficient mice. Interestingly, 60 % of PACAP KO mice developed colorectal tumors with aggressive-appearing pathology [96].

As both studies showed, PACAP-deficient mice receiving normal water without DSS displayed no signs of inflammation (and of colitis), suggesting that the lack of endogenous PACAP does not result in immune depression leading to spontaneous development of colitis. This is in agreement with our own observations that mice lacking endogenous PACAP do not spontaneously develop increased inflammatory status (not shown), but based on the above-described reports, the lack of endogenous PACAP leads to increased vulnerability to inflammation and inflammation-associated cancer development in the colon.

In summary, PACAP is widely distributed in the gastrointestinal tract and exerts several different effects. Based on the studies reviewed here in detail, PACAP seems to be an important endogenous protective factor in ischemic and inflammatory bowel diseases and exogenous PACAP treatment can counteract changes in ischemic and inflammatory models. Human data indicate that PACAP levels change in inflammatory conditions the clinical relevance of which needs further investigation.

Acknowledgements This study was supported by OTKA K104984, 119759 Arimura Foundation, TAMOP 4.2.4.A/2-11-1-2012-0001 "National Excellence Program," Bolyai Scholarship of the Hungarian Academy of Sciences, PTE AOK KA Research Grant, PTE-KA 300019 and New National Excellence Program (UNKP). This work is dedicated to the 650th anniversary of the University of Pecs.

References

1. Arimura A, Somogyvari-Vigh A, Weill C, Fiore RC, Tatsuno I, Bay V, et al. PACAP functions as a neurotrophic factor. Ann N Y Acad Sci. 1994;739:228–43.
2. Somogyvari-Vigh A, Reglodi D. Pituitary adenylate cyclase activating polypeptide: a potential neuroprotective peptide. Curr Pharm Des. 2004;10:2861–89.
3. Lee EH, Seo SR. Neuroprotective roles of pituitary adenylate cyclase-activating polypeptide in neurodegenerative diseases. BMB Rep. 2014;47:369–75.
4. Reglodi D, Kiss P, Lubics A, Tamas A. Review on the protective effects of PACAP in models of neurodegenerative diseases in vitro and in vivo. Curr Pharm Des. 2011;17:962–72.
5. Shioda S, Nakamachi T. PACAP as a neuroprotective factor in ischemic neuronal injuries. Peptides. 2015;72:202–7.
6. Fujimiya M, Inui A. Peptidergic regulation of gastrointestinal motility in rodents. Peptides. 2000;21:1565–82.
7. Holmberg A, Schwerte T, Pelster B, Holmgren S. Ontogeny of the gut motility control system in zebrafish Danio rerio embryos and larvae. J Exp Biol. 2004;207:4085–94.
8. Farre R, Auli M, Lecea B, Martinez E, Clave P. Pharmacologic characterization of intrinsic mechanisms controlling tone and relaxation of porcine lower esophageal sphincter. J Pharmacol Exp Ther. 2006;316:1238–48.
9. Lindestrom LM, Ekblad E. Origins and projections of nerve fibres in rat pyloric sphincter. Auton Neurosci. 2002;97:73–82.
10. Rattan S, Chakder S. Sites of actions of contractile and relaxant effects of pituitary adenylate cyclase activating peptide (PACAP) in the internal anal sphincter smooth muscle. Ann N Y Acad Sci. 1998;865:503–11.
11. Al-Qudah M, Alkahtani R, Akbarali HI, Murthy KS, Grider JR. Stimulation of synthesis and release of brain-derived neurotropic factor from intestinal smooth muscle cells by substance P and pituitary adenylate cyclase-activating peptide. Neurogastroenterol Motil. 2015;27:1162–74.

12. Wu MJ, Kee KH, Na J, Kim SW, Bae Y, Shin DH, et al. Pituitary adenylate cyclase-activating polypeptide inhibits pacemaker activity of colonic interstitial cells of Cajal. Korean J Physiol Pharmacol. 2015;19:435–40.
13. Ferencz A, Szanto Z, Borsiczky B, Kiss K, Kalmar-Nagy K, Telek G, et al. The effects of preconditioning on the oxidative stress in small bowel autotransplantation. Surgery. 2002;132:877–84.
14. Mittal A, Phillips ARJ, Loveday B, Windsor JA. The potential role for xanthine oxidase inhibition in major intra-abdominal surgery. World J Surg. 2008;32:288–95.
15. Reglodi D, Somogyvari-Vigh A, Vigh S, Kozicz T, Arimura A. Delayed systemic administration of PACAP38 is neuroprotective in transient middle cerebral artery occlusion in rat. Stroke. 2000;31:1411–7.
16. Uchida D, Arimura A, Somogyvari-Vigh A, Shioda S, Banks WA. Prevention of ischemia-induced death of hippocampal neurons by pituitary adenylate cyclase activating polypeptide. Brain Res. 1996;736:280–6.
17. Brifault C, Gras M, Liot D, May V, Vaudry D, Wurtz O. Delayed pituitary adenylate cyclase-activating polypeptide delivery after brain stroke improves functional recovery by inducing m2 microglia/macrophage polarization. Stroke. 2015;46:520–8.
18. Chen Y, Samal B, Hamelink CR, Xiang CC, Chen Y, Chen M, et al. Neuroprotection by endogenous and exogenous PACAP following stroke. Regul Pept. 2006;137:4–19.
19. Lenti L, Zimmermann A, Kis D, Olah O, Toth GK, Hegyi O, et al. PACAP and VIP differentially preserve neurovascular reactivity after global cerebral ischemia in newborn pigs. Brain Res. 2009;1283:50–7.
20. Tamas A, Reglodi D, Szanto Z, Borsiczky B, Nemeth J, Lengvari I. Comparative neuroprotective effects of preischemic PACAP and VIP administration in permanent occlusion of the middle cerebral artery in rats. Neuro Endocrinol Lett. 2002;23:249–54.
21. Atlasz T, Szabadfi K, Kiss P, Tamas A, Toth G, Reglodi D, et al. Evaluation of the protective effects of PACAP with cell-specific markers in ischemia-induced retinal degeneration. Brain Res Bull. 2010;81:497–504.
22. Danyadi B, Szabadfi K, Reglodi D, Mihalik A, Danyadi T, Kovacs Z, et al. PACAP application improves functional outcome of chronic retinal ischemic injury in rats-evidence from electroretinographic measurements. J Mol Neurosci. 2014;54:293–9.
23. Seki T, Itoh H, Nakamachi T, Endo K, Wada Y, Nakamura K, et al. Suppression of rat retinal ganglion cell death by PACAP following transient ischemia induced by high intraocular pressure. J Mol Neurosci. 2011;43:30–4.
24. Ji H, Zhang Y, Shen XD, Gao F, Huang CY, Abad C, et al. Neuropeptide PACAP in mouse liver ischemia and reperfusion injury: immunomodulation by the cAMP-PKA pathway. Hepatology. 2013;57:1225–37.
25. Horvath G, Racz B, Reglodi D, Kovacs K, Kiss P, Gallyas Jr F, et al. Effects of PACAP on mitochondrial apoptotic pathways and cytokine expression in rats subjected to renal ischemia/reperfusion. J Mol Neurosci. 2010;42:411–8.
26. Laszlo E, Kiss P, Horvath G, Szakaly P, Tamas A, Reglodi D. The effects of pituitary adenylate cyclase-activating polypeptide in renal ischemia/reperfusion. Acta Biol Hung. 2014;65:369–78.
27. Laszlo E, Varga A, Kovacs K, Jancso G, Kiss P, Tamas A, et al. Ischemia/reperfusion-induced kidney injury in heterozygous PACAP-deficient mice. Transplant Proc. 2015;47:2210–5.
28. Szakaly P, Kiss P, Lubics A, Magyarlaki T, Tamas A, Racz B, et al. Effects of PACAP on survival and renal morphology in rats subjected to renal ischemia/reperfusion. J Mol Neurosci. 2008;36:89–96.
29. Roth E, Weber G, Kiss P, Horvath G, Toth G, Gasz B, et al. Effect of PACAP and preconditioning against ischemia/reperfusion-induced cardiomyocyte apoptosis in vitro. Ann N Y Acad Sci. 2009;1163:512–6.
30. Ferencz A, Racz B, Tamas A, Reglodi D, Lubics A, Nemeth J, et al. Influence of PACAP on oxidative stress and tissue injury following small-bowel autotransplantation. J Mol Neurosci. 2009;37:168–76.

31. Ferencz A, Racz B, Tamas A, Nedvig K, Nemeth J, Kalmar-Nagy K, et al. Changes and effect of PACAP-38 on intestinal ischemia-reperfusion and autotransplantation. Transplant Proc. 2009;41:57–9.
32. Ferencz A, Reglodi D, Kalmar-Nagy K, Horvath OP, Roth E, Weber G, et al. Influence of pituitary adenylate cyclase-activating polypeptide on the activation of mitogen activated protein kinases following small bowel cold preservation. Transplant Proc. 2009;41:60–2.
33. Nedvig K, Weber G, Nemeth J, Kovacs K, Reglodi D, Kemeny A, et al. Changes of PACAP immunoreactivities and cytokine levels after PACAP-38 containing intestinal preservation and autotransplantation. J Mol Neurosci. 2012;48:788–94.
34. Seaborn T, Masmoudi-Kouli O, Fournier A, Vaudry H, Vaudry D. Protective effects of pituitary adenylate cyclase-activating polypeptide (PACAP) against apoptosis. Curr Pharm Des. 2011;17:204–14.
35. Brown D, Tamas A, Reglodi D, Tizabi Y. PACAP protects against inflammatory-mediated toxicity in dopaminerg SH-SY5Y cells: implication for Parkinson's disease. Neurotox Res. 2014;26:230–9.
36. Dejda A, Seaborn T, Bourgault S, Touzani O, Fournier A, Vaudry H, et al. PACAP and a novel stable analog protect rat brain from ischemia: insight into the mechanisms of action. Peptides. 2011;32:1207–16.
37. Rozzi SJ, Borelli G, Ryan K, Steiner JP, Reglodi D, Mocchetti I, et al. PACAP27 is protective against tat-induced neurotoxicity. J Mol Neurosci. 2014;54:485–93.
38. Racz B, Horvath G, Reglodi D, Gasz B, Kiss P, Gallyas Jr F, et al. PACAP ameliorates oxidative stress in the chicken inner ear: an in vitro study. Regul Pept. 2010;160:91–8.
39. Szabadfi K, Szabo A, Kiss P, Reglodi D, Setalo Jr G, Kovacs K, et al. PACAP promotes neuron survival in early experimental diabetic retinopathy. Neurochem Int. 2014;64:84–91.
40. Banki E, Kovacs K, Nagy D, Juhasz T, Degrell P, Csanaky K, et al. Molecular mechanisms underlying the nephroprotective effects of PACAP in diabetes. J Mol Neurosci. 2014;54:300–9.
41. Gasz B, Racz B, Roth E, Borsiczky B, Ferencz A, Tamas A, et al. Pituitary adenylate cyclase-activating polypeptide protects cardiomyocytes against oxidative stress-induced apoptosis. Peptides. 2006;27:87–94.
42. Mori H, Nakamachi T, Ohtaki H, Yofu S, Sato A, Endo K, et al. Cardioprotective effect of endogenous pituitary adenylate cyclase-activating polypeptide on doxorubicin-induced cardiomyopathy in mice. Circ J. 2010;74:1183–90.
43. Miyamoto K, Tsumuraya T, Ohtaki H, Dohi K, Satoh K, Xu Z, et al. PACAP38 suppresses cortical damage in mice with traumatic brain injury by enhancing antioxidant activity. J Mol Neurosci. 2014;54:370–9.
44. Masmoudi-Kouki O, Douiri S, Hamdi Y, Kaddour H, Bahdoudi S, Vaudry D, et al. Pituitary adenylate cyclase-activating polypeptide protects astroglial cells against oxidative stress-induced apoptosis. J Neurochem. 2011;117:403–11.
45. Horvath G, Brubel R, Kovacs K, Reglodi D, Opper B, Ferencz A, et al. Effects of PACAP on oxidative stress-induced cell death in rat kidney and human hepatocyte cells. J Mol Neurosci. 2011;43:67–75.
46. Reglodi D, Fabian Z, Tamas A, Lubics A, Szeberenyi J, Alexy T, et al. Effects of PACAP on in vitro and in vivo neuronal cell death, platelet aggregation, and production of reactive oxygen radicals. Regul Pept. 2004;123:51–9.
47. Delgado M, Genea D. Inhibition of endotoxin-induced macrophage chemokine production by vasoactive intestinal peptide and pituitary adenylate cyclise activating polypeptide in vitro and in vivo. J Immunol. 2001;167:966–75.
48. Abad C, Waschek JA. Immunomodulatory roles of VIP and PACAP in models of multiple sclerosis. Curr Pharm Des. 2011;17:1025–35.
49. Ganea D, Rodriguez R, Delgado M. Vasoactive intestinal peptide and pituitary adenylate cyclase-activating polypeptide: players in innate and adaptive immunity. Cell Mol Biol (Noisy-le-grand). 2003;49:127–42.

50. Delgado M, Jonakait GM, Ganea D. Vasoactive intestinal peptide and pituitary adenylate cyclase-activating polypeptide inhibit chemokine production in activated microglia. Glia. 2002;39:148–61.
51. Delgado M, Leceta J, Ganea D. Vasoactive intestinal peptide and pituitary adenylate cyclase-activating polypeptide inhibit the production of inflammatory mediators by activated microglia. J Leukoc Biol. 2003;73:155–64.
52. Wainwright DA, Xin J, Sanders VM, Jones KJ. Differential actions of pituitary adenylyl cyclase-activating polypeptide and interferon gamma on Th2- and Th1-associated chemokine expression in cultured murine microglia. J Neurodegener Regen. 2008;1:31–4.
53. Wada Y, Nakamachi T, Endo K, Seki T, Ohtaki H, Tsuchikawa D, et al. PACAP attenuates NMDA-induced retinal damage in association with modulation of the microglia/macrophage status into an acquired deactivation subtype. J Mol Neurosci. 2013;51:493–502.
54. Banki E, Degrell P, Kiss P, Kovacs K, Kemeny A, Csanaky K, et al. Effect of PACAP treatment on kidney morphology and cytokine expression in rat diabetic nephropathy. Peptides. 2013;42:125–30.
55. Szabo A, Danyadi B, Bognar E, Szabadfi K, Fabian E, Kiss P, et al. Effect of PACAP on MAP kinases, Akt and cytokine expressions in rat retinal hypoperfusion. Neurosci Lett. 2012;523:93–8.
56. Ferencz A, Weber G, Helyes Z, Hashimoto H, Baba A, Reglodi D. Presence of endogenous PACAP-38 ameliorated intestinal cold preservation tissue injury. J Mol Neurosci. 2010;42:428–34.
57. Ferencz A, Kiss P, Weber G, Helyes Z, Shintani N, Baba A, et al. Comparison of intestinal warm ischemic injury in PACAP knockout and wild-type mice. J Mol Neurosci. 2010;42:435–42.
58. Ferencz A, Nedvig K, Fekecs T, Racz B, Weber G, Hashimoto H, et al. Comparison of intestinal cold preservation injury on pituitary adenylate cyclase-activating polypeptide in knockout and wild-type mice. Transplant Proc. 2010;42:2290–2.
59. Reglodi D, Kiss P, Szabadfi K, Atlasz T, Gabriel R, Horvath G, et al. PACAP is an endogenous protective factor-insights from PACAP-deficient mice. J Mol Neurosci. 2012;48:482–92.
60. Ohtaki H, Nakamachi T, Dohi K, Aizawa Y, Takaki A, Hodoyama K, et al. Pituitary adenylate cyclase-activating polypeptide (PACAP) decreases ischemic neuronal cell death in association with IL-6. Proc Natl Acad Sci U S A. 2006;103:7488–93.
61. Nakamachi T, Ohtaki H, Yofu S, Dohi K, Watanabe J, Mori H, et al. Endogenous pituitary adenylate cyclase activating polypeptide is involved in suppression of edema in the ischemic brain. Acta Neurochir Suppl. 2010;106:43–6.
62. Szabadfi K, Atlasz T, Kiss P, Danyadi B, Tamas A, Helyes Z, et al. Mice deficient in pituitary adenylate cyclase-activating polypeptide (PACAP) are more susceptible to retinal ischemic injury in vivo. Neurotox Res. 2012;21:41–8.
63. Szakaly P, Laszlo E, Kovacs K, Racz B, Horvath G, Ferencz A, et al. Mice deficient in pituitary adenylate cyclase activating polypeptide (PACAP) show increased susceptibility to in vivo renal ischemia/reperfusion injury. Neuropeptides. 2011;45:113–21.
64. Elekes K, Sandor K, Moricz A, Kereskai L, Kemeny A, Szoke E, et al. Pituitary adenylate cyclase-activating polypeptide plays an anti-inflammatory role in endotoxin-induced airway inflammation: In vivo study with gene-deleted mice. Peptides. 2011;32:1439–46.
65. Kemeny A, Reglodi D, Cseharovszky R, Hashimoto H, Baba A, Szolcsanyi J, et al. Pituitary adenylate cyclase-activating polypeptide deficiency enhances oxazolone-induced allergic contact dermatitis in mice. J Mol Neurosci. 2010;42:443–9.
66. Maasz G, Pirger Z, Reglodi D, Petrovics D, Schmidt J, Kiss P, et al. Comparative protein composition of the brains of PACAP-deficient mice using mass spectrometry-based proteomic analysis. J Mol Neurosci. 2014;54:310–9.
67. Watson MB, Nobuta H, Abad C, Lee SK, Bala N, Zhu C, et al. PACAP deficiency sensitizes nigrostriatal dopaminergic neurons to paraquat-induced damage and modulates central and peripheral inflammatory activation in mice. Neuroscience. 2013;240:277–86.
68. Abraham C, Cho JH. Inflammatory bowel disease. N Engl J Med. 2009;19:2066–78.

69. Ekelund M, Ekblad E. Intestinal adaptation in atrophic rat ileum is accompanied by supersensitivity to vasoactive intestinal peptide, pituitary adenylate cyclase-activating peptide and nitric oxide. Scand J Gastroenterol. 2001;36:251–7.
70. Lazar Z, Shahbazian A, Benko R, Toth G, Penke B, Bartho L, et al. PACAP-(6-38) inhibits the effects of vasoactive intestinal polypeptide, but not PACAP, on the small intestinal circular muscle. Eur J Pharmacol. 2001;431:259–64.
71. Mao YK, Wang YF, Moogk C, Fox-Threlkeld JE, Xiao Q, McDonald TJ, et al. Locations and molecular forms of PACAP and sites and characteristics of PACAP receptors in canine ileum. Am J Physiol. 1998;274:G217–25.
72. Pirone A, Baoan D, Piano I, Della Santina L, Baglini A, Lenzi C. Pituitary adenylate cyclase-activating peptide (PACAP) immunoreactivity distribution in the small intestine of the adult New Hampshire chicken. Acta Histochem. 2011;113:477–83.
73. Zizzo MG, Mulé F, Serio R. Mechanisms underlying the inhibitory effects induced by pituitary adenylate cyclase-activating peptide in mouse ileum. Eur J Pharmacol. 2005;521:133–8.
74. Belai A, Boulos PB, Robson T, Burnstock G. Neurochemical coding in the small intestine of patients with Crohn's disease. Gut. 1997;40:767–74.
75. He CL, Soffer EE, Ferris CD, Walsh RM, Szurszewski JH, Farrugia G. Loss of interstitial cells of cajal and inhibitory innervation in insulin-dependent diabetes. Gastroenterology. 2001;121:427–34.
76. Heimesaat MM, Dunay IR, Schulze S, Fischer A, Grundmann U, Alutis M, et al. Pituitary adenylate cyclase-activating polypeptide ameliorates experimental acute ileitis and extra-intestinal sequelae. PLoS One. 2014;9:e108389.
77. Leung PS, So SC, Lam SY, Tsang LL, Chung YW, Chan HC. Local regulation of anion secretion by pituitary adenylate cyclase-activating polypeptide in human colonic T84 cells. Cell Biol Int. 2001;25:123–9.
78. Rettenbacher M, Reubi JC. Localization and characterization of neuropeptide receptors in human colon. Naunyn Schmiedebergs Arch Pharmacol. 2001;364:291–304.
79. Miampamba M, Germano PM, Arli S, Wong HH, Scott D, Tache Y, et al. Expression of pituitary adenylate cyclase-activating polypeptide and PACAP type 1 receptor in the rat gastric and colonic myenteric neurons. Regul Pept. 2002;105:145–54.
80. Uyttebroek L, Shepherd IT, Harrisson F, Hubens G, Blust R, Timmermans JP, et al. Neurochemical coding of enteric neurons in adult and embryonic zebrafish (Danio rerio). Cancer Res. 2012;72:1705–16.
81. Gonkowski S, Calka J. Changes in pituitary adenylate cyclase-activating peptide 27-like immunoreactive nervous structures in the porcine descending colon during selected pathological processes. J Mol Neurosci. 2012;48:777–87.
82. Gonkowski S, Kaminska B, Landowski P, Calka J. Immunohistochemical distribution of cocaine- and amphetamine-regulated transcript peptide-like immunoreactive (CART-LI) nerve fibers and various degree of co-localization with other neuronal factors in the circular muscle layer of human descending colon. Histol Histopathol. 2013;28:851–8.
83. Wojtkiewicz J, Rowniak M, Gonkowski S, Crayton R, Majewski M, Robak A, et al. Proliferative enteropathy (PE)-induced changes in the calbindin-immunoreactive (CB-IR) neurons of inferior mesenteric ganglion supplying the descending colon in the pig. J Mol Neurosci. 2012;48:757–65.
84. Kun J, Szitter I, Kemeny A, Perkecz A, Kereskai L, Pohoczky K, et al. Upregulation of the transient receptor potential ankyrin 1 ion channel in the inflamed human and mouse colon and its protective roles. PLoS One. 2014;9:e108164.
85. Vinuesa AG, Sancho R, Garcia-Limones C, Behrens A, Ten Dijke P, Calzado MA, et al. Vanilloid receptor-1 regulates neurogenic inflammation in colon and protects mice from colon cancer. J Comp Neurol. 2010;518:4419–38.
86. Szanto Z, Sarszegi Z, Reglodi D, Nemeth J, Szabadfi K, Kiss P, et al. PACAP immunoreactivity in human malignant tumor samples and cardiac diseases. J Mol Neurosci. 2012;48:667–73.
87. Tamas A, Javorhazy A, Reglodi D, Sarlos DP, Banyai D, Semjen D, et al. Examination of PACAP-like immunoreactivity in urogenital tumor samples. J Mol Neurosci. 2016;59:177–83.

88. Godlewski J, Lakomy IM. Changes in vasoactive intestinal peptide, pituitary adenylate cyclase-activating polypeptide and neuropeptide Y-ergic structures of the enteric nervous system in the carcinoma of the human large intestine. Folia Histochem Cytobiol. 2010;48:208–16.
89. Simpson J, Sundler F, Humes DJ, Jenkins D, Scholefield JH, Spiller RC. Post inflammatory damage to the enteric nervous system in diverticular disease and its relationship to symptoms. Neurogastroenterol Motil. 2009;21:847–e58.
90. Kaminska B, Landowski P, Gonkowski S, Majewski M, Renke J, Korzon M. Changes in the number of neuroprotective transmitter containing mucosal nerve fibres in children with ulcerative colitis. Med Wieku Rozwoj. 2006;10:483–91.
91. Kaminska B, Landowski P, Gonkowski S, Szlagatys-Sidorkiewitz A, Majewski M, Dobosz M, et al. Analysis of enteral nervous system in children with drug resistant ulcerative colitis. Med Wieku Rozwoj. 2007;11:117–22.
92. Shen Z, Larsson LT, Malmfors G, Absood A, Håkanson R, Sundler F. A novel neuropeptide, pituitary adenylate cyclase-activating polypeptide (PACAP), in human intestine: evidence for reduced content in Hirschsprung's disease. Cell Tissue Res. 1992;269:369–74.
93. Jakab B, Reglodi D, Jozsa R, Hollosy T, Tamas A, Lubics A, et al. Distribution of PACAP-38 in the central nervous system of various species determined by a novel radioimmunoassay. J Biochem Biophys Methods. 2004;61:189–98.
94. Nemeth J, Jakab B, Jozsa R, Hollosy T, Tamas A, Lubics A, et al. PACAP-27 radioimmunoassay: description and application of a novel method. J Radioanal Nucl Chem. 2007;273:327–32.
95. Azuma YT, Hagi K, Shintani N, Kuwamura M, Nakajima H, Hashimoto H, et al. PACAP provides colonic protection against dextran sodium sulfate induced colitis. J Cell Physiol. 2008;216:111–9.
96. Nemetz N, Abad C, Lawson G, Nobuta H, Chhith S, Doung L, et al. Induction of colitis and rapid development of colorectal tumors in mice deficient in the neuropeptide PACAP. Int J Cancer. 2008;122:1803–9.

Chapter 18
Renoprotective Effects of Pituitary Adenylate Cyclase-Activating Polypeptide 38 (PACAP38)

M-Altaf Khan and Vecihi Batuman

Abstract Pituitary adenylate-cyclase activating polypeptide 38 (PACAP38) is a naturally occurring pleiotropic neuropeptide with immunomodulatory actions. Research over the past two decades have revealed potent renoprotective properties of this peptide in diverse renal disorders including myeloma kidney, ischemia–reperfusion injury (IRI), nephrotoxic injury mediated by cisplatin, aminoglycosides, calcineurin inhibitors, iodinated radiocontrast dyes, and others. The kidney expresses both PACAP and PACAP receptors, and the renoprotective properties of PACAP are mediated through modulation of inflammatory pathways within the kidney as well as circulation derived cells. In this chapter we review the data on the effects of PACAP38 on immune pathways in various kidney injury models. In general, PACAP38 restrains heightened immune responses by inhibiting phosphorylation of mitogen-activated protein kinases (MAPK), and inhibits nuclear transcription factors including NF-κB. PACAP38 blocks hyperactivation of innate immunity related genes and reduces the expression of Toll-like receptors (TLRs) as well as their adaptor proteins. PACAP38 may be a promising therapy in a broad range of renal disorders.

Keywords PACAP38 • Acute kidney injury • Myeloma kidney • Ischemia–reperfusion injury • Cisplatin • Cyclosporine A • Gentamicin • Diabetic nephropathy • Contrast-induced nephropathy • Toll-like receptors

---

M.-A. Khan
Department of Medicine, Section of Nephrology & Hypertension, School of Medicine, Tulane University, New Orleans, LA 70112, USA

V. Batuman (✉)
Department of Medicine, Section of Nephrology & Hypertension, School of Medicine, Tulane University, New Orleans, LA 70112, USA

Tulane Cancer Center, School of Medicine, Tulane University, New Orleans, LA 70112, USA

Department of Veterans Affairs, Southeast Louisiana Veterans Health Care System, New Orleans, LA 70112, USA

Nephrology Section-8545, School of Medicine, Tulane University, 1430 Tulane Avenue, New Orleans, LA 70112-2632, USA
e-mail: vbatuma@tulane.edu

D. Reglodi, A. Tamas (eds.), *Pituitary Adenylate Cyclase Activating Polypeptide — PACAP*, Current Topics in Neurotoxicity 11,
DOI 10.1007/978-3-319-35135-3_18

Abbreviations

AKI	Acute kidney injury
BUN	Blood urea nitrogen
CIN	Contrast-induced nephropathy
CKD	Chronic kidney disease
CM	Contrast media
CsA	Cyclosporine A
ECM	Extracellular matrix
EMT	Epithelial–mesenchymal transformation
FCAs	Flow cytometry assays
H&E	Hematoxylin and eosin
IL	Interleukin
IRI	Ischemia–reperfusion injury
KIM-1	Kidney injury molecule 1
LC	Light chains
LDH	Lactate dehydrogenase
MAPK	Mitogen-activated protein kinase
MM	Multiple myeloma
MyD88	Myeloid differentiation primary response gene 88
NF-κB	Nuclear factor κB
PACAP38	Pituitary adenylate cyclase-activating polypeptide 38, PAS, periodic acid-Schiff
RPTECs	Renal proximal tubule epithelial cells
STZ	Streptozotocin
TLRs	Toll-like receptors
TNF-α	Tumor necrosis factor α
TRIF	Toll/interleukin 1 receptor domain-containing adapter inducing interferon-β
TUNEL	Terminal deoxynucleotidyl transferase dUTP nick end labeling
VIP	Vasoactive intestinal peptide

Introduction

Pituitary adenylate cyclase-activating polypeptide 38 (PACAP38) was isolated in 1989 from ovine hypothalamus at Tulane University during a screen for novel hypophysiotropic factors [1–3], and later on it was found to be a pleiotropic peptide with potent anti-inflammatory, immune modulatory, and antioxidant properties and its potential cytoprotective effects were recognized [4, 5]. The cytoprotective effects of PACAP have been extensively studied especially in the brain and were extended to investigations with other organ systems including the kidney [6–26]. PACAP38 exerts immunomodulatory effects on various immune responses by suppressing the production of proinflammatory cytokines, modulating MAPK, and NF-κB signaling pathways [6, 27]. PACAP38 has been investigated as potential therapy for a variety of autoimmune and inflammatory diseases, such as rheumatoid arthritis, septic shock, asthma, Crohn's disease, primary pulmonary hypertension, and multiple sclerosis [5–7, 28, 29].

PACAP38 has also been administered to healthy human volunteers by investigators in several laboratories and by us to a single patient with multiple myeloma (MM) under a single-patient protocol approved by the FDA without any untoward side effects [5, 30]. The anti-apoptotic properties of PACAP38 have been shown in the nervous system [28, 29, 31] and the kidney [16, 19]. We have reported that PACAP38 is renoprotective against toxic myeloma light chains (LC), IRI-induced acute kidney injury (AKI), iodinated radiocontrast nephrotoxicity, cyclosporine A, cisplatin, and also that it can reverse epithelial–mesenchymal transformation (EMT) in kidney cells, in vitro [12–14, 16–20].

We previously reported the inhibitory effects of PACAP38 on reactive oxygen species (ROS)- and TLRs-mediated NF-κB activation in hypoxia/ischemia models of AKI [12, 19]. More recently, we have reported the renoprotective effects of PACAP38 as an anti-inflammatory agent modulating TLRs and as an antioxidant by suppressing genes encoding Nox enzymes in hypertensive (eNOS$^{-/-}$) mice against radiocontrast-induced nephropathy (CIN) [14]. PACAP38 protects renal proximal tubule epithelial cells (RPTECs) in vitro and also protects the structure and function of the kidney in vivo against the deleterious effects of nephrotoxin-induced injury, diabetes, stress, and hypoxia/ischemia–reperfusion related AKI. These observations indicate that PACAP38 could be a potential therapeutic agent for the treatment of AKI and chronic kidney disease (CKD). The detailed discussion about the renoprotective effects of PACAP38 against various types of kidney injuries is described as follows.

Protective Effects of PACAP38 in Myeloma Kidney

AKI in MM is generally induced by the toxic and inflammatory effects of monoclonal free LC on kidney proximal tubule cells and by the formation of intratubular casts through interaction with Tamm–Horsfall proteins. Production of excessive quantities of LC is seen in most cases of LC-associated kidney injury, although a direct relation between quantity and nephrotoxicity does not exist, due to variable toxic effects of different types of LC species [32]. LC-induced toxic effects include inhibition of transport functions in the kidney sometimes resulting in Fanconi syndrome, generation of ROS, cytoskeletal abnormalities, and apoptosis and necrosis in proximal tubule cells [27, 32, 33]. The inflammatory pathways that are activated as a result of LC toxicity also could explain how acute injury can progress into chronic tubulointerstitial nephritis in patients with myeloma kidney and identify attractive opportunities for novel therapeutic interventions [27, 32]. Although the pathophysiological mechanisms have been elucidated, there is no curative treatment for MM, and myeloma kidney, apart from the limited use of steroids, thalidomide and its

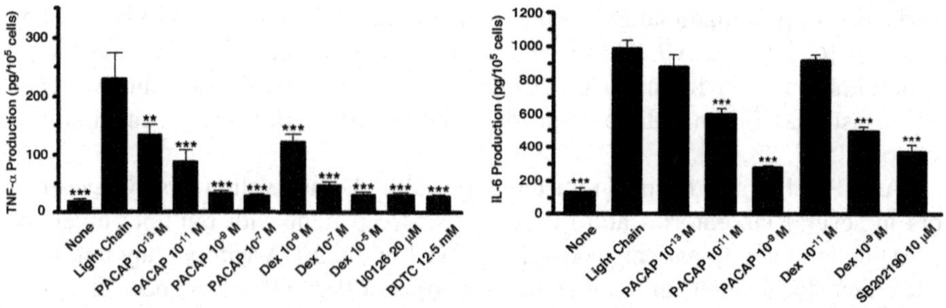

Fig. 18.1 Effect of PACAP38 or dexamethasone on myeloma LC-induced TNF-α and IL-6 production in cultured human RPTECs. Effects of 20 μM U0126, a MEK1/2 inhibitor; 12.5 mM PDTC, a NF-κB inhibitor; and 10 μM SB202190, a p38 MAPK inhibitor, are also shown. ***$P<0.001$ and **$P<0.01$, as compared with respective LC-stimulated value. (Adopted from Arimura et al. 2006)

derivatives, and the newer proteasome inhibitor bortezomib [34]. Since PACAP38 is known to reduce cytokine production through suppression of NF-KB and p38 MAPK, pathways that participate in myeloma LC nephrotoxicity, we tested the effects PACAP38 on myeloma kidney injury [6, 7, 20]. In these studies, human immunoglobulin κ-LC was purified from the urine of a MM patient with known κ-type myeloma. Human RPTECs were exposed to 50 μM of κ-LC for 72 h in the presence or absence of PACAP38 or dexamethasone at various concentrations. PACAP38 was more effective than dexamethasone in suppressing LC-induced inflammatory cytokines IL-6 and TNF-α in RPTECs in a dose dependent manner (Fig. 18.1) [6].

Morphologically, κ-LC also induced considerable damage in cultured RPTECs causing detachment of the cells from the cultured plate and aggregation and necrosis of the cells [6]. PACAP38 nearly completely prevented the cell injury resulted from the κ-LC (Fig. 18.2) [6]. κ-LC-induced cytokine production is mediated by activation of MAPKs, especially p38 MAPK. PACAP38 suppressed LC-stimulated p38 MAPK phosphorylation, but not the extracellular signal regulated kinase (ERK) activation. PACAP38 also suppressed LC-stimulated activation of p50 subunit of NF-κB. P65 subunit of NF-κB was activated in cultured RPTECs before addition of LC, but it was also suppressed by PACAP38. These changes including the morphological alterations were prevented when PACAP38 was added together with LC. Suppression of LC-stimulated TNF-α production by PACAP38 was attenuated by either M65 (a PAC1-specific inhibitor) or a VPAC1-specific inhibitor. Vasoactive intestinal peptide (VIP), a paralog of PACAP38, also suppressed LC-induced TNF-α production, but to a lesser extent than PACAP-38, and VIP-induced suppression was greatly reduced by a VPAC1 inhibitor. These findings suggest that both PAC1- and VPAC1-receptors are involved in PACAP-mediated suppression of cytokine production [7].

To examine the renoprotective effects of PACAP38 in vivo, rats were administered intravenously κ-LC either alone or with PACAP38 for 72 h, and then TNF-α level in the kidney was determined. K-LC administration increased TNF-α content in the kidney several fold, but simultaneous injection of PACAP38 reduced the TNF-α level to nearly the control level. This suggests that administration of PACAP38 reduces cytokine production stimulated by κ-LC in the rat kidney in vivo

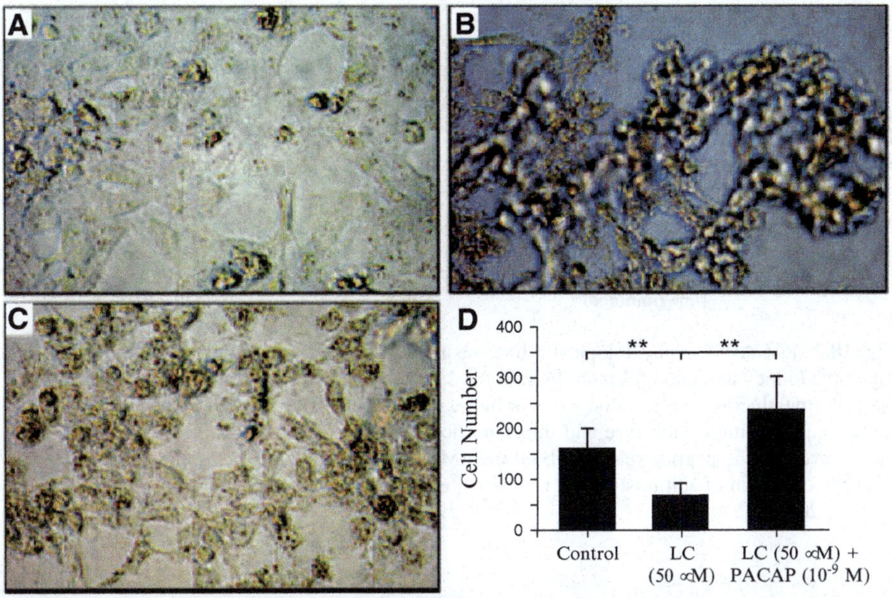

Fig. 18.2 Photomicrograph of cultured human RPTECs. (**a**) Untreated RPTECs; (**b**) RPTECs cultured with 50 μM LC for 72 h; (**c**) RPTECs cultured with 50 μM LC and 1 nM PACAP38 for 72 h; (**d**) Number of healthy cells in each treatment group (Control, LC 50 μM, LC 50 μM+PACAP 10^{-9} M). Mean and SE of cells in 10 visual fields was used in each group. (Adopted from Arimura et al. 2006)

[6, 7]. PACAP38 is also known to stimulate growth of some cancer cells. Thus, if PACAP38 were to stimulate the growth of myeloma cells, its renoprotective effects would be greatly compromised. Therefore, effects of PACAP38 on myeloma cells were examined in cell culture. When four lines of LC producing human myeloma cells were cultured in the non-inactivating serum containing media, addition of PACAP38 to the media dose-dependently suppressed the growth of myeloma cells, as did dexamethasone. However, myeloma cells do not grow in peripheral blood, but grow in the bone marrow and adhere to the stromal cells. Adhesion of myeloma cells stimulates the release of growth factors such as IL-6 from stromal cells, enhancing the growth of myeloma cells. PACAP38 markedly suppressed the production of IL-6 induced by adhesion of myeloma cells to bone marrow stromal cells. These results clearly show that PACAP38 suppresses growth of myeloma cells directly and also by affecting their internal milieu and point out to a potential antitumor effect of PACAP38 on reticuloendothelial system-derived cancers [7].

CKD in patients with MM is characterized by extensive tubulointerstitial fibrosis often along with cast deposition. We observed that toxic myeloma LC isolated from patients with myeloma kidney can induce EMT possibly contributing to the interstitial fibrosis seen in such patients [18]. To determine the role of EMT as a potential mechanism contributing to the characteristic tubulointerstitial renal fibrosis in MM, RPTECs were exposed to κ-LC (50 μM) in vitro for periods up to 72 h [18]. κ-LC caused marked cellular and morphological alterations in RPTECs, accompanied

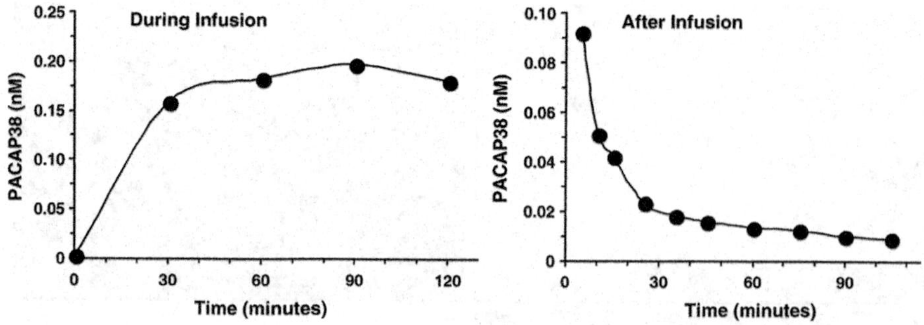

Fig. 18.3 PACAP38 in physiological saline was infused IV to a male patient with active MM and myeloma kidney at a rate of 4 pmol/kg/min for 2 h. The continuous IV infusion did not appear to alter plasma glucose levels, blood gases or blood pressure and heart rate. The subject experienced a transient flushing on the face and upper portion of the body, but no unpleasant feelings. The blood levels of the peptide reached about 0.2 nM during the infusion, and rapidly declined soon after the cessation of administration. The half-life in the blood was about 10 min. (Adopted from Li et al. 2007)

with increased expression of profibrotic TGF-β1, FSP-1 and extracellular matrix (ECM) components (E-cadherin and α-SMA). The use of bone morphogenetic protein-7 (BMP-7) or PACAP38 reversed EMT by promoting the formation of cell aggregates, and the reacquisition of E-cadherin expression and renal proximal tubule epithelial morphology within the confluent cell monolayer during and after LC exposure. PACAP38 was as effective as BMP-7 in reversing EMT [18].

In a case study, an 81-year-old male patient with active MM and myeloma kidney was infused intravenously with synthetic human PACAP38 at a rate of 4 pmol/kg/min for 120 min (Fig. 18.3) [30]. The continuous infusion increased the level of PACAP38 in blood, and reached a plateau at about 0.2 nM during the infusion. The level of PACAP38 in the blood rapidly declined after the cessation of administration with a half-life of about 5–10 min. The continuous infusion did not significantly alter the basal glucose level, blood gases, or blood pressure. There was a large reduction in free λ-LC excretion in urine after the start of the treatment with PACAP38, which, however, was coincidental with the simultaneous dexamethasone treatment, which the patient was receiving at the time. Perhaps more significant, the patient reported marked improvement in his bone pain, which may have been mediated by the anti-TNF-α effects of PACAP38. These studies showed that PACAP38 can be safely used in humans and suggest that it has the potential to be used as a novel therapeutic agent for the treatment of MM and myeloma kidney [30].

Dissecting the cellular and molecular events has identified novel avenues for potential treatment for myeloma kidney. Among these, PACAP38 has emerged as a promising novel therapy. PACAP38 prevented LC-induced renal injury patterns in vitro, and in experimental animals in vivo. Further clinical studies are necessary to establish the usefulness of these novel therapies in preserving the kidney function in patients with MM and myeloma kidney [20, 35].

Protective Effects of PACAP38 in Renal IRI

IRI is an important cause of AKI. Some surgical interventions, kidney transplantation and AKI seen in diverse settings including shock may involve IRI [36]. Microvascular perfusion failure, no-reflow and tissue hypoxia play a causative role in the pathophysiology of kidney damage despite reperfusion and reoxygenation later on. Reperfusion following ischemia induces an inflammatory response predominantly through activation of the innate immunity through TLRs [37]. Inflammatory cells (such as leukocytes, macrophages, and neutrophils) taking part in this response produce cytokines, chemokines, oxygen free radicals and lipid mediators that contribute to the development of the injury. Apoptosis and necrosis are also involved in this process. There are several studies reported establishing the protective effects of PACAP38 against renal IRI. The protective effects of PACAP38 in renal IRI are achieved through complex mechanisms, including modulation of TLRS and the anti-apoptotic, antioxidant, and anti-inflammatory actions of the peptide [12, 15, 21].

Riera et al. reported that treatment with PACAP38 enhances cAMP and protects the rat kidney from IRI by modulation of inflammatory responses [38]. The observation that continuous administration of a high dose of PACAP38 after bilateral renal IRI minimized the occurrence of post-IRI renal insufficiency, reduced the inflammatory responses, and ameliorated the kidney structural damage supports this. More recently Szakaly et al. reported that mice deficient in PACAP38 are more susceptible to renal IRI in vivo, further supporting its renoprotective role in IRI [25]. PACAP38 treatment was effective not only when its continuous administration began immediately before reperfusion, but also when its treatment was delayed to 6 h after IRI. Data from this study supported the hypothesis that PACAP38 exerted its beneficial effect on the kidney by acting through not the parenchyma but through circulating cells rather in renal cells. First, it enhanced plasma but not renal cAMP levels. Second, PACAP38 infusion reduced serum IL-6 levels. Third, the lower myeloperoxidase (MPO) activity and number of CD45+ (a marker for leukocytes) cells in ischemic kidneys from rats treated with PACAP38 reflects a reduction in the post-ischemic renal inflammatory leukocytes infiltration. However, these observations need to be reconciled with the findings supporting cytoprotective effects of PACAP38 directly at the site of injury mediated by its receptors present in the kidney [6, 8], and suggest that renoprotective effects of PACAP38 are mediated through both its systemic and local effects. In IRI animals without PACAP38 treatment, renal MPO activity increased at 6 h after IRI and slightly declined at 24 h after IRI. Continuous infusion of PACAP38 reduced the MPO activity at 6 h after IRI and produced a further decline at 24 h. The number of CD45+ cells had a similar pattern, i.e., 24 h after IRI, the number of CD45+ cells were significantly lower in PACAP38–treated kidneys than in untreated kidneys further substantiating the role of inflammatory cells in this experiment. Thus, enhancement of endogenous circulating cAMP with PACAP38 modulates post-IRI inflammatory responses including inflammatory cell infiltration and strongly protects from IRI-related AKI, even when administration is delayed for 6 h after injury [38].

Szakaly et al. showed that PACAP38 treatment prolonged the renal ischemic time, ameliorated renal morphological damage, and significantly increased survival rate after renal IRI in rats [24]. A remarkable difference was observed in the 60-min IRI group rats as more control IRI-subjected rats died prematurely, while almost all PACAP38-treated IRI rats survived until the time of sacrifice [24].

To explore possible mechanisms underlying the renoprotective effects of PACAP38 against hypoxic/ischemic injury, Li et al. first performed studies using primary RPTECs derived from wild-type (C57BL6/J, MyD88$^{+/+}$) and MyD88$^{-/-}$ mice, in vitro [19]. During IRI, the generation of inflammatory cascade involved the activation of TLRs, the principal mediators of innate immunity within the kidney. The myeloid differentiation primary response gene 88 (MyD88) is essential for signaling by TLRs, resulting in the production of pro-inflammatory cytokines [39–42]. Thus, MyD88 is a central adaptor protein for the majority of TLRs, acting as a link between the receptors and downstream pathways [43, 44]. In this study, hypoxic conditions were induced by immersing the RPTECs monolayers in mineral oil for 1.5 h at 37 °C in the presence or absence of PACAP38 (10^{-8} M) 1 h before or 1.5 h after hypoxia. Control cell monolayers remained in serum-free medium. After washing with PBS, cells were re-incubated with fresh serum-free medium for 20 h to allow reoxygenation. There were significant increases in protein levels (determined by ELISA) of inflammatory chemokines and cytokines (MCP-1, IL-6, and MIP-2) after hypoxic injury and 20 h following reoxygenation in both MyD88$^{+/+}$ and MyD88$^{-/-}$ RPTECs. The baseline levels of cytokines in MyD88$^{-/-}$ cells were significantly lower than the levels observed in MyD88$^{+/+}$ RPTECs. PACAP38 was effective at both time intervals, 1 h before and 1.5 h after hypoxia, and resulted in a significant suppression of cytokine production in both MyD88$^{+/+}$ and MyD88$^{-/-}$ RPTECs. PACAP38 also significantly suppressed apoptosis and mRNA expressions of TLR4, MCP-1, and IL-6 induced by hypoxia in both MyD88$^{+/+}$ and MyD88$^{-/-}$ RPTECs. These studies demonstrate that the administration of PACAP38 to MyD88$^{-/-}$ RPTECs further ameliorates the hypoxic renal injury beyond the improvement conferred by MyD88-deficient status alone [19].

Li et al. evaluated the protective effects of PACAP38 against IRI on renal function and histology using wild-type (C57BL6/J) male mice in vivo [19]. Mice were subjected to 45 min of bilateral renal ischemia followed by 10 days of reperfusion. In the treatment group mice received PACAP38 (20 μg/100 μl of saline) intraperitoneally (IP) 1 h before IRI and additional doses were given at 12 h and 5 days after ischemia. The control group mice were sham operated and received equal volumes of saline IP. Mice subjected to IRI only, had significantly higher levels of serum creatinine, blood urea nitrogen (BUN), and serum and kidney tissue TNF-α compared to control sham-operated mice. Treatment with PACAP38 significantly reduced levels of serum creatinine, BUN and TNF-α in IRI mice. Histopathology and immunofluorescence studies also demonstrated that PACAP38 significantly ameliorated renal tubular damage, apoptosis and neutrophil infiltration induced by IRI in mice kidney. PACAP38 significantly downregulated the mRNA expression of

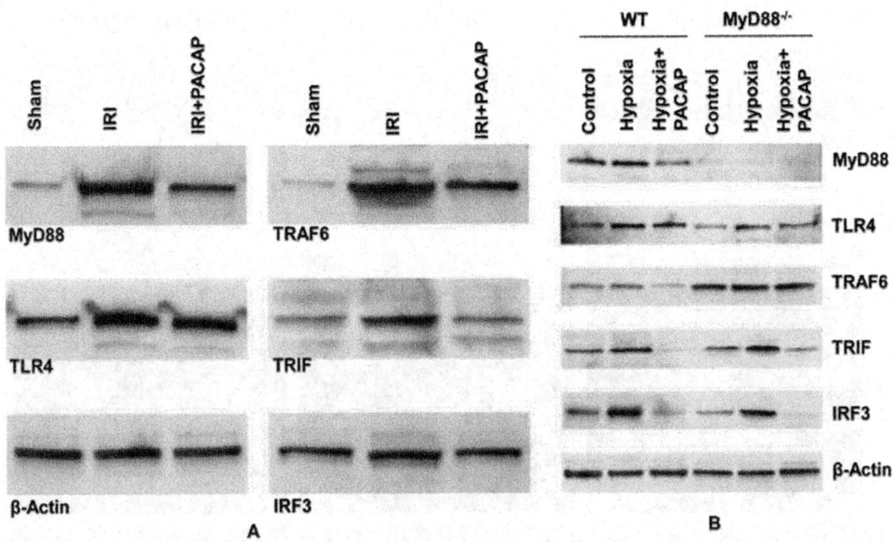

Fig. 18.4 Western blots showing changes in TLR4, MyD88, and the co-effector molecules in the mouse kidney subjected to IRI (**a**), and RPTECs prepared from WT and MyD88$^{-/-}$ mice subjected to hypoxia and reoxygenation (**b**). (**a**) Treatment with PACAP38 significantly inhibited the expression of MyD88 and TLR4 as well as TRAF6, TRIF, and IRF3 expression in the renal cortex from WT mice following renal IRI. In primary RPTECs of WT cells, exposure to hypoxia led to increases in MyD88, TLR4, TRAF6, TRIF, and IRF3 expression, and PACAP38 treatment reversed these responses. In MyD88$^{-/-}$ RPTECs, immunoblots showed reduced basal levels of TLR4, TRIF, and IRF3 expression; responses were similar except for TRAF6 expression, probably due to the compensatory action of TRAF6 for MyD88 deficiency in the MyD88$^{-/-}$ RPTECs subjected to hypoxia and reoxygenation. WT = Wild type. (Adopted from Li et al. 2010b)

TLR4, MyD88, IL-6, and MCP-1 that was significantly upregulated by IRI in mice kidney. Western blots showed that PACAP38 significantly reduced protein levels of TLR4, TRAF6, and IRF3 in mice kidney that were substantially elevated after IRI (Fig. 18.4). Interestingly, protein levels of both adaptor proteins (MyD88 and TRIF) for TLRs were significantly increased by renal IRI in mice and by hypoxia in RPTECs cultures and PACAP38 significantly reduced their levels close to the controls (Fig. 18.4). This indicates that IRI related AKI may involve both MyD88-dependent and MyD88-independent pathways and treatment with PACAP38 is effective against AKI signaled by either MyD88-dependent or TRIF-dependent pathways that activate inflammatory responses [19].

In another study, Khan et al. evaluated the delayed effects of PACAP38 administered 24 h after IRI in mice kidney [12]. Most of the therapeutic strategies that have been tested in AKI models generally have focused on prevention of renal IRI and there is not much information about the potential effect of PACAP38 to reverse established IRI [45]. To explore the possible regenerative property of PACAP38, wild-type male mice were subjected to 45 min of bilateral renal ischemia and reperfusion for 72 h. In this study, reperfusion time was reduced from 10 days as in previous study Li et al. to 72 h to minimize the effects of natural healing of damaged

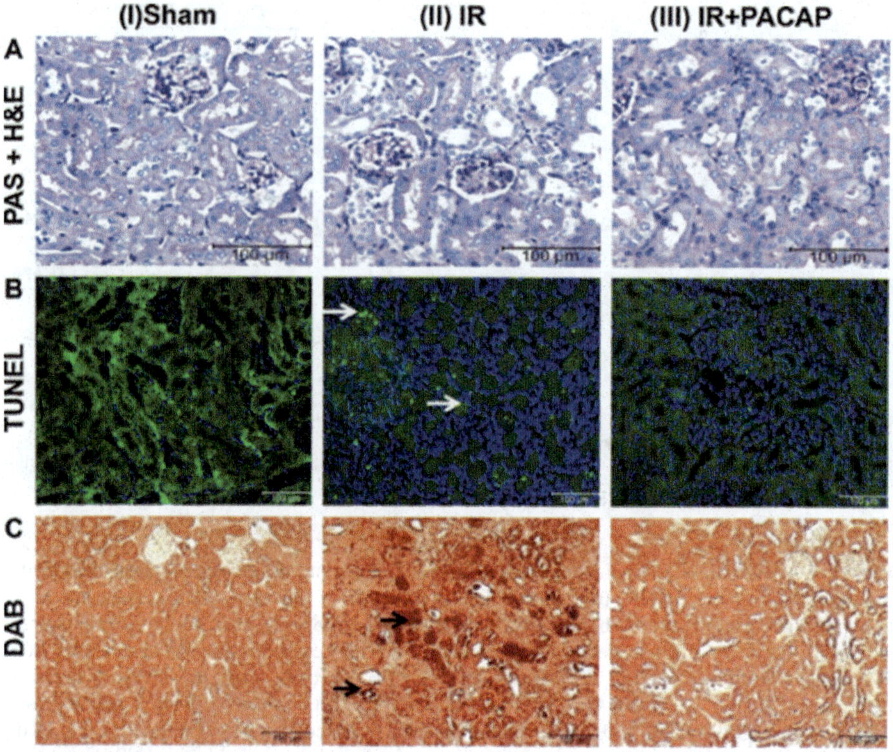

Fig. 18.5 Representative images of kidney sections showing the effects of PACAP38 on histopathology in mouse kidney after IRI. Treatment with PACAP38 preserved tubular morphology (**a**, PAS + H&E staining), reduced apoptosis (**b**, TUNEL assay), and reduced neutrophil infiltration (**c**, immunohistochemistry (3,3_-diaminobenzidine substrate method)) in mouse kidney proximal tubular cells. Scale bar = 100 μM in **a**, **b** and **c**. (Adopted from Khan et al. 2012)

kidney tubules after IRI and to detect the actual effects of PACAP38 [12, 19]. Treatment groups received two doses of PACAP38 (20 μg/100 μl of saline) IP at 24 and 48 h after IRI. The control sham-operated group and IRI only mice received same volume of saline on the same schedule. In these experiments PACAP38 reversed the mRNA expression of TLR-associated genes that were hyper-activated during IRI in the mouse kidney even when the treatment was given 24 h after IRI. PACAP38 significantly improved renal function as assessed by serum creatinine measurements, significantly downregulated the mRNA expression of AKI biomarkers (KIM-1, Nogo-B1, and netrin-1), reduced levels of TNF-α and decreased the intensity of mouse kidney tubule cell structural damage, apoptosis, and neutrophil infiltration as assessed by histology and immunohistochemistry (Fig. 18.5). TLR-specific PCR arrays showed that IRI in the mouse kidney induced the largest increase of the TLR2 expression followed by TLR6 triggering a cascade of downstream inflammatory pathways leading to the activation of NF-κB and its associated genes, similar to observations reported by Leemans et al. [46]. TLR2 is an important initiator of inflammatory responses leading to AKI and renal dysfunction in IRI.

PACAP38 treatment after IRI also significantly reduced the protein levels of proinflammatory cytokines and chemokines such as MCP-1 and IFN-γ toward baseline determined by ELISA. The levels of IL-1β, IL-2, and IL-6 were also significantly increased by IRI injury and PACAP38 treatment significantly reduced the levels of these cytokines except IL-2. These results indicate that PACAP38 can inhibit both the expression (mRNA levels) and protein levels of the inflammatory cytokines in the mouse kidney subjected to IRI. These experiments also showed that PACAP38 is a powerful modulator of innate immunity: IRI stimulated 71 % of assayed genes related to TLRs signaling pathways in mouse kidney subjected to IRI. PACAP38 significantly reversed 67 % of these genes towards healthier baseline levels even when treatment was started 24 h after IRI. These changes were associated with significant improvement in kidney function, as serum creatinine concentrations, proximal tubule cell apoptosis and neutrophil infiltration were reduced [12].

The renoprotective effects of endogenous PACAP38 were confirmed in renal cells derived from PACAP-deficient and wild type mice subjected to hypoxia in vitro [47]. Similarly, PACAP-deficient mice subjected to renal IRI in vivo had worse renal outcomes compared to wild type mice [25]. In another study, in rats subjected to renal IRI, PACAP38 countered the IRI induced decrease of the anti-apoptotic Bcl-2, both after 45 and 60 min of IRI [48]. PACAP38 also demonstrated significant antioxidant effects in rat kidney cell cultures through interaction with the endogenous scavenger system [10]. These studies showed that endogenous PACAP protects against hypoxic injury in the kidney and exogenous PACAP38 could attenuate tissue injury through both anti-inflammatory and anti-apoptotic effects. In summary, there is extensive evidence that both the exogenous and the endogenous PACAP have protective effects against renal IRI in both in vitro and in vivo. The antiapoptotic, anti-inflammatory, and antioxidant actions of the PACAP38 can all play an important role in the treatment of IRI-associated AKI [15].

Protective Effects of PACAP38 Against Cisplatin-Induced Kidney Injury

Cisplatin and other platinum derivatives are the most widely used chemotherapeutic agents to treat solid tumors including ovarian, head and neck, and testicular germ cell tumors. However, it has dose-limiting toxic side effects on the kidney, cochlea, and nerves. Nephrotoxicity is the most common and clinically important toxicity due to cisplatin constraining its therapeutic utility. Studies have shown that several mechanisms, including oxidative stress, DNA damage, and inflammatory responses, are major players in cisplatin-induced nephrotoxicity. Even though the establishment of cisplatin-induced nephrotoxicity can be alleviated by diuretics and pre-hydration, the prevalence of cisplatin nephrotoxicity is still high, occurring in approximately one-third of patients who have undergone cisplatin therapy [49].

Several apoptotic pathways have been implicated in cisplatin-induced renal epithelial cell death. These include the extrinsic pathway activated through the TNF and Fas, cell death receptors, as well as the intrinsic mitochondrial and the endoplasmic reticulum stress pathways [49]. Cisplatin induces renal cell death by activation of p53, and p53 tumor suppression induces cell cycle arrest or apoptosis in response to DNA damage, oncogene activation, and hypoxia [50, 51]. Thus, pharmacological or genetic intervention can inhibit p53 activation by reducing the activation of caspases, induction of apoptosis, and eventually renal injury by cisplatin [52, 53].

Li et al. identified a pivotal role of PACAP38 in ameliorating cisplatin nephrotoxicity by suppressing p53 and inflammatory cytokine (TNF-α) both in vitro and in vivo [16, 17]. For studies in vitro, human RPTECs were exposed to 50 μM cisplatin for 0–72 h in the presence or absence of PACAP38 (10^{-10}–10^{-6} M) or p53 siRNA (50 nM). Exposure of RPTECs to cisplatin for 24-h significantly increased the secretion of TNF-α in the culture medium, and activated p53 transcription in the nuclear extracts. Treatment with PACAP38 significantly reduced TNF-α secretion and suppressed p53 activation in a dose-dependent manner. The protective effect of PACAP38 on cisplatin-induced tubular cell apoptosis was also examined by TUNEL assay and flow cytometry assays (FCAs) analysis (Fig. 18.6). The number of apoptotic cells was significantly increased in RPTECs exposed to cisplatin from 0.6 to 42.2 % and was reduced markedly to 7.4 % or 3.4 % by exogenous p53 siRNA or PACAP38 treatment, respectively (Fig. 18.6).

P53 transcriptionally controls the caspases that are involved in the initiation and execution phases of apoptosis in response to cisplatin exposure [53–55]. In this study, the upregulation of p53 induced the activation of caspase-7 and PARP-1, which in turn reduced the endogenous levels of essential DNA repair protein APE-1 after cisplatin exposure. Inhibiting cisplatin-induced p53 activation with PACAP38 prevented the subsequent activation of these executioners of apoptosis, suggesting that inhibition of these key nuclear enzymes could provide marked renoprotection. Treatment with PACAP38 also reduced the levels of proapoptotic proteins, including Bax, and increased the levels of the anti-apoptotic molecules Bcl-2 and Bcl-xL on the mitochondrial surface, which should antagonize the proapoptotic molecules that trigger the release of cytochrome c and, subsequently, lead to the activation of the proinflammatory cytokine TNF-α. Cisplatin exposure also altered RPTECs integrins and PACAP38 partially restored the collagen-binding integrins (fibronectin and collagen IV). These studies partially revealed the mechanisms whereby cisplatin causes cell death in RPTECs and protection mechanism by PACAP38.

For in vivo studies [17], male (C57BL/6J) mice were given a single IP. injection of cisplatin (20 mg/kg bw). One treatment group received PACAP38 (20 μg/100 μl saline) 2 h before cisplatin with three additional doses at 24, 48, and 72 h after cisplatin injections. In other groups PACAP38 treatment started either at 24, 48, or 72 h after cisplatin injection with subsequent additional doses. Control group received the same volume of saline on the same schedule. All mice were euthanized

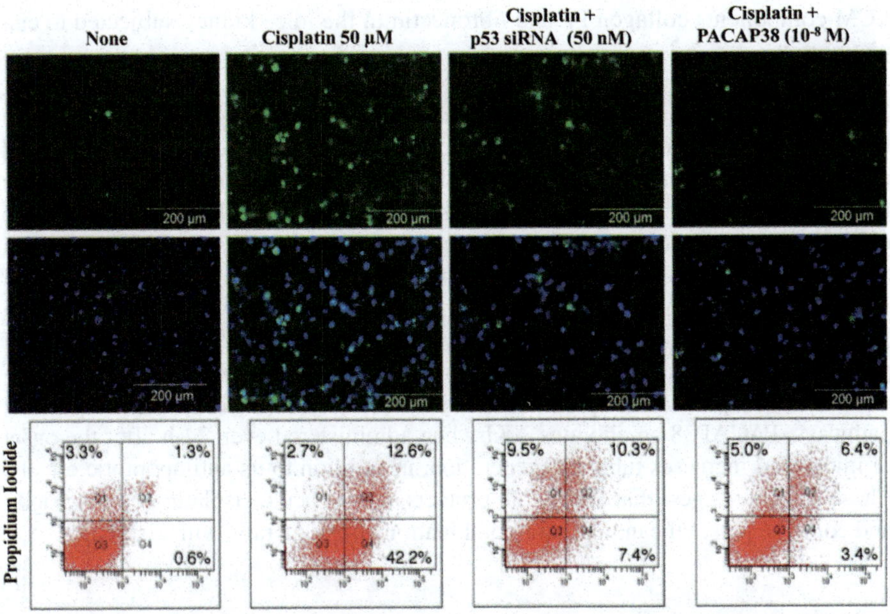

Fig. 18.6 Effects of PACAP38 on cisplatin-induced apoptotic/necrotic cell death in human RPTECs. The dose-dependent inhibitory effect of PACAP38 on apoptotic cell death was assessed by TUNEL assay and FCAs assays. PACAP38 had a significantly greater inhibitory potency than p53 siRNA against cisplatin-induced apoptotic cell death. The number of apoptotic cells was significantly increased in cells exposed to cisplatin and protection was observed in the cells treated with PACAP38 or p53 siRNA. Similarly, the fraction of FITC–annexin V cells was increased significantly by cisplatin exposure and was reduced sharply by p53 siRNA or PACAP38 treatment. (Adopted from Li et al. 2010a)

at 96 h post-cisplatin injection. PACAP38 protected kidneys from cisplatin-induced nephrotoxicity when treatment started either pre- or post-cisplatin injections. PACAP38 treatment started 2 h before or 24 h after cisplatin injection was significantly more effective compared to 48 h or 72 h treatment start time. PACAP38 significantly decreased the levels of BUN and serum creatinine in mice subjected to cisplatin injury. PACAP38 also demonstrated renoprotective effects by significantly reducing the levels of TNF-α in plasma and the kidney. Cisplatin caused extensive damaged in mice kidney structurally and morphologically and PACAP38 significantly prevented or reversed kidney injury even when the treatment was started 24 h after cisplatin injection. PACAP38 also significantly reversed p53 activation and apoptotic cell death in the mouse kidney as demonstrated by Western blots, immunohistology, and FCAs. Octamer-4, vimentin, and PCNA are key factors involved in cell regeneration and differentiation process. PACAP38 was also involved in renal tubular cell repair/regeneration mechanism after cisplatin injury as demonstrated by upregulation in the expression of Octamer 4 and vimentin, and PCNA levels by immunohistology after PACAP38 treatment. PACAP38 also partially restored the

ECM components collagen IV and fibronectin in the mice kidney subjected to cisplatin injury.

These studies, showed that PACAP38 significantly reduces renal tubular injury, inhibits the apoptotic cascade, increases the expression of collagen IV, and decreases fibronectin to maintain the structural integrity of the tubular basement membrane and prevents tubulointerstitial fibrosis, and reduces the levels of BUN and creatinine in mice with cisplatin nephrotoxicity [16, 17]. Treatment with PACAP38 was more effective than p53 silencing in blocking the activation of caspases and PARP-1 with the decline in renal TNF-α preceding the restoration of integrin-mediated cell–ECM interactions for repairing tubular cell dysfunction. The renoprotection with PACAP38 in cisplatin nephrotoxicity may be associated with the suppression of p53-mediated upregulation of APE-1 leading to decreased apoptosis and TNF-α production after exposure to cisplatin, possibly by both p53-dependent and -independent pathways. PACAP38 ameliorates AKI when administered even 24-h after the onset of injury and increases tubular regeneration, in addition to its anti-apoptotic effects. These studies suggest that PACAP38 protects against acute cisplatin nephrotoxicity and could be a useful candidate for combination chemotherapy with cisplatin.

Protective Effects of PACAP38 Against Cyclosporine A (CsA) and Gentamicin Induced Nephrotoxicity

CsA, a calcineurin inhibitor (CNI), is used to improve organ allograft transplantation and patient survival. In addition, CsA has also been used to treat many autoimmune disorders, such as psoriasis, rheumatoid arthritis, systemic lupus erythematosus, steroid-resistant nephrotic syndrome, and inflammatory bowel disease. Despite the impressive reduction in the number of acute rejection episodes after the use of CsA in clinical practice, its long-term therapy is typically associated with CsA-induced nephrotoxicity [56]. The mechanisms underlying CsA-induced nephropathy have been studied thoroughly now as having a dose-dependent multifactorial cause including vasorelaxant-vasoconstrictor disequilibrium, oxidative stress, apoptosis and proliferation-fibrosis [57, 58]. At present, the main focus is to find the most adequate duration for CsA exposure and the proper moment to promote dose reduction and/or replacement to other less nephrotoxic drug, in order to maintain renoprotection without compromising the graft by rejection [56].

Khan et al. evaluated the protective effects of PACAP38 on CsA-induced nephrotoxicity in human RPTECs and in intact mice [13]. Human RPTECs were exposed to CsA (25–50 μM) for 24 h in the presence or absence of PACAP38 or VIP at a concentration of (10^{-10}–10^{-6} M). For in vivo studies, male BALB/c mice were injected a single IP dose of CsA (5 mg/kg bw) and the treatment group received PACAP38 (20 μg/100 μl of saline) 2 h before CsA injection and additional doses were given daily for 10 days. RPTECs exposed to 50 μM CsA for 24 h showed discernible morphological changes by displaying elongated fibroblast-like morphology with a large degree of cell detachment, significant release of lactate dehydrogenase (LDH) and pro-fibrotic TGF-β1 compared to untreated RPTECs. Treatment with PACAP38 (10^{-8} M) 1-h before exposure to CsA prevented the development of fibroblast like alterations in RPTECs, preserved their morphology, and significantly reduced the release of LDH and TGF-β1 compared to RPTECs treated with PACAP38 1-h after CsA exposure. Apoptosis was significantly less in cells exposed to CsA in the presence of PACAP38 than in the cells exposed to CsA alone. VIP (a related peptide in the PACAP family) was much less effective than PACAP38 in ameliorating the CsA-induced cytotoxic changes. PACAP38 exhibited a dose-dependent inhibitory effect on CsA-induced apoptosis after a 24-h exposure as determined by cytoplasmic histone-associated DNA fragmentation. Although VIP at 10^{-6} M also ameliorated DNA fragmentation, PACAP38 was significantly more potent as the median effective concentration (EC50) for PACAP38 was significantly less than for VIP.

For in vivo studies, renal function was evaluated by measuring serum creatinine levels and by quantifying morphological abnormalities in kidneys of mice after CsA administration. The improvement in renal function was reflected in kidney morphology. Exposure to CsA caused significant increases in serum creatinine, tubular injury and interstitial fibrosis and cast formation compared to saline treated control mice. PACAP38 administration significantly reduced serum creatinine level, tubulointerstitial damage and fibrosis compared to mice treated with CsA only. PACAP38 also significantly ameliorated the production of TGF-β1 (by ELISA) and apoptosis (by TUNEL assay) induced by CsA in mice kidney. TGF-β1 has been widely proposed as a major mediator of CsA-induced renal fibrosis and EMT in various types of cells [59, 60]. This study confirmed the stimulatory effects of CsA on TGF-β1 secretion and increased production in vitro and in vivo providing further evidence that TGF-β1 may be a key mediator of nonhemodynamically mediated CsA nephrotoxicity.

In this study, a number of key EMT markers such as E-cadherin, ZO-1, α-SMA, and E2A were significantly altered by CsA, and PACAP38 partially reversed these changes [13]. Decreased levels of E-cadherin and ZO-1 are inversely correlated with increased production of α-SMA and E2A in the kidney [61, 62]. PACAP38 partially restored the production of ZO-1 while it had no effect on E-cadherin in kidneys of mice treated with CsA. In untreated control mouse kidney cortex, α-SMA, a phenotypic marker for myofibroblasts that is not expressed in epithelial cells, was detected at very low levels in the basal state, as expected, but was markedly increased after treatment with CsA, indicating a myofibroblastic transformation. PACAP38 reduced CsA-induced α-SMA production to basal levels. Changes in E2A protein levels were also noted after CsA exposure in kidneys that paralleled the increased levels of α-SMA. PACAP38 again reduced the production of E2A protein to basal levels.

In an earlier study Duymelinck et al. showed that CsA alters ECM turnover in the kidney of CsA-treated animals [63]. These authors observed that CsA significantly upregulated mRNA expression of ECM components such as collagen III, collagen IV, laminin, and fibronectin in mouse kidneys. Khan et al. found that in mice treated with both PACAP38 and CsA, the expression of these ECM components was significantly suppressed compared to the mice treated with CsA only [13]. The EMT-associated genes and the regulation of ECM can also be regulated by ROS generation

[64]. Experimental and clinical data strongly suggest that CsA-induced renal injury also involves increased production of free radicals such as H_2O_2 [65, 66]. As expected, CsA exposure resulted in the production of the ROS in tubule cells. PACAP38 suppressed the levels of CsA-induced superoxide below control levels in the mouse kidney and close to the control level in RPTECs [13]. These results suggest that PACAP38 contributes to the prevention of the nephrotoxic effects of CsA-induced ROS via inhibition of NOX-2 enzyme. These observations suggest that PACAP has the potential to be an effective renoprotective adjunct in patients on cyclosporine.

Gentamicin, an aminoglycoside antibiotic, is used for the treatment of many types of bacterial infection, especially those involving Gram-negative bacteria. However, gentamicin therapy is associated with two forms of toxicity, nephrotoxicity and ototoxicity, as side effects. Gentamicin-induced nephrotoxicity has been reported to appear during 10–25 % of therapeutic courses, and the typical clinical manifestation is renal transport and electrolyte abnormalities, such as magnesium and potassium wasting and non-oliguric renal insufficiency [67, 68]. Li et al. showed that PACAP38 can ameliorate gentamicin nephrotoxicity and reduce the production of TNF-α in gentamicin-treated rats [20].

Protective Effects of PACAP38 Against Diabetic Nephropathy

Diabetic nephropathy contributes to nearly 50 % of end-stage renal disease worldwide. Despite recent improvements in glycemic control and the use of inhibitors of the renin-angiotensin-aldosterone system for slowing progression of diabetic nephropathy, the burden of disease continues to increase globally. Therefore further mechanistic studies are needed to understand the pathogenesis of diabetic nephropathy and to develop new therapeutic strategies [69]. Recent studies have shown that increased expression of TGF-β1 plays a key role in the pathogenesis of diabetic nephropathy as it promotes accumulation of ECM components, apoptosis, dedifferentiation of podocytes, and EMT of proximal tubules, all of which are thought to facilitate a decline in nephron number and renal function [69, 70].

In a pilot study to evaluate the renoprotective effects of PACAP38 against diabetic nephropathy, rats received streptozotocin (STZ) intravenous (IV) at a dose of 65 mg/kg bw [7, 20]. All of these rats developed hyperglycemia within a few days and showed proteinuria within a week. PACAP38 was administered IV for 2 weeks using osmotic mini pumps. Administration of PACAP38 significantly decreased STZ-induced profibrotic cytokine TGF-β1 both in plasma and kidney tissue and also significantly suppressed the production of proinflammatory cytokine TNF-α in the rat kidney. Treatment with PACAP38 also reversed the tubular and glomerular structural damage induced by STZ as evaluated histologically. PACAP38 also ameliorated proteinuria, polyuria and hyperglycemia in diabetic rats.

In more recent studies, Banki et al. also demonstrated effects of PACAP38 to ameliorate STZ-induced diabetic nephropathy in rats [8, 9]. In these studies, histological analysis revealed severe diabetic changes in kidneys of STZ-treated animals (glomerular PAS-positive area expansion, tubular damage, known as the Armanni–Ebstein phenomenon). PACAP38 treatment significantly reversed these structural alterations in the kidney. Diabetic kidneys also showed significant cytokine activation compared to their healthy controls and PACAP38 downregulated the levels of several cytokines including CINC-1, TIMP-1, LIX, MIG, and s-ICAM [8]. In another study, PACAP38 treatment countered the STZ-induced increases in the levels of the proapoptotic p38MAPK and cleaved caspase-3, and also decreased the p60 subunit of NF-κB. Although the levels of the antiapoptotic factors pAkt, and pERK1/2, increased slightly in the diabetic kidneys, PACAP38 treatment resulted in a more robust elevation of these proteins. PCR and Western blot revealed downregulation of fibrotic markers, like collagen IV and TGF-β1 in the kidney. PACAP38 treatment also resulted in increased expression of the antioxidant glutathione [9].

Sakamoto et al. localized VPAC1, one of the PACAP38 receptors, to glomerular podocytes, which also reportedly may have crucial roles not only in normal glomerular physiology but also in pathology [23]. PACAP38 was effective in the downregulation of proinflammatory cytokines, such as MCP-1 and IL-6, which had been induced by the activation of TLRs with lipopolysaccharide. PACAP38 downregulated the expression of MCP-1 through the protein kinase A signaling pathway; this led to the attenuation of the activation of extracellular signal-regulated kinase and NF-κB signaling. These studies suggest that PACAP38 could be a possible treatment option for diabetic nephropathy through its anti-fibrotic and anti-inflammatory effects on glomerular podocytes and epithelial tubular cells.

Protective Effects of PACAP38 Against CIN

Every year, almost 8 million patients with acute chest pain are admitted to emergency departments for evaluation using contrast-enhanced imaging at an estimated cost of more than $10 billion [71]. The majority of these patients are older individuals with comorbidities, and are at risk of CIN at rates as high as 50 %, especially in elderly diabetic patients with cardiovascular disorders. Patients who develop CIN after CM administration sustain an increase in both short- and long-term morbidity and mortality [72]. Overall, the economic burden associated with CIN is high because the average 1-year cost of treating a single patient with CIN is $11,812 [73]. Despite considerable research, an effective and reliable treatment to prevent CIN is still not available. Thus there is a need for novel therapies that can reduce the incidence of CIN and its overall economic burden [73].

The pathophysiology of CIN is a combination of hypoxic and toxic renal tubular damage associated with renal endothelial dysfunction and altered intrarenal microcirculation. Oxidative stress due to ROS is thought to be a critical factor in the development of diabetic nephropathy and its complications [74–77]. The mechanisms responsible for the pathogenesis of CIN are attributed to a combination of the direct tubular toxicity of CM, reduction in medullary blood flow, and excess generation of

ROS [78]. Innate immunity, mediated through TLRs, contribute to ROS generation [79] and play a major role in the pathogenesis of hypoxia/ischemia-induced and other types of AKI [12, 19, 45, 80]. TLRs are a family of evolutionarily conserved transmembrane pattern/damage-recognition receptors that transduce intracellular signals in response to diverse ligands, such as pathogen-associated molecular patterns, cytokines, and molecules released from dying cells, ie, histones and environmental stressors [81, 82]. Stimulation of TLRs initiates a cascade of signaling events that lead to the production of myriad cytokines and effector molecules [83]. The integrated role of ROS generation with TLRs activation in the development of CIN is not well understood, essential for the identification of molecular targets and the development of novel therapeutics.

Several pharmacologic agents have been tested to reduce the risk of CIN in patients with preexisting renal disease. However, none of these strategies have shown any consistent benefit. ROS scavenger N-acetylcysteine (NAC), the acetylated variant of the amino acid cysteine, is an excellent source of sulfhydryl groups. By increasing glutathione synthesis, NAC promotes detoxification, thus acting directly as a free radical scavenger [84, 85]. NAC is widely used as an adjunctive prophylactic agent to prevent CIN. However, there are conflicting data on the efficacy of NAC in humans undergoing contrast procedures and it has not yet received approval from the FDA for this indication. In order to develop a reliable therapeutic strategy for CIN, Khan et al. investigated the therapeutic effects of PACAP38 to mitigate CIN using both in vitro and in vivo models [14].

For in vitro studies, human RPTECs were exposed to ionic (Urografin) or non-ionic (iohexol) CM at concentrations of 25–100 mg iodine/ml for 24 h to evaluate cytotoxic effects of CM. To determine the dose–response curves for inhibition of contrast induced cell injury, RPTECs were treated with 10^{-9}–10^{-6} M of PACAP38 or VIP for 1 h prior to exposing them to CM at an optimal concentration of 50 mg iodine/mL for 24 h. After 24 h exposure, Urografin was more cytotoxic than iohexol. Incubation of RPTECs with CM significantly increased the release of LDH and KIM-1 into the culture medium, induced apoptosis, and suppressed cell proliferation compared to the untreated cells. Treatment of CM-exposed RPTECs with either PACAP38 or VIP inhibited the release of LDH and KIM-1, prevented apoptosis, and restored cell proliferation in a dose-dependent manner. Both PACAP38 and VIP at a dose range of 10^{-8}–10^{-6} M exhibited significant renoprotective effects compared to CM only treated RPTECs, but PACAP38 was significantly more potent than VIP at each of these dosage points.

For in vivo studies, endothelial nitric oxide synthase deficient (eNOS$^{-/-}$) mice were used because eNOS$^{-/-}$ mice are hypertensive and more prone to develop CIN than wild type mice. Male homozygous eNOS$^{-/-}$ mice were water deprived for 24 h and then given an IV injection of CM Urografin (1.85 g iodine/kg). The CM-injected group received PACAP38 (10 μg/100 μl) IP 1 h before and 12 h after the injection of Urografin. Control mice received an equal volume of saline on the same schedule, and all mice were euthanized at 72 h after CM injection [14]. Exposure of eNOS$^{-/-}$ mice to Urografin caused significant elevations in the serum levels of creatinine, cystatin C, and showed extensive renal tubule cell injury. PACAP38

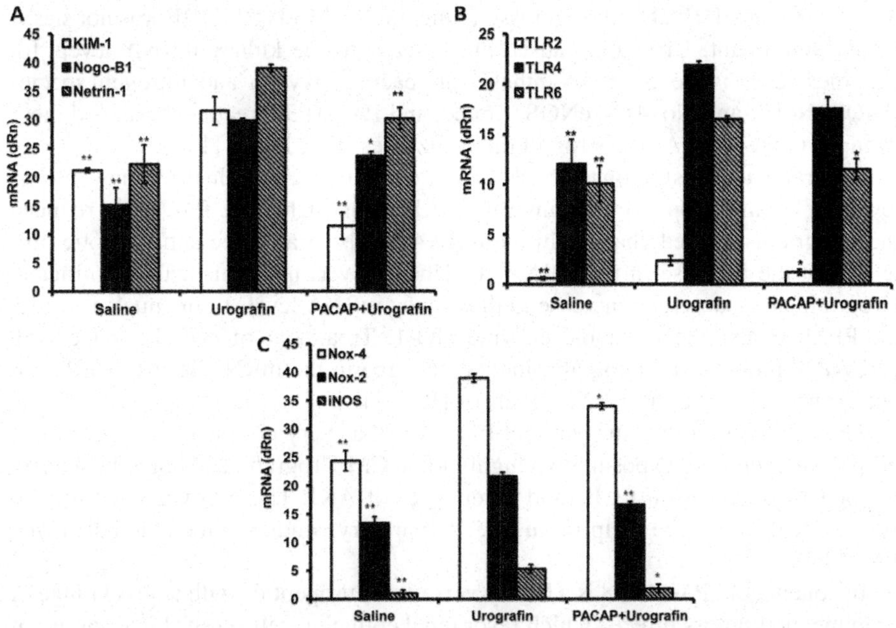

Fig. 18.7 Reduction of contrast-induced kidney injury biomarkers, TLRs, and ROS by PACAP38. Administration of Urografin to eNOS$^{-/-}$ mice caused an increase in the kidney mRNA levels of the kidney injury biomarkers (KIM-1, Nogo-B1, and Netrin-1) (**a**), TLRs (TLR2, TLR4, and TLR6) (**b**), and ROS generating enzymes (Nox-4, Nox-2, and iNOS) (**c**). Treatment with PACAP38 ameliorated the expression of all these kidney injury inducing factors. *$p<0.05$ and **$p<0.01$ compared to the mice treated only with Urografin. Bars represent the mean ± SE. (Adopted from Khan et al. 2013)

prevented renal insufficiency by significantly reducing the elevated serum levels of creatinine, cystatin C, and reduced the extent of kidney injury caused by Urografin administration in eNOS$^{-/-}$ mice. Treatment with PACAP38 also markedly reduced levels of apoptosis and neutrophil infiltration induced by Urografin as shown by immunohistology. PACAP38 also significantly reduced levels of Urografin-induced pro-inflammatory (TNF-α, MCP-1, IFN-γ, and IL-6) and pro-fibrotic cytokines (TGF-β1 and fibronectin) both at protein and mRNA expression levels. Administration of Urografin to eNOS$^{-/-}$ mice caused an increase in the kidney mRNA levels of the kidney injury biomarkers KIM-1, Nogo B1, and netrin-1. Treatment of eNOS-deficient mice with PACAP38 reduced the expression of all the three kidney injury biomarkers (Fig. 18.7a). In eNOS$^{-/-}$ mice, Urografin also caused an increase in the kidney mRNA levels of TLR2, TLR4, and TLR6. PACAP38 reduced these changes in the mRNA levels of all the three TLRs although the decrease in the mRNA level for TLR4 was not statistically significant (Fig. 18.7b). Urografin administration to eNOS$^{-/-}$ mice also caused an increase in the kidney mRNA levels of the TLR adaptor proteins MyD88 and TRIF. Treatment of eNOS$^{-/-}$ mice with PACAP38 reduced these Urografin-induced changes in the mRNA levels for MyD88 and TRIF, but the decrease in the mRNA levels for TRIF was not statistically significant. Urografin caused an increase in the kidney mRNA levels for enzymes responsible for the synthesis of reactive oxygen and nitrogen species (iNOS, Nox2, and Nox4) in eNOS$^{-/-}$ mice, and PACAP38 treatment reduced these changes in the mRNA levels for all three enzymes (Fig. 18.7c) [14].

Urografin administration to eNOS$^{-/-}$ mice caused a decrease in the kidney mRNA levels of the antiapoptotic proteins Bcl-2, APE1, and Ube2V1. PACAP38 reduced the Urografin-induced changes in the mRNA levels of all three antiapoptotic proteins, but the increase in the mRNA for Ube2V1 was not statistically significant. Urografin also caused an increase in the kidney mRNA levels of the nuclear receptor PPAR-α and the inducible enzyme HO-1. Treatment of eNOS$^{-/-}$ mice with PACAP38 reduced the Urografin-induced change in the mRNA levels of PPAR-α but further increased the mRNA levels of HO-1 [14].

These studies further suggest that PACAP38 has very promising renoprotective properties even after exposure to a highly toxic CM Urografin and could be a novel therapeutic agent against CIN and other types of AKI. These in vitro and in vivo studies will likely be helpful in the preparatory studies aimed at advancing PACAP38 in clinical trials.

In conclusion, PACAP38 is a highly versatile neuropeptide with a broad range of immune modulatory effects, which favor renal epithelial cell survival in a variety of diverse insults, ranging from oxidant injury to nephrotoxin-induced injury and pathologic states like myeloma kidney and diabetic nephropathy. PACAP has protective effects on kidney and other organs through its systemic effects and through its receptors present in the affected organs. It is surprising that this highly promising immune modulator has not been yet introduced to clinical applications, and it is time that translational studies to be initiated for this purpose.

Acknowledgements The authors would like to dedicate this chapter to the memory of the late Dr. Akira Arimura.

References

1. Miyata A, Arimura A, Dahl RR, Minamino N, Uehara A, Jiang L, et al. Isolation of a novel 38 residue-hypothalamic polypeptide which stimulates adenylate cyclase in pituitary cells. Biochem Biophys Res Commun. 1989;164:567–74.
2. Miyata A, Jiang L, Dahl RD, Kitada C, Kubo K, Fujino M, et al. Isolation of a neuropeptide corresponding to the N-terminal 27 residues of the pituitary adenylate cyclase activating polypeptide with 38 residues (PACAP38). Biochem Biophys Res Commun. 1990;170:643–8.
3. Miyata A, Sano H, Li M, Matsuda Y, Kaiya H, Sato K, et al. Genomic organization and chromosomal localization of the mouse pituitary adenylate cyclase activating polypeptide (PACAP) gene. Ann N Y Acad Sci. 2000;921:344–8.
4. Ganea D, Delgado M. Vasoactive intestinal peptide (VIP) and pituitary adenylate cyclase-activating polypeptide (PACAP) as modulators of both innate and adaptive immunity. Crit Rev Oral Biol Med. 2002;13:229–37.

5. Vaudry D, Falluel-Morel A, Bourgault S, Basille M, Burel D, Wurtz O, et al. Pituitary adenylate cyclase-activating polypeptide and its receptors: 20 years after the discovery. Pharmacol Rev. 2009;61:283–357.
6. Arimura A, Li M, Batuman V. Potential protective action of pituitary adenylate cyclase-activating polypeptide (PACAP38) on in vitro and in vivo models of myeloma kidney injury. Blood. 2006;107:661–8.
7. Arimura A, Li M, Batuman V. Treatment of renal failure associated with multiple myeloma and other diseases by PACAP-38. Ann N Y Acad Sci. 2006;1070:1–4.
8. Banki E, Degrell P, Kiss P, Kovacs K, Kemeny A, Csanaky K, et al. Effect of PACAP treatment on kidney morphology and cytokine expression in rat diabetic nephropathy. Peptides. 2013;42:125–30.
9. Banki E, Kovacs K, Nagy D, Juhasz T, Degrell P, Csanaky K, et al. Molecular mechanisms underlying the nephroprotective effects of PACAP in diabetes. J Mol Neurosci. 2014;54:300–9.
10. Horvath G, Brubel R, Kovacs K, Reglodi D, Opper B, Ferencz A, et al. Effects of PACAP on oxidative stress-induced cell death in rat kidney and human hepatocyte cells. J Mol Neurosci. 2011;43:67–75.
11. Horvath G, Mark L, Brubel R, Szakaly P, Racz B, Kiss P, et al. Mice deficient in pituitary adenylate cyclase activating polypeptide display increased sensitivity to renal oxidative stress in vitro. Neurosci Lett. 2010;469:70–4.
12. Khan AM, Li M, Abdulnour-Nakhoul S, Maderdrut JL, Simon EE, Batuman V. Delayed administration of pituitary adenylate cyclase-activating polypeptide 38 ameliorates renal ischemia/reperfusion injury in mice by modulating Toll-like receptors. Peptides. 2012;38: 395–403.
13. Khan AM, Li M, Brant E, Maderdrut JL, Majid DS, Simon EE, et al. Renoprotection with pituitary adenylate cyclase-activating polypeptide in cyclosporine A-induced nephrotoxicity. J Investig Med. 2011;59:793–802.
14. Khan AM, Maderdrut JL, Li M, Toliver HL, Coy DH, Simon EE, et al. Pituitary adenylate cyclase-activating polypeptide prevents contrast-induced nephropathy in a novel mouse model. Physiol Rep. 2013;1:e00163.
15. Laszlo E, Kiss P, Horvath G, Szakaly P, Tamas A, Reglodi D. The effects of pituitary adenylate cyclase activating polypeptide in renal ischemia/reperfusion. Acta Biol Hung. 2014;65: 369–78.
16. Li M, Balamuthusamy S, Khan AM, Maderdrut JL, Simon EE, Batuman V. Pituitary adenylate cyclase-activating polypeptide ameliorates cisplatin-induced acute kidney injury. Peptides. 2010;31:592–602.
17. Li M, Balamuthusamy S, Khan AM, Maderdrut JL, Simon EE, Batuman V. Pituitary adenylate cyclase-activating polypeptide prevents cisplatin-induced renal failure. J Mol Neurosci. 2011;43:58–66.
18. Li M, Hering-Smith KS, Simon EE, Batuman V. Myeloma light chains induce epithelial-mesenchymal transition in human renal proximal tubule epithelial cells. Nephrol Dial Transplant. 2008;23:860–70.
19. Li M, Khan AM, Maderdrut JL, Simon EE, Batuman V. The effect of PACAP38 on MyD88-mediated signal transduction in ischemia-/hypoxia-induced acute kidney injury. Am J Nephrol. 2010;32:522–32.
20. Li M, Maderdrut JL, Lertora JJ, Arimura A, Batuman V. Renoprotection by pituitary adenylate cyclase-activating polypeptide in multiple myeloma and other kidney diseases. Regul Pept. 2008;145:24–32.
21. Reglodi D, Kiss P, Horvath G, Lubics A, Laszlo E, Tamas A, et al. Effects of pituitary adenylate cyclase activating polypeptide in the urinary system, with special emphasis on its protective effects in the kidney. Neuropeptides. 2012;46:61–70.
22. Reglodi D, Kiss P, Szabadfi K, Atlasz T, Gabriel R, Horvath G, et al. PACAP is an endogenous protective factor-insights from PACAP-deficient mice. J Mol Neurosci. 2012;48:482–92.

23. Sakamoto K, Kuno K, Takemoto M, He P, Ishikawa T, Onishi S, et al. Pituitary adenylate cyclase-activating polypeptide protects glomerular podocytes from inflammatory injuries. J Diabetes Res. 2015;2015:727152.
24. Szakaly P, Kiss P, Lubics A, Magyarlaki T, Tamas A, Racz B, et al. Effects of PACAP on survival and renal morphology in rats subjected to renal ischemia/reperfusion. J Mol Neurosci. 2008;36:89–96.
25. Szakaly P, Laszlo E, Kovacs K, Racz B, Horvath G, Ferencz A, et al. Mice deficient in pituitary adenylate cyclase activating polypeptide (PACAP) show increased susceptibility to in vivo renal ischemia/reperfusion injury. Neuropeptides. 2011;45:113–21.
26. Wang X, Wei H, Zhao T, Zhu X, Yang X, Chen D, et al. Evidence for pituitary adenylate cyclase-activating peptide as a direct immunoregulator in teleost head kidney. Fish Shellfish Immunol. 2013;34:265–72.
27. Batuman V. Proximal tubular injury in myeloma. Contrib Nephrol. 2007;153:87–104.
28. Arimura A. Perspectives on pituitary adenylate cyclase activating polypeptide (PACAP) in the neuroendocrine, endocrine, and nervous systems. Jpn J Physiol. 1998;48:301–31.
29. Vaudry D, Pamantung TF, Basille M, Rousselle C, Fournier A, Vaudry H, et al. PACAP protects cerebellar granule neurons against oxidative stress-induced apoptosis. Eur J Neurosci. 2002;15:1451–60.
30. Li M, Maderdrut JL, Lertora JJ, Batuman V. Intravenous infusion of pituitary adenylate cyclase-activating polypeptide (PACAP) in a patient with multiple myeloma and myeloma kidney: a case study. Peptides. 2007;28:1891–5.
31. Seaborn T, Masmoudi-Kouli O, Fournier A, Vaudry H, Vaudry D. Protective effects of pituitary adenylate cyclase-activating polypeptide (PACAP) against apoptosis. Curr Pharm Des. 2011;17:204–14.
32. Batuman V. The pathogenesis of acute kidney impairment in patients with multiple myeloma. Adv Chronic Kidney Dis. 2012;19:282–6.
33. Khan AM, Li M, Balamuthusamy S, Maderdrut JL, Simon EE, Batuman V. Myeloma light chain-induced renal injury in mice. Nephron Exp Nephrol. 2010;116:e32–41.
34. Moreau P, Attal M, Facon T. Frontline therapy of multiple myeloma. Blood. 2015;125:3076–84.
35. Sengul S, Li M, Batuman V. Myeloma kidney: toward its prevention--with new insights from in vitro and in vivo models of renal injury. J Nephrol. 2009;22:17–28.
36. Bonventre JV. Pathophysiology of acute kidney injury: roles of potential inhibitors of inflammation. Contrib Nephrol. 2007;156:39–46.
37. Paulus P, Rupprecht K, Baer P, Obermuller N, Penzkofer D, Reissig C, et al. The early activation of toll-like receptor (TLR)-3 initiates kidney injury after ischemia and reperfusion. PLoS One. 2014;9:e94366.
38. Riera M, Torras J, Cruzado JM, Lloberas N, Liron J, Herrero I, et al. The enhancement of endogenous cAMP with pituitary adenylate cyclase-activating polypeptide protects rat kidney against ischemia through the modulation of inflammatory response. Transplantation. 2001;72:1217–23.
39. Han J. MyD88 beyond Toll. Nat Immunol. 2006;7:370–1.
40. Pulskens WP, Teske GJ, Butter LM, Roelofs JJ, van der Poll T, Florquin S, et al. Toll-like receptor-4 coordinates the innate immune response of the kidney to renal ischemia/reperfusion injury. PLoS One. 2008;3:e3596.
41. Shigeoka AA, Holscher TD, King AJ, Hall FW, Kiosses WB, Tobias PS, et al. TLR2 is constitutively expressed within the kidney and participates in ischemic renal injury through both MyD88-dependent and -independent pathways. J Immunol. 2007;178:6252–8.
42. Wolfs TG, Buurman WA, van Schadewijk A, de Vries B, Daemen MA, Hiemstra PS, et al. In vivo expression of Toll-like receptor 2 and 4 by renal epithelial cells: IFN-gamma and TNF-alpha mediated up-regulation during inflammation. J Immunol. 2002;168:1286–93.
43. Goldstein DR, Tesar BM, Akira S, Lakkis FG. Critical role of the Toll-like receptor signal adaptor protein MyD88 in acute allograft rejection. J Clin Invest. 2003;111:1571–8.

44. O'Neill LA. The role of MyD88-like adapters in Toll-like receptor signal transduction. Biochem Soc Trans. 2003;31:643–7.
45. Arumugam TV, Okun E, Tang SC, Thundyil J, Taylor SM, Woodruff TM. Toll-like receptors in ischemia-reperfusion injury. Shock. 2009;32:4–16.
46. Leemans JC, Stokman G, Claessen N, Rouschop KM, Teske GJ, Kirschning CJ, et al. Renal-associated TLR2 mediates ischemia/reperfusion injury in the kidney. J Clin Invest. 2005;115:2894–903.
47. Horvath G, Racz B, Szakaly P, Kiss P, Laszlo E, Hau L, et al. Mice deficient in neuropeptide PACAP demonstrate increased sensitivity to in vitro kidney hypoxia. Transplant Proc. 2010;42:2293–5.
48. Horvath G, Racz B, Reglodi D, Kovacs K, Kiss P, Gallyas Jr F, et al. Effects of PACAP on mitochondrial apoptotic pathways and cytokine expression in rats subjected to renal ischemia/reperfusion. J Mol Neurosci. 2010;42:411–8.
49. Oh GS, Kim HJ, Shen A, Lee SB, Khadka D, Pandit A, et al. Cisplatin-induced kidney dysfunction and perspectives on improving treatment strategies. Electrolyte Blood Press. 2014;12:55–65.
50. Bassett EA, Wang W, Rastinejad F, El-Deiry WS. Structural and functional basis for therapeutic modulation of p53 signaling. Clin Cancer Res. 2008;14:6376–86.
51. Wei Q, Dong G, Yang T, Megyesi J, Price PM, Dong Z. Activation and involvement of p53 in cisplatin-induced nephrotoxicity. Am J Physiol Renal Physiol. 2007;293:F1282–91.
52. Molitoris BA, Dagher PC, Sandoval RM, Campos SB, Ashush H, Fridman E, et al. siRNA targeted to p53 attenuates ischemic and cisplatin-induced acute kidney injury. J Am Soc Nephrol. 2009;20:1754–64.
53. Seth R, Yang C, Kaushal V, Shah SV, Kaushal GP. p53-dependent caspase-2 activation in mitochondrial release of apoptosis-inducing factor and its role in renal tubular epithelial cell injury. J Biol Chem. 2005;280:31230–9.
54. Wesierska-Gadek J, Gueorguieva M, Wojciechowski J, Tudzarova-Trajkovska S. In vivo activated caspase-3 cleaves PARP-1 in rat liver after administration of the hepatocarcinogen N-nitrosomorpholine (NNM) generating the 85 kDa fragment. J Cell Biochem. 2004;93:774–87.
55. Yang C, Kaushal V, Haun RS, Seth R, Shah SV, Kaushal GP. Transcriptional activation of caspase-6 and -7 genes by cisplatin-induced p53 and its functional significance in cisplatin nephrotoxicity. Cell Death Differ. 2008;15:530–44.
56. Sereno J, Vala H, Nunes S, Rocha-Pereira P, Carvalho E, Alves R, et al. Cyclosporine A-induced nephrotoxicity is ameliorated by dose reduction and conversion to sirolimus in the rat. J Physiol Pharmacol. 2015;66:285–99.
57. Hesselink DA, Bouamar R, van Gelder T. The pharmacogenetics of calcineurin inhibitor-related nephrotoxicity. Ther Drug Monit. 2010;32:387–93.
58. Reis FN. The unsolved cyclosporine-induced kidney injury: is paricalcitol a feasible new renoprotective option? Kidney Int. 2010;77:1055–7.
59. Bottinger EP, Bitzer M. TGF-beta signaling in renal disease. J Am Soc Nephrol. 2002;13:2600–10.
60. Campistol JM, Inigo P, Larios S, Bescos M, Oppenheimer F. Role of transforming growth factor-beta1 in the progression of chronic allograft nephropathy. Nephrol Dial Transplant. 2001;16 Suppl 1:114–6.
61. Slattery C, Campbell E, McMorrow T, Ryan MP. Cyclosporine A-induced renal fibrosis: a role for epithelial-mesenchymal transition. Am J Pathol. 2005;167:395–407.
62. Yang J, Liu Y. Dissection of key events in tubular epithelial to myofibroblast transition and its implications in renal interstitial fibrosis. Am J Pathol. 2001;159:1465–75.
63. Duymelinck C, Deng JT, Dauwe SE, De Broe ME, Verpooten GA. Inhibition of the matrix metalloproteinase system in a rat model of chronic cyclosporine nephropathy. Kidney Int. 1998;54:804–18.
64. Wu WS. The signaling mechanism of ROS in tumor progression. Cancer Metastasis Rev. 2006;25:695–705.

65. Djamali A, Vidyasagar A, Adulla M, Hullett D, Reese S. Nox-2 is a modulator of fibrogenesis in kidney allografts. Am J Transplant. 2009;9:74–82.
66. Navarro-Antolin J, Hernandez-Perera O, Lopez-Ongil S, Rodriguez-Puyol M, Rodriguez-Puyol D, Lamas S. CsA and FK506 up-regulate eNOS expression: role of reactive oxygen species and AP-1. Kidney Int Suppl. 1998;68:S20–4.
67. Abdeen A, Sonoda H, El-Shawarby R, Takahashi S, Ikeda M. Urinary excretion pattern of exosomal aquaporin-2 in rats that received gentamicin. Am J Physiol Renal Physiol. 2014;307:F1227–37.
68. Lopez-Novoa JM, Quiros Y, Vicente L, Morales AI, Lopez-Hernandez FJ. New insights into the mechanism of aminoglycoside nephrotoxicity: an integrative point of view. Kidney Int. 2011;79:33–45.
69. Zheng X, Bhalla V. The missing link: studying the alternative TGF-beta pathway provides a unifying theory for different components of diabetic nephropathy. Diabetes. 2015;64:1898–900.
70. Hathaway CK, Gasim AM, Grant R, Chang AS, Kim HS, Madden VJ, et al. Low TGFbeta1 expression prevents and high expression exacerbates diabetic nephropathy in mice. Proc Natl Acad Sci U S A. 2015;112:5815–20.
71. Espinosa G, Annapragada A. Cost-effectiveness of a novel blood-pool contrast agent in the setting of chest pain evaluation in an emergency department. AJR Am J Roentgenol. 2013;201:710–9.
72. Finn WF. The clinical and renal consequences of contrast-induced nephropathy. Nephrol Dial Transplant. 2006;21:i2–10.
73. Subramanian S, Tumlin J, Bapat B, Zyczynski T. Economic burden of contrast-induced nephropathy: implications for prevention strategies. J Med Econ. 2007;10:119–34.
74. Brezniceanu ML, Lau CJ, Godin N, Chenier I, Duclos A, Ethier J, et al. Reactive oxygen species promote caspase-12 expression and tubular apoptosis in diabetic nephropathy. J Am Soc Nephrol. 2010;21:943–54.
75. Eid S, Maalouf R, Jaffa AA, Nassif J, Hamdy A, Rashid A, et al. 20-HETE and EETs in diabetic nephropathy: a novel mechanistic pathway. PLoS One. 2013;8:e70029.
76. Ha H, Lee HB. Reactive oxygen species as glucose signaling molecules in mesangial cells cultured under high glucose. Kidney Int Suppl. 2000;77:S19–25.
77. Kuroki T, Isshiki K, King GL. Oxidative stress: the lead or supporting actor in the pathogenesis of diabetic complications. J Am Soc Nephrol. 2003;14:S216–20.
78. Heyman SN, Rosen S, Khamaisi M, Idee JM, Rosenberger C. Reactive oxygen species and the pathogenesis of radiocontrast-induced nephropathy. Invest Radiol. 2010;45:188–95.
79. Lim H, Kim D, Lee SJ. Toll-like receptor 2 mediates peripheral nerve injury-induced NADPH oxidase 2 expression in spinal cord microglia. J Biol Chem. 2013;288:7572–9.
80. Amura CR, Renner B, Lyubchenko T, Faubel S, Simonian PL, Thurman JM. Complement activation and toll-like receptor-2 signaling contribute to cytokine production after renal ischemia/reperfusion. Mol Immunol. 2012;52:249–57.
81. Allam R, Scherbaum CR, Darisipudi MN, Mulay SR, Hagele H, Lichtnekert J, et al. Histones from dying renal ells aggravate kidney injury via TLR2 and TLR4. J Am Soc Nephrol. 2012;23:1375–88.
82. Kawai T, Akira S. TLR signaling. Semin Immunol. 2007;19:24–32.
83. El-Achkar TM, Dagher PC. Renal Toll-like receptors: recent advances and implications for disease. Nat Clin Pract Nephrol. 2006;2:568–81.
84. Poletti PA, Platon A, De Seigneux S, Dupuis-Lozeron E, Sarasin F, Becker CD, et al. N-acetylcysteine does not prevent contrast nephropathy in patients with renal impairment undergoing emergency CT: a randomized study. BMC Nephrol. 2013;14:119.
85. Wang HZ, Peng ZY, Wen XY, Rimmele T, Bishop JV, Kellum JA. N-acetylcysteine is effective for prevention but not for treatment of folic acid-induced acute kidney injury in mice. Crit Care Med. 2011;39:2487–94.

Chapter 19
Neuroplasticity of PACAP Expression and Function in Micturition Reflex Pathways

Eric J. Gonzalez, Beatrice Girard, Karen M. Braas, Victor May, and Margaret A. Vizzard

Abstract Micturition, the storage and periodic elimination of urine, requires a complex neural control system that coordinates the activities of the smooth muscle of the urinary bladder and urethra and the smooth and striated muscle of the urethral sphincters. The lower urinary tract (LUT) reflex mechanisms, organized at the level of the lumbosacral spinal cord, are modulated predominantly by supraspinal controls. Complex neural organization is necessary for the coordination of the reciprocal functions of the urinary bladder, urethra, and urethral sphincters to result in normal micturition function. Injury or diseases of the nervous system, as well as disorders of the peripheral organs, can produce LUT dysfunction. Numerous neuropeptide/receptor systems are expressed in central and peripheral nervous system pathways that regulate the LUT and expression can also be found in both neural and non-neural (e.g., urothelium) components. Pituitary adenylate cyclase-activating polypeptide (PACAP; *Adcyap1*) and its cognate receptor, PAC1 (*Adcyap1r1*), have tissue-specific distributions in diverse systems including the LUT. PACAP and associated receptors exhibit neurophenotypic changes with neural injury, inflammation, stress, and disease of the LUT. Changes in the balance of the PACAP/receptor system in central and peripheral bladder reflex pathways may underlie and/or contribute to LUT dysfunction including urinary urgency, increased voiding frequency, nocturia, urinary incontinence, detrusor dyssynergia, and/or pain. The PACAP/receptor system in LUT pathways may thus represent a potential target for therapeutic intervention.

Keywords Lower urinary tract • Spinal cord • Dorsal root ganglia • Urinary bladder • Neurochemistry • Urothelium • Detrusor smooth muscle • Nerve growth factor • Knockout mice • Cystitis • Spinal cord injury

---

E.J. Gonzalez • B. Girard • K.M. Braas • V. May • M.A. Vizzard (✉)
Department of Neurological Sciences, University of Vermont College of Medicine,
D405A Given Research Building, 149 Beaumont Avenue, Burlington, VT 05405, USA
e-mail: margaret.vizzard@uvm.edu

D. Reglodi, A. Tamas (eds.), *Pituitary Adenylate Cyclase Activating Polypeptide — PACAP*, Current Topics in Neurotoxicity 11,
DOI 10.1007/978-3-319-35135-3_19

Abbreviations

ATP	Adenosine triphosphate
BPS/IC	Bladder pain syndrome/interstitial cystitis
CGRP	Calcitonin gene-related peptide
CYP	Cyclophosphamide
DCM	Dorsal commissure
DH	Dorsal horn
DRG	Dorsal root ganglia
h	Hours
IR	Immunoreactivity
L	Lumbar
LCP	Lateral collateral pathway
LUT	Lower urinary tract
MPG	Major pelvic ganglia
NGF-OE	Nerve growth factor-overexpression
nNOS	Neuronal nitric oxide synthase
p	Phosphorylated
PACAP	Pituitary adenylate cyclase-activating polypeptide
PC	Pheochromocytoma
PCR	Polymerase chain reaction
pCREB	Phosphorylated cAMP-response element binding protein
PLC	Phospholipase C
SCI	Spinal cord injury
SP	Substance P
SPN	Sacral parasympathetic nucleus
T	Thoracic
Trk	Receptor tyrosine kinase
VIP	Vasoactive intestinal polypeptide
WT	Wild type

Introduction

Micturition Reflex Pathways to the Urogenital Tract

Micturition is organized between two modes of operation: storage and elimination. During storage, somatosympathetic excitatory inputs to the urethral sphincters and sympathetic inputs to the bladder wall are tonically active [1, 2]. In contrast, during reflexive or voluntary elimination, parasympathetic inputs to the urinary bladder wall are active whereas somatosympathetic inputs to the bladder wall and urethral sphincters are inhibited [3]. Although spinal reflexes underlie most of the storage phase, reflexive or voluntary micturition reflex mechanisms are modulated by supraspinal regulation in the pontine micturition center [4].

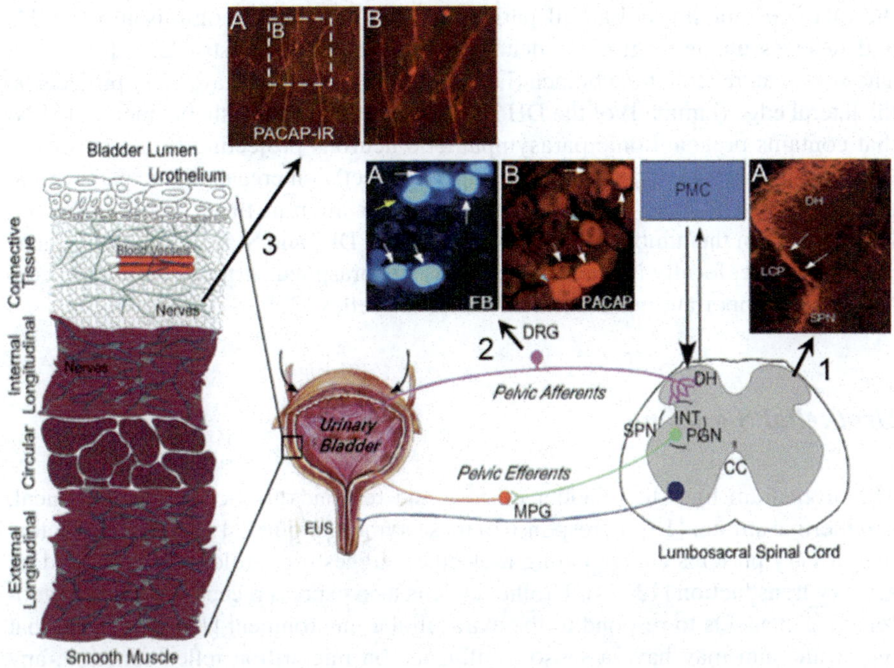

Fig. 19.1 Wiring diagram of micturition reflexes emphasizing the reflex elements that express PACAP-IR. The PACAP/receptor system has been identified in micturition reflex pathways with contributions to normal LUT function as well as that after neural injury, disease, or inflammation. PACAP-IR is expressed in normal LUT pathways but expression is dramatically increased following injury, disease, and inflammation of the urinary bladder (see text for details). Robust PACAP-IR is expressed in lumbosacral DRG (2), spinal cord (1), including the superficial laminae of the DH (1) and the LCP of Lissauer's tract (*white arrows*, 1). Bladder afferent cells in the DRG, retrogradely labeled with the conventional tracer, Fast Blue (FB), express PACAP-IR (*white arrows*, 2). Not all PACAP-IR cells in the lumbosacral DRG are bladder afferent cells (*blue arrows*, 2). Not all presumptive bladder afferent cells expressing FB, also exhibit PACAP-IR (*yellow arrow*, 2). PACAP-IR is also present in the urinary bladder including expression in urothelial cells that line the urinary bladder and in nerve fibers of the suburothelial nerve plexus (3). Postganglionic neurons in the major pelvic ganglia (MPG) also express PACAP-IR (not shown). *INT* interneurons, *PGN* preganglionic neurons, *SPN* sacral parasympathetic nucleus, *CC* central canal, *PMC* pontine micturition center, *EUS* external urethral sphincter

Slowly adapting mechanoreceptors in the urinary bladder wall underlie the switch from storage to elimination [5]. The thinly myelinated Aδ afferent fibers of the hypogastric and pelvic nerves increase their activity as hydrostatic pressure rises [6]. Bladder afferent nerves that terminate peripherally in the urinary bladder may also signal through unmyelinated C-fibers that respond to nociceptive stimulation by chemicals, inflammation, and elevated intravesical pressures [7–9]. Although C-fibers are quiescent during normal bladder filling, their activation may contribute to the development of lower urinary tract (LUT) symptoms and functional disorders of the urinary bladder [10, 11].

Bladder afferent fibers from the pelvic nerve project into Lissauer's tract where collateral branches extend along the superficial laminae of the dorsal horn (DH) [2, 8, 12] (Fig. 19.1). The ventromedial collateral branch follows the medial edge of the DH (i.e., medial collateral pathways) into the dorsal commissure (DCM) and receives inputs from the pudendal nerve and urogenital structures [2, 8, 12]. The ventrolateral collateral branch (i.e., lateral collateral pathway, LCP) projects on the lateral edge (lamina I) of the DH into the sacral parasympathetic nucleus (SPN) that contains preganglionic parasympathetic neurons projecting to the periphery [2, 8, 12] (Fig. 19.1). In addition to synapsing directly on preganglionic parasympathetic neurons in some species, primary bladder afferent fibers also synapse on interneurons in the lumbosacral DCM, superficial DH, and SPN [2, 6]. These interneurons project locally in the spinal cord or to supraspinal cortical modulatory centers and are important in the normal micturition reflex [2, 6].

Urothelial Signaling

The urothelium lines the bladder mucosa and responds to mechanical, chemical, and thermal stimuli [13]. In response to these stimuli, urothelial cells secrete factors like urinary proteins and signaling molecules suggesting a role in urinary bladder sensory transduction [13, 14]. Urothelial cells also express receptors and mechanosensitive channels to respond to the extracellular environment [15–18]. Given that the urothelium may have a sensory influence on micturition reflex function, any disruption to urothelial signaling mechanisms and/or the underlying neural network may contribute to pathological conditions of the urinary bladder [13].

Neurochemistry of Micturition Pathways

Bladder afferent fibers contain a variety of neuropeptides, including calcitonin-gene related peptide (CGRP), substance P (SP), neurokinin A, neurokinin B, vasoactive intestinal polypeptide (VIP), pituitary adenylate cyclase-activating polypeptide (PACAP), cholecystokinin, and enkephalins [12, 19–24] (Fig. 19.1). With the exception of CGRP, all of these substances are predominantly expressed in small diameter (presumably C-fiber) afferents [12, 19, 20, 25–32]. The administration of capsaicin, which acts selectively on small-diameter afferents to deplete neurotransmitter stores, reduces the levels of SP, neurokinin A, and CGRP but not VIP or enkephalin within the pelvic viscera [33]. These findings are consistent with SP, CGRP, and related tachykinins expression in afferent pathways to the pelvic viscera [33]. The following sections focus on the expression, distribution, and functional plasticity of members of the VIP–secretin–glucagon family of hormones, PACAP and VIP in micturition reflex pathways (Fig. 19.1). The contributions of other peptides to micturition reflex pathways have recently been described [24].

Pituitary Adenylate Cyclase-Activating Polypeptide (PACAP) and VPAC/PAC$_1$ Receptor Signaling

PACAP belongs to the VIP/secretin/glucagon family of bioactive peptides and was isolated from hypothalami based on its stimulation of anterior pituitary adenylyl cyclase (AC) activity [34, 35]. The rat PACAP precursor protein consists of 175 amino acid residues with posttranslational processing resulting in two α-amidated forms, PACAP38 and PACAP27 [34, 36–39]. PACAP38 has 38 amino acid residues, whereas PACAP27 has the carboxyl terminus truncated and exhibits 68 % homology to VIP [35, 37, 40]. The distribution of these two forms is tissue-specific with PACAP38 typically predominating expression in most tissues [34, 41]. PACAP38 remains identical among mammalian species suggesting similar physiologically important roles such as signaling modulation and trophic functions in the nervous and endocrine systems [34, 42].

There are three distinct G-protein-coupled receptors for PACAP and VIP: PAC1, VPAC1, and VPAC2 [43–48] (Fig. 19.2). PAC1 receptors exhibit high affinity for PACAP and display unique patterns of AC and phospholipase C (PLC) activation for PACAP27 or PACAP38 [48–52]. The potency of PACAP27 and PACAP38 to PAC1 receptors is affected by alternative splicing to receptor transcripts resulting in the presence (short) or absence (very-short) of a 21-residue insert into the amino-terminal extracellular domain [53] (Fig. 19.2). Other variants from the alternative splicing of two 84 base pair HIP and HOP cassettes result in the unique patterns of AC and PLC activation [48] (Fig. 19.2). VPAC1 and VPAC2 receptors, on the other hand, exhibit high affinity for both PACAP and VIP and are solely coupled to AC [48]. The expression of PAC1 and VPAC receptors is tissue- and cell type-specific. It was previously shown that rat superior cervical ganglia sympathetic neurons express PAC1(short) HOP1 while VPAC receptors were sparsely expressed in ganglion non-neuronal cells [49–51, 54–60].

PACAP and PAC$_1$ Receptor Neuronal Functions in the LUT

PACAP peptides have diverse functions in endocrine, nervous, gastrointestinal, and cardiovascular systems and are expressed in many central nervous system neurons and sensory and autonomic ganglia [34, 36, 41, 42, 61–76]. PACAP facilitates neuronal calcium influx, induces depolarization of the membrane, activates AC and PLC, and stimulates neurotransmitter secretion [49, 50, 54, 55, 58–60, 77–79]. Widespread PACAP-immunoreactivity (IR) has been demonstrated in nerve fibers within the urinary bladder smooth muscle, suburothelial plexus and surrounding blood vessels [80] (Fig. 19.1). Neonatal capsaicin treatment significantly reduced PACAP suggesting these fibers are derived from sensory neurons [80]. These results are consistent with the expression of PACAP in DRG and its neurochemical plasticity following nerve injury or inflammation [42, 81–83] (Fig. 19.1).

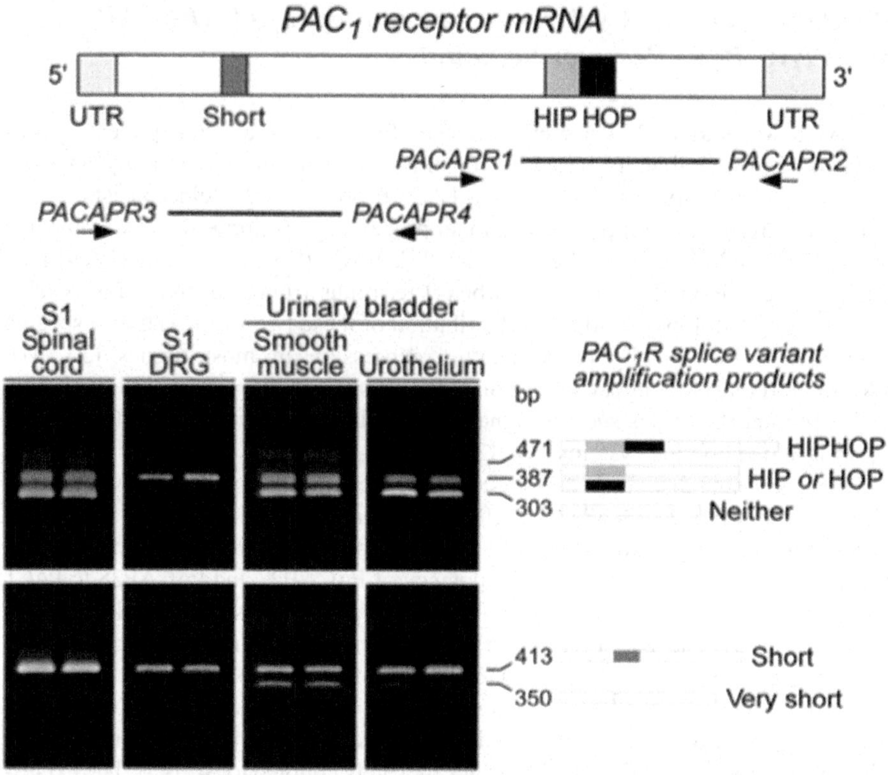

Fig. 19.2 LUT tissues express PAC1 receptor variants. Complementary DNA templates were prepared from rat S1 spinal cord, S1 DRG, and bladder detrusor and urothelium total RNA. The region spanning the alternative splice site for the HIP and HOP exons within the third cytoplasmic loop was amplified using PACAPR1/2 oligonucleotide primers. Six third cytoplasmic loop isoform fragments containing neither, one or both HIP and HOP cassettes can potentially be amplified with these primers. LUT tissues express PAC1 receptor isoforms in a tissue-specific manner. S1 DRG express predominantly the one cassette isoform; other tissues possess both the null and the one cassette variant. Schematic shading: *Dark grey*, *short* region containing exons 4 and 5; *light grey*, HIP exon cassette; *black*, HOP cassette. Thick line, region amplified using PACAPR1/2 primers. LUT tissue expression of PAC1 receptor isoforms also results from alternative splicing in amino-terminal extracellular domain. Complementary DNA templates from LUT samples described above were amplified using primers PACAPR3/4, which flank the amino-terminal extracellular domain splice site. The amplified fragments of indicated sizes represent isoforms with both (*short*) or neither (*very short*) exons 4 and 5. All LUT tissues express the *short* variant; urinary detrusor smooth muscle also demonstrates *very short* PAC1 receptor expression. *Shading* in schematic denotes alternatively spliced exons. *Thick line*, region amplified using primers PACAR3/4. Figure modified from ref. [98]

PACAP- and VIP-Mediated Effects on Urothelium and Detrusor Smooth Muscle

The urothelium acts as a selective barrier to prevent urinary constituents from penetrating the underlying tissue [84, 85]. A disruption to the properties of barrier function may occur through trauma, infection or disorders affecting the bladder like bladder pain syndrome/interstitial cystitis (BPS/IC) or spinal cord injury (SCI) [86, 87]. It has been suggested that this loss of barrier integrity contributes to the altered sensory processing observed in cystitis. Recent studies have demonstrated that the urothelium expresses PAC1 receptors that upon stimulation release ATP to stimulate receptors on underlying sensory nerve fibers [88] (Fig. 19.2). ATP release was evoked by PACAP27, PACAP38, and VIP application to cultured urothelial cells with PACAP27 and PAC1 receptor antagonism blocking ATP release [88]. These results suggest PACAP and PAC1 signaling may regulate micturition reflex function at the level of the urothelium [88] (Figs. 19.1 and 19.2).

PACAP and VIP have direct effects on smooth muscle cells. PACAP or VIP elicit relaxation of guinea pig stomach, rat ileum, rabbit iris sphincter and dilator muscles, cat and human esophageal sphincter, and human and guinea pig airways but elicit contraction in guinea pig ileum and gall bladder [89]. The effects of PACAP on urinary bladder smooth muscle have not been well described despite PACAP, PAC1, and VPAC expression in nerve fibers within the detrusor [80] (Figs. 19.1 and 19.2). PACAP27 has a small effect on isolated bladder smooth muscle strips even though it facilitates micturition in conscious, open outlet rats [90]. These studies, however, did not take into account PAC1 and VPAC receptor cross talk and the peptide selectivity of various PAC1 isoforms.

Unlike PACAP, VIP is expressed in postganglionic efferent neurons of the major pelvic ganglia (MPG) and minimally innervates the urinary bladder [80, 91–93]. VIP administration to the detrusor smooth muscle had no effect on spontaneous or carbachol-induced bladder contractions but intrathecal or intra-arterial VIP administration facilitated micturition [94]. These conflicting roles may result from VIP receptor distribution varying across species and target tissue [94–97]. Taken together, it appears that PACAP/receptor signaling has more influence on micturition reflex function than VIP/receptor signaling [94–98].

PACAP and VIP Expression and Effects on MPG Neurons

PACAP/receptor expression has been demonstrated in the MPG, ganglia that supply autonomic (sympathetic and parasympathetic) innervation to the urinary bladder, following neuronal injury. PAC1, VPAC1, and VPAC2 transcripts were reported in the MPG in cell culture for 4 h [99], with only VPAC2 transcript significantly increased by day 3 [100]. PACAP transcript and PACAP-IR similarly increased in the MPG by day 3 in culture [100]. Unlike PACAP, VIP transcript expression remained unchanged in a 3-day culture [100]. Furthermore, the application of VIP, PACAP, and maxadilan, a PAC1-selective agonist, increased neuronal excitability and decreased after-hyperpolarization in the MPG [99]. Taken together, these studies suggest PACAP/receptor signaling in the MPG may have a role in micturition reflex function following injury (Fig. 19.1).

PACAP or VIP Knockout Mice Exhibit Altered Micturition Reflexes

Bladder dysfunction and altered somatic sensation have previously been demonstrated in mice with a genetic disruption or deletion to PACAP or VIP. PACAP (+/–) and PACAP (–/–) mice display less somatic sensitivity to mechanical stimuli in the pelvic and hindpaw regions relative to controls [101]. On the other hand, VIP (–/–) mice display increased somatic sensitivity in the pelvis and decreased paw pressure threshold following inflammation [102]. These dissimilarities may reflect distinct roles for VIP and PACAP in bladder sensory function. In contrast to the observed mechanosensitivity differences, both knockout mice exhibited an increase in bladder mass with hypertrophy specific to the lamina propria and detrusor smooth muscle in PACAP (–/–) mice and only the detrusor smooth muscle in VIP (–/–) mice [101, 103].

Functionally, PACAP (–/–) mice have increased bladder capacity, void volume, and longer intercontraction intervals relative to controls [101]. It has been argued that these mice exhibit partial outlet obstruction because of extended and incomplete emptying of the bladder [101]. VIP (–/–) mice, however, do not functionally present with changes in the basal tone of the bladder but do have increased void volume, longer intercontraction intervals, and complete emptying of the bladder [102]. Bladder dysfunction in VIP (–/–) mice may result from functional changes within the bladder itself or neuroplasticity among bladder afferent cells. Along with the increase in tissue mass, the urinary bladders of VIP (–/–) mice have increased urea permeability, increased basal expression of NGF and an exaggerated proinflammatory response to inflammation [102–104]. Additionally, there are elevated basal levels of phosphorylated cAMP response-element binding protein (pCREB) in lumbosacral (L1, L2, L5–S1) DRG and in L6 and S1 afferent neurons projecting directly from the bladder of VIP (–/–) mice suggesting elevated afferent activity of the urinary bladder [103].

Neuroplasticity of PACAP/Receptor Expression and Function with Cystitis

The regulation of transcript and peptide expression of PACAP and its receptors previously demonstrated with cyclophosphamide (CYP)-induced cystitis has been suggested to underlie the development of urinary bladder dysfunction. Following a downregulation in transcript expression after acute (4 h) CYP-induced cystitis, PACAP transcript expression is dramatically upregulated in the urothelium and L6 and S1 DRG after intermediate (48 h) or chronic (10 day) CYP-induced cystitis [88]. Similarly, transcript expression of the PAC1 receptor is down-regulated in the urothelium, detrusor smooth muscle and L6 and S1 DRG after acute (4 h) CYP-induced cystitis, but upregulated in the urothelium and detrusor smooth muscle after intermediate (48 h) or chronic (10 day) CYP-induced cystitis [88]. VPAC1 and

VPAC2 transcript expression, however, remains upregulated in the urothelium and detrusor smooth muscle from acute (4 h) CYP-induced cystitis to intermediate (48 h) CYP-induced cystitis with the down-regulation of VPAC2 transcript expression occurring after chronic (10 day) CYP-induced cystitis [88]. PACAP-IR in the spinal cord is restricted to nerve fibers and is increased in micturition reflex associated regions after intermediate (48 h) or chronic (10 day) CYP-induced cystitis [22, 105]. Following intermediate or chronic CYP-induced cystitis, PACAP-IR is increased in the superficial laminae (I-II) of the DH, medial to lateral extent of the dorsal horn, LCP of Lissauer, SPN, and S1 spinal segments [22, 105] (Fig. 19.1). Additionally, the percentage of Fast Blue labeled bladder afferent cells positive for PACAP increased in L1-L2, L6, and S1 DRG following chronic (10 day) CYP-induced cystitis [22] (Fig. 19.1).

The aforementioned regulation of PACAP and its receptors in areas associated with the micturition reflex suggests this neuropeptide may have a role in bladder dysfunction with inflammation (Fig. 19.2). In support of this, the intrathecal (L6-S1) or intravesical administration of a PAC1 receptor antagonist, PACAP6-38, was able to increase bladder capacity but not intravesical pressure with intermediate (48 h) CYP-induced cystitis [98]. The different routes of administration with similar functional effects suggest PACAP6-38 may have multiple sites of action. Administration of PACAP6-38 at the level of the spinal cord may be acting on superficial DH neurons to block PACAP release from C-fiber afferents, whereas, PACAP at the level of the urinary bladder may be acting on urothelial, suburothelial or detrusor smooth muscle cells [98] (Fig. 19.1). Despite not yet knowing its specific site of action, the inhibition of aberrant PACAP signaling seems to be a promising target to reduce voiding frequency with cystitis.

PACAP Expression in LUT with CYP-Induced Cystitis in PACAP Promoter-Dependent EGFP BAC Transgenic Mice

We previously demonstrated an upregulation of PACAP expression in rodent micturition pathways following CYP-induced cystitis [98]. We subsequently examined the effects of CYP-induced cystitis (4 h, 48 h, chronic) in PACAP promoter-dependent EGFP BAC transgenic mice [106]. We induced bladder inflammation in adult mice by injecting CYP intraperitoneally to produce acute (150 mg/kg; 4 h), intermediate (150 mg/kg; 48 h), and chronic (75 mg/kg; every third day for 10 days) cystitis. In control (no inflammation) animals, low basal expression of PACAP-EGFP+ fibers was present in the superficial DH at all segmental levels examined (L1, L2, L4–S1). Dorsal root ganglia (DRG; L1, L2, L6, S1) from control animals also exhibited PACAP-EGFP+ cells. After CYP-induced cystitis, PACAP-EGFP+ cells increased dramatically in spinal segments and DRG (L1, L2, L6, and S1) involved in micturition reflexes. Small diameter, PACAP-EGFP+ DRG cells co-localized with TRPV1- and TRPV4-IR [106]. The density of PACAP-EGFP+ nerve fibers was increased in the superficial laminae (I–II) of the L1, L2, L6, and S1 DH. No changes in PACAP-EGFP+ nerve fibers were observed in the L4–L5 segments. PACAP-EGFP+ nerve fibers also increased in the lateral collateral pathway in L6–S1 spinal cord. Following CYP-induced cystitis, PACAP-EGFP+ urothelial cells were observed and the number of PACAP-EGFP+ urothelial cells increased with duration of cystitis. PACAP-EGFP+ urothelial cells were co-localized with TRPV4-IR [106]. Changes in PACAP expression in LUT pathways after cystitis may play a role in altered visceral sensation (allodynia) and/or increased voiding frequency in the chronic inflammatory pain syndrome, interstitial cystitis/bladder pain syndrome.

Neuroplasticity of PACAP/Receptor Expression and Function with Spinal Cord Injury (SCI)

SCI has been demonstrated to regulate the transcript and peptide expression of PACAP and its receptors within the spinal cord and urinary bladder. An increase in PACAP and PAC1 receptor transcript expression is observed in the spinal cord following a moderate compression model of SCI [107]. PACAP- and PAC1-immunoreactive cells are also increased around the site of injury and co-localized with NeuN-positive cells (i.e., neuronal marker) [107]. In other studies utilizing spinal cord (Thoracic (T)7–T9) transection, PACAP-IR is increased 6 weeks after SCI in micturition reflex associated regions [108]. Within the upper lumbar (L1–L2) spinal cord, PACAP-IR increased in the superficial laminae (I–II) of the DH, medial to lateral extent of the DH, and a fiber bundle extending laterally from Lissauer's tract [108]. Similarly within the lumbosacral (L6–S1) spinal cord, PACAP-IR increased in the DH, DCM, SPN, and LCP of Lissauer [108]. Increased PACAP-IR is not limited to the spinal cord, but the percentage of PACAP-positive bladder afferent cells labeled with Fast Blue are also increased in L1–L2, L6, and S1 DRG 48 h to 6 weeks after SCI [108]. Unlike the increased PACAP-IR observed at the level of the spinal cord and DRG, the urinary bladder has decreased PACAP-IR in the urothelium and detrusor smooth muscle from 5 days to 3 weeks after SCI [108].

The regulation of PACAP and its receptors around the site of SCI and micturition reflex regions suggests this neuropeptide may have a protective role to help facilitate bladder function. In support of this, the intrathecal administration of PACAP-38 following transection of the T8–T9 spinal cord resulted in large amplitude and long duration bladder contractions under isovolumetric conditions [109]. Additionally, intrathecal administration of a PAC1 receptor antagonist, PACAP6-38, following transection of the T8–T10 spinal cord reduced filling, threshold and peak micturition pressures, number and amplitude of non-voiding contractions, and had shorter intercontraction intervals [110]. Taken together, these studies suggest PACAP may act on parasympathetic efferent pathways at the level of the spinal cord and/or DRG [109]. Aside from its possible role in bladder function, PACAP may also be protecting from a loss of motor function. PACAP (+/–) mice showed a greater injury volume surrounding SCI and also exhibited lower Basso Mouse Scale motor scores on days 3, 7, and 14 suggesting impaired motor function [107]. These studies demonstrate the significance of PACAP regulation following SCI and argue for a role of PACAP in both bladder and somatomotor function with injury.

Role of Nerve Growth Factor (NGF) and Associated Receptors in LUT Plasticity

Cytokines and growth factors, including NGF, are upregulated at the site of tissue injury, inflammation, and/or target organ hypertrophy [111–115]. Following noxious peripheral stimulation, for example, levels of neuroactive compounds (e.g., enkephalin [112], dynorphin [116], CGRP [115, 117, 118], SP [23, 112, 116, 118], neuropeptide Y [112]; neuronal nitric oxide synthase (nNOS) [28, 119, 120] and PACAP [22, 121]) have been demonstrated to increase in DRG and spinal cord neurons. NGF, in particular, is also released from the target organ for tyrosine kinase receptor (Trk) type 1 (TrkA) binding and retrograde transport in DRG afferent neurons [122]. The subsequent increase in NGF expression within the DRG neurons may induce increased production of neuropeptides (i.e., SP, CGRP, and PACAP) and alter sensory transduction [115, 117, 118]. In addition, a large percentage of pelvic visceral afferent neurons express neurotrophic factor receptors, including Trk for NGF and related substances [123–126]. Following cystitis or SCI, neurotrophic factor receptors exhibit neuroplastic increases in TrkA- and TrkB-IR and Trk phosphorylation in bladder afferent neurons [125, 126].

CYP-Induced Cystitis

Increases in the number of TrkA-immunoreactive cell profiles were detected in the L1 and L6 DRG (fourfold) and the S1 DRG (1.5-fold) but not in the L2, L4, and L5 DRG with CYP-induced cystitis of acute and chronic duration compared with control rats [125]. The number of TrkB-IR cell profiles increased in the L1 and L2 DRG (L1: 2.6-fold; L2: 1.4-fold) and in the L6 and S1 DRG (L6: 2.2-fold; S1: 1.3-fold) only after acute CYP treatment (8 h) [125]. After CYP treatment, the percentage of bladder afferent cell profiles expressing TrkA-IR (~50 %) increased in L1 and L6 DRG. The percentage of bladder afferent cell profiles expressing TrkB-IR (~45 %) in L1, L2, L6, and S1 DRG also increased compared with control cell profiles [125]. The increase in TrkA-IR in bladder afferent cells occurred 8 h after CYP treatment and was maintained in L1 DRG with chronic (10 days) CYP-induced cystitis. However, the increase in bladder afferent cells expressing TrkB-IR only occurred at the most acute time point examined (8 h). TrkA-IR and TrkB-IR cell profiles also demonstrated phosphorylated Trk-IR with acute and/or chronic CYP-induced cystitis [125].

Spinal Cord Injury (SCI)

After SCI, a significant increase in the number of TrkB-immunoreactive cells was also detected in the L6–S1 DRG and in the L1–L2 DRG but not in the L4–L5 DRG compared with control rats [127]. After SCI, the percentage of FB-labeled cells expressing TrkA- or TrkB-IR in L1 and L6 DRG significantly increased compared with control DRG. After SCI, the percentage of TrkA-immunoreactive cells expressing phosphorylated (p)-Trk-IR significantly increased (1.5- to 2.3-fold increase) in the L1, L6, and S1 DRG. The percentage of TrkB-immunoreactive cells expressing p-Trk-IR after SCI also increased (1.3-fold increase) in the L1 and L6 DRG [127]. These results demonstrate that (1) TrkA- and TrkB-IR is increased in bladder afferent cells after SCI and (2) TrkA and TrkB receptors are phosphorylated in DRG after SCI. Neuroplasticity of LUT reflexes after SCI may be mediated by both NGF and brain-derived neurotrophic factor in target tissues [127].

NGF and PACAP Interactions

Recent reports have demonstrated reciprocal regulatory interactions between NGF and PACAP in rat pheochromocytoma (PC)12 cells and in DRG cells. Recent studies [128] have implicated NGF as a positive regulator of PACAP expression in nociceptive DRG cells. In rat PC12 cells, both NGF and PACAP can induce PC differentiation into a neuronal phenotype [129]. Upon PC12 transfection of a PACAP promoter-luciferase construct, exogenously applied PACAP and NGF, added either alone or in combination, upregulated PACAP gene expression [130, 131]. In addition, the neurotrophins can also facilitate expression of the PACAP-selective PAC1 receptor. NGF upregulated the PAC1 receptor promoter in PC12 cells; both NGF and BDNF induced PAC1 receptor promoter activity and mRNA expression in cerebellar granule cells [132].

Conversely, PACAP has also been shown to upregulate TrkA and TrkB receptor expression and/or phosphorylation in PC12 cells and hippocampal neurons, respectively in a Src-dependent manner [133]. Studies with sympathetic neuroblasts also demonstrated that PACAP can augment TrkA and TrkC expression in the neuronal differentiation process [134]. The ability for the PACAP and NGF signaling pathways to demonstrate reciprocal regulatory processes may be a primary example of an important feed-forward mechanism to amplify a trophic survival or differentiation response during neuronal development or regeneration. In bladder inflammation or other pathophysiological events, the same feed-forward mechanism may present complications and exacerbate dysfunction.

Transgenic Mouse Model with Chronic Urothelial Overexpression of NGF (NGF-OE)

Our laboratory has characterized a transgenic mouse model of urothelium-specific, NGF-OE that represents a novel approach to exploring the role of NGF in urinary bladder inflammation and sensory function [135]. Functionally, NGF-OE mice exhibit urinary bladder hyperreflexia with frequent urination and the presence of non-voiding bladder contractions as well as referred somatic pelvic hypersensitivity [135]. No changes in the electrical properties of the MPG neurons of NGF-OE mice were detected using intracellular recording, suggesting that the urinary bladder phenotype in NGF-OE mice is not influenced by changes in the efferent limb of the micturition reflex. NGF-OE mice may represent a useful animal model of BPS/IC because the changes observed in the urinary bladders of these mice are consistent with certain changes observed in this syndrome. Pleiotropic changes, subsequent to NGF-OE, including changes in the expression of growth factors, neuroactive compounds, and ion channels (e.g., transient receptor potential (TRP) channels) [136, 137] can also directly modulate pain and bladder/visceral sensory function and could contribute to altered urinary bladder function in NGF-OE mice [11, 137–139].

Recent studies also demonstrate changes in PACAP/VIP and receptor expression in micturition pathways in NGF-OE mice [140]. Results demonstrate upregulation of PAC1 receptor transcript and PAC1-IR in urothelium of NGF-OE mice whereas PACAP transcript and PACAP-IR were decreased in urothelium of NGF-OE mice [140]. In contrast, VPAC1 receptor transcript was decreased in both urothelium and detrusor smooth muscle of NGF-OE mice [140]. VIP transcript expression and VIP-IR was not altered in urinary bladder of NGF-OE mice [140]. Changes in PACAP, VIP and associated receptors transcripts and peptide expression in micturition pathways resemble some, but not all, changes observed after induction of urinary bladder inflammation known to involve NGF production.

Contributions of PACAP/Receptor Signaling to Increased Voiding Frequency and Somatic Sensitivity in NGF-OE

Given the presence of PAC1-IR fibers, the expression of PAC1 receptor expression in bladder tissues, and the abilities of PACAP to facilitate detrusor contractility, whether PACAP/receptor signaling contributes to bladder hyperreflexia and somatic sensitivity was recently evaluated [141]. Intravesical administration of PACAP6-38 (300 nM) significantly increased bladder capacity (2.0-fold), intercontraction interval and void volume in NGF-OE mice. Intravesical instillation of PACAP6-38 also decreased filling and peak micturition pressure in NGF-OE mice [141]. PACAP6-38 had no effects on WT mice. Intravesical administration of PACAP6-38 (300 nM) significantly reduced pelvic sensitivity in NGF-OE mice but was without effect in WT mice. PACAP/receptor signaling contributes to the increased voiding frequency and pelvic sensitivity observed in NGF-OE mice [141].

We have extended these studies to address the contribution of target-derived NGF in combination with CYP-induced cystitis to determine whether additional changes in neuropeptides/receptors are observed in micturition reflex pathways due to the presence of additional inflammatory mediators in the urinary bladder [142]. Quantitative polymerase chain reaction (PCR) was used to determine PACAP/ VIP, SP, galanin, and receptor transcript expression in the urinary bladder (urothelium, detrusor) in NGF-OE mice and wild type (WT) mice with CYP-induced cystitis (4 h, 48 h, and chronic) [142]. With CYP-induced cystitis (4 h), WT and NGF-OE mice exhibited similar changes in galanin transcript expression in the urothelium (30-fold increase) and detrusor (threefold increase). In contrast, PACAP, VIP, and SP transcripts exhibited differential changes in WT and NGF-OE with CYP-induced cystitis. PAC1, VPAC1, and VPAC2 transcript expression also exhibited differential responses in NGF-OE mice that were tissue (urothelium vs. detrusor) and CYP-induced cystitis duration-dependent [142]. Using conscious cystometry, NGF-OE mice treated with CYP exhibited significant increases in voiding frequency above that observed in control NGF-OE mice [142]. These studies are consistent with target-derived NGF and other inflammatory mediators affecting neurochemical plasticity and the reflex function of micturition pathways.

We now have determined whether additional changes in neuropeptides/receptors and growth factor/receptors are observed in the urinary bladder (urothelium, detrusor) and lumbosacral dorsal root ganglia (DRG) involved in micturition reflexes in NGF-OE mice with CYP-induced cystitis (Girard and Vizzard, unpublished observations). Quantitative PCR was used to determine NGF, BDNF, VEGF, and receptors (TrkA, TrkB, $p75^{NTR}$) and PACAP/VIP and receptors (PAC1, VPAC1, VPAC2) transcripts expression in tissues from NGF-OE and wild type (WT) mice with CYP-induced cystitis (4 h, 48 h, and chronic). As expected in urothelium of control NGF-OE mice, NGF mRNA was significantly increased. Urothelial expression of NGF mRNA in NGF-OE mice treated with CYP (4 h 48 h, and chronic) was not further increased but maintained with all durations of CYP treatment evaluated. In contrast, CYP-induced cystitis (4 h 48 h, and chronic) in NGF-OE mice demonstrated significant regulation in BDNF, VEGF mRNA, TrkA, TrkB, and $P75^{NTR}$ in urothelium and detrusor smooth muscle. Similarly, CYP-induced cystitis (4 h 48 h, and chronic) in NGF-OE mice resulted in significant differential changes in WT and NGF-OE in transcript expression for NGF, BDNF, and receptors (TrkA, TrkB, $p75^{NTR}$) and PACAP/VIP and receptors (PAC1, VPAC1, VPAC2) in lumbosacral DRG that was also CYP-induced cystitis duration-dependent. These studies are consistent with target-derived NGF and other inflammatory mediators affecting neurochemical plasticity and contributing to reflex function of micturition pathways.

Perspectives and Significance

PACAP (*Adcyap1*) and its cognate receptor, PAC1 (*Adcyap1r1*), have tissue-specific distributions in diverse systems including micturition reflex pathways including expression in both neural and non-neural (e.g., urothelium) components. PACAP and associated receptors exhibit neuroplastic changes in expression and function with neural injury, inflammation, and diseases of the LUT. Changes in the PACAP/receptor system in micturition pathways may underlie and/or contribute to LUT dysfunction including the symptoms of urinary urgency, increased voiding frequency, nocturia, urinary incontinence, detrusor dyssynergia, and/or pain. The PACAP/receptor system in micturition reflexes may represent a potential target for therapeutic intervention.

Acknowledgements The authors thank current and former members of the Vizzard laboratory who have contributed to the studies described within including: Mary Beth Klinger, Susan Malley, Abbey Peterson, Kimberly Corrow, Katarina Zvarova, Peter Zvara, Li-ya Qiao, and Bopaiah P. Cheppudira. Gratitude is expressed to Dr. James A. Waschek, Dept. of Psychiatry and Behavioral Sciences, David Geffen School of Medicine, University of California Los Angeles for providing the PACAP-EGFP mice. The authors thank Dr. Jan Fahrenkrug, University of Copenhagen, Copenhagen, Denmark, for his generosity in supplying the monoclonal PACAP antisera used in the studies described.

Grants Research from the Vizzard laboratory described herein was funded by the National Institutes of Health (NIH) grants DK051369 (MAV), DK060481 (MAV). Additional support was also provided by grants from the National Center for Research Resources (5 P30 RR 032135) and the National Institute of General Medical Sciences (8 P30 GM 103498) from the NIH.

References

1. Kuru M. Nervous control of micturition. Physiol Rev. 1965;45:425–94.
2. de Groat WC, Kruse MN. Central processing and morphological plasticity in lumbosacral afferent pathways from the lower urinary tract. In: Mayer EA, Raybould HE, editors. Basic and clinical aspects of chronic abdominal pain research and clinical management, vol. 9. Amsterdam: Elsevier Science Publishers; 1993. p. 219–35.
3. Middleton JW, Keast JR. Artificial autonomic reflexes: using functional electrical stimulation to mimic bladder reflexes after injury or disease. Auton Neurosci. 2004;113:3–15.
4. Griffiths DJ. The pontine micturition centres. Scand J Urol Nephrol Suppl. 2002;210:21–6.
5. Kingsley RE, Gable SR, Kingsley TR, Saint Joseph Medical Center (South Bend Ind.). Magnetic Resonance Imaging Center. Concise text of neuroscience. Baltimore: Williams & Wilkins; 1996. viii, 564 p. 335–434.
6. Fowler CJ, Griffiths D, de Groat WC. The neural control of micturition. Nat Rev Neurosci. 2008;9:453–66.
7. Habler HJ, Janig W, Koltzenburg M. Activation of unmyelinated afferent fibres by mechanical stimuli and inflammation of the urinary bladder in the cat. J Physiol. 1990;425:545–62.
8. Andersson KE. Bladder activation: afferent mechanisms. Urology. 2002;59(5 Suppl 1):43–50.
9. Mazieres L, Jiang C, Lindstrom S. The C fibre reflex of the cat urinary bladder. J Physiol. 1998;513(Pt 2):531–41.

10. de Groat WC, Booth AM, Yoshimura N. Neurophysiology of micturition and its modification in animal models of human disease. In: Maggi CA, editor. The autonomic nervous system. London: Harwood Academic Publishers; 1993. p. 227–90.
11. Yoshimura N, Seki S, Chancellor MB, de Groat WC, Ueda T. Targeting afferent hyperexcitability for therapy of the painful bladder syndrome. Urology. 2002;59(5 Suppl 1):61–7.
12. Donovan MK, Winternitz SR, Wyss JM. An analysis of the sensory innervation of the urinary system of the rat. Brain Res Bull. 1983;11:321–4.
13. Birder L, Andersson KE. Urothelial signaling. Physiol Rev. 2013;93:653–80.
14. Birder L, Wyndaele JJ. From urothelial signalling to experiencing a sensation related to the urinary bladder. Acta Physiol (Oxf). 2013;207:34–9.
15. Kullmann FA, Shah MA, Birder LA, de Groat WC. Functional TRP and ASIC-like channels in cultured urothelial cells from the rat. Am J Physiol Renal Physiol. 2009;296:F892–901.
16. Moro C, Tajouri L, Chess-Williams R. Adrenoceptor function and expression in bladder urothelium and lamina propria. Urology. 2013;81:211e1–7.
17. Wang EC, Lee JM, Ruiz WG, Balestreire EM, von Bodungen M, Barrick S, et al. ATP and purinergic receptor-dependent membrane traffic in bladder umbrella cells. J Clin Invest. 2005;115:2412–22.
18. Bschleipfer T, Schukowski K, Weidner W, Grando SA, Schwantes U, Kummer W, et al. Expression and distribution of cholinergic receptors in the human urothelium. Life Sci. 2007;80:2303–7.
19. de Groat WC, Vizzard MA, Araki I, Roppolo J. Spinal interneurons and preganglionic neurons in sacral autonomic reflex pathways. Prog Brain Res. 1996;107:97–111.
20. de Groat WC, Kawatani M, Hisamitsu T, Booth AM, Roppolo JR, Thor K, et al. Neural control of micturition: the role of neuropeptides. J Auton Nerv Syst. 1986;Suppl:369–87.
21. Keast JR. Patterns of co-existence of peptides and differences of nerve fibre types associated with noradrenergic and non-noradrenergic (putative cholinergic) neurons in the major pelvic ganglion of the male rat. Cell Tissue Res. 1991;266:405–15.
22. Vizzard MA. Up-regulation of pituitary adenylate cyclase-activating polypeptide in urinary bladder pathways after chronic cystitis. J Comp Neurol. 2000;420:335–48.
23. Vizzard MA. Alterations in neuropeptide expression in lumbosacral bladder pathways following chronic cystitis. J Chem Neuroanat. 2001;21:125–38.
24. Arms L, Vizzard MA. Neuropeptides in lower urinary tract function. Handbook Exp Pharmacol. 2011;202:395–423.
25. Ek A, Alm P, Andersson KE, Persson K. Adrenergic and cholinergic nerves of the human urethra and urinary bladder. Acta Physiol Scand. 1977;99:345–52.
26. Keast JR, de Groat WC. Segmental distribution and peptide content of primary afferent neurons innervating the urogenital organs and colon of male rats. J Comp Neurol. 1992;319:615–23.
27. Su HC, Polak JM, Mulderry PK, Ghatei MA, Gibson SJ, Terenghi G, et al. Calcitonin gene-related peptide immunoreactivity in afferent neurons supplying the urinary tract: combined retrograde tracing and immunohistochemistry. Neuroscience. 1986;18:727–47.
28. Vizzard MA, de Groat WC. Increased expression of neuronal nitric oxide synthase (NOS) in bladder afferent pathways following chronic bladder irritation. J Comp Neurol. 1996;370:191–202.
29. Vizzard MA, Erdman SL, de Groat WC. Localization of NADPH-diaphorase in bladder afferent and postganglionic efferent neurons of the rat. J Auton Nerv Syst. 1993;44:85–90.
30. Vizzard MA, Erdman SL, de Groat WC. Localization of NADPH-diaphorase in pelvic afferent and efferent pathways of the rat. Neurosci Lett. 1993;152:72–6.
31. Vizzard MA, Erdman SL, de Groat WC. Increased expression of neuronal nitric oxide synthase in dorsal root ganglion neurons after systemic capsaicin administration. Neuroscience. 1995;67:1–5.
32. Vizzard MA, Erdman SL, Förstermann U, de Groat WC. Differential distribution of nitric oxide synthase in neural pathways to the urogenital organs (urethra, penis, urinary bladder) of the rat. Brain Res. 1994;646:279–91.

33. de Groat WC. Neuropeptides in pelvic afferent pathways. Experientia. 1987;43:801–13.
34. Arimura A. Perspectives on pituitary adenylate cyclase activating polypeptide (PACAP) in the neuroendocrine, endocrine, and nervous systems. Jpn J Physiol. 1998;48:301–31.
35. Ogi K, Kimura C, Onda H, Arimura A, Fujino M. Molecular cloning and characterization of cDNA for the precursor of rat pituitary adenylate cyclase activating polypeptide (PACAP). Biochem Biophys Res Commun. 1990;173:1271–9.
36. Braas KM, May V, Harakall SA, Hardwick JC, Parsons RL. Pituitary adenylate cyclase-activating polypeptide expression and modulation of neuronal excitability in guinea pig cardiac ganglia. J Neurosci. 1998;18:9766–79.
37. Kimura C, Ohkubo S, Ogi K, Hosoya M, Itoh Y, Onda H, et al. A novel peptide which stimulates adenylate cyclase: molecular cloning and characterization of the ovine and human cDNAs. Biochem Biophys Res Commun. 1990;166:81–9.
38. Ohkubo S, Kimura C, Ogi K, Okazaki K, Hosoya M, Onda H, et al. Primary structure and characterization of the precursor to human pituitary adenylate cyclase activating polypeptide. DNA Cell Biol. 1992;11:21–30.
39. Okazaki K, Itoh Y, Ogi K, Ohkubo S, Onda H. Characterization of murine PACAP mRNA. Peptides. 1995;16(7):1295–9.
40. Miyata A, Jiang L, Dahl RD, Kitada C, Kubo K, Fujino M, et al. Isolation of a neuropeptide corresponding to the N-terminal 27 residues of the pituitary adenylate cyclase activating polypeptide with 38 residues (PACAP38). Biochem Biophys Res Commun. 1990;170(2):643–8.
41. Arimura A, Somogyvari-Vigh A, Miyata A, Mizuno K, Coy DH, Kitada C. Tissue distribution of PACAP as determined by RIA: highly abundant in the rat brain and testes. Endocrinology. 1991;129:2787–9.
42. Moller K, Reimer M, Ekblad E, Hannibal J, Fahrenkrug J, Kanje M, et al. The effects of axotomy and preganglionic denervation on the expression of pituitary adenylate cyclase activating peptide (PACAP), galanin and PACAP type 1 receptors in the rat superior cervical ganglion. Brain Res. 1997;775:166–82.
43. Hashimoto H, Ishihara T, Shigemoto R, Mori K, Nagata S. Molecular cloning and tissue distribution of a receptor for pituitary adenylate cyclase-activating polypeptide. Neuron. 1993;11:333–42.
44. Hosoya M, Onda H, Ogi K, Masuda Y, Miyamoto Y, Ohtaki T, et al. Molecular cloning and functional expression of rat cDNAs encoding the receptor for pituitary adenylate cyclase activating polypeptide (PACAP). Biochem Biophys Res Commun. 1993;194:133–43.
45. Inagaki N, Yoshida H, Mizuta M, Mizuno N, Fujii Y, Gonoi T, et al. Cloning and functional characterization of a third pituitary adenylate cyclase-activating polypeptide receptor subtype expressed in insulin-secreting cells. Proc Natl Acad Sci U S A. 1994;91:2679–83.
46. Ishihara T, Shigemoto R, Mori K, Takahashi K, Nagata S. Functional expression and tissue distribution of a novel receptor for vasoactive intestinal polypeptide. Neuron. 1992;8:811–9.
47. Lutz EM, Sheward WJ, West KM, Morrow JA, Fink G, Harmar AJ. The VIP2 receptor: molecular characterisation of a cDNA encoding a novel receptor for vasoactive intestinal peptide. FEBS Lett. 1993;334:3–8.
48. Spengler D, Waeber C, Pantaloni C, Holsboer F, Bockaert J, Seeburg PH, et al. Differential signal transduction by five splice variants of the PACAP receptor. Nature. 1993;365:170–5.
49. Braas KM, May V. Pituitary adenylate cyclase-activating polypeptides directly stimulate sympathetic neuron neuropeptide Y release through PAC(1) receptor isoform activation of specific intracellular signaling pathways. J Biol Chem. 1999;274:27702–10.
50. Braas KM, May V. Pituitary adenylate cyclase-activating polypeptides, PACAP-38 and PACAP-27, regulation of sympathetic neuron catecholamine, and neuropeptide Y expression through activation of type I PACAP/VIP receptor isoforms. Ann N Y Acad Sci. 1996;805:204–16.
51. Braas KM, May V. Novel activation and interaction of sympathetic neuron signal transduction pathways by pituitary adenylate cyclase activating polypeptides. Soc Neurosci Abstr. 1995;21:1845.

52. Deutsch PJ, Sun Y. The 38-amino acid form of pituitary adenylate cyclase-activating polypeptide stimulates dual signaling cascades in PC12 cells and promotes neurite outgrowth. J Biol Chem. 1992;267:5108–13.
53. Pantaloni C, Brabet P, Bilanges B, Dumuis A, Houssami S, Spengler D, et al. Alternative splicing in the N-terminal extracellular domain of the pituitary adenylate cyclase-activating polypeptide (PACAP) receptor modulates receptor selectivity and relative potencies of PACAP-27 and PACAP-38 in phospholipase C activation. J Biol Chem. 1996;271: 22146–51.
54. Beaudet MM, Parsons RL, May V. Electrophysiological actions and signal transduction mechanisms of pituitary adenylate cyclase activating polypeptide (PACAP) in the rat superior cervical ganglion. Soc Neurosci Abstr. 1998;24:2050.
55. Beaudet MM, Parsons RL, Braas KM, May V. Mechanisms mediating pituitary adenylate cyclase-activating polypeptide depolarization of rat sympathetic neurons. J Neurosci. 2000;20:7353–61.
56. DiCicco-Bloom E, Deutsch PJ, Maltzman J, Zhang J, Pintar JE, Zheng J, et al. Autocrine expression and ontogenetic functions of the PACAP ligand/receptor system during sympathetic development. Dev Biol. 2000;219:197–213.
57. Lu N, Zhou R, DiCicco-Bloom E. Opposing mitogenic regulation by PACAP in sympathetic and cerebral cortical precursors correlates with differential expression of PACAP receptor (PAC1-R) isoforms. J Neurosci Res. 1998;53:651–62.
58. May V, Braas KM. Pituitary adenylate cyclase-activating polypeptide (PACAP) regulation of sympathetic neuron neuropeptide Y and catecholamine expression. J Neurochem. 1995;65:978–87.
59. May V, Beaudet MM, Parsons RL, Hardwick JC, Gauthier EA, Durda JP, et al. Mechanisms of pituitary adenylate cyclase activating polypeptide (PACAP)-induced depolarization of sympathetic superior cervical ganglion (SCG) neurons. Ann N Y Acad Sci. 1998;865:164–75.
60. May V, Braas KM. Pituitary adenylate cyclase activating polypeptide (PACAP) stimulation of the sympathetic neuron mitogen-activated protein kinase (MAPK) pathway. Soc Neurosci Abstr. 1998;24:116.
61. Brandenburg CA, May V, Braas KM. Expression, secretion and plasticity of endogenous pituitary adenylate cyclase activating polypeptides (PACAP) in rat superior cervical ganglion neurons. Soc Neurosci Abstr. 1995;21:1598.
62. Brandenburg CA, May V, Braas KM. Induction of novel pituitary adenylate cyclase activating polypeptide (PACAP) precursor messenger RNA in rat superior cervical ganglion neurons. Soc Neurosci Abstr. 1996;22:1764.
63. Brandenburg CA, May V, Braas KM. Identification of endogenous sympathetic neuron pituitary adenylate cyclase-activating polypeptide (PACAP): depolarization regulates production and secretion through induction of multiple propeptide transcripts. J Neurosci. 1997;17:4045–55.
64. Ghatei MA, Takahashi K, Suzuki Y, Gardiner J, Jones PM, Bloom SR. Distribution, molecular characterization of pituitary adenylate cyclase-activating polypeptide and its precursor encoding messenger RNA in human and rat tissues. J Endocrinol. 1993;136:159–66.
65. Holgert H, Holmberg K, Hannibal J, Fahrenkrug J, Brimijoin S, Hartman BK, et al. PACAP in the adrenal gland—relationship with choline acetyltransferase, enkephalin and chromaffin cells and effects of immunological sympathectomy. Neuroreport. 1996;8:297–301.
66. Klimaschewski L, Hauser C, Heym C. PACAP immunoreactivity in the rat superior cervical ganglion in comparison to VIP. Neuroreport. 1996;7(15–17):2797–801.
67. Koves K, Arimura A, Somogyvari-Vigh A, Vigh S, Miller J. Immunohistochemical demonstration of a novel hypothalamic peptide, pituitary adenylate cyclase-activating polypeptide, in the ovine hypothalamus. Endocrinology. 1990;127:264–71.
68. Koves K, Arimura A, Gorcs TG, Somogyvari-Vigh A. Comparative distribution of immunoreactive pituitary adenylate cyclase activating polypeptide and vasoactive intestinal polypeptide in rat forebrain. Neuroendocrinology. 1991;54:159–69.

69. Masuo Y, Suzuki N, Matsumoto H, Tokito F, Matsumoto Y, Tsuda M, et al. Regional distribution of pituitary adenylate cyclase activating polypeptide (PACAP) in the rat central nervous system as determined by sandwich-enzyme immunoassay. Brain Res. 1993;602:57–63.
70. Moller K, Zhang YZ, Hakanson R, Luts A, Sjolund B, Uddman R, et al. Pituitary adenylate cyclase activating peptide is a sensory neuropeptide: immunocytochemical and immunochemical evidence. Neuroscience. 1993;57:725–32.
71. Moller K, Reimer M, Hannibal J, Fahrenkrug J, Sundler F, Kanje M. Pituitary adenylate cyclase-activating peptide (PACAP) and PACAP type 1 receptor expression in regenerating adult mouse and rat superior cervical ganglia in vitro. Brain Res. 1997;775:156–65.
72. Nogi H, Hashimoto H, Hagihara N, Shimada S, Yamamoto K, Matsuda T, et al. Distribution of mRNAs for pituitary adenylate cyclase-activating polypeptide (PACAP), PACAP receptor, vasoactive intestinal polypeptide (VIP), and VIP receptors in the rat superior cervical ganglion. Neurosci Lett. 1997;227:37–40.
73. Portbury AL, McConalogue K, Furness JB, Young HM. Distribution of pituitary adenylyl cyclase activating peptide (PACAP) immunoreactivity in neurons of the guinea-pig digestive tract and their projections in the ileum and colon. Cell Tissue Res. 1995;279:385–92.
74. Shiotani Y, Kimura S, Ohshige Y, Yanaihara C, Yanaihara N. Immunohistochemical localization of pituitary adenylate cyclase-activating polypeptide (PACAP) in the adrenal medulla of the rat. Peptides. 1995;16:1045–50.
75. Sundler F, Ekblad E, Hannibal J, Moller K, Zhang YZ, Mulder H, et al. Pituitary adenylate cyclase-activating peptide in sensory and autonomic ganglia: localization and regulation. Ann N Y Acad Sci. 1996;805:410–26.
76. Tatsuno I, Somogyvari-Vigh A, Arimura A. Developmental changes of pituitary adenylate cyclase activating polypeptide (PACAP) and its receptor in the rat brain. Peptides. 1994;15:55–60.
77. May V, Beaudet MM, Parsons RL, Braas KM. PACAP modulates rat sympathetic neuron depolarization through IP3. Ann N Y Acad Sci. 2000;921:186–94.
78. Murase T, Kondo K, Otake K, Oiso Y. Pituitary adenylate cyclase-activating polypeptide stimulates arginine vasopressin release in conscious rats. Neuroendocrinology. 1993;57:1092–6.
79. Tatsuno I, Yada T, Vigh S, Hidaka H, Arimura A. Pituitary adenylate cyclase activating polypeptide and vasoactive intestinal peptide increase cytosolic free calcium concentration in cultured rat hippocampal neurons. Endocrinology. 1992;131:73–81.
80. Fahrenkrug J, Hannibal J. Pituitary adenylate cyclase activating polypeptide immunoreactivity in capsaicin-sensitive nerve fibres supplying the rat urinary tract. Neuroscience. 1998;83:1261–72.
81. Larsen JO, Hannibal J, Knudsen SM, Fahrenkrug J. Expression of pituitary adenylate cyclase-activating polypeptide (PACAP) in the mesencephalic trigeminal nucleus of the rat after transection of the masseteric nerve. Mol Brain Res. 1997;46:109–17.
82. Zhang Q, Shi TJ, Ji RR, Zhang YZ, Sundler F, Hannibal J, et al. Expression of pituitary adenylate cyclase-activating polypeptide in dorsal root ganglia following axotomy: time course and coexistence. Brain Res. 1995;705:149–58.
83. Zhang Y-Z, Hannibal J, Zhao Q, Moller K, Danielsen N, Fahrenkrug J, et al. Pituitary adenylate cyclase activating peptide expression in the rat dorsal root ganglia: up-regulation after peripheral nerve injury. Neuroscience. 1996;74:1099–110.
84. Negrete HO, Lavelle JP, Berg J, Lewis SA, Zeidel ML. Permeability properties of the intact mammalian bladder epithelium. Am J Physiol. 1996;271(4 Pt 2):F886–94.
85. Zeidel ML. Low permeabilities of apical membranes of barrier epithelia: what makes watertight membranes watertight? Am J Physiol. 1996;271(2 Pt 2):F243–5.
86. Lavelle JP, Meyers SA, Ruiz WG, Buffington CAT, Zeidel ML, Apodaca G. Urothelial pathophysiological changes in feline interstitial cystitis: a human model. Am J Physiol Renal Physiol. 2000;278:F540–53.
87. Apodaca G, Kiss S, Ruiz WG, Meyers S, Zeidel M, Birder L. Disruption of bladder epithelial function after spinal cord injury. Am J Physiol. 2003;284:F966–76.

88. Girard BM, Wolf-Johnston A, Braas KM, Birder LA, May V, Vizzard MA. PACAP-mediated ATP release from rat urothelium and regulation of PACAP/VIP and receptor mRNA in micturition pathways after cyclophosphamide (CYP)-induced cystitis. J Mol Neurosci. 2008;36:310–20.
89. Fahrenkrug J. VIP and autonomic neurotransmission. Pharmacol Therap. 1989;41:515–34.
90. Ishizuka O, Alm P, Larsson B, Mattiasson A, Andersson KE. Facilitatory effect of pituitary adenylate cyclase-activating polypeptide on micturition in normal, conscious rats. Neuroscience. 1995;66:1009–14.
91. Chapple CR, Milner P, Moss HE, Burnstock G. Loss of sensory neuropeptides in the obstructed human bladder. Br J Urol. 1992;70:373–81.
92. Smet PJ, Moore KH, Jonavicius J. Distribution and colocalization of calcitonin gene-related peptide, tachykinins, and vasoactive intestinal peptide in normal and idiopathic unstable human urinary bladder. Lab Invest. 1997;77:37–49.
93. Wanigasekara Y, Kepper ME, Keast JR. Immunohistochemical characterisation of pelvic autonomic ganglia in male mice. Cell Tissue Res. 2003;311:175–85.
94. Igawa Y, Persson K, Andersson KE, Uvelius B, Mattiasson A. Facilitatory effect of vasoactive intestinal polypeptide on spinal and peripheral micturition reflex pathways in conscious rats with and without detrusor instability. J Urol. 1993;149:884–9.
95. Erol K, Ulak G, Donmez T, Cingi MI, Alpan RS, Ozdemir M. Effects of vasoactive intestinal polypeptide on isolated rat urinary bladder smooth muscle. Urol Int. 1992;49:151–3.
96. Hernandez M, Barahona MV, Recio P, Benedito S, Martinez AC, Rivera L, et al. Neuronal and smooth muscle receptors involved in the PACAP- and VIP-induced relaxations of the pig urinary bladder neck. Br J Pharmacol. 2006;149:100–9.
97. Uckert S, Stief CG, Lietz B, Burmester M, Jonas U, Machtens SA. Possible role of bioactive peptides in the regulation of human detrusor smooth muscle—functional effects in vitro and immunohistochemical presence. World J Urol. 2002;20:244–9.
98. Braas KM, May V, Zvara P, Nausch B, Kliment J, Dunleavy JD, et al. Role for pituitary adenylate cyclase activating polypeptide in cystitis-induced plasticity of micturition reflexes. Am J Physiol Regul Integr Comp Physiol. 2006;290:R951–62.
99. Tompkins JD, Girard BM, Vizzard MA, Parsons RL. VIP and PACAP effects on mouse major pelvic ganglia neurons. J Mol Neurosci. 2010;42:390–6.
100. Girard BM, Galli JR, Young BA, Vizzard MA, Parsons RL. PACAP expression in explant cultured mouse major pelvic ganglia. J Mol Neurosci. 2010;42:370–7.
101. May V, Vizzard MA. Bladder dysfunction and altered somatic sensitivity in PACAP-/- mice. J Urol. 2010;183:772–9.
102. Studeny S, Cheppudira BP, Meyers S, Balestreire EM, Apodaca G, Birder LA, et al. Urinary bladder function and somatic sensitivity in vasoactive intestinal polypeptide (VIP)(-/-) mice. J Mol Neurosci. 2008;36:175–87.
103. Jensen DG, Studeny S, May V, Waschek J, Vizzard MA. Expression of phosphorylated cAMP response element binding protein (p-CREB) in bladder afferent pathways in VIP-/- mice with cyclophosphamide (CYP)-induced cystitis. J Mol Neurosci. 2008;36:299–309.
104. Girard BM, Malley SE, Braas KM, Waschek JA, May V, Vizzard MA. Exaggerated expression of inflammatory mediators in vasoactive intestinal polypeptide knockout (VIP-/-) mice with cyclophosphamide (CYP)-induced cystitis. J Mol Neurosci. 2008;36:188–99.
105. Herrera GM, Braas KM, May V, Vizzard MA. PACAP enhances mouse urinary bladder contractility and is upregulated in micturition reflex pathways after cystitis. Ann N Y Acad Sci. 2006;1070:330–6.
106. Mathews ME, Malley, S, Girard, BM, Braas, KM, Wascek, JA, May, V, Vizzard, MA. Pituitary adenylate cyclase-activating polypeptide (PACAP) expression in lower urinary tract pathways (LUT) with cyclophosphamide (CYP)-induced cystitis in PACAP promoter-dependent EGFP BAC transgenic mice. Soc Neurosci Abstr Planner. 2015. In press.
107. Tsuchikawa D, Nakamachi T, Tsuchida M, Wada Y, Hori M, Farkas J, et al. Neuroprotective effect of endogenous pituitary adenylate cyclase-activating polypeptide on spinal cord injury. J Mol Neurosci. 2012;48:508–17.

108. Zvarova K, Dunleavy JD, Vizzard MA. Changes in pituitary adenylate cyclase activating polypeptide expression in urinary bladder pathways after spinal cord injury. Exp Neurol. 2005;192:46–59.
109. Yoshiyama M, de Groat WC. Effects of intrathecal administration of pituitary adenylate cyclase activating polypeptide on lower urinary tract functions in rats with intact or transected spinal cords. Exp Neurol. 2008;211:449–55.
110. Zvara P, Braas KM, May V, Vizzard MA. A role for pituitary adenylate cyclase activating polypeptide (PACAP) in detrusor hyperreflexia after spinal cord injury (SCI). Ann N Y Acad Sci. 2006;1070:622–8.
111. Dray A. Inflammatory mediators of pain. Br J Anaesth. 1995;75:125–31.
112. Lewin GR, Mendell L. Nerve growth factor and nociception. Trends Neurosci. 1993;16:353–9.
113. Lindholm D, Heumann R, Meyer M, Thoenen H. Interleukin-1 regulates synthesis of nerve growth factor in non-neuronal cells of the rat sciatic nerve. Nature. 1987;330:658–9.
114. Meller ST, Cummings CP, Traub RJ, Gebbhart GF. The role of nitric oxide in the development and maintenance of the hyperalgesia produced by intraplantar injection of carrageenan in the rat. Neuroscience. 1994;60:367–74.
115. Woolf CJ, Allchorne A, Safieh-Garabedian B, Poole S. Cytokines, nerve growth factor and inflammatory hyperalgesia: the contribution of tumour necrosis factor. Br J Pharmacol. 1997;121:417–24.
116. Ruda MA, Iadarola MJ, Cohen LV, Yound II WS. In situ hybridization, histochemistry and immunocytochemistry reveal an increase in spinal dynorphin biosynthesis in a rat model of peripheral inflammation and hyperalgesia. Proc Natl Acad Sci U S A. 1988;85:622–6.
117. Donnerer J, Schuligoi R, Stein C. Increased content and transport of substance P and calcitonin gene-related peptide in sensory nerves innervating inflamed tissue: evidence for a regulatory function of nerve growth factor in vivo. Neuroscience. 1992;49:693–8.
118. Gary MB, Hargreaves KM. Enhanced release of immunoreactive CGRP and substance P from spinal dorsal horn slices occurs during carrageenan inflammation. Brain Res. 1992;582:139–42.
119. Vizzard MA, Erdman SL, de Groat WC. Increased expression of neuronal nitric oxide synthase (NOS) in visceral neurons after nerve injury. J Neurosci. 1995;15:4033–45.
120. Vizzard MA. Increased expression of neuronal nitric oxide synthase in bladder afferent and spinal neurons following spinal cord injury. Dev Neurosci. 1997;19:232–46.
121. Jongsma H, Danielsen N, Sundler F, Kanje M. Alteration of PACAP distribution and PACAP receptor binding in the rat sensory nervous system following sciatic nerve transection. Brain Res. 2000;853:186–96.
122. Johnson Jr EM, Taniuchi M, Clark HB, Springer JE, Koh S, Tayrien MW, et al. Demonstration of the retrograde transport of nerve growth factor receptor in the peripheral and central nervous system. J Neurosci. 1987;7:923–9.
123. McMahon SB, Armanini MP, Ling LH, Phillips HS. Expression and coexpression of Trk receptors in subpopulations of adult primary sensory neurons projecting to identified peripheral targets. Neuron. 1994;12:1161–71.
124. Wright DE, Snyder WD. Neurotrophin receptor mRNA expression defines distinct populations of neurons in rat dorsal root ganglia. J Comp Neurol. 1995;351:329–38.
125. Qiao LY, Vizzard MA. Cystitis-induced upregulation of tyrosine kinase (TrkA, TrkB) receptor expression and phosphorylation in rat micturition pathways. J Comp Neurol. 2002;454:200–11.
126. Qiao LY, Vizzard MA. Up-regulation of tyrosine kinase (Trka, Trkb) receptor expression and phosphorylation in lumbosacral dorsal root ganglia after chronic spinal cord (T8-T10) injury. J Comp Neurol. 2002;449:217–30.
127. Qiao LY, Vizzard MA. Spinal cord injury-induced expression of TrkA, TrkB, phosphorylated CREB, and c-Jun in rat lumbosacral dorsal root ganglia. J Comp Neurol. 2005;482:142–54.
128. Jongsma Wallin H, Danielsen N, Johnston JM, Gratto KA, Karchewski LA, Verge VMK. Exogenous NT-3 and NGF differentially modulate PACAP expression in adult sensory neurons, suggesting distinct roles in injury and inflammation. Eur J Neurosci. 2001;14:267–82.

129. Grumolato L, Louiset E, Alexandre D, AitAli D, Turquier V, Fournier A, et al. PACAP and NGF regulate common and distinct traits of the sympathoadrenal lineage: effects on electrical properties, gene markers and transcription factors in differentiating PC12 cells. Eur J Neurosci. 2003;17:71–82.
130. Hashimoto H, Hagihara N, Koga K, Yamamoto K, Shintani N, Tomimoto S, et al. Synergistic induction of pituitary adenylate cyclase-activating polypeptide (PACAP) gene expression by nerve growth factor and PACAP in PC12 cells. J Neurochem. 2000;74:501–7.
131. Yamamoto K, Hashimoto H, Hagihara N, Nishino A, Fujita T, Matsuda T, et al. Cloning and characterization of the mouse pituitary adenylate cyclase-activating polypeptide (PACAP) gene. Gene. 1998;211:63–9.
132. Jamen F, Bouschet T, Laden JC, Bockaert J, Brabet P. Up-regulation of the PACAP type-1 receptor (PAC1) promoter by neurotrophins in rat PC12 cells and mouse cerebellar granule cells via the Ras/mitogen-activated protein kinase cascade. J Neurochem. 2002;82:1199–207.
133. Lee FS, Rajagopal R, Kim AH, Chang PC, Chao MV. Activation of Trk neurotrophin receptor signaling by pituitary adenylate cyclase-activating polypeptides. J Biol Chem. 2002;277: 9096–102.
134. DiCiccoBloom E, Deutsch PJ, Maltzman J, Zhang JW, Pintar JE, Zheng J, et al. Autocrine expression and ontogenetic functions of the PACAP ligand/receptor system during sympathetic development. Develop Biol. 2000;219:197–213.
135. Schnegelsberg B, Sun TT, Cain G, Bhattacharya A, Nunn PA, Ford AP, et al. Overexpression of NGF in mouse urothelium leads to neuronal hyperinnervation, pelvic sensitivity, and changes in urinary bladder function. Am J Physiol Reg Integr Comp Physiol. 2010;298:R534–47.
136. Allen SJ, Dawbarn D. Clinical relevance of the neurotrophins and their receptors. Clin Sci (Lond). 2006;110:175–91.
137. Pezet S, McMahon SB. Neurotrophins: mediators and modulators of pain. Ann Rev Neurosci. 2006;29:507–38.
138. Szallasi A, Cortright DN, Blum CA, Eid SR. The vanilloid receptor TRPV1: 10 years from channel cloning to antagonist proof-of-concept. Nat Rev Drug Discov. 2007;6:357–72.
139. Ford AP, Gever JR, Nunn PA, Zhong Y, Cefalu JS, Dillon MP, et al. Purinoceptors as therapeutic targets for lower urinary tract dysfunction. Br J Pharmacol. 2006;147 Suppl 2:S132–43.
140. Girard BM, Malley SE, Braas KM, May V, Vizzard MA. PACAP/VIP and receptor characterization in micturition pathways in mice with overexpression of NGF in urothelium. J Mol Neurosci. 2010;42:378–89.
141. Girard BM, Malley, S, Mathews, ME, Vizzard, MA. Contributions of pituitary adenylate cyclase-activating polypeptide (PACAP)/receptor signaling to increased voiding frequency and somatic sensitivity in mice with urothelium-specific overexpression (OE) of nerve growth factor (NGF) in the urinary bladder. Soc Neurosci Abstr Planner. 2015. In press.
142. Girard BM, Malley SE, Vizzard MA. Neurotrophin/receptor expression in urinary bladder of mice with overexpression of NGF in urothelium. Am J Physiol Renal Physiol. 2011;300:F345–55.

Part VI
Functions of PACAP in the Skeletal System

Chapter 20
Role of PACAP and VIP Signalling in Regulation of Chondrogenesis and Osteogenesis

Tamas Juhasz, Andrea Tamas, Roza Zakany, and Dora Reglodi

Abstract Pituitary adenylate cyclase activating polypeptide (PACAP) and vasoactive intestinal peptide (VIP) are multifunctional proteins that can regulate diverse physiological processes. These are also regarded as neurotrophic and anti-inflammatory substances in the CNS, and PACAP is reported to prevent harmful effects of oxidative stress. In the last decade more and more data accumulated on the similar function of PACAP in various tissues, but its cartilage- and bone-related presence and functions have not been widely investigated yet. In this summary we plan to verify the presence and function of PACAP and VIP signalling tool kit during cartilage differentiation and bone formation. We give evidence about the protective function of PACAP in cartilage regeneration with oxidative or mechanically stress and also with the modulation of PACAP signalling in vitro in osteogenic cells. Our observations imply the therapeutic perspective that PACAP might be applicable as a natural agent exerting protecting effect during joint inflammation and/or may promote cartilage regeneration during degenerative diseases of articular cartilage.

Keywords PKA • Mechanical stress • Hedgehog • BMP • Runx2 • Oxidative stress

Abbreviations

ALP	Alkaline phosphatase
BMP	Bone morphogenetic protein
cAMP	Cyclic adenosine monophosphate

T. Juhasz (✉) • R. Zakany
Department of Anatomy, Histology and Embryology, Faculty of Medicine,
University of Debrecen, Nagyerdei krt. 98, Debrecen 4032, Hungary
e-mail: juhaszt@anat.med.unideb.hu

A. Tamas • D. Reglodi
Department of Anatomy, MTA-PTE "Lendület" PACAP Research Team, Medical School,
University of Pecs, Szigeti út 12, Pecs 7624, Hungary

D. Reglodi, A. Tamas (eds.), *Pituitary Adenylate Cyclase Activating Polypeptide — PACAP*, Current Topics in Neurotoxicity 11,
DOI 10.1007/978-3-319-35135-3_20

CREB	cAMP response element-binding protein
ECM	Extracellular matrix
HH	Hedgehog
IHH	Indian Hedgehog
MAPK	Mitogen-activated protein kinase
NFAT	Nuclear factor of activated T cells
PAC1	Pituitary adenylate cyclase-activating polypeptide type I receptor
PACAP	Pituitary adenylate cyclase polypeptide
PKA	Protein kinase A
PKC	Protein kinase C
PP2A	Protein phosphatase 2A
PP2B	Protein phosphatase 2B
PTHrP	Parathyroid hormone related peptide
Runx2	Runt-related transcription factor 2
SHH	Sonic Hedgehog
TGFβ	Transforming growth factor-β
VIP	Vasoactive intestinal polypeptide
VPAC	Vasoactive intestinal peptide receptor

Various signalling pathways influence the proper limb development and activating signals for many of them may arrive from the outer environment, partly via transmission by the surrounding extracellular matrix (ECM). Regulatory molecules or physical stimuli are able to induce the activation of specific receptors which can determine the cell fate during the differentiation cycle. Chemical signals reaching the tissue or directly the cells may originate from blood or interstitial fluid; in the latter case autocrine and paracrine signalling are the most frequent in developing organisms [1]. The best known mechanotransducer of developing cells is the primary cilium [2]. Changes of the intracellular ion concentration as the consequence of the activation of various ion channels can alter the resting membrane potential of cells and may influence proliferation and differentiation [3]. All of the processes mentioned above have impact on the cellular differentiation program of skeletal elements. Hyaline cartilage especially articular cartilage is an avascular and aneural tissue with a uniquely organized extracellular matrix. The sequential receptor activation by paracrine or autocrine ways plays crucial role in the differentiation of cartilage tissue, although many details of these mechanisms are still not well explored. The significance of the better understanding of cartilage formation is underscored by the fact, that production of proper long bone architecture requires a cartilage template and involves time and growth factor dependent activation of precisely defined regulating mechanisms and signalling cascade systems [4].

PACAP and VIP

VIP and PACAP are small hormone like peptides which belong to the VIP-secretin-growth hormone releasing hormone (GHRH)-glucagon superfamily. PACAP was first extracted from ovine hypothalamus, and was thought to be an important neurohormone regulating various processes in the CNS [5]. It is produced by a variety of cells and tissues in addition to neuronal cells. Among others, specific cells of the intestinal system can produce VIP along with some immune and endocrine cells. VIP has important functions in neuronal development and both in innate and acquired immunity [6].

PACAP has two bioactive forms: a shorter, 27 amino acid (PACAP 27) and a longer 38 amino acid (PACAP 38) variant [7]. The N-terminal region of the polypeptide is evolutionary conserved and shows a high homology with that of VIP [5]. In the last decade, increasing amount of evidence has emerged regarding the important roles of PACAP in peripheral organs such as uterus [8], ovary [9], and teeth [10]; moreover, its presence has been proven in human milk [11]. Nonetheless, only sporadic data exist about its function in skeletal elements [12–15].

Three main G protein coupled receptors of these peptides have been identified such as PAC1, VPAC1, and VPAC2 from which PACAP binding to PAC1 has the highest affinity, while the latter two attract PACAP and VIP with equal affinity [16]. Alteration in the conformation of these receptors results in the elevation of intracellular cAMP level leading to the activation of protein kinase A (PKA) [5]. The so-called "canonical" signalling activation may lead to the nuclear translocation of CREB or Sox9 transcription factors and consequent activation of the expression of various genes. PAC1 receptor activation is also able to trigger the phosphorylation of key elements in MAPK pathways, such as ERK and p38 kinases [5], subsequently regulating cellular division or apoptotic program induction [17]. The versatility of PACAP/VIP receptor induced signal transduction mechanisms indicates its multifactorial regulation, implying a vast array of signalling connections. Activation of IP_3 receptors inducing the release of Ca^{2+} from endoplasmic reticulum (ER) can be involved as the result of PACAP binding [18]. The elevation of intracellular Ca^{2+} concentration activates various Ca^{2+} dependent signalling molecules such as classical PKCs, MAPK [19] or protein phosphatases like PP2B [20]. G-protein coupled receptors (GPCR) have been proven to have communication with other signalling cascades, implying that PACAP receptor activation may cross talk with WNT β-catenin [21], TGFβ [22], BMP [23], Hedgehog [24], and Notch signal transduction [25]. Multifactorial or pleiotropic effects of PACAP have been investigated in several biological processes and it has been proven to prevent apoptosis, ischemic conditions, inflammation and oxidative stress [21, 26–29].

PACAP and VIP Regulates Chondrogenic Differentiation

The complete network of signalling cascades governing chondrogenic differentiation is not fully discovered. As articular cartilage has very poor regeneration capacity and no effective or curative treatment is available currently for degenerative cartilage diseases such as osteoarthritis the identification of new pharmacological targets for reconstruction of cartilage is very important. Hormone like peptides are not in the focus of research as blood vessels are not found in the articular cartilage and hormones can only be delivered by diffusion from the synovial fluid into the joint cartilage. Low level of nutrition and poor oxygenation also are consequences of the avascular nature of this tissue. Indeed, it can be a question of interest if the differentiating chondroprogenitor cells or chondrocytes are able to release small regulatory peptides and can influence their own differentiation via autocrine or paracrine ways.

Cartilage differentiation in vivo starts with rapid proliferation and aggregation of mesenchymal-like chondroprogenitor cells resulting in the formation of precartilage nodules, in which the final commitment of chondroprogenitor cells can be fulfilled. Chondrogenic nodules and cartilage specific extracellular matrix production are both required for proper hyaline cartilage structure formation [30]. The ECM of articular cartilage, composed mostly of high molecular mass proteoglycans (PG) such as aggrecan, glucosaminoglycans (GAG) like hyaluronan form a highly organized network attracting high amount of water. The extremely hydrated nature of the cartilage is essential to proper mechanical functions and offers substantial mechanical stability. Moreover, the negatively charged PGs and GAGs also function as an extracellular ion pool which can influence the normal lifecycle of chondrogenic cells [30]. Several transcription factors and receptors have been proven to play role in the matrix synthesis. One of these transcription factor families is the SoxE from which Sox5, Sox6, and Sox9 are essential for the induction of mRNA expression of cartilage matrix-specific proteins (e.g., COL2A1, aggrecan core protein). Sox9 is one of the pivotal signalling elements of chondrogenesis and its regulation by reversible phosphorylation is a key momentum of the proper differentiation cycle [31]. Also CREB transcription factor regulates cartilage-specific matrix production directly and it plays an important role in lubricin secretion of articular cartilage [32]. Moreover, Sox9 promoter is known to be regulated by the CREB that binds to a CRE site upstream of Sox9 [33]. Our laboratory has demonstrated that Sox9 and CREB transcription factors are phosphorylated by PKA during cartilage formation [34, 35]. Nuclear translocation of the phosphorylated forms of these transcription factors enhances the matrix production of chondrocytes. Moreover, a complex regulatory mechanism and synergism between Sox9 function and the cAMP-PKA-CREB pathway including connections to the BMP signalling was published in both mature and differentiating chondrocytes [34, 36]. We have shown that the activation of signalling elements phosphorylated by PKA can be equilibrated by Ser/Thr protein phosphatases such as PP2A and PP2B [37–39]. Our group also demonstrated that PP2A is a negative regulator of chondrogenesis and was involved in the mechanotransduction [34, 38]. On the other hand, we showed that PP2B was a positive regulator of chondrogenesis and influenced the reversible phosphorylation of ERK1/2 and NFAT4 [37, 39]. In one of our previous works, we presented data that the changes in cytosolic free Ca^{2+} concentration showed a characteristic developmental stage dependent pattern in chondrogenic cells [40]. Additionally, differentiating stage dependent Ca^{2+} oscillations were identified in chondrogenic cells [41]. A wide range of receptors with substantial Ca^{2+} permeability have been described in the last decade as part of the mapping of chondrocyte specific channelome [3, 42].

Since the regulation of these signalling pathways which can modulate cartilage formation and chondrocyte biology is cAMP or Ca^{2+} dependent it implicates the importance of PACAP/VIP neuropeptides in proper hyaline cartilage formation. A potential role of VIP has been demonstrated in bone remodelling and it is known to have important functions in inflammatory diseases [43]. Moreover, this neuropeptide is able to regulate Ca^{2+} release of neonatal osteoblasts. VIP exerted a positive effect in rheumatoid arthritis indicating that its application in the therapy of this disorder can be promising [44]. Although the articular cartilage is aneural, the surrounding synovial membrane is rich in nerve endings, which release VIP into the synovial fluid and subsequently induce anti-inflammatory processes [45]. About the functions of PACAP in the adult joints we still have exiguous knowledge despite the fact that PACAP-positive nerve endings have been described in cartilage canals of porcine epiphyseal cartilage more than 15 years ago [46]. We have demonstrated that the mRNAs of preproPACAP as well as PAC1, VPAC1 and VPAC2 receptors are expressed in chicken “high density” chondrogenic cell cultures. Additionally, we have shown the expression of the PAC1 receptor protein in chondroprogenitor cells and an increased extracellular matrix synthesis was detected during PACAP administration suggesting the supportive effect of this neuropeptide in cartilage development [13]. Our findings implied the presence of PACAP-related autocrine and/or paracrine effects in cartilage itself, reflecting on a possible new signalling mechanism in the regeneration of hyaline cartilage [47, 48]. In the light of these data Giunta and coworkers identified PACAP positive chondrocytes in the superficial zone of articular cartilage and in synovial fluid [12]. Moreover, the reduction of the neurohormone level was demonstrated during osteoarthritis in the articulating cartilage [12]. Although the expression of the VIP receptors has been detected in chondrogenic cells, there are some data pointing out that VIP probably does not influence matrix production of chondrocytes and synovial cells [49]. As classical downstream targets of PAC1 receptor, Gs and/or Gq proteins become activated triggering the elevation of intracellular cAMP concentration or the intracellular release of Ca^{2+} ions [5]. These events initiate the activation of certain protein kinases such as PKA, PKC, and MAPK which all play essential, but probably distinct roles in chondrogenesis [35, 37, 39]. During proper chondrogenesis PKA is proved to phosphorylate CREB and Sox9 transcription factors [31, 35] by which induce the secretion of cartilage specific ECM components (Fig. 20.1). PACAP administration into the medium of chondrogenic cell cultures increased the phosphorylation both of Sox9 and CREB, and enhanced matrix production of the differentiating cells by effecting on both the expression of aggrecan, collagen type II (Fig. 20.1) and the expression

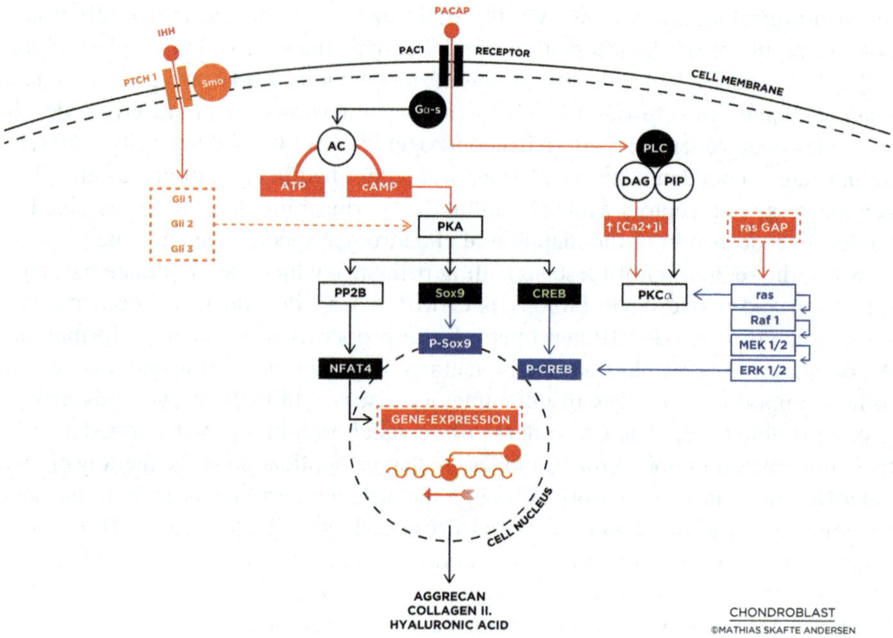

Fig. 20.1 Effects of PACAP on chondrogenic signalling mechanisms. The increased concentration of cAMP level elevates PKA activity. PKA phosphorylates CREB and Sox9 which translocate into the nucleus of chondrogenic cells and induce the gene expression of collagen type II, aggrecan, and various GAGs such as hyaluronic acid. Activation of PAC1 receptor can also elevate the intracellular Ca^{2+} concentration leading to increased PP2B, PKC, or MAPK signalling activity. The elevated activity of PP2B regulates NFAT4 being responsible for the augmented matrix production. PACAP has a connection with the IHH signalling pathway by decreasing the activity of Gli transcription factors

of matrix synthesizing enzymes such as Chst11, Extl1, or HAS enzymes [13]. PAC1 receptor activation also can be responsible for the elevation of intracellular Ca^{2+} concentration which in turn can activate the Ca^{2+} dependent phosphatase PP2B (also known as calcineurin). Therefore, we investigated the involvement of this Ser/Thr phosphatase in PACAP signalling pathways and connection between PP2B activity and PACAP signalling was proven [13] (Fig. 20.1). Similar observations were found in chromaffin cells [50]. These in vitro results indicate that the presence of PACAP is essential for proper cartilage formation although the phenotype of PACAP KO mice did not show any dramatic macroscopic morphological alteration of the skeleton [51]. Although the analysis of the genetically modified animals has not been completed yet, our initial observations suggested alterations in the composition of the cartilage extracellular matrix and in the expression of various signalling molecules in the knee joints of PACAP KO mice (our unpublished data). In the reproductory system of these mice, the lack of PACAP gene resulted in reduced fertility and altered mating behavior of females [52], moreover the maturation [53] and the morphology [54] of gonadal cells showed notable differences. Further evidences support the idea of the regulation effect of PACAP on differentiation of various cells, as alterations have been found in tooth formation of PACAP KO mice [10]. The complex phenotypic changes raise the possibility of multiple cross talk of PACAP signalling with developmental pathways connected to various morphogenes, as well as certain compensatory mechanisms of PACAP signalling cascades. For instance MAPK and Wnt signalling both play important roles in the proper cartilage formation and tissue patterning [55] and a PACAP-independent PAC1 receptor activation has been directly linked to the regulation of Wnt/β-catenin pathways [21]. Notch signalling activation plays a crucial role in chondrogenesis [56] and exerts modulatory function in osteoarthritis [57]. Recently, a cross talk of G protein coupled receptors and Notch signalling has been reported in bacterial LPS induced macrophages [58]. SHH pathway is another essential positive chondroregulatory pathway [59] and it can be inhibited by PACAP activation [60].

Besides the direct influence of PACAP on chondrogenesis, we have demonstrated a chondro-protective effect of this neuropeptide in chondrogenic cell cultures where its administration compensated the harmful effects of oxidative stress. Similar phenomenon has been observed in ischemia or oxidative stress induced apoptosis in the central nervous system [61] or in diabetic kidney [62]. Moreover, PACAP deficient mice showed higher sensitivity to injury during retinal ischemic conditions, axonal lesion, intestinal inflammation or oxidative stress of the kidneys [63]. The activation of PACAP/VIP system had positive effect in rheumatoid arthritis [64, 65] and decreased expression of PACAP was identified in osteoarthritic knee joint [12]. PACAP is known to exert cytoprotective effects in several other peripheral organs and in the CNS, for example cardioprotective effects of these peptides have been demonstrated [66] and a positive effect in Parkinson's disease also have been shown [67]. On the basis of these observations, a protective and/or ameliorating effect of PACAP was likely in diseases of the articulating cartilage or skeletal stress situations. Indeed, the addition of PACAP1-38 during oxidative stress prevented the inhibition of cartilage matrix production by normalizing the phosphorylation of Sox9 and CREB in chicken chondrogenic cells [13]. Recently, the involvement of PACAP or VIP signalling activation in mechanotransduction of developing articular cartilage has been proved by our group [68] and the importance of PKA in mechanical cellular response was proposed [34]. Mechanical load of in vitro chondrogenic cell cultures resulted in an increased PAC1 receptor and PACAP expression, and supported the undifferentiated stage of chondroblasts. Mechanical load of in vitro chondrogenic cell cultures resulted in increased PAC1 receptor and PACAP expression and the activation of IHH (Fig. 20.1) inducing the elevated expression of collagen type X which was normalized by PACAP 1-38 addition [68]. In these experiments, we found that PACAP administration was able to reduce the expression of matrix metalloproteinases (MMP) during oxidative stress in chondrogenic cell cultures (our unpublished data). Similar results have been published in alveolar cells where both VIP and PACAP were able to decrease the expression of certain MMPs and reduced the activation and expression of caspase3 [69]. It is also important to note that VIP and its receptors in synovial fibroblasts [70] are able to regulate inflammatory factors release [71]. These data all strongly suggest that PACAP is a promising future therapeutic agent in inflammatory and degenerative joint diseases [72].

Bone Formation Under the Control of VIP and PACAP Signalling Cascades

Development of long bones is determined and organized by a cartilage template. Invasion of the calcified cartilage by the osteoprogenitor cells and their precisely regulated differentiation are required for the proper osteogenesis [4]. Development of bone tissue is supported by complex bone specific developing mechanisms and signalling. Differentiation of osteoblasts from osteoprogenitors is followed by an initial deposition of a bone specific organic ECM abundant in collagen type I completed with another bone specific matrix components such as osteocalcin or osteonectin [4]. Calcification of bone matrix also requires osteoblast activity, calcium hydroxyapatite crystals accumulate and deposit into the collagen rich bone matrix. Various signalling cascades are considered as bone specific regulatory mechanisms, such as BMP, WNT, Notch, and Hedgehog signalling pathways [73–77]. Thyroid hormones are crucial components of bone remodelling and differentiation and neuronal connections are also found to be involved in bone production [74]. Consequently, the activation of signalling mechanisms mentioned above by hormones and neuropeptides may derive from different sources in the surrounding tissues. The presence and precise timing of the expression of growth factors such as BMPs are essential for proper bone production. These factors bind to their own receptors; to BMPRs. Subsequently, their activation induces the phosphorylation of Smad1/5 and with the help of Smad4 the complex is translocated into the nuclei of osteogenic cells and initiates expression of bone specific genes (Fig. 20.2.) such as the transcription factor osterix, alkaline phosphatase (ALP), or collagen type I [78, 79]. Sequential release of BMPs is regulated by complex networks of signalling cascades including CREB, one of the transcription factors activated via PKA signalling pathways [79] and activate the gene expression of the proteins mentioned previously (Fig. 20.2). On the other hand a well-balanced expression of hedgehog signalling elements governed by another bone specific transcription factor, Runx2 is also essential for proper long bone formation [80] (Fig. 20.2). Runx2 can be directly phosphorylated by PKA [81] and subsequently activates the expression of bone specific signalling elements or ECM components (Fig. 20.2). FGFs are also essential for proper bone formation. FGF2 decreases the ALP activity, Runx2 activation, and collagen secretion and elevates the proliferation of osteoblasts [82]. This complex signalization involves broad spectrum cross talk opportunities with the PACAP/VIP signalization, further highlighting the significance of neuropeptide signalling in bone formation and regeneration.

At the beginning of enchondral ossification, blood vessels invade into the bone template, in a process regulated by VEGFs. The expression of these growth factors is also under the control of PKA–Runx2 axis [83]. Formation and invasion of the blood vessels into the calcified cartilage matrix result in the appearance of osteoprogenitor cells and differentiation of osteoblasts producing the bone matrix. This process can also be regulated by neuropeptides [84] or invading neuronal elements. During the elongation of long bones PACAP positive nerve fibers have been shown

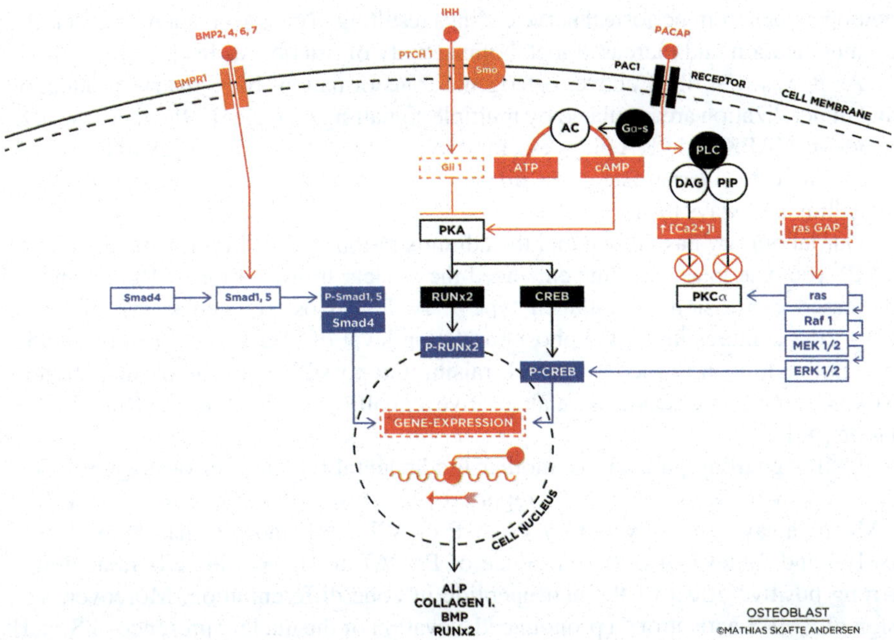

Fig. 20.2 Diverse connections of PACAP signalling pathways in osteogenic cells. PACAP binding to its receptors elevates the intracellular cAMP concentration and activates PKA in osteoblast cells. CREB is partly phosphorylated by PKA while the nuclear translocation of Runx2 is regulated by the kinase. The presence of PACAP does not induce intracellular Ca^{2+} release, subsequently the Ca^{2+} dependent signalling pathways are not activated (*arrows* crossed by *red lines*). The neuropeptide trigger the expression of BMPs simultaneously activated Smad transcription factors and cross talk with Runx2. SHH or IHH binding to PTCH1 receptor can induce the nuclear translocation of Gli1 transcription factor. Activation of Gli1 is inhibited by the continuous activation of PKA

penetrating the bone matrix [46], and VIP positive sympathetic nerve endings have also been identified releasing these neuropeptides [85]. As an interesting observation, receptor composition and effects of VIP exhibited differences in cells of bones developed in different ways, i.e., via membranous or endochondral ossification. Moreover, the direct communication of sympathetic nerve fibers with osteoblasts showed an embryonic origin dependent response and signalization, suggesting that the innervation of periosteum by peptidergic fibers plays important function both in bone regeneration and formation [86].

It has been proven that in calvaria derived cell lines, in which the receptors and proteins of PACAP and/or VIP signalling are present, PKA activation can be induced by PACAP or VIP addition [14, 87]. Accumulation of cAMP in osteoblasts is proved to be a result of combined activation of PACAP/VIP and regulates diverse signalling pathways influencing osteoblast differentiation. In line with this, presence of certain neuropeptides was shown to be elevated after bone fracture, indicating their importance in successful regeneration [88]. A recent report demonstrated release of various neuropeptides from periosteal nerve endings resulting in enhancement of intercellular communication and increased metabolic activity of osteoblasts [89].

As it was described above, osteogenic transformation, bone matrix production and mineralization are regulated by multiple signalling cascades [90], where the activation of MAPK and PKA plays essential roles. Runx2 is one of the key transcription factors and it governs osteoblast differentiation [91] and it is regulated by PKA signalling pathways [92].

Our laboratory has proven that the administration of PACAP into the medium of UMR-106 osteoblast cell line enhanced the nuclear translocation of Runx2 and an increased expression of collagen type I, ALP and osterix genes was observed (Fig. 20.2). Interestingly, the phosphorylation level of CREB transcription factors was not significantly increased [14], raising the possibility of the participation of MAPK signalling cascade as an alternative signalling pathway for the Runx2 regulation [93].

BMP signalling pathway is another fundamental regulator of osteogenesis and cross talk with Runx2 has been reported [94] (Fig. 20.2). Moreover, the TGFβ/BMP pathways are activated by PACAP or VIP [23]. Indeed, the expression of BMP6 and 7 elevated in the presence of PACAP in UMR-106 cells indicating a strong positive effect of the neuropeptide in bone differentiation. Moreover, as a sign of BMPR activation, a pronounced elevation of the nuclear presence of Smad1 transcription factor was detected under the effect of PACAP administration [14]. TGFβ/BMP signalling pathways are involved in the regulation and activation of VIP signalling cascades. Exogenous VIP may also result in the activation of Smads [95] suggesting a complex signalling with numerous alternative routes during bone development [14].

The involvement of MAPK system has been reported in bone development and in fracture healing. The administration of PACAP and VIP is proven to upregulate the activation of ERK 1/2 in osteoblasts [96, 97] or adipogenic models [98]. The two neurohormones have negative effect on JNK and p38 phosphorylation in monocytes in vivo and in vitro [99]. Additionally, intracellular Ca^{2+} concentration can be elevated by PACAP [100] or VIP [101], resulting in an activation of classical PKCs and ERK both influencing osteoblast differentiation [102]. Nonetheless, in UMR-106 cell line we were not able to detect significant Ca^{2+} concentration change and no significant alteration was detected in activation of classical PKCs such as PKCα [14] (Fig. 20.2). Ca^{2+} influx can be evoked by PACAP [101] PACAP and VIP are able to decrease the Ca^{2+} entry via L- and N-type calcium channels in neurons [102]. It is known that the administration of PACAP affects Ca^{2+} oscillation [103] and alters the Ca^{2+} related vesicular transport of chromaffin cells [104]. Besides this dynamic alteration of intracellular Ca-homeostasis, PACAP also exerts effects on matrix mineralization. We have reported that the inorganic matrix components of UMR-106 cell line can be elevated by PACAP addition [14]. Moreover, an altered mineralization was detected during tooth formation of PACAP deficient mice [10], suggesting a yet unknown connection between PACAP and mineralized tissue formation. As a possible mechanism for PACAP induced extracellular Ca^{2+} accumulation during osteogenesis, calcitonin gene-related peptide was proven to influence osteoclast function [105] and the presence of PACAP decreased the matrix-resorption and consequent Ca-release by these cells [90, 106].

Hedgehog signalling is of key importance amongst the regulatory mechanisms of bone and cartilage development [80]. A well-defined balance between Indian hedgehog (IHH) and parathyroid hormone related peptide (PTHrP) is essential for proper long bone formation, regulation of proliferation, and matrix production of osteoblasts via the activation of Runx2 transcription factor [107]. PTHrP directly communicates with PKA signalling inducing the activation of CREB and NFAT factors in osteoblasts [108]. In UMR-106 cells the application of PACAP elevated the expression of PTHrP without altering the IHH expression [14]. Sonic hedgehog (SHH) pathway is known to be regulated by PACAP signalling [60] and the activation of PKA downregulates the function of Gli1, which consequently decreases the proliferation [24] (Fig. 20.2). In PACAP KO mice, enhanced SHH signalling was detected during tooth development [10]. On the contrary, exogenous administration of PACAP elevated the expression of SHH and a more pronounced nuclear presence of Gli1 was found in rat UMR-106 cells [14]. This contradiction may stem from the osteosarcoma origin of UMR cells, as malignant cells can exhibit alterations of various signalling mechanisms. Another hypothesis is that the cellular presence of the repressor form of Gli2 or Gli3 transcription factors are also increased in UMR-106 cells which upregulate the activation of SHH signalling pathways in this tumor cell line. Nevertheless, we have only sparse data about the function of VIP in bone development, but it is a suppressor of bone resorption and has very similar effect as mechanical load in MC3T3 osteoblastic cell line [43].

Conclusion

Articular cartilage is optimally able to dissipate the mechanical stress, which loads the articular surface and transmits it to the subchondral bone. Chondrocytes represent the only type of cells in cartilage with a very limited capacity to reproduce themselves even in a healthy adult human. Therefore, they cannot regenerate the cartilage tissue in case of traumatic injury or loss of joint surface, which can be the ultimate consequence of any type of inflammatory joint diseases. Protection of this tissue during joint inflammation, stimulation of its poor regeneration capacity or production and implantation of artificial cartilage all are major challenges of the modern reparative medicine. PACAP was originally described as a hormone-like product of neurons, able to reduce the harmful consequences of various brain injuries, as well as playing regulatory roles during brain development. On the basis of the data presented and summarized above, PACAP and VIP are important modulators of the physiological differentiating processes of articular cartilage and may influence endochondral ossification. As PACAP and VIP neuropeptides are naturally synthetized signal molecules of humans, they can be good candidates for application via intraarticular injection as chondroprotective agents or they can be interesting candidates for the regulation of callus formation.

Acknowledgements The authors are grateful to Mrs. Krisztina Bíró for excellent technical assistance and to Ms. Andrea Pálfi, Ms. Solveig Lind Helgadottir, Ms. Katarina Karadottir, and Mr. Vince Szegecki medical students for their skillful help during the studies and for Mathias Skafte Andersen for the graphical figures. This work was supported by grants from Akira Arimura Foundation Research Grant, the Hungarian Science Research Fund (OTKA 119759 and OTKA K 104984), PTE AOK KA Research Grant, Bolyai Scholarship, and the Hungarian Academy of Sciences (MTA "Lendület" Program) and from the New Széchenyi Plan (TÁMOP-4.2.2.A-11/1/KONV-2012-0053; The project is co-financed by the European Union and the European Social Fund). This research and T.J. was supported by Szodoray Lajos and Magyary Zoltán Funds by the European Union and the State of Hungary, co-financed by the European Social Fund in the framework of TÁMOP 4.2.4. A/2-11-1-2012-0001 "National Excellence Program" New National Excellence Program (UNKP). This work is dedicated to the 650th anniversary of the University of Pecs. T.J. and R.Z. are supported by GOP-1.1.1-11-2012-0197 financed by the Hungarian government and the EU. This work was supported by the University of Debrecen (RH/751/2015).

References

1. Yan JY, Tian FM, Wang WY, Cheng Y, Song HP, Zhang YZ, et al. Parathyroid hormone (1-34) prevents cartilage degradation and preserves subchondral bone micro-architecture in guinea pigs with spontaneous osteoarthritis. Osteoarthritis Cartilage. 2014;22:1869–77.
2. Rais Y, Reich A, Simsa-Maziel S, Moshe M, Idelevich A, Kfir T, et al. The growth plate's response to load is partially mediated by mechano-sensing via the chondrocytic primary cilium. Cell Mol Life Sci. 2015;72:597–615.
3. Barrett-Jolley R, Lewis R, Fallman R, Mobasheri A. The emerging chondrocyte channelome. Front Physiol. 2010;1:135.
4. Paiva KB, Granjeiro JM. Bone tissue remodeling and development: focus on matrix metalloproteinase functions. Arch Biochem Biophys. 2014;561C:74–87.
5. Vaudry D, Falluel-Morel A, Bourgault S, Basille M, Burel D, Wurtz O, et al. Pituitary adenylate cyclase-activating polypeptide and its receptors: 20 years after the discovery. Pharmacol Rev. 2009;61:283–357.
6. Moody TW, Hill JM, Jensen RT. VIP as a trophic factor in the CNS and cancer cells. Peptides. 2003;24:163–77.
7. Miyata A, Arimura A, Dahl RR, Minamino N, Uehara A, Jiang L, et al. Isolation of a novel 38 residue-hypothalamic polypeptide which stimulates adenylate cyclase in pituitary cells. Biochem Biophys Res Commun. 1989;164:567–74.
8. Reglodi D, Tamas A, Koppan M, Szogyi D, Welke L. Role of PACAP in female fertility and reproduction at gonadal level—recent advances. Front Endocrinol (Lausanne). 2012;3:155.
9. Koppan M, Varnagy A, Reglodi D, Brubel R, Nemeth J, Tamas A, et al. Correlation between oocyte number and follicular fluid concentration of pituitary adenylate cyclase-activating polypeptide (PACAP) in women after superovulation treatment. J Mol Neurosci. 2012;48:617–22.
10. Sandor B, Fintor K, Felszeghy S, Juhasz T, Reglodi D, Mark L, et al. Structural and morphometric comparison of the molar teeth in pre-eruptive developmental stage of PACAP-deficient and wild-type mice. J Mol Neurosci. 2014;54:331–41.
11. Csanaky K, Banki E, Szabadfi K, Reglodi D, Tarcai I, Czegledi L, et al. Changes in PACAP immunoreactivity in human milk and presence of PAC1 receptor in mammary gland during lactation. J Mol Neurosci. 2012;48:631–7.
12. Giunta S, Castorina A, Marzagalli R, Szychlinska MA, Pichler K, Mobasheri A, et al. Ameliorative effects of PACAP against cartilage degeneration. Morphological, immunohistochemical and biochemical evidence from in vivo and in vitro models of rat osteoarthritis. Int J Mol Sci. 2015;16:5922–44.

13. Juhasz T, Matta C, Katona E, Somogyi C, Takacs R, Gergely P, et al. Pituitary adenylate cyclase activating polypeptide (PACAP) signalling exerts chondrogenesis promoting and protecting effects: implication of calcineurin as a downstream target. PLoS One. 2014;9:e91541.
14. Juhasz T, Matta C, Katona E, Somogyi C, Takacs R, Hajdu T, et al. Pituitary adenylate cyclase-activating polypeptide (PACAP) signalling enhances osteogenesis in UMR-106 cell line. J Mol Neurosci. 2014;54:555–73.
15. Kovacs CS, Chik CL, Li B, Karpinski E, Ho AK. Pituitary adenylate cyclase-activating peptide stimulates cyclic AMP accumulation in UMR 106 osteoblast-like cells. J Endocrinol. 1996;149:287–95.
16. Gourlet P, Vandermeers A, Vertongen P, Rathe J, De NP, Cnudde J, et al. Development of high affinity selective VIP1 receptor agonists. Peptides. 1997;18:1539–45.
17. Dohi K, Mizushima H, Nakajo S, Ohtaki H, Matsunaga S, Aruga T, et al. Pituitary adenylate cyclase-activating polypeptide (PACAP) prevents hippocampal neurons from apoptosis by inhibiting JNK/SAPK and p38 signal transduction pathways. Regul Pept. 2002;109:83–8.
18. Tanaka K, Shibuya I, Uezono Y, Ueta Y, Toyohira Y, Yanagihara N, et al. Pituitary adenylate cyclase-activating polypeptide causes Ca2+ release from ryanodine/caffeine stores through a novel pathway independent of both inositol trisphosphates and cyclic AMP in bovine adrenal medullary cells. J Neurochem. 1998;70:1652–61.
19. Szabo A, Danyadi B, Bognar E, Szabadfi K, Fabian E, Kiss P, et al. Effect of PACAP on MAP kinases, Akt and cytokine expressions in rat retinal hypoperfusion. Neurosci Lett. 2012;523:93–8.
20. Schuhmann K, Romanin C, Baumgartner W, Groschner K. Intracellular Ca2+ inhibits smooth muscle L-type Ca2+ channels by activation of protein phosphatase type 2B and by direct interaction with the channel. J Gen Physiol. 1997;110:503–13.
21. Yu R, Cui Z, Li M, Yang Y, Zhong J. Dimer-dependent intrinsic/basal activity of the class B G protein-coupled receptor PAC1 promotes cellular anti-apoptotic activity through Wnt/beta-catenin pathways that are associated with dimer endocytosis. PLoS One. 2014;9:e113913.
22. Oka H, Jin L, Kulig E, Scheithauer BW, Lloyd RV. Pituitary adenylate cyclase-activating polypeptide inhibits transforming growth factor-beta1-induced apoptosis in a human pituitary adenoma cell line. Am J Pathol. 1999;155:1893–900.
23. Pavelock KA, Girard BM, Schutz KC, Braas KM, May V. Bone morphogenetic protein down-regulation of neuronal pituitary adenylate cyclase-activating polypeptide and reciprocal effects on vasoactive intestinal peptide expression. J Neurochem. 2007;100:603–16.
24. Niewiadomski P, Zhujiang A, Youssef M, Waschek JA. Interaction of PACAP with Sonic hedgehog reveals complex regulation of the hedgehog pathway by PKA. Cell Signal. 2013;25:2222–30.
25. Lu Q, Tong B, Luo Y, Sha L, Chou G, Wang Z, et al. Norisoboldine suppresses VEGF-induced endothelial cell migration via the cAMP-PKA-NF-kappaB/Notch1 pathway. PLoS One. 2013;8:e81220.
26. Botz B, Bolcskei K, Kereskai L, Kovacs M, Nemeth T, Szigeti K, et al. Differential regulatory role of pituitary adenylate cyclase-activating polypeptide in the serum-transfer arthritis model. Arthritis Rheumatol. 2014;66:2739–50.
27. Fabian E, Reglodi D, Mester L, Szabo A, Szabadfi K, Tamas A, et al. Effects of PACAP on intracellular signaling pathways in human retinal pigment epithelial cells exposed to oxidative stress. J Mol Neurosci. 2012;48:493–500.
28. Heimesaat MM, Dunay IR, Schulze S, Fischer A, Grundmann U, Alutis M, et al. Pituitary adenylate cyclase-activating polypeptide ameliorates experimental acute ileitis and extra-intestinal sequelae. PLoS One. 2014;9:e108389.
29. Rozzi SJ, Borelli G, Ryan K, Steiner JP, Reglodi D, Mocchetti I, et al. PACAP27 is protective against tat-induced neurotoxicity. J Mol Neurosci. 2014;54:485–93.
30. Goldring MB, Tsuchimochi K, Ijiri K. The control of chondrogenesis. J Cell Biochem. 2006;97:33–44.

31. de Crombrugghe B, Lefebvre V, Behringer RR, Bi W, Murakami S, Huang W. Transcriptional mechanisms of chondrocyte differentiation. Matrix Biol. 2000;19:389–94.
32. Ogawa H, Kozhemyakina E, Hung HH, Grodzinsky AJ, Lassar AB. Mechanical motion promotes expression of Prg4 in articular cartilage via multiple CREB-dependent, fluid flow shear stress-induced signaling pathways. Genes Dev. 2014;28:127–39.
33. Piera-Velazquez S, Hawkins DF, Whitecavage MK, Colter DC, Stokes DG, Jimenez SA. Regulation of the human SOX9 promoter by Sp1 and CREB. Exp Cell Res. 2007;313:1069–79.
34. Juhasz T, Matta C, Somogyi C, Katona E, Takacs R, Soha RF, et al. Mechanical loading stimulates chondrogenesis via the PKA/CREB-Sox9 and PP2A pathways in chicken micromass cultures. Cell Signal. 2014;26:468–82.
35. Zakany R, Szucs K, Bako E, Felszeghy S, Czifra G, Biro T, et al. Protein phosphatase 2A is involved in the regulation of protein kinase A signaling pathway during in vitro chondrogenesis. Exp Cell Res. 2002;275:1–8.
36. Zhao L, Li G, Zhou GQ. SOX9 directly binds CREB as a novel synergism with the PKA pathway in BMP-2-induced osteochondrogenic differentiation. J Bone Miner Res. 2009;24:826–36.
37. Juhasz T, Matta C, Mészár Z, Nagy G, Szijgyarto Z, Molnar Z, et al. Optimalized transient transfection of chondrogenic primary cell cultures. CEJB. 2010;5:572–84.
38. Zakany R, Bako E, Felszeghy S, Hollo K, Balazs M, Bardos H, et al. Okadaic acid-induced inhibition of protein phosphatase 2A enhances chondrogenesis in chicken limb bud micromass cell cultures. Anat Embryol (Berl). 2001;203:23–34.
39. Zakany R, Szijgyarto Z, Matta C, Juhasz T, Csortos C, Szucs K, et al. Hydrogen peroxide inhibits formation of cartilage in chicken micromass cultures and decreases the activity of calcineurin: implication of ERK1/2 and Sox9 pathways. Exp Cell Res. 2005;305:190–9.
40. Matta C, Fodor J, Szijgyarto Z, Juhasz T, Gergely P, Csernoch L, et al. Cytosolic free Ca2+ concentration exhibits a characteristic temporal pattern during in vitro cartilage differentiation: a possible regulatory role of calcineurin in Ca-signalling of chondrogenic cells. Cell Calcium. 2008;44:310–23.
41. Fodor J, Matta C, Olah T, Juhasz T, Takacs R, Toth A, et al. Store-operated calcium entry and calcium influx via voltage-operated calcium channels regulate intracellular calcium oscillations in chondrogenic cells. Cell Calcium. 2013;54:1–16.
42. Somogyi CS, Matta C, Foldvari Z, Juhasz T, Katona E, Takacs AR, et al. Polymodal transient receptor potential vanilloid (TRPV) ion channels in chondrogenic cells. Int J Mol Sci. 2015;16:18412–38.
43. Yoo YM, Kwag JH, Kim KH, Kim CH. Effects of neuropeptides and mechanical loading on bone cell resorption in vitro. Int J Mol Sci. 2014;15:5874–83.
44. Sethi V, Rubinstein I, Kuzmis A, Kastrissios H, Artwohl J, Onyuksel H. Novel, biocompatible, and disease modifying VIP nanomedicine for rheumatoid arthritis. Mol Pharm. 2013;10:728–38.
45. Konttinen YT, Tiainen VM, Gomez-Barrena E, Hukkanen M, Salo J. Innervation of the joint and role of neuropeptides. Ann N Y Acad Sci. 2006;1069:149–54.
46. Strange-Vognsen HH, Arnbjerg J, Hannibal J. Immunocytochemical demonstration of pituitary adenylate cyclase activating polypeptide (PACAP) in the porcine epiphyseal cartilage canals. Neuropeptides. 1997;31:137–41.
47. Ahmed N, Dreier R, Gopferich A, Grifka J, Grassel S. Soluble signalling factors derived from differentiated cartilage tissue affect chondrogenic differentiation of rat adult marrow stromal cells. Cell Physiol Biochem. 2007;20:665–78.
48. Gelse K, Brem M, Klinger P, Hess A, Swoboda B, Hennig F, et al. Paracrine effect of transplanted rib chondrocyte spheroids supports formation of secondary cartilage repair tissue. J Orthop Res. 2009;27:1216–25.
49. Rahman S, Dobson PR, Bunning RA, Russell RG, Brown BL. The regulation of connective tissue metabolism by vasoactive intestinal polypeptide. Regul Pept. 1992;37:111–21.

50. Lee HW, Hahm SH, Hsu CM, Eiden LE. Pituitary adenylate cyclase-activating polypeptide regulation of vasoactive intestinal polypeptide transcription requires Ca2+ influx and activation of the serine/threonine phosphatase calcineurin. J Neurochem. 1999;73:1769–72.
51. Hattori S, Takao K, Tanda K, Toyama K, Shintani N, Baba A, et al. Comprehensive behavioral analysis of pituitary adenylate cyclase-activating polypeptide (PACAP) knockout mice. Front Behav Neurosci. 2012;6:58.
52. Shintani N, Mori W, Hashimoto H, Imai M, Tanaka K, Tomimoto S, et al. Defects in reproductive functions in PACAP-deficient female mice. Regul Pept. 2002;109:45–8.
53. Barberi M, Di Paolo V, Latini S, Guglielmo MC, Cecconi S, Canipari R. Expression and functional activity of PACAP and its receptors on cumulus cells: effects on oocyte maturation. Mol Cell Endocrinol. 2013;375:79–88.
54. Brubel R, Kiss P, Vincze A, Varga A, Varnagy A, Bodis J, et al. Effects of pituitary adenylate cyclase activating polypeptide on human sperm motility. J Mol Neurosci. 2012;48:623–30.
55. Zhang Y, Pizzute T, Pei M. A review of crosstalk between MAPK and Wnt signals and its impact on cartilage regeneration. Cell Tissue Res. 2014;358:633–49.
56. Serrano MJ, So S, Hinton RJ. Roles of notch signalling in mandibular condylar cartilage. Arch Oral Biol. 2014;59:735–40.
57. Sassi N, Gadgadi N, Laadhar L, Allouche M, Mourali S, Zandieh-Doulabi B, et al. Notch signaling is involved in human articular chondrocytes de-differentiation during osteoarthritis. J Recept Signal Transduct Res. 2014;34:48–57.
58. Sangphech N, Osborne BA, Palaga T. Notch signaling regulates the phosphorylation of Akt and survival of lipopolysaccharide-activated macrophages via regulator of G protein signaling 19 (RGS19). Immunobiology. 2014;219:653–60.
59. Kwon HJ. ATP oscillations mediate inductive action of FGF and Shh signalling on prechondrogenic condensation. Cell Biochem Funct. 2013;31:75–81.
60. Waschek JA, Cicco-Bloom E, Nicot A, Lelievre V. Hedgehog signaling: new targets for GPCRs coupled to cAMP and protein kinase A. Ann N Y Acad Sci. 2006;1070:120–8.
61. Sanchez A, Chiriva-Internati M, Grammas P. Transduction of PACAP38 protects primary cortical neurons from neurotoxic injury. Neurosci Lett. 2008;448:52–5.
62. Banki E, Kovacs K, Nagy D, Juhasz T, Degrell P, Csanaky K, et al. Molecular mechanisms underlying the nephroprotective effects of PACAP in diabetes. J Mol Neurosci. 2014;54:300–9.
63. Reglodi D, Kiss P, Szabadfi K, Atlasz T, Gabriel R, Horvath G, et al. PACAP is an endogenous protective factor-insights from PACAP-deficient mice. J Mol Neurosci. 2012;48:482–92.
64. Hernanz A, Medina S, de Miquel E, Martin-Mola E. Effect of calcitonin gene-related peptide, neuropeptide Y, substance P, and vasoactive intestinal peptide on interleukin-1beta, interleukin-6 and tumor necrosis factor-alpha production by peripheral whole blood cells from rheumatoid arthritis and osteoarthritis patients. Regul Pept. 2003;115:19–24.
65. Pulsatelli L, Dolzani P, Silvestri T, De GR, Salvarani C, Macchioni P, et al. Synovial expression of vasoactive intestinal peptide in polymyalgia rheumatica. Clin Exp Rheumatol. 2006;24:562–6.
66. Dvorakova MC. Cardioprotective role of the VIP signaling system. Drug News Perspect. 2005;18:387–91.
67. Reglodi D, Kiss P, Lubics A, Tamas A. Review on the protective effects of PACAP in models of neurodegenerative diseases in vitro and in vivo. Curr Pharm Des. 2011;17:962–72.
68. Juhasz T, Szentleleky E, Somogyi CS, Takacs R, Dobrosi N, Engler M, et al. Pituitary adenylate cyclase activating polypeptide (PACAP) pathway is induced by mechanical load and reduces the activity of hedgehog signaling in chondrogenic micromass cell cultures. Int J Mol Sci. 2015;16:17344–67.
69. Onoue S, Ohmori Y, Endo K, Yamada S, Kimura R, Yajima T. Vasoactive intestinal peptide and pituitary adenylate cyclase-activating polypeptide attenuate the cigarette smoke extract-induced apoptotic death of rat alveolar L2 cells. Eur J Biochem. 2004;271:1757–67.

70. Juarranz Y, Gutierrez-Canas I, Santiago B, Carrion M, Pablos JL, Gomariz RP. Differential expression of vasoactive intestinal peptide and its functional receptors in human osteoarthritic and rheumatoid synovial fibroblasts. Arthritis Rheum. 2008;58:1086–95.
71. Carrion M, Perez-Garcia S, Jimeno R, Juarranz Y, Gonzalez-Alvaro I, Pablos JL, et al. Inflammatory mediators alter interleukin-17 receptor, interleukin-12 and -23 expression in human osteoarthritic and rheumatoid arthritis synovial fibroblasts: immunomodulation by vasoactive intestinal peptide. Neuroimmunomodulation. 2013;20:274–84.
72. Mobasheri A. The future of osteoarthritis therapeutics: emerging biological therapy. Curr Rheumatol Rep. 2013;15:385.
73. Chen G, Deng C, Li YP. TGF-beta and BMP signaling in osteoblast differentiation and bone formation. Int J Biol Sci. 2012;8:272–88.
74. Gogakos AI, Duncan Bassett JH, Williams GR. Thyroid and bone. Arch Biochem Biophys. 2010;503:129–36.
75. Kim JH, Liu X, Wang J, Chen X, Zhang H, Kim SH, et al. Wnt signaling in bone formation and its therapeutic potential for bone diseases. Ther Adv Musculoskelet Dis. 2013;5:13–31.
76. Pan A, Chang L, Nguyen A, James AW. A review of hedgehog signaling in cranial bone development. Front Physiol. 2013;4:61.
77. Yavropoulou MP, Yovos JG. The role of Notch signaling in bone development and disease. Hormones (Athens). 2014;13:24–37.
78. Wang L, Park P, La MF, Than K, Rahman S, Lin CY. Bone formation induced by BMP-2 in human osteosarcoma cells. Int J Oncol. 2013;43:1095–102.
79. Zhang R, Edwards JR, Ko SY, Dong S, Liu H, Oyajobi BO, et al. Transcriptional regulation of BMP2 expression by the PTH-CREB signaling pathway in osteoblasts. PLoS One. 2011;6:e20780.
80. Ehlen HW, Buelens LA, Vortkamp A. Hedgehog signaling in skeletal development. Birth Defects Res C Embryo Today. 2006;78:267–79.
81. Jonason JH, Xiao G, Zhang M, Xing L, Chen D. Post-translational regulation of Runx2 in bone and cartilage. J Dent Res. 2009;88:693–703.
82. Hughes-Fulford M, Li CF. The role of FGF-2 and BMP-2 in regulation of gene induction, cell proliferation and mineralization. J Orthop Surg Res. 2011;6:8.
83. Yang J, Zhao L, Tian W, Liao Z, Zheng H, Wang G, et al. Correlation of WWOX, RUNX2 and VEGFA protein expression in human osteosarcoma. BMC Med Genomics. 2013;6:56.
84. Lerner UH, Persson E. Osteotropic effects by the neuropeptides calcitonin gene-related peptide, substance P and vasoactive intestinal peptide. J Musculoskelet Neuronal Interact. 2008;8:154–65.
85. Hohmann EL, Elde RP, Rysavy JA, Einzig S, Gebhard RL. Innervation of periosteum and bone by sympathetic vasoactive intestinal peptide-containing nerve fibers. Science. 1986;232:868–71.
86. Bataille C, Mauprivez C, Hay E, Baroukh B, Brun A, Chaussain C, et al. Different sympathetic pathways control the metabolism of distinct bone envelopes. Bone. 2012;50:1162–72.
87. Suzuki A, Kotoyori J, Oiso Y, Kozawa O. Pituitary adenylate cyclase-activating polypeptide induces cAMP production independently from vasoactive intestinal polypeptide in osteoblast-like cells. Cell Signal. 1994;6:11–6.
88. Onuoha GN. Circulating sensory peptide levels within 24 h of human bone fracture. Peptides. 2001;22:1107–10.
89. Ma W, Zhang X, Shi S, Zhang Y. Neuropeptides stimulate human osteoblast activity and promote gap junctional intercellular communication. Neuropeptides. 2013;47:179–86.
90. Lundberg P, Lundgren I, Mukohyama H, Lehenkari PP, Horton MA, Lerner UH. Vasoactive intestinal peptide (VIP)/pituitary adenylate cyclase-activating peptide receptor subtypes in mouse calvarial osteoblasts: presence of VIP-2 receptors and differentiation-induced expression of VIP-1 receptors. Endocrinology. 2001;142:339–47.
91. Okura H, Sato S, Kishikawa S, Kaneto S, Nakashima T, Yoshida N, et al. Runx2-I isoform contributes to fetal bone formation even in the absence of specific N-terminal amino acids. PLoS One. 2014;9:e108294.

92. Li TF, Dong Y, Ionescu AM, Rosier RN, Zuscik MJ, Schwarz EM, et al. Parathyroid hormone-related peptide (PTHrP) inhibits Runx2 expression through the PKA signaling pathway. Exp Cell Res. 2004;299:128–36.
93. Duan Y, Wang Z, Zhang H, He Y, Fan R, Cheng Y, et al. Extremely low frequency electromagnetic field exposure causes cognitive impairment associated with alteration of the glutamate level, MAPK pathway activation and decreased CREB phosphorylation in mice hippocampus: reversal by procyanidins extracted from the lotus seedpod. Food Funct. 2014;5:2289–97.
94. Pitts RL, Wang S, Jones EA, Symes AJ. Transforming growth factor-beta and ciliary neurotrophic factor synergistically induce vasoactive intestinal peptide gene expression through the cooperation of Smad, STAT, and AP-1 sites. J Biol Chem. 2001;276:19966–73.
95. Persson E, Lerner UH. The neuropeptide VIP regulates the expression of osteoclastogenic factors in osteoblasts. J Cell Biochem. 2011;112:3732–41.
96. Li D, Jiao J, Shatos MA, Hodges RR, Dartt DA. Effect of VIP on intracellular [Ca2+], extracellular regulated kinase 1/2, and secretion in cultured rat conjunctival goblet cells. Invest Ophthalmol Vis Sci. 2013;54:2872–84.
97. Mustafa T, Grimaldi M, Eiden LE. The hop cassette of the PAC1 receptor confers coupling to Ca2+ elevation required for pituitary adenylate cyclase-activating polypeptide-evoked neurosecretion. J Biol Chem. 2007;282:8079–91.
98. Maeda Y, Sekiguchi F, Yamanaka R, Sugimoto R, Yamasoba D, Tomita S, et al. Mechanisms for proteinase-activated receptor 1-triggered prostaglandin E2 generation in mouse osteoblastic MC3T3-E1 cells. Biol Chem. 2015;396:153–62.
99. Lv B, Tang Y, Chen F, Xiao X. Vasoactive intestinal peptide and pituitary adenylate cyclase-activating polypeptide inhibit tissue factor expression in monocyte in vitro and in vivo. Shock. 2009;31:185–91.
100. Miraoui H, Oudina K, Petite H, Tanimoto Y, Moriyama K, Marie PJ. Fibroblast growth factor receptor 2 promotes osteogenic differentiation in mesenchymal cells via ERK1/2 and protein kinase C signaling. J Biol Chem. 2009;284:4897–904.
101. May V, Clason TA, Buttolph TR, Girard BM, Parsons RL. Calcium influx, but not intracellular calcium release, supports PACAP-mediated ERK activation in HEK PAC1 receptor cells. J Mol Neurosci. 2014;54:342–50.
102. Hayashi K, Endoh T, Shibukawa Y, Yamamoto T, Suzuki T. VIP and PACAP inhibit L-, N- and P/Q-type Ca2+ channels of parasympathetic neurons in a voltage independent manner. Bull Tokyo Dent Coll. 2002;43:31–9.
103. Harfi I, Sariban E. Mechanisms and modulation of pituitary adenylate cyclase-activating protein-induced calcium mobilization in human neutrophils. Ann N Y Acad Sci. 2006;1070:322–9.
104. Payet MD, Bilodeau L, Breault L, Fournier A, Yon L, Vaudry H, et al. PAC1 receptor activation by PACAP-38 mediates Ca2+ release from a cAMP-dependent pool in human fetal adrenal gland chromaffin cells. J Biol Chem. 2003;278:1663–70.
105. Akopian A, Demulder A, Ouriaghli F, Corazza F, Fondu P, Bergmann P. Effects of CGRP on human osteoclast-like cell formation: a possible connection with the bone loss in neurological disorders? Peptides. 2000;21:559–64.
106. Winding B, Wiltink A, Foged NT. Pituitary adenylyl cyclase-activating polypeptides and vasoactive intestinal peptide inhibit bone resorption by isolated rabbit osteoclasts. Exp Physiol. 1997;82:871–86.
107. Franceschi RT, Xiao G. Regulation of the osteoblast-specific transcription factor, Runx2: responsiveness to multiple signal transduction pathways. J Cell Biochem. 2003;88:446–54.
108. Park HJ, Baek K, Baek JH, Kim HR. The cooperation of CREB and NFAT is required for PTHrP-induced RANKL expression in mouse osteoblastic cells. J Cell Physiol. 2015;230:667–79.

Chapter 21
Role of PACAP/VIP in Bone and Joint Physiology and Pathophysiology

Balint Botz and Zsuzsanna Helyes

Abstract Pituitary adenylate cyclase-activating polypeptide (PACAP) and vasoactive intestinal polypeptide (VIP) are both important neuropeptides released primarily from the innervating sensory nerves and a variety of different non-neural cellular sources of the skeletal system. This review summarizes the data available about the expression and function of these peptides and their receptors in skeletal tissues with special emphasis on their disease conditions. Results obtained in animal models and clinical observations are collected, their relevance and importance are discussed. There are a lot of data about the effects of PACAP/VIP on cellular signaling of skeletal tissue cells in vitro, but their physiological/pathophysiological in vivo importance is not well understood. Furthermore, since these peptides are potent immunomodulators, they exert substantial effects on the progress of inflammatory/degenerative joint diseases. Additionally, as key elements of nociceptive signaling, they also contribute to the development of chronic pain arising from arthritis/osteoarthritis. In summary, the (1) direct effects of PACAP/VIP on skeletal tissues, (2) their immunoregulatory properties, and (3) their pivotal importance in nociception and peripheral/central sensitization all have potential clinical relevance.

Keywords PACAP • VIP • Bone physiology • Joint physiology • Rheumatoid arthritis • Osteoarthritis • Pain

B. Botz
Department of Pharmacology and Pharmacotherapy, Molecular Pharmacology Research Team, János Szentágothai Research Centre, University of Pecs, Szigeti u. 12. and Ifjúság ú. 20, Pecs 7624, Hungary

Department of Radiology, University of Pecs, Pecs, Hungary

Z. Helyes (✉)
Department of Pharmacology and Pharmacotherapy, Molecular Pharmacology Research Team, János Szentágothai Research Centre, University of Pecs, Szigeti u. 12. and Ifjúság ú. 20., Pecs 7624, Hungary

MTA-PTE NAP-B Chronic Pain Research Group, Pecs, Hungary
e-mail: zsuzsanna.helyes@aok.pte.hu

D. Reglodi, A. Tamas (eds.), *Pituitary Adenylate Cyclase Activating Polypeptide — PACAP*, Current Topics in Neurotoxicity 11,
DOI 10.1007/978-3-319-35135-3_21

Introduction

It has been clearly demonstrated during the last decades that the physiological homeostasis of the skeletal system is not only influenced by a number of hormones and local mediators, but also by elements of the nervous system. This is supported by the fact, that almost all skeletal tissues, including the bone marrow and cortex, the periosteum, and the synovium, are innervated by sensory and sympathetic nerves [1–3] that produce and release neuropeptides [4–6]. Owing to their distribution, these peptidergic fibers have been implicated in the normal physiology of skeletal tissues and also in the pathogenesis of the diseases affecting them, such as autoimmune or degenerative arthritides [1, 7]. The periosteum is particularly densely innervated by peptidergic fibers, which strongly correlates with its high sensitivity to painful stimuli [8]. A great proportion of peptide-containing nerves are involved in nociceptive signaling, but the locally released mediators can also influence pain, tissue remodeling and turnover, as well as immune cell recruitment. PACAP and VIP were shown to influence bone and cartilage development besides balancing immune functions and pain during neurogenic and non-neurogenic inflammatory conditions of the joints, such as autoimmune and degenerative arthritides. Furthermore, there is evidence for their regulatory role in skeletal tissues and cellular signaling under both normal and disease conditions [9, 10].

This chapter gives an overview about the presence, signaling pathways, and most relevant functions of these peptides, summarizing the evidence obtained from in vitro and in vivo experimental conditions, as well as clinical studies. We also discuss the current perspectives of the PACAP/VIP system as a target for drug development against certain diseases of the skeletal system with special emphasis put on arthritic conditions and notably joint pain.

Expression of PACAP/VIP and Their Receptors in Skeletal Tissues

Even before the discovery of PACAP, the presence of the closely related VIP was demonstrated in nerve fibers of the rat bone. VIP immunoreactivity was found to be widespread in the epiphysis and periosteum, whilst being moderate around blood vessels [5, 11]. VIP immunopositivity was markedly diminished in the periosteum of the rat following chemical sympathectomy suggesting that most VIP-containing fibers in the bone are sympathetic [6]. The presence of both PACAP and VIP-positive nerve fibers was demonstrated in various parts of the rat temporomandibular joint (synovial membrane, articular disc, joint capsule) [12]. VIP-immunopositive fibers were also identified in the synovial connective tissue of the mouse knee joint, particularly in the highly vascularized regions [13].

The immunopositivity for PACAP was also demonstrated in the cartilage canals of newborn and juvenile pigs. PACAP-immunoreactive fibers were found to innervate blood vessels, with a high degree of co-localization with calcitonin gene-related peptide (CGRP) and substance P (SP), but—interestingly—not with VIP [14]. The expression of PACAP was recently confirmed in differentiating chicken chondrocytes as well [15]. In summary, the marked expression (and release) of PACAP/VIP in osseous and chondrogenic tissues and innervating fibers is well-substantiated by current evidence.

The presence of VPAC receptors on both osteoclasts and a significant proportion (20 %) of stromal cells/osteoblasts of the rat was first demonstrated by measuring their ability to bind VIP by atomic force microscopy [16]. The expression of the yet not characterized VIP receptors was previously confirmed on the SaOs-2 human osteosarcoma cell line [17]. Indeed, according to later studies, VPAC1 receptor, but not VPAC2 or PAC1 is expressed on osteoblastic cells derived from human periosteum (SaM-I) and osteosarcoma cell lines (HOS, SaOS-2, MG-63). In addition, VIP mRNA was not found in the osteoblastic cells, excluding the possibility of autocrine signaling by VIP in osteoblasts [18]. In contrast, undifferentiated primary calvary osteoblasts of the mouse express primarily the VPAC2 receptor, while VPAC1 expression is only induced during their differentiation [19]. The monoclonal mouse preosteoblast cell line MC3T3-EI also expresses VPAC2 [20]. In contrast, mouse osteoclasts express primarily VPAC1 and PAC1 receptors [21]. Differentiating chicken chondrocytes express all known receptors of VIP/PACAP; furthermore, PAC1 expression peaks at the time of the final commitment to chondrogenic cells [15]. VPAC2, but not VPAC1 receptor expression was confirmed in cultured synovial cells of patients suffering from rheumatoid arthritis (RA) [22]. Based on this, we can conclude that the presence of all known VIP/PACAP receptors has been confirmed on numerous skeletal tissue cell types; however, the receptorial expression pattern seems to vary according to both the degree of differentiation and the origin of tissues/cells investigated.

Influence of PACAP/VIP on Cellular Signaling in Skeletal Tissues

As early as in 1983 it was shown that VIP stimulates bone resorption in mouse calvarial organ culture by a cyclic adenosine monophosphate (cAMP)-dependent mechanism [23]. Furthermore, VIP triggers a rapid cAMP increase in SaOs-2 human osteosarcoma cells [17]. A similar cAMP-increasing effect of VIP was also demonstrated in human osteoblast-like cells, but not in articular chondrocytes or synovial cells [24]. PACAP-27 and PACAP-38 also stimulate cAMP accumulation in the mouse calvaria and isolated bone cells [25]. PACAP increases cAMP-accumulation, but not protein kinase C (PKC) activity or Ca^{2+} influx in the osteoblast-like MC3T3-EI cells, and importantly, PACAP was found to be more potent than VIP in this respect [26]. PACAP-27 and PACAP-38 and VIP all stimulate cAMP elevation in the UMR 106 osteoblast-like tumor cell line. PACAP-27 was found to be slightly more potent than PACAP-38, and both are tenfold more potent when compared to VIP. The antagonists PACAP6-27 and PACAP6-38 are able to block this cAMP-increasing effect, while VIP antagonists are ineffective [27]. A later investigation on mouse calvarial osteoblasts showed that PACAP-38 is a tenfold more potent stimulator of cAMP-production than VIP, and it is also more potent than PACAP-27. Radioligand binding also demonstrated a stronger specific binding of PACAP-38. PACAP6-38 inhibits the effect of both PACAP and VIP in this model too. These findings would suggest that the PAC1 receptor might be pivotal to this cAMP-increasing effect, but both the authors and others (see earlier) found that osteoblasts express predominantly VPAC2 receptors, not PAC1. Therefore, it was concluded that PACAP-preferring VPAC2 receptors are likely to be responsible for the increased potency of PACAP [19].

VIP, PACAP-27, and PACAP-38 all potently inhibit the formation of tartrate-resistant acid phosphatase-positive multinuclear cells in mouse bone marrow cultures stimulated by 1,25(OH)2-vitamin D3 (D3). Furthermore, VIP and PACAP-38 decrease the stimulatory effect of D3 on receptor activator of NF-kappaB (RANK) and RANK ligand (RANKL) expression. Since both RANK and RANKL are important for osteoclastogenesis, this was suggested as the primary mechanism responsible for the inhibitory effect of VIP and PACAP on osteoclast differentiation. Furthermore, they abolish the inhibitory effect of D3 on osteoprotegerin (OPG) expression—providing another anti-osteoclastogenic mechanism. Based on these observations, it was suggested that VIP and PACAP may regulate osteoclast differentiation by affecting the expression of certain key proteins such as RANK, RANKL, and OPG [28]. However, both VIP and PACAP were found to stimulate RANKL mRNA expression in mouse calvarial osteoblasts, while they inhibited the expression of OPG and macrophage colony-stimulating factor (M-CSF). VPAC2 receptor activation and consequent cAMP accumulation coupled with ERK phosphorylation was suggested to be responsible for the increased RANKL/OPG ratio [29]. These observations demonstrate that activation of VPAC2 receptors in osteoblasts enhances the RANKL/OPG ratio by mechanisms mediated by cAMP and ERK pathways suggesting an important role for VIP/PACAP in bone remodeling. Culturing neonatal mouse calvarial osteoblasts in the presence of VIP-stimulates alkaline phosphatase (ALP) activity and Ca^{2+} accumulation within bone noduli without affecting cell proliferation. This raised the possibilty that VIP-ergic signaling might represent a neurohormonal control mechanism of the anabolic processes of the bone [30]. VIP exerts a bidirectional influence on the activity of rat osteoclasts in vitro: initially, it elicits an acute rise of Ca^{2+}, cell contraction and diminished motility, with consequently decreased resorbing activity. However, prolonged experiments showed that this effect is transient, and chronic VIP administration stimulates bone resorption [16]. Other studies also revealed that PACAP is able to inhibit thyroid hormone-induced osteocalcin-synthesis in osteoblasts by suppressing the activation of p38 MAP kinases [31]. It was also demonstrated on the MC3T3-EI cell line that VPAC2-activation by PACAP/VIP activates the adenylate cyclase pathway by facilitating cAMP accumulation, thereby inhibiting osteoblastic differentiation by suppressing ALP production. PACAP/VIP also facilitate the production of the proinflammatory cytokine IL-6, which plays a pivotal role in bone resorption. VPAC2 downregulation by siRNA blocks the effects of PACAP/VIP, while the PAC1-selective agonist maxadilan has no effect [20]. Others observed a similar effect of PACAP/VIP in mouse calvarial osteoblasts and the MC3T3-E1 osteoblastic cell line, where both peptides potentiate IL-1β-induced IL-6 production. Since the potency of PACAP and VIP is similar, and as secretin (also an agonist on VPAC1) has no effect, VPAC2 receptor activation was suggested to be the possible pathway responsible for this phenomenon [32]. Further studies showed increased activation of the cAMP-PKA-CREB pathway and the transcription factor C/EBP, but diminished activation of AP-1, as the primary downstream signaling mechanisms of this effect [33]. This facilitation of IL-6 release by PACAP/VIP implicates these peptides as stimulators of osteoclast activity and consequently bone remodeling. In summary, VPAC2-receptor signaling and consequent cAMP-increase seems to be the central element of VIP/PACAP signaling in osteoblasts and osteoclasts, but not in chondrocytes or synovial cells.

Role of PACAP/VIP in Bone and Joint Physiology

PACAP/VIP has been determined to be important regulators of chondrogenic and osteogenic differentiation [10]. The remarkable presence of VIP-positive fibers in the bone, indirectly implicated a role for this peptide in bone physiology relatively early [11]. As PACAP-distribution in the cartilage canals is the highest around blood vessels, it was suggested to be a regulator of blood flow in the bone [14]. VIP was found to be able to stimulate the production of the proinflammatory mediator PGE2 in human articular chondrocytes, osteoblast-like, and synovial cells—an effect which can be blocked by VIP antagonist. Thus, it was hypothesized that locally released VIP may facilitate bone resorption by inducing PGE2-production. Furthermore, VIP significantly stimulates metalloprotease activity of human articular chondrocytes [24]. VIP, PACAP-27, and PACAP-38 all inhibit the bone resorption activity of isolated rabbit osteoclasts [34]. Recently it has been demonstrated on the UMR 106 osteoblast cell line, that interestingly both PACAP and the PAC1 antagonist PACAP6-38 promote osteogenesis, cell proliferation, and mineralisation in vitro suggesting the presence of a not yet identified receptor and/or signaling mechanism in these cells. They also facilitate the expression of key proteins, such as collagen type I, ALP, and osterix. The expression of vascular endothelial growth factor (VEGF), several bone morphogenetic proteins (BMP 2, 4, 6, 7) and other regulatory factors are also elevated. Thus, it was proposed, that PACAP might facilitate bone formation and fracture healing in vivo [35]. A similar protective effect of PACAP was observed on chondrogenic cell cultures, where mechanical stress facilitates PACAP-signaling and PAC1 expression, while diminishing the presence VPAC1 and VPAC2 receptors. Application of exogenous PACAP acts against the hypertrophic differentiation of chondrocytes induced by mechanical load by diminishing the activation of hedgehog signaling and collagen type X expression. Thus, PACAP was proposed as a potential chondroprotective therapeutic agent in osteoarthritis [36].

Despite these pronounced in vitro effects, studies using global various PACAP, VIP, PAC1, or VPAC1 or VPAC2 gene-deficient mice reported no remarkable difference in bone and joint phenotype of any knockouts, but growth retardation due to intestinal or endocrine dysfunctions was commonly observed [37–42]. Recently a slightly, but not significantly increased bone density was found in the distal tibia of PACAP gene-deficient mice generated on the outbred CD1 background [43].

It was demonstrated in vitro that the PAC1 agonist PACAP-38 and the traditionally PAC1 antagonist PACAP6-38 both stimulate chondrocyte proliferation and thereby cartilage formation, and increase the expression of PAC1 and the Sox9 and CREB transcription factors which are pivotal for chondrogenesis. Both PACAP-38 and PACAP6-38 increase the expression of the Ca^{2+}-dependent serine-threonine phosphatase calcineurin, thus it was proposed that endogenous PACAP promotes chondrogenesis via this pathway. Furthermore, an in vitro chondroprotective effect of PACAP was demonstrated by its ability to partially rescue chondrogenesis during H_2O_2-induced oxidative stress, whilst conspicuously the conventionally PAC1 antagonist PACAP6-38 proved to be even more potent in this respect. Thus, distinct PAC1 receptor isoforms, splice variants or a presently not identified other target might be responsible for this chondroprotective effect. Inhibition of calcineurin can antagonize this action of PACAP [15]. As a consequence, PAC1-signaling seems to be predominantly responsible for the manifold trophic and protective effects of PACAP in cartilaginous tissues.

Immunomodulatory Effects of PACAP/VIP in Arthritis

Rheumatoid arthritis is a chronic systemic inflammatory condition of autoimmune etiology. The articular pain and joint swelling are the hallmark features of this disease, which are accompanied by degenerative symptoms in the later stage of the disease. It affects 0.5–1 % of the population worldwide, thereby being one of the most common forms of autoimmune diseases. The paramount importance of joint innervation and neuropeptides in inflammatory arthritides is now supported by a large number of preclinical studies [44, 45]. Furthermore, clinical case reports also demonstrated that local nerve damage or CNS lesions both prevent the development of autoimmune arthritis in a site-specific manner [46–48]. Thus, the exact role of neurogenic peptide mediators including VIP and PACAP generated a great interest, as these peptides modulate not only the inflammation, but also the central sensitisation and the development/aggravation of chronic pain.

Since the activation of the innate and adaptive immune system is a crucial step in autoimmune joint inflammation, it is important to briefly recapitulate that both VIP and PACAP have manifold immunomodulatory effects on a plethora of immune cells which emphasizes their role in this disease [9]. In vitro both peptides stimulate the release of the potent inflammatory mediators, histamine and serotonin [49–52]. PACAP and VIP also influence the recruitment, activity, cytokine, and chemokine production of lymphoid cells, macrophages and polymorphonuclear cells [9].

It was first suggested on the basis of the profound stimulatory effect of VIP on the production of the proinflammatory mediator PGE2 that these neuropeptides might be involved in the modulation of inflammation in rheumatoid arthritis. It was also demonstrated that VIP facilitates metalloprotease enzyme activity of articular chondrocytes, thus it was suggested to influence extracellular matrix breakdown in the cartilage [24]. It was also noted that VIP immunopositivity of periosteal nerve fibers is diminished in the vicinity of joint inflammation during experimental arthritis [13]. Topical application of VIP onto the capsular surface of the rat knee joint triggers a dose-dependent increase of synovial blood flow, which can be blocked by the VPAC1/2 antagonist VIP6-28, suggesting the VPAC1/2-dependent effect of VIP/PACAP on the synovial microvasculature. This response is decreased in acutely inflamed joints, but can be reversed by local denervation. Chronically inflamed joints exhibit the same degree of vasodilatory response, which is mast cell-independent [53].

In the collagen II-induced arthritis (CIA) model of the mouse it was demonstrated that PACAP acts as a potent immunomodulator. Daily i.p. treatment with a relatively high dosage (~5 nmol/animal) of PACAP considerably diminishes paw edema and clinical disease severity. Production of numerous proinflammatory factors (TNF-α, IL-1α, IL-1β, IL-6, IL-12, IL-18, and iNOS) and chemokines (RANTES, MCP-1, MIP-1α, MIP-1β, and MIP-2) decreased, whereas the levels of anti-inflammatory factors (IL-10, IL-1Ra) increased following PACAP treatment. Production of some isoenzymes of the matrix metalloproteinase (MMP) family were found to be inhibited by PACAP (MMP-2), while the level of other MMPs (MMP-9 and -13) is unaffected. Furthermore, PACAP ameliorates cellular infiltration, cartilage damage, and the occurrence of bone erosions. In vitro PACAP inhibits the chemotaxis of synovial T cells and macrophages isolated from CIA mice, and modulates the balance of Th1 and Th2 cells in arthritis by facilitating a Th2 response instead of the generation of Th1 cells highly characteristic of CIA [54]. Interestingly, another study by the same authors demonstrated that treatment with VIP in using the same dosage abrogates CIA and influences cytokine levels in the perfectly same manner. This interesting identical effect of VIP and PACAP might indicate that only the VPAC receptors, but not PAC1 plays a role in the aforementioned effect of PACAP [55]. VIP also abrogates disease severity in the rat CIA model, facilitated $CD4^+$ $CD25^+$ regulatory T cells (Treg), and diminishes Th17- and Th1 responses [56]. Despite the convincing effects of these peptides in this murine models, this approach cannot easily be transformed into clinical therapy, as according to later studies in the same (CIA) model, these large VIP doses (~5 nmol/animal) also elicit a massive, 50 % blood pressure drop in mice [57]. This makes treatment of rheumatoid arthritis (or basically any disease) with high doses of free VIP/PACAP an unrealistic goal due to their powerful vasoactive properties, especially in case of PACAP as it is considerably more vasoactive than VIP [58]. In contrast to free VIP, a tenfold lower dose of micellar VIP (0.5 nmol/animal) has been shown to ameliorate arthritis severity more effectively, without having a dramatic hypotensive side effect [57]. Besides its marked effect on inflammatory parameters, treatment of CIA mice/rats with VIP was found to decrease the RANK-RANKL/OPG ratio and to induce an anti-inflammatory CD4+, CD25+, Foxp3+ regulatory T cell (Treg) response [56, 59, 60]. VIP also modulates the activity of important transcription factors (NFkB and AP-1). Furthermore, levels of mediators stimulating bone resorption (IL-1β, IL-6, IL-11, IL-17, TNFα, PGE, NO) are diminished, whereas the levels of protective cytokines (IL-4 and -10) are increased upon VIP-treatment [59]. Additionally, treatment with dendritic cells cultured ex vivo in the presence of VIP ameliorates disease severity in the CIA model [61]. In a rat osteoarthritis (OA) model decreased synovial levels of PACAP were found. In vitro treatment with PACAP also exerts anti-inflammatory effects (decreased apoptosis, iNOS, and COX-2) on isolated rat chondrocytes exposed to IL-1β [62]. In the K/BxN serum-induced passive transfer model of autoimmune arthritis, global PACAP gene-deficiency results in a mixed picture. PACAP knockout mice show diminished disease severity, vascular hyperpermeability, macrophage activity, along with an impaired neutrophil activity in the acute phase of the disease. However, they display increased neutrophil free radical production, increased synovial hyperplasia, and a pronounced arthritic bone spur formation in the later phase of the disease [43]. The observed differences in contrast with the CIA model are likely due to the fact that this passive autoantibody-transfer model activates primarily myeloid lineages, and develops even if B/T cells are entirely absent [63]. Furthermore, the K/BxN serum transfer model relies on an initial rapidly occurring microvascular permeability increase, which is neurogenically driven and also requires mast cell activation [64]. Both PACAP and VIP have been corroborated as key modulators of neurogenic inflammation by eliciting vasodilation, and permeability increase, and they also degranulate mast cells [49, 51, 58, 65, 66]. Thus, the direct effects of PACAP/VIP on mast cells, myeloid cells, and vascular responses might be dominant in the K/BxN serum-transfer model, whilst their immunoregulatory effects on the lymphocytes (notably T cell responses) cannot be detected. Altogether, the above results show that VIP/PACAP ameliorate structural remodeling and damage of cartilage and bone in RA/OA.

It also has to be pointed out that the differences between PACAP-27 and PACAP-38 signaling received little attention in neuropeptide research. Although the latter is the dominant isoform, expression of PACAP-27 is also not negligible in peripheral tissues [67]. However, this could be the causing factor of some conflicting results gathered about the role of the endogenous and exogenously administered peptides (which almost exclusively relied on PACAP-38). It has been demonstrated that key differences exist in their signaling, as PACAP-27, but not PACAP-38 acts as an agonists on the formyl peptide receptor like 1 (FPRL1) receptor by initiating Ca^{2+}-mobilization and ERK phosphorylation. Additionally, PACAP-27 facilitated CD11b upregulation and chemotaxis of neutrophils in vitro [68] in contrast to PACAP-38 and VIP, which were found to inhibit it [69]. According to later studies, not only PACAP-27, but also VIP was found to ho have an agonistic action on the FPRL1 receptor of isolated human monocytes, rising intracellular Ca^{2+} and eliciting pro-inflammatory effects [70, 71]. Furthermore, the FPRL1 receptor has profound implications in RA being expressed on FLS and most inflammatory cells. FPRL1 agonists in vitro increase the secretion of key MMP enzymes by RA fibroblast-like synoviocytes (FLS) (MMP-1, 3) [72]. FPRL1-activation induces IL-1β, TNF-α, and

IL-8 production of human neutrophils [73, 74], furthermore it facilitates synovitis, neovascularization and bone/joint damage in RA [75]. Thus, the possible effect of PACAP-27 via FPRL1-activation might also be involved in the pathophysiological mechanisms of inflammatory arthritides. The in vivo results about the role of VIP/PACAP are summarized in Table 21.1.

Table 21.1 Experimental in vivo studies on the role of VIP/PACAP in inflammatory/degenerative arthritides

Peptide	Model	Observed role	Species	Reference
VIP	Methylated BSA-induced arthritis	Diminution of VIP-positive joint innervation	Mouse	[13]
	Kaolin/carrageenan and adjuvant-induced monoarthritis—topical administration of VIP	VIP induced synovial hyperemia is diminished in acute but not chronic arthritis. Responsiveness can be reestablished by surgical denervation	Rat	[53]
	CIA—systemic treatment with VIP	Decreased disease severity and Th1/Th17 response, decreased RANK-RANKL/OPG ratio	Rat	[56]
	CIA—systemic treatment with tolerogenic dendritic cells generated in the presence of VIP	Decreased disease severity due to the induction of regulatory T cells and the consequent production of TGFβ and IL-10	Mouse	[61]
	CIA—systemic treatment with micellar, low dose VIP	Decreased disease severity without the hypotensive side effect of free VIP	Mouse	[57]
	CIA—systemic treatment with VIP	Decreased disease severity, joint damage, cytokine and chemokine expression, decreased Th1 response, decreased RANK-RANKL/OPG ratio, induction of Treg cells	Mouse	[55, 59, 60]
PACAP	CIA—systemic treatment with PACAP	Decreased disease severity, joint damage, cytokine and chemokine expression, decreased Th1 response	Mouse	[54]
	K/BxN serum-transfer Arthritis—PACAP gene-deficient mice	Increased disease severity, vascular responses, and macrophage activity. Biphasic effect on neutrophil activity. Increased synovial hyperplasia and osteophyte formation	Mouse	[43]
	Anterior cruciate ligament transection model of osteoarthritis	Decreased level of PACAP in the synovial fluid, weaker PACAP immunopositivity on osteoarthritic cartilage	Rat	[62]

PACAP and VIP: Key Mediators of Chronic Joint Pain

In addition to the immunoregulatory/chondroprotective effects of PACAP/VIP, these neuropeptides are also important during the development of chronic arthritic pain, influencing both peripheral and central sensitization processes [76]. The earliest indirect observations suggesting the importance of these peptides showed that the number of PACAP-mRNA-expressing neurons increases significantly in the 5th lumbar dorsal root ganglia (L5 DRG) of rats in the early, but not in the late phase of paw inflammation elicited by intraplantar Freund's complete adjuvant (CFA) injection [77]. In the same CFA model systemic treatment with anti-nerve growth factor (a-NGF) diminishes PACAP expression, which implies that NGF is a positive endogenous modulator of neural PACAP expression during inflammation [78]. Intraarticular injection of VIP into the knee joints of healthy rats elicits increased afferent firing rate and nociception, which can be abolished by co-treatment with the VPAC1/2 antagonist VIP6-28. Furthermore, local VIP6-28 treatment in the sodium monoiodoacetate-induced osteoarthritis model of the rat diminishes nociception. Similarly to VIP, in a knee joint inflammation model of the rat local PACAP-38 injection increases rotation-induced afferent firing—a clear sign of peripheral sensitization. This can be again blocked by VIP6-28 suggesting that VPAC1/2 activation is the central element of VIP/PACAP-induced peripheral sensitization [79–81]. In accordance with these findings, PACAP gene deficiency also results in diminished hyperalgesia in the serum-transfer arthritis model of the mouse [43]. However, taking into consideration the known proinflammatory effects of VIP/PACAP, it is very likely that not a direct action of the peptides *per se*, but rather secondary messengers are responsible for the peripheral sensitization (e.g., via mast cell-derived mediators). Others provided compelling additional evidence about the role of PACAP in central sensitization during chronic arthritic pain, as elevated levels of the peptide have been found in the DRG and spinal cord of mice in the chronic phase of the CIA model [82]. As a conclusion, in contrast to the perplexing complexity and divergent effect of these peptides on the inflammatory components, there is a well-substantiated evidence about the overall algogenic effect of both VIP and PACAP in joint inflammation. This strongly points towards the VPAC receptors, and therefore, selective VPAC-antagonism could provide a novel, peripheral approach to treat joint pain in both inflammatory and degenerative arthritides. A schematic representation of the role of VIP/PACAP in RA is outlined in Fig. 21.1.

Clinical Results About PACAP/VIP and Their Relevance in Rheumatologic Conditions

As early as 1986, it was shown, that VIP-levels are elevated in the synovial fluid, but not in the serum of patients suffering of inflammatory joint disease, which is less prominent in those who receive anti-inflammatory medication (corticosteroids). Furthermore, intra-articular corticosteroid injection was found to diminish

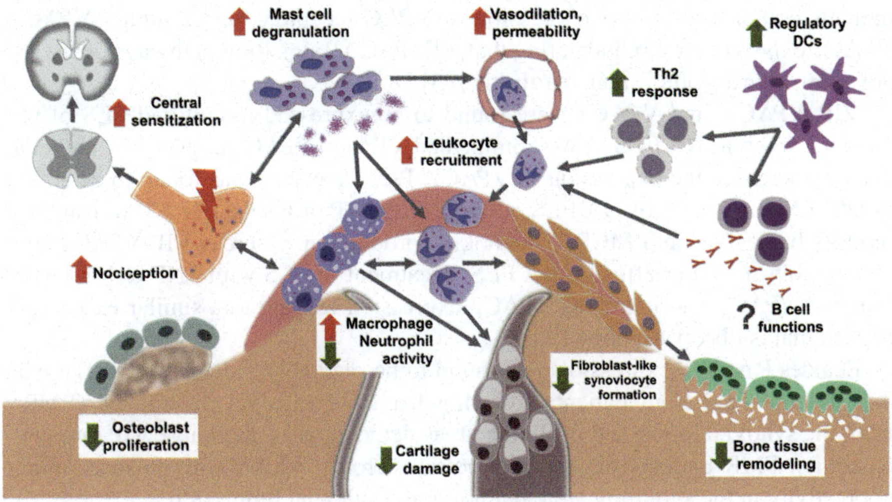

Fig. 21.1 The divergent effects of PACAP/VIP on various aspects of inflammatory/degenerative arthritides: immune components, skeletal tissue cells and nociceptive signaling

VIP-levels dramatically in the synovial fluid [83]. Contradictorily, others found only sporadical presence of VIP in the synovial fluid of both RA patients and control subjects [84]. VIP-immunopositivity was also reported in the sublining area of the synovium in RA patients [85]. Additionally, weak expression of VIP was reported in the temporomandibular joint without any correlation to the disease of the joint (disc derangement or RA) [86]. Recent studies however found slightly, but not significantly decreased serum VIP levels in patients suffering from RA or spondyloarthritis (SpA). Serum VIP levels also showed a weak inverse correlation with the disease activity. Based on this, VIP was proposed as a early prognostic biomarker of RA/SpA, but the reported small differences and considerable variations of serum VIP levels would pose a challenge that remains to be solved [87, 88].

In vitro VIP diminishes the proliferation, proinflammatory cytokine (IL-6, IL-8) and MMP-2-production of isolated whole blood cells of RA patients, and also the activity of the transcription factor CREB which is important for the upregulation of proinflammatory cytokine (e.g., IL-1β) production [22]. VIP was also shown to increase the production of several cytokines (IL-1β, IL-6, and TNF-α) in isolated human leukocytes. This effect was significantly more prominent in samples obtained from RA patients than from healthy or OA subjects [89]. In contrast, in human synovial cells and cultured FLS of OA and RA patients VIP downregulates the production of numerous proinflammatory chemokines (CCL2, CXCL8), cytokines (IL-6, IL-12), and cytokine receptors (IL-17RA) [90, 91]. It was also shown that VIP decreases Toll-like receptor (TLR-4) expression and signaling in RA FLS cultures [92]. It was reported recently, that VIP downregulates the expression of a variety of pathogenic markers (IL-2, IL-21, and STAT3) in cultured memory Th cells of patient suffering from early RA, further underlining that VIP may abrogate RA primarily by dampening Th1/Th17-responses. Furthermore, memory Th cells of RA patients demonstrate lower expression of VPAC receptors, and a higher VPAC2/VPAC1 expression ratio, indicating that VIP/PACAP signaling pathways on T cells are altered during autoimmune arthritis [93].

VIP, VPAC1, and VPAC2 were found to be expressed in cultured FLS of OA patients, whereas RA FLS shows diminished VIP and VPAC1 expression, and conversely increased the expression of VPAC2. PAC1 receptor was found to be absent in all FLS. Treatment of OA FLS cultures with VIP or a selective VPAC1 agonist inhibits IL-6, IL-8, and MCP-1 chemokine production, whereas VIP/VPAC2 agonist exhibits a similar effect in RA FLS. Treatment of FLS with TNF-α diminishes VIP and VPAC1 and increases VPAC2 expression leading to a similar expression pattern that is observed in RA [94].

Besides RA, VIP expression was found to be markedly induced in the synovial biopsies obtained from patients suffering from polymyalgia rheumatica (PMR). Both the synovial lining and sublining areas demonstrate remarkable VIP immunoreactivity. PMR patients treated with corticosteroids exhibit a diminished immunopositivity in the sublining area, but not in the synovial lining [85].

In conclusion, existing clinical data do not provide a clear picture about the dynamics of PACAP/VIP expression during inflammatory joint diseases. A number of confounding factors might cause these inconclusive/conflicting results, such as the lack of properly controlled studies, pooling data coming from different, uncharacterized disease subtypes and stages, known or unknown comorbidities of patients, and the unaccounted effects of ongoing anti-inflammatory medication such as disease-modifying antirheumatic drugs (DMARD) or biologics. Therefore, their early prognostic biomarker value in these clinical conditions is difficult to determine at this stage, and further studies are needed to conclude on that.

Conclusions

PACAP and VIP receptors are present on a broad range of skeletal and immune cells, and the innervating sensory/sympathetic fibers are primarily responsible for the production of these peptide mediators. A growing body of evidence supports the marked involvement of the VIP/PACAP system in the physiological functions of osteoblasts, osteoclasts, and cartilage development, but it is yet to be determined whether they can be utilized as a therapeutic approach to treat/prevent disorders of the skeletal system. PACAP/VIP exhibit interesting and complex immunomodulatory properties during inflammatory/degenerative arthritis, but establishing their clinical significance and relation to existing therapeutic agents needs further investigations. This would be a particularly pertinent question as already existing therapeutic modalities (DMARDs, biologics) might possess an ability to influence the local levels of these peptide mediators. Methotrexate (MTX) for example has been shown to facilitate the dipeptidyl peptidase IV (DPPIV) activity of stimulated macrophages. Decreased DPPIV activity was observed in peripheral mononuclear cells in preclinical arthritis models [95, 96]. Since VIP and PACAP-38 (but not

PACAP-27) are degraded primarily via DPPIV [97–99], it is feasible that MTX or other DMARDs might restore the normal turnover of VIP/PACAP or other peptidergic mediators in diseased tissues, ultimately contributing to their therapeutic effects. In summary, further studies are needed to determine the potential diagnostic and/or therapeutic value of the VIP/PACAP system in inflammatory and degenerative joint diseases. Finally, due to their algogenic properties and importance during peripheral/central sensitization, targeting PACAP and VIP-signaling should be considered as an alternative approach to prevent or treat chronic, persistent, therapy-resistant, poorly manageable joint/bone pain.

References

1. Hukkanen M, Konttinen YT, Rees RG, Santavirta S, Terenghi G, Polak JM. Distribution of nerve endings and sensory neuropeptides in rat synovium, meniscus and bone. Int J Tissue React. 1992;14:1–10.
2. Ackermann PW, Finn A, Ahmed M. Sensory neuropeptidergic pattern in tendon, ligament and joint capsule. A study in the rat. Neuroreport. 1999;10:2055–60.
3. Lerner UH, Persson E. Osteotropic effects by the neuropeptides calcitonin gene-related peptide, substance P and vasoactive intestinal peptide. J Musculoskelet Neuronal Interact. 2008;8:154–65.
4. Bjurholm A, Kreicbergs A, Brodin E, Schultzberg M. Substance P- and CGRP-immunoreactive nerves in bone. Peptides. 1988;9:165–71.
5. Bjurholm A, Kreicbergs A, Terenius L, Goldstein M, Schultzberg M. Neuropeptide Y-, tyrosine hydroxylase- and vasoactive intestinal polypeptide-immunoreactive nerves in bone and surrounding tissues. J Auton Nerv Syst. 1988;25:119–25.
6. Hill EL, Elde R. Distribution of CGRP-, VIP-, D beta H-, SP-, and NPY-immunoreactive nerves in the periosteum of the rat. Cell Tissue Res. 1991;264:469–80.
7. Bring DK, Heidgren ML, Kreicbergs A, Ackermann PW. Increase in sensory neuropeptides surrounding the Achilles tendon in rats with adjuvant arthritis. J Orthop Res. 2005;23:294–301.
8. Grönblad M, Liesi P, Korkala O, Karaharju E, Polak J. Innervation of human bone periosteum by peptidergic nerves. Anat Rec. 1984;209:297–9.
9. Vaudry D, Falluel-Morel A, Bourgault S, Basille M, Burel D, Wurtz O, et al. Pituitary adenylate cyclase-activating polypeptide and its receptors: 20 years after the discovery. Pharmacol Rev. 2009;61:283–357.
10. Juhász T, Helgadottir SL, Tamás A, Reglődi D, Zákány R. PACAP and VIP signaling in chondrogenesis and osteogenesis. Peptides. 2015;66:51–7.
11. Hohmann EL, Elde RP, Rysavy JA, Einzig S, Gebhard RL. Innervation of periosteum and bone by sympathetic vasoactive intestinal peptide-containing nerve fibers. Science. 1986;232:868–71.
12. Uddman R, Grunditz T, Kato J, Sundler F. Distribution and origin of nerve fibers in the rat temporomandibular joint capsule. Anat Embryol (Berl). 1998;197:273–82.
13. Buma P, Elmans L, Van Den Berg WB, Schrama LH. Neurovascular plasticity in the knee joint of an arthritic mouse model. Anat Rec. 2000;260:51–61.
14. Strange-Vognsen HH, Arnbjerg J, Hannibal J. Immunocytochemical demonstration of pituitary adenylate cyclase activating polypeptide (PACAP) in the porcine epiphyseal cartilage canals. Neuropeptides. 1997;31:137–41.
15. Juhász T, Matta C, Katona É, Somogyi C, Takács R, Gergely P, et al. Pituitary adenylate cyclase activating polypeptide (PACAP) signalling exerts chondrogenesis promoting and protecting effects: implication of calcineurin as a downstream target. PLoS One. 2014;9:e91541.

16. Lundberg P, Lie A, Bjurholm A, Lehenkari PP, Horton MA, Lerner UH, et al. Vasoactive intestinal peptide regulates osteoclast activity via specific binding sites on both osteoclasts and osteoblasts. Bone. 2000;27:803–10.
17. Hohmann EL, Tashjian Jr AH. Functional receptors for vasoactive intestinal peptide on human osteosarcoma cells. Endocrinology. 1984;114:1321–7.
18. Togari A, Arai M, Mizutani S, Mizutani S, Koshihara Y, Nagatsu T. Expression of mRNAs for neuropeptide receptors and beta-adrenergic receptors in human osteoblasts and human osteogenic sarcoma cells. Neurosci Lett. 1997;233:125–8.
19. Lundberg P, Lundgren I, Mukohyama H, Lehenkari PP, Horton MA, Lerner UH. Vasoactive intestinal peptide (VIP)/pituitary adenylate cyclase-activating peptide receptor subtypes in mouse calvarial osteoblasts: presence of VIP-2 receptors and differentiation-induced expression of VIP-1 receptors. Endocrinology. 2001;142:339–47.
20. Nagata A, Tanaka T, Minezawa A, Poyurovsky M, Mayama T, Suzuki S, et al. cAMP activation by PACAP/VIP stimulates IL-6 release and inhibits osteoblastic differentiation through VPAC2 receptor in osteoblastic MC3T3 cells. J Cell Physiol. 2009;221:75–83.
21. Ransjö M, Lie A, Mukohyama H, Lundberg P, Lerner UH. Microisolated mouse osteoclasts express VIP-1 and PACAP receptors. Biochem Biophys Res Commun. 2000;274(2):400–4.
22. Takeba Y, Suzuki N, Kaneko A, Asai T, Sakane T. Evidence for neural regulation of inflammatory synovial cell functions by secreting calcitonin gene-related peptide and vasoactive intestinal peptide in patients with rheumatoid arthritis. Arthritis Rheum. 1999;42:2418–29.
23. Hohmann EL, Levine L, Tashjian Jr AH. Vasoactive intestinal peptide stimulates bone resorption via a cyclic adenosine 3′,5′-monophosphate-dependent mechanism. Endocrinology. 1983;112:1233–9.
24. Rahman S, Dobson PR, Bunning RA, Russell RG, Brown BL. The regulation of connective tissue metabolism by vasoactive intestinal polypeptide. Regul Pept. 1992;37:111–21.
25. Lerner UH, Lundberg P, Ransjö M, Persson P, Håkanson R. Helodermin, helospectin, and PACAP stimulate cyclic AMP formation in intact bone, isolated osteoblasts, and osteoblastic cell lines. Calcif Tissue Int. 1994;54:284–9.
26. Suzuki A, Kotoyori J, Oiso Y, Kozawa O. Pituitary adenylate cyclase-activating polypeptide induces cAMP production independently from vasoactive intestinal polypeptide in osteoblast-like cells. Cell Signal. 1994;6:11–6.
27. Kovacs CS, Chik CL, Li B, Karpinski E, Ho AK. Pituitary adenylate cyclase-activating peptide stimulates cyclic AMP accumulation in UMR 106 osteoblast-like cells. J Endocrinol. 1996;149:287–95.
28. Mukohyama H, Ransjö M, Taniguchi H, Ohyama T, Lerner UH. The inhibitory effects of vasoactive intestinal peptide and pituitary adenylate cyclase-activating polypeptide on osteoclast formation are associated with upregulation of osteoprotegerin and downregulation of RANKL and RANK. Biochem Biophys Res Commun. 2000;271:158–63.
29. Persson E, Lerner UH. The neuropeptide VIP regulates the expression of osteoclastogenic factors in osteoblasts. J Cell Biochem. 2011;112:3732–41.
30. Lundberg P, Boström I, Mukohyama H, Bjurholm A, Smans K, Lerner UH. Neuro-hormonal control of bone metabolism: vasoactive intestinal peptide stimulates alkaline phosphatase activity and mRNA expression in mouse calvarial osteoblasts as well as calcium accumulation mineralized bone nodules. Regul Pept. 1999;85:47–58.
31. Kanno Y, Ishisaki A, Yoshida M, Nakajima K, Tokuda H, Numata O, et al. Adenylyl cyclase-cAMP system inhibits thyroid hormone-stimulated osteocalcin synthesis in osteoblasts. Mol Cell Endocrinol. 2005;229:75–82.
32. Persson E, Lerner UH. The neuropeptide VIP potentiates IL-6 production induced by proinflammatory osteotropic cytokines in calvarial osteoblasts and the osteoblastic cell line MC3T3-E1. Biochem Biophys Res Commun. 2005;335:705–11.
33. Persson E, Voznesensky OS, Huang YF, Lerner UH. Increased expression of interleukin-6 by vasoactive intestinal peptide is associated with regulation of CREB, AP-1 and C/EBP, but not NF-kappaB, in mouse calvarial osteoblasts. Bone. 2005;37:513–29.
34. Winding B, Wiltink A, Foged NT. Pituitary adenylyl cyclase-activating polypeptides and vasoactive intestinal peptide inhibit bone resorption by isolated rabbit osteoclasts. Exp Physiol. 1997;82:871–86.

35. Juhász T, Matta C, Katona É, Somogyi C, Takács R, Hajdú T, et al. Pituitary adenylate cyclase-activating polypeptide (PACAP) signalling enhances osteogenesis in UMR-106 cell line. J Mol Neurosci. 2014;54:555–73.
36. Juhász T, Szentléleky E, Somogyi CS, Takács R, Dobrosi N, Engler M, et al. Pituitary adenylate cyclase activating polypeptide (PACAP) pathway is induced by mechanical load and reduces the activity of hedgehog signaling in chondrogenic micromass cell cultures. Int J Mol Sci. 2015;16:17344–67.
37. Jamen F, Persson K, Bertrand G, Rodriguez-Henche N, Puech R, Bockaert J, et al. PAC1 receptor-deficient mice display impaired insulinotropic response to glucose and reduced glucose tolerance. J Clin Invest. 2000;105:1307–15.
38. Goetzl EJ, Voice JK, Shen S, Dorsam G, Kong Y, West KM, et al. Enhanced delayed-type hypersensitivity and diminished immediate-type hypersensitivity in mice lacking the inducible VPAC(2) receptor for vasoactive intestinal peptide. Proc Natl Acad Sci U S A. 2001;98: 13854–9.
39. Asnicar MA, Köster A, Heiman ML, Tinsley F, Smith DP, Galbreath E, et al. Vasoactive intestinal polypeptide/pituitary adenylate cyclase-activating peptide receptor 2 deficiency in mice results in growth retardation and increased basal metabolic rate. Endocrinology. 2002;143:3994–4006.
40. Colwell CS, Michel S, Itri J, Rodriguez W, Tam J, Lelievre V, et al. Disrupted circadian rhythms in VIP- and PHI-deficient mice. Am J Physiol Regul Integr Comp Physiol. 2003;285:R939–49.
41. Hashimoto H, Shintani N, Baba A. New insights into the central PACAPergic system from the phenotypes in PACAP- and PACAP receptor-knockout mice. Ann N Y Acad Sci. 2006;1070:75–89.
42. Fabricius D, Karacay B, Shutt D, Leverich W, Schafer B, Takle E, et al. Characterization of intestinal and pancreatic dysfunction in VPAC1-null mutant mouse. Pancreas. 2011;40: 861–71.
43. Botz B, Bölcskei K, Kereskai L, Kovács M, Németh T, Szigeti K, et al. Differential regulatory role of pituitary adenylate cyclase-activating polypeptide in the serum-transfer arthritis model. Arthritis Rheumatol. 2014;66:2739–50.
44. Levine JD, Khasar SG, Green PG. Neurogenic inflammation and arthritis. Ann N Y Acad Sci. 2006;1069:155–67.
45. Borbély É, Botz B, Bölcskei K, Kenyér T, Kereskai L, Kiss T, et al. Capsaicin-sensitive sensory nerves exert complex regulatory functions in the serum-transfer mouse model of autoimmune arthritis. Brain Behav Immun. 2015;45:50–9.
46. Bland JH, Eddy WM. Hemiplegia and rheumatoid hemiarthritis. Arthritis Rheum. 1968;11:72–80.
47. Kane D, Lockhart JC, Balint PV, Mann C, Ferrell WR, McInnes IB. Protective effect of sensory denervation in inflammatory arthritis (evidence of regulatory neuroimmune pathways in the arthritic joint). Ann Rheum Dis. 2005;64:325–7.
48. Kim CW, Kim MJ, Park SB, Han SH. A case of rheumatoid arthritis with unilateral knee synovial hypertrophy in hemiplegia. Ann Rehabil Med. 2012;36:144–7.
49. Lowman MA, Benyon RC, Church MK. Characterization of neuropeptide-induced histamine release from human dispersed skin mast cells. Br J Pharmacol. 1988;95:121–30.
50. Seebeck J, Kruse ML, Schmidt-Choudhury A, Schmidtmayer J, Schmidt WE. Pituitary adenylate cyclase activating polypeptide induces multiple signaling pathways in rat peritoneal mast cells. Eur J Pharmacol. 1998;352:343–50.
51. Schmidt-Choudhury A, Furuta GT, Galli SJ, Schmidt WE, Wershil BK. Mast cells contribute to PACAP-induced dermal oedema in mice. Regul Pept. 1999;82:65–9.
52. Schmidt-Choudhury A, Meissner J, Seebeck J, Goetzl EJ, Xia M, Galli SJ, et al. Stem cell factor influences neuro-immune interactions: the response of mast cells to pituitary adenylate cyclase activating polypeptide is altered by stem cell factor. Regul Pept. 1999;83:73–80.
53. McDougall JJ, Barin AK. The role of joint nerves and mast cells in the alteration of vasoactive intestinal peptide (VIP) sensitivity during inflammation progression in rats. Br J Pharmacol. 2005;145:104–13.

54. Abad C, Martinez C, Leceta J, Gomariz RP, Delgado M. Pituitary adenylate cyclase-activating polypeptide inhibits collagen-induced arthritis: an experimental immunomodulatory therapy. J Immunol. 2001;167:3182–9.
55. Delgado M, Abad C, Martinez C, Leceta J, Gomariz RP. Vasoactive intestinal peptide prevents experimental arthritis by downregulating both autoimmune and inflammatory components of the disease. Nat Med. 2001;7:563–8.
56. Deng S, Xi Y, Wang H, Hao J, Niu X, Li W, et al. Regulatory effect of vasoactive intestinal peptide on the balance of Treg and Th17 in collagen-induced arthritis. Cell Immunol. 2010;265:105–10.
57. Sethi V, Rubinstein I, Kuzmis A, Kastrissios H, Artwohl J, Onyuksel H. Novel, biocompatible, and disease modifying VIP nanomedicine for rheumatoid arthritis. Mol Pharm. 2013;10:728–38.
58. Warren JB, Donnelly LE, Cullen S, Robertson BE, Ghatei MA, Bloom SR, et al. Pituitary adenylate cyclase-activating polypeptide: a novel, long-lasting, endothelium-independent vasorelaxant. Eur J Pharmacol. 1991;197:131–4.
59. Juarranz Y, Abad C, Martinez C, Arranz A, Gutierrez-Cañas I, Rosignoli F, et al. Protective effect of vasoactive intestinal peptide on bone destruction in the collagen-induced arthritis model of rheumatoid arthritis. Arthritis Res Ther. 2005;7:R1034–45.
60. Gonzalez-Rey E, Fernandez-Martin A, Chorny A, Delgado M. Vasoactive intestinal peptide induces CD4+, CD25+ T regulatory cells with therapeutic effect in collagen-induced arthritis. Arthritis Rheum. 2006;54:864–76.
61. Chorny A, Gonzalez-Rey E, Fernandez-Martin A, Pozo D, Ganea D, Delgado M. Vasoactive intestinal peptide induces regulatory dendritic cells with therapeutic effects on autoimmune disorders. Proc Natl Acad Sci U S A. 2005;102:13562–7.
62. Giunta S, Castorina A, Marzagalli R, Szychlinska MA, Pichler K, Mobasheri A, et al. Ameliorative effects of PACAP against cartilage degeneration. morphological, immunohistochemical and biochemical evidence from in vivo and in vitro models of rat osteoarthritis. Int J Mol Sci. 2015;16:5922–44.
63. Korganow AS, Ji H, Mangialaio S, Duchatelle V, Pelanda R, Martin T, et al. From systemic T cell self-reactivity to organ-specific autoimmune disease via immunoglobulins. Immunity. 1999;10:451–61.
64. Stangenberg L, Burzyn D, Binstadt BA, Weissleder R, Mahmood U, Benoist C, et al. Denervation protects limbs from inflammatory arthritis via an impact on the microvasculature. Proc Natl Acad Sci U S A. 2014;111:11419–24.
65. Kulka M, Sheen CH, Tancowny BP, Grammer LC, Schleimer RP. Neuropeptides activate human mast cell degranulation and chemokine production. Immunology. 2008;123:398–410.
66. Schytz HW, Holst H, Arendt-Nielsen L, Olesen J, Ashina M. Cutaneous nociception and neurogenic inflammation evoked by PACAP38 and VIP. J Headache Pain. 2010;11:309–16.
67. Arimura A, Somogyvári-Vigh A, Miyata A, Mizuno K, Coy DH, Kitada C. Tissue distribution of PACAP as determined by RIA: highly abundant in the rat brain and testes. Endocrinology. 1991;129:2787–9.
68. Kim Y, Lee BD, Kim O, Bae YS, Lee T, Suh PG, et al. Pituitary adenylate cyclase-activating polypeptide 27 is a functional ligand for formyl peptide receptor-like 1. J Immunol. 2006;176:2969–75.
69. Kinhult J, Uddman R, Laan M, Lindén A, Cardell LO. Pituitary adenylate cyclase-activating peptide inhibits neutrophil chemotaxis. Peptides. 2001;22:2151–4.
70. El Zein N, Badran B, Sariban E. The neuropeptide pituitary adenylate cyclase activating polypeptide modulates Ca2+ and pro-inflammatory functions in human monocytes through the G protein-coupled receptors VPAC-1 and formyl peptide receptor-like 1. Cell Calcium. 2008;43:270–84.
71. El Zein N, Badran B, Sariban E. VIP differentially activates beta2 integrins, CR1, and matrix metalloproteinase-9 in human monocytes through cAMP/PKA, EPAC, and PI-3K signaling pathways via VIP receptor type 1 and FPRL1. J Leukoc Biol. 2008;83:972–81.

72. O'Hara R, Murphy EP, Whitehead AS, FitzGerald O, Bresnihan B. Local expression of the serum amyloid A and formyl peptide receptor-like 1 genes in synovial tissue is associated with matrix metalloproteinase production in patients with inflammatory arthritis. Arthritis Rheum. 2004;50:1788–99.
73. Furlaneto CJ, Campa A. A novel function of serum amyloid A: a potent stimulus for the release of tumor necrosis factor-alpha, interleukin-1beta, and interleukin-8 by human blood neutrophil. Biochem Biophys Res Commun. 2000;268:405–8.
74. He R, Sang H, Ye RD. Serum amyloid A induces IL-8 secretion through a G protein-coupled receptor, FPRL1/LXA4R. Blood. 2003;101:1572–81.
75. Szekanecz Z, Besenyei T, Paragh G, Koch AE. Angiogenesis in rheumatoid arthritis. Autoimmunity. 2009;42:563–73.
76. Tajti J, Tuka B, Botz B, Helyes Z, Vecsei L. Role of pituitary adenylate-cyclase-activating polypeptide in nociception and migraine. CNS Neurol Disord Drug Targets. 2015;14:540–53.
77. Zhang Y, Danielsen N, Sundler F, Mulder H. Pituitary adenylate cyclase-activating peptide is upregulated in sensory neurons by inflammation. Neuroreport. 1998;9:2833–6.
78. Jongsma Wallin H, Pettersson LM, Verge VM, Danielsen N. Effect of anti-nerve growth factor treatment on pituitary adenylate cyclase activating polypeptide expression in adult sensory neurons exposed to adjuvant induced inflammation. Neuroscience. 2003;120:325–31.
79. Schuelert N, McDougall JJ. Electrophysiological evidence that the vasoactive intestinal peptide receptor antagonist VIP6-28 reduces nociception in an animal model of osteoarthritis. Osteoarthritis Cartilage. 2006;14:1155–62.
80. McDougall JJ, Watkins L, Li Z. Vasoactive intestinal peptide (VIP) is a modulator of joint pain in a rat model of osteoarthritis. Pain. 2006;123:98–105.
81. Sándor K, Bölcskei K, McDougall JJ, Schuelert N, Reglodi D, Elekes K, et al. Divergent peripheral effects of pituitary adenylate cyclase-activating polypeptide-38 on nociception in rats and mice. Pain. 2009;141:143–50.
82. Sándor K, Bas DB, Petterson J, Gregory J, Christianson C, Codeluppi S, et al. Collagen antibody-induced arthritis (CAIA) evokes transient inflammation but persistent allodynia. Ann Rheum Dis. 2011;70:A90.
83. Lygren I, Ostensen M, Burhol PG, Husby G. Gastrointestinal peptides in serum and synovial fluid from patients with inflammatory joint disease. Ann Rheum Dis. 1986;45:637–40.
84. Larsson J, Ekblom A, Henriksson K, Lundeberg T, Theodorsson E. Concentration of substance P, neurokinin A, calcitonin gene-related peptide, neuropeptide Y and vasoactive intestinal polypeptide in synovial fluid from knee joints in patients suffering from rheumatoid arthritis. Scand J Rheumatol. 1991;20:326–35.
85. Pulsatelli L, Dolzani P, Silvestri T, De Giorgio R, Salvarani C, Macchioni P, et al. Synovial expression of vasoactive intestinal peptide in polymyalgia rheumatica. Clin Exp Rheumatol. 2006;24:562–6.
86. Holmlund A, Ekblom A, Hansson P, Lind J, Lundeberg T, Theodorsson E. Concentrations of neuropeptides substance P, neurokinin A, calcitonin gene-related peptide, neuropeptide Y and vasoactive intestinal polypeptide in synovial fluid of the human temporomandibular joint. A correlation with symptoms, signs and arthroscopic findings. Int J Oral Maxillofac Surg. 1991;20:228–31.
87. Martínez C, Ortiz AM, Juarranz Y, Lamana A, Seoane IV, Leceta J, et al. Serum levels of vasoactive intestinal peptide as a prognostic marker in early arthritis. PLoS One. 2014;9:e85248.
88. Seoane IV, Tomero E, Martínez C, Garcia-Vicuña R, Juarranz Y, Lamana A, et al. Vasoactive intestinal peptide in early spondyloarthritis: low serum levels as a potential biomarker for disease severity. J Mol Neurosci. 2015;25. Epub ahead of print.
89. Hernanz A, Medina S, de Miguel E, Martín-Mola E. Effect of calcitonin gene-related peptide, neuropeptide Y, substance P, and vasoactive intestinal peptide on interleukin-1beta, interleukin-6 and tumor necrosis factor-alpha production by peripheral whole blood cells from rheumatoid arthritis and osteoarthritis patients. Regul Pept. 2003;115:19–24.
90. Juarranz MG, Santiago B, Torroba M, Gutierrez-Cañas I, Palao G, Galindo M, et al. Vasoactive intestinal peptide modulates proinflammatory mediator synthesis in osteoarthritic and rheumatoid synovial cells. Rheumatology (Oxford). 2004;43:416–22.

91. Carrión M, Pérez-García S, Jimeno R, Juarranz Y, González-Álvaro I, Pablos JL, et al. Inflammatory mediators alter interleukin-17 receptor, interleukin-12 and -23 expression in human osteoarthritic and rheumatoid arthritis synovial fibroblasts: immunomodulation by vasoactive intestinal Peptide. Neuroimmunomodulation. 2013;20:274–84.
92. Gutiérrez-Cañas I, Juarranz Y, Santiago B, Arranz A, Martinez C, Galindo M, et al. VIP down-regulates TLR4 expression and TLR4-mediated chemokine production in human rheumatoid synovial fibroblasts. Rheumatology (Oxford). 2006;45:527–32.
93. Jimeno R, Gomariz RP, Garín M, Gutiérrez-Cañas I, González-Álvaro I, Carrión M, et al. The pathogenic Th profile of human activated memory Th cells in early rheumatoid arthritis can be modulated by VIP. J Mol Med (Berl). 2015;93:457–67.
94. Juarranz Y, Gutiérrez-Cañas I, Santiago B, Carrión M, Pablos JL, Gomariz RP. Differential expression of vasoactive intestinal peptide and its functional receptors in human osteoarthritic and rheumatoid synovial fibroblasts. Arthritis Rheum. 2008;58:1086–95.
95. Olivo RA, Nascimento NG, Teixeira CF, Silveira PF. Methotrexate and cyclosporine treatments modify the activities of dipeptidyl peptidase IV and prolyl oligopeptidase in murine macrophages. Clin Dev Immunol. 2008;2008:794050.
96. Yamasaki SC, Murari-do-Nascimento S, Silveira PF. Neutral aminopeptidase and dipeptidyl peptidase IV in the development of collagen II-induced arthritis. Regul Pept. 2012;173:47–54.
97. Lambeir AM, Durinx C, Proost P, Van Damme J, Scharpé S, De Meester I. Neutral aminopeptidase and dipeptidyl peptidase IV in the development of collagen II-induced arthritis. FEBS Lett. 2001;507:327–30.
98. Zhu L, Tamvakopoulos C, Xie D, Dragovic J, Shen X, Fenyk-Melody JE, et al. The role of dipeptidyl peptidase IV in the cleavage of glucagon family peptides: in vivo metabolism of pituitary adenylate cyclase activating polypeptide-(1-38). J Biol Chem. 2003;278:22418–23.
99. Bourgault S, Vaudry D, Botia B, Couvineau A, Laburthe M, Vaudry H, et al. Novel stable PACAP analogs with potent activity towards the PAC1 receptor. Peptides. 2008;29:919–32.

Part VII
PACAP in the Regulation of the Reproductive System

Chapter 22
Role of PACAP and Its PACAP Type I Receptor in the Central Control of Reproductive Hormones

Haruhiko Kanasaki, Aki Oride, Mijiddorj Tselmeg, Unurjargal Sukhbaatar, and Satoru Kyo

Abstract The regulation of sexual maturation and reproductive function requires precise cooperation of hormonal regulation at hypothalamic, pituitary, and gonadal levels. Gonadotropin-releasing hormone (GnRH), which is released from the hypothalamus in a pulsatile manner, regulates synthesis and secretion of the pituitary gonadotropins: luteinizing hormone (LH) and follicle-stimulating hormone (FSH). Gonadotropins, in turn, are secreted into the systemic circulation and act on the gonads to regulate follicular development, steroidogenesis, and gametogenesis. GnRH and gonadotropins are key players in controlling female reproductive functions. In addition, it has gradually become clear that kisspeptin neurons in the hypothalamus activate GnRH neurons through G protein-coupled receptor 54 (GPR54), and many investigations thus far have advanced our understanding of the physiological control of the hypothalamic–pituitary–gonadal axis.

Pituitary adenylate cyclase-activating polypeptide (PACAP) type I receptor (PAC1R) is a G protein-coupled receptor that is activated by PACAP. PACAP is present in two bioactive amidated forms, PACAP38 and PACAP28, and was first isolated as a hypothalamic activator of cAMP production in pituitary cells. PAC1R and PACAP are expressed not only in the central nervous system, but also in peripheral organs, and function in the hypothalamus and anterior pituitary alone or in cooperation with other hypothalamic neuropeptides.

In this review, we summarize the current state of knowledge on how PACAP and PAC1R affect the central control of reproductive functions. The effect of PACAP and its PAC1R on pituitary lactotrophs, gonadotrophs, and GnRH producing neurons is mainly described.

Keywords PACAP • PAC1 receptor • TRH • GnRH • Prolactin • Gonadotropins

---

H. Kanasaki (✉) • A. Oride • M. Tselmeg • U. Sukhbaatar • S. Kyo
Department of Obstetrics and Gynecology, School of Medicine, Shimane University, 89-1 Enya Cho, Izumo, Shimane Prefecture 693-8501, Japan
e-mail: kanasaki@med.shimane-u.ac.jp

D. Reglodi, A. Tamas (eds.), *Pituitary Adenylate Cyclase Activating Polypeptide — PACAP*, Current Topics in Neurotoxicity 11,
DOI 10.1007/978-3-319-35135-3_22

Introduction

Pituitary adenylate cyclase-activating polypeptide (PACAP) was first isolated from ovine hypothalamus extract as a novel regulatory peptide that stimulates adenylate cyclase activity in anterior pituitary cells [1]. Because hypothalamic neurons containing PACAP project toward the median eminence and terminate in the perivascular region of the hypothalamo-pituitary portal circulation, PACAP was initially viewed as a hypothalamic-hypophysiotropic neuropeptide. Subsequently, however, it was shown that PACAP and its receptors are expressed not only in the central nervous system, but also in most peripheral organs where PACAP exerts its effects on various tissues and physiological functions.

Distribution of PACAP

Miyata and his coworkers [1] in Arimura's laboratory isolated a peptide in pure form that markedly increased cAMP accumulation in pituitary cells, and sequencing of the peptide revealed that it comprised 38 amino acid residues (PACAP38). PACAP27, which corresponds to the N-terminal 27 amino acids of PACAP38, was subsequently discovered as another peptide capable of stimulating adenylate cyclase activity in adenohypophyseal cells [2]. Because the sequence of PACAP27 shares 68 % identity with vasoactive intestinal polypeptide (VIP), PACAP is recognized as a highly conserved member of the VIP–secretin–glucagon peptide superfamily [3]. Although radioimmunoassays revealed that the highest concentration of PACAP is found in the hypothalamic area [4], PACAP-containing neurons are not restricted to the hypothalamic area but are widely distributed in various brain regions including the cerebral cortex, amygdala, hippocampus, pineal gland, substantia nigra, cerebellum, and pons [5–7]. In the hypothalamus, PACAP-immunoreactive neurons are located primarily in the parvocellular and magnocellular neurons of the paraventricular (PVN) and supraoptic nuclei [8, 9]. Furthermore, PACAP mRNA is expressed in the PVN and arcuate nucleus (ARC) [7]. PACAP-containing neuronal fibers are connected to the anterior pituitary through the hypophyseal portal blood vessels and work as a hypothalamic-pituitary releasing factor [10–12]. PACAP released into the blood circulation exhibits poor stability with a half-life of 5–10 min [13]. PACAP is variably expressed in peripheral tissues including the peripheral nervous system [14], eye [15], pituitary [16, 17], adrenal gland [18], pancreas [19], mammary gland [20], gastrointestinal tract [21], lymphoid tissues [22], gonads [23, 24], placenta [25], and urinary tract and uterus [23, 26]. The ratio of PACAP27 and PACAP38 varies between the different organs [4]. In contrast to the central nervous system, PACAP and VIP are often found to be coexpressed in the same cells in peripheral organs [27, 28].

Receptors for PACAP and Its Distribution

Three PACAP receptors have been cloned in vertebrates thus far and have been termed the PAC1, VPAC1, and VPAC2 receptors according to their relative affinity for PACAP and VIP. The PAC1 receptor (PAC1R) is relatively specific for PACAP38 and PACAP27 with similar affinities, but VIP does not efficiently bind PAC1R. In contrast, PACAP38, PACAP27, and VIP bind VPAC1 and VPAC2 receptors with equal affinity. The distribution and relative density of PACAP receptor mRNAs have been clearly demonstrated. In the central nervous system, PAC1R expression is generally abundant and widely distributed compared with VPAC1 and VPAC2 receptors. Receptors for PACAP have also been identified with widespread distribution in peripheral organs and cumulative knowledge regarding PACAP and its receptors were described in a comprehensive review [29].

These observations and cumulative knowledge regarding PACAP were described in a comprehensive review [29]. PACAP initiates activation of intracellular signaling pathways including the phospholipase C (PLC)/protein kinase C (PKC)/calcium and adenylate cyclase/protein kinase A (PKA) pathways. PACAP also activates extracellular signal-regulated kinase (ERK) [30, 31]. The predominant PAC1R mainly couples with Gs protein and induces cAMP accumulation, which activates PKA [1].

Action of PACAP and Its Receptor in Anterior Pituitary Cells

Evidence that PACAP acts as a hypothalamic-pituitary releasing factor includes the observation that PACAP is localized in the hypothalamus, the physical connection between PACAP-expressing neuronal fibers and the median eminence as well as pituitary stalks, and the existence of its receptors in anterior pituitary cells [16, 32]. Indeed, PACAP stimulates the release of growth hormone, adrenocorticotropin, gonadotropins (luteinizing hormone; LH, follicle-stimulating hormone; FSH), prolactin, and thyroid-stimulating hormone [1, 33]. In addition, PACAP is produced within the anterior pituitary gland, and is expressed by gonadotrophs and folliculo-stellate cells, but not by other hormone-producing cells [17, 34–36].

PACAP in Prolactin-Producing Cells

Prolactin is released from lactotrophs or somatolactotrophs in the anterior pituitary gland. Although the synthesis and secretion of prolactin are predominantly under the inhibitory control of dopamine through dopamine D_2 receptors [37], prolactin synthesis and secretion is also regulated by several hypothalamic hormones such as thyrotropin-releasing hormone (TRH). TRH is the primary secretagogue for thyrotropin-stimulating hormone from thyrotrophs and regulates prolactin secretion, although it is unclear at present whether TRH is a physiological regulator of prolactin

[38]. VIP and peptide histidine also participate in prolactin regulation [39, 40]. In fact, PACAP has a stimulatory effect on prolactin secretion in vivo. A previous study demonstrated that intravenous injection of PACAP into rats induces an increase in plasma prolactin concentration [41]. Consistent with this notion, PACAP-knockout animals were shown to have decreased circulating prolactin levels [42]. In contrast, however, studies investigating the effect of PACAP on pituitary cells have led to controversial results. PACAP was reported to be devoid of prolactin-releasing activity in cultured rat adenohypophyseal cells [43] and in ovine [44] and bovine [45] pituitary cultures. It is possible that PACAP may also regulate prolactin by modulating various hypothalamic factors. However, PACAP receptors are indeed expressed in all cell types in the rat adenohypophysis [16], and it is also evident that pituitary prolactin-producing cells express PAC1R in rats [46]. Our previous study using GH3 cells, a rat somatolactotroph cell line, demonstrated that PACAP has a stimulatory effect on these cells and accelerates prolactin synthesis. In GH3 cells, PACAP increased the transcription rate of prolactin mRNA in a dose-dependent manner; however, its effect was very limited, probably because the expression level of PACAP receptors was relatively low [30]. In the same study, the stimulatory effect of PACAP on prolactin expression was dramatically increased by PAC1R-overexpression in GH3 cells. In addition, we found that PACAP itself increased endogenous PAC1R expression [30].

TRH and PACAP in Prolactin Producing Cells

TRH, a known secretagogue for prolactin, binds to TRH receptors within lactotrophs and stimulates the synthesis and release of prolactin. After binding to its receptor, TRH stimulates inositol phospholipid metabolism by activating receptors and accelerates its signaling cascades, which include PKC activation and calcium release from storage sites [47]. Extracelluar signal-regulated kinase (ERK) activation by TRH is mediated by both PKC-dependent and -independent pathways [48, 49]. There is much evidence supporting the involvement of ERK in prolactin gene expression [49–51]. In experiments using GH3 cells overexpressing PAC1R, PACAP dramatically increased prolactin promoter activity with a concomitant increase in ERK phosphorylation. In addition, transfection with a higher amount of PAC1R expression vector led to a further increase in prolactin promoter activity in response to PACAP. Moreover, combined treatment with PACAP and TRH further increased prolactin promoter activation compared with that stimulated by each factor alone [30]. These results strongly suggest that PACAP is a regulator of prolactin in lactotrophs and works cooperatively with TRH.

Interaction Between PAC1R and TRH Receptor

Lactotrophs express both the TRH receptor and PAC1R. In experiments using GH3 cells, PACAP enhanced the stimulatory action of TRH on the prolactin promoter. PACAP has been found to decrease the activity of pyroglutamyl peptidase II [52], a

TRH-specific ectoenzyme that cleaves the pyroglutamyl-histidyl peptide bond of TRH [53]. TRH is a secretagogue for prolactin; thus, the inhibition of pyroglutamyl peptidase II induced by PACAP may cause indirect stimulation of prolactin release through a reduction in TRH degradation. While an interaction between TRH and PACAP almost certainly exists, an interaction between the TRH receptor and PAC1R within lactotrophs is somewhat ambiguous. TRH and PACAP receptors share signaling pathways such as ERK and cAMP/PKA [30, 54]. It is a well-known phenomenon within the pituitary gonadotrophs that continuous exposure to gonadotropin-releasing hormone (GnRH) leads to desensitization of GnRH receptors and eliminates the effect of GnRH on gonadotropin secretion [55]. Similar to the case of the GnRH receptor, it was reported that prolonged stimulation with TRH decreases the number of TRH receptors [56]. Using GH3 cells, we examined how the functions of the TRH receptor and PAC1R change after prolonged stimulation with TRH or PACAP. In experiments using GH3 cells overexpressing PAC1R, prolonged treatment with TRH abolished the effect of TRH on the prolactin promoter. In addition, the response of the prolactin promoter to PACAP was eliminated by this treatment. ERK as well as cAMP/PKA signaling activated by TRH or PACAP was also disrupted by prolonged PACAP or TRH treatment. We therefore hypothesized that sustained stimulation with TRH or PACAP desensitizes their own and each other's receptor functions [54].

PACAP in Gonadotrophs

Gonadotropin synthesis and release is predominantly regulated by GnRH, but there is evidence that PACAP acts on gonadotrophs as well. Receptors for PACAP exist in all endocrine cell types as well as in folliculostellate cells of the adenohypophysis [16]. In addition, PACAP mRNA is expressed in gonadotrophs and folliculostellate cells [17, 35, 36]. These observations suggest that although PACAP was initially viewed as a hypothalamic-hypophysiotropic factor, it may act within the pituitary as an autocrine-paracrine factor. Injection of PACAP increases plasma levels of LH and FSH in rats [43]. PACAP stimulates expression and secretion of LH and FSH either alone or synergistically with GnRH in pituitary cultures [57]. The gonadotropin-stimulating effect of PACAP occurs with a concomitant increase in cytosolic calcium concentration [58]. In addition, it was reported that PACAP expression within the pituitary changes markedly depending on gender, time of day and stage of estrous cycle. Indeed, there are relatively few PACAP-secreting cells during diestrus, but their numbers increase substantially in proestrus [17, 59, 60].

GnRH Receptor and PAC1R in Gonadotrophs

PACAP has been shown to stimulate the expression of gonadotrophin subunits in rat pituitary cells as well as in gonadotroph cell lines such as αT3-1 and LβT2 [31, 61, 62]. In addition, it is known that PACAP increases the expression of the GnRH receptor and its own PAC1R in gonadotrophs [63–65]. In contrast, GnRH itself affects the expression of the GnRH receptor [66], PACAP and PAC1R within gonadotrophs [65]. We speculate that the functional interaction between GnRH and PACAP, or between the GnRH receptor and PAC1R, participates in GnRH pulse-frequency-dependent gonadotropin subunit gene expression. It is generally agreed that hypothalamic GnRH is released from the hypothalamus in a pulsatile manner, and the pattern of the GnRH pulse varies during the reproductive cycle [67], thereby determining the dominance of the output of the gonadotropins LH and FSH from pituitary gonadotrophs. In experiments using ovariectomized rhesus monkeys, Wildt and coworkers [68] demonstrated that administration of a higher frequency of GnRH pulses increases LH secretion, whereas lower frequencies decrease LH but increase FSH secretion. GnRH pulse-frequency-dependent specific regulation of LH and FSH was also observed in the gonadotroph cell line LβT2. In these cells, expression of the LHβ-subunit is optimally stimulated by a GnRH pulse interval of 30 min, whereas that of the FSHβ-subunit is preferentially stimulated by a longer GnRH pulse interval of 2 h [69, 70]. The detailed mechanism by which the GnRH pulse frequency determines the specificity of LH and FSH expression remains unknown. In experiments using LβT2 cells, we found that when cells were exposed to low-frequency GnRH pulses, levels of PACAP and PAC1R within the gonadotrophs were increased compared to high-frequency GnRH pulses [65]. On the other hand, it was known that the number of cell surface GnRH receptors is increased at higher frequencies of GnRH pulse [71]. Expression of LHβ-subunit is optimally stimulated by relatively high cell-surface density of GnRH receptors, whereas expression of FSHβ-subunit is favored at a lower density of GnRH receptor [69, 72]. It is known that increasing concentrations of PACAP or increasing densities of PAC1R by overexpression of PAC1R increase the effect of PACAP on gonadotropin subunit expression [73]. Details are still remain unknown, but the changes in PACAP and PAC1R expression within the gonadotrophs under the low frequency of GnRH pulse might contribute to the differential regulation of LHβ- and FSHβ-subunit expression as an autocrine/paracrine factor (Fig. 22.1). In addition, we have observed that PAC1R expression in gonadotrophs potentiates the effect of GnRH on gonadotropin subunit gene expression with concomitant potentiation of GnRH-mediated increases in ERK phosphorylation [74]. Our observations suggest that functional crosstalk exists between GnRH's and PACAP's actions, and that these peptides induce intracellular crosstalk between the GnRH receptor and PAC1Rs.

Effect of PACAP in GnRH-Producing Neurons

PACAP is abundantly expressed in the central nervous system including the hypothalamus [11]. In addition, binding sites that display relatively high affinity for PACAP are distributed in the hypothalamic areas. Because GnRH-producing neurons exist in this area, PACAP may regulate gonadotropins by modulation of GnRH

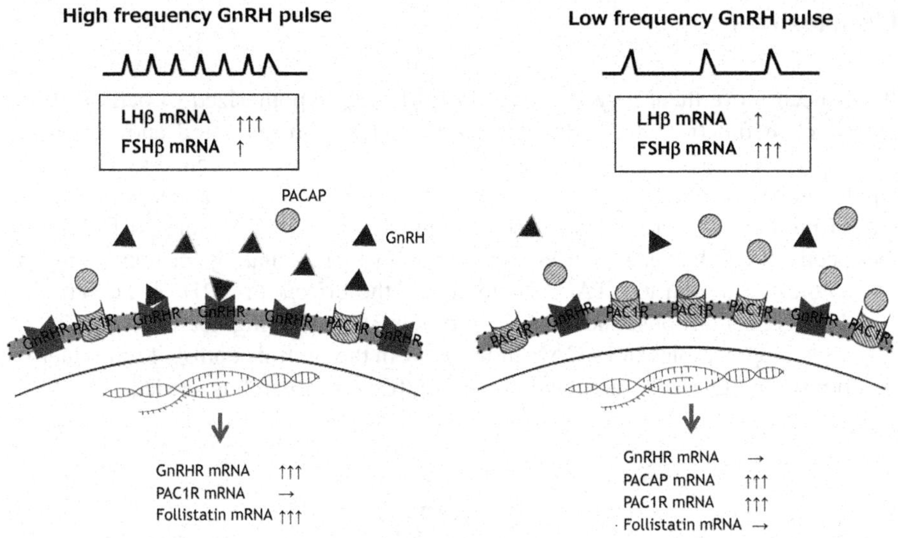

Fig. 22.1 A schematic summary of the involvement of PACAP and Pac1R in GnRH pulse frequency-dependent specific regulation of LHβ- and FSHβ-subunit expression. Under the high frequency GnRH stimulation, GnRH receptor (GnRHR) expression was predominantly observed in gonadotrophs. On the other hand, PACAP and PAC1R expression responded preferentially tp low frequency GnRH pulses without GnRHR induction. Follistatin expression within gonadotrophs was dependent on high frequency GnRH pulse

neuronal activity. Previous studies demonstrated that administration of PACAP by intracerebroventricular injection or direct injection into the medial basal hypothalamus, which contains GnRH-neurons, suppresses GnRH pulsatility in ewes [44, 75]. In contrast, PACAP was shown to increase GnRH mRNA expression levels in male rats [76]. In addition, a number of studies have implicated PACAP in the pubertal process via alteration of GnRH neuronal activity [77, 78]. In support of a direct action PACAP on GnRH neuronal activity, receptors for PACAP have been demonstrated in an immortalized GnRH cell line [79]. We have previously demonstrated that PACAP stimulates intracellular signaling cascades including ERK and cAMP-dependent kinase pathways with a concomitant increase in GnRH receptor expression in the immortalized GnRH cell line GT1-7 [80]. Although GnRH is a crucial component of the hypothalamic-pituitary gonadal axis, it has gradually become obvious that kisspeptin-secreting neurons activate GnRH neurons through G protein-coupled receptor 54 (GPR54) [81]. At present, it remains largely unknown how PACAP interacts with kisspeptin at the receptor level in GnRH neurons, although we have observed that kisspeptin-induced activation of ERK signaling was prevented by PACAP, whereas kisspeptin-stimulated cAMP/PKA signaling (which was determined by cAMP response element (CRE)-mediated transcriptional activity) was potentiated in the presence of PACAP [80]. These observations suggest that the effects of kisspeptin in GnRH neurons were differently modulated by PACAP according to its signaling pathways.

Conclusions

It has been more than 20 years since PACAP was hypothesized to participate in reproductive functions [82]. PACAP exerts its function in central and peripheral organs by itself, and also modifies many functions that are controlled by their principal regulators. In this review, we focused on lactotroph-, gonadotroph-, and GnRH-producing neurons, and describe the action of PACAP and its PAC1R in these cells. PACAP exerts its effects in these cells by modulating hormone or receptor synthesis. In addition, PACAP modulates the effects of TRH on lactotrophs, GnRH on gonadotrophs, and kisspeptin on GnRH-producing neurons (Fig. 22.2). The physiological roles of PACAP and PAC1R in the central control of reproductive functions remain largely unknown; therefore, further research is needed.

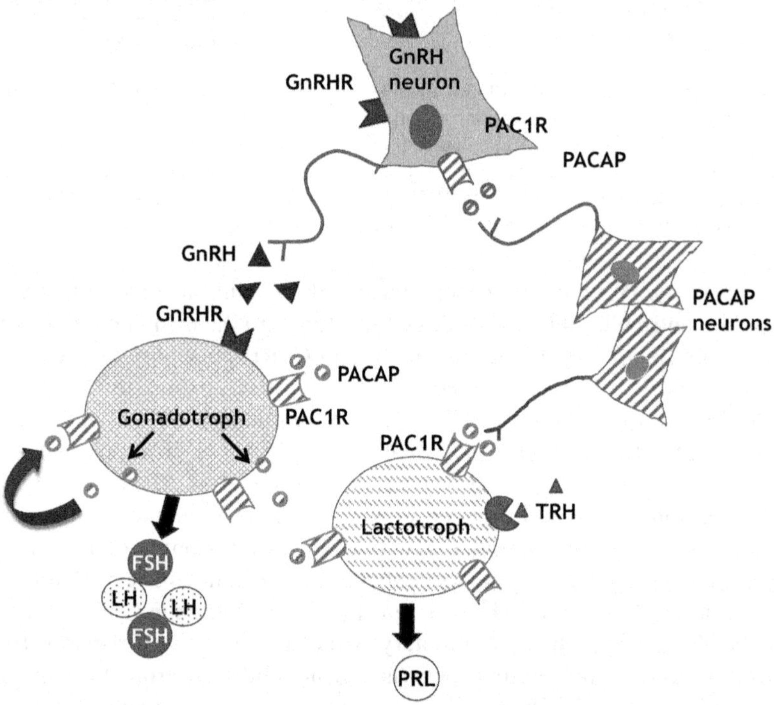

Fig. 22.2 A schematic summary of this review. PACAP acts as a hypothalamic factor to stimulate gonadotropin and prolactin (PRL) production in gonadotrophs and lactotrophs. PACAP also modulates GnRH- and TRH-induced gonadotropin and PRL expression. In addition, PACAP acts on GnRH neurons to modulate their functions

References

1. Miyata A, Arimura A, Dahl RR, Minamino N, Uehara A, Jiang L, et al. Isolation of a novel 38 residue-hypothalamic polypeptide which stimulates adenylate cyclase in pituitary cells. Biochem Biophys Res Commun. 1989;164(1):567–74.
2. Miyata A, Jiang L, Dahl RD, Kitada C, Kubo K, Fujino M, et al. Isolation of a neuropeptide corresponding to the N-terminal 27 residues of the pituitary adenylate cyclase activating polypeptide with 38 residues (PACAP38). Biochem Biophys Res Commun. 1990;170(2):643–8.
3. Segre GV, Goldring SR. Receptors for secretin, calcitonin, parathyroid hormone (PTH)/PTH-related peptide, vasoactive intestinal peptide, glucagonlike peptide 1, growth hormone-releasing hormone, and glucagon belong to a newly discovered G-protein-linked receptor family. Trends Endocrinol Metab. 1993;4(10):309–14.
4. Arimura A, Somogyvari-Vigh A, Miyata A, Mizuno K, Coy DH, Kitada C. Tissue distribution of PACAP as determined by RIA: highly abundant in the rat brain and testes. Endocrinology. 1991;129(5):2787–9.
5. Gonzalez BJ, Basille M, Vaudry D, Fournier A, Vaudry H. Pituitary adenylate cyclase-activating polypeptide. Ann Endocrinol. 1998;59(5):364–405. Pituitary adenylate cyclase-activating polypeptide.
6. Ghatei MA, Takahashi K, Suzuki Y, Gardiner J, Jones PM, Bloom SR. Distribution, molecular characterization of pituitary adenylate cyclase-activating polypeptide and its precursor encoding messenger RNA in human and rat tissues. J Endocrinol. 1993;136(1):159–66.
7. Hannibal J. Pituitary adenylate cyclase-activating peptide in the rat central nervous system: an immunohistochemical and in situ hybridization study. J Comp Neurol. 2002;453(4):389–417.
8. Koves K, Arimura A, Somogyvari-Vigh A, Vigh S, Miller J. Immunohistochemical demonstration of a novel hypothalamic peptide, pituitary adenylate cyclase-activating polypeptide, in the ovine hypothalamus. Endocrinology. 1990;127(1):264–71.
9. Koves K, Arimura A, Gorcs TG, Somogyvari-Vigh A. Comparative distribution of immunoreactive pituitary adenylate cyclase activating polypeptide and vasoactive intestinal polypeptide in rat forebrain. Neuroendocrinology. 1991;54(2):159–69.
10. Dow RC, Bennie J, Fink G. Pituitary adenylate cyclase-activating peptide-38 (PACAP)-38 is released into hypophysial portal blood in the normal male and female rat. J Endocrinol. 1994;142(1):R1–4.
11. Hannibal J, Mikkelsen JD, Clausen H, Holst JJ, Wulff BS, Fahrenkrug J. Gene expression of pituitary adenylate cyclase activating polypeptide (PACAP) in the rat hypothalamus. Regul Pept. 1995;55(2):133–48.
12. Kimura S, Ohshige Y, Lin L, Okumura T, Yanaihara C, Yanaihara N, et al. Localization of pituitary adenylate cyclase-activating polypeptide (PACAP) in the hypothalamus-pituitary system in rats: light and electron microscopic immunocytochemical studies. J Neuroendocrinol. 1994;6(5):503–7.
13. Li M, Maderdrut JL, Lertora JJ, Batuman V. Intravenous infusion of pituitary adenylate cyclase-activating polypeptide (PACAP) in a patient with multiple myeloma and myeloma kidney: a case study. Peptides. 2007;28(9):1891–5.
14. Mulder H, Uddman R, Moller K, Zhang YZ, Ekblad E, Alumets J, et al. Pituitary adenylate cyclase activating polypeptide expression in sensory neurons. Neuroscience. 1994;63(1):307–12.
15. Seki T, Shioda S, Izumi S, Arimura A, Koide R. Electron microscopic observation of pituitary adenylate cyclase-activating polypeptide (PACAP)-containing neurons in the rat retina. Peptides. 2000;21(1):109–13.
16. Vigh S, Arimura A, Gottschall PE, Kitada C, Somogyvari-Vigh A, Childs GV. Cytochemical characterization of anterior pituitary target cells for the neuropeptide, pituitary adenylate cyclase activating polypeptide (PACAP), using biotinylated ligands. Peptides. 1993;14(1):59–65.
17. Koves K, Kantor O, Scammell JG, Arimura A. PACAP colocalizes with luteinizing and follicle-stimulating hormone immunoreactivities in the anterior lobe of the pituitary gland. Peptides. 1998;19(6):1069–72.

18. Frodin M, Hannibal J, Wulff BS, Gammeltoft S, Fahrenkrug J. Neuronal localization of pituitary adenylate cyclase-activating polypeptide 38 in the adrenal medulla and growth-inhibitory effect on chromaffin cells. Neuroscience. 1995;65(2):599–608.
19. Filipsson K, Sundler F, Hannibal J, Ahren B. PACAP and PACAP receptors in insulin producing tissues: localization and effects. Regul Pept. 1998;74(2–3):167–75.
20. Skakkebaek M, Hannibal J, Fahrenkrug J. Pituitary adenylate cyclase activating polypeptide (PACAP) in the rat mammary gland. Cell Tissue Res. 1999;298(1):153–9.
21. Hannibal J, Ekblad E, Mulder H, Sundler F, Fahrenkrug J. Pituitary adenylate cyclase activating polypeptide (PACAP) in the gastrointestinal tract of the rat: distribution and effects of capsaicin or denervation. Cell Tissue Res. 1998;291(1):65–79.
22. Gaytan F, Martinez-Fuentes AJ, Garcia-Navarro F, Vaudry H, Aguilar E. Pituitary adenylate cyclase-activating peptide (PACAP) immunolocalization in lymphoid tissues of the rat. Cell Tissue Res. 1994;276(2):223–7.
23. Fahrenkrug J, Hannibal J. Pituitary adenylate cyclase activating polypeptide innervation of the rat female reproductive tract and the associated paracervical ganglia: effect of capsaicin. Neuroscience. 1996;73(4):1049–60.
24. Shioda S, Legradi G, Leung WC, Nakajo S, Nakaya K, Arimura A. Localization of pituitary adenylate cyclase-activating polypeptide and its messenger ribonucleic acid in the rat testis by light and electron microscopic immunocytochemistry and in situ hybridization. Endocrinology. 1994;135(3):818–25.
25. Koh PO, Kwak SD, Kim HJ, Roh G, Kim JH, Kang SS, et al. Expression patterns of pituitary adenylate cyclase activating polypeptide and its type I receptor mRNAs in the rat placenta. Mol Reprod Dev. 2003;64(1):27–31.
26. Spencer F, Chi L, Zhu M. Temporal relationships among uterine pituitary adenylate cyclase-activating polypeptide, decidual prolactin-related protein and progesterone receptor mRNAs expressions during decidualization and gestation in rats. Comp Biochem Physiol C Toxicol Pharmacol. 2001;129(1):25–34.
27. Uddman R, Luts A, Absood A, Arimura A, Ekelund M, Desai H, et al. PACAP, a VIP-like peptide, in neurons of the esophagus. Regul Pept. 1991;36(3):415–22.
28. Luts L, Sundler F. Peptide-containing nerve fibers in the parathyroid glands of different species. Regul Pept. 1994;50(2):147–58.
29. Vaudry D, Falluel-Morel A, Bourgault S, Basille M, Burel D, Wurtz O, et al. Pituitary adenylate cyclase-activating polypeptide and its receptors: 20 years after the discovery. Pharmacol Rev. 2009;61(3):283–357.
30. Mijiddorj T, Kanasaki H, Purwana IN, Oride A, Miyazaki K. Stimulatory effect of pituitary adenylate-cyclase activating polypeptide (PACAP) and its PACAP type I receptor (PAC1R) on prolactin synthesis in rat pituitary somatolactotroph GH3 cells. Mol Cell Endocrinol. 2011;339(1–2):172–9.
31. Harada T, Kanasaki H, Mutiara S, Oride A, Miyazaki K. Cyclic adenosine 3',5'monophosphate/protein kinase A and mitogen-activated protein kinase 3/1 pathways are involved in adenylate cyclase-activating polypeptide 1-induced common alpha-glycoprotein subunit gene (Cga) expression in mouse pituitary gonadotroph LbetaT2 cells. Biol Reprod. 2007;77(4):707–16.
32. Rawlings SR, Hezareh M. Pituitary adenylate cyclase-activating polypeptide (PACAP) and PACAP/vasoactive intestinal polypeptide receptors: actions on the anterior pituitary gland. Endocr Rev. 1996;17(1):4–29.
33. Okada R, Yamamoto K, Ito Y, Mochida H, Tonon MC, Fournier A, et al. VIP and PACAP stimulate TSH release from the bullfrog pituitary. Peptides. 2007;28(9):1784–9.
34. Grafer CM, Thomas R, Lambrakos L, Montoya I, White S, Halvorson LM. GnRH stimulates expression of PACAP in the pituitary gonadotropes via both the PKA and PKC signaling systems. Mol Endocrinol. 2009;23(7):1022–32. Pubmed Central PMCID: 2703603.
35. Radleff-Schlimme A, Leonhardt S, Wuttke W, Jarry H. Evidence for PACAP to be an autocrine factor on gonadotrope cells. Ann N Y Acad Sci. 1998;865:486–91.

36. Jin L, Tsumanuma I, Ruebel KH, Bayliss JM, Lloyd RV. Analysis of homogeneous populations of anterior pituitary folliculostellate cells by laser capture microdissection and reverse transcription-polymerase chain reaction. Endocrinology. 2001;142(5):1703–9.
37. Ben-Jonathan N, Hnasko R. Dopamine as a prolactin (PRL) inhibitor. Endocr Rev. 2001;22(6):724–63.
38. Yamada M, Saga Y, Shibusawa N, Hirato J, Murakami M, Iwasaki T, et al. Tertiary hypothyroidism and hyperglycemia in mice with targeted disruption of the thyrotropin-releasing hormone gene. Proc Natl Acad Sci U S A. 1997;94(20):10862–7. Pubmed Central PMCID: 23510.
39. Vijayan E, Samson WK, Said SI, McCann SM. Vasoactive intestinal peptide: evidence for a hypothalamic site of action to release growth hormone, luteinizing hormone, and prolactin in conscious ovariectomized rats. Endocrinology. 1979;104(1):53–7.
40. Muratori M, Romano C, Gambino G, Faglia G. Prolactin responsiveness to peptide histidine methionine-27 in normal subjects and hyperprolactinemic patients. Horm Res. 1994;42(6):257–61.
41. Yamauchi K, Murakami Y, Nishiki M, Tanaka J, Koshimura K, Kato Y. Possible involvement of vasoactive intestinal polypeptide in the central stimulating action of pituitary adenylate cyclase-activating polypeptide on prolactin secretion in the rat. Neurosci Lett. 1995;189(2):131–4.
42. Isaac ER, Sherwood NM. Pituitary adenylate cyclase-activating polypeptide (PACAP) is important for embryo implantation in mice. Mol Cell Endocrinol. 2008;280(1–2):13–9.
43. Hart GR, Gowing H, Burrin JM. Effects of a novel hypothalamic peptide, pituitary adenylate cyclase-activating polypeptide, on pituitary hormone release in rats. J Endocrinol. 1992;134(1):33–41.
44. Sawangjaroen K, Anderson ST, Curlewis JD. Effects of pituitary adenylate cyclase-activating polypeptide (PACAP) and vasoactive intestinal polypeptide (VIP) on hormone secretion from sheep pituitary cells in vitro. J Neuroendocrinol. 1997;9(4):279–86.
45. Hashizume T, Soliman EB, Kanematsu S. Effects of pituitary adenylate cyclase-activating polypeptide (PACAP), prostaglandin E2 (PGE2) and growth hormone releasing factor (GRF) on the release of growth hormone from cultured bovine anterior pituitary cells in vitro. Domest Anim Endocrinol. 1994;11(4):331–7.
46. Vertongen P, Velkeniers B, Hooghe-Peters E, Robberecht P. Differential alternative splicing of PACAP receptor in pituitary cell subpopulations. Mol Cell Endocrinol. 1995;113(2):131–5.
47. Gershengorn MC. Mechanism of thyrotropin releasing hormone stimulation of pituitary hormone secretion. Annu Rev Physiol. 1986;48:515–26.
48. Ohmichi M, Sawada T, Kanda Y, Koike K, Hirota K, Miyake A, et al. Thyrotropin-releasing hormone stimulates MAP kinase activity in GH3 cells by divergent pathways. Evidence of a role for early tyrosine phosphorylation. J Biol Chem. 1994;269(5):3783–8.
49. Wang YH, Maurer RA. A role for the mitogen-activated protein kinase in mediating the ability of thyrotropin-releasing hormone to stimulate the prolactin promoter. Mol Endocrinol. 1999;13(7):1094–104.
50. Kanasaki H, Fukunaga K, Takahashi K, Miyazaki K, Miyamoto E. Mitogen-activated protein kinase activation by stimulation with thyrotropin-releasing hormone in rat pituitary GH3 cells. Biol Reprod. 1999;61(1):319–25.
51. Kanasaki H, Yonehara T, Yamamoto H, Takeuchi Y, Fukunaga K, Takahashi K, et al. Differential regulation of pituitary hormone secretion and gene expression by thyrotropin-releasing hormone. A role for mitogen-activated protein kinase signaling cascade in rat pituitary GH3 cells. Biol Reprod. 2002;67(1):107–13.
52. Vargas MA, Bourdais J, Sanchez S, Uriostegui B, Moreno E, Joseph-Bravo P, et al. Multiple hypothalamic factors regulate pyroglutamyl peptidase II in cultures of adenohypophyseal cells: role of the cAMP pathway. J Neuroendocrinol. 1998;10(3):199–206.
53. Charli JL, Vargas MA, Cisneros M, de Gortari P, Baeza MA, Jasso P, et al. TRH inactivation in the extracellular compartment: role of pyroglutamyl peptidase II. Neurobiology (Bp). 1998;6(1):45–57.

54. Mijiddorj T, Kanasaki H, Unurjargal S, Oride A, Purwana I, Miyazaki K. Prolonged stimulation with thyrotropin-releasing hormone and pituitary adenylate cyclase-activating polypeptide desensitize their receptor functions in prolactin-producing GH3 cells. Mol Cell Endocrinol. 2013;365(2):139–45.
55. Knobil E. Patterns of hypophysiotropic signals and gonadotropin secretion in the rhesus monkey. Biol Reprod. 1981;24(1):44–9.
56. Hinkle PM, Tashjian Jr AH. Thyrotropin-releasing hormone regulates the number of its own receptors in the GH3 strain of pituitary cells in culture. Biochemistry. 1975;14(17):3845–51.
57. Tsujii T, Ishizaka K, Winters SJ. Effects of pituitary adenylate cyclase-activating polypeptide on gonadotropin secretion and subunit messenger ribonucleic acids in perifused rat pituitary cells. Endocrinology. 1994;135(3):826–33.
58. Canny BJ, Rawlings SR, Leong DA. Pituitary adenylate cyclase-activating polypeptide specifically increases cytosolic calcium ion concentration in rat gonadotropes and somatotropes. Endocrinology. 1992;130(1):211–5.
59. Koves K, Kantor O, Molnar J, Heinzlmann A, Szabo E, Szabo F, et al. The role of PACAP in gonadotropic hormone secretion at hypothalamic and pituitary levels. J Mol Neurosci. 2003;20(2):141–52.
60. Szabo E, Nemeskeri A, Arimura A, Koves K. Effect of PACAP on LH release studied by cell immunoblot assay depends on the gender, on the time of day and in female rats on the day of the estrous cycle. Regul Pept. 2004;123(1–3):139–45.
61. Mutiara S, Kanasaki H, Harada T, Miyazaki K. Dopamine D(2) receptor expression and regulation of gonadotropin alpha-subunit gene in clonal gonadotroph LbetaT2 cells. Mol Cell Endocrinol. 2006;259(1–2):22–9.
62. Burrin JM, Aylwin SJ, Holdstock JG, Sahye U. Mechanism of action of pituitary adenylate cyclase-activating polypeptide on human glycoprotein hormone alpha-subunit transcription in alphaT3-1 gonadotropes. Endocrinology. 1998;139(4):1731–7.
63. Pincas H, Laverriere JN, Counis R. Pituitary adenylate cyclase-activating polypeptide and cyclic adenosine 3',5'-monophosphate stimulate the promoter activity of the rat gonadotropin-releasing hormone receptor gene via a bipartite response element in gonadotrope-derived cells. J Biol Chem. 2001;276(26):23562–71.
64. Oride A, Kanasaki H, Purwana IN, Miyazaki K. Possible involvement of mitogen-activated protein kinase phosphatase-1 (MKP-1) in thyrotropin-releasing hormone (TRH)-induced prolactin gene expression. Biochem Biophys Res Commun. 2009;382(4):663–7.
65. Kanasaki H, Purwana IN, Mijiddorj T, Oride A, Miyazaki K. Possible involvement of PACAP and PACAP type 1 receptor in GnRH-induced FSH beta-subunit gene expression. Regul Pept. 2011;167(2–3):227–32.
66. Kaiser UB, Conn PM, Chin WW. Studies of gonadotropin-releasing hormone (GnRH) action using GnRH receptor-expressing pituitary cell lines. Endocr Rev. 1997;18(1):46–70.
67. Crowley Jr WF, Filicori M, Spratt DI, Santoro NF. The physiology of gonadotropin-releasing hormone (GnRH) secretion in men and women. Recent Prog Horm Res. 1985;41:473–531.
68. Wildt L, Hausler A, Marshall G, Hutchison JS, Plant TM, Belchetz PE, et al. Frequency and amplitude of gonadotropin-releasing hormone stimulation and gonadotropin secretion in the rhesus monkey. Endocrinology. 1981;109(2):376–85.
69. Bedecarrats GY, Kaiser UB. Differential regulation of gonadotropin subunit gene promoter activity by pulsatile gonadotropin-releasing hormone (GnRH) in perifused L beta T2 cells: role of GnRH receptor concentration. Endocrinology. 2003;144(5):1802–11.
70. Kanasaki H, Bedecarrats GY, Kam KY, Xu S, Kaiser UB. Gonadotropin-releasing hormone pulse frequency-dependent activation of extracellular signal-regulated kinase pathways in perifused LbetaT2 cells. Endocrinology. 2005;146(12):5503–13.
71. Ahren B, Filipsson K. Pituitary adenylate cyclase activating polypeptide stimulates insulin secretion in a glucose-dependent manner in vivo. Ann N Y Acad Sci. 1998;865:466–70.
72. Kaiser UB, Sabbagh E, Katzenellenbogen RA, Conn PM, Chin WW. A mechanism for the differential regulation of gonadotropin subunit gene expression by gonadotropin-releasing hormone. Proc Natl Acad Sci U S A. 1995;92(26):12280–4. Pubmed Central PMCID: 40340.

73. Purwana IN, Kanasaki H, Oride A, Mijiddorj T, Shintani N, Hashimoto H, et al. GnRH-induced PACAP and PAC1 receptor expression in pituitary gonadotrophs: a possible role in the regulation of gonadotropin subunit gene expression. Peptides. 2010;31(9):1748–55.
74. Purwana IN, Kanasaki H, Oride A, Mijiddorj T, Miyazaki K. Expression of the pituitary adenylate cyclase-activating polypeptide (PACAP) type 1 receptor (PAC1R) potentiates the effects of GnRH on gonadotropin subunit gene expression. Biochem Biophys Res Commun. 2011;410(2):295–300.
75. Anderson ST, Sawangjaroen K, Curlewis JD. Pituitary adenylate cyclase-activating polypeptide acts within the medial basal hypothalamus to inhibit prolactin and luteinizing hormone secretion. Endocrinology. 1996;137(8):3424–9.
76. Li S, Grinevich V, Fournier A, Pelletier G. Effects of pituitary adenylate cyclase-activating polypeptide (PACAP) on gonadotropin-releasing hormone and somatostatin gene expression in the rat brain. Brain Res Mol Brain Res. 1996;41(1–2):157–62.
77. Szabo F, Horvath J, Heinzlmann A, Arimura A, Koves K. Neonatal PACAP administration in rats delays puberty through the influence of the LHRH neuronal system. Regul Pept. 2002;109(1–3):49–55.
78. Choi EJ, Ha CM, Kim MS, Kang JH, Park SK, Choi WS, et al. Central administration of an antisense oligodeoxynucleotide against type I pituitary adenylate cyclase-activating polypeptide receptor suppresses synthetic activities of LHRH-LH axis during the pubertal process. Brain Res Mol Brain Res. 2000;80(1):35–45.
79. Olcese J, McArdle CA, Middendorff R, Greenland K. Pituitary adenylate cyclase-activating peptide and vasoactive intestinal peptide receptor expression in immortalized LHRH neurons. J Neuroendocrinol. 1997;9(12):937–43.
80. Kanasaki H, Mijiddorj T, Sukhbaatar U, Oride A, Miyazaki K. Pituitary adenylate cyclase-activating polypeptide (PACAP) increases expression of the gonadotropin-releasing hormone (GnRH) receptor in GnRH-producing GT1-7 cells overexpressing PACAP type I receptor. Gen Comp Endocrinol. 2013;193:95–102.
81. Kauffman AS, Clifton DK, Steiner RA. Emerging ideas about kisspeptin- GPR54 signaling in the neuroendocrine regulation of reproduction. Trends Neurosci. 2007;30(10):504–11.
82. McArdle CA. Pituitary adenylate cyclase-activating polypeptide: a key player in reproduction? Endocrinology. 1994;135(3):815–7.

Chapter 23
Occurrence and Functions of PACAP in the Placenta

Gabriella Horvath, Jozsef Nemeth, Reka Brubel, Balazs Opper, Miklos Koppan, Andrea Tamas, Laszlo Szereday, and Dora Reglodi

Abstract Pituitary adenylate cyclase activating polypeptide (PACAP) is an endogenous neuropeptide with a widespread distribution both in the nervous system and peripheral organs. The peptide is also present in the female gonadal system, indicating its role in reproductive functions. While a lot of data are known on PACAP-induced effects in oogenesis and in the regulation of gonadotropin secretion at pituitary level, its placental effects are somewhat neglected in spite of the documented implantation deficit in mice lacking endogenous PACAP. The aim of the present review is to give a brief summary on the occurrence and actions of PACAP and its receptors in the placenta. Radioimmunoassay (RIA) measurements revealed increased serum PACAP levels during the third trimester and several changes in placental PACAP content in obstetrical pathological conditions, further supporting the function of PACAP during pregnancy.

Both the peptide and its receptors have been shown in different parts of the placenta and the umbilical cord. PACAP influences blood vessel and smooth muscle contractility of the uteroplacental unit and is involved in regulation of local hormone secretion. The effects of PACAP on trophoblast cells have been mainly studied in vitro. Effects of PACAP on cell survival, angiogenesis and invasion/proliferation have been described in different trophoblast cell lines. PACAP increases proliferation and decreases invasion in proliferative extravillous trophoblast cells, but not in primary trophoblast cells, where PACAP decreased the

G. Horvath (✉) • R. Brubel • B. Opper • A. Tamas • D. Reglodi
Department of Anatomy, MTA-PTE "Lendulet" PACAP Research Team, University of Pecs, Pecs, Hungary
e-mail: gabriella.horvath@aok.pte.hu

J. Nemeth
Department of Pharmacology and Pharmacotherapy, University of Debrecen, Debrecen, Hungary

M. Koppan
Department of Obstetrics and Gynecology, University of Pecs, Pecs, Hungary

L. Szereday
Department of Medical Microbiology and Immunology, University of Pecs, Pecs, Hungary

D. Reglodi, A. Tamas (eds.), *Pituitary Adenylate Cyclase Activating Polypeptide — PACAP*, Current Topics in Neurotoxicity 11,
DOI 10.1007/978-3-319-35135-3_23

secretion of various angiogenic markers. PACAP pretreatment enhances survival of non-tumorous primary trophoblast cells exposed to oxidative stress, but it does not influence the cell death-inducing effects of methotrexate in proliferative extravillous cytotrophoblast cells. Interestingly, PACAP has pro-apoptotic effect in choriocarcinoma cells suggesting that the effect of PACAP depends on the type of trophoblast cells. These data strongly support that PACAP plays a role in normal and pathological pregnancies and our review provides an overview of currently available experimental data worth to be further investigated to elucidate the exact role of this peptide in the placenta.

Keywords Pregnancy • Trophoblast • Proliferation • Migration • Placenta • Human

Introduction

Pituitary adenylate cyclase activating polypeptide (PACAP) was first described as a hypothalamic neuropeptide acting on the pituitary [1]. Numerous subsequent studies have described its regulatory effects in the hypothalamo-hypophyseal-endocrine gland axis at all levels. Shortly after its discovery it became evident that PACAP occurs at high levels in several peripheral organs, especially in the gonads. Arimura and coworkers showed that after the hypothalamus, highest PACAP levels are found in the testis [2]. This drew the attention to the peptide as a regulator of male fertility and reproduction. Indeed, PACAP was found to influence spermatogenesis at various levels [3–6].

PACAP is also involved in female reproductive functions. Although our knowledge on PACAP in reproductive functions is still limited, currently available data clearly indicate that the neuropeptide plays an important regulatory role in female reproductive physiology and pathology (rev. [7]). Briefly, PACAP, at the hypothalamic level, influences receptive behavior in female rodents, in association with gonadotropin releasing hormone and steroids [8], and plays an important modulatory role in pituitary hormone production. The role of PACAP in the hypothalamo-pituitary-gonadal axis has been reviewed several times previously [9–13] and is reviewed in the present book in two chapters (11. and 12.). PACAP is present in the ovary, in the ovarian follicular fluid and plays an important role in oocyte maturation [7, 14, 15]. In humans, the level of immunoreactive PACAP in the follicular fluid of hyperstimulated women is correlated with the number of retrieved oocytes [16]. PACAP also plays a role in the muscle contraction of the vaginal wall as well as that of the uterus and uterine tube [17–19]. Decreased immunoreactivity was shown in vaginal wall diseases and the plasticity of the PACAPergic system was demonstrated after vaginal reconstructive surgery [20, 21]. The PACAPergic innervation of the female genital tract was also described and has been associated with nerves originating from the paracervical ganglia [22, 23].

During pregnancy, the serum level of PACAP increases in the third trimester in healthy pregnant women and it markedly decreases during delivery, reaching pre-birth levels 3 days after delivery [24]. Winters and colleagues indicated a possible difference between cesarian and vaginal births regarding PACAP levels in the fetal cord blood [25]. In pregnant rats, Papka et al. [26] detected PACAP immunoreactivity in cervical nerves, lumbosacral dorsal root ganglia and in the spinal cord. Immunoreactivity showed changes during pregnancy, indicating that sensory nerve-derived PACAP is involved in the innervation of the cervix and may play a role in cervical ripening [26]. The role of PACAP in reproduction and offspring care does not seem to finish with birth: PACAP plays a complex regulatory role in breastfeeding. The action of PACAP on prolactin synthesis and release is complex and influenced by several factors [27, 28]. The high concentrations of PACAP in the milk suggest that the peptide plays a role in the development of the newborn [29–31]. Equally possible function of PACAP in the milk is a local regulatory action of milk production and/or mammary gland development [32, 33]. (This is reviewed elsewhere in this book by Tamas and co-workers 49.). In addition, PACAP was implicated to play a role in maternal behavior as showed by the decreased maternal crouching behavior of PACAP knockout mice [34].

It seems that placental functions of the neuropeptide are somewhat neglected in spite of the findings of Isaac and Sherwood [35], who described that the reproductive rate of PACAP knockout mice is lower due to implantation insufficiency, clearly indicating a placental role of endogenous PACAP. The present review briefly summarizes the occurrence and actions of PACAP and its receptors in the placenta.

Occurrence of PACAP and PACAP Receptors in the Placenta

Occurrence of PACAP and its receptors are summarized in a schematic drawing (Fig. 23.1). The gene encoding VPAC receptors was found to be weakly expressed in human placenta at relative prevalent levels comparable to that in the testis, kidney and thymus [36]. Another study also confirmed these findings [37].

Subsequent studies gave further insight into the occurrence and distribution of PACAP and its receptors in the placenta. Radioimmunoassay and immunocytochemistry first confirmed the expression of PACAP27 and PACAP38 in human placentas [38]. PACAP levels in the placenta were compared to those in the isthmic region of the uterus and in the umbilical cord. Both forms of PACAP could be detected in the examined specimens of the uteroplacental unit. PACAP38 concentration was higher than PACAP27 levels in all examined regions. Uterus and placenta showed similar levels of immunoreactivity, while intensity in the umbilical cord was much weaker [38]. These authors found no immunoreactive nerve fibers in the placenta or umbilical cord, immunoreactive nerve fibers were only present in the uterus, with isthmic region and nonpregnant myometrium showing stronger immunoreactivity than pregnant uterus. PACAP immunoreactivity by radioimmunoassay was confirmed later by Brubel et al. [39]: both forms of the peptide could be detected with PACAP38

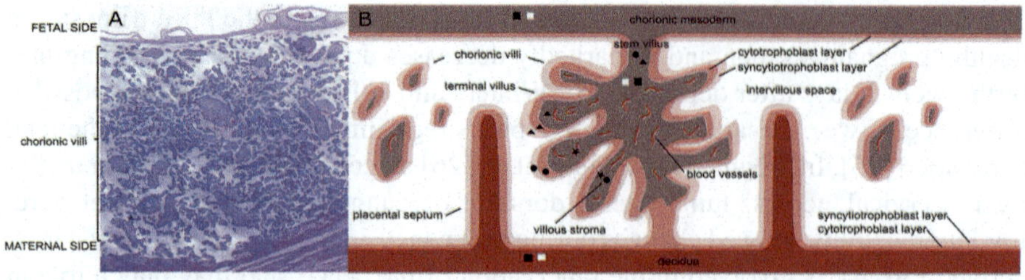

Fig. 23.1 Distribution of PACAP and PACAP receptors in the human placenta. (**A**). Representative photomicrograph of the human placenta indicating the parts depicted in the (**B**) schematic illustration of the human placenta. *Star:* PACAP38 immunohistochemistry, *filled circle:* PACAP mRNA, *filled triangle:* PAC1 mRNA, *filled square:* PACAP38 radioimmunoassay, *open square:* PACAP27 radioimmunoassay. Exact localization of VPAC receptors is not known

showing stronger immunoreactivity, in agreement with the above earlier observations. Furthermore, it was also found that different parts of the human placenta contained similar levels of PACAPs, such as central maternal, peripheral maternal, central fetal, and peripheral fetal parts. The umbilical cord showed very weak immunoreactivity. In addition, Brubel et al. [39] also compared the level of immunoreactivity in the 1st trimester samples and full-term placentas. Markedly stronger immunoreactivity for PACAP38 was found in full-term placentas on both the maternal and fetal sides. In contrast, PACAP27-immunoreactivity only increased on the maternal side, while it did not change on the fetal side towards the end of pregnancy.

Scaldaferri and collagues [40] studied PACAP and PAC1 receptor expression by means of Northern blot analysis, polymerase chain reaction (PCR) and immunohistochemistry. The authors detected the presence of PACAP and PAC1 receptor in both rat and full-term human placentas. In human placentas, strong immunohistochemical staining was observed in stromal cells surrounding blood vessels and weaker signal was detected in vessel walls in stem villi. In terminal villi, stromal cells expressed PACAP38 immunoreactivity. In stem villi, the stromal immunoreactivity was restricted to the periphery, while this spatial distribution pattern was lacking in terminal villi, where immunostaining was dispersed throughout the stroma. In rat placentas, several immunostained cells were observed in the labyrinth and in the villous-like structures of the intraplacental yolk sac, structures derived from yolk sac extensions into placental disc during late pregnancy [40].

Isoforms of the PAC1 receptor were also studied by RT-PCR. Different isoform expression was revealed in rat and human placentas. Rat placenta was shown to express 3 isoforms: the short, hip or hop variant and the hip-hop variant. In contrast, human placenta only expressed the SV2 form, homologous to the rat hop form. Radioligand receptor binding assay revealed that the relative potencies of PACAP-related peptides were PACAP27, PACAP38 with comparable strong binding (almost equipotent, PACAP27 slightly stronger) and VIP with weaker binding (10 times less potent). Growth hormone releasing hormone and unrelated peptides, such as beta-endorphin and corticotropin-releasing hormone, did not bind to the receptor [40].

PACAP and PAC1 receptor mRNAs were investigated in rat placenta also by Koh and coworkers [41]. Rat placenta consists of decidua basalis, junctional and labyrinth zones. Expression of PACAP and PAC1 receptor mRNA was detected by in situ hybridization in decidual cells, and in chorionic vessels and stromal cells of the labyrinth zone [41]. In decidual cells, signals were strongest on day 13.5 of gestation, then decreased with more advanced stages. No signal could be detected in the junctional zone. In contrast, signals were gradually increasing with advancing pregnancy in the labyrinth zone, in the branching villi, stem villi and also in chorionic vessels.

Koh et al. [42] investigated the expression of PACAP and PAC1 receptor mRNA from human legal abortions of 6–7 weeks, from induced abortions of 14–24 weeks (second trimester) and term placentas by cesarian section or normal vaginal delivery. In situ hybridization revealed expression of PACAP and PAC1 receptor mRNAs in stem villi and terminal villi. In 7- and 14-week-old samples, PACAP mRNA was detected in stroma cells surrounding blood vessels within stem villi, with moderate expression level, while stronger expression of PACAP mRNA was found at later stages [42]. PACAP mRNA was only weakly expressed in cytotrophoblast and syncytiotrophoblast cells. PAC1 receptor expression was detected in the same areas: stronger expression was described in stroma cells of the villi, while weaker expression in the trophoblast cells. Similar pattern of VIP immunoreactivity was detected by Marzioni and coworkers [43]. Immunostaining was present in both trophoblast layers and in the endothelium of the fetal vessels.

The gradual increase of the mRNA expression for both PACAP and its specific receptor implies a role of PACAP in placental growth. The findings with RIA also confirmed these increasing levels in late placentas compared to early placentas [4].

Expression of PACAP in Pathological Pregnancies

Butadiene diepoxide is a reactive metabolite of 1,3-butadiene that is an important industrial chemical and causes a dose-dependent inhibition of deciduoma development in rats [44, 45]. Placental expression of PACAP mRNA significantly decreased in rats pretreated with 1,3-butadiene, a chemical toxin for reproduction in rats [44, 46]. The decrease was more drastic on gestation day 12 (63 %) than on day 9 (48 %).

Our recent preliminary investigations have focused on the levels of PACAP38 and PACAP27 in cases of different pathological situations. These measurements were done in order to show possible changes of PACAP expression caused by maternal smoking during pregnancy or fetal distress or hypoxia leading to presence of meconium in the amniotic fluid. Human placentas were collected from full-term placentas. Samples were taken from the chorionic villi (fetal side), the decidua (maternal side) and the umbilical cord. Four different groups were examined: (1) normal pregnancy and birth; (2) amniotic fluid with meconium—premature birth (36–38 weeks); (3) premature birth (31–32 weeks) with smoking during pregnancy; (4) post term birth with smoking during pregnancy ($n=3$ in all groups).

The procedure used was in accordance with protocols approved by the ethical committee (no. 2784,3117, University of Pecs; 8-28/92 009-10 I 8EKU, ETT TUKEB, Ministry of Health, Hungary). Tissue samples were weighed and homogenized in ice-cold distilled water. The homogenate was centrifuged (12,000 rpm, 4 °C, 30 min), and the supernatant was further processed for RIA analysis of PACAP38 and PACAP27 contents, as previously described [47, 48].

Briefly, the conditions were as follows: antisera: PACAP38 "88111-3" (working dilution, 1:10,000) and PACAP27 "88123" (dilution: 1:45,000); tracer: mono-125I-labelled ovine PACAP24-38 and mono-125I-labelled ovine PACAP27 prepared in our laboratory (5000 cpm/tube); standard: ovine PACAP38 and PACAP27 ranging from 0 to 1000 fmol/ml; buffer: assay prepared in 1 ml of 0.05 mol/l (pH 7.4) phosphate buffer containing 0.1 mol/l sodium chloride, 0.25 % (wt/vol) bovine serum albumin, and 0.05 % (wt/vol) sodium azide; incubation time: 48–72 h of incubation at 4 °C; separation solution: charcoal/dextran/milk powder (10:1:0.2 g in 100 ml of distilled water). Results are given as fmol/mg PACAP38-like immunoreactivity and PACAP27-like immunoreactivity in the tissue samples. Differences between PACAP contents were assessed by ANOVA test.

As sample sizes were low, no definite statistical comparison was possible, but some conclusions on tendencies can be drawn based on these preliminary results. We found detectable differences in levels of PACAP38 between pathological and physiological pregnancies suggesting that PACAP expression may be disturbed or upregulated during pathological events related to pregnancy (Fig. 23.2). PACAP27 levels did not show pronounced alterations in any examined condition. In accordance with previous results, levels of PACAP27 were significantly lower than PACAP38 levels in each sample. Under hypoxic condition, as indicated by the presence of meconium in the amniotic fluid, PACAP levels did not change in most samples. Only a slight decrease was observed in the chorionic villi and in the umbilical cord. Samples from premature births of smoking mothers showed marked increases in all regions examined except for the central decidua, where a slight decrease was observed. Decreases were also detected from samples derived from post term births of smoking mothers. Although our results are preliminary from limited number of clinical cases, some tendencies are promising as it seems that PACAP levels change in some pathological conditions. PACAP alterations have been observed in several diseases in both tissue samples and body fluids including plasma, cerebrospinal fluid and follicular fluid [16, 49]. Recent studies have shown that PACAP levels were lower in lung cancer, colon and kidney tumor samples compared to healthy tissue, while higher in prostate cancer samples compared with samples from benign prostatic hyperplasia [50, 51]. As PACAP is indicated as a potential biomarker for various conditions by several studies, it would be important to conduct a clinical study including enough pathological placenta samples to draw final conclusions (clinical review: see Reglodi et al. in this book, chapter 2.).

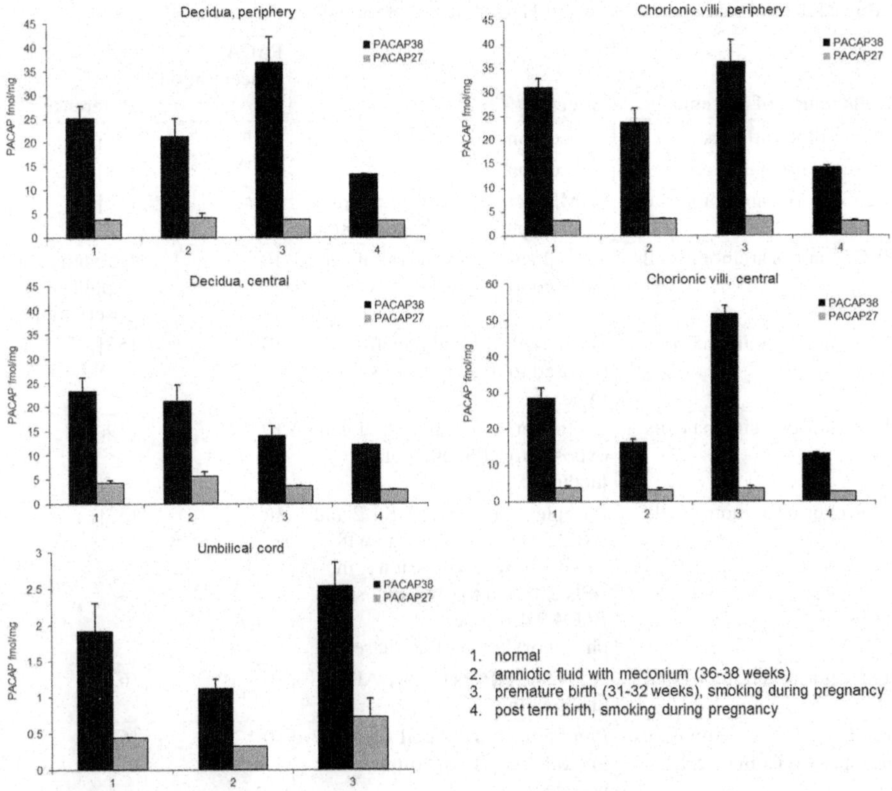

Fig. 23.2 Levels of PACAP27 and PACAP38 in placentas from healthy and pathological pregnancies. Results are given as mean fmol/mg tissue PACAP ± SEM. Differences between levels of PACAP38 and PACAP27-like immunoreactivities were significant in all cases

Effects of PACAP in Vessels of the Placenta

Effects of PACAP on different components and cells of the placenta are summarized in Table 23.1. Steenstrup et al. [38] investigated the effects of PACAP on vessels and smooth muscle contractility in the uteroplacental unit. They found that preincubation of the vessels with PACAPs and VIP produced a significant and concentration-dependent inhibition of the norepinephrine-induced contraction on the intramyometrial and stem villus arteries. The high concentration needed for significant relaxation indicates that the local release of the peptides is necessary to achieve this effect in vivo. These results show that PACAP causes relaxation of the placental vessels. In contrast, no effect was observed on either the amplitude, tone or frequency of strips of spontaneously contracted myometrial smooth muscle obtained from pregnant women [38]. These observations indicate that PACAP may be involved in the regulation of the uteroplacental blood flow. The time-related localization of endometrial/

Table 23.1 Summary of the effects of PACAP in the placenta

Cells/region of placenta	Effect	PACAP concentration (M)	Reference
Stem villous arteries	Relaxation	10^{-10}–10^{-6}	[38]
Intramyometrial arteries	Relaxation	10^{-10}–10^{-6}	[38]
JEG3 choriocarcinoma cells	cAMP increase, IL-6 secretion, alpha-subunit gene transcription	10^{-7}	[52]
JEG3 choriocarcinoma cells	No influence on survival of cells exposed to oxidative stress	10^{-7}	Horvath et al. unpublished observation
JAR choriocarcinoma cells	Decreased survival in cells exposed to oxidative stress or hypoxia	10^{-7}	[53]
JAR choriocarcinoma cells	No influence on survival of cells exposed to LPS, ethanol, methotrexate	10^{-9}–10^{-7}	[39, 53]
JAR choriocarcinoma cells	Phosphorylation of ERK1/2 and JNK↑; Akt, GSK-3β, and p38 MAPK↓, Bax expression↓; in cells exposed to oxidative stress, PACAP decreased phosphorylation of all these	10^{-7}	[53]
JAR choriocarcinoma cells	Agonistic effects of PACAP6-38 on signaling	10^{-7}	[62]
HTR-8/SVneo nontumorous primary trophoblast cells	Pretreatment increased survival in oxidative stress, co-treatment no effect	10^{-8}–10^{-7}	[61]
HTR-8/SVneo nontumorous primary trophoblast cells	No effect on invasion	10^{-7}	[61]
HTR-8/SVneo nontumorous primary trophoblast cells	Reduced levels of angiogenic factors active A, ADAMTS-1, angiogenin, angiopoietin-1, endocrine gland-derived vascular endothelial growth factor, and endoglin	10^{-6}	[61]
HIPEC65 proliferative extravillous cytotrophoblast cells	Induced proliferation, but no effect on methotrexate-induced cell death decreased invasion	10^{-7}	[61]
Decidual and peripheral mononuclear cells from early pregnancies	No effect on secreted angiogenic molecules or inflammatory cytokine production	10^{-6}	[61]

uterine PACAP was described by Spencer et al. [45]. PACAP mRNA pattern showed similarity with that of the progesterone receptor during decidualization, but it was consistently lower during gestation. Furthermore, both uterine and placental mRNA expression pattern of decidual prolactin-related protein corresponded to PACAP and progesterone receptor mRNA levels. They suggest that PACAP could be important in facilitating endometrial blood flow and increase availability of metabolic substrates to the developing deciduoma or embryo [45].

Effects on Hormonal Secretion of the Placenta

Using JEG3 choriocarcinoma cells PACAP was found to induce a 12-fold increase in cAMP secretion [52]. This action of PACAP was rapid, cAMP increase started already after 30 min. cAMP in the placenta is known to stimulate alpha-subunit expression. Alpha-subunit of the hypophyseal hormones LH, FSH, and TSH is also present in the placental hormone human chorionic gonadotropin. PACAP38 was found to positively regulate alpha-gene transcription in JEG3 cells, with maximal effect at 100 nM concentration. The time course of this effect showed that PACAP effect started after 8 h. Similar effects were observed with the homolog peptide, VIP, but the effects of VIP an alpha-gene transcription started only after 24 h [52]. These findings show that PACAP may be involved in placental hormone secretion, and the similar effects of PACAP and VIP suggest that these effects are mediated by the shared VPAC receptors.

Effects of PACAP on Survival of Trophoblast Cells

Effects of PACAP on trophoblast cell survival were studied in JAR human choriocarcinoma cells [53]. PACAP treatment alone did not influence the survival rate. Cells exposed to oxidative stress induced by H_2O_2 showed decreased survival rate, which was further decreased by PACAP. A similar effect was observed in cells undergoing chemically induced hypoxia by $CoCl_2$. No effects on survival were observed in cells exposed to lipopolysaccharide (LPS), methotrexate or ethanol [39, 53]. These findings were contradicting the general survival-promoting effect of PACAP observed in many different cell lines and tissues both in vitro and in vivo [54]. Examining the signaling pathway revealed that PACAP treatment alone slightly increased phosphorylation of ERK1/2 (extracellular signal-regulated kinase) and JNK (c-Jun N-terminal kinase) but decreased that of Akt (protein kinase B), MAPK (mitogen-activated protein kinase), GSK-3β (glycogen synthase kinase 3 beta) and the expression of Bax. Oxidative stress alone increased phosphorylation of JNK and slightly decreased that of Akt, ERK, and GSK-3β, while no changes were observed in the expression of phospho-p38 and Bax. In cells exposed to oxidative stress, PACAP treatment decreased phosphorylation of all of these examined signaling molecules compared to H_2O_2-treated controls. These results indicate that PACAP sensitizes the cells to some stressors, like oxidative stress and in vitro hypoxia, while it does not affect the deleterious effects of other stressors [53]. The reason for this is not known at the moment, and may or may not reflect the physiological role of the peptide, since these experiments were performed in choriocarcinoma cells, in which the receptor expression and signaling induced by PACAP may be significantly altered.

Similar results were described in retinoblastoma cells, where PACAP treatment induced cell death [55] in spite of the well-known protective effects of PACAP in the retina [56, 57]. PACAP also inhibited the growth of the neuronal tumor medulloblastoma in spite of the well-known neuroprotective effects of the peptide [58–60]. An equally possible explanation is that the sensitizing effect of PACAP to stressors such as hypoxia and oxidative stress may be involved in the adaptation promoting effect of PACAP in pregnancy and under pathological conditions. The first explanation is supported by later findings in non-tumorous primary trophoblast cells (HTR-8/Svneo cells), where PACAP pretreatment led to a significant increase in survival measured by MTT test in cells exposed to oxidative stress by H_2O_2, while co-treatment had no effect [61]. However, PACAP treatment did not influence the cell death-inducing effect of methotrexate in HIPEC65 proliferative extravillous cytotrophoblast cell line, but induced proliferation of these cells when treated alone with PACAP [61]. These results show that the effects of PACAP on proliferation and survival of trophoblast cells depend on the type of stressor, the timing of treatment and the type of cell.

Another interesting finding regarding survival and signaling effects in this cell line was that the generally accepted PAC1/VPAC2 antagonist PACAP6-38 exerted agonistic effects in JAR cells [62]. This does not seem to be specific for this cell line, since agonistic effects were found for example on rat trachea neuropeptide release, retinoblastoma cell survival and cartilage and bone development [55, 62–64]. The reason for this might be the expression of a yet unknown splice variant of the receptor or the tumorous nature of the JAR cells.

Effects of PACAP on Invasion of Trophoblast Cells

PACAP treatment did not influence the invasion of HTR-8/Svneo human first trimester extravillous primary trophoblast cells, while it decreased the invasion of HIPEC invasive, proliferative extravillous cytotrophoblast cells [61].

Effects of PACAP on Trophoblast Angiogenesis

Effects of PACAP on angiogenic factors were investigated in HTR-8/Svneo cells using an angiogenesis array method. Levels of several angiogenic markers were markedly decreased in the cell culture supernatant after 24 h of PACAP treatment.

Secreted levels of active A, ADAMTS-1, angiogenin, angiopoietin-1, endocrine gland-derived vascular endothelial growth factor, and endoglin were reduced [61]. In human peripheral blood and decidual mononuclear cells obtained from healthy pregnant women undergoing elective termination of apparently normal pregnancies no effect on levels of secreted angiogenic molecules was found. Similarly, PACAP treatment had no effect on the inflammatory cytokine production of these cells [61].

Observations in PACAP Knockout Mice, Effects on Implantation

It has been reported several times that the reproductive rate of PACAP and PAC1 receptor knockout mice is lower than that of wild types [34, 65]. The reason for this is not elucidated yet, several mechanisms may be responsible for this effect. For example, PACAP is involved in spermatogenesis and sperm motility [4], in steroid hormone synthesis, in ovarian folliculogenesis, and in reproductive behavior [34, 35]. The small litter can also be due to the premature intrauterine death and early postnatal death due to defects in breathing and temperature regulation [66, 67]. In addition to this multifactorial mechanism, it seems that placental defects are also partially responsible for the lower reproductive rate of PACAP knockout mice. Isaac and Sherwood [35] observed that while the puberty onset, estrous cycle and seminal plugs of PACAP knockout mice were normal, significantly fewer PACAP null females gave birth following mating than wild types. The authors found no defect in ovulation, ovarian histology or fertilization of released eggs, only 13 % had implanted embryos 6.5 days after mating compared to 81 % in wild types. Levels of prolactin and progesterone were significantly lower in PACAP knockout females. These observations suggest that impaired implantation is involved in the observed decreased fertility, the details of which need further clarification [35].

In summary, we give a brief review of data supporting a role of PACAP in normal and pathological pregnancies. The currently available experimental data are worth to be further investigated to elucidate the exact role in the placenta and evaluate the potential biomarker value of PACAP in reproductive pathology.

Acknowledgements This study was supported by OTKA K104984, 115874, 119759, K104960, K119529, Arimura Foundation, "Lendület" Program, and Bolyai Scholarship of the Hungarian Academy of Sciences, TAMOP 4.2.4.A/2-11-1-2012-0001 "National Excellence Program", PTE AOK Research Grant and New National Excellence Program (UNKP). This work is dedicated to the 650th anniversary of the University of Pecs.

References

1. Miyata A, Arimura A, Dahl RR, Minamino N, Uehara A, Jiang L, et al. Isolation of a novel 38 residue-hypothalamic polypeptide which stimulates adenylate cyclase in pituitary cells. Biochem Biophys Res Commun. 1989;164:567–74.

2. Arimura A, Somogyvari-Vigh A, Miyata A, Mizuno K, Coy DH, Kitada C. Tissue distribution of PACAP as determined by RIA: highly abundant in the rat brain and testes. Endocrinology. 1991;129:2787–9.
3. Agnese M, Valiante S, Angelini F, Laforgia V, Andreuccetti P, Prisco M. Pituitary adenylate cyclase-activating polypeptide and its receptor PAC1 in the testis of Triturus carnifex and Podarcis sicula. Gen Comp Endocrinol. 2010;168:256–61.
4. Brubel R, Kiss P, Vincze A, Varga A, Varnagy A, Bodis J, et al. Effects of pituitary adenylate cyclase activating polypeptide on human sperm motility. J Mol Neurosci. 2012;48:623–30.
5. Li M, Arimura A. Neuropeptides of the pituitary adenylate cyclase-activating polypeptide/vasoactive intestinal polypeptide/growth hormone-releasing hormone/secretin family in testis. Endocrine. 2003;20:201–14.
6. Shuto Y, Somogyvari-Vigh A, Shioda S, Onda H, Arimura A. Effect of hypophysectomy on pituitary adenylate cyclase-activating polypeptide gene expression in the rat testis. Peptides. 1995;16:1039–44.
7. Reglodi D, Tamas A, Koppan M, Szogyi D, Welke L. Role of PACAP in female fertility and reproduction at gonadal level—recent advances. Front Endocrinol (Lausanne). 2012;3:155.
8. Apostolakis EM, Riherd DN, O'Malley BW. PAC1 receptors mediate pituitary adenylate cyclase-activating polypeptide- and progesterone-facilitated receptivity in female rats. Mol Endocrinol. 2005;19:2798–811.
9. Counis R, Laverriere JN, Garrel-Lazayres G, Cohen-Tannoudji J, Lariviere S, Bleux C, et al. What is the role of PACAP in gonadotrope function? Peptides. 2007;28:1797–804.
10. Koves K, Kantor O, Lakatos A, Szabo E, Kirilly E, Heinzlmann A, et al. Advent and recent advances in research on the role of pituitary adenylate cyclase-activating polypeptide (PACAP) in the regulation of gonadotropic hormone secretion of female rats. J Mol Neurosci. 2014;54:494–511.
11. Sherwood NM, Adams BA, Isaac ER, Wu S, Fradinger EA. Knocked down and out: PACAP in development, reproduction and feeding. Peptides. 2007;28:1680–7.
12. Sherwood NM, Krueckl SL, McRory JE. The origin and function of the pituitary adenylate cyclase activating polypeptide (PACAP)/glucagon superfamily. Endocr Rev. 2000;21:619–70. Review.
13. Vaudry D, Falluel-Morel A, Bourgault S, Basille M, Burel D, Wurtz O, et al. Pituitary adenylate cyclase-activating polypeptide and its receptors: 20 years after the discovery. Pharmacol Rev. 2009;61:283–357.
14. Barberi M, Di Paolo V, Latini S, Guglielmo MC, Cecconi S, Canipari R. Expression and functional activity of PACAP and its receptors on cumulus cells: effects on oocyte maturation. Mol Cell Endocrinol. 2013;375:79–88.
15. Brubel R, Reglodi D, Jambor E, Koppan M, Varnagy A, Biro Z, et al. Investigation of pituitary adenylate cyclase activating polypeptide in human gynecological and other biological fluids by using MALDI TOF mass spectrometry. J Mass Spectrom. 2011;46:189–94.
16. Koppan M, Varnagy A, Reglodi D, Brubel R, Nemeth J, Tamas A, et al. Correlation between oocyte number and follicular fluid concentration of pituitary adenylate cyclase-activating polypeptide (PACAP) in women after superovulation treatment. J Mol Neurosci. 2012;48:617–22.
17. Aughton KL, Hamilton-Smith K, Gupta J, Morton JS, Wayman CP, Jackson VM. Pharmacological profiling of neuropeptides on rabbit vaginal wall and vaginal artery smooth muscle in vitro. Br J Pharmacol. 2008;155:236–43.
18. Giraldi A, Alm P, Werkstrom V, Myllymaki L, Wagner G, Andersson KE. Morphological and functional characterization of a rat vaginal smooth muscle sphincter. Int J Impot Res. 2002;14:271–82.
19. Steenstrup BR, Alm P, Hannibal J, Jorgensen JC, Palle C, Junge J, et al. Pituitary adenylate cyclase activating polypeptide: occurrence and relaxant effect in female genital tract. Am J Physiol. 1995;269:E108–17.
20. Hong X, Huang L, Song Y. Role of vasoactive intestinal peptide and pituitary adenylate cyclase activating polypeptide in the vaginal wall of women with stress urinary incontinence and pelvic organ prolapse. Int Urogynecol J Pelvic Floor Dysfunct. 2008;19:1151–7.

21. Radziszewski P, Majewski M, Baranowski W, Czaplicki M, Bossowska A, Dobronski P, et al. Re-innervation pattern of the "neovagina" created from the bladder flap in patients with Mayer-Rokitanski-Kistner-Hauser syndrome: an immunochemical study. Gynecol Endocrinol. 2009;25:362–71.
22. Fahrenkrug J, Hannibal J. Pituitary adenylate cyclase activating polypeptide innervation of the rat female reproductive tract and the associated paracervical ganglia: effect of capsaicin. Neuroscience. 1996;73:1049–60.
23. Fahrenkrug J, Steenstrup BR, Hannibal J, Alm P, Ottesen B. Role of PACAP in the female reproductive organs. Ann N Y Acad Sci. 1996;805:394–407.
24. Reglodi D, Gyarmati J, Ertl T, Borzsei R, Bodis J, Tamas A, et al. Alterations of pituitary adenylate cyclase activating polypeptide (PACAP)-like immunoreactivity in the human plasma during pregnancy and after birth. J Endocrinol Invest. 2010;33:443–5.
25. Winters SJ, King JC, Brees CK, Moore Jr JP. Pituitary adenylate cyclase-activating polypeptide (PACAP) in fetal cord blood. Early Hum Dev. 2014;90:451–3.
26. Papka RE, Workley M, Usip S, Mowa CN, Fahrenkrug J. Expression of pituitary adenylate cyclase activating peptide in the uterine cervix, lumbosacral dorsal root ganglia and spinal cord of rats during pregnancy. Peptides. 2006;27:743–52.
27. Mijiddorj T, Kanasaki H, Unurjargal S, Oride A, Purwana I, Miyazaki K. Prolonged stimulation with thyrotropin-releasing hormone and pituitary adenylate cyclase-activating polypeptide desensitize their receptor functions in prolactin-producing GH3 cells. Mol Cell Endocrinol. 2013;365:139–45.
28. Tohei A, Ikeda M, Hokao R, Shinoda M. The different effects of i.c.v. injection of pituitary adenylate cyclase activating polypeptide (PACAP) on prolactin secretion in adult male and lactating rats. Exp Anim. 2009;58:489–95.
29. Borzsei R, Mark L, Tamas A, Bagoly T, Bay C, Csanaky K, et al. Presence of pituitary adenylate cyclase activating polypeptide-38 in human plasma and milk. Eur J Endocrinol. 2009;160:561–5.
30. Csanaky K, Banki E, Szabadfi K, Reglodi D, Tarcai I, Czegledi L, et al. Changes in PACAP immunoreactivity in human milk and presence of PAC1 receptor in mammary gland during lactation. J Mol Neurosci. 2012;48:631–7.
31. Csanaky K, Reglodi D, Banki E, Tarcai I, Mark L, Helyes Z, et al. Examination of PACAP38-like immunoreactivity in different milk and infant formula samples. Acta Physiol Hung. 2013;100:28–36.
32. Csanaky K, Doppler W, Tamas A, Kovacs K, Toth G, Reglodi D. Influence of terminal differentiation and PACAP on the cytokine, chemokine, and growth factor secretion of mammary epithelial cells. J Mol Neurosci. 2014;52:28–36.
33. McFarland-Mancini M, Hugo E, Loftus J, Ben-Jonathan N. Induction of prolactin expression and release in human preadipocytes by cAMP activating ligands. Biochem Biophys Res Commun. 2006;344:9–16.
34. Shintani N, Mori W, Hashimoto H, Imai M, Tanaka K, Tomimoto S, et al. Defects in reproductive functions in PACAP-deficient female mice. Regul Pept. 2002;109:45–8.
35. Isaac ER, Sherwood NM. Pituitary adenylate cyclase-activating polypeptide (PACAP) is important for embryo implantation in mice. Mol Cell Endocrinol. 2008;280:13–9.
36. Sreedharan SP, Huang JX, Cheung MC, Goetzl EJ. Structure, expression, and chromosomal localization of the type I human vasoactive intestinal peptide receptor gene. Proc Natl Acad Sci U S A. 1995;92:2939–43.
37. Adamou JE, Aiyar N, Van Horn S, Elshourbagy NA. Cloning and functional characterization of the human vasoactive intestinal peptide (VIP)-2 receptor. Biochem Biophys Res Commun. 1995;209:385–92.
38. Steenstrup BR, Jorgensen JC, Alm P, Hannibal J, Junge J, Fahrenkrug J, et al. Pituitary adenylate cyclase activating polypeptide (PACAP): occurrence and vasodilatory effect in the human uteroplacental unit. Regul Pept. 1996;61:197–204.
39. Brubel R, Boronkai A, Reglodi D, Racz B, Nemeth J, Kiss P, et al. Changes in the expression of pituitary adenylate cyclase-activating polypeptide in the human placenta during pregnancy and its effects on the survival of JAR choriocarcinoma cells. J Mol Neurosci. 2010;42:450–8.

40. Scaldaferri ML, Modesti A, Palumbo C, Ulisse S, Fabbri A, Piccione E, et al. Pituitary adenylate cyclase activating polypeptide (PACAP) and PACAP-receptor type 1 expression in rat and human placenta. Endocrinology. 2000;141:1158–67.
41. Koh PO, Kwak SD, Kim HJ, Roh G, Kim JH, Kang SS, et al. Expression patterns of pituitary adenylate cyclase activating polypeptide and its type I receptor mRNAs in the rat placenta. Mol Reprod Dev. 2003;64:27–31.
42. Koh PO, Won CK, Noh HS, Cho CJ, Choi WS. Expression of pituitary adenylate cyclase activating polypeptide and its type I receptor mRNAs in human placenta. J Vet Sci. 2005;6:1–5.
43. Marzioni D, Fiore G, Giordano A, Nabissi M, Florio P, Verdenelli F, et al. Placental expression of substance P and vasoactive intestinal peptide: evidence for a local effect on hormone release. J Clin Endocrinol Metab. 2005;90:2378–83.
44. Spencer F, Chi L, Zhu M. A mechanistic assessment of 1,3-butadiene diepoxide-induced inhibition of uterine deciduoma proliferation in pseudopregnant rats. Reprod Toxicol. 2001;15:253–60.
45. Spencer F, Chi L, Zhu M. Temporal relationships among uterine pituitary adenylate cyclase-activating polypeptide, decidual prolactin-related protein and progesterone receptor mRNAs expressions during decidualization and gestation in rats. Comp Biochem Physiol C Toxicol Pharmacol. 2001;129:25–34.
46. Chi L, Nixon E, Spencer F. Uterine-ovarian biochemical and developmental interactions to the postimplantation treatment with a butadiene metabolite, diepoxybutane, in pregnant rats. J Biochem Mol Toxicol. 2002;16:147–53.
47. Jakab B, Reglodi D, Jozsa R, Hollosy T, Tamas A, Lubics A, et al. Distribution of PACAP-38 in the central nervous system of various species determined by a novel radioimmunoassay. J Biochem Bioph Meth. 2004;61:189–98.
48. Nemeth J, Jakab B, Jozsa R, Hollosy T, Tamas A, Lubics A, et al. PACAP-27 radioimmunoassay: description and application of a novel method. J Radioanal Nucl Chem. 2007;273:327–32.
49. Bukovics P, Czeiter E, Amrein K, Kovacs N, Pal J, Tamas A, et al. Changes of PACAP level in cerebrospinal fluid and plasma of patients with severe traumatic brain injury. Peptides. 2014;60:18–22.
50. Szanto Z, Sarszegi Z, Reglodi D, Nemeth J, Szabadfi K, Kiss P, et al. PACAP immunoreactivity in human malignant tumor samples and cardiac diseases. J Mol Neurosci. 2012;48:667–73.
51. Tamas A, Javorhazy A, Reglodi D, Sarlos DP, Banyai D, Semjen D, et al. Examination of PACAP-like immunoreactivity in urogenital tumor samples. J Mol Neurosci. 2016; 59:177–83.
52. Desai BJ, Burrin JM. PACAP-38 positively regulates glycoprotein hormone alpha-gene expression in placental cells. Mol Cell Endocrinol. 1994;99:31–7.
53. Boronkai A, Brubel R, Racz B, Tamas A, Kiss P, Horvath G, et al. Effects of pituitary adenylate cyclase activating polypeptide on the survival and signal transduction pathways in human choriocarcinoma cells. Ann N Y Acad Sci. 2009;1163:353–7.
54. Somogyvari-Vigh A, Reglodi D. Pituitary adenylate cyclase activating polypeptide: a potential neuroprotective peptide. Curr Pharm Des. 2004;10:2861–89.
55. Wojcieszak J, Zawilska JB. PACAP38 and PACAP6-38 exert cytotoxic activity against human retinoblastoma Y79 cells. J Mol Neurosci. 2014;54:463–8.
56. D'Amico AG, Maugeri G, Reitano R, Bucolo C, Saccone S, Drago F, et al. PACAP modulates expression of hypoxia-inducible factors in streptozotocin-induced diabetic rat retina. J Mol Neurosci. 2015;57:501–9.
57. Danyadi B, Szabadfi K, Reglodi D, Mihalik A, Danyadi T, Kovacs Z, et al. PACAP application improves functional outcome of chronic retinal ischemic injury in rats-evidence from electroretinographic measurements. J Mol Neurosci. 2014;54:293–9.
58. Cohen JR, Resnick DZ, Niewiadomski P, Dong H, Liau LM, Waschek JA. Pituitary adenylyl cyclase activating polypeptide inhibits gli1 gene expression and proliferation in primary medulloblastoma derived tumorsphere cultures. BMC Cancer. 2010;10:676.

59. Miyamoto K, Tsumuraya T, Ohtaki H, Dohi K, Satoh K, Xu Z, et al. PACAP38 suppresses cortical damage in mice with traumatic brain injury by enhancing antioxidant activity. J Mol Neurosci. 2014;54:370–9.
60. Reglodi D, Kiss P, Lubics A, Tamas A. Review on the protective effects of PACAP in models of neurodegenerative diseases in vitro and in vivo. Curr Pharm Des. 2011;17:962–72.
61. Horvath G, Reglodi D, Brubel R, Halasz M, Barakonyi A, Tamas A, et al. Investigation of the possible functions of PACAP in human trophoblast cells. J Mol Neurosci. 2014;54:320–30.
62. Reglodi D, Borzsei R, Bagoly T, Boronkai A, Racz B, Tamas A, et al. Agonistic behavior of PACAP6-38 on sensory nerve terminals and cytotrophoblast cells. J Mol Neurosci. 2008;36:270–8.
63. Juhasz T, Matta C, Katona E, Somogyi C, Takacs R, Hajdu T, et al. Pituitary adenylate cyclase-activating polypeptide (PACAP) signalling enhances osteogenesis in UMR-106 cell line. J Mol Neurosci. 2014;54:555–73.
64. Juhasz T, Szentleleky E, Somogyi CS, Takacs R, Dobrosi N, Engler M, et al. Pituitary adenylate cyclase activating polypeptide (PACAP) pathway is induced by mechanical load and reduces the activity of hedgehog signaling in chondrogenic micromass cell cultures. Int J Mol Sci. 2015;16:17344–67.
65. Jamen F, Rodriguez-Henche N, Pralong F, Jegou B, Gaillard R, Bockaert J, et al. PAC1 null females display decreased fertility. Ann N Y Acad Sci. 2000;921:400–4.
66. Gray SL, Cummings KJ, Jirik FR, Sherwood NM. Targeted disruption of the pituitary adenylate cyclase-activating polypeptide gene results in early postnatal death associated with dysfunction of lipid and carbohydrate metabolism. Mol Endocrinol. 2001;15:1739–47.
67. Gray SL, Yamaguchi N, Vencovakoh P, Sherwood NM. Temperature-sensitive phenotype in mice lacking pituitary adenylate cyclase-activating polypeptide. Endocrinology. 2002;143:3946–54.

Chapter 24
PACAP in the Reproductive System

Rita Canipari, Virginia Di Paolo, Marzia Barberi, and Sandra Cecconi

Abstract The production of a mature oocyte that is capable of being fertilized and undergoing embryo development is the major function of the female reproductive system. This process depends on the close association between germ cells and somatic cells, i.e., granulosa and theca cells. The growth of the mammalian oocyte is finely regulated by the coordinate expression of autocrine and paracrine factors. The initial phases of folliculogenesis are independent from gonadotropins, but at later stages, it is strictly dependent on follicle-stimulating hormone (FSH) and luteinizing hormone (LH). However, increasing amounts of paracrine/autocrine growth factors participate in the modulation of gonadotropin effects. Pituitary adenylate cyclase-activating peptide (PACAP) is an extremely conserved regulatory peptide that was initially found in the central nervous system. However, PACAP has been found together with its receptors in many organs, tissues and cell types, including lung, testis, adrenal gland, and ovary. In the ovary, PACAP expression has been found mainly in preovulatory follicles after the LH surge. However, PACAP is constantly expressed through the ovarian cycle in the nerve fibers in the hilus region and in the interstitial tissue in the proximity of the primary follicles. In the ovary, PACAP promotes steroidogenesis and cAMP production, reduces the rate of apoptosis in granulosa cells, and accelerates oocyte meiotic maturation.

Keywords PACAP • VIP • PAC_1-R • $VPAC_1$-R • $VPAC_2$-R • Oocyte • Ovary • Ovulation

Introduction

The production of a mature developmental-competent oocyte is a complex process during which germ cells and somatic cells maintain a close association. This bidirectional oocyte–follicular cell communication is essential for both oocyte and follicular

R. Canipari (✉) • V. Di Paolo • M. Barberi
Department of Anatomy, Histology, Forensic Medicine and Orthopedic, Section of Histology, Sapienza University of Rome, Via A. Scarpa 16, Rome 00161, Italy
e-mail: rita.canipari@uniroma1.it

S. Cecconi
Department of Health Sciences, University of L'Aquila, L'Aquila, Italy

D. Reglodi, A. Tamas (eds.), *Pituitary Adenylate Cyclase Activating Polypeptide — PACAP*, Current Topics in Neurotoxicity 11,
DOI 10.1007/978-3-319-35135-3_24

somatic cell function and development [1]. In addition, the growth of the mammalian oocyte is finely regulated by the coordinate expression of autocrine and paracrine factors, which are produced according to an ordered series of maturation and differentiation steps that occur in both germ cells and somatic cells. The initial phases of folliculogenesis are gonadotropin independent, and a number of intraovarian factors that are produced either by theca or granulosa cells (GCs) or by the oocyte itself or are derived from plasma play an important role during this process. In contrast, at later stages, folliculogenesis is strictly dependent on follicle-stimulating hormone (FSH) and luteinizing hormone (LH), though an increasing number of paracrine/autocrine growth factors participate in the modulation of gonadotropin effects.

We are interested in the study of the intraovarian pituitary adenylate cyclase-activating peptide (PACAP)/vasoactive intestinal polypeptide (VIP) system in female reproduction.

The name of PACAP27 and 38 reflects their potent action in stimulating cAMP production in anterior pituitary cells [2, 3]. PACAP is an extremely conserved regulatory peptide and has been studied in several species, such as mammals, fish and other non-mammalian species [4–9]. Based on sequence similarity, PACAP belongs to the PACAP–VIP–secretin–GHRH–glucagon superfamily [10, 11]. The biological effects of PACAP are mediated through the binding to three G-protein-coupled transmembrane receptors PAC_1-R, $VPAC_1$-R, and $VPAC_2$-R. Whereas PAC_1-R is a PACAP-specific receptor, $VPAC_1$-R and $VPAC_2$-R are shared by PACAP and VIP and have similar affinities for PACAP and VIP [11, 12]. These three receptors are coupled with the adenylate cyclase (AC) pathway, but only PAC_1-R is coupled with phospholipase C (PLC). Moreover, because of alternative splicing, PAC_1-R exists in five different splice variant forms, which either contain or lack each of the two alternative exons, named "hip" and "hop" [13].

In addition to the central nervous system, PACAP, VIP, and their receptors have been found in various organs and peripheral tissues, such as lung, testis, adrenal glands, and ovary [for review see refs. 11, 14, 15] and increase cAMP accumulation in various cell types in culture [2, 16–19].

This wide distribution suggests that such peptides may have multifunctional roles, including that of a hypophysiotropic hormone, neuromodulator, vasoregulator, and regulator of secretion of several factors, hormones, and enzymes [11].

PACAP/VIP and Their Receptors in the Reproductive System

Gonads

Interestingly, very high levels of PACAP are found in both female and male gonads

Testis

In the testis, histological studies have shown that PACAP immunoreactivity is mainly present in the developing acrosome of rat spermatids [20–23]. PACAP mRNA and protein are expressed at high levels in germ cells in a stage-specific manner during spermatogenesis in rodent [24, 25] and human testis [26]. Moreover, this expression locally activates cAMP-coupled receptors that are located in somatic Sertoli cells [21]. PACAP binding sites have been found in germ cells, Leydig cells, and Sertoli cells, and these cells respond to PACAP in vitro by increasing the cAMP levels [27]. The stage-dependent expression of PACAP and consequent cAMP production may act as a critical modulator of the spermatogenic cycle and participate in sperm maturation [21].

Ovary

PACAP, VIP and their receptors have been found in the ovary of several species, from fish to human [28–34], and their distribution in the organ depends on the studied species [35, 36].

Zebrafish

In the zebrafish ovary, two forms of PACAP ($PACAP_{38}$-1 and $PACAP_{38}$-2) and three PACAP receptors (PAC_1-R, $VPAC_1$-R, and $VPAC_2$-R) are present [31, 32]. PACAP and PACAP receptors are differentially expressed in the oocyte and in follicle cells, with $PACAP_{38}$-1 expression restricted to the oocyte and $PACAP_{38}$-2 to both oocytes and follicle cells. Concerning the receptors, PAC_1-R and $VPAC_2$-R are expressed in the oocyte, whereas $VPAC_1$-R is exclusively located in follicle cells. During folliculogenesis, a different pattern of expression has been observed: $PACAP_{38}$-1 and the two oocyte-associated receptors have high expression in the early stages of folliculogenesis, while a decrease occurs when the follicles approach the final stage of growth. In contrast, an increased expression of $VPAC_1$-R has been observed in the follicle cells prior to maturation [31].

Rodents

In the mouse, the presence of PACAP and two isoforms of PAC1-Rs has been shown as early as 11.5 days post coitum (dpc) in mouse gonadal ridges in both primordial germ cells (PGCs) and gonadal somatic cells. Mostly PACAP-immunostaining was found on the PGC surface, but its origin remains to be established [37]. In contrast, PAC_1-R mRNA is present in both PGCs and gonadal somatic cells [37].

In rodents, PACAP expression has been found mainly in preovulatory follicles after the LH surge [28, 38–40]. However, PACAP is constantly expressed through the ovarian cycle both in the nerve fibers in the hilus region [41] and in the interstitial tissue in the proximity of the primary follicles [39, 42].

In the preovulatory follicle, PACAP is transiently expressed both in the mural granulosa cells lining the antrum and in the cumulus cells (CCs) surrounding the oocyte but not in theca cells [28, 39]. Interestingly, after ovulation, PACAP expression is maintained in the corpus luteum [43, 44], and its expression increases with pregnancy (R), which suggests that it has a role in late gestation.

In contrast, VIP has been detected by immunohistochemistry in neonatal rat ovaries as early as 2 days after birth, but no mRNA has been found in the ovary [45], with the exception of data that were reported by Gozes and Tsafriri [46]. In that report, the authors showed the presence of VIP mRNA in rat ovaries, suggesting the local production of VIP in the ovaries. However, the transection of the nerve fibers that reach the ovary eliminates radioimmunoassayable VIP levels, which indicates that ovarian VIP might originate from the extrinsic innervations of the gland [47, 48].

PACAP/VIP receptors have been shown in the rat and mouse ovary during the first wave of folliculogenesis [30, 42, 49]. In particular, PAC_1-R is the predominant form in the ovary and is mainly expressed in GCs of growing follicles, increases with increasing age [30, 50], decreases after PMSG stimulation [30, 51], and is transiently stimulated by LH in preovulatory follicles [30, 52]. Moreover, PAC1-R has been found on fully grown GV oocytes and disappears during meiotic maturation [30]. $VPAC_1$-R is primarily expressed in theca cells and decreases with increasing age [30]. This decrease may be only apparent as a consequence of the relative increase in the GC population with respect to theca cells. Interestingly, $VPAC_1$-R has been localized by immunofluorescence in association with the blood vessel wall and in the stroma near follicles [30]. $VPAC_2$-R is present in both GCs and theca cells.

In the ovary, the different distribution of the receptors in the different compartments is evidenced by the different second messengers that are elicited by these two peptides.

In theca cells, where only two VPAC receptors are present, treatment with PACAP or VIP does not significantly affect IP production, even at high concentrations (10^{-6} M), while both PACAP and VIP produce statistically significant increases in cAMP levels. In contrast, in GCs that were treated with PACAP, an increased level of IPs was observed, as well as an increased level of cAMP, in response to both PACAP and VIP [30], confirming the presence of active PAC_1-R.

Monovulatory Species

The presence of PACAP and its receptors has been demonstrated in the bovine ovary [33] and in human GCs [29]. In the human ovary, the mRNAs for PAC_1- $VPAC_1$- and $VPAC_2$-R were present in non-luteinized GCs. In contrast, in granulosa-luteal (GL) cells, PAC_1-R mRNA was barely detectable, but the receptor was still present as a protein [29]. PACAP was found in human follicular fluid (FF) [53], and its presence in FF was inversely correlated with the number of retrieved oocytes after superovulation treatment [54]. The lower number of oocytes that were obtained in the presence of a higher concentration of PACAP in the FF (>290 fmol/ml) suggested that at such high concentrations, PACAP may override other intraovarian regulatory mechanisms [54].

The fact that PACAP and VIP receptors are differently distributed in the ovaries suggests distinct functional roles for these two peptides, as discussed later in this review.

Uterus

PACAP and PAC_1-R are present in both human and rat placenta at term, suggesting they may have a role in placental function. Both are co-expressed in decidual cells, chorionic vessels, and stromal cells of the chorionic villi of the rat placenta, though they show regional and temporal differences [55–57]. PACAP stimulates VEGF release in pituitary folliculostellate cells [58], and in the placenta, vasculogenesis and angiogenesis are important processes [59]. Indeed, in rat chorionic vessels, PACAP significantly increases with gestation, which suggests that PACAP may act as a tropic factor and promote vessel growth through VEGF activity [57].

PACAP and Primordial Germ Cells

Among the growth factors that are essential for PGCs in culture, none seems to be able to directly stimulate PGC proliferation. Agents that increase intracellular cAMP (i.e., dibutyryl cAMP and forskolin) markedly stimulate PGC proliferation in culture [60]. PACAP-38 and -27 bind PGCs and activate adenylate cyclase (AC) in these cells and stimulate their proliferation, which suggests that PACAP might be an important regulator of PGC proliferation and that its action is mediated by cAMP [37].

Functions of PACAP in the Seminiferous Tubule

The presence of high levels of PACAP in the testis [22] and of PACAP receptors in the different cell populations of the testis [11] provides evidence that the peptide may operate as a local regulator of gonadal activity.

PACAP, produced by germ cells [22, 23], is able to modulate functions of both germ cells and somatic cells. PACAP activates or inhibits protein synthesis in spermatocytes or spermatids, respectively [61], induces a concentration-dependent stimulation of testosterone secretion from isolated rat Leydig cells [62–64] and, in cultured Sertoli cells, PACAP increases cAMP concentration and stimulates estradiol and inhibin secretion [16].

Moreover, in rat epididymal epithelium, PACAP stimulates chloride secretion, which is important for sperm activation and storage [65]. In the golden hamster, inhibition of PACAP with a specific PACAP receptor antagonist inhibits sperm motility in vitro in a dose-dependent manner [66]. Conversely, PACAP increases sperm motility in human [67] and mouse [68] spermatozoa.

Interestingly, male germ cells express a testis-specific prohormone convertase (PC4), at approximately the same time when PACAP immunoreactivity appears [69]. PC4 has been demonstrated to be the best candidate enzyme for processing pro-PACAP to generate PACAP38 and PACAP27 in the gonads [70]. Therefore, the generation of PACAP active form appears to be finely regulated by PC4 at specific stages of seminiferous tubule, strongly suggesting that PACAP may play a role in the stage-dependent germ cell development.

Moreover, PACAP, expressed in a stage dependent manner [71], may direct Sertoli cell functions by cyclical modulation of cAMP levels, depending on the stage of the seminiferous epithelial cycle, thus providing an optimal microenvironment for spermatogenesis [72, 73].

Functions of PACAP in the Production of a Mature Egg

As mentioned above, the different localization of PACAP and VIP and their receptors in the various compartments of the female reproductive tract and the activation of different second messengers suggest a different but synergistic role for these two peptides in the reproductive process.

PACAP Regulation of Follicle Recruitment and Growth

The number of follicles is determined early in life. Primordial follicle formation occurs when the oocytes that survive the process of germ cell cluster breakdown are individually surrounded by squamous (i.e., flattened) pre-GCs. The majority of primordial follicles exist in a quiescent state, but periodically, a cohort of these follicles is recruited for growth, differentiation and maturation. One of the initial events in primordial follicle development (i.e., the primordial to primary transition) is characterized by a change in GCs from a squamous to cuboidal morphology [74] and the expression of markers of cell proliferation [75]. As soon as the follicle begins to grow, it is possible to observe the recruitment of theca cells from the stromal-interstitial cell population and the subsequent proliferation of both granulosa and theca cells. Follicular recruitment is under the influence of positive and negative modulators, which must be accurately balanced to avoid the premature exhaustion of the follicle pool, which can, in turn, cause reproductive senescence in aging animals or menopause in the case of humans [1, 76].

The presence of PACAP in the GCs of preovulatory follicles and in the interstitial tissue in the proximity of the primary follicles after the LH surge and the presence of PACAP/VIP receptors in the ovary of immature mice and rats [42, 49] suggest a role for this peptide in the first stages of folliculogenesis [39, 42, 77, 78]. Gras et al. [77] suggested that PACAP stimulation might increase the growing potential of immature rat follicles because GCs that are primed with PACAP have a higher production of estradiol in response to FSH. However, in contrast to their data, we could not demonstrate FSH-R mRNA stimulation by PACAP. Moreover, the addition of PACAP to organ cultures of 2-day-old rat ovaries resulted in fewer growing follicles compared with control cultures [42]. PACAP was also able to inhibit FSH-dependent follicle growth in rats and mice whole follicle cultures [42, 49] by inhibiting cell proliferation and not influencing cell viability. In fact, PACAP significantly inhibited FSH-stimulated cyclin-D2 and PCNA expression [42]. Considering these results, it is possible to hypothesize that PACAP, similarly to anti-Mullerian hormone (AMH) [79], may negatively regulate follicle selection by inhibiting GC proliferation. However, the low production of PACAP before the LH surge suggests that PACAP, differently from AMH, might not have a function in follicle assembly or during the first wave of folliculogenesis and may act only in cycling ovaries after the LH surge. Indeed, PACAP is produced by preovulatory follicles and may modulate the growth of primary and preantral follicles in close proximity. Therefore, PACAP may contribute, together with many other factors, to the control of the premature exhaustion of the follicle pool by a transient inhibition of their growth.

PACAP and VIP Functions in Preovulatory Follicles

The involvement of PACAP in final oocyte maturation and ovulation is evidenced by the observations that PACAP exerts positive effects on several parameters that are related to follicle maturation and ovulation. PACAP stimulates steroidogenesis in GCs [30, 80, 81], contributes to the survival of GCs in several species [28, 29, 52, 82] and accelerates meiotic maturation in rat and mouse oocytes [82, 83]. Moreover, the production of PACAP and its receptors is stimulated by gonadotropins in preovulatory follicles [32, 33, 50, 52].

PACAP/VIP and Blood Flow

The presence of both VIP receptors and VIP-containing nerve fibers around blood vessels and theca layers of follicles of several species [45, 84–86] supports an important role for this peptide in blood flow regulation. VIP increases cerebral blood flow in several species [87] and has a relaxant effect on rabbit ovarian arteries [88]. Therefore, it is possible that VIP can contribute to the increased blood flow around preovulatory follicles that is observed after the LH surge [89]. In fact, an increased rate of blood flow is developmentally important for the generation of a healthy follicle and a competent oocyte [90], whereas a decrease in vascularization can be often associated with follicular atresia. The increased ovarian stromal blood flow may lead to a greater delivery of gonadotropins to the GCs of the preovulatory follicles [91]. In turn, gonadotropin induces the production of PACAP in the preovulatory follicle.

It is interesting to note the absence of fibers containing VIP around arteries and arterioles and a decrease in PACAP-IR nerve terminals in ovaries of patients with polycystic ovaries [92]. Wojtkiewicz and colleagues suggested a decreased effect of these bioactive agents on blood flow through these vessels in the cystic gonads and that the altered distribution and/or number of nerve fibers may be important in this disease [92].

PACAP and Oocyte Meiotic Maturation

In the mouse, PAC_1-R has also been found on fully grown GV oocytes [30], and both PACAP and PAC_1-R are transiently induced in cumulus cells (CCs) in response to the LH surge [82].

Mammalian oocytes are arrested at the dictyate stage until shortly before ovulation. It is only after the preovulatory surge of LH that the oocyte resumes meiotic maturation and becomes fertilizable. The inhibition of oocyte maturation is mediated by the cells of the cumulus oophorus, and the relief of inhibition is a result of the effect of gonadotropins on the CCs and not on the oocyte itself. In addition to gonadotropins, a number of other factors have been reported to regulate oocyte maturation. Most of these factors do not directly affect the oocyte but instead exert their activity via CCs [93]. Among these factors are EGF-like factors amphiregulin (AREG), epiregulin (EPI), and beta-cellulin (BTC), which are produced in the preovulatory follicle and play an important role in CC functions [94]. The facts that AREG expression in CCs precedes PACAP expression, PAC1-R is stimulated by EGF, and PACAP prevents CC apoptosis more efficiently than FSH and EGF suggest a potential cooperation among these factors in the physiological events that are correlated with mouse oocyte maturation [82].

In the rat, we showed that PACAP exerts a positive stimulation of meiotic maturation [95].

In the mouse, PACAP not only positively affected oocyte meiotic maturation but was also able to improve oocyte developmental potential. In fact, PACAP reverses the blockade of hypoxanthine-mediated spontaneous meiotic maturation, accelerates the completion of meiotic maturation with an increasing number of oocytes reaching the first polar body stage after 12 h of culture and efficiently promotes the decondensation of the sperm head and the formation of the male pronucleus [82]. Met II oocytes that matured for 12 h in the presence of PACAP displayed a correct arrangement of cortical granules (CGs) distributed in the sub-cortical position with a clearly visible CG-free domain overlaying the spindle in Met II oocytes [82]. Interestingly, we were also able to demonstrate that PACAP acted directly on denuded rat oocytes because PACAP delayed meiotic maturation by modulating their intracellular cAMP levels [83]. This result was not due to a cytotoxic effect because the inhibition that was obtained with PACAP was reversible once the peptide was removed from the medium. Therefore, our data indirectly suggest that rat oocytes possess specific PACAP receptors, and the lack of action of VIP on oocytes strongly indicates PAC_1-R [83]. In fact, more recently, we were able to detect, by RT-PCR, mRNA for PAC_1-R in denuded pronase-treated rat oocytes [30]. Moreover, to further support the presence of PAC_1-R, PACAP was able to increase in denuded oocytes not only cAMP levels but also the intracellular levels of calcium [30].

Cumulus Mucification and CC Apoptosis

The expanded cumulus provides the proper environment for oocyte development and successful fertilization [96, 97]. The CC number, cumulus compactness and CC apoptosis are critical parameters in determining oocyte developmental competence, embryo quality, fertilization rate and pregnancy outcome [98–102]. In response to the LH surge and soluble factors that are produced by the oocyte, CCs secrete a muco-elastic matrix and undergo a dramatic expansion. A few hours after ovulation, this matrix undergoes degradation, and a gradual shedding of CCs parallels the aging of oocytes [103, 104]. PACAP added to COCs in vitro exhibited only a moderate effect on cumulus expansion but very efficiently inhibited apoptosis. Interestingly, PAC_1-R in CCs is also maintained in ovulated COCs, which suggests a role for these receptors at later times. After ovulation, the destabilization of the matrix occurs also in the presence of FSH. In contrast, PACAP prevented the precocious disassembly of the cumulus after ovulation and maintained cumulus integrity and cell viability, either in combination with FSH or alone [82].

PACAP is present on the acrosome of epididymal spermatozoa [68]. PACAP that is released by the spermatozoa can interact with PAC_1-R on CCs at the site of fertilization, which causes the release of a soluble factor by CC itself and thereby promotes fertilization [68].

PACAP/VIP and Proteolytic Enzymes That Are Involved in Ovulation

PACAP and VIP play an important role at the time of ovulation by inducing the production of plasminogen activator (PA) in the preovulatory follicle. Plasminogen activators (PAs) are proteolytic enzymes that are involved in numerous biological processes [105, 106]. Two forms of PA have been detected in the ovary: the tissue-type (tPA) and the urokinase-type (uPA), both of which seem to be involved in gonadotropin-induced ovulation, as their synthesis is regulated by gonadotropins [107–109]. This role is supported by the finding that gonadotropin-induced ovulation is inhibited by the injection of serine protease inhibitors in rat ovarian bursa [110–112]. In addition to gonadotropins, a number of other factors play a key role in the regulation of the PA production in the ovary [113–115]. PACAP and VIP are among these factors. In agreement with the presence of different receptors in the different ovarian compartments [30], PACAP stimulates PA production in rat GCs that were cultured either as a whole follicle or as isolated cells [95]; in contrast, VIP can stimulate GC-PA production, but only when GCs are cultured as whole follicles.

PACAP and Progesterone

Even if the LH surge is the signal for the rapid preovulatory increase in progesterone production, progesterone levels remain elevated despite the decreased gonadotropin levels. PACAP stimulates progesterone secretion in GCs [81, 83, 116, 117], and the immunoneutralization of endogenous PACAP significantly reduces acute progesterone accumulation and impairs subsequent luteinization, which suggest that PACAP plays an important role in LH-induced progesterone production during the periovulatory period [117, 118]. Therefore, it can be speculated that the transient LH induction of PACAP and PAC_1-R at the time of ovulation may support progesterone production at later times.

The stimulatory effect of PACAP on progesterone secretion has been found also in cells that were isolated from human corpora lutea from nonpregnant patients [116].

The fact that PACAP is not essential at the time of ovulation is evident when observing PACAP−/− animals. These animals have normal follicular development and ovulation but show impaired fertility at later stages of the establishment of pregnancy [119]. In fact, only 13 % of these females show implantation 6.5 days after mating. The normal efficiency of ovulation in these animals suggests that PACAP might cooperate with other factors or that the loss of PACAP can be functionally complemented by alternative or compensatory mechanisms. Associated with the decrease in implantation, the prolactin and progesterone levels are significantly lower in females lacking PACAP. These results suggest an important role for PACAP in the establishment of pregnancy, with the progesterone that is produced at the ovarian level causing changes in the uterus as necessary to support embryo implantation [119].

Conclusions

In summary, the demonstration of PACAP expression in the ovary and its multiple effects on ovarian physiology indicate that PACAP is more than a hypothalamic peptide. The timing of expression and the different localization of PACAP and its receptors suggest intriguing possibilities for a role of this peptide in the complex ovarian physiology around the periovulatory period.

This role could be of importance for in vitro maturation (IVM) protocols of human oocytes, which is a promising safe technique to avert many side effects of gonadotropin stimulation protocols for in vitro fertilization (IVF). Unfortunately, the pregnancy rates from oocytes that are matured in vitro are still very low [120, 121]. When IVM techniques do not provide the correct environment for oocytes, although they may ensure nuclear maturation and some degree of fertilization after IVF, the result is a low rate of poor-quality blastocysts. Hence, considering the beneficial effect of PACAP on both nuclear and cytoplasmic maturation, it may be hypothesized that PACAP supplementation in IVM oocytes might reconstruct a physiological milieu for the oocyte to sustain proper maturation and preimplantation development.

References

1. Canipari R, Cellini V, Cecconi S. The ovary feels fine when paracrine and autocrine networks cooperate with gonadotropins in the regulation of folliculogenesis. Curr Pharm Des. 2012;18:245–55.
2. Miyata A, Arimura A, Dahl RR, Minamino N, Uehara A, Jiang L, et al. Isolation of a novel 38 residue-hypothalamic polypeptide which stimulates adenylate cyclase in pituitary cells. Biochem Biophys Res Commun. 1989;164:567–74.

3. Arimura A. PACAP: the road to discovery. Peptides. 2007;28:1617–9.
4. McRory JE, Parker DB, Ngamvongchon S, Sherwood NM. Sequence and expression of cDNA for pituitary adenylate cyclase activating polypeptide (PACAP) and growth hormone-releasing hormone (GHRH)-like peptide in catfish. Mol Cell Endocrinol. 1995;108:169–77.
5. Parker DB, Coe IR, Dixon GH, Sherwood NM. Two salmon neuropeptides encoded by one brain cDNA are structurally related to members of the glucagon superfamily. Eur J Biochem. 1993;215:439–48.
6. Parker DB, Power ME, Swanson P, Rivier J, Sherwood NM. Exon skipping in the gene encoding pituitary adenylate cyclase-activating polypeptide in salmon alters the expression of two hormones that stimulate growth hormone release. Endocrinology. 1997;138:414–23.
7. Fradinger EA, Sherwood NM. Characterization of the gene encoding both growth hormone-releasing hormone (GRF) and pituitary adenylate cyclase-activating polypeptide (PACAP) in the zebrafish. Mol Cell Endocrinol. 2000;165:211–9.
8. Hu Z, Lelievre V, Tam J, Cheng JW, Fuenzalida G, Zhou X, et al. Molecular cloning of growth hormone-releasing hormone/pituitary adenylyl cyclase-activating polypeptide in the frog Xenopus laevis: brain distribution and regulation after castration. Endocrinology. 2000;141:3366–76.
9. McRory JE, Parker RL, Sherwood NM. Expression and alternative processing of a chicken gene encoding both growth hormone-releasing hormone and pituitary adenylate cyclase-activating polypeptide. DNA Cell Biol. 1997;16:95–102.
10. Kimura C, Ohkubo S, Ogi K, Hosoya M, Itoh Y, Onda H, et al. A novel peptide which stimulates adenylate cyclase: molecular cloning and characterization of the ovine and human cDNAs. Biochem Biophys Res Commun. 1990;166:81–9.
11. Vaudry D, Falluel-Morel A, Bourgault S, Basille M, Burel D, Wurtz O, et al. Pituitary adenylate cyclase-activating polypeptide and its receptors: 20 years after the discovery. Pharmacol Rev. 2009;61:283–357.
12. Lutz EM, Sheward WJ, West KM, Morrow JA, Fink G, Harmar AJ. The VIP2 receptor: molecular characterisation of a cDNA encoding a novel receptor for vasoactive intestinal peptide. FEBS Lett. 1993;334:3–8.
13. Spengler D, Waeber C, Pantaloni C, Holsboer F, Bockaert J, Seeburg PH, et al. Differential signal transduction by five splice variants of the PACAP receptor. Nature. 1993;365:170–5.
14. Arimura A. Pituitary adenylate cyclase activating polypeptide (PACAP): discovery and current status of research. Regul Pept. 1992;37:287–302.
15. Arimura A. Receptors for pituitary adenylate cyclase-activating polypeptide: comparison with vasoactive intestinal peptide receptors. Trends Endocrinol Metab. 1992;3:288–94.
16. Heindel JJ, Powell CJ, Paschall CS, Arimura A, Culler MD. A novel hypothalamic peptide, pituitary adenylate cyclase activating peptide, modulate Sertoli cells function in vitro. Biol Reprod. 1992;47:800–6.
17. Chedeville A, Mirossay L, Chastre E, Hurbain-Kosmath I, Lopez M, Gespach C. Interaction of VIP, PACAP and related peptides in normal and leukemic human monocytes and macrophages. FEBS Lett. 1993;319:171–6.
18. Deutsch PJ, Sun Y. The 38-amino acid form of pituitary adenylate cyclase-activating polypeptide stimulates dual signaling cascades in PC12 cells and promotes neurite outgrowth. J Biol Chem. 1992;267:5108–13.
19. Suzuki A, Kotoyori J, Oiso Y, Kozawa O. Pituitary adenylate cyclase-activating polypeptide induces cAMP production independently from vasoactive intestinal polypeptide in osteoblast-like cells. Cell Signal. 1994;6:11–6.
20. Arimura A, Somogyvari Vigh A, Miyata A, Mizuno K, Coy DH, Kitada C. Tissue distribution of PACAP as determined by RIA: highly abundant in the rat brain and testes. Endocrinology. 1991;129:2787–9.
21. Li M, Arimura A. Neuropeptides of the pituitary adenylate cyclase-activating polypeptide/vasoactive intestinal polypeptide/growth hormone-releasing hormone/secretin family in testis. Endocrine. 2003;20:201–14.

22. Shioda S, Legradi G, Leung WC, Nakajo S, Nakaya K, Arimura A. Localization of pituitary adenylate cyclase-activating polypeptide and its messenger ribonucleic acid in the rat testis by light and electron microscopic immunocytochemistry and in situ hybridization. Endocrinology. 1994;135:818–25.
23. Li M, Funahashi H, Mbikay M, Shioda S, Arimura A. Pituitary adenylate cyclase activating polypeptide-mediated intracrine signaling in the testicular germ cells. Endocrine. 2004;23:59–75.
24. Konen J, Paavola M, Penttila TL, Parvinen M, Pelto-Huijko M. Stage-specific expression of pituitary adenylate cyclase-activating polypeptide (PACAP) mRNA in the rat seminiferous tubules. Endocrinology. 1994;135:2291–4.
25. Hannibal J, Fahrenkrug J. Expression of pituitary adenylate cyclase activating polypeptide (PACAP) gene by rat spermatogenic cells. Regul Pept. 1995;55:111–5.
26. Nakamura K, Nakamachi T, Endo K, Ito K, Machida T, Oka T, et al. Distribution of pituitary adenylate cyclase-activating polypeptide (PACAP) in the human testis and in testicular germ cell tumors. Andrologia. 2014;46:465–71.
27. Vaudry D, Gonzalez BJ, Basille M, Pamantung TF, Fournier A, Vaudry H. PACAP acts as a neurotrophic factor during histogenesis of the rat cerebellar cortex. Ann N Y Acad Sci. 2000;921:293–9.
28. Barberi M, Muciaccia B, Morelli MB, Stefanini M, Cecconi S, Canipari R. Expression localisation and functional activity of pituitary adenylate cyclase-activating polypeptide, vasoactive intestinal polypeptide and their receptors in mouse ovary. Reproduction. 2007;134: 281–92.
29. Morelli MB, Barberi M, Gambardella A, Borini A, Cecconi S, Coticchio G, et al. Characterization, expression, and functional activity of pituitary adenylate cyclase-activating polypeptide and its receptors in human granulosa-luteal cells. J Clin Endocrinol Metab. 2008;93:4924–32.
30. Vaccari S, Latini S, Barberi M, Teti A, Stefanini M, Canipari R. Characterization and expression of different pituitary adenylate cyclase-activating polypeptide/vasoactive intestinal polypeptide receptors in rat ovarian follicles. J Endocrinol. 2006;191:287–99.
31. Zhou R, Tsang AH, Lau SW, Ge W. Pituitary adenylate cyclase-activating polypeptide (PACAP) and its receptors in the zebrafish ovary: evidence for potentially dual roles of PACAP in controlling final oocyte maturation. Biol Reprod. 2011;85:615–25.
32. Wang Y, Wong AO, Ge W. Cloning, regulation of messenger ribonucleic acid expression, and function of a new isoform of pituitary adenylate cyclase-activating polypeptide in the zebrafish ovary. Endocrinology. 2003;144:4799–810.
33. Sayasith K, Brown KA, Sirois J. Gonadotropin-dependent regulation of bovine pituitary adenylate cyclase-activating polypeptide in ovarian follicles prior to ovulation. Reproduction. 2007;133:441–53.
34. Gobbetti A, Zerani M, Miano A, Bramucci M, Murri O, Amici D. Presence of pituitary adenylate cyclase-activating polypeptide 38-immuno-like material in the brain and ovary of the female crested newt, Triturus carnifex: its involvement in the ovarian synthesis of prostaglandins and steroids. J Endocrinol. 1997;152:141–6.
35. Thomas RL, Crawford NM, Grafer CM, Halvorson LM. Pituitary adenylate cyclase-activating polypeptide (PACAP) in the hypothalamic-pituitary-gonadal axis: a review of the literature. Reprod Sci. 2013;20:857–71.
36. Reglodi D, Tamas A, Koppan M, Szogyi D, Welke L. Role of PACAP in female fertility and reproduction at gonadal level—recent advances. Front Endocrinol (Lausanne). 2012;3:155.
37. Pesce M, Canipari R, Ferri JL, Siracusa G, De Felici M. Pituitary adenylate cyclase-activating polypeptide (PACAP) stimulates adenylate cyclase and promotes proliferation of mouse primordial germ cells. Development. 1996;122:215–21.
38. Ko C, In YH, Park-Sarge OK. Role of progesterone receptor activation in pituitary adenylate cyclase activating polypeptide gene expression in rat ovary. Endocrinology. 1999;140:5185–94.
39. Gras S, Hannibal J, Georg B, Fahrenkrug J. Transient periovulatory expression of pituitary adenylate cyclase activating peptide in rat ovarian cells. Endocrinology. 1996;137:4779–85.

40. Scaldaferri L, Arora K, Lee SH, Catt KJ, Moretti C. Expression of PACAP and its type-I receptor isoforms in the rat ovary. Mol Cell Endocrinol. 1996;117:227–32.
41. Steenstrup BR, Alm P, Hannibal J, Jorgensen JC, Palle C, Junge J, et al. Pituitary adenylate cyclase-activating polypeptide: occurrence and relaxant effect in female genital tract. Am J Physiol. 1995;269:E108–17.
42. Latini S, Chiarpotto M, Muciaccia B, Vaccari S, Barberi M, Guglielmo MC, et al. Inhibitory effect of pituitary adenylate cyclase activating polypeptide on the initial stages of rat follicle development. Mol Cell Endocrinol. 2010;320:34–44.
43. Kotani E, Usuki S, Kubo T. Rat corpus luteum expresses both PACAP and PACAP type IA receptor mRNAs. Peptides. 1997;18:1453–5.
44. Kotani E, Usuki S, Kubo T. Effect of pituitary adenylate cyclase-activating polypeptide (PACAP) on progestin biosynthesis in cultured granulosa cells from rat ovary and expression of mRNA encoding PACAP type IA receptor. J Reprod Fertil. 1998;112:107–14.
45. Ahmed CE, Dees WL, Ojeda SR. The immature rat ovary is innervated by vasoactive intestinal peptide (VIP)-containing fibers and responds to VIP with steroid secretion. Endocrinology. 1986;118:1682–9.
46. Gozes I, Tsafriri A. Detection of vasoactive intestinal peptide-encoding messenger ribonucleic acid in the rat ovaries. Endocrinology. 1986;119:2606–10.
47. Dees WL, Ahmed CE, Ojeda SR. Substance P- and vasoactive intestinal peptide-containing fibers reach the ovary by independent routes. Endocrinology. 1986;119:638–41.
48. Advis JP, Ahmed CE, Ojeda SR. Direct hypothalamic control of vasoactive intestinal peptide (VIP) levels in the developing rat ovary. Brain Res Bull. 1989;22:605–10.
49. Cecconi S, Rossi G, Barberi M, Scaldaferri L, Canipari R. Effect of pituitary adenylate cyclase-activating polypeptide and vasoactive intestinal polypeptide on mouse preantral follicle development in vitro. Endocrinology. 2004;145:2071–9.
50. Park HJ, Lee J, Wang L, Park JH, Kwon HB, Arimura A, et al. Stage-specific expression of pituitary adenylate cyclase-activating polypeptide type I receptor messenger ribonucleic acid during ovarian follicle development in the rat. Endocrinology. 2000;141:702–9.
51. Ko C, Park-Sarge OK. Progesterone receptor activation mediates LH-induced type-I pituitary adenylate cyclase activating polypeptide receptor (PAC(1)) gene expression in rat granulosa cells. Biochem Biophys Res Commun. 2000;277:270–9.
52. Lee J, Park HJ, Choi HS, Kwon HB, Arimura A, Lee BJ, et al. Gonadotropin stimulation of pituitary adenylate cyclase-activating polypeptide (PACAP) messenger ribonucleic acid in the rat ovary and the role of PACAP as a follicle survival factor. Endocrinology. 1999;140: 818–26.
53. Brubel R, Reglodi D, Jambor E, Koppan M, Varnagy A, Biro Z, et al. Investigation of pituitary adenylate cyclase activating polypeptide in human gynecological and other biological fluids by using MALDI TOF mass spectrometry. J Mass Spectrom. 2011;46:189–94.
54. Koppan M, Varnagy A, Reglodi D, Brubel R, Nemeth J, Tamas A, et al. Correlation between oocyte number and follicular fluid concentration of pituitary adenylate cyclase-activating polypeptide (PACAP) in women after superovulation treatment. J Mol Neurosci. 2012;48: 617–22.
55. Scaldaferri ML, Modesti A, Palumbo C, Ulisse S, Fabbri A, Piccione E, et al. Pituitary adenylate cyclase-activating polypeptide (PACAP) and PACAP-receptor type 1 expression in rat and human placenta. Endocrinology. 2000;141:1158–67.
56. Brubel R, Boronkai A, Reglodi D, Racz B, Nemeth J, Kiss P, et al. Changes in the expression of pituitary adenylate cyclase-activating polypeptide in the human placenta during pregnancy and its effects on the survival of JAR choriocarcinoma cells. J Mol Neurosci. 2010;42:450–8.
57. Koh PO, Kwak SD, Kim HJ, Roh G, Kim JH, Kang SS, et al. Expression patterns of pituitary adenylate cyclase activating polypeptide and its type I receptor mRNAs in the rat placenta. Mol Reprod Dev. 2003;64:27–31.
58. Gloddek J, Pagotto U, Paez PM, Arzt E, Stalla GK, Renner U. Pituitary adenylate cyclase-activating polypeptide, interleukin-6 and glucocorticoids regulate the release of vascular endothelial growth factor in pituitary folliculostellate cells. J Endocrinol. 1999;160:483–90.

59. Demir R, Kaufmann P, Castellucci M, Erbengi T, Kotowski A. Fetal vasculogenesis and angiogenesis in human placental villi. Acta Anat (Basel). 1989;136:190–203.
60. De Felici M, Dolci S, Pesce M. Proliferation of mouse primordial germ cells in vitro: a key role for cAMP. Dev Biol. 1993;157:277–80.
61. West AP, McKinnell C, Sharpe RM, Saunders PT. Pituitary adenylate cyclase activating polypeptide can regulate testicular germ cell protein synthesis in vitro. J Endocrinol. 1995;144:215–23.
62. Romanelli F, Fillo S, Isidori A, Conte D. Pituitary adenylate cyclase-activating polypeptide regulates rat Leydig cell function in vitro. Neuropeptides. 1997;31:311–7.
63. Rossato M, Nogara A, Gottardello F, Bordon P, Foresta C. Pituitary adenylate cyclase activating polypeptide stimulates rat Leydig cell steroidogenesis through a novel transduction pathway. Endocrinology. 1997;138:3228–35.
64. El-Gehani F, Zhang FP, Pakarinen P, Rannikko A, Huhtaniemi I. Gonadotropin-independent regulation of steroidogenesis in the fetal rat testis. Biol Reprod. 1998;58:116–23.
65. Zhou WL, Leung PS, Wong TP, Dun NJ, Wong PY, Chan HC. Local regulation of epididymal anion secretion by pituitary adenylate cyclase-activating polypeptide. J Endocrinol. 1997;154:389–95.
66. Gozes I, Perl O, Zamostiano R, Rubinraut S, Fridkin M, Shochat L, et al. Multiple actions of a hybrid PACAP antagonist: neuronal cell killing and inhibition of sperm motility. Ann N Y Acad Sci. 1998;865:266–73.
67. Brubel R, Kiss P, Vincze A, Varga A, Varnagy A, Bodis J, et al. Effects of pituitary adenylate cyclase activating polypeptide on human sperm motility. J Mol Neurosci. 2012;48:623–30.
68. Tanii I, Aradate T, Matsuda K, Komiya A, Fuse H. PACAP-mediated sperm-cumulus cell interaction promotes fertilization. Reproduction. 2011;141:163–71.
69. Seidah NG, Day R, Hamelin J, Gaspar A, Collard MW, Chretien M. Testicular expression of PC4 in the rat: molecular diversity of a novel germ cell-specific Kex2/subtilisin-like proprotein convertase. Mol Endocrinol. 1992;6:1559–70.
70. Basak A, Toure BB, Lazure C, Mbikay M, Chretien M, Seidah NG. Enzymic characterization in vitro of recombinant proprotein convertase PC4. Biochem J. 1999;343(Pt 1):29–37.
71. Kononen J, Paavola M, Penttila TL, Parvinen M, Pelto-Huikko M. Stage-specific expression of pituitary adenylate cyclase-activating polypeptide (PACAP) mRNA in the rat seminiferous tubules. Endocrinology. 1994;135:2291–4.
72. Toppari J, Kangasniemi M, Kaipia A, Mali P, Huhtaniemi I, Parvinen M. Stage- and cell-specific gene expression and hormone regulation of the seminiferous epithelium. J Electron Microsc Tech. 1991;19:203–14.
73. Daniel PB, Habener JF. Pituitary adenylate cyclase-activating polypeptide gene expression regulated by a testis-specific promoter in germ cells during spermatogenesis. Endocrinology. 2000;141:1218–27.
74. Cecconi S, Nottola SA, Canipari R. Regulation of ovarian follicle assembly and development in mammals In: Gonzales-Bulnes Antonio, editor. Novel concepts in ovarian endocrinology. Madrid: 2007. p. 43–81.
75. Wandji SA, Srsen V, Voss AK, Eppig JJ, Fortune JE. Initiation in vitro of growth of bovine primordial follicles. Biol Reprod. 1996;55:942–8.
76. Edson MA, Nagaraja AK, Matzuk MM. The mammalian ovary from genesis to revelation. Endocr Rev. 2009;30:624–712.
77. Gras S, Host E, Fahrenkrug J. Role of pituitary adenylate cyclase-activating peptide (PACAP) in the cyclic recruitment of immature follicles in the rat ovary. Regul Pept. 2005;128:69–74.
78. Koh PO, Kwak SD, Kang SS, Cho GJ, Chun SY, Kwon HB, et al. Expression of pituitary adenylate cyclase activating polypeptide (PACAP) and PACAP type I A receptor mRNAs in granulosa cells of preovulatory follicles of the rat ovary. Mol Reprod Dev. 2000;55:379–86.
79. Durlinger AL, Visser JA, Themmen AP. Regulation of ovarian function: the role of anti-Mullerian hormone. Reproduction. 2002;124:601–9.
80. Heindel JJ, Sneeden J, Powell CJ, Davis B, Culler MD. A novel hypothalamic peptide, pituitary adenylate cyclase-activating peptide, regulates the function of rat granulosa cells in vitro. Biol Reprod. 1996;54:523–30.

81. Zhong Y, Kasson BG. Pituitary adenylate cyclase-activating polypeptide stimulates steroidogenesis and adenosine 3',5'-monophosphate accumulation in cultured rat granulosa cells. Endocrinology. 1994;135:207–13.
82. Barberi M, Paolo VD, Latini S, Guglielmo MC, Cecconi S, Canipari R. Expression and functional activity of PACAP and its receptors on cumulus cells: effects on oocyte maturation. Mol Cell Endocrinol. 2013;375:79–88.
83. Apa R, Lanzone A, Mastrandrea M, Miceli F, Macchione E, Fulghesu AM, et al. Effect of pituitary adenylate cyclase-activating peptide on meiotic maturation in follicle-enclosed, cumulus-enclosed, and denuded rat oocytes. Biol Reprod. 1997;57:1074–9.
84. Hulshof SC, Dijkstra G, Van der Beek EM, Bevers MM, Figueiredo JR, Beckers JF, et al. Immunocytochemical localization of vasoactive intestinal peptide and neuropeptide Y in the bovine ovary. Biol Reprod. 1994;50:553–60.
85. Kannisto P, Ekblad E, Helm G, Owman C, Sjoberg NO, Stjernquist M, et al. Existence and coexistence of peptides in nerves of the mammalian ovary and oviduct demonstrated by immunocytochemistry. Histochemistry. 1986;86:25–34.
86. Jorgensen JC, Kannisto P, Liedberg F, Ottesen B, Owman C, Schmidt G. The influence of neuropeptide Y and norepinephrine on ovulation in the rat ovary. Peptides. 1991;12:975–82.
87. Grant S, Lutz EM, McPhaden AR, Wadsworth RM. Location and function of VPAC1, VPAC2 and NPR-C receptors in VIP-induced vasodilation of porcine basilar arteries. J Cereb Blood Flow Metab. 2006;26:58–67.
88. Jorgensen JC. Interaction between norepinephrine, NPY and VIP in the ovarian artery. Peptides. 1991;12:831–7.
89. Acosta TJ, Hayashi KG, Ohtani M, Miyamoto A. Local changes in blood flow within the preovulatory follicle wall and early corpus luteum in cows. Reproduction. 2003;125:759–67.
90. Van Blerkom J. Intrafollicular influences on human oocyte developmental competence: perifollicular vascularity, oocyte metabolism and mitochondrial function. Hum Reprod. 2000;15 Suppl 2:173–88.
91. Redmer DA, Reynolds LP. Angiogenesis in the ovary. Rev Reprod. 1996;1:182–92.
92. Wojtkiewicz J, Jana B, Kozlowska A, Crayton R, Majewski M, Zalecki M, et al. Innervation pattern of polycystic ovaries in the women. J Chem Neuroanat. 2014;61–62:147–52.
93. Canipari R. Oocyte—granulosa cell interactions. Hum Reprod Update. 2000;6:279–89.
94. Park JY, Su YQ, Ariga M, Law E, Jin SL, Conti M. EGF-like growth factors as mediators of LH action in the ovulatory follicle. Science. 2004;303:682–4.
95. Apa R, Lanzone A, Miceli F, Vaccari S, Macchione E, Stefanini M, et al. Pituitary adenylate cyclase-activating polypeptide modulates plasminogen activator expression in rat granulosa cell. Biol Reprod. 2002;66:830–5.
96. Chen L, Russell PT, Larsen WJ. Functional significance of cumulus expansion in the mouse: roles for the preovulatory synthesis of hyaluronic acid within the cumulus mass. Mol Reprod Dev. 1993;34:87–93.
97. Hizaki H, Segi E, Sugimoto Y, Hirose M, Saji T, Ushikubi F, et al. Abortive expansion of the cumulus and impaired fertility in mice lacking the prostaglandin E receptor subtype EP(2). Proc Natl Acad Sci U S A. 1999;96:10501–6.
98. Blondin P, Sirard MA. Oocyte and follicular morphology as determining characteristics for developmental competence in bovine oocytes. Mol Reprod Dev. 1995;41:54–62.
99. Boni R, Cuomo A, Tosti E. Developmental potential in bovine oocytes is related to cumulus-oocyte complex grade, calcium current activity, and calcium stores. Biol Reprod. 2002;66:836–42.
100. Host E, Mikkelsen AL, Lindenberg S, Smidt-Jensen S. Apoptosis in human cumulus cells in relation to maturation stage and cleavage of the corresponding oocyte. Acta Obstet Gynecol Scand. 2000;79:936–40.
101. Lee KS, Joo BS, Na YJ, Yoon MS, Choi OH, Kim WW. Cumulus cells apoptosis as an indicator to predict the quality of oocytes and the outcome of IVF-ET. J Assist Reprod Genet. 2001;18:490–8.

102. Ruvolo G, Bosco L, Pane A, Morici G, Cittadini E, Roccheri MC. Lower apoptosis rate in human cumulus cells after administration of recombinant luteinizing hormone to women undergoing ovarian stimulation for in vitro fertilization procedures. Fertil Steril. 2007;87:542–6.
103. Salustri A, Camaioni A, D'Alessandris C. Endocrine and paracrine regulation of cumulus expansion. Zygote. 1996;4:313–5.
104. Russell DL, Salustri A. Extracellular matrix of the cumulus-oocyte complex. Semin Reprod Med. 2006;24:217–27.
105. Dano K, Andreasen A, Grondahl-Hansen J, Kristensen P, Nielsen LS, Skriver L. Plasminogen activators, tissue degradation and cancer. Adv Cancer Res. 1985;44:139–266.
106. Blasi F, Vassalli JD, Dano K. Urokinase-type plasminogen activator proenzyme, receptor and inhibitors. J Cell Biol. 1987;104:801–4.
107. Canipari R, Strickland S. Studies on the hormonal regulation of plasminogen activator production in the rat ovary. Endocrinology. 1986;118:1652–9.
108. Canipari R, O'Connell ML, Meyer G, Strickland S. Mouse ovarian granulosa cells produce urokinase-type plasminogen activator, whereas the corresponding rat cells produce tissue-type plasminogen activator. J Cell Biol. 1987;105:977–81.
109. O'Connell ML, Canipari R, Strickland S. Hormonal regulation of tissue plasminogen activator secretion and mRNA levels in rat granulosa cells. J Biol Chem. 1987;262:2339–44.
110. Macchione E, Epifano O, Stefanini M, Belin D, Canipari R. Urokinase redistribution from the secreted to the cell-bound fraction in granulosa cells of rat preovulatory follicles [In Process Citation]. Biol Reprod. 2000;62:895–903.
111. Reich R, Miskin R, Tsafriri A. Follicular plasminogen activator: involvement in ovulation. Endocrinology. 1985;116:516–23.
112. Tsafriri A, Bicsak TA, Cajander SB, Ny T, Hsueh AJ. Suppression of ovulation rate by antibodies to tissue-type plasminogen activator and alpha 2-antiplasmin. Endocrinology. 1989;124:415–21.
113. Too CK, Weiss TJ, Bryant-Greenwood GD. Relaxin stimulates plasminogen activator secretion by rat granulosa cells in vitro. Endocrinology. 1982;111:1424–6.
114. Wang C. Luteinizing hormone releasing hormone stimulates plasminogen activator production by rat granulosa cells. Endocrinology. 1983;112:1130–2.
115. Hsueh AJ, Liu YX, Cajander S, Peng XR, Dahl K, Kristensen P, et al. Gonadotropin-releasing hormone induces ovulation in hypophysectomized rats: studies on ovarian tissue-type plasminogen activator activity, messenger ribonucleic acid content, and cellular localization. Endocrinology. 1988;122:1486–95.
116. Apa R, Lanzone A, Mastrandrea M, Miceli F, de Feo D, Caruso A, et al. Control of human luteal steroidogenesis: role of growth hormone- releasing hormone, vasoactive intestinal peptide, and pituitary adenylate cyclase-activating peptide. Fertil Steril. 1997;68:1097–102.
117. Gras S, Hannibal J, Fahrenkrug J. Pituitary adenylate cyclase-activating polypeptide is an auto/paracrine stimulator of acute progesterone accumulation and subsequent luteinization in cultured periovulatory granulosa/lutein cells. Endocrinology. 1999;140:2199–205.
118. Fahrenkrug J. PACAP—a multifaceted neuropeptide. Chronobiol Int. 2006;23:53–61.
119. Isaac ER, Sherwood NM. Pituitary adenylate cyclase-activating polypeptide (PACAP) is important for embryo implantation in mice. Mol Cell Endocrinol. 2008;280:13–9.
120. Suikkari AM, Soderstrom-Anttila V. In-vitro maturation of eggs: is it really useful? Best Pract Res Clin Obstet Gynaecol. 2007;21:145–55.
121. Chian RC, Lim JH, Tan SL. State of the art in in-vitro oocyte maturation. Curr Opin Obstet Gynecol. 2004;16:211–9.

Part VIII
PACAP in Biological Barriers and Barrier-Forming Cells

Chapter 25
Transport of Pituitary Adenylate Cyclase Activating Polypeptide Across the Blood–Brain Barrier: Consequences for Disease States and Therapeutic Effects

William A. Banks

Abstract The pituitary adenylate cyclase activating polypeptides (PACAP38 and PACAP27) are pluripotent peptides that exert many effects directly on brain. These direct effects on brain often occur even when PACAP is administered peripherally, especially PACAP38. This ability of peripheral PACAP38 to directly affect the brain is largely due to its transport across the blood–brain barrier (BBB) by peptide transport system-6 (PTS-6). PTS-6 is actually a family of transporters that transport PACAP38 from blood to brain (influx) and brain to blood (efflux) and PACAP27 from brain to blood. The efflux transporter for PACAP27 is beta-1 ATPase and its inhibition results in a fourfold increase in the accumulation by brain and an increase in the therapeutic effect of peripherally administered PACAP27. The activity of the PACAP38 influx component of PTS-6 is not static but can increase or decrease after injuries to the CNS. Because of PTS-6 influx, very low doses of PACAP38 administered peripherally can have profound therapeutic effects on brain. However, the short half-life and rapid degradation of PACAP27 and especially PACAP38 in blood complicates peripheral administration. Analogs of PACAP38, such as conjugation with a TAT peptide or acylation, can have improved therapeutic effects. Bypassing the BBB by administering PACAP38 intranasally at the level of the cribriform plate is also promising.

Keywords PACAP • Blood–brain barrier • Stroke • Brain • Spinal cord injury • Transport • Antisense • Acylation • Intranasal • Central nervous system • Transcellular diffusion

---

W.A. Banks (✉)
Geriatric Research Education and Clinical Center, Veterans Affairs Puget Sound Health Care System, Bldg 1/Rm810a, VAPSHCS, 1660 S. Columbian Way, Seattle, WA 98108, USA

Division of Gerontology and Geriatric Medicine, Department of Medicine, University of Washington School of Medicine, Seattle, WA, USA
e-mail: wabanks1@uw.edu

D. Reglodi, A. Tamas (eds.), *Pituitary Adenylate Cyclase Activating Polypeptide — PACAP*, Current Topics in Neurotoxicity 11,
DOI 10.1007/978-3-319-35135-3_25

Introduction

Pituitary adenylate cyclase activating polypeptide (PACAP), discovered by Akira Arimura approximately 25 years ago [1], is a pluripotent and very powerful peptide. Among its remarkable characteristics is the saturable blood-to-brain transport of its 38 amino acid form (PACAP38). This transport activity has been reified as peptide transport system-6 (PTS-6) and because of this transporter, very small, peripherally administered doses of PACAP can have remarkable effects on the brain.

Many peptides cross the BBB by saturable mechanisms as well as by transcellular diffusion. The latter refers to a non-saturable mechanism by which the physicochemical characteristics of a substance, most notably its lipid solubility and molecular weight, determine the extent to which that substance can cross the cellular membranes of the blood–brain barrier (BBB). Although peptides tend to be water soluble, many can cross enough by this mechanism to affect brain function [2, 3]. As discussed in more detail below, PACAP27 crosses the BBB by the mechanism of transmembrane diffusion.

Major roadblocks to the therapeutic use of peptides, whether their target is central or peripheral, is their rapid degradation and clearance from blood. This means that a peripherally administered dose of a peptide does not circulate for long in the blood stream and therefore is not presented to the BBB for the opportunity of entering the central nervous system.

As discussed in more detail below, these BBB and pharmacokinetic characteristic of the PACAPs underlie their abilities to access the brain and to exert effects on the CNS. These characteristics dictate study design and strategies for those who want to examine the effects of the PACAPs on brain or to improve on the parent molecules' abilities to access the brain after peripheral administration.

PACAP may also play a role in the maintenance and protection of the BBB. Brain endothelial cells grown in culture express the mRNA for the VPAC1 and VPAC2 receptors, but not for the PAC1 receptor [4]. PACAP38 increases tight junction protein levels through a posttranscriptional mechanism and increases the electrical resistance in an in vitro model of the BBB. PACAP38 also protected the in vitro BBB tight junctions and electrical resistance from deterioration when the model was exposed to glucose deprivation or oxidative stress. PACAP38 did not, however, protect brain endothelial cells from apoptosis induced by glucose deprivation or oxidative stress [4].

Characteristics of PACAP Transport

The characteristics of the BBB transport of both PACAP38 and PACAP27 were described over two decades ago [5]. The results showed a complex system of BBB transporters for the PACAPs (Fig. 25.1), reified as peptide transport system-6 (PTS-6). PACAP38 is transported in by a saturable transport system at a unidirectional influx rate initially measured to be about 2.9 μl/g-min. However, PACAP38

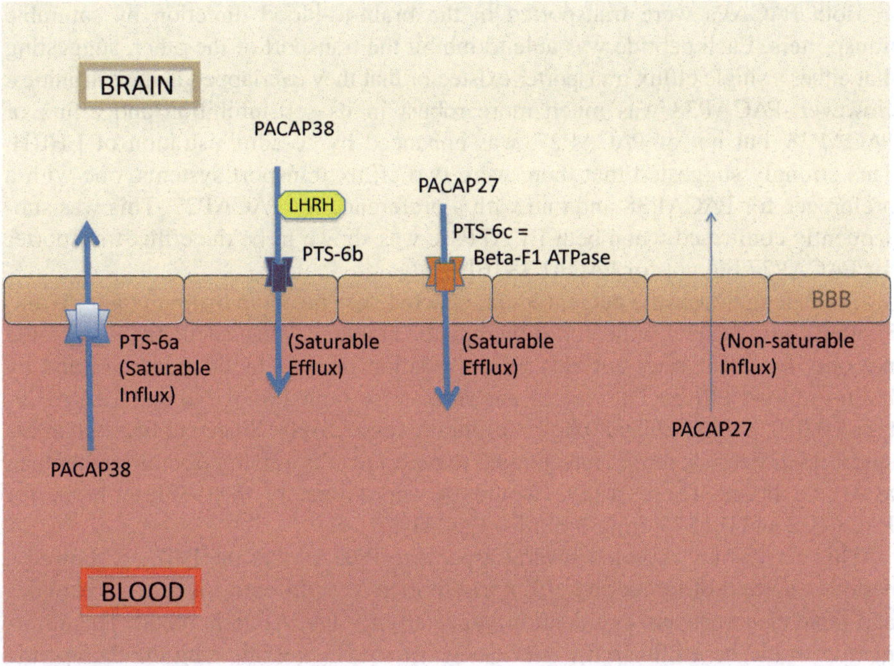

Fig. 25.1 Peptide Transport System-6 (PTS-6) at the blood–brain barrier. PTS-6 is here divided into its three functional components: saturable blood-to-brain transport of PACAP38 (PTS-6a); saturable brain-to-blood transport of PACAP38 (PTS-6b); saturable brain-to-blood transport of PACAP27 (PTS-6c). PACAP27 blood-to-brain influx is not saturable, but dependent on transcellular diffusion. The efflux of PACAP27 is not modulated by LHRH and the transporter protein is known to be beta-F1 ATPase. Efflux of PACAP38 is not mediated by this protein and is modulated by LHRH

is rapidly degraded in brain, but especially so in blood. Levels of PACAP38 corrected for degradation indicated a half-time clearance from blood of less than one minute and a unidirectional influx rate of about 15 μl/g-min. Results using the brain perfusion method which eliminates peripheral degradation found a similar rate of 16 μl/g-min. This is a very rapid transport rate for a peptide and indicates a robust transport system across the BBB for PACAP38. It also indicates that major blocks to the use of PACAP38 for therapeutic purposes is its rapid clearance from blood and degradation by peripheral tissues.

PACAP27 also crossed the BBB, but its transport was not saturable. PACAP27 is smaller than PACAP38 and about seven times more lipid soluble. If transport of the PACAPs were exclusively by the transcellular route, then the transport of PACAP27 should be greater than that of PACAP28, but it was about 2.1 μl/g-min, similar to the uncorrected rate of transport for PACAP38. PACAP27 was also rapidly degraded, but whereas it was more stable in blood than PACAP38, it was less stable in brain.

Both PACAPs were transported in the brain-to-blood direction by saturable transporters. Each peptide was able to inhibit the transport of the other, suggesting that either a single efflux transporter existed or that they overlapped in their affinities. However, PACAP38 was much more robust in its self-inhibition and efflux of PACAP38, but not of PACAP27, was enhanced by co-administration of LHRH. This strongly suggested that there were two efflux transport systems, one with a preference for PACAP38 and one with a preference for PACAP27. This was subsequently confirmed when beta-F1 ATPase was shown to be the efflux transporter for PACAP27 but not for PACAP38 [6].

The calculation of the percent of the injected dose taken up by brain (%Inj/brain) is often a more useful indicator of therapeutic potential. This calculation is affected not only by influx rate, but also by degradation rates in blood and brain and by brain-to-blood efflux. The values are similar for both PACAP38 and PACAP27, being 0.053 %Inj/brain and 0.067 %Inj/brain, respectively. Since publication of this study, it has become more conventional to report results as per g of brain rather than as whole brain. These results would be approximately 0.119 %Inj/g-brain for PACAP38 and 0.151 %Inj/g-brain for PACAP27.

This early study revealed several aspects to PACAP and the BBB. It showed a complex system of influx and efflux transporters that differed for the two peptides and rapid clearance and degradation, especially for PACAP38 in blood. This information would be useful in the later design of studies investigating the therapeutic potential of the PACAPs.

Other studies have found that PACAP38 is transported throughout the CNS, including the cervical, thoracic, and lumbar regions of the spinal cord; only the pons-medulla of young, healthy, male CD-1 mice showed no transport of PACAP38 [7]. Uptake by the hypothalamus and hippocampus is especially high, being about 7 and 3 times higher, respectively, than for whole brain [8].

Alterations in Transporter Function in Disease States and CNS Injuries

The PACAP transporters are not static, but vary in response to disease. Studies with spinal cord injury are illustrative of this. An early study severed the spinal cord in mice between lumbar vertebrae 2 and 3 and measured responses of PTS-6 and BBB integrity from 30 min to 53 days post lesion [7]. BBB disruption was limited to brain starting 7 days after injury and the lumbar spinal cord distal to injury between 16 and 30 days post injury. By contrast, the changes in PTS-6 activity as measured by the blood-to-brain transport of PACAP38 was widespread and varied. Overall, two distinct patterns were observed regarding changes in PTS-6 activity. There was an early phase (days 3–7) in which PACAP38 uptake was greatly reduced by brain and the cervical spinal cord, but not in the lumbar spinal cord. A late phase (days 7–30) of increased PACAP38 uptake occurred primarily in the spinal cord.

PTS-6 activity also changes following stroke in a complex pattern that is unrelated to BBB disruption [9]. Four h after unilateral middle cerebral artery occlusion, PTS-6 activity showed an arithmetic increase throughout the brain, although this increase reached statistical significance only in the striatum. PTS-6 activity then decreased significantly throughout the brain, reaching the lowest level of activity 24 h after injury. By 48 after injury, PTS-6 activity had recovered. As the authors of that study pointed out, alterations in PTS-6 activity will likely affect the amount of PACAP that must be administered peripherally in order to achieve a therapeutic level in brain.

PTS-6 activity was examined in brain and spinal cord between 6 h and 7 days after cardiac arrest [10]. No changes were seen except for a small increase in PACAP uptake by the thoracic spinal cord 24 h after cardiac arrest and resuscitation.

PTS-6 activity is greatly reduced in young and aged SAMP8 mice in comparison to that of young CD-1 mice. The SAMP8 mouse develops cognitive impairments with aging that are reversed with treatments that lower amyloid-beta peptide. In many of its characteristics, the young SAMP8 is similar to the young CD-1, but with aging the SAMP8 mouse develops several phenotypic characteristics of Alzheimer's disease, including those related to the BBB. We found that the aged SAMP8 mouse had a statistically significant decrease in the uptake of PACAP38 by the whole brain, hippocampus, olfactory bulb, and hypothalamus; in most cases, the values for these regions in young mice were intermediate between that of young CD-1 and aged SAMP8 mice [8]. These results suggest that the age-related phenotype of the SAMP8 is associated with a decreased ability of circulating PACAP to access the CNS. Whether this plays a part in the cognitive deterioration of the SAMP8 mouse is yet to be determined.

In comparison to the several conditions that impact PACAP transport across the BBB, activation of the innate immune system and the induction of inflammation by injecting lipopolysaccharide ip did not change either the efflux or the influx of PACAP38 [11].

CNS Effects After Peripheral Administration

Given the ability of the PACAPs to readily cross the BBB, it would be postulated that they should be effective after their peripheral administration in the treatment of various conditions or diseases of the brain. An effect on the CNS after the peripheral administration of a drug is not proof that the drug's effect requires it to cross the BBB as there are other mechanism by which drugs can exert their effects on the CNS that do not require BBB transport. However, BBB transport is one mechanism by which a drug can exert its effects on the CNS.

That peripheral PACAP38 can influence the development of the CNS was shown by Reglodi et al. [12]. They administered PACAP38 subcutaneously to neonatal rats for the first 14 days of life. They found several aspects of development were accelerated by PACAP38, including eye opening, incisor eruption, ear unfolding, and ear twitch and eyelid reflexes. PACAP38 also had positive effects on negative geotaxis, forelimb placing, hindlimb grasping, gait, and auditory startle. Although it is widely believed that the rodent neonatal BBB is leaky, studies clearly show that it is as much a barrier to proteins as is the adult BBB [13–15]. Therefore, these effects of peripheral PACAP38 on CNS development are consistent with the presence of PTS-6 activity during the neonatal period.

One of the first studies that used the characteristics of PACAP BBB transport in a therapeutic design was that of Uchida who examined the effects of PACAP38 in 4 vessel stroke in rats [16]. Because of the short half-live of PACAP38 in the circulation, an intravenous infusion was used to administer PACAP38. Because the transport of PACAP38 across the BBB was robust, low infusion rates of 16 and 160 pmol/h were used. Both doses were effective in preserving CA1 hippocampal neurons, even when the iv infusions were begun 24 h after the four vessel stroke.

This study also performed a dose response curve for PACAP38 administered by the intracerebroventricular (icv) route. It found that the optimal dose was an infusion of 1 pmol/min and gave results superior to 10 pmol/min. By comparison, the 16 pmol/min iv dose was more effective than the 160 pmol/min iv dose. Thus, the iv route required only 16 times more drug than the icv route. This is remarkably favorable for the iv route as for many peptides the iv/icv ratio is 100:1 or 1000:1. This favorable iv/icv ratio may have been caused by the rapid uptake of peripherally administered PACAP38 by the hippocampus, which is about 3 times faster than the uptake by whole brain [8]. The icv administration of radioactive PACAP38 showed that about 2 %Inj/g-brain was taken up by the hippocampus. As reviewed above, the original transport study of PACAP38 [5] found the uptake by whole brain to be about 0.119 %Inj/g-brain after the iv administration of PACAP38. The icv–iv ratio for %Inj/g-brain is 17 (2/0.119 = 16.8). Thus, the rapid uptake of PACAP by brain reflects the potent therapeutic effect of PACAP.

Another study confirmed that peripheral administration of a low dose PACAP38 (30 ng/kg) was protective in stroke and also in oxygen-glucose deprivation with reoxygenation [17]. These studies showed that PACAP once in the CNS acts both directly through its receptors but also by stimulating brain-derived neurotropic factor and its pathways and by down regulating nerve growth factor and neurite outgrowth factor receptors.

Traumatic brain injury has emerged as the signature injury of the early twenty-first century in both military and civilian life. Both CSF and plasma levels of PACAP38 increase and do so in tandem, consistent with BBB transport activities [18]. Furthermore, blood levels of PACAP38 are significantly higher in those patients who eventually die of their injuries. Interestingly, CSF levels of PACAP38 were not different between those who died and those who survived their injuries. This means that those who died had not only an elevation in plasma levels of PACAP38, but also lower CSF/plasma ratios. A low CSF/plasma ratio could be caused by an impairment in PTS-6 in those most severely injured. Such an impairment could result in subtherapeutic levels of PACAP being delivered to the brain, thus putting these patients at increased risk of death. Several studies have shown that icv administration of PACAP38 improves outcomes in animal models of traumatic brain injury. However, only one study has investigated the effectiveness of peripherally administered PACAP38 [19]. This study used the low infusion rate of 16 pmol/h and began therapy immediately after injury. PACAP38 decreased several measures of oxidative stress while elevating levels of the antioxidant enzymes superoxide dismutase and glutathione peroxidase. PACAP38 also reduced neurodegeneration as measured by fluoro-jade staining, injury area, and injury volume.

A slow (1 mg/kg over 7 days) subcutaneous infusion of PACAP38 was found to be protective against methamphetamine neurotoxicity induced 28 days after the PACAP infusion had finished [20]. As in the study my Miyamoto et al, PACAP38 reduced measures of oxidative stress. It also reduced evidence for astrogliosis and microgliosis by reducing elevations in GFAP and GLUT5 staining, respectively.

Approaches for Enhanced Delivery to the CNS

Based on the studies above, several strategies present themselves for the improved delivery of the PACAPs to the brain. These include synthesis of PACAP analogs that are more enzymatically resistant or better able to cross the BBB, inhibition of efflux transporters, or bypassing the BBB. All these strategies have been reported.

One approach attached to the PACAP38 C-terminus an 11-amino acid portion of the HIV-1 TAT protein (PACAP-TAT). The PACAP-TAT molecule retained full activity at the PAC1 receptor and crossed the BBB 2.5 times better than native PACAP38 [21] and was more potent after peripheral administration in the suppression of food intake [22]. An acylated version of PACAP was also effective at a very low dose after its intraperitoneal administration in an model of stroke [23].

Intranasal administration is often a useful way to bypass the BBB and deliver substances directly to the CNS [24]. Substances are absorbed by the olfactory and trigeminal nerves and can be delivered to various regions of the brain. We found that PACAP38 could be delivered to the brain by intranasal administration with about 0.25–0.5 %Inj/g of the administered dose being taken up by whole brain [25]. PACAP38 displays an unusual pattern in that the brain region with highest uptake is not the olfactory bulb, but the occipital cortex and the striatum. A very low dose of intranasal PACAP38 (0.01 μg/mouse) improved memory in the aged SAMP8 mouse, an animal model of Alzheimer's disease.

Co-administering the PACAP38 with beta-cyclodextrin increased uptake of PACAP38 in several brain regions with the increase by the hypothalamus being particularly high [25]. For some regions of the brain, such as the olfactory bulb and the striatum, combining beta-cyclodextrin with PACAP38 did not change uptake. Thus, beta-cyclodextrin had a targeting effect on brain distribution. This was seen with other cyclodextrins as well, although different patterns of brain distribution were produced. For example, alpha-cyclodextrin improved uptake of PACAP38 by the olfactory bulb while decreasing uptake by occipital cortex and striatum, whereas (2-hydroxypropyl)-beta-cyclodextrin improved thalamic uptake while decreasing uptake by striatum. These results show that enough PACAP38 can be delivered to brain by intranasal administration to affect the brain. The results also show that it may be possible to direct PACAP to certain regions of the brain or away from other regions, and thus modify the type of brain effect induced, by combining PACAP with various cyclodextrins.

An approach to improve the therapeutic delivery of PACAP27 to brain involved directing antisense molecules against the brain-to-blood efflux transporter for PACAP27, beta-F1 ATPase [6]. Efflux transporters generally limit the ability of their ligands to access and be retained by the CNS and their inhibition can greatly increase the therapeutic effects of their ligands. Naked oligophoshorothioate molecules are readily taken up by the BBB and transported across the BBB, exerting profound effects on CNS function [26]. Peripheral injection of a cocktail of oligophosphorothioate antisense molecules directed against beta-F1 ATPase selectively inhibited the efflux of PACAP27, but not of PACAP38, from brain resulting in a fourfold increase in brain uptake [6]. This resulted in an ability of peripherally administered PACAP27 to improve cognition in the SAMP8, an animal model of Alzheimer's disease. In comparison, treatment with the antisense cocktail alone was as effective as antisense plus PACAP in decreasing infarct volume 24 h after induction of middle cerebral artery occlusion in mice. The latter finding suggests that inhibition of the efflux of PACAP27 endogenous to brain may be sufficient to improve outcomes.

Summary

The PACAPs are potent molecules that can exert powerful effects on the central nervous system after their systemic administration. A large part of this effect resides in their abilities to cross the blood–brain barrier. However, the relation between the blood–brain barrier and the PACAPs is complex. This complexity occurs both because of the presence of influx and efflux mechanisms and because these mechanisms change in magnitude with insults to the central nervous system. The short half-life of the PACAPs in circulation limits their presentation to the blood–brain barrier and hence to the central nervous system. Several strategies appear promising in making the PACAPs more efficient therapeutics, including their administration by the intranasal route, inhibition of the efflux systems, acylation, and attachment to penetrating peptides.

References

1. Miyata A, Arimura A, Dahl RR, Minamino N, Uehara A, Jiang L, et al. Isolation of a novel 38 residue-hypothalamic polypeptide which stimulates adenylate cyclase in pituitary cells. Biochem Biophys Res Commun. 1989;164:567–74.
2. Banks WA, Kastin AJ, Coy DH. Evidence that [125 I]N-Tyr-delta sleep-inducing peptide crosses the blood-brain barrier by a non-competitive mechanism. Brain Res. 1984;301: 201–7.

3. Banks WA, Kastin AJ. Peptides and the blood-brain barrier: lipophilicity as a predictor of permeability. Brain Res Bull. 1985;15:287–92.
4. Wilhelm I, Fazakas C, Tamas A, Toth G, Reglodi D, Krizbai IA. PACAP enhances barrier properties of cerebral microvessels. J Mol Neurosci. 2014;54:469–76.
5. Banks WA, Kastin AJ, Komaki G, Arimura A. Passage of pituitary adenylate cyclase activating polypeptide 1-27 and pituitary adenylate cyclase activating polypeptide 1-38 across the blood-brain barrier. J Pharmacol Exp Therap. 1993;267:690–6.
6. Dogrukol-Ak D, Kumar VB, Ryerse JS, Farr SA, Verma S, Nonaka K, et al. Isolation of peptide transport system-6 from brain endothelial cells: therapeutic effects with antisense inhibition in Alzheimer's and stroke models. J Cereb Blood Flow Metab. 2009;29:411–22.
7. Banks WA, Kastin AJ, Arimura A. Effect of spinal cord injury on the permeability of the blood-brain and blood-spinal cord barriers to the neurotropin PACAP. Exp Neurol. 1998;151:116–23.
8. Nonaka N, Banks WA, Mizushima H, Shioda S, Morley JE. Regional differences in PACAP transport across the blood-brain barrier in mice: A possible influence of strain, amyloid beta protein, and age. Peptides. 2002;23:2197–202.
9. Somogyvari-Vigh A, Pan W, Reglodi D, Kastin AJ, Arimura A. Effect of middle cerebral artery occulsion on the passage of pituitary adenylate cyclase activating polypeptide across the blood-brain barrier in the rat. Regul Pept. 2000;91:89–95.
10. Mizushima H, Banks WA, Dohi K, Shioda S, Matsumoto H, Matsumoto K. The effect of cardiac arrest on the permeability of the mouse blood-brain and blood-spinal cord barrier to pituitary adenylate cyclase activating polypeptide (PACAP). Peptides. 1999;20:1337–40.
11. Nonaka N, Shioda S, Banks WA. Effect of lipopolysaccharide on the transport of pituitary adenylate cyclase activating polypeptide across the blood-brain barrier. Exp Neurol. 2005;191:137–44.
12. Reglodi D, Kiss P, Tamas A, Lengvari I. The effects of PACAP and PACAP antagonist on the neurobehavioral development of newborn rats. Behav Brain Res. 2003;140:131–9.
13. Daneman R, Zhou L, Kebede AA, Barres BA. Pericytes are required for blood-brain barrier integrity during embryogenesis. Nature. 2010;468:562–6.
14. Dziegielewska KM, Hinds LA, Mollgard K, Reynolds ML, Saunders NR. Blood-brain, blood-cerebrospinal fluid and cerebrospinal fluid-brain barriers in a marsupial (Macropus eugenii) during development. J Physiol. 1988;403:307–88.
15. Moos T, Mollgard K. Cerebrovascular permeability to azo dyes and plasma proteins in rodents of different ages. Neuropathol Appl Neurobiol. 1993;19:120–7.
16. Uchida D, Arimura A, Somogyvari-Vigh A, Shioda S, Banks WA. Prevention of ischemia-induced death of hippocampal neurons by pituitary adenylate cyclase activating polypeptide. Brain Res. 1996;736:280–6.
17. Lazarovici P, Cohen G, Arien-Zakay H, Chen J, Zhang C, Chopp M, et al. Multimodal neuroprotection induced by PACAP38 in oxygen-glucose deprivation and middle cerebral artery occlusion stroke models. J Mol Neurosci. 2012;48:526–40.
18. Bukovics P, Czeiter E, Amrein K, Kovacs N, Pal J, Tamas A, et al. Changes in PACAP level in cerebrospinal fluid and plasma of patients with severe traumatic brain injury. Peptides. 2014;60:18–22.
19. Miyamoto K, Tsumuraya T, Ohtaki H, Dohi K, Satoh K, Xu Z, et al. PACAP38 suppresses cortical damage in mice with traumatic brain injury by enhancing antioxidant activity. J Mol Neurosci. 2014;54:370–9.
20. Guillot TS, Richardson JR, Wang MZ, Li YJ, Taylor TN, Ciliaz BX, et al. PACAP38 increases vesicular monoamine transporter 2 (VMAT2) expression and attenuates methamphetamine toxicity. Neuropeptides. 2008;42:423–34.
21. Yu R, Guo X, Huang L, Zeng Z, Zhang H. The novel peptide PACAP-TAT with enhanced traversing ability attenuates the severe lung injury by repeated smoke inhalation. Peptides. 2012;38:142–9.

22. Yu R, Zeng Z, Guo X, Zhang H, Liu X, Ding Y, et al. The TAT peptide endows PACAP with an enhanced ability to traverse bio-barriers. Neurosci Lett. 2012;527:1–5.
23. Takeda N, Murozono M, Watanabe S, Isshiki A, Watanabe Y. Neuroprotective effects of novel derivatives of vasoactive intestinal peptide and pituitary adenylate cyclase-activating peptide in two brain ischemic models of mice. Masui. 2005;54:240–8.
24. Lochhead JJ, Thorne RG. Intranasal delivery of biologics to the central nervous system. Adv Drug Deliv Rev. 2012;64:614–28.
25. Nonaka N, Farr SA, Nakamachi T, Morley JE, Nakamura M, Shioda S, et al. Intranasal administration of PACAP: Uptake by brain and brain region targeting with cyclodextrins. Peptides. 2012;36:168–75.
26. Banks WA, Farr SA, Butt W, Kumar VB, Franko MW, Morley JE. Delivery across the blood-brain barrier of antisense directed against amyloid beta: reversal of learning and memory deficits in mice overexpressing amyloid precursor protein. J Pharmacol Exp Therap. 2001;297:1113–21.

Chapter 26
Effects of PACAP on Biological Barriers

Imola Wilhelm and Istvan A. Krizbai

Abstract Biological barriers are formed by epithelial or specialized endothelial monolayers that separate and regulate the transport between compartments of different chemical compositions. The unique barrier properties of epithelial and brain endothelial cells (forming the blood–brain barrier) are primarily connected to the tight junctions that seal the intercellular way of transport. Here we review the functions of the different biological barriers, the regulation of tight junctions and the effects of pituitary adenylate cyclase-activating polypeptide (PACAP) on epithelia and the cerebral endothelium. PACAP and PACAP receptors are widely expressed by these cells and also by immune cells, smooth muscle cells, and nerve endings associated to them. The protective effect of PACAP on kidney, lung, retinal pigment, and other epithelial cells in different pathological conditions is well documented; however, in brain endothelial cells no pro-survival effect of PACAP has been observed. Moreover, little is known about the effects of PACAP on tight junctions, except for the outer blood–retina barrier and the blood–brain barrier, where it helps to maintain the integrity of the barrier in pathological conditions.

Keywords Barrier • Tight junctions • Epithelium • Endothelium • Blood–brain barrier • Blood–retinal barrier • PACAP • Survival • Protective

Abbreviations

BBB	Blood–brain barrier
BRB	Blood–retinal barrier
cAMP	Cyclic adenosine monophosphate
CNS	Central nervous system
EMT	Epithelial–mesenchymal transition
MDCK	Madin-Darby canine kidney

I. Wilhelm • I.A. Krizbai (✉)
Biological Research Centre, Hungarian Academy of Sciences, Institute of Biophysics, Temesvári krt. 62, Szeged 6726, Hungary
e-mail: krizbai.istvan@brc.mta.hu

D. Reglodi, A. Tamas (eds.), *Pituitary Adenylate Cyclase Activating Polypeptide — PACAP*, Current Topics in Neurotoxicity 11,
DOI 10.1007/978-3-319-35135-3_26

PKA	Protein kinase A
RBEC	Rat brain endothelial cell
RPE	Retinal pigment epithelium
TJ	Tight junction
ZO	Zonula occludens

Structure and Function of Biological Barriers

In multicellular organisms selective permeable interfaces separate and interconnect tissue compartments of different chemical compositions. The barrier between the distinct fluid compartments is formed by specialized cell types, mainly epithelial cells continuously interconnected by tight junctions (TJs). Endothelia—although from histological point of view are specialized epithelia—usually do not form tight barriers. However, in case of the blood-brain barrier (BBB) microvascular endothelial cells form the barrier. Biological barriers not only separate the different tissues within the organism, but delimitate the organism from the external environment as well.

The unique barrier properties of epithelial and brain endothelial cells are mainly given by the TJs that seal the paracellular way of transport (Fig. 26.1a–c), forcing most molecular and even cellular traffic to take the transcellular route. Here however, the luminal secretions of the cells, the plasma membrane and specific transport systems regulate the passage of molecules.

Tight Junctions (TJs)

TJs are molecular complexes of intercellular adhesion, characteristic to barrier forming cells. They constitute a barrier for and regulate the passage of ions, water and water soluble molecules between adjacent cells (gate function) and also block the free diffusion of proteins and lipids between the apical and basolateral domains of the plasma membrane (fence function). On electron micrographs TJs are viewed as points of cell contact where the exterior membrane leaflets of the neighboring cells appear to fuse (kissing points), while on freeze fracture images a branching network of strands are visible on the protoplasmic (P-) face with complementary grooves on the exoplasmic (E-) face. By immunofluorescence, TJ proteins display a cell border distribution (chicken wire-like pattern) observed from above (Fig. 26.1c). From the side, dots concentrated at the uppermost portion of the lateral membrane are seen.

Molecular Composition of TJs

TJs are composed of transmembrane proteins responsible for mediation of cell–cell adhesion, and cytoplasmic (plaque) proteins which link the integral membrane proteins to the actin cytoskeleton (Fig. 26.1b). Moreover, several regulatory signaling molecules are located in close proximity of the TJs.

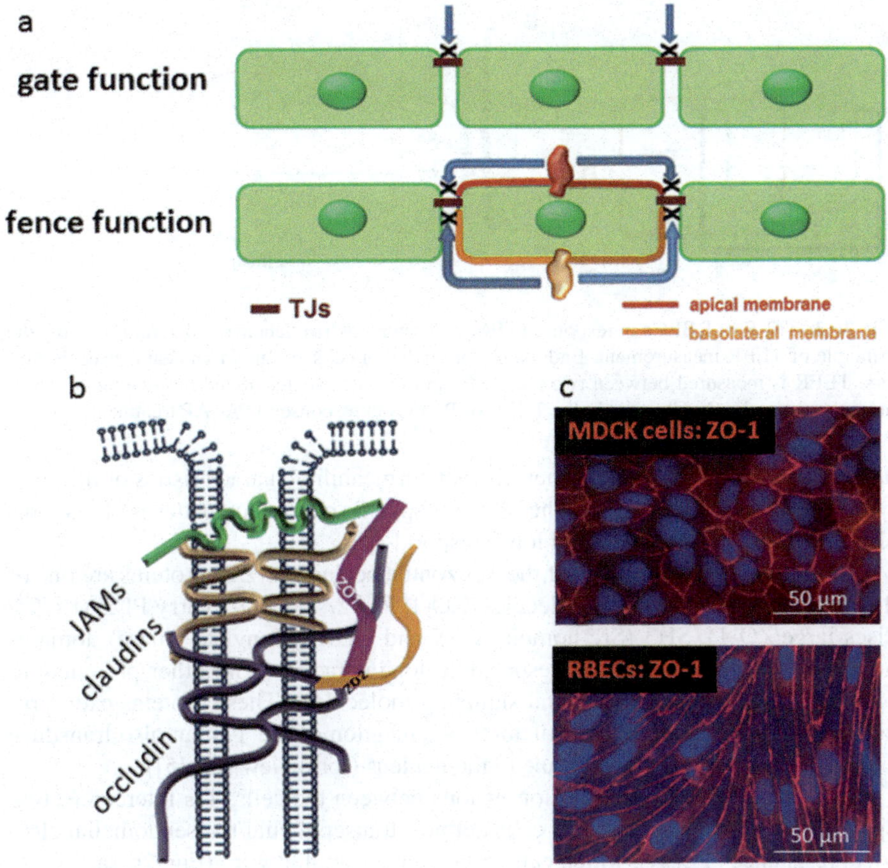

Fig. 26.1 Tight junctions of barrier forming cells. (**a**) Tight junctions prevent the free intercellular passage of molecules (gate function) and inhibit the diffusion of proteins between the apical and basolateral domains of the plasma membrane (fence function). (**b**) Schematic diagram showing the main molecular components of the TJs. (**c**) Immunofluorescence staining of the TJ protein ZO-1 in epithelial (MDCK) and rat brain endothelial cells (RBECs)

Transmembrane proteins of the TJs are: tetraspan proteins (occludin, tricellulin/marvelD2, marvelD3, claudins), single-span proteins of the immunoglobulin superfamily (junctional adhesion molecules/JAMs, coxsackievirus-adenovirus receptor/CAR, endothelial cell-selective adhesion molecule/ESAM), and other single-span proteins (Crumbs3/CRB3, blood vessel epicardial substance/Bves) (reviewed in ref. [1]). The first described transmembrane protein of the TJs was occludin [2], but the backbone of the TJs is formed by claudins, discovered a few years later [3]. The claudin family presently consists of more than 30 members, which are small 20–27 kDa proteins able to interact in a homophilic and heterophilic manner in *cis* and *trans* (same and different types of claudins interacting in the membrane of the same cell or with claudins of the adjacent cell). The different claudins may exhibit varying barrier-forming or pore-forming properties, and therefore the tissue-specific

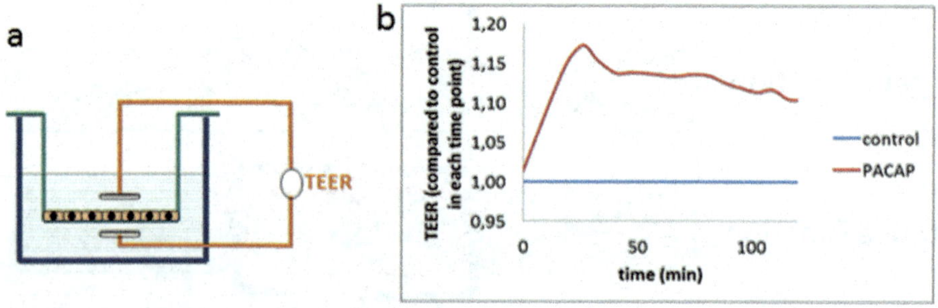

Fig. 26.2 TEER of RBECs in response to PACAP treatment. (**a**) Schematic diagram showing the principle of TEER measurement. Endothelial or epithelial cells are grown on semipermeable filters. TEER is measured between two electrodes placed into the apical and basolateral compartment, respectively. (**b**) Changes in the TEER of RBECs in response to PACAP treatment

distribution of claudins determines distinct permeability characteristics of different types of epithelia. For example the segment-specific expression patterns of claudins along the nephron regulate renal ion transport [4].

Among the plaque proteins of the TJs zonula occludens (ZO) proteins are one of the most important scaffold molecules. ZO-1, ZO-2, and ZO-3 carry PDZ (Psd95/discs large/ZO-1), SH3 (Src homology 3), and GUK (guanylate kinase) domains and a proline rich region—all responsible for interaction with other proteins (TJ proteins, cytoskeletal proteins and signaling molecules). These proteins exert crucial functions not only in the structural organization of the TJs but also transduce signals from the plasma membrane to the nucleus (for review see: [5]).

By preventing the free diffusion of ions between the cells, TJs interconnecting epithelial or cerebral endothelial cells define a transepithelial/transendothelial electrical resistance (TEER), which can be easily measured in vitro (Fig. 26.2a).

Regulation of TJs: Role of cAMP

TJ proteins can be regulated at the mRNA level, e.g., glucocorticoids upregulate, while Snail/Slug downregulate their gene expression. Moreover, TJ proteins are post-transcriptionally regulated by signaling molecules localized to the cell membrane. The most important regulators are Ca^{2+}, cyclic nucleotides, different kinases and phosphatases, and G-proteins. In addition, TJ proteins themselves have regulatory roles, e.g., ZO-1 and ZO-2 can shuttle to the nucleus to regulate transcription factors [5].

Crosstalk of TJs with signaling pathways has been comprehensively summarized elsewhere [6]; here we would like to underline the role of cAMP and protein kinase A (PKA) which are the main signaling pathways mediating the effect of PACAP. In brain endothelial cells cAMP stabilizes the TJs, increases TEER, decreases permeability and increases the complexity of junctional strands (for review see: [7]). Cyclic AMP-dependent PKA phosphorylates claudin-5 at Thr207, which is accompanied by increased barrier function and induces the expression of several TJ proteins, including claudin-5, occludin and ZO proteins [8]. Moreover, cAMP elevating agents, like adrenomedullin or cilostazol have also been shown to stabilize the TJs of cerebral endothelial cells [9, 10].

In epithelial cells the picture is more diverse or even controversial. In kidney (MDCK/Madin-Darby canine kidney) or intestinal epithelia (T84 cells) PKA signaling is important for the assembly of functional TJs and prevents junctional disassembly [11]. On the other hand, PKA has been shown to be involved in the disruption of TJs in MDCK and other epithelial cells [12].

Epithelial and Endothelial Barriers of Different Organs

Barriers Separating the Organism from the External Environment

Protection of the organism from the external environment is given by the skin, the intestine and the respiratory tract. In the skin, TJs of the stratum granulosum of keratinocytes form the barrier, where claudin-1 and claudin-4 are expressed [13].

The content of the intestine (external environment) is separated from the internal body compartment by epithelial cells covering the lumen of the gastrointestinal tract. The gastrointestinal epithelium is responsible for the absorption of nutrients and also for providing a physical barrier to the permeation of pathogens, toxins and antigens (reviewed in ref. [14]). Claudin expression in the different segments of the gastrointestinal tract is tightly regulated [15].

Bronchial and alveolar epithelial cells separate airspace from blood, acting as a barrier to pathogens and foreign particles.

Barriers of the Brain and Retina

The central nervous system (CNS) has three barriers that regulate the molecular exchange between the blood and the neural tissue: the arachnoid epithelium, the choroid plexus epithelium and the brain endothelium forming the BBB (for review see: [16]). Besides having an endothelial phenotype, brain endothelial cells have special barrier characteristics which distinguish them from endothelia of peripheral tissues (for review see: [17]). In this respect the role of the continuous line of TJs interconnecting brain endothelial cells is indisputable (Fig. 26.1c). TJs restrict the paracellular movement of molecules, while the transcellular barrier is given by the low level of endocytosis and transcytosis. Besides, a complex set of enzymes provides an enzymatic barrier and the cerebral endothelium expresses a large number of efflux transporters as well (reviewed in ref. [18]). Moreover, the low level of leukocyte adhesion molecules limits the entry of immune cells into the CNS. The barrier phenotype of cerebral endothelial cells is induced and maintained by the neural environment, i.e., pericytes and astrocytes coming in close contact with endothelial cells. Besides the barrier function, cerebral endothelial cells possess an important carrier function, represented mainly by solute-like carrier transporters. This is responsible for the transport of nutrients to the brain and the removal of metabolites.

The blood-retina barrier (BRB) is composed of an inner and an outer barrier. The outer barrier is formed by the retinal pigment epithelium (RPE), while the inner barrier is given by endothelial cells surrounded by pericytes and glial cells (Müller cells)—similar to the BBB. Both barrier forming cell types in the retina are equipped with a well-developed junctional complex mediating a highly selective diffusion of molecules between the blood and the retina.

Other Biological Barriers

Hepatocytes and bile duct epithelial cells separate the blood and the bile. Therefore, the apical membranes of hepatocytes delimitate bile canaliculi, while the basolateral sides are in contact with the blood. The two membrane parts are equipped with different transport systems, and are separated by TJs.

In the kidney, epithelial cells along the nephron separate urine from renal parenchyma. Different segments of the nephron have different paracellular transport characteristics due to the differences in the distribution of junctional proteins. Claudin composition of TJs highly determines the different steps of urine formation.

Besides forming a barrier, epithelial cells may have secretory functions as well, mainly in exocrine glands. Mammary ducts, for example, are lined by a single layer of columnar or cuboidal epithelial cells interconnected by a continuous belt of TJs. The structure of the mammary ductal epithelial barrier is almost identical to that of other epithelia, with the exception that the cells are located on contractile myoepithelial cells and not directly on the basal lamina [19].

The blood-testis barrier is formed by TJs connecting Sertoli cells of the seminiferous tubules. It controls the environment in which germ cells develop. The placental barrier controls the exchange of nutrients, gases, waste, hormones, and other endogenous and foreign molecules between maternal and fetal circulations. It is formed by syncytiotrophoblast cells, an epithelial layer covering embryonic placental villi.

Sensory tissues of the nose, the ears, and the eyes—although retain an epithelial organization—are very different from classical barrier-forming epithelia. In these epithelial layers sensory cells are embedded, which are neurons or neuron-like cells (olfactory sensory neurons, auditory hair cells, and photoreceptors, respectively).

Targets of PACAP in Epithelial and Endothelial Barriers

Pituitary adenylate cyclase-activating polypeptide (PACAP) and its receptors (PAC1, VPAC1, and VPAC2) are widely expressed in various tissues, including epithelia and endothelia. These receptors are coupled to the adenylate cyclase/cAMP system. By modulating cAMP levels, PACAP may regulate barrier properties, which besides its well-known pro-survival and anti-inflammatory effects may be an important factor in the regulation of barrier functions. In this chapter we focus on the expression of the molecules of the PACAP system and effects of PACAP on epithelial and endothelial cells and their junctions.

Expression of PACAP, PACAP Receptors and Transporters in Barrier Forming Tissues

PACAP and its receptors are widely expressed in barrier forming tissues both in healthy and pathological conditions (Table 26.1).

In brain endothelial cells the presence of PACAP, and both VPAC1 and VPAC2 mRNA was detected; however, PAC1 was absent [20]. Moreover, the peptide transport system-6 (PTS-6) has also been shown to be expressed by brain endothelial cells, which is responsible for the bidirectional transport of PACAP-38 and for the efflux of PACAP-27 through the BBB [21] (reviewed in Chapter 25 by Banks).

In the retina RPE cells express the mRNA of PACAP, PAC1, VPAC1, and VPAC2 as well [22], while PACAP receptor expression of retinal endothelial cells has not been studied so far. In addition, PACAP immunopositivity was found in the neurons of the retina [23], where selective and non-selective PACAP receptors are also expressed [24], suggesting a complex role of PACAP in the regulation of retinal function.

In the digestive system VPAC1 is expressed in epithelia lining the human biliary tree [25], while hepatocytes express all three PACAP receptors [26]. In colon epithelial cells no or weak VPAC1 expression was found [27, 28], and its disappearance was seen in severe ulcerative colitis [28]. Moreover, PACAP and its receptors are expressed in immune cells and nerve endings associated to the alimentary tract. The importance of PACAP in gastrointestinal and hepatic physiology is discussed in Chapter 16 by Benhammou et al.

Table 26.1 Expression of PACAP receptors in barrier forming cells

Cell type	PAC1	VPAC1	VPAC2
Brain endothelial cells	–	+	+
RPE cells	+	+	+
Colon epithelial cells		+/–	
Biliray epithelial cells		+	
Hepatocytes	+	+	+
Lung epithelial cells	+	+	+
Kidney epithelial cells	+	+	+
Mammary epithelial cells	+	+	+
Keratinocytes	+		
Sertoli cells	+		
Placenta	+		

In lung, kidney, prostate, and mammary epithelia the mRNA expression of all three classes of PACAP/VIP receptors has been shown [29–32]. In the skin mainly non-epithelial cells express PACAP and PAC1, the level of which may increase in psoriasis [33] or inflammation [34]. VPAC2 mRNA was found in keratinocytes of the basal layer and in glandular cells surrounded by VIP-immunoreactive nerve fibers [35]. In Sertoli cells an alternative splice variant of PAC1 receptor has been described, named PAC1R(3a) [36]. The placenta expresses PACAP and PAC1 genes [37].

All these data underline the wide distribution of PACAP and its receptors in epithelia. Not only epithelial cells themselves express PACAP, VIP or their receptors, but immune cells, smooth muscle cells and nerve endings associated to them—these might also substantially affect epithelial functions.

Effects of PACAP on TJs

TJs are critical components of biological barriers and regulate the permeability of different substances between tissue compartments. Increase in the permeability of different epithelial or endothelial barriers may lead to severe consequences and it usually reflects a change in the TJs. One of the most spectacular outcomes of barrier breakdown is edema, which may occur in the skin, but also in the lungs, the brain, or other organs. Depending on the localization, type and severity of this change, a diverse range of disease states connected to the TJs may occur. These include infections, diverse acute and chronic inflammatory disorders, inherited monogenic and polygenic diseases, and carcinomas.

Tightening the junctions may become an important approach in the treatment of some disorders affecting epithelia or endothelia, e.g., inflammatory conditions or metastases. Cyclic AMP is known to strengthen the TJs of the cerebral endothelium and several epithelia. Nevertheless, cAMP/PKA is the most important downstream pathway of all three receptors of PACAP (PAC1, VPAC1, and VPAC2), which are extensively expressed in epithelia and endothelia. However, little is known about the effects of PACAP on the TJs in physiological and pathological conditions.

Effects of PACAP on the TJs of the BBB and the BRB

Disruption of the BBB has an important impact on the outcome of several neurological diseases, including stroke, brain trauma, epilepsy, neurodegenerative disorders, primary and metastatic brain tumors, inflammatory disorders and CNS infections [40]. In all these diseases extrinsic or intrinsic damaging factors lead to TJ disruption and permeability increase, which initiate or aggravate the disease.

In endothelial cells of the BBB PACAP was shown to induce an increase in the TEER (Fig. 26.2b) and to protect against glucose-deprivation- and oxidative stress-induced junctional damage [20]. In control conditions apical administration of PACAP induced an increase in the junctional staining of claudin-5, occludin, and ZO-1 in rat and human brain endothelial cells, which was reflected by a marked increase in the TEER. Moreover, PACAP could significantly improve glucose deprivation- and/or oxidative stress-induced junctional disassembly, i.e., partly or almost totally maintaining the continuity of the junctions, without affecting the expression of the junctional proteins. This suggests that PACAP may help to maintain the integrity of the BBB in pathological conditions, like ischemia/reoxygenation.

In the retina PACAP and VIP have been shown to protect the TJs of the outer BRB in the presence of high glucose and IL-1beta [22]. Hyperglycemia and inflammation are major factors in diabetic macular edema, which is a severe disease characterized by the breakdown of the TJs of RPE cells. Treatment with PACAP or VIP significantly reduced the FITC-dextran hyperpermeability induced by high glucose and IL-1beta, and increased TEER values. PACAP was able to restore claudin-1 and ZO-1 loss as well [22]. These data suggest that PACAP is not only retinoprotective, but can also protect against the breakdown of the outer BRB in diabetic conditions.

Effects of PACAP on the TJs of the Digestive System

Inflammatory disorders of the digestive system are often associated with changes in barrier permeability. Inflammatory cytokines have been shown to alter the structure of the TJs of diverse epithelia and endothelia (reviewed in ref. [41]). This is an important mechanism in several inflammatory diseases, including inflammatory bowel diseases (Crohn's disease, ulcerative colitis, etc.). The most important proinflammatory cytokines that increase TJ permeability are IFN-gamma, TNF-alpha, IL-1beta, IL-6, and TGF-beta. Inflammatory cytokines are able of directly modifying TJ composition and structure by two mechanisms: alteration of TJ protein composition by replacing barrier forming with those with "leaky" properties, and internalization of TJ proteins. Furthermore, microorganisms can also contribute to TJ dysfunction (described in detail in ref. [42]). Certain bacterial toxins or microorganisms, for example, target the TJs of intestinal epithelia leading to diarrhea.

Little is known about the direct role of PACAP in the regulation of gastrointestinal epithelial TJs; however, the general cAMP increasing effect may be beneficial regarding barrier function. In addition, PACAP is a potent anti-inflammatory agent in several inflammatory conditions of different epithelia which can also contribute to barrier maintenance. For example it ameliorates *Toxoplasma gondii*-induced acute ileitis [43]. However, in inflammatory conditions immune cells and/or nerve endings and not directly epithelial cells may be the main targets of PACAP.

Effects of PACAP on the TJs of Other Epithelia

A direct effect of the PACAP on the TJs has not been demonstrated so far; however, it is noteworthy that in the skin PACAP has anti-inflammatory effects in allergic contact dermatitis [44] and in the lungs it ameliorates endotoxin-induced inflammation [45]. Here again however, immune cells and/or nerve endings and not epithelial cells are the main targets of PACAP.

A more direct role for PACAP was found in the stria vascularis of the rat cochlea, where it has been suggested that PACAP may modulate TJ integrity. PAC-1 receptor immunoreactivity was found to be localized to the TJs between marginal cells in all cochlear turns, close to the apical surface of the stria vascularis [46].

Anti-apoptotic and Pro-survival Effects of PACAP on Epithelia and Endothelia

PACAP has anti-apoptotic effects in various tissues and organs against various insults. Several in vitro and in vivo results indicate that this protective effect applies for diverse epithelial cells as well.

PACAP proved to be renoprotective against several insults, e.g., in ischemia/hypoxia, in cisplatin-, cyclosporine A-, or contrast-induced injury of proximal tubule epithelial cells [47–50]. In addition, kidney cells of PACAP knockout mice are more vulnerable to hypoxia [51, 52]. The renoprotective effects of PACAP are discussed in detail in [53] and Chapter 18 by Khan and Batuman.

In the lung, PACAP-TAT—a peptide with increased traversing ability through the BBB, the blood-air barrier and the blood-testis barrier (see Chapter 10 by R. Yu)—proved to be protective against smoke-induced lung injury [54]. It has also been shown that PACAP expression is decreased in lung and colon cancers [55]. In the small intestine endogenous PACAP is protective against intestinal warm and cold ischemia [56, 57].

In sensory epithelia PACAP may affect receptor cells and supporting epithelial cells. PACAP protects cochlear sensory epithelium with supporting cells and sensory hair cells against oxidative stress-induced apoptosis [58]. PACAP was also protective against kanamycin-induced hair cell toxicity [59]. PACAP also protects olfactory sensory neurons against cell death induced by axotomy or TNF-alpha [60].

PACAP supports survival of neurons in the retina in several injuries (diabetic, ischemic, etc.) (reviewed in ref. [61]). Moreover, PACAP has anti-apoptotic role in RPE cells of the outer BRB [62]. It has been suggested that PACAP released from retinal neural cells upregulates semaphorin 4A expression in RPE cells and thereby contributes to the maintenance of retinal structure and function [63].

Interestingly, in brain endothelial cells the lack of pro-survival effect of PACAP was observed in the presence of oxidative stress and/or glucose deprivation [20]. In contrast to brain endothelia, PACAP has been shown to have protective effects in microvascular endothelia of non-cerebral origin. PACAP was shown to have anti-apoptotic effects in mouse hemangioendothelioma cells challenged by oxidative stress [64] and to inhibit hyperglycemia-induced proliferation of H5V murine microvascular endothelial cells [65].

Role of PACAP in Carcinomas

Not only healthy epithelial cells express PACAP receptors, but carcinomas as well. The most frequently occurring human tumors, including breast, prostate, pancreas, lung, colon, stomach, liver, and urinary bladder carcinomas abundantly express PACAP receptors, predominantly the VPAC1 type [38]. In colon epithelial carcinoma cells VPAC1 mRNA was found to be expressed whereas VPAC2 mRNA was not detected [39]; moreover, VPAC1 overexpression was shown to be associated with poor differentiation of colon cancer [27]. Unfortunately, anti-VIP/PACAP therapy in tumors is limited by the presence of these receptors in normal tissues.

Malignant transformation of epithelia leads to the development of different carcinomas. An important step in this process is the epithelial–mesenchymal transition (EMT), during which epithelial cells gradually loose barrier characteristics and acquire an invasive phenotype. Despite its possible regulatory capacity on the junctions, it is not known how PACAP may affect this process. The well-established anti-apoptotic effect of PACAP, however, may influence carcinogenesis as well. The complex role of PACAP in cancers is subject of a separate chapter (Chapter 47 by Moody and Jensen).

Other Effects of PACAP on Brain Endothelia

PACAP may have special effects on cerebral microvascular endothelial cells. It has been recently shown that PACAP promotes endothelial tube formation which together with its anti-apoptotic effect defines a significant pro-angiogenic effect. Furthermore, it has been suggested that age-related dysregulation of autocrine PACAP signaling in cerebral endothelial cells may contribute to impaired angiogenic capacity in ageing [66].

Conclusions

Fluid compartments of multicellular organisms are separated by biological barriers, which are formed by specialized cell types, epithelial or endothelial cells continuously interconnected by TJs. PACAP and its receptors are widely expressed in epithelial and endothelial barriers both in healthy and pathological conditions. Not only epithelial cells themselves express PACAP receptors, but immune cells, smooth muscle cells, and nerve endings associated to them—these might also substantially affect epithelial functions.

PACAP has direct or indirect (other cell type-mediated) protective and anti-inflammatory effects in epithelia. Interestingly, in brain endothelial cells the lack of pro-survival effect of PACAP was observed in oxidative stress and/or glucose deprivation.

Increase in the permeability through the TJs may aggravate a large number of diseases; therefore, it is of great importance to understand possible mechanisms of TJ modulation. Cyclic AMP (the most important effector of PACAP receptors) is known to strengthen the TJs of the cerebral endothelium and of several epithelia. PACAP can protect the TJs of the outer blood–retina barrier and of the blood–brain barrier; however, there is no experimental evidence on the effects of PACAP on other TJ barriers. Considering the clinical importance of biological barriers, it would be important to understand the possible protective impact of PACAP on epithelial TJs in health and disease.

References

1. Bauer HC, Traweger A, Zweimueller-Mayer J, Lehner C, Tempfer H, Krizbai I, et al. New aspects of the molecular constituents of tissue barriers. J Neural Transm. 2011;118:7–21.
2. Furuse M, Hirase T, Itoh M, Nagafuchi A, Yonemura S, Tsukita S, et al. Occludin: a novel integral membrane protein localizing at tight junctions. J Cell Biol. 1993;123:1777–88.
3. Furuse M, Fujita K, Hiiragi T, Fujimoto K, Tsukita S. Claudin-1 and -2: novel integral membrane proteins localizing at tight junctions with no sequence similarity to occludin. J Cell Biol. 1998;141:1539–50.
4. Kiuchi-Saishin Y, Gotoh S, Furuse M, Takasuga A, Tano Y, Tsukita S. Differential expression patterns of claudins, tight junction membrane proteins, in mouse nephron segments. J Am Soc Nephrol. 2002;13:875–86.
5. Traweger A, Toepfer S, Wagner RN, Zweimueller-Mayer J, Gehwolf R, Lehner C, et al. Beyond cell-cell adhesion: emerging roles of the tight junction scaffold ZO-2. Tissue Barriers. 2013;1:e25039.
6. Gonzalez-Mariscal L, Tapia R, Chamorro D. Crosstalk of tight junction components with signaling pathways. Biochim Biophys Acta. 1778;2008:729–56.
7. Krizbai IA, Deli MA. Signalling pathways regulating the tight junction permeability in the blood-brain barrier. Cell Mol Biol (Noisy-le-Grand). 2003;49:23–31.
8. Ishizaki T, Chiba H, Kojima T, Fujibe M, Soma T, Miyajima H, et al. Cyclic AMP induces phosphorylation of claudin-5 immunoprecipitates and expression of claudin-5 gene in blood-brain-barrier endothelial cells via protein kinase A-dependent and -independent pathways. Exp Cell Res. 2003;290:275–88.
9. Kis B, Deli MA, Kobayashi H, Abraham CS, Yanagita T, Kaiya H, et al. Adrenomedullin regulates blood-brain barrier functions in vitro. Neuroreport. 2001;12:4139–42.
10. Horai S, Nakagawa S, Tanaka K, Morofuji Y, Couraud PO, Deli MA, et al. Cilostazol strengthens barrier integrity in brain endothelial cells. Cell Mol Neurobiol. 2013;33:291–307.
11. Kohler K, Zahraoui A. Tight junction: a co-ordinator of cell signalling and membrane trafficking. Biol Cell. 2005;97:659–65.
12. Klingler C, Kniesel U, Bamforth SD, Wolburg H, Engelhardt B, Risau W. Disruption of epithelial tight junctions is prevented by cyclic nucleotide-dependent protein kinase inhibitors. Histochem Cell Biol. 2000;113:349–61.
13. Sugawara T, Iwamoto N, Akashi M, Kojima T, Hisatsune J, Sugai M, et al. Tight junction dysfunction in the stratum granulosum leads to aberrant stratum corneum barrier function in claudin-1-deficient mice. J Dermatol Sci. 2013;70:12–8.
14. Suzuki T. Regulation of intestinal epithelial permeability by tight junctions. Cell Mol Life Sci. 2013;70:631–59.
15. Gunzel D, Yu AS. Claudins and the modulation of tight junction permeability. Physiol Rev. 2013;93:525–69.

16. Abbott NJ, Ronnback L, Hansson E. Astrocyte-endothelial interactions at the blood-brain barrier. Nat Rev Neurosci. 2006;7:41–53.
17. Wilhelm I, Krizbai IA. In vitro models of the blood-brain barrier for the study of drug delivery to the brain. Mol Pharm. 2014;11:1949–63.
18. Wilhelm I, Fazakas C, Krizbai IA. In vitro models of the blood-brain barrier. Acta Neurobiol Exp (Wars). 2011;71:113–28.
19. Owens MB, Hill AD, Hopkins AM. Ductal barriers in mammary epithelium. Tissue Barriers. 2013;1:e25933.
20. Wilhelm I, Fazakas C, Tamas A, Toth G, Reglodi D, Krizbai IA. PACAP enhances barrier properties of cerebral microvessels. J Mol Neurosci. 2014;54:469–76.
21. Dogrukol-Ak D, Kumar VB, Ryerse JS, Farr SA, Verma S, Nonaka N, et al. Isolation of peptide transport system-6 from brain endothelial cells: therapeutic effects with antisense inhibition in Alzheimer and stroke models. J Cereb Blood Flow Metab. 2009;29:411–22.
22. Scuderi S, D'Amico AG, Castorina A, Imbesi R, Carnazza ML, D'Agata V. Ameliorative effect of PACAP and VIP against increased permeability in a model of outer blood retinal barrier dysfunction. Peptides. 2013;39:119–24.
23. Seki T, Shioda S, Ogino D, Nakai Y, Arimura A, Koide R. Distribution and ultrastructural localization of a receptor for pituitary adenylate cyclase activating polypeptide and its mRNA in the rat retina. Neurosci Lett. 1997;238:127–30.
24. Nilsson SF, De Neef P, Robberecht P, Christophe J. Characterization of ocular receptors for pituitary adenylate cyclase activating polypeptide (PACAP) and their coupling to adenylate cyclase. Exp Eye Res. 1994;58:459–67.
25. Chignard N, Mergey M, Barbu V, Finzi L, Tiret E, Paul A, et al. VPAC1 expression is regulated by FXR agonists in the human gallbladder epithelium. Hepatology. 2005;42:549–57.
26. Robberecht P, Gourlet P, Cauvin A, Buscail L, De Neef P, Arimura A, et al. PACAP and VIP receptors in rat liver membranes. Am J Physiol. 1991;260:G97–102.
27. Liu S, Zeng Y, Li Y, Guo W, Liu J, Ouyang N. VPAC1 overexpression is associated with poor differentiation in colon cancer. Tumour Biol. 2014;35:6397–404.
28. Jonsson M, Norrgard O, Forsgren S. Epithelial expression of vasoactive intestinal peptide in ulcerative colitis: down-regulation in markedly inflamed colon. Dig Dis Sci. 2012;57:303–10.
29. Busto R, Carrero I, Guijarro LG, Solano RM, Zapatero J, Noguerales F, et al. Expression, pharmacological, and functional evidence for PACAP/VIP receptors in human lung. Am J Physiol. 1999;277:L42–8.
30. Brubel R, Horvath G, Reglodi D, Lubics A, Tamas A, Kiss P, et al. Presence of pituitary adenylate cyclase activating polypeptide and its type I receptor in the rat kidney. Transplant Proc. 2011;43:1297–9.
31. Nagakawa O, Junicho A, Akashi T, Koizumi K, Matsuda T, Fuse H, et al. Vasoactive intestinal peptide and pituitary adenylate cyclase activating polypeptide stimulate interleukin-6 production in prostate cancer cells and prostatic epithelial cells. Oncol Rep. 2005;13:1217–21.
32. Garcia-Fernandez MO, Collado B, Bodega G, Cortes J, Ruiz-Villaespesa A, Carmena MJ, et al. Pituitary adenylate cyclase-activating peptide/vasoactive intestinal peptide receptors in human normal mammary gland and breast cancer tissue. Gynecol Endocrinol. 2005;20:327–33.
33. Steinhoff M, McGregor GP, Radleff-Schlimme A, Steinhoff A, Jarry H, Schmidt WE. Identification of pituitary adenylate cyclase activating polypeptide (PACAP) and PACAP type 1 receptor in human skin: expression of PACAP-38 is increased in patients with psoriasis. Regul Pept. 1999;80:49–55.
34. Helyes Z, Kun J, Dobrosi N, Sandor K, Nemeth J, Perkecz A, et al. Pituitary adenylate-cyclase activating polypeptide is up-regulated in murine skin inflammation and mediates transient receptor potential vanilloid-1-induced neurogenic edema. J Invest Dermatol. 2015;35:2209–18.
35. Fischer TC, Dinh QT, Peiser C, Loser C, Fischer A, Groneberg DA. Simultaneous detection of receptor mRNA and ligand protein in human skin tissues. J Cutan Pathol. 2002;29:65–71.
36. Daniel PB, Kieffer TJ, Leech CA, Habener JF. Novel alternatively spliced exon in the extracellular ligand-binding domain of the pituitary adenylate cyclase-activating polypeptide (PACAP)

type 1 receptor (PAC1R) selectively increases ligand affinity and alters signal transduction coupling during spermatogenesis. J Biol Chem. 2001;276:12938–44.
37. Scaldaferri ML, Modesti A, Palumbo C, Ulisse S, Fabbri A, Piccione E, et al. Pituitary adenylate cyclase-activating polypeptide (PACAP) and PACAP-receptor type 1 expression in rat and human placenta. Endocrinology. 2000;141:1158–67.
38. Reubi JC, Laderach U, Waser B, Gebbers JO, Robberecht P, Laissue JA. Vasoactive intestinal peptide/pituitary adenylate cyclase-activating peptide receptor subtypes in human tumors and their tissues of origin. Cancer Res. 2000;60:3105–12.
39. Anderson CM, Mendoza ME, Kennedy DJ, Raldua D, Thwaites DT. Inhibition of intestinal dipeptide transport by the neuropeptide VIP is an anti-absorptive effect via the VPAC1 receptor in a human enterocyte-like cell line (Caco-2). Br J Pharmacol. 2003;138:564–73.
40. Rosenberg GA. Neurological diseases in relation to the blood-brain barrier. J Cereb Blood Flow Metab. 2012;32:1139–51.
41. Forster C. Tight junctions and the modulation of barrier function in disease. Histochem Cell Biol. 2008;130:55–70.
42. Guttman JA, Finlay BB. Tight junctions as targets of infectious agents. Biochim Biophys Acta. 1788;2009:832–41.
43. Heimesaat MM, Dunay IR, Schulze S, Fischer A, Grundmann U, Alutis M, et al. Pituitary adenylate cyclase-activating polypeptide ameliorates experimental acute ileitis and extra-intestinal sequelae. PLoS One. 2014;9:e108389.
44. Kemeny A, Reglodi D, Cseharovszky R, Hashimoto H, Baba A, Szolcsanyi J, et al. Pituitary adenylate cyclase-activating polypeptide deficiency enhances oxazolone-induced allergic contact dermatitis in mice. J Mol Neurosci. 2010;42:443–9.
45. Elekes K, Sandor K, Moricz A, Kereskai L, Kemeny A, Szoke E, et al. Pituitary adenylate cyclase-activating polypeptide plays an anti-inflammatory role in endotoxin-induced airway inflammation: in vivo study with gene-deleted mice. Peptides. 2011;32:1439–46.
46. Abu-Hamdan MD, Drescher MJ, Ramakrishnan NA, Khan KM, Toma VS, Hatfield JS, et al. Pituitary adenylyl cyclase-activating polypeptide (PACAP) and its receptor (PAC1-R) in the cochlea: evidence for specific transcript expression of PAC1-R splice variants in rat microdissected cochlear subfractions. Neuroscience. 2006;140:147–61.
47. Li M, Khan AM, Maderdrut JL, Simon EE, Batuman V. The effect of PACAP38 on MyD88-mediated signal transduction in ischemia-/hypoxia-induced acute kidney injury. Am J Nephrol. 2010;32:522–32.
48. Li M, Balamuthusamy S, Khan AM, Maderdrut JL, Simon EE, Batuman V. Pituitary adenylate cyclase-activating polypeptide prevents cisplatin-induced renal failure. J Mol Neurosci. 2011;43:58–66.
49. Khan AM, Li M, Brant E, Maderdrut JL, Majid DS, Simon EE, et al. Renoprotection with pituitary adenylate cyclase-activating polypeptide in cyclosporine A-induced nephrotoxicity. J Investig Med. 2011;59:793–802.
50. Khan AM, Maderdrut JL, Li M, Toliver HL, Coy DH, Simon EE, et al. Pituitary adenylate cyclase-activating polypeptide prevents contrast-induced nephropathy in a novel mouse model. Physiol Rep. 2013;1:e00163.
51. Horvath G, Racz B, Szakaly P, Kiss P, Laszlo E, Hau L, et al. Mice deficient in neuropeptide PACAP demonstrate increased sensitivity to in vitro kidney hypoxia. Transplant Proc. 2010;42:2293–5.
52. Szakaly P, Laszlo E, Kovacs K, Racz B, Horvath G, Ferencz A, et al. Mice deficient in pituitary adenylate cyclase activating polypeptide (PACAP) show increased susceptibility to in vivo renal ischemia/reperfusion injury. Neuropeptides. 2011;45:113–21.
53. Reglodi D, Kiss P, Szabadfi K, Atlasz T, Gabriel R, Horvath G, et al. PACAP is an endogenous protective factor-insights from PACAP-deficient mice. J Mol Neurosci. 2012;48:482–92.
54. Yu R, Guo X, Huang L, Zeng Z, Zhang H. The novel peptide PACAP-TAT with enhanced traversing ability attenuates the severe lung injury induced by repeated smoke inhalation. Peptides. 2012;38:142–9.

55. Szanto Z, Sarszegi Z, Reglodi D, Nemeth J, Szabadfi K, Kiss P, et al. PACAP immunoreactivity in human malignant tumor samples and cardiac diseases. J Mol Neurosci. 2012;48:667–73.
56. Ferencz A, Kiss P, Weber G, Helyes Z, Shintani N, Baba A, et al. Comparison of intestinal warm ischemic injury in PACAP knockout and wild-type mice. J Mol Neurosci. 2010;42: 435–42.
57. Ferencz A, Weber G, Helyes Z, Hashimoto H, Baba A, Reglodi D. Presence of endogenous PACAP-38 ameliorated intestinal cold preservation tissue injury. J Mol Neurosci. 2010;42: 428–34.
58. Racz B, Horvath G, Reglodi D, Gasz B, Kiss P, Gallyas Jr F, et al. PACAP ameliorates oxidative stress in the chicken inner ear: an in vitro study. Regul Pept. 2010;160:91–8.
59. Nemeth A, Szabadfi K, Fulop B, Reglodi D, Kiss P, Farkas J, et al. Examination of calcium-binding protein expression in the inner ear of wild-type, heterozygous and homozygous pituitary adenylate cyclase-activating polypeptide (PACAP)-knockout mice in kanamycin-induced ototoxicity. Neurotox Res. 2014;25:57–67.
60. Kanekar S, Gandham M, Lucero MT. PACAP protects against TNFalpha-induced cell death in olfactory epithelium and olfactory placodal cell lines. Mol Cell Neurosci. 2010;45:345–54.
61. Atlasz T, Szabadfi K, Kiss P, Racz B, Gallyas F, Tamas A, et al. Pituitary adenylate cyclase activating polypeptide in the retina: focus on the retinoprotective effects. Ann N Y Acad Sci. 2010;1200:128–39.
62. Fabian E, Reglodi D, Mester L, Szabo A, Szabadfi K, Tamas A, et al. Effects of PACAP on intracellular signaling pathways in human retinal pigment epithelial cells exposed to oxidative stress. J Mol Neurosci. 2012;48:493–500.
63. Ko JA, Hirata J, Yamane K, Sonoda KH, Kiuchi Y. Up-regulation of semaphorin 4A expression in human retinal pigment epithelial cells by PACAP released from cocultured neural cells. Cell Biochem Funct. 2015;33:29–36.
64. Racz B, Gasz B, Borsiczky B, Gallyas Jr F, Tamas A, Jozsa R, et al. Protective effects of pituitary adenylate cyclase activating polypeptide in endothelial cells against oxidative stress-induced apoptosis. Gen Comp Endocrinol. 2007;153:115–23.
65. Castorina A, Giunta S, Mazzone V, Cardile V, D'Agata V. Effects of PACAP and VIP on hyperglycemia-induced proliferation in murine microvascular endothelial cells. Peptides. 2010;31:2276–83.
66. Banki E, Sosnowska D, Tucsek Z, Gautam T, Toth P, Tarantini S, et al. Age-related decline of autocrine pituitary adenylate cyclase-activating polypeptide impairs angiogenic capacity of rat cerebromicrovascular endothelial cells. J Gerontol A Biol Sci Med Sci. 2015;70:665–74.

Chapter 27
Role of PACAP in Astrocytes and Astrocytic Tumors

Tomoya Nakamachi

Abstract In the central nervous system (CNS), astrocytes are the most numerous among glial cells and have diverse physiological functions. These cells also play an important role in many CNS disorders and pathologies. Pituitary adenylate cyclase activating polypeptide (PACAP) is abundantly expressed in the CNS, acting as a neuroprotectant against various neurological threats. PACAP also has been studied as an astrocytic regulator from diversified aspects. PACAP receptor expression has dynamically changed in pathological condition. In this chapter, we summarize the expression and function of PACAP and PACAP receptors in astrocytes and astrocytic tumors, and discuss the role of PACAP in physiological and pathological conditions.

Keywords PACAP • PACAP receptors • Astrocyte • Astrocytoma • Pathology • Proliferation • Anti-apoptosis • Growth factor • Cytokine • Diversity • Neuroinflammation

Astrocytes

Astrocyte is a type of glial cells, found in the central nervous system (CNS). Michael von Lenhossek coined the term "astrocyte" to describe the star-shaped cells he discovered in brain sections using silver staining techniques in 1893. Glial cells are the most abundant cell type in the CNS, with astrocytes accounting for 20–40 % of the total number of mammalian brain cells [1]. Although astrocytes are classically classified according to their morphology into fibrous and protoplasmic astrocytes, several types of specialized astrocytes have been identified [2, 3]. Originally, astrocytes were considered as simple supporting cells or "brain glue" in the CNS. However, recent studies revealed that astrocytes are heterogeneous, with several functions, such as metabolic support of neurons, maintenance of synapses, and control of the

T. Nakamachi (✉)
Laboratory of Regulatory Biology, Graduate School of Science and Engineering, University of Toyama, 3190-Gofuku, Toyama-shi, Toyama 930-8555, Japan
e-mail: nakamachi@sci.u-toyama.ac.jp

D. Reglodi, A. Tamas (eds.), *Pituitary Adenylate Cyclase Activating Polypeptide — PACAP*, Current Topics in Neurotoxicity 11,
DOI 10.1007/978-3-319-35135-3_27

brain's water homeostasis in normal conditions. Moreover, astrocytes are pathophysiologically associated with neurodegenerative diseases [4] and neuroinflammations [5]. It has been reported that neurotransmitters or neuromodulators are important factors for astrocyte activation [6, 7]. Pituitary adenylate cyclase activating polypeptide (PACAP) has been studied as an astrocytic regulator from diverse aspects [8, 9]. This chapter summarizes the expression and function of PACAP and PACAP receptors in astrocytes and astrocytic tumors.

PACAP and PACAP Receptor Expression in Astrocytes

PACAP is a bioactive neuropeptide identified from ovine hypothalamus extracts using cAMP elevation as an indicator in cultured pituitary cells [10]. PACAP exists in two forms, PACAP27 and PACAP38 (27 and 38 amino acids, respectively) [11]. PACAP belongs to the vasoactive intestinal polypeptide (VIP)/secretin/glucagon family, and its closest paralog is VIP. PACAP and VIP share three types of G protein-coupled receptors: PAC1 receptor (PAC1R), VPAC1 receptor (VPAC1R), and VPAC2 receptor (VPAC2R) [12, 13]. The binding affinity of PACAP to PAC1R is approximately 1000 times greater than that of PACAP to VPAC1R and VPAC2R, suggesting that PAC1R is a relatively selective receptor for PACAP, whereas VPAC1R and VPAC2R are common receptors for PACAP and VIP [14].

The possibility of PACAP receptor expression in astrocytes was first indicated using binding assay on cultured astrocyte membrane. High and low affinity binding sites for [125I]PACAP27 were found on the rat astrocyte membrane with approximately 1000 times difference in affinity between the sites [15]. Furthermore, PACAP binds to a 57-kDa membrane protein with a high affinity, and VIP, at a much higher concentration, binds to the same protein [16]. These data imply that at least two types of PACAP receptors with different affinities to PACAP are expressed in rat astrocytes. Grimaldi and Cavallaro reported that PAC1R, VPAC1R, and VPAC2R mRNA were detected in primary cultures of rat astrocytes using reverse transcription-polymerase chain reaction [17, 18]. Another study reported that PAC1R mRNA levels are relatively higher than other VIP receptor mRNA levels in cultured rat astrocytes [19]. In lower vertebrates, PACAP and PAC1R mRNA were identified in cultured tilapia astrocytes [20]. PACAP/VIP receptor immunoreactivity has also been observed in astrocytes in the intact rodent brain [21, 22]. These reports suggest that PACAP directly affects astrocytes and is involved in their function.

PACAP receptor expression in astrocytes is affected by stress or damage. In an in vitro study, the intensity of PAC1R immunoreactivity strongly increased 2 days after scratch injury in cultured mouse astrocytes [23], whereas in an in vivo study, PAC1R immunoreactivity was increased in astrocytes 5 days after injury in a mouse model of cortical stab wound injury [24]. In a mouse spinal cord injury model, PAC1R-immunopositive astrocytes appeared around the injury site between days 7 and 14 after injury [25]. In a mouse bilateral common carotid artery occlusion model, which caused delayed neuronal death in the hippocampal region from days 3 to 7 after ischemia, PAC1R mRNA levels significantly increased on day 7 while VPAC1R mRNA levels significantly decreased on day 3 after ischemia [26]. In the same animal model few PAC1R-immunopositive astrocytes were detected in the hippocampal area until the 3rd day after ischemia, but PAC1R-positive astrocytes were widely distributed in the hippocampus between days 7 and 14 and were converging around the damaged hippocampal area by day 28 [27]. In a mouse model of cold-induced brain injury, VPAC2R-immunopositive astrocytes mainly appeared in the cerebral cortex on post-injury day 7, and PAC1R and VPAC2R mRNA expression increased in dbcAMP-stimulated cultured astrocytes [28]. These results suggest that PACAP and PACAP receptors, particularly PAC1R expressed in the reactive astrocytes, play an important role in the reactive astrocytes rather than in the normal astrocytes after brain and spinal cord injury. We describe the functions and underlying mechanisms of PACAP in astrocytes in the following sections.

Proliferative Effect of PACAP in Astrocytes

The proliferative effect of PACAP in astrocytes was first identified in a study using rat primary cultures. Tatsuno et al. reported that PACAP administration significantly increased astrocyte proliferation with a peak at a concentration of 10^{-10} M and stimulated intracellular cAMP levels at a concentration of 10^{-10} M or higher in a dose dependent manner [29]. VIP also stimulated intracellular cAMP levels in astrocytes, but it required approximately 10,000 times higher concentration compared to PACAP [29]. In addition, PACAP stimulates MAPK in both a PKA- and PKC-independent manner in cultured rat astrocytes [30]. Using similar primary astrocyte cultures, Hashimoto et al. demonstrated that PACAP stimulated astrocytic proliferation with a peak at a concentration of 10^{-9} M and that this response was blocked by a cAMP antagonist or an ERK inhibitor but not by a PKA inhibitor [19]. PACAP38 increased the cAMP and phosphorylated ERK levels in astrocytes. PACAP38-induced ERK phosphorylation and thymidine incorporation were completely inhibited by the cAMP antagonist but not the PKA inhibitor, implying that a cAMP-dependent but PKA-independent mechanism was involved in the increased proliferation [19]. In primary cultures of rat Müller cells (the astrocyte-like glial cells in the retina), treatment with PACAP38 at a concentration of 10^{-9} to 10^{-7} M increased bromodeoxyuridine incorporation, which is commonly used for the detection of cell proliferation [31]. PACAP treatment in scratch-injured mouse astrocyte cultures significantly increased the number of astrocytes expressing the proliferating cell marker Ki67. This reaction was suppressed by pretreatment with PACAP6-38, a PAC1R, and VPAC2R antagonist [23]. These results suggest that PACAP plays an important role in astrocyte proliferation via PACAP/VIP receptors.

Anti-apoptotic Effect of PACAP on Astrocytes

PACAP has a considerable potential to protect neuronal cells against various neurotoxic agents in vitro and to suppress neuronal damage and death in different neuropathological models in vivo [32–36]. However, few studies report that PACAP affects the survival of astrocyte and neurons.

Addition of PACAP (10^{-9} M) markedly suppressed the toxic effect of H_2O_2 (100–300 μM) on rat cortical astrocytes cultures. Co-incubation with H_2O_2 (300 μM) and graded concentrations of PACAP (10^{-16} to 10^{-6} M) resulted in a dose-dependent increase in the number of surviving astrocytes [37]. Pre-incubation with the PACAP receptor antagonist PACAP6-38 (10^{-6} M), the PKA inhibitor H89, the PLC inhibitor U73122, the PKC inhibitor chelerythrine, or the mitogen-activated protein (MAP)-kinase kinase (MEK) inhibitor U0126 did not affect H_2O_2-induced cell death but abolished the protective effect of PACAP on astroglial cell survival [37]. In primary cultures of rat astrocytes subjected to oxygen-glucose deprivation, PACAP38 at a concentration range of 10^{-3} to 10^{-6} M stimulated cAMP levels, but a cell protective effect was not detected [38]. At present, there is not enough information on the astrocyte-protective effect of PACAP, and further studies need to confirm the above mentioned results and clarify their functional significance in injured nervous tissues.

Effect of PACAP on Humoral Factors Released by Astrocytes

The neuroprotective potential of PACAP has been studied in both in vivo and in vitro models [33, 35, 39]. In the LPS-induced neurotoxicity model, PACAP suppressed neuronal death at nanomolar or higher concentrations in a neuron-rich culture, whereas the neuroprotective effect was bimodal at subpicomolar and nanomolar concentrations in a neuron–glia mixed culture [40, 41]. These results imply that PACAP directly affects neurons and indirectly affects glial cells to suppress neuronal damage. Astrocytes play an important role in inflammation, neuronal protection, and repair by releasing a wide variety of neurotrophic factors, cytokines, and chemokines [42].

Interleukin (IL)-6 has been considered to be a proinflammatory cytokine; however, intracerebroventricular injection of IL-6 decreases infarct volume following ischemia [43], suggesting that IL-6 has a neuroprotective function in nervous tissues. Tatsuno et al. reported that PACAP treatment stimulates IL-6 secretion from folliculostellate cells, which are astrocyte-like cells present in the pituitary [44]. PACAP also stimulates IL-6 production in folliculostellate cell lines, TtT/GF cells [45, 46], and cultured Müller cells [31]. In primary cultures of rat astrocytes, PACAP38 was significantly more potent in stimulating IL-6 compared with VIP. This effect was synergic with that of IL-1 [29, 47]. In an in vivo model, IL-6-immunopositive astrocytes appeared around the damaged hippocampal area 7 days after the common carotid artery occlusion. These astrocytes merged with PAC1R immunoreactivity [26]. Intracerebroventricular injection of PACAP increased the expression of IL-6 mRNA in the mouse cerebral cortex, whereas the neuroprotective effect of PACAP in middle cerebral artery occlusion ischemia was abolished in IL-6 null mice [48]. These reports suggest that PACAP is a potent stimulator of IL-6 secretion by astrocyte and astroglial-like cells and that PACAP prevents neuronal cell death via IL-6 signaling.

PACAP also stimulates astrocytes to release chemokines [32, 49]. PACAP treatment stimulated the release of macrophage inflammatory protein 1-alpha (MIP-1α) and regulated-on-activation, normal T expressed and secreted (RANTES) at a concentration of 10^{-6} and 10^{-13} M, respectively, in primary astrocyte cultures [50]. Furthermore, co-treatment with a RANTES neutralizing antibody decreased PACAP38-mediated neuroprotection, but co-treatment with a MIP-1α neutralizing antibody did not affect PACAP38-mediated neuroprotection against sodium nitroprusside-induced toxicity [51] and HIV envelope protein gp120 toxicity [50]. These results indicate that PACAP38 induces neuroprotection, in part, through the release of RANTES. These results imply that PACAP is a regulator of chemokine expression in astrocytes.

Activity-dependent neuroprotective protein (ADNP) was identified as a VIP-regulated gene, which is secreted by astrocytes [52, 53]. The octapeptide called NAPVSIPQ or NAP, an active fragment of ADNP, potently protects the nervous system in vivo and cultured neuronal cells in vitro [54, 55]. Li et al. [41] have reported that PACAP38 treatment increased the ADNP mRNA levels with a peak at a concentration of 10^{-13} M and that the neuroprotective effect of PACAP38 in rat neuron–glia cocultures was attenuated by co-treatment with the antisense oligonucleotide of ADNP. In mouse neuron–glia cocultures, ADNP mRNA expression was biphasically elevated by PACAP38 treatment at a concentration of 10^{-13} and 10^{-9} M mediated by the inositol 1,4,5-triphosphate (IP3)/phospholipase C (PLC) pathway and the PKA and IP3/PLC pathways, respectively [56]. In the mouse brain, ADNP-immunoreactive cells in the cerebral cortex and cerebellum were co-localized with the astrocyte marker GFAP and PAC1R immunoreactivities [22]. These observations suggest that ADNP is mainly expressed in astrocytes and may contribute to the neuroprotective effect of PACAP.

PACAP also potently stimulates the production and secretion of endozepines (endogenous benzodiazepine ligands), octadecaneuropeptide (ODN), and diazepam-binding inhibitor (DBI), which act as neuroprotectants. PACAP (10^{-13} to 10^{-6} M) treatment of cultured rat astrocytes induces a dose-dependent stimulation of ODN release, whereas VIP is at least 1000-fold less potent, and PACAP6-38 completely suppresses the stimulatory effects of both PACAP and VIP [57]. PACAP also stimulates both DBI gene expression and DBI secretion in a dose-dependent manner through PAC1R [58]. Thus, PACAP is an effective inducer of cytokines, chemokines, and neurotrophic factors, which are related to the neuroprotective potential of PACAP.

PACAP in Astrocytomas

Astrocytomas (astrocytic tumors) are the most common primary CNS neoplasms. Astrocytomas are forms of primary tumors arising from astrocytes and comprise pilocytic astrocytomas (grade I), low-grade astrocytomas (grade II), anaplastic astrocytomas (grade III), and glioblastomas (glioblastoma multiforme, grade IV) [59]. The most aggressive astrocytoma, glioblastomas, have a potential to grow rapidly. The median survival time is only 12–15 months following diagnosis, even if the patient receives optimal medical treatment [60]. Effective drugs have yet to be developed to treat high grade astrocytomas.

PACAP receptor expressions and the proliferative effect of PACAP have been reported in astrocytomas. In binding and adenylate cyclase assays using human astrocytoma membranes, PACAP27 and PACAP38 show approximately 100–300 times higher affinity and are more potent on cAMP elevation compared to VIP, indicating that astrocytomas have at least two subclasses of PACAP binding sites [61]. PAC1R and VPAC2R mRNAs were detected in human astrocytoma cell lines by reverse transcription-polymerase chain reaction [62, 63]. PACAP and VIP stimulated the proliferation of the rat astrocytoma C6 cells at concentrations of 10^{-15} to 10^{-8} M [64]. More recently, in four human astrocytoma cell lines, derived from grade III to grade IV astrocytomas, PAC1-R immunoreactivity and PAC1R and VPAC1R mRNAs were detected in all cells with the greatest intensity in the grade IV astrocytoma cells YH-13; however, VPAC2R mRNA was not detected in any of the astrocytoma cell lines tested [65]. When PACAP38 (10^{-13} to 10^{-7} M) was applied to the astrocytoma cells, proliferation was significantly stimulated in only YH-13 cells at a PACAP38 concentration of 10^{-11} M [65]. These reports suggest that PACAP has a proliferative potential in astrocytomas, and PACAP receptor expression is correlated with the malignant and proliferative potential of astrocytoma cell lines. Further studies on PACAP and PACAP receptor expression in clinical tumor specimens will clarify the relation between the PACAP system and the malignancy level of astrocytomas in humans.

Conclusion

In over 20 years of research in astrocytic biology, the knowledge on the role of PACAP has progressed. In particular, animal and culture studies have shown that PACAP is a key regulator of astrocytes, affecting its proliferation, survival, and humoral factor secretion (Fig. 27.1). Thus, astrocytes are important targets for PACAP in neuroprotection. At present, how PACAP regulates the diverse astrocyte functions and how its regulative effects are related to pathology remains unclear. Heterogeneity of astrocyte physiology has recently been reported in detail and may explain and lead to diverse astrocyte functions [66, 67]. Furthermore, single-cell gene expression profiling revealed that reactive astrocytes induced by ischemic

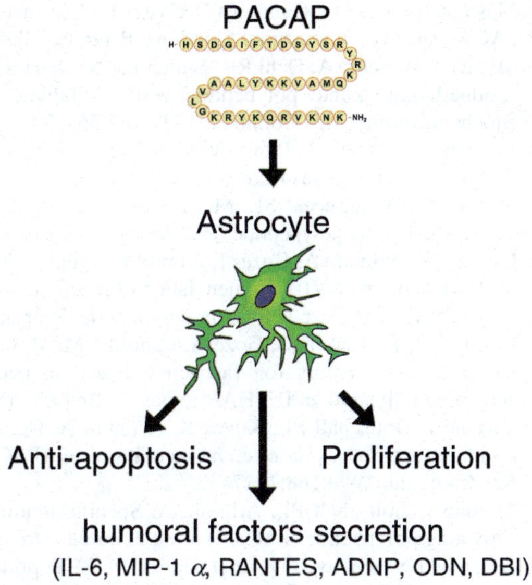

Fig. 27.1 Summary of regulative effect of PACAP on astrocytes. *IL-6* Interleukin-6, *MIP-1α* macrophage inflammatory protein 1-alpha, *RANTES* regulated-on-activation, normal T expressed and secreted, *ADNP* activity-dependent neuroprotective protein, *ODN* octadecaneuropeptide, *DBI* diazepam-binding inhibitor

damage could be classified in at least two subgroups [68], implying that functional heterogeneity may be associated with pathological conditions. Further accumulation of knowledge related to astrocytic heterogeneity could help elucidate the mechanisms of astrocyte regulation by PACAP and clarify the role of PACAP in physiological and pathological conditions in the CNS.

Acknowledgments This work was supported by JSPS KAKENHI 23249079, 15K15670, 15H04394 and by a research grant from Toyama First Bank Scholarship Foundation.

References

1. Herculano-Houzel S. The glia/neuron ratio: how it varies uniformly across brain structures and species and what that means for brain physiology and evolution. Glia. 2014;62:1377–91.
2. Sofroniew MV, Vinters HV. Astrocytes: biology and pathology. Acta Neuropathol (Berl). 2010;119:7–35.
3. Sovrea AS, Bosca AB. Astrocytes reassessment—an evolving concept part one: embryology, biology, morphology and reactivity. J Mol Psychiatry. 2013;1:18.
4. Phatnani H, Maniatis T. Astrocytes in neurodegenerative disease. Cold Spring Harb Perspect Biol. 2015;7(6).
5. Jha MK, Suk K. Management of glia-mediated neuroinflammation and related patents. Recent Pat Inflamm Allergy Drug Discov. 2014;8:118–24.
6. Bekar LK, He W, Nedergaard M. Locus coeruleus alpha-adrenergic-mediated activation of cortical astrocytes in vivo. Cereb Cortex. 2008;18:2789–95.
7. Caruso C, Carniglia L, Durand D, Scimonelli TN, Lasaga M. Astrocytes: new targets of melanocortin 4 receptor actions. J Mol Endocrinol. 2013;51:R33–50.
8. Nakamachi T, Farkas J, Watanabe J, Ohtaki H, Dohi K, Arata S, et al. Role of PACAP in neural stem/progenitor cell and astrocyte—from neural development to neural repair. Curr Pharm Des. 2011;17:973–84.

9. Masmoudi-Kouki O, Gandolfo P, Castel H, Leprince J, Fournier A, Dejda A, et al. Role of PACAP and VIP in astroglial functions. Peptides. 2007;28:1753–60.
10. Miyata A, Arimura A, Dahl RR, Minamino N, Uehara A, Jiang L, et al. Isolation of a novel 38 residue-hypothalamic polypeptide which stimulates adenylate cyclase in pituitary cells. Biochem Biophys Res Commun. 1989;164:567–74.
11. Arimura A, Shioda S. Pituitary adenylate cyclase activating polypeptide (PACAP) and its receptors: neuroendocrine and endocrine interaction. Front Neuroendocrinol. 1995;16:53–88.
12. Sherwood NM, Krueckl SL, McRory JE. The origin and function of the pituitary adenylate cyclase-activating polypeptide (PACAP)/glucagon superfamily. Endocr Rev. 2000;21:619–70.
13. Harmar AJ, Arimura A, Gozes I, Journot L, Laburthe M, Pisegna JR, et al. International Union of Pharmacology. XVIII. Nomenclature of receptors for vasoactive intestinal peptide and pituitary adenylate cyclase-activating polypeptide. Pharmacol Rev. 1998;50:265–70.
14. Harmar AJ, Fahrenkrug J, Gozes I, Laburthe M, May V, Pisegna JR, et al. Pharmacology and functions of receptors for vasoactive intestinal peptide and pituitary adenylate cyclase-activating polypeptide: IUPHAR review 1. Br J Pharmacol. 2012;166:4–17.
15. Tatsuno I, Gottschall PE, Koves K, Arimura A. Demonstration of specific binding sites for pituitary adenylate cyclase activating polypeptide (PACAP) in rat astrocytes. Biochem Biophys Res Commun. 1990;168:1027–33.
16. Tatsuno I, Gottschall PE, Arimura A. Specific binding sites for pituitary adenylate cyclase activating polypeptide (PACAP) in rat cultured astrocytes: molecular identification and interaction with vasoactive intestinal peptide (VIP). Peptides. 1991;12:617–21.
17. Grimaldi M, Cavallaro S. Functional and molecular diversity of PACAP/VIP receptors in cortical neurons and type I astrocytes. Eur J Neurosci. 1999;11:2767–72.
18. Grimaldi M, Cavallaro S. Expression and coupling of PACAP/VIP receptors in cortical neurons and type I astrocytes. Ann N Y Acad Sci. 2000;921:312–6.
19. Hashimoto H, Kunugi A, Arakawa N, Shintani N, Fujita T, Kasai A, et al. Possible involvement of a cyclic AMP-dependent mechanism in PACAP-induced proliferation and ERK activation in astrocytes. Biochem Biophys Res Commun. 2003;311:337–43.
20. Chi-Wei L, Chang SL, Weng CF. Pituitary adenylate cyclase-activating polypeptide (PACAP) regulates the expression of PACAP in cultured tilapia astrocytes. Exp Biol Med. 2007;232: 262–76.
21. Joo KM, Chung YH, Kim MK, Nam RH, Lee BL, Lee KH, et al. Distribution of vasoactive intestinal peptide and pituitary adenylate cyclase-activating polypeptide receptors (VPAC1, VPAC2, and PAC1 receptor) in the rat brain. J Comp Neurol. 2004;476:388–413.
22. Nakamachi T, Ohtaki H, Yofu S, Dohi K, Watanabe J, Hayashi D, et al. Pituitary adenylate cyclase-activating polypeptide (PACAP) type 1 receptor (PAC1R) co-localizes with activity-dependent neuroprotective protein (ADNP) in the mouse brains. Regul Pept. 2008;145:88–95.
23. Nakamachi T, Nakamura K, Oshida K, Kagami N, Mori H, Watanabe J, et al. Pituitary adenylate cyclase-activating polypeptide (PACAP) stimulates proliferation of reactive astrocytes in vitro. J Mol Neurosci. 2011;43:16–21.
24. Suzuki R, Arata S, Nakajo S, Ikenaka K, Kikuyama S, Shioda S. Expression of the receptor for pituitary adenylate cyclase-activating polypeptide (PAC1-R) in reactive astrocytes. Brain Res Mol Brain Res. 2003;115:10–20.
25. Tsuchikawa D, Nakamachi T, Tsuchida M, Wada Y, Hori M, Farkas J, et al. Neuroprotective effect of endogenous pituitary adenylate cyclase-activating polypeptide on spinal cord injury. J Mol Neurosci. 2012;48:508–17.
26. Nakamachi T, Tsuchida M, Kagami N, Yofu S, Wada Y, Hori M, et al. IL-6 and PACAP receptor expression and localization after global brain ischemia in mice. J Mol Neurosci. 2012;48:518–25.
27. Nakamachi T, Farkas J, Kagami N, Wada Y, Hori M, Tsuchikawa D, et al. Expression and distribution of pituitary adenylate cyclase-activating polypeptide receptor in reactive astrocytes induced by global brain ischemia in mice. Acta Neurochir Suppl. 2013;118:55–9.
28. Nishimoto M, Miyakawa H, Wada K, Furuta A. Activation of the VIP/VPAC2 system induces reactive astrocytosis associated with increased expression of glutamate transporters. Brain Res. 2011;1383:43–53.

29. Tatsuno I, Morio H, Tanaka T, Uchida D, Hirai A, Tamura Y, et al. Pituitary adenylate cyclase-activating polypeptide (PACAP) is a regulator of astrocytes: PACAP stimulates proliferation and production of interleukin 6 (IL-6), but not nerve growth factor (NGF), in cultured rat astrocyte. Ann N Y Acad Sci. 1996;805:482–8.
30. Moroo I, Tatsuno I, Uchida D, Tanaka T, Saito J, Saito Y, et al. Pituitary adenylate cyclase activating polypeptide (PACAP) stimulates mitogen-activated protein kinase (MAPK) in cultured rat astrocytes. Brain Res. 1998;795:191–6.
31. Nakatani M, Seki T, Shinohara Y, Taki C, Nishimura S, Takaki A, et al. Pituitary adenylate cyclase-activating peptide (PACAP) stimulates production of interleukin-6 in rat Muller cells. Peptides. 2006;27:1871–6.
32. Dejda A, Sokolowska P, Nowak JZ. Neuroprotective potential of three neuropeptides PACAP. VIP PHI Pharmacol Rep. 2005;57:307–20.
33. Ohtaki H, Nakamachi T, Dohi K, Shioda S. Role of PACAP in ischemic neural death. J Mol Neurosci. 2008;36:16–25.
34. Nakamachi T, Ohtaki H, Yofu S, Dohi K, Watanabe J, Mori H, et al. Endogenous pituitary adenylate cyclase activating polypeptide is involved in suppression of edema in the ischemic brain. Acta Neurochir Suppl. 2010;106:43–6.
35. Somogyvari-Vigh A, Reglodi D. Pituitary adenylate cyclase activating polypeptide: a potential neuroprotective peptide. Curr Pharm Des. 2004;10:2861–89.
36. Endo K, Nakamachi T, Seki T, Kagami N, Wada Y, Nakamura K, et al. Neuroprotective effect of PACAP against NMDA-induced retinal damage in the mouse. J Mol Neurosci. 2011;43:22–9.
37. Masmoudi-Kouki O, Douiri S, Hamdi Y, Kaddour H, Bahdoudi S, Vaudry D, et al. Pituitary adenylate cyclase-activating polypeptide protects astroglial cells against oxidative stress-induced apoptosis. J Neurochem. 2011;117:403–11.
38. Jozwiak-Bebenista M, Kowalczyk E, Nowak JZ. The cyclic AMP effects and neuroprotective activities of PACAP and VIP in cultured astrocytes and neurons exposed to oxygen-glucose deprivation. Pharmacol Rep. 2015;67:332–8.
39. Shioda S, Ohtaki H, Nakamachi T, Dohi K, Watanabe J, Nakajo S, et al. Pleiotropic functions of PACAP in the CNS: neuroprotection and neurodevelopment. Ann N Y Acad Sci. 2006;1070:550–60.
40. Kong LY, Maderdrut JL, Jeohn GH, Hong JS. Reduction of lipopolysaccharide-induced neurotoxicity in mixed cortical neuron/glia cultures by femtomolar concentrations of pituitary adenylate cyclase-activating polypeptide. Neuroscience. 1999;91:493–500.
41. Li M, David C, Kikuta T, Somogyvari-Vigh A, Arimura A. Signaling cascades involved in neuroprotection by subpicomolar pituitary adenylate cyclase-activating polypeptide 38. J Mol Neurosci. 2005;27:91–105.
42. Sofroniew MV. Multiple roles for astrocytes as effectors of cytokines and inflammatory mediators. Neuroscientist. 2014;20:160–72.
43. Loddick SA, Turnbull AV, Rothwell NJ. Cerebral interleukin-6 is neuroprotective during permanent focal cerebral ischemia in the rat. J Cereb Blood Flow Metab. 1998;18:176–9.
44. Tatsuno I, Somogyvari-Vigh A, Mizuno K, Gottschall PE, Hidaka H, Arimura A. Neuropeptide regulation of interleukin-6 production from the pituitary: stimulation by pituitary adenylate cyclase activating polypeptide and calcitonin gene-related peptide. Endocrinology. 1991;129:1797–804.
45. Nagashima AC, Giacomini D, Castro CP, Pereda MP, Renner U, Stalla GK, et al. Transcriptional regulation of interleukin-6 in pituitary folliculo-stellate TtT/GF cells. Mol Cell Endocrinol. 2003;201:47–56.
46. Matsumoto H, Koyama C, Sawada T, Koike K, Hirota K, Miyake A, et al. Pituitary folliculo-stellate-like cell line (TtT/GF) responds to novel hypophysiotropic peptide (pituitary adenylate cyclase-activating peptide), showing increased adenosine 3′,5′-monophosphate and interleukin-6 secretion and cell proliferation. Endocrinology. 1993;133:2150–5.
47. Gottschall PE, Tatsuno I, Arimura A. Regulation of interleukin-6 (IL-6) secretion in primary cultured rat astrocytes: synergism of interleukin-1 (IL-1) and pituitary adenylate cyclase activating polypeptide (PACAP). Brain Res. 1994;637:197–203.

48. Ohtaki H, Nakamachi T, Dohi K, Aizawa Y, Takaki A, Hodoyama K, et al. Pituitary adenylate cyclase-activating polypeptide (PACAP) decreases ischemic neuronal cell death in association with IL-6. Proc Natl Acad Sci U S A. 2006;103:7488–93.
49. Brenneman DE. Neuroprotection: a comparative view of vasoactive intestinal peptide and pituitary adenylate cyclase-activating polypeptide. Peptides. 2007;28:1720–6.
50. Brenneman DE, Hauser JM, Spong C, Phillips TM. Chemokine release is associated with the protective action of PACAP-38 against HIV envelope protein neurotoxicity. Neuropeptides. 2002;36:271–80.
51. Sanchez A, Tripathy D, Grammas P. RANTES release contributes to the protective action of PACAP38 against sodium nitroprusside in cortical neurons. Neuropeptides. 2009;43:315–20.
52. Bassan M, Zamostiano R, Davidson A, Pinhasov A, Giladi E, Perl O, et al. Complete sequence of a novel protein containing a femtomolar-activity-dependent neuroprotective peptide. J Neurochem. 1999;72:1283–93.
53. Gozes I, Divinsky I, Pilzer I, Fridkin M, Brenneman DE, Spier AD. From vasoactive intestinal peptide (VIP) through activity-dependent neuroprotective protein (ADNP) to NAP: a view of neuroprotection and cell division. J Mol Neurosci. 2003;20:315–22.
54. Gozes I. Activity-dependent neuroprotective protein: from gene to drug candidate. Pharmacol Ther. 2007;114:146–54.
55. Gozes I. NAP (davunetide) provides functional and structural neuroprotection. Curr Pharm Des. 2011;17:1040–4.
56. Nakamachi T, Li M, Shioda S, Arimura A. Signaling involved in pituitary adenylate cyclase-activating polypeptide-stimulated ADNP expression. Peptides. 2006;27:1859–64.
57. Masmoudi O, Gandolfo P, Leprince J, Vaudry D, Fournier A, Patte-Mensah C, et al. Pituitary adenylate cyclase-activating polypeptide (PACAP) stimulates endozepine release from cultured rat astrocytes via a PKA-dependent mechanism. FASEB J. 2003;17:17–27.
58. Masmoudi-Kouki O, Gandolfo P, Leprince J, Vaudry D, Pelletier G, Fournier A, et al. PACAP stimulates biosynthesis and release of endozepines from rat astrocytes. Ann N Y Acad Sci. 2006;1070:411–6.
59. Kleihues P, Louis DN, Scheithauer BW, Rorke LB, Reifenberger G, Burger PC, et al. The WHO classification of tumors of the nervous system. J Neuropathol Exp Neurol. 2002;61:215–25.
60. Ahmed R, Oborski MJ, Hwang M, Lieberman FS, Mountz JM. Malignant gliomas: current perspectives in diagnosis, treatment, and early response assessment using advanced quantitative imaging methods. Cancer Manag Res. 2014;6:149–70.
61. Robberecht P, Woussen-Colle MC, Vertongen P, De Neef P, Hou X, Salmon I, et al. Expression of pituitary adenylate cyclase activating polypeptide (PACAP) receptors in human glial cell tumors. Peptides. 1994;15:661–5.
62. Vertongen P, Camby I, Darro F, Kiss R, Robberecht P. VIP and pituitary adenylate cyclase activating polypeptide (PACAP) have an antiproliferative effect on the T98G human glioblastoma cell line through interaction with VIP2 receptor. Neuropeptides. 1996;30:491–6.
63. Sharma A, Walters J, Gozes Y, Fridkin M, Brenneman D, Gozes I, et al. A vasoactive intestinal peptide antagonist inhibits the growth of glioblastoma cells. J Mol Neurosci. 2001;17:331–9.
64. Dufes C, Alleaume C, Montoni A, Olivier JC, Muller JM. Effects of the vasoactive intestinal peptide (VIP) and related peptides on glioblastoma cell growth in vitro. J Mol Neurosci. 2003;21:91–102.
65. Nakamachi T, Sugiyama K, Watanabe J, Imai N, Kagami N, Hori M, et al. Comparison of expression and proliferative effect of pituitary adenylate cyclase-activating polypeptide (PACAP) and its receptors on human astrocytoma cell lines. J Mol Neurosci. 2014;54:388–94.
66. Matyash V, Kettenmann H. Heterogeneity in astrocyte morphology and physiology. Brain Res Rev. 2010;63:2–10.
67. Oberheim NA, Goldman SA, Nedergaard M. Heterogeneity of astrocytic form and function. Methods Mol Biol. 2012;814:23–45.
68. Rusnakova V, Honsa P, Dzamba D, Stahlberg A, Kubista M, Anderova M. Heterogeneity of astrocytes: from development to injury—single cell gene expression. PLoS One. 2013;8:e69734.

Chapter 28
Multiple Actions of Pituitary Adenylate Cyclase-Activating Polypeptide (PACAP) in Schwann Cell Biology

Alessandro Castorina

Abstract The neuropeptide pituitary adenylate cyclase activating polypeptide (PACAP) is a well-recognized endogenously produced pleiotropic molecule that elicits a broad range of biological functions in various cell types, including in Schwann cells (SCs). Its main actions, which have been primarily associated to neuroprotection both in the central and peripheral nervous system (CNS and PNS), are currently being extended, and new intriguing roles in regulating several aspects of CNS and PNS physiology are emerging. The overall goal of the present chapteris to provide an update of the recent advances that have been made to elucidate the multiple actions elicited by PACAP in relationship to the complex biology of SCs, whose seminal role is to provide myelin ensheathment to nerve fibers forming peripheral nerves. After reviewing the main functions of this specific type of neuroglia, a major goal will be to describe PACAP regulatory roles in several biological processes in SCs, including myelin generation, cell protection against injury and its secretagogue activity on proteolytic enzymes. Furthermore, a discussion on the impact that PACAP-mediated changes may have on peripheral axonal myelination and repair processes after nerve injury will be opened. The final aim will be to illustrate the potential benefits of PACAP treatment in chronic pain. In this section, particular attention will be given to those findings inferring on SCs involvement in the development of neuropathic pain and the key modulatory role elicited by PACAP. The contents herein summarized could offer a new perspective of PACAP to aid in the future development of more stable peptide analogues in the scenario of peripheral neuropathies and provide a comprehensive tool that will help to dissect novel PACAP-regulated functions in SCs.

Keywords Schwann cells • PACAP • Myelin • Nerve repair • Peripheral nervous system • Tissue plasminogen activator • Schwannoma cells • Trophic factor • Neuropeptide • Regeneration

---

A. Castorina (✉)
Department of Biomedical and Biotechnological Sciences, Section of Human Anatomy and Histology, University of Catania, Via S. Sofia, 87, Catania 95123, Italy
e-mail: alecasto@unict.it

D. Reglodi, A. Tamas (eds.), *Pituitary Adenylate Cyclase Activating Polypeptide — PACAP*, Current Topics in Neurotoxicity 11,
DOI 10.1007/978-3-319-35135-3_28

Introduction

Pituitary adenylate cyclase-activating polypeptide (PACAP) is a hypophysiotropic molecule that was discovered in 1989 by Miyata and coworkers [1]. Since its discovery, owing to its pleiotropic activities and the ubiquitous distribution in several tissues, it has attracted more than three thousand publications in literature, and many more are still arising. PACAP is detectable is many cell types of the central and peripheral nervous system (CNS and PNS), including neuronal and glial cell populations, where for the most it elicits neuroprotective and trophic functions [2–5]. Intriguingly, over the last decade, studies from our laboratories and other research groups have contributed to the localization of the peptide in SCs [2, 6], the myelin-producing glia whose seminal role is to provide electric insulation to peripheral nerves. These studies have highlighted novel interesting roles of PACAP in this cell type, hence opening a new scenario for potential therapeutic implications of PACAP to treat several pathologies in which the affected cell population is mainly composed of SCs, including peripheral neuropathies such as Charcot–Marie–Tooth disease (CMT), Guillain–Barrè Syndrome (GBS), and chronic idiopathic demyelinating polyneuropathy (CIDP) among others. Furthermore, given the ability of SCs to interact with neurons through the reciprocal exchange of trophic factors and the activation of several intracellular cascades, uncovering novel regulatory functions of the peptide in SCs has begun to attract the attention of scholars interested in the field of regenerative medicine and those studying pain mechanisms, as PACAP could aid in the development of targeted therapies to promote both peripheral nerve recovery after injury and the generation of new drugs to counteract or prevent the genesis of neuropathic pain.

The present chapter is envisioned with the aim to provide an overview of the main biological activities elicited by PACAP in SCs that have been identified so far and possibly, offer new insights to unravel more potential activities of PACAP in the PNS. The chapter is structured in sections, each describing PACAP regulatory functions on specific aspects of SC biology, including pro-myelinating, protective activities, as well as other interesting functions of recent discovery. All this with the hope to provide a more clear view of the underlying events through which such an endogenous molecule takes part in the control of peripheral nerve homeostasis, repair and regeneration processes, but also in the reestablishment of nerve conduction after injury. Finally, a concluding section is dedicated to describing the molecular mechanisms underpinning how the proposed PACAP activities in SCs may have consequences in the modulation of neuropathic pain.

The topics covered in this chapter are gathered from experimental evidences in our laboratories and cumulative literature sources that pose PACAP as an important regulatory agent, which hides the potential to control many critical aspects of SCs biology that are beyond myelin formation, and that could therefore serve as a useful tool to expand our current knowledge of the multiple PACAP activities in this specific PNS-associated glial cell type.

The Dichotomy of SCs: Myelin vs. Non-myelin Forming Cells

SCs, named after the physiologist Theodor Schwann, are the most common cell type in the PNS. These cells, derived from the neural crest, are part of the peripheral nerve neuroglia and are therefore specific to the PNS [7]. SCs phenotype is determined by a complex array of factors, including environmental stimuli, release of growth factors, inflammatory cytokines and injury—all of which can profoundly modify the intrinsic cellular processes of the cell. This notion in particular has attracted scientific interest, especially soon after the discovery that SCs possess a plastic and reversible phenotype, which has been principally correlated to the tight interaction existing between these type of glial cells and the physiological/pathological conditions of neighboring neurons. Indeed, findings have demonstrated that SCs may reversibly shift to either myelin-forming [8] or non-myelin-forming cells in the PNS [9], and are present in the form of perineuronal satellite cells (PSC) both in the autonomic nervous system [10], in perisynaptic areas [11], in dorsal root ganglia [12] and autonomic ganglia [13].

Myelin-forming SCs are very large cells that are highly specialized and that actively interact with axons for normal nerve function, maintenance and repair. Some ensheath nociceptor fibers with relatively few compact myelin lamellae, while others form hundreds of spirals of compact myelin lamellae around the larger diameter sensory fibers. Of note, the axonal diameter is directly correlated to the number of compact myelin lamellae formed, which in turn positively correlates with nerve conduction velocity (i.e., the more lamellae, the faster the nerve conduction). Importantly, while the typical hallmark of a fast-conduction myelinated nerve has generally been extrapolated from the number of compact myelin lamellae as indicated above, it now appears clear that some "non-compact" regions of the inner and outer mesaxon, namely the paranodal "loops," the Schmidt Lanterman incisures, and the transverse processes and nodal microvilli also seem to play a pivotal role in maintaining an organization that is critical for a fast and efficient nerve conduction [14]. So, when the myelinated nerve is damaged, despite the loss of compact myelin, the inability of SCs to maintain its complex architecture and a functional interaction with the ensheathed axon also has to be taken into account. Other elements to be included in the complex scenario here described are that SCs also play a role as facultative antigen presenting cells [15], and thereby are able to respond to both self and foreign antigens. When triggered, their immune activation can result in progressive demyelination and halted nerve conduction, even in the absence of axonal damage. Myelin-forming SCs also play active roles in immunomodulation. They constitutively express many receptors associated to immune response [16], and can regulate their surface expression of MHC class I and class II antigens [15].

In a study, Griffin et al. [9] has estimated that approximately 80 % of peripheral nerves are made up of unmyelinated nerve fibers, although the presence of SCs is ascertained. The reason is that most of SCs, although ensheathing axons, do not form compact myelin, but each cell has many axons embedded within grooves formed in its cytoplasmic membrane. Of notice, studies have identified that these population of unmyelinated SCs express P2 nucleotide receptors, through which ATP is released and taken up by axonal P2Y receptors, resulting in increased axonal excitability of nociceptors. By contrast, the purine nucleoside adenosine exerts exactly the opposite role in unmyelinating SCs, primarily by downregulating this excitability, which has a dampening effect on pain neurotransmission [17, 18]. Therefore, the presence of non-myelin forming SCs is essentially justified by the fact that this particular SC phenotype elicits supportive functions in the PNS, such as the release of trophic molecules, promotes cell-to-cell communication, exerts scavenging activities when triggered (i.e. after nerve injury), and provides replacement to dead SCs.

PACAP, PAC1, and VPAC Receptor Subtypes

The polypeptide PACAP (chemical formula: $C_{203}H_{331}N_{63}O_{53}S$) was originally isolated from an ovine hypothalamus extract by Miyata and coworkers in 1989 [1]. Since its discovery, two peptide isoforms have been identified, a 38-amino acid form(His-Ser-Asp-Gly-Ile-Phe-Thr-Asp-Ser-Tyr-Ser-Arg-Tyr-Arg-Lys-Gln-Met-Ala-Val-Lys-Lys-Tyr-Leu-Ala-Ala-Val-Leu-Gly-Lys-Arg-Tyr-Lys-Gln-Arg-Val-Lys-Asn-Lys-NH2) named PACAP38 and a C-terminally truncated 27-amino acid form (PACAP27) [19]. PACAP belongs to a superfamily of peptides that includes secretin, glucagon, and the peptide histidine–isoleucine (PHI). The peptide sequence is highly conserved among species, raising the suspect that its role could be vital for normal cellular biology. Its high affinity receptors are G protein-coupled receptors, and can be separated into two main groups, PAC1 and VPAC types (which comprise the VPAC1 and VPAC2 subtypes). Both PAC1 and VPAC receptors can activate either adenylate cyclase (AC) or phospho-lipase C with consequent stimulation of cAMP production and IP accumulation [20, 21]. A peptide with about 68 % structural homology to PACAP is the vasoactive intestinal peptide (VIP), but its ability to activate AC has been shown to be about 1000-fold lower than PACAP [20, 21].

Evidences have demonstrated that PACAP activity is strictly dependent on the subtype of receptor being expressed on the cell surface, which in turn may couple to different G proteins, or by the alternative splicing of PAC1 receptor. According to different studies, PAC1 receptor, which has the highest affinity to PACAP with respect to the homolog VIP, may trigger different signaling pathways to fine-tune the biological activity of cells [21]. Whether and how PAC1 receptor switches a signaling pathway or another seems to be regulated by alternative splicing mechanisms during exon assembly, which determines the alternate insertion/deletion of mRNAs encoding for specific amino acid cassettes (null, Hip, Hop, or Hip-Hop cassettes, respectively) within the third intracellular loop of the seven transmembrane receptor [22]. For instance, as elegantly illustrated by Blechman and Levkowitz [23] and others [24, 25], the coordinated expression of PAC1 receptor splicing isoforms in brain regions regulate many physiological processes in the developing brain and retina. In a study, using a SC-like cultures, we have shown the coexistence of the PAC1null (i.e., the PAC1 receptor lacking the cassette insertion) and the PAC1Hip variants [2], both of which are considered the predominant isoforms present in the brain and that are mainly coupled to AC activation [23]. Other factors that can affect PACAP activity, including peptide concentration, should also be considered, as PACAP-mediated signaling pathways seem to be distinctly activated at different concentration ranges, following an inverted U shape [26].

Among PACAP functions, the one that mostly emerges is its strong protective action, promoting survival in many types of neuronal [5, 27, 28] and non-neuronal cells and tissues, including chondrocytes, endothelial cells, SCs, retinas, lungs, and ovaries [2, 29–36]. Further evidences of the protective role elicited by PACAP have been obtained using genetically engineered mice lacking PACAP gene [37, 38]. In vitro and in vivo studies have shown that the peptide acts through PAC1 receptors to stimulate various downstream executors of the protein kinase A and C (PKA and PKC) pathways [39]. It also activates ion channels, β-arrestin signaling, and mitogen-activated protein (MAP) kinase in some tissues [40–42]. At the subcellular level it has been demonstrated that PACAP is detectable in close proximity to the plasma membrane, localizes to areas corresponding to the rough endoplasmic reticulum, the Golgi's apparatus, and the cytoplasm [43, 44]. The almost ubiquitous expression of PACAP in tissues, along with the variety of factors contributing to trigger the activation of diverse intracellular cascades increases the complexity of PACAP regulatory mechanisms. On the other hand, it helps to explain the wide spectrum of biological functions it elicits as well as its potential therapeutic validity as a drug candidate to treat or ameliorate many pathological conditions of the PNS and CNS.

PACAP as a Pro-myelinating Factor in SCs

The major hurdles encountered when performing studies on primary SCs are the difficulties associated to the reproduction in vitro of an environment that reliably mimics that observed in vivo and therefore, that fully respects SCs physiological requirements for proper functioning. For instance, it is accepted that primary cultures of SCs do not normally express myelin genes at levels comparable to normal myelinating cells when expanded in vitro, unless cells are cocultured with neurons and supplemented with appropriate trophic factors [45, 46]. It is hence clear that a bidirectional dialog between axons and SCs is fundamental for the formation of myelin during development, its maintenance during adult life, and during the remyelination process of regenerating axons after injury. In fact, during embryonic development, axonal signalling is critical for SC migration, survival and proliferation, since this ensures that the number of SCs and the axons with which they are associated are matched [47]. Conversely, SCs release a number of trophic factors needed to support the growth of immature neurons and the development of their axon properties, particularly during the myelination process. In the PNS, PACAP and the structurally homolog VIP are known to elicit neuro- and glio-protective activities against a broad range of insults [2–4], but these endogenous trophic factors have also been endowed with pro-regenerating effects after peripheral nerve injuries in vivo, where PACAP expression is induced at the level of the injury site [48]. In some studies it has also been proposed that local administration of the PACAP homolog VIP may promote the early myelination and growth of regenerating axons after sciatic nerve transection [49]. Other support in this direction has come from studies using PACAP-deficient mice. These animals exhibit impaired nerve repair mechanisms and an enhanced local neuroinflammatory profile after experimental nerve injury [50]. Additional evidences are from studies using monkey trigeminal nerve dissociated in culture, which have demonstrated that both neurons and SCs express PAC1 receptors and that the addition of PACAP to culture media promoted axonal outgrowth [6].

The functional repair process following peripheral nerve injury consists of a sequence of well-orchestrated steps that culminate in axonal regrowth, *de novo* synthesis of myelin components and subsequent myelin ensheathment of the rejuvenated axon. As mentioned above, SCs can reversibly switch to a proliferating non-myelin producing phenotype immediately after nerve injury. Such plasticity of SCs seems to be reliant on the release by damaged axons of trophic molecules such as neuregulin-1, which binds to SCs surface receptors to first induce proliferation and subsequently to dictate axon's final ensheathment fate [51–53]. Notably, some studies have suggested that in vitro, a partial SC differentiation can also be obtained by removing growth factors from culture media [54, 55]. By taking advantage of these insights, our research group had been able to confirm the occurrence of a partial differentiation in Schwann cell-like cultures (RT4D6P2T cell line), an immortalized cell line that has consistently proved to possess many of the biochemical features of primary SCs [46, 56–58]. In that occasion, we found that serum starvation induced significant morphological changes in these cells, which partly acquired the typical spindle-like shape resembling that of terminally differentiated SCs (Fig. 28.1), with the exception of very rare and

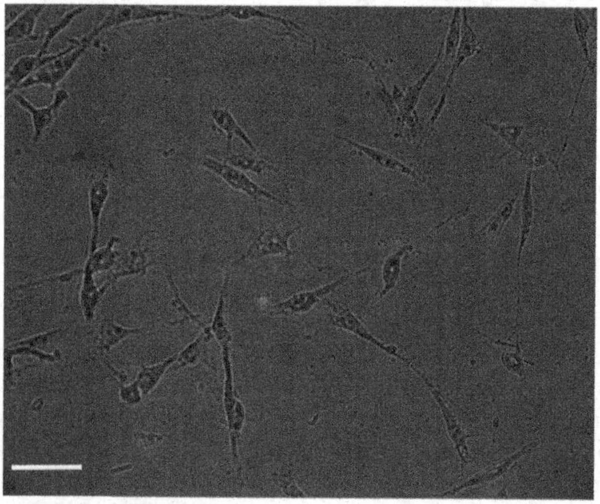

Fig. 28.1 Representative phase-contrast photomicrograph of the RT4-D6P2T Schwann cell (SC) line. Depicted is a representative photomicrograph showing the morphological appearance of the SC line used in our in vitro studies [2–4, 54, 55]. The typical spindle-like shape resembling that of differentiated SCs is appreciable. Cells were plated on glass cover slips, fixed in 4 % *para*-formaldehyde and observed in phase contrast microscopy using the Axiovert 40 microscope (Carl Zeiss Inc.) equipped with a digital camera (AxioCam MRc5; Carl Zeiss). Original magnification was 40X. Scale bar = 50 μm

isolated cells that presented a peculiar astrocytic-like morphology (a representative photomicrograph is shown in Fig. 28.2), possibly as a consequence of the immortalization process itself [59]. However, we found that endogenous PACAP expression, as well as that of PAC1 and VPAC2 receptors, was remarkably increased in SC-like cultures after serum deprivation [56]. In addition, we discovered that some of the key myelin components (i.e., myelin basic protein, myelin associated glycoprotein, and P0 protein) and the phosphatidylinositide-3-kinase (PI3K) substrate protein kinase B (aka Akt) were also significantly upregulated after removal of serum, and even more after exogenous supplementation of PACAP [56]. These changes could partly be prevented by pretreatment with PACAP6-38, a nonspecific PACAP receptor antagonist, and by wortmannin, a specific PI3K inhibitor. We concluded that PACAP and VIP promoted the expression of these myelin-associated genes in such tumor-derived SC line, probably through a mechanism that involves the PI3K/Akt signaling cascade. Unfortunately, in that occurrence we were unable to measure the effective production of myelin, nor to provide clear evidence that cells were successfully differentiated. With this knowledge, a more accurate explanation for the outcomes of the study could

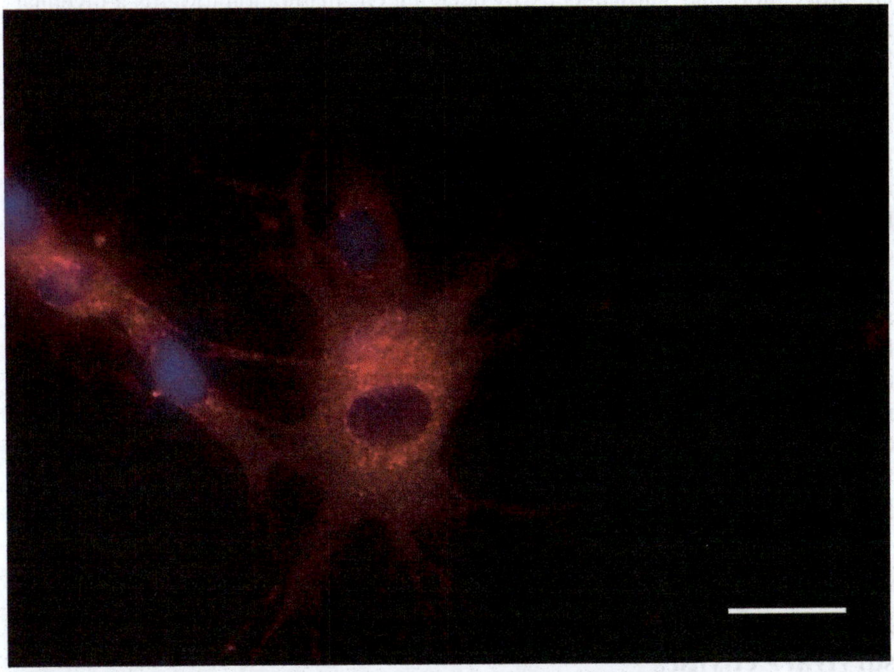

Fig. 28.2 Immunofluorescent S-100 staining of RT4-D6P2T cells showing an isolated astrocytic-shaped cell. RT4-D6P2T cells (ATCC n° CRL-2768) obtained from American Type Culture Collection (Rockville, MD, USA) were cultured on glass cover slips and fixed in 4 % *para*-formaldehyde, permeabilized with 0.2 % Triton X-100, blocked with 0.1 % BSA in PBS and then probed with a rabbit S-100 antibody (1:100 dilution; Santa Cruz Biotechnology, cat no. sc-28533). Detection was then performed using an Alexa fluor 594-conjugated secondary antibody. Nuclei were counterstained with DAPI (#940110 Vector Laboratories). Immunofluorescent images were captured using an Axiovert 40 fluorescence microscope (Carl Zeiss Inc.). Original magnification 40X. Scale bar = 30 μm

be that, independently of the source, PACAP may act as a predisposing factor able to switch in SCs those intracellular events that precede the assembly of new myelin, potentially by increasing the expression of genes encoding essential myelin components. However, the limited number of mechanistic observations in merit poses a limit that impedes us to draw out further conclusions. Additional evidences from both in vitro and in vivo studies could certainly help to fill in the existing gap.

Regarding the effects of PACAP in the CNS in relationship to its role in myelin generation, peculiar differences have been found when reporting how endogenous PACAP affects oligodendrocyte (OL) proliferation *vs.* maturation to a fully myelinating phenotype. Some evidences in vivo have demonstrated that CNS myelination, as opposite to the PNS, is a process that is functionally delayed by the endogenous peptide, possibly as a physiological attempt to impede the release of factors by the myelin sheath that could block axonal outgrowth, thereby creating a permissive environment for efficient development of synapses and formation of novel neural networks [57]. The underlying mechanisms to explain such an atypical behavior of OLs remains partly unresolved. From the few available data, PACAP seems to act at two different stages of OL development, firstly by stimulating the proliferation of OL progenitors and, subsequently by delaying OL maturation towards a myelinating phenotype [59]. The mechanisms to regulate such distinct events also seem to be dependent on two distinct molecular routes: (1) PACAP potentiates fibroblast growth factor 2 mitogenic properties on OL progenitors and (2) it efficaciously antagonizes a molecular switch, the morphogenetic factor sonic hedgehog, to disengage cells from the naturally occurring differentiation process [60]. It is noteworthy to highlight that PACAP effects on SCs and OLs are completely opposite, with PACAP promoting differentiation in SCs and inhibiting it in OLs. The reason for the discrepancy is still unknown. It is possible that the different microenvironment (i.e., neighboring axons or other glial cells) release pro-differentiating factors in the PNS and not in the CNS. Alternatively, it is possible that OLs respond to a tightly regulated developmental switch in which PACAP firstly promotes axonal sprouting at the expenses of myelination, and only after neuronal branching has been completed, it is downregulated to facilitate the reinstatement of myelination. This mechanism could be transcriptionally regulated by fine-tuning the alternative splicing of the PACAP-preferring PAC1 receptor isoforms, to activate different signaling cascades [22, 23], but this option still needs to be verified.

Protective Roles of PACAP in SCs: Direct and Indirect Pro-survival Activities

The use of the RT4D6P2T cell line to investigate PACAP activities has proven to be a suitable tool to determine whether the peptide provides benefits to SCs exposed to different types of insults. Data retrieved from our studies using this tumor-derived cell line converged towards a strong protective role of PACAP in this type of neuroglia. In 2008, when our research group started testing the effects of PACAP and VIP in this SC-like culture, we showed that both endogenous molecules elicited a sustained protective effect against nutrient deprivation (PACAP > VIP), which surprisingly, was partially maintained even after 3 continuous days of serum starvation [2]. That protective outcome was associated to reduced apoptosis, as formation of oligonucleosomes and the expression of the pro-apoptotic protein BAX were both strikingly decreased after PACAP application. The remarkable result soon prompted us to investigate further in the attempt to identify novel underlying mechanisms that could be implicated in regulating the resilience of SCs to nutrient starvation and how these mechanisms were affected by PACAP treatment.

SCs are known to be a regenerative cell type of the PNS. Many of the cellular functions of these cells are dictated through the activation of multiple signaling pathways, including the RAS-MAPK. Neurofibromatosis type 1 gene (NF1) encodes a RAS GTPase-activating protein called neurofibromin and is one of several genes whose functional inactivation (due to gene mutations or lowered expression) affects the RAS-MAPK signaling, causing a group of related diseases collectively known as RASopathies. The interesting aspect regarding RASopathies is that they commonly include SCs disorders characterized by somatic NF1 mutations, which are typically associated to cellular resistance to therapy and death [61]. In addition, because the NF1 gene also governs most of the cellular processes of SCs related to cell cycle entry [62], tumor suppressor activity [63] and more recently resistance to apoptosis [64], analyzing whether this key player involved in such a number of growth-related biological activities could be deregulated after nutrient deprivation and affected by PACAP was considered an aspect worthy of consideration.

The studies revealed that RT4D6P2T cells expressed very low NF1 levels under normal nutrient conditions, but starvation significantly induced NF1 gene and protein [3]. Treatment with either PACAP or VIP succeeded to lower NF1 expression to basal levels. These data, in line with Shapira's studies [64], supported a pro-apoptotic function of this gene in SCs. Moreover, PACAP appeared to be a negative regulator of the putative pro-apoptotic gene expression, a phenomenon that we restricted to the specific type of insult SCs received and that, undeniably, is likely to involve other molecular targets [4].

Several studies have consistently reported that PACAP, but also VIP, stimulate the endogenous production of additional trophic factors involved in cell protection, such as the activity dependent neuroprotective protein (ADNP) [4, 26, 65, 66] endozepines, including the octadecaneuropeptide (ODNP) [67, 68], and brain derived neurotrophic factor (BDNF) [60, 69–72]. There are therefore compelling reasons to believe that, among the myriad of other endogenous pro-survival molecules potentially activated by PACAP in SCs, at least two may in part account for the indirect prosurvival/antiapoptotic responses triggered by PACAP following insults. Based on these convergent data, it is proposed that PACAP plays a pivotal role in protection against cell death, which may be the summation of direct activities over genes related to cell survival and/or indirect effects triggered by the intricate interaction with other trophic molecules. Whether other intracellular systems are also implicated in PACAP-mediated protection in SCs remains to be established, but given the number of substrates triggered and the different responsiveness of these cells to external *stimuli*, the possibility to identify new routes or activated molecules is more likely to become a probability.

PACAP as a Secretagogue for Plasminogen Activators in SCs

Despite the many advances in regenerative medicine, effective strategies to guarantee the appropriate regeneration of peripheral nerves after injury have still demonstrated to be unsatisfactory. Among the many unresolved questions, those that have mostly limited the successful progress in nerve repair therapies have to be searched at the axonal/cellular level. According to the available information on the normal physiological response of peripheral nerves to injury, following damage, the axon is interrupted and forms two segments: a proximal stump (the portion of the axon that remains attached to the soma) and a distal one. After a relatively brief latent period, the separated distal stump rapidly degenerates (after 24–36 h) in a process termed "Wallerian degeneration" (WD), characterized by axonal membrane disintegration, degradation of the myelin sheath and infiltration of macrophages and SCs [73]. Interestingly, during WD, part of the axon's neurolemma is preserved and the perineural glia may be recruited to form a "bridge" to eventually guide the regrowth of the proximal axon stump towards the denervated terminal nerve ending. Unfortunately, this naturally occurring process is in great part dependent on the ability of infiltrating cells to release trophic factors (to support nerve regrowth) and to effectively clear cellular components and myelin debris that accumulates in the path of the artificial bridge formed in the gap between the proximal and distal stump, which impedes proper axon sprouting. Infiltrating macrophages, perineural glial cells, as well as neurites and transdifferentiated SCs (please see below for details), are recruited to also take part in the clearance process, but in most cases are unable to withstand with these requests, resulting in an ineffective nerve regrowth, especially when nerve stumps are physically too distant for the development of the new axonal connection [74]. These events strongly limit the regenerative capacities of peripheral nerves after damage.

Effective phagocytic activity of residual myelin and other cell debris needs to be preceded by local release of proteolytic enzymes. These enzymes digest the remaining extracellular matrix (ECM), myelin droplets, and any other cellular products associated to WD, to promote the sprouting of the rejuvenating axon. Plasminogen activators (PAs) are proteases that are capable of degrading most of ECM and dissolving adhesive contacts to promote growth cone migration and axonal outgrowth. The PAs, which include tissue PA (tPA) and urokinase PA (uPA), convert the inactive plasminogen to its active serine protease plasmin. In turn, plasmin has broad activity and digests most ECM proteins as well as some cell surface molecules, such as neural cell adhesion molecule (NCAM) [75], and activates other proteases, including matrix metalloproteases [76–79].

When a peripheral nerve is damaged SCs recapitulate many of the activities that are triggered during development to insure axon regrowth. Such injury-induced ability to revert back from a myelinating phenotype to a nondifferentiated, proliferative phenotype is referred to as "transdifferentiation." Transdifferentiation is a transient reversible cell passage that initiates in SCs the intracellular accumulation of a series of substrates that are necessary to support debris clearance in the first instance, and myelin regeneration during axon regrowth. Unfortunately, the mechanisms that regulate SCs transdifferentiation are not well understood. Fortunately, recent insights from Raoult and coworkers have opened a new venue of investigation, since they demonstrated that PACAP stimulates the expression of tPA in cerebellar-derived neuronal cells [80]. In that study, the authors contextualized their results to the migratory capacities of cerebellar granule neurons (CGNs), and did not consider any further implications in terms of nerve regenerative capacities. A couple of years later, we shifted Raoult's study to our SCs model, and found that PACAP behaved similarly as in CGNs, being a potent stimulator of tPA expression and activity [69]. With interest, we noticed that PACAP-driven tPA induction was mediated by the PI3K/Akt signaling, the same intracellular pathway activated for the induction of myelin components, but had a diverse kinetic. Indeed, tPA induction peaked at 4 h and progressively dropped up to 12 h after PACAP treatment. Conversely, PACAP-driven expression of myelin components grew after 24–48 h [56]. Moreover, we noticed that brain derived neurotrophic factor (BDNF), a powerful trophic factor, also strongly induced tPA activity, mostly by boosting endogenous PACAP levels. Our interpretation of these findings, contextualized to a model of nerve injury, was that PACAP (and also BDNF), released at the injury site by neurons and SCs, may elicit temporally distinct actions in SCs. It is supposed that PACAP acts as a secretagogue for plasminogens to promote SCs-mediated enzymatic digestion of cellular debris during the earliest phase of nerve injury (i.e., when SCs transdifferentiate) and subsequently trigger the intracellular accumulation of myelin substrates to support remyelination of the newly regrown axon (for a schematic illustration of the proposed mechanism please refer to Fig. 28.3). These conclusions, although need to be supported by further evidences, certainly add a new piece of knowledge on the increasing number of biological activities elicited by SCs in which PACAP plays a central role to support nerve repair after injury.

SCs as a Potential Target of PACAP for the Treatment for Neuropathic Pain of Peripheral Origin

Neuropathic pain is manifested as a painful response to a normally non-painful stimulus (allodynia) or as an overly exuberant response to a painful stimulus. Neuropathic pain may result from disorders of the PNS or the CNS. Thus, it may be divided into peripheral neuropathic pain, central neuropathic pain, or combined (peripheral and central). In this section the role of PACAP will be discussed solely in the context of peripheral neuropathic pain, where SCs appear to play an important role.

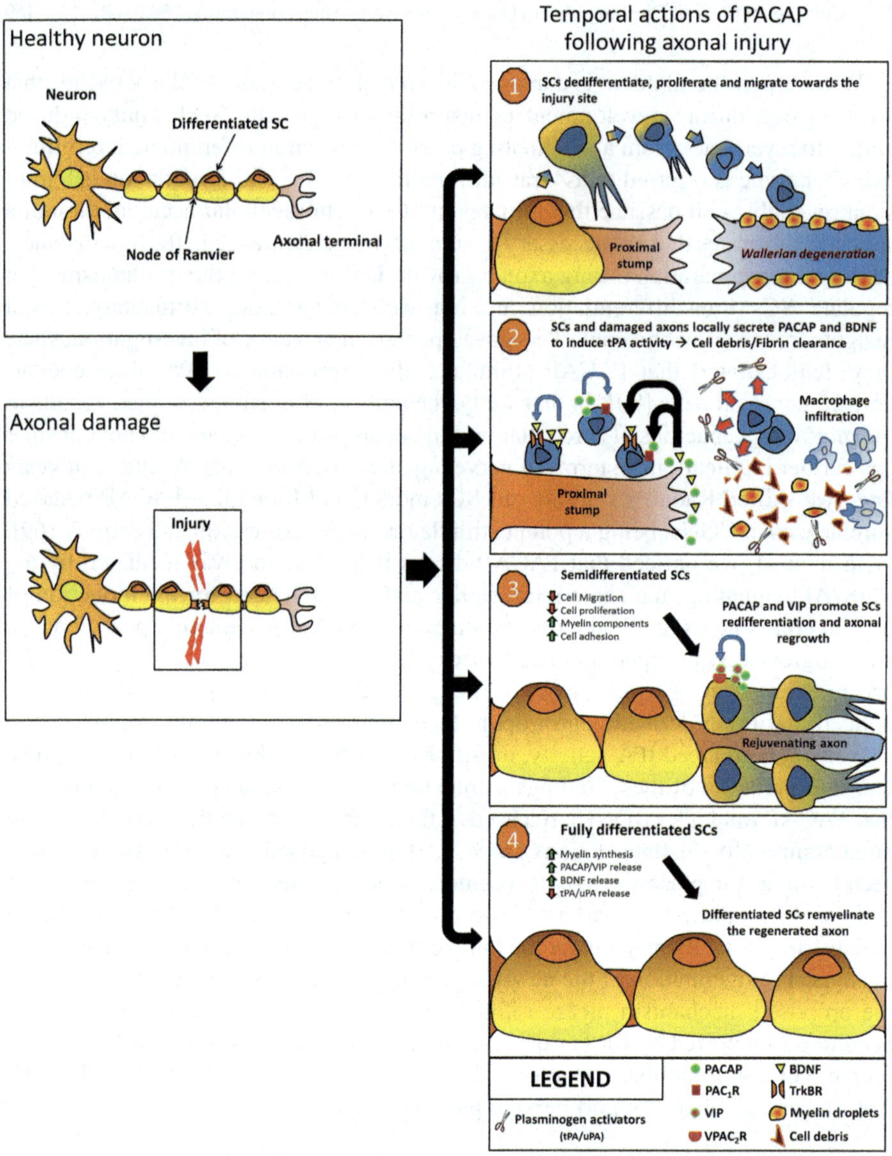

Fig. 28.3 Proposed model illustrating the temporal actions of PACAP and BDNF in relationship to Schwann cells differentiating and proteolytic activities following peripheral nerve injury. Injured axons of peripheral nerves regenerate spontaneously, and various factors seem to contribute to this regenerative ability. After nerve damage, the distal stump of the injured axon undergoes Wallerian degeneration and Schwann cells (SCs) rapidly dedifferentiate, proliferate, and migrate towards the injury site to create a permissive environment for axonal regeneration (1). The actions of PACAP and BDNF on these early processes are unclear. Once SCs reach the injured site, local release of PACAP and BDNF at the proximal axon stump or by SCs (possibly through an autocrine/paracrine mechanism) seems to be a determinant in promoting the production and release of scavenging molecules (tPA and uPA), and augmented synthesis of myelin components, while inhibiting SCs migratory capacities (2 and 3). Upon clearance of cell debris/fibrin deposits, PACAP and/or BDNF promote axonal regrowth, and SCs aligned along the rejuventating neurite begin to differentiate (most likely through a PACAP-VIP/VPAC2-mediated mechanism [72]) towards a fully myelinating phenotype (3). Finally, differentiated SCs wrap around the newly formed axon and produce myelin sheaths to promote functional recovery (4). *PAC1* PACAP-preferring receptor. *TrkBR* Tropomyosin related kinase B receptor. *VPAC2* PACAP/VIP receptor

It is now accepted that many of the central responses in neuropathic pain arise from the effects of extracellular mediators that are released at the injury site and transported retrogradely up to the afferent neuron. To support this, blockade of retrograde transport has shown to effectively dampen many abnormal pain behaviors in rat models of chronic pain [81, 82]. Thus, the pathophysiology of the injured nerve is of extreme importance in the generation of neuropathic pain states. It is well known that the nerve injury site becomes rich in extracellular mediators, including cytokines, growth factors and proteases. SCs play a key role in regulating this microenvironment. In healthy nerves, SCs provide the necessary myelin ensheathment of neurites and paracrine trophic support to nerve; however, in injured nerves, transdifferentiation of SCs is rapidly triggered and cells undergo dramatic phenotypic modulation, regaining capacity to proliferate, migrate, and secrete numerous factors that control WD and nerve regeneration. With the understanding of SC physiology, the response of these cells to cytokines and extracellular mediators, and the capacity to express cytokines in a regulated manner emerges as a central problem in the pain field.

Experimental evidence from animal models of neuropathic pain has revealed that therapeutics aimed at preventing SC atrophy, loss of basal lamina, and promote SCs survival may be effective in counteracting the development of painful peripheral neuropathies [83]. Converging studies point to the existence of distinct neuroinflammatory signatures following the development of a neuropathic pain condition, and these signatures seem to correlate to the appearance and persistence of associated behavioral comorbidities [84–89]. In relationship to SCs, the utmost cytokines involved seem to be the tumor necrosis factor α (TNFα) [90], interleukin-1α (IL-1α) and IL-1β [91], and IL-6 [92].

PACAP exerts divergent effects in acute somatic and visceral inflammatory pain models, being an excitatory factor and causing neurogenic vasodilation during pain transmission originating from both exteroceptive and interoceptive areas and anti-nociceptive, anti-hyperalgesic and with anti-allodynic effects in visceral pain [93–95]. PACAP has been ascribed as a potent pro-survival factor in many types of glia, including SCs [2, 28]. In addition, as reviewed by Delgado and colleagues in 2007 [93], PACAP and VIP have consistently demonstrated to also play significant immunomodulatory activities, acting at diverse levels of the immune system, but primarily by reducing the levels of pro-inflammatory cytokines in macrophages and microglia [94, 95] and by increasing the expression of anti-inflammatory molecules, such as IL-10 [96]. Even though such abilities have never been tested directly in SCs, recent data has proven that this glial cell type also expresses functional receptors for pro-inflammatory cytokines, which upon binding, seem to be able to reduce SCs myelination, as demonstrated in studies using SCs/neuron cocultures [97]. Therefore, taking into consideration our previous findings and the new emerging reports, it can be concluded that PACAP might contribute to dampen or prevent the occurrence of three key events that are critical for the genesis of neuropathic pain: (1) progressive SCs death, (2) release of pro-inflammatory cytokines by SCs and other infiltrating cells at the site of injury and (3) the consequent hyperactivity of nociceptive transmission caused by peripheral nerve inflammation, SCs and axonal death.

Conclusions and Future Directions

SCs elicit key regulatory functions on the physiology of peripheral nerves both in healthy and disease conditions. It is now ascertained that the role of SCs is not merely limited to wrapping around axons of motor and sensory neurons to form a myelin sheath to provide electrical insulation and allow nerve conduction, but also to maintain axons viable and facilitate the recovery of peripheral nerves after injury. SCs achieve this goal by providing trophic support, release chemoattractants and proteolytic factors. Being aware that this cell type is a target of PACAP raises important queries on the underlying potential encompassed by this peptide in regards to its therapeutic utilization in regenerative medicine, but also for the treatment of chronic pain conditions, and/or to impede the relentless progression of certain peripheral neuropathies, such as CMT, GBS, and CIDP. PACAP and the cognate peptide VIP emerge as two important endogenous molecules in this context, and their multiple activities in SCs reflect only part of the entire picture (for a summary of PACAP-stimulated SCs functions please refer to Fig. 28.4). The ubiquitous expression of the neurotrophic molecule and of its binding receptors in several vital organs and cell types seem to be just a cue of the many therapeutic possibilities hidden behind this peptide of relatively recent discovery. Unluckily, despite the large number of existing experimental/preclinical investigations, which have certainly contributed to gain insights on the beneficial functions of PACAP, the current efforts by biotechnology and investigators to developing stable forms of the peptide to maximize its utility in therapy have achieved only marginal results. As recently reported by Marzagalli et al. [34], PACAP is theoretically a good candidate molecule for drug discovery purposes, given its ability to cross the blood brain barrier and circulate in the brain through saturable and non-saturable systems. But when injected systemically in its native form, it has low bioavailability, particularly in the brain. The reason is that PACAP is rapidly hydrolyzed by the ubiquitous enzyme dipeptidyl-peptidase IV (DPP IV) to form, PACAP (3–38) or PACAP (5–38), two shorter forms of the peptide with an antagonist activity on PAC1 receptors [98]. The degradation by DPP IV in the blood stream also results in poor metabolic stability and short half-life (which is estimated to be between 2 and 10 min after PACAP is injected into mice or humans) [99].

Independent research groups are currently working to develop a more stable form of PACAP, which can be suitable for clinical use. To date, a novel cyclopeptide has been obtained from the cyclization of PACAP (1–5), and that possesses potent activity towards PAC1 [98, 99], suggesting that efforts are still being put by scientists to overcome the issues associated with low peptide bioavailability. In this regard, a very recent publication by Fournier's research group [100] appears remarkable, because of the potential positive impact this study holds. A novel PACAP-derivative, termed Ac-[Phe(pI)^{6}, Nle^{17}]PACAP(1–27), has been developed and characterized in a rodent model of Parkinson's disease. This new PACAP analogue retains all the beneficial activities of PACAP, appears not to be degraded by DPP IV, and most surprisingly, is devoid of adverse effects in the cardiovascular system. If the analogue will result successful in further experimental testing, it promises to become a leading molecule for the development of PACAP-derived drugs by the pharmaceutical industry.

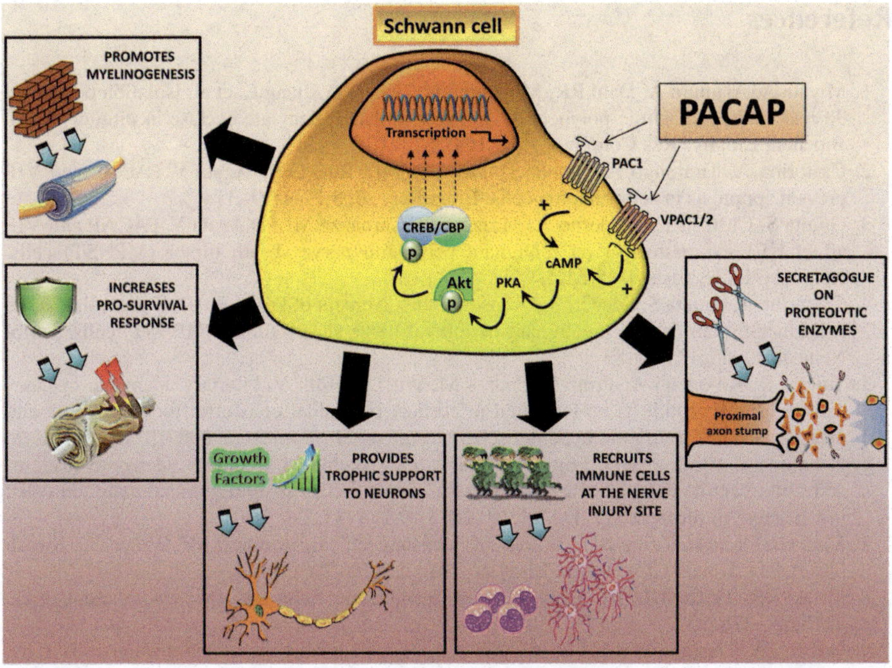

Fig. 28.4 Exemplified view of PACAP-stimulated biological activities in SCs. SCs take part in the regulation of numerous biological functions, which are differently triggered by PACAP according to the concurrent stimuli received through the surrounding microenvironment and/or in specific contexts (i.e., injury, nutrient deprivation, etc.). Exogenous or endogenous PACAP stimulation (through autocrine/paracrine loops) activates either PAC1 and VPAC1/2 receptors to potently increase intracellular cAMP accumulation, thereby triggering the canonical cAMP/PKA/Akt intracellular signaling cascade, among others. Akt-dependent and -independent priming of different downstream targets (including the transcription factor CREB [55]) may then trigger the transcriptional machinery for the induction of multiple genes. Such events lead to several biological responses, like the accumulation of myelin components [54], the increased resilience to cell death [2–4] or the augmented trophic support to ensheathed neurons [4]. After axonal injury, PACAP also stimulates SCs and other resident glia to aid in the recruitment of infiltrating immune cells at the site of injury [85], but also to release proteolytic factors capable to promote the clearance of ECM/cell/myelin debris and accelerate nerve regeneration [55, 72]. *PKA* protein kinase A. *Akt* V-Akt murine thymoma viral oncogene homolog 1. *CREB* cAMP responsible element binding protein. *CBP* CREB binding protein

Acknowledgments The economical sources for the realization of this book chapter are from the Department of Biomedical and Biotechnological Sciences, School of Medicine, University of Catania, Italy (National Grant–PiTecnoBio PON 01_02464). I would like to thank Dr. Rubina Marzagalli for her help in retrieving literature information and for the preparation of the iconographic material. A further thanks goes to Mr. P. Asero for his technical support.

Authors' disclosures The author declares that there is no conflict of interests regarding the publication of this chapter.

References

1. Miyata A, Arimura A, Dahl RR, Minamino N, Uehara A, Jiang L, et al. Isolation of a novel 38 residue-hypothalamic polypeptide which stimulates adenylate cyclase in pituitary cells. Biochem Biophys Res Commun. 1989;164:567–74.
2. Castorina A, Tiralongo A, Giunta S, Carnazza ML, Rasi G, D'Agata V. PACAP and VIP prevent apoptosis in schwannoma cells. Brain Res. 2008;1241:29–35.
3. Giunta S, Castorina A, Adorno A, Mazzone V, Carnazza ML, D'Agata V. PACAP and VIP affect NF1 expression in rat malignant peripheral nerve sheath tumor (MPNST) cells. Neuropeptides. 2010;44:45–51.
4. Castorina A, Giunta S, Scuderi S, D'Agata V. Involvement of PACAP/ADNP signaling in the resistance to cell death in malignant peripheral nerve sheath tumor (MPNST) cells. J Mol Neurosci. 2012;48:674–83.
5. Stumm R, Kolodziej A, Prinz V, Endres M, Wu DF, Höllt V. Pituitary adenylate cyclase-activating polypeptide is up-regulated in cortical pyramidal cells after focal ischemia and protects neurons from mild hypoxic/ischemic damage. J Neurochem. 2007;103:1666–81.
6. Nakajima E, Walkup RD, Fujii A, Shearer TR, Azuma M. Pituitary adenylate cyclase-activating peptide induces neurite outgrowth in cultured monkey trigeminal ganglion cells: involvement of receptor PAC1. Mol Vis. 2013;19:174–83.
7. Kidd GJ, Ohno N, Trapp BD. Biology of Schwann cells. In: Aminoff MJ, Boller FO, Swaab DF, editors. Handb Clin Neurol. Elsevier; 2013;115:55–79.
8. Monk KR, Feltri ML, Taveggia C. New insights on Schwann cell development. Glia. 2015;63:1376–93.
9. Griffin JW, Thompson WJ. Biology and pathology of nonmyelinating Schwann cells. Glia. 2008;56:1518–31.
10. Yamazaki S, Ema H, Karlsson G, Yamaguchi T, Miyoshi H, Shioda S, et al. Nonmyelinating Schwann cells maintain hematopoietic stem cell hibernation in the bone marrow niche. Cell. 2011;147:1146–58.
11. Ko CP, Sugiura Y, Feng Z. The biology of perisynaptic (terminal) Schwann cells. In: Armati PJ, editor. The biology of Schwann cells. Cambridge: Cambridge University Press; 2007. p. 72–99.
12. Zhou XF, Deng YS, Chie E, Xue Q, Zhong JH, McLachlan EM, et al. Satellite cell-derived nerve growth factor and neurotrophin-3 are involved in noradrenergic sprouting in the dorsal root ganglia following peripheral nerve injury in the rat. Eur J Neurosci. 1999;11:1711–22.
13. Schwyn RC. An autoradiographic study of satellite cells in autonomic ganglia. Am J Anat. 1967;121:727–40.
14. Arroyo EJ, Scherer S. The molecular organisation of myelinating Schwann cells. In: Armati PJ, editor. The biology of Schwann cells. Cambridge: Cambridge University Press; 2007. p. 37–54.
15. Armati PJ, Pollard JD, Gatenby P. Rat and human Schwann cells in vitro can synthesize and express MHC molecules. Muscle Nerve. 1990;13:106–16.
16. Nagano S, Takeda M, Ma L, Soliven B. Cytokine-induced cell death in immortalized Schwann cells: roles of nitric oxide and cyclic AMP. J Neurochem. 2001;77:1486–95.
17. Irnich D, Burgstahler R, Bostock H, Grafe P. ATP affects both axons and Schwann cells of unmyelinated C fibres. Pain. 2001;92:343–50.
18. Armati PJ, Mathey EK. An update on Schwann cell biology—immunomodulation, neural regulation and other surprises. J Neurol Sci. 2013;333:68–72.
19. Miyata A, Jiang L, Dahl RD, Kitada C, Kubo K, Fujino M, et al. Isolation of a neuropeptide corresponding to the N-terminal 27 residues of the pituitary adenylate cyclase activating polypeptide with 38 residues (PACAP38). Biochem Biophys Res Commun. 1990;170: 643–8.
20. Dickson L, Finlayson K. VPAC and PAC receptors: from ligands to function. Pharmacol Ther. 2009;121:294–316.

21. Lutz EM, Ronaldson E, Shaw P, Johnson MS, Holland PJ, Mitchell R. Characterization of novel splice variants of the PAC1 receptor in human neuroblastoma cells: consequences for signaling by VIP and PACAP. Mol Cell Neurosci. 2006;31:193–209.
22. Holighaus Y, Mustafa T, Eiden LE. PAC1hop, null and hip receptors mediate differential signaling through cyclic AMP and calcium leading to splice variant-specific gene induction in neural cells. Peptides. 2011;32:1647–55.
23. Blechman J, Levkowitz G. Alternative splicing of the pituitary adenylate cyclase-activating polypeptide receptor PAC1: mechanisms of fine tuning of brain activity. Front Endocrinol (Lausanne). 2013;4:55.
24. Lakk M, Szabó B, Völgyi B, Gábriel R, Dénes V. Development-related splicing regulates pituitary adenylate cyclase-activating polypeptide (PACAP) receptors in the retina. Invest Ophthalmol Vis Sci. 2012;53:7825–32.
25. Dénes V, Czotter N, Lakk M, Berta G, Gábriel R. PAC1-expressing structures of neural retina alter their PAC1 isoform splicing during postnatal development. Cell Tissue Res. 2014;355:279–88.
26. Li M, David C, Kikuta T, Somogyvari-Vigh A, Arimura A. Signaling cascades involved in neuroprotection by subpicomolar pituitary adenylate cyclase-activating polypeptide 38. J Mol Neurosci. 2005;27:91–105.
27. Botia B, Jolivel V, Burel D, Le Joncour V, Roy V, Naassila M, et al. Neuroprotective effects of PACAP against ethanol-induced toxicity in the developing rat cerebellum. Neurotox Res. 2011;19:423–34.
28. Jóźwiak-Bębenista M, Kowalczyk E, Nowak JZ. The cyclic AMP effects and neuroprotective activities of PACAP and VIP in cultured astrocytes and neurons exposed to oxygen-glucose deprivation. Pharmacol Rep. 2015;67:332–8.
29. Juhász T, Matta C, Katona É, Somogyi C, Takács R, Gergely P, et al. Pituitary adenylate cyclase activating polypeptide (PACAP) signalling exerts chondrogenesis promoting and protecting effects: implication of calcineurin as a downstream target. PLoS One. 2014;9:e91541.
30. Giunta S, Castorina A, Marzagalli R, Szychlinska MA, Pichler K, Mobasheri A, et al. Ameliorative effects of PACAP against cartilage degeneration. Morphological, immunohistochemical and biochemical evidence from in vivo and in vitro models of rat osteoarthritis. Int J Mol Sci. 2015;16:5922–44.
31. Castorina A, Giunta S, Mazzone V, Cardile V, D'Agata V. Effects of PACAP and VIP on hyperglycemia-induced proliferation in murine microvascular endothelial cells. Peptides. 2010;31:2276–83.
32. Scuderi S, D'Amico AG, Castorina A, Imbesi R, Carnazza ML, D'Agata V. Ameliorative effect of PACAP and VIP against increased permeability in a model of outer blood retinal barrier dysfunction. Peptides. 2013;39:119–24.
33. Giunta S, Castorina A, Bucolo C, Magro G, Drago F, D'Agata V. Early changes in pituitary adenylate cyclase-activating peptide, vasoactive intestinal peptide and related receptors expression in retina of streptozotocin-induced diabetic rats. Peptides. 2012;37:32–9.
34. Marzagalli R, Scuderi S, Drago F, Waschek JA, Castorina A. Emerging role of PACAP as a new potential therapeutic target in major diabetes complications. Int J Endocrinol. 2015;2015:160928.
35. Ji H, Zhang Y, Shen XD, Gao F, Huang CY, Abad C, et al. Neuropeptide PACAP in mouse liver ischemia and reperfusion injury: immunomodulation by the cAMP-PKA pathway. Hepatology. 2013;57:1225–37.
36. Lee J, Park HJ, Choi HS, Kwon HB, Arimura A, Lee BJ, et al. Gonadotropin stimulation of pituitary adenylate cyclase-activating polypeptide (PACAP) messenger ribonucleic acid in the rat ovary and the role of PACAP as a follicle survival factor. Endocrinology. 1999;140: 818–26.
37. Waschek JA. VIP and PACAP: neuropeptide modulators of CNS inflammation, injury, and repair. Br J Pharmacol. 2013;169:512–23.

38. Reglodi D, Kiss P, Szabadfi K, Atlasz T, Gabriel R, Horvath G, et al. PACAP is an endogenous protective factor-insights from PACAP-deficient mice. J Mol Neurosci. 2012;48:482–92.
39. Shioda S, Ohtaki H, Nakamachi T, Dohi K, Watanabe J, Nakajo S, et al. Pleiotropic functions of PACAP in the CNS: neuroprotection and neurodevelopment. Ann N Y Acad Sci. 2006;1070:550–60.
40. Broca C, Quoyer J, Costes S, Linck N, Varrault A, Deffayet PM, et al. Beta-Arrestin 1 is required for PAC1 receptor-mediated potentiation of long-lasting ERK1/2 activation by glucose in pancreatic beta-cells. J Biol Chem. 2009;284:4332–42.
41. Moroo I, Tatsuno I, Uchida D, Tanaka T, Saito J, Saito Y, et al. Pituitary adenylate cyclase activating polypeptide (PACAP) stimulates mitogen-activated protein kinase (MAPK) in cultured rat astrocytes. Brain Res. 1998;795:191–6.
42. Juhász T, Helgadottir SL, Tamás A, Reglődi D, Zákány R. PACAP and VIP signaling in chondrogenesis and osteogenesis. Peptides. 2015;66:51–7.
43. Izumi S, Seki T, Shioda S, Zhou CJ, Arimura A, Koide R. Ultrastructural localization of PACAP immunoreactivity in the rat retina. Ann N Y Acad Sci. 2000;921:317–20.
44. Muroi M, Shioda S, Yada T, Zhou CJ, Nakai Y, Nakajo S, et al. Distribution and ultrastructural localization of PACAP receptors in the rat pancreatic islets. Ann N Y Acad Sci. 1998;865:438–40.
45. Jesuraj NJ, Nguyen PK, Wood MD, Moore AM, Borschel GH, Mackinnon SE, et al. Differential gene expression in motor and sensory Schwann cells in the rat femoral nerve. J Neurosci Res. 2012;90:96–104.
46. Hai M, Muja N, DeVries GH, Quarles RH, Patel PI. Comparative analysis of Schwann cell lines as model systems for myelin gene transcription studies. J Neurosci Res. 2002;69:497–508.
47. Jessen KR, Mirsky R. The origin and development of glial cells in peripheral nerves. Nat Rev Neurosci. 2005;6:671–82.
48. Tamas A, Reglodi D, Farkas O, Kovesdi E, Pal J, Povlishock JT, et al. Effect of PACAP in central and peripheral nerve injuries. Int J Mol Sci. 2012;13:8430–48.
49. Zhang QL, Liu J, Lin PX, Webster HD. Local administration of vasoactive intestinal peptide after nerve transection accelerates early myelination and growth of regenerating axons. J Peripher Nerv Syst. 2002;7:118–27.
50. Armstrong BD, Abad C, Chhith S, Cheung-Lau G, Hajji OE, Nobuta H, et al. Impaired nerve regeneration and enhanced neuroinflammatory response in mice lacking pituitary adenylyl cyclase activating peptide. Neuroscience. 2008;151:63–73.
51. Garratt AN, Voiculescu O, Topilko P, Charnay P, Birchmeier C. A dual role of erbB2 in myelination and in expansion of the Schwann cell precursor pool. J Cell Biol. 2000;148:1035–46.
52. Michailov GV, Sereda MW, Brinkmann BG, Fischer TM, Haug B, Birchmeier C, et al. Axonal neuregulin-1 regulates myelin sheath thickness. Science. 2004;304:700–3.
53. Taveggia C, Zanazzi G, Petrylak A, Yano H, Rosenbluth J, Einheber S, et al. Neuregulin-1 type III determines the ensheathment fate of axons. Neuron. 2005;47:681–94.
54. Cheng L, Esch FS, Marchionni MA, Mudge AW. Control of Schwann cell survival and proliferation: autocrine factors and neuregulins. Mol Cell Neurosci. 1998;12:141–56.
55. Syroid DE, Maycox PR, Burrola PG, Liu N, Wen D, Lee KF, et al. Cell death in the Schwann cell lineage and its regulation by neuregulin. Proc Natl Acad Sci U S A. 1996;93:9229–34.
56. Castorina A, Scuderi S, D'Amico AG, Drago F, D'Agata V. PACAP and VIP increase the expression of myelin-related proteins in rat schwannoma cells: involvement of PAC1/VPAC2 receptor-mediated activation of PI3K/Akt signaling pathways. Exp Cell Res. 2014;322:108–21.
57. Vincze A, Reglodi D, Helyes Z, Hashimoto H, Shintani N, Abrahám H. Role of endogenous pituitary adenylate cyclase activating polypeptide (PACAP) in myelination of the rodent brain: lessons from PACAP-deficient mice. Int J Dev Neurosci. 2011;29:923–35.

58. Bansal R, Pfeiffer SE. Regulated galactolipid synthesis and cell surface expression in Schwann cell line D6P2T. J Neurochem. 1987;49:1902–11.
59. Lee M, Lelievre V, Zhao P, Torres M, Rodriguez W, Byun JY, et al. Pituitary adenylyl cyclase-activating polypeptide stimulates DNA synthesis but delays maturation of oligodendrocyte progenitors. J Neurosci. 2001;21:3849–59.
60. Lelievre V, Ghiani CA, Seksenyan A, Gressens P, de Vellis J, Waschek JA. Growth factor-dependent actions of PACAP on oligodendrocyte progenitor proliferation. Regul Pept. 2006;137:58–66.
61. Ratner N, Miller SJ. A RASopathy gene commonly mutated in cancer: the neurofibromatosis type 1 tumour suppressor. Nat Rev Cancer. 2015;15:290–301.
62. Eccleston PA. Regulation of Schwann cell proliferation: mechanisms involved in peripheral nerve development. Exp Cell Res. 1992;199:1–9.
63. Uhlmann EJ, Gutmann DH. Tumor suppressor gene regulation of cell growth: recent insights into neurofibromatosis 1 and 2 gene function. Cell Biochem Biophys. 2001;34:61–78.
64. Shapira S, Barkan B, Friedman E, Kloog Y, Stein R. The tumor suppressor neurofibromin confers sensitivity to apoptosis by Ras-dependent and Ras-independent pathways. Cell Death Differ. 2007;14:895–906.
65. Zusev M, Gozes I. Differential regulation of activity-dependent neuroprotective protein in rat astrocytes by VIP and PACAP. Regul Pept. 2004;123:33–41.
66. Nakamachi T, Li M, Shioda S, Arimura A. Signaling involved in pituitary adenylate cyclase-activating polypeptide-stimulated ADNP expression. Peptides. 2006;27:1859–64.
67. Masmoudi O, Gandolfo P, Leprince J, Vaudry D, Fournier A, Patte-Mensah C, et al. Pituitary adenylate cyclase-activating polypeptide (PACAP) stimulates endozepine release from cultured rat astrocytes via a PKA-dependent mechanism. FASEB J. 2003;17:17–27.
68. Masmoudi-Kouki O, Gandolfo P, Leprince J, Vaudry D, Pelletier G, Fournier A, et al. PACAP stimulates biosynthesis and release of endozepines from rat astrocytes. Ann N Y Acad Sci. 2006;1070:411–6.
69. Castorina A, Waschek JA, Marzagalli R, Cardile V, Drago F. PACAP interacts with PAC1 receptors to induce tissue plasminogen activator (tPA) expression and activity in Schwann cell-like cultures. PLoS One. 2015;10:e0117799.
70. Andero R, Ressler KJ. Fear extinction and BDNF: translating animal models of PTSD to the clinic. Genes Brain Behav. 2012;11:503–12.
71. Pettersson LM, Geremia NM, Ying Z, Verge VM. Injury-associated PACAP expression in rat sensory and motor neurons is induced by endogenous BDNF. PLoS One. 2014;9:e100730.
72. Ogata K, Shintani N, Hayata-Takano A, Kamo T, Higashi S, Seiriki K, et al. PACAP enhances axon outgrowth in cultured hippocampal neurons to a comparable extent as BDNF. PLoS One. 2015;10:e0120526.
73. Conforti L, Gilley J, Coleman MP. Wallerian degeneration: an emerging axon death pathway linking injury and disease. Nat Rev Neurosci. 2014;15:394–409.
74. Lewis GM, Kucenas S. Perineurial glia are essential for motor axon regrowth following nerve injury. J Neurosci. 2014;34:12762–77.
75. Endo A, Nagai N, Urano T, Ihara H, Takada Y, Hashimoto K, et al. Proteolysis of highly polysialylated NCAM by the tissue plasminogen activator-plasmin system in rats. Neurosci Lett. 1998;246:37–40.
76. Keski-Oja J, Lohi J, Tuuttila A, Tryggvason K, Vartio T. Proteolytic processing of the 72,000-Da type IV collagenase by urokinase plasminogen activator. Exp Cell Res. 1992;202:471–6.
77. Murphy G, Docherty AJ. The matrix metalloproteinases and their inhibitors. Am J Respir Cell Mol Biol. 1992;7:120–5.
78. Siconolfi LB, Seeds NW. Mice lacking tPA, uPA, or plasminogen genes showed delayed functional recovery after sciatic nerve crush. J Neurosci. 2001;21:4348–55.
79. Siconolfi LB, Seeds NW. Mice lacking tissue plasminogen activator and urokinase plasminogen activator genes show attenuated matrix metalloproteases activity after sciatic nerve crush. J Neurosci Res. 2003;74:430–4.

80. Raoult E, Roussel BD, Bénard M, Lefebvre T, Ravni A, Ali C, et al. Pituitary adenylate cyclase-activating polypeptide (PACAP) stimulates the expression and the release of tissue plasminogen activator (tPA) in neuronal cells: involvement of tPA in the neuroprotective effect of PACAP. J Neurochem. 2011;119:920–31.
81. Yamamoto T, Yaksh TL. Effects of colchicine applied to the peripheral nerve on the thermal hyperalgesia evoked with chronic nerve constriction. Pain. 1993;55:227–33.
82. Yamamoto T, Takahara A. Recent updates of N-type calcium channel blockers with therapeutic potential for neuropathic pain and stroke. Curr Top Med Chem. 2009;9:377–95.
83. Höke A. Mechanisms of disease: what factors limit the success of peripheral nerve regeneration in humans? Nat Clin Pract Neurol. 2006;2:448–54.
84. Redford EJ, Hall SM, Smith KJ. Vascular changes and demyelination induced by the intraneural injection of tumour necrosis factor. Brain. 1995;118:869–78.
85. Wagner R, Myers RR. Endoneurial injection of TNF-alpha produces neuropathic pain behaviors. Neuroreport. 1996;7:2897–901.
86. Watkins LR, Maier SF. Glia: a novel drug discovery target for clinical pain. Nat Rev Drug Discov. 2003;2:973–85.
87. Mor D, Bembrick AL, Austin PJ, Keay KA. Evidence for cellular injury in the midbrain of rats following chronic constriction injury of the sciatic nerve. J Chem Neuroanat. 2011;41:158–69.
88. Austin PJ, Bembrick AL, Denyer GS, Keay KA. Injury-dependent and disability-specific lumbar spinal gene regulation following sciatic nerve injury in the rat. PLoS One. 2015;10:e0124755.
89. Austin PJ, Berglund AM, Siu S, Fiore NT, Gerke-Duncan MB, Ollerenshaw SL, et al. Evidence for a distinct neuro-immune signature in rats that develop behavioural disability after nerve injury. J Neuroinflammation. 2015;12:96.
90. Campana WM, Li X, Shubayev VI, Angert M, Cai K, Myers RR. Erythropoietin reduces Schwann cell TNF-alpha, Wallerian degeneration and pain-related behaviors after peripheral nerve injury. Eur J Neurosci. 2006;23:617–26.
91. Shamash S, Reichert F, Rotshenker S. The cytokine network of Wallerian degeneration: tumor necrosis factor-alpha, interleukin-1alpha, and interleukin-1beta. J Neurosci. 2002;22:3052–60.
92. Frieboes LR, Palispis WA, Gupta R. Nerve compression activates selective nociceptive pathways and upregulates peripheral sodium channel expression in Schwann cells. J Orthop Res. 2010;28:753–61.
93. Sándor K, Bölcskei K, McDougall JJ, Schuelert N, Reglodi D, Elekes K, et al. Divergent peripheral effects of pituitary adenylate cyclase-activating polypeptide-38 on nociception in rats and mice. Pain. 2009;141:143–50.
94. Sándor K, Kormos V, Botz B, Imreh A, Bölcskei K, Gaszner B, et al. Impaired nocifensive behaviours and mechanical hyperalgesia, but enhanced thermal allodynia in pituitary adenylate cyclase-activating polypeptide deficient mice. Neuropeptides. 2010;44:363–71.
95. Botz B, Imreh A, Sándor K, Elekes K, Szolcsányi J, Reglődi D, et al. Role of pituitary adenylate-cyclase activating polypeptide and Tac1 gene derived tachykinins in sensory, motor and vascular functions under normal and neuropathic conditions. Peptides. 2013;43:105–12.
96. Delgado M, Abad C, Martinez C, Juarranz MG, Leceta J, Ganea D, et al. PACAP in immunity and inflammation. Ann N Y Acad Sci. 2003;992:141–57.
97. Delgado M, Jonakait GM, Ganea D. Vasoactive intestinal peptide and pituitary adenylate cyclase-activating polypeptide inhibit chemokine production in activated microglia. Glia. 2002;39:148–61.
98. Delgado M, Ganea D. Inhibition of endotoxin-induced macrophage chemokine production by VIP and PACAP in vitro and in vivo. Arch Physiol Biochem. 2001;109:377–82.
99. Delgado M, Munoz-Elias EJ, Gomariz RP, Ganea D. Vasoactive intestinal peptide and pituitary adenylate cyclase-activating polypeptide enhance IL-10 production by murine macrophages: in vitro and in vivo studies. J Immunol. 1999;162:1707–16.

100. Stettner M, Lohmann B, Wolffram K, Weinberger JP, Dehmel T, Hartung HP, et al. Interleukin-17 impedes Schwann cell-mediated myelination. J Neuroinflammation. 2014;11:63.
101. Yu R, Wang J, Li J, Wang Y, Zhang H, Chen J, et al. A novel cyclopeptide from the cyclization of PACAP(1-5) with potent activity towards PAC1 attenuates STZ-induced diabetes. Peptides. 2010;31:1062–7.
102. Cheng H, Ding Y, Yu R, Chen J, Wu C. Neuroprotection of a novel cyclopeptide C*HSDGIC* from the cyclization of PACAP (1-5) in cellular and rodent models of retinal ganglion cell apoptosis. PLoS One. 2014;9:e108090.
103. Lamine A, Létourneau M, Doan ND, Maucotel J, Couvineau A, Vaudry H, et al. Characterizations of a synthetic pituitary adenylate cyclase-activating polypeptide analog displaying potent neuroprotective activity and reduced in vivo cardiovascular side effects in a Parkinson's disease model. Neuropharmacology. 2015.

Part IX
PACAP in Sensory Systems

Chapter 29
Sniffing Out a Role for PACAP in the Olfactory System

Mary T. Lucero

Abstract In mammals, the olfactory system comprises a main olfactory epithelium containing olfactory sensory neurons located in the dorsal recesses of the nasal cavity, olfactory bulbs, and olfactory cortex which includes a number of discrete brain regions such as those associated with emotions (lateral amygdala) and memory (entorhinal cortex). An accessory olfactory system of vomeronasal organ and accessory olfactory bulbs connect to the hypothalamic/gonadotrophic axis. PACAP and the high affinity PAC1 receptor are found at high levels throughout these pathways. Recent studies in the developing olfactory epithelium show that functional PACAP signaling is important for proliferation and neuroprotection. Physiological studies of rodent olfactory epithelial slices reveal that PACAP elicits increases in intracellular calcium in olfactory sensory neurons and protects olfactory slices and primary cultures from injury-related cell death. PACAP is also found in the olfactory ensheathing cells that wrap the sensory axons as they leave the olfactory epithelium to synapse in the olfactory bulb. Within olfactory bulb, PACAP is expressed in mitral and tufted cells, which synapse with PAC1 receptor-expressing granule cells and provide the output to higher olfactory centers. Importantly, PACAP promotes release of glutamate and GABA in the developing olfactory bulb network. Physiological studies using PACAP and GABA stimulation show that as early as postnatal days 2–5, the responses of subsets of olfactory granule cells shift from immature to mature profiles. These interesting and robust effects of PACAP make the olfactory system an excellent model for further studies of the roles of PACAP in neurogenesis, neuromodulation, neuroprotection, and neuroregeneration.

Keywords Pituitary adenylate cyclase activating polypeptide • Olfactory bulb • Pleiotropic • Confocal imaging • Calcium oscillations

M.T. Lucero (✉)
Department of Neuroscience and Physiology, American University of the Caribbean School of Medicine, One University Drive at Jordan Road, Cupecoy, Dutch Lowlands, Sint Maarten, The Netherlands
e-mail: mlucero@aucmed.edu

D. Reglodi, A. Tamas (eds.), *Pituitary Adenylate Cyclase Activating Polypeptide — PACAP*, Current Topics in Neurotoxicity 11,
DOI 10.1007/978-3-319-35135-3_29

Introduction

The visual and auditory sensory systems develop elaborate structures (eyes, ears) for collecting specific signals (light, sound) and for delivering these signals to specialized sensory epithelia where the information is transduced from photons or vibrations to the chemical and electrical language used by the brain. In both sensory systems, the sensory epithelium is protected from the external environment so that the majority of the delicate sensory cells (rods and cones, hair cells) will survive a lifetime. By comparison, the olfactory system also develops an elaborate structure for collecting specific odorant signals (nose) and has a sensory epithelium for transducing odorant signals to neural correlates, however the delicate sensory neurons of the olfactory system are separated from the external environment by only a thin layer of mucus. Rather than lasting a lifetime, individual olfactory sensory neurons (OSNs) survive on the order of months and at any given time, about 10 % of the neurons in the olfactory epithelium (OE) are being replaced [1]. The constant and dynamic changes in OSNs due to environmental damage at the level of the OE are matched by the constant turnover of interneurons in the olfactory bulb, the brain region where OSNs make their first synapse. Thus, the olfactory sensory system has been studied for understanding how we detect and process our odorant environment as well as for understanding the basic mechanisms of adult neuronal regeneration and neuroprotection. As expected, no quintessential molecule for regeneration has been identified because the process is complex and requires multifactorial coordinated molecular interactions and levels of activity. An emerging theme is that molecules and signaling systems important in early development are also found in regeneration. Pituitary adenylate cyclase activating polypeptide (PACAP) is a molecule that has been widely studied in the development and neuroprotection of the hippocampus and cerebellum, but has received relatively little attention in the olfactory system. This is surprising given that among the highest levels of PACAP-specific PAC1 receptors are found in olfactory bulb [2, 3]. However, study of PACAP in OE and olfactory bulb is progressing and providing interesting insights into the contributions that PACAP makes in early development, neuroprotection, and regeneration of the olfactory system.

PACAP and PAC1 Receptors

PACAP is a highly conserved molecule found in organisms ranging from invertebrates to man [4–6]. The expression of closely related PACAP peptides have been reported in tunicates [5], frogs [7], fishes [8], chicken [9], reptiles [10], ovine [11], rodents and humans [12]. In all organisms studied, this small (27 or 38 amino acid) alpha-amidated peptide is present at high levels during early development and persists in regions of the nervous system that show high levels of plasticity in adulthood including the olfactory system, and centers for learning and memory.

There are three types of PACAP receptors, PAC1, VPAC1, and VPAC2 [13]. PACAP binds with high affinity (K_D in the low nM range) to all three of these class 2G-protein coupled receptors and the closely related vasoactive intestinal peptide (VIP) binds equally well to VPAC 1 and 2, but the PAC1 receptor has low affinity (μM) for VIP. The PAC1 receptor has extensive alternative splice variants that can be largely grouped into four classes: N-terminal deletions which affect ligand binding and 2nd messenger coupling; intracellular loop variations (hip, hop, null cassettes) which affect ligand binding, transmembrane 4 variations that affect receptor function; and 5′UTR splicing that may affect mRNA expression (for reviews see refs. [14, 15]). Both forms of PACAP (27 and 38 aa) and multiple splice variants of PAC1 receptors are found in the olfactory system [16–19]. This review focuses on PACAP and PAC1 receptors in mammals; however, there is a need for future studies and review of VIP and VPAC receptors in the olfactory system.

Components of the Olfactory System

The mammalian olfactory system is comprised of the peripheral OE within the nasal cavity and the central brain structures which include the olfactory bulb, lateral olfactory tract, anterior olfactory nucleus, olfactory tubercle, piriform cortex, amygdala, and entorhinal cortex. In rats, the OE originates from an olfactory placode at E12 and PACAP immunoreactivity has been visualized in the developing OE as early as E13 [16]. The OE develops into a pseudostratified structure with two types of progenitor basal cells (horizontal and globose), immature sensory neurons, mature sensory neurons, microvillar cells, and glial-like sustentacular cells which span the epithelium from the basement membrane to the apical surface (Fig. 29.1). The olfactory neuroepithelium is avascular but contains the ducts of Bowman's glands that produce the mucus which covers the surface of the epithelium. Just below the basement membrane is the lamina propria which contains the capillaries, Bowman's glands, and the nerve bundles containing axons that emanate from each of the OSNs. Immediately as they penetrate the basement membrane, OSN axons are wrapped by olfactory ensheathing cells (OECs) which support the developing axons to reach their targets in the olfactory bulb. Extrinsic innervation originating from the superior cervical ganglion as well as trigeminal nerve also passes through the lamina propria, penetrates the olfactory basement membrane, and terminates within the OE [20]. This extrinsic innervation affects not only blood flow and mucus secretion [20], but also releases neurotransmitters such as dopamine and norepinephrine [21] which modulate the sensitivity of OSNs to odorant stimulation [22–24].

In addition to the olfactory system which mainly detects volatile odorants, a vomeronasal system is present in most mammals which includes paired vomeronasal organs in the anterior nasal septum containing sensory neurons that project to the accessory olfactory bulb, and is used for detecting aqueous odorants and pheromones such as those found in urine [25]. Although overlaps exist, the vomeronasal

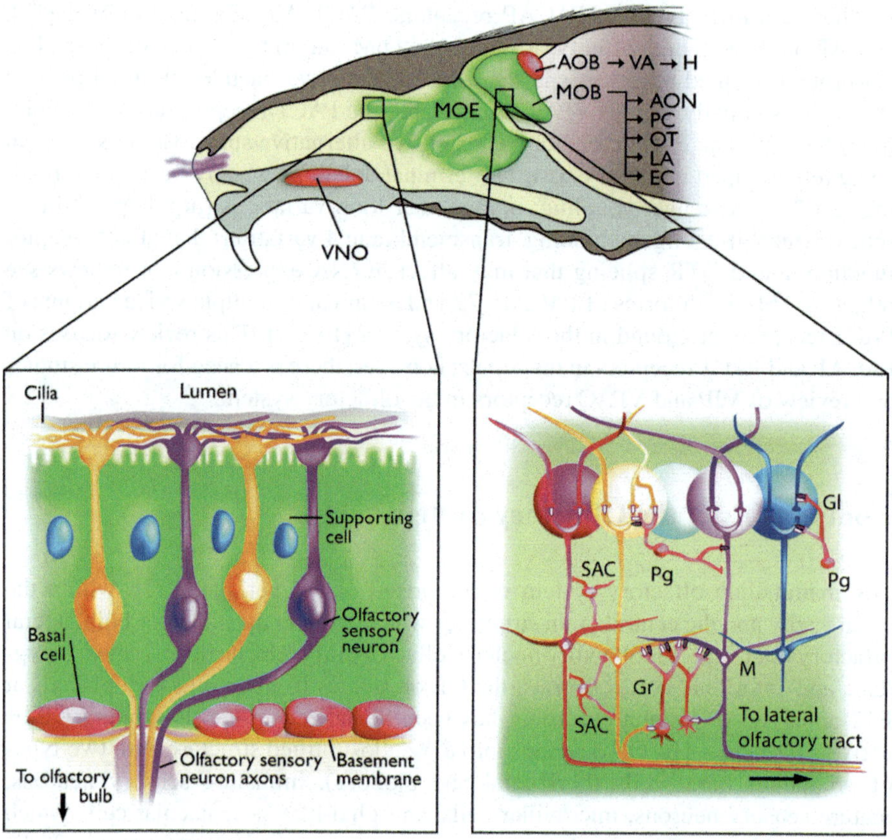

Fig. 29.1 The olfactory system in rodent. *Top:* a simplified sagittal view of the rat head. The main olfactory system is highlighted in *green*, and the accessory olfactory system appears in *red*. The presence of turbinates in the main olfactory epithelium (MOE) increases the surface area of the sensory organ. Axons of sensory neurons in the MOE project to the main olfactory bulb (MOB) and axons of sensory neurons in the vomeronasal organ (VNO) project to the accessory olfactory bulb (AOB). Information provided by pheromone signals is then transmitted to the vomeronasal amygdala (VA) before reaching specific nuclei of the hypothalamus (H). Output projections of the MOB target the primary olfactory cortex that include the anterior olfactory nucleus (AON), the piriform cortex (PC), the olfactory tubercle (OT), the lateral part of the cortical amygdala (LA), and the entorhinal cortex (EC). *Left inset:* schematic illustration of the olfactory epithelium showing the three major cell types and olfactory ensheathing cells (OECs) wrapping axons as they penetrate the basement membrane. Note that there are actually two types of basal cells and that microvillar cells are not shown. *Right inset:* basic circuitry of the main olfactory bulb. Olfactory sensory neurons that express the same odorant receptor gene project their axons to either of two glomeruli (GL) in the olfactory bulb. Four populations of sensory neurons each expressing a different odorant receptor gene are depicted by different colors. Their axons converge on specific glomeruli where they synapse with the dendrite of local interneurons (periglomerular neurons (Pg) and second order neurons (mitral cells, M). The lateral dendrites of mitral cells contact the apical dendrites of granule cells (Gr). Short axon cells (SAC) are bulbar interneurons that contact both apical and lateral dendrites of mitral cells. Modified with permission from Lledo P-M, Gheusi G, Vincent J-D. Information Processing in the Mammalian Olfactory System. *Physiol Rev* 85: 282, Fig. 1, 2005 [46]

system is considered more important for detecting pheromonal cues while the olfactory system detects a wide range of odorants [26, 27]. However, many social behaviors such as mate recognition are influenced by the main olfactory system (OE and olfactory bulb) [28–31].

PACAP and PAC1 Receptors Are Present in Olfactory Neurons, Stem Cells, and Olfactory Ensheathing Cells

PACAP mRNA expression or immunolabeling has been observed in the OE of rodents ranging in age from E13 to adult [16]. PACAP mRNA expression appears to be distributed throughout the OE of E13 mice and more localized to basal cells, neuronal cells and olfactory ensheathing cells in neonatal or adult OE [16, 32–34]. Although the stimulus for release of PACAP within the developing or mature OE has not been identified, physiological studies of PACAP effects on postnatal day 0 (P0) to P4 OE slices showed that functional PAC1 receptors are present on OSNs, and that PACAP elicits release of Ca^{2+} from intracellular stores (Fig. 29.2 [34]).

The antibody labeling of PAC1 receptors in adult rodent shows staining in both sensory neurons and basal cells of the OE [16] suggesting that PACAP may have both autocrine and paracrine effects within the OE. Expression studies using rtPCR reveal multiple splice variants of the PAC1 receptor from both neonates and adults. Specifically, PAC1 splice variants with the normal (N), short (S) and very short (VS) N-terminal domains and one (Hop) or no cassette (R) in the intracellular loop were found in both neonatal and adult OE [16]. Further, rtPCR of clonally derived olfactory placode cell lines show that there seems to be a switch as the cells mature with more immature olfactory neuronal precursors (OP27) expressing three splice variants (N/R,N/Hop1, S/Hop2) and the more mature immediate neuronal precursors (OP6) expressing the same three variants plus S/R, S/Hop1, and N/Hop2 [17].

PACAP Promotes Proliferation and Neuroprotection of Developing OE

The role of PACAP and other amidated peptides in embryonic development of the OE has been studied by examining the expression of PACAP mRNA as well as the enzyme responsible for PACAP's carboxy-terminal amidation called peptidylglycine α-amidating monooxygenase (PAM) [16]. The presence of both PACAP and PAM in the developing OE of E13 mice suggests a role for PACAP in early development. Block of PACAP amidation by inhibiting PAM results in reduced survival of neonatal OE cultures and a thinner OE due to reductions in the immature olfactory neurons [16]. In primary cultures of P2 rat, PACAP38 significantly increased both proliferation and survival [16], while in OE slices from neonatal mice, PACAP protected against tumor necrosis factor α (TNFα)-induced cell death [17]. In primary

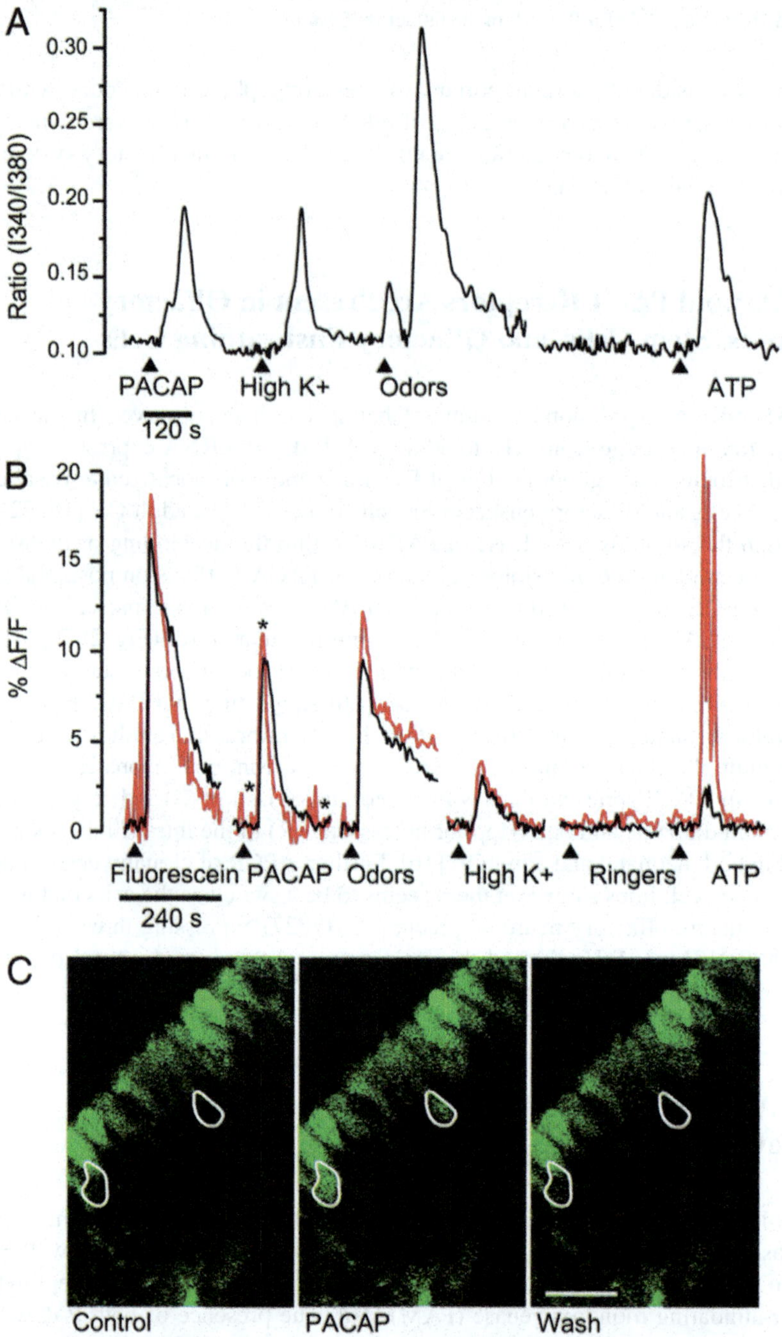

Fig. 29.2 PACAP-elicited calcium responses in adult cultures and neonatal slices of mouse olfactory epithelium. (**a**) Ratiometric fura imaging experiments were performed on adult mouse ORN cultures using six coverslips obtained from three different mice. A representative trace shows increases in intracellular $[Ca^{2+}]_i$ in response to 100 nM PACAP, high K^+ Ringer, 25 μM n-amyl acetate, 50 μM R-carvone (Odors), and 10 μM ATP. (**b**) Two traces obtained from a slice of P4 mouse OE loaded with fluo-4 AM are shown. Test solutions (1:1000 fluorescein; 100 nM PACAP; 25 μM n-amyl acetate; 50 μM R-carvone(Odors); high K^+ Ringer; Ringer control; 10 μM ATP) were loop

cultures of adult rat and mouse OE, PACAP protected against axotomy-induced cell death via inhibition of expression of A-type K^+ channels in OSNs [33, 35]. Thus PACAP has both proliferative and neuro-protective roles in developing and adult OE. Interestingly, PACAP mRNA expression and immunolabeling was also found in the olfactory ensheathing cells (OECs) that envelope the axons of OSNs and PACAP may contribute to the unique regenerative properties of OECs [16, 32–34]. PACAP release from OECs has also been proposed to play a role in maintaining the balance between TNFα and interleukin 6 during bacterial infection of a compromised OE [36].

Collectively, the expression of PACAP and multiple PAC1 receptor splice variants in the OE suggests that PACAP plays a role in the early development and adult maintenance of olfactory structures. However, initial studies with PACAP knockout mice show that at P3, the OE of PACAP$^{-/-}$ mice appears similar to WT although both GAP43 staining (an indicator of immature OSNs) and BRDU-labeling (an indicator of proliferation) are significantly reduced [37]. By P7, there are no longer significant differences suggesting that PACAP is not absolutely required for development of the OE but likely supports the initial efforts of other growth/transcription factors that have been identified as required for OE development [38–42]. Application of chemical ablation (methimazole [43] or methyl bromide gas [44] or olfactory bulbectomy [1, 45] methods for inducing synchronized regeneration of adult OE to PACAP or PAC1 receptor knockout mice would greatly advance understanding of the role of PACAP in the survival and continual regeneration of adult OE. The distinct layering of developmental cell types from basal stem cells to apical adult neurons make the OE an excellent model for further studies investigating the cell types expressing the PAC1 receptor splice variants in neuronal development, neuroprotection, and regeneration.

PACAP and PAC1 Receptors Are Highly Expressed in Olfactory Bulb

The OSNs make their first synapses onto dendrites of both projection neurons (mitral and tufted cells) and interneurons within glomeruli of the olfactory bulb (Fig. 29.1) [46, 47]. Interestingly, PACAP and especially PAC1 receptors are found at high levels in the olfactory bulbs of vertebrates ranging from the most primitive

Fig. 29.2 (continued) injected into bath flow (15- to 20-s delay) at the arrowheads. There was a 4-min wash between test substance applications (data not shown). The data are expressed as a percent change in fluorescence over the baseline (%ΔF/F). Traces were obtained from the ORNs outlined in **c**. (**c**) Pseudocolor confocal images taken at the times indicated by the asterisks on the PACAP trace in **b**. Control (*left*) was before PACAP application, PACAP (*middle*) was during the peak of the response, and Wash (*right*) was following superfusion of PACAP onto the slice. Fluorescent cells along the surface are sustentacular cells that have loaded with fluo-4AM. Scale bar, 50 μm. Reprinted with permission from Hegg CC, Au E, Roskams AJ, Lucero MT. PACAP is present in the olfactory system and evokes calcium transients in olfactory receptor neurons. *J Neurophysiol* 90: 2716 Fig. 5, 2003 [34]

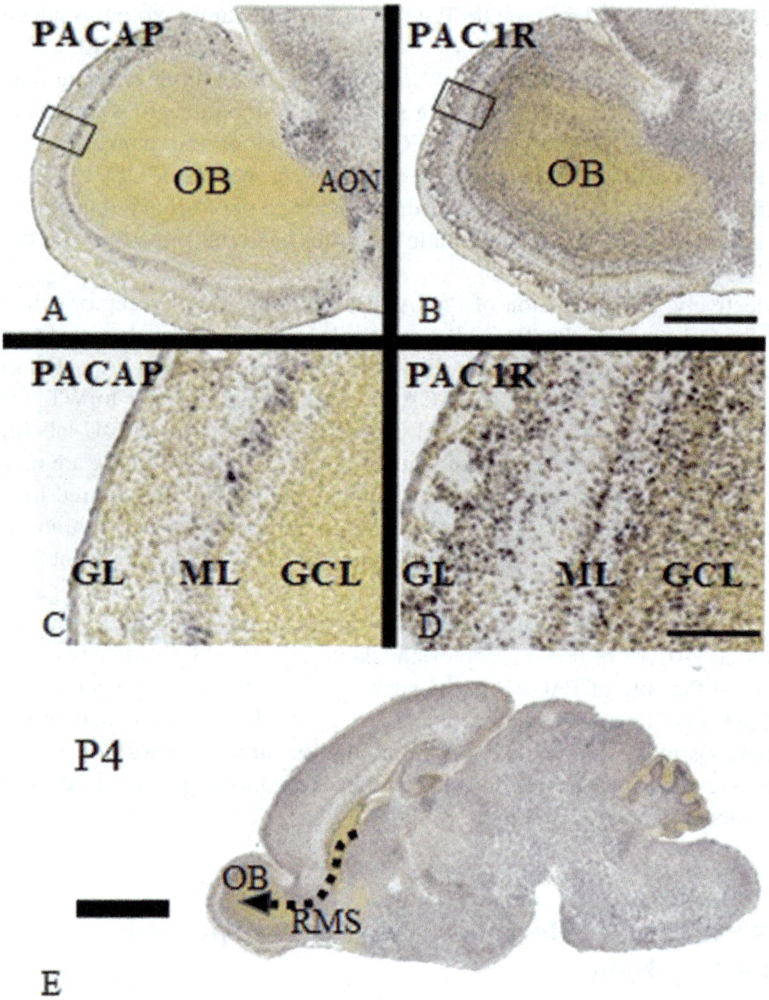

Fig. 29.3 In situ hybridization (ISH) of PACAP and PAC1 receptor expression in P4 mouse brain. (**a**, **c**) In situ hybridization shows PACAP expression in mitral cell layer (ML) and anterior olfactory nucleus (AON). (**b**, **d**) In situ hybridization shows PAC1 receptor expression in glomerular (GL), ML, and granule cell layers (GCL). (**e**) Sagittal section of P4 mouse brain showing migratory path from subventricular zone to olfactory bulb (OB) via the rostral migratory stream (RMS). Scale bars: **b** = 500 μm; **d** = 250 μm, **e** = 1 cm. Modified with permission from the [49]

jawless fish, [5] to mammals [15]. The highest PACAP expression in the olfactory system of mammals is found in the anterior olfactory nucleus which sends efferent projections to the region of extremely high PAC1 receptor expression, the granule cell layer of olfactory bulb (Fig. 29.3; [48–51]). In addition, PACAP is expressed in the mature mitral and tufted projection neurons within the olfactory bulb as well as scattered cells in the granule cell layer [49]. The presence of both PACAP and PAC1 receptors in mitral cells suggests both paracrine and autocrine signaling within the mitral cell layer (Fig. 29.3).

PACAP KO Mice Show Deficits in Olfactory/Vomeronasal Mediated Social Behaviors

Behavioral tests of both PAC1$^{-/-}$ and PACAP$^{-/-}$ mice reveal that loss of PACAP signaling increased motor activity and reduced anxiety behavior leading to increased novelty seeking behaviors [52, 53] while enhanced PACAP signaling in humans is thought to increase anxiety and be associated with post-traumatic stress disorder [54]. Mice in which PAC1 receptors are knocked out are still capable of odorant detection suggesting that PACAP is not an absolute requirement for a functioning olfactory system [53]. However, a number of olfactory and vomeronasal mediated behaviors, especially those dealing with social and sexual recognition are markedly abnormal without PACAP signaling [52, 53, 55]. Specifically, female PAC1$^{-/-}$ mice show reduced affiliative behavior towards males and low fertility [53]. Male PAC1$^{-/-}$ mice show reduced intruder responses, abnormal gender identification, and decreased investigation of female urine [53]. Although our unpublished observations found that PACAP is present in the vomeronasal organ, and the expression profile of PACAP and PAC1 receptors in the accessory olfactory bulb is similar to the main olfactory bulb [49], there are as yet no physiological studies on PACAP in the vomeronasal system.

Physiological and Anatomical Studies of PACAP in Developing Olfactory Bulb

Surprisingly, despite the high levels of PAC1 receptors in olfactory bulb [2, 3, 7, 9, 48–51, 56], and impairment of olfactory/vomeronasal pathways in knockout models, there have been very few physiological studies of PACAP in the olfactory bulb. During brain development, PACAP is expressed in the postmitotic neurons and PAC1 receptors are expressed in regions of ongoing proliferation as well as in later stages of migrating neuroblasts to promote the switch from migration to maturation [50]. Specifically, in migrating cerebellar granule cells, PACAP slows migration and promotes neurite outgrowth via activation of cAMP pathways and release of tissue-type plasminogen activator (tPA) [57, 58]. Likewise, immunohistochemistry studies in developing olfactory bulb, revealed that the PAC1 receptor protein is present before cells begin to migrate from the subventricular zone, is absent in the rostral migratory stream (RMS) and reappears as migrating neuroblasts enter the olfactory bulb [56]. The first biochemical studies of PACAP in olfactory bulb were those using PACAP-induced cAMP signaling as a positive control for cAMP activation and used

PAC1 receptors as a site of modulation for examination of the transduction pathway used by $GABA_B$ receptors in olfactory bulb interneurons [59–61]. Presumably, PACAP was used because of its robust and reproducible cAMP response profile. More recently, a physiological study of PACAP in developing mouse olfactory bulb has confirmed that PACAP and PAC1 receptors are functional in the early postnatal mouse olfactory bulb [62]. Dynamic confocal calcium imaging of genetically identified migrating precursor neurons within P2-P5 slices of mouse olfactory bulb were used to test the hypothesis that PACAP triggers developmentally important intracellular Ca^{2+} concentration ($[Ca^{2+}]_i$) oscillations that help establish the complex circuitry of the olfactory bulb. By recording PACAP-induced calcium responses in the presence and absence of neurotransmission, the authors showed that one third of the PACAP-responsive granule cells are directly activated by PACAP while the majority are recruited through PACAP effects on the developing network via PACAP-induced neurotransmitter release of both glutamate and GABA [62]. Consistent with this functional observation, immunostaining of PAC1 receptors labeled one third of the GABAergic cells of the same age range [62].

In a second series of experiments, the authors used both functionally and genetically identified granule cells to study the time course for development of PACAP responsiveness in mouse olfactory bulb. The functional experiments took advantage of the fact that immature migrating neuroblasts contain high concentrations of intracellular Cl^-, depolarize with GABAergic stimulation, and open voltage-gated Ca^{2+} channels while mature granule cells hyperpolarize to GABA and do not increase $[Ca^{2+}]_i$ [63, 64]. Thus, maturation state and PACAP responsiveness were measured in the same genetically identified cells (based on DLX2 or GAD65 promoter driven expression of td-Tomato) using Ca^{2+} imaging. Consistent with immunological studies of PAC1 receptors in developing olfactory bulb [56], three populations of genetically identified GABAergic cells in slices of P2-P5 mice were identified: (1) immature migrating neuroblasts that do not express functional PAC1 receptors (unresponsive to PACAP) but showed excitatory Ca^{2+} responses to GABA, (2) maturing granule cells that express both functional PAC1 receptors and show excitatory GABA responses, and (3) mature granule cells that respond to PACAP but not GABA. The balance of each cell type shifted over the 4 day window from a majority of immature migrating neuroblasts to a majority of mature granule cells. Interestingly, not all migrating GABAergic neuroblasts showed the same developmental shift in PACAP responsiveness. GABAergic neurons in olfactory bulb express, either GAD65 or GAD67 with some cells expressing both. Genetic labeling of the GABAergic neurons with GFP revealed that during the early postnatal window, only the GAD67 neurons which take up early residency in the granule cell layer of the olfactory bulb expressed PAC1 receptors and switched to mature neurons. The GAD65 neurons which migrate through the granule cell layer until they reach the glomerular layer did not change their developmental profile by P5. These observations further support the findings in cerebellum that PACAP plays a role in the inhibition of migration and neuroblast maturation [57, 65] however, the mechanism for delayed expression of PAC1 receptors in subsets of migrating olfactory bulb neuroblasts has yet to be determined. Based on Allen Developing Mouse Brain

Expression profiles, it appears that once migrating GABAergic neuroblasts reach the glomerular layer, PAC1 receptors are expressed (Fig. 29.3b, d).

The initial studies of PACAP in the development of the OE and olfactory bulb confirm that PACAP signaling in early olfactory development is functional and robust. However, based on its neuroprotective and neuroregenerative properties, PACAP may play an even more important role in protection and regeneration of the olfactory neurons in adults. The insights gained from study of the developmental pathways especially in terms of technical approaches for identifying cell types and functional maturity, lay the foundation for future studies.

Future Studies of PACAP in the Olfactory System

The Ca^{2+} imaging of PACAP responses in olfactory bulb revealed complex calcium signaling that ranged from single transient or sustained increases in $[Ca^{2+}]_i$ to Ca^{2+} oscillations that lasted from minutes to hours (Fig. 29.4). How might these Ca^{2+} oscillations affect migration of olfactory bulb neuroblasts? In cerebellar neurons, reduction of Ca^{2+} oscillation amplitudes was associated with decreased migration and increased differentiation [65]. Do spontaneous oscillations in granule cells of olfactory bulb reduce amplitude in the presence of PACAP? Given the importance of PACAP in the suprachiasmatic nucleus [66] and the observation that PACAP or PAC1 knock out mice have altered circadian rhythms [67–73], what role might the PACAP-induced long term changes in Ca^{2+} oscillations play in the circadian clock intrinsic to the olfactory bulb? PACAP/PAC1 knockout mice will be useful for examining abnormalities in olfactory bulb circuitry especially in terms of migration and integration of GABAergic neuroblasts that continually move from the subventricular zone, through the rostral migratory stream to the olfactory bulb (Fig. 29.3e). As the availability of mice with genetically identified olfactory bulb neuronal subtypes becomes more available, the cell specificity and significance of PACAP signaling in developing and adult olfactory bulb networks will be easier to tease out. In addition, the physiology and circuitry of olfactory bulbs in nonmammalian organisms such as frog [74, 75], catfish [76], and zebrafish [77] are well studied and could become excellent models for physiological studies of PACAP. Some questions that might be addressed in the future include the role of PACAP in signaling between the anterior olfactory nucleus which expresses high levels of PACAP (Fig. 29.3a, c) and the granule cell layer of olfactory bulb which expresses PAC1 receptors (Fig. 29.3b, d).

The observation that PACAP promotes both GABA and glutamate release in olfactory bulb suggests a modulatory effect of PACAP on network excitability [62]. What are the effects of PACAP on modulating release of other neurotransmitters such as norepinephrine, acetylcholine, dopamine, or serotonin? In vivo optogenetic studies could visualize modulatory PACAP effects by optical stimulation of efferent brain regions (locus coeruleus for norepinephrine) and record Ca^{2+} responses from granule cells of olfactory bulb in the presence and absence of PAC1 receptor agonists or antagonists. Similar experiments could be performed in PACAP or PAC1

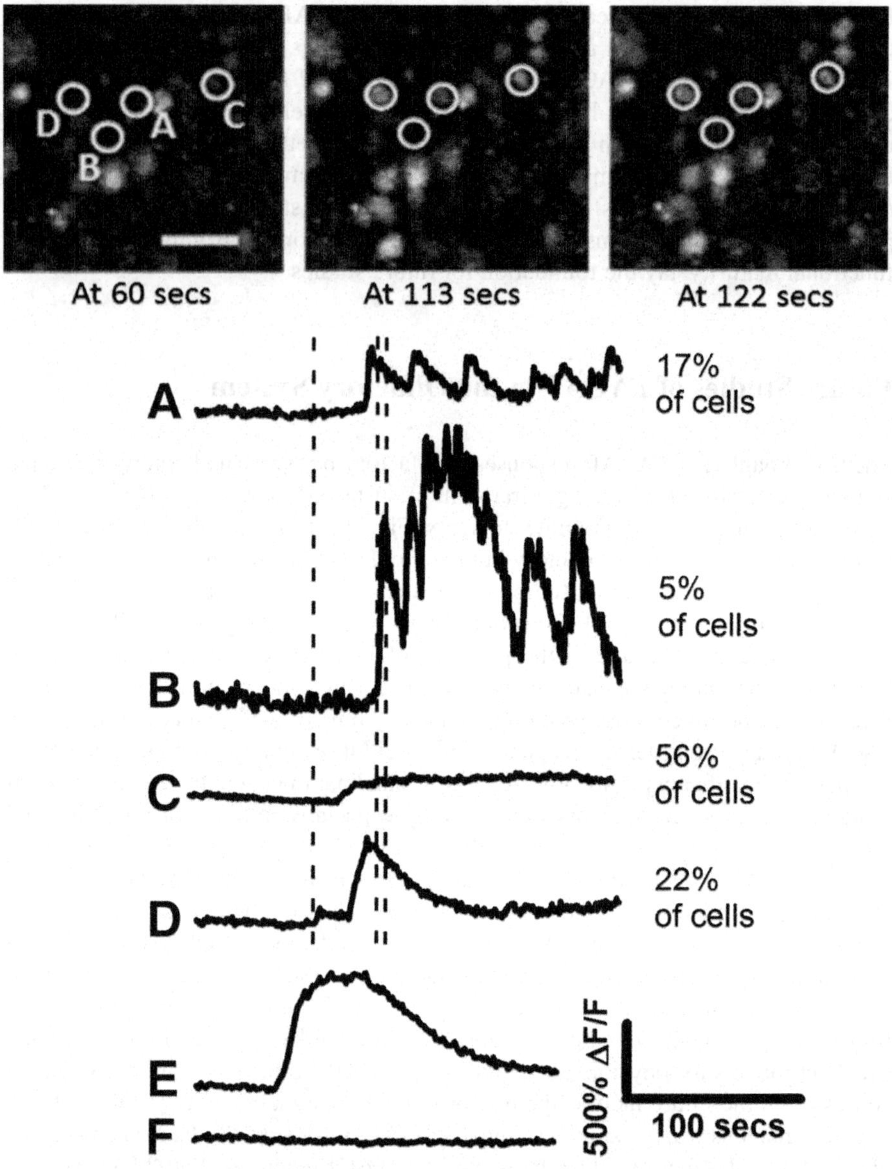

Fig. 29.4 PACAP-induced Ca^{2+} responses in neonatal olfactory bulb (OB) are heterogeneous. A time series of confocal images from a fluo-4-loaded slice of postnatal day (P) 4 mouse OB shows the changes in $[Ca^{2+}]_i$ in response to 40 nM PACAP. Traces in (**A–D**) reflect the % fluorescence intensity changes (%ΔF/F) over time for the regions of interest (ROIs) marked (**A–D**) at top. *Dashed lines* match the images to the time points on the traces. The PACAP-induced calcium oscillations of 1413 cells (from 37 pups) were categorized into four main groups: (**A**) 17 % of cells; slow saw tooth oscillation, (**B**) 5 % of cells; fast saw tooth oscillation, (**C**) 56 % of cells; sustained response, (**D**) 22 % of cells; single transient (returns to baseline within 100 s), (**E**) HK (50 mM K^+) used to confirm the vitality of the cells, and (**F**) 0.1 % BSA is the PACAP vehicle control. Modified from Irwin M, Greig A, Tvrdik P, Lucero MT. PACAP modulation of calcium ion activity in developing granule cells of the neonatal mouse olfactory bulb. *J Neurophysiol* 113: 1237, Fig. 2 2015 [62]

knockout mice to further explore the modulatory effects of PACAP on network excitability. The olfactory bulb is ideally situated for such recordings [78–81].

The abnormal social recognition of PAC1 knockout mice poses interesting questions regarding the links between PACAP signaling in the hypothalamic/pituitary system, the immune system, and the olfactory system. Recently it has been shown that an important component of social recognition involves major urinary proteins (MUPs) excreted in the urine and receptors present in the vomeronasal organ [82]. Sensitivity to social cues can be up or down regulated at the level of sensory neurons. For example, non-estrous female rodents which have high progesterone are not able to smell male odors as well as estrous females with low progesterone due to inhibitory progesterone effects on sensory neurons [83]. Does PACAP somehow interface between its effects on hypothalamic gonadotropin release [84], ovarian progestin production [85] and its effects on odor sensitivity either through altering intracellular Ca^{2+} or cAMP in OSNs or vomeronasal neurons?

The growing interest in understanding the development, regenerative capabilities, circuitry, coding, and modulation of the olfactory system combined with the studies showing that PACAP and PAC1 receptors are abundant and functional in the olfactory system from development to adulthood makes it clear that more studies of PACAP and its receptors should focus on the olfactory system.

References

1. Graziadei PP, Graziadei GA. Neurogenesis and neuron regeneration in the olfactory system of mammals. I. Morphological aspects of differentiation and structural organization of the olfactory sensory neurons. J Neurocytol. 1979;8:1–18.
2. Hashimoto H, Ishihara T, Shigemoto R, Mori K, Nagata S. Molecular cloning and tissue distribution of a receptor for pituitary adenylate cyclase-activating polypeptide. Neuron. 1993;11:333–42.
3. Hashimoto H, Nogi H, Mori K, Ohishi H, Shigemoto R, Yamamoto K, et al. Distribution of the mRNA for a pituitary adenylate cyclase-activating polypeptide receptor in the rat brain: an in situ hybridization study. J Comp Neurol. 1996;371:567–77.
4. Cardoso JCR, Pinto VC, Vieira FA, Clark MS, Power DM. Evolution of secretin family GPCR members in the metazoa. BMC Evol Biol. 2006;6:108.
5. Ng SYL, Chow BKC, Kasamatsu J, Kasahara M, Lee LTO. Agnathan VIP, PACAP and their receptors: ancestral origins of today's highly diversified forms. PLoS One. 2012;7:e44691.
6. Sherwood N, Krueckl SL, McRory JE. The origin and function of the pituitary adenylate cyclase-activating polypeptide (PACAP)/glucagon superfamily. Endocr Rev. 2000;21:619–70.
7. Hu Z, Lelievre V, Rodriguez WI, Cheng JW, Waschek JA. Comparative distributions of pituitary adenylyl cyclase-activating polypeptide and its selective type I receptor mRNA in the frog (Xenopus laevis) brain. Regul Pept. 2002;109:15–26.
8. Mathieu M, Girosi L, Vallarino M, Tagliafierro G. PACAP in developing sensory and peripheral organs of the zebrafish, Danio rerio. Eur J Histochem. 2005;49:167–78.
9. Peeters K, Gerets HH, Princen K, Vandesande F. Molecular cloning and expression of a chicken pituitary adenylate cyclase-activating polypeptide receptor. Mol Brain Res. 1999;71:244–55.
10. Valiante S, Prisco M, Capaldo A, Zambrano I, De Falco M, Andreuccetti P, et al. Molecular characterization and gene expression of the pituitary adenylate cyclase-activating polypeptide (PACAP) in the lizard brain. Brain Res. 2007;1127:66–75.

11. Miyata A, Arimura A, Dahl RR, Minamino N, Uehara A, Jiang L, et al. Isolation of a novel 38 residue-hypothalamic polypeptide which stimulates adenylate cyclase in pituitary cells. Biochem Biophys Res Commun. 1989;164:567–74.
12. Kimura C, Ohkubo S, Ogi K, Hosoya M, Itoh Y, Onda H, et al. A novel peptide which stimulates adenylate cyclase: molecular cloning and characterization of the ovine and human cDNAs. Biochem Biophys Res Commun. 1990;166:81–9.
13. McCulloch DA, MacKenzie CJ, Johnson MS, Robertson DN, Holland PJ, Ronaldson E, et al. Additional signals from VPAC/PAC family receptors. Biochem Soc Trans. 2002;30:441–6.
14. Blechman J, Levkowitz G. Alternative splicing of the pituitary adenylate cyclase-activating polypeptide receptor PAC1: mechanisms of fine tuning of brain activity. Front Endocrinol (Lausanne). 2013;4:1–19.
15. Vaudry D, Falluel-morel A, Bourgault S, Basille M, Burel D, Wurtz O, et al. Pituitary adenylate cyclase-activating polypeptide and its receptors : 20 years after the discovery. Pept Res. 2009;61:283–357.
16. Hansel DE, May V, Eipper BA, Ronnett GV. Pituitary adenylyl cyclase-activating peptides and alpha-amidation in olfactory neurogenesis and neuronal survival in vitro. J Neurosci. 2001;21:4625–36.
17. Kanekar S, Gandham M, Lucero MT. PACAP protects against TNFα-induced cell death in olfactory epithelium and olfactory placodal cell lines. Mol Cell Neurosci. 2010;45:345–54.
18. Peeters K, Gerets HHJ, Arckens L, Vandesande F. Distribution of pituitary adenylate cyclase-activating polypeptide and pituitary adenylate cyclase-activating polypeptide type I receptor mRNA in the chicken brain. J Comp Neurol. 2000;423:66–82.
19. Zhou CJ, Kikuyama S, Shibanuma M, Hirabayashi T, Nakajo S, Arimura A, et al. Cellular distribution of the splice variants of the receptor for pituitary adenylate cyclase-activating polypeptide (PAC1-R) in the rat brain by in situ RT-PCR. Mol Brain Res. 2000;75:150–8.
20. Getchell ML, Getchell TV. Fine structural aspects of secretion and extrinsic innervation in the olfactory mucosa. Microsc Res Tech. 1992;23:111–27.
21. Lucero MT, Squires A. Catecholamine concentrations in rat nasal mucus are modulated by trigeminal stimulation of the nasal cavity. Brain Res. 1998;807:234–6.
22. Hegg CC, Lucero MT. Dopamine reduces odor- and elevated-K(+)-induced calcium responses in mouse olfactory receptor neurons in situ. J Neurophysiol. 2004;91:1492–9.
23. Lucero MT. Peripheral modulation of smell: fact or fiction? Semin Cell Dev Biol. 2013;24:58–70.
24. Vargas G, Lucero MT. Dopamine modulates inwardly rectifying hyperpolarization-activated current (I H) in cultured rat olfactory receptor neurons. J Neurophysiol. 1999;81:149–58.
25. Chamero P, Leinders-Zufall T, Zufall F. From genes to social communication: molecular sensing by the vomeronasal organ. Trends Neurosci. 2012;35:597–606.
26. Eisthen HL. Phylogeny of the vomeronasal system and of receptor cell types in the olfactory and vomeronasal epithelia of vertebrates. Microsc Res Tech. 1992;23:1–21.
27. Kelliher KR. The combined role of the main olfactory and vomeronasal systems in social communication in mammals. Horm Behav. 2007;52:561–70.
28. Keller M, Baum MJ, Brock O, Brennan PA, Bakker J. The main and the accessory olfactory systems interact in the control of mate recognition and sexual behavior. Behav Brain Res. 2009;200:268–76.
29. Restrepo D, Arellano J, Oliva AM, Schaefer ML, Lin W. Emerging views on the distinct but related roles of the main and accessory olfactory systems in responsiveness to chemosensory signals in mice. Horm Behav. 2004;46:247–56.
30. Spehr M, Kelliher KR, Li X-H, Boehm T, Leinders-Zufall T, Zufall F. Essential role of the main olfactory system in social recognition of major histocompatibility complex peptide ligands. J Neurosci. 2006;26:1961–70.
31. Spehr M, Spehr J, Ukhanov K, Kelliher KR, Leinders-Zufall T, Zufall F. Parallel processing of social signals by the mammalian main and accessory olfactory systems. In: Cellular and molecular life sciences. 2006; p. 1476–84.

32. Au E, Roskams AJ. Olfactory ensheathing cells of the lamina propria in vivo and in vitro. Glia. 2003;41:224–36.
33. Han P, Lucero MT. Pituitary adenylate cyclase activating polypeptide reduces A-type K+ currents and caspase activity in cultured adult mouse olfactory neurons. Neuroscience. 2005;134:745–56.
34. Hegg CC, Au E, Roskams AJ, Lucero MT. PACAP is present in the olfactory system and evokes calcium transients in olfactory receptor neurons. J Neurophysiol. 2003;90:2711–9.
35. Han P, Lucero MT. Pituitary adenylate cyclase activating polypeptide reduces expression of Kv1.4 and Kv4.2 subunits underlying A-type K^+ current in adult mouse olfactory neuroepithelia. Neuroscience. 2006;138:411–9.
36. Herbert RP, Harris J, Chong K, Chapman J, West AK, Chuah M. Cytokines and olfactory bulb microglia in response to bacterial challenge in the compromised primary olfactory pathway. J Neuroinflammation. 2012;9:109.
37. Stamm ML. In vivo characterization of pituitary adenylate cyclase-activating polypeptide in mouse olfactory epithelium. Univ Utah. 2011.
38. Carson C, Murdoch B, Roskams AJ. Notch 2 and Notch 1/3 segregate to neuronal and glial lineages of the developing olfactory epithelium. Dev Dyn. 2006;235:1678–88.
39. Elmas C, Erdoğan D, Ozoğul C. Expression of growth factors in fetal human olfactory mucosa during development. Growth Dev Aging. 2003;67:11–25.
40. Hansel DE, Eipper BA, Ronnett GV. Regulation of olfactory neurogenesis by amidated neuropeptides. J Neurosci Res. 2001;66:1–7.
41. Mackay-Sim A, Chuah MI. Neurotrophic factors in the primary olfactory pathway. Prog Neurobiol. 2000;62:527–59.
42. Marks C, Belluscio L, Ibáñez CF. Critical role of GFRα1 in the development and function of the main olfactory system. J Neurosci. 2012;32:17306–20.
43. Genter MB, Deamer NJ, Blake BL, Wesley DS, Levi PE. Olfactory toxicity of methimazole: dose-response and structure-activity studies and characterization of flavin-containing monooxygenase activity in the Long-Evans rat olfactory mucosa. Toxicol Pathol. 1995;23:477–86.
44. Schwob JE, Youngentob SL, Mezza RC. Reconstitution of the rat olfactory epithelium after methyl bromide-induced lesion. J Comp Neurol. 1995;359:15–37.
45. Graziadei PP, Levine RR, Graziadei GA. Regeneration of olfactory axons and synapse formation in the forebrain after bulbectomy in neonatal mice. Proc Natl Acad Sci. 1978;75:5230–4.
46. Lledo P-M, Gheusi G, Vincent J-D. Information processing in the mammalian olfactory system. Physiol Rev. 2005;85:281–317.
47. Shepherd GM, Chen WR, Greer CA. Olfactory bulb. In: Shepherd GM, editor. The synaptic organization of the brain. New York: Oxford University Press; 2004. p. 165–216.
48. Alexandre D, Vaudry H, Jégou S, Anouar Y. Structure and distribution of the mRNAs encoding pituitary adenylate cyclase-activating polypeptide and growth hormone-releasing hormone-like peptide in the frog, Rana ridibunda. J Comp Neurol. 2000;421:234–46.
49. Allen Developing Mouse Brain Atlas [Online]. 2015. http://developingmouse.brain-map.org.
50. Jaworski DM, Proctor MD. Developmental regulation of pituitary adenylate cyclase-activating polypeptide and PAC1 receptor mRNA expression in the rat central nervous system. Dev Brain Res. 2000;120:27–39.
51. Shioda S, Shuto Y, Somogyvari-Vigh A, Legradi G, Onda H, Coy DH, et al. Localization and gene expression of the receptor for pituitary adenylate cyclase-activating polypeptide in the rat brain. Neurosci Res. 1997;28:345–54.
52. Hattori S, Takao K, Tanda K, Toyama K, Shintani N, Baba A, et al. Comprehensive behavioral analysis of pituitary adenylate cyclase-activating polypeptide (PACAP) knockout mice. Front Behav Neurosci. 2012;6:1–18.
53. Nicot A, Otto T, Brabet P, DiCicco-Bloom EM. Altered social behavior in pituitary adenylate cyclase-activating polypeptide type I receptor-deficient mice. J Neurosci. 2004;24:8786–95.
54. Ressler KJ, Mercer KB, Bradley B, Jovanovic T, Mahan A, Kerley K, et al. Post-traumatic stress disorder is associated with PACAP and the PAC1 receptor. Nature. 2011;470:492–7.

55. Gaszner B, Kormos V, Kozicz T, Hashimoto H, Reglodi D, Helyes Z. The behavioral phenotype of pituitary adenylate-cyclase activating polypeptide-deficient mice in anxiety and depression tests is accompanied by blunted c-Fos expression in the bed nucleus of the stria terminalis, central projecting Edinger-Westphal nucleus. Neuroscience. 2012;202:283–99.
56. Matsuno R, Ohtaki H, Nakamachi T, Watanabe J, Yofu S, Hayashi D, et al. Distribution and localization of pituitary adenylate cyclase-activating polypeptide-specific receptor (PAC1R) in the rostral migratory stream of the infant mouse brain. Regul Pept. 2008;145:80–7.
57. Cameron DB, Raoult E, Galas L, Jiang Y, Lee K, Hu T, et al. Role of PACAP in controlling granule cell migration. Cerebellum. 2009;8:433–40.
58. Falluel-Morel A, Vaudry D, Aubert N, Galas L, Benard M, Basille M, et al. PACAP and ceramides exert opposite effects on migration, neurite outgrowth, and cytoskeleton remodeling. Ann N Y Acad Sci. 2006;1070:265–70.
59. Olianas MC, Onali P. Stimulation of guanosine 5'-O-(3-[35S] thiotriphosphate) binding by cholinergic muscarinic receptors in membranes of rat olfactory bulb. J Neurochem. 1996;67:2549–56.
60. Olianas MC, Onali P. Impairment of muscarinic stimulation of adenylyl cyclase by heparin in rat olfactory bulb. Life Sci. 1997;61:515–22.
61. Olianas MC, Onali P. GABA(B) receptor-mediated stimulation of adenylyl cyclase activity in membranes of rat olfactory bulb. Br J Pharmacol. 1999;126:657–64.
62. Irwin M, Greig A, Tvrdik P, Lucero MT. PACAP modulation of calcium ion activity in developing granule cells of the neonatal mouse olfactory bulb. J Neurophysiol. 2015;113:1234–48.
63. Dzhala VI, Kuchibhotla KV, Glykys JC, Kahle KT, Swiercz WB, Feng G, et al. Progressive NKCC1-dependent neuronal chloride accumulation during neonatal seizures. J Neurosci. 2010;30:11745–61.
64. Mejia-Gervacio S, Murray K, Lledo P-M. NKCC1 controls GABAergic signaling and neuroblast migration in the postnatal forebrain. Neural Dev. 2011;6:4.
65. Cameron DB, Galas L, Jiang Y, Raoult E, Vaudry D, Komuro H. Cerebellar cortical-layer-specific control of neuronal migration by pituitary adenylate cyclase-activating polypeptide. Neuroscience. 2007;146:697–712.
66. Michel S, Itri J, Han JH, Gniotczynski K, Colwell CS. Regulation of glutamatergic signalling by PACAP in the mammalian suprachiasmatic nucleus. BMC Neurosci. 2006;7:15.
67. Colwell CS, Michel S, Itri J, Rodriguez W, Tam J, Lelièvre V, et al. Selective deficits in the circadian light response in mice lacking PACAP. Am J Physiol Regul Integr Comp Physiol. 2004;287:R1194–201.
68. Faluhelyi N, Reglodi D, Csernus V. Development of the circadian melatonin rhythm and its responsiveness to PACAP in the embryonic chicken pineal gland. Ann N Y Acad Sci. 2005;1040:305–9.
69. Hannibal J. Roles of PACAP-containing retinal ganglion cells in circadian timing. Int Rev Cytol. 2006;251:1–39.
70. Kawaguchi C, Tanaka K, Isojima Y, Shintani N, Hashimoto H, Baba A, et al. Changes in light-induced phase shift of circadian rhythm in mice lacking PACAP. Biochem Biophys Res Commun. 2003;310:169–75.
71. Mertens I, Husson SJ, Janssen T, Lindemans M, Schoofs L. PACAP and PDF signaling in the regulation of mammalian and insect circadian rhythms. Peptides. 2007;28:1775–83.
72. Nagy AD, Csernus VJ. The role of PACAP in the control of circadian expression of clock genes in the chicken pineal gland. Peptides. 2007;28:1767–74.
73. Racz B, Horvath G, Faluhelyi N, Nagy AD, Tamas A, Kiss P, et al. Effects of PACAP on the circadian changes of signaling pathways in chicken pinealocytes. J Mol Neurosci. 2008;36:220–6.
74. Manzini I, Brase C, Chen T-W, Schild D. Response profiles to amino acid odorants of olfactory glomeruli in larval Xenopus laevis. J Physiol. 2007;581:567–79.
75. Zelles T, Boyd JD, Hardy AB, Delaney KF. Branch-specific Ca^{2+} influx from Na^{+}-dependent dendritic spikes in olfactory granule cells. J Neurosci. 2006;26:30–40.

76. Nikonov AA, Caprio J. Electrophysiological evidence for a chemotopy of biologically relevant odors in the olfactory bulb of the channel catfish. J Neurophysiol. 2001;86:1869–76.
77. Edwards JG, Michel WC. Pharmacological characterization of ionotropic glutamate receptors in the zebrafish olfactory bulb. Neuroscience. 2003;122:1037–47.
78. Bhalla US, Bower JM. Multiday recordings from olfactory bulb neurons in awake freely moving rats: spatially and temporally organized variability in odorant response properties. J Comput Neurosci. 1997;4:221–56.
79. Homma R, Kovalchuk Y, Konnerth A, Cohen LB, Garaschuk O. In vivo functional properties of juxtaglomerular neurons in the mouse olfactory bulb. Front Neural Circuits. 2013;7:23.
80. Rubin BD, Katz LC. Optical imaging of odorant representations in the mammalian olfactory bulb. Neuron. 1999;23:499–511.
81. Wachowiak M, Economo MN, Díaz-Quesada M, Brunert D, Wesson DW, White JA, et al. Optical dissection of odor information processing in vivo using GCaMPs expressed in specified cell types of the olfactory bulb. Ann Intern Med. 2013;158:5285–8300.
82. Janotová K, Stopka P. Mechanisms of chemical communication: the role of major urinary proteins. Folia Zool. 2009;58:41–55.
83. Dey S, Chamero P, Pru JK, Chien M-S, Ibarra-Soria X, Spencer KR, et al. Cyclic regulation of sensory perception by a female hormone alters behavior. Cell. 2015;161:1334–44.
84. Petersen B, Buchfelder M, Fahlbusch R, Adams EF. Pituitary adenylate cyclase-activating polypeptide directly stimulates LH and FSH secretion by human pituitary gonadotrophinomas. Exp Clin Endocrinol Diabetes. 1996;104:250–5.
85. Kotani E, Usuki S, Kubo T. Effect of pituitary adenylate cyclase-activating polypeptide (PACAP) on progestin biosynthesis in cultured granulosa cells from rat ovary and expression of mRNA encoding PACAP type IA receptor. J Reprod Fertil. 1998;112:107–14.

Chapter 30
Protective Effects of PACAP in the Retina

Tamas Atlasz, Alexandra Vaczy, Dora Werling, Peter Kiss, Andrea Tamas, Krisztina Kovacs, Eszter Fabian, Timea Kvarik, Barbara Mammel, Bese Danyadi, Emese Lokos, and Dora Reglodi

Abstract Pituitary adenylate cyclase activating polypeptide (PACAP) is a widespread neuropeptide that is well known for its general cytoprotective effects in different neuronal injuries, such as traumatic brain and spinal cord injury, models of neurodegenerative diseases, and cerebral ischemia. PACAP and its receptors also occur in the retina. In this review, we summarize the retinoprotective effects of PACAP. In vitro, PACAP is protective against glutamate, thapsigargin, anisomycin, oxidative stress, UV light, high glucose, inflammation, and anoxia. Both the neural retina and the pigment epithelial cells can be protected by PACAP in various experimental paradigms. In vivo, the protective effects of intravitreal PACAP treatment have been shown in the following models in rats and mice: excitotoxic injury induced by glutamate, *N*-methyl-D-aspartate (NMDA) or kainate, ischemic injury induced by carotid artery ligation and high intraocular pressure, degeneration caused by UV-A light, optic nerve transection, and streptozotocin-induced diabetic retinopathy as well as retinopathy of prematurity. Molecular biological methods have revealed that PACAP activates anti-apoptotic, while inhibits pro-apoptotic signaling pathways, and it also stimulates an anti-inflammatory environment in the retina. Altogether, PACAP is suggested to be a potential therapeutic retinoprotective agent in various retinal diseases.

Keywords Retina • Diabetic retinopathy • Ischemia • Excitotoxic injury • Retinopathy of prematurity

---

T. Atlasz (✉)
Department of Anatomy, MTA-PTE "Lendulet" PACAP Research Team, University of Pecs, Pecs, Hungary

Department of Sportbiology, University of Pecs, Pecs, Hungary

Janos Szentagothai Research Center, University of Pecs, Pecs, Hungary
e-mail: attam33@gmail.com

A. Vaczy • D. Werling • P. Kiss • A. Tamas • E. Fabian • T. Kvarik • B. Mammel
B. Danyadi • E. Lokos • D. Reglodi
Department of Anatomy, MTA-PTE "Lendulet" PACAP Research Team, University of Pecs, Pecs, Hungary

K. Kovacs
Department of Biochemistry and Medical Chemistry, University of Pecs, Pecs, Hungary

D. Reglodi, A. Tamas (eds.), *Pituitary Adenylate Cyclase Activating Polypeptide — PACAP*, Current Topics in Neurotoxicity 11,
DOI 10.1007/978-3-319-35135-3_30

Abbreviations

6-OHDA	6-hydroxydopamine
AIF	Apoptosis inducing factor
AMPA receptor	α-amino-3-hydroxy-5-methyl-4-isoxazolepropionic acid receptor
Bad	Bcl-2-associated death promoter
Bax	Bcl-2-associated X protein
BBB	Blood–brain barrier
BCCAO	Bilateral common carotid artery occlusion
Bcl-2	B-cell lymphoma 2
Bcl-xl	B-cell lymphoma-extra large
cAMP/PKA/CREB	Cyclic AMP/protein-kinase A/cAMP response element-binding protein
CIAP-1	Cellular inhibitor of apoptosis protein-1
CINC	Cytokine-induced neutrophil chemoattractant
CNS	Central nervous system
CNTF	Ciliary neurotrophic factor
CtBP2	C-terminal binding protein 2
DR	Diabetic retinopathy
eNOS	Endothelial nitric oxide synthase
ERG	Electroretinography
ERK	Extracellular signal-regulated kinase
Fas TNFR	Fas tumor necrosis factor receptor
FITC	Fluorescein isothiocyanate
GCL	Ganglion cell layer
GFAP	Glial fibrillary acidic protein
HIF1α, 2α, 3α	Hypoxia-inducible factor 1-alpha, 2-alpha, 3-alpha
HO-1	Heme oxygenase-1
HSP-27	Heat shock protein-27
IL-1β	Interleukin-1 beta
INL	Inner nuclear layer
IP3	Inositol trisphosphate
IPL	Inner plexiform layer
JNK	c-Jun N-terminal kinase
LPS	Lipopolysaccharide
MAPK	Mitogen-activated protein kinase
MIP-1	Macrophage inflammatory protein 1
MSG	Monosodium glutamate
NBL	Neuroblast layer
NF-$_{\kappa}$B	Nuclear factor-$_{\kappa}$B
NFL	Nerve fiber layer
NMDA	*N*-methyl-D-aspartate
NPY	Neuropeptide Y

OLM-ILM	Outer limiting membrane-inner limiting membrane
ONL	Outer nuclear layer
OPL	Outer plexiform layer
Ops	Oscillatory potentials
PAC1	PACAP receptor 1
PACAP	Pituitary adenylate cyclase activating polypeptide
PACAP KO	PACAP knockout
PKCα	Protein kinase C alpha
PLC gamma-1	Phospholipase C gamma-1
RANTES	Regulated on activation normal T cell expressed and secreted
RGCs	Retinal ganglion cells
ROP	Retinopathy of prematurity
ROS	Reactive oxygen species
RPE	Retinal pigment epithelial
RSK1/2	Ribosomal s6 kinase 1, 2
sICAM	Soluble intercellular adhesion molecule
STAT4	Signal transducer and activator of transcription 4
TGF-β	Transforming growth factor beta
TH	Tyrosine hydroxylase
TIMP-1	Tissue inhibitor metallopeptidase inhibitor-1
Trail R2 DR5	TRAIL receptor 2—death receptor 5
TUNEL	Terminal deoxynucleotidyl transferase dUTP nick end labeling
UV-A, B	Ultraviolet A, ultraviolet B
VEGF	Vascular endothelial growth factor
VGAT	Vesicular GABA transporter
VGLUT1	Vesicular glutamate transporter 1
VIP	Vasoactive intestinal peptide
VPAC1	Vasoactive intestinal peptide (VIP) receptor type 1
VPAC2	Vasoactive intestinal peptide (VIP) receptor type 2
ZO-1	Zona occludens protein 1

Pituitary Adenylate Cyclase Activating Polypeptide (PACAP)

Pituitary adenylate cyclase activating polypeptide (PACAP) was isolated from ovine hypothalamus based on its stimulating effect on adenylate cyclase enzyme in anterior pituitary cells [1, 2]. PACAP has two isoforms: PACAP1-27 and PACAP1-38, containing 27 and 38 amino acids, respectively. PACAP is a member of the vasoactive intestinal peptide (VIP)/secretin/glucagon superfamily [3]. Its receptors can be divided into two groups, PAC1 receptor and VPAC receptors (VPAC1 and VPAC2) [4]. The biological effects of PACAP are diverse [4], but one of the most studied actions of the peptide is its potent neuroprotective effect. Earlier studies have shown that PACAP protects neurons in vitro against various toxic agents (glutamate, 6-OHDA, oxidative stress, etc.). The neuroprotective role of PACAP has been also demonstrated in vivo in different animal models, such as cerebral ischemia, traumatic brain injury, spinal cord injury, and Parkinson's disease. These neuroprotective effects have been summarized in several reviews by different authors [4–11].

PACAP is also protective in sensory organs, including the auditory and visual systems [12–14]. The retina is a special tissue, an extension of the central nervous system (CNS) containing different neurons, and it also provides an excellent model to measure neuronal degenerations. PACAP has been shown to protect the different neurons of the retina in numerous in vitro and in vivo models of retinal degenerations. The purpose of the present review is to summarize the current knowledge of the retinoprotective functions of PACAP. The retinoprotective effects of PACAP provide a new, exciting area of research, and several novel data have been published since the appearance of the first two extensive reviews on the retinoprotective effects of PACAP [12, 15]. Therefore, we briefly summarize retinoprotective effects until 2010 and discuss findings in the last 5 years more in detail.

Distribution of PACAP in the Retina

Several studies have described the presence of PACAP and PACAP receptors in the rat retina. PACAP immunopositive cell bodies were found in amacrine, horizontal cells in the inner nuclear layer (INL), in the ganglion cell layer (GCL) and also in the Muller glial cells. PACAP positive nerve fibers were also present in the nerve fiber layer (NFL), and also in the inner plexiform layer (IPL), but no PACAP positivity was found in the outer nuclear layer (ONL) or the retinal pigment epithelium (RPE) layer [16–20]. The presence of PACAP was also found in the mouse [21], teleost [22], turtle [23], and chicken retina [24]. PACAP appearance showed in a subset of retinal ganglion cells, which cells are photosensitive due to expression of melanopsin. The PACAP expression in this subset of ganglion cells is partly controlled by dopamine, while it is not affected by photoreceptor degeneration [25, 26]. Melanopsin-containing retinal ganglion cells are known to be more resistant in various types of degeneration, such as axonal injury [27], and it has been proven by numerous studies that these melanopsin-containing cells almost all co-store PACAP, which might contribute to the highly resistant nature of these cells [28–30]. PACAP- and melanopsin-containing retinal ganglion cells play an important role in several light-activated non-image-forming functions of the retina and the complimentary signaling due to PACAP neurotransmission has been shown to be important in addition to the glutamatergic signaling [31, 32].

PACAP immunopositivity was also found in the limboretinal pathway [33]. PACAP and its receptors were also described in the developing retina of mammals [34]. In developing chicken retina PACAP was expressed in the INL from embryonic day 8 (E8) [35]. In zebrafish during the pharyngula period, at 24 h of post fertilization PACAP-like immunoreactivity (PACAP-LI) was detected in the superficial layer of the retina, and during the hatching period, at 72 h of post fertilization in the GCL [36]. PACAP-LI was absent from the retina in the zebrafish at day 13 of larval development, suggesting that PACAP may have a role in the control of cell differentiation and proliferation [36, 37]. In the rat, PACAP mRNA was expressed in the GCL in the developing retina at E20 [38].

Distribution of PACAP Receptors in the Retina

Earlier studies showed the distribution and localization of PACAP receptors in the retina. Nilsson et al. [39] characterized the receptors of PACAP and their coupling to adenylate cyclase in albino rat eye, and D'Agata and Cavallaro also described the expression of PACAP/VIP receptor variants and their coupling to phospholipase C in the rat retina [16]. In chicken retina at early stage (E6) PAC1 receptor and its mRNA can be found [35]. The mRNA of PACAP receptors and the protein expression have also been shown in all the layers of the neonatal rat retina [40]. Seki and coworkers described the localization and strong expression of PAC1 receptor and its mRNA in different retinal layers (NFL, INL, GCL), weaker expression in the ONL, OPL, IPL, and the outer segments of the photoreceptors in the rat retina [41, 42]. Moreover, different PAC1 receptor mRNA variants (short, hop) were found in the cell bodies and the processes of ganglion and amacrine cells [41]. Finally, PAC1 receptor has been described in Muller glial cells [19, 43].

PACAP also plays a role in retinal development. The expression of the specific PAC1 receptor and VPAC receptors has been confirmed in the developing mammalian retina, including the retinal progenitor cells [44]. A shift in PAC1 receptor isoform expression (null, hip, hop1, hiphop1) during development might explain the different roles exerted during differentiation of retinal neural elements [45, 46].

PACAP controls the proliferation of retinal progenitor cells through the activation of all 3 receptors, and the cAMP/PKA/CREB pathway. This activation resulted in downregulation of cyclin D1 and the transcription factor Klf4, while $p27^{kip1}$ protein content did not change [44, 47]. The overall effect of these changes was an antiproliferative effect.

Protective Effects of PACAP in the Retina In Vitro

The in vitro protective effects are summarized in Table 30.1.

In Vitro Protective Effects in the Neural Retina

The first in vitro studies presented that VIP, a related peptide to PACAP [48], or PACAP1-27 and PACAP1-38 [49] have retinoprotective effect against glutamate toxicity using cultured retinal neurons. PACAP1-27, PACAP1-38, and VIP increased

Table 30.1 Summary of the effects of PACAP in the retina in vitro

Cell type	Treatments	Observed effect	Reference
Retinal cells from pups	Glutamate-induced toxicity, in vitro PACAP1-27 and PACAP1-38 (10, 100 nmol–1 μmol, dose dependent), cultured	↑: Intracellular Ca^{2+} and no effect by PACAP1-27, PACAP1-38 ↑: cAMP by VIP ↑: MAPK and cAMP by PACAPs, antagonized by PACAP6-38 (PACAP antagonist) and H-89 (PKA inhibitor)	[49]
Retinal explants from newborn	Cell death by anisomycin (1 μg/ml) within the neuroblastic layer (NBL) 1–10 nM PACAP1-38	↓: TUNEL positive and pyknotic cells by PACAP1-38, PACAP1-27, or maxadilan (PAC1 receptor agonist), blocked by PACAP6-38 or Maxd.4 (PAC1 receptor antagonist) ↑: pCREB by PACAP1-38; cAMP but not IP3 by PACAP, pyknotic profiles blocked by Rp-CAMPS (PKA inhibitor)	[40]
Photoreceptor cells	Cell death by thapsigargin (10 nmol PACAP1-38)	↓: Dead cells	[40]
Turtle horizontal cells	Anoxic condition (PACAP1-38, 0.165 μmol)	↑: Light response time dependent	[50]
Developing retinal explant	PACAP1-38, 10 nmol	↑: cAMP, p-CREB, nuclear Klf4 ↓: cyclin D1 No change in $p27^{kip1}$	[44, 47]
RGC-5 retinal ganglion cells	PACAP1-5 (1–100 nmol); UV-B-induced injury	↑: Viability, bcl-2 ↓: Bax	[73]
RGC-5 retinal ganglion cells	Cyclic PACAP1-5 (1–100 μmol), UV-B-induced injury	↑: Viability, bcl-2 Maintained structural integrity ↓: Apoptosis, ROS, bax, caspase-3	[73, 91]
ARPE19, human	PACAP1-38 0.1–1 μmol (IL-1β stimulation)	↓: The IL-1beta stimulated IL-6, IL-8, and MCP-1 mRNA and protein l, NF-kB	[51]
ARPE19, human	Oxidative stress by H_2O_2 (PACAP1-38: 10 pmol–1 μmol)	↑: Viable cells, blocked by PACAP6-38 and reduced by PI3K/Akt inhibitor but not MAPK inhibitors ↓: Apoptosis	[52]
ARPE19, human	Oxidative stress by H_2O_2 PACAP1-38 10–100 nM	↑: pERK/CREB; Akt; Lyn; Yes; Src; ↓: p-p38; p-JNK;Bad; Bax; Trail R2 DR5; FADD; Fas TNFR; HIF1-alpha; HO-1; cIAP-1; HSP-27; SMAC/Diablo; p53; eNOS, Paxillin, PLCgamma-1, STAT4, RSK1/2	[53]

(continued)

Table 30.1 (continued)

Cell type	Treatments	Observed effect	Reference
ARPE19, human	Normal and high glucose and Il-β1, + 100 nmol PACAP1-38 or VIP	↑: Transendothelial resistance, claudin and Zo-1 tight junctional protein expression ↓: Blood–retinal barrier permeability	[55]
ARPE19, human	PACAP1-38 1 and 10 μmol PACAP6-38 1 and 10 μmol	↑: Semaphorin4a mRNA, protein ↑: pERK PACAP6-38 blocked these effects and ↑: IL-6	[54]
Retinoblastoma Y79 cells, human	PACAP1-38 and PACAP6-38 nmol and 1–5 μmol	Nanomolar: no effect on cell viability Micromolar: ↓: Cell viability, further ↓: with PACAP6-38; no effect: PACAP1-27 and maxadilan	[115]

cAMP, and the protective effects of VIP or PACAP isoforms were antagonized by VIP6-28 (VIP antagonist) or PACAP6-38 (PACAP antagonist), respectively. In addition, H-89 (PKA inhibitor) also decreased the protective effects of VIP or PACAP1-27, PACAP1-38.

Silveira et al. [40] found that anisomycin (1 μg/ml) inhibited protein synthesis and induced cell death in the neuroblastic layer of retinal explants from newborn rats. PACAP1-27 or PACAP1-38 dose-dependently prevented the anisomycin-induced cell death. One nanomolar PACAP1-27 or PACAP1-38 administration in parallel significantly reduced the number of dead cells with the terminal deoxynucleotidyl transferase dUTP nick end labeling (TUNEL) procedure and the number of pyknotic profiles per mm^2 in the neuroblast layer (NBL). A similar retinoprotective effect was observed with PAC1 receptor agonist maxadilan, but not with glucagon. Neuroprotection by PACAP1-38 was also measured in photoreceptor cell death caused by thapsigargin in the ONL [40]. In the retinal explant model, PACAP6-38 and the specific PAC1 receptor antagonist Maxd.4 antagonized the neuroprotective effect. Molecular and immunohistochemical analysis demonstrated that the retinoprotective effects are most probably mediated by the PAC1 receptor, which is expressed in all the retinal layers of the neonatal retina. Treatment of PACAP1-38 induced CREB immunoreactivity in the NBL. PACAP1-38 administration also increased cAMP, but not IP3, and a cAMP-dependent protein kinase inhibitor Rp-cAMPS could block the retinoprotective effects of PACAP1-38.

In adult turtles, horizontal cells showed light responses 18, 22, 42, and 46 h after removal of the eyes. Rabl et al. [50] demonstrated that the amplitudes of light responses were larger in slices at all time-points in PACAP1-38-containing solution (0.165 μM PACAP) compared to control retinal slices in anoxic condition. These results provide evidence that PACAP1-38 has neuroprotective effects in hypoxic/anoxic conditions in the adult turtle retina in vitro.

In Vitro Protective Effects in the Retinal Pigment Epithelial Cells

Retinal pigment epithelial (RPE) cells play an essential role in photoreceptor survival and metabolism. Several diseases are associated with malfunction of the pigment cells. The first results concerning the effects of PACAP came from Zhang et al. [51]. Presence of mRNA for PAC1 and VPAC1 receptors was confirmed in unstimulated human RPE cells (ARPE19). VPAC2 mRNA was expressed in cells stimulated with IL-1β. PACAP treatment at 0.1–1 μM concentrations inhibited the IL-1 β-stimulated IL-6, IL-8, and MCP-1 mRNA and protein levels. The dense immunofluorescence of NF-kB in the nucleus after stimulation with IL-1β was decreased by PACAP [51]. Studies following these first results in pigment cells revealed several other effects of PACAP in these retinal cells.

ARPE19 cells exposed to PACAP1-38 could be rescued against oxidative stress induced by H_2O_2 in a dose-dependent manner [52]. Highest efficacy was observed at 100 nM PACAP concentration, but it was effective in a range 10 pM–1 μM, while not effective at 1 pM. Using flow cytometric and JC-1 assay, PACAP treatment was proven to reduce apoptotic cell death. PACAP1-38 and 6-38 alone had no effect on cell survival. The survival-promoting effect was inhibited by PI3K/Akt inhibitors, but not affected by MAPK inhibitors [52]. A subsequent study investigated the signaling pathways in oxidative stress-induced pigment epithelial cell damage [53]. Western blot results revealed that PACAP treatment could activate the protective ERK/CREB and Akt pathways, while it inhibited the phosphorylation of p38 MAPK and JNK. Several cytokines and other apoptotic markers were altered upon oxidative stress, as measured by cytokine and phospho-kinase arrays. PACAP administration counteracted several of these changes. 10 and 100 nM PACAP decreased the oxidative stress-induced increases of the proinflammatory and/or proapoptotic factors Bad; Bax; Trail R2 DR5; FADD; Fas TNFR; HIF-1α; HO-1; cIAP-1; HSP-27; SMAC/Diablo; p53; eNOS; Paxillin; PLCgamma-1; STAT4; RSK1/2. In contrast, PACAP could increase the protective Lyn; Yes; and Src.

A more recent study has reported that PACAP regulates semaphorin4A expression from RPE cells [54]. The semaphorin protein family has members that influence brain and retinal development, by providing neural guidance. Semaphorin4A is expressed in the GCL, INL and RPE cell during the developmental period of contact-building between photoreceptors and RPE cells. Human ARPE cells were cocultured with PC12 cells (prototype neuronal cells) based on a collagen vitrigel membrane. The authors found that in the presence of neural cells the protein and mRNA levels of semaphorin4A (but not other semaphorins) were increased, an effect mimicked by PACAP (but not VIP, NPY, or enkephalin). Furthermore, this effect was also associated with ERK phosphorylation (but not that of JNK or p38 MAPK). All these effects of PACAP were blocked by the PACAP antagonist PACAP6-38. Furthermore, PACAP6-38 downregulated the cytokine IL-6 expression, indicating that endogenous PACAP participates in the neural cell-induced IL-6 production (other cytokines out of the tested ones were not influenced by the neuronal co-culture). These results suggest that the PACAP secreted by neural cells regulates several factors secreted by RPE cells, important for retinal development and homeostasis [54]. The local secretion of PACAP necessary for these effects is supported by the high concentration used in this study (1 μM and 10 μM PACAP, lower concentrations were not reported), a concentration much higher than the ones reported to exert protective effects against oxidative stress [52, 53].

Another study investigated the potential of PACAP and VIP against the disruption of tight junctions in the retina, a major factor in macular edema developing in diabetic retinopathy [55]. ARPE19 cells were cultured either in normal glucose or in high glucose in combination with the pro-inflammatory IL-1β. Effects on permeability were evaluated by measuring both apical-to-basolateral movements of fluorescein isothiocyanate (FITC) dextran and transepithelial electrical resistance [55]. Results of these experiments demonstrated that normal glucose + IL-1β and, to a greater extent, high glucose + IL-1β significantly increased FITC-dextran diffusion, paralleled by decreased electrical resistance. PACAP or VIP reversed these effects. Furthermore, high glucose + IL-1β-induced reduction of claudin-1 and ZO-1 expression was reversed by PACAP and VIP. Occludin expression was not affected in any of the conditions tested. Altogether, these findings show that both peptides counteract high glucose + IL-1β-induced damage in ARPE19 cells, suggesting that they might be relevant to the maintenance of outer blood–retinal barrier function in edema accompanying diabetic retinopathy [55]. These results are in accordance with our earlier observations in an in vitro blood–brain barrier (BBB) model [56]. We found that PACAP treatment improved the barrier properties of the brain endothelium. PACAP induced an increase in the transendothelial electrical resistance, which is the most important marker of the tightness of the tight junctions. Moreover, PACAP had a protective role against glucose deprivation- and oxidative stress-induced junctional damage in microvascular brain endothelial cells [56].

Protective Effects of PACAP in the Retina In Vivo

The in vivo protective effects are summarized in Table 30.2.

Excitotoxic Injury in the Retina

Glutamate, a nonessential amino acid, acts as an excitatory neurotransmitter in the retina under normal conditions. However, glutamate in high concentrations can be toxic to the retinal cells by overstimulation of the glutamate receptors. Monosodium glutamate (MSG) administration given systemically passes the BBB of newborn rats and so it can, can injure the retina [57] and lead to loss of the entire inner retinal layers [58]. The excitotoxic damage of the retina is a main factor in retinal diseases, such as glaucoma and ischemia-induced retinopathy. We have previously shown in

Table 30.2 Summary of the effects of PACAP in the retina in vivo

Lesion	Treatment	Observed effect	Reference
Excitotoxicity			
Glutamate induced excitotoxic injury in newborn rats	PACAP1-38 (1–100 pmol/5 μl) intravitreal	↑: Thickness of the different layers by 100 pmol PACAP1-38	[58, 59, 61]
	PACAP1-38 (100 pmol/5 μl)	↓: Caspase-3, JNK, cytochrome C, AIF ↑: Phospho-Bad, p-ERK1/2, p-CREB, phospho-PKA, Bcl-xl, 14-3-3 by PACAP1-38 ↑: Cytochrome C, caspase-3, JNK, phospho-Bad, phospho-PKA, Bcl-xl, 14-3-3 by the antagonist PACAP6-38	[65–67]
	PACAP1-38 100 pmol/5 μl intravitreal	PACAP1-38 (not subtype specific) effects: ↑: VGLUT1, TH, GABA, parvalbumin, calbindin, calretinin, PKCα expression	[64]
	PACAP1-27 (100 pmol/5 μl) PACAP6-38, PACAP6-27 (1000 pmol/5 μl) intravitreal	↑: Thickness of the different layers by PACAP1-27 ↓: Thickness of the different layers by PACAP6-38 or PACAP6-27	[60]
	100 pmol PACAP 1–38 intravitreal	↓:Caspase-3, caspase-9	[68]
	PACAP1-38 100 pmol/5 μl intravitreal	Functional improvement	[75]
Kainic acid induced injury in adult rat retina	PACAP1-38 10 pmol pretreatment, intravitreal	↓: Loss of ganglion cells	[43]
NMDA rats	–	↓: VIP immunoreactivity No change: PACAP immunoreactivity	[72]
NMDA rats	Cyclic PACAP1-5, 100 fmol–10 μmol	↓: Ganglion cell loss, apoptosis	[73]
NMDA mice	PACAP1-38 10^{-10} M	↓: Ganglion cell loss, apoptosis ↓: Ganglion cell loss in PACAP heterozygous knockout mice (Increased cell loss in knockout animals)	[107]

Lesion	Treatment	Observed effect	Reference
NMDA mice	PACAP1-38 10^{-8-12} M	↓: Ganglion cell loss ↑: Number of microglia/macrophages, deactivation phenotype of microglia/macrophages, IL-10 mRNA, TGFβ mRNA (attenuated by PACAP6-38)	[74]
Ischemia			
(BCCAO) permanent occlusion of common carotid arteries in adult rat	PACAP1-38 (100 pmol/5 μl) Intravitreal	↑: General cytoprotective mechanism; thickness of retinal layers by PACAP1-38; ↑: VGLUT1, TH, GABA, parvalbumin, calbindin, calretinin, PKCα, GFAP	[13, 76]
BCCAO in rats	PACAP1-38 (100 pmol/5 μl) Intravitreal	Functional improvement	[84]
BCCAO in rats	PACAP1-38 (100 pmol/5 μl) Intravitreal	↓: p-p38 MAPK, p-JNK ↑: p-Akt, p-ERK, CINC, CNTF, fractalkine, sICAM, IL-1, LIX, Selectin, MIP-1, RANTES TIMP-1, VEGF, thymus chemokine	[83]
BCCAO in rats	PACAP1-38 (100 pmol/5 μl) Intravitreal	PACAP38 is the most effective form compared to other fragments and related peptides	[80, 81]
BCCAO in rats	PACAP1-27 (1 μmol) Eye drop	↑: Thickness of retinal layers ↓: GFAP immunoreactivity	[82]
BCCAO in mice	PACAP1-38 in PACAP knockout mice (100 pmol/3 μl)	↑: Thickness of retinal layers, ↓: Ganglion cell loss in PACAP homozygous knockout mice (Increased cell loss in knockout animals)	[106]
High intraocular pressure in rats	1 fmol–1000 pmol PACAP, intravitreal	↓: Ganglion cell death ↑: Light aversion, ERG changes	[79]
Ex vivo retinal explant, sodium azide treatment	100 nmol PACAP1-38	↓: Cell death, apoptosis, vascular endothelial growth factor overexpression, glutamate release, glutamine, caspase-3 mRNA, oxidized glutathione, cysteine, accumulation of peroxidized lipids, inflammatory mediators, nitric oxide metabolites, inositol, purine metabolism, lactate ↑: Glutathione, gamma-glutamyl cysteine, cAMP, NADPH	[85]

(continued)

Table 30.2 (continued)

Lesion	Treatment	Observed effect	Reference
Optic nerve injury			
Optic nerve injury in Adult rat	PACAP1-38 (10 and 100 pmol)	↓: Ganglion cell death	[88]
UV light			
UV-A induced retinal degeneration in adult rat	PACAP1-38 (100 pmol/5 µl)	↓: Pyknotic cells, empty cell body shapes by intravitreal PACAP1-38 treatment; ↑: the cell number in the ONL, INL and GCL and the whole retinal thickness by posttraumatic PACAP1-38 treatment	[89]
Focused UV-A irradiation	PACAP1-38 (100 pmol/5 µl)	Protective effects in ONL, IPL, PL	[90]
Diabetic retinopathy			
Streptozotocin-induced diabetes	100 pmol/5 µl 3×	↓: Degeneration in arborization and synapses of dopaminergic amacrine cells	[12, 13]
	100 pmol/5 µl 3×	↓: Müller glial cell activation, degeneration of ganglion cells, photoreceptor outer segments and terminals	[80, 92, 104, 106]
	100 pmol/5 µl 3×	↑: p-Akt, p-ERK1, p-ERK2, PKC, Bcl-2, ↓: p-p38MAPK, caspase 3, 8, 12, ↓:apoptosis	[93]
	100 pmol/5 µl 3×	↑: Intact ribbon bipolar synapses ↓: Degeneration of retinal pigment epithelium more retained ultrastructure after PACAP treatment	[94]
	100 µmol 1×	↓: HIF-1alpha, HIF-2alpha ↑: HIF-3alpha	[95]
Retinopathy of prematurity-oxygen-induced retinopathy in rats	3× intravitreal PACAP1-38 (100 pmol/5 µl)	↓: Avascular area, inflammatory cytokines, no changes in neovascular tufts	[103]

a number of studies that PACAP is protective against excitotoxicity induced by systemic monosodium-glutamate lesion. We showed that, three times intravitreal MSG treatment on postnatal days 1, 5 and 9 caused severe neurodegeneration in the inner retinal layers in newborn rat pups [58, 59]. Three weeks after the MSG administration the thickness of the total retina was decreased, the distance of the outer limiting membrane-inner limiting membrane (OLM-ILM) was only half of the normal retina. In addition, the IPL almost disappeared, and the fusion of the INL and GCL could be observed. The number of the cells per 100 μm in the GCL was also decreased [58–61]. Although PACAP1-38 is able to cross the BBB, systemic PACAP1-38 treatment caused insignificant improvement of the retinal morphology following MSG-induced excitotoxic injury [62]. In contrast, intravitreally injected PACAP1-38 (1–100 pmol) resulted in significant amelioration of this degeneration [58, 59, 61]. While 1 pmol PACAP1-38 was slightly effective, treatment with 100 pmol PACAP1-38 was able to significantly attenuate the MSG-induced toxicity and the entire retina seemed to be normal. Similar protection could be seen by PACAP1-27 treatment [60]. The application of the two widely used PACAP antagonists (PACAP6-38, PACAP6-27) led to a more pronounced degeneration, indicating that, endogenously present PACAP plays a key role against retinal toxicity [60]. A similar degree of retinoprotection was found using enriched environment, but combining these two neuroprotective strategies (PACAP and enriched environment) did not result in increased neuroprotection in excitotoxic retinal injury [63].

Cell-type specific effects of PACAP were shown in MSG-induced retinal degeneration by means of immunohistochemistry. In the MSG-treated retinas, the cell bodies and processes in the INL, IPL, and GCL layers displayed less immunoreactivity for vesicular glutamate transporter 1 (VGLUT1), tyrosine hydroxylase (TH), vesicular GABA transporter (VGAT), parvalbumin, and calretinin. After MSG treatment the calbindin-positive horizontal cells did not appear to be affected. In MSG+PACAP1-38 co-treated retinas, the immunoreactivity patterns were similar compared to the normal eyes. These data suggest that PACAP has a general, but not subtype-specific protective mechanism against glutamate-induced excitotoxic injury in the rat retina [64].

We also studied potential protective signaling mechanisms in MSG treated retinas. We found that intravitreal administration of PACAP preceding the MSG treatments induced significant increases in the phosphorylation, that is, the activation of ERK1/2 and its downstream target, CREB, 12 h after the treatment compared to the contralateral untreated eye during the first two treatments, with no further elevations 24 h after treatments. These results demonstrate that the degenerative effect of MSG and the protective effect of PACAP involve complex kinase signaling pathways and are related to cAMP/ERK/CREB activation [65]. Furthermore, after MSG treatment proapoptotic signaling proteins such as caspase-3, JNK, apoptosis inducing factor (AIF) and cytochrome-c increased, and some anti-apoptotic proteins (phospho-PKA, phospho-Bad, Bcl-xL, 14-3-3) decreased. Co-treatment with PACAP1-38 counteracted the MSG-induced apoptotic effects, the level of pro-apoptotic signals was significantly decreased, while the anti-apoptotic proteins were significantly increased [65–67]. A follow-up study confirmed the anti-apoptotic effect of PACAP in MSG-induced retinal lesion of the newborns by reinforcing that

PACAP decreased caspase-3 levels and also demonstrating the decreased level of caspase-9 [68]. In addition, we showed that the PACAP antagonist PACAP6-38 increased cytochrome-c release, caspase-3, JNK activity and decreased phospho-Bad, phospho PKA, Bcl-xL, 14-3-3 protein activity [66, 67]. These in vivo observations, similarly to the results of Atlasz et al. [60], indicating that endogenously present PACAP has retinoprotective effects in glutamate-induced excitotoxicity.

We studied the short-term functional consequences of MSG treatment in the mouse retina. Spontaneous and light-evoked spikes of retinal ganglion cells (RGCs) from wild type mice were recorded using a 60-channel multielectrode array. MSG remarkably elevated the free intracellular calcium (Ca^{2+}) concentration and also increased the spontaneous spiking 4–5 min after drug application. During this time, spike correlations between RGC pairs were reduced. However, after 10–15 min of MSG application, the spontaneous activity of most RGCs was dramatically reduced or totally eliminated. Pretreatment with PACAP1-38 prevented the MSG effects as indicated by little or no change in the spontaneous spiking patterns during the course of recordings (up to 60 min). Moreover, the Ca^{2+} influx was noticeably decreased by PACAP1-38. In addition, MSG blocked the light-evoked responses of all recorded cells. Based on these data, application of PACAP1-38, rescued RGCs from the short-term MSG-induced insults [69].

Kainic acid is a glutamate receptor agonist, which leads to excitotoxic cell death in the CNS, including the retina [43]. It was found that 5 nmol kainic acid caused an extensive loss of AMPA/kainate receptors expressing ganglion cells [70]. Previously we showed that microiontophoretically applied PACAP1-38 could block the excitatory effects of kainic acid in several brain areas in vivo [71]. In the rat retina, pretreatment with 10 pmol intravitreal PACAP1-38 was neuroprotective against kainic-acid induced neurotoxicity, as shown by the attenuated ganglion cell loss [43].

N-methyl-D-aspartate (NMDA) treatment also leads to similar lesions. A radioimmunoassay analysis showed that in contrast to VIP, retinal PACAP levels do not change after intravitreal NMDA injection, which might be caused by a compensatory upregulation [72]. A recent study has confirmed that PACAP-mediated pathways are protective in NMDA-induced retinal lesion. Cheng et al. [73] showed that intravitreal NMDA injection led to a decrease in light aversion, and decreases in the a and b-waves in the electroretinogram accompanied by a photopic negative response. These alterations were reversed by cyclic PACAP1-5 treatment. This treatment also resulted in decreased ganglion cell loss and apoptosis assessed by TUNEL-staining [73].

A similar protective effect was described in mice with PACAP1-38, at a concentration of 100 pmol. The PACAP-attenuated NMDA-induced retinal damage in mice was associated with modulation of the microglia/macrophage status into an acquired deactivation subtype [74]. NMDA injection led to an increase in microglia/macrophage number, while PACAP co-administration attenuated this effect in addition to the diminished ganglion cell loss. The co-injection of the antagonist PACAP6-38 counteracted these protective effects and the beneficial effects were not observed in IL-10 knockout mice. PACAP administration also led to elevated expression of the mRNA for the protective cytokines IL-10 and TGF-β [74]. In summary, these results show that acquired deactivation type of the microglia/macrophage cells favors neuroprotection in the retina and this is stimulated by PACAP.

The question was raised whether these apparent morphological improvements by PACAP administration also lead to functional amelioration in MSG-induced retinal damage [75]. Electroretinographic recordings revealed a marked decrease in both the b- and a-wave values after MSG treatment, reflecting the function of the inner retinal layers and the photoreceptors, respectively. In retinas receiving intravitreal PACAP treatment, these values were significantly increased. Thus, the functional outcome, although not parallel with the morphology, was significantly improved after PACAP treatment. These observations are important from the clinical point of view, showing, that PACAP treatment is able to improve the functional properties of the retina in excitotoxic damage [75].

Ischemic Injury in the Retina

Retina is one of the most sensitive tissues to hypoxia in the human body. Several diseases such as artery occlusion, diabetic retinopathy and high intraocular pressure cause hypoperfusion in the retina. Depending on the rat strain, permanent bilateral common carotid artery occlusion (BCCAO) leads to reduction in the cerebral blood flow, causes hypoperfusion-induced biochemical and behavioral changing and also produces retinal degeneration [76, 77]. Our research group showed in a series of experiments that intravitreal PACAP is highly protective in ischemic retinal lesion. We found that BCCAO caused severe damage in all layers of the retina. Two weeks after permanent carotid artery ligation, the thickness of the plexiform layers was markedly decreased, and as a consequence, the distance between the OLM-ILM was significantly less compared to the control retina. In several treated animals, in the IPL some dots of about 1 μm in diameter were seen. Based on the relatively even distribution, these dots seemed to be degenerating bipolar cell terminals. In addition, the outer segments of the photoreceptors became shorter, and their structure was also damaged after the insult [76]. Intravitreal PACAP1-38 treatment following BCCAO led to a nearly intact appearance of the retinal layers, which is well reflected by the morphometric analysis. The thickness of all retinal layers, but not the number of cells in the GCL, was preserved in PACAP1-38 treated animals. PACAP antagonist PACAP6-38 significantly attenuated the protective effects of PACAP1-38.

In 2010, we provided immunohistochemical description of several retinal cell types that were damaged by BCCAO and could be partially or fully rescued by intravitreal PACAP1-38 treatment. Using the cell type specific markers (VGLUT1, TH, VGAT, parvalbumin, calbindin, calretinin, PKCα, and GFAP) marked changes were detected after BCCAO, and the alterations observed in the immunostaining for all examined antisera were counteracted by intravitreal PACAP1-38 treatment [13]. These studies provided the basis for the promising therapeutic potential of PACAP in ischemic retinal diseases [78].

Subsequent studies confirmed these original observations. Seki et al. [79] showed that intravitreal PACAP treatment protected ganglion cells against a different kind of ischemic lesion induced by high intraocular pressure. In this model, PACAP exerted a bimodal protective effect with peaks at 10 fM and 10–100 pM. The protective effect was blocked by cAMP antagonist at both high and low protective doses, while a MAP kinase inhibitor only blocked the effects of PACAP at a low dose, suggesting that the protective effects of PACAP are mediated by different signaling at low and high doses [79].

In a series of studies, we confirmed these findings and compared the efficacy of several analogs and related peptides to PACAP1-38 [80, 81]. We investigated the effects of the following PACAP fragments: PACAP 4-13, 4-22, 6-10, 6-15, 11-15, and 20-31 and three related peptides (VIP, secretin, glucagon) in BCCAO-induced ischemic retinopathy. We confirmed that the most effective molecule is PACAP1-38, while the other fragments or related peptides did not lead to retinoprotection [81]. The only exception was VIP, which showed retinoprotection in a dose ten times higher than that of PACAP [80]. In addition, our preliminary results demonstrate that even in form of eye-drops, PACAP can lead to retinoprotective effects in ischemic retinopathy, a route of application with major possible clinical relevance [82].

In order to elucidate some of the molecular mechanisms, we studied MAP kinases and Akt expression as well as cytokines in BCCAO-induced retinal ischemia in rats [83]. We revealed that PACAP treatment alone did not influence the phosphorylation of Akt or the MAPKs, but decreased the hypoperfusion-induced activation of both p38MAPK and JNK and increased the activation of the protective Akt and ERK1/2 in hypoperfused retinas. The cytokine profile dramatically changed after BCCAO, with most cytokines and chemokines (such as CINC, CNTF, fractalkine, sICAM, IL-1, LIX, Selectin, MIP-1, RANTES, and TIMP-1) increased, a result attenuated by PACAP. In addition, PACAP increased the expression of vascular endothelial growth factor (VEGF) and thymus chemokine. These results provided further insight into the neuroprotective mechanism induced by PACAP in ischemic retinal injuries, showing that PACAP ameliorated hypoperfusion injury involving Akt, MAPK pathways and anti-inflammatory actions [83].

Whether the morphological amelioration is accompanied by functional improvement was tested using electroretinography (ERG) in another set of experiments [84]. Retinal damage and protective effects of PACAP were quantified by the changes in the waveforms and amplitudes. BCCAO-induced functional retinal degeneration was already observed on postoperative days 2 and lasted throughout the entire observation period. Intravitreal injection of PACAP immediately after BCCAO resulted in a more retained retinal function as assessed by average ERG waveforms compared to the BCCAO-operated groups on both postoperative 2nd and 14th days. ERGs of PACAP-treated ischemic eyes were similar to the intact controls in contrast to the ERGs of saline-treated BCCAO retinas. PACAP treatment significantly counteracted the ischemia-induced alterations in the amplitudes of both the "a" and "b" waves of the ERG on postoperative day 14. The elapse time of the major oscillatory potentials (OPs) was reduced in the BCCAO-ischemic group, but PACAP treatment led to significant protection. The results confirm that the previously described morphological protection induced by PACAP treatment is reflected in functional improvement in ischemic retinal lesions [84].

A recent study has revealed several metabolomics changes induced by PACAP in ex vivo retinal ischemia induced by sodium azide treatment for various periods of time [85]. PACAP induced several changes in addition to the reduction of ischemia-induced cell death, VEGF overexpression, and glutamate release. Apoptosis, as evaluated by TUNEL staining, was reduced in the ONL, INL, and GCL. Elevated levels of caspase-3 mRNA were also reversed by PACAP treatment. Glutamine, a precursor of glutamate, was significantly elevated in ischemic conditions, and reversed by PACAP administration. Ischemia is associated with oxidative injury and so the authors evaluated the oxidative status of the retinal explants. It was found that PACAP had a positive effect on glutathione while it reduced the oxidized glutathione levels, and counteracted the increased gamma-glutamyl cysteins levels with accompanying increase of cysteine. The accumulation of pro-inflammatory factors and peroxidized lipids was prevented by PACAP. Three nitric oxide metabolites were found to be upregulated in ischemic retinas: arginine, ornithine and citrulline. Treatment with PACAP reversed these changes. The significantly decreased level of cAMP was restored by PACAP treatment and glycolytic flues were normalized preventing the over-accumulation of lactate or promoting the downregulation of the glyoxalate antioxidant system. Upregulated inositol levels were also abolished. PACAP seemed to promote a shift towards pentose-phosphate pathway while maintaining higher levels of NADPH compared to ischemic controls. It also led to a downregulation of purine metabolism. This study concluded that peptidergic systems, such as PACAP, deserve attention in treating retinal ischemia [85].

Thanks to all these results, PACAP is now on the list of potential neuroprotective molecules in combating ischemic retinopathy [78, 86].

Optic Nerve Injury in the Retina

Optic nerve transection is a model of apoptotic neuronal cell death in the adult CNS [87]. Seki et al. [88] reported that optic nerve trauma resulted approximately 55 % loss of cells in the GCL. In this model, the number of cells in the GCL in animals treated before the operation with PACAP1-38 at 10 and 100 pmol in 3 μl saline (dose-dependent) intravitreally, was significantly increased compared with the control group on the 14th day after surgery [88], providing evidence that PACAP is protective in this injury model.

UV Radiation Injury in the Retina

We showed, for the first time, that PACAP protected against two different kinds (diffused and focused) of ultraviolet (UV)-A induced degenerations. Thickness of all layers as well as the number of cells in the ONL and INL and in the GCL was evaluated by standard morphological and morphometrical analyses. After diffuse UV-A irradiation, significant decrease of the number of the cells in the GCL was observed. In addition, severe degeneration in the ONL and INL was seen, and the IPL was swollen. PACAP1-38 treatment after the UV-A irradiation significantly ameliorated the light-induced degeneration, the ONL and INL thickness seemed to be normal, and the number of cells in INL, ONL, and GCL was significantly increased [89]. In a model of focused UV-A irradiation, the shape of the laser focus was observed on the retina, tissue gaps were found in the ONL and in the photoreceptor layer, and the IPL was reduced. The total thickness of the retina was significantly less than in unaffected areas. We observed protective effects of PACAP in all of the different retinal layers [90].

Succeeding studies confirmed this effect in several different models. In an in vitro model of UV-B irradiation, PACAP was shown to protect RGC-5 retinal ganglion cells expressing PAC1 receptors [73, 91]. This effect was mimicked also by a cyclic PACAP derivative, cyclic PACAP1-5 [73]. Furthermore, pretreatment with this short derivative effectively blocked apoptosis, maintained structural integrity, and decreased production of reactive oxygen species (ROS). Both PACAP and cyclic PACAP1-5 counteracted the UV-B-induced decrease in the protective Bcl-2 and the increase in the deleterious Bax protein levels [73].

Diabetic Retinopathy in the Retina

Our recent studies have shown that PACAP is also protective in diabetic retinopathy (DR), a major microangiopathy affecting the vision of diabetic patients. Our first results demonstrated that in the early stages of diabetic retinopathy, amacrine cells underwent characteristic degeneration, especially the dopaminergic amacrine cells. PACAP prevented the typical alterations induced in 1-month diabetes: the degeneration in arborization and synapses of tyrosine-hydroxylase positive dopaminergic amacrine cells [13]. In a rat model of streptozotocin-induced diabetic retinopathy, early signs do not include overall changes in thickness of retinal layers, but the decrease in the number of ganglion cells and the upregulation of GFAP as a sign of Müller glial cell activation, can be observed. These changes were counteracted by three times 100 pmol/5 μl intravitreal PACAP administration [92]. Furthermore, the degeneration of the photoreceptor outer segments and terminals was attenuated in PACAP-treated rats. PACAP increased the levels of PAC1-receptor and TH determined with molecular biological methods. These early results suggested a therapeutic potential in DR.

Subsequently, our research group and others confirmed these early findings [93]. Intraocular PACAP injection markedly attenuated diabetic retinal injury: increased levels of the anti-apoptotic p-Akt, p-ERK1, p-ERK2, PKC, Bcl-2, while decreased levels of the pro-apoptotic p-p38MAPK and activated caspases (8, 3, 12) were detected. The number of apoptotic cells increased in all nuclear layers of diabetic retinas, but significantly decreased after PACAP treatment. Our results clearly demonstrated that the protective effects of PACAP are mediated, at least partly, by attenuating apoptosis, including also that of the dopaminergic amacrine cells [93]. Our most recent study shed further light on this PACAP-induced retinoprotection. Electron microscopic analysis revealed that retinal pigment epithelium, the ribbon synapses and other synaptic profiles suffered alterations in diabetes. However, in PACAP-treated diabetic retinas more bipolar ribbon synapses were found intact in the inner plexiform layer than in DR animals. The ribbon synapse was marked with C-terminal binding protein 2 (CtBP2)/Bassoon and formed horseshoe-shape ribbons, which were more retained in PACAP-treated diabetic retinas than in DR rats. These results are supported by molecular biological data. The selective degeneration of related structures such as bipolar and ganglion cells could be ameliorated by PACAP treatment. In summary, intravitreal administration of PACAP may have therapeutic potential in streptozotocin-induced DR through maintaining synapse integrity in the vertical pathway [94].

D'Amico and coworkers [95] analyzed the expression of hypoxia-induced factors (HIF) in DR 3 weeks after the induction of diabetes with streptozotocin. They found that the expression of HIF-1α and HIF-2α significantly increased in diabetic rats, but not after 1 intravitreal PACAP treatment. Conversely, the expression of HIF-3α was significantly downregulated in retinas of diabetic rats, and increased after PACAP treatment. These results suggest that the retinoprotective effects of PACAP are partially mediated by interfering with the expression of HIFs that play an important role in pathological vasculogenesis occurring in DR [95].

These studies, together with the results demonstrating the effects of PACAP on the blood–retinal barrier properties in vitro (see above), show that PACAP has therapeutic potential in DR [96–98]. Several experiments support the efficacy of PACAP treatment in animal models of diabetes-related diseases, such as DR, another major microangiopathic complication of diabetes [99, 100]. Thus, PACAP is among the emerging molecules to fight diabetic complications, similarly to VEGF antagonists, antioxidants, anti-inflammatory agents, and other neuropeptides [96, 101, 102].

Retinopathy of Prematurity (ROP)

ROP is a major complication in premature newborns. It is a vasoproliferative disorder, the main features of which are well depicted in a rat model of oxygen-induced retinopathy. Our preliminary studies have shown that PACAP is able to reverse some of the deleterious effects of this retinopathy. Intravitreal PACAP1-38 treatment ameliorated the vascular extent of the retinopathic eyes without alterations in vessel morphology. Moreover, cytokine expression profiles indicate that PACAP may block the new vessel formation and have a positive influence on the remodeling of the tissue [103].

Studies in PACAP and Receptor Gene Modified Animals

We have described that although PACAP deficient mice have normal retinas at young age, aging signs appear earlier than in their wild-type mates [104]. As it has been shown in numerous models, PACAP knockout (KO) mice are more vulnerable to different types of tissue injuries from ischemia to oxidative stress and even trauma [105]. In our experiments, PACAP KO mice that underwent 10 min of BCCAO followed by 2-week reperfusion period had significantly greater retinal damage, as shown by the thickness of the whole retina, the morphometric analysis of the individual retinal layers, and the cell numbers in the INL and GCL. Exogenous PACAP administration could partially protect against this retinal degeneration in the KO mice. These results clearly show that endogenous PACAP reacts as a stress-response peptide that is necessary for endogenous protection against different retinal insults [106].

The higher vulnerability of PACAP deficient mice against lesions was confirmed by another study by Endo et al. [107]. The authors found, in accordance with our results, that under normal circumstances, there was no difference in the number of retinal ganglion cells in heterozygous PACAP deficient mice and wild-type counterparts [107]. However, when these mice were subjected to NMDA-induced toxicity, PACAP heterozygous mice reacted with increased loss of ganglion cells [107]. The number of apoptotic cells, as measured by TUNEL staining, peaked on day 1 in heterozygous mice, while the peak was on day 3 in wild types, suggesting that apoptotic cell death began earlier in mice deficient in PACAP. These effects were reversed by simultaneous injection of PACAP in heterozygous mice. These results confirm that endogenous PACAP is protective in the retina and prove that even partial loss of PACAP-driven neuroprotection decreases resistance against harmful effects [107]. Our recent, preliminary results also confirm that endogenous PACAP is protective in lipopolysaccharide (LPS)-induced retinal inflammation, as shown by the increased inflammatory status of PACAP KO. Twenty-four hours after LPS treatment several cytokines and chemokines showed significant increase in the PACAP knockout mice compared to the normal, wild types [108].

Little is known about retinal morphology or function in mice with PAC1 receptor abnormalities. Engelund and coworkers [109] described that pupillary light reflex was altered in PAC1 receptor knockout mice. The finding that PACAPergic signaling is essential for normal retinal development is also confirmed by studies with transgenic mice. It was described by Lang et al. [110] during development that overexpression of the PAC1 receptor led to an early exit from retinal proliferation, reduced signaling of GABAergic neurons. Furthermore, this led to decreased visual function [110]. These data show that a proper balance of PACAP signaling is required for appropriate retinogenesis and visual acuity.

Studies in Tumor Cells

PACAP can exert different effects in cells derived from various sources and especially from tumors. While in most cell types PACAP is protective, in some tumors it exerts toxic effects [111]. Furthermore, PACAP6-38 is in most cell types an antagonist, it can also behave as an agonist in some, tumorous and even non-tumorous, cells [112, 113]. Retinoblastoma is a malignant tumor of retinal progenitor cells. Human retinoblastoma cells are known to express PACAP receptors [114]. In Y79 retinoblastoma cells, PACAP1-38 did not influence cell survival at nanomolar concentrations, but at 1–5 μM concentrations, it reduced survival [115]. Interestingly, PACAP1-27 and maxadilan had negligible effects, while two membrane-penetrating analogs of PACAP also decreased cell viability. The authors also revealed that this cytotoxic effect might be via an interaction of p38, MEK1/2, and JNK signaling, because inhibitors of these pathways led to an even more expressed cytotoxicity [115]. These results suggest that high concentrations of PACAP1-38 might lead to a transmembrane penetration and lead to direct cytotoxicity at sites other than the conventional PACAP receptor-mediated signaling.

Acknowledgements This study was supported by OTKA K104984, 119759 Arimura Foundation, TAMOP 4.2.4.A/2-11-1-2012-0001 "National Excellence Program," MTA-PTE "Lendulet" Program, Bolyai Scholarship of the Hungarian Academy of Sciences, PTE AOK KA Research Grant, National Brain Research Programme KTIA_13_NAP-A-III/5 and New National Excellence Program (UNKP). This work is dedicated to the 650th anniversary of the University of Pecs.

References

1. Arimura A. PACAP: the road to discovery. Peptides. 2007;28:1617–9.
2. Miyata A, Arimura A, Dahl RR, Minamino N, Uehara A, Jiang L, et al. Isolation of a novel 38 residue-hypothalamic polypeptide which stimulates adenylate cyclase in pituitary cells. Biochem Biophys Res Commun. 1989;164:567–74.
3. Arimura A, Shioda S. Pituitary adenylate cyclase polypeptide (PACAP) and its receptors: neuroendocrine and endocrine interactions. Front Neuroendocrinol. 1995;16:53–8.
4. Vaudry D, Falluel-Morel A, Bourgault S, Basille M, Burel D, Wurtz O, et al. Pituitary adenylate cyclase activating polypeptide and its receptors: 20 years after the discovery. Pharmacol Rev. 2009;61:283–357.
5. Ohtaki H, Nakamachi T, Dohi K, Shioda S. Role of PACAP in ischemic neural death. J Mol Neurosci. 2008;36:16–25.
6. Reglodi D, Kiss P, Lubics A, Tamas A. Review of the protective effects of PACAP in models of neurodegenerative diseases in vitro and in vivo. Curr Pharm Des. 2011;17:962–72.
7. Reglodi D, Renaud J, Tamas A, Tizabi Y, Socias B, Del-Bel E, et al. Novel tactics for neuroprotection in Parkinson's disease: role of antibiotics, polyphenols and neuropeptides. Prog Neurobiol. doi: 10.1016/j.pneurobio.2015.10.004. Accessed 2 Nov 2015.
8. Shioda S, Ohtaki H, Nakamachi T, Dohi K, Watanabe J, Nakajo S, et al. Pleiotropic functions of PACAP in the CNS: neuroprotection and neurodevelopment. Ann N Y Acad Sci. 2006;1070:550–60.
9. Shioda S, Nakamachi T. PACAP as a neuroprotective factor in ischemic neuronal injuries. Peptides. 2015;72:202–7.
10. Somogyvari-Vigh A, Reglodi D. Pituitary adenylate cyclase activating polypeptide: a potential neuroprotective peptide. Rev Curr Pharm Des. 2004;10:2861–89.
11. Waschek JA. Multiple actions of pituitary adenylyl cyclase activating peptide in nervous system development and regeneration. Dev Neurosci. 2002;24:14–23.
12. Atlasz T, Szabadfi K, Kiss P, Racz B, Gallyas F, Tamas A, et al. Pituitary adenylate cyclase activating polypeptide in the retina: focus on the retinoprotective effects. Ann N Y Acad Sci. 2010;1200:128–39.
13. Atlasz T, Szabadfi K, Kiss P, Tamas A, Toth G, Reglodi D, et al. Evaluation of the protective effects of PACAP with cell-specific markers in ischemia-induced retinal degeneration. Brain Res Bull. 2010;81:497–504.

14. Nemeth A, Szabadfi K, Fulop B, Reglodi D, Kiss P, Farkas J, et al. Examination of calcium-binding protein expression in the inner ear of wild type, heterozygous and homozygous pituitary adenylate cyclase activating polypeptide (PACAP)-knockout mice in kanamycin-induced ototoxicity. Neurotox Res. 2014;25:57–67.
15. Nakamachi T, Matkovits A, Seki T, Shioda S. Distribution and protective function of pituitary adenylate cyclase-activating polypeptide in the retina. Front Endocrinol (Lausanne). 2012;3:145.
16. D'Agata V, Cavallaro S. Functional and molecular expression of PACAP/VIP receptors in the rat retina. Mol Brain Res. 1998;54:161–4.
17. Hannibal J, Fahrenkrug J. Target areas innervated by PACAP-immunoreactive retinal ganglion cells. Cell Tissue Res. 2004;316:99–113.
18. Izumi S, Seki T, Shioda S, Zhou CJ, Arimura A, Koide R. Ultrastructural localization of PACAP immunoreactivity in the rat retina. Ann NY Acad Sci. 2000;921:317–20.
19. Kubrusly RC, da Cunha MC, Reis RA, Soares H, Ventura AL, Kurtenbach E, et al. Expression of functional receptors and transmitter enzymes in cultured Muller cells. Brain Res. 2005;1038:141–9.
20. Seki T, Shioda S, Izumi S, Arimura A, Koide R. Electron microscopic observation of pituitary adenylate cyclase activating polypeptide (PACAP)-containing neurons in the rat retina. Peptides. 2000;21:109–13.
21. Mathis U, Schaeffel F. Glucagon-related peptides in the mouse retina and the effects of deprivation of form vision. Graefe's Arch Clin Exp Ophthalmol. 2007;245:267–75.
22. Grone BP, Zhao S, Chen CC, Fernald RD. Localization and diurnal expression of melanopsin, vertebrate ancient opsin, and pituitary adenylate cyclase activating peptide mRNA in a teleost retina. J Biol Rhythms. 2007;22:558–61.
23. Reglodi D, Somogyvari-Vigh A, Vigh J, Li M, Lengvari I, Arimura A. Pituitary adenylate cyclase activating polypeptide is highly abundant in the nervous system of anoxia-tolerant turtle, Pseudemys scripta elegans. Peptides. 2001;22:873–8.
24. Jozsa R, Somogyvari-Vigh A, Reglodi D, Hollosy T, Arimura A. Distribution and daily variations of PACAP in the chicken brain. Peptides. 2001;22:1371–7.
25. Sakamoto K, Liu C, Kasamatsu M, Pozdeyev NV, Iuvone PM, Tosini G. Dopamine regulates melanopsin mRNA expression in intrinsically photosensitive retinal ganglion cells. Eur J Neurosci. 2005;22:3129–36.
26. Wan J, Zheng H, Hu BY, Xiao HL, She ZJ, Chen ZL, et al. Acute photoreceptor degeneration down-regulates melanopsin expression in adult rat retina. Neurosci Lett. 2006;400:48–52.
27. Perez de Sevilla Muller L, Sargoy A, Rodriguez AR, Brecha NC. Melanopsin ganglion cells are the most resistant retinal ganglion cell type to axonal injury in the rat retina. PLoS One. 2014;9:e93274.
28. Hannibal J, Kankipati L, Strang CE, Peterson BB, Dacey D, Gamlin PD. Central projections of intrinsically photosensitive retinal ganglion cells in the macaque monkey. J Comp Neurol. 2014;522:2231–48.
29. La Morgia C, Ross-Cisneros FN, Hannibal J, Montagna P, Sadun AA, Carelli V. Melanopsin-expressing retinal ganglion cells: implications for human diseases. Vision Res. 2011;51:296–302.
30. Langel JL, Smale L, Esquiva G, Hannibal J. Central melanopsin projections in the diurnal rodent. Arvicanthis niloticus. Front Neuroanat. 2015;9:93.
31. Delwig A, Majumdar S, Ahern K, LaVail MM, Edwards R, Hnaski TS, et al. Glutamatergic neurotransmission from melanopsin retinal ganglion cells is required for neonatal photoaversion but not adult pupillary light reflex. PLoS One. 2013;8:e83974.
32. Purrier N, Engeland WC, Kofuji P. Mice deficient of glutamatergic signaling from intrinsically photosensitive retinal ganglion cells exhibit abnormal circadian photoentrainment. PLoS One. 2014;9:e111449.
33. Vereczki V, Koves K, Csaki A, Grosz K, Hoffman GE, Fiskum G. Distribution of hypothalamic, hippocampal and other limbic peptidergic neuronal cell bodies giving rise to retinopetal fibers: anterograde and retrograde tracing and neuropeptide immunohistochemical studies. Neuroscience. 2006;140:1089–100.

34. Bagnoli P, Dal Monte M, Casini G. Expression of neuropeptides and their receptors in the developing retina of mammals. Histol Histopathol. 2003;18:1219–42.
35. Borba JC, Henze IP, Silveira MS, Kubrusly RC, Gardino PF, de Mello MC, et al. Pituitary adenylate cyclase activating polypeptide (PACAP) can act as a determinant of the tyrosine hydroxylase phenotype of dopaminergic cells during retina development. Dev Brain Res. 2005;156:193–201.
36. Mathieu M, Girosi L, Vallarino M, Tagliafierro G. PACAP in developing sensory and peripheral organs of the zebrafish, Danio rerio. Eur J Histochem. 2005;49:167–78.
37. Mathieu M, Ciarlo M, Trucco N, Griffero F, Damonte G, Salis A, et al. Pituitary adenylate cyclase activating polypeptide in the brain, spinal cord and sensory organs of the zebrafish, Danio rerio, during development. Brain Res Dev Brain Res. 2004;151:169–85.
38. Skoglosa Y, Takei N, Lindholm D. Distribution of pituitary adenylate cyclase activating polypeptide mRNA in the developing rat brain. Mol Brain Res. 1999;65:1–13.
39. Nilsson SF, De Neef P, Robberecht P, Christophe J. Characterization of ocular receptors for pituitary adenylate cyclase activating polypeptide (PACAP) and their coupling to adenylate cyclase. Exp Eye Res. 1994;58:459–67.
40. Silveira MS, Costa MR, Bozza M, Linden R. Pituitary adenylate cyclase activating polypeptide prevents induced cell death in retinal tissue through activation of cyclic AMP-dependent protein kinase. J Biol Chem. 2002;277:16075–80.
41. Seki T, Shioda S, Ogino D, Nakai Y, Arimura A, Koide R. Distribution and ultrastructural localization of a receptor for pituitary adenylate cyclase activating polypeptide and its mRNA in the rat retina. Neurosci Lett. 1997;238:127–30.
42. Seki T, Izumi S, Shioda S, Zhou CJ, Arimura A, Koide R. Gene expression for PACAP receptor mRNA in the rat retina by in situ hybridization and in situ RT-PCR. Ann N Y Acad Sci. 2000;921:366–9.
43. Seki T, Nakatani M, Taki C, Shinohara Y, Ozawa M, Nishimura S, et al. Neuroprotective effects of PACAP against kainic acid-induced neurotoxicity in rat retina. Ann N Y Acad Sci. 2006;1070:531–4.
44. Njaine B, Martins RA, Santiago MF, Linden R, Silveira MS. Pituitary adenylyl cyclase-activating polypeptide controls the proliferation of retinal progenitor cells through downregulation of cyclin D1. Eur J Neurosci. 2010;32:311–21.
45. Denes V, Czotter N, Lakk M, Berta G, Gabriel R. PAC1-expressing structures of neural retina alter their PAC1 isoform splicing during postnatal development. Cell Tissue Res. 2014;355:279–88.
46. Lakk M, Szabo B, Volgyi B, Gabriel R, Denes V. Development-related splicing regulates pituitary adenylate cyclase-activating polypeptide (PACAP) receptors in the retina. Invest Ophthalmol Vis Sci. 2012;53:7825–32.
47. Njaine B, Rocha-Martins M, Vieira-Vieira CH, De-Melo LD, Linden R, Braas K, et al. Pleiotropic functions of pituitary adenylyl cyclase-activating polypeptide on retinal ontogenesis: involvement of KLF4 in the control of progenitor cell proliferation. J Mol Neurosci. 2014;54:430–42.
48. Shoge K, Mishima HK, Saitoh T, Ishihara K, Tamura Y, Shiomi H, et al. Protective effects of vasoactive intestinal peptide against delayed glutamate neurotoxicity in cultured retina. Brain Res. 1998;809:127–36.
49. Shoge K, Mishima HK, Saitoh T, Ishihara K, Tamura Y, Shiomi H, et al. Attenuation by PACAP of glutamate-induced neurotoxicity in cultured retinal neurons. Brain Res. 1999;839:66–73.
50. Rabl K, Reglodi D, Banvolgyi T, Somogyvari-Vigh A, Lengvari I, Gabriel R, et al. PACAP inhibits anoxia-induced changes in physiological responses in horizontal cells in the turtle retina. Regul Pept. 2002;109:71–4.
51. Zhang XY, Hayasaka S, Chi ZL, Cui HS, Hayasaka Y. Effect of pituitary adenylate cyclase-activating polypeptide (PACAP) on IL-6, IL-8, and MCP-1 expression in human retinal pigment epithelial cell line. Curr Eye Res. 2005;30:1105–11.
52. Mester L, Kovacs K, Racz B, Solti I, Atlasz T, Szabadfi K, et al. Pituitary adenylate cyclase-activating polypeptide is protective against oxidative stress in human retinal pigment epithelial cells. J Mol Neurosci. 2011;43:35–43.

53. Fabian E, Reglodi D, Mester L, Szabo A, Szabadfi K, Tamas A, et al. Effects of PACAP on intracellular signaling pathways in human retinal pigment epithelial cells exposed to oxidative stress. J Mol Neurosci. 2012;48:493–500.
54. Ko JA, Hirata J, Yamane K, Sonoda KH, Kiuchi Y. Up-regulation of semaphorin 4A expression in human retinal pigment epithelial cells by PACAP released from cocultured neural cells. Cell Biochem Funct. 2015;33:29–36.
55. Scuderi S, D'Amico AG, Castorina A, Imbesi R, Carnazza ML, D'Agata V. Ameliorative effect of PACAP and VIP against increased permeability in a model of outer blood retinal barrier dysfunction. Peptides. 2013;39:119–24.
56. Wilhelm I, Fazakas C, Tamas A, Toth G, Reglodi D, Krizbai IA. PACAP enhances barrier properties of cerebral microvessels. J Mol Neurosci. 2014;54:469–76.
57. Szabadfi K, Atlasz T, Horvath G, Kiss P, Hamza L, Farkas J, et al. Early postnatal enriched environment decreases retinal degeneration induced by monosodium glutamate treatment. Brain Res. 2009;1259:107–12.
58. Babai N, Atlasz T, Tamas A, Reglodi D, Toth G, Kiss P, et al. Search for the optimal monosodium glutamate treatment schedule to study the neuroprotective effects of PACAP in the retina. Ann N Y Acad Sci. 2006;1070:149–55.
59. Babai N, Atlasz T, Tamas A, Reglodi D, Kiss P, Gabriel R. Degree of damage compensation by various PACAP treatments in monosodium glutamate-induced retina degeneration. Neurotox Res. 2005;8:227–33.
60. Atlasz T, Szabadfi K, Reglodi D, Kiss P, Tamas A, Toth G, et al. Effects of pituitary adenylate cyclase activating polypeptide (PACAP1-38) and its fragments on retinal degeneration induced by neonatal MSG treatment. Ann N Y Acad Sci. 2009;1163:348–52.
61. Tamas A, Gabriel R, Racz B, Denes V, Kiss P, Lubics A, et al. Effects of pituitary adenylate cyclase activating polypeptide in retinal degeneration induced by monosodium-glutamate. Neurosci Lett. 2004;372:110–3.
62. Kiss P, Tamas A, Lubics A, Lengvari I, Szalai M, Hauser D, et al. Effects of systemic PACAP treatment in monosodium glutamate-induced behavioral changes and retinal degeneration. Ann N Y Acad Sci. 2006;1070:365–70.
63. Kiss P, Atlasz T, Szabadfi K, Horvath G, Griecs M, Farkas J, et al. Comparison between PACAP- and enriched environment-induced retinal protection in MSG-treated newborn rats. Neurosci Lett. 2011;487:400–5.
64. Atlasz T, Szabadfi K, Kiss P, Babai N, Koszegi Z, Tamas A, et al. PACAP-mediated neuroprotection of neurochemically identified cell types in MSG-induced retinal regeneration. J Mol Neurosci. 2008;36:97–104.
65. Racz B, Tamas A, Kiss P, Toth G, Gasz B, Borsiczky B, et al. Involvement of ERK and CREB signaling pathways in the protective effect of PACAP in monosodium glutamate-induced retinal lesion. Ann N Y Acad Sci. 2006;1070:507–11.
66. Racz B, Gallyas Jr F, Kiss P, Toth G, Hegyi O, Gasz B, et al. The neuroprotective effects of PACAP in monosodium glutamate-induced retinal lesion involves inhibition of proapoptotic signaling pathways. Regul Pept. 2006;137:20–6.
67. Racz B, Gallyas Jr F, Kiss P, Tamas A, Lubics A, Lengvari I, et al. Effects of pituitary adenylate cyclase activating polypeptide (PACAP) on the PKA-Bad-14-3-3 signaling pathway in glutamate-induced retinal injury in neonatal rats. Neurotox Res. 2007;12:95–104.
68. Lakk M, Denes V, Gabriel R. Pituitary adenylate cyclase-activating polypeptide receptors signal via phospholipase C pathway to block apoptosis in newborn rat retina. Neurochem Res. 2015;40:1402–9.
69. Atlasz T, Balogh M, Werling D, Zhang Y, Reglodi D, Bloomfield S, et al. Pituitary adenylate cyclase-activating polypeptide (PACAP) prevents monosodium glutamate (MSG) induced functional disturbances in the mouse retina. Presented at 9th FENS Forum of Neuroscience July 5–9, 2014, Milan, Italy.
70. Zhang D, Sucher NJ, Lipton SA. Co-expression of AMPA/kainate receptor-operated channels with high and low Ca^{2+} permeability in single rat retinal ganglion cells. Neuroscience. 1995;67:177–88.

71. Atlasz T, Koszegi Z, Babai N, Tamas A, Reglodi D, Kovacs P, et al. Microiontophoretically applied PACAP blocks excitatory effects of kainic acid in vivo. Ann N Y Acad Sci. 2006;1070:143–8.
72. Teuchner B, Dimmer A, Humpel C, Amberger A, Fischer-Colbrie R, Nemeth J, et al. VIP, PACAP-38, BDNF and ADNP in NMDA-induced excitotoxicity in the rat retina. Acta Ophthalmol. 2011;89:670–5.
73. Cheng H, Ding Y, Yu R, Chen J, Wu C. Neuroprotection of a novel cyclopeptide C*HSDGIC* from the cyclization of PACAP (1-5) in cellular and rodent models of retinal ganglion cell apoptosis. PLoS One. 2014;9, e108090.
74. Wada Y, Nakamachi T, Endo K, Seki T, Ohtaki H, Tsuchikawa D, et al. PACAP attenuates NMDA-induced retinal damage in association wit modulation of the microglia/macrophage status into an acquired deactivation subtype. J Mol Neurosci. 2013;51:493–502.
75. Varga B, Szabadfi K, Kiss P, Fabian E, Tamas A, Griecs M, et al. PACAP improves functional outcome in excitotoxic retinal lesion: an electroretinographic study. J Mol Neurosci. 2011;43:44–50.
76. Atlasz T, Babai N, Kiss P, Reglodi D, Tamas A, Szabadfi K, et al. Pituitary adenylate cyclase activating polypeptide is protective in bilateral carotid occlusion-induced retinal lesion in rats. Gen Comp Endocrinol. 2007;153:108–14.
77. Mester L, Szabo A, Atlasz T, Szabadfi K, Reglodi D, Kiss P, et al. Protection against chronic hypoperfusion-induced retinal neurodegeneration by PARP inhibition via activation of PI3-kinase Akt pathway and suppression of JNK and p38 MAP kinases. Neurotox Res. 2009;16:68–76.
78. Szabadfi K, Mester L, Reglodi D, Kiss P, Babai N, Racz B, et al. Novel neuroprotective strategies in ischemic retinal lesions. Int J Mol Sci. 2010;11:544–61.
79. Seki T, Itoh H, Nakamachi T, Endo K, Wada Y, Nakamura K, et al. Suppression of rat retinal ganglion cell death by PACAP following transient ischemia induced by high intraocular pressure. J Mol Neurosci. 2011;43:30–4.
80. Szabadfi K, Danyadi B, Kiss P, Tamas A, Fabian E, Gabriel R, et al. Protective effects of vasoactive intestinal peptide (VIP) in ischemic retinal degeneration. J Mol Neurosci. 2012;48:501–7.
81. Werling D, Reglodi D, Kiss P, Toth G, Szabadfi K, Tamas A, et al. Investigation of PACAP fragments and related peptides in chronic retinal hypoperfusion. J Ophthalmol. 2014:563812.
82. Werling D, Kvarik T, Varga R, Nagy N, Mayer F, Vaczy A, et al. Investigating the retinoprotective effects of PACAP eye-drop in ischemic retinopathy. Presented at 12th International Symposium on VIP, PACAP and Related Peptides, September 21–26, 2015, Cappadocia, Turkey.
83. Szabo A, Danyadi B, Bognar E, Szabadfi K, Fabian E, Kiss P, et al. Effect of PACAP on MAP kinases, Akt and cytokine expressions in rat retinal hypoperfusion. Neurosci Lett. 2012;523:93–8.
84. Danyadi B, Szabadfi K, Reglodi D, Mihalik A, Danyadi T, Kovacs Zs, et al. PACAP application improves functional outcome of chronic retinal ischemic injury in rats—evidence from electroretinographic measurements. J Mol Neurosci. 2014;54:293–9.
85. D'Alessandro A, Cervia D, Catalani E, Gevi F, Zolla L, Casini G. Protective effects of the neuropeptides PACAP, substance P and the somatostatin analogue octreotide in retinal ischemia: a metabolomic analysis. Mol Biosyst. 2014;10:1290–304.
86. Cervia D, Casini G. The neuropeptide systems and their potential role in the treatment of mammalian retinal ischemia: a developing story. Curr Neuropharmacol. 2013;11:95–101.
87. Quigley HA, Nickells RW, Kerrigan LA, Pease ME, Thibault DJ, Zack DJ. Retinal ganglion cell death in experimental glaucoma and after axotomy occurs by apoptosis. Invest Ophthalmol Vis Sci. 1995;36:774–86.
88. Seki T, Itoh H, Nakamachi T, Shioda S. Suppression of ganglion cell death by PACAP following optic nerve transection in the rat. J Mol Neurosci. 2008;36:57–60.
89. Atlasz T, Szabadfi K, Kiss P, Marton Z, Griecs M, Hamza L, et al. Effects of PACAP in UV-A radiation-induced retinal degeneration models in rats. J Mol Neurosci. 2011;43:51–7.

90. Atlasz T, Szabadfi K, Molnar A, Kiss P, Reglodi D, Marton ZS, et al. PACAP protects rat retina from UV-A radiation-induced degeneration. Presented at the Satellite Symposium of the 9th International Symposium on VIP, PACAP and Related Peptides, October 2–3, 2009, Yakushima, Japan.
91. Ding Y, Cheng H, Yu R, Tang C, Liu X, Chen J. Effects of cyclopeptide C*HSDGIC* from the cyclization of PACAP (1-5) on the proliferation and UVB-induced apoptosis of the retinal ganglion cell line RGC-5. Peptides. 2012;36:280–5.
92. Szabadfi K, Atlasz T, Kiss P, Reglodi D, Szabo A, Kovacs K, et al. Protective effects of the neuropeptide PACAP in diabetic retinopathy. Cell Tissue Res. 2012;348:37–46.
93. Szabadfi K, Szabo A, Kiss P, Reglodi D, Setalo Jr G, Kovacs K, et al. PACAP promotes neuron survival in early experimental diabetic retinopathy. Neurochem Int. 2014;64:84–91.
94. Szabadfi K, Reglodi D, Szabo A, Szalontai B, Valasek A, Setalo Gy Jr et al. Pituitary adenylate cyclase activating polypeptide, a potential therapeutic agent for diabetic retinopathy in rats: focus on the vertical information processing pathway. Neurotox Res. 2016;29:432–46.
95. D'Amico AG, Maugeri G, Reitano R, Bucolo C, Saccone S, Drago F, et al. PACAP modulates expression of hypoxia-inducible factors in streptozotocin-induced diabetic rat retina. J Mol Neurosci. 2015;57:501–9.
96. Gabriel R. Neuropeptides and diabetic retinopathy. Br J Clin Pharmacol. 2013;75:1189–201.
97. Marzagalli R, Scuderi S, Drago F, Waschek JA, Castorina A. Emerging role of PACAP as a new potential therapeutic target in major diabetes complications. Int J Endocrinol. 2015;2015:160928.
98. Szabadfi K, Pinter E, Reglodi D, Gabriel R. Neuropeptides, trophic factors, and other substances providing morphofunctional and metabolic protection in experimental models of diabetic retinopathy. Int Rev Cell Mol Biol. 2014;311:1–121.
99. Banki E, Degrell P, Kiss P, Kovacs K, Kemeny A, Csanaky K, et al. Effect of PACAP treatment on kidney morphology and cytokine expression in rat diabetic nephropathy. Peptides. 2013;42C:125–30.
100. Banki E, Kovacs K, Nagy D, Juhasz T, Degrell P, Csanaky K, et al. Molecular mechanisms underlying the nephroprotective effects of PACAP in diabetes. J Mol Neurosci. 2014;54:300–9.
101. Ola MS, Ahmed MM, Ahmad R, Abuohashish HM, Al-Rejaie SS, Alhomida AS. Neuroprotective effects of rutin in streptozotocin-induced diabetic rat retina. J Mol Neurosci. 2015;56:440–8.
102. Scuderi S, D'Amico AG, Federico C, Saccone S, Magro G, Bucolo C, et al. Different retinal expression patterns of IL-1α, IL-1β, and their receptors in a rat model of type 1 STZ-induced diabetes. J Mol Neurosci. 2015;56:431–9.
103. Kvarik T, Mammel B, Reglodi D, Kovacs K, Werling D, Bede B, Vaczy A, Fabian E, Toth G, Kiss P, Tamas A, Ertl T, Gyarmati J, Atlasz T. PACAP is protective in a rat model of retinopathy of prematurity. J Mol Neurosci 2016 in press. DOI:10.1007/s12031-016-0797-5.
104. Szabadfi K, Kiss P, Reglodi D, Tamas A, Hashimoto H, Helyes ZS, et al. Neurochemical differences between wild type and PACAP KO mice in adult and aging retina. Presented at the 8th FENS Forum of Neuroscience, July 14–18, 2012, Barcelona, Spain.
105. Reglodi D, Kiss P, Szabadfi K, Atlasz T, Gabriel R, Horvath G, et al. PACAP is an endogenous protective factor—insights from PACAP deficient mice. J Mol Neurosci. 2012;48:482–92.
106. Szabadfi K, Atlasz T, Kiss P, Danyadi B, Tamas A, Helyes ZS, et al. Mice deficient in pituitary adenylate cyclase activating polypeptide (PACAP) are more susceptible to retinal ischemic injury in vivo. Neurotox Res. 2012;21:41–8.
107. Endo K, Nakamachi T, Seki T, Kagami N, Wada Y, Nakamura K, et al. Neuroprotective effect of PACAP against NMDA-induced retinal damage in the mouse. J Mol Neurosci. 2011;43:22–9.
108. Atlasz T, Váczy A, Kovács K, Lőkös E, Kvárik T, Werling D, et al. The protective role of the endogenous PACAP in LPS-induced inflammation in the retina. Presented at 12th International Symposium on VIP, PACAP and Related Peptides, September 21–26, 2015, Cappadocia, Turkey.

109. Engelund A, Fahrenkrug J, Harrison A, Luuk H, Hannibal J. Altered pupillary light reflex in PACAP receptor-1 deficient mice. Brain Res. 2012;1453:17–25.
110. Lang B, Zhao L, McKie L, Forrester JV, McCaig CD, Jackson IJ, et al. GABAergic amacrine cells and visual function are reduced in PAC1 transgenic mice. Neuropharmacology. 2010;58:215–25.
111. Brubel R, Boronkai A, Reglodi D, Racz B, Nemeth J, Kiss P, et al. Changes in the expression of pituitary adenylate cyclase activating polypeptide (PACAP) in the human placenta during pregnancy and its effects on survival of JAR choriocarcinoma cells. J Mol Neurosci. 2010;42:450–8.
112. Juhasz T, Helgadottir SL, Reglodi D, Tamas A, Zakany R. Signalisation of VIP and PACAP in chondrogenesis and osteogenesis. Peptides. 2015;66:51–7.
113. Saghy E, Payrits M, Helyes ZS, Reglodi D, Banki E, Toth G, et al. Stimulatory effect of pituitary adenylate-cyclase activating polypeptide 6-38, M65 and vasoactive intestinal polypeptide 6-28 on trigeminal sensory neurons. Neuroscience. 2015;308:144–56.
114. Olianas MC, Ennas MG, Lampis G, Onali P. Presence of pituitary adenylate cyclase-activating polypeptide receptors in Y-79 human retinoblastoma cells. J Neurochem. 1996;67:1803–9.
115. Wojcieszak J, Zawilska JB. PACAP38 and PACAP6-38 exert cytotoxic activity against human retinoblastoma Y79 cells. J Mol Neurosci. 2014;54:463–8.

Chapter 31
Pituitary Adenylate Cyclase-Activating Polypeptide in the Auditory System

Balazs D. Fulop, Dora Reglodi, Adrienn Nemeth, and Andrea Tamas

Abstract Pituitary adenylate cyclase-activating polypeptide (PACAP) is a neuropeptide with well-known neurotrophic, neuroprotective, antioxidant, and antiapoptotic effects. It also exerts protective effects in sensory organs, such as visual and olfactory system. In this review we present data regarding the localization and effects of PACAP in the auditory system. PACAP and its specific receptor (PAC1-R) are present in the organ of Corti in hair cells, supporting cells, and different nerve fibers. They are also present in the spiral ganglion showing co-localization with efferent fibers of glutamatergic and adrenergic pathways, probably directly affecting the efferent signal transduction in the inner ear. PACAP and its specific receptor also occur in the stria vascularis suggesting a role in endolymph production; furthermore, they are present in central pathways of the auditory system such as the cochlear nuclei, superior olivary complex, inferior colliculus, and medial geniculate body. PAC1-receptor is also present in the inner ear in PACAP-deficient mice, occurring at the same localization but with altered expression compared to wild-type mice. PACAP protects hair cells from H_2O_2-induced apoptosis in chicken inner ear cell cultures in vitro, and the lack of PACAP affects the Ca^{2+}-binding protein expression in hair cells in PACAP-deficient mice under control circumstances and after ototoxic drug treatment.

Keywords PACAP • PAC1-R • Inner ear • Auditory system • Ototoxicity

B.D. Fulop • D. Reglodi • A. Tamas (✉)
Department of Anatomy, MTA-PTE "Lendulet" PACAP Research Team, Medical School, University of Pecs, Szigeti u. 12, Pecs 7624, Hungary
e-mail: andreatamassz@gmail.com

A. Nemeth
Department of Anatomy, MTA-PTE "Lendulet" PACAP Research Team, Medical School, University of Pecs, Szigeti u. 12, Pecs 7624, Hungary

Department of Oto-rhino-laryngology, University of Pecs, Pecs, Hungary

D. Reglodi, A. Tamas (eds.), *Pituitary Adenylate Cyclase Activating Polypeptide — PACAP*, Current Topics in Neurotoxicity 11,
DOI 10.1007/978-3-319-35135-3_31

Abbreviations

AC	Adenylate cyclase
ApoD	Apolipoprotein D
ApoJ	Apolipoprotein J
Bad	bcl-2-associated death promoter
Bax	bcl-2-like protein 4
Bcl	B-cell lymphoma
ChAT	Choline acetyltransferase
CREB	cAMP response element-binding protein
DBH	Dopamine β-hydroxylase
HZ	Heterozygous PACAP-deficient mice
ERK	Extracellular-signal-regulated kinase
GluR2/3	Glutamate receptor 2/3
JC-1	5,5′,6,6′-Tetrachloro-1,1′,3,3′-tetraethylbenzimi-dazolylcarbocyanine iodide
JNK	c-Jun N-terminal kinase
LW	Lateral wall
MNTB	Medial nucleus of trapezoid body
MTT assay	Colorimetric assay for assessing cell metabolic activity
nNOS	Neuronal nitric oxide synthase
OC	Organ of Corti
p38MAPK	p38 mitogen-activated protein kinases
P5	5th postnatal day
P7	7th postnatal day
PACAP	Pituitary adenylate cyclase-activating polypeptide
PACAP KO	Homozygous PACAP-deficient mice
PACAP38	1-38 Amino-acid isoform of PACAP
PAC1-R	PAC1-receptor
PKA	Protein kinase A
PLC	Phospholipase C
ROS	Reactive oxygen species
RT-PCR	Reverse transcription-polymerase chain reaction
SG	Spiral ganglion
SOC	Superior olivary complex
VIP	Vasoactive intestinal polypeptide
WT	Wild type mice

Introduction

Pituitary adenylate cyclase-activating polypeptide (PACAP) is known as a general cytoprotective factor. It protects neuronal and non-neuronal cells against various injuries and also exerts protective effects in sensory organs (Chaps. 29 and 30). In this review we summarize the experiments on the effects of PACAP in the auditory system. We describe the occurrence of PACAP, and its specific receptor, PAC1-receptor (PAC1-R), in the cochlea [1–3] and in the central pathways of the auditory system [4–6]. We also present the changes in PAC1-receptor expression in the inner ear of PACAP-deficient mice [7]. Thereafter, we discuss the functional changes in the inner ear related to PACAP. We summarize the effects of PACAP on the endolymph composition [8], on apoptosis in chicken inner ear cell cultures in vitro [9], and its effects on Ca^{2+}-binding protein expression in hair cells [7, 10].

PACAP and PAC1-R in the Cochlea

Kawano et al. [3] studied the expression of PACAP mRNA in the cochlea of Wistar rats using RT-PCR and in situ hybridization techniques. The RT-PCR primers designed against PACAP mRNA yielded the expected 553 bp RNA sequence. In situ hybridization returned cytoplasmic labeling, in most cells of the spiral ganglion and in the marginal cells of stria vascularis in the lateral wall.

Abu-Hamdan et al. [1] investigated the expression of PACAP and different splice variants of PAC1-R in rat cochlea. They divided the cochlea into three subfractions, such as organ of Corti (OC), spiral ganglion (SG), and lateral wall (LW) (Fig. 31.1).

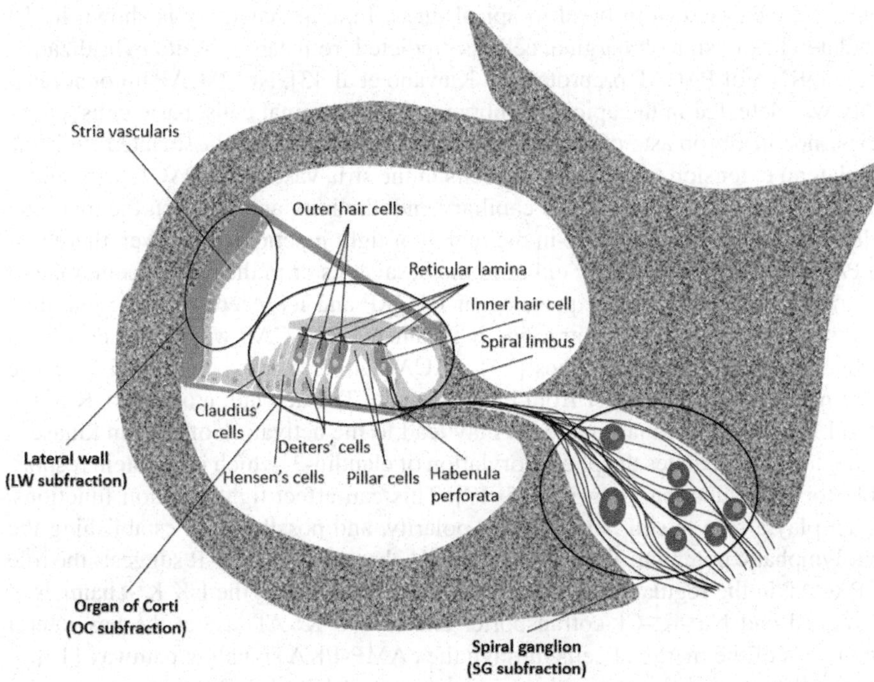

Fig. 31.1 Schematic drawing of the cochlea showing lateral wall (LW), organ of Corti (OC), and spiral ganglion (SG) subfractions

They were working with ACI Black Agouti (Harlan Sprague-Dawley) rats of both sexes. First they examined the expression of PACAP in the cochlea of 20-day-old rats using RT-PCR. Amplification products of predicted size were present in all three cochlear subfractions. For PAC1-R all the three subfractions contained mRNA for the short splice variant (~842 bp) of the third intracellular loop. Additional amplification products were found in the LW fraction, which represents other splice variants such as formerly known hop1, hop2, hip, and a new hop1 variant with a 193 bp deletion in the carboxyl-terminal corresponding region. Western blot technique confirmed the presence of PAC1-R short splice variant in 6-week-old rats with a ~53 kDa band in all three subfractions and a ~49 kDa in the LW fraction, which might represent the novel hop1 variant. Hip-hop1 or hip-hop2 variants were not present in the cochlear subfractions based on sequencing results. The primers designed against the amino terminus corresponding region showed the presence of one amplification product at ~500 bp in all three subfractions in accordance with previously described sequences [11]. This was also confirmed by Western blot, showing a band at ~70 kDa in all three subfractions. Additionally they found a new variant in the OC subfraction with a 7-amino acid deletion. There were no amplification products indicating the presence of other, previously described splice variants [12, 13]. Further Western blot examinations suggested that PAC1-R splice variants are glycosylated in LW and OC subfractions.

They also used immunohistology in the LW subfraction in 2-month-old rats where PACAP and PAC1-R were found mainly in the stria vascularis. The immunoreactivity increased from basal to apical turns. Immunoreactivity is shown in the basolateral extension of marginal cells as predicted from earlier in situ hybridization of the mRNA of PACAP preprotein by Kawano et al. [3]. No PACAP immunoreactivity was detected in the apical membrane of the marginal cells, basal cells, capillaries, nor in fibroblasts of the spiral ligament. PAC1-R was associated with the basolateral extension of the marginal cells in the stria vascularis. PAC1-R immunoreactivity was also present in the capillary endothelium, and between the marginal cells close to the apical surface in the region of tight junctions. However, there was no PAC1-R immunoreactivity either in the basal cells or in the apical membrane of the marginal cells. The co-expression of PACAP and its receptor on the marginal cells suggests an autocrine/intracrine mechanism of PACAP which was described earlier in several other cell types [14]. PACAP either might be secreted from the marginal cells or could enter from blood plasma. The activation of PAC1-R in the lateral membranes of marginal cells may lead to the activation of protein kinase A (PKA), which mediates the phosphorylation of claudin-3, which is a protein responsible for tight-junction integrity [15–17]. This can affect tight junction functions, which play a role in cell permeability, polarity, and possibly also establishing the endolymphatic potential. The proximity of K^+ channels to PAC1-R suggests the role of PACAP in the regulation of the K^+ transport by affecting the IsK K^+-channels at the apical and Na^+-K^+-Cl^--cotransporter and/or Na^+-K^+-ATPase at the basolateral extensions of the marginal cells through the cAMP-PKA signaling pathway [15].

Drescher et al. [2] examined the localization of PACAP, PAC1-R, and several afferent and efferent neuronal markers [choline acetyltransferase (ChAT), gluta-

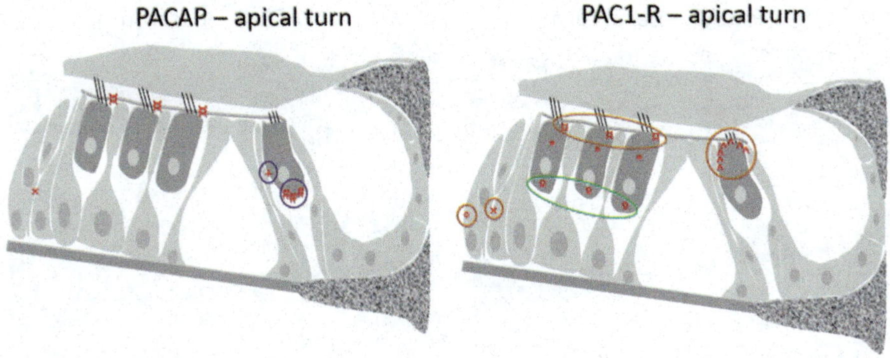

Fig. 31.2 Localization of PACAP and PAC1-R in the organ of Corti in the apical turn of the cochlea. PACAP immunoreactivity: cup under the inner hair cell (#), between inner hair cell and inner pillar cell (+), reticular lamina at the outer hair cell region (¤), Hensen's cell region (×). PAC1-R immunoreactivity: sites at the apical part of the outer hair cells (*), inverted V form at the apical site of the inner hair cell (^), base of the outer hair cells (o̲), reticular lamina in the outer and inner hair cell regions (¤), Hensen's cell region (x), Claudius' cell region (o). *green circle*: colocalization with GluR2/3; *purple circle*: colocalization with ChAT; *brown circle*: colocalization with DBH

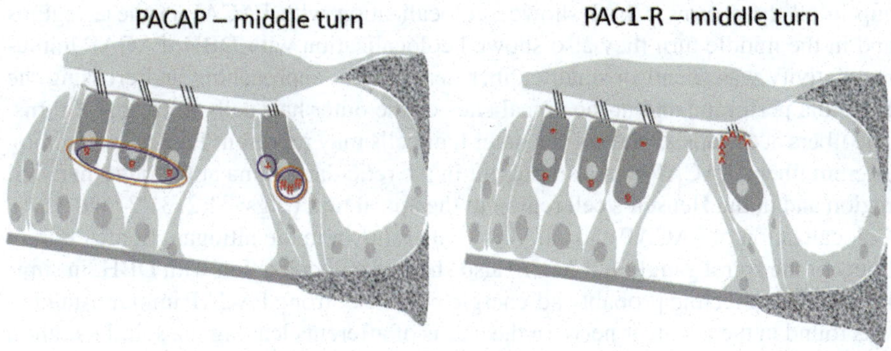

Fig. 31.3 Localization of PACAP and PAC1-R in the organ of Corti in the middle turn of the cochlea. PACAP immunoreactivity: cup under the inner hair cell (#), between inner hair cell and inner pillar cell (+), base of the outer hair cells (o̲). PAC1-R immunoreactivity: sites at the apical part of the outer hair cells (*), base of the outer hair cells (o̲), inverted V form at the apical site of the inner hair cell (^). *purple circle*: colocalization with ChAT; *brown circle*: colocalization with DBH

mate receptor 2/3 (GluR2/3), dopamine β-hydroxylase (DBH)] with immunohistochemistry in the cochlea of 2-month-old ACI Black Agouti rats. Their aim was to show the localization of PACAP and PAC1-R in the organ of Corti (Figs. 31.2, 31.3, and 31.4) and spiral ganglion, and elucidate the role of PACAP in the afferent and efferent signaling of the inner ear.

PACAP was found at the base of outer hair cells in the basal and middle turns, in small-caliber nerve fibers under and lateral of inner hair cells forming an efferent

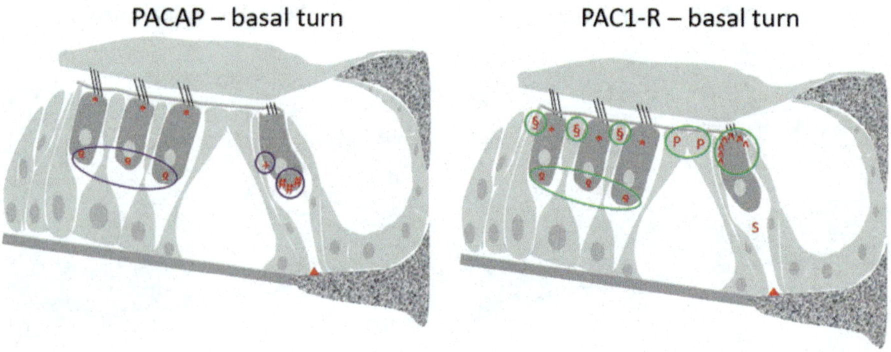

Fig. 31.4 Localization of PACAP and PAC1-R in the organ of Corti in the basal turn of the cochlea. PACAP immunoreactivity: cup under the inner hair cell (#), lateral from inner hair cell (+), base of the outer hair cells (o), apical part of the outer hair cells (*), habenula perforata (*filled triangle*). PAC1-R immunoreactivity: sites at the apical part of the outer hair cells (*), inverted V form at the apical site of the inner hair cell (^), habenula perforata (*filled triangle*), base of the outer hair cells (o), apical site of Deiters' cells (§), head of the inner and outer pillar cells (P), diagonal fibers under the inner hair cell (S). *green circle*: colocalization with GluR2/3; *purple circle*: colocalization with ChAT

cup in all three turns. ChAT showed colocalization with PACAP in these regions and in the middle turn they also showed colocalization with DBH. PACAP immunoreactivity was found in small-caliber nerve fibers approaching and crossing the habenula perforata, and at the apical sites of the outer hair cells in the basal turns. The fibers at the apical sites of the outer hair cells may represent PACAP containing efferent fibers. PACAP was also present in the reticular lamina at the outer hair cell region and in the Hensen's cell region in the apical turn (Figs. 31.2, 31.3, and 31.4). Colocalization of PACAP and ChAT was also shown in the intraganglionic efferent fibers of the spiral ganglion. PACAP also showed colocalization with DBH, in similar regions, reflecting probably adrenergic neurons. Strong PACAP immunostaining was found in the cochlear nerve in the axons of afferents leaving the spiral ganglion (Fig. 31.1), corresponding to previous PACAP preprotein mRNA findings of Kawano et al. [3]. Although type I axons had high immunoreactivity the cell bodies showed low density of immunoreactivity which suggests afferent axonal transport of PACAP in these cells, presumably to modulate excitatory amino acid release at the axon terminals of the afferent cells in the cochlear nucleus. The efferent intraganglionic spiral bundle showed also PACAP immunoreactivity. These small-caliber nerve fibers were overlapping type I/II afferent cell bodies and showing colocalization with ChAT.

PAC1-R immunoreactivity was found in small-caliber nerve fibers approaching the habenula perforata in the basal turn. It was also found as an inverted V form at the apical site of the inner hair cells in all three turns, showing colocalization with GluR2/3 in the basal and with DBH in the apical turns. PAC1-R was also found at the base and at the apical part of the outer hair cells in all three turns, at the base colocalizing with Glu2/3 in the basal and apical turns. It was also found in nerve fibers at the head of the inner and outer pillar cells in the basal turn also colocalizing with GluR2/3. It was also found in the reticular lamina in the outer hair cell regions in the apical turn, showing colocalization with DBH, and in diagonal fibers under the inner hair cell in the basal turn. Regarding the supporting cells there was PAC-1R immunopositivity at the apical site of Deiters' cells in the basal turn (GluR2/3 colocalization), and in the Hensen's and Claudius' cell region in the apical turns colocalizing with DBH (Figs. 31.2, 31.3 and 31.4). PAC1-R immunoreactivity was found in type I afferent cells in the spiral ganglion, in small-caliber nerve fibers in the spiral limbus, and in the cochlear nerve.

PACAP showed colocalization with ChAT and DBH in the efferent cholinergic and adrenergic fibers suggesting that PACAP acts as an efferent neuromodulator in the inner ear. It is known that type I afferents of the inner hair cells get cholinergic and partially dopaminergic afferentation from lateral olivocochlear fibers [18, 19]. Dopamine acts as a protective neurotransmitter in case of excitotoxicity for type I afferents [20, 21] and PACAP is able to upregulate dopaminergic functions [22]. Drescher et al. [2] showed more evidence for the previous findings of Spoendlin [23], that adrenergic efferent innervation is also present in different sites of the organ of Corti, such as supranuclear sites of inner and outer hair cells, and Deiters' and Hensen's cell regions. PACAP is associated with adrenergic nerve terminals [24] and takes part in the modification of adrenergic functions [25].

Glutamate is the primary neurotransmitter in the cochlea [18, 26] and the release of glutamate in acoustic trauma/ischemia may lead to cell death and destruction of the afferent nerve endings [27, 28]. Morio et al. [29] showed that in neuronal cells PACAP suppressed glutamate-induced cell death. Drescher et al. [2] demonstrated the presence of PAC1-R in cell bodies and dendrites of type I afferent cells in the spiral ganglion. It also showed colocalization with GluR2/3 under the outer hair cells in the organ of Corti suggesting that PACAP interacts directly with glutamate-mediated signal transduction in the auditory system. Based on these finding PACAP may prevent glutamate-induced cell death in the inner ear upon various insults.

Our research group examined the inner ear of wild type (WT), heterozygous (HZ) and homozygous PACAP-deficient (KO) mice [7]. The gross anatomical morphology of PACAP-deficient mice is similar to the WT mice, but they maintain an increased vulnerability against various insults and have an altered behavior, decreased reproductive function and several biochemical alterations [30]. Partial lack of PACAP also leads to higher vulnerability of heterozygous PACAP-deficient mice [31–35]. We used hematoxylin-eosin staining to show the microscopic structure of the inner ear, and immunofluorescent staining against PAC1-R. The morphology of the inner ear showed no differences between the homozygous PACAP-deficient and WT mice. Our research group found PAC1-R immunopositivity in both inner and outer hair cells, outer phalangeal cells (Deiters' cells), and pillar cells in accordance with the previous results of Kawano et al. [3], Abu-Hamdan et al. [1] and Drescher et al. [2]. The localization of PAC1-R immunoreactivity did not show any differences between homozygous PACAP-deficient and WT mice, but in PACAP-deficient mice the hair cells and outer phalangeal cells showed lower intensity of PAC1-R immunoreactivity compared to WT mice.

PACAP in the Brainstem and Thalamus

Hannibal [4] used mouse monoclonal antibody to show PACAP immunoreactivity in the adult rat brain (Fig. 31.5). The cell bodies and nerve fibers of the cochlear nuclei (especially the ventral cochlear nuclei) showed moderate PACAP expression. Furthermore the root of the cochlear nerve showed also PACAP immunoreactivity. The deep layer of inferior colliculus contained also weak PACAP positive cell bodies and nerve fibers. Nerve fibers showed weak PACAP immunoreactivity in the medial geniculate body (Fig. 31.5). Palkovits et al. [36] also found PACAP in the inferior colliculus and medial geniculate nucleus in human brain samples with radioimmunoassay examination.

Kausz et al. [5] found PACAP immunoreactive cell bodies in the dorsal and ventral cochlear nucleus, superior olivary complex (SOC), and trapezoid nucleus (Fig. 31.5). Reuss et al. [6] described the localization of PACAP, vasoactive intestinal polypeptide

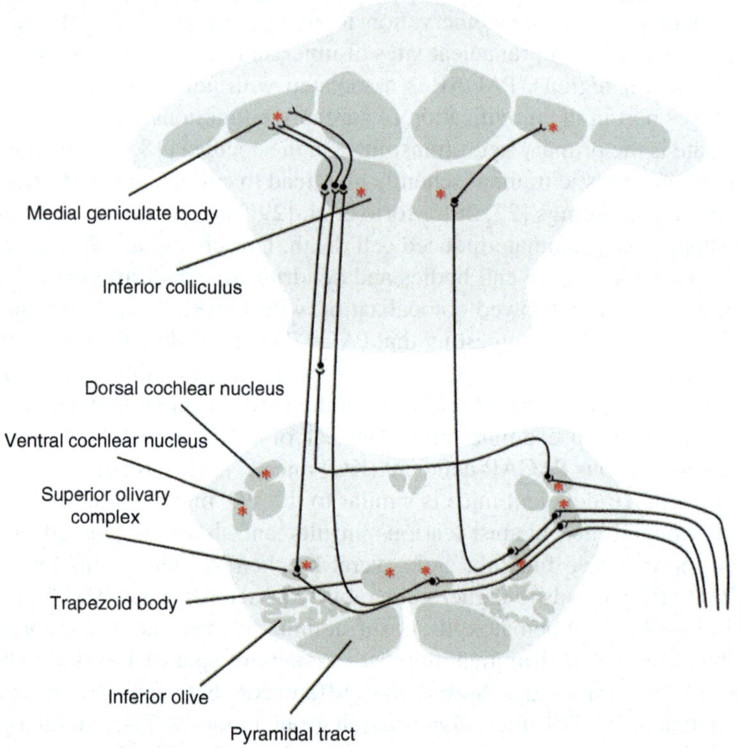

Fig. 31.5 Schematic picture of the auditory pathway showing the occurrence of PACAP in the brainstem and thalamus. PACAP immunoreactivity was present in cell bodies and nerve fibers of dorsal and ventral cochlear nuclei. It was also found in several parts of the superior olivary complex (SOC), such as the medial nucleus of the trapezoid body (MNTB), medial superior olive, ventral nucleus of the trapezoid body. PACAP immunopositive cell bodies were also found in the inferior colliculus, and PACAP immunoreactive nerve fibers in the medial geniculate body

(VIP), neuronal nitric oxide synthase (nNOS) and other neurotransmitters in the SOC in adult male Djungarian hamsters (phodopus sungorus) using immunohistochemistry and retrograde tracing. PACAP immunoreactivity was found in approximately one third of all neurons, whereas VIP was not present in the neurons of the SOC. PACAP was present in many large neurons of the medial nucleus of the trapezoid body (MNTB) and in the medial superior olive. These cells are suggested to project to the lateral superior olive, periolivary regions, and inferior colliculus [37]. The large PACAP-immunopositive neurons of the medial superior olive were also nNOS positive. One fourth of all neurons contained PACAP in the lateral superior olive and in the periolivary regions. 40 % of the olivocochlear neurons, which were positive with fluorochrome after the retrograde tracing in MNTB and ventral nucleus of the trapezoid body, were also positive for PACAP, such as some periolivary neurons (Fig. 31.5). Therefore, these cells could serve as the origin of PACAP found in the cochlea by Kawano et al. [3]. PACAP was found in the medial olivocochlear efferents, which target the outer hair cells, but not in the lateral olivocochlear efferents, which target the inner hair cells. This further substantiates the theory of Drescher et al. [2] that PACAP plays a greater role in the efferent signaling of the cochlea, and through it an indirect effect on the afferent signaling.

Functional Studies

Effects of PACAP on the Endolymph Composition

Our research group examined the protein composition of chicken endolymph after PACAP38 treatment [8]. We treated 1-day-old chicken with intraperitoneal PACAP38 injection. Endolymph was collected 1, 6, and 24 h after PACAP38 injection. Then we removed the membranous labyrinth and the endolymph was squeezed to a sterile paper strip. Then we utilized microchip electrophoresis using protein 230 Plus LabChip Kit evaluated with Agilent 2100 Bioanalyzer System [8].

Endolymph, which fills the membranous labyrinth, is produced in the stria vascularis, it is rich in potassium and poor in sodium. However, perilymph, which fills the scala tympani and vestibuli, is rich in sodium and poor in potassium. Although endolymph is derived from the plasma, its composition resembles more the composition of the intracellular fluid. Endolymph affects several physiological and pathological functions of the labyrinth [38, 39]. Several neuropeptides have an influence on endolymph secretion, e.g., substance P, vasopressin, and somatostatin [40, 41]. Abu-Hamdan et al. [1] described the presence of PACAP and PAC1-R in the stria vascularis. The expression of the peptide and its receptor increases from basal to apical turns. Basal-apical gradient of other components, such as the endolymphatic potential [42] and endolymphatic K^+ and Cl^- concentrations [43] in the cochlea had already been described. The gradient present in both PACAP and PAC1-R might be partly the cause for these gradients, as they could contribute to the maintenance of endolymphatic potential through regulating the tight junction permeability in the stria vascularis. Based on these results we aimed at elucidating the role of PACAP in endolymph production.

However, there are no data for the protein composition of chicken endolymph, it is well described in other species. It has been studied mostly in human and guinea pigs samples. It yields a complex protein composition containing albumin, α-chymotrypsin, α-antitrypsin, transferrin, apolipoprotein D (ApoD), apolipoprotein J (ApoJ), fetuin, and other proteins. The composition of these proteins is similar to the perilymph, but the level of proteins is approximately the 15–20 % of the perilymph levels. That could be based on the mechanism of production, such as the proteins of the endolymph are filtered from the perilymph through the basal and marginal cells of the spiral ligament. Endolymph is rich in ApoD and ApoJ proteins which are at higher levels in the perilymph and liquor but low in the blood plasma. Their function is presumably to protect the integrity of the extracellular membranes bordering the endolymphatic space.

In our experiment we found several protein peaks between 14 and 80 kDa, the range where the majority of the above mentioned proteins are located. There was no significant difference between control and PACAP-treated groups at any given time. Based on these results we can conclude that under the given circumstances there was no change on the composition of the endolymph proteins after PACAP treatment. However, these negative findings can be a consequence of the chosen circumstances. We could not exclude that other species, route of PACAP administration or applied dose would cause significant changes in the protein composition of the endolymph.

Effects of PACAP on Apoptosis in an Inner Ear Cell Culture In Vitro

Although physiological apoptosis is essential in the development of the inner ear, increased apoptosis, caused by toxins, acoustic overstimulation, oxidative stress, infections, etc., leads to permanent cell damage and loss of hearing [44, 45]. Oxidative stress is highly involved in these pathological conditions, hence antioxidative and antiapoptotic drugs, among others PACAP [46–48], could be used as inner ear protecting factors [49, 50]. Our research group studied the effects of PACAP on apoptosis in a chicken in vitro cell culture model [9] which contained both sensory hair cells and supporting cells. We used 6 series of 40 newly hatched chickens for this study. Chicken cochlea is simpler and easier to dissect, therefore, a favorite model for cochlear experiments, since the reactive oxygen species (ROS) generation and apoptotic reaction to toxic agents (e.g., aminoglycoside) is similar to mammals [51–53]. We used H_2O_2 as a ROS generating molecule for causing cell death via oxidative stress. Treatment lasted 2 hours, and evaluation followed immediately thereafter.

MTT assay showed that there were more than 90 % living cells in the control group and PACAP treatment alone did not result in significant alterations. It also showed that H_2O_2 treatment led to a significant drop in cell viability and an increase in the number of apoptotic and necrotic cells. Cell viability significantly increased when H O -treated cells were co-incubated with PACAP. JC-1 assay is for the detection of apoptosis, and this staining also showed higher cell viability in case of H_2O_2

and PACAP co-administration compared to H_2O_2 treatment. AnnexinV and propidium iodide co-staining verified these results: co-administration of PACAP with H_2O_2 resulted in a significant increase of living cells, and significant decrease of apoptotic cells compared to the H_2O_2-treated group. There was also a decrease in the number of necrotic cells compared to the H_2O_2-treated group, but the difference was not significant. The protected cells are suggested to be primarily sensory hair cells, as the used H_2O_2 concentration causes damage to the hair cells with no injury to other neuroepithelial cells [54].

These antiapoptotic effects of PACAP were verified by the regression of the activation of the caspase-3 apoptotic protein in case of PACAP and H_2O_2 co-treatment compared to the H_2O_2-treated cells. The protective effect of PACAP is based on several molecular signaling pathways. Most of these effects is mediated by PAC1-R [55, 56]. PAC1-R activates adenylate cyclase (AC) and phospholipase C (PLC). AC activation leads to the activation of the PKA signaling pathways resulting in the elevation of the antiapoptotic extracellular-signal-regulated kinase (ERK), phosphorylation of cAMP response element-binding protein (CREB), and inhibition of proapoptotic c-Jun N-terminal kinase (JNK) and p38 mitogen-activated protein kinases (p38MAPK). PACAP also affects the intrinsic mitochondrial pathway of apoptosis, through activation of the antiapoptotic members of the Bcl peptide family (Bcl-2, Bcl-xL) and inhibition of the proapoptotic Bad, Bax. These effects are moderated partly through the 14-3-3 protein, which phosphorylates Bad, therefore, it is not able to bind and inactivate Bcl-2 and Bcl-xL. These effects result in decreased release of cytochrome *c* from the mitochondria, which also results in the inhibition of apoptosis. PLC activation leads to decreased caspase activation through inositol trisphosphate and diacylglycerol pathways. The antiapoptotic effects and the inhibition of caspase-3 activation could provide the molecular background, how PACAP can protect the inner ear cells against glutamatergic excitotoxicity, aminoglycoside, or other drug effects, during acoustic overstimulation or in other pathological conditions.

Effects of PACAP on Ca^{2+}-Binding Protein Expression of Hair Cells

We investigated the effects of endogenous PACAP on hair cell Ca^{2+}-binding protein composition under control circumstances and in kanamycin induced ototoxicity [7, 10]. The Ca^{2+}-concentration in the endolymph and the intracellular Ca^{2+} concentration in the hair cells are strictly regulated and important for normal hearing processes [57]. Several pathological conditions lead to elevated Ca^{2+} concentration in endolymph and hair cells, such as acoustic overstimulation [58, 59], drug induced ototoxicity [60], vestibular labyrinth destruction [57], and surgically induced endolymphatic hydrops [61]. The subsequent high intracellular Ca^{2+} concentration is toxic to cells [62, 63]. Ca^{2+}-buffering proteins (calretinin, parvalbumin, calbindin, etc.) presumably play an essential role in regulating the Ca^{2+} level of hair cells and may protect them against the toxic effects of high intracellular Ca^{2+} concentration [64–68].

We used 5-day-old (P5) and 7-day-old (P7) WT, HZ, and PACAP KO mice in the experiments. We chose this age group because the development of the inner ear of mice still goes on during this period of time and all three Ca^{2+}-binding proteins are present in the inner and outer hair cells on these postnatal days based on previous studies in various animal species [65]. To model toxic insults to the inner ear we chose kanamycin as an ototoxic agent. It is a member of the aminoglycoside antibiotics, which are widely used in the clinical practice together with the platinum based chemotherapeutics although possibly causing permanent cochlear damage [69]. Kanamycin causes hair cell damage mainly through ROS production; hence antioxidant and antiapoptotic drugs (such as PACAP) can be possibly used as otoprotective agents [69, 70].

We used immunofluorescent staining against Ca^{2+}-binding proteins on sections produced from the cochlea on P5 and P7. Mice sacrificed on P5 did not receive any treatment, mice sacrificed on P7 were treated with one dose of kanamycin (1 mg/g body weight) or physiological saline as a control on P5. The localization of parvalbumin, calretinin, and calbindin was in accordance with previously described results and did not show any differences between the groups. However, we found significant alterations in the Ca^{2+}-binding protein density. All three Ca^{2+}-binding proteins showed stronger immunopositivity in the hair cells of homozygous PACAP-deficient mice, than in WT mice on P5. We found similar results for parvalbumin (Fig. 31.6a, e) and calretinin (Fig. 31.7:a, e) on P7 in the physiological saline treated groups (calbindin was not examined on P7). Parvalbumin and calretinin immunopositivity was especially pronounced in the hair bundles of the hair cells. Heterozygous PACAP-deficient animals in the physiological saline group also showed higher expression of calretinin (Fig. 31.7a, c) in inner and outer hair cells, and higher expression of parvalbumin in the outer hair cells compared to WT mice (Fig. 31.6a, c) on P7.

Parvalbumin expression in the hair cells of both WT and heterozygous PACAP-deficient mice was elevated after kanamycin treatment on P7, and the expression was significantly higher in the heterozygous PACAP-deficient mice, compared to WT mice (Fig. 31.6a–d). The baseline stronger parvalbumin immunopositivity of homozygous PACAP-deficient mice did not increase further upon kanamycin treatment (Fig. 31.6e, f). Calretinin immunolabeling yielded similar results in WT and homozygous PACAP-deficient mice (Fig. 31.7a, b, e, f) on P7 as parvalbumin. Heterozygous PACAP-deficient mice, similarly to homozygous PACAP-deficient mice showed high immunoreactivity of calretinin under control circumstances; therefore, after kanamycin treatment this higher baseline immunofluorescence did not increase further (Fig. 31.7c, d).

Apoptosis is caused by ROS generation in aminoglycoside induced ototoxicity, and leads to cell death through an extrinsic, death-receptor-mediated pathway and an intrinsic, mitochondria-mediated pathway [69, 71]. In case of aminoglycoside-induced ototoxicity the intrinsic pathway has a greater role by activating G proteins followed by stress-activated protein kinase activation (JNK). Subsequently cytochrome *c* release, caspase-8 and 9 activation and elevation of intracellular Ca^{2+} concentration follows [62, 63]. The elevation of Ca^{2+} concentration and the activation of proapoptotic pathways can be prevented by increased expression of Ca^{2+}-

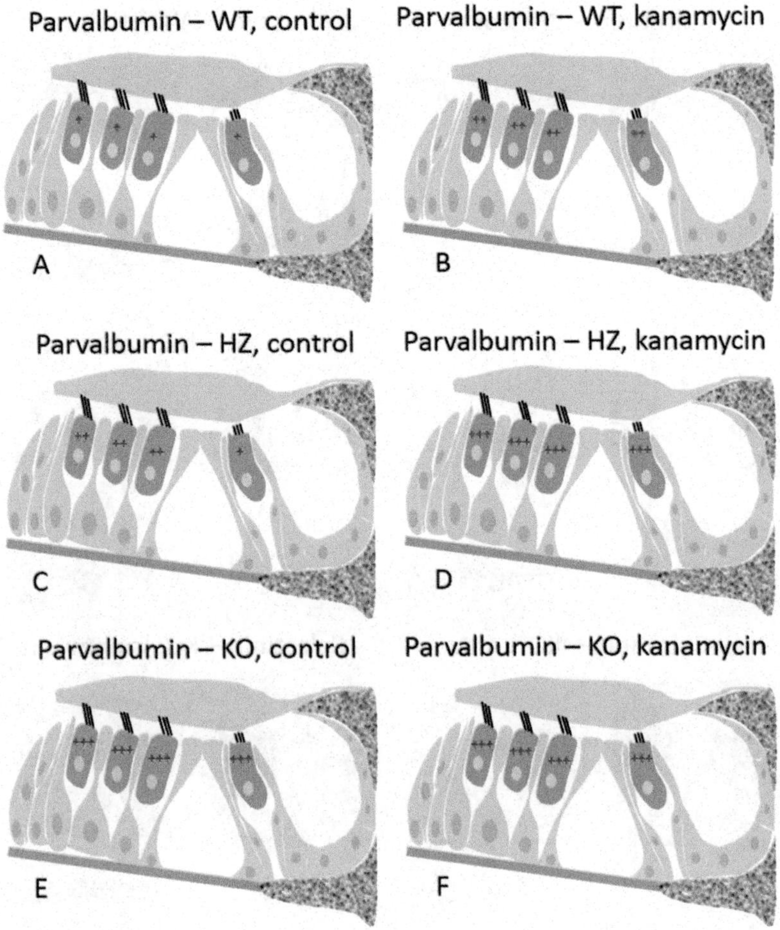

Fig. 31.6 Parvalbumin expression in inner and outer hair cells in wild-type (WT), heterozygous (HZ), and homozygous PACAP-deficient (KO) mice under control circumstances and after kanamycin treatment. + mild, ++ moderate, +++ strong immunoreactivity

binding proteins. When the buffer capacity of the Ca^{2+}-binding proteins reaches its maximum the intracellular Ca^{2+} concentration increases and leads to the activation of apoptotic pathways [72].

Although the direct effect of PACAP is to elevate the intracellular Ca^{2+} concentration through PLC signaling [73] it was also shown in hippocampal neurons that PACAP can protect the cells against the glutamate-induced toxic Ca^{2+} concentration elevation [74]. Our results show that in the lack of PACAP the Ca^{2+}-binding protein expression increases in the inner and outer hair cells, which is a probable endogenous compensatory mechanism against the elevated Ca^{2+} level in the hair cells caused by the lack of PACAP. However, the elevated Ca^{2+}-binding protein level cannot be further upregulated upon ototoxic insults, as the Ca^{2+}-binding protein expression reaches its maximum level.

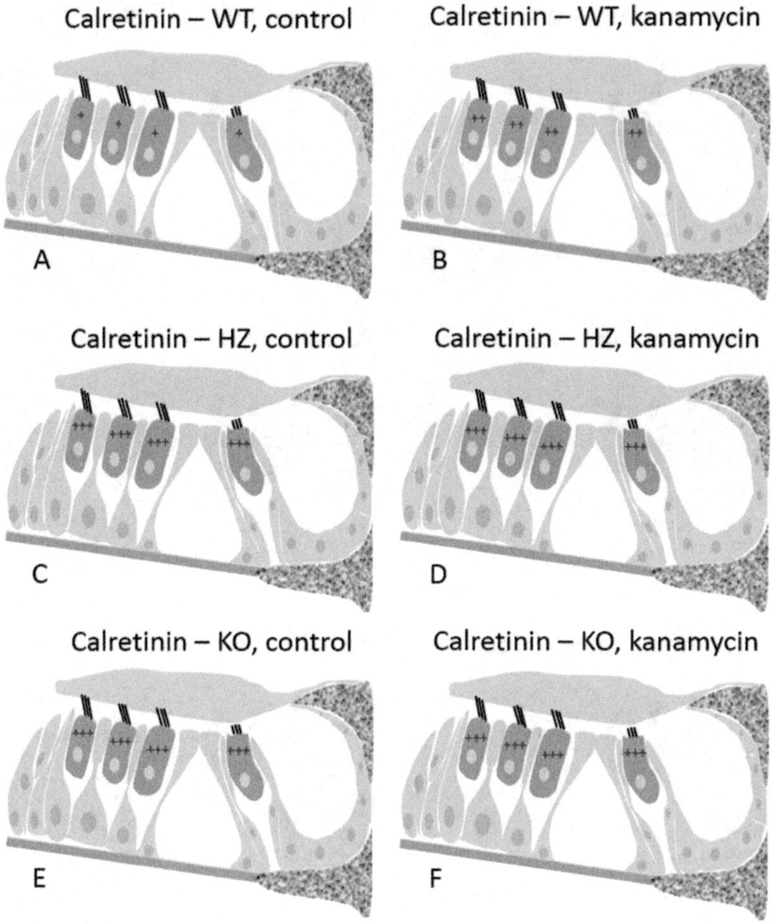

Fig. 31.7 Calretinin expression in inner and outer hair cells in wild-type (WT), heterozygous (HZ), and homozygous PACAP-deficient (KO) mice under control circumstances and after kanamycin treatment. + mild, ++ moderate, +++ strong immunoreactivity

Discussion

PACAP and PAC1-R were described in several parts of the auditory system. They are present in the organ of Corti in hair cells, supporting cells, and different nerve fibers. They were also showed in spiral ganglion and stria vascularis. Their presence in the stria vascularis suggests a role of PACAP affecting endolymph production which we tried to prove by studying the endolymph protein composition in chicken cochlea after PACAP treatment but we did not find any differences in this experimental setup. Colocalization of PACAP and PAC1-R with GluR2/3 and DBH in several structures in the cochlea, mostly efferent nerve fibers, suggests that PACAP could theoretically modulate glutamatergic, dopaminergic, cholinergic, and adrenergic efferents of hair cells, whereby they would also take place in the regulation of the afferent signaling. The presence of PACAP and its receptor PAC1-R was shown in several parts of the auditory pathway (cochlear nuclei, SOC, inferior colliculus, medial geniculate body). PACAP was found in the olivocochlear neurons in the SOC, which could serve as the origin of PACAP found in the cochlea.

PACAP protects the hair cells from H_2O_2 toxicity in vitro, increasing the number of living cells mostly through decreasing apoptosis. In PACAP-deficient mice there is a higher concentration of Ca^{2+}-binding proteins in the hair cells, which is probably a protective mechanism against the hypothesized higher intracellular Ca^{2+} concentration caused by the pathological conditions in the lack of PACAP. Kanamycin treatment causes also Ca^{2+}-binding protein level elevation in WT and HZ PACAP-deficient mice, but there is no further elevation in homozygous PACAP-deficient mice, where the baseline Ca^{2+}-binding protein level is already elevated.

In summary, PACAP and PAC1-R are present and cause functional changes in the cochlea. With further examination we hope to prove the protective effects of the exogenous PACAP in ototoxicity in vivo and examine its role throughout various clinical studies.

Acknowledgement MTA-PTE "Lendulet" Program, OTKA K104984, 119759, Arimura Foundation, TAMOP 4.2.4.A/2-11-1-2012-0001 "National Excellence Program," New National Excellence Program (UNKP), Janos Bolyai Research Scholarship of Hungarian Academy of Sciences. This work is dedicated to the 650th anniversary of the University of Pecs.

References

1. Abu-Hamdan MD, Drescher MJ, Ramakrishnan NA, Khan KM, Toma VS, Hatfield JS, et al. Pituitary adenylyl cyclase-activating polypeptide (PACAP) and its receptor (PAC1- R) in the cochlea: evidence for specific transcript expression of PAC1-R splice variants in rat microdissected cochlear subfractions. Neuroscience. 2006;140:147–61.
2. Drescher MJ, Drescher DG, Khan KM, Hatfield JS, Ramakrishnan NA, Abu-Hamdan MD, et al. Pituitary adenylyl cyclase-activating polypeptide (PACAP) and its receptor (PAC1- R) are positioned to modulate afferent signaling in the cochlea. Neuroscience. 2006;142:139–64.
3. Kawano H, Shimozono M, Tono T, Miyata A, Komune S. Expression of pituitary adenylate cyclase-activating polypeptide mRNA in the cochlea of rats. Mol Brain Res. 2001;94:200–3.
4. Hannibal J. Pituitary adenylate cyclase-activating peptide in the rat central nervous system: an immunohistochemical and in situ hybridization study. J Comp Neurol. 2002;453:389–417.
5. Kausz M, Murai Z, Arimura A, Koves K. Distribution of pituitary adenylate cyclase activating polypeptide (PACAP) immunoreactive elements in the brain stem of rats studied by immunohistochemistry. Neurobiology. 1999;7:19–31.
6. Reuss S, Disque-Kaiser U, Antoniou-Lipfert P, Gholi MN, Riemann E, Riemann R. Neurochemistry of olivocochlear neurons in the hamster. Anat Rec (Hoboken). 2009;292:461–71.
7. Tamas A, Szabadfi K, Nemeth A, Fulop B, Kiss P, Atlasz T, et al. Comparative examination of inner ear in wild type and pituitary adenylate cyclase activating polypeptide (PACAP)-deficient mice. Neurotox Res. 2012;21:435–44.
8. Gaal V, Mark L, Kiss P, Kustos I, Tamas A, Kocsis B, et al. Investigation of the effects of PACAP on the composition of tear and endolymph proteins. J Mol Neurosci. 2008;36:321–9.

9. Racz B, Horvath G, Reglodi D, Gasz B, Kiss P, Gallyas Jr F, et al. PACAP ameliorates oxidative stress in the chicken inner ear: an in vitro study. Regul Pept. 2010;160:91–8.
10. Nemeth A, Szabadfi K, Fulop B, Reglodi D, Kiss P, Farkas J, et al. Examination of calcium-binding protein expression in the inner ear of wild-type, heterozygous and homozygous pituitary adenylate cyclase-activating polypeptide (PACAP)-knockout mice in kanamycin-induced ototoxicity. Neurotox Res. 2014;25:57–67.
11. Spengler D, Waeber C, Pantaloni C, Holsboer F, Boekaert J, Seeburg PH, et al. Differential signal transduction by five splice variants of the PACAP receptor. Nature. 1993;365:170–5.
12. Chatterjee TK, Sharma RV, Fisher RA. Molecular cloning of a novel variant of the pituitary adenylate cyclase-activating polypeptide (PACAP) receptor that stimulates calcium influx by activation of L-type calcium channels. J Biol Chem. 1996;271:32226–32.
13. Dautzenberg FM, Mevenkamp G, Wille S, Hauger RL. N-terminal splice variants of the type I PACAP receptor: isolation, characterization and ligand binding/selectivity determinants. J Neuroendocrinol. 1999;11:941–9.
14. Shintani N, Suetake S, Hashimoto H, Koga K, Kasai A, Kawaguchi C, et al. Neuroprotective action of endogenous PACAP in cultured rat cortical neurons. Regul Pept. 2005;126:123–8.
15. Drescher MJ, Khan KM, Hatfield JS, Shakir AH, Drescher DG. Immunohistochemical localization of adenylyl cyclase isoforms in the lateral wall of the rat cochlea. Mol Brain Res. 2000;76:289–98.
16. Florian P, Amasheh S, Lessidrensky M, Todt I, Bloedow A, Ernst A, et al. Claudins in the tight junctions of stria vascularis marginal cells. Biochem Biophys Res Commun. 2003;304:5–10.
17. Kitajiri S, Furuse M, Morita K, Saishin-Kiuchi Y, Kido H, Ito J, et al. Expression patterns of claudins tight junction adhesion molecules in the inner ear. Hear Res. 2004;187:25–34.
18. Eybalin M. Neurotransmitters and neuromodulators of the mammalian cochlea. Physiol Rev. 1993;73:309–73.
19. Morley BJ, Li H-S, Hiel H, Drescher DG, Elgoyhen AB. Identification of the subunits of the nicotinic cholinergic receptors in the rat cochlea using RT-PCR and in situ hybridization. Mol Brain Res. 1998;53:78–87.
20. D'Aldin C, Puel JL, Leducq R, Crambes O, Eybalin M, Pujol R. Effects of a dopaminergic agonist in the guinea pig cochlea. Hear Res. 1995;90:202–11.
21. Pujol R, Puel JL. Excitotoxicity, synaptic repair, and functional recovery in the mammalian cochlea: a review of recent findings. Ann N Y Acad Sci. 1999;884:249–54.
22. Anderson ST, Kusters DHL, Clarke IJ, Pow DV, Curlewis JD. Expression of pituitary adenylate cyclase activating polypeptide type 1 receptor (PAC1R) in the ewe hypothalamus: distribution and colocalization with tyrosine hydroxylase-immunoreactive neurones. J Neuroendocrinol. 2005;17:298–305.
23. Spoendlin H. Primary neurons and synapses. In: Friedmann I, Ballantyne J, editors. Ultrastructural atlas of the inner ear. London: Butterworths; 1984. p. 133–64.
24. Shioda S, Yada T, Muroya S, Uramura S, Nakajo S, Ohtaki H, et al. Functional significance of colocalization of PACAP and catecholamine in nerve terminals. Ann N Y Acad Sci. 2000;921:211–7.
25. May V, Beaudet MM, Parsons RL, Hardwick JC, Gauthier EA, Durda JP, et al. Mechanisms of pituitary adenylate cyclase activating polypeptide (PACAP)-induced depolarization of sympathetic superior cervical ganglion (SCG) neurons. Ann N Y Acad Sci. 1998;865:164–75.
26. Matsubara A, Laake JH, Davanger S, Usami S, Ottersen OP. Organization of AMPA receptor subunits at a glutamate synapse: a quantitative immunogold analysis of hair cell synapses in the rat organ of Corti. J Neurosci. 1996;16:4457–67.
27. Puel JL. Chemical synaptic transmission in the cochlea. Prog Neurobiol. 1995;47:449–76.
28. Pujol R, Puel JL, Gervais d'Aldin C, Eybalin M. Pathophysiology of the glutamatergic synapses in the cochlea. Acta Otolaryngol Stockh. 1993;113:330–4.
29. Morio H, Tatsuno I, Hirai A, Tamura Y, Saito Y. Pituitary adenylate cyclase-activating polypeptide protects rat-cultured cortical neurons from glutamate-induced cytotoxicity. Brain Res. 1996;741:82–8.

30. Reglodi D, Kiss P, Szabadfi K, Atlasz T, Gabriel R, Horvath G, et al. PACAP is an endogenous protective factor-insights from PACAP-deficient mice. J Mol Neurosci. 2012;48:482–92.
31. Endo K, Nakamachi T, Seki T, Kagami N, Wada Y, Nakamura K, et al. Neuroprotective effect of PACAP against NMDA-induced retinal damage in the mouse. J Mol Neurosci. 2011;43:22–9.
32. May V, Vizzard MA. Bladder dysfunction and altered somatic sensitivity inPACAP-/- mice. J Urol. 2010;183:772–9.
33. Nakamachi T, Ohtaki H, Yofu S, Watanabe J, Mori H, Sato A, et al. Endogenous pituitary adenylate cyclase activating polypeptide is involved in suppression of edema in the ischemic brain. Acta Neurochir Suppl. 2010;106:43–6.
34. Ohtaki H, Nakamachi T, Dohi K, Aizawa Y, Takaki A, Hodoyama K, et al. Pituitary adenylate cyclase-activating polypeptide (PACAP) decreases ischemic neuronal cell death in association with IL-6. Proc Natl Acad Sci U S A. 2006;103:7488–93.
35. Tsuchikawa D, Nakamachi T, Tsuchida M, Wada Y, Hori M, Farkas J, et al. Neuroprotective effect of endogenous pituitary adenylate cyclase-activating polypeptide on spinal cord injury. J Mol Neurosci. 2012;48:508–17.
36. Palkovits M, Somogyvari-Vigh A, Arimura A. Concentrations of pituitary adenylate cyclase activating polypeptide (PACAP) in human brain nuclei. Brain Res. 1995;699:116–20.
37. Schwartz IR. The superior olivary complex and lateral lemniscal nuclei. In: Webster DB, Popper AN, Fay RF, editors. The mammalian auditory pathway: neuroanatomy. New York: Springer; 1992. p. 117–67.
38. Fukushima K, Takeda T, Kakigi A, Takeda S, Sawada S, Nishioka R, et al. Effects of lithium on endolymph homeostasis and experimentally induced endolymphatic hydrops. ORL J Otorhinolaryngol Relat Spec. 2005;67:282–8.
39. Hanner P, Jennische E, Lange S, Lonnroth I, Wahlstrom B. Increased antisecretory factor reduces vertigo in patients with Meniere's disease: a pilot study. Hear Res. 2004;190:31–6.
40. Ferrary E, Sterkers O. Mechanisms of endolymph secretion. Kidney Int Suppl. 1998;65:S98–103.
41. Kitano H, Suzuki M, Kitanishi T, Yazawa Y, Kitajima K, Isono T, et al. Regulation of inner ear fluid in the rat by vasopressin. Neuroreport. 1999;10:1205–7.
42. Misrahy GA, Hildreth KM, Shinabarger EW, Gannon WJ. Electrical properties of wall of endolymphatic space of the cochlea (guinea pig). Am J Physiol. 1958;194:396–402.
43. Sterkers O, Ferrary E, Amiel C. Inter- and intracompartmental osmotic gradients within the rat cochlea. Am J Physiol. 1984;247:F602–6.
44. Huang T, Cheng AG, Stupak H, Liu W, Kim A, Staecker H, et al. Oxidative stress-induced apoptosis of cochlear sensory cells: otoprotective strategies. Int J Dev Neurosci. 2000;18:259–70.
45. Yehoash R. Cochlear pathology, sensory cell death and regeneration. Br Med Bull. 2002;63:25–38.
46. Ohtaki H, Nakamachi T, Dohi K, Shioda S. Role of PACAP in ischemic neural death. J Mol Neurosci. 2008;36:16–25.
47. Reglodi D, Fabian ZS, Tamas A, Lubics A, Szeberenyi J, Alexy T, et al. Effects of PACAP on in vitro and in vivo neuronal cell death, platelet aggregation, and production of reactive oxygen radicals. Regul Pept. 2004;123:51–9.
48. Reglodi D, Kiss P, Lubics A, Tamas A. Review on the protective effects of PACAP in models of neurodegenerative diseases in vitro and in vivo. Curr Pharm Des. 2011;17:962–72.
49. Kopke RD, Liu W, Gabaiyadeh R, Jacono A, Feghali J, Spray D, et al. Use of organotypic cultures of Corti's organ to study the protective effects of antioxidant molecules on cisplatininduced damage of auditory hair cells. Am J Otol. 1997;18:559–71.
50. Seidman MD, Vivek P. Intratympanic treatment of hearing loss with novel and traditional agents. Otolaryngol Clin North Am. 2004;37:973–90.
51. Cheng AG, Cunningham LL, Rubel EW. Hair cell death in the avian basilar papilla: characterization of the in vitro model and caspase activation. J Assoc Res Otolaryngol. 2003;4:91–105.
52. Hirose K, Westrum LE, Stone JS, Zirpel L, Rubel EW. Dynamic studies of ototoxicity in mature avian auditory epithelium. Ann N Y Acad Sci. 1999;884:389–409.

53. Matsui JI, Gale JE, Warchol ME. Critical signaling events during the aminoglycoside-induced death of sensory hair cells in vitro. J Neurobiol. 2004;61:250–66.
54. Dehne N, Lautermann J, ten Cate WJ, Rauen U, de Grott H. In vitro effects of hydrogen peroxide on the cochlear neurosensory epithelium of guinea pig. Hear Res. 2000;143:162–70.
55. Dejda A, Jolivel V, Bourgault S, Seaborn T, Fournier A, Vaudry H, et al. Inhibitory effect of PACAP on caspase activity in neuronal apoptosis: a better understanding towards therapeutic applications in neurodegenerative diseases. J Mol Neurosci. 2008;36:26–37.
56. Pantaloni C, Brabet P, Bilanges B, Dumuis A, Houssami S, Spengler D, et al. Alternative splicing in the N-terminal extracellular domain of the pituitary adenylate cyclaseactivating polypeptide (PACAP) receptor modulates receptor selectivity and relative potencies of PACAP-27 and PACAP-38 in phospholipase C activation. J Biol Chem. 1996;271:22146–51.
57. Ikeda R, Nakaya K, Oshima T, Kawase T, Kobayashi T. Calcium concentration in cochlear endolymph after vestibular labyrinth injury. Neuroreport. 2010;21:651–5.
58. Ikeda K, Kusakari J, Takasaka T. Ionic changes in cochlear endolymph of the guinea pig induced by acoustic injury. Hear Res. 1988;32:103–10.
59. Lendvai B, Halmos GB, Polony G, Kapocsi J, Horvath T, Aller M, et al. Chemical neuroprotection in the cochlea: the modulation of dopamine release from lateral olivocochlear efferents. Neurochem Int. 2011;59:150–8.
60. Li Y, Ding D, Jiang H, Fu Y, Salvi R. Co-administration of cisplatin and furosemide causes rapid and massive loss of cochlear hair cells in mice. Neurotox Res. 2011;20:307–19.
61. Salt AN, DeMott J. Endolymph calcium increases with time after surgical induction of hydrops in guinea-pigs. Hear Res. 1994;74:115–21.
62. Orrenius S, McCabe Jr MJ, Nicotera P. Ca(2+)-dependent mechanisms of cytotoxicity and programmed cell death. Toxicol Lett. 1992;64–65:357–64.
63. Trump BF, Berezesky IK. The role of altered [Ca2+]i regulation in apoptosis, oncosis, and necrosis. Biochim Biophys Acta. 1996;1313:173–8.
64. Gilbert RJ, Pothoulakis C, LaMont JT, Yakubovich M. Clostridium difficile toxin B activates calcium influx required for actin disassembly during cytotoxicity. Am J Physiol. 1995;268:G487–95.
65. Hackney CM, Mahendrasingam S, Penn A, Fettiplace R. The concentrations of calcium buffering proteins in mammalian cochlear hair cells. J Neurosci. 2005;25:7867–75.
66. Heizmann CW. Calcium-binding proteins: basic concepts and clinical implications. Gen Physiol Biophys. 1992;11:411–25.
67. Iacopino A, Christakos S, German D, Sonsalla PK, Altar CA. Calbindin-D28K-containing neurons in animal models of neurodegeneration:possible protection from excitotoxicity. Brain Res Mol Brain Res. 1992;13:251–61.
68. Idrizbegovic E, Salman H, Niu X, Canlon B. Presbyacusis and calcium-binding protein immunoreactivity in the cochlear nucleus of BALB/c mice. Hear Res. 2006;216–17:198–206.
69. Rybak LP, Whitworth CA. Ototoxicity: therapeutic opportunities. Drug Discov Today. 2005;10:1313–21.
70. Poirrier AL, Van den Ackerveken P, Kim TS, Vandenbosch R, Nguyen L, Lefebvre PP, et al. Ototoxic drugs: difference in sensitivity between mice and guinea pigs. Toxicol Lett. 2010;193:41–9.
71. Huth ME, Ricci AJ, Cheng AG. Mechanisms of aminoglycoside ototoxicity and targets of hair cell protection. Int J Otolaryngol. 2011;937861.
72. Tombal B, Denmeade SR, Gillis JM, Isaacs JT. A supramicromolar elevation of intracellular free calcium ([Ca(2+)](i)) is consistently required to induce the execution phase of apoptosis. Cell Death Differ. 2002;9:561–73.
73. Vaudry D, Falluel-Morel A, Bourgault A, Basille M, Burel D, Wurtz O, et al. Pituitary adenylate cyclase activating polypeptide and its receptors: 20 years after the discovery. Pharm Rev. 2009;61:283–357.
74. Dong Y, Tang TS, Lu CL, He C, Dong JB, Huang XY, et al. Pituitary adenylate cyclase activating polypeptide ameliorates the damage and inhibits the increase of intracellular calcium concentration in cultured hippocampal neurons induced by glutamate. Sheng Li Xue Bao. 2000;52:402–6.

Part X
Neuroprotective Effects of PACAP

Chapter 32
PACAP Signaling in Neuroprotection

Destiny-Love Manecka, Loubna Boukhzar, Anthony Falluel-Morel, Isabelle Lihrmann, and Youssef Anouar

Abstract Pituitary adenylate cyclase-activating polypeptide (PACAP) exerts its neuroprotective effects mainly through its PACAP-preferring receptor PAC1. Upon binding to this receptor and stimulation of adenylate cyclase and phospholipase C through G proteins, PACAP activates multiple signaling cascades to promote its neuroprotective actions. The neuroprotective function of PACAP has been demonstrated in vitro and in vivo during neuronal differentiation, in neurodegenerative disease models, and after administration of toxic molecules. PACAP neuroprotective activity mainly involves different kinases including protein kinase A, extracellular regulated kinase, AMP kinase, or phosphatidylinositol-3′OH kinase which in turn activate several transcription factors such as the cAMP response element-binding protein, activating transcription factor 2, or nuclear factor kappa B to regulate different genes involved in cell survival. PACAP also induces neuroprotection via indirect pathways involving other neurotrophic factors, cytokines or chemokines. These different pathways activated by PACAP converge to inhibit caspases and therefore apoptosis during neurodevelopment or in various pathological conditions of the central nervous system.

Keywords Cell signaling • Protein kinase • Neuron survival • Neuroprotection • G protein-coupled receptor

Introduction

Pituitary adenylate cyclase-activating polypeptide (PACAP) exerts its various effects through three seven-transmembrane G protein-coupled receptors [1–4]. These receptors have been classified into two categories according to their relative

D.-L. Manecka • L. Boukhzar • A. Falluel-Morel • I. Lihrmann • Y. Anouar (✉)
INSERM U982, Laboratory of Neuronal and Neuroendocrine Differentiation and Communication, Institute for Research and Innovation in Biomedicine, University of Rouen, Mont-Saint-Aignan 76821, France

Normandy University, Caen, France
e-mail: youssef.anouar@univ-rouen.fr

D. Reglodi, A. Tamas (eds.), *Pituitary Adenylate Cyclase Activating Polypeptide — PACAP*, Current Topics in Neurotoxicity 11,
DOI 10.1007/978-3-319-35135-3_32

affinity for PACAP and VIP and the signaling pathways they stimulate [5]. Type one PACAP receptor (PAC1) has an affinity for PACAP 100- to 10,000-fold greater than for VIP. This receptor is positively coupled to adenylate cyclase (AC) and phospholipase C (PLC) whereas type 2 receptors (VPAC1 and VPAC2) bind PACAP and VIP with the same affinity and are coupled to AC but are ineffective on PLC. The ubiquitous distribution of PACAP and its receptors is consistent with a pleiotropic role of the peptide in many physiological functions. This review summarizes briefly the neurotrophic and neuroprotective effects of PACAP and describes the signaling pathways that trigger its neuroprotective effects.

PACAP Neurotrophic and Neuroprotective Actions

Neurotrophic Effects

It has been shown that PACAP exerts multiple trophic effects on neurons and glial cells in sensory, sympathetic, and central nervous systems. Evidence indicates that PACAP can both stimulate and inhibit nerve cell proliferation. Indeed, PACAP stimulates astrocyte, sympathetic neuroblast, and stem cell proliferation [6–8], while it inhibits cerebral cortical precursor and neural tube cell proliferation [6, 9–12]. These opposite effects may involve the expression of different variants of PAC1 receptor [6, 13]. Furthermore, PACAP is able to induce neuronal-like differentiation in vitro of different types of stem cells [14, 15] and cell lines derived from neuroblastoma [16], cortical precursors [6], immature cerebellar granular cells [17], neuroepithelial cells [18, 19], and pheochromocytoma [20, 21].

Neuroprotective Effects

Besides its trophic function, PACAP exerts neuroprotective effects in physiological and pathological conditions. For instance, PACAP inhibits programmed cell death in cerebellar granular cells, which naturally occurs during cerebellar ontogenesis [22–26]. PACAP also protects cerebellar cells from apoptosis induced by 4-hydroxynonenal [27], ethanol [28–30], ceramides [31, 32], cisplatin [33], or hydrogen peroxide-induced oxidative stress [34–36].

PACAP is also neuroprotective in many neurodegenerative disease models and experimental injuries. In these conditions, it has been shown that PACAP concentration is increased in neuronal cells, indicating that this peptide could be implicated in protective mechanisms [37, 38]. For example, intracerebroventricular or intravenous injection of PACAP in rat before or after permanent occlusion of middle cerebral artery significantly reduces infarct size [39, 40]. In fact, the blood–brain barrier is temporarily more permeable to PACAP after ischemia, a condition which might facilitate the effect of the peptide [41].

In a rat Parkinson's disease model, PACAP protects dopaminergic cells from apoptosis, reduces neurologic deficits, and improves behavioral impairments [42, 43]. In vitro, PACAP inhibits apoptosis induced by several compounds that trigger neurotoxicity and that are used to model neurodegenerative diseases such as rotenone or 6-hydroxydopamine [44, 45] for Parkinson's disease or β-amyloid peptide for Alzheimer's disease [46]. In a mouse model of Alzheimer's disease, PACAP has been shown to ameliorate cognitive deficits [47].

PACAP Signaling in Neuroprotection

PACAP protective effects involve mainly the PAC1 receptor and its variants. The G-protein (Gs and Gq)-coupled PAC1 receptor is positively linked to different cAMP-regulated protein kinases and to PLC, and both pathways mediate the neuroprotective effects of PACAP (Fig. 32.1).

Signaling Through cAMP-Regulated Kinases

Stimulation of AC by PACAP receptors produces cAMP and thus activates cAMP-dependent protein kinase (PKA). In many cell types, it has been demonstrated that PACAP neuroprotective function can be mimicked by cAMP analogs and abolished by PKA inhibitors, implying that the cAMP/PKA pathway is key to PACAP-induced neuroprotection [25, 45, 46, 48, 49]. In fact, studies performed in many cell models showed that the neuroprotective effects of PACAP are mainly exerted through cAMP-activated PKA and mitogen-activated protein kinases (MAPK) [26, 48, 50, 51]. Indeed, analysis of apoptosis induced by potassium withdrawal in primary cultures of rat cerebellar granule cells and by rotenone in PC12 cells showed that PACAP activates ERK signaling via cAMP stimulation in order to protect neuronal cells [26, 45, 52]. ERK activation is achieved through the recruitment of two small GTPases within the plasma membrane: Ras and Rap 1 [53–56]. These small GTPases can be activated by PKA or by a protein called Epac via a PKA-independent pathway [57]. Ras and Rap1 initiate the activation of the MAPK cascade and lead to the stimulation of the MAPK kinase kinase Raf which in turn phosphorylates the MAPK kinase MEK which is ultimately responsible for ERK phosphorylation. Activated phospho-ERK is able to regulate the activity of specific transcription factors involved in neuroprotection [26, 45, 52]. Upon ERK activation, PACAP is able to stimulate anti-apoptotic proteins via the regulation of specific transcription factors. Thus, PACAP protects rat cerebellar granule cell cultures from C2-ceramide-induced apoptosis via ERK-mediated increase of activator protein-1 (AP-1) DNA binding. PACAP increases both c-Fos gene expression and the proportion of c-Fos-containing AP-1 dimers that stimulate the expression of the proto-oncogene B-cell lymphoma protein 2 (Bcl-2) in cerebellar granule neurons [33, 58, 59]. Bcl-2 is a

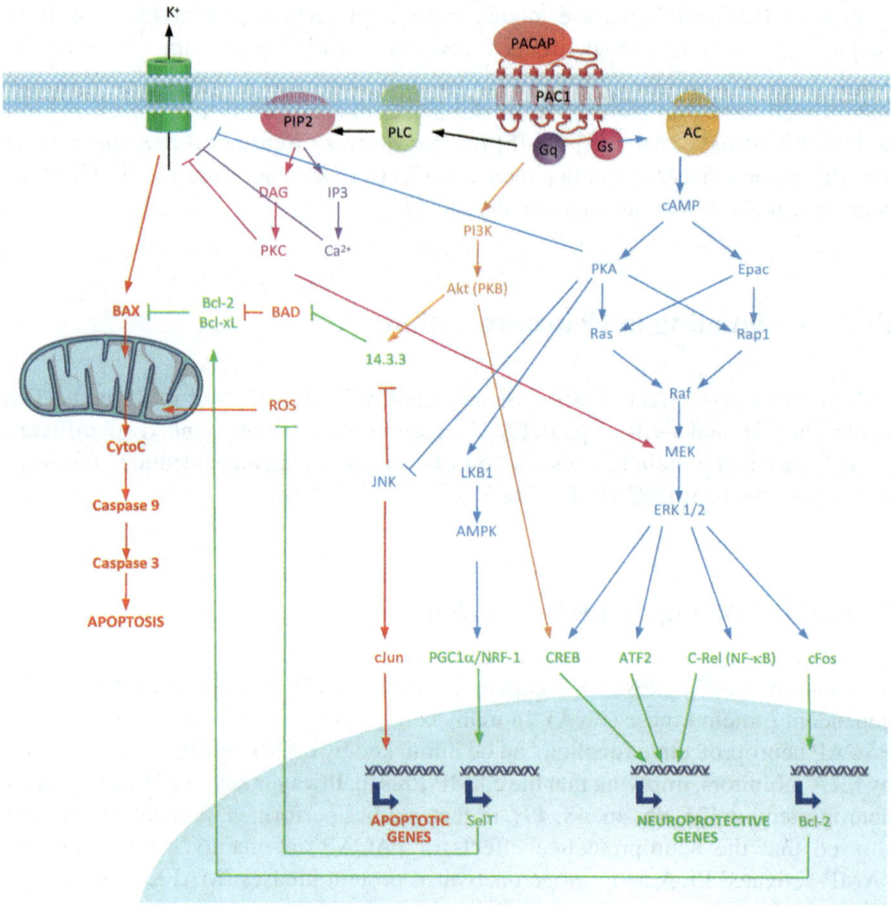

Fig. 32.1 Schematic representation of the main intracellular signaling pathways that seem to be involved in the neuroprotective functions of PACAP in different cell types. *AC* adenylyl cyclase, *Akt* enzyme member of the serine/threonine-specific protein kinase family, *AMPK* adenosine monophosphate-activated protein kinase, *ATF2* activating transcription factor-2, *Bad* Bcl-2-associated death promoter, *Bax* Bcl-2-associated X protein, *Bcl-2* B-cell lymphoma 2, *Bcl-xL* B-cell lymphoma-extra large, Ca^{2+} calcium, *cAMP* cyclic adenosine monophosphate, *cFos* Finkel Biskis Jinkins osteosarcoma-related oncogene, *cJun* jun oncogene, *CREB* cAMP response element-binding protein, *caspase* cysteinyl-aspartate-cleaving proteases, *cytoC* cytochrome c, *DAG* diacylglycerol, *Epac* exchange factor directly activated by cAMP, *ERK* extracellular signal-regulated kinase, *Gs* class s GTP-binding protein, *Gq* class q GTP-binding protein, *IP3* inositol 1,4,5-trisphosphate, *JNK* c-jun amino-terminal protein kinase, K^+ potassium ion, *LKB1* serine–threonine liver kinase B1, *MEK* mitogen-activated protein kinase ERK kinase (MEK), *NF-kB* nuclear factor of kappa light chain in B cell, *NRF-1* nuclear respiratory factor 1, *PACAP* pituitary adenylate cyclase-activating polypeptide, *PAC* PACAP-specific receptor, *PGC1α* peroxisome proliferator-activated receptor-γ-coactivator-1α, *PI3K* phosphatidylinositol-3′OH kinase, *PIP2* phosphatidylinositol bisphosphate, *PKA* protein kinase A, *PKB* protein kinase B, *PKC* protein kinase C, *PLC* phospholipase C, *Raf* Raf proto-oncogene serine/threonine-protein kinase, *Rap* small GTPase of the RAS oncogene family, *Ras* retrovirus-associated DNA sequences, *ROS* reactive oxygen species, *SelT* selenoprotein T

member of a family of proteins named Bcl-2 family, which are implicated in the regulation of apoptosis. Bcl-2 inhibits Bcl-2-associated X protein (Bax), a pro-apoptotic Bcl-2 family member [60]. Bcl-2 has been shown to be implicated in PACAP-induced neuroprotection after focal ischemia in mice and after alcohol administration in rats [29, 61]. Other regulators of Bcl-2 family members are regulated by PACAP in order to inhibit apoptosis. Thus, during glutamate-induced apoptosis, PACAP promotes cell survival via the stimulation of Bcl-2-associated death promoter (Bad) protein and 14-3-3 proteins in a PKA/Bad/14-3-3- and Akt/Bad/14-3-3-dependent cascades that increase the expression of Bcl-2 and the Bcl-2 homolog B-cell lymphoma-extra large (Bcl-xL) [62, 63]. The transcription factor c-Jun is another AP-1 family member whose phosphorylation by c-Jun N-terminal kinase (JNK), also called stress-activated kinases (SAPK), is implicated in neuronal apoptosis induced by various factors or conditions including ethanol [29], ischemia [64, 65], and glutamate [62] in animals, or growth factor withdrawal in PC12 cells [66]. It has been shown that neuroprotective/anti-apoptotic effects exerted by PACAP in these models involve the inhibition of JNK and the activation of c-jun. Although the signaling pathways recruited by PACAP to reduce JNK activity have not been clearly established yet, it is thought that PKA stimulation could be responsible for JNK down-regulation. In fact, it has been shown in PC12 cells that PKA activity is reduced in apoptotic cells and that the induction of PKA activity by the cAMP agonist dibutyryl cAMP reduces JNK activity, leading to inhibition of apoptosis [67]. Neuroprotection through JNK inhibition could be part of the PKA/Bad/14-3-3 neuroprotective cascade since, after ischemia, apoptosis is induced by inhibition of Bad by JNK [68, 69].

There is now good evidence that potassium (K^+) channels play an important role in the regulation of apoptosis [70, 71]. Indeed, various studies have shown that programmed cell death is accompanied by potassium ion outflow in many cell types including neurons [72–76], suggesting that enhancement of K^+ currents may induce cell death while reduction of K^+ outflow would promote cell survival. In cultured rat cerebellar granule cells, PACAP-induced cAMP/PKA pathway reduces potassium efflux via the inhibition of the delayed rectifier K current (I_K). This inhibition of I_K current reduces caspase activity and participates in PACAP-protective effects in these cells [49].

Gene expression analyses of PACAP-induced neuronal differentiation using microarray data revealed the activation of precedently unsuspected signaling components recruited by PACAP to promote neuroprotection. We have recently identified selenoprotein T (SelT) as a PACAP-regulated gene via cAMP/PKA pathway stimulation, which protects neuronal cells against oxidative stress in vitro and in vivo (77, 78, and unpublished results). Interestingly, PACAP stimulates the gene expression of SelT by activating the AMP kinase/peroxisome proliferator activated receptor y coactivator-1α (PGC-1α) in a cAMP/PKA-dependent manner (manuscript in preparation). Since PGC-1α is a key regulator of mitochondrial gene transcription, this finding highlights the importance of PACAP and the underlying signaling pathways for neuronal cell metabolism and survival.

At the transcriptional level, several studies showed that PACAP neuroprotective action involves the activation of the cAMP response element-binding protein (CREB), a transcription factor whose activity is stimulated by PKA [79, 80]. Analysis of PACAP-protective effect in glutamate-induced retinal degeneration revealed that ERK activity is responsible for CREB activation [81]. Following transient global ischemia in rat, PACAP reduces oxidative DNA damage and hippocampal neuronal death via the stimulation of apurinic/apyrimidinic endonuclease 1 (APE1), an enzyme involved in the DNA base excision repair pathway. The stimulation of APE1 gene expression by PACAP requires the activation of CREB and activating transcription factor-2 via PKA and p38 stimulation [82]. Finally, it has been shown that activation of the PI3K-dependent serine/threonine protein kinase Akt, also known as protein kinase B (PKB), by Gαq protein following PACAP stimulation also leads to increased neuronal survival [28, 66], probably by activating CREB as it has been shown in Schwann cells [66, 83]. Besides CREB, we showed in a recent study that PACAP promotes PC12 cell survival by stimulating both the gene expression and nuclear translocation of the anti-apoptotic c-Rel and p52/p105 subunits of the NF-kB transcription factor. This effect is mediated by ERK activation and involves several components of the NF-kB pathway [84].

Signaling Through PLC

Activation of PLC by PAC1 receptor induces the hydrolysis of the plasma membrane lipid phosphatidylinositol 4,5-bisphosphate (PIP_2) into two second messengers, inositol 1,4,5-trisphosphate (IP_3) and diacylglycerol (DAG). Cytosolic IP_3 triggers the release of calcium (Ca^{2+}) from intracellular stores through the IP_3 receptor (IP_3R), whereas membrane DAG activates protein kinase C (PKC).

PKC-Mediated Signaling

Controversial results have been reported regarding the implication of PKC in PACAP-mediated neuroprotection. For instance, PACAP-promoted PC12 cell survival in the presence of rotenone or the prion protein fragments 106–126 has been shown to be independent of PKC [45, 85]. On the contrary, other studies showed that PACAP and PAC1 receptor activation of PKC stimulates ERK in PC12 cells, which is known to promote the survival of these cells [26, 45, 52, 53, 84, 86]. Furthermore, PKC has been linked to PACAP-induced neuroprotection in rat cerebellar granule cell cultures [25].

As mentioned above, K^+ outflow is implicated in apoptosis mechanisms. Like the cAMP/PKA pathway, PLC/PKC signaling plays a role in PACAP-mediated inhibition of K^+ outflows. Indeed, PLC and both downstream PKC and Ca^{2+} inhibit A-type K^+ currents and participate in neuroprotective effects of PACAP on olfactory receptor neurons [87, 88].

Calcium-Mediated Signaling

PACAP-mediated increase in intracellular Ca^{2+} has also been described as a signaling messenger for neuroprotection, although only a few studies have addressed this issue. Thus, it has been shown that PACAP induces a rapid and long-lasting PKA-dependent increase in action potential firing associated to Ca^{2+} transients, which are essential for the anti-apoptotic actions of PACAP [78]. It has also been shown that Ca^{2+} is involved in ERK-mediated activation of the NF-kB pathway, which leads to PC12 cell survival [70].

Indirect Neuroprotective Pathways

PACAP regulates other effectors like growth factors, cytokines, or chemokines to achieve neuroprotection (Fig. 32.2). The most described indirect neuroprotective effect of PACAP is through the regulation of brain-derived neurotrophic factor (BDNF). Indeed, it has been reported that the protective effect of PACAP is partially due to the stimulation of the gene expression and secretion of BDNF, which is reduced after injury. Thus, in primary cultures of rat cerebral cortex cells, the excitotoxic NMDA and serum deprivation reduce BDNF expression. In both conditions, BDNF inhibition is prevented by PACAP, indicating that the neuroprotective effect of PACAP is at least partially mediated by the stimulation of BDNF expression [89]. Similarly, BDNF is down-regulated in response to the toxicity induced by inflammatory mediators or by oligomeric Aβ42 peptides in human neuroblastoma cells, and PACAP treatment attenuates these effects and subsequent CREB

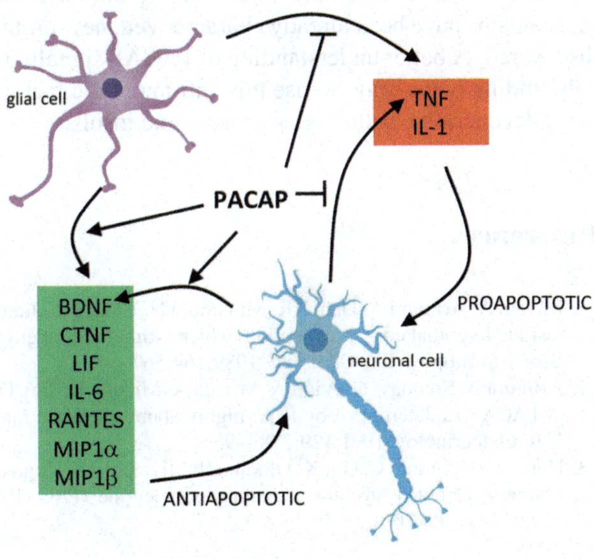

Fig. 32.2 Schematic representation of PACAP indirect neuroprotection. PACAP stimulates release of trophic factors BDNF, CTNF, and LIF, cytokines IL-6, RANTES, MIP-1α, and MI1β and inhibits release of deleterious cytokines TNF and IL-1. PACAP also regulates neuronal release of TNF and BDNF

down-regulation [47, 90]. In an Alzheimer's disease transgenic mouse model, PACAP administration increases the expression of protective proteins including BDNF, and enhances the non-amyloidogenic pathway of APP processing [47]. It has been shown that PC12 cells express and release the pro-inflammatory cytokine tumor necrosis factor (TNF), which can act in an autocrine manner to inhibit cell viability via its death receptor TNFR1. In these cells, PACAP-mediated survival is associated to inhibition of TNF mRNA expression and protein release [91]. In vivo, it has been shown that PACAP exerts neuroprotective effects through the regulation of glial cell activity and neuronal microenvironment. Indeed, PACAP protects cortical cells from sodium nitroprusside and the envelope protein gp120 of the human immunodeficiency virus at least in part through the induction of the release of chemokines such as regulated on activation normal T expressed and secreted (RANTES) and macrophage inflammatory protein 1 (MIP)-1 α and β from astrocytes [92, 93]. After ischemia-induced apoptosis of neurons in mice hippocampus, administration of PACAP inhibits the release of deleterious factors like interleukin (IL)-1 and TNF by microglial cells, and enhances the release of the protective factors IL-6, ciliary neurotrophic factor (CNTF), and leukemia inhibitory factor (LIF) from astrocytes which in turn, akin to PACAP, can protect neurons via the inhibition of JNK or the activation of ERK [65, 94].

Conclusion

PACAP promotes neuronal survival by directly or indirectly inhibiting apoptosis in both physiological and pathological conditions. Currently, several signal transduction pathways activated by PACAP and leading to cell survival in normal or pathological conditions have been identified. Although complex signaling transduction mechanisms have been already characterized, new protagonists are continuing to be discovered. A better understanding of PACAP signaling pathways will undoubtedly help finding better ways to use this anti-apoptotic factor as a potential treatment for neurodegenerative pathologies or ischemic insults.

References

1. Miyata A, Arimura A, Dahl RR, Minamino N, Uehara A, Jiang L, et al. Isolation of a novel 38 residue-hypothalamic polypeptide which stimulates adenylate cyclase in pituitary cells. Biochem Biophys Res Commun. 1989;164:567–74.
2. Arimura A, Somogyvári-Vigh A, Miyata A, Mizuno K, Coy DH, Kitaba C. Tissue distribution of PACAP as determined by RIA: highly abundant in the rat brain and testes: endocrinology 129. Endocrinoloy. 1991;129:2787–9.
3. Hosoya M, Kimura C, Ogi K, Ohkubo S, Miyamoto Y, Kugoh H, et al. Structure of the human pituitary adenylate cyclase activating polypeptide (PACAP) gene. Biochim Biophys Acta. 1992;1129:199–206.

4. Miyata A, Jiang L, Dahl RR, Kitaba C, Kubo K, Fujino M, et al. Isolation of a neuropeptide corresponding to the N-terminal 27 residues of the pituitary adenylate cyclase activating polypeptide with 38 residues (PACAP38). Biochem Biophys Res Commun. 1990;170:643–8.
5. Harmar AJ, Arimura A, Gozes I, Journot L, Laburthe M, Pisegna JR, et al. International Union of Pharmacology. XVIII. Nomenclature of receptors for vasoactive intestinal peptide and pituitary adenylate cyclase-activating polypeptide. Pharmacol Rev. 1998;50:265–70.
6. Lu N, Zhou R, DiCicco-Bloom E. Opposing mitogenic regulation by PACAP in sympathetic and cerebral cortical precursors correlates with differential expression of PACAP receptor (PAC1-R) isoforms. J Neurosci Res. 1998;53:651–62.
7. Mercer A, Ronnholm H, Holmberg J, Lundh H, Heidrich J, Zachrisson O, et al. PACAP promotes neural stem cell proliferation in adult mouse brain. J Neurosci Res. 2004;76:205–15.
8. Tatsuno I, Morio H, Tanaka T, Uchida D, Hirai A, Tamura Y, et al. Pituitary adenylate cyclase-activating polypeptide (PACAP) is a regulator of astrocytes: PACAP stimulates proliferation and production of interleukin 6 (IL-6), but not nerve growth factor (NGF), in cultured rat astrocytes. Ann N Y Acad Sci. 1996;805:482–8.
9. DiCicco-Bloom E, Lu N, Pintar JE, Zhang J. The PACAP ligand/receptor system regulates cerebral cortical neurogenesis. Ann N Y Acad Sci. 1998;865:274–89.
10. Lu N, DiCicco-Bloom E. Pituitary adenylate cyclase-activating polypeptide is an autocrine inhibitor of mitosis in cultured cortical precursor cells. Proc Natl Acad Sci. 1997;94:3357–62.
11. Suh J, Lu N, Nicot A, Tatsuno I, DiCicco-Bloom E. PACAP is an anti-mitogenic signal in developing cerebral cortex. Nat Neurosci. 2001;4:123–4.
12. Waschek JA, Casillas RA, Nguyen TB, DiCicco-Bloom EM, Carpenter EM, Rodriguez WI. Neural tube expression of pituitary adenylate cyclase-activating peptide (PACAP) and receptor: potential role in patterning and neurogenesis. Proc Natl Acad Sci. 1998;95:9602–7.
13. Nicot A, DiCicco-Bloom E. Regulation of neuroblast mitosis is determined by PACAP receptor isoform expression. Proc Natl Acad Sci. 2001;98:4758–63.
14. Cazillis M, Gonzalez BJ, Billardon C, Lombet A, Fraichard A, Samarut J, et al. VIP and PACAP induce selective neuronal differentiation of mouse embryonic stem cells. Eur J Neurosci. 2004;19:798–808.
15. Ohno F, Watanabe J, Sekihara H, Hirabayashi T, Arata S, Kikuyama S, et al. Pituitary adenylate cyclase-activating polypeptide promotes differentiation of mouse neural stem cells into astrocytes. Regul Pept. 2005;126:115–22.
16. Hoshino M, Li M, Zheng LQ, Suzuki M, Mochizuki T, Yanaihara N. Pituitary adenylate cyclase activating peptide and vasoactive intestinal polypeptide: differentiation effects on human neuroblastoma NB-OK-1 cells. Neurosci Lett. 1993;159:35–8.
17. Gonzalez BJ, Basille M, Vaudry D, Fournier A, Vaudry H. Pituitary adenylate cyclase-activating polypeptide promotes cell survival and neurite outgrowth in rat cerebellar neuroblasts. Neuroscience. 1997;78:419–30.
18. Zhou CJ, Shioda S, Shibanuma M, Nakajo S, Funahashi H, Nakai Y, et al. Pituitary adenylate cyclase-activating polypeptide receptors during development: expression in the rat embryo at primitive streak stage. Neuroscience. 1999;93:375–91.
19. Zhou C-J, Yada T, Kohno D, Kikuyama S, Suzuki R, Mizushima H, et al. PACAP activates PKA, PKC and Ca2+ signaling cascades in rat neuroepithelial cells. Peptides. 2001;22:1111–7.
20. Deutsch PJ, Sun Y. The 38-amino acid form of pituitary adenylate cyclase-activating polypeptide stimulates dual signaling cascades in PC12 cells and promotes neurite outgrowth. J Biol Chem. 1992;267:5108–13.
21. Ghzili H, Grumolato L, Thouënnon E, Tanguy Y, Turquier V, Vaudry H, et al. Role of PACAP in the physiology and pathology of the sympathoadrenal system. Front Neuroendocrinol. 2008;29:128–41.
22. Allais A, Burel D, Isaac ER, Gray SL, Basille M, Ravni A, et al. Altered cerebellar development in mice lacking pituitary adenylate cyclase-activating polypeptide: Altered cerebellar development in PACAP-/- mice. Eur J Neurosci. 2007;25:2604–18.

23. Vaudry D, Gonzalez BJ, Basille M, Fournier A, Vaudry H. Neurotrophic activity of pituitary adenylate cyclase-activating polypeptide on rat cerebellar cortex during development. Proc Natl Acad Sci. 1999;96:9415–20.
24. Vaudry D, Gonzalez BJ, Basille M, Pamantung TF, Fournier A, Vaudry H. PACAP acts as a neurotrophic factor during histogenesis of the rat cerebellar cortex. Ann N Y Acad Sci. 2000;921:293–9.
25. Vaudry D, Gonzalez BJ, Basille M, Pamantung TF, Fontaine M, Fournier A, et al. The neuroprotective effect of pituitary adenylate cyclase-activating polypeptide on cerebellar granule cells is mediated through inhibition of the CED3-related cysteine protease caspase-3/CPP32. Proc Natl Acad Sci. 2000;97:13390–5.
26. Villalba M, Bockaert J, Journot L. Pituitary adenylate cyclase-activating polypeptide (PACAP-38) protects cerebellar granule neurons from apoptosis by activating the mitogen-activated protein kinase (MAP kinase) pathway. J Neurosci. 1997;17:83–90.
27. Ito Y, Arakawa M, Ishige K, Fukuda H. Comparative study of survival signal withdrawal-and 4-hydroxynonenal-induced cell death in cerebellar granule cells. Neurosci Res. 1999;35:321–7.
28. Bhave SV, Hoffman PL. Phosphatidylinositol 3′-OH kinase and protein kinase A pathways mediate the anti-apoptotic effect of pituitary adenylyl cyclase-activating polypeptide in cultured cerebellar granule neurons: modulation by ethanol: mechanisms of PACAP neuroprotection and effects of ethanol. J Neurochem. 2003;88:359–69.
29. Botia B, Jolivel V, Burel D, Le Joncour V, Roy V, Naassila M, et al. Neuroprotective effects of PACAP against ethanol-induced toxicity in the developing rat cerebellum. Neurotox Res. 2011;19:423–34.
30. Vaudry D, Rousselle C, Basille M, Falluel-Morel A, Pamantung TF, Fontaine M, et al. Pituitary adenylate cyclase-activating polypeptide protects rat cerebellar granule neurons against ethanol-induced apoptotic cell death. Proc Natl Acad Sci. 2002;99:6398–403.
31. Falluel-Morel A, Aubert N, Vaudry D, Basille M, Fontaine M, Fournier A, et al. Opposite regulation of the mitochondrial apoptotic pathway by C2-ceramide and PACAP through a MAP-kinase-dependent mechanism in cerebellar granule cells: PACAP and C2-ceramide control mitochondrial apoptotic pathway. J Neurochem. 2004;91:1231–43.
32. Vaudry D, Falluel-Morel A, Basille M, Pamantung TF, Fontaine M, Fournier A, et al. Pituitary adenylate cyclase-activating polypeptide prevents C2-ceramide-induced apoptosis of cerebellar granule cells. J Neurosci Res. 2003;72:303–16.
33. Aubert N, Vaudry D, Falluel-Morel A, Desfeux A, Fisch C, Ancian P, et al. PACAP prevents toxicity induced by cisplatin in rat and primate neurons but not in proliferating ovary cells: Involvement of the mitochondrial apoptotic pathway. Neurobiol Dis. 2008;32:66–80.
34. Rácz B, Gasz B, Borsiczky B, Gallyas F, Tamás A, Józsa R, et al. Protective effects of pituitary adenylate cyclase activating polypeptide in endothelial cells against oxidative stress-induced apoptosis. Gen Comp Endocrinol. 2007;153:115–23.
35. Vaudry D, Pamantung TF, Basille M, Rousselle C, Fournier A, Vaudry H, et al. PACAP protects cerebellar granule neurons against oxidative stress-induced apoptosis. Eur J Neurosci. 2002;15:1451–60.
36. Vaudry D, Hamelink C, Damadzic R, Eskay RL, Gonzalez B, Eiden LE. Endogenous PACAP acts as a stress response peptide to protect cerebellar neurons from ethanol or oxidative insult. Peptides. 2005;26:2518–24.
37. Waschek JA. Multiple actions of pituitary adenylyl cyclase activating peptide in nervous system development and regeneration. Dev Neurosci. 2002;24:14–23.
38. Zigmond RE. LIF, NGF, and the cell body response to axotomy. Neuroscientist. 1997;3:176–85.
39. Reglodi D, Somogyvari-Vigh A, Vigh S, Kozicz T, Arimura A. Delayed systemic administration of PACAP38 is neuroprotective in transient middle cerebral artery occlusion in the rat. Stroke. 2000;31:1411–7.
40. Reglodi D, Tamas A, Somogyvari-Vigh A, Szanto Z, Kertes E, Lenard L, et al. Effects of pretreatment with PACAP on the infarct size and functional outcome in rat permanent focal cerebral ischemia. Peptides. 2002;23:2227–34.

41. Somogyvári-Vigh A, Pan W, Reglodi D, Kastin AJ, Arimura A. Effect of middle cerebral artery occlusion on the passage of pituitary adenylate cyclase activating polypeptide across the blood–brain barrier in the rat. Reg Pept. 2000;91:89–95.
42. Reglődi D, Lubics A, Tamás A, Szalontay L, Lengvári I. Pituitary adenylate cyclase activating polypeptide protects dopaminergic neurons and improves behavioral deficits in a rat model of Parkinson's disease. Behav Brain Res. 2004;151:303–12.
43. Reglődi D, Tamás A, Lubics A, Szalontay L, Lengvári I. Morphological and functional effects of PACAP in 6-hydroxydopamine-induced lesion of the substantia nigra in rats. Reg Pept. 2004;123:85–94.
44. Deguil J, Jailloux D, Page G, Fauconneau B, Houeto J-L, Philippe M, et al. Neuroprotective effects of pituitary adenylate cyclase–activating polypeptide (PACAP) in MPP+-induced alteration of translational control in Neuro-2a neuroblastoma cells. J Neurosci Res. 2007;85:2017–25.
45. Wang G, Qi C, Fan G-H, Zhou H-Y, Chen S-D. PACAP protects neuronal differentiated PC12 cells against the neurotoxicity induced by a mitochondrial complex I inhibitor, rotenone. FEBS Lett. 2005;579:4005–11.
46. Onoue S, Endo K, Ohshima K, Yajima T, Kashimoto K. The neuropeptide PACAP attenuates β-amyloid (1–42)-induced toxicity in PC12 cells. Peptides. 2002;23:1471–8.
47. Rat D, Schmitt U, Tippmann F, Dewachter I, Theunis C, Wieczerzak E, et al. Neuropeptide pituitary adenylate cyclase-activating polypeptide (PACAP) slows down Alzheimer's disease-like pathology in amyloid precursor protein-transgenic mice. FASEB J. 2011;25:3208–18.
48. Kienlen Campard P, Crochemore C, René F, Monnier D, Koch B, Loeffler JP. PACAP type I receptor activation promotes cerebellar neuron survival through the cAMP/PKA signaling pathway. DNA Cell Biol. 1997;16:323–33.
49. Mei YA, Vaudry D, Basille M, Castel H, Fournier A, Vaudry H, et al. PACAP inhibits delayed rectifier potassium current via a cAMP/PKA transduction pathway: evidence for the involvement of IK in the anti-apoptotic action of PACAP. Eur J Neurosci. 2004;19:1446–58.
50. Takei N, Skoglösa Y, Lindholm D. Neurotrophic and neuroprotective effects of pituitary adenylate cyclase-activating polypeptide (PACAP) on mesencephalic dopaminergic neurons. J Neurosci Res. 1998;54:698–706.
51. Tanaka J, Koshimura K, Murakami Y, Sohmiya M, Yanaihara N, Kato Y. Neuronal protection from apoptosis by pituitary adenylate cyclase-activating polypeptide. Reg Pept. 1997;72:1–8.
52. Journot L, Villalba M, Bockaert J. PACAP-38 protects cerebellar granule cells from apoptosis. Ann N Y Acad Sci. 1998;865:100–10.
53. Barrie AP, Clohessy AM, Buensuceso CS, Rogers MV, Allen JM. Pituitary adenylyl cyclase-activating peptide stimulates extracellular signal-regulated kinase 1 or 2 (ERK1/2) activity in a Ras-independent, mitogen-activated protein Kinase/ERK kinase 1 or 2-dependent manner in PC12 cells. J Biol Chem. 1997;272:19666–71.
54. Bouschet T, Perez V, Fernandez C, Bockaert J, Eychene A, Journot L. Stimulation of the ERK pathway by GTP-loaded Rap1 requires the concomitant activation of Ras, protein kinase C, and protein kinase A in neuronal cells. J Biol Chem. 2003;278:4778–85.
55. Jamen F, Bouschet T, Laden J-C, Bockaert J, Brabet P. Up-regulation of the PACAP type-1 receptor (PAC1) promoter by neurotrophins in rat PC12 cells and mouse cerebellar granule cells via the Ras/mitogen-activated protein kinase cascade. J Neurochem. 2002;82:1199–207.
56. Obara Y, Horgan AM, Stork PJS. The requirement of Ras and Rap1 for the activation of ERKs by cAMP, PACAP, and KCl in cerebellar granule cells. J Neurochem. 2007;101:470–82.
57. Emery AC, Eiden LE. Signaling through the neuropeptide GPCR PAC1 induces neuritogenesis via a single linear cAMP- and ERK-dependent pathway using a novel cAMP sensor. FASEB J. 2012;26:3199–211.
58. Aubert N, Falluel-Morel A, Vaudry D, Xifro X, Rodriguez-Alvarez J, Fisch C, et al. PACAP and C2-ceramide generate different AP-1 complexes through a MAP-kinase-dependent pathway: involvement of c-Fos in PACAP-induced Bcl-2 expression. J Neurochem. 2006;99:1237–50.
59. Vaudry D, Gonzalez BJ, Basille M, Anouar Y, Fournier A, Vaudry H. Pituitary adenylate cyclase-activating polypeptide stimulates both c-fos gene expression and cell survival in rat

cerebellar granule neurons through activation of the protein kinase A pathway. Neuroscience. 1998;84:801–12.
60. Youle RJ, Strasser A. The BCL-2 protein family: opposing activities that mediate cell death. Nat Rev Mol Cell Biol. 2008;9:47–59.
61. Ohtaki H, Nakamachi T, Dohi K, Aizawa Y, Takaki A, Hodoyama K, et al. Pituitary adenylate cyclase-activating polypeptide (PACAP) decreases ischemic neuronal cell death in association with IL-6. Proc Natl Acad Sci. 2006;103:7488–93.
62. Rácz B, Gallyas F, Kiss P, Tóth G, Hegyi O, Gasz B, et al. The neuroprotective effects of PACAP in monosodium glutamate-induced retinal lesion involve inhibition of proapoptotic signaling pathways. Reg Pept. 2006;137:20–6.
63. Rácz B, Gasz B, Gallyas F, Kiss P, Tamás A, Szántó Z, et al. PKA-Bad-14-3-3 and Akt-Bad-14-3-3 signaling pathways are involved in the protective effects of PACAP against ischemia/reperfusion-induced cardiomyocyte apoptosis. Reg Pept. 2008;145:105–15.
64. Dohi K, Mizushima H, Nakajo S, Ohtaki H, Matsunaga S, Aruga T, et al. Pituitary adenylate cyclase-activating polypeptide (PACAP) prevents hippocampal neurons from apoptosis by inhibiting JNK/SAPK and p38 signal transduction pathways. Reg Pept. 2002;109:83–8.
65. Shioda S, Ozawa H, Dohi K, Mizushima H, Matsumoto K, Nakajo S, et al. PACAP protects hippocampal neurons against apoptosis: involvement of JNK/SAPK signaling pathway. Ann N Y Acad Sci. 2006;865:111–7.
66. May V, Lutz E, MacKenzie C, Schutz KC, Dozark K, Braas KM. Pituitary adenylate cyclase-activating polypeptide (PACAP)/PAC1HOP1 receptor activation coordinates multiple neurotrophic signaling pathways: Akt activation through phosphatidylinositol 3-kinase and vesicle endocytosis for neuronal survival. J Biol Chem. 2010;285:9749–61.
67. Hur KC. Protein kinase A functions as a negative regulator of C-jun N-terminal kinase but not of p38 mitogen-activated protein kinase in PC12 cells. Integr Biosci. 2005;9:173–9.
68. Fan J, Xu G, Nagel DJ, Hua Z, Zhang N, Yin G. A model of ischemia and reperfusion increases JNK activity, inhibits the association of BAD and 14-3-3, and induces apoptosis of rabbit spinal neurocytes. Neurosci Lett. 2010;473:196–201.
69. Kamada H, Nito C, Endo H, Chan PH. Bad as a converging signaling molecule between survival PI3-K/Akt and death JNK in neurons after transient focal cerebral ischemia in rats. J Cereb Blood Flow Metab. 2007;27:521–33.
70. Lang F, Gulbins E, Szabo I, Lepple-Wienhues A, Huber SM, Duranton C, et al. Cell volume and the regulation of apoptotic cell death. J Mol Recognit. 2004;17:473–80.
71. Shieh CC, Coghlan M, Sullivan JP, Gopalakrishnan M. Potassium channels: molecular defects, diseases, and therapeutic opportunities. Pharmacol Rev. 2000;52:557–93.
72. Colom LV, Diaz ME, Beers DR, Neely A, Xie W-J, Appel SH. Role of potassium channels in amyloid-induced cell death. J Neurochem. 1998;70:1925–34.
73. Wang Z. Roles of K+channels in regulating tumour cell proliferation and apoptosis. Eur J Physiol. 2004;448:274–86.
74. Yu SP, Yeh C-H, Sensi SL, Gwag BJ, Canzoniero MT, Farhangrazi ZS, et al. Mediation of neuronal apoptosis by enhancement of outward potassium current. Science. 1997;278:114–7.
75. Yu SP, Yeh C-H, Gottron F, Wang X, Grabb MC, Choi DW. Role of outward delayed rectifier K+ current in ceramide-induced caspases activation and apoptosis in cultures cortical neurons. J Neurochem. 1999;73:933–41.
76. Yu SP, Yeh C-H, Strasser U, Tian M, Choi DW. NMDA receptor-mediated K+ efflux and neuronal apoptosis. Science. 1999;284:336–9.
77. Grumolato L, Ghzili H, Montero-Hadjadje M, Gasman S, Lesage J, Tanguy Y, et al. Selenoprotein T is a PACAP-regulated gene involved in intracellular Ca^{2+} mobilization and neuroendocrine secretion. FASEB J. 2008;22:1756–68.
78. Castex MT, Arabo A, Bénard M, Roy V, Le Joncour V, Prévost G, Bonnet JJ, Anouar Y, Falluel-Morel A. Selenoprotein T deficiency leads to neurodevelopmental abnormalities and hyperactive behavior in mice. Mol Neurobiol. In press.
79. Baxter PS, Martel M-A, McMahon A, Kind PC, Hardingham GE. Pituitary adenylate cyclase-activating peptide induces long-lasting neuroprotection through the induction of activity-

dependent signaling via the cyclic AMP response element-binding protein-regulated transcription co-activator 1: activity-dependent neuroprotection by PACAP signaling. J Neurochem. 2011;118:365–78.
80. McIlvain HB, Baudy A, Sullivan K, Liu D, Pong K, Fennell M, et al. Pituitary adenylate cyclase-activating peptide (PACAP) induces differentiation in the neuronal F11 cell line through a PKA-dependent pathway. Brain Res. 2006;1077:16–23.
81. Racz B, Tamas A, Kiss P, Toth G, Gasz B, Borsiczky B, et al. Involvement of ERK and CREB signaling pathways in the protective effect of PACAP in monosodium glutamate-induced retinal lesion. Ann N Y Acad Sci. 2006;1070:507–11.
82. Stetler RA, Gao Y, Zukin RS, Vosler PS, Zhang L, Zhang F, et al. Apurinic/apyrimidinic endonuclease APE1 is required for PACAP-induced neuroprotection against global cerebral ischemia. Proc Natl Acad Sci. 2010;107:3204–9.
83. Castorina A, Waschek JA, Marzagalli R, Cardile V, Drago F. PACAP interacts with PAC1 receptors to induce tissue plasminogen activator (tPA) expression and activity in Schwann cell-like cultures. PLoS One. 2015;10:e0117799.
84. Manecka D-L, Mahmood SF, Grumolato L, Lihrmann I, Anouar Y. Pituitary adenylate cyclase-activating polypeptide (PACAP) promotes both survival and neuritogenesis in PC12 cells through activation of nuclear factor B (NF- B) pathway: involvement of extracellular signal-regulated kinase (ERK), calcium and c-REL. J Biol Chem. 2013;288:14936–48.
85. Onoue S, Ohshima K, Endo K, Yajima T, Kashimoto K. PACAP protects neuronal PC12 cells from the cytotoxicity of human prion protein fragment 106–126. FEBS Lett. 2002;522:65–70.
86. Lazarovici P, Jiang H, Fink D. The 38-amino-acid form of pituitary adenylate cyclase-activating polypeptide induces neurite outgrowth in PC12 cells that is dependent on protein kinase C and extracellular signal-regulated kinase but not on protein kinase A, nerve growth factor receptor tyrosine kinase, p21ras G protein, and pp60c-src cytoplasmic tyrosine kinase. Mol Pharmacol. 1998;54:547–58.
87. Han P, Lucero MT. Pituitary adenylate cyclase-activating polypeptide reduces A-type K+ currents and caspase activity in cultured adult mouse olfactory neurons. Neuroscience. 2005;134:745–56.
88. Han P, Lucero MT. Pituitary adenylate cyclase activating polypeptide reduces expression of Kv1.4 and Kv4.2 subunits underlying A-type K+ current in adult mouse olfactory neuroepithelia. Neuroscience. 2006;138:411–9.
89. Frechilla D, Garcia-Osta A, Palacios S, Cenarruzabeitia E, Del Rio J. BDNF mediates the neuroprotective effect of PACAP-38 on rat cortical neurons. Neuropharmacol Neurotoxicol. 2001;12:919–23.
90. Brown D, Tamas A, Reglodi D, Tizabi Y. PACAP Protects against inflammatory-mediated toxicity in dopaminergic SH-SY5Y cells: implication for Parkinson's disease. Neurotox Res. 2014;26:230–9.
91. Manecka D-L, Lelièvre V, Anouar Y. Inhibition of constitutive TNF production is associated with PACAP-mediated differentiation in PC12 cells. FEBS Lett. 2014;588:3008–14.
92. Brenneman DE, Hauser JM, Spong CY, Phillips TM. Chemokine release is associated with the protective action of PACAP-38 against HIV envelope protein neurotoxicity. Neuropeptides. 2002;36:271–80.
93. Sanchez A, Tripathy D, Grammas P. RANTES release contributes to the protective action of PACAP38 against sodium nitroprusside in cortical neurons. Neuropeptides. 2009;43:315–20.
94. Shioda S. Pleiotropic functions of PACAP in the CNS: neuroprotection and neurodevelopment. Ann N Y Acad Sci. 2006;1070:550–60.

Chapter 33
PACAP Expression and Plasticity in the Peripheral Nervous System

Karen M. Braas, Rodney L. Parsons, Margaret A. Vizzard, James A. Waschek, and Victor May

Abstract Pituitary adenylate cyclase activating polypeptide (*Adcyap1*; PACAP) is a well-studied neural and endocrine pleiotropic peptide important in development and the homeostatic regulation of many physiological systems. Accordingly, PACAP and its cognate PAC1 receptor (*Adcyap1r1*) are localized in many central nervous system (CNS) regions and widely distributed in the peripheral nervous system (PNS). PACAP has been identified in a population of small nociceptive neurons in the dorsal root ganglion (DRG) targeting many peripheral tissues. PACAP-immunoreactive preganglionic sympathetic and parasympathetic nerve fibers densely innervate a variety of autonomic and enteric ganglia. Furthermore, a small fraction of the autonomic and enteric postganglionic neurons also endogenously express PACAP. Notably, PACAP belongs to a cohort of neuroplasticity peptides, and PNS insult invariably induces PACAP expression in the affected neurons. Axotomy, nerve crush, inflammation, and neural tissue explant paradigms can dramatically induce neuronal PACAP transcripts, immunoreactivity, content, and cell numbers. This plasticity in PACAP expression has been best appreciated in studies using the PACAP-EGFP transgenic mouse line. The mechanisms driving PACAP expression after insults are not well understood, but likely reflect the aggregate effects of cytokine/inflammatory activation signals at the injury site and an abrogation of retrograde signaling from the target tissues. The few comparative studies suggest that the mechanistic signatures for peptide induction may differ among the

K.M. Braas • R.L. Parsons • M.A. Vizzard • V. May (✉)
Department of Neurological Sciences, University of Vermont College of Medicine,
149 Beaumont Avenue, Burlington, VT 05405, USA
e-mail: victor.may@uvm.edu

J.A. Waschek
Department of Psychiatry and Biobehavioral Sciences, David Geffen School of Medicine,
University of California Los Angeles,
635 South Charles E. Young Drive NRB 345, Los Angeles, CA 90095, USA

D. Reglodi, A. Tamas (eds.), *Pituitary Adenylate Cyclase Activating Polypeptide — PACAP*, Current Topics in Neurotoxicity 11,
DOI 10.1007/978-3-319-35135-3_33

various neural systems. This induction of PACAP expression in physiological insults has been implicated to participate in fiber regeneration and anti-inflammatory responses, but PACAP neuroplasticity may also have broader roles in maintaining neurocircuit homeostasis.

Keywords PACAP • PAC1 receptor • Sensory • Dorsal root ganglion (DRG) • Autonomic • Sympathetic • Parasympathetic • Superior cervical ganglion (SCG) • Cardiac ganglion • Phenotypic plasticity • Injury • Inflammation

Introduction

Many studies have comprehensively examined the distribution and expression of pituitary adenylate cyclase activating polypeptide (*Adcyap1*; PACAP) in the central nervous system. These surveys can be accessed in a number of excellent primary investigations and reviews [1–4], and from inspection of the Allen Brain Atlas (http://www.brain-map.org/). There are fewer critical studies of PACAP expression and plasticity in the peripheral nervous system and this chapter aims to provide an annotated overview of that literature supplemented with primary data from our laboratories.

In examining the publications reporting PACAP distribution, we have been mindful of several issues. For many peptidergic systems, including PACAP, the immunocytochemical localizations often identified only neuronal fibers, varicosities, and terminals, without significant signal detection in neuronal soma. These localization patterns likely reflected the preferential antibody binding to the fully post-translationally processed mature peptides *en route* to the terminals, rather than the peptide precursor protein molecules in the cell bodies. Consequently, the origins of tissue PACAP fibers were sometimes misinterpreted until clarification by in situ hybridization of neuronal PACAP transcripts. Finally, while the endogenous expression of PACAP in many neuronal tissues is unequivocal, some of the early studies were performed in colchicine-treated animals. Though colchicine disrupted microtubule polymerization and blocked axonal vesicular transport, facilitating peptide immunocytochemical localization in the cell soma, colchicine likely instilled cellular stress- and injury-induced responses. The resulting induction of PACAP expression may have altered distribution estimates from both immunocytochemistry and in situ hybridization analyses. Hence the sections below will not contain exhaustive information but attempt to generalize consensus data from the many studies. From the many investigations, PACAP has been shown to be widely distributed in sensory ganglia, participating in nociception, and autonomic/enteric neural systems, regulating diverse functions including cardiac and respiratory output, stress responses, gastrointestinal secretion and motility, and urogenital activities. With a number of other peptides, PACAP also demonstrates neurophenotypic plasticity in responses to a variety of physiological challenges.

Sensory Systems

PACAP has long been appreciated to be a sensory peptide. Under normal basal conditions, PACAP immunoreactivity and transcripts were identified in a subpopulation (approximately 10 %) of dorsal root ganglion (DRG) neurons, which appeared to represent predominantly small caliber unmyelinated C-nociceptors [5–8]. PACAP was largely co-localized with substance P- and calcitonin gene-related peptide (CGRP)-expressing sensory neurons which has been corroborated in recent single cell RNA sequencing studies [9]. However, the overall fraction of neurons expressing PACAP was considerably smaller than that for either substance P or CGRP [6, 10]. In the spinal cord, the PACAPergic DRG central fibers were localized predominantly to nociceptive targets in laminae I and II of the dorsal horn [5, 11]. Further, sensory neuron degeneration upon capsaicin treatments diminished PACAP immunoreactivity in both central and peripheral axons, and decreased PACAP content in parallel with CGRP depletion, further supporting PACAP expression in CGRP-containing nociceptive C-fibers [5]. As in DRG, PACAP immunoreactivity and/or transcripts were also identified in subpopulations of trigeminal and nodose ganglia neurons with CGRP [5, 6, 12, 13]. Accordingly, sensory fibers co-expressing PACAP and CGRP have been shown to be widely distributed in peripheral tissues including skin, cerebral vessels, pineal gland, eye, submandibular salivary ducts, tongue, trachea, larynx, urinary bladder and urogenital organs, and the gastrointestinal tract [5, 14–18].

Autonomic Nervous System

Sympathetic Nervous System

As observed in sensory ganglia, PACAP immunoreactivity and precursor protein transcripts were identified in autonomic ganglia and fibers during late embryonic development (E14–E18) [19]. Immunocytochemical processing for PACAP in the superior cervical ganglion (SCG) at the head of the paravertebral sympathetic trunk revealed intense staining of dense fiber networks circumscribing the postganglionic neurons, suggesting the presence of PACAP in sympathetic preganglionic neurons and fibers [20]. Although the sources of the SCG fibers containing PACAP were initially unclear, in situ hybridization coupled with retrograde fluorogold tracing from the sympathetic trunk demonstrated unequivocally that sympathetic preganglionic PACAP-expressing neurons in the intermediolateral cell column of the thoracic cord project to sympathetic ganglia [19, 21, 22]. As nearly all of the SCG postganglionic neurons express PAC1 receptors and PACAP potently and efficaciously stimulated SCG catecholamine and neuropeptide Y (NPY) secretion [20, 23, 24], PACAP represents one of the preganglionic noncholinergic regulators of sympathetic function. Preganglionic PACAP fibers were also found to innervate stellate, celiac, and mesenteric ganglia [25, 26]. In good agreement, PACAP was identified in preganglionic abdominal splanchnic nerves that target the adrenal medulla for epinephrine release [27, 28]. In aggregate, the high levels of PACAP expression in sympathetic systems implicate roles in stress responses, which has been supported in physiological and transgenic animal studies.

A small population of postganglionic sympathetic ganglia neurons also express PACAP endogenously under normal physiological states [20, 29]. Although PACAP immunoreactivity was detected in developing SCG neurons, and PACAP transcript and peptide levels were reported in adult SCG by quantitative PCR and radioimmunoassay, respectively, SCG neurons with endogenous PACAP identified by in situ hybridization analyses represented only a very small fraction of the total population (1–3 %). To obviate non-neuronal sources of SCG PACAP production, highly enriched SCG neurons free of contaminating background cells were maintained in serum-free cultures. Even under these conditions, SCG PACAP transcripts and peptides were readily measured, supporting endogenous neuronal SCG PACAP synthesis. But as in many neuronal systems, endogenous PACAP expression exhibits plasticity and can be dramatically induced under different physiological challenges (see below).

Parasympathetic Nervous System

PACAP also participates in parasympathetic function. For these investigations, the accessibility of the guinea pig cardiac ganglia model has been invaluable for detailed cellular, molecular, and electrophysiological studies. As for SCG, immunocytochemical inspection of cardiac ganglia whole mount preparations demonstrated dense pericellular PACAP networks surrounding the postganglionic neurons [30]. The PACAP-immunoreactive fibers degenerated upon culture of cardiac ganglion explants demonstrating their extrinsic nature. These PACAP-containing fibers were co-localized with choline-acetyltransferase (ChAT)-immunoreactivity, implicating vagal preganglionic origins, but not with either NPY- or substance P-immunoreactivity, which served as markers for sympathetic and sensory fibers, respectively. Furthermore, in the brainstem, in situ hybridization analyses identified PACAP transcript expression in neurons of the nucleus ambiguus and dorsal motor nucleus of the vagus. Hence, despite PACAP identification in nodose sensory ganglia, these results in aggregate demonstrated that PACAP innervation of guinea pig cardiac ganglia neurons was largely via vagal preganglionic parasympathetic fibers. These observations implicated vagal PACAP regulation of cardiac output and function [31–33], and inferred PACAP vagal regulation of other visceral parasympathetic ganglia in respiratory, gastric, pancreatic, and gastrointestinal physiology. PACAP fiber networks have also been described in the cranial ciliary, sphenopalatine, submandibular, and otic parasympathetic ganglia, which were consistent with PACAP/PAC1 receptor-specific effects in the head and neck.

Similar to sympathetic ganglia, a small population of parasympathetic cardiac ganglion neurons (4 %) expressed PACAP endogenously. Intrinsic PACAP was also identified in all four cranial parasympathetic ganglia by immunocytochemistry and in situ hybridization [12, 19]. Noteworthy, unlike sensory or other autonomic ganglia where endogenous PACAP-expressing neurons represented a small fraction of the population, nearly all of the postganglionic parasympathetic neurons in the otic and sphenopalatine ganglia expressed PACAP under normal baseline conditions [12]. The functional implications of prevalent postganglionic parasympathetic innervation of head and neck structures are unclear but recently, sphenopalatine PACAP fiber innervation of the middle meningeal artery has been implicated in migraine.

Mixed Sympathetic and Parasympathetic Ganglion

The major pelvic ganglion (MPG) is a mixed parasympathetic and sympathetic ganglion identified by a composite of cholinergic and catecholaminergic neurons, respectively, that provide major autonomic input to urogenital organs and the lower bowel. Although low levels of MPG PACAP transcripts were identified in normal control tissues by PCR, no PACAP-immunoreactive fibers or neurons were identified in the ganglion [34]. This contrasted with the prevalent expression of vasoactive intestinal peptide (VIP) neurons and fibers in MPG parasympathetic cholinergic neurons. However, unlike VIP in which the levels appeared relatively stable, PACAP expression was induced in some MPG neurons following experimental manipulations.

Enteric Nervous System

The enteric nervous system of the gastrointestinal tract is composed of ganglionated fiber networks in the myenteric plexus and the submucosal plexus. PACAP-immunoreactive fibers were identified in both myenteric and submucosal plexi, often co-localized with CGRP and other sensory peptides implicating extrinsic PACAP origins from DRG or vagal projections. However, capsaicin-induced neurodegeneration or extrinsic denervation only diminished gut PACAP content and immunoreactive fibers, implicating some endogenous PACAP production from intrinsic sources. Accordingly, both immunocytochemistry and in situ hybridization studies identified endogenous PACAP expression in the myenteric plexus throughout the gastrointestinal tract and in the submucosal plexus of the small and large intestine [15].

PACAP Neuroplasticity

Many neuronal systems exhibit phenotypic neuroplasticity under a variety of experimental paradigms in vitro and in vivo that can sometimes result in rather dramatic changes in neurotransmitter and neuropeptide expression. These neurophenotypic plasticity responses are important in development and in adult tissues to maintain neurocircuit homeostasis, which can impact a broad range of activities including behavior [35–38]. For peripheral transmitters, the perturbations can alter the relative levels of tissue catecholaminergic- versus cholinergic-producing neurons [39, 40]. For neuropeptides, the physiological challenges in neuronal injuries or inflammation can reduce the levels of some peptides, but dramatically induce the expression of others from levels that might have otherwise appeared negligible or absent [41, 42]. VIP and PACAP fall into the latter category, and are invariably highly induced in sensory, autonomic, and motor neurons. The plasticity in PACAP expression is illustrated in the following few examples.

Sensory Neurons

The dynamics in DRG PACAP expression have been well described. As noted earlier, approximately 10 % of DRG neurons express PACAP immunoreactivity and transcripts, mostly in small (<35 μm diameter) C-fiber nociceptors. A variety of nerve injury (transection, crush, or chronic constriction injury) and inflammation (adjuvant, cyclophosphamide) models dramatically increased DRG PACAP transcripts, immunoreactivity, and neurons in a time-dependent manner [7, 8, 11, 43–45]. From the small population of DRG PACAP neurons under baseline conditions, the increase in PACAP-expressing DRG neurons after sciatic nerve transection, for example, could be detected within a day after physiological challenge, reaching 50 % after 3 days, and a maximal sustained peak of more than 75 % after 10 days [7]. Notably, the DRG populations demonstrating PACAP induction appeared to depend on the nature of the injury. The increase in DRG PACAP neurons after sciatic nerve transection largely reflected heightened PACAP expression in large DRG neurons (>35 μm diameter), which was accompanied by a modest PACAP reduction in the small caliber DRG neuronal population (<35 μm diameter) [7, 11]. These observations were mirrored by increased PACAP-immunoreactive fiber density in the deep laminae (large DRG neuronal projections) of the spinal cord dorsal horn and decreased central PACAP fibers to the superficial layers (small C-fiber nociceptor projections). The induction of DRG PACAP-expressing neurons following inflammatory insult was confined to the small-sized neuronal population [46] with corresponding increases in PACAP-immunoreactive fiber staining in the superficial layers of the dorsal horn [45], whereas nerve compression increased PACAP levels in both small and large DRG neuronal populations [43]. Transection of the masseteric nerve produced a similar temporal increase in PACAP expression in the trigeminal ganglion [47].

Sympathetic Neurons

Injury to other peripheral neuronal tissues elicited the same PACAP induction responses. From the 1 to 3 % of postganglionic sympathetic SCG neurons that express PACAP under basal states, axotomy of the internal and/or external carotid nerves [20] or placing SCG ganglia in explant cultures augmented the SCG PACAP neuronal population to nearly 40 % (>10-fold increase) which was matched by a 30- to 50-fold increase in SCG PACAP transcript levels and a 30-fold augmentation of ganglionic PACAP peptide content (Fig. 33.1). Similar peptide induction responses were produced following chemical sympathectomy with 6-hydroxydopamine [20, 48] or nerve growth factor (NGF) neutralizing antisera treatments [49]. By contrast, the induction of PACAP and other peptides was far lower following SCG deafferentation or decentralization injury, after transection of sympathetic preganglionic fibers in the sympathetic trunk, from damage to the few sympathetic postganglionic axons that backtracked into the sympathetic chain [20, 50] (Fig. 33.1).

Parasympathetic Neurons

Placing postganglionic parasympathetic cardiac ganglia in explant cultures for 2–4 days changed the number of PACAP-expressing neurons from 1 to 4 % to a maximal increase of 25–35 %, in parallel with a 15-fold induction in PACAP precursor protein transcripts [30, 51]. Compared to other peripheral ganglia and peptides, in which axotomy or tissue explantation can produce double-digit fold changes, the PACAP transcript changes in the mixed autonomic major pelvic ganglia were smaller (<5 %) in explants and appeared absent following transection of the cavernous nerve [52].

PACAP Neuroplasticity in PACAP-EGFP Mice

The recent availability of a PACAP-EGFP transgenic mouse line has allowed studies of PACAP neuroplasticity with high resolution. In this model, expression of the EGFP cassette, inserted upstream of the PACAP ATG initiation codon and under PACAP gene (*Adcyap1*) promoter control, has allowed detailed central and peripheral nervous system PACAP regulation and induction studies under a variety of physiological paradigms (Fig. 33.2). Under baseline conditions, native PACAP-EGFP fluorescence was low or absent in a variety of peripheral tissues. Placing the SCG and MPG into explant cultures also dramatically induced PACAP-EGFP expression (Fig. 33.2C–F); the number of MPG PACAP-expressing neurons appeared greater than levels indicated in previous PCR analyses. Comparable to previous studies, chronic constriction injury by partial sciatic nerve ligation dramatically induced the number of PACAP-expressing neurons in the ipsilateral DRG, compared to the contralateral sham control ganglion (Fig. 33.2A, B); the increases in DRG PACAP expression were reflected by increased PACAP-EGFP fiber fluorescence in the sciatic nerve, proximal to the ligation, and in the DRG central axons to the ipsilateral dorsal horn (Fig. 33.2G). Similar to previous studies, the dense injury-induced PACAP-immunoreactive fibers were prevalent in the ipsilateral gracile fasciculus and deeper laminae (II–IV) of the dorsal horn (Fig. 33.2G) which implicated preferential PACAP induction in the large (>35 μm diameter) DRG neurons. There were no apparent changes in the number of second order PACAP-EGFP

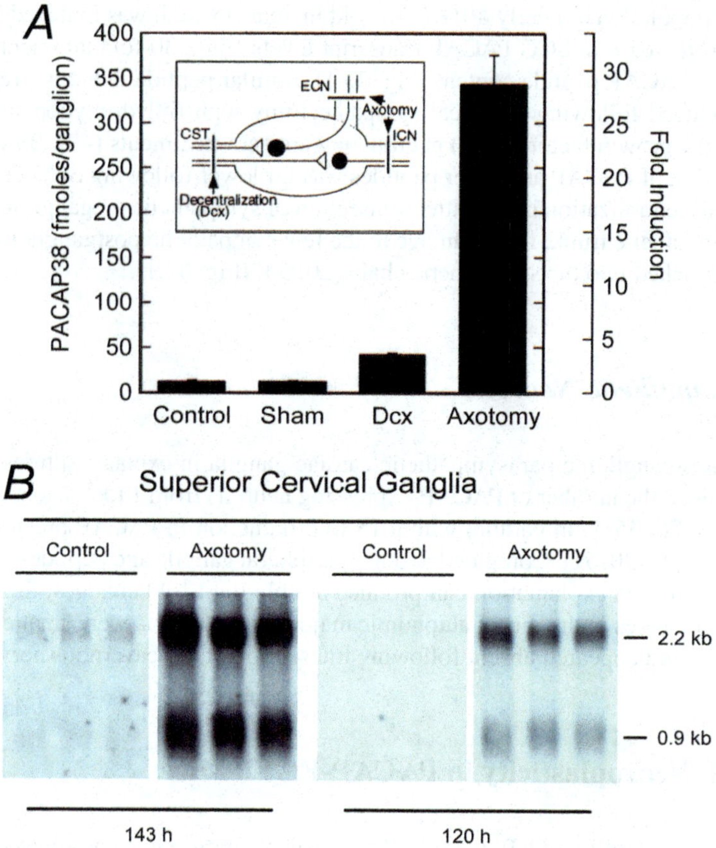

Fig. 33.1 Superior cervical ganglion (SCG) axotomy increases PACAP transcript expression. Adult male rat SCGs were axotomized by transection of the internal and external carotid nerves (ICN and ECN, respectively) or decentralized (Dcx) by transection of the sympathetic trunk (schematic in inset). Control SCGs were from unoperated animals and sham ganglia were surgically exposed but untouched. Following 7 days postsurgical recovery, the rats were euthanized by decapitation and the harvested SCG immediately frozen on dry ice. (**A**) The SCGs were homogenized in 5N acetic acid with protease inhibitors; the extracts were lyophilized and reconstituted in assay buffer for PACAP38 radioimmunoassay [29]. Axotomy increased SCG PACAP content nearly 30-fold; PACAP induction following decentralization was more modest (threefold). Data expressed either as mean fmol/ganglion (*left axis*) or fold change from control (*right axis*) ± SEM. *n*=5 per group. (**B**) SCG RNA was also prepared for Northern analysis as previously described [29]. Axotomy dramatically induced SCG PACAP mRNA levels. The blots were exposed to film with intensifying screens for either 143 h (*left panel*) or 120 h (*right panel*), demonstrating PACAP mRNA levels in control SCG and relative induction of the 2.2 and 0.9 kb PACAP transcripts after axotomy. Each lane represents RNA from an individual ganglion

Fig. 33.2 (continued) ipsilateral dorsal horn (DH) and gracile fasciculus (GF, panel **G**, right half of spinal cord). Notably, CCI also induced PACAP-EGFP expression in motor neurons of the ventral horn (VH). (**C, D**) Compared to control SCGs which were frozen immediately after harvest for cryosectioning (**C**), explantation of intact PACAP-EGFP SCG in NGF-free medium for 24 h increased PACAP expression (**D**). (**E**, F) Similarly, explant culture of the mixed sympathetic/parasympathetic major pelvic ganglion (MPG) induced PACAP-EGFP expression (**F**) compared to control ganglion (**E**)

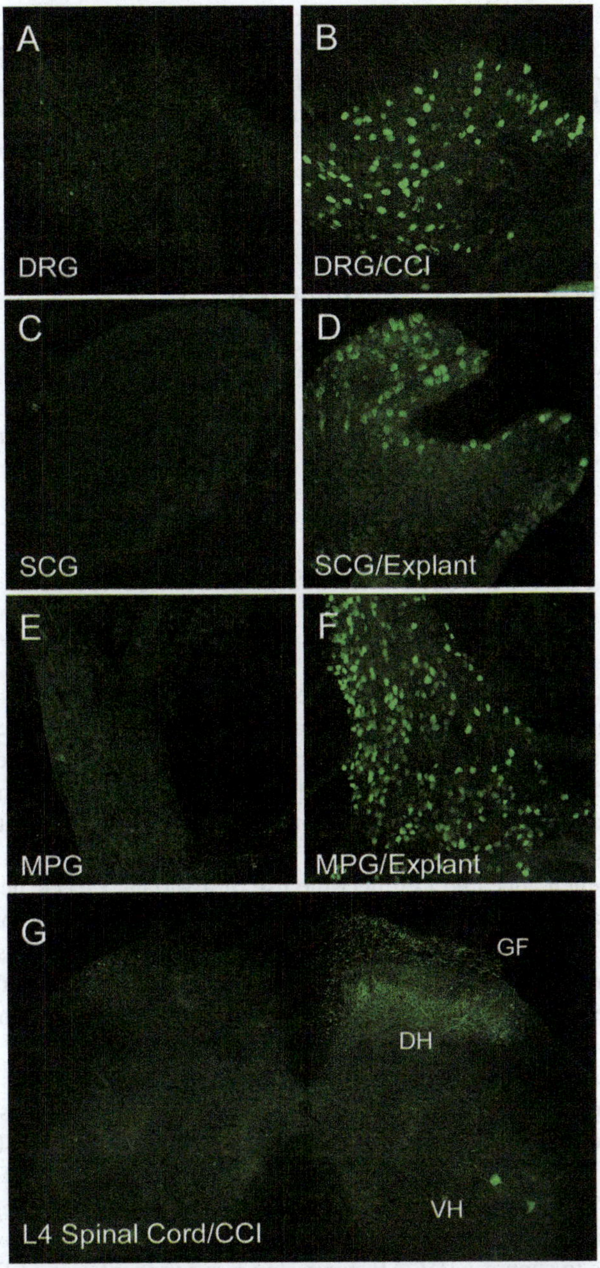

Fig. 33.2 Peripheral ganglia PACAP expression in transgenic PACAP-EGFP mice is induced after nerve injury or explantation. PACAP-EGFP mice received unilateral partial sciatic nerve ligation for chronic constriction injury (CCI, 14 days). The mice were deeply anesthetized and euthanized by 4% paraformaldehyde perfusion fixation, and tissue cryosections were prepared for native EGFP fluorescence microscopy. (**A**, **B**) Compared to sham control L4 dorsal root ganglion (DRG, **A**), CCI induced dramatically the number of PACAP-EGFP neurons in the ipsilateral DRG (**B**). The increase in DRG PACAP-EGFP was reflected in the increased PACAP fiber density in the

neurons in the superficial (lamina I) layer of the dorsal horn compared to the contralateral sham control side. In agreement with sciatic nerve transection results [22], chronic constriction injury also increased PACAP transcript expression in ventral horn motor neurons (Fig. 33.2G). These studies demonstrate the utility of PACAP-EGFP mice to unambiguously identify induced PACAPergic neurons and pathways under different physiological states.

Mechanisms of PACAP Neuroplasticity

The regulatory mechanisms underlying PACAP expression and injury-induced plasticity are not completely understood. As in other peptidergic systems, multiple factors and signaling mechanisms undoubtedly participate in the endogenous baseline and induced levels of PACAP expression.

Neuronal Activity, Trophic Factors, and Transcriptional Regulators of PACAP Expression

Neuronal activity and the resulting changes in calcium influx can regulate neuronal transmitter and peptide identity and respecification [36, 37]. The expression of galanin, VIP, substance P, and PACAP, for example, can be dramatically induced upon depolarization of cultured neurons [29, 53, 54]. The PACAP mRNA in nervous tissues is typically 2.2 kb in size and interestingly, depolarization not only augmented the 2.2 kb PACAP transcript but induced the expression of a smaller 0.9 kb form which appeared to reflect the use of a variant 3' UTR hexamer signal sequence for alternative transcript polyadenylation [55]. SCG axotomy similarly produced a shortened form of the PACAP transcript (Fig. 33.1). The short PACAP transcript variant appeared to demonstrate enhanced stability, but whether the ~1 kb deletion sequence also harbors transcriptional regulatory elements and/or impacts PACAP mRNA folding to regulate expression is currently unclear.

A variety of other factors influence PACAP expression. The glass bottom boat class of bone morphogenetic proteins (BMP), especially BMP6, can suppress neuronal PACAP expression, but reciprocally upregulate VIP levels [56]. The effects of neurotrophins on PACAP expression are complex. Signaling by NGF through the high affinity TrkA receptor can stimulate PC12 and DRG nociceptor PACAP expression [46, 57, 58] but repress SCG PACAP levels (see below). Neurotrophin-3 (NT-3) appears to suppress PACAP expression in a subpopulation of DRG neurons following injury [57]. The PACAP gene has two neural-restrictive silencer-like elements (NRSLE) [59, 60], suggesting that it can be repressed by RE-1 silencing transcriptional factor (REST). Other regulators include the gp130 cytokines [61] (see below), thyroid transcriptional factor-1 (TTF-1) [62], and T-cell leukemia homeobox protein 1/3 (Tlx1/3) [63]. The roles of Tlx1/3 are notable as Tlx1/3 coordinates PACAP and glutaminergic neuronal identities.

Activation Mechanisms in Injury/Inflammation-Induced PACAP Expression

Roles for activation and repressive retrograde signals have been postulated in models regulating neurophenotypic plasticity elicited by injury or inflammation. Each model was championed at various times, but both mechanisms undoubtedly contribute to the plasticity process. The activation model has emphasized the roles of the neuropoietic cytokines, including leukemia inhibitory factor (LIF), ciliary neurotrophic factor (CNTF), interleukin-6 (IL-6), interleukin-11 (IL-11), oncostatin M, and cardiotrophin-1, with the canonical gp130 receptor—JAK/STAT signaling pathway in neuropeptide induction. This process has been well reviewed [61]. SCG axotomy increased ganglia LIF and IL-6 expression in the supporting satellite and Schwann cells without effects on SCG gp130 transcript levels [64, 65]. Furthermore, conditioned medium stimulation of sympathetic culture VIP expression was blocked by preabsorption with anti-LIF [66], and SCG axotomy of LIF–/– animals attenuated VIP, galanin, and neurokinin A expression [67]. However, in vivo SCG exposure to LIF or axotomy of the facial motor nucleus in LIF–/– or IL-6–/– mice had no effects on galanin or VIP/PACAP induction, respectively [49, 68]. In accord, direct application of LIF, CNTF, or IL-6 to primary sympathetic cultures had no or only modest (<2-fold) effects on PACAP expression. These results implicated roles for other gp130 cytokines; however to obviate the testing of each potential mediator, a transgenic model was used in which $gp130^{fl/fl}$ animals were crossed with dopamine-beta-hydroxylase (DBH)-cre mice to delete gp130 from catecholaminergic neurons, including gp130 receptors in the SCG. Axotomy of the $gp130^{DBHcre}$ mouse SCG failed to demonstrate injury-induced VIP or PACAP expression [65], but paradoxically, cholecystokinin transcript levels were still augmented despite the previous demonstration of regulation by LIF. The cytokines may not be derived solely from endogenous ganglia non-neuronal cells. In injuries, immune cells can release a variety of cytokines and mediators generating an inflammatory milieu that can influence neuronal function and phenotypic responses [69, 70]. In support of this concept, axotomy of the facial motor nucleus failed to induce PACAP expression in the immunodeficient SCID mouse; however, VIP induction was not affected. The abrogated PACAP injury response in the SCID mouse could be reversed upon infusion of splenocytes, consistent with roles for inflammatory agents [68]. These studies in aggregate suggest that cytokines and inflammatory factors, from endogenous ganglia or exogenous immune cell origins, can participate in injury-induced peptidergic plasticity.

Abrogation of Retrograde Mechanisms in Injury/Inflammation-Mediated PACAP Expression

Other studies have suggested additional modes of peptide induction in injury. Using the SCG as example, numerous studies have shown that only axotomy of the internal or external carotid nerves are able to dramatically increase SCG VIP, PACAP, and galanin expression; SCG decentralization, as described above, produced only modest changes from axotomy of the few postganglionic SCG fibers traveling in the sympathetic trunk. If the SCG peptidergic responses were dependent on satellite/Schwann cells or immune responses, it is unclear how the cellular mediators would distinguish the two forms of injury. Further, it is unclear how long distance axonal or terminal damage can generate responses back at the soma. As examples, mid-thigh sciatic nerve contusions, catecholaminergic nerve terminal injuries with 6-hydroxydopamine, or TRPV1-mediated sensory nerve terminal damage with capsaicin or the agonist resiniferatoxin dramatically induces peptide expression and plasticity—yet how do the corresponding SCG or DRG neuronal cell bodies recognize the injuries to mount plasticity responses? One mechanism relates to neuronal connectivity with the cognate peripheral target tissues and the retrograde transport of target-derived neurotrophic factors. The retrograde transport of NGF via TrkA receptors at sympathetic nerve terminals back to the soma has been well studied to generate survival signals via CREB signaling [71–74]. In congruence, immunosympathectomy with NGF neutralizing antisera to unoperated SCG recapitulated many of the peptide induction responses observed in axotomy [49]. These observations suggest that retrograde target tissue signaling represents a repressive neurotrophic signal and that axotomy, in relieving that neurotrophic support, activates a stress signal for PACAP induction.

Our own PACAP studies with sympathetic neuronal cultures have supported many of these observations. Primary sympathetic neurons, free of contaminating support cells, can be cultured in defined medium supplemented with NGF. The low levels of endogenous PACAP expression under these culture conditions could be induced over eightfold upon NGF withdrawal in a rapid time-dependent manner, and suppressed again upon the reintroduction of NGF into the culture medium (Fig. 33.3). The levels of SCG PACAP expression were exquisitely sensitive to NGF concentration and could be blocked with NGF neutralizing antisera or with the Trk receptor signaling inhibitor K252a (Fig. 33.4A). NGF/TrkA signaling takes place not only at the plasma membrane but continues in signaling endosomes upon endocytosis. Hence, the blockade of either clathrin- or caveolae-mediated internalization mechanisms also stimulated PACAP transcript expression (Fig. 33.4A). The roles of NGF/TrkA retrograde signaling were tested in compartmentalized Campenot chambers in which the SCG soma were separated from the axons, and in this model the withdrawal of NGF from the axonal compartment alone was sufficient to induce PACAP transcript expression in the cell bodies (Fig. 33.4C–D). NGF withdrawal placed the cultures into modes of cellular stress as reflected by an increase in neuronal JNK phosphorylation; suppression of JNK activation with the inhibitor SP600125 completely blocked the induction in PACAP expression during NGF withdrawal (Fig. 33.4B). By contrast, NGF had a lesser effect on VIP expression; withdrawal of cultured sympathetic neuron NGF augmented VIP transcript levels approximately 50 %. These VIP levels were markedly different from those after SCG axotomy in which VIP induction was typically equal or greater than that for PACAP. Consistent with other studies, cultured SCG neuron treatment with LIF, CNTF, or IL-6 had no effects on PACAP expression.

Despite the apparent roles of NGF in regulation of SCG PACAP levels, NGF may not be the principal retrograde neuroplasticity signal for all peptides or neuronal

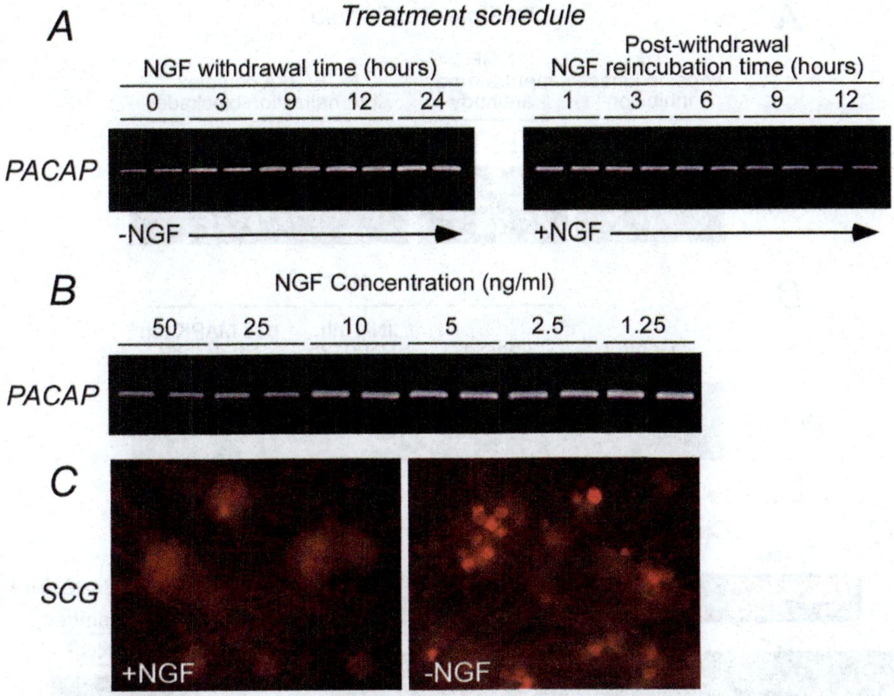

Fig. 33.3 NGF withdrawal from SCG cultures induces neuronal PACAP expression. Primary neonatal rat SCGs were enzymatically dissociated and seeded into 24-well culture plates; the cultures were treated with arabinoside cytosine to remove background cells and maintained in defined serum-free medium containing 50 ng/ml NGF as previously described [29]. (**A**) After 7 days in vitro, NGF was withdrawn from the sympathetic cultures and the cells were harvested at the times shown for RNA extraction and semi-quantitative PCR assessments of PACAP transcript levels (*left panel*). After 24 h in NGF-free medium, NGF was returned back to the replicate plates and the cultures harvested at the times shown after growth factor add-back (*right panel*). NGF withdrawal increased SCG culture PACAP transcript expression in a time-dependent manner; conversely NGF add-back gradually diminished PACAP transcript levels over 12 h. (**B**) SCG neurons were cultured in 50 ng/ml NGF and after 7 days, the defined medium was replaced with fresh medium containing the indicated NGF concentrations; the neurons were maintained in culture for an additional 12 h before harvesting for PCR analyses. Decreasing NGF concentrations increased culture PACAP transcript levels in a concentration-dependent manner. Each lane represents an individual replicate culture well. (**C**) SCG cultures were maintained in NGF-supplemented (+NGF) or NGF-free (-NGF) defined medium for 24 h before fixation and processing for PACAP immunocytochemistry using a Cy3-conjugated secondary antibody. NGF withdrawal increased the number of PACAP-immunoreactive neurons in the SCG cultures

systems. Neonatal sympathetic neurons are sensitive to NGF levels for trophic support, and accordingly the cultures are more responsive to cellular stress following neurotrophin withdrawal to initiate phenotypic responses. Adult sympathetic neurons may be more resilient to changes in NGF levels, and the relative contribution of abrogated retrograde NGF/TrkA signaling to other signals (i.e., inflammatory cytokines) may be different. Notably, one study demonstrated that the concurrent addi-

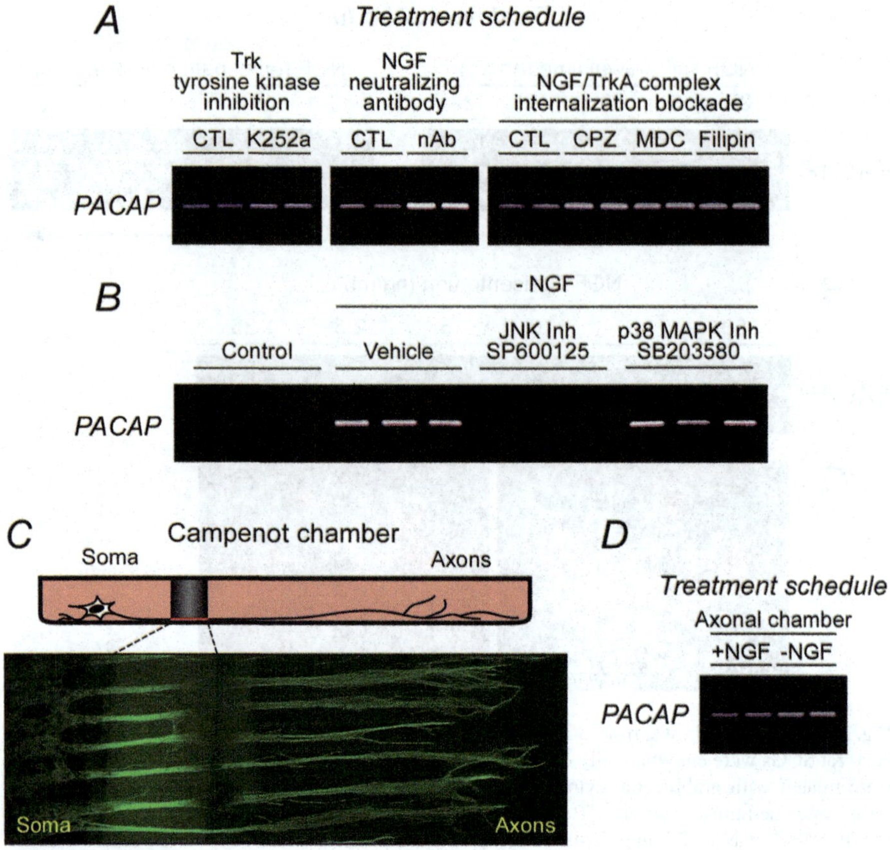

Fig. 33.4 Abrogation of retrograde NGF signaling and JNK activation increases SCG PACAP expression. (**A**) SCG neurons were cultured in defined medium containing 50 ng/ml NGF for 7 days before addition of the neurotrophin Trk receptor tyrosine kinase inhibitor K252a (100 nM) or NGF neutralizing antibody (nAb, 1 μg/ml) for 24 h. Replicate cultures were also treated with endocytosis inhibitors chlorpromazine (CPZ, 10 μg/ml, clathrin endocytosis inhibitor), monodansylcadaverine (MDC, 200 μM, clathrin endocytosis inhibitor), or filipin (2 μg/ml, caveolae endocytosis inhibitor). Blockade of NGF/Trk signaling and endocytosis/trafficking induced SCG neuronal PACAP transcript expression examined by PCR analyses. (**B**) NGF withdrawal from sympathetic cultures stimulated JNK activation [76]. To assess the roles of JNK/c-jun signaling in PACAP expression, SCG cultures were withdrawn from NGF, and treated simultaneously with the JNK inhibitor SP600125 or an unrelated p38 MAPK inhibitor SB203580. NGF removal induced culture PACAP transcripts (Vehicle) compared to control NGF-supplemented cultures (Control). The addition of the JNK inhibitor to the NGF-free cultures blocked PACAP induction; the unrelated p38 MAPK inhibitor had no effects on PACAP induction. (**C**) Schematic of SCG Campenot compartmentalized culture preparation. The SCG neurons were seeded onto a 1 mm wide compartment (Soma, *left*) and the axons were allowed to tunnel through a vacuum grease-sealed teflon barrier to reach the NGF-saturated axonal compartment (Axons, *right*). Hence, the SCG neuronal somas were compartmentalized from the axons. *Dotted lines* depict the width of the teflon barrier. (**D**) Sympathetic neuron somal PACAP mRNA levels from Campenot cultures maintained with NGF (50 ng/ml) in the axonal compartment (+NGF) were compared to those in which NGF was withdrawn from the axonal compartment for 24 h (-NGF). The neuronal soma compartments contained 10 ng/ml NGF (minimal survival dose) in all conditions. NGF withdrawal from the axonal compartment and resulting abrogation of NGF/TrkA retrograde neurotrophin signaling were sufficient to induce neuronal PACAP transcript expression

tion of anti-NGF and LIF to unoperated SCG induced peptide expression levels beyond those of anti-NGF alone, suggesting that dual activation signals and loss of target tissue factors are synergistic for neuropeptide plasticity [49]. Unlike sympathetic neurons, neonatal DRG neurons can be maintained in culture without NGF and by contrast, NGF treatments facilitated DRG PACAP expression in vitro, which appeared consistent with the stimulatory effects of NGF on DRG PACAP in vivo [46, 57, 75]. Indeed, NGF-induced DRG peptidergic plasticity has been associated with small caliber C-nociceptor hyperalgesia in tissue inflammation [75]. These results suggest NGF has different mechanisms of action on neuropeptide plasticity in different tissues; NGF suppresses PACAP expression in sympathetic neurons through neurotrophic signaling, but augments PACAP levels in DRG sensory neurons. For DRG and other autonomic tissues, the abrogation of different retrograde target tissue factors, in combination with activation mechanisms, may be important for axotomy-induced peptidergic plasticity. Described earlier, DRG PACAP expression is sensitive to NT-3/TrkC signaling [57]. In agreement, the induction of PACAP expression in guinea pig cardiac ganglion explants was not suppressed with NGF treatments but by the glial cell line-derived neurotrophic factor neurturin, implicating glial-derived neurotrophic factor (GDNF) receptor alpha2 (GFRalpha2) signaling in the regulation of cardiac neuron PACAP expression [51]. These observations suggest that there may not be a unifying mechanism underlying neurotransmitter/neuropeptide phenotypic plasticity. Rather, these studies in aggregate illustrate that neuropeptidergic plasticity mechanisms are complex; there are different mechanisms underlying injury- and inflammation-induced neurophenotypic plasticity, and the induction of each peptide may have a unique mechanistic signature. Accordingly, the totality of the peripheral nervous system peptide plasticity responses to the many physiological challenges represents an aggregate of these mechanisms.

Functional Roles of PACAP in Neuropeptidergic Plasticity

The induction of specific peptides in neuroplasticity has long been suggested to represent a reprogramming of cellular functions from neurotransmission to neuroprotection and regeneration. Many of the induced peptidergic systems, including PACAP, can engage neurotrophic signaling pathways [2, 76]; however, unequivocal evidence for these functions has proven somewhat elusive. Sciatic nerve transection in galanin−/− mice delayed nerve regeneration as assessed by toe spreading [77]. Axotomy of the facial nerve in PACAP−/− mice had no apparent effects on neuronal survival and demonstrated modest impairments in nerve regeneration, but enhanced inflammatory responses at the injury site [78]. Among other potential roles, these studies implicated injury-induced PACAP expression for anti-inflammatory processes.

However, injury- and inflammation-induced neuropeptidergic plasticity may also be viewed broadly as intrinsic neuronal mechanisms to maintain neural system homeostasis [36]. The balance of excitatory and inhibitory neurocircuit signaling has little margin for error, which is essential for proper activities in developing and mature networks. Under system perturbations such as stress, injury, or disease, the abilities of subpopulations of neurons to undergo neurochemical dynamics and plasticity may be a means of maintaining system homeostatic responsiveness. Congruently, PACAP signaling has clear roles in neuronal excitability [32, 79–81]. Noteworthy, JNK/c-jun regulation of the homeobox gene *tlx3* may coordinate PACAP and glutaminergic phenotypic identities in some neuronal systems [63, 82], and PACAP can modulate glutaminergic signaling by regulating cell surface NMDA and mGlu receptor expression. Hence, one potential function of DRG PACAP induction in injury and inflammation may be the initiation of adaptive mechanisms to maintain homeostatic signaling and responsiveness in the dorsal horn. Whether PACAP neuroplasticity has comparable homeostatic functions in other neuronal systems remains to be studied.

Conclusions

PACAP is widely distributed in sensory, autonomic, and enteric neurons. The population of endogenous PACAP-expressing neurons in these ganglia can be increased dramatically from the low number under basal conditions in parallel with heightened levels of PACAP transcripts, immunoreactivity, and content after experimental neuronal injury or inflammation. The mechanisms underlying the neuronal plasticity of PACAP and other peptides have not been fully elucidated, but likely involve multiple signals from the abrogation of retrograde target tissue regulators and an augmentation of stimulatory cytokine/inflammatory factors. The differential induction of PACAP, VIP, and other peptides may require distinct mechanistic signatures for the various neural tissues. Among roles including neurotrophic and anti-inflammatory responses, PACAP neuroplasticity may be an intrinsic mechanism required to maintain neurocircuit homeostasis.

Acknowledgments We thank Galen Missig, Beatrice M. Girard, Anton Kurtz, and Kristin C. Schutz for technical support in some of the studies. This work was supported by grants NS-01636 (KMB), HD-27468 and MH-096764 (VM), DK-051369, DK-060481 and DK-065989 (MAV), NS-23978 and HL-65481 (RLP), and funds from the University of Vermont (UVM) Center of Biomedical Research Excellence (Neuroscience COBRE, NCRR P30 RR032135/ NIGMS P30 GM103498).

References

1. Hannibal J. Pituitary adenylate cyclase-activating peptide in the rat central nervous system: an immunohistochemical and in situ hybridization study. J Comp Neurol. 2002;453:389–417.
2. Vaudry D, Falluel-Morel A, Bourgault S, Basille M, Burel D, Wurtz O, et al. Pituitary adenylate cyclase-activating polypeptide and its receptors: 20 years after the discovery. Pharmacol Rev. 2009;61:283–357.
3. Hashimoto H, Nogi H, Mori K, Ohishi H, Shigemoto R, Yamamoto K, et al. Distribution of the mRNA for a pituitary adenylate cyclase-activating polypeptide receptor in the rat brain: an in situ hybridization study. J Comp Neurol. 1996;371:567–77.

4. Jaworski DM, Proctor MD. Developmental regulation of pituitary adenylate cyclase-activating polypeptide and PAC(1) receptor mRNA expression in the rat central nervous system. Dev Brain Res. 2000;120:27–39.
5. Moller K, Zhang YZ, Hakanson R, Luts A, Sjolund B, Uddman R, et al. Pituitary adenylate cyclase activating peptide is a sensory neuropeptide: immunocytochemical and immunochemical evidence. Neuroscience. 1993;57:725–32.
6. Mulder H, Uddman R, Moller K, Zhang YZ, Ekblad E, Alumets J, et al. Pituitary adenylate cyclase activating polypeptide expression in sensory neurons. Neuroscience. 1994;63:307–12.
7. Zhang Q, Shi TJ, Ji RR, Zhang YZ, Sundler F, Hannibal J, et al. Expression of pituitary adenylate cyclase-activating polypeptide in dorsal root ganglia following axotomy: time course and coexistence. Brain Res. 1995;705:149–58.
8. Zhang YZ, Hannibal J, Zhao Q, Moller K, Danielsen N, Fahrenkrug J, et al. Pituitary adenylate cyclase activating peptide expression in the rat dorsal root ganglia: up-regulation after peripheral nerve injury. Neuroscience. 1996;74:1099–110.
9. Usoskin D, Furlan A, Islam S, Abdo H, Lonnerberg P, Lou D, et al. Unbiased classification of sensory neuron types by large-scale single-cell RNA sequencing. Nat Neurosci. 2015;18:145–53.
10. Noguchi K, Dubner R, De Leon M, Senba E, Ruda MA. Axotomy induces preprotachykinin gene expression in a subpopulation of dorsal root ganglion neurons. J Neurosci Res. 1994;37:596–603.
11. Jongsma H, Danielsen N, Sundler F, Kanje M. Alteration of PACAP distribution and PACAP receptor binding in the rat sensory nervous system following sciatic nerve transection. Brain Res. 2000;853:186–96.
12. Mulder H, Uddman R, Moller K, Elsas T, Ekblad E, Alumets J, et al. Pituitary adenylate cyclase activating polypeptide is expressed in autonomic neurons. Regul Pept. 1995;59:121–8.
13. Rytel L, Palus K, Całka J. Co-expression of PACAP with VIP, SP and CGRP in the porcine nodose ganglion sensory neurons. Anat Histol Embryol. 2015;44:86–91.
14. Zvarova K, Dunleavy JD, Vizzard MA. Changes in pituitary adenylate cyclase activating polypeptide expression in urinary bladder pathways after spinal cord injury. Exp Neurol. 2005;192:46–59.
15. Hannibal J, Ekblad E, Mulder H, Sundler F, Fahrenkrug J. Pituitary adenylate cyclase activating polypeptide (PACAP) in the gastrointestinal tract of the rat: distribution and effects of capsaicin or denervation. Cell Tissue Res. 1998;291:65–79.
16. Edvinsson L, Elsas T, Suzuki N, Shimizu T, Lee TJ. Origin and co-localization of nitric oxide synthase, CGRP, PACAP, and VIP in the cerebral circulation of the rat. Microsc Res Tech. 2001;53:221–8.
17. Fahrenkrug J, Hannibal J. Pituitary adenylate cyclase activating polypeptide innervation of the rat female reproductive tract and the associated paracervical ganglia: effect of capsaicin. Neuroscience. 1996;73:1049–60.
18. Wang ZY, Alm P, Hakanson R. Distribution and effects of pituitary adenylate cyclase-activating peptide in the rabbit eye. Neuroscience. 1995;69:297–308.
19. Nielsen HS, Hannibal J, Fahrenkrug J. Embryonic expression of pituitary adenylate cyclase-activating polypeptide in sensory and autonomic ganglia and in spinal cord of the rat. J Comp Neurol. 1998;394:403–15.
20. Moller K, Reimer M, Ekblad E, Hannibal J, Fahrenkrug J, Kanje M, et al. The effects of axotomy and preganglionic denervation on the expression of pituitary adenylate cyclase activating peptide (PACAP), galanin and PACAP type 1 receptors in the rat superior cervical ganglion. Brain Res. 1997;775:166–82.
21. Beaudet MM, Braas KM, May V. Pituitary adenylate cyclase activating polypeptide (PACAP) expression in sympathetic preganglionic projection neurons to the superior cervical ganglion. J Neurobiol. 1998;36:325–36.
22. Pettersson LM, Heine T, Verge VM, Sundler F, Danielsen N. PACAP mRNA is expressed in rat spinal cord neurons. J Comp Neurol. 2004;471:85–96.
23. Braas KM, May V. Pituitary adenylate cyclase-activating polypeptides directly stimulate sympathetic neuron neuropeptide Y release through PAC(1) receptor isoform activation of specific intracellular signaling pathways. J Biol Chem. 1999;274:27702–10.

24. May V, Braas KM. Pituitary adenylate cyclase-activating polypeptide (PACAP) regulation of sympathetic neuron neuropeptide Y and catecholamine expression. J Neurochem. 1995;65:978–87.
25. Zalecki M. Localization and neurochemical characteristics of the extrinsic sympathetic neurons projecting to the pylorus in the domestic pig. J Chem Neuroanat. 2012;43:1–13.
26. Ermilov LG, Schmalz PF, Miller SM, Szurszewski JH. PACAP modulation of the colon-inferior mesenteric ganglion reflex in the guinea pig. J Physiol. 2004;560:231–47.
27. Stroth N, Kuri BA, Mustafa T, Chan SA, Smith CB, Eiden LE. PACAP controls adrenomedullary catecholamine secretion and expression of catecholamine biosynthetic enzymes at high splanchnic nerve firing rates characteristic of stress transduction in male mice. Endocrinology. 2013;154:330–9.
28. Hamelink C, Tjurmina O, Damadzic R, Young WS, Weihe E, Lee HW, et al. Pituitary adenylate cyclase-activating polypeptide is a sympathoadrenal neurotransmitter involved in catecholamine regulation and glucohomeostasis. Proc Natl Acad Sci U S A. 2002;99:461–6.
29. Brandenburg CA, May V, Braas KM. Identification of endogenous sympathetic neuron pituitary adenylate cyclase-activating polypeptide (PACAP): depolarization regulates production and secretion through induction of multiple propeptide transcripts. J Neurosci. 1997;17:4045–55.
30. Calupca MA, Vizzard MA, Parsons RL. Origin of pituitary adenylate cyclase-activating polypeptide (PACAP)-immunoreactive fibers innervating guinea pig parasympathetic cardiac ganglia. J Comp Neurol. 2000;423:26–39.
31. Braas KM, May V, Harakall SA, Hardwick JC, Parsons RL. Pituitary adenylate cyclase-activating polypeptide expression and modulation of neuronal excitability in guinea pig cardiac ganglia. J Neurosci. 1998;18:9766–79.
32. Tompkins JD, Lawrence YT, Parsons RL. Enhancement of Ih, but not inhibition of IM, is a key mechanism underlying the PACAP-induced increase in excitability of guinea pig intrinsic cardiac neurons. Am J Physiol Regul Integr Comp Physiol. 2009;297:R52–9.
33. Hoover DB, Tompkins JD, Parsons RL. Differential activation of guinea pig intrinsic cardiac neurons by the PAC1 agonists maxadilan and pituitary adenylate cyclase-activating polypeptide 27 (PACAP27). J Pharmacol Exp Ther. 2009;331:197–203.
34. Girard BM, Galli JR, Young BA, Vizzard MA, Parsons RL. PACAP expression in explant cultured mouse major pelvic ganglia. J Mol Neurosci. 2010;42:370–7.
35. Demarque M, Spitzer NC. Neurotransmitter phenotype plasticity: an unexpected mechanism in the toolbox of network activity homeostasis. Dev Neurobiol. 2012;72:22–32.
36. Spitzer NC. Activity-dependent neurotransmitter respecification. Nat Rev Neurosci. 2012;18:94–106.
37. Spitzer NC, Borodinsky LN, Root CM. Homeostatic activity-dependent paradigm for neurotransmitter specification. Cell Calcium. 2005;37:417–23.
38. Dulcis D, Jamshidi P, Leutgeb S, Spitzer NC. Neurotransmitter switching in the adult brain regulates behavior. Science. 2013;340:449–53.
39. Francis NJ, Landis SC. Cellular and molecular determinants of sympathetic neuron development. Annu Rev Neurosci. 1999;22:541–66.
40. Walicke PA, Campenot RB, Patterson PH. Determination of transmitter function by neuronal activity. Proc Natl Acad Sci U S A. 1977;74:5767–71.
41. Zigmond RE, Hyatt-Sachs H, Mohney RP, Schreiber RC, Shadiack AM, Sun Y, et al. Changes in neuropeptide phenotype after axotomy of adult peripheral neurons and the role of leukemia inhibitory factor. Perspect Dev Neurobiol. 1996;4:75–90.
42. Hokfelt T, Zhang X, Wiesenfeld-Hallin Z. Messenger plasticity in primary sensory neurons following axotomy and its functional implications. Trends Neurosci. 1994;17:22–30.
43. Pettersson LM, Dahlin LB, Danielsen N. Changes in expression of PACAP in rat sensory neurons in response to sciatic nerve compression. Eur J Neurosci. 2004;20:1838–48.
44. Zhang Y, Danielsen N, Sundler F, Mulder H. Pituitary adenylate cyclase-activating peptide is upregulated in sensory neurons by inflammation. Neuroreport. 1998;9:2833–6.
45. Vizzard MA. Up-regulation of pituitary adenylate cyclase-activating polypeptide in urinary bladder pathways after chronic cystitis. J Comp Neurol. 2000;420:335–48.

46. Jongsma Wallin H, Pettersson LM, Verge VM, Danielsen N. Effect of anti-nerve growth factor treatment on pituitary adenylate cyclase activating polypeptide expression in adult sensory neurons exposed to adjuvant induced inflammation. Neuroscience. 2003;120:325–31.
47. Larsen JO, Hannibal J, Knudsen SM, Fahrenkrug J. Expression of pituitary adenylate cyclase-activating polypeptide (PACAP) in the mesencephalic trigeminal nucleus of the rat after transection of the masseteric nerve. Mol Brain Res. 1997;46:109–17.
48. Hyatt-Sachs H, Bachoo M, Schreiber R, Vaccariello SA, Zigmond RE. Chemical sympathectomy and postganglionic nerve transection produce similar increases in galanin and VIP mRNA but differ in their effects on peptide content. J Neurobiol. 1996;30:543–55.
49. Shadiack AM, Vaccariello SA, Sun Y, Zigmond RE. Nerve growth factor inhibits sympathetic neurons' response to an injury cytokine. Proc Natl Acad Sci U S A. 1998;95:7727–30.
50. Mohney RP, Siegel RE, Zigmond RE. Galanin and vasoactive intestinal peptide messenger RNAs increase following axotomy of adult sympathetic neurons. J Neurobiol. 1994;25:108–18.
51. Girard BM, Young BA, Buttolph TR, White SL, Parsons RL. Regulation of neuronal pituitary adenylate cyclase-activating polypeptide expression during culture of guinea-pig cardiac ganglia. Neuroscience. 2007;146:584–93.
52. Girard BM, Galli JR, Vizzard MA, Parsons RL. Galanin expression in the mouse major pelvic ganglia during explant culture and following cavernous nerve transection. J Mol Neurosci. 2012;48:713–20.
53. Klimaschewski L. Regulation of galanin in rat sympathetic neurons in vitro. Neurosci Lett. 1997;234:87–90.
54. Sun Y, Rao MS, Landis SC, Zigmond RE. Depolarization increases vasoactive intestinal peptide- and substance P-like immunoreactivities in cultured neonatal and adult sympathetic neurons. J Neurosci. 1992;12:3717–28.
55. Harakall SA, Brandenburg CA, Gilmartin GA, May V, Braas KM. Induction of multiple pituitary adenylate cyclase activating polypeptide (PACAP) transcripts through alternative cleavage and polyadenylation of proPACAP precursor mRNA. Ann N Y Acad Sci. 1998;865:367–74.
56. Pavelock KA, Girard BM, Schutz KC, Braas KM, May V. Bone morphogenetic protein down-regulation of neuronal pituitary adenylate cyclase-activating polypeptide and reciprocal effects on vasoactive intestinal peptide expression. J Neurochem. 2007;100:603–16.
57. Jongsma Wallin H, Danielsen N, Johnston JM, Gratto KA, Karchewski LA, Verge VM. Exogenous NT-3 and NGF differentially modulate PACAP expression in adult sensory neurons, suggesting distinct roles in injury and inflammation. Eur J Neurosci. 2001;14:267–82.
58. Hashimoto H, Hagihara N, Koga K, Yamamoto K, Shintani N, Tomimoto S, et al. Synergistic induction of pituitary adenylate cyclase-activating polypeptide (PACAP) gene expression by nerve growth factor and PACAP in PC12 cells. J Neurochem. 2000;74:501–7.
59. Sugawara H, Inoue K, Iwata S, Shimizu T, Yamada K, Mori N, et al. Neural-restrictive silencers in the regulatory mechanism of pituitary adenylate cyclase-activating polypeptide gene expression. Regul Pept. 2004;123:9–14.
60. Sugawara H, Tominaga A, Inoue K, Takeda Y, Yamada K, Miyata A. Functional characterization of neural-restrictive silencer element in mouse pituitary adenylate cyclase-activating polypeptide (PACAP) gene expression. J Mol Neurosci. 2014;54:526–34.
61. Zigmond RE. gp130 cytokines are positive signals triggering changes in gene expression and axon outgrowth in peripheral neurons following injury. Front Mol Neurosci. 2012;4:62.
62. Kim MS, Hur MK, Son YJ, Park JI, Chun SY, D'Elia AV, et al. Regulation of pituitary adenylate cyclase-activating polypeptide gene transcription by TTF-1, a homeodomain-containing transcription factor. J Biol Chem. 2002;277:36863–71.
63. Guo Z, Zhao C, Huang M, Huang T, Fan M, Xie Z, et al. Tlx1/3 and Ptf1a control the expression of distinct sets of transmitter and peptide receptor genes in the developing dorsal spinal cord. J Neurosci. 2012;32:8509–20.

64. Banner LR, Patterson PH. Major changes in the expression of the mRNAs for cholinergic differentiation factor/leukemia inhibitory factor and its receptor after injury to adult peripheral nerves and ganglia. Proc Natl Acad Sci U S A. 1994;91:7109–13.
65. Habecker BA, Sachs HH, Rohrer H, Zigmond RE. The dependence on gp130 cytokines of axotomy induced neuropeptide expression in adult sympathetic neurons. Dev Neurobiol. 2009;69:392–400.
66. Sun Y, Rao MS, Zigmond RE, Landis SC. Regulation of vasoactive intestinal peptide expression in sympathetic neurons in culture and after axotomy: the role of cholinergic differentiation factor/leukemia inhibitory factor. J Neurobiol. 1994;25:415–30.
67. Rao MS, Sun Y, Escary JL, Perreau J, Tresser S, Patterson PH, et al. Leukemia inhibitory factor mediates an injury response but not a target-directed developmental transmitter switch in sympathetic neurons. Neuron. 1993;11:1175–85.
68. Armstrong BD, Hu Z, Abad C, Yamamoto M, Rodriguez WI, Cheng J, et al. Lymphocyte regulation of neuropeptide gene expression after neuronal injury. J Neurosci Res. 2003;74:240–7.
69. Costigan M, Scholz J, Woolf CJ. Neuropathic pain: a maladaptive response of the nervous system to damage. Annu Rev Neurosci. 2009;32:1–32.
70. Basbaum AI, Bautista DM, Scherrer G, Julius D. Cellular and molecular mechanisms of pain. Cell. 2009;139:267–84.
71. Kuruvilla R, Zweifel LS, Glebova NO, Lonze BE, Valdez G, Ye H, et al. A neurotrophin signaling cascade coordinates sympathetic neuron development through differential control of TrkA trafficking and retrograde signaling. Cell Calcium. 2004;118:243–55.
72. Ye H, Kuruvilla R, Zweifel LS, Ginty DD. Evidence in support of signaling endosome-based retrograde survival of sympathetic neurons. Neuron. 2003;39:57–68.
73. Tsui-Pierchala BA, Ginty DD. Characterization of an NGF-P-TrkA retrograde-signaling complex and age-dependent regulation of TrkA phosphorylation in sympathetic neurons. J Neurosci. 1999;19:8207–18.
74. Riccio A, Pierchala BA, Ciarallo CL, Ginty DD. An NGF-TrkA-mediated retrograde signal to transcription factor CREB in sympathetic neurons. Science. 1997;277:1097–100.
75. Woolf CJ. Phenotypic modification of primary sensory neurons: the role of nerve growth factor in the production of persistent pain. Philos Trans R Soc Lond B Biol Sci. 1996;351:441–8.
76. May V, Lutz E, MacKenzie C, Schutz KC, Dozark K, Braas KM. Pituitary adenylate cyclase-activating polypeptide (PACAP)/PAC1HOP1 receptor activation coordinates multiple neurotrophic signaling pathways: Akt activation through phosphatidylinositol 3-kinase and vesicle endocytosis for neuronal survival. J Biol Chem. 2010;285:9749–61.
77. Holmes FE, Mahoney S, King VR, Bacon A, Kerr NC, Pachnis V, et al. Targeted disruption of the galanin gene reduces the number of sensory neurons and their regenerative capacity. Proc Natl Acad Sci U S A. 2000;97:11563–8.
78. Armstrong BD, Abad C, Chhith S, Cheung-Lau G, Hajji OE, Nobuta H, et al. Impaired nerve regeneration and enhanced neuroinflammatory response in mice lacking pituitary adenylyl cyclase activating peptide. Neuroscience. 2008;151:63–73.
79. Merriam LA, Baran CN, Girard BM, Hardwick JC, May V, Parsons RL. Pituitary adenylate cyclase 1 receptor internalization and endosomal signaling mediate the pituitary adenylate cyclase activating polypeptide-induced increase in guinea pig cardiac neuron excitability. J Neurosci. 2013;33:4614–22.
80. Tompkins JD, Hardwick JC, Locknar SA, Merriam LA, Parsons RL. Ca2+ influx, but not Ca2+ release from internal stores, is required for the PACAP-induced increase in excitability in guinea pig intracardiac neurons. J Neurophysiol. 2006;95:2134–42.
81. Tompkins JD, Merriam LA, Girard BM, May V, Parsons RL. Nickel suppresses the PACAP-induced increase in guinea pig cardiac neuron excitability. Am J Physiol Cell Physiol. 2015;308:C857–66.
82. Marek KW, Kurtz LM, Spitzer NC. cJun integrates calcium activity and tlx3 expression to regulate neurotransmitter specification. Nat Neurosci. 2010;13:944–50.

Chapter 34
The Neuropeptide PACAP, a Potent Disease Modifier Candidate for Brain Stroke Treatment

Coralie Brifault, David Vaudry, and Olivier Wurtz

Abstract Despite years of intense research, acute ischemic stroke remains a leading cause of death and long-term disabilities worldwide. Although preclinical studies lead to the identification of over 1000 potential neuroprotective compounds, the current treatments for brain ischemia only rely on clot thrombolysis through injection of recombinant-tissue plasminogen activator (r-tPA) or mechanical revascularization, which benefit to less than 10 % of stroke victims due to a narrow therapeutical time window and side effects. Consequently, there is a crucial need for the identification of new molecules and the development of other strategies that could target later phases of the pathophysiological cascade of mechanisms following stroke. Indeed, because stroke initiates a complex series of pathophysiological events evolving both in time and location, putative therapeutic molecules need to be effective on several of the biochemical processes evoked by stroke. Pituitary adenylate cyclase-activating polypeptide (PACAP) has been reported to decrease infarct volume and improve functional recovery in several models of global and focal brain ischemia. The unique particularity of PACAP relies on its ability to act on various pathological processes of cerebral ischemia. PACAP can counteract excitotoxicity,

C. Brifault
Department of Pathology, University of California San Diego, 9500 Gilman Drive, La Jolla, CA 92093, USA

D. Vaudry • O. Wurtz (✉)
Institut National de la Santé et de la Recherche Médicale (Inserm) U982—DC2N Laboratory, University of Rouen, 2 Place E. Blondel, Mont-Saint-Aignan, Cedex 76821, France

Institute for Research and Innovation in Biomedicine (IRIB), Normandy University, Rouen, France

Laboratory of Neuronal and Neuroendocrine Differentiation and Communication, Rouen University, Mont-Saint-Aignan, Cedex 76821, France
e-mail: olivier.wurtz@univ-rouen.fr

D. Reglodi, A. Tamas (eds.), *Pituitary Adenylate Cyclase Activating Polypeptide — PACAP*, Current Topics in Neurotoxicity 11,
DOI 10.1007/978-3-319-35135-3_34

inhibit apoptosis, reduce oxidative stress, modulate inflammation, and promote brain repair mechanisms. Nevertheless, due to several limitations, the pertinence of a potential therapeutical use of PACAP is still under investigation to successfully ensure the bench to bedside continuum.

Keywords PACAP • Brain stroke • Ischemia • Neuroprotection • Therapeutical drug • Inflammation

Abbreviations

AEPO	Asialo-erythropoietin
AIF	Apoptosis inducing factor
AIS	Acute ischemic stroke
BBB	Blood brain barrier
CBF	Cerebral blood flow
CNS	Central nervous system
CRMP2	Collapsing response mediator protein 2
DPP-IV	Dipeptidyl peptidase IV
EPO	Erythropoietin
GHRH	Growth hormone-releasing hormone
GPx-1	Glutathione peroxidase-1
HIF-1α	Hypoxic inducible factor-1α
i.c.v.	Intracerebroventricular
i.p.	Intraperitoneal
i.v.	Intravenous
IL	Interleukin
MCA	Middle cerebral artery
MCAO	Middle cerebral artery occlusion
NOS2	Nitric oxide synthase 2
PACAP	Pituitary adenylate cyclase-activating polypeptide
PEG	Polyethylene glycol
PLA	Polylactic acid
PLGA	Poly(lactic-co-glycolic acid) copolymer
PTS-6	Peptide transporter system 6
r-tPA	Recombinant-tissue plasminogen activator
SOD2	Superoxide dismutase-2
TNF	Tumor necrosis factor
VEGF	Vascular endothelial growth factor
VIP	Vasoactive intestinal peptide

Brain Stroke

Definition and Epidemiology

Stroke is a cerebrovascular accident caused by an abrupt reduction of cerebral blood flow (CBF) generating irreversible cerebral tissue damages. Although stroke can sometimes be due to intracerebral hemorrhage, the ischemic origin accounts for 87 % of the cases and results from the occlusion of an arterial cerebral vessel responsible for sudden focal neurological deficits. According to the World Health Organization, stroke encompasses the transient ischemic attack defined as neurological deficits disappearing spontaneously in less than 1 h without neither sequel nor brain imaging abnormalities, and the acute ischemic stroke (AIS) corresponding to clinical symptoms persisting beyond 24 h which will be in the scope of this review article.

In the United States, 795,000 Americans experience a new or recurrent stroke with 610,000 first-ever cases each year. It represents one stroke victim every 40 s and it is responsible for one death every 4 min. We assume that 6.6 million Americans already experienced a stroke event, an estimated stroke prevalence of 2.6 % on the 2009–2012 period. Stroke is considered as the third cause of death after heart diseases and cancer, the second cause of dementia after Alzheimer's disease, and the first cause of nontraumatic acquired disability in adults. The consequences after cerebral ischemia vary, but are often dramatic. Mortality rate was 20 % at 1 month and from 32 to 60 % after 3 years. While 25 % of patients recover without apparent sequel, 75 % of stroke survivors have various physical and psychological handicaps depending on the location and severity of the lesions. Among people bearing irreversible deficits, 30 % will not be autonomous in their daily activities and 40 % will not be able to resume their work. The management of stroke victims represents therefore a significant societal cost. In Europe, about 3–4 % of public health expenditures are dedicated to stroke, which represents a total of 64 billion € each year [1, 2]. The incidence and prevalence of stroke being closely associated with the aging of the population in industrialized countries, the projections for 2030 foresee a 20.5 % increase in prevalence of stroke in the United States [3]. Thus, it becomes obvious that stroke victims will represent a major public health and economic concerns.

Pathophysiology

Representing 2 % of the total body weight but consuming 20 % of cardiac output, the brain is particularly dependent on the continuous supply of oxygen and nutrients. The transient or permanent decrease in CBF observed in stroke results in 30 % of cases from occlusion of the middle cerebral artery (MCA) or one of its collaterals.

The consecutive energetic depletion initiates a complex series of pathophysiological events, which evolve in a spatiotemporal fashion and lead to a massive and progressive neurodegeneration responsible for disabling functional deficits [4].

From the spatial point of view, the hypoperfusion gradient from the periphery to the ischemic core delimits two distinct zones. The ischemic core, where energy supply is insufficient to allow cell survival, is the location of rapid and irreversible cell death, whereas the penumbra area, surrounding the ischemic core, is defined as an electrically silent but metabolically active tissue, thanks to the presence of arterial collaterals sufficient to support cell survival at least for a short period of time.

From the temporal point of view, the cascade of pathophysiological events responsible for the extension of the lesions, spreading from the ischemic core toward the penumbra, can be seen as a set of entangled processes taking place in two phases: an acute phase occurring during the first hours after the occlusion which mainly involves excitotoxic and oxidative stress processes, and a chronic phase developing over days and weeks after the stroke onset which includes apoptotic and inflammatory mechanisms. All these events lead to cerebral infarction whose size and location determine the nature and severity of the neurological deficits among stroke's victims.

Despite years of intense research and preclinical identification of numerous potential neuroprotective compounds, the current treatments for brain ischemia only rely on clot thrombolysis through injection of recombinant-tissue plasminogen activator (r-tPA) or mechanical revascularization. However, these treatments benefit to less than 10 % of stroke victims due to a narrow therapeutical time window and side effects [5]. Consequently, there is a crucial need for the development of other strategies that could target the overall pathophysiological cascade of events following stroke in the acute and chronic phases.

The PACAPergic System, an Endogenous Mechanism for Cell Protection After Stroke

The PACAPergic System

Pituitary adenylate cyclase-activating polypeptide (PACAP) was first isolated by Arimura and his colleagues from ovine hypothalamus extracts on the basis of its capacity to stimulate adenylate cyclase activity [6]. This neuropeptide exists in two forms, with a 38-amino acid peptide (PACAP38) and a truncated 27-residue fragment (PACAP27). PACAP, which exhibits 68 % sequence identity with vasoactive intestinal peptide (VIP), was classified as a member of the secretin/glucagon/growth hormone-releasing hormone (GHRH) superfamily [7]. Interestingly, the biologically active region of PACAP, i.e., its N-terminal part, has been highly conserved during evolution, from fish to mammals, suggesting that this neuropeptide must exert essential biological functions.

PACAP binds to three different seven-transmembrane G protein-coupled receptors named by the International Union of Pharmacology PAC1, VPAC1, and VPAC2 receptors according to their respective affinity toward PACAP and VIP [8]. Indeed PAC1 receptor exhibits a high affinity for both PACAP isoforms and a much lower affinity for VIP [9] whereas VPAC1 and VPAC2 receptors show similar affinity for PACAP and VIP [7]. More detailed information on PACAP and its receptors has already been published and is available in other published reviews [7, 10].

Several studies have reported the wide distribution of PACAP and its receptors in the central nervous system (CNS) and in peripheral tissues, supporting the hypothesis that PACAP exerts important biological activities. In fact, in the developing brain, PACAP exerts neurotrophic effects by modulating cell proliferation [11, 12], enhancing neuronal differentiation [13, 14] and blocking apoptosis [15–17]. In peripheral organs, PACAP has a strong dilatory activity on the vascular system [18] and bronchi [19]. Moreover, in vitro and in vivo studies have revealed that PACAP acts as a potent regulator of the immune system, exerting a general anti-inflammatory effect on innate [20, 21] and adaptive responses [21–23]. Altogether, these data indicate that PACAP is a unique neuropeptide which acts through a complex interplay of cellular and molecular mechanisms to inhibit cell death, promote neurotrophic activities, and regulate inflammatory responses.

Endogenous PACAP Expression After Cerebral Ischemia

Several studies have shown that PACAP expression is rapidly upregulated after brain artery occlusion, suggesting that the neuropeptide may play a role in the early pathophysiological events of ischemia. In a permanent model of MCA occlusion (MCAO), Stumm et al. have identified a biphasic regulation of PACAP in rat cortical pyramidal cells. They demonstrated that PACAP mRNA level in the frontoparietal cortex was increased by 95 % 6 h after artery occlusion but was reduced by 74 % 2 days after ischemia as compared with sham-operated animals [24]. A similar pattern of expression was also observed after transient MCAO in mice with an increase of PACAP mRNA by 251 % after 6 h [24], suggesting that the rapid upregulation of PACAP is independent of the stroke model. In accordance with this hypothesis, an increase of PACAP transcript is also observed in a mouse model of global ischemia, with a very strong elevation of PACAP mRNA expression in the granule cell layer of the dentate gyrus of the hippocampus 1 and 3 days after stroke [25]. According to these reports, PACAP mRNAs are essentially localized in cortical and pyramidal neurons [24, 25], suggesting that neurons are the main source of PACAP after stroke. Nevertheless, recent data revealed that in the penumbral region, PACAP upregulation is not restricted to neurons but can also be observed in astrocytes and endothelial cells. Interestingly, some blood-derived $CD34^+$ cells may also produce the neuropeptide upon infiltration in ischemic brain regions and can contribute to the elevated PACAP expression observed in human brain samples from patients who died 1–3 days after stroke [26]. Finally, the importance of endogenous PACAP expression in stroke models has been highlighted in PACAP-deficient mice which exhibit increased infarct volumes and functional deficits when compared to their control littermates [27, 28].

Although the molecular mechanisms of PACAP upregulation are still unclear, it is likely that PACAP may be expressed in response to stress-related environmental factors appearing in ischemic conditions. Stumm et al. showed that PACAP upregulation is mediated by the activation of the NMDA receptor and excessive glutamatergic neurotransmission, a hallmark of the initial ischemic cascade [24]. Another mechanism by which PACAP expression may increase is through activation of the hypoxic inducible factor-1α (HIF-1α) transcription factor whose transcriptional activity is strongly enhanced in hypoxic conditions. Indeed, no increased PACAP expression can be observed in HIF-1α deficient mice or after chemical inhibition during experimental ischemia [26]. In parallel, Zhang et al. demonstrated that PACAP expression is upregulated in sensory neurons in a model of adjuvant-induced inflammation in rat paw, suggesting that inflammatory conditions stimulate PACAP expression in neuronal cells [29].

Thus, these observations lead to the conclusion that in pathological conditions, endogenous PACAP expression is triggered in response to stress-related cues such as cell death, tissue lesion, or local inflammation, suggesting that the PACAPergic system could be part of an endogenous mechanism of protection aimed to support cell survival and to modulate the inflammatory response.

Expression of PACAP Receptors After Stroke

If less data are available regarding expression of PACAP receptors after stroke, they also support the idea that the endogenous PACAPergic system contributes to a mechanism of protection initiated after brain injury. Indeed, from day 1 to day 3 after ischemia, a transient increase of the PAC1 mRNA expression has been reported in neurons at the level of the post-ischemic fronto-parietal cortex, primary olfactory cortex, and caudate putamen in models of focal cerebral ischemia [24, 30, 31]. Similar results are observed in a mouse model of global ischemia where numerous PAC1 positive neurons are detected in the hippocampus 1 day after transient MCAO. In contrast, transcriptomic analysis revealed a significant decrease of VPAC1 mRNA at day 3, while VPAC2 mRNA levels remained unchanged over time [32]. From day 3, the PAC1 receptor is no longer detectable in neurons but its expression gradually increases in astrocytes. Indeed, an increase of PAC1 mRNA was observed in astrocytes of the corpus callosum in a permanent MCAO model [24]. Nakamachi and collaborators have also reported an elevation of PAC1 mRNA in hippocampal astrocytes surrounding the damaged CA1 pyramidal 7 days after ischemia [32], suggesting that the PACAPergic system may play a role in the reactive astrogliosis occurring after brain injury.

Altogether, these data reinforce the hypothesis of a participation of the PACAPergic system in an endogenous protective mechanism through the targeting at different levels of many pathophysiological processes in order to promote cell survival.

PACAP's Therapeutical Potential in Stroke Models: Mechanisms of Action

Neuroprotective Action of Exogenous PACAP's Delivery After Ischemia

Since its discovery, numerous studies using different approaches have demonstrated the neuroprotective effects of PACAP against ischemic lesions in various animal models of brain ischemia (Table 34.1).

Table 34.1 Reported PACAP's neuroprotective effects in experimental models of cerebral ischemia

Species	Model	Route	Doses	Delay	Results	References
Rat	4VO	*i.c.v*	0.1–10 pmol/h	0 h	↓ Infarct volume	[31]
		i.v.	5 nmol/kg + 16 or 160 pmol/h	24 h	↓ Infarct volume; ↓ JNK-SAPK; ↑IL-6	
Rat	CA	*i.c.v.*	1 pmol/h	0 h	↓ Infarct volume	[34]
Rat	tMCAO	*i.v.*	5 nmol/kg + 160 pmol/h	4, 8 and 12 h	↓ Infarct volume ↓ Functional deficits	[36]
Rat	tMCAO	*i.v.*	20 nmol/kg + 160 pmol/h	4 h	↓ Infarct volume	[37]
Rat	pMCAO	*i.c.v.*	450 pmol	Pre-ischemic	↓ Infarct volume	[97]
Rat	CA	*i.c.v.*	1 pmol/h	Pre-ischemic	↓ Infarct volume ↓ TUNEL ↓ p38-JNK-SAPK	[33]
Rat	pMCAO	*i.c.v.*	0.25–4 μg 2 μg or 2 μg × 7 days	Pre-ischemic	↓ Infarct volume and functional deficits	[39]
Rat	pMCAO	*i.c.v.*	2 μg	Pre-ischemic	↓ TUNEL	[43]
Mice (PACAP KO)	pMCAO	*i.c.v.*	1 pmol	0 h	↑ Infarct volume and functional deficits ↑ Cytochrome c; ↓ Bcl-2	[28]
Mice (IL-6 KO)	tMCAO	*i.v.*	5 nmol/kg + 16 pmol/h	0 h	No effect of PACAP	[28]
Mice (PACAP KO)	pMCAO	*i.v.*	0.75 nmol	1 h	↑ Infarct volume and functional deficits	[27]
		i.c.v.	40 pmol	1 h		

(continued)

Table 34.1 (continued)

Species	Model	Route	Doses	Delay	Results	References
Rat	tMCAO	*i.v.*	0.002–20 μg/kg	1 h	↓ Infarct volume and functional deficits ↓ TNF-α; TNFR1; iNOS; MIP-1α; NF-kB	[35]
Mice	pMCAO	*i.c.v.*	1 pmol	0 h	↑ IL-22; IGF-1; CTRAM	[38]
Mice	pMCAO	*i.c.v.*	1 pmol	0 h	↑ S100a5; IL-1β; Ccl2; IL-11; Cxcr2; FGF-21 and CRMP2 ↓ Prlr	[58]
Mice	pMCAO	*i.c.v.*	Constitutive secretion by ESC	3 Days	↓ Functional deficits ↓ Pro-inflammatory responses ↑ M2 microglia	[40]
Rat	tMCAO	*i.p.*	0.1, 1 or 10 μg/kg × 5 days	4 h	↓ Infarct volume and functional deficits ↑ $CD34^+$ cells homing ↑ Angiogenesis and CBF	[26]

4VO: 4 vessels occlusion, *CA* cardiac arrest, *tMCAO* transient middle cerebral artery occlusion, *pMCAO* permanent cerebral artery occlusion, *KO* knock-out, *i.c.v.* intracerebroventricular, *i.v.* intravenous, *i.p.* intraperitoneal, *ESC* embryonic stem cells, *CBF* cerebral blood flow

The therapeutic benefit of exogenous PACAP administration for the treatment of cerebral ischemia was reported for the first time in 1996 in a rat model of transient global ischemia. A chronic intracerebroventricular (*i.c.v.*) infusion of PACAP, 1 pmol/h for 7 days, started immediately after induction of the ischemic lesion significantly reduced hippocampal neuronal cell death [31]. Other studies, using global ischemia models, confirmed that *i.c.v.* administration of PACAP before or during the ischemia induction correlates with a reduction of infarct volumes [33, 34]. Similarly, in animal models of transient focal cerebral ischemia, the intravenous (*i.v.*) injection of PACAP induces beneficial effects when administered before the induction of ischemia and up to 4 h after arterial occlusion, including a reduction of lesion volume and functional deficits [28, 35–37]. Finally, the therapeutic potential of PACAP was also confirmed in permanent models of cerebral ischemia, in which PACAP was administrated either *i.v.* or *i.c.v.* up to 1 h after stroke [27, 28, 38, 39].

These data highlight that PACAP is a potent neuroprotective agent when administered during the acute phase of brain ischemia, specifically within the first hours after artery occlusion. Nevertheless, a study also reports that a delayed delivery of PACAP beyond the acute phase can also improve functional recovery 7 and 14 days after stroke. Such effect was obtained using a stem cell-based strategy to deliver PACAP in the vicinity of the infarct area 3 days after the stroke onset in a mouse model of permanent MCAO and indicates that PACAP may represent an interesting therapeutical candidate to counteract the delayed pathophysiological processes occurring during the chronic phase of cerebral ischemia and therefore constitutes a novel alternative for stroke treatment in a currently unexploited therapeutical time window [40].

Antiapoptotic Role of PACAP

The impairment of blood supply after vessel occlusion in a brain territory initiates a pathophysiological cascade of events leading to a massive neuronal loss. If cell death responsible for ischemic lesions has firstly been considered only of necrotic nature, it is now established that part of the cells also die by apoptosis [41]. Apoptotic processes preferentially develop within the penumbra and participate in the extension of the lesions from the infarct core toward the ischemic penumbra, therefore representing a potentially important therapeutic target [42]. Several in vitro and in vivo studies on ischemia report that the protective effect of PACAP is based on its well-recognized antiapoptotic properties as illustrated in rats by a 50 % reduction of apoptotic cell death after *i.c.v.* administration of PACAP (2 μg) immediately before permanent MCA occlusion [43].

The antiapoptotic property of the neuropeptide relies on its ability to interfere at multiple levels of the neurotoxic cascade that ends to cell apoptosis. The O_2 and glucose withdrawal impairs the activity of critical ATP-dependent ionic pumps such as ATP-dependent Na^+/K^+ and Ca^{2+} pumps disrupting ionic homeostasis and transmembrane gradients resulting in a massive release of excitatory amino acids, especially glutamate [44]. Lucas and Newhouse in 1957 and then Olney in 1969 have established a correlation between extracellular glutamate concentration and cytotoxicity, and shown that glutamate neurotoxicity is due to its excitatory effect, leading to the concept of excitotoxicity [45–47]. The binding of glutamate to its ionotropic NMDA, AMPA, and kainate receptors increases the permeability of these channel receptors to Cl^-, Na^+, and Ca^{2+} ions. Furthermore, the intracellular accumulation of Ca^{2+} triggers molecular events leading to the activation of apoptotic processes. At high concentration (above 10 nM), PACAP inhibits glutamate receptors transmission, from NMDA- as well as AMPA-type receptors [48, 49] and increases glial glutamate uptake and metabolism in astrocytes by promoting the expression of GLT-1, GLAST transporters, and glutamine synthetase [50]. Thus, PACAP can interfere in a first step with this initial excitotoxic cascade by modulating glutamate signaling (Fig. 34.1).

Consecutively to energetic depletion and associated excitotoxicity, the increased cytosolic Ca^{2+} concentration initiates the uncoupling of the mitochondrial oxidative phosphorylation, leading to deleterious free radicals production. In oxidative stress

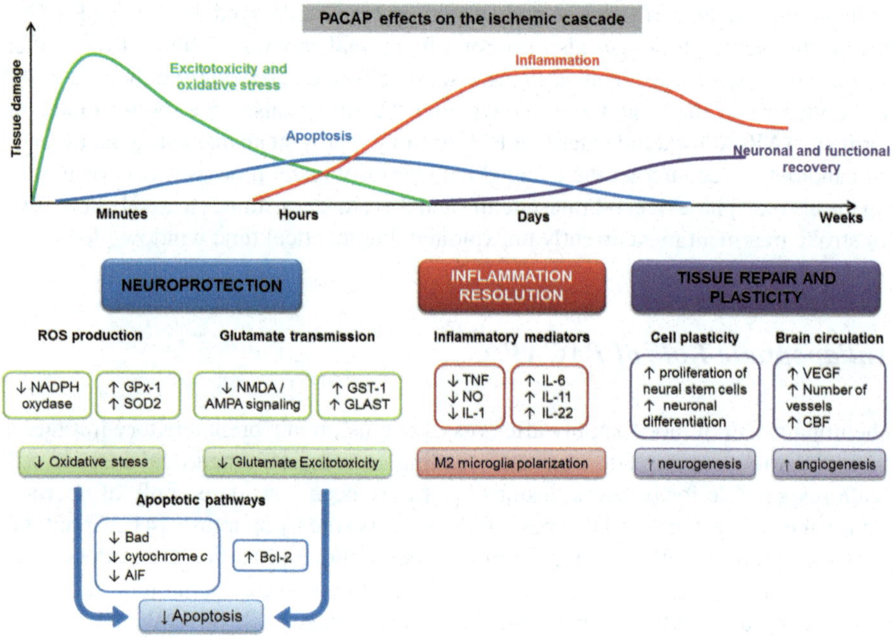

Fig. 34.1 Summary of PACAP's neuroprotective effects during acute ischemic stroke. PACAP acts on the different physiopathological events of the ischemic cascade developing in the early as well as the late phase of brain stroke. During the acute phase of ischemia, PACAP prevents neuronal cell death directly by regulating the caspase-dependent apoptotic pathways and indirectly by reducing oxidative stress and glutamate excitotoxicity. Moreover, PACAP affects the local inflammatory response in both phases of ischemia by modulating the balance between pro- and anti-inflammatory mediators and promoting the redirection of the microglial cell response toward a neuroprotective M2 phenotype. In the late phase, the PACAP-dependent redirection of the microglial response combined to the neurotrophic properties of the peptide on neural progenitors amplifies the post-ischemic neurogenesis and neoangiogenesis processes improving the endogenous tissue repair and functional recovery

conditions, such as after hydrogen peroxide exposure [17], lipopolysaccharide-induced dopaminergic neurotoxicity [51] or controlled cortical impact [52], PACAP reduces the production of reactive oxygen species by inhibiting the activity of NADPH oxidase and inducing the expression of Glutathione Peroxidase-1 (GPx-1) and Superoxide Dismutase-2 (SOD-2) antioxidant enzymes. Moreover, in a model of global ischemia, PACAP counteracts hippocampal CA1 neuronal death consecutive to oxidative DNA damages *via* induction of apurinic/apyrimidinic endonuclease 1 DNA repair enzyme [53]. Therefore, it appears that PACAP's anti-apoptotic effect results not only from the modulation of glutamatergic transmission but also depends on a powerful antioxidant activity, two of the initial processes responsible for the activation of the apoptotic cascade (Fig. 34.1).

Besides the inhibition of these initial steps of the neurotoxic cascade, the investigation of the mechanisms involved in the protective action of PACAP revealed its ability to inhibit directly the caspase-dependent mitochondrial apoptotic pathway. Western blot analysis performed in PACAP-deficient mice reported a rapid decrease, within 6 h after permanent focal ischemia, of Bcl-2 expression levels in the mitochondrial fraction and a concomitant cytosolic accumulation of cytochrome *c* release [28]. PACAP can also regulate the expression of other members of the Bcl-2 family. In particular, in an in vitro model of cortical neurons submitted to oxygen/glucose deprivation, PACAP promotes cell survival through the phosphorylation and therefore the inactivation of Bad [24]. Similarly, the neuroprotective effect of PACAP against monosodium glutamate-induced retinal degeneration is mediated by inhibition of pro-apoptotic signaling pathways and relies on the ability of PACAP to phosphorylate Bad but also to decrease cytochrome *c* and apoptosis inducing factor (AIF) release [54] (Fig. 34.1).

Altogether, these observations demonstrate that PACAP has a potent antiapoptotic activity due to combined inhibitory actions on the initiation phase as well as the effector phase of the apoptotic pathway.

Immunomodulatory Effects of PACAP

The post-ischemic inflammation is a complex process initiated early after the stroke onset, sustained over weeks, exerting both detrimental and beneficial roles and controlled by microglia, a unique immunocompetent cell population in the brain parenchyma. A correlation between the neuroprotective effect of PACAP and its immunomodulatory role was first proposed in 1998 by Shioda and coworkers. In a rodent model of global cerebral ischemia, *i.c.v.* infusion of PACAP for 48 h before ischemia causes an increase in interleukin (IL)-6 levels in the cerebrospinal fluid of animals correlated to a decrease of brain lesions [34]. Immunofluorescence studies highlighting IL-6 production in PAC1-expressing cortical neurons and absence of PACAP-mediated neuroprotection in IL-6 knock-out mice following MCAO suggest that PACAP's neuroprotective properties rely on the PACAP-dependent IL-6 production (Fig. 34.1). Although the mechanisms underlying the protective effects of PACAP through the production of IL-6 are not clearly identified to date, such results illustrate the close link between modulation of the inflammatory response and neuroprotection.

Microarray-based transcriptomic studies showed in a mouse model of permanent focal ischemia that *i.c.v.* administration of the neuropeptide PACAP immediately after MCA occlusion modulates the expression of numerous genes encoding inflammatory mediators 6 and 24 h after PACAP injection. Among these genes, *Il-22* and *Il-11* were found upregulated after PACAP treatment whereas *tumor necrosis factor (tnf)* was downregulated [38] (Fig. 34.1). Similarly, *i.v.* administration of very low quantity of PACAP (0.02 μg/kg) in a rat model of transient ischemia negatively regulates the expression of pro-inflammatory compounds such as TNF-α, the nitric oxide synthase (NOS)-2, the chemokine MIP-1α, and the transcription factor NF-kB (Fig. 34.1). Importantly, the modulation of the local inflammatory response seems to correlate with a reduction in lesion volume and neurological deficits in animals suggesting that the anti-inflammatory properties of the neuropeptide PACAP could account for a large part of its neuroprotective effects [35]. Interestingly, these results have been confirmed in a model of delayed PACAP delivery 3 days after the stroke onset [40]. In this study, the observed modulation of the local inflammatory response after delayed PACAP delivery reveals a PACAP-dependent shift of the inflammatory response toward a resolutive phenotype, supporting cell survival and tissue repair, that correlates with decreased neurological deficits in animals. This PACAP-dependent skewing of the inflammatory response relies, at least partly, on direct or indirect neuropeptide's effects on microglial cells (Fig. 34.1). Indeed, we report a PACAP-dependent polarization of microglial cell phenotype toward a neuroprotective M2-like phenotype as illustrated by an increased number of Arginase-1 positive microglial cells in the peri-infarct zone [40].

Altogether, these studies highlight that therapeutical strategies aimed at targeting the inflammatory process in the early phase as well as in the late phase of stroke and more specifically by skewing the microglial response toward a neuroprotective response rather than simply inhibiting inflammation could improve functional recovery in stroke victims.

Neurotrophic Properties of PACAP

PACAP exerts other important biological activities capable of interfering with the delayed pathophysiological events following stroke. More particularly, PACAP could mobilize different endogenous mechanisms of tissue repair including modulation of inflammation but also processes such as neurogenesis and angiogenesis, which in concert improve tissue homeostasis, neuroplasticity, and neuronal functions.

Following stroke, neurogenesis processes, which promote the generation of new parenchymal cells from neural stem cells and progenitor cells, are increased in the subventricular zone and in the dentate gyrus of the hippocampus. However, only 0.2 % of dead neurons are efficiently replaced by newborn cells [55], which is insufficient to achieve a complete functional recovery. Although there is no direct demonstration of a beneficial effect of PACAP in the neurogenic processes in in vivo models of ischemia, evidence from in vivo and in vitro studies using neural cells indicates that PACAP, due to its neurotrophic properties, may represent a promising pharmacological tool to improve such mechanisms. Indeed, besides an in vitro pro-mitogenic effect of PACAP on various cell types including neural stem/progenitor cells [56, 57], *i.c.v.* PACAP administration in naive mice as well as delayed PACAP delivery 3 days after ischemia (unpublished observations) increase the number of BrdU-positive cells in the subventricular zone and in the dentate gyrus [12, 56], indicating that PACAP enhances in vivo adult neurogenesis. Although the molecular mechanisms involved are not clearly identified so far, the pro-mitogenic effect of PACAP seems mediated through the PAC1 receptor which is expressed in the neurogenic areas [56].

Confirming further the potential of PACAP to enhance neurogenesis, several studies also report that PACAP signaling can guide stem/progenitor cell differentiation toward a neuronal phenotype. In particular, in the presence of PACAP, embryonic stem (ES) cells grown as embryoid bodies present an increased number of neuritic extensions compared to control cells, reflecting the neurogenic effect of the neuropeptide in the earliest stages of development. Furthermore, electrophysiological data demonstrate that PACAP enhances the amplitude of outward currents in these embryoid bodies-derived cells after 2 weeks of treatment [13], illustrating the functionality of the newly formed neurons. Furthermore, recent work revealed that in a murine stroke model, PACAP increases the expression of the *collapsing response mediator protein 2* (CRMP2), known to participate in axonal growth and neuronal differentiation, suggesting that PACAP could promote neuroregeneration through CRMP2 induction [58]. The fact that migration and differentiation of neural cells contribute to the mechanisms associated with endogenous neurogenesis to improve functional recovery [59, 60] suggests that PACAP could enhance post-stroke neurogenesis and consequently improve neurological and functional outcomes (Fig. 34.1).

Another mechanism that plays a critical role in improving long-term recovery after stroke is the reestablishment of a functional microvasculature in the ischemic penumbra. Indeed, neuronal function is critically dependent on regulation of CBF [61]. Among the numerous factors regulating angiogenesis and vascular maturation, the vascular endothelial growth factor (VEGF) appears to be a key player. In rodents, VEGFc and its receptors are overexpressed during the first 7–10 days after transient MCA occlusion and would be involved in post-ischemic revascularization processes [62, 63]. Previous investigations have demonstrated that PACAP stimulates VEGF production in rat pituitary folliculostellate cells [64], suggesting that it may play an important role in angiogenesis. Such pro-angiogenic properties of PACAP have recently been confirmed in a transient model of MCAO in rat where intraperitoneal (*i.p.*) injections of PACAP for 5 consecutive days after MCA ligation lead to a higher cerebral microvascular perfusion and to an increased number of cerebral blood vessels in the penumbral area compared with control animals, as visualized with FITC-dextran. Moreover, a significant elevation of CBF in the ischemic cortex of PACAP-treated animals was observed 1 week after ischemia [26], confirming that PACAP may be considered as a potent angiogenic factor after stroke (Fig. 34.1).

PACAP Therapeutical Potential for Stroke Treatment: Limitations and Possibilities

All the above-mentioned observations indicate that PACAP could be useful as a therapeutic agent to reduce cell loss and promote neurological and functional recovery after ischemic stroke. However, its clinical use may be compromised by a poor bioavailability, a short half-life, and potential side effects as discussed below.

Time Window for PACAP Administration

Although PACAP represents a very interesting candidate for stroke treatment, its ability to induce tissue protection and functional recovery has mainly been investigated early after administration of the peptide, i.e., within the first 4 h after the stroke onset [33, 35, 36, 38]. It should be remembered that a narrow time window of only 4.5 h is one of the most important limitations of the current available treatments for brain ischemia which are only given to about 10 % of stroke victims. As a consequence, the development of therapeutical strategies with enlarged time window is a priority and in this context, our recent findings of a beneficial effect of PACAP 3 days after the stroke onset offer new avenues for the use of PACAP as a therapeutic drug with unexploited therapeutical window that could benefit to a larger number of stroke patients [40].

PACAP Bioavailability and Administration Route

Numerous studies using rodent models of cerebral ischemia demonstrate promising results after an intravenous injection of PACAP, as illustrated in Table 34.1. In human however, strategies based on a systemic administration of pharmacological agents are restricted by the difficulties to access the damaged cerebral tissues especially after cerebrovascular disruption. Interestingly, while the passage of numerous molecules is restricted by the blood brain barrier (BBB), PACAP38 crosses the BBB *via* the specific peptide transporter system 6 (PTS-6) [65]. Moreover, in two different models of brain ischemia, one caused by cardiac arrest and the other by MCAO, a rapid transient increase of PACAP influx into the brain within 4 h after *i.v.* injection has been shown [66, 67]. Although the exact mechanisms involved in the observed influx of PACAP have not been directly demonstrated, these changes of PACAP permeability are more likely attributable to an enhanced PTS-6 activity across the BBB rather than the ischemia-induced increased BBB permeability. This ability of PACAP to rapidly enter the brain may account for the neuroprotective effects observed early after the stroke onset [35–37]. Nevertheless, treatment with the native peptide starting 8 or 12 h after occlusion did not result in reduced infarct or improve functional recovery [35–37], which could be explained by the decreased entry of PACAP beginning 6 h after ischemia [66]. Thus it appears that while intravenous PACAP injection at early time points after ischemia may represent a relevant approach for neuroprotection, after the first 4 h, *i.v.* PACAP injections do not reach efficiently the brain parenchyma to promote neurological improvement and functional recovery. Thus, alternative strategies to improve the delivery of PACAP in the vicinity of the infarct area, such as stem cell-expressing PACAP injected 3 days after stroke [40], may be of great interest. Moreover, by enhancing the rate of entry of PACAP in the CNS, neuroprotection would be achieved with lower doses of the peptide, leading to a reduction of the adverse side effects associated with an *i.v.* injection.

PACAP Stability

In the systemic circulation, PACAP, as most natural peptides, is rapidly degraded by plasmatic enzymes including carboxypeptidase and endopeptidase, greatly reducing its biological half-life and therefore restricting its use as a therapeutical agent. Furthermore, in vitro and in vivo studies revealed that PACAP is rapidly metabolized by dipeptidyl peptidase-IV (DPP-IV), a ubiquitous aminopeptidase that releases dipeptides from the N-terminal fragment of diverse peptides. Thus, after intravenous injection to mice, the half-life of PACAP in plasma is less than 2 min [68]. In human, a case study in an elderly male patient estimated PACAP blood half-life between 5 and 10 min [69], while in healthy volunteers PACAP exhibits a mean plasma half-life of 3.5 ± 1.3 min [70]. Moreover, the cleavage of PACAP by DPP-IV generates two metabolites PACAP(3-38) and PACAP(5-38) [68] that behave as PAC1 antagonists [71], reducing further the biological effects of PACAP. As a consequence, obtaining an effective therapeutical dose of PACAP in the ischemic territory may be hampered by this low metabolic stability.

Potential Side Effects

PACAP and its receptors are widely distributed throughout the CNS and peripheral tissues. Such a large distribution implies that this neuropeptide exerts numerous biological activities [7], some of them being potential side effects when injected systemically, therefore preventing the safe therapeutic use of the peptide. For instance, PACAP and its receptors are expressed in the cardiovascular system [72–74]. Considering that cerebral ischemia results from vascular impairment, potential effects of PACAP on the cardiovascular system have to be considered very carefully. In fact, in healthy volunteers, intravenous infusion of PACAP (10 pmol/kg/min) for 20 min significantly increases heart rate [70]. As PACAP was reported to be a potent and long-lasting vasodilator, this tachycardia probably results from the peripheral vasodilatation caused by the peptide [75, 76]. It has also been showed that PACAP in a dose-dependent manner can cause tachycardia, bradycardia, and auricular fibrillation [77], various events that may lead to the occurrence of ischemic stroke. Moreover, it is worth noting that in human, all subjects who received an *i.v.* injection of PACAP experienced a sustained skin flushing on the face and chest [70] and some of them were subjected to headache [78]. Because such physiological responses may impede the use of PACAP as a safe therapeutical drug for stroke treatment, the development of approaches to limit PACAP side effects represents an important goal. As discussed below, different strategies may be considered to counteract the limitations associated with PACAP poor bioavailability and side effects, including the design of stable analogs of PACAP with improved selectivity and the development of strategies to enhance its specific brain delivery.

Development of Stable Analogs

One possibility to increase PACAP metabolic stability is to develop PACAP stable analogs presenting a reduced sensitivity to plasmatic endopeptidase and DPP-IV. In this attempt, chemical modifications at potential peptidase cleavage sites into the peptide sequence can drastically improve its resistance to enzymatic degradation. For instance, the sequence of PACAP contains two potential cleavage sites for plasmatic endopeptidases represented by two dibasic pairs Arg^{14}-Lys^{15} and Lys^{20}-Lys^{21} [79]. Therefore, the replacement of the Lysine amino acid by an Alanine residue confers an increased resistance to enzymatic digestion. Similarly, N-terminal acetylation of the peptide abolished its degradation by DPP-IV. Structure-activity relationship studies led to the development of two metabolically stable PACAP analogs, i.e., acetyl-[Ala^{15}, Ala^{20}]PACAP38-propylamide and acetyl-PACAP27-propylamide that exhibit a complete resistance to DPP-IV degradation and an increased plasmatic half-life while remaining highly active [79]. It has been shown in a rat model of transient cerebral ischemia that PACAP38 and its analog acetyl-[Ala^{15}, Ala^{20}] PACAP38-propylamide, when administered intravenously at very low dose (0.02 μg/kg), present similar potency and act through same mechanisms to reduce neuronal cell loss and functional deficits [35]. These results indicate that picomolar doses of PACAP and analogs are sufficient to exert maximal neuroprotective activities in case of stroke and it is tempting to speculate that at such low doses, side effects associated with the systemic administration of PACAP would be considerably reduced.

Nevertheless, PACAP analogs have also shown detrimental effects in animal stroke models. As a matter of fact, the administration of a VPAC2 agonist worsens neuronal damages after stroke [80]. Therefore, development of Ac-[Phe(*p*I)^{6}, Nle^{17}]PACAP(1-27) which activates PAC1 and VPAC1 receptors but not VPAC2 could be a new tool to reduce PACAP side effects [81]. Actually, in an in vivo model of Parkinson disease, this analog was as potent as PACAP to restore tyrosine hydroxylase expression and to modulate the inflammatory response in the *substantia nigra* but had significantly less effects on mean arterial pressure when compared to PACAP [82].

Enhancement of Brain Delivery

Another possible approach to counteract the limitations associated with the plasmatic instability of PACAP and its potential peripheral side effects is the development of strategies aiming at enhancing the specific delivery of PACAP within the ischemic brain.

Intranasal Delivery

During the last few decades, the nasal route for drug delivery into the brain has been extensively investigated. The "nose-to-brain" delivery is of great interest because this route is noninvasive and can deliver drugs directly into the brain, bypassing the

BBB which is impermeable for a high percentage of molecules. Additionally, such approach avoids rapid metabolism in the systemic circulation as well as peripheral side effects, therefore representing a promising hope for counteracting the limitations associated with the intravenous injection of PACAP. Several therapeutic proteins such as granulocyte-colony stimulating factor, transforming growth factor-beta, erythropoietin (EPO), and insulin-growth factor-1 have been successfully delivered into the brain of rodents in different models of stroke [83–85], indicating that intranasal drug delivery is a promising method to specifically provide therapeutical compounds into the ischemic brain. Moreover, different formulations including simple drop (nanogel) or nasal spray (insoluble powder or emulsions) can be designed to maximize the convenience and comfort of patients. Interestingly, nasal application of PACAP can reach the brain rapidly and was effective to slow down Alzheimer's disease-like pathology in a mice transgenic model [86]. Noteworthy, such administration of the peptide has also been reported in human and did not affect heart rate and blood pressure in patients [86], indicating that using the "nose-to-brain" route for PACAP delivery can efficiently avoid the side effects of PACAP on the cardiovascular system.

Colloidal Carriers

The use of colloidal carriers including liposomes and nanoparticles could represent another interesting avenue for a selective and prolonged delivery of PACAP and its derivatives into the ischemic brain. Colloidal carriers are drug vectors, particles, or vesicles in the nanometer size range used for the transport of loaded drugs to a target site. These systems are biocompatible, nontoxic, and biodegradable and therefore represent a promising tool for a clinical use. They can carry various therapeutic agents (hydrophilic, lipophilic, and amphoteric drug molecules) which are generally encapsulated, entrapped, adsorbed, or chemically attached to the carrier surface [87]. These encapsulation systems offer the advantage to protect drug molecules from enzymatic degradation in the blood circulation, significantly increasing their bioavailability. The biocompatibility and longevity of such particles can be achieved through surface modifications involving a coating of biodegradable polymers such as chitosan, PLGA (poly(lactic-co-glycolic acid) copolymer), PLA (polylactic acid), and PEG (polyethylene glycol) [88]. In particular, PEGylation of drug-loaded liposomes is very efficient to enhance the longevity of particles in the systemic bloodstream [89]. Such surface modifications are also used to achieve tissue specificity. For example, the biopolymer polysorbate, PLGA, and PEG have been successful in delivering therapeutics into the brain parenchyma [90, 91]. Other ligands can also be added to specifically target the brain including insulin, transferrin, lactoferrin, glutathione, and apolipoprotein [92]. It is worth mentioning that such nanocarrier formulations also exist for nose-to-brain delivery [93].

Colloidal vectors, in particular polymeric nanoparticles and liposomes, have been used with success to transport therapeutically relevant agents into the brain parenchyma in various animal models of brain ischemia. For instance, in a transient model of MCAO, chitosan-nanoparticles loaded with the basic fibroblast growth factor and/or with an inhibitor of caspase-3 have been administrated systematically just before the induction of ischemia. In this study the authors found that nanoparticles rapidly transport the active molecules to the brain parenchyma in their active form, which results in reduced infarct volume and functional impairment [94]. PEGylated liposomes administrated within 3 h after the stroke onset also accumulate in the ischemic penumbral area in a time-dependent manner with a peak of accumulation at 24 h [95]. Hence, the long-lasting accumulation of the nanocarriers in the vicinity of the infarct area could lead to a prolonged drug delivery and an increased therapeutic efficacy of PACAP. Supporting this hypothesis, asialoerythropoietin (AEPO, a metabolite of EPO) loaded liposomes present higher protective effect compared to native AEPO [95]. Furthermore, magnetic resonance imaging studies revealed that 100 nm-liposomes, injected intravenously 1 or 2 h after permanent MCAO in rats, accumulated in the ipsilateral parenchyma [96]. These results raise the possibility that liposome-based strategies may be possible even in the absence of reperfusion and could therefore benefit to patients who are not eligible for r-tPA treatment. All these observations indicate that application of such technologies to PACAP could not only enhance its therapeutic efficacy to improve neurological and functional ischemia-induced deficits, but also enable a reduction of the dose of peptide which must be administered to achieve a therapeutic response, thus minimizing its adverse side effects. Remarkably, encapsulation systems may also present the advantage of targeting PACAP delivery at the cellular level. As discussed previously a possible mechanism underlying the beneficial effect of PACAP during cerebral ischemia relies on its immunoregulatory activity on microglial cells. Thus, incorporating molecules specifically recognized by microglial cells in the surface of liposomes or nanoparticles would be of great interest. For instance, it would be interesting to evaluate the therapeutical potential of PACAP-loaded particles coated with molecules known to activate microglia phagocytic function without triggering a pro-inflammatory program such as the agonist of the TREM2 receptor Hsp60 [97].

Altogether, these approaches represent alternative strategies for the stabilization and development of pharmaceutical formulation of PACAP and its derivatives to benefit from its neuroprotective properties while overcoming its limitations in the context of brain stroke.

Conclusion

In conclusion, there is now clear evidence that PACAP exerts a potent neuroprotective activity and improves functional recovery in various models of cerebral ischemia. PACAP acts at very low doses on the main pathophysiological events developing in the early and in the late phases of brain stroke, making PACAP a potential therapeutical candidate for stroke treatment even in still unexploited therapeutical windows. Interestingly, several pathophysiological processes modulated by PACAP in stroke models are known to be involved in many other neurodegenerative diseases among which Parkinson's disease, multiple sclerosis, spinal cord injury, or Alzheimer's disease, suggesting that PACAP could represent also a novel alternative for the treatment of these neurological troubles. Nevertheless, because of the wide distribution of PACAP and its receptors, a systemic delivery of the peptide would have many peripheral actions responsible for numerous side effects limiting the PACAP's therapeutical use. Therefore, it is now necessary to develop selective analogs and efficient route of administration to specifically and locally target the signaling mechanisms and cell responses relaying the protective effects of PACAP without affecting peripheral organs.

References

1. Evers SMAA, Struijs JN, Ament AJHA, van Genugten MLL, Jager JHC, van den Bos GAM. International comparison of stroke cost studies. Stroke. 2004;35:1209–15.
2. Olesen J, Gustavsson A, Svensson M, Wittchen H-U, Jönsson B. The economic cost of brain disorders in Europe. Eur J Neurol. 2012;19:155–62.
3. Mozaffarian D, Benjamin EJ, Go AS, Arnett DK, Blaha MJ, Cushman M, et al. Heart disease and stroke statistics-2015 update: a report from the American Heart Association. Circulation. 2015;131(4):e29–322.
4. Dirnagl U, Iadecola C, Moskowitz MA. Pathobiology of ischaemic stroke: an integrated view. Trends Neurosci. 1999;22:391–7.
5. Donnan GA, Davis SM, Parsons MW, Ma H, Dewey HM, Howells DW. How to make better use of thrombolytic therapy in acute ischemic stroke. Nat Rev Neurol. 2011;7:400–9.
6. Miyata A, Arimura A, Dahl RR, Minamino N, Uehara A, Jiang L, et al. Isolation of a novel 38 residue-hypothalamic polypeptide which stimulates adenylate cyclase in pituitary cells. Biochem Biophys Res Commun. 1989;164:567–74.
7. Vaudry D, Falluel-Morel A, Bourgault S, Basille M, Burel D, Wurtz O, et al. Pituitary adenylate cyclase-activating polypeptide and its receptors: 20 years after the discovery. Pharmacol Rev. 2009;61:283–357.
8. Harmar AJ, Arimura A, Gozes I, Journot L, Laburthe M, Pisegna JR, et al. International Union of Pharmacology. XVIII. Nomenclature of receptors for vasoactive intestinal peptide and pituitary adenylate cyclase-activating polypeptide. Pharmacol Rev. 1998;50:265–70.
9. Lutz EM, MacKenzie CJ, Johnson M, West K, Morrow JA, Harmar AJ, et al. Domains determining agonist selectivity in chimaeric VIP2 (VPAC2)/PACAP (PAC1) receptors. Br J Pharmacol. 1999;128:934–40.
10. Harmar AJ, Fahrenkrug J, Gozes I, Laburthe M, May V, Pisegna JR, et al. Pharmacology and functions of receptors for vasoactive intestinal peptide and pituitary adenylate cyclase-activating polypeptide: IUPHAR review 1. Br J Pharmacol. 2012;166:4–17.
11. Nicot A, DiCicco-Bloom E. Regulation of neuroblast mitosis is determined by PACAP receptor isoform expression. Proc Natl Acad Sci U S A. 2001;98:4758–63.
12. Ohta S, Gregg C, Weiss S. Pituitary adenylate cyclase-activating polypeptide regulates forebrain neural stem cells and neurogenesis in vitro and in vivo. J Neurosci Res. 2006;84:1177–86.
13. Chafai M, Louiset E, Basille M, Cazillis M, Vaudry D, Rostène W, et al. PACAP and VIP promote initiation of electrophysiological activity in differentiating embryonic stem cells. Ann N Y Acad Sci. 2006;1070:185–9.
14. Gonzalez BJ, Basille M, Vaudry D, Fournier A, Vaudry H. Pituitary adenylate cyclase-activating polypeptide promotes cell survival and neurite outgrowth in rat cerebellar neuroblasts. Neuroscience. 1997;78:419–30.

15. Masmoudi-Kouki O, Douiri S, Hamdi Y, Kaddour H, Bahdoudi S, Vaudry D, et al. Pituitary adenylate cyclase-activating polypeptide protects astroglial cells against oxidative stress-induced apoptosis. J Neurochem. 2011;117:403–11.
16. Vaudry D, Falluel-Morel A, Basille M, Pamantung TF, Fontaine M, Fournier A, et al. Pituitary adenylate cyclase-activating polypeptide prevents C2-ceramide-induced apoptosis of cerebellar granule cells. J Neurosci Res. 2003;72:303–16.
17. Vaudry D, Pamantung TF, Basille M, Rousselle C, Fournier A, Vaudry H, et al. PACAP protects cerebellar granule neurons against oxidative stress-induced apoptosis. Eur J Neurosci. 2002;15:1451–60.
18. Huang M, Shirahase H, Rorstad OP. Comparative study of vascular relaxation and receptor binding by PACAP and VIP. Peptides? 14:755–62.
19. Lindén A, Yoshihara S, Chan B, Nadel JA. Inhibition of bronchoconstriction by pituitary adenylate cyclase activating polypeptide (PACAP 1-27) in guinea-pigs in vivo. Br J Pharmacol. 1995;115:913–6.
20. Delgado M, Miller Jonakait G, Ganea D. Vasoactive intestinal peptide and pituitary adenylate cyclase-activating polypeptide inhibit chemokine production in activated microglia. Glia. 2002;39:148–61.
21. Delgado M, Munoz-Elias EJ, Gomariz RP, Ganea D. VIP and PACAP inhibit IL-12 production in LPS-stimulated macrophages. Subsequent effect on IFNgamma synthesis by T cells. J Neuroimmunol. 1999;96:167–81.
22. Tan Y-V, Abad C, Lopez R, Dong H, Liu S, Lee A, et al. Pituitary adenylyl cyclase-activating polypeptide is an intrinsic regulator of Treg abundance and protects against experimental autoimmune encephalomyelitis. Proc Natl Acad Sci U S A. 2009;106:2012–7.
23. Tan Y-V, Abad C, Wang Y, Lopez R, Waschek JA. VPAC2 (vasoactive intestinal peptide receptor type 2) receptor deficient mice develop exacerbated experimental autoimmune encephalomyelitis with increased Th1/Th17 and reduced Th2/Treg responses. Brain Behav Immun. 2015;44:167–75.
24. Stumm R, Kolodziej A, Prinz V, Endres M, Wu DF, Höllt V. Pituitary adenylate cyclase-activating polypeptide is up-regulated in cortical pyramidal cells after focal ischemia and protects neurons from mild hypoxic/ischemic damage. J Neurochem. 2007;103:1666–81.
25. Riek-Burchardt M, Kolodziej A, Henrich-Noack P, Reymann KG, Höllt V, Stumm R. Differential regulation of CXCL12 and PACAP mRNA expression after focal and global ischemia. Neuropharmacology. 2010;58:199–207.
26. Lin C-H, Chiu L, Lee H-T, Chiang C-W, Liu S-P, Hsu Y-H, et al. PACAP38/PAC1 signaling induces bone marrow-derived cells homing to ischemic brain. Stem Cells. 2015;33:1153–72.
27. Chen Y, Samal B, Hamelink CR, Xiang CC, Chen Y, Chen M, et al. Neuroprotection by endogenous and exogenous PACAP following stroke. Regul Pept. 2006;137:4–19.
28. Ohtaki H, Nakamachi T, Dohi K, Aizawa Y, Takaki A, Hodoyama K, et al. Pituitary adenylate cyclase-activating polypeptide (PACAP) decreases ischemic neuronal cell death in association with IL-6. Proc Natl Acad Sci U S A. 2006;103:7488–93.
29. Zhang Y, Danielsen N, Sundler F, Mulder H. Pituitary adenylate cyclase-activating peptide is upregulated in sensory neurons by inflammation. Neuroreport. 1998;9:2833–6.
30. Gillardon F, Hata R, Hossmann K-A. Delayed up-regulation of Zac1 and PACAP type I receptor after transient focal cerebral ischemia in mice. Mol Brain Res. 1998;61:207–10.
31. Uchida D, Arimura A, Somogyvári-Vigh A, Shioda S, Banks WA. Prevention of ischemia-induced death of hippocampal neurons by pituitary adenylate cyclase activating polypeptide. Brain Res. 1996;736:280–6.
32. Nakamachi T, Tsuchida M, Kagami N, Yofu S, Wada Y, Hori M, et al. IL-6 and PACAP receptor expression and localization after global brain ischemia in mice. J Mol Neurosci. 2012;48:518–25.
33. Dohi K, Mizushima H, Nakajo S, Ohtaki H, Matsunaga S, Aruga T, et al. Pituitary adenylate cyclase-activating polypeptide (PACAP) prevents hippocampal neurons from apoptosis by inhibiting JNK/SAPK and p38 signal transduction pathways. Regul Pept. 2002;109:83–8.

34. Shioda S, Ozawa H, Dohi K, Mizushima H, Matsumoto K, Nakajo S, et al. PACAP protects hippocampal neurons against apoptosis: involvement of JNK/SAPK signaling pathway. Ann N Y Acad Sci. 1998;865:111–7.
35. Dejda A, Seaborn T, Bourgault S, Touzani O, Fournier A, Vaudry H, et al. PACAP and a novel stable analog protect rat brain from ischemia: Insight into the mechanisms of action. Peptides. 2011;32:1207–16.
36. Reglodi D, Somogyvari-Vigh A, Vigh S, Kozicz T, Arimura A. Delayed systemic administration of PACAP38 is neuroprotective in transient middle cerebral artery occlusion in the rat. Stroke. 2000;31:1411–7.
37. Reglodi D, Somogyvari-Vigh A, Vigh S, Maderdrut JL, Arimura A. Neuroprotective effects of PACAP38 in a rat model of transient focal ischemia under various experimental conditions. Ann N Y Acad Sci. 2000;921:119–28.
38. Hori M, Nakamachi T, Rakwal R, Shibato J, Ogawa T, Aiuchi T, et al. Transcriptomics and proteomics analyses of the PACAP38 influenced ischemic brain in permanent middle cerebral artery occlusion model mice. J Neuroinflammation. 2012;9:256.
39. Reglodi D, Tamás A, Somogyvári-Vigh A, Szántó Z, Kertes E, Lénárd L, et al. Effects of pretreatment with PACAP on the infarct size and functional outcome in rat permanent focal cerebral ischemia. Peptides. 2002;23:2227–34.
40. Brifault C, Gras M, Liot D, May V, Vaudry D, Wurtz O. Delayed pituitary adenylate cyclase—activating polypeptide delivery after brain stroke improves functional recovery by inducing m2 microglia/macrophages polarization. Stroke. 2015;46:520–8.
41. Linnik MD, Zobrist RH, Hatfield MD. Evidence supporting a role for programmed cell death in focal cerebral ischemia in rats. Stroke. 1993;24:2002–8.
42. Charriaut-Marlangue C, Margaill I, Represa A, Popovici T, Plotkine M, Ben-Ari Y. Apoptosis and necrosis after reversible focal ischemia: an in situ DNA fragmentation analysis. J Cereb Blood Flow Metab. 1996;16:186–94.
43. Reglodi D, Fábián Z, Tamás A, Lubics A, Szeberényi J, Alexy T, et al. Effects of PACAP on in vitro and in vivo neuronal cell death, platelet aggregation, and production of reactive oxygen radicals. Regul Pept. 2004;123:51–9.
44. Katsura K, Kristián T, Siesjö BK. Energy metabolism, ion homeostasis, and cell damage in the brain. Biochem Soc Trans. 1994;22:991–6.
45. Lucas DR, Newhouse JP. The toxic effect of sodium L-glutamate on the inner layers of the retina. AMA Arch Ophthalmol. 1957;58:193–201.
46. Olney JW. Glutamate-induced retinal degeneration in neonatal mice. Electron microscopy of the acutely evolving lesion. J Neuropathol Exp Neurol. 1969;28(3):455–74.
47. Olney JW, Adamo NJ, Ratner A. Monosodium glutamate effects. Science. 1971;172(3980):294.
48. Costa L, Santangelo F, Li Volsi G, Ciranna L. Modulation of AMPA receptor-mediated ion current by pituitary adenylate cyclase-activating polypeptide (PACAP) in CA1 pyramidal neurons from rat hippocampus. Hippocampus. 2009;19:99–109.
49. Liu GJ, Madsen BW. PACAP38 modulates activity of NMDA receptors in cultured chick cortical neurons. J Neurophysiol. 1997;78:2231–4.
50. Figiel M, Engele J. Pituitary adenylate cyclase-activating polypeptide (PACAP), a neuron-derived peptide regulating glial glutamate transport and metabolism. J Neurosci. 2000;20:3596–605.
51. Yang S, Yang J, Yang Z, Chen P, Fraser A, Zhang W, et al. Pituitary adenylate cyclase-activating polypeptide (PACAP) 38 and PACAP4-6 are neuroprotective through inhibition of NADPH oxidase: potent regulators of microglia-mediated oxidative stress. J Pharmacol Exp Ther. 2006;319:595–603.
52. Miyamoto K, Tsumuraya T, Ohtaki H, Dohi K, Satoh K, Xu Z, et al. PACAP38 suppresses cortical damage in mice with traumatic brain injury by enhancing antioxidant activity. J Mol Neurosci. 2014;54:370–9.
53. Stetler RA, Gao Y, Zukin RS, Vosler PS, Zhang L, Zhang F, et al. Apurinic/apyrimidinic endonuclease APE1 is required for PACAP-induced neuroprotection against global cerebral ischemia. Proc Natl Acad Sci U S A. 2010;107:3204–9.

54. Rácz B, Gallyas F, Kiss P, Tóth G, Hegyi O, Gasz B, et al. The neuroprotective effects of PACAP in monosodium glutamate-induced retinal lesion involve inhibition of proapoptotic signaling pathways. Regul Pept. 2006;137(1-2):20–6.
55. Arvidsson A, Collin T, Kirik D, Kokaia Z, Lindvall O. Neuronal replacement from endogenous precursors in the adult brain after stroke. Nat Med. 2002;8:963–70.
56. Mercer A, Rönnholm H, Holmberg J, Lundh H, Heidrich J, Zachrisson O, et al. PACAP promotes neural stem cell proliferation in adult mouse brain. J Neurosci Res. 2004;76(2):205–15.
57. Hori M, Nakamachi T, Shibato J, Rakwal R, Tsuchida M, Shioda S, et al. PACAP38 differentially effects genes and CRMP2 protein expression in ischemic core and penumbra regions of permanent middle cerebral artery occlusion model mice brain. Int J Mol Sci. 2014;15:17014–34.
58. Zhang R, Zhang L, Zhang Z, Wang Y, Lu M, Lapointe M, et al. A nitric oxide donor induces neurogenesis and reduces functional deficits after stroke in rats. Ann Neurol. 2001;50:602–11.
59. Zhao Y, Guan Y-F, Zhou X-M, Li G-Q, Li Z-Y, Zhou C-C, et al. Regenerative neurogenesis after ischemic stroke promoted by nicotinamide phosphoribosyltransferase-nicotinamide adenine dinucleotide cascade. Stroke. 2015;46:1966–74.
60. Pratt PF, Medhora M, Harder DR. Mechanisms regulating cerebral blood flow as therapeutic targets. Curr Opin Investig Drugs. 2004;5(9):952–6.
61. Shin S-H, Lee J, Lim KS, Rhim T, Lee SK, Kim Y-H, et al. Sequential delivery of TAT-HSP27 and VEGF using microsphere/hydrogel hybrid systems for therapeutic angiogenesis. J Control Release. 2013;166:38–45.
62. Shin Y-J, Choi J-S, Lee J-Y, Choi J-Y, Cha J-H, Chun M-H, et al. Differential regulation of vascular endothelial growth factor-C and its receptor in the rat hippocampus following transient forebrain ischemia. Acta Neuropathol. 2008;116:517–27.
63. Gloddek J, Pagotto U, Paez Pereda M, Arzt E, Stalla GK, Renner U. Pituitary adenylate cyclase-activating polypeptide, interleukin-6 and glucocorticoids regulate the release of vascular endothelial growth factor in pituitary folliculostellate cells. J Endocrinol. 1999;160:483–90.
64. Dogrukol-Ak D, Kumar VB, Ryerse JS, Farr SA, Verma S, Nonaka N, et al. Isolation of peptide transport system-6 from brain endothelial cells: therapeutic effects with antisense inhibition in Alzheimer and stroke models. J Cereb Blood Flow Metab. 2009;29:411–22.
65. Mizushima H, Banks WA, Dohi K, Shioda S, Matsumoto H, Matsumoto K. The effect of cardiac arrest on the permeability of the mouse blood-brain and blood-spinal cord barrier to pituitary adenylate cyclase activating polypeptide (PACAP). Peptides. 1999;20(11):1337–40.
66. Somogyvári-Vigh A, Pan W, Reglödi D, Kastin AJ, Arimura A. Effect of middle cerebral artery occlusion on the passage of pituitary adenylate cyclase activating polypeptide across the blood–brain barrier in the rat. Regul Pept. 2000;91:89–95.
67. Zhu L, Tamvakopoulos C, Xie D, Dragovic J, Shen X, Fenyk-Melody JE, et al. The role of dipeptidyl peptidase IV in the cleavage of glucagon family peptides: in vivo metabolism of pituitary adenylate cyclase activating polypeptide-(1-38). J Biol Chem. 2003;278:22418–23.
68. Li M, Maderdrut JL, Lertora JJL, Batuman V. Intravenous infusion of pituitary adenylate cyclase-activating polypeptide (PACAP) in a patient with multiple myeloma and myeloma kidney: a case study. Peptides. 2007;28:1891–5.
69. Birk S, Sitarz JT, Petersen KA, Oturai PS, Kruuse C, Fahrenkrug J, et al. The effect of intravenous PACAP38 on cerebral hemodynamics in healthy volunteers. Regul Pept. 2007;140:185–91.
70. Robberecht P, Gourlet P, De Neef P, Woussen-Colle MC, Vandermeers-Piret MC, Vandermeers A, et al. Structural requirements for the occupancy of pituitary adenylate-cyclase-activating-peptide (PACAP) receptors and adenylate cyclase activation in human neuroblastoma NB-OK-1 cell membranes. Discovery of PACAP(6-38) as a potent antagonist. Eur J Biochem. 1992;207:239–46.
71. Cardell LO, Uddman R, Luts A, Sundler F. Pituitary adenylate cyclase activating peptide (PACAP) in guinea-pig lung: distribution and dilatory effects. Regul Pept. 1991;36:379–90.
72. Miyata A, Sato K, Hino J, Tamakawa H, Matsuo H, Kangawa K. Rat aortic smooth-muscle cell proliferation is bidirectionally regulated in a cell cycle-dependent manner via PACAP/VIP Type 2 receptor. Ann N Y Acad Sci. 1998;865:73–81.

73. Shigyo M, Aizawa H, Inoue H, Matsumoto K, Takata S, Hara N. Pituitary adenylate cyclase activating peptide regulates neurally mediated airway responses. Eur Respir J. 1998;12:64–70.
74. Dorner GT, Wolzt M, Schmetterer L, Eichler HG. Effect of pituitary adenylate cyclase activating polypeptide 1–27 on ocular, cerebral and skin blood flow in humans. Naunyn Schmiedebergs Arch Pharmacol. 1998;358:657–62.
75. Warren JB, Cockcroft JR, Larkin SW, Kajekar R, Macrae A, Ghatei MA, et al. Pituitary adenylate cyclase activating polypeptide is a potent vasodilator in humans. J Cardiovasc Pharmacol. 1992;20:83–7.
76. Chang Y, Lawson LJ, Hancock JC, Hoover DB. Pituitary adenylate cyclase-activating polypeptide: localization and differential influence on isolated hearts from rats and guinea pigs. Regul Pept. 2005;129:139–46.
77. Schytz HW, Birk S, Wienecke T, Kruuse C, Olesen J, Ashina M. PACAP38 induces migraine-like attacks in patients with migraine without aura. Brain. 2009;132:16–25.
78. Bourgault S, Vaudry D, Botia B, Couvineau A, Laburthe M, Vaudry H, et al. Novel stable PACAP analogs with potent activity towards the PAC1 receptor. Peptides. 2008;29:919–32.
79. Darsalia V, Mansouri S, Wolbert P, Barde S, Sjöholm A, Patrone C. The specific VPAC2 agonist Bay 55-9837 increases neuronal damage and hemorrhagic transformation after stroke in type 2 diabetic rats. Neuropeptides. 2013;47:133–7.
80. Doan N-D, Bourgault S, Dejda A, Létourneau M, Detheux M, Vaudry D, et al. Design and in vitro characterization of PAC1/VPAC1-selective agonists with potent neuroprotective effects. Biochem Pharmacol. 2011;81:552–61.
81. Lamine A, Létourneau M, Doan ND, Maucotel J, Couvineau A, Vaudry H, et al. Characterizations of a synthetic pituitary adenylate cyclase-activating polypeptide analog displaying potent neuroprotective activity and reduced in vivo cardiovascular side effects in a Parkinson's disease model. Neuropharmacology. 2016;108:440–50.
82. Fletcher L, Kohli S, Sprague SM, Scranton RA, Lipton SA, Parra A, et al. Intranasal delivery of erythropoietin plus insulin-like growth factor-I for acute neuroprotection in stroke. Laboratory investigation. J Neurosurg. 2009;111:164–70.
83. Guerra-Crespo M, Sistos A, Gleason D, Fallon JH. Intranasal administration of PEGylated transforming growth factor-alpha improves behavioral deficits in a chronic stroke model. J Stroke Cerebrovasc Dis. 2010;19:3–9.
84. Sun B, He M, Han X, Sun J, Yang M, Yuan H, et al. Intranasal delivery of granulocyte colony-stimulating factor enhances its neuroprotective effects against ischemic brain injury in rats. Mol Neurobiol. 2014. In press.
85. Kinhult J, Adner M, Uddman R, Cardell LO. Pituitary adenylate cyclase-activating polypeptide, effects in the human nose. Clin Exp Allergy. 2003;33:942–9.
86. Denora N, Trapani A, Laquintana V, Lopedota A, Trapani G. Recent advances in medicinal chemistry and pharmaceutical technology—strategies for drug delivery to the brain. Curr Top Med Chem. 2009;9:182–96.
87. Dikpati A, Madgulkar AR, Kshirsagar SJ, Chahal AS. Targeted drug delivery to CNS using nanoparticles. J Adv Pharm Sci. 2012;2:179–91.
88. Malik DK, Baboota S, Ahuja A, Hasan S, Ali J. Recent advances in protein and peptide drug delivery systems. Curr Drug Deliv. 2007;4(2):141–51.
89. Kreuter J, Ramge P, Petrov V, Hamm S, Gelperina SE, Engelhardt B, et al. Direct evidence that polysorbate-80-coated poly(butylcyanoacrylate) nanoparticles deliver drugs to the CNS via specific mechanisms requiring prior binding of drug to the nanoparticles. Pharm Res. 2003;20:409–16.
90. Simşek S, Eroğlu H, Kurum B, Ulubayram K. Brain targeting of atorvastatin loaded amphiphilic PLGA-b-PEG nanoparticles. J Microencapsul. 2013;30:10–20.
91. Alyautdin R, Khalin I, Nafeeza MI, Haron MH, Kuznetsov D. Nanoscale drug delivery systems and the blood-brain barrier. Int J Nanomed. 2014;9:795–811.
92. Gartziandia O, Herran E, Pedraz JL, Carro E, Igartua M, Hernandez RM. Chitosan coated nanostructured lipid carriers for brain delivery of proteins by intranasal administration. Colloids Surf B Biointerfaces. 2015;134:304–13.

93. Yemisci M, Caban S, Gursoy-Ozdemir Y, Lule S, Novoa-Carballal R, Riguera R, et al. Systemically administered brain-targeted nanoparticles transport peptides across the blood-brain barrier and provide neuroprotection. J Cereb Blood Flow Metab. 2015;35:469–75.
94. Ishii T, Asai T, Oyama D, Fukuta T, Yasuda N, Shimizu K, et al. Amelioration of cerebral ischemia-reperfusion injury based on liposomal drug delivery system with asialo-erythropoietin. J Control Release. 2012;160:81–7.
95. Ishii T, Fukuta T, Agato Y, Oyama D, Yasuda N, Shimizu K, et al. Nanoparticles accumulate in ischemic core and penumbra region even when cerebral perfusion is reduced. Biochem Biophys Res Commun. 2013;430:1201–5.
96. Neumann H, Takahashi K. Essential role of the microglial triggering receptor expressed on myeloid cells-2 (TREM2) for central nervous tissue immune homeostasis. J Neuroimmunol. 2007;184:92–9.
97. Tamás A, Reglõdi D, Szántó Z, Borsiczky B, Németh J, Lengvári I. Comparative neuroprotective effects of preischemic PACAP and VIP administration in permanent occlusion of the middle cerebral artery in rats. Neuro Endocrinol Lett. 2002;23:249–54.
98. Scharf E, May V, Braas KM, Shutz KC, Mao-Draayer Y. Pituitary adenylate cyclase-activating polypeptide (PACAP) and vasoactive intestinal peptide (VIP) regulate murine neural progenitor cell survival, proliferation, and differentiation. J Mol Neurosci. 2008;36:79–88.

Part XI
PACAP in the Pathomechanism of Migraine and Pain

Chapter 35
Pituitary Adenylate Cyclase Activating Polypeptide (PACAP) in Migraine Pathophysiology

Lars Edvinsson

Abstract Pituitary adenylate cyclase activating peptide (PACAP) can be found in the trigeminal (TG) and the spenopalatine ganglia (SPG), and in the trigeminocervical complex. In the SPG it is co-stored with vasoactive intestinal peptide (VIP) and nitric oxide synthase, while it is co-stored with calcitonin gene-related peptide (CGRP) in the trigeminal system. PACAP has been shown to trigger migraine in patients without marked vasodilator effects. Recently, it has been found that PACAP levels are elevated in migraineurs during headache compared to interictal samples in the external jugular vein. The elevation significantly reduced after sumatriptan treatment. In addition, the PACAP receptors (PAC_1, $VPAC_1$, and $VPAC_2$) are present in sensory neurons and in vascular smooth muscle related to the trigeminovascular system. Since PACAP is elevated in cats and humans following trigeminovascular activation, and this change is reduced when headache is treated, PACAP or its receptors may be a promising target for migraine therapeutics.

Keywords PACAP • PACAP receptors • Trigeminal ganglion • Migraine • CGRP • VIP • NOS • Otic ganglion • Sphenopalatine ganglion • Trigeminal nucleus caudalis

Abbreviations

BBB	Blood–brain barrier
CGRP	Calcitonin gene-related peptide
PACAP	Pituitary adenylate cyclase-activating peptide
VIP	Vasoactive intestinal peptide

L. Edvinsson (✉)
Department of Medicine, Institute of Clinical Sciences, Lund University, Lund 221 84, Sweden
e-mail: lars.edvinsson@med.lu.se

D. Reglodi, A. Tamas (eds.), *Pituitary Adenylate Cyclase Activating Polypeptide — PACAP*, Current Topics in Neurotoxicity 11,
DOI 10.1007/978-3-319-35135-3_35

Introduction

Migraine is a complex disorder that is influenced by genetic and environmental factors; it is among the most common disorders in the world and the most common cause of neurological disability worldwide [1]. The pathophysiology behind migraine is still unclear, but current view strongly suggests that it is primarily a brain disorder [2, 3] associated with several extracerebral aspects such as sensitization at peripheral and central sensory nerve endings [4]. Although the acute attack starts in the central nervous system (CNS) aspects of the peripheral nervous system (PNS), in particular via the sensory trigeminal ganglion (TG), are important for the complete attack with for example pain. It is here the cranial autonomic and sensory nervous systems come into play, in particular the parasympathetic sphenopalatine (SPG) and otic ganglia (OTG), and the sensory TG [5]. Extensive neuroanatomical mapping and selective denervation studies have shown that the TG contains calcitonin gene-related peptide (CGRP), amylin, neurokinins, pituitary adenylate cyclase activating peptide (PACAP), and nitric oxide synthase (NOS), while the SPG and the OTG store vasoactive intestinal peptide (VIP), PACAP, and NOS, with a minor contribution of acetylcholine [5, 6]. Their central connections are to the trigeminocervical complex and the superior and inferior salivatory nuclei, respectively.

We have previously shown that direct activation of the trigeminovascular system causes the release of the vasodilator neuropeptide CGRP and substance P [7]. CGRP has been shown to be elevated in humans during spontaneous migraine attacks [8] while cluster headache attacks show release of both CGRP and VIP [9]. Interestingly, treatment of migraine or cluster headache attacks with sumatriptan results in reduction in headache and in normalization of the elevated CGRP levels [9, 10]. CGRP is also elevated during nitroglycerine-triggered migraine attacks [11] and reduced with sumatriptan treatment [12]. CGRP is reported elevated in chronic migraine [13] while systemic CGRP infusion can trigger migraine-like attacks [14]. Thus, translational efforts from identification of the transmitter CGRP, demonstration of its neuroanatomical localization and release in genuine migraine attacks formed the basis for the development of CGRP receptor antagonists and demonstration of their role as a target for novel migraine therapeutics [15].

Peripheral sensory neurons serve a dual role as afferent neurons, conveying sensory information from the periphery to the CNS, and as efferent effectors mediating, e.g., neurogenic inflammation. In this context neuropeptides are crucial for both these functions in primary sensory neurons. In the afferent functions, they act as messengers and modulators in addition to a principal transmitter. Upon release at peripheral nerve terminals, they induce an efferent response, "neurogenic inflammation," which comprises vasodilatation, plasma extravasation, and recruitment of immune cells. In this context PACAP, which is a VIP-resembling peptide, has been established as a sensory neuropeptide with localization in TG and in dorsal root ganglia [16]. In relation to migraine related tissues PACAP has been found in different parts of the human trigeminocervical complex [17, 18] and the SPG and OTG [19, 20].

PACAP in TG and SPG

PACAP is a member of a superfamily of structurally related peptides which share many functions, the VIP-glucagon-growth hormone releasing factor-secretin superfamily of signaling peptides [21]. PACAP is encoded by the ADCYAP1 gene, resulting in a propeptide of 175 amino acids. Following truncation, PACAP is expressed in one of two forms containing either 27 or 38 amino acids, PACAP-27 and PACAP-38, respectively. PACAP-38 is more prevalent, representing about 90 % of the PACAP forms in mammalian tissues [22]. PACAP is expressed throughout the CNS, as well as in peripheral organs and glands [23].

Early studies in relation to pain structures revealed PACAP immunoreactive nerve fibers in the superficial layer of the dorsal horns of the spinal cord, in nerve cell bodies (mostly of small size) of the spinal ganglia and the TG [16]. In addition, PACAP mRNA was found in the nerve cell bodies. Interestingly the PACAP labeled neurons constituted 10 % of the total number of nerve cell bodies in the dorsal root ganglia while the population of CGRP mRNA labeled nerve cell bodies constituted 46 % [24]. This agrees well with later immunocytochemical work on the TG [5]. Furthermore, PACAP mRNA was seen in a small number of single nerve cell bodies of small size that also were labeled by the CGRP mRNA probe [24]. Neurograms obtained following stimulation of the ipsilateral peroneal nerve or the sciatic nerve suggested that intrathecal PACAP may act as a neurotransmitter or modulator in sensory C-fibers and possibly with a contribution of A-fibers [25]. Interestingly PACAP-27 was more potent than PACAP-38 or VIP at this function. Further support for a sensory role of PACAP in the rat spinal cord came from capsaicin administration in vivo [26]. Capsaicin administration resulted in elevated cisternal concentrations of PACAP-27 by 177 %, PACAP-38 by 93 % and CGRP by 692 % [26], supporting their role in nociception.

PACAP has been found in intracranial migraine related tissues such as nerve fibers in intracranial blood vessels, in SPG and in TG, in the trigeminal nucleus caudalis (TNC) [17–20]. Most of the neurons in the OTG and SPG co-store VIP/PACAP and NOS; and, a small population of cells also contain choline acetyltransferase, the enzyme that synthesizes acetylcholine [5]. Denervation studies revealed that the main part of perivascular PACAP originates in the parasympathetic ganglia [27]. The TG contains a subpopulation of neurons that store PACAP (about 10 %), but not VIP [19, 20]. The TNC and the dorsal roots at C1-2 contain the central fibers which also store PACAP in the superficial layers [17, 28]. Interestingly, sensory CGRP-containing C- fibers (probably from the TG) project to the cranial parasympathetic ganglia, and there are CGRP receptors on satellite glial cells (SGC) in the SPG [29]. The SGC are closely attached to the neurons via gap junctions which provide possibilities for interactions. These findings show morphological evidence for an interaction between the intracranial parasympathetic and sensory systems.

Involvement of PACAP in Migraine

PACAP has a widespread distribution in the brain, the brainstem, and spinal cord as well as in dorsal root ganglia [30, 31]. These include several regions of interest in the discussions of migraine pathophysiology [2] such as midbrain, hypothalamus and thalamus, as well as raphe nuclei and locus coeruleus and TNC [31]. Intrathecal administration of PACAP elicited a dose-dependent decrease in the tail-flick latency, while higher doses resulted in biting and scratching; these behaviors were interpreted as pain-like syndrome and supported a role of PACAP as a sensory neurotransmitter involved in nociception [32].

A role for PACAP in migraine has been suggested by four main observations; (1) PACAP levels are increased in the cranial circulation of the cat upon nociceptive stimulation of the superior sagittal sinus [33, 34] (2). PACAP levels in the external jugular vein are reduced with amelioration of migraine headache when subjects were treated with sumatriptan, and lower levels of PACAP occur between attacks when compared with attacks [34] (3). The plasma levels of PACAP-38 are altered in ictal and interictal periods of migraine patients [35] (4). The intravenous infusion of PACAP-38 in subjects known to have migraine results in migraine-like headache [36]. The obvious, but unanswered, questions are; what are the sources of endogenous PACAP and how can systemic PACAP induce migraine? There are at present two possibilities; (1) PACAPs do not pass the wall of cerebral arteries as studied in isolated perfused middle cerebral arteries. This may agree with the view that the peptide is a large molecule with vasodilator effects but does not pass the blood–brain barrier (BBB) [37, 38]. (2) However, Banks suggests that it does pass the BBB and provides evidence in his chapter in this book (see below). Tuka and colleagues [38a] reported that stimulation of the TG or systemic nitroglycerol injection increases the level of PACAP in the TNC. In addition PACAP-38 was elevated in plasma after TG stimulation. Thus, the relevant target of systemically administered PACAP is still not settled. It is my view that the elevated plasma level of PACAP reported in several studies probably originates in structures located outside the BBB but it needs to be clarified. Recently it was observed that infusion of PACAP-38 resulted in marked dilatation of extracranial arteries (branch of the main trunk of the maxillary artery) but not the intracranial middle cerebral arteries [39]. In support, this response preceded the onset of migraine-like headache and was sensitive to triptan administration [39, 40].

PACAP Receptors

When released, PACAP may bind to three different G-protein coupled receptors which also can bind VIP as well. VIP, PACAP-38 and PACAP-27 all have equal affinity in binding to the VIP/PACAP receptors $VPAC_1$ and $VPAC_2$ receptors. The third receptor, PAC_1, has high affinity for both forms of PACAP, but has 100- to 1000-fold lower affinity for VIP [41]. The principal effect of $VPAC_1$/$VPAC_2$ or PAC_1

receptor stimulation is an increase in cAMP through adenylyl cyclase activation. Activation of other second messenger systems, including phospholipase C and phospholipase D, has been implicated and may occur along with adenylyl cyclase activation. Clinical experiments have shown that PACAP-38 but not VIP elicits a migraine-like headache [36], thus implicating the PAC_1 receptor as a putative anti-migraine target. Systemic administration of nitroglycerol or PACAP-38 produced photophobia and meningeal vasodilatation in wild-type but not in PACAP-knockout mice [42]. Those effects and an increased activation 4 h later in the TG and in the TNC were attributed to stimulation of the peripheral terminals of PACAP-ergic trigeminal sensory nerves in the meningeal region in the wild-type but not in the PACAP-knockout mice.

All three PACAP receptors have been found in migraine-related SPG and TG ganglia and on intracranial blood vessels [37, 38, 43]. In support for a more specific role of PACAP was suggested by Walker and colleagues who quantified PACAP-related signaling in primary cell cultures from TG (neurons and glia) and showed that PACAP, but not VIP, induced cAMP production [44]. The response was antagonized by PACAP(6-38), which is consistent with PAC_1 receptor pharmacology [45]. Interestingly, PACAP-38, but not PACAP-27 or VIP, in addition caused ERK1/2 activation in cultured glial cells but not in neurons. The task of in-depth analysis of PAC_1 receptor pharmacology and development of specific receptor blockers will not be easy because ten splice variants displaying different intracellular coupling have been reported [46]. The development of a specific PAC1 receptor antagonist with or without BBB permeability might provide novel insights into the pathophysiology of migraine and provide a novel target for drug development.

Acknowledgements The study was supported by the Swedish Research Council (grant no 5958), the Heart-Lund Foundation, and a regional ALF grant.

Competing Interests The author declares no competing interests; however, he has served on the board of TEVA Pharma and Eli Lilly Pharma.

References

1. Murray CJ, Vos T, Lozano R, Naghavi M, Flaxman AD, Michaud C, et al. Disability-adjusted life years (DALYs) for 291 diseases and injuries in 21 regions, 1990-2010: a systematic analysis for the Global Burden of Disease Study 2010. Lancet. 2012;380:2197–223.
2. Akerman S, Holland P, Goadsby PJ. Diencephalic and brainstem mechanisms in migraine. Nat Rev Neurosci. 2011;12:570–84.
3. Charles A. Migraine: a brain state. Curr Opin Neurol. 2013;26:235–9.
4. Olesen J, Burstein R, Ashina M, Tfelt-Hansen P. Origin of pain in migraine: evidence for peripheral sensitisation. Lancet Neurol. 2009;8:679–90.
5. Edvinsson L, Uddman R. Neurobiology in primary headaches. Brain Res Rev. 2005;48:438–56.
6. Edvinsson L. Tracing neural connections to pain pathways with relevance to primary headaches. Cephalalgia. 2011;31:737–47.
7. Zagami AS, Goadsby PJ, Edvinsson L. Stimulation of the superior sagittal sinus in the cat causes release of vasoactive peptides. Neuropeptides. 1990;16:69–75.

8. Goadsby PJ, Edvinsson L, Ekman R. Vasoactive peptide release in the extracerebral circulation of humans during migraine headache. Ann Neurol. 1990;28:183–7.
9. Goadsby PJ, Edvinsson L. Human in vivo evidence for trigeminovascular activation in cluster headache. Brain. 1994;117:427–34.
10. Goadsby PJ, Edvinsson L. The trigeminovascular system and migraine: studies characterizing cerebrovascular and neuropeptide changes seen in humans and cats. Ann Neurol. 1993;33:48–56.
11. Juhasz G, Zsombok T, Modos EA, Olajos S, Jakab B, Nemeth J, et al. NO-induced migraine attack: strong increase in plasma calcitonin gene-related peptide (CGRP) concentration and negative correlation with platelet serotonin release. Pain. 2003;106:461–70.
12. Juhasz G, Zsombok T, Jakab B, Nemeth J, Szolcsanyi J, Bagdy G. Sumatriptan causes parallel decrease in plasma calcitonin gene-related peptide (CGRP) concentration and migraine headache during nitroglycerin induced migraine attack. Cephalalgia. 2005;25:179–83.
13. Cernuda-Morollon E, Larrosa D, Ramon C, Vega J, Martinez-Camblor P, Pascual J. Interictal increase of CGRP levels in peripheral blood as a biomarker for chronic migraine. Neurology. 2013;81:1191–6.
14. Lassen LH, Haderslev PA, Jacobsen VB, Iversen HK, Sperling B, Olesen J. CGRP may play a causative role in migraine. Cephalalgia. 2002;22:54–61.
15. Ho TW, Edvinsson L, Goadsby PJ. CGRP and its receptors provide new insights into migraine pathophysiology. Nat Rev Neurol. 2010;6:761–6.
16. Moller K, Zhang YZ, Håkanson R, Luts A, Sjölund B, Uddman R, et al. Pituitary adenylate cyclase activating peptide is a sensory neuropeptide: immunocytochemical and immunochemical evidence. Neuroscience. 1993;57:725–32.
17. Uddman R, Tajti J, Hou M, Sundler F, Edvinsson L. Neuropeptide expression in the human trigeminal nucleus caudalis and in the cervical spinal cord C1 and C2. Cephalalgia. 2002;22:112–6.
18. Uddman R, Goadsby PJ, Jansen I, Edvinsson L. PACAP, a VIP-like peptide: immunohistochemical localization and effect upon cat pial arteries and cerebral blood flow. J Cereb Blood Flow Metab. 1993;13:291–7.
19. Uddman R, Tajti J, Moller S, Sundler F, Edvinsson L. Neuronal messengers and peptide receptors in the human sphenopalatine and otic ganglia. Brain Res. 1999;826:193–9.
20. Csati A, Tajti J, Kuris A, Tuka B, Edvinsson L, Warfvinge K. Distribution of vasoactive intestinal polypeptide, pituitary adenylate cyclase activating peptide, nitric oxide synthase and their receptors in human and rat sphenopalatine ganglion. Neuroscience. 2012;202:158–68.
21. Vaudry D, Gonzalez BJ, Basille M, Yon L, Fournier A, Vaudry H. Pituitary adenylate cyclase-activating polypeptide and its receptors: from structure to functions. Pharmacol Rev. 2000;52:269–324.
22. Bourgault S, Vaudry D, Botia B, Couvineau A, Laburthe M, Vaudry H, et al. Novel stable PACAP analogs with potent activity towards the PAC1 receptor. Peptides. 2008;29:919–32.
23. Fahrenkrug J, Hannibal J. Neurotransmitters co-existing with VIP or PACAP. Peptides. 2004;25:393–401.
24. Mulder H, Uddman R, Möller K, Zhang Y-Z, Ekblad E, Alumets J, et al. Pituitary adenylate cyclase activating polypeptide expression in sensory neurons. Neuroscience. 1994;63:307–12.
25. Zhang Y-Z, Sjölund B, Möller K, Håkanson R, Sundler F. Pituitary adenylate cyclase activating peptide produces a marked and long-lasting depression of a C-fibre-evoked flexion reflex. Neuroscience. 1993;57:733–7.
26. Zhang Y, Malmberg AB, Yaksh TL, Sjölund B, Sundler F, Håkanson R. Capsaicin-evoked release of pituitary adenylate cyclase activating peptide (PACAP) and calcitonin gene-related peptide (CGRP) from rat spinal cord in vivo. Regul Pept. 1997;69:83–7.
27. Edvinsson L, Elsås T, Suzuki N, Shimizu T, Lee TJ. Origin and co-localization of nitric oxide synthase, CGRP, PACAP, and VIP in the cerebral circulation of the rat. Microsc Res Tech. 2001;53:221–8.
28. Christiansen T, Bruun A, Knight YE, Goadsby PJ, Edvinsson L. Immunoreactivity of NOS, CGRP, PACAP, SP and VIP in the trigeminal nucleus caudalis and in the cervical spinal cord C1 and C2 of the cat. J Headache Pain. 2003;4:156–63.

29. Csati A, Tajti J, Tuka B, Edvinsson L, Warfvinge K. Calcitonin gene-related peptide and its receptor components in the human sphenopalatine ganglion—interaction with the sensory system. Brain Res. 2012;1435:29–39.
30. Masuo Y, Tokito F, Matsumoto Y, Shimamoto N, Fujino M. Ontogeny of pituitary adenylate cyclase-activating polypeptide (PACAP) and its binding sites in the rat brain. Neurosci Lett. 1994;170:43–6.
31. Palkovits M, Somogyvári-Vigh A, Arimura A. Concentrations of pituitary adenylate cyclase activating polypeptide (PACAP) in human brain nuclei. Brain Res. 1995;699:116–20.
32. Narita M, Dun SL, Dun NJ, Tseng LF. Hyperalgesia induced by pituitary adenylate cyclase-activating polypeptide in the mouse spinal cord. Eur J Pharmacol. 1996;311:121–6.
33. Zagami AS, Edvinsson L, Hoskin KL, Goadsby PJ. Stimulation of the superior sagittal sinus causes extracranial release of PACAP. Cephalalgia. 1995;15 Suppl 14:109.
34. Zagami AS, Edvinsson L, Goadsby PJ. Pituitary adenylate cyclase activating polypeptide and migraine. Ann Clin Transl Neurol. 2014;1:1036–40.
35. Tuka B, Helyes Z, Markovics A, Bagoly T, Szolcsányi J, Szabó N, et al. Alterations in PACAP-38-like immunoreactivity in the plasma during ictal and interictal periods of migraine patients. Cephalalgia. 2013;33:1085–95.
36. Schytz HW, Birk S, Wienecke T, Kruuse C, Olesen J, Ashina M. PACAP38 induces migraine-like attacks in patients with migraine without aura. Brain. 2009;132:16–25.
37. Erdling A, Sheykhzade M, Maddahi A, Bari F, Edvinsson L. VIP/PACAP receptors in cerebral arteries of rat: characterization, localization and relation to intracellular calcium. Neuropeptides. 2013;47:85–92.
38. Syed AU, Koide M, Braas KM, May V, Wellman GC. Pituitary adenylate cyclase-activating polypeptide (PACAP) potently dilates middle meningeal arteries: implications for migraine. J Mol Neurosci. 2013;48:574–83.
38a. Tuka B, Helyes Z, Markovics A, Bagloy T, Németh J, Márk L, et al. Peripheral and central alterations of pituitary adenylate cyclase activating polypeptide-like immunoreactivity in the rat in response to activation of the trigeminovascular system. Peptides. 2012;33:307–16.
39. Amin FM, Asghar MS, Guo S, Hougaard A, Hansen AE, Schytz HW, et al. Headache and prolonged dilatation of the middle meningeal artery by PACAP38 in healthy volunteers. Cephalalgia. 2012;32:140–9.
40. Amin FM, Hougaard A, Schytz HW, Asghar MS, Lundholm E, Parvaiz AI, et al. Investigation of the pathophysiological mechanisms of migraine attacks induced by pituitary adenylate cyclase-activating polypeptide-38. Brain. 2014;137:779–94.
41. Dickson L, Finlayson K. VPAC and PAC receptors: from ligands to function. Pharmacol Ther. 2009;121:294–316.
42. Markovics A, Kormos V, Gaszer B, Lashgarara A, Szoke E, Sandor K, et al. Pituitary adenylate cyclase-activating polypeptide plays a key role in nitroglycerol-induced trigeminovascular activation in mice. Neurobiol Dis. 2012;45:633–44.
43. Knutsson M, Edvinsson L. Distribution of mRNA for VIP and PACAP receptors in human cerebral arteries and cranial ganglia. Neuroreport. 2002;13:507–9.
44. Walker C, Sundrum T, Hay DL. PACAP receptor pharmacology and agonist bias: analysis in primary neurons and glia from the trigeminal ganglia and transfected cells. Br J Pharmacol. 2014;171:1521–33.
45. Harmar AJ, Fahrenkrug J, Gozes I, Laburthe M, May V, Pisegna JR, et al. Pharmacology and functions of receptors for vasoactive intestinal peptide and pituitary adenylate cyclase-activating polypeptide: IUPHAR review I. Br J Pharmacol. 2012;166:4–17.
46. Blechman J, Levkowitz G. Alternative splicing of the pituitary adenylate cyclase-activating polypeptide receptor PAC1: mechanisms of fine tuning of brain activity. Front Endocrinol. 2013;4:55.

Chapter 36
PACAP Regulation of Vascular Tone: Differential Mechanism Among Vascular Beds

Arsalan U. Syed, Masayo Koide, Victor May, and George C. Wellman

Abstract The homeostatic control of cranial and cerebral blood flow is essential for normal brain function. Brain surface (cerebral) and meningeal (cranial) arteries/arterioles are densely innervated by sensory and autonomic fibers containing a variety of vasoactive peptides including pituitary adenylate cyclase activating polypeptide (PACAP, *Adcyap1*) which has been shown to exert potent vasodilatory effects in a variety of vascular beds. Recent studies have demonstrated that PACAP potency, activation of PACAP receptor subtypes, and signaling to potassium channels to mediate vasodilatory responses are different between cerebral and cranial arteries, such as the cerebellar artery and the middle meningeal artery (MMA). PACAP demonstrates exquisite picomolar potency at the PACAP-selective PAC1 receptor (*Adcyap1r1*) to activate ATP-sensitive potassium (K_{ATP}) channels of the MMA causing vasodilation. Although PACAP frequently colocalizes with the vasodilatory calcitonin gene related peptide (CGRP) in the fiber networks innervating the MMA, PACAP potency is nearly three orders of magnitude greater than that of CGRP. CGRP is equipotent in the MMA and cerebral arteries. In contrast, PACAP is a less potent vasodilator of cerebral arteries compared to the MMA, with nanomolar concentrations required to activate VPAC receptors leading to K_{ATP} and large-conductance calcium-dependent potassium (BK) channel activation and cerebellar artery dilation. Further, PACAP's effects on the brain are multidimensional. PACAP signaling also exerts direct neurotrophic and neuroprotective effects in the central nervous system against degenerative processes or physiological insults. Thus, PACAP regulation of cerebral blood flow helps sustains brain metabolic demands, maintains

---

A.U. Syed • M. Koide • G.C. Wellman (✉)
Department of Pharmacology, University of Vermont College of Medicine,
149 Beaumont Avenue, Burlington, VT 05405, USA
e-mail: george.wellman@uvm.edu

V. May
Department of Pharmacology, University of Vermont College of Medicine,
149 Beaumont Avenue, Burlington, VT 05405, USA

Department of Neurological Sciences, University of Vermont College of Medicine,
149 Beaumont Avenue, Burlington, VT 05405, USA

D. Reglodi, A. Tamas (eds.), *Pituitary Adenylate Cyclase Activating Polypeptide — PACAP*, Current Topics in Neurotoxicity 11,
DOI 10.1007/978-3-319-35135-3_36

neural physiological functions—including learning and memory, and protects neurons from neurodegeneration in disease and trauma. Coherent with CGRP vascular responses and functions, the vasodilatory effects of PACAP in the MMA also implicate a role in genesis of migraine disorders.

Keywords PACAP • PAC1 receptor • Middle meningeal artery (MMA) • Cerebral arteries • Cranial vessels • ATP-sensitive potassium channels (K_{ATP}) • Large-conductance calcium-activated potassium channels (BK) • Vascular smooth muscle • Migraine

Introduction

Many regulatory peptides, including vasopressin, endothelin, calcitonin gene-related peptide (CGRP), angiotensin peptides, natriuretic peptides, substance P, cholecystokinin, neuropeptide Y (NPY), and the vasoactive intestinal peptide (VIP)/pituitary adenylate cyclase-activating polypeptide (PACAP) family of related peptides, have central roles in maintaining systemic and cerebrovascular homeostasis [1, 2]. Of these peptides, PACAP has been identified in sensory and autonomic neuronal fibers along diverse arterial beds in numerous species [2–6]. Direct PACAP G protein-coupled PAC1 and VPAC receptor signaling within vascular smooth muscle cells has been shown to unvaryingly induce vasodilation with high potency and efficacy [6–11]. Among peripheral organs, PACAP fibers have been found to innervate vascular beds of the skin, eye, salivary glands, respiratory airways, cardiac tissues, gastrointestinal tract, mesentery, pancreas, kidney, testis, corpus cavernosum, ovary, and placenta [11–26]. From this wide distribution, it is clear that PACAP plays an important role in maintaining vascular function throughout a broad spectrum of physiological systems.

In addition, there is an increasing appreciation for the roles of PACAP in the regulation of cerebral hemodynamics. Due to the high metabolic demands coupled with limited energy storage within the CNS, the blood supply to the brain must be continuous and operate within a narrow range to maintain structural and functional integrity of the brain [27–29]. In this regard, the brain is like no other organ and highly susceptible to irreversible damage even upon transient ischemic events. Hence, a series of complex regulatory mechanisms preserve cerebrovascular homeostasis. Upon systemic vascular challenges, the brain, through neural and humoral control, can redirect and redistribute blood flow from the peripheral circulatory system to cerebral circulation. Complementary to these changes in blood flow dynamics, the cerebrovasculature also has autoregulatory mechanisms that can respond to the perturbations in arterial pressure that typically occur in the course of normal physiological activities. In addition, cerebral blood flow is differentially modulated by sensory/autonomic innervation, and astrocytic signaling along the length of the cerebral artery [30, 31]. Pial, or brain surface arteries are densely innervated by sensory and autonomic sympathetic and parasympathetic fibers containing diverse regulatory transmitters and peptides for vasodilation (i.e., acetylcholine, CGRP,

VIP, PACAP) and vasoconstriction (norepinephrine, NPY) activities [1, 3, 23]. However, as vessels pass through the Virchow-Robin space and enter the brain parenchyma, extrinsic perivascular nerves disappear and intracerebral arterioles come under the regulatory control of astrocytic endfeet [30, 31]. Hence, cerebral arterial tone is regulated in part by sensory and autonomic peptides. The homeostatic balance of cerebral blood flow is delicate but critical for normal brain function, and altered cerebrovascular activities from dysregulation, disease or trauma have been shown not only to precipitate stroke but contribute in other pathologies including neurodegenerative disorders, behavioral abnormalities, and migraine [32, 33]. The following sections highlight PACAP signaling mechanisms drawing data from our work in the middle meningeal and cerebellar arteries to provide insights into their potential roles in the regulation of cerebral and cranial blood flow [9, 10].

PACAP and PACAP Receptors in Cerebral and Cranial Vessels

The PACAP precursor molecule is endoproteolytically processed to PACAP27 or PACAP38 bioactive peptides, which share significant homology with VIP [6, 34–36]. Accordingly, PACAP and VIP share receptor subtypes. PACAP peptides bind selectively at PAC1 receptors, but both VIP and PACAP bind with near equal affinity at VPAC1 and VPAC2 receptors. PAC1 and VPAC receptors have been identified in cerebral and cranial arteries by immunocytochemistry, ligand binding studies, transcript analyses, and physiological assays [1, 3, 8, 37, 38]. Further, PACAP infusions have been shown to facilitate cerebral artery vasodilation in cranial window [39] and isolated vessel preparations [7]. However, PACAP signaling mechanisms underlying vasodilation can differ among cerebral and cranial vascular beds (see below). There are several sources that can supply PACAP to both cerebral and cranial vessels. One is the trigeminal ganglion in which subpopulations of sensory neurons express a variety of vasoregulatory peptides including PACAP, VIP or CGRP [1, 23]. As in dorsal root ganglion (DRG), PACAP is expressed in approximately 10 % of the trigeminal neurons and frequently colocalized with CGRP for synergistic functions. The other potential sources of PACAP include the parasympathetic sphenopalatine and otic ganglia in the head; but the relative abundance of PACAP neurons in these ganglia has been highly variable ranging from 5 to 6 % to nearly all neurons in the population [3–5]. The reasons underlying the discrepancy are unclear but may reflect immunocytochemical vs. in situ hybridization methods and detection of low PACAP-expressing neurons. Retrograde tracing studies have suggested that the majority of the PACAP fibers in cerebral arteries are derived from peripheral trigeminal sensory axons [3]. Our immunocytochemical studies demonstrating prevalent dual PACAP and CGRP colocalization in neuronal fibers along the middle meningeal artery are consistent with that interpretation (Fig. 36.1), but many fibers also demonstrate PACAP-immunoreactivity alone suggesting that some may also be of postganglionic parasympathetic origins.

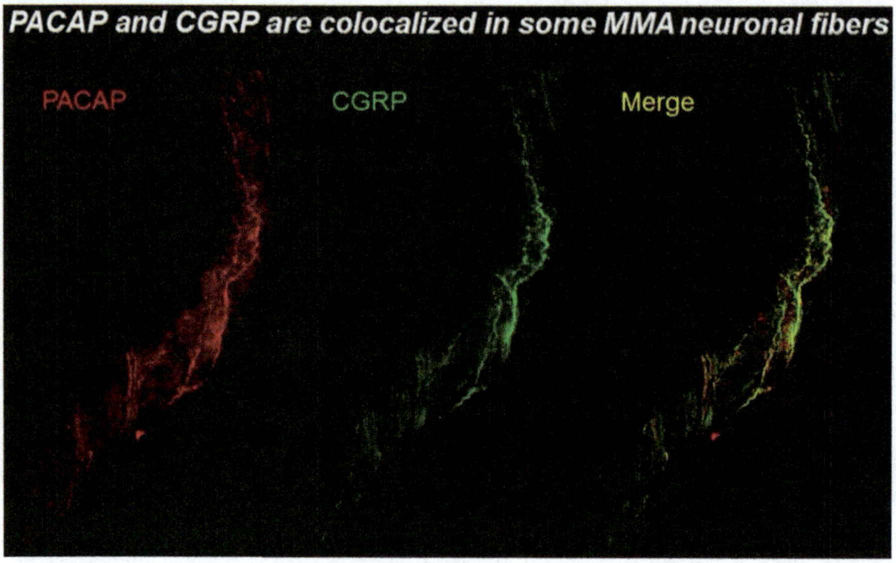

Fig. 36.1 PACAP and CGRP immunoreactivities are colocalized in fibers innervating the MMA. Rat MMA was dissected from the dura, perfusion fixed in 4 % paraformaldehyde, and immunocytochemically dually processed for PACAP (1:10, Jen Hannibal, Bisperg Hospital, Copenhagen Denmark) and CGRP [1–17] (1:1500, Ian Dickerson), Univ Rochester) for visualization with Cy3 (PACAP, *red*, *left panel*) and AlexaFluor 488 (CGRP, *green*, *middle panel*), respectively. The fluorescent signals were merged (*yellow*, *left panel*) to illustrate high levels of colocalization; the distribution patterns suggested that the fibers were sensory in origin from the trigeminal ganglion. Occasional fibers expressed only PACAP or CGRP

Both PAC1 and VPAC receptors have been described in cerebral and cranial vessels based on ligand binding, transcript analyses and pharmacological assessments. The relative abundance of each receptor subtypes in cerebral and cranial arteries has not been well characterized but these blood vessels undoubtedly express multiple PACAP receptor classes and all are likely to have physiological roles [1, 8–10, 23, 39]. PACAP peptides bind to all three receptors with high affinity and hence are relevant ligands regardless of receptor subtype expression. The vasodilatory effects of PACAP appear to be direct on arterial smooth muscle and largely endothelium independent [11, 37, 40].

PACAP Mediates Middle Meningeal Artery (MMA) Vasodilation via PAC1 Receptor Activation and K_{ATP} Channels

Unlike many previous studies, we used an ex vivo pressurized arteriograph system to examine the vasodilatory effects of PACAP on the MMA to obviate pharmacological approaches to pre-constrict the arteries. Myogenic tone (pressure-induced constriction) is an intrinsic property of resistance arteries and largely dependent on the intraluminal pressure on the arterial walls. Hence, ex vivo-pressurized arteriography utilizes this intrinsic property of the resistance arteries to mimic physiological conditions. For these studies, the MMA was dissected from adult male rats and cannulated onto glass micropipettes as described previously [10]. The artery was bathed continuously with artificial cerebral spinal fluid (aCSF) at 37 °C and the intraluminal pressure increased to 40 mmHg to allow development of myogenic tone. Upon artery constriction and stabilization, the peptides and drugs were added to the bathing solution to assess changes in vessel diameter, and at the end of the treatments, the maximum passive diameter of the artery was assessed in Ca^{2+}-free aCSF containing vasodilator diltiazem (100 μM) and forskolin (1 μM).

Using this approach, we found that PACAP38 potently dilated the MMA in a concentration dependent manner with an EC_{50} of 1 pM; VIP by contrast was approximately 1000-fold less potent and induced dilation in MMA with EC_{50} of 1.4 nM [10], which suggested that the PACAP effects were mediated by the PACAP selective PAC1 receptor. MMA pretreatments with a PAC1/VPAC2 receptor antagonist PACAP6-38 (100 nM) completely abolished the vasodilation induced by 3 pM PACAP38. However, 100 nM PACAP6-38 attenuated the vasodilatory effects of 3 nM PACAP38 by 50 %, which may have reflected pharmacological mass action effects or potential roles for VPAC1 receptor signaling in the MMA. Even though VPAC1 receptors have been described in cerebral arteries [8], semi quantitative PCR analyses of rat MMA cDNA only detected prominent PAC1 and VPAC2 receptor transcript expression. These PCR results were consistent with the peptide potency studies in our pressurized vessel preparations, but whether MMA also demonstrates very low levels of VPAC1 receptor remains to be evaluated. Further, more detailed diagnostic restriction digest analyses of the PAC1 amplicon revealed that the MMA PAC1 receptor was predominantly the PAC1Hop1 receptor variant [10]. This observation is notable and suggests that the MMA PAC1 receptor signaling is capable of engaging multiple intracellular second messenger cascades to elicit vascular responses.

Potassium channels play a crucial role in arterial tone via regulation of vascular smooth muscle membrane potential. Potassium channel activation causes vascular smooth muscle membrane hyperpolarization, leading to decreased L-type voltage-gated calcium channel (VDCC) open probability and a reduction of intracellular calcium to facilitate smooth muscle relaxation and vasodilation (Fig. 36.3, modified from [9]). From many studies, the PKA-dependent phosphorylation and activation of several potassium channels, including ATP-sensitive potassium (K_{ATP}) channels [41], large conductance calcium-activated potassium (BK) channels [42] and voltage-gated potassium (K_V) channels [43] can participate in the cerebral vasodilatory process. Accordingly, the ability for PAC1Hop1 receptors to engage the adenylyl cyclase/cAMP/PKA pathway for potassium channel phosphorylation and activation represents a principal mechanism to stimulate MMA dilation.

Among the different potassium channels, our studies have shown that PACAP and PAC1 receptor activation of K_{ATP} channels is sufficient to induce MMA vasodilation (Fig. 36.2a–c, f). For these studies, the isolated MMA segments were again pressurized to 40 mmHg and allowed to develop myogenic tone, and 3 nM PACAP produced a 56.2 ± 7.6 % increase in dilation. However, in the presence of the K_{ATP}

channel inhibitor glibenclamide (10 μM) or a vascular selective K_{ATP} channel inhibitor PNU37883 (10 μM) [44], the PACAP effects were completely abolished. By contrast, MMA treatment with a BK channel inhibitor, paxilline (1 μM), failed to attenuate the PACAP vasodilatory responses (Fig. 36.2d, f). Furthermore, pretreat-

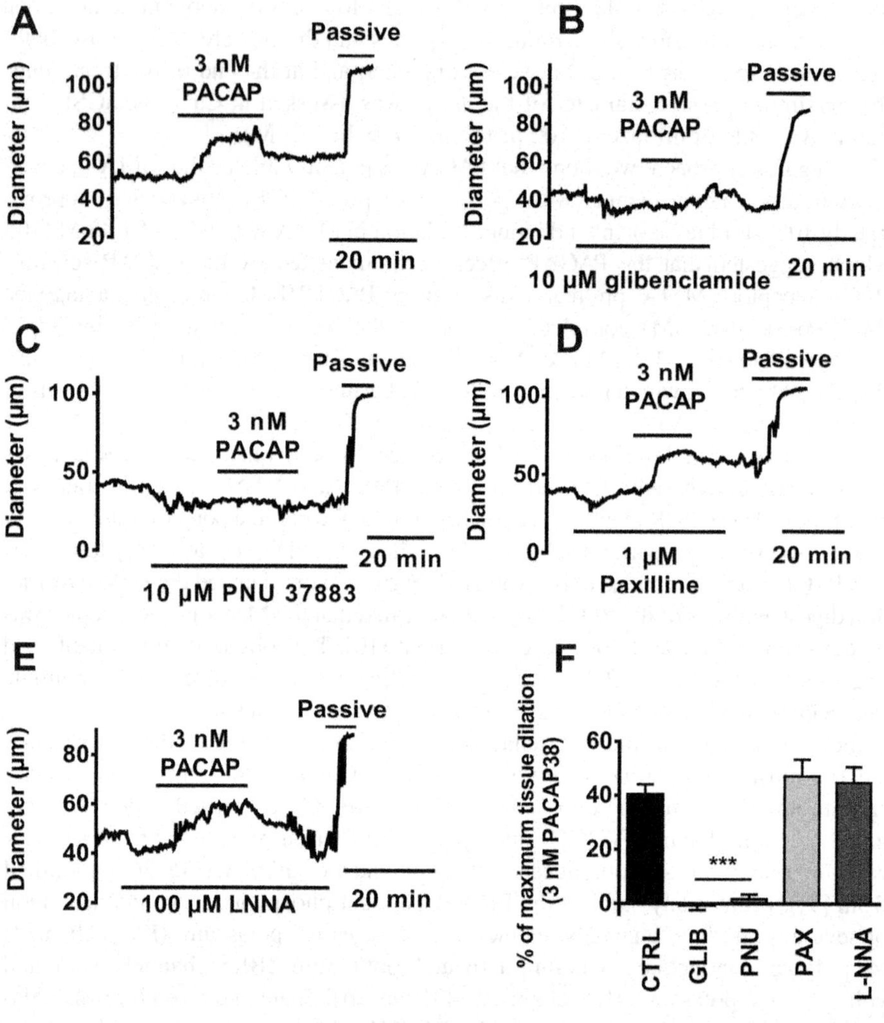

Fig. 36.2 PACAP-induces MMA vasodilation through K_{ATP} channel activation. Segments of the MMA were cannulated and pressurized for superfusion experiments as described previously [10]. MMA perfusions with 3 nM PACAP produced vasodilation (**a**; $n=8$). The addition of K_{ATP} channel inhibitor glibenclamide (**b**; GLIB, 10 μM, $N=4$) or the vascular selective K_{ATP} channel inhibitor PNU37883 (**c**; PNU, 10 μM, $n=4$) blocked the PACAP vasodilatory effect. By contrast, treatments with the BK channel inhibitor paxilline (**d**; PAX, 1 μM, $n=4$) or the nitric oxide synthase inhibitor L-NNA (**e**; 100 μM, $n=4$) had no effect. The data are summarized in (**f**) represent as mean ± SE. ***, significantly different from PACAP alone (CTRL, control) at $p<0.001$. One-way ANOVA followed by post hoc comparisons of the means using Tukey test

ment with the competitive nitric oxide synthase inhibitor nitro-L-arginine (L-NNA, 100 μM) had no effect on PACAP-mediated MMA vasodilation (Fig. 36.2e, f). Together, these results suggest that PACAP/PAC1 receptor-mediated activation of cAMP/PKA and K_{ATP} channels can regulate MMA tone exclusively without engaging BK channel functionor nitric oxide (NO) production. The spatial and temporal regulation of K_{ATP} channels by PKA requires further investigation but A-kinase-anchoring proteins (AKAPs) have been postulated to play crucial roles in conferring the specific PKA interactions needed for channel regulatory fidelity [45].

PACAP and CGRP Demonstrate Differential Potencies in Stimulating MMA Vasodilation

Described earlier, PACAP and CGRP are frequently colocalized in the fiber networks around the cerebral arteries including the MMA (Fig. 36.1). CGRP also has vasodilatory functions via G protein-coupled calcitonin receptor-like receptors (CALCR). We have begun to examine whether there are differences between PACAP and CGRP in their vasodilatory activities with respect to potency, efficacy or signaling mechanisms. Using the same ex vivo pressurized MMA vessel preparation, concentration-dependence studies demonstrated that the EC_{50} for CGRP-induced MMA dilation was 1 nM, which was comparable to the potency of VIP, but again 1000-fold lower than that for PACAP-mediated vasodilation (EC_{50} 1 pM). These results appeared to agree with previously published data in which PAC1 receptor-stimulated increase in cAMP production was 500-fold more potent than that of CGRP [20]. The efficacy of PACAP and CGRP in MMA vasodilation appeared comparable; the mechanisms underlying the differences in potency are under study.

PACAP Mediates Cerebellar Artery Vasodilation via Different Receptors and Channels

Unlike the responses in MMA, PACAP and VIP dilated cerebellar arteries with equal potency. Using the same ex vivo procedures to perfuse the cannulated cerebellar arterial segments, PACAP38 and VIP induced vasodilation with an EC_{50} of 11 and 6 nM, respectively [9, 10]. All three VIP/PACAP receptor subtypes have been identified in cerebral arteries but distinct from the MMA in which PAC1 receptor signaling was implicated, the near equal potency of VIP and PACAP in eliciting cerebellar arterial dilation indicated that the responses were largely mediated by VPAC receptors.

In cerebellar arteries, PACAP-mediated stimulation of the cAMP/PKA pathway not only activated K_{ATP} channels but also BK channels to increase spontaneous

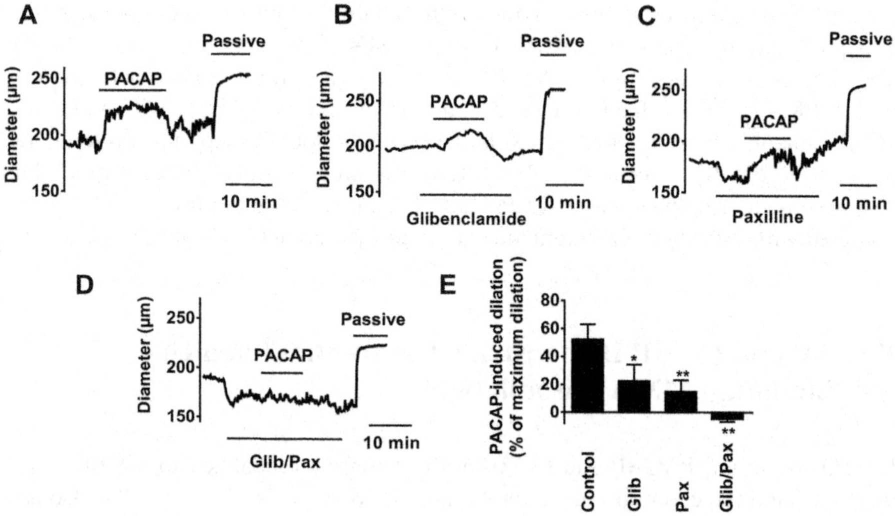

Fig. 36.3 PACAP-induced dilation in isolated cerebellar artery is mediated via both K_{ATP} and BK channels. (**a**) Representative trace of 3 nM PACAP-induced vasodilation of cerebellar artery. (**b**, **c**) Representative traces of 3 nM PACAP-induced vasodilation in the presence of K_{ATP} channel inhibitor, glibenclamide (Glib) (10 μM) (B) and BK channel inhibitor, paxilline (Pax) (1 μM) (**c**). The vasodilation was significantly attenuated but not abolished. (**d**) Representative trace of 3 nM PACAP-induced vasodilation in the presence of 10 μM glibenclamide and 1 μM paxilline combined. The combined treatment with glibenclamide and paxilline abolished 3 nM PACAP-induced vasodilation in cerebellar artery. (**e**) Summary data of PACAP-induced vasodilation of cerebellar artery in the absence and presence of glibenclamide and paxilline (n=4–7). Data are expressed as mean ± SE. * $P < 0.05$, ** $P < 0.01$ vs. PACAP-induced dilation in the absence of K^+ channel blockers (control). Reproduced and modified with permission from the Journal of Molecular Neuroscience (Koide et al. 2014)

transient outward currents (STOCs) for vasodilation [9, 46–49]. The BK channels, like K_{ATP} channels, can be activated by direct PKA phosphorylation [48, 49], but characteristically, BK channels are regulated by local increases in intracellular calcium (Ca^{2+} sparks). PACAP signaling and stimulation of PKA can potentially activate ryanodine receptors (RyRs) on the sarcoplasmic reticulum (SR) for intracellular calcium release, resulting in a localized transient increase in cytosolic calcium up to 10 μM, termed a calcium spark [27, 47, 48]. These calcium spark events in turn can lead to localized BK channel activation on the plasma membrane. Additionally, PKA can also phosphorylate phospholamban (PLB), a pentameric protein that regulates the sarcoplasmic reticulum calcium-ATPase (SERCA) pump, critical for the management of intracellular calcium. PLB in the unphosphorylated state inhibits SERCA activity; PACAP activation of PKA and downstream phosphorylation of PLB relieve that inhibition to increase SERCA activity and sarcoplasmic reticulum calcium load, resulting in increased calcium spark frequency and BK channel activity [27, 47, 48]. Unlike the MMA studies, the addition of the K_{ATP} channel blocker glibenclamide or the BK channel blocker paxilline only attenuated and did not abolish

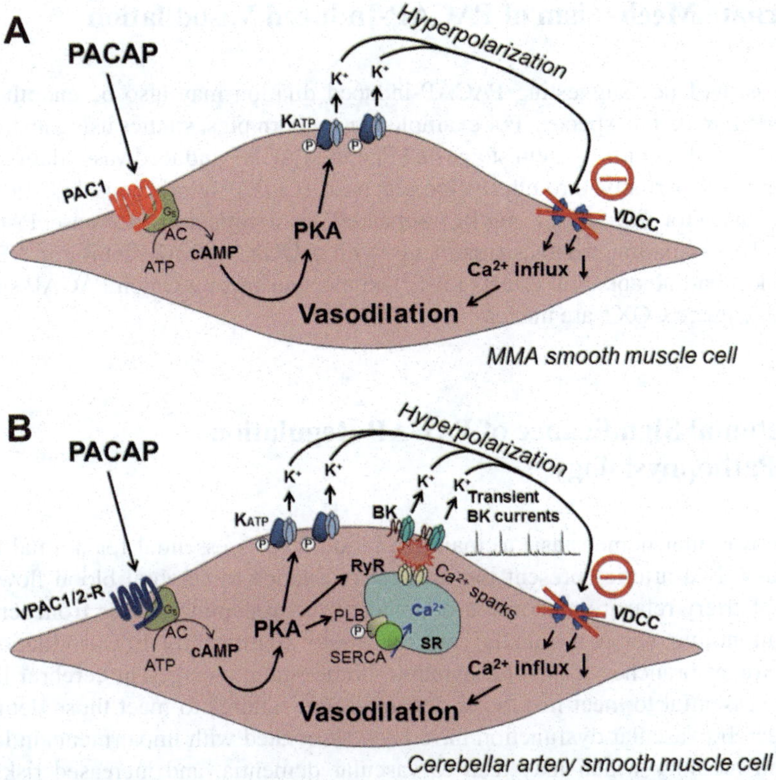

Fig. 36.4 Schematic of PACAP mechanisms in vascular smooth muscle vasodilation. (**a**) PACAP binding to PAC1 G protein-coupled receptors stimulates cAMP/PKA pathways, resulting in K_{ATP} channel phosphorylation and activation. The ensuing hyperpolarization closes voltage-dependent calcium channels (VDCC) and reduces intracellular calcium to facilitate smooth muscle relaxation and vasodilation. This is the principal PACAP vasodilatory mechanism in MMA. (**b**) In cerebellar arteries, PACAP-mediated cAMP/PKA signaling by binding to VPAC1/2-R, can also activate calcium-sensitive potassium (BK) channels by direct channel phosphorylation (not shown), activation of ryanodine receptors (RyR) or disinhibition of phospholamban (PLB) on the sarcoplasmic reticulum ATPase pump (SERCA). The latter mechanisms increase sarcoplasm reticulum calcium release and spark generation to activate BK channels. The BK mechanisms are synergistic with those via K_{ATP} for cerebellar artery vasodilation. Reproduced and modified with permission from the Journal of Molecular Neuroscience (Koide et al. 2014)

the PACAP-induced vasodilation in cerebellar arteries (Fig. 36.3b–c, e, modified from [9]). However, the concurrent addition of glibenclamide and paxilline fully eliminated the PACAP vasodilatory responses (Fig. 36.3d, e, modified from [9]). These findings suggest that similar to coronary and pulmonary arteries [13, 14], PACAP regulates cerebral artery tone via both K_{ATP} and BK channels, but in contradistinction, PACAP only regulates K_{ATP} channels for MMA vasodilation (Fig. 36.4).

Alternate Mechanism of PACAP-Induced Vasodilation

There is evidence suggesting PACAP-induced dilation may also be endothelium dependent in certain species. For example in new born pigs, studies using an in vivo closed cranial window system suggested that the PACAP-induced vasodilation may be mediated partially through cyclooxygenase (COX). Intravenous injections of COX1 inhibitor SC-560 (1 mg/kg) appeared to significantly decrease PACAP-induced vasodilation whereas treatments with a COX2-specific inhibitor NS-398 (1 mg/kg) had no apparent effects [50]. The mechanisms by which PACAP signaling can engage COX1 are unclear.

Functional Significance of PACAP Regulation and Pathophysiology

Cerebrovascular homeostasis to maintain blood flow is essential for normal brain activities. Pial arteries present the greatest resistance to cerebral blood flow and cerebral artery relaxation after neurotransmitter/neuropeptide release from sensory and autonomic nerve terminals can contribute significantly to vasodilation in downstream branches [51]. The dynamic homeostatic changes in cerebral blood flow is essential to meet metabolic demands, and failures to meet those demands from cerebrovascular dysfunction have been associated with impairments in learning and memory which may lead to vascular dementia, and increased risks for Alzheimer's and related neurodegenerative diseases [52]. PACAP receptor activation can engage neurotrophic signaling pathways for neuroprotection against injuries, but the abilities for PACAP to regulate vasodilation and cerebral blood flow may offer means to mitigate ischemic damage [15, 17, 53].

Recently, there has been interest in PACAP signaling in the MMA and migraine. Migraine is a complex and often debilitating neurological disorder characterized by intense unilateral pulsating headache, frequently accompanied with aura, photophobia, phonophobia, nausea, and related symptoms [32, 54]. The causes of migraine are not understood and are likely to be multifactorial. One mechanism that has been debated for decades implicates MMA vasodilation from trigeminal sensory and/or sphenopalatine parasympathetic fiber innervation and signaling. A different and more recent concept suggests that migraine results from cortical spreading depression [55, 56]. But regardless whether MMA vasodilation plays a causal or contributory role in the pathophysiology of migraine, there is good evidence that vasodilatory peptides participate in migraine development. As described in previous work, CGRP is localized to sensory fibers innervating the MMA, facilitates MMA vasodilation, and can induce migraine-like headaches similar to spontaneous migraine attacks. Magnetic resonance angiography imaging correlated MMA vasodilation with migraine onset upon exogenous CGRP infusions and notably, CGRP receptor antagonists can ameliorate migraine headaches [32, 57]. However, clinical trials examining the efficacy of the CGRP receptor antagonists were halted because of liver toxicity [58]. More recently, plasma PACAP levels have been reported to be elevated in patients with migraine [59]. Further, PACAP is associated with post-traumatic stress disorder (PTSD) which is highly comorbid with migraine, suggesting an intersection of PACAP mechanisms and neural pathways [60]. As described above, PACAP is frequently colocalized with CGRP in sensory fibers innervating the MMA and similar to CGRP, exogenous PACAP administration can initiate migraine ipsilateral to the infusion site coincident with MMA vasodilation [61, 62]. Specific small molecule PACAP receptor antagonists have yet to be identified but blocking MMA PACAP signaling, either alone or in combination with CGRP receptor antagonists, may offer means for migraine therapeutics. From these examples, a better understanding of PACAP and VIP neurocircuits and signaling mechanisms regulating cerebral arterial tone may be clinically relevant for a number of neurophysiological disorders and states.

Conclusion

The regulation of cerebrovascular resistance is essential to meet brain metabolic demands and function. PACAP has potent vasodilatory activities in a variety of vascular systems including the cerebral and cranial arteries. The PACAP fibers innervating the cerebral arteries are sensory or autonomic in origin, and interestingly, PACAP potency, the PACAP receptor subtypes and the downstream signaling mechanisms mediating the vasodilatory effects appear to vary for cranial blood vessels. PACAP signaling via the PAC1 and VPAC receptors in the MMA activates exclusively K_{ATP} channels for vasodilation; the PACAP effects on cerebellar arteries appear to be primarily mediated by VPAC receptor regulation of K_{ATP} and BK channels. Thus the roles of PACAP in the brain are multidimensional. In addition to its direct trophic and protective activities on neurons, the roles of PACAP in regulating blood vessel diameter may be important in maintaining learning and memory, and mitigating brain damage from ischemic events. Additionally, the vasodilatory effects of PACAP on the MMA may have implications in migraine disorders.

References

1. Edvinsson L, Elsas T, Suzuki N, Shimizu T, Lee TJ. Origin and co-localization of nitric oxide synthase, CGRP, PACAP, and VIP in the cerebral circulation of the rat. Microsc Res Tech. 2001;53:221–8.
2. Goadsby PJ. Autonomic nervous system control of the cerebral circulation. Handb Clin Neurol. 2013;117:193–201.
3. Baeres FM, Moller M. Origin of PACAP-immunoreactive nerve fibers innervating the subarachnoidal blood vessels of the rat brain. J Cereb Blood Flow Metab. 2004;24:628–35.
4. Mulder H, Uddman R, Moller K, Zhang YZ, Ekblad E, Alumets J, et al. Pituitary adenylate cyclase activating polypeptide expression in sensory neurons. Neuroscience. 1994;63:307–12.

5. Mulder H, Uddman R, Moller K, Elsas T, Ekblad E, Alumets J, et al. Pituitary adenylate cyclase activating polypeptide is expressed in autonomic neurons. Regul Pept. 1995;59:121–8.
6. Sherwood NM, Krueckl SL, McRory JE. The origin and function of the pituitary adenylate cyclase-activating polypeptide (PACAP)/glucagon superfamily. Endocr Rev. 2000;21:619–70.
7. Anzai M, Suzuki Y, Takayasu M, Kajita Y, Mori Y, Seki Y, et al. Vasorelaxant effect of PACAP-27 on canine cerebral arteries and rat intracerebral arterioles. Eur J Pharmacol. 1995;285:173–9.
8. Baun M, Hay-Schmidt A, Edvinsson L, Olesen J, Jansen-Olesen I. Pharmacological characterization and expression of VIP and PACAP receptors in isolated cranial arteries of the rat. Eur J Pharmacol. 2011;670(1):186–94.
9. Koide M, Syed AU, Braas KM, May V, Wellman GC. Pituitary adenylate cyclase activating polypeptide (PACAP) dilates cerebellar arteries through activation of large-conductance Ca^{2+}-activated (BK) and ATP-sensitive K_{ATP} K^+ channels. J Mol Neurosci. 2014;54:443–50.
10. Syed AU, Koide M, Braas KM, May V, Wellman GC. Pituitary adenylate cyclase-activating polypeptide (PACAP) potently dilates middle meningeal arteries: implications for migraine. J Mol Neurosci. 2012;48:574–83.
11. Wilson AJ, Warren JB. Adenylate cyclase-mediated vascular responses of rabbit aorta, mesenteric artery and skin microcirculation. Br J Pharmacol. 1993;110:633–8.
12. Banki E, Hajna Z, Kemeny A, Botz B, Nagy P, Bolcskei K, et al. The selective PAC1 receptor agonist maxadilan inhibits neurogenic vasodilation and edema formation in the mouse skin. Neuropharmacology. 2014;85:538–47.
13. Bruch L, Bychkov R, Kastner A, Bulow T, Ried C, Gollasch M, et al. Pituitary adenylate-cyclase-activating peptides relax human coronary arteries by activating K_{ATP} and K_{Ca} channels in smooth muscle cells. J Vasc Res. 1997;34:11–8.
14. Bruch L, Rubel S, Kastner A, Gellert K, Gollasch M, Witt C. Pituitary adenylate cyclase activating peptides relax human pulmonary arteries by opening of K_{ATP} and K_{Ca} channels. Thorax. 1998;53:586–7.
15. Danyadi B, Szabadfi K, Reglodi D, Mihalik A, Danyadi T, Kovacs Z, et al. PACAP application improves functional outcome of chronic retinal ischemic injury in rats-evidence from electroretinographic measurements. J Mol Neurosci. 2014;54:293–9.
16. Dorner GT, Wolzt M, Eichler HG, Schmetterer L. Effect of pituitary adenylate cyclase activating polypeptide 1-27 on ocular, cerebral and skin blood flow in humans. Naunyn Schmiedebergs Arch Pharmacol. 1998;358:657–62.
17. Elsas T, White LR. Evidence for a possible synergism between pituitary adenylate cyclase activating polypeptide and calcitonin gene-related peptide in porcine ophthalmic artery. Acta Ophthalmol Scand. 1997;75:159–61.
18. Graf AH, Schiechl A, Hacker GW, Hauser-Kronberger C, Steiner H, Arimura A, et al. Helospectin and pituitary adenylate cyclase activating polypeptide in the human vagina. Regul Pept. 1995;55:277–86.
19. Hedlund P, Alm P, Ekstrom P, Fahrenkrug J, Hannibal J, Hedlund H, et al. Pituitary adenylate cyclase-activating polypeptide, helospectin, and vasoactive intestinal polypeptide in human corpus cavernosum. Br J Pharmacol. 1995;116:2258–66.
20. Lerner EA, Ribeiro JM, Nelson RJ, Lerner MR. Isolation of maxadilan, a potent vasodilatory peptide from the salivary glands of the sand fly Lutzomyia longipalpis. J Biol Chem. 1991;266:11234–6.
21. Steenstrup BR, Jorgensen JC, Alm P, Hannibal J, Junge J, Fahrenkrug J, et al. Pituitary adenylate cyclase activating polypeptide (PACAP): occurrence and vasodilatory effect in the human uteroplacental unit. Regul Pept. 1996;61:197–204.
22. Uddman R, Luts A, Arimura A, Sundler F. Pituitary adenylate cyclase-activating peptide (PACAP), a new vasoactive intestinal peptide (VIP)-like peptide in the respiratory tract. Cell Tissue Res. 1991;265:197–201.
23. Uddman R, Goadsby PJ, Jansen I, Edvinsson L. PACAP, a VIP-like peptide: immunohistochemical localization and effect upon cat pial arteries and cerebral blood flow. J Cereb Blood Flow Metab. 1993;13:291–7.

24. Vuckovic A, Rondelet B, Brion JP, Naeije R. Expression of vasoactive intestinal peptide and related receptors in overcirculation-induced pulmonary hypertension in piglets. Pediatr Res. 2009;66:395–9.
25. Warren JB, Larkin SW, Coughlan M, Kajekar R, Williams TJ. Pituitary adenylate cyclase activating polypeptide is a potent vasodilator and oedema potentiator in rabbit skin in vivo. Br J Pharmacol. 1992;106:331–4.
26. Yao W, Sheikh SP, Ottesen B, Jorgensen JC. Vascular effects and cyclic AMP production produced by VIP, PHM, PHV, PACAP-27, PACAP-38, and NPY on rabbit ovarian artery. Peptides. 1996;17:809–15.
27. Brayden JE, Nelson MT. Regulation of arterial tone by activation of calcium-dependent potassium channels. Science. 1992;256:532–5.
28. Li Y, Baylie RL, Tavares MJ, Brayden JE. TRPM4 channels couple purinergic receptor mechanoactivation and myogenic tone development in cerebral parenchymal arterioles. J Cereb Blood Flow Metab. 2014;34:1706–14.
29. Nelson MT, Quayle JM. Physiological roles and properties of potassium channels in arterial smooth muscle. Am J Physiol. 1995;268:C799–822.
30. Dunn KM, Hill-Eubanks DC, Liedtke WB, Nelson MT. TRPV4 channels stimulate Ca^{2+}-induced Ca^{2+} release in astrocytic endfeet and amplify neurovascular coupling responses. Proc Natl Acad Sci U S A. 2013;110:6157–62.
31. Filosa JA, Morrison HW, Iddings JA, Du W, Kim KJ. Beyond neurovascular coupling, role of astrocytes in the regulation of vascular tone. Neuroscience. 2016;323:96–109.
32. Asghar MS, Hansen AE, Amin FM, van der Geest RJ, Koning P, Larsson HB, et al. Evidence for a vascular factor in migraine. Ann Neurol. 2011;69:635–45.
33. Longden TA, Dabertrand F, Hill-Eubanks DC, Hammack SE, Nelson MT. Stress-induced glucocorticoid signaling remodels neurovascular coupling through impairment of cerebrovascular inwardly rectifying K^+ channel function. Proc Natl Acad Sci U S A. 2014;111:7462–7.
34. Seidah NG, Chretien M, Day R. The family of subtilisin/kexin like pro-protein and prohormone convertases: divergent or shared functions. Biochimie. 1994;76:197–209.
35. Seidah NG, Chretien M. Proprotein and prohormone convertases: a family of subtilases generating diverse bioactive polypeptides. Brain Res. 1999;848:45–62.
36. Vaudry D, Falluel-Morel A, Bourgault S, Basille M, Burel D, Wurtz O, et al. Pituitary adenylate cyclase-activating polypeptide and its receptors: 20 years after the discovery. Pharmacol Rev. 2009;61:283–357.
37. Chan KY, Baun M, de Vries R, van den Bogaerdt AJ, Dirven CM, Danser AH, et al. Pharmacological characterization of VIP and PACAP receptors in the human meningeal and coronary artery. Cephalalgia. 2011;31:181–9.
38. Erdling A, Sheykhzade M, Maddahi A, Bari F, Edvinsson L. VIP/PACAP receptors in cerebral arteries of rat: characterization, localization and relation to intracellular calcium. Neuropeptides. 2013;47:85–92.
39. Tong S, Parfenova H, Shibata M, Zuckerman S, Armstead WM, Leffler CW. Pituitary adenylate cyclase-activating polypeptide dilates cerebral arterioles of newborn pigs. Proc Soc Exp Biol Med. 1993;203:343–7.
40. Huang M, Shirahase H, Rorstad OP. Comparative study of vascular relaxation and receptor binding by PACAP and VIP. Peptides. 1993;14:755–62.
41. Wellman GC, Quayle JM, Standen NB. ATP-sensitive K^+ channel activation by calcitonin gene-related peptide and protein kinase A in pig coronary arterial smooth muscle. J Physiol. 1998;507:117–29.
42. Kleppisch T, Nelson MT. ATP-sensitive K^+ currents in cerebral arterial smooth muscle: pharmacological and hormonal modulation. Am J Physiol. 1995;269:H1634–40.
43. Brignell JL, Perry MD, Nelson CP, Willets JM, Challiss RA, Davies NW. Steady-state modulation of voltage-Gated K^+ channels in rat arterial smooth muscle by cyclic AMP-dependent protein kinase and protein phosphatase 2B. PLoS One. 2015;10:e0121285.

44. Wellman GC, Barrett-Jolley R, Koppel H, Everitt D, Quayle JM. Inhibition of vascular K_{ATP} channels by U-37883A: a comparison with cardiac and skeletal muscle. Br J Pharmacol. 1999;128:909–16.
45. Kapiloff MS, Chandrasekhar KD. A-kinase anchoring proteins: temporal and spatial regulation of intracellular signal transduction in the cardiovascular system. J Cardiovasc Pharmacol. 2011;58:337–8.
46. Brayden JE, Li Y, Tavares MJ. Purinergic receptors regulate myogenic tone in cerebral parenchymal arterioles. J Cereb Blood Flow Metab. 2013;33:293–9.
47. Jaggar JH, Wellman GC, Heppner TJ, Porter VA, Perez GJ, Gollasch M, et al. Ca^{2+} channels, ryanodine receptors and Ca^{2+}-activated K^+ channels: a functional unit for regulating arterial tone. Acta Physiol Scand. 1998;164:577–87.
48. Wellman GC, Santana LF, Bonev AD, Nelson MT. Role of phospholamban in the modulation of arterial Ca^{2+} sparks and Ca^{2+}-activated K^+ channels by cAMP. Am J Physiol Cell Physiol. 2001;281:C1029–37.
49. Wellman GC, Nelson MT. Signaling between SR and plasmalemma in smooth muscle: sparks and the activation of Ca^{2+}-sensitive ion channels. Cell Calcium. 2003;34:211–29.
50. Lenti L, Domoki F, Kis D, Hegyi O, Toth GK, Busija DW, et al. Pituitary adenylate cyclase-activating polypeptide induces pial arteriolar vasodilation through cyclooxygenase-dependent and independent mechanisms in newborn pigs. Brain Res. 2007;1165:81–8.
51. Iadecola C. Neurovascular regulation in the normal brain and in Alzheimer's disease. Nat Rev Neurosci. 2004;5:347–60.
52. Claassen JA. New cardiovascular targets to prevent late onset Alzheimer disease. Eur J Pharmacol. 2015;763:131–4.
53. Banki E, Kovacs K, Nagy D, Juhasz T, Degrell P, Csanaky K, et al. Molecular mechanisms underlying the nephroprotective effects of PACAP in diabetes. J Mol Neurosci. 2014;54:300–9.
54. Edvinsson L. PACAP and its receptors in migraine pathophysiology. Br J Pharmacol. 2014;1:1036–40.
55. Ayata C. Cortical spreading depression triggers migraine attack: pro. Headache. 2010;50:725–30.
56. Cui Y, Kataoka Y, Watanabe Y. Role of cortical spreading depression in the pathophysiology of migraine. Neurosci Bull. 2014;30:812–22.
57. Asghar MS, Hansen AE, Kapijimpanga T, van der Geest RJ, van der Koning P, Larsson HB, et al. Dilation by CGRP of middle meningeal artery and reversal by sumatriptan in normal volunteers. Neurology. 2010;75:1520–6.
58. Bigal ME, Walter S. Monoclonal antibodies for migraine: preventing calcitonin gene-related peptide activity. CNS Drugs. 2014;28:389–99.
59. Tuka B, Helyes Z, Markovics A, Bagoly T, Szolcsanyi J, Szabo N, et al. Alterations in PACAP-38-like immunoreactivity in the plasma during ictal and interictal periods of migraine patients. Cephalalgia. 2013;33:1085–95.
60. Bukovics P, Czeiter E, Amrein K, Kovacs N, Pal J, Tamas A, et al. Changes of PACAP level in cerebrospinal fluid and plasma of patients with severe traumatic brain injury. Peptides. 2014;60:18–22.
61. Amin FM, Asghar MS, Guo S, Hougaard A, Hansen AE, Schytz HW, et al. Headache and prolonged dilatation of the middle meningeal artery by PACAP38 in healthy volunteers. Cephalalgia. 2012;32:140–9.
62. Amin FM, Hougaard A, Schytz HW, Asghar MS, Lundholm E, Parvaiz AI, et al. Investigation of the pathophysiological mechanisms of migraine attacks induced by pituitary adenylate cyclase-activating polypeptide-38. Brain. 2014;137:779–94.

Chapter 37
PACAP Circuits Mediating the Sensory and Behavioral Consequences of Pain

Galen Missig, Rodney L. Parsons, Margaret A Vizzard, Sayamwong E. Hammack, Karen M. Braas, and Victor May

Abstract The co-prevalence of stress-related psychopathologies with chronic pain has implicated common mechanistic underpinnings. PACAP has been associated with chronic stress-related behaviors and disorders; the expression and regulation of PACAP and PAC1 receptors in sensory pathways have implicated the PACAPergic system in nociceptive signaling. Although conflicting effects of PACAP in mediating anti-nociception or pro-nociception have been reported in peripheral vs. central peptide infusion experiments, neuropathic and inflammatory pain studies using PACAP or PAC1 receptor knockout mice have consistently illustrated the pro-nociceptive action of PACAP signaling. PACAP has recently been identified in the parabrachioamygdaloid tract which transmits nociceptive information from the dorsal horn of the spinal cord to the pontine lateral parabrachial nucleus (LPBn), where PACAP neurons project to the lateral capsular division of the central amygdala (CeLC), or the nociceptive amygdala. The amygdala is the principal integration center for emotional responses; hence these observations implicate a neurocircuit in which PACAP nociceptive signaling can impact stress-related behaviors. PACAP expression was induced along the spino-parabrachioamygdaloid tract in a partial sciatic nerve ligation chronic constriction injury (CCI) model of neuropathic pain. Amygdala PACAP infusions increased stress-related behaviors concomitant with heightened thermal nociceptive sensitivity, and conversely, CCI-induced behavioral and pain responses were attenuated upon acute amygdala PACAP receptor antagonist administration. Chronic pain increased amygdala ERK activation which could be mimicked by PACAP; notably, inhibition of PACAP receptor-mediated ERK signaling with drugs that block either MEK or endosomal signaling mitigated PACAP-mediated nociceptive hypersensitivity. Accordingly, PACAP signaling and

G. Missig • R.L. Parsons • M.A. Vizzard • K.M. Braas • V. May (✉)
Department of Neurological Sciences, University of Vermont College of Medicine, 149 Beaumont Avenue, Burlington, VT 05405, USA
e-mail: victor.may@uvm.edu

S.E. Hammack
Department of Psychological Science, University of Vermont, 110 Colchester Avenue, Burlington, VT 05405, USA

D. Reglodi, A. Tamas (eds.), *Pituitary Adenylate Cyclase Activating Polypeptide — PACAP*, Current Topics in Neurotoxicity 11,
DOI 10.1007/978-3-319-35135-3_37

function in the parabrachioamygdaloid tract may be a mediator of the adverse emotional consequences of chronic pain.

Keywords PACAP • PAC1 receptor • Stress-related behavior • Nociception • Lateral parabrachial nucleus (LPBn) • Central nucleus of the amygdala (CeA) • Lateral capsular division of central amygdala (CeLC) • Spino-parabrachioamygdaloid tract • Extracellular signal-regulated kinase (ERK)

Introduction

A variety of bioactive neuropeptides participate in the formation, transmission, modulation, and perception of pain. Substance P and neurokinin A of the tachykinin family of peptides, and calcitonin-gene related peptide (CGRP) for example, have been well studied in pain due to their expression patterns along nociceptive pathways, and abilities to initiate and modulate nociceptive transmission. Although appreciated as a sensory peptide within a few years after discovery, the recent accumulation of evidence has generated renewed interest in pituitary adenylate cyclase activating polypeptide (PACAP, *Adcyap1*) as a nociceptive peptide critical in mediating the development of chronic pain and pain-related behavioral responses. As other chapters will have discussed, the PACAP precursor molecule can be alternatively post-translationally processed to two alpha-amidated bioactive peptides—PACAP27 or PACAP38—which belong to the VIP/secretin/glucagon family of peptides [1, 2]. PACAP has high binding affinity to three G protein-coupled receptor subtypes. PACAP binds selectively at the PAC1 receptor (*Adcyap1r1*); both PACAP and VIP bind to VPAC1 and VPAC2 (*Vipr1/Vipr2*) with high affinity [3–5]. Uniquely, the PAC1 receptor has multiple receptor isoforms, resulting from alternative splicing in the gene transcripts corresponding to the third cytoplasmic domain (Hip and/or Hop variants), which exhibit differential activation of multiple signal transduction pathways in a tissue-specific manner [5–7]. The abilities of PACAP and PAC1 receptor signaling to potently transduce both rapid acute and long-term responses provide means for PACAP to effect neurotransmitter signaling and trophic support for cellular proliferation, differentiation, and neuroplasticity. These attributes of PACAP/PAC1 receptor signaling may be central for neurocircuit dynamics to maintain homeostasis and yet, sustained PACAP signaling may contribute to neural maladaptations leading to physiological and behavioral pathologies, including pain- and stress-related disorders.

Nociceptive Pathways

Nociception is the process by which noxious stimuli are detected by an organism to allow for adaptive physiological and behavioral responses essential for survival. The cellular and molecular mechanisms of nociception have been well reviewed

[8–11] and are briefly summarized here to facilitate discussions in later sections. The detection of noxious stimuli is performed by primary afferent neurons from the trigeminal ganglia that carry information from the face and neck, and the dorsal root ganglia (DRG) which transduce signals from the rest of the body. Between the two major nociceptor classes, myelinated A(delta) medium fibers are fast-conducting and transmit well-localized sharp pain while unmyelinated small caliber C-fibers are largely polymodal and convey more delocalized prolonged slow pain that may be related to nociceptive sensitization processes. Specialized channels on these nociceptors are critical to this process, including several members of the transient receptor potential (TRP) ion channel family such as TRPV1 and TRPA, the degenerin/epithelia sodium channels (DEG/ENaC), and the piezo2 mechanosensitive channels [9, 12–14]. C-fibers also have been characterized as heterogeneous in their neurochemical profiles as either peptidergic or nonpeptidergic; the peptidergic population has been well described to express CGRP, substance P, and nerve growth factor (NGF)-selective TrkA receptors, and the nonpeptidergic population can be distinguished by c-RET neurotrophin receptor expression, the presence of the Mrg family of G protein-coupled receptors, or IB4 isolectin binding [9].

The resulting transduction of noxious mechanical, chemical, or thermal stimuli at the afferent terminals activates a variety of voltage-gated sodium and potassium channels for action potential generation and central axon synaptic release of glutamate/peptide onto second order neurons. In the spinal cord, the DRG A(delta) and peptidergic C-fiber nociceptors terminate primarily onto neurons in the superficial layers laminae I and II of the dorsal horn where the integration of complex descending and local signals modulate pain transmission to higher centers. The A(delta) and large caliber mechanical (touch) A(beta) fibers also converge onto deeper laminae III–V. The main outputs from the secondary projection neurons form organized tracts to the thalamus (spinothalamic tract), medulla, and brainstem (spinoreticular, spinoparabrachial, and spinomesencephalic tracts) [11], and within the brain, the intersection of networked regions or a pain matrix drives nociceptive processing and pain experience. The matrix is not a rigidly defined set of structures as pain is a multifactorial subjective experience, but acute pain commonly accesses the primary and secondary somatosensory cortices, insular cortex, anterior cingulate cortex, and thalamus [15]. From diversity of pain experience and circumstance, the number of regions can be enlarged to include the hippocampus, amygdala, anterior insula, entorhinal complex, and prefrontal, temporal and parietal cortices, resulting in the integration of pain and emotional information. The associations between stress-related behaviors and chronic pain are well appreciated. Unlike acute stress which can produce analgesia, chronic stress can heighten pain experiences leading to hyperalgesia [16]. Conversely, chronic pain is stressful and amplifies stress-associated behaviors. The high comorbidity of chronic pain and stress-related psychopathologies can adversely impact both disorders complicating clinical management [17, 18]. The neuroanatomical pathways, neurochemistry, and mechanisms underlying persistent pain/nociceptive hypersensitivity and stress-related disorders are not well understood but thought to stem from interrelated maladaptive processes. The recent appreciation of PACAP signaling in chronic stress and pain circuits suggests that the PACAPergic system may participate in that process.

PACAP and PAC1 Receptor Expression and Plasticity in Peripheral Nociceptive Pathways

The initial evidence that PACAP participates in nociception stemmed from its distribution and expression patterns within the peripheral nervous system. Complementing other sensory peptides, PACAP immunoreactivity and mRNA expression have been identified in both DRG and trigeminal ganglion neurons from immunocytochemical and in situ hybridization histochemical studies, respectively [19, 20]. Under normal physiological conditions, PACAP immunoreactivity in DRG neuronal soma and peripheral axons has been identified in small to medium-sized unmyelinated capsaicin-sensitive C-fiber nociceptor afferents with other sensory peptides including CGRP and substance P. PACAP expression within a defined subset of peptidergic DRG neurons has been confirmed in recent classifications using single-cell RNA sequencing [21]. In the spinal cord, the central axons of PACAPergic DRG neurons are largely confined to lamina I and II of the dorsal horn corresponding to projections important for nociceptive transmission [20, 22, 23]. The expression of PACAP in DRG sensory neurons is detailed in a different chapter in this volume. In addition to DRG, there is also a population of PACAP-expressing neurons in lamina I and II of the spinal cord dorsal horn presenting the possibility that PACAP may be expressed in second order neurons in the nociceptive pathway as well [24, 25]. Based on in vitro receptor autoradiography and in situ hybridization, PAC1 receptors are densely expressed in laminae I and II of the dorsal horn in correspondence with PACAP DRG central axon projections. While the distribution of PACAP fibers, PACAP neurons, and PAC1 receptors in the superficial layers of the dorsal horn is suggestive, the potential functional PACAP to PACAP connectivity between DRG and second order dorsal horn PACAPergic neurons is still unclear. From ultrastructural studies, PACAP signaling on PACAP-expressing neurons has been suggested in the enteric nervous systems [26]. Only a few isolated neurons in the ventral horn appear to express PAC1 receptors [24]; PAC1 receptors are not apparent in DRG neurons implying that PACAP does not act in an autocrine or paracrine manner in the ganglion or presynaptically in the dorsal horn [22].

Among several sensory peptides, PACAP demonstrates phenotypic plasticity in various peripheral models of injury- and inflammation-induced pain. Across different experimental paradigms including axotomy, nerve compression, and adjuvant treatments, DRG PACAP transcripts, peptide levels, and cell numbers can be dramatically induced [22, 23, 27–31]. Notably, depending on the nature of insult, there appears to be an induction of PACAP within select DRG neuronal populations with concurrent changes in central and peripheral axon peptide immunoreactivity. PACAP is normally identified to a subpopulation of small and medium-sized nociceptive cells and following inflammatory insult the induction of PACAP appears to be confined to the same small-sized neuronal population [29]. Accordingly, inflammatory cyclophosphamide-induced cystitis augments DRG PACAP neuronal numbers and immunoreactive fiber density in the superficial layers of the dorsal horn, consistent with projections from DRG small neuron induction of PACAP [32].

By contrast, axotomy shifts PACAP expression in the different DRG populations resulting in apparent decreased levels in small DRG neurons and increased peptide expression in the medium and large-sized DRG neurons [22, 23]. Large DRG neurons project to deeper layers of the dorsal horn and in coherence with axotomy-mediated induction patterns, immunoreactive PACAP fibers appear reduced in the superficial layers of the dorsal horn but enhanced in the deeper laminae. However, whether the decrease in PACAP fiber immunoreactivity in the superficial dorsal horn laminae reflected heightened C-fiber PACAP secretion has not been determined. Nerve compression increases PACAP levels in both small and large neuronal populations [30]. The mechanisms underlying the various PACAP induction patterns to different injuries and the consequences of the dynamics in fiber projections in pain remain unclear but are supported in recent transgenic animal studies (see below). Whether the second order PACAP neurons in laminae I and II of the dorsal horn also demonstrate plasticity under the different injury models is unknown although no overt changes were observed following axotomy [24]. In contrast to DRG PACAP inductions, PACAP binding in the dorsal horn after injury was diminished without apparent changes in PAC1 receptor transcript levels [22]. Although the expression patterns for PACAP and PAC1 receptors exhibit an inverse relationship in some studies, the loss of PACAP binding may reflect higher PAC1 receptor internalization and turnover following heightened signaling [33, 34]. Likewise, VPAC1 receptor expression is decreased but VPAC2 receptors are increased following neuropathic pain [35]. While the changes in PACAP expression in the multiple experimental models may be related to enhanced nociceptive neurotransmission, the interpretations are complicated by cellular stress-induced plasticity responses to the various injury challenges. PACAP/PAC1 receptor activation can engage neurotrophic pro-survival signals to promote regeneration [3]; hence the induction in DRG PACAP expression in the neuropathic and inflammatory pain paradigms may have distinct, dual or overlapping activities in nociception and trophic support.

PACAP and PAC1 Receptors in Nociception

From PACAP and PAC1 receptor expression, distribution, and plasticity observed in experimental injury models, the PACAPergic system was implicated to facilitate nociceptive responses. While seemingly straightforward, the PACAP infusion studies were in fact equivocal as to whether PACAP was pro- or anti-nociceptive. At peripheral nerve terminals the actions of PACAP appeared largely anti-nociceptive. While intraplantar PACAP injections alone had no effect on thermal or mechanical sensitivity in naïve animals, intraplantar PACAP injections proved anti-allodynic, anti-nociceptive, and anti-hyperalgesic in experimental models of somatic and visceral inflammatory pain [36]. However, PACAP at knee joint afferents resulted in increased mechanical sensitivity [36]. Intrathecal PACAP injections inhibited spinal and inflammatory nociceptive responses [37–39] whereas others reported PACAP administration was anti-nociceptive in the early phase of formalin-induced pain, but transitioned to pro-nociception in the late phase of the inflammatory response [40].

The pro-nociceptive actions of PACAP, however, were compelling. Intrathecal PACAP infusions to naïve rats produced hyperalgesia, as revealed in thermal hypersensitivity and tail flick latency tests, and amplified pain neurotransmission to the dorsal horn via NMDA mechanisms [41, 42]. The intrathecal nociceptive effects of PACAP were gradual but long lasting, which were in contradistinction to the rapid and transient effects of substance P [40]. Demonstrating a direct effect of PACAP signaling, PACAP application to spinal cord neurons increased excitability of multireceptive cells in lamina III-V of the dorsal horn [43]. Importantly, in comparable studies, blockade of PACAP signaling with the PAC1/VPAC2 receptor antagonist PACAP(6-38) or neutralizing PACAP antibodies attenuated the thermal hypersensitivity and nocifensive responses in a variety of neuropathic and inflammatory pain models [41, 44]. Further, while PACAP(6-38) had no effects alone or upon nonnoxious stimulation, the receptor antagonist blocked the increased excitation of dorsal horn neurons to noxious stimuli [35]. The effects of C-fiber stimulation on spinal nociceptive reflex responses were facilitated by PACAP administration and inhibited with a specific PAC1 receptor antagonist [45, 46]. The causes for the observed discrepancies in the PACAP nociceptive effects are not well understood but may be related to dose and temporal parameters, and route or site of PACAP administration, especially after pain initiation in the various experimental models. Under specific circumstances, PACAP may have activated autoregulatory or descending inhibitory pathways or stimulated anti-inflammatory responses by blocking immune cell cytokine release into the peripheral milieu of pain mediators to produce anti-nociceptive effects. Based on PACAP and PAC1 receptor expression and distribution in the sensory pathways, and the preponderance of electrophysiological and behavioral data, however, the central effects of PACAP in injury appear to result in system sensitization and are pro-nociceptive.

Nociception Studies in PACAP/PAC1 Receptor Knockout Mice

The most convincing evidence for PACAP involvement in pain stems from studies using transgenic PACAP (PACAP–/–) and PAC1 receptor (PAC1R–/–) knockout mice which have been coherent in demonstrating the facilitatory roles of PACAP signaling in chronic pain. PACAP–/– mice display a range of physiological and neuropsychiatric phenotypes including decreased locomotor activities, decreased feeding behaviors, altered memory performance, and attenuated stress responses reflecting the multifaceted roles of PACAP [47–49]. In several experimental models, PACAP–/– mice exhibit important deficits in neuropathic pain development. Under control conditions, naïve PACAP–/– mice showed unaltered or slight decreases in sensitivity responses to thermal or mechanical stimuli [31, 50, 51]. However, in chronic pain models with intraplantar noxious stimuli,

PACAP−/− mice displayed a marked loss in the induction of mechanical and thermal hypersensitivity, and nocifensive behaviors [31, 51]. Similarly, PACAP−/− mice failed to develop thermal or mechanical hypersensitivity in response to spinal nerve transection or sciatic nerve ligation, and demonstrated substantially attenuated writhing activities in response to intraperitoneal acetic acid injection [31, 51, 52]. The diminished nociceptive responses in the PACAP−/− mice to either formalin or acetic acid treatments were accompanied by decreased c-Fos expression in the somatosensory cortex and periaqueductal grey (PAG), indicating a tangible diminution in nociceptive transmission rather than an absence of behavioral responses [51].

PACAP activation of multiple different receptor subtypes and PAC1 receptor-mediated intracellular signaling appear central to nociceptive mechanisms. This was implied as PACAP nociceptive responses were recapitulated with PAC1 receptor selective agonist maxadilan and blocked by the specific receptor antagonist max.d.4 [46]. Accordingly, as in PACAP−/− animals, mice with PAC1 receptor deficiency (PAC1R−/−) under naïve conditions also exhibited normal responses to acute thermal or mechanical stimuli, but demonstrated reduced nocifensive responses to intraplantar formalin administration and decreased abdominal responses to intraperitoneal acetic acid injection [53, 54]. The knockout studies conducted to date have not addressed the different potential sites of PACAP/PAC1 receptor action mediating the nociceptive responses; however, $PAC1^{CamKCre2}$ mice with forebrain-specific deletions of the PAC1 receptor (PAC1 receptor deletions in the forebrain cortical areas, hippocampus and olfactory bulb) did not demonstrate diminished chemical and visceral pain responses [53]. Thus, the nociceptive actions of PACAP likely resided within the peripheral pathways, spinal cord and brainstem, or possibly in combination of these regions [53]. The PACAP knockout studies do not exclude possible roles for VIP or PACAP on VPAC1/VPAC2 receptor signaling in pain responses as VIP is an important mediator of inflammatory processes, and VIP administration is often potently anti-inflammatory [55]. Nevertheless, these studies in aggregate implicated PACAP and PAC1 receptor involvement in the development of nociceptive hypersensitivity across several modalities of chronic pain.

PACAP Sensory Pathways PACAP-EGFP Mice

The recent availability of the PACAP-EGFP mouse, in which neuronal EGFP expression is under the control of the PACAP promoter, has enabled a unique means of examining PACAP expression and plasticity in sensory pathways. Under basal conditions, little native EFGP fluorescence was observed in peripheral DRG neurons and fibers which may parallel the inherent low levels of PACAP expression. However, in recapitulation of prior immunocytochemistry and in situ hybridization transcript studies, chronic constriction injury (CCI) and persistent neuropathic pain elicited by unilateral partial sciatic nerve ligation in the PACAP-EGFP mice dramatically induced PACAP-EGFP neuronal expression in the ipsilateral lumbar DRGs with concomitant increases in EGFP fluorescence in both peripheral sensory fibers in the sciatic nerve proximal to the ligation and central axons projecting into the dorsal horn (see Chap. 33). The contralateral DRGs and fibers were devoid or demonstrated little basal EGFP signals. Notably, central axons projecting to the deeper laminae II–IV of the dorsal horn and gracile fasciculus exhibited induced PACAP-EGFP fluorescence, which appeared consistent with the enhanced PACAP staining patterns in the spinal cord following sciatic nerve transection [22]. The fibers were typically large caliber A(delta) mechanosensory (touch) neurons, but how these contribute to chronic pain signaling is still unclear. The second order DRG sensory neurons have multiple ascending projections including the spinoparabrachial tract which relays nociceptive signals to the lateral parabrachial nucleus (LPBn) for subsequent upstream signaling to forebrain targets such as the nociceptive lateral capsular division of the central amygdala (CeLC, Fig. 37.1) [8, 56–58]. Hence CCI increases PACAP expression in several neural elements in sensory tracts, including elements within the spino-parabrachioamygdaloid pathway impacting the behavioral consequences of chronic pain.

Central PACAP Signaling at the Intersection of Stress and Chronic Pain

Altered PACAP/PAC1 receptor expression and signaling participate in stress-related psychopathologies including posttraumatic stress disorder (PTSD), and given the high comorbidity of stress- and pain-related disorders, PACAP could be part of the common mechanistic underpinnings linking these disorders [17, 18, 59]. To date, most studies have emphasized PACAP roles in pain signaling at the peripheral nociceptor and spinal cord. From expression patterns and function in the nociceptive tracts described above, the PACAPergic systems appear contributory to the maladaptive neuroplasticity that can facilitate the adverse emotional consequences of chronic pain.

The CeLC is often termed the nociceptive amygdala and recent studies have identified dense PACAP-immunoreactive fibers in this nucleus (Fig. 37.2) [60]. From anterograde tracing and excitotoxic lesion studies, the CeLC PACAP fibers were shown to represent axonal projections from the LPBn; notably, the bed nucleus of the stria terminalis (BNST) also received significant PACAP innervation from the LPBn (Fig. 37.3). As the LPBn receives sensory projections from the dorsal horn, these results implicated PACAP nociceptive signaling to the amygdala via the spino-parabrachioamygdaloid tract impacting stress- and pain-related behaviors. In congruence, CCI induced LPBn PACAP transcript levels and increased transgenic mice PACAP-EGFP+ cell number, coinciding with increased PACAP fiber density in the CeLC (data not shown). To assess the roles of PACAP signaling in the CeLC, direct PACAP infusion into the CeA of rats resulted in both heightened emotional behaviors shown by increased anxiety-like behavior on the elevated plus maze, and amplified pain-related responses as indicated by thermal hypersensitivity (Fig. 37.4) [60]. The thermal hypersensitivity responses were recapitulated following CeA infusions

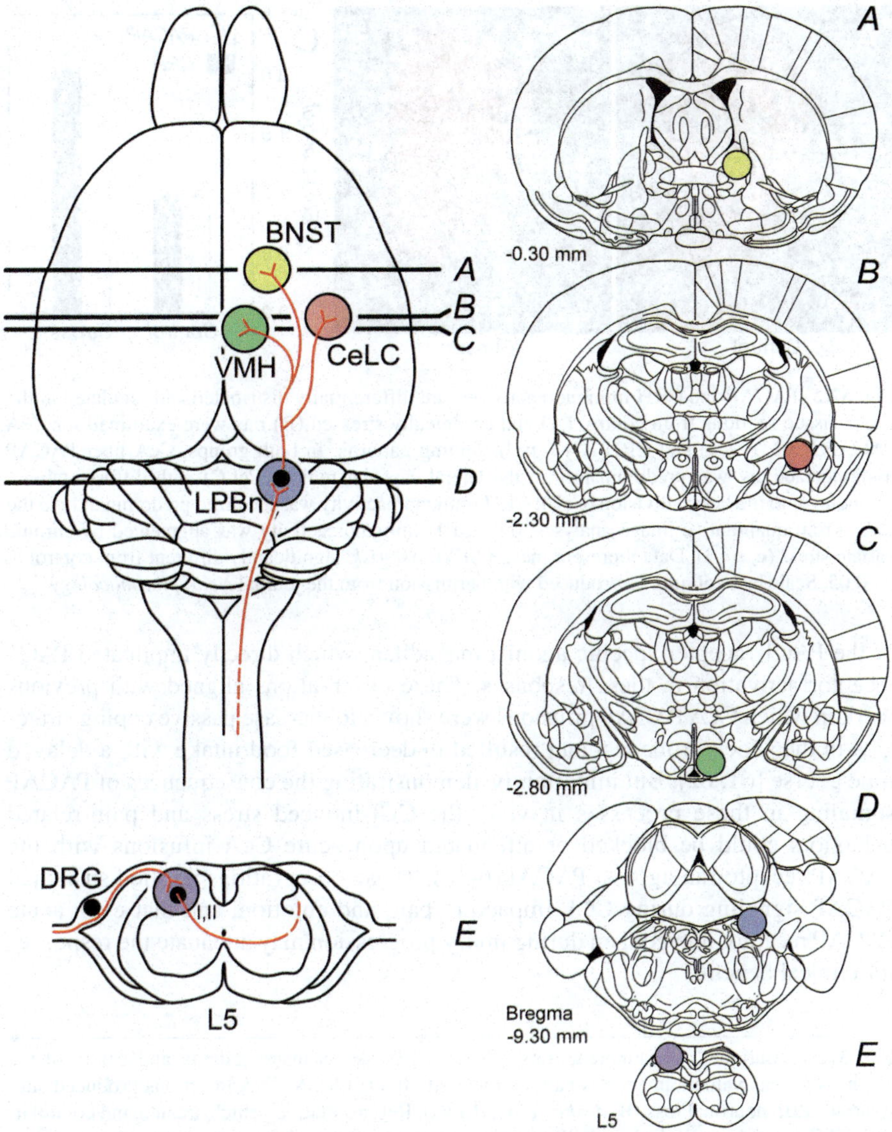

Fig. 37.1 Schematic of PACAP in the spino-parabrachioamygdaloid tract. *Lower left* (**E**), representative L5 spinal cord segment illustrating dorsal root ganglion (DRG) PACAP neurons projecting to the dorsal horn, where second order neuronal fibers from laminae I/II travel in the dorsolateral fasciculus to target the LPBn in the brainstem. The third order LPBn PACAP neurons project to primary targets in the lateral capsular division of the central amygdala (CeLC, nociceptive amygdala), bed nucleus of the stria terminalis (BNST), and ventromedial hypothalamus (VMH). Rat brain levels for the BNST (**A**), CeLC (**B**), VMH (**C**), and LPBn (**D**) are shown with the corresponding coronal sections on the *right*. Numbers below each coronal denote distance from Bregma. *Black dots*, neuronal soma. *Colored circles*, terminal fields in rat brain levels and corresponding coronal section. These PACAP circuits provide means for chronic pain impact emotional behavior and feeding. It is unclear whether the second order fibers are also PACAPergic; the PACAP primary afferents also travel in the gracile fasciculus of the spinal cord

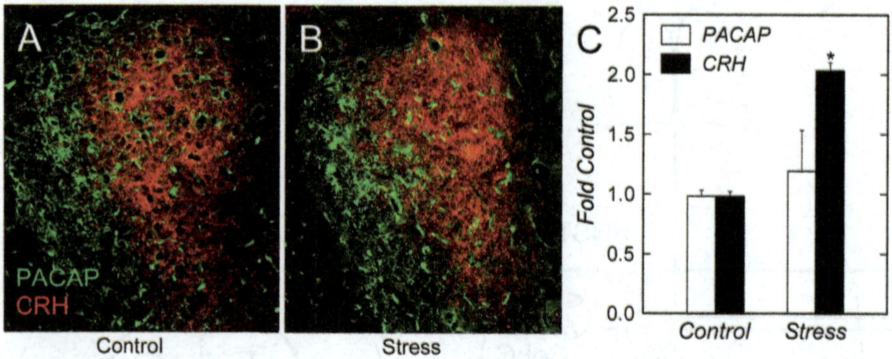

Fig. 37.2 PACAP and CRH immunoreactivities are differentially distributed and regulated in the CeA. Tissue sections from control (**A**) and chronically stressed (**B**) rats were examined for CeA PACAP (Cy2, *green*) and CRH (Cy3, *red*) staining patterns. In both groups, CeA fiber PACAP immunoreactivity was predominantly in the lateral capsular region (CeLC) with diffuse staining extending into the lateral division (CeL); CRH immunoreactivity was localized predominantly to the CeL. From quantitative image analyses, only CRH immunoreactivity was augmented by chronic variate stress (**c**, $n = 3$). Data represent mean ± SEM. *Asterisk*, significantly different from control at $p < 0.05$. Scale bar, 250 μm. Reproduced with permission from the journal *Neuropharmacology*

of the PAC1 receptor specific agonist maxadilan, which directly implicated PAC1 receptor activation in these responses. These observations aligned with previous work in which CeA PACAP infusions were shown to increase passive coping strategies in the shock probe test and resulted in decreased food intake with a delayed time course [61, 62]. But importantly, demonstrating the consequences of PACAP signaling in these processes in vivo, the CCI-induced stress and pain-related behaviors could be blocked or attenuated upon acute CeA infusions with the PACAP receptor antagonist PACAP(6-38). These observations implied sustained PACAP signaling during CCI, impacting pain and emotion, and that even acute PACAP receptor antagonism during injury progression may attenuate the responses for clinical benefits.

Fig. 37.3 (continued) immunoreactivity (Cy3, *red*) to assess the specificity and extent of the lesion. Whereas vehicle injections had no apparent effects (**A**), NMDA injections produced substantial LPBn neuronal loss (**B**, *dashed circled* area). Representative vehicle treated and contralateral NMDA excitotoxic lesioned PBn in the same animal are shown; the lesioned image was flipped to facilitate comparison. CeA and BNST tissue sections from the NMDA excitotoxic lesioned animals were processed for dual PACAP and CGRP immunocytochemical localizations. CGRP has roles in stress and pain, and is prominently expressed in the PBn. Tissue sections ipsilateral to LPBn—vehicle injections (*left panels*) demonstrated substantial PACAP (Cy3, *red*) and CGRP (AlexaFluor 488, *green*) colocalization in the CeLC (**C**, **E**) and BNST (**G**, **I**); colocalization in merged micrographs illustrated in *yellow*. By contrast, the same CeLC and BNST regions in the contralateral half that received LPBn NMDA excitotoxic lesion (PBn—lesion) demonstrated marked decreases in both PACAP and CGRP immunoreactivities (**D**, **F**, **H**, **J**). Again, micrographs from the stained CeLC and BNST regions from the PBn—lesioned side was flipped for comparisons with the control vehicle—injected side from the same animals to facilitate comparisons. *scp* superior cerebellar peduncle, *CPu* caudate putamen. Representative figures from three separate animals. Scale bar, 200 μm in corresponding tissues. Reproduced with permission from the journal *Neuropharmacology*

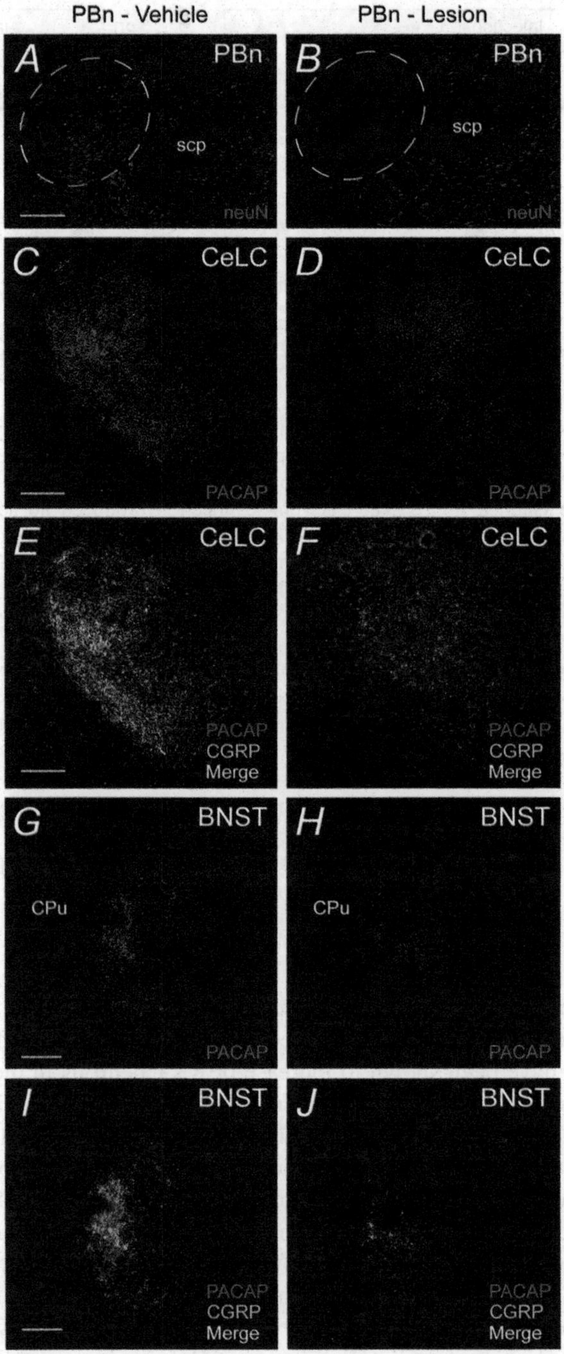

Fig. 37.3 Excitotoxic LPBn lesions diminish PACAP and CGRP fiber immunoreactivities in the CeLC and BNST. The LPBn was unilaterally lesioned with NMDA; the contralateral LPBn received vehicle. After 7 days, the PBn sections were processed for neuron-specific nuclear NeuN

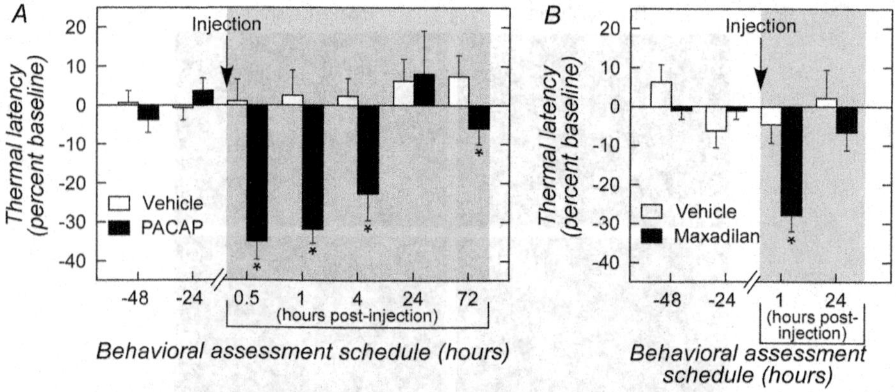

Fig. 37.4 CeA PACAP/PAC1 receptor signaling increases thermal sensitivity. (**A**) Rats were habituated in Hargreave's thermal sensitivity apparatus with 2 days of baseline assessments (24 and 48 h). PACAP was subsequently infused into the CeA (single injection) for thermal testing at the indicated time (*shaded area*). Whereas vehicle injection produced no apparent responses changes compared to baseline (*white bars*), CeA PACAP infusions consistently decreased thermal latency responses (*black bars*) up to 4 h post-treatment. The responses dissipated by 24 h; the small but significant decrease in thermal latency at 72 h may reflect latent plasticity events. $n=6$–8 per group, data represent mean response ± SEM, *, different from corresponding vehicle control, $p<0.025$. (**B**) The PACAP-induced decrease in thermal latency was mirrored in CeA infusions with the PAC1 receptor specific agonist maxadilan. The maxadilan responses observed at 1 h was again dissipated by 24 h. $n=7$–8 per group, data represent mean response ± SEM, *, different from corresponding vehicle control, $p<0.002$. Reproduced with permission from the journal *Neuropharmacology*

Nociceptive PACAP/PAC1 Receptor Signaling Through ERK Activation

A signature of nociceptive signaling is extracellular signaling regulated kinase (ERK) activation which participates in the neuroplasticity that promotes the manifestation of chronic pain and stress-related disorders [63]. Both inflammation and axotomy injury have been shown to increase pERK+ neurons in the DRG; during inflammatory or neuropathic pain, there are increased pERK levels in lamina I and II neurons of the spinal cord and the ensuing hypersensitivity development can be abrogated upon blockade of ERK phosphorylation by intrathecal application of an immediate upstream mitogen-activated protein kinase kinase (MEK) inhibitor [64, 65]. As in other chronic pain models, CCI can increase the number of pERK+ neurons in the CeLC. ERK signaling has been shown to contribute to pain-related enhancement of PBn-CeLC synaptic neurotransmission and similarly, inhibition of CeA ERK activation attenuates pain-related behavioral hypersensitivity [63, 66, 67]. PACAP and PAC1 receptor signaling can stimulate and sustain ERK activation potently and efficaciously [33, 68] and CeA PACAP infusions recapitulated the

ERK activation and stress/pain-related behaviors associated with CCI, with concomitant increases in the neuronal activity marker c-Fos. The relationship between PACAP and pERK in nociception was enhanced further by the close apposition of PACAP-immunoreactive fibers with CeA pERK+ neurons in CCI, and the abilities of MEK inhibitors to attenuate PACAP-mediated ERK activation and thermal nociceptive sensitivity. There are multiple intracellular PAC1 receptor effector mechanisms that activate ERK including PKA and PKC [33, 69, 70], but more recently, PAC1 receptor internalization and endosomal signaling have been suggested to be a means to sustain cellular ERK levels [33, 34]. In accord, CeA pretreatment with the endocytosis inhibitor Pitstop 2 prevented PACAP-induced ERK phosphorylation and thermal hypersensitivity. These results may be examples of how G protein-coupled receptor (GPCR) internalization can produce divergent, long term and targeted signaling for physiological function and may not always be associated with degradation and desensitization. In contrast to the transient effects of other sensory peptides, PAC1 receptor endosomal ERK signaling may be one mechanism underlying the gradual but sustained actions of PACAP in pain.

PACAP and Glutamate Signaling

In addition to stimulation of ERK-mediated neuroplasticity, PACAP signaling may also regulate postsynaptic neuronal function by modulating glutamatergic neurotransmission. PACAP is coexpressed with glutamate in a variety of systems including retina, suprachiasmatic nucleus, central amygdala, and BNST [71, 72]. Furthermore, in the developing dorsal horn of the spinal cord, the same transcription factors that determine glutamatergic cell fate also appear to control PACAP expression [73]. The co-release of PACAP with glutamate may function to modulate excitatory neurotransmission, as NMDA receptor blockade in the ventromedial hypothalamus (VMH) diminished PACAP-induced hypophagia [74]. The attenuation of fear conditioning by intra-BLA PACAP(6-38) administration was mediated through altered NMDA signaling [75] and intrathecal PACAP-mediated pain resulted in a dose-dependent enhancement of NMDA-induced aversive behaviors and potentiated NMDA currents in dorsal horn neurons [41]. In addition, transgenic PACAP−/− mice failed to develop mechanical allodynia to NMDA, but allodynia could be restored by co-infusion of PACAP with NMDA [31]. There are multiple mechanisms by which PACAP could potentially modulate glutamatergic signaling. In the dorsal horn, there is evidence that PACAP may promote the functional coupling of nitric oxide synthase to NMDA receptors [31]. In the hippocampus, PACAP was found to enhance synaptic NMDA receptor trafficking and surface expression through Gq, PKC, and Src signaling mechanisms [76–78]. In the amygdala, PACAP resulted in potentiation of the BLA-CeA transmission through a postsynaptic mechanism involving synaptic targeting of GluR1 subunit-containing AMPA receptors [79]. Alternatively, PACAP may enhance glutamate signaling by regulating mGluR function [80].

Conclusions

Pain is a multidimensional experience comprising both somatosensory and affective components. There are a number of lines of evidence indicating a role for PACAP in pain, including expression within nociceptive pathways, upregulation following injury or noxious stimulation, lack of pain development in genetic deletion models, and the ability to induce pain-related behaviors upon CeA peptide administration. While the role of PACAP in primary sensory neurons in pain has been established, recent evidence suggests that this role may extend centrally along nociceptive pathways including parabrachial nucleus to amygdala projections to mediate the emotional components of pain.

Acknowledgments This work was supported by grants MH-97988 (SEH/VM), MH-080935 (SEH), MH-096764 (VM), DK-051369, DK-060481 and DK-065989 (MAV), NS-23978 and HL-65481 (RLP), and funds from the University of Vermont (UVM) Center of Biomedical Research Excellence (Neuroscience COBRE, NCRR P30 RR032135/NIGMS P30 GM103498).

References

1. Miyata A, Arimura A, Dahl RR, Minamino N, Uehara A, Jiang L, et al. Isolation of a novel 38 residue-hypothalamic polypeptide which stimulates adenylate cyclase in pituitary cells. Biochem Biophys Res Commun. 1989;16:567–74.
2. Miyata A, Jiang L, Dahl RD, Kitada C, Kubo K, Fujino M, et al. Isolation of a neuropeptide corresponding to the N-terminal 27 residues of the pituitary adenylate cyclase activating polypeptide with 38 residues (PACAP38). Biochem Biophys Res Commun. 1990;31:643–8.
3. Vaudry D, Falluel-Morel A, Bourgault S, Basille M, Burel D, Wurtz O, et al. Pituitary adenylate cyclase-activating polypeptide and its receptors: 20 years after the discovery. Pharmacol Rev. 2009;61:283–357.
4. Harmar AJ, Fahrenkrug J, Gozes I, Laburthe M, May V, Pisegna JR, et al. Pharmacology and functions of receptors for vasoactive intestinal peptide and pituitary adenylate cyclase-activating polypeptide: IUPHAR review 1. Br J Pharmacol. 2012;166:4–17.
5. Blechman J, Levkowitz G. Alternative splicing of the pituitary adenylate cyclase-activating polypeptide receptor PAC1: mechanisms of fine tuning of brain activity. Front Endocrinol. 2013;4:55.
6. Spengler D, Waeber C, Pantaloni C, Holsboer F, Bockaert J, Seeburg PH, et al. Differential signal transduction by five splice variants of the PACAP receptor. Nature. 1993;365:170–5.
7. Braas KM, May V. Pituitary adenylate cyclase-activating polypeptides directly stimulate sympathetic neuron neuropeptide Y release through PAC(1) receptor isoform activation of specific intracellular signaling pathways. J Biol Chem. 1999;274:27702–10.
8. Gauriau C, Bernard JF. Pain pathways and parabrachial circuits in the rat. Exp Physiol. 2002;87:251–8.
9. Basbaum AI, Bautista DM, Scherrer G, Julius D. Cellular and molecular mechanisms of pain. Cell. 2009;139:267–84.
10. von Hehn CA, Baron R, Woolf CJ. Deconstructing the neuropathic pain phenotype to reveal neural mechanisms. Neuron. 2012;73:638–52.
11. Todd AJ. Neuronal circuitry for pain processing in the dorsal horn. Nat Rev Neurosci. 2010;11:823–36.

12. Ben-Shahar Y. Sensory functions for degenerin/epithelial sodium channels (DEG/ENaC). Adv Genet. 2011;76:1–26.
13. Julius D. TRP channels and pain. Annu Rev Cell Dev Biol. 2013;29:355–84.
14. Ranade SS, Woo SH, Dubin AE, Moshourab RA, Wetzel C, Petrus M, et al. Piezo2 is the major transducer of mechanical forces for touch sensation in mice. Nature. 2014;516:121–5.
15. Tracey I, Mantyh PW. The cerebral signature for pain perception and its modulation. Neuron. 2007;55:377–91.
16. McEwen BS, Kalia M. The role of corticosteroids and stress in chronic pain conditions. Metabolism. 2010;59 Suppl 1:S9–15.
17. Scioli-Salter ER, Forman DE, Otis JD, Gregor K, Valovski I, Rasmusson AM. The shared neuroanatomy and neurobiology of comorbid chronic pain and PTSD: therapeutic implications. Clin J Pain. 2015;31:363–74.
18. Asmundson GJ, Katz J. Understanding the co-occurrence of anxiety disorders and chronic pain: state-of-the-art. Depress Anxiety. 2009;26:888–901.
19. Mulder H, Uddman R, Moller K, Zhang YZ, Ekblad E, Alumets J, et al. Pituitary adenylate cyclase activating polypeptide expression in sensory neurons. Neuroscience. 1994;63:307–12.
20. Moller K, Zhang YZ, Hakanson R, Luts A, Sjolund B, Uddman R, et al. Pituitary adenylate cyclase activating peptide is a sensory neuropeptide: immunocytochemical and immunochemical evidence. Neuroscience. 1993;57:725–32.
21. Usoskin D, Furlan A, Islam S, Abdo H, Lonnerberg P, Lou D, et al. Unbiased classification of sensory neuron types by large-scale single-cell RNA sequencing. Nat Neurosci. 2015;18:145–53.
22. Jongsma H, Danielsen N, Sundler F, Kanje M. Alteration of PACAP distribution and PACAP receptor binding in the rat sensory nervous system following sciatic nerve transection. Brain Res. 2000;853:186–96.
23. Zhang YZ, Hannibal J, Zhao Q, Moller K, Danielsen N, Fahrenkrug J, et al. Pituitary adenylate cyclase activating peptide expression in the rat dorsal root ganglia: up-regulation after peripheral nerve injury. Neuroscience. 1996;74:1099–110.
24. Pettersson LM, Heine T, Verge VM, Sundler F, Danielsen N. PACAP mRNA is expressed in rat spinal cord neurons. J Comp Neurol. 2004;471:85–96.
25. Beaudet MM, Braas KM, May V. Pituitary adenylate cyclase activating polypeptide (PACAP) expression in sympathetic preganglionic projection neurons to the superior cervical ganglion. J Neurobiol. 1998;36:325–36.
26. Nagahama M, Tsuzuki M, Mochizuki T, Iguchi K, Kuwahara A. Light and electron microscopic studies of pituitary adenylate cyclase-activating peptide (PACAP)--immunoreactive neurons in the enteric nervous system of rat small and large intestine. Anat Embryol. 1998;198:341–52.
27. Zhang Y, Danielsen N, Sundler F, Mulder H. Pituitary adenylate cyclase-activating peptide is upregulated in sensory neurons by inflammation. Neuroreport. 1998;9:2833–6.
28. Zhang Q, Shi TJ, Ji RR, Zhang YZ, Sundler F, Hannibal J, et al. Expression of pituitary adenylate cyclase-activating polypeptide in dorsal root ganglia following axotomy: time course and coexistence. Brain Res. 1995;705:149–58.
29. Jongsma Wallin H, Pettersson LM, Verge VM, Danielsen N. Effect of anti-nerve growth factor treatment on pituitary adenylate cyclase activating polypeptide expression in adult sensory neurons exposed to adjuvant induced inflammation. Neuroscience. 2003;120:325–31.
30. Pettersson LM, Dahlin LB, Danielsen N. Changes in expression of PACAP in rat sensory neurons in response to sciatic nerve compression. Eur J Neurosci. 2004;20:1838–48.
31. Mabuchi T, Shintani N, Matsumura S, Okuda-Ashitaka E, Hashimoto H, Muratani T, et al. Pituitary adenylate cyclase-activating polypeptide is required for the development of spinal sensitization and induction of neuropathic pain. J Neurosci. 2004;24:7283–91.
32. Vizzard MA. Up-regulation of pituitary adenylate cyclase-activating polypeptide in urinary bladder pathways after chronic cystitis. J Comp Neurol. 2000;420:335–48.

33. May V, Buttolph TR, Girard BM, Clason TA, Parsons RL. PACAP-induced ERK activation in HEK cells expressing PAC1 receptors involves both receptor internalization and PKC signaling. Am J Physiol Cell Physiol. 2014;306:C1068–79.
34. Merriam LA, Baran CN, Girard BM, Hardwick JC, May V, Parsons RL. Pituitary adenylate cyclase 1 receptor internalization and endosomal signaling mediate the pituitary adenylate cyclase activating polypeptide-induced increase in guinea pig cardiac neuron excitability. J Neurosci. 2013;33:4614–22.
35. Dickinson T, Fleetwood-Walker SM. VIP and PACAP: very important in pain? Trends Pharmacol Sci. 1999;20:324–9.
36. Sandor K, Bölcskei K, McDougall JJ, Schuelert N, Reglodi D, Elekes K, et al. Divergent peripheral effects of pituitary adenylate cyclase-activating polypeptide-38 on nociception in rats and mice. Pain. 2009;141:143–50.
37. Zhang YZ, Sjolund B, Moller K, Hakanson R, Sundler F. Pituitary adenylate cyclase activating peptide produces a marked and long-lasting depression of a C-fibre-evoked flexion reflex. Neuroscience. 1993;57:733–7.
38. Yamamoto T, Tatsuno I. Antinociceptive effect of intrathecally administered pituitary adenylate cyclase activating polypeptide (PACAP) on the rat formalin test. Neurosci Lett. 1995;184:32–5.
39. Zhang Y, Malmberg AB, Sjolund B, Yaksh TL. The effect of pituitary adenylate cyclase activating peptide (PACAP) on the nociceptive formalin test. Neurosci Lett. 1996;207:187–90.
40. Shimizu T, Katahira M, Sugawara H, Inoue K, Miyata A. Diverse effects of intrathecal pituitary adenylate cyclase-activating polypeptide on nociceptive transmission in mice spinal cord. Regul Pept. 2004;123:117–22.
41. Ohsawa M, Brailoiu GC, Shiraki M, Dun NJ, Paul K, Tseng LF. Modulation of nociceptive transmission by pituitary adenylate cyclase activating polypeptide in the spinal cord of the mouse. Pain. 2002;100:27–34.
42. Narita M, Dun SL, Dun NJ, Tseng LF. Hyperalgesia induced by pituitary adenylate cyclase-activating polypeptide in the mouse spinal cord. Eur J Pharmacol. 1996;311:121–6.
43. Dickinson T, Fleetwood-Walker SM, Mitchell R, Lutz EM. Evidence for roles of vasoactive intestinal polypeptide (VIP) and pituitary adenylate cyclase activating polypeptide (PACAP) receptors in modulating the responses of rat dorsal horn neurons to sensory inputs. Neuropeptides. 1997;31:175–85.
44. Davis-Taber R, Baker S, Lehto SG, Zhong C, Surowy CS, Faltynek CR, et al. Central pituitary adenylate cyclase 1 receptors modulate nociceptive behaviors in both inflammatory and neuropathic pain states. J Pain. 2008;9:449–56.
45. Xu XJ, Wiesenfeld-Hallin Z. Intrathecal pituitary adenylate cyclase activating polypeptide facilitates the spinal nociceptive flexor reflex in the rat. Neuroscience. 1996;72:801–4.
46. Sakashita Y, Kurihara T, Uchida D, Tatsuno I, Yamamoto T. Involvement of PACAP receptor in primary afferent fibre-evoked responses of ventral roots in the neonatal rat spinal cord. Br J Pharmacol. 2001;132:1769–76.
47. Girard BA, Lelievre V, Braas KM, Razinia T, Vizzard MA, Ioffe Y, et al. Noncompensation in peptide/receptor gene expression and distinct behavioral phenotypes in VIP- and PACAP-deficient mice. J Neurochem. 2006;99:499–513.
48. Hattori S, Takao K, Tanda K, Toyama K, Shintani N, Baba A, et al. Comprehensive behavioral analysis of pituitary adenylate cyclase-activating polypeptide (PACAP) knockout mice. Front Behav Neurosci. 2012;6:58.
49. Hashimoto H, Shintani N, Tanaka K, Mori W, Hirose M, Matsuda T, et al. Altered psychomotor behaviors in mice lacking pituitary adenylate cyclase-activating polypeptide (PACAP). Proc Natl Acad Sci U S A. 2001;98:13355–60.
50. May V, Vizzard MA. Bladder dysfunction and altered somatic sensitivity in PACAP-/- mice. J Urol. 2010;183:772–9.
51. Sandor K, Kormos V, Botz B, Imreh A, Bolcskei K, Gaszner B, et al. Impaired nocifensive behaviours and mechanical hyperalgesia, but enhanced thermal allodynia in pituitary adenylate cyclase-activating polypeptide deficient mice. Neuropeptides. 2010;44:363–71.

52. Botz B, Imreh A, Sandor K, Elekes K, Szolcsanyi J, Reglodi D, et al. Role of pituitary adenylate-cyclase activating polypeptide and Tac1 gene derived tachykinins in sensory, motor and vascular functions under normal and neuropathic conditions. Peptides. 2013;43:105–12.
53. Martin M, Otto C, Santamarta MT, Torrecilla M, Pineda J, Schutz G, et al. Morphine withdrawal is modified in pituitary adenylate cyclase-activating polypeptide type I-receptor-deficient mice. Mol Brain Res. 2003;110:109–18.
54. Jongsma H, Pettersson LM, Zhang YZ, Reimer MK, Kanje M, Waldenstrom A, et al. Markedly reduced chronic nociceptive response in mice lacking the PAC1 receptor. Neuroreport. 2001;12:2215–9.
55. Delgado M, Pozo D, Ganea D. The significance of vasoactive intestinal peptide in immunomodulation. Pharmacol Rev. 2004;56:249–90.
56. Veinante P, Yalcin I, Barrot M. The amygdala between sensation and affect: a role in pain. J Mol Psychiatry. 2013;1:9.
57. Rouwette T, Vanelderen P, Roubos EW, Kozicz T, Vissers K. The amygdala, a relay station for switching on and off pain. Eur J Pain. 2012;16:782–92.
58. Bernard JF, Bester H, Besson JM. Involvement of the spino-parabrachio-amygdaloid and -hypothalamic pathways in the autonomic and affective emotional aspects of pain. Prog Brain Res. 1996;107:243–55.
59. Ressler KJ, Mercer KB, Bradley B, Jovanovic T, Mahan A, Kerley K, et al. Post-traumatic stress disorder is associated with PACAP and the PAC1 receptor. Nature. 2011;470:492–7.
60. Missig G, Roman CW, Vizzard MA, Braas KM, Hammack SE, May V. Parabrachial nucleus (PBn) pituitary adenylate cyclase activating polypeptide (PACAP) signaling in the amygdala: implication for the sensory and behavioral effects of pain. Neuropharmacology. 2014;86:38–48.
61. Legradi G, Das M, Giunta B, Hirani K, Mitchell EA, Diamond DM. Microinfusion of pituitary adenylate cyclase-activating polypeptide into the central nucleus of amygdala of the rat produces a shift from an active to passive mode of coping in the shock-probe fear/defensive burying test. Neural Plast. 2007;2007:79102.
62. Iemolo A, Ferragud A, Cottone P, Sabino V. Pituitary adenylate cyclase-activating peptide in the central amygdala causes anorexia and body weight loss via the melanocortin and the TrkB systems. Neuropsychopharmacology. 2015;40:1846–55.
63. Ji RR, Gereau 4th RW, Malcangio M, Strichartz GR. MAP kinase and pain. Brain Res Rev. 2009;60:135–48.
64. Ji RR, Baba H, Brenner GJ, Woolf CJ. Nociceptive-specific activation of ERK in spinal neurons contributes to pain hypersensitivity. Nat Neurosci. 1999;2:1114–9.
65. Obata K, Yamanaka H, Dai Y, Tachibana T, Fukuoka T, Tokunaga A, et al. Differential activation of extracellular signal-regulated protein kinase in primary afferent neurons regulates brain-derived neurotrophic factor expression after peripheral inflammation and nerve injury. J Neurosci. 2003;23:4117–26.
66. Carrasquillo Y, Gereau RW. Activation of the extracellular signal-regulated kinase in the amygdala modulates pain perception. J Neurosci. 2007;27:1543–51.
67. Cheng SJ, Chen CC, Yang HW, Chang YT, Bai SW, Chen CC, et al. Role of extracellular signal-regulated kinase in synaptic transmission and plasticity of a nociceptive input on capsular central amygdaloid neurons in normal and acid-induced muscle pain mice. J Neurosci. 2011;31:2258–70.
68. May V, Lutz E, MacKenzie C, Schutz KC, Dozark K, Braas KM. Pituitary adenylate cyclase-activating polypeptide (PACAP)/PAC1HOP1 receptor activation coordinates multiple neurotrophic signaling pathways: Akt activation through phosphatidylinositol 3-kinase and vesicle endocytosis for neuronal survival. J Biol Chem. 2010;285:9749–61.
69. Bouschet T, Perez V, Fernandez C, Bockaert J, Eychene A, Journot L. Stimulation of the ERK pathway by GTP-loaded Rap1 requires the concomitant activation of Ras, protein kinase C, and protein kinase A in neuronal cells. J Biol Chem. 2003;278:4778–85.
70. Barrie AP, Clohessy AM, Buensuceso CS, Rogers MV, Allen JM. Pituitary adenylyl cyclase-activating peptide stimulates extracellular signal-regulated kinase 1 or 2 (ERK1/2) activity in

a Ras-independent, mitogen-activated protein kinase/ERK kinase 1 or -dependent manner in PC12 cells. J Biol Chem. 1997;272:19666–71.

71. Engelund A, Fahrenkrug J, Harrison A, Hannibal J. Vesicular glutamate transporter 2 (VGLUT2) is co-stored with PACAP in projections from the rat melanopsin-containing retinal ganglion cells. Cell Tissue Res. 2010;340:243–55.
72. Hannibal J, Moller M, Ottersen OP, Fahrenkrug J. PACAP and glutamate are co-stored in the retinohypothalamic tract. J Comp Neurol. 2000;418:147–55.
73. Guo Z, Zhao C, Huang M, Huang T, Fan M, Xie Z, et al. Tlx1/3 and Ptf1a control the expression of distinct sets of transmitter and peptide receptor genes in the developing dorsal spinal cord. J Neurosci. 2012;32:8509–20.
74. Resch JM, Maunze B, Phillips KA, Choi S. Inhibition of food intake by PACAP in the hypothalamic ventromedial nuclei is mediated by NMDA receptors. Physiol Behav. 2014;133:230–5.
75. Schmidt SD, Myskiw JC, Furini CR, Schmidt BE, Cavalcante LE, Izquierdo I. PACAP modulates the consolidation and extinction of the contextual fear conditioning through NMDA receptors. Neurobiol Learn Mem. 2015;118:120–4.
76. Macdonald DS, Weerapura M, Beazely MA, Martin L, Czerwinski W, Roder JC, et al. Modulation of NMDA receptors by pituitary adenylate cyclase activating peptide in CA1 neurons requires G alpha q, protein kinase C, and activation of Src. J Neurosci. 2005;25:11374–84.
77. Chowdhury D, Marco S, Brooks IM, Zandueta A, Rao Y, Haucke V, et al. Tyrosine phosphorylation regulates the endocytosis and surface expression of GluN3A-containing NMDA receptors. J Neurosci. 2013;33:4151–64.
78. Trepanier CH, Jackson MF, MacDonald JF. Regulation of NMDA receptors by the tyrosine kinase Fyn. FEBS J. 2012;279:12–9.
79. Cho JH, Zushida K, Shumyatsky GP, Carlezon WAJ, Meloni EG, Bolshakov VY. Pituitary adenylate cyclase-activating polypeptide induces postsynaptically expressed potentiation in the intra-amygdala circuit. J Neurosci. 2012;32:14165–77.
80. Kammermeier PJ. Endogenous homer proteins regulate metabotropic glutamate receptor signaling in neurons. J Neurosci. 2008;28:8560–7.

Part XII
Role of PACAP in Inflammatory Processes

Chapter 38
PACAP Modulation of CNS and Peripheral Inflammation

Yukio Ago, Michael C. Condro, Abha K. Rajbhandari, Christina Van, Bhavaani Jayaram, Victor May, and James A. Waschek

Abstract Initially defined as a defense against infection, the immune system is known to function in injury and repair and in age-related neurodegenerative diseases. Because of the capacity of immune cells to execute highly aggressive responses, precise control mechanisms are required to prevent collateral damage to healthy cells and to facilitate repair. PACAP is a neuropeptide that has evolved to function in these latter capacities and appears to do so at multiple levels. While PACAP is known to act directly on peripheral immune cells to modify their activities in vitro and to strongly modify inflammation and clinical symptomatology in rodent models of inflammatory diseases (see Chap. 7), considerable evidence indicates that PACAP is upregulated and acts within the compromised brain to regulate inflammatory and regenerative functions of astrocytes and microglia. Moreover, PACAP acts as a neurotransmitter/neuromodulator in the hypothalamus and peripheral nervous system and may thereby regulate inflammation in the periphery via its regulation of the hypothalamic pituitary adrenal axis (HPA) and autonomic control of immune cells in lymph nodes and other immune organs. While many details remain to be elucidated, we will describe in this chapter potential neural circuitries and mechanisms by which PACAP might regulate CNS and peripheral inflammation and immune cell function.

Y. Ago
Department of Psychiatry, David Geffen School of Medicine, University of California at Los Angeles, 635 Charles E Young Drive South, NRB 345, Los Angeles, CA 90095, USA

Laboratory of Medicinal Pharmacology, Graduate School of Pharmaceutical Sciences, Osaka University, Osaka 565-0871, Japan

M.C. Condro • A.K. Rajbhandari • C. Van • B. Jayaram • J.A. Waschek (✉)
Department of Psychiatry, David Geffen School of Medicine, University of California at Los Angeles, 635 Charles E Young Drive South, NRB 345, Los Angeles, CA 90095, USA
e-mail: jwaschek@mednet.ucla.edu

V. May
Department of Neurological Sciences, University of Vermont College of Medicine, 149 Beaumont Avenue, Burlington, VT 05405, USA

D. Reglodi, A. Tamas (eds.), *Pituitary Adenylate Cyclase Activating Polypeptide — PACAP*, Current Topics in Neurotoxicity 11,
DOI 10.1007/978-3-319-35135-3_38

Keywords Pituitary adenylyl cyclase activating peptide (PACAP) • Vasoactive intestinal peptide (VIP) • Inflammation • Neuroinflammation • Astrocyte • Microglia • Sympathetic • Autonomic • Stress • Blood–brain barrier (BBB)

Abbreviations

BBB	Blood–brain barrier
BNST	Bed nucleus of stria terminalis
COX-2	Cyclooxygenase-2
CRF	Corticotropin releasing factor
HPA	Hypothalamic pituitary adrenal axis
IFN-γ	Interferon-γ
IL	Interleukin
IML	Intermediolateral
LPS	Lipopolysaccharide
MEK	Mitogen-activated protein kinase kinase
NO	Nitric oxide
PACAP	Pituitary adenylyl cyclase activating peptide
PKA	Protein kinase A
PSNS	Parasympathetic nervous system
PVN	Paraventricular nucleus of hypothalamus
SNpc	Substantia nigra pars compacta
SNS	Sympathetic nervous system
TJ	Tight junctions
TNF-α	Tumor necrosis factor-α
VIP	Vasoactive intestinal peptide

PACAP Regulation of Systemic Inflammation

Introduction

Like the closely related peptide vasoactive intestinal peptide (VIP), PACAP is well known to potently inhibit inflammatory activities of immune cells (see chapter by Delgado et al.) and to exert powerful anti-inflammatory activities when given exogenously to animal models of system inflammatory disease including as collagen-induced arthritis [1] and endotoxic shock [2]. If one presumes that the actions of exogenous PACAP reflect the activities of the endogenous neuropeptide, what are the potential neuronal/hormonal circuits by which PACAP acts? As discussed later, PACAP is abundantly expressed in stress circuitries known to regulate immune function in the periphery. Although still unproven, we propose that PACAP acts as a classical neurotransmitter/neuromodulator within these circuits, and may also be locally released from sympathetic nerve terminals in immune organs such as lymph nodes and spleen.

General Model of CNS Regulation of Systemic Inflammation

The brain regulates nearly all aspects of physiology, so it is not surprising that the brain actively regulates peripheral inflammation. Neural regulation of inflammation, whether it be a response to infection, injury, or chronic inflammatory disease in a peripheral organ, can be viewed in the general brain organization of stress responses (Fig. 38.1). In this model of neural circuitry, the brain responds to different kinds of stress, such as hemodynamic, metabolic, and emotional, using some common pathways. Stress signals in the periphery first reach the CNS by way of ascending neuronal pathways in the spinal cord and brainstem [3]. With respect to inflammatory stress, a well-described example of an ascending pathway is that utilized by peripheral cytokines, which signal to the brain by way of vagal afferents to the nucleus tractus solitarius in the brainstem [4]. In an alternative humoral pathway, cytokines released into the blood may directly enter the brain circumventricular organs (which exhibit a relatively porous blood–brain barrier). Stress signals reaching the hindbrain are transmitted to the hypothalamus either directly or routed indirectly through the cortex or limbic system, where they regulate the

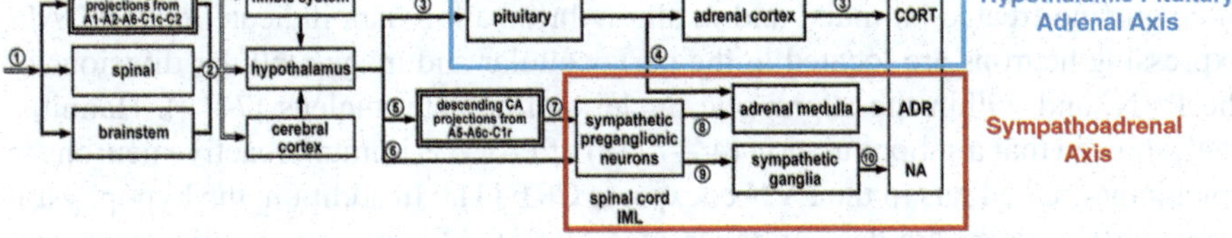

Fig. 38.1 A block diagram for the illustration of the central organization of stress responses (adapted from Kvetnansky et al. [3]) (1) Viscero- and somatosensory stressful stimuli; (2) ascending pathways from spinal and medullary noncatecholamine (CA) and medullary CA neurons to integrative centers in the hypothalamus, limbic system, and the cerebral cortex; (3) hypothalamo–pituitary–adrenal (HPA) axis, ultimately releasing corticosterone (CORT); (4) circulating adrenocorticotropic hormone influences adrenaline (ADR) synthesis in adrenal medullary chromaffin cells; (5) hypothalamic projections to descending CA neurons in the pons-medulla; (6) premotor sympathetic projections from the hypothalamus; (7) premotor sympathetic neurons from the descending CA cell groups to the intermediolateral cell column (IML); (8) sympatho-adrenomedullary system; (9) sympatho-neural system; (10) varicosities in the sympathetic postganglionic nerve terminals. The output of the HPA axis is boxed in *blue*; the sympathoadrenal axis in *red*. The latter includes the sympathetic neurons and the catecholaminergic cells of the adrenal medulla. The catecholaminergic cells of the adrenal medulla are regulated in a manner analogous to sympathetic neurons, i.e., by the equivalent of sympathetic preganglionic innervation

hypothalamic–pituitary–adrenal and the sympathoadrenal axes. These are arguably the most critical systems that respond to systemic homeostatic challenges. The HPA and the adrenal component of the sympathoadrenal axis are well known to regulate inflammation in the periphery, most prominently through the release of cortisol and catecholamines, respectively, while the sympathetic nervous system (SNS) regulates peripheral inflammation by way of its innervation of lymphoid organs, including the thymus, lymph nodes, bone marrow, spleen, and Peyer's patches. Considerable evidence indicates that PACAP is a major regulator of both the HPA and sympathoadrenal axes, and as discussed later, may regulate inflammation in the periphery in this manner.

PACAP Regulation of the HPA Axis and Potential Impact on Inflammation

The HPA axis plays an important role in mediating stress responses via release of the corticotropin releasing factor (CRF) from the hypothalamus and extrahypothalamic regions such as the amygdala. CRF is a well-studied 41-amino acid signaling peptide that is activated in response to highly arousing or stressful situations. CRF acts in the pituitary to release adrenocorticotropic hormone, which acts in the adrenal cortex to release glucocorticoids. The paraventricular nucleus of hypothalamus (PVN) is known to be the principal source of CRF in the brain. Several types of evidence support that PACAP is an important regulator of stress responses, and interacts with, and in some cases, controls CRF expression and activity. First, the hypothalamus is a major site of PACAP synthesis, exhibiting the highest concentration of PACAP in the rat brain [5, 6]. Previous immunohistochemical studies (most in colchicine-treated animals) and in situ hybridization data indicate that PACAP-expressing neurons are located in the parvocellular and magnocellular divisions of the PVN, and within the supraoptic nuclei and arcuate nucleus [7–14]. Hannibal et al. showed that a subpopulation (44–77 %) of PACAP-immunoreactive neurons in colchicine-treated rats in the PVN coexpress CRF [11]. In addition, the hypophysiotropic CRF neurons are densely innervated by PACAP nerve fibers [15]. Receptors for PACAP are also highly expressed in the PVN [16, 17].

Gene knockout and pharmacological studies demonstrate that PACAP and the PACAP-selective PAC1 receptors play critical roles in regulating the HPA axis, and do so at least in part via actions on the CRF system. For example, restraint stress-induced inductions of hypothalamic CRF mRNA and serum corticosterone were shown to be reduced in mice that lack PACAP [18, 19]. Moreover, under conditions of longer term restraint (14–21 days), elevations in corticosterone and weight loss were shown to be reduced in both PACAP- and PAC1 receptor-deficient mice [19]. In agreement, intracerebroventricular injection of PACAP increased CRF mRNA levels, an effect that was blocked by the PACAP receptor antagonist PACAP6-38 [20]. In a different strain of PACAP knockout mice, diminished corticosterone responses to restraint stress, and also to open field exposure were observed [21]. In this latter work, stress-induced inductions in cFos levels in CRF-containing neurons were reduced in PACAP-deficient mice. Overall the data suggest that PACAP acts as an autocrine/paracrine factor on CRF neurons to induce CRF gene expression and enhance and/or prolong the activity of the HPA axis. It follows that this particular PACAP function is also anti-inflammatory due to the increased release of systemic glucocorticoids, which generally act to suppress inflammation.

Evidence suggests that PACAP also acts via extrahypothalamic regions such as the bed nucleus of stria terminalis (BNST), amygdala, and possibly hippocampus to modulate responses to stress. For example, chronic stress induces PACAP and PAC1 receptor gene expression in the BNST [22, 23], where this signaling system has been shown to be important for regulating stress responses [24–27]. Furthermore, PACAP injection into the posterior BNST produces chronic stress responses such as anorexia and weight loss [25], and increased serum corticosterone levels [24]. These effects appear to be mediated by PAC1 receptors [27]. In another model, a 10-shock fear-conditioning protocol significantly induced PAC1 mRNA levels in the amygdala [28]. The effect of microinjection of PACAP into the central nucleus of the amygdala was examined in a shock-probe fear protocol [29]. Time spent withdrawn in the opposite corner of the probe, exploration, locomotion, distance, and velocity was all reduced in PACAP injected shock-challenged, but not shock-unchallenged rats, suggesting an enhanced fear response. PACAP in the hippocampus and/or amygdala also appear to be important for fear learning, because global and forebrain-specific PAC1-knockout mice showed decreased freezing in contextual fear conditioning without effects on other task performances [30, 31].

Interestingly, the bacterial endotoxin lipopolysaccharide (LPS) and interleukin (IL)-1β have been shown to induce PACAP gene expression in neurons of the medial parvocellular part of the PVN [32]. Although the impact of hypothalamic PACAP on the immune system is unclear, PACAP actions in these extrahypothalamic sites in stress responses are potentially transmitted to the HPA and sympathoadrenal axes by the general pathways described in Fig. 38.1.

PACAP in the Autonomic Nervous System Regulation of Inflammation

In addition to its regulation of the HPA axis, PACAP may also influence inflammation through signaling in the autonomic nervous system (reviewed in [33]). This system can influence the immune system primarily through direct innervation and release of neuromodulators in primary lymphoid tissues, such as the bone marrow and thymus, and secondary tissues, such as lymph nodes and the spleen (reviewed in [4, 34, 35]). The autonomic nervous system is typically divided into two pathways: the sympathetic nervous system (SNS), mediated by ganglia in the thoracic spinal segments that provide noradrenergic and peptidergic innervation to target organs, and the parasympathetic nervous system (PSNS), mediated by ganglia within the head and neck and by smaller ganglia located near or within target tissues, and utilizing acetylcholine and other neuroactive molecules. The SNS, rather than the PSNS, is believed to constitute the primary descending pathway of the so-called inflammatory reflex [4]. In this reflex circuit, inflammation is sensed by the peripheral nervous system, and then relayed through autonomic neurons in the hindbrain and spinal cord to augment or (usually) suppress inflammation by release of neuromodulators in lymphoid tissues. While PACAP is found in parasympathetic neurons [36], anatomical studies to date have found no conclusive direct parasympathetic innervation of lymphoid organs despite acetylcholine receptors present on immune cells. For years the question of the source of the acetylcholine in lymphoid tissues has puzzled the field. Recent evidence suggests that vagus nerve stimulation of celiac or other sympathetic ganglia cause the release of noradrenalin in lymphoid organs that results in the release of acetylcholine from specialized T cells (reviewed in [4, 35]). Thus, acetylcholine in lymphoid organs may come from T cells rather than from a direct autonomic source. There is, however, ample evidence for SNS innervation of lymphoid organs [34], and therefore this system has been the focus of studies of autonomic control of immune responses.

An abundance of studies have examined the general influence of the SNS on peripheral inflammation and the role of catecholamines in particular. Most of these studies, as well as studies in which immune cells were treated with norepinephrine receptor analogs in vitro, suggest that norepinephrine acts primarily via β-2 adrenergic receptors, and mainly inhibits innate inflammatory responses, and either promotes or inhibits adaptive immunity [34, 37]. However, almost nothing is known about peptidergic effects in this system. PACAP is expressed by preganglionic neurons originating in the intermediolateral (IML) region of the spinal cord that project to sympathetic ganglia [38, 39]. In fact, a majority of the IML projections to the superior cervical ganglia (SCG) express PACAP mRNA [38]. PACAP application to sympathetic nerves in vivo results in increased electrical activity [40] and release of catecholamines [41]. PACAP action through PAC1 receptors was shown to upregulate the expression of enzymes involved in the synthesis of catecholamines, such as tyrosine hydroxylase expression in the SCG, and to stimulate release of catecholamines and neuropeptides from SGC neurons [41]. Moreover, PACAP stimulation of SGC cultured neurons elicits gene expression and/or release of such factors at potencies two orders of magnitude higher than with VIP [41, 42], again implicating PAC1 receptors. Additional evidence for PACAP involvement in sympathetic regulation comes from studies involving knockout mice. PACAP-deficient mice have pathological phenotypes related to abnormal catecholamine release from the SNS, such as impaired thermoregulation [43]. Finally, PACAP exposure to cultured postganglionic sympathetic neurons induces expression of both PACAP and VIP, among other secreted factors [44]. These results suggest that PACAP release onto SCG neurons results in further downstream release of PACAP and VIP, which potentially act directly via their receptors on immune cells in lymphoid tissues. A proposed model of PACAP regulation of the SNS is presented in Fig. 38.2. However, considerable further studies will be needed to more closely examine the role of PACAP in the SNS and how it may function to alter immune responses.

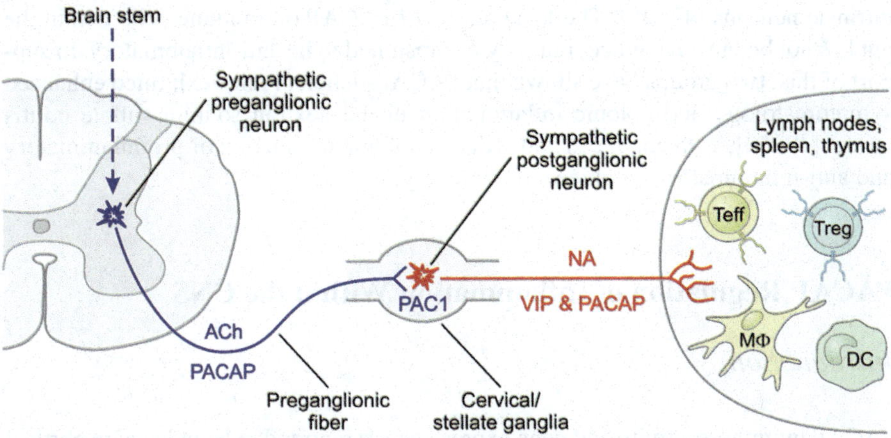

Fig. 38.2 Potential neural circuitry by which PACAP and VIP modulate peripheral immune cell activity during inflammatory stress (adapted from Waschek [33]). Neurons in the brain stem and hypothalamus (not shown) sense inflammation and activate preganglionic sympathetic neurons in the spinal cord interomediolateral (IML) column. PACAP is expressed with acetylcholine (ACh) in the preganglionic neurons in thoracic spinal cord IML neurons. When released during stress or inflammation, PACAP acts via PAC1 receptors expressed on sympathetic neurons in cervical/stellate ganglia to alter noradrenaline (NA) synthesis and/or activity, and to increase the production of VIP and/or PACAP. The latter neuropeptides are released along with NA in peripheral immune sites such as the thymus, lymph nodes, and spleen. Once released in these target organs, VIP or PACAP may then act directly with VPAC1, VPAC2, or PAC1 receptors on immune cells, or indirectly via action on stromal cells expressing these receptors. *Teff* T effector cell (Th1, Th2, and Th17), *Treg* regulatory T cell), *DC* dendritic cells, *MΦ* macrophage

In addition to the SNS, PACAP regulates the related sympathoadrenal axis by way of its expression in cholinergic neurons that innervate the adrenal medulla (see Fig. 38.1). For example, PACAP is detected in the innervation of the adrenal cortex and has been shown to stimulate secretion of catecholamines from the adrenal medulla [45]. Furthermore, PACAP-deficient mice exhibit prolonged hypoglycemia upon insulin challenge [46], a process dependent on sympathetic noradrenalin. Upon fasting and administration of insulin, these mice exhibit significantly lower levels of plasma epinephrine than wild-type controls. Although it is not completely clear if circulating catecholamines can regulate systemic inflammation, most immune cells express adrenergic receptors and respond to catecholamines in vitro. Mirroring the actions in the SNS, PACAP may thus affect the immune system indirectly by regulating the synthesis and release of catecholamines from the adrenal medulla.

PACAP Regulation of Inflammation in the Gastrointestinal Tract

Due to the prevalence of inflammatory bowel disease in the human population, it is of interest that PACAP is expressed in the extrinsic innervation of the intestine, most likely in sensory and vagal afferents, and in both myenteric and submucosal intrinsic neurons [47, 48]. The local action of PACAP on immune cells within the gut has not been well studied, but may be presumed to be anti-inflammatory. In support of this, two groups have shown that PACAP-deficient mice exhibited enhanced symptomatology and colonic inflammation in the dextran sodium sulfate colitis model [49, 50], with enhanced and reduced colonic production of proinflammatory and anti-inflammatory cytokines, respectively [49].

PACAP Regulation of Inflammation Within the CNS

Introduction

PACAP immunoreactivity and gene expression are reported to be induced in cortical and other CNS neurons in stroke and traumatic injury models [51–55], suggesting that PACAP might act locally within the brain to protect neurons and/or promote repair following injury. In this respect, PACAP administration in several models of CNS injury has been found to decrease both neuron loss and neurological deficits (see other chapters). Moreover, in at least one case, PACAP administration resulted in a pronounced diminution of inflammatory changes within the brain at the level of gene expression [56]. While PACAP is well known to act directly on neurons in culture to promote their survival, the latter finding suggests that PACAP might act indirectly on other cell types within the brain to modify inflammation, thereby reducing injury and promoting repair. As discussed later, an abundance of data suggests that PACAP acts on at least two types of glia, microglia and astrocytes, to regulate their inflammatory state and promote processes involved in protection and repair.

Local Immunomodulatory Action of PACAP on Astrocytes During Pathological Stress and Injury

Astrocytes behave as one of the innate immune effector cells in the CNS, amplifying the effects of inflammation, but they can also provide support and provide protection for neurons and other glial cells [57, 58]. Astrocytes also provide functional support to neurons by maintaining local ion and pH homeostasis, storing CNS glycogen, and clearing neuronal waste [59, 60]. Astrocyte dysfunction and apoptosis may thus contribute to pathogenesis of many acute and chronic neurodegenerative disorders, such as ischemia, Alzheimer's disease, Parkinson's disease, and multiple sclerosis [61–65].

All PACAP/VIP receptors, PAC1, VPAC1, and VPAC2, are reportedly expressed in cultured rat astrocytes [66–70]. The presence of PACAP/VIP receptors has also been described in both resting and reactive brain astrocytes [71–73], suggesting that some of the actions of PACAP and VIP in the brain may be mediated through astroglial cells. The following two sections will address putative immunomodulatory and neuroprotective actions of PACAP on astrocytes.

PACAP Regulation of Reactive Astrogliosis

Although reactive astrogliosis is best known as a process that can impede repair, considerable evidence indicates that it evolved as a protective function in CNS. Hallmarks of reactive astrogliosis include proliferation, induction of glial fibrillary acidic protein (GFAP) expression, and a morphology change to a hypertrophic state. Interestingly, PACAP38 was shown to stimulate the proliferation of astrocytes in vitro [69, 74], an effect which was completely abolished by the cyclic AMP antagonist Rp-cAMP, whereas the protein kinase A (PKA) inhibitor H89 had no effect [69]. These results suggest that PACAP-mediated stimulation of proliferation of astrocytes may involve a cAMP-dependent, but PKA-independent, pathway. The proliferative effect of PACAP on reactive astrocytes has also been reported in a scratch injury model using primary astrocyte cultures [75]. The percentage of Ki67 (a marker of proliferating cells) and glial fibrillary acidic protein double-positive cells increased in groups treated with PACAP (0.1–10 pM), particularly in the peri-injured area where many reactive astrocytes were found. The PACAP receptor antagonist PACAP6-38 suppressed the PACAP-induced proliferation of reactive astrocytes. In other studies, Masmoudi-Kouki et al. [76] have shown that subnanomolar concentrations of PACAP38 dose dependently protects cultured rat astrocytes from apoptosis induced by H_2O_2. PACAP6-38 totally suppressed the protective effect of PACAP. The protective action of PACAP38 was also blocked by the H89, the protein kinase C inhibitor chelerythrine and the mitogen-activated protein kinase kinase (MEK) inhibitor U0126. PACAP38 stimulated glutathione formation and blocked H_2O_2-evoked reactive oxygen species accumulation and glutathione content reduction. Additionally, PACAP38 prevented the decrease of mitochondrial activity and caspase 3 activation induced by H_2O_2. Together, PACAP exerts a potent protective effect against oxidative stress in astrocytes.

Astrocytes are known to release numerous cytokines and growth factors into the extracellular space [77, 78], and considerable evidence indicates that astrocyte-conditioned media support the survival of neurons in vitro [79–84]. It is of interest that treatment of cultured cerebellar granule cells with supernatants from astrocytes treated with 1 nM PACAP provoked a significant reduction of neuron death induced by H_2O_2 [76]. Pretreatment with PACAP6-38 attenuated the protective action of conditioned medium from PACAP-exposed astrocytes on granule cells. PACAP has also been shown to stimulate glial cells to generate various factors such as activity-dependent neurotrophic factor, IL-1 and IL-6, macrophage inflammatory protein, neurotrophin-3, protease nexin-1, and CCL5/RANTES (regulated on activation, normal T cell expressed and secreted) (reviewed in [85]). For example, in cultured rat astrocytes, very low concentrations of PACAP (0.1 pM) were able to stimulate the production of activity-dependent neuroprotective protein, an effect which was blocked by a PAC1 receptor antagonist [86]. PACAP38 was significantly more potent in stimulating IL-6 than VIP, and this effect was synergic with that of IL-1 [87, 88]. PACAP has also been shown to stimulate IL-6 production in cultured rat Müller cells, which are the predominant glial element in the retina, and this effect was blocked by PACAP6-38 [89, 90]. Although IL-6 is generally considered to be a proinflammatory cytokine, several studies indicate its actions on neurons are antiapoptotic due its induction of the STAT3 pathway. In support of this hypothesis, intracerebroventricular injection of recombinant IL-6 in rats has been demonstrated to decrease infarct volume following ischemia [91], whereas administration of IL-6 receptor antibody produced the opposite effect [92]. These reports suggest that IL-6 can have a neuroprotective function in CNS injury. PACAP administration to wild-type mice subject to the middle cerebral artery occlusion ischemia provided neuroprotection associated with induced IL-6 production, and the neuroprotective was abolished in IL-6 deficient mice [93]. These findings indicate that PACAP prevents neuronal cell death at least in part via IL-6 signaling.

Increased Glutamate Uptake and Metabolism by PACAP in Astrocytes

Glutamate is the major excitatory neurotransmitter in the mammalian CNS and is known to be involved in complex processes such as memory formation. However, excess levels of extracellular glutamate are considered detrimental in several neurodegenerative disorders [94]. Astrocytes perform the majority of glutamate uptake in the brain [95]. Following injury, astrocytes upregulate expression of glutamate transporters and glutamine synthetase, which allow them to sequester glutamate and convert it to glutamine, thus limiting neuronal and oligodendrocyte damage [96]. Downregulation of astrocytic glutamate transporter expression with antisense oligonucleotides leads to neurodegeneration and progressive paralysis [95]. Reduced expression of GLAST (EAAT1) and/or GLT-1 (EAAT2) has been observed after brain injury as well as in patients suffering from amyotrophic lateral sclerosis, Alzheimer's disease, and Huntington's disease and is thought to contribute in part to the neuronal cell death associated with these pathological conditions [97].

PACAP38 was shown to induce a robust increase in mRNA and protein levels of both GLT-1 and GLAST in cultured astrocytes derived from the cerebral cortex [98] and striatum [99]. The PACAP-induced increase in glutamate transporter levels correlated with an increase in the maximal velocity of the glutamate uptake [98]. PACAP also increased the K_m of the uptake process. In addition, PACAP38 robustly induced the expression of glutamine synthetase. Because the effects of PACAP38 on glial glutamate transporter expression were attenuated in the presence of PACAP6-38, PACAP promotes glial glutamate turnover via the PAC1 or VPAC2 receptor. Interestingly, the PKA inhibitor H89 almost completely abolished the PACAP-induced expression of GLAST and also decreased, although to a lesser extent, the PACAP induction of GLT-1 expression [98]. On the other hand, the PKC inhibitor Gö6976 did not interfere with the effects of PACAP on GLAST expression but resulted in a robust inhibition of PACAP-induced GLT-1 expression. Together, these findings suggest that PACAP differentially affects the expression of GLAST and GLT-1 via PKA and PKC in astrocytes. Importantly, PAC1 receptor expression is upregulated in astrocytes in several CNS injury models such as global ischemia [100], contusion spinal cord injury [54], and cortical stab injury [73]. These findings suggest that PAC1 receptors in these cells might provide some protective functions via astrocytes.

PACAP Regulation of Microglia Function

Just as macrophages respond to injury in peripheral tissues, microglia—the resident macrophages of the nervous system—become activated in the injured brain. While this is important for protection and repair, overactive microglial release of proinflammatory cytokines, including tumor necrosis factor-α (TNF-α), interleukin (IL)-1β, IL-6, IL-12, cyclooxygenase-2 (COX-2), and other factors such as nitric oxide (NO), reactive oxygen species, prostaglandin E2 (PGE2), and superoxide, is believed to contribute to the manifestation of neurodegenerative diseases, stroke, and CNS traumatic injury [101, 102]. Like for macrophages in non-CNS sites (see Chap. 7), PACAP is reported to inhibit the production of proinflammatory factors in activated microglia in culture. For example, PACAP was found to inhibit LPS-induced microglial production of TNF-α, IL-6, IL-1 β, NO, and COX-2 in a dose- and time-dependent manner [103, 104]. This anti-inflammatory activity was reported to be mediated by both VPAC1 [103] and PAC1 receptors [105]. Likewise, in an oxygen and glucose deprivation/reoxygenation microglia cell line model, PACAP was found to inhibit TNF-α in association with a reduction in NF-kB signaling [106]. In other studies, PACAP was found to inhibit interferon-γ (IFN-γ)-induced Jak1/2 and STAT1 phosphorylation in cultured microglia and to lower mRNA levels of IFN regulatory factor-1 [103]. PACAP also inhibits LPS-induced microglia costimulatory activation molecule CD40 expression by increasing IL-10 production [107]. The production of chemotactic factors has also been shown to be regulated by PACAP. For example, interaction of PACAP with the VPAC1 receptor in microglia was shown to inhibit the LPS-induced production of the CXC chemokines MIP-2 and KC, and the CC chemokines MIP-1α, MIP-1β, MCP-1, and RANTES [108]. Overall, the in vitro studies suggest that PACAP can regulate multiple immune processes via action on microglia.

In vivo studies also support a role of PACAP in microglia activation. In a focal ischemia mouse stroke model, mice given a stereotaxic intracerebraventricular injection of embryonic stem (ES) cells expressing PACAP had decreased levels of proinflammatory TNF-α in the infarcted hemisphere, and increased levels of anti-inflammatory IL-10, as compared to mice injected with saline or control ES cells [56]. Furthermore, mice that had received PACAP-producing ES cells exhibited microglia with reduced cell body size, increased number of fiber processes, and increased expression of arginase-1, a well-established marker of a "M2" anti-inflammatory macrophage phenotype [56]. Interestingly, in naïve PACAP-deficient mice, microglia in the substantia nigra pars compacta (SNpc, the main location of the dopaminergic vulnerable cells in Parkinson's Disease patients), exhibited an altered morphology, suggestive of a preactivated state [109]. When these mice were challenged with paraquat, an environmental pesticide associated with increased Parkinson's Disease risk, tyrosine hydroxylase staining in the SNpc was significantly reduced compared to paraquat-treated wild-type controls, and this was associated with increased expression of TNF-α gene expression in this brain region [109]. Finally, in a facial nerve crush model, microglia activation was more pronounced in the facial motor nucleus of PACAP-deficient vs. control mice [110]. This activation was associated with increased gene expression of TNF-α, IL-6, and IFN-γ RNA; a reduction in IL-4 RNA levels; and a delay in axon regeneration. These studies provide in vivo evidence that PACAP promotes an anti-inflammatory microglia phenotype with a potential impact on nerve regeneration.

PACAP Regulation of the Blood–Brain Barrier

The blood–brain barrier [BBB] is the major physical barrier that contributes to the immune privileged environment of the CNS. Composed of microvascular endothelial cells in close association with pericytes and astrocytic end feet, the BBB acts as a molecular sieve that prevents leukocytes and unwanted molecules from entering the CNS. Cerebral endothelial cells are characterized by tight junctions that block the passage of macromolecules from the blood into the brain. Tight junctions (TJ) regulate paracellular permeability and also contribute to the transendothelial electrical resistance. While a saturable mechanism has been identified that regulates PACAP transit across the BBB ([111] and see Chap. 2), there is a paucity of studies to investigate the effects of PACAP or the related peptide VIP on the transit of leukocytes and molecules across the BBB. PACAP nerve fibers are found surrounding the blood vessels in sheep brain [112] and in adventitia and adventitia-media border of cat cerebral arteries [113]. PACAP has been shown to significantly increase transendothelial electrical resistance, and was reported to increase expression of junctional proteins in cerebral microvessels [114]. Scuderi et al. demonstrated that PACAP counteracted the damaging effects of high glucose and cytokines by augmenting/restoring the expression of TJ proteins like ZO-1 and caludin-1 in a model of outer blood retinal barrier dysfunction [115]. The ability of PACAP to regulate chemokine production in microglia and astrocytes (see above and depicted in Fig. 38.3) is also likely to impact the transmigration of leukocytes into brain, and might be involved in homing microglia to sites of CNS injury or neurodegeneration. The BBB and immune cell flux in the brain has great importance in CNS injury, multiple sclerosis, and many neurologic diseases, so further investigation of PACAP in these specific processes seem warranted.

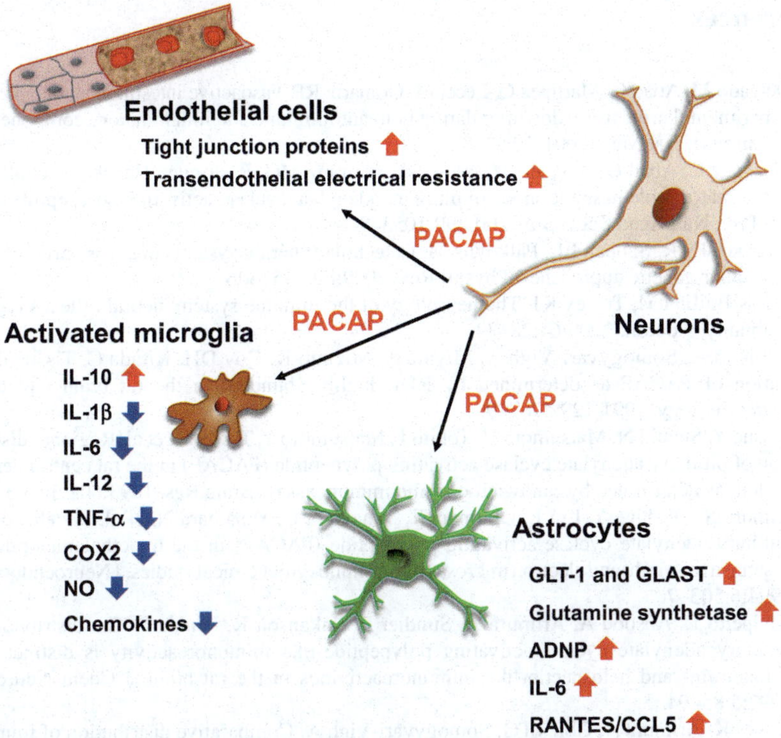

Fig. 38.3 Local anti-inflammatory actions of PACAP within the brain. PACAP acts within the brain on at least three target cell types to modify the immune response: microglia, astrocytes, and vascular endothelial cells. Reported actions on each of these cell types are indicated (see text for details and abbreviations)

Conclusions

PACAP and it receptors are expressed in neural circuits that mediate different types of stress, including inflammatory stress. While an abundance of data indicate that PACAP can act regulate inflammatory and the mechanisms involved remain unclear. However, PACAP expression is upregulated in multiple types of neurons in various nervous system injury models and by inflammation stimuli. This chapter posits that PACAP acts to control systemic inflammation by way of both the HPA and the sympathoadrenal axis, and controls local inflammation within the brain, primarily by way of high affinity receptors expressed on microglia and astrocytes.

Acknowledgements This work was funded by the Program for Advancing Strategic International Networks to Accelerate the Circulation of Talented Researchers (Grant No. S2603, Japan), the National Multiple Sclerosis Society (RG4859), Department of Defense, Congressionally Directed Medical Research Award MS120120 (W81XWH), and the National Institutes of Health (MH098506).

References

1. Delgado M, Abad C, Martinez C, Leceta J, Gomariz RP. Vasoactive intestinal peptide prevents experimental arthritis by downregulating both autoimmune and inflammatory components of the disease. Nat Med. 2001;7:563–8.
2. Martinez C, Abad C, Delgado M, Arranz A, Juarranz MG, Rodriguez-Henche N, et al. Anti-inflammatory role in septic shock of pituitary adenylate cyclase-activating polypeptide receptor. Proc Natl Acad Sci U S A. 2002;99:1053–8.
3. Kvetnansky R, Sabban EL, Palkovits M. Catecholaminergic systems in stress: structural and molecular genetic approaches. Physiol Rev. 2009;89:535–606.
4. Rosas-Ballina M, Tracey KJ. The neurology of the immune system: neural reflexes regulate immunity. Neuron. 2009;64:28–32.
5. Arimura A, Somogyvari-Vigh A, Miyata A, Mizuno K, Coy DH, Kitada C. Tissue distribution of PACAP as determined by RIA: highly abundant in the rat brain and testes. Endocrinology. 1991;129:2787–9.
6. Masuo Y, Suzuki N, Matsumoto H, Tokito F, Matsumoto Y, Tsuda M, et al. Regional distribution of pituitary adenylate cyclase activating polypeptide (PACAP) in the rat central nervous system as determined by sandwich-enzyme immunoassay. Brain Res. 1993;602:57–63.
7. Kimura S, Ohshige Y, Lin L, Okumura T, Yanaihara C, Yanaihara N, et al. Localization of pituitary adenylate cyclase-activating polypeptide (PACAP) in the hypothalamus-pituitary system in rats: light and electron microscopic immunocytochemical studies. J Neuroendocrinol. 1994;6:503–7.
8. Kivipelto L, Absood A, Arimura A, Sundler F, Hakanson R, Panula P. The distribution of pituitary adenylate cyclase-activating polypeptide-like immunoreactivity is distinct from helodermin- and helospectin-like immunoreactivities in the rat brain. J Chem Neuroanat. 1992;5:85–94.
9. Koves K, Arimura A, Gorcs TG, Somogyvari-Vigh A. Comparative distribution of immunoreactive pituitary adenylate cyclase activating polypeptide and vasoactive intestinal polypeptide in rat forebrain. Neuroendocrinology. 1991;54:159–69.
10. Hannibal J, Fahrenkrug J. Expression of pituitary adenylate cyclase activating polypeptide (PACAP) gene by rat spermatogenic cells. Regul Pept. 1995;55:111–5.
11. Hannibal J, Mikkelsen JD, Fahrenkrug J, Larsen PJ. Pituitary adenylate cyclase-activating peptide gene expression in corticotropin-releasing factor-containing parvicellular neurons of the rat hypothalamic paraventricular nucleus is induced by colchicine, but not by adrenalectomy, acute osmotic, ether, or restraint stress. Endocrinology. 1995;136:4116–24.
12. Piggins HD, Stamp JA, Burns J, Rusak B, Semba K. Distribution of pituitary adenylate cyclase activating polypeptide (PACAP) immunoreactivity in the hypothalamus and extended amygdala of the rat. J Comp Neurol. 1996;376:278–94.
13. Das M, Vihlen CS, Legradi G. Hypothalamic and brainstem sources of pituitary adenylate cyclase-activating polypeptide nerve fibers innervating the hypothalamic paraventricular nucleus in the rat. J Comp Neurol. 2007;500:761–76. PubMed PMID: 17154257.
14. Murase T, Kondo K, Arima H, Iwasaki Y, Ito M, Miura Y, et al. The expression of pituitary adenylate cyclase-activating polypeptide (PACAP) mRNA in rat brain: possible role of endogenous PACAP in vasopressin release. Neurosci Lett. 1995;185:103–6. PubMed PMID: 7746497.
15. Legradi G, Hannibal J, Lechan RM. Pituitary adenylate cyclase-activating polypeptide-nerve terminals densely innervate corticotropin-releasing hormone-neurons in the hypothalamic paraventricular nucleus of the rat. Neurosci Lett. 1998;246:145–8.
16. Hashimoto H, Nogi H, Mori K, Ohishi H, Shigemoto R, Yamamoto K, et al. Distribution of the mRNA for a pituitary adenylate cyclase-activating polypeptide receptor in the rat brain: an in situ hybridization study. J Comp Neurol. 1996;371:567–77.
17. Nomura M, Ueta Y, Serino R, Kabashima N, Shibuya I, Yamashita H. PACAP type I receptor gene expression in the paraventricular and supraoptic nuclei of rats. Neuroreport. 1996;8:67–70.

18. Stroth N, Eiden LE. Stress hormone synthesis in mouse hypothalamus and adrenal gland triggered by restraint is dependent on pituitary adenylate cyclase-activating polypeptide signaling. Neuroscience. 2010;165:1025–30.
19. Mustafa T, Jiang SZ, Eiden AM, Weihe E, Thistlethwaite I, Eiden LE. Impact of PACAP and PAC1 receptor deficiency on the neurochemical and behavioral effects of acute and chronic restraint stress in male C57BL/6 mice. Stress. 2015;18(4):408–18.
20. Grinevich V, Fournier A, Pelletier G. Effects of pituitary adenylate cyclase-activating polypeptide (PACAP) on corticotropin-releasing hormone (CRH) gene expression in the rat hypothalamic paraventricular nucleus. Brain Res. 1997;773:190–6.
21. Tsukiyama N, Saida Y, Kakuda M, Shintani N, Hayata A, Morita Y, et al. PACAP centrally mediates emotional stress-induced corticosterone responses in mice. Stress. 2011;14:368–75.
22. Hammack SE, Cheung J, Rhodes KM, Schutz KC, Falls WA, Braas KM, et al. Chronic stress increases pituitary adenylate cyclase-activating peptide (PACAP) and brain-derived neurotrophic factor (BDNF) mRNA expression in the bed nucleus of the stria terminalis (BNST): roles for PACAP in anxiety-like behavior. Psychoneuroendocrinology. 2009;34:833–43.
23. Lezak KR, Roman CW, Braas KM, Schutz KC, Falls WA, Schulkin J, et al. Regulation of bed nucleus of the stria terminalis PACAP expression by stress and corticosterone. J Mol Neurosci. 2014;54:477–84.
24. Lezak KR, Roelke E, Harris OM, Choi I, Edwards S, Gick N, et al. Pituitary adenylate cyclase-activating polypeptide (PACAP) in the bed nucleus of the stria terminalis (BNST) increases corticosterone in male and female rats. Psychoneuroendocrinology. 2014;45:11–20.
25. Kocho-Schellenberg M, Lezak KR, Harris OM, Roelke E, Gick N, Choi I, et al. PACAP in the BNST produces anorexia and weight loss in male and female rats. Neuropsychopharmacology. 2014;39:1614–23.
26. Hammack SE, Roman CW, Lezak KR, Kocho-Shellenberg M, Grimmig B, Falls WA, et al. Roles for pituitary adenylate cyclase-activating peptide (PACAP) expression and signaling in the bed nucleus of the stria terminalis (BNST) in mediating the behavioral consequences of chronic stress. J Mol Neurosci. 2010;42:327–40.
27. Roman CW, Lezak KR, Hartsock MJ, Falls WA, Braas KM, Howard AB, et al. PAC1 receptor antagonism in the bed nucleus of the stria terminalis (BNST) attenuates the endocrine and behavioral consequences of chronic stress. Psychoneuroendocrinology. 2014;47:151–65.
28. Ressler KJ, Mercer KB, Bradley B, Jovanovic T, Mahan A, Kerley K, et al. Post-traumatic stress disorder is associated with PACAP and the PAC1 receptor. Nature. 2011;470:492–7.
29. Legradi G, Das M, Giunta B, Hirani K, Mitchell EA, Diamond DM. Microinfusion of pituitary adenylate cyclase-activating polypeptide into the central nucleus of amygdala of the rat produces a shift from an active to passive mode of coping in the shock-probe fear/defensive burying test. Neural Plast. 2007;2007:79102.
30. Otto C, Kovalchuk Y, Wolfer DP, Gass P, Martin M, Zuschratter W, et al. Impairment of mossy fiber long-term potentiation and associative learning in pituitary adenylate cyclase activating polypeptide type I receptor-deficient mice. J Neurosci. 2001;21:5520–7.
31. Otto C, Martin M, Wolfer DP, Lipp HP, Maldonado R, Schutz G. Altered emotional behavior in PACAP-type-I-receptor-deficient mice. Mol Brain Res. 2001;92:78–84.
32. Hannibal J, Jessop DS, Fahrenkrug J, Harbuz MS, Larsen PJ. PACAP gene expression in neurons of the rat hypothalamo-pituitary-adrenocortical axis is induced by endotoxin and interleukin-1beta. Neuroendocrinology. 1999;70:73–82.
33. Waschek JA. VIP and PACAP: neuropeptide modulators of CNS inflammation, injury, and repair. Br J Pharmacol. 2013;169:512–23.
34. Nance DM, Sanders VM. Autonomic innervation and regulation of the immune system (1987-2007). Brain Behav Immun. 2007;21:736–45.
35. Tracey KJ. Reflex control of immunity. Nat Rev Immunol. 2009;9:418–28.
36. Hannibal J. Pituitary adenylate cyclase-activating peptide in the rat central nervous system: an immunohistochemical and in situ hybridization study. J Comp Neurol. 2002;453: 389–417.
37. Bellinger DL, Millar BA, Perez S, Carter J, Wood C, ThyagaRajan S, et al. Sympathetic modulation of immunity: relevance to disease. Cell Immunol. 2008;252:27–56.

38. Beaudet MM, Braas KM, May V. Pituitary adenylate cyclase activating polypeptide (PACAP) expression in sympathetic preganglionic projection neurons to the superior cervical ganglion. J Neurobiol. 1998;36:325–36.
39. Pettersson LM, Heine T, Verge VM, Sundler F, Danielsen N. PACAP mRNA is expressed in rat spinal cord neurons. J Comp Neurol. 2004;471:85–96.
40. Tanida M, Shintani N, Morita Y, Tsukiyama N, Hatanaka M, Hashimoto H, et al. Regulation of autonomic nerve activities by central pituitary adenylate cyclase-activating polypeptide. Regul Pept. 2010;161:73–80.
41. May V, Braas KM. Pituitary adenylate cyclase-activating polypeptide (PACAP) regulation of sympathetic neuron neuropeptide Y and catecholamine expression. J Neurochem. 1995;65:978–87.
42. Braas KM, May V. Pituitary adenylate cyclase-activating polypeptides, PACAP-38 and PACAP-27, regulation of sympathetic neuron catecholamine, and neuropeptide Y expression through activation of type I PACAP/VIP receptor isoforms. Ann N Y Acad Sci. 1996;805:204–16. discussion 17–8.
43. Gray SL, Yamaguchi N, Vencova P, Sherwood NM. Temperature-sensitive phenotype in mice lacking pituitary adenylate cyclase-activating polypeptide. Endocrinology. 2002;143:3946–54.
44. Braas KM, Schutz KC, Bond JP, Vizzard MA, Girard BM, May V. Microarray analyses of pituitary adenylate cyclase activating polypeptide (PACAP)-regulated gene targets in sympathetic neurons. Peptides. 2007;28:1856–70.
45. Isobe K, Nakai T, Takuwa Y. Ca(2+)-dependent stimulatory effect of pituitary adenylate cyclase-activating polypeptide on catecholamine secretion from cultured porcine adrenal medullary chromaffin cells. Endocrinology. 1993;132:1757–65.
46. Hamelink C, Tjurmina O, Damadzic R, Young WS, Weihe E, Lee HW, et al. Pituitary adenylate cyclase-activating polypeptide is a sympathoadrenal neurotransmitter involved in catecholamine regulation and glucohomeostasis. Proc Natl Acad Sci U S A. 2002;99:461–6.
47. Nagahama M, Tsuzuki M, Mochizuki T, Iguchi K, Kuwahara A. Light and electron microscopic studies of pituitary adenylate cyclase-activating peptide (PACAP)--immunoreactive neurons in the enteric nervous system of rat small and large intestine. Anat Embryol (Berl). 1998;198:341–52.
48. Hannibal J, Ekblad E, Mulder H, Sundler F, Fahrenkrug J. Pituitary adenylate cyclase activating polypeptide (PACAP) in the gastrointestinal tract of the rat: distribution and effects of capsaicin or denervation. Cell Tissue Res. 1998;291:65–79.
49. Azuma YT, Hagi K, Shintani N, Kuwamura M, Nakajima H, Hashimoto H, et al. PACAP provides colonic protection against dextran sodium sulfate induced colitis. J Cell Physiol. 2008;216:111–9.
50. Nemetz N, Abad C, Lawson G, Nobuta H, Chhith S, Duong L, et al. Induction of colitis and rapid development of colorectal tumors in mice deficient in the neuropeptide PACAP. Int J Cancer. 2008;122:1803–9.
51. Stumm R, Kolodziej A, Prinz V, Endres M, Wu DF, Hollt V. Pituitary adenylate cyclase-activating polypeptide is up-regulated in cortical pyramidal cells after focal ischemia and protects neurons from mild hypoxic/ischemic damage. J Neurochem. 2007;103:1666–81.
52. Skoglosa Y, Lewen A, Takei N, Hillered L, Lindholm D. Regulation of pituitary adenylate cyclase activating polypeptide and its receptor type 1 after traumatic brain injury: comparison with brain-derived neurotrophic factor and the induction of neuronal cell death. Neuroscience. 1999;90:235–47.
53. Armstrong BD, Hu Z, Abad C, Yamamoto M, Rodriguez WI, Cheng J, et al. Induction of neuropeptide gene expression and blockade of retrograde transport in facial motor neurons following local peripheral nerve inflammation in severe combined immunodeficiency and BALB/C mice. Neuroscience. 2004;129:93–9.
54. Tsuchikawa D, Nakamachi T, Tsuchida M, Wada Y, Hori M, Farkas J, et al. Neuroprotective effect of endogenous pituitary adenylate cyclase-activating polypeptide on spinal cord injury. J Mol Neurosci. 2012;48:508–17.

55. Zhou X, Rodriguez WI, Casillas RA, Ma V, Tam J, Hu Z, et al. Axotomy-induced changes in pituitary adenylate cyclase activating polypeptide (PACAP) and PACAP receptor gene expression in the adult rat facial motor nucleus. J Neurosci Res. 1999;57:953–61.
56. Brifault C, Gras M, Liot D, May V, Vaudry D, Wurtz O. Delayed pituitary adenylate cyclase-activating polypeptide delivery after brain stroke improves functional recovery by inducing m2 microglia/macrophage polarization. Stroke. 2015;46:520–8.
57. Minagar A, Shapshak P, Fujimura R, Ownby R, Heyes M, Eisdorfer C. The role of macrophage/microglia and astrocytes in the pathogenesis of three neurologic disorders: HIV-associated dementia, Alzheimer disease, and multiple sclerosis. J Neurol Sci. 2002;202:13–23.
58. Nair A, Frederick TJ, Miller SD. Astrocytes in multiple sclerosis: a product of their environment. Cell Mol Life Sci. 2008;65:2702–20.
59. Brown AM, Ransom BR. Astrocyte glycogen and brain energy metabolism. Glia. 2007;55:1263–71.
60. Nedergaard M, Ransom B, Goldman SA. New roles for astrocytes: redefining the functional architecture of the brain. Trends Neurosci. 2003;26:523–30.
61. Braak H, Sastre M, Del Tredici K. Development of alpha-synuclein immunoreactive astrocytes in the forebrain parallels stages of intraneuronal pathology in sporadic Parkinson's disease. Acta Neuropathol. 2007;114:231–41.
62. Giffard RG, Swanson RA. Ischemia-induced programmed cell death in astrocytes. Glia. 2005;50:299–306.
63. Seifert G, Schilling K, Steinhauser C. Astrocyte dysfunction in neurological disorders: a molecular perspective. Nat Rev Neurosci. 2006;7:194–206.
64. Takuma K, Ago Y, Matsuda T. The glial sodium-calcium exchanger: a new target for nitric oxide-mediated cellular toxicity. Curr Prot Pept Sci. 2013;14:43–50.
65. Takuma K, Baba A, Matsuda T. Astrocyte apoptosis: implications for neuroprotection. Prog Neurobiol. 2004;72:111–27.
66. Ashur-Fabian O, Giladi E, Brenneman DE, Gozes I. Identification of VIP/PACAP receptors on rat astrocytes using antisense oligodeoxynucleotides. J Mol Neurosci. 1997;9:211–22.
67. Grimaldi M, Cavallaro S. Functional and molecular diversity of PACAP/VIP receptors in cortical neurons and type I astrocytes. Eur J Neurosci. 1999;11:2767–72.
68. Grimaldi M, Cavallaro S. Expression and coupling of PACAP/VIP receptors in cortical neurons and type I astrocytes. Ann N Y Acad Sci. 2000;921:312–6.
69. Hashimoto H, Kunugi A, Arakawa N, Shintani N, Fujita T, Kasai A, et al. Possible involvement of a cyclic AMP-dependent mechanism in PACAP-induced proliferation and ERK activation in astrocytes. Biochem Biophys Res Commun. 2003;311:337–43.
70. Jaworski DM. Expression of pituitary adenylate cyclase-activating polypeptide (PACAP) and the PACAP-selective receptor in cultured rat astrocytes, human brain tumors, and in response to acute intracranial injury. Cell Tissue Res. 2000;300:219–30.
71. Joo KM, Chung YH, Kim MK, Nam RH, Lee BL, Lee KH, et al. Distribution of vasoactive intestinal peptide and pituitary adenylate cyclase-activating polypeptide receptors (VPAC1, VPAC2, and PAC1 receptor) in the rat brain. J Comp Neurol. 2004;476:388–413.
72. Nakamachi T, Ohtaki H, Yofu S, Dohi K, Watanabe J, Hayashi D, et al. Pituitary adenylate cyclase-activating polypeptide (PACAP) type 1 receptor (PAC1R) co-localizes with activity-dependent neuroprotective protein (ADNP) in the mouse brains. Regul Pept. 2008;145:88–95.
73. Suzuki R, Arata S, Nakajo S, Ikenaka K, Kikuyama S, Shioda S. Expression of the receptor for pituitary adenylate cyclase-activating polypeptide (PAC1-R) in reactive astrocytes. Mol Brain Res. 2003;115:10–20.
74. Tatsuno I, Arimura A. Pituitary adenylate cyclase-activating polypeptide (PACAP) mobilizes intracellular free calcium in cultured rat type-2, but not type-1, astrocytes. Brain Res. 1994;662:1–10.
75. Nakamachi T, Nakamura K, Oshida K, Kagami N, Mori H, Watanabe J, et al. Pituitary adenylate cyclase-activating polypeptide (PACAP) stimulates proliferation of reactive astrocytes in vitro. J Mol Neurosci. 2011;43:16–21.

76. Masmoudi-Kouki O, Douiri S, Hamdi Y, Kaddour H, Bahdoudi S, Vaudry D, et al. Pituitary adenylate cyclase-activating polypeptide protects astroglial cells against oxidative stress-induced apoptosis. J Neurochem. 2011;117:403–11.
77. Martin DL. Synthesis and release of neuroactive substances by glial cells. Glia. 1992; 5:81–94.
78. Ota Y, Zanetti AT, Hallock RM. The role of astrocytes in the regulation of synaptic plasticity and memory formation. Neural Plast. 2013;2013:185463.
79. Dhandapani KM, Hadman M, De Sevilla L, Wade MF, Mahesh VB, Brann DW. Astrocyte protection of neurons: role of transforming growth factor-beta signaling via a c-Jun-AP-1 protective pathway. J Biol Chem. 2003;278:43329–39.
80. Eriksen JL, Druse MJ. Astrocyte-mediated trophic support of developing serotonin neurons: effects of ethanol, buspirone, and S100B. Brain Res Dev Brain Res. 2001;131:9–15.
81. Hailer NP, Wirjatijasa F, Roser N, Hischebeth GT, Korf HW, Dehghani F. Astrocytic factors protect neuronal integrity and reduce microglial activation in an in vitro model of N-methyl-D-aspartate-induced excitotoxic injury in organotypic hippocampal slice cultures. Eur J Neurosci. 2001;14:315–26.
82. Lin MS, Hung KS, Chiu WT, Sun YY, Tsai SH, Lin JW, et al. Curcumin enhances neuronal survival in N-methyl-d-aspartic acid toxicity by inducing RANTES expression in astrocytes via PI-3K and MAPK signaling pathways. Prog Neuropsychopharmacol Biol Psychiatry. 2011;35:931–8.
83. Takarada T, Nakamichi N, Kawagoe H, Ogura M, Fukumori R, Nakazato R, et al. Possible neuroprotective property of nicotinic acetylcholine receptors in association with predominant upregulation of glial cell line-derived neurotrophic factor in astrocytes. J Neurosci Res. 2012;90:2074–85.
84. Yamamuro A, Ago Y, Takuma K, Maeda S, Sakai Y, Baba A, et al. Possible involvement of astrocytes in neuroprotection by the cognitive enhancer T-588. Neurochem Res. 2003;28:1779–83.
85. Masmoudi-Kouki O, Gandolfo P, Castel H, Leprince J, Fournier A, Dejda A, et al. Role of PACAP and VIP in astroglial functions. Peptides. 2007;28:1753–60.
86. Nakamachi T, Li M, Shioda S, Arimura A. Signaling involved in pituitary adenylate cyclase-activating polypeptide-stimulated ADNP expression. Peptides. 2006;27:1859–64.
87. Gottschall PE, Tatsuno I, Arimura A. Regulation of interleukin-6 (IL-6) secretion in primary cultured rat astrocytes: synergism of interleukin-1 (IL-1) and pituitary adenylate cyclase activating polypeptide (PACAP). Brain Res. 1994;637:197–203.
88. Tatsuno I, Morio H, Tanaka T, Uchida D, Hirai A, Tamura Y, et al. Pituitary adenylate cyclase-activating polypeptide (PACAP) is a regulator of astrocytes: PACAP stimulates proliferation and production of interleukin 6 (IL-6), but not nerve growth factor (NGF), in cultured rat astrocyte. Ann N Y Acad Sci. 1996;805:482–8.
89. Nakatani M, Seki T, Shinohara Y, Taki C, Nishimura S, Takaki A, et al. Pituitary adenylate cyclase-activating peptide (PACAP) stimulates production of interleukin-6 in rat Muller cells. Peptides. 2006;27:1871–6.
90. Seki T, Hinohara Y, Taki C, Nakatani M, Ozawa M, Nishimura S, et al. PACAP stimulates the release of interleukin-6 in cultured rat Muller cells. Ann N Y Acad Sci. 2006;1070:535–9.
91. Loddick SA, Turnbull AV, Rothwell NJ. Cerebral interleukin-6 is neuroprotective during permanent focal cerebral ischemia in the rat. J Cereb Blood Flow Metab. 1998;18:176–9.
92. Yamashita T, Sawamoto K, Suzuki S, Suzuki N, Adachi K, Kawase T, et al. Blockade of interleukin-6 signaling aggravates ischemic cerebral damage in mice: possible involvement of Stat3 activation in the protection of neurons. J Neurochem. 2005;94:459–68.
93. Ohtaki H, Nakamachi T, Dohi K, Aizawa Y, Takaki A, Hodoyama K, et al. Pituitary adenylate cyclase-activating polypeptide (PACAP) decreases ischemic neuronal cell death in association with IL-6. Proc Natl Acad Sci U S A. 2006;103:7488–93.
94. Mehta A, Prabhakar M, Kumar P, Deshmukh R, Sharma PL. Excitotoxicity: bridge to various triggers in neurodegenerative disorders. Eur J Pharmacol. 2013;698:6–18.

95. Rothstein JD, Dykes-Hoberg M, Pardo CA, Bristol LA, Jin L, Kuncl RW, et al. Knockout of glutamate transporters reveals a major role for astroglial transport in excitotoxicity and clearance of glutamate. Neuron. 1996;16:675–86.
96. Liberto CM, Albrecht PJ, Herx LM, Yong VW, Levison SW. Pro-regenerative properties of cytokine-activated astrocytes. J Neurochem. 2004;89:1092–100.
97. Sims KD, Robinson MB. Expression patterns and regulation of glutamate transporters in the developing and adult nervous system. Crit Rev Neurobiol. 1999;13:169–97.
98. Figiel M, Engele J. Pituitary adenylate cyclase-activating polypeptide (PACAP), a neuron-derived peptide regulating glial glutamate transport and metabolism. J Neurosci. 2000;20:3596–605.
99. Schluter K, Figiel M, Rozyczka J, Engele J. CNS region-specific regulation of glial glutamate transporter expression. Eur J Neurosci. 2002;16:836–42.
100. Nakamachi T, Farkas J, Kagami N, Wada Y, Hori M, Tsuchikawa D, et al. Expression and distribution of pituitary adenylate cyclase-activating polypeptide receptor in reactive astrocytes induced by global brain ischemia in mice. Acta Neurochir Suppl. 2013;118:55–9.
101. Block ML, Hong JS. Microglia and inflammation-mediated neurodegeneration: multiple triggers with a common mechanism. Prog Neurobiol. 2005;76:77–98.
102. Hill KE, Zollinger LV, Watt HE, Carlson NG, Rose JW. Inducible nitric oxide synthase in chronic active multiple sclerosis plaques: distribution, cellular expression and association with myelin damage. J Neuroimmunol. 2004;151:171–9.
103. Delgado M. Inhibition of interferon (IFN) gamma-induced Jak-STAT1 activation in microglia by vasoactive intestinal peptide: inhibitory effect on CD40, IFN-induced protein-10, and inducible nitric-oxide synthase expression. J Biol Chem. 2003;278:27620–9.
104. Gonzalez-Rey E, Delgado M. Vasoactive intestinal peptide inhibits cyclooxygenase-2 expression in activated macrophages, microglia, and dendritic cells. Brain Behav Immun. 2008;22:35–41.
105. Kim WK, Kan Y, Ganea D, Hart RP, Gozes I, Jonakait GM. Vasoactive intestinal peptide and pituitary adenylyl cyclase-activating polypeptide inhibit tumor necrosis factor-alpha production in injured spinal cord and in activated microglia via a cAMP-dependent pathway. J Neurosci. 2000;20:3622–30.
106. Qin X, Sun ZQ, Dai XJ, Mao SS, Zhang JL, Jia MX, et al. Toll-like receptor 4 signaling is involved in PACAP-induced neuroprotection in BV2 microglial cells under OGD/reoxygenation. Neurol Res. 2012;34:379–89.
107. Kim WK, Ganea D, Jonakait GM. Inhibition of microglial CD40 expression by pituitary adenylate cyclase-activating polypeptide is mediated by interleukin-10. J Neuroimmunol. 2002;126:16–24.
108. Delgado M, Jonakait GM, Ganea D. Vasoactive intestinal peptide and pituitary adenylate cyclase-activating polypeptide inhibit chemokine production in activated microglia. Glia. 2002;39:148–61.
109. Watson MB, Nobuta H, Abad C, Lee SK, Bala N, Zhu C, et al. PACAP deficiency sensitizes nigrostriatal dopaminergic neurons to paraquat-induced damage and modulates central and peripheral inflammatory activation in mice. Neuroscience. 2013;240:277–86. Pubmed Central PMCID: 3637876.
110. Armstrong BD, Abad C, Chhith S, Cheung-Lau G, Hajji OE, Nobuta H, et al. Impaired nerve regeneration and enhanced neuroinflammatory response in mice lacking pituitary adenylyl cyclase activating peptide. Neuroscience. 2008;151:63–73.
111. Banks WA, Kastin AJ, Komaki G, Arimura A. Passage of pituitary adenylate cyclase activating polypeptide1-27 and pituitary adenylate cyclase activating polypeptide1-38 across the blood-brain barrier. J Pharmacol Exp Therap. 1993;267:690–6.
112. Koves K, Arimura A, Somogyvari-Vigh A, Vigh S, Miller J. Immunohistochemical demonstration of a novel hypothalamic peptide, pituitary adenylate cyclase-activating polypeptide, in the ovine hypothalamus. Endocrinology. 1990;127(1):264–71.

113. Uddman R, Goadsby PJ, Jansen I, Edvinsson L. PACAP, a VIP-like peptide: immunohistochemical localization and effect upon cat pial arteries and cerebral blood flow. J Cereb Blood Flow Metab. 1993;13:291–7.
114. Wilhelm I, Fazakas C, Tamas A, Toth G, Reglodi D, Krizbai IA. PACAP enhances barrier properties of cerebral microvessels. J Mol Neurosci. 2014;54:469–76.
115. Scuderi S, D'Amico AG, Castorina A, Imbesi R, Carnazza ML, D'Agata V. Ameliorative effect of PACAP and VIP against increased permeability in a model of outer blood retinal barrier dysfunction. Peptides. 2013;39:119–24.

Chapter 39
PACAP Regulation of Inflammatory and Free Radical Networks in Neuronal and Nonneuronal Diseases

Hirokazu Ohtaki and Seiji Shioda

Abstract Experimental and clinical evidence suggests that under conditions of acute and chronic inflammation, many organs and tissues are subjected to considerable oxidative stress induced by oxygen- and nitrogen-derived free radicals and/or reactive oxygen species (ROS). However, free radicals and ROS are essential in the first line of defense tools against infective agents, including bacteria and viruses. Moreover, a free radical, nitric oxide (NO) acts as an intra- and extracellular signal transducer. Thus, depending on the situation, free radicals and ROS can protect or damage cells and cellular processes. Pituitary adenylate cyclase-activating polypeptide (PACAP) has been shown to suppress inflammation by suppressing proinflammatory cytokines in diseases of the central nervous system (CNS) and other tissues and organs, with most of these diseases involving the oxidative stress. Recent studies have suggested that PACAP may regulate oxidative stress. This chapter summarizes free radicals and ROS and then the current knowledge on the effects of PACAP on CNS and non-CNS diseases that involve oxidative stress and inflammatory response.

Keywords PACAP • Oxidative stress • Free radicals • Reactive oxygen species • Inflammation • Inflammatory cytokines • Superoxide dismutase (SOD) • Superoxide anion • Central nervous system • Mesenchymal stem/progenitor cells (MSCs)

H. Ohtaki (✉)
Department of Anatomy, Showa University School of Medicine,
1-5-8 Hatanodai, Shinagawa-ku, Tokyo 142-8555, Japan
e-mail: taki@med.showa-u.ac.jp

S. Shioda
Peptide Drug Innovation, Global Research Center for Innovative Life Science, Hoshi University School of Pharmacy and Pharmaceutical Sciences, Tokyo 142-8501, Japan

D. Reglodi, A. Tamas (eds.), *Pituitary Adenylate Cyclase Activating Polypeptide — PACAP*, Current Topics in Neurotoxicity 11,
DOI 10.1007/978-3-319-35135-3_39

Introduction

Experimental and clinical evidence suggests that under conditions of acute and chronic inflammation, many organs and tissues are subjected to considerable oxidative stress induced by oxygen- and nitrogen-derived free radicals and/or reactive oxygen species (ROS). However, free radicals and ROS are essential in the first line of defense tools against infective agents, including bacteria and viruses. Moreover, a free radical, nitric oxide (NO) acts as an intra- and extracellular signal transducer. Thus, depending on the situation, free radicals and ROS can protect or damage cells and cellular processes.

Pituitary adenylate cyclase-activating polypeptide (PACAP) is a neuropeptide involved in cytokine networks during immune/inflammatory responses. PACAP has been shown to suppress inflammation by decreasing proinflammatory cytokines in diseases of the central nervous system (CNS), including ischemia, traumatic brain injury (TBI), spinal cord injury (SCI), and neurodegenerative diseases. Moreover, PACAP has been reported to suppress inflammation in diseases of other tissues and organs, with most of these diseases involving oxidative stress. Recent studies have suggested that PACAP may regulate oxidative stress. This chapter summarizes current knowledge on the effects of PACAP on CNS and non-CNS diseases that involve oxidative stress.

Free Radicals and Reactive Oxygen Species (ROS)

To understand role of PACAP in free radical networks, it is important to define free radical and ROS, terms frequently used interchangeably. ROS is a generic description of oxygen species more reactive than ordinary oxygen molecules (O_2); these include singlet oxygen ($^1O_2$), hydrogen peroxide (H_2O_2), superoxide anion ($O_2^{\bullet-}$), and hydroxyl radical ($HO^\bullet$). Free radicals are chemical species with an unpaired electron. Thus, of the four ROS, $O_2^{\bullet-}$ and $HO^\bullet$ are free radicals, whereas H_2O_2 and $^1O_2$ are not. Free radicals also include other oxygen species, including lipid peroxides such as alkoxyl radical ($LO^\bullet$) and peroxyl radical ($LOO^\bullet$), reactive nitrogen species (RNS) such as NO (more correctly, $NO^\bullet$) and peroxynitrite ($ONOO^-$), as well as carbon, sulfur, and halogens [1] (Fig. 39.1).

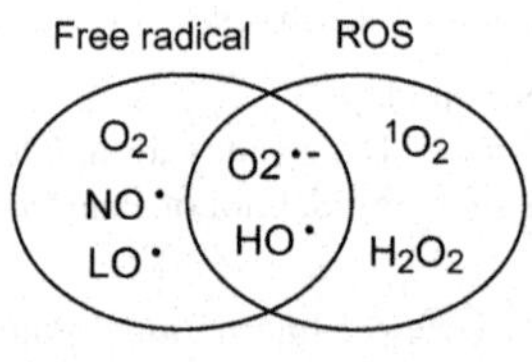

Fig. 39.1 Free radicals and ROS

Metabolism of Free Radicals and Reactive Oxygen Species

In eukaryotes, more than 90 % of the O_2 consumed under aerobic conditions is reduced directly to water by cytochrome oxidase, situated on the internal mitochondrial membrane, in an electron-transport chain involving four electrons [1–3]. This electron-transport chain is coupled with oxidative phosphorylation to produce energy in the form of ATP. The remaining O_2, however, amounting to less than 10 %, undergoes a one electron reduction, yielding $O_2^{\bullet-}$, the primary free radical molecule. It is impossible to completely prevent the production of free radicals in body. The $O_2^{\bullet-}$ is formed by other metabolic processes, including hydroxylation by NADPH cytochrome P450 reductase in the endoplasmic reticulum. During defense responses, neutrophils and macrophages produce $O_2^{\bullet-}$ by activating NADPH oxidase on the plasma membrane. Although most of the $O_2^{\bullet-}$ generated is released into phagocytic vehicles for germicidal activity, some $O_2^{\bullet-}$ is released extracellularly and may enhance inflammation [4]. The metabolism of $O_2^{\bullet-}$ within the body is illustrated in Fig. 39.2. Although $O_2^{\bullet-}$ is not a strong oxidant, it can easily react with NO to produce the oxidative metabolite, $ONOO^-$, a much stronger oxidant and highly toxic reactive nitrogen species (RNS). The production of $ONOO^-$ and the overmetabolism of NO can be prevented by superoxide dismutase (SOD), which metabolizes $O_2^{\bullet-}$ to H_2O_2. H_2O_2, in turn, undergoes autolysis due to the low binding energy between its oxygen atoms, yielding $HO^\bullet$ and lipid peroxides. $HO^\bullet$ is also a very strong oxidant, which reacts with and oxidizes components within the body, including sugars, proteins, lipids, and DNA, impairing their functions. Because the reactions of $HO^\bullet$ are extremely rapid, there are no specific radical scavengers for this molecule in the body. However, the proform of $HO^\bullet$, H_2O_2, is hydrolyzed to water (H_2O) by catalase and glutathione peroxidases (GPx). Therefore, catalase and GPx suppress $HO^\bullet$ toxicity. Oxidation of sugars, proteins, lipids, and DNA by $ONOO^-$ and $HO^\bullet$ reduces cell and tissue activity and may impair bodily homeostasis [1].

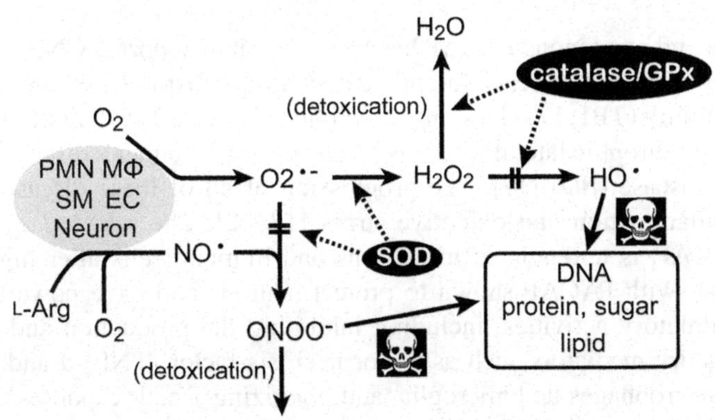

Fig. 39.2 Schematic illustration of reactive oxygen metabolism. *PNM* polymorphonuclear leukocyte, *Mϕ* macrophages, *SM* smooth muscles, *EC* endothelial cells, *L-Arg* L-Arginine

Regulation of Oxidative Stress in the Body

Many antioxidative molecules and enzymes can inhibit the toxicity of constitutively generated free radicals and ROS. Antioxidative molecules include glutathione, vitamin C (L-ascorbic acid), and vitamin E (D-α-tocopherol) and its analogs. The major antioxidative enzymes include SOD, catalase, and GPx, which contribute to the $O_2^{\bullet-}$ metabolic pathway and ultimately produce H_2O. Under healthy condition, bodies and cells maintain a homeostatic balance between ROS and free radical synthesis and scavenging. However, aging and unusual conditions, including ischemia/reperfusion, inflammation, exposure to X-ray and ultraviolet radiation, and exposure to toxins, can alter bodily homeostasis, resulting in ROS and free radical synthesis dominating over scavenging. This imbalance is defined as oxidative stress [1].

Inflammation and Oxidative Stress

PACAP has been reported to suppress tissue damage in various physically and chemically induced pathological models that result from inflammatory responses. Inflammation consists of three principal components: (1) increased blood supply to the infected/damaged area, (2) increased permeability of blood vessels, and (3) migration of leukocytes from the venules to the surrounding tissues. These migrating leukocytes, which include immune system cells such as macrophages and neutrophils, as well as parenchymal cells, respond and cross-talk with each other via pathways mediated by cytokines, chemokines, and release free radicals.

CNS

Endogenous and exogenous PACAP has been shown to suppress CNS neural damage resulting from brain ischemia and ischemia/reperfusion [5–8] and from traumatic brain injury (TBI) [9–11], spinal cord injury (SCI) [12–14], multiple sclerosis (MS) [15], amyotrophic lateral sclerosis (ALS) [16], Alzheimer's disease (AD), and Parkinson's disease (PD) [17]. The progression of all of these diseases involves increased inflammation and oxidative stress [18–22]. The *adcyap1* gene, which encodes PACAP, is upregulated in neurons and immune cells after injury and/or inflammation, with PACAP shown to protect neurons and have powerful *in vivo* immunomodulatory activities, including inhibiting the production and release of proinflammatory mediators such as tumor necrosis factor (TNF)-α and interferon (IFN)-γ in macrophages and microglia, and polarizing T-cell responses away from Type 1 helper (Th1) and Th17 phenotypes and toward a Th2 phenotype [23].

Cerebral Ischemia

The neuroprotective effect of PACAP on the CNS was first reported in ischemic diseases [8]. However, the contribution of PACAP-associated anti-inflammation to the suppression of neural damage is unclear, because PACAP has shown antiapoptotic activity. On the other hands, PACAP has been suggested to involve in cytokine network in immune and inflammatory cells [24, 25]. For example, in the presence of IL-1β, PACAP enhanced IL-6 secretion from astrocytes culture [26]. Indeed, exogenous PACAP did not have neuroprotective effects against ischemia of IL-6 gene deficient (KO) mice, suggesting that IL-6 production is a neurotrophic pathway of PACAP [5]. More recently, injection of PACAP and its analog was shown to have direct anti-inflammatory activity in ischemia, with both molecules suppressing the expression of genes encoding macrophage inflammatory protein-1α (MIP-1α, *Ccl3*), inducible NO synthase (iNOS, *Nos2*), TNFα (*Tnf*), while enhancing the expression of IL-6 (*Il6*) [27].

TBI

The immunoreactivity of PACAP has been reported in postmortem traumatized human brains [28]. PACAP27 and 38 immunoreactions were reduced in the core-region of the cortex, but were increased in the pericontusional cortex, after TBI. PACAP immunoreactions were observed in neurons and in astrocytes positive for glial fibrillary acidic protein (GFAP). Conversely, 2′,3′-cyclic-nucleotide 3′-phosphodiesterase (CNPase)-positive oligodendrocytes temporary decreased PACAP immunoreactions [28]. PACAP treatment of rats after the induction of TBI resulted in a significant reduction in brain [9–11] and axonal [29] damage. Pretreatment with PACAP significantly increased CD4-positive and decreased CD8-positive T-lymphocytes in the blood and spleen [9]. Pretreatment with PACAP suppressed toll-like receptor 4 (TLR4) upregulation, which was induced by TBI, and prevented intracellular NF-kB signaling by downregulating MyD88, phosphorylated IkB, and NF-kB. PACAP also suppressed increases in proinflammatory cytokines, such as IL-1β and TNFα, in the brain parenchyma around the injured regions [10].

We directly demonstrated that PACAP increased the levels of antioxidative enzymes, reducing oxidative stress. Intravenous injection of PACAP reduced the region positive for injury, accompanied by the $ONOO^-$ oxidative metabolite, 3-nitrotyrosine. PACAP increased manganese-SOD (Mn-SOD) and GPx-1 in neocortical regions, consistent with an increase in $O_2^{\bullet-}$ and prior to oxidative damage in the traumatized region (Fig. 39.3) [30].

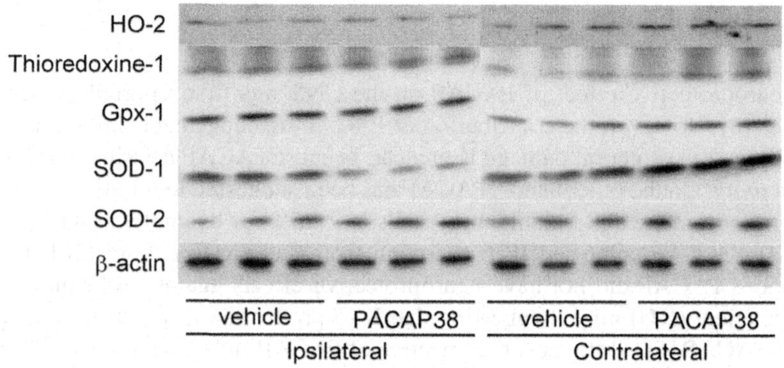

Fig. 39.3 PACAP enhancement of antioxidant activity after TBI [11]. Western blotting image of 3 h TBI at ipsilateral and contralateral hemisphere. PACAP38 increased Mn-SOD and Gpx-1 compared with vehicle administration at both ipsilateral and contralateral hemisphere

SCI

The expression of *adcyap1* was increased in the spinal cord after SCI [13, 14, 31, 32]. Treatment with PACAP suppressed the injury [12, 13], whereas PACAP heterozygous KO mice showed an increase in injured region after SCI [14, 33]. These findings suggested that endogenous and exogenous PACAP play suppressive roles after SCI.

Recently, two different groups, including ours, have shown that PACAP was involved in the attenuation of SCI by human mesenchymal stem/stromal cells (hMSCs) [33, 34]. Bone marrow derived hMSCs have been shown to suppress CNS injury, including brain ischemia [35, 36], TBI [37], and SCI [33, 38]. Injection of immortalized hMSCs or PACAP suppressed SCI, whereas their combination acted synergistically, increasing the activity of the antioxidative enzymes, Mn-SOD and peroxiredoxin-1/6 (Prxs, Prx-1 and -6) [34]. These results suggested that PACAP and hMSCs suppressed SCI by enhancing antioxidative activity. In contrast, we found that, although hMSCs transplanted into wild-type spinal cord after SCI significantly reduce the size of the injury and the impairment of motor function, transplantation of hMSCs into PACAP heterozygous KO mice did not have protective effects [33]. Transplantation of hMSCs reduced the expression of the *il1b* (IL-1β), *tnf*, *Il10* and *tgfb* (TGFβ) gene, and increased the expression of the IL-4 gene (*il4*). However, transplantation of hMSCs into PACAP KO mice abolished the effect of PACAP on the expression of the *il1b*, *tgfb*, and *il4*. Naive hMSCs in vitro did not express *ADCYAP1* and *ADCYAP1R1* (which encodes PAC1), but did express *ADCYAP1* and *ADCYAP1R1* after treatment with IFNγ. Therefore, hMSCs may have suppressed some inflammatory responses mediated by PACAP [33].

Neurodegenerative Diseases

Neuroinflammation is a major risk factor in PD. PACAP treatment ameliorated the symptoms of PD in rodent models [39, 40]. PACAP27 reduced dopaminergic neuron loss and motor deficits induced by Prostaglandin J2 (PGJ2) injection, without preventing microglial activation [40]. PGJ2 is derived from PGD2, the principal product of cyclooxygenase, which is synthesized by nonenzymatic dehydration in the mammalian CNS. Dopaminergic neurons in the substantia nigra pars compact (SNpc) of PACAP KO mice were found to be vulnerable against paraquat-induced damage. Relative to control mice, the SNpc in PACAP KO mice showed greater *tnf* expression and a number of Th17 cells but impaired induction of Treg cells [41].

Postmortem examination of patients with AD showed reduced PACAP expression [42, 43], whereas treatment with PACAP ameliorated AD symptoms [17]. Because AD accompanies inflammatory responses, the neuroprotective effects of PACAP may involve anti-inflammatory and antioxidative mechanisms. However, the direct contribution of PACAP to these mechanisms is unclear.

ALS is a neurodegenerative disease, which induces the death of motor neurons and results in the impairment of muscular functions. The gene encoding an antioxidative enzyme SOD1 (Zn/Cu-SOD, *SOD1*) is mutated in about 20 % of patients with familial ALS. Hemizygous transgenic mice containing the G93A mutant (glycine 93 changed to alanine) form of SOD1 (SOD1(G93A)) developed ALS. PACAP-deficient SOD1(G93A) mice showed greater preganglionic parasympathetic neuronal loss than SOD1(G93A) mice, and microglia that switched from an amoeboid to a hypertrophic phenotype were observed in SOD1(G93A) mice. However, these alterations in phenotype were not observed in PACAP-deficient SOD1(G93A) mice. Taken together, the findings suggest a protective role of endogenous PACAP, along with the suppression of microglial activation [16].

Experimental autoimmune encephalomyelitis (EAE) is the most commonly used experimental model of the human inflammatory demyelinating disease, multiple sclerosis (MS). Immunization of PACAP KO mice with myelin oligodendrocyte glycoprotein (MOG) exacerbated the clinical and pathological symptoms of EAE compared with wild-type mice. The increased sensitivity was accompanied by enhanced expression in the spinal cord of genes encoding the proinflammatory cytokines [*tnf*, *il6*, *ifng* (IFNγ), *il12a* (IL-12p35), *il23a* (IL-23p19), and *il17*], chemokines [*ccl2* (MCP-1/CCL2), *ccl3* (MIP-1alpha/CCL3), and *ccl5* (RANTES/CCL5)], and chemotactic factor receptors, along with the downregulation of genes encoding anti-inflammatory cytokines (*il4*, *il10*, and *tgfb*) [15]. PACAP injection ameliorated both the clinical and pathological manifestations of EAE. Ex vivo examination showed that the MOG35-55-specific Th1 response was significantly inhibited in mice treated with PACAP. In vitro analysis revealed that PACAP suppressed the production of proinflammatory cytokines, including *tnf*, *il1b*, and *il12*, and the costimulatory factor B7-2 (*cd86*) by macrophages and microglia, which may function as antigen-presenting cells (APC) in the CNS [44].

Human Immunodeficiency Virus Type-1 (HIV) Infection

HIV infection of the CNS promotes neuronal injury and apoptosis, culminating in HIV-associated neurocognitive disorders (HAND). Viral proteins, such as transactivator of transcription (Tat) and gp120, have emerged as leading candidates to explain HIV-mediated neurotoxicity. PACAP was the most potent secretagogue for RANTES on astrocytes cultures. Gp120-mediated neuronal cell death was prevented by cotreatment with PACAP38 [45]. Tat induced the production of ROS, and other indices of mitochondrial destabilization, within a few hours. In addition, Tat induced DNA double-stranded breaks (DSBs). Injection of PACAP27 inhibited all Tat-mediated toxic effects, including DNA DSBs. Importantly, PACAP27 prevented the induction of neuronal loss induced by Tat. The neuroprotective effect of PACAP27 correlated with its ability to release RANTES, which acts as an antiapoptotic chemokine [46].

Microglial Activation

PACAP and VIP (PACAP/VIP) have anti-inflammatory effects, both in innate and adaptive immunity. In innate immunity, VIP/PACAP inhibited the production of proinflammatory cytokines and chemokines by macrophages, microglia, and dendritic cells. In addition, VIP/PACAP reduced the expression of costimulatory molecules (particularly CD80 and CD86) on antigen-presenting cells (APC), thereby suppressing the stimulation of antigen-specific CD4-positive T-lymphocytes [25].

IFNγ activates microglia to generate the classically activating (M1) phenotype, resulting in the production of proinflammatory cytokines and ROS, mediated by the JAK/STAT1 pathway [4, 47, 48]. PACAP deactivated M1 microglia and inhibited JAK/STAT phosphorylation. STAT1 binds to the IFNγ activated site motif in IFN regulatory factor-1 and the CD40 promoter and to the IFN-stimulated response element motif of the IP-10 promoter. PACAP modulated expression of the genes encoding IP-10, CD40, and iNOS, with these effects of PACAP mediated through the specific receptor VPAC1 and the cAMP/protein kinase A transduction pathway [49]. PACAP inhibited TNFα, IL-1β, IL-6, and NO production by lipopolysaccharide (LPS)-activated microglia. LPS is a ligand of toll-like receptor 4 (TRL4) and induces the M1 phenotype in microglia/macrophages, an induction mediated by the NFkB pathway. PACAP was found to regulate proinflammatory factors by inhibiting p65 nuclear translocation and NFkB-DNA binding [50]. PACAP also inhibited the expression of the macrophage-derived CXC chemokines macrophage inflammatory protein-2 (MIP-2) and KC (IL-8), and of the CC chemokines MIP-1α, MIP-1β, MIP-1, and RANTES in vivo and in vitro. Its inhibition of chemokine gene expression correlated with the PACAP inhibition of NFkB binding and transactivating activity [51]. These findings were supported by in vivo study evidence in mice with ischemia and SCI, in that microglial activation was affected by PACAP injection or PACAP KO [27, 33, 52].

Peripheral Nervous System (PNS)

Peripheral nerve inflammation frequently accompanies nerve injury and occurs in several diseases, including Guillain–Barre syndrome, and acute and chronic inflammatory demyelinating polyneuropathy. The effect of PACAP on the PNS was determined using a mouse model of facial nerve injury [53–55]. Although neural injury did not differ significantly in PACAP KO and wild-type mice, axon regeneration after crush injury was significantly delayed in PACAP KO mice. This impaired regeneration was associated with increased expression of genes encoding the proinflammatory cytokines, including *tnf*, *ifng*, *il6*, and decreased expression of the gene encoding the anti-inflammatory cytokine *il4* at the site of injury. Similar changes in cytokine expression and increased microglial responses were observed in the brainstem facial motor nucleus [53]. *Adcyap1* expression in motor neurons in the facial motor nucleus increased after the injury; however, the number of motor neurons expressing *adcyap1* after the injury was significantly reduced in mice with severe combined immunodeficiency (SCID). *Adcyap1* expression was fully reversed by an infusion of normal splenocytes, suggesting that *adcyap1* induction requires inflammatory mediators [54, 55]. *Adcyap1* induction, however, was observed in mice lacking leukemia inhibitory factor (LIF) and IL-6, and in mice lacking both receptors for TNFα. These results suggested that an inflammatory response, probably involving T lymphocytes, was necessary for the axotomy-induced increase in PACAP but not in VIP [55].

Retina

PACAP also ameliorated retinal damage in rodents induced by injection of *N*-methyl-D-aspartate (NMDA) [56–58]. The NMDA receptor is an ionotropic glutamate receptor that mediates excitotoxicity. Cell protection by PACAP was not observed when PACAP was injected into IL-10 KO mice, an effect that may contribute to microglia/macrophage activation [58]. PACAP and PACAP fragments 1–5 increased the viability of the retinal ganglion cell line RGC-5 exposed to ultraviolet irradiation. PACAP fragments 1–5 also reduced ROS production in situ [56], and PACAP and PACAP fragments 1–5 reduced Bax and increased Bcl-2 expression.

PACAP was found to protect human pigment epithelial cells (ARPE-19 cells) against H_2O_2-induced injury. Oxidative stress activated several markers of apoptosis, including Bad, Bax, HIF-1α, several heat shock proteins, TNF-related apoptosis-inducing ligand, and Fas-associated protein with death domain, with all of these effects reversed by PACAP treatment. Furthermore, PACAP activated the protective Akt pathway, as well as reversing the effects of oxidative stress on several other signaling molecules, such as Chk2, Yes, Lyn, paxillin, p53, PLC, STAT4, and RSK [59].

Nonneural Cell Inflammation and PACAP

Aging

Adcyap1 expression was found to decrease with age in rat cerebromicrovascular endothelial cells (CMVECs), along with a reduction in formation of capillary-like structures and increases in $O_2^{\bullet-}$ and apoptosis. Knock down of *adcyap1* by RNA interference also impaired tube formation in young CMVECs [60].

The level of reactive oxygen metabolites (ROMs) in mouse serum increased with physiological aging. PACAP KO mice were more susceptible to age-associated increases in ROMs but, conversely, exhibited lower biological antioxidative potential (BAP) with aging (Fig. 39.4). While PACAP did not directly affect radical scavenging activity in vitro, intravenous PACAP treatment of wild-type mice decreased ROMs and increased serum BAP within a few hours in a dose-dependent fashion, with ROMs and BAP returning to basal levels at 24 h. These effects were not

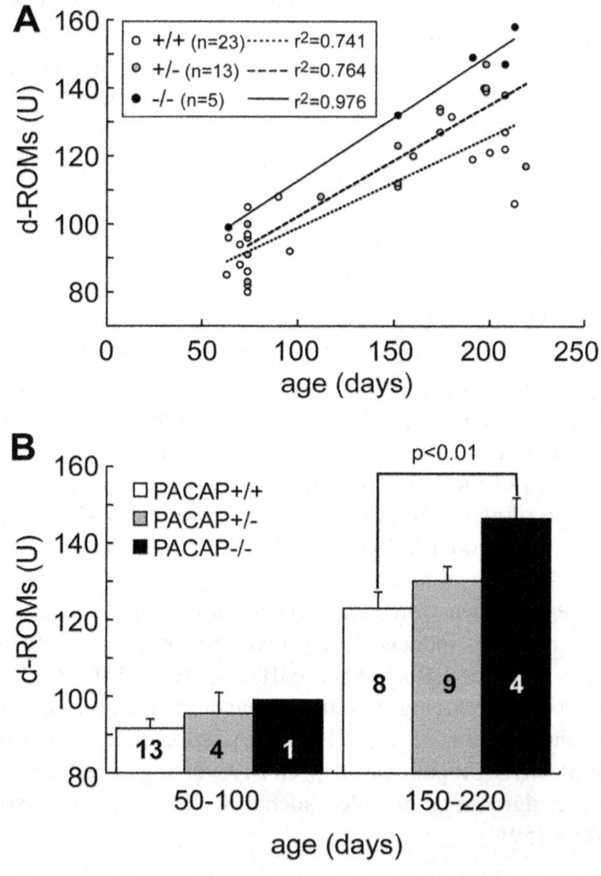

Fig. 39.4 The plasma of aged PACAP KO mice displayed increased oxidative metabolites [61]. Plasma samples of PACAP $^{+/+}$, $^{+/-}$ and $^{-/-}$ mice were analyzed for oxidative metabolites. (**a**) The ROMs value was highly correlated with age. (**b**) Mean plasma ROMs in younger (50–100 days old) or older (150–220 days old) mice revealed significantly greater oxidative stress in the older PACAP$^{-/-}$ animals than in their wild-type (PACAP$^{+/+}$) littermates

observed with VIP or PACAP 6–38, suggesting that PACAP suppressed oxidative stress through de novo synthesis mediated by PAC1 receptor [61].

Levels of advanced glycation end products (AGEs) have been shown to increase with age, especially in individuals with diabetes. The recognition of receptor for AGEs (RAGE) is involved in inflammation and oxidative stress including diabetic inflammation and AD. Therefore, RAGE shedding is considered a therapeutic target. PACAP27 was found to induce RAGE shedding in vitro, a process mediated by Ca^{2+} signaling, PKCα/PKCβ1, CAMKII, PI3 kinase, and MAP kinase. Moreover, ADAM10 and matrix metalloproteinase-9 (MMP-9) were shown to play important roles in PACAP-induced RAGE shedding [62].

PACAP was found to delay testicular aging [63]. PACAP KO mice showed impaired steroidogenesis. Interestingly, aged (15-month old) PACAP KO mice showed decreased $ONOO^-$ in germ cells, had low number of apoptotic cells, and had well conserved testicular structure.

Systemic Inflammation Such as Shock

PACAP was found to inhibit the release and cytokine activity of high mobility group box 1 (HMGB1) and to improve survival during lethal endotoxemia. PACAP also suppressed HMGB1 release induced by TNFα or IFNγ [64]. PACAP also inhibited the expression of LPS-induced tissue factor (TF), which is expressed on the surface of activated monocytes and is the major procoagulant that initiates thrombus formation in sepsis. VIP and PACAP blocked both the migration of c-Rel/p65 and the phosphorylation of p38 and JNK, as demonstrated by immunoblotting [65]. PACAP decreased LPS-induced IL-6 production, neutrophil infiltration and the expression of intercellular adhesion molecule-1 (ICAM-1), vascular cell adhesion molecule-1 (VCAM-1), and fibrinogen through PAC1 receptor [66].

PACAP protected mice from lethal endotoxemia by inhibiting IL-6 and TNF [67]. PAC1KO mice exhibited high mortality, which could not be reversed by injection of VIP or PACAP. The PAC1 receptor acted in vivo as an anti-inflammatory receptor, at least in part, by attenuating LPS-induced IL-6 production [68, 69].

Nephropathy

PACAP has been reported to ameliorate nephropathy [70–76]. PACAP attenuated contrast agent-induced nephropathy in HK-2 human renal proximal tubule epithelial cells and in endothelial NOS gene (*nos3*) deficient mice. PACAP decreased the expression of proinflammatory cytokines such as TNFα and IFNγ, as well as macrophage markers and *nos2* [75, 76]. Primary kidney cells obtained from PACAP KO mice showed reduced viability following exposure to oxidative stress induced by

H_2O_2 and hypoxia induced by $CoCl_2$, effects reversed by exogenous PACAP [72, 73]. PACAP injection into mice subjected to renal ischemic/reperfusion increased mitochondrial expression of the antiapoptotic factor bcl-2 and suppressed the expression of anti-inflammatory cytokines such as TIMP-1 and RANTES [77]. PACAP and the VPAC1 gene, but not the PAC1 gene, were expressed in kidneys. TLRs, the principal mediators of innate immunity, were localized to podocytes of glomeruli and endothelial cells. LPS-induced expression of IL-6 and MCP-1 was significantly attenuated in the presence of PACAP [78].

PACAP38 inhibited myeloma light chain-induced proinflammatory cytokine expression with greater potency than dexamethasone, and attenuated the resulting damage to renal proximal tubule epithelial cells. These results indicated that the effects of PACAP38 are mediated through the inhibition of p38 MAPK phosphorylation and nuclear translocation of the p50 subunit of NFkB via both PAC1 and VPAC1 receptors. PACAP was also shown to be effective in other in vivo animal models of kidney hypertrophy, including streptozotocin-induced diabetic nephropathy and gentamicin-induced nephrotoxicity [79].

Heart

PACAP enhanced the viability of cardiomyocytes subjected to H_2O_2-induced oxidative stress by suppressing mitochondrial apoptotic signaling [80, 81]. Compared with wild-type mice, PACAP (+/−) KO mice with doxorubicin-induced cardiomyopathy had higher serum ROM levels, and their hearts had higher numbers of TUNEL-positive apoptotic cells and 8-hydroxy-2′-deoxyguanosine (8-OHdG), an oxidative metabolite of DNA [82]. All of these effects, however, were prevented by exogenous PACAP.

Skin

PACAP and PAC1R mRNA and proteins are expressed in plantar skin, dorsal paw, and ears and are increased during neurogenic inflammatory responses induced by capsaicin and transient receptor potential vanilloid 1 (TRPV1) receptor agonist [83]. Mice deficient in the PACAP gene had significantly smaller areas of capsaicin-induced neurogenic paw edema than wild-type mice, but not of complete Freund's adjuvant (CFA)-induced edema, suggesting that PACAP may influence TRPV1-mediated acute neurogenic edema. However, the selective PAC1 receptor agonist maxadilan inhibited neurogenic vasodilation and edema formation in the ear due to mustard oil-induced TRPA1 receptor activation [84].

Joints

Osteoarthritis (OA) is the most common form of degenerative joint disease and has been associated with inflammation and hypocellularity resulting from chondrocyte apoptosis. Changes in PACAP distribution in cartilage and in synovial fluid (SF) were assessed by immunohistochemistry and ELISA. PACAP abundance in cartilage tissue and SF was high in healthy controls. Rats with anterior cruciate ligament transection (ACLT)-induced OA showed increased IL-1β and decreased PACAP levels in SF. In vitro, PACAP prevented IL-1β-induced chondrocyte apoptosis, and reduced expression of apoptosis-related proteins, iNOS and cyclooxygenase 2 (COX2), suggesting that PACAP has anti-inflammatory effects [85]. PACAP greatly reduced the frequency and severity of collagen-induced arthritis in mice, improving clinical symptoms and ameliorating joint damage, by blocking inflammatory cytokines and chemokines such as TNF-α, IL-6, IL-12, iNOS, IL-18, IL-1β, IL-1α, RANTES, MCP-1, MIP-1α, MIP-1β, and MIP-2. PACAP also altered the Th1/Th2 balance, resulting in a Th2 dominant response.

PACAP KO mice with experimental arthritis induced by K/BxN serum transfer showed lower levels of edema formation, mechanical hyperalgesia and macrophage-derived superoxide activity levels in joints than wild-type mice [86]. PACAP deficiency reduced early accumulation of neutrophils by slowing their extravasation from the vessels, but facilitated their function in later phases. In addition, PACAP reduced activity and ROS production and promoted inflammation-induced pathologic bone neoformation. Taken together, these findings suggested that PACAP has complex inflammatory regulatory function in joints.

Digestive Tissues

PACAP ameliorated acute ileitis induced by *T. gondii* infection in mice, a model of inflammatory bowel disease. PACAP treated mice showed lower levels of ileal T-lymphocytes, neutrophils, monocytes, and macrophages, accompanied by lower levels of proinflammatory cytokines such as IL-23p19, IL-22, IFNγ, and MCP-1, and higher levels of the anti-inflammatory cytokine IL-4 in spleen and serum. The PACAP-induced reduction of intestinal proinflammatory responses was also revealed by the lower IFN-γ and NO concentrations in mesenteric lymph nodes (MLNs). NO levels in MLNs were lower following PACAP than PLC treatment [87]. Hepatocellular damage after ischemia and reperfusion of the liver, mediated by the cAMP-PKA pathway, was exacerbated in PACAP KO mice, but these effects were inhibited by treatment with PACAP27 or 38. PACAP treatment resulted in neutrophil/macrophage infiltration and activation, reduced hepatocyte necrosis/apoptosis, and augmented hepatic IL-10 expression [88]. PACAP (–/–) KO mice had significantly poorer histopathological scores upon challenge with the colitis-inducing agent, dextran sulfate sodium (DSS). DSS significantly upregulated the production of IL-1β and IL-6 in the proximal colon, as well as significantly increasing the production of IFNγ, IL-1β, IL-6, IL-12, and KC (CXCL1) in the distal colon, of PACAP (−/−) KO compared with wild-type mice [89].

Over-expression of PACAP aggravated cerulein-induced acute pancreatitis in mice. PACAP-transgenic (Tg) mice exhibited more severe pathophysiological signs of cerulein-induced pancreatitis, including higher serum amylase and lipase levels, accompanied by the exacerbation of pancreatic edema, necrosis, and inflammation. Cerulein treatment resulted in a similar magnitude of increase in mRNA expression of several proinflammatory cytokines (*tnf*, *il1b*, and *il6*) in wild-type and PACAP-Tg mice. In addition, the mRNA and protein levels of regenerating gene III beta (RegIIIβ), a key factor in the pancreatic response to acute pancreatitis, were upregulated by cerulein in wild-type mice, but were attenuated in PACAP-Tg mice [90].

Neuroinflammation driven by the vanilloid-type ion channel receptor transient receptor potential vanilloid type 1 (TRPV-1) is thought to play a role in the pathophysiology of inflammatory bowel diseases. TPRV-1-deficient mice showed defects in the expression of the anti-inflammatory neuropeptides, VIP and PACAP, contributing to the generation of a local proinflammatory environment [91].

Respiratory System

LPS significantly enhanced intranasal endotoxin-induced subacute airway inflammation in PACAP KO mice, accompanied by more severe inflammatory histopathological changes. Polymorphonuclear leucocytes (PNMs) were increased in PACAP (−/−) KO mice. In contrast, IL-1β concentrations were similar in PACAP (−/−) KO and wild-type mice [92].

Increased *adcyap1r1* expression in lung tissue was present under inflammatory conditions, and *adcyap1r1* expression was detected on DCs. In ovalbumin-induced asthma models, PAC1R deficiency resulted in inflammatory effects, while agonistic stimulation resulted in anti-inflammatory effects. Neither the gene-depletion nor the pharmacologic studies showed effects on long function [93].

Neutrophils

In most inflammatory models, PACAP has been reported to suppress inflammation. However, in monocytes and neutrophils, PACAP may play a role in VPAC1-mediated proinflammatory activity dependent on transient increases in Ca^{2+} [94–97]. This PACAP-induced transient increase in Ca^{2+} enhanced H_2O_2 production and upregulated CD11b.

Reduced PMN response to PACAP was observed after cells were exposed to inhibitors of the cAMP/protein kinase A, protein kinase C, and PI3K pathways; to pertussis toxin and genistein; and after chelation of intracellular calcium or depletion of extracellular calcium. In PACAP-treated PMNs, ROS production and CD11b membrane upregulation, in contrast to lactoferrin release, were dependent on both intra- and extracellular calcium concentrations, whereas MMP-9 release was unaffected by extracellular calcium depletion. These results indicate that both extracellular and intracellular calcium play key roles in the proinflammatory activities of PACAP [98].

References

1. Inoue M, Sato EF, Nishikawa M, Park AM, Kira Y, Imada I, et al. Mitochondrial generation of reactive oxygen species and its role in aerobic life. Curr Med Chem. 2003;10:2495–505.
2. Barnham KJ, Masters CL, Bush AI. Neurodegenerative diseases and oxidative stress. Nat Rev Drug Discov. 2004;3:205–14.
3. D'Autréaux B, Toledano MB. ROS as signalling molecules: mechanisms that generate specificity in ROS homeostasis. Nat Rev Mol Cell Biol. 2007;8:813–24.
4. Dohi K, Ohtaki H, Nakamachi T, Yofu S, Satoh K, Miyamoto K, et al. Gp91phox (NOX2) in classically activated microglia exacerbates traumatic brain injury. J Neuroinflammation. 2010;7:41.
5. Ohtaki H, Nakamachi T, Dohi K, Aizawa Y, Takaki A, Hodoyama K, et al. Pituitary adenylate cyclase-activating polypeptide (PACAP) decreases ischemic neuronal cell death in association with IL-6. Proc Natl Acad Sci U S A. 2006;103:7488–93.
6. Ohtaki H, Nakamachi T, Dohi K, Shioda S. Role of PACAP in ischemic neural death. J Mol Neurosci. 2008;36:16–25.
7. Reglodi D, Tamás A, Somogyvári-Vigh A, Szántó Z, Kertes E, Lénárd L, et al. Effects of pretreatment with PACAP on the infarct size and functional outcome in rat permanent focal cerebral ischemia. Peptides. 2002;23:2227–34.
8. Uchida D, Arimura A, Somogyvári-Vigh A, Shioda S, Banks WA. Prevention of ischemia-induced death of hippocampal neurons by pituitary adenylate cyclase activating polypeptide. Brain Res. 1996;736:280–6.
9. Hua R, Mao SS, Zhang YM, Chen FX, Zhou ZH, Liu JQ. Effects of pituitary adenylate cyclase activating polypeptide on CD4(+)/CD8(+) T cell levels after traumatic brain injury in a rat model. World J Emerg Med. 2012;3:294–8.
10. Mao SS, Hua R, Zhao XP, Qin X, Sun ZQ, Zhang Y, et al. Exogenous administration of PACAP alleviates traumatic brain injury in rats through a mechanism involving the TLR4/MyD88/NF-kB pathway. J Neurotrauma. 2012;29:1941–59.
11. Miyamoto K, Tsumuraya T, Ohtaki H, Dohi K, Satoh K, Xu Z, et al. PACAP38 suppresses cortical damage in mice with traumatic brain injury by enhancing antioxidant activity. J Mol Neurosci. 2014;54:370–9.
12. Chen WH, Tzeng SF. Pituitary adenylate cyclase-activating polypeptide prevents cell death in the spinal cord with traumatic injury. Neurosci Lett. 2005;384:117–21.
13. Tsuchida M, Nakamachi T, Sugiyama K, Tsuchikawa D, Watanabe J, Hori M, et al. PACAP stimulates functional recovery after spinal cord injury through axonal regeneration. J Mol Neurosci. 2014;54:380–7.
14. Tsuchikawa D, Nakamachi T, Tsuchida M, Wada Y, Hori M, Farkas J, et al. Neuroprotective effect of endogenous pituitary adenylate cyclase-activating polypeptide on spinal cord injury. J Mol Neurosci. 2012;48:508–17.

15. Tan YV, Abad C, Lopez R, Dong H, Liu S, Lee A, et al. Pituitary adenylyl cyclase-activating polypeptide is an intrinsic regulator of Treg abundance and protects against experimental autoimmune encephalomyelitis. Proc Natl Acad Sci U S A. 2009;106:2012–7.
16. Ringer C, Büning LS, Schäfer MK, Eiden LE, Weihe E, Schütz B. PACAP signaling exerts opposing effects on neuroprotection and neuroinflammation during disease progression in the SOD1(G93A) mouse model of amyotrophic lateral sclerosis. Neurobiol Dis. 2013;54:32–42.
17. Reglodi D, Kiss P, Lubics A, Tamas A. Review on the protective effects of PACAP in models of neurodegenerative diseases in vitro and in vivo. Curr Pharm Des. 2011;17:962–72.
18. Barber SC, Shaw PJ. Oxidative stress in ALS: key role in motor neuron injury and therapeutic target. Free Radic Biol Med. 2010;48:629–41.
19. Conti A, Miscusi M, Cardali S, Germanò A, Suzuki H, Cuzzocrea S, et al. Nitric oxide in the injured spinal cord: synthases cross-talk, oxidative stress and inflammation. Brain Res Rev. 2007;54:205–18.
20. Hajieva P, Behl C. Antioxidants as a potential therapy against age-related neurodegenerative diseases: amyloid Beta toxicity and Alzheimer's disease. Curr Pharm Des. 2006;12:699–704.
21. Miyamoto K, Ohtaki H, Dohi K, Tsumuraya T, Song D, Kiriyama K, et al. Therapeutic time window for edaravone treatment of traumatic brain injury in mice. Biomed Res Int. 2013;2013:379206.
22. Ohtaki H, Takeda T, Dohi K, Yofu S, Nakamachi T, Satoh K, et al. Increased mitochondrial DNA oxidative damage after transient middle cerebral artery occlusion in mice. Neurosci Res. 2007;58:349–55.
23. Waschek JA. VIP and PACAP: neuropeptide modulators of CNS inflammation, injury, and repair. Br J Pharmacol. 2013;169:512–23.
24. Delgado M, Abad C, Martinez C, Juarranz MG, Leceta J, Ganea D, et al. PACAP in immunity and inflammation. Ann N Y Acad Sci. 2003;992:141–57.
25. Gonzalez-Rey E, Varela N, Chorny A, Delgado M. Therapeutical approaches of vasoactive intestinal peptide as a pleiotropic immunomodulator. Curr Pharm Des. 2007;13:1113–9.
26. Gottschall PE, Tatsuno I, Arimura A. Regulation of interleukin-6 (IL-6) secretion in primary cultured rat astrocytes: synergism of interleukin-1 (IL-1) and pituitary adenylate cyclase activating polypeptide (PACAP). Brain Res. 1994;637:197–203.
27. Dejda A, Seaborn T, Bourgault S, Touzani O, Fournier A, Vaudry H, et al. PACAP and a novel stable analog protect rat brain from ischemia: insight into the mechanisms of action. Peptides. 2011;32:1207–16.
28. van Landeghem FK, Weiss T, Oehmichen M, von Deimling A. Cellular localization of pituitary adenylate cyclase-activating peptide (PACAP) following traumatic brain injury in humans. Acta Neuropathol. 2007;113:683–93.
29. Tamás A, Zsombok A, Farkas O, Reglödi D, Pál J, Büki A, et al. Postinjury administration of pituitary adenylate cyclase activating polypeptide (PACAP) attenuates traumatically induced axonal injury in rats. J Neurotrauma. 2006;23:686–95.
30. Miyamoto K, Ohtaki H, Dohi K, Tsumuraya T, Nakano H, Kiriyama K, et al. Edaravone increases regional cerebral blood flow after traumatic brain injury in mice. Acta Neurochir Suppl. 2013;118:103–9.
31. Pettersson LM, Heine T, Verge VM, Sundler F, Danielsen N. PACAP mRNA is expressed in rat spinal cord neurons. J Comp Neurol. 2004;471:85–96.
32. Zvarova K, Dunleavy JD, Vizzard MA. Changes in pituitary adenylate cyclase activating polypeptide expression in urinary bladder pathways after spinal cord injury. Exp Neurol. 2005;192:46–59.
33. Tsumuraya T, Ohtaki H, Song D, Sato A, Watanabe J, Hiraizumi Y, et al. Human mesenchymal stem/stromal cells suppress spinal inflammation in mice with contribution of pituitary adenylate cyclase-activating polypeptide (PACAP). J Neuroinflammation. 2015;12:35.
34. Fang KM, Chen JK, Hung SC, Chen MC, Wu YT, Wu TJ, et al. Effects of combinatorial treatment with pituitary adenylate cyclase activating peptide and human mesenchymal stem cells on spinal cord tissue repair. PLoS One. 2010;5:e15299.

35. Honmou O, Onodera R, Sasaki M, Waxman SG, Kocsis JD. Mesenchymal stem cells: therapeutic outlook for stroke. Trends Mol Med. 2012;18:292–7.
36. Ohtaki H, Ylostalo JH, Foraker JE, Robinson AP, Reger RL, Shioda S, et al. Stem/progenitor cells from bone marrow decrease neuronal death in global ischemia by modulation of inflammatory/immune responses. Proc Natl Acad Sci U S A. 2008;105:14638–43.
37. Watanabe J, Shetty AK, Hattiangady B, Kim DK, Foraker JE, Nishida H, et al. Administration of TSG-6 improves memory after traumatic brain injury in mice. Neurobiol Dis. 2013;59:86–99.
38. Saito F, Nakatani T, Iwase M, Maeda Y, Hirakawa A, Murao Y, et al. Spinal cord injury treatment with intrathecal autologous bone marrow stromal cell transplantation: the first clinical trial case report. J Trauma. 2008;64:53–9.
39. Reglődi D, Tamás A, Lengvári I, Toth G, Szalontay L, Lubics A. Comparative study of the effects of PACAP in young, aging, and castrated males in a rat model of Parkinson's disease. Ann N Y Acad Sci. 2006;1070:518–24.
40. Shivers KY, Nikolopoulou A, Machlovi SI, Vallabhajosula S, Figueiredo-Pereira ME. PACAP27 prevents Parkinson-like neuronal loss and motor deficits but not microglia activation induced by prostaglandin J2. Biochim Biophys Acta. 1842;2014:1707–19.
41. Watson MB, Nobuta H, Abad C, Lee SK, Bala N, Zhu C, et al. PACAP deficiency sensitizes nigrostriatal dopaminergic neurons to paraquat-induced damage and modulates central and peripheral inflammatory activation in mice. Neuroscience. 2013;240:277–86.
42. Han P, Caselli RJ, Baxter L, Serrano G, Yin J, Beach TG, et al. Association of pituitary adenylate cyclase-activating polypeptide with cognitive decline in mild cognitive impairment due to Alzheimer disease. JAMA Neurol. 2015;72:333–9.
43. Han P, Tang Z, Yin J, Maalouf M, Beach TG, Reiman EM, et al. Pituitary adenylate cyclase-activating polypeptide protects against β-amyloid toxicity. Neurobiol Aging. 2014;35:2064–71.
44. Kato H, Ito A, Kawanokuchi J, Jin S, Mizuno T, Ojika K, et al. Pituitary adenylate cyclase-activating polypeptide (PACAP) ameliorates experimental autoimmune encephalomyelitis by suppressing the functions of antigen presenting cells. Mult Scler. 2004;10:651–9.
45. Brenneman DE, Hauser JM, Spong C, Phillips TM. Chemokine release is associated with the protective action of PACAP-38 against HIV envelope protein neurotoxicity. Neuropeptides. 2002;36(4):271–80.
46. Rozzi SJ, Borelli G, Ryan K, Steiner JP, Reglodi D, Mocchetti I, et al. PACAP27 is protective against tat-induced neurotoxicity. J Mol Neurosci. 2014;54:485–93.
47. Sato A, Ohtaki H, Tsumuraya T, Song D, Ohara K, Asano M, et al. Interleukin-1 participates in the classical and alternative activation of microglia/macrophages after spinal cord injury. J Neuroinflammation. 2012;9:65.
48. Song D, Ohtaki H, Tsumuraya T, Miyamoto K, Shibato J, Rakwal R, et al. The anti-inflammatory property of human bone marrow-derived mesenchymal stem/stromal cells is preserved in late-passage cultures. J Neuroimmunol. 2013;263(1–2):55–63.
49. Delgado M. Inhibition of interferon (IFN) gamma-induced Jak-STAT1 activation in microglia by vasoactive intestinal peptide: inhibitory effect on CD40, IFN-induced protein-10, and inducible nitric-oxide synthase expression. J Biol Chem. 2003;278:27620–9.
50. Delgado M, Leceta J, Ganea D. Vasoactive intestinal peptide and pituitary adenylate cyclase-activating polypeptide inhibit the production of inflammatory mediators by activated microglia. J Leukoc Biol. 2003;73:155–64.
51. Delgado M, Ganea D. Inhibition of endotoxin-induced macrophage chemokine production by vasoactive intestinal peptide and pituitary adenylate cyclase-activating polypeptide in vitro and in vivo. J Immunol. 2001;167:966–75.
52. Brifault C, Gras M, Liot D, May V, Vaudry D, Wurtz O. Delayed pituitary adenylate cyclase-activating polypeptide delivery after brain stroke improves functional recovery by inducing m2 microglia/macrophage polarization. Stroke. 2015;46:520–8.
53. Armstrong BD, Abad C, Chhith S, Cheung-Lau G, Hajji OE, Nobuta H, et al. Impaired nerve regeneration and enhanced neuroinflammatory response in mice lacking pituitary adenylyl cyclase activating peptide. Neuroscience. 2008;151:63–73.

54. Armstrong BD, Hu Z, Abad C, Yamamoto M, Rodriguez WI, Cheng J, et al. Induction of neuropeptide gene expression and blockade of retrograde transport in facial motor neurons following local peripheral nerve inflammation in severe combined immunodeficiency and BALB/C mice. Neuroscience. 2004;129:93–9.
55. Armstrong BD, Hu Z, Abad C, Yamamoto M, Rodriguez WI, Cheng J, et al. Lymphocyte regulation of neuropeptide gene expression after neuronal injury. J Neurosci Res. 2003;74:240–7.
56. Cheng H, Ding Y, Yu R, Chen J, Wu C. Neuroprotection of a novel cyclopeptide C*HSDGIC* from the cyclization of PACAP (1-5) in cellular and rodent models of retinal ganglion cell apoptosis. PLoS One. 2014;9:e108090.
57. Endo K, Nakamachi T, Seki T, Kagami N, Wada Y, Nakamura K, et al. Neuroprotective effect of PACAP against NMDA-induced retinal damage in the mouse. J Mol Neurosci. 2011;43:22–9.
58. Wada Y, Nakamachi T, Endo K, Seki T, Ohtaki H, Tsuchikawa D, et al. PACAP attenuates NMDA-induced retinal damage in association with modulation of the microglia/macrophage status into an acquired deactivation subtype. J Mol Neurosci. 2013;51:493–502.
59. Fabian E, Reglodi D, Mester L, Szabo A, Szabadfi K, Tamas A, et al. Effects of PACAP on intracellular signaling pathways in human retinal pigment epithelial cells exposed to oxidative stress. J Mol Neurosci. 2012;48:493–500.
60. Banki E, Sosnowska D, Tucsek Z, Gautam T, Toth P, Tarantini S, et al. Age-related decline of autocrine pituitary adenylate cyclase-activating polypeptide impairs angiogenic capacity of rat cerebromicrovascular endothelial cells. J Gerontol A Biol Sci Med Sci. 2015;70:665–74.
61. Ohtaki H, Satoh A, Nakamachi T, Yofu S, Dohi K, Mori H, et al. Regulation of oxidative stress by pituitary adenylate cyclase-activating polypeptide (PACAP) mediated by PACAP receptor. J Mol Neurosci. 2010;42:397–403.
62. Metz VV, Kojro E, Rat D, Postina R. Induction of RAGE shedding by activation of G protein-coupled receptors. PLoS One. 2012;7:e41823.
63. Lacombe A, Lelievre V, Roselli CE, Salameh W, Lue YH, Lawson G, et al. Delayed testicular aging in pituitary adenylate cyclase-activating peptide (PACAP) null mice. Proc Natl Acad Sci U S A. 2006;103:3793–8.
64. Tang Y, Lv B, Wang H, Xiao X, Zuo X. PACAP inhibit the release and cytokine activity of HMGB1 and improve the survival during lethal endotoxemia. Int Immunopharmacol. 2008;8:1646–51.
65. Lv B, Tang Y, Chen F, Xiao X. Vasoactive intestinal peptide and pituary adenylate cyclase-activating polypeptide inhibit tissue factor expression in monocyte in vitro and in vivo. Shock. 2009;31:185–91.
66. Martínez C, Arranz A, Juarranz Y, Abad C, García-Gómez M, Rosignoli F, et al. PAC1 receptor: emerging target for septic shock therapy. Ann N Y Acad Sci. 2006;1070:405–10.
67. Delgado M, Martinez C, Pozo D, Calvo JR, Leceta J, Ganea D, et al. Vasoactive intestinal peptide (VIP) and pituitary adenylate cyclase-activation polypeptide (PACAP) protect mice from lethal endotoxemia through the inhibition of TNF-alpha and IL-6. J Immunol. 1999;162:1200–5.
68. Martinez C, Abad C, Delgado M, Arranz A, Juarranz MG, Rodriguez-Henche N, et al. Anti-inflammatory role in septic shock of pituitary adenylate cyclase-activating polypeptide receptor. Proc Natl Acad Sci U S A. 2002;99:1053–8.
69. Martínez C, Delgado M, Pozo D, Leceta J, Calvo JR, Ganea D, et al. Vasoactive intestinal peptide and pituitary adenylate cyclase-activating polypeptide modulate endotoxin-induced IL-6 production by murine peritoneal macrophages. J Leukoc Biol. 1998;63:591–601.
70. Arimura A, Li M, Batuman V. Potential protective action of pituitary adenylate cyclase-activating polypeptide (PACAP38) on in vitro and in vivo models of myeloma kidney injury. Blood. 2006;107:661–8.
71. Horvath G, Brubel R, Kovacs K, Reglodi D, Opper B, Ferencz A, et al. Effects of PACAP on oxidative stress-induced cell death in rat kidney and human hepatocyte cells. J Mol Neurosci. 2011;43:67–75.

72. Horvath G, Mark L, Brubel R, Szakaly P, Racz B, Kiss P, et al. Mice deficient in pituitary adenylate cyclase activating polypeptide display increased sensitivity to renal oxidative stress in vitro. Neurosci Lett. 2010;469:70–4.
73. Horvath G, Racz B, Szakaly P, Kiss P, Laszlo E, Hau L, et al. Mice deficient in neuropeptide PACAP demonstrate increased sensitivity to in vitro kidney hypoxia. Transplant Proc. 2010;42:2293–5.
74. Khan AM, Li M, Abdulnour-Nakhoul S, Maderdrut JL, Simon EE, Batuman V. Delayed administration of pituitary adenylate cyclase-activating polypeptide 38 ameliorates renal ischemia/reperfusion injury in mice by modulating Toll-like receptors. Peptides. 2012;38:395–403.
75. Khan AM, Li M, Brant E, Maderdrut JL, Majid DS, Simon EE, et al. Renoprotection with pituitary adenylate cyclase-activating polypeptide in cyclosporine A-induced nephrotoxicity. J Investig Med. 2011;59:793–802.
76. Khan AM, Maderdrut JL, Li M, Toliver HL, Coy DH, Simon EE, et al. Pituitary adenylate cyclase-activating polypeptide prevents contrast-induced nephropathy in a novel mouse model. Physiol Rep. 2013;1, e00163.
77. Horvath G, Racz B, Reglodi D, Kovacs K, Kiss P, Gallyas F, et al. Effects of PACAP on mitochondrial apoptotic pathways and cytokine expression in rats subjected to renal ischemia/reperfusion. J Mol Neurosci. 2010;42:411–8.
78. Sakamoto K, Kuno K, Takemoto M, He P, Ishikawa T, Onishi S, et al. Pituitary adenylate cyclase-activating polypeptide protects glomerular podocytes from inflammatory injuries. J Diabetes Res. 2015;2015:727152.
79. Li M, Maderdrut JL, Lertora JJ, Arimura A, Batuman V. Renoprotection by pituitary adenylate cyclase-activating polypeptide in multiple myeloma and other kidney diseases. Regul Pept. 2008;145:24–32.
80. Gasz B, Rácz B, Roth E, Borsiczky B, Ferencz A, Tamás A, et al. Pituitary adenylate cyclase activating polypeptide protects cardiomyocytes against oxidative stress-induced apoptosis. Peptides. 2006;27:87–94.
81. Gasz B, Rácz B, Röth E, Borsiczky B, Tamás A, Boronkai A, et al. PACAP inhibits oxidative stress-induced activation of MAP kinase-dependent apoptotic pathway in cultured cardiomyocytes. Ann N Y Acad Sci. 2006;1070:293–7.
82. Mori H, Nakamachi T, Ohtaki H, Yofu S, Sato A, Endo K, et al. Cardioprotective effect of endogenous pituitary adenylate cyclase-activating polypeptide on doxorubicin-induced cardiomyopathy in mice. Circ J. 2010;74:1183–90.
83. Helyes Z, Kun J, Dobrosi N, Sándor K, Németh J, Perkecz A, et al. Pituitary adenylate cyclase-activating polypeptide is upregulated in murine skin inflammation and mediates transient receptor potential vanilloid-1-induced neurogenic edema. J Invest Dermatol. 2015;135:2209–18.
84. Banki E, Hajna Z, Kemeny A, Botz B, Nagy P, Bolcskei K, et al. The selective PAC1 receptor agonist maxadilan inhibits neurogenic vasodilation and edema formation in the mouse skin. Neuropharmacology. 2014;85:538–47.
85. Giunta S, Castorina A, Marzagalli R, Szychlinska MA, Pichler K, Mobasheri A, et al. Ameliorative effects of PACAP against cartilage degeneration. Morphological, immunohistochemical and biochemical evidence from in vivo and in vitro models of rat osteoarthritis. Int J Mol Sci. 2015;16:5922–44.
86. Botz B, Bölcskei K, Kereskai L, Kovács M, Németh T, Szigeti K, et al. Differential regulatory role of pituitary adenylate cyclase-activating polypeptide in the serum-transfer arthritis model. Arthritis Rheumatol. 2014;66:2739–50.
87. Heimesaat MM, Dunay IR, Schulze S, Fischer A, Grundmann U, Alutis M, et al. Pituitary adenylate cyclase-activating polypeptide ameliorates experimental acute ileitis and extra-intestinal sequelae. PLoS One. 2014;9:e108389.
88. Ji H, Zhang Y, Shen XD, Gao F, Huang CY, Abad C, et al. Neuropeptide PACAP in mouse liver ischemia and reperfusion injury: immunomodulation by the cAMP-PKA pathway. Hepatology. 2013;57:1225–37.

89. Azuma YT, Hagi K, Shintani N, Kuwamura M, Nakajima H, Hashimoto H, et al. PACAP provides colonic protection against dextran sodium sulfate induced colitis. J Cell Physiol. 2008;216:111–9.
90. Hamagami K, Sakurai Y, Shintani N, Higuchi N, Ikeda K, Hashimoto H, et al. Over-expression of pancreatic pituitary adenylate cyclase-activating polypeptide (PACAP) aggravates cerulein-induced acute pancreatitis in mice. J Pharmacol Sci. 2009;110:451–8.
91. Vinuesa AG, Sancho R, García-Limones C, Behrens A, ten Dijke P, Calzado MA, et al. Vanilloid receptor-1 regulates neurogenic inflammation in colon and protects mice from colon cancer. Cancer Res. 2012;72:1705–16.
92. Elekes K, Sandor K, Moricz A, Kereskai L, Kemeny A, Szoke E, et al. Pituitary adenylate cyclase-activating polypeptide plays an anti-inflammatory role in endotoxin-induced airway inflammation: in vivo study with gene-deleted mice. Peptides. 2011;32:1439–46.
93. Lauenstein HD, Quarcoo D, Plappert L, Schleh C, Nassimi M, Pilzner C, et al. Pituitary adenylate cyclase-activating peptide receptor 1 mediates anti-inflammatory effects in allergic airway inflammation in mice. Clin Exp Allergy. 2011;41:592–601.
94. El Zein N, Badran B, Sariban E. The neuropeptide pituitary adenylate cyclase activating polypeptide modulates Ca^{2+} and pro-inflammatory functions in human monocytes through the G protein-coupled receptors VPAC-1 and formyl peptide receptor-like 1. Cell Calcium. 2008;43:270–84.
95. El Zein N, Badran BM, Sariban E. The neuropeptide pituitary adenylate cyclase activating protein stimulates human monocytes by transactivation of the Trk/NGF pathway. Cell Signal. 2007;19:152–62.
96. Harfi I, D'Hondt S, Corazza F, Sariban E. Regulation of human polymorphonuclear leukocytes functions by the neuropeptide pituitary adenylate cyclase-activating polypeptide after activation of MAPKs. J Immunol. 2004;173:4154–63.
97. Harfi I, Sariban E. Mechanisms and modulation of pituitary adenylate cyclase-activating protein-induced calcium mobilization in human neutrophils. Ann N Y Acad Sci. 2006;1070:322–9.
98. Harfi I, Corazza F, D'Hondt S, Sariban E. Differential calcium regulation of proinflammatory activities in human neutrophils exposed to the neuropeptide pituitary adenylate cyclase-activating protein. J Immunol. 2005;175:4091–102.

Chapter 40
Immunobiology of the Pituitary Adenylate Cyclase-Activating Peptide

Mario Delgado

Abstract The immune system is faced with the daunting job of defending the organism against invading pathogens, while at the same time maintaining tolerance to the body's own tissues, thereby preserving its integrity. Loss of immune tolerance compromises immune homeostasis and results in the onset of autoimmune disorders. Identification of the endogenous factors that control immune tolerance and inflammation is a key goal in the field of immunology. Evidences from the last two decades indicate that the neuropeptide pituitary adenylate cyclase-activating peptide (PACAP) is one of these endogenous factors. PACAP is produced by immune cells, and through its binding to specific receptors, it exerts potent anti-inflammatory actions and participates in the maintenance of immune tolerance at multiple levels, especially in immunological disorders. Therefore, PACAP is a key element in the bidirectional communication that exists between the neuroendocrine and immune systems in order to maintain body's homeostasis.

Keywords Immune system • Inflammation • Autoimmunity • Lymphocyte • Macrophage • Microglia

Introduction: When Neuroendocrine and Immune Systems Talk Each Other

The immune system responds to pathogen invasion with two temporarily separate but physically linked responses, mediated by a variety of cells. The first response, named innate immunity, is rapid and involves neutrophils, monocytes/macrophages, and dendritic cells (DCs) in the periphery and microglia in the central nervous system (CNS). In contrast, the second response, termed adaptive immunity, occurs later, following activation of T and B lymphocytes through specific receptors.

M. Delgado (✉)
Instituto de Parasitologia y Biomedicina Lopez-Neyra, CSIC,
Avd. Conocimiento, PT Ciencias de la Salud, Granada 18016, Spain
e-mail: mdelgado@ipb.csic.es

D. Reglodi, A. Tamas (eds.), *Pituitary Adenylate Cyclase Activating Polypeptide — PACAP*, Current Topics in Neurotoxicity 11,
DOI 10.1007/978-3-319-35135-3_40

In contrast to the innate response, adaptive immunity leads to the development of memory for a specific antigen. The successful elimination of most pathogens requires crosstalk between the innate and adaptive arms of the immune system. The innate immune system recognizes conserved pathogen-associated molecular patterns through pattern-recognition receptors, such as Toll-like receptors (TLRs), which induce the release of proinflammatory cytokines, chemokines and free radicals, recruitment of inflammatory cells to the site of infection, and lysis of infected host cells by cytotoxic T lymphocytes and natural killer cells. Following the elimination of the injurious agent, inflammation has to be resolved, and cells participating in innate and adaptive immunity have to be deactivated or eliminated to reestablish homeostasis. The consequences of uncontrolled activation of innate and adaptive immunity and of sustained production of inflammatory mediators are deleterious for the host [1]. Acute inflammation can lead to chronic inflammation, scarring and tissue destruction, and eventually to organ failure. A further damage arises from potential autoimmune responses occurring during chronic inflammatory responses, in which the immune mediators that respond to pathogen-derived antigens can also react to self-antigens.

Although there is substantial information on the molecular basis of the initiation of an inflammatory process, the mechanisms that resolve inflammatory responses are much more unknown. It has been recognized that inflammation is self-controlled by endogenous anti-inflammatory mediators secreted by cells of the host immune system during an ongoing process, and that the ability to control an inflammatory state depends on the local balance between pro- and anti-inflammatory factors [1]. Moreover, the immune response against self-antigens is normally prevented and controlled by long-term tolerance, established and regulated by central clonal deletion of self-reactive T cells in the thymus, induction of anergy in self-reactive T cells in the periphery and generation of antigen-specific regulatory T cells (Treg) which suppress the activity of self-reactive effector T cells [2]. Only very recently, we have started unrevealing the cells and factors involved in the induction and maintenance of immune tolerance. A number of traditional immunosuppressive and anti-inflammatory cytokines, such as IL-10, IL-13, and TGF-b1, have been shown to also play an important role in tolerance, particularly in the generation of Treg cells [2].

In the last three decades, it became evident that the immune response is not solely mediated by immune cells, and that cells and factors belonging to other integrative systems cooperate to regulate the complex immune homeostasis. Thus, a bidirectional connection exists between the neuroendocrine and immune systems, that serves to mount a variety of coordinated responses to external/internal danger. The immune system acts as a sixth sense signaling the brain to respond to pathogen infection, inflammation and tissue injury, leading to changes in febrile and behavioral responses [3]. Conversely, the immune system is controlled by the CNS, mainly in response to environmental stress. This intimate network is supported on a mutual biochemical language, that involves shared ligands (neuropeptides, hormones, cytokines) and the respective receptors. For example, glucocorticoids and norepinephrine are endogenous immunoregulatory agents produced by the hypothalamic-pituitary-adrenal axis and the sympathetic nervous system, respectively, in response to stress or systemic inflammation [3]. Moreover, various neuropeptides are released at the peripheral peptidergic endings of sensory and efferent nerves in close proximity to immune cells in response to various invasive stimuli [3]. The discovery that some neuropeptides are also produced by diverse immune cells [4] is probably the most exciting finding added to this crosstalk. After that, it was reasonable to assume that these neuropeptides could contribute, in a cytokine-like manner, to the regulation of the immune response. From the growing list of neuropeptides currently thought to posses immunomodulatory actions (approximately 50), the family comprising the vasoactive intestinal peptide (VIP) and the pituitary adenylate cyclase-activating peptide (PACAP) emerged probably as one of the most relevant for various reasons. First, they are produced by immune cells during inflammatory conditions or following antigenic stimulation. Second, they exert their effect through various G-protein-coupled receptors (GPCRs) expressed in various immune cells. Third, they signal mainly through the activation of cAMP/protein kinase A (PKA) and downregulate the activation of several transduction pathways and transcription factors essential for the expression of most of the inflammatory mediators. In this chapter, I will review the discovery of PACAP in the immune system, will focus on recent developments regarding the effects of PACAP on immune tolerance and resolution of inflammation, and will highlight the effectiveness of using PACAP in treating several preclinical models of inflammatory and autoimmune disorders.

PACAP Is Produced by Immune Cells

Five years after its discovery by Arimura in the ovine pituitary, Gaytan and coworkers described PACAP-like immunoreactivity in immune cells, apparently with lymphocyte morphology, in samples isolated from rat thymus, bone marrow, spleen, and lymph nodes [5]. By using updated molecular techniques, we later confirmed that PACAP was produced by different subsets of lymphocytes, including CD4+CD8+ thymocytes and CD4+ and CD8+ T cells of central (thymus) and peripheral lymphoid organs (spleen and lymph nodes) [6]. The preferential immune localization of PACAP in lymphocytes, but not in macrophages, corresponded exactly with those described previously for VIP (reviewed in [7]), suggesting that the two peptides of the family could play a major role in these cells, as it was confirmed later. VIP was then described to be specifically produced by Th2 cells, but not by Th1 cells, in response to antigen stimulation in vitro and in vivo [8]. However, this interesting fact has not been confirmed for PACAP. Moreover, it is still unsolved whether lymphocytes preferentially express any of the two forms of PACAP (PACAP27 or PACAP38), as occurs in other tissues and cells. In any case, as discussed later, evidence indicates that endogenous PACAP plays a major role in the control of inflammatory and self-reactive immune responses.

PACAP Receptors in Cells of the Immune System

The two alternatively processed forms of PACAP, as well as VIP, elicit their biological action in immune system through binding to a subset of three specific membrane receptors belonging to the large family B of GPCRs: VPAC1 (also known as VIP1, VIP/PACAP type II, or PVR 2), VPAC2 (also known as VIP2, VIP/PACAP type III, or PVR 3), and PAC1 (also known as VIP/PACAP type I receptor, or PVR 1) [9]. In general, PACAP receptors in the immune system share the same molecular basis of ligand–receptor interaction as in other cells and tissues, with an important exception [7]. Thus, VPAC1 and VPAC2 bind VIP and PACAP with equal affinity ($Kd \approx 0.6$ nM). However, the PAC1 isoform expressed in immune cells binds VIP and PACAP with similar affinities ($Kd \approx 100$ nM), clearly differing from the PAC1 isoforms expressed in the CNS, that show a high affinity for PACAP-27 and PACAP-38 ($Kd \approx 0.5$ nM), but a much lower affinity for VIP ($Kd > 500$ nM) [10]. The expression of fully functional VIP/PACAP receptors in the immune system was first claimed in human peripheral blood lymphocytes three decades ago through binding techniques (using ^{125}I-VIP as a ligand) and adenylyl cyclase measurements [11]. Later on, VIP/PACAP binding sites were identified in human peripheral lymphocytes and monocytes and in murine lymphocytes and macrophages [7]. The cloning and molecular characterization of the three VIP/PACAP receptor genes allowed the study of the pattern of expression of PACAP receptors in immune cells (Table 40.1). In general, it is accepted that VPAC1 is constitutively expressed in lymphocytes (including thymocytes, CD4 and CD8 T cells in peripheral lymphoid organs), macrophages, monocytes, DCs, microglia, and mast cells, and that VPAC2 is scarcely expressed in these cells when they are in a naïve or resting state, but its expression is induced following stimulation in lymphocytes, monocytes, and macrophages (Table 40.1). PAC1 is only expressed by cells of the macrophage/monocyte lineage, including microglia and osteoclasts. Interestingly, the expression of PAC1 is lost during differentiation of monocytes or bone marrow precursors to DCs. Similar to other tissues and cells, VPAC1, VPAC2, and PAC1 are coupled to adenylate cyclase activation and subsequent activation of PKA pathway in immune cells [7]. Moreover, VIP binding to PAC1 activates phospholipase C and protein kinase C (PKC) in macrophages and monocytes.

Most of the studies using specific agonists and antagonists for the different VIP receptors have established that VPAC1 is the major mediator in the immunomodulatory effects of PACAP, both in vitro and in vivo, with a moderate involvement of VPAC2, and minimal or none for PAC1 [7, 12, 13]. However, the development of mice deficient for VPAC2 or PAC1 revealed that both receptors must participate in the effects of PACAP on the immune system, since increased susceptibility for inflammatory disorders has been described in both PAC1-KO and VPAC2-KO mice [14–19]. Although VPAC1-deficient mice are not presently available due to the crucial role played by this receptor during the embryonic development, a critical role for VPAC1 in the regulation of immune disorders has been reported in humans. Reduced expression of VPAC1 has been associated with reduced responses to its ligands (VIP and PACAP) in immune cells of patients with autoimmune and chronic inflammatory disorders such as ankylosing spondylitis, rheumatoid arthritis and

Table 40.1 PACAP receptors on immune cells

Immune cells	Receptors VPAC1	VPAC2	PAC1	Comments
Lymphocytes				
Thymocytes:				
Murine	+	–(+)	–	VPAC1 expressed on DP and SP $CD4^+$ and $CD8^+$. VPAC2 inducible upon stimulation
Murine	+	+	–	VPAC2 involved in SP $CD4^+$ differentiation
Human	+	+	–	VPAC2 mainly expressed
CD4 T cells	+	-(+)	–	Spleen and lymph node cells. VPAC2 induced upon TCR/antigen stimulation and inflammation
CD8 T cells				
Human PBLs	+	–(+)	–	VPAC1 decreases after stimulation. VPAC2 low levels
B cells	+	–	ND	Controversial. VPAC1 mRNA detected on rat, but not on mouse. Few VIP-binding sites described on human B cells
Mast cells	+	+	–	VPAC2 decreases after stimulation and in atopic dermatitis
Granulocytes				
Neutrophils	ND	ND	ND	Suggested VIP-binding sites by functional studies, although with pharmacological VIP doses (>>*K*d of VPACs)
Macrophages				
Monocytes	+	–(+)	+	VPAC2 induced after inflammatory stimulation
Alveolar MΦ	+	+	ND	VPAC1 increases after lung inflammation
Peritoneal MΦ	+	–(+)	+	VPAC2 induced after stimulation in cell lines
Microglia	+	–	+	VPAC2 unresponsive to inflammatory stimulation
Osteoclasts	+	–	+	
Dendritic cells				
Langerhans cell	+	+	–	
BM-DCs	+	+	–	

DP double positive, *SP* single positive, *BM-DCs* bone marrow-derived DCs, *MΦ* macrophages, *PBLs* peripheral blood lymphocytes, *ND* not determined. For references see [7]

osteoarthritis [20–23]. Interestingly, the reduced expression of VPAC1 in patients with arthritis was genetically associated with a polymorphism found the 3′UTR region of the VPAC1 gene [20, 22, 23], and the microRNA 525-5p, which specifically targets the 3′UTR region of the VPAC1 gene, decreased the expression of VPAC1 transcripts in activated monocytes [24, 25]. Moreover, it has been reported a decrease in VPAC2 expression in Th1 cells of patients with multiple sclerosis, although no associations with genetic polymorphisms were found in this case [26]. In agreement, deficiency in VPAC2 gene predisposed mice to suffer a more severe autoimmune encephalomyelitis, a preclinical model of multiple sclerosis [19]. These data suggest that defects in the VIP/PACAP receptor/signaling system predisposes to a higher frequency of inflammatory and autoimmune disorders.

Effects of PACAP on Immune Cells

PACAP, together with VIP, are two of the best studied immunoregulatory neuropeptides. Many studies indicate that PACAP affects both innate and adaptive immune responses (Fig. 40.1), and acts as a major anti-inflammatory factor in experimental models of inflammatory and autoimmune diseases, and evidence identified PACAP as a very attractive factor for the design of new therapeutic strategies.

Effects of PACAP on the Innate Immune Response

Early data indicated that PACAP could exert a dual role in regulating innate immune response depending on the state of activation of the cell and its environment. Thus, PACAP is able to stimulate the phagocytic activity, the adhesion and mobility and the release of free radicals, and the inflammatory cytokine IL-6 by resting macrophages [27–29]. This effect could be related with a primary effect on killing pathogens, the immediate function of these cells. In agreement, recent data has identified PACAP as a potential natural antimicrobial peptide. Based on an amphipathic and cationic structure, PACAP binds to the negatively charged surface of pathogens and kills them through various mechanisms that involve loss of membrane integrity and energetic collapse of the pathogen [30].

Fig. 40.1 (continued) of Treg cells with suppressive effects on activated T cells through two mechanisms that imply indirect actions on the generation of tolerogenic DCs and probably direct effects on naïve T cells and expansion of Treg cells. The impairment on Th17 responses has been only observed in vivo. As a consequence, the PACAP effects on the innate and adaptive immune responses result in diminished inflammation and organ-specific autoimmunity and in restoration of immune tolerance, which could be therapeutically exploited on autoimmune and inflammatory diseases

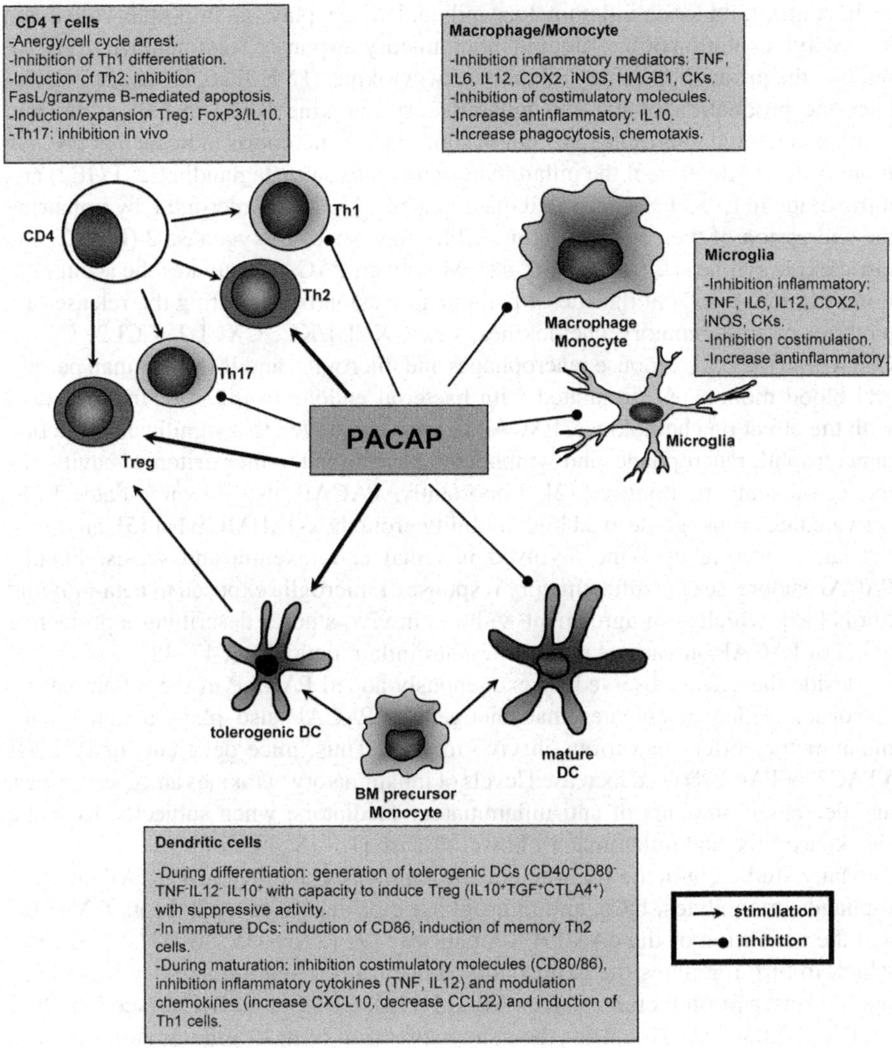

Fig. 40.1 PACAP regulates key events of the innate and adaptive immune responses. PACAP is released in the context of an immune response through different sources: nerves release PACAP as a neuropeptide in lymphoid organs, lymphocytes secrete PACAP under stimulation, and blood contains variable amounts of PACAP as a consequence of endocrine secretion or a therapeutic systemic administration of the peptide. PACAP can impair the inflammatory response directly acting on macrophages, monocytes, and microglia by inhibiting the production of inflammatory mediators (cytokines, chemokines CKs, lipids, and free radicals) and by inducing the production of the anti-inflammatory cytokine IL-10. In addition, PACAP deactivates the costimulatory response of macrophages and DCs on T cells, limiting the clonal expansion of Th1 cells under inflammatory conditions. On the other hand, PACAP regulates the adaptive response in different ways. First, PACAP inhibits the differentiation of Th1 cells and favors the expansion of Th2 cells through various nonexcluding mechanisms that involve regulation of DC functions, Th1-differentiating factors, chemokines, and apoptosis. B cell responses and switching IgG isotype are indirectly affected by this PACAP effect vias Th2-responses. Second, PACAP induces the emergence

In contrast, under an inflammatory milieu, PACAP plays an important role in the successful resolution of the ongoing inflammatory response. Treatment with PACAP inhibits the production of the inflammatory cytokines TNF, IL-6, IL-12, and stimulates the production of the anti-inflammatory cytokine IL-10 in LPS-stimulated macrophages and microglia [31–39]. In addition, recent reports indicate that PACAP inhibits the production of the inflammatory mediators prostaglandin E2 (PGE2) and nitric oxide in LPS/IFNγ-activated macrophages, DCs, and microglia, by impairing the expression of their respective inducible enzymes cycloxygenase 2 (COX2) and nitric oxide synthase 2 (iNOS) [40, 41]. Moreover, PACAP regulates the accumulation of immune cells at the site of pathogen invasion by inhibiting the release of a plethora of inflammatory chemokines, i.e., CXCL1/KC, CXCL2, CCL2, CCL3, CCL4, and CCL5, by mouse macrophages and microglia, and IL-8 in human peripheral blood monocytes stimulated with bacterial endotoxin [42–44]. In agreement with the effect on chemokines, PACAP administration led to a significant reduction in neutrophil, macrophage, and lymphocyte recruitment to the peritoneal cavity in a model of acute peritonitis [42]. Importantly, PACAP also downregulates LPS-activated macrophage-derived high mobility group box-1 (HMGB1) [45], an essential late-occurring cytokine involved in lethal endotoxemia and sepsis. Finally, PACAP suppresses the inflammatory response of microglia exposed to beta-amyloid fibrils [46], which is in agreement with the in vivo studies describing a protective effect of PACAP on various models of neuroinflammation [32, 47, 48].

Beside the effect observed by exogenously added PACAP in the inflammatory response, evidence indicates that endogenous PACAP also plays a major anti-inflammatory effect in various in vivo models. Thus, mice deficient for PACAP, VPAC2, or PAC1 showed increased levels of inflammatory cytokines and chemokines, and decreased amounts of anti-inflammatory mediators, when subjected to septic shock, arthritis, and inflammatory bowel disease [16–18, 49–51].

Many studies indicate that the anti-inflammatory activity of PACAP on macrophages, monocytes, DCs, and microglia are primarily exerted through VPAC1 and the activation of the cAMP/PKA pathway [7, 12, 31–35, 40–42, 44, 52–54], which in turn regulates the expression and/or transactivating activity of a plethora of transcription factors such as AP-1, NFkB, CREB, and IRF-1 (see Fig. 40.2 for details) that are critical for the gene expression of most inflammatory mediators [31, 33, 34, 40, 42, 44, 46, 52, 55–58]. In addition, a PKA-independent pathway also participates in the PACAP deactivation of macrophages and monocytes by inhibiting the nuclear translocation of the transcription factor NFkB [59], although it does not seem to be involved in the effect of PACAP on microglia and DCs. In addition to the inhibition all these downstream signaling pathways, PACAP also reduces the capacity of monocyte/macrophages to detect TLR-mediated signaling by inducing membrane shedding of CD14, an essential accessory molecule for TLR signaling [60]. Finally, the participation of PAC1 and PKC pathway has been demonstrated only in the stimulatory effect of PACAP on adherence, migration, phagocytosis, superoxide production, and IL-6 production in resting macrophages [27–29].

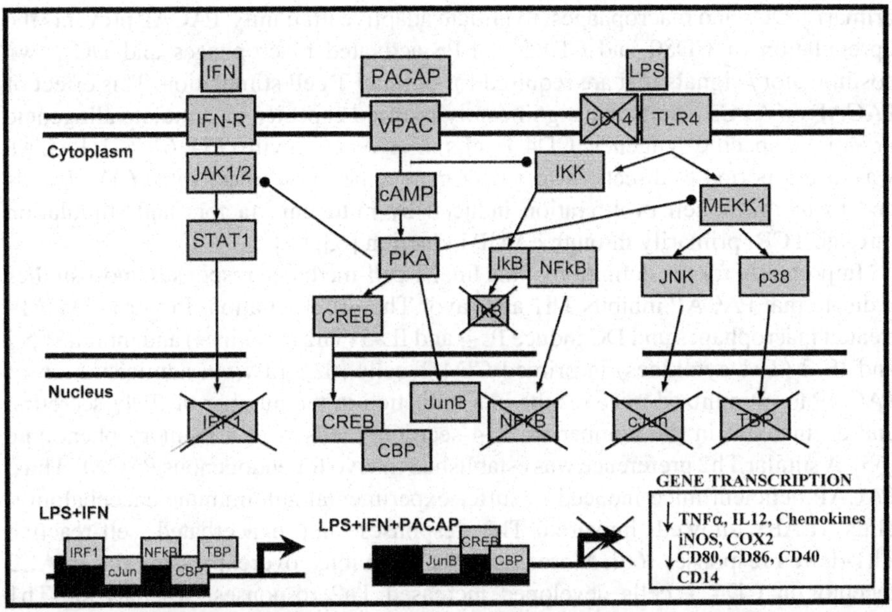

Fig. 40.2 Molecular mechanisms and transcription factors involved in the PACAP signaling during inflammatory response (macrophage, microglia, and DCs). Binding of PACAP to VPAC1 induces cAMP and activates PKA, and exerts several effects: (1) inhibition of IFNγ-induced Jak1/Jak2 phosphorylation, STAT1 activation, and binding to promoters of inflammatory genes (CD40, CXCL10, iNOS, COX2) which are dependent of the regulation by IRF-1; (2) inhibition of various MAPK cascades, initiated with the suppression of MEKK1/MEK4 and of MEKK1/MEK6 activities, and subsequent inhibition of Jun-kinase (JNK) and p38 MAPK activities, respectively. Consequently there is a change in the composition of transcription factor AP-1 with c-Jun being replaced by JunB, and inhibition of TATA-box binding protein (TBP) phosphorylation and nuclear translocation. Moreover, through a PKA-independent mechanism, PACAP–VPAC1 interaction inhibits IkB-kinase activity and suppresses nuclear translocation and activation of the transcription factor NFkB. AP1, TATA-box protein, and NFkB act in concert to activate gene transcription of most of inflammatory cytokines and chemokines as well as costimulatory molecules. In parallel, PACAP-induced PKA activation stimulates cAMP-responsive element binding (CREB) factor to compete with NFkB for coactivators, such as p300-CBP, required for transcription of inflammatory genes. LPS, bacterial lipopolysaccharide. *Arrows* indicate activation. *Dotted-end lines* indicate inhibition

Effects of PACAP on the Adaptive Immune T Cell Response

In addition to their immediate function in killing pathogens, macrophages, microglia, and especially DCs have the capacity to process naïve antigen and initiate an adaptive immune response, through stimulatory and costimulatory contacts with naïve T cells bearing the appropriate TCR. Following recognition of antigenic peptides complexed to MHC class II, CD4 T lymphocytes proliferate and differentiate into Th1, Th2, Th17, and inducible peripheral Treg effectors. CD4 T cells are major targets for PACAP regulation, with PACAP affecting both activation of naïve T cells and their differentiation into effector cells (Fig. 40.1). This effect on CD4 T cells is partially mediated by regulating the capacity of antigen-presenting cells (APCs), primarily DCs and macrophages, to initiate adaptive immunity. PACAP prevents the upregulation of CD80 and CD86 in LPS-activated macrophages and DCs, two costimulatory signals that are required for optimal T cell stimulation. This effect of PACAP on APCs results in a significantly reduced capacity to stimulate allogeneic or antigen-specific syngeneic CD4 T cells in vivo and in vitro [54, 61, 62]. PACAP was also reported to directly affect IL-2 gene expression in activated CD4 T cells and to inhibit T cell proliferation induced by mitogenic factors and stimulation through TCR, primarily through cAMP induction [63, 64].

Importantly for the definition of the final T cell-mediated responses, most studies indicate that PACAP inhibits Th1 and favor Th2 differentiation. In vitro, PACAP-treated macrophages and DC induce IL-4 and IL-5 (Th2 cytokines) and inhibit IFNγ and IL-2 (Th1 cytokines) in primed CD4 T cells [62]. In vivo, administration of PACAP to immunized mice results in a reduction in the number of IFNγ-secreting and an increase in the number of IL-4-secreting cells with a memory phenotype [65]. A similar Th2 preference was established in vivo for endogenous PACAP. Thus, PACAP-deficient mice induced to suffer experimental autoimmune encephalomyelitis (EAE) showed impaired Th2 responses and exacerbated self-reactive Th1-driven responses [66]. Moreover, transgenic mice overexpressing the VPAC2 receptor in CD4 T cells developed increased Th2 responses, whereas the Th1 response prevailed in VPAC2-deficient mice [14, 67, 68]. A number of nonexcluding mechanisms contribute to the PACAP-induced Th2 bias. PACAP affects Th1/Th2 generation indirectly by inhibiting IL-12 production in activated APCs, and directly by blocking IL-12 signaling through the inhibition of JAK2/STAT1 phosphorylation [31, 36]. Finally, PACAP also preferentially supports the survival of Th2, but not of Th1 effectors through a mechanism that involves the inhibition of FasL and granzyme B expression [69–72]. In addition to promoting Th2 differentiation and survival, PACAP also affects Th1/Th2 migration in a differential manner. PACAP promotes Th2 and inhibits Th1 migration, by promoting antigen-stimulated DC production of the Th2-attracting chemokine CCL22 and inhibiting the release of CXCL10, a Th1-attracting chemokine [53, 73]. In agreement with the in vitro studies, intraperitoneal administration of PACAP-treated DC pulsed with antigen led to the preferential accumulation of Th2 effectors in the peritoneal cavity [53].

Regarding Th17 cells, it remains unsolved whether PACAP affects in vitro the activation or differentiation of this lineage of CD4 effector T cells that play a major role in autoimmunity [74]. However, evidence in experimental of autoimmunity supports that PACAP reduces the generation of Th17 cells. Thus, mice deficient in PACAP or VPAC2 subjected to autoimmune disorders showed increased numbers of Th17 cells in peripheral lymphoid organs and affected tissues [19, 66].

PACAP and the Induction of Immune Tolerance

Probably one of the most important pieces added recently to the puzzle of immunobiology of PACAP was that identified this neuropeptide as a critical factor involved in the induction of immune tolerance. Diverse in vivo studies showed that the administration of PACAP to animals suffering autoimmune and inflammatory disorders increased the number of CD4+FoxP3+ Treg cells concomitant with disease amelioration [75, 76]. Similarly, PACAP- and VPAC2-deficient mice suffering autoimmunity have decreased levels of Treg cells in thymus, spleen, and lymph nodes [19, 66, 77]. The PACAP-mediated increase in Treg cells during the ongoing autoimmune response would regulate the activity of self-reactive Th1 and Th17 clones and subsequent inflammatory destructive responses. In addition, this is a mechanism to induced long-lasting therapeutic responses, even in the absence of PACAP.

Whether PACAP is able to induce directly this Treg cell population is still unknown. However, various recent studies described the capacity of VIP to induce de novo the generation of human and mouse IL-10 producing CD4+FoxP3+ Treg cells from a naïve population [78, 79]. Even in patients with sarcoidosis, the administration of nebulized VIP resulted in increased numbers of $CD4^{+}FoxP3^{+}CD127^{-}CD25^{+}$ Treg in the bronchoalveolar lavage [79]. Because the effects of VIP on the generation of Treg cells are mediated through the VPAC1-cAMP-PKA pathway [78, 80], it is quite plausible that PACAP would exert a similar action. Interestingly, it has been recently observed that thymic and splenic Treg cells isolated from PACAP-deficient mice showed impaired capacity to expand upon stimulation [77]. This finding differs from that observed for VIP that showed no effect on the expansion of already generated Treg cells [78, 79].

In any case, beside its potential capacity to generate or expand directly Treg cells, evidence indicates that PACAP is able to increase the population of Treg cells indirectly through its action on DCs. Although DCs are essential for the stimulation of antigen-specific T cells, they also function to establish and maintain tolerance. DCs regulate the response of effector T cells by causing anergy or deletion of stimulated T cells, and by inducing Treg [81]. Similar to Treg, tolerogenic DC can be divided into natural and inducible tolerogenic DCs. The later are of major interest since they can be induced with biological and pharmacological agents and manipulated to present specific autoantigens. Together with galectin 1, vitamin D3, IL-10, and TNF, PACAP belongs to the group of biological agents that induce tolerogenic DCs [81]. This, differentiation of murine bone marrow-derived DC in the presence of PACAP led to the development of DC with a tolerogenic phenotype, i.e., low expression of CD40/80/86, reduced production of TNF and IL-12, and increased secretion of IL-10 following LPS stimulation. The DCs generated in the presence of PACAP in vitro and in vivo induced IL-10-producing $CD4^{+}Foxp3^{+}$ Treg which inhibited the proliferation of allogeneic or syngeneic Ag-specific T cells and were able to transfer tolerance to naïve recipients [82, 83]. Although it remains to be confirmed for PACAP, similar results were obtained with human blood monocyte-derived DC differentiated in the presence of VIP. VIP led to the generation of human tolerogenic DC which in turn induced IL-10 producing human CD4 and CD8 Treg. Both types of human Treg suppressed the proliferation/activation of antigen-specific Th1 cells [84], extending the potential application of these cells to human disorders.

Potential of PACAP in the Treatment of Inflammatory and Autoimmune Diseases

The evidence reviewed in the previous section indicates that, through its potent and diverse anti-inflammatory and/or immunosuppressive actions, PACAP is part of a feedback circuit that limits ongoing inflammatory and immune responses. Recent studies proved the relevance of PACAP to human health. Treatment with PACAP decreases the frequency and severity of various experimental models of sepsis, ileitis, hepatitis, respiratory inflammatory disorders, neurodegenerative disorders, rheumatoid arthritis, inflammatory bowel disease and multiple sclerosis [15, 76, 85–90]. In disorders characterized by an exacerbated inflammatory response, such as sepsis, inflammatory pulmonary disease and hepatitis, the beneficial effect of PACAP is exerted through the downregulation of a wide spectrum of inflammatory cytokines (mainly TNF, IL-6, IL-12 and IFNγ), chemokines and mediators of oxidative stress, at both systemic and local levels. PACAP have been proven effective in neurodegenerative diseases characterized by inflammation, such as spinal cord injury, brain trauma and Parkinson's disease [38, 91–93], by diminishing neuroinflammation and neurodegeneration, partially through its effect on microglia.

In the case of autoimmune disorders, the therapeutic effects of PACAP are associated with the reduction of both early events that are associated with the initiation and establishment of autoimmunity, and of later phases that are associated with an evolving immune and destructive inflammatory response. PACAP impairs the development of self-reactive Th1 and Th17 cells, their entry into the target organs, the release of proinflammatory cytokines and chemokines, and the subsequent recruitment and activation of macrophages and neutrophils. Although PACAP can directly regulate the activation of macrophages and T cells, PACAP also regulates the immune response through the modulation at multiple levels of the differentiation and activation of DCs and through the induction of Treg cells.

Despite the effectiveness of PACAP in experimental models of inflammation and autoimmunity, no clinical trials have been initiated to date to treat immune disorders. In the case of VIP, only two clinical trials have been performed. A phase I clinical trial sponsored by the National Institutes of Health (NCT00004494 clinical trial, http://www.ClinicalTrials.gov) demonstrated the effectiveness of systemic VIP administration in patients with sepsis and pulmonary distress syndrome. Moreover, in a recent open label phase II clinical trial, VIP has been proven clinically effective (decrease of inflammatory markers in the lung and increase of Treg cell numbers) in patients with sarcoidosis, a systemic disease of unknown etiology characterized by the formation of granulomas especially in the lung that courses with both inflammatory and autoimmune components [79]. One of the major obstacles for the translation of PACAP-based treatments into viable clinic therapies is related to its sensitivity to degradation by peptidases. Many general strategies have been proposed to increase the VIP half-life (residue substitutions, combination with endopeptidase inhibitors) and to improve its targeted tissue delivery (gene-based therapy, insertion in nanoparticles or micelles) with certain success in several experimental models of immune disorders [94], which should be also extended to PACAP in a near future. We must also keep in mind the possibility of using PACAP to generate ex vivo antigen-specific tolerogenic DCs and Treg cells for treatments related to immune diseases.

In summary, PACAP emerges as a key endogenous regulator of the immune homeostasis with a great potential to be included in the design of new therapeutic strategies to treat immune disorders that course with exacerbated inflammatory and autoimmune responses. However, its fully translation to the clinical practice will depend on the development of new tools that improve its stability and/or delivery.

References

1. Goodnow CC. Multistep pathogenesis of autoimmune diseases. Cell. 2007;130:25–35.
2. Sakaguchi S, Miyara M, Costantino CM, Hafler DA. FOXP3+ regulatory T cells in the human immune system. Nat Rev Immunol. 2010;10:490–500.
3. Sternberg EM. Neural regulation of innate immunity: a coordinated nonspecific host response to pathogens. Nat Rev Immunol. 2006;6:318–28.
4. Gonzalez-Rey E, Chorny A, Delgado M. Regulation of immune tolerance by anti-inflammatory neuropeptides. Nat Rev Immunol. 2007;7:52–63.
5. Gaytan F, Martinez-Fuente AJ, Gracia-Navarro F, Vaudry H, Aguilar E. Pituitary adenylate cyclase-activating peptide (PACAP) immunolocalization in lymphoid tissues of the rat. Cell Tissue Res. 1994;276:223–7.
6. Abad C, Martinez C, Leceta J, Juarranz MG, Delgado M, Gomariz RP. Pituitary adenylate-cyclase-activating polypeptide expression in the immune system. Neuroimmunomodulation. 2002;10:177–86.
7. Delgado M, Pozo D, Ganea D. The significance of vasoactive intestinal peptide in immunomodulation. Pharmacol Rev. 2004;56:249–90.
8. Delgado M, Ganea D. Cutting edge: is vasoactive intestinal peptide a type 2 cytokine? J Immunol. 2001;166:2907–12.
9. Harmar A, Arimura A, Gozes I, Journot L, Laburthe M, Pisegna J, et al. International Union of Pharmacology. XVIII. Nomenclature of receptors for vasoactive intestinal peptide and pituitary adenylate cyclase-activating polypeptide. Pharmacol Rev. 1998;50:265–70.
10. Pozo D, Delgado M, Martinez C, Gomariz RP, Guerrero JM, Calvo JR. Functional characterization and mRNA expression of pituitary adenylate cyclase activating polypeptide (PACAP) type I receptors in rat peritoneal macrophages. Biochim Biophys Acta. 1997;1359:250–62.
11. Guerrero J, Prieto J, Elorza F, Ramirez R, Goberna R. Interaction of vasoactive intestinal peptide with human blood mononuclear cells. Mol Cell Endocrinol. 1981;21:151–60.
12. Delgado M, Gomariz RP, Martinez C, Abad C, Leceta J. Anti-inflammatory properties of the type 1 and type 2 vasoactive intestinal peptide receptors: role in lethal endotoxic shock. Eur J Immunol. 2000;30:3236–46.
13. Gonzalez-Rey E, Delgado M. Anti-inflammatory neuropeptide receptors: new therapeutic targets for immune disorders? Trends Pharmacol Sci. 2007;28:482–91.
14. Goetzl EJ, Voice JK, Shen S, Dorsam G, Kong Y, West KM, et al. Enhanced delayed-type hypersensitivity and diminished immediate-type hypersensitivity in mice lacking the inducible VPAC(2) receptor for vasoactive intestinal peptide. Proc Natl Acad Sci U S A. 2001;98:13854–9.
15. Lauenstein HD, Quarcoo D, Plappert L, Schleh C, Nassimi M, Pilzner C, et al. Pituitary adenylate cyclase-activating peptide receptor 1 mediates anti-inflammatory effects in allergic airway inflammation in mice. Clin Exp Allergy. 2011;41:592–601.
16. Martinez C, Abad C, Delgado M, Arranz A, Juarranz MG, Rodriguez-Henche N, et al. Anti-inflammatory role in septic shock of pituitary adenylate cyclase-activating polypeptide receptor. Proc Natl Acad Sci U S A. 2002;99:1053–8.

17. Martínez C, Juarranz Y, Abad C, Arranz A, Miguel BG, Rosignoli F, et al. Analysis of the role of the PAC1 receptor in neutrophil recruitment, acute-phase response, and nitric oxide production in septic shock. J Leukoc Biol. 2005;77:729–38.
18. Samarasinghe AE, Hoselton SA, Schuh JM. The absence of VPAC2 leads to aberrant antibody production in Aspergillus fumigatus sensitized and challenged mice. Peptides. 2011;32:131–7.
19. Tan YV, Abad C, Wang Y, Lopez R, Waschek JA. VPAC2 (vasoactive intestinal peptide receptor type 2) receptor deficient mice develop exacerbated experimental autoimmune encephalomyelitis with increased Th1/Th17 and reduced Th2/Treg responses. Brain Behav Immun. 2015;44:167–75.
20. Delgado M, Robledo G, Rueda B, Varela N, O'Valle F, Hernandez-Cortes P, et al. Genetic association of vasoactive intestinal peptide receptor with rheumatoid arthritis: altered expression and signal in immune cells. Arthritis Rheum. 2008;58:1010–9.
21. Juarranz Y, Gutierrez-Canas I, Santiago B, Carrion M, Pablos JL, Gomariz RP. Differential expression of vasoactive intestinal peptide and its functional receptors in human osteoarthritic and rheumatoid synovial fibroblasts. Arthritis Rheum. 2008;58:1086–95.
22. Paladini F, Cocco E, Cauli A, Cascino I, Vacca A, Belfiore F, et al. A functional polymorphism of the vasoactive intestinal peptide receptor 1 gene correlates with the presence of HLA-B*2705 in Sardinia. Genes Immun. 2008;9:659–67.
23. Paladini F, Belfiore F, Cocco E, Carcassi C, Cauli A, Vacca A, et al. HLA-E gene polymorphism associates with ankylosing spondylitis in Sardinia. Arthritis Res Ther. 2009;11:R171.
24. Cocco E, Paladini F, Macino G, Fulci V, Fiorillo MT, Sorrentino R. The expression of vasoactive intestinal peptide receptor 1 is negatively modulated by microRNA 525-5p. PLoS One. 2010;5:e12067.
25. Paladini F, Porciello N, Camilli G, Tuncer S, Cocco E, Fiorillo MT, et al. Single nucleotide polymorphisms in the 3′UTR of VPAC-1 cooperate in modulating gene expression and impact differently on the interaction with miR525-5p. PLoS One. 2014;9:e112646.
26. Sun W, Hong J, Zang YC, Liu X, Zhang JZ. Altered expression of vasoactive intestinal peptide receptors in T lymphocytes and aberrant Th1 immunity in multiple sclerosis. Int Immunol. 2006;18:1691–700.
27. Delgado M, Garrido E, de la Fuente M, Gomariz RP. Pituitary adenylate cyclase-activating polypeptide (PACAP-38) stimulates rat peritoneal macrophage functions. Peptides. 1996;17:1097–105.
28. Garrido E, Delgado M, Martínez C, Gomariz RP, De la Fuente M. Pituitary adenylate cyclase-activating polypeptide (PACAP38) modulates lymphocyte and macrophage functions: stimulation of adherence and opposite effect on mobility. Neuropeptides. 1996;30:583–95.
29. Martínez C, Delgado M, Pozo D, Leceta J, Calvo JR, Ganea D, et al. VIP and PACAP enhance IL-6 release and mRNA levels in resting peritoneal macrophages: in vitro and in vivo studies. J Neuroimmunol. 1998;85:155–67.
30. Delgado M, Anderson P, Garcia-Salcedo JA, Caro M, Gonzalez-Rey E. Neuropeptides kill African trypanosomes by targeting intracellular compartments and inducing autophagic-like cell death. Cell Death Differ. 2009;16:406–16.
31. Delgado M, Ganea D. Vasoactive intestinal peptide and pituitary adenylate cyclase-activating polypeptide inhibit interleukin-12 transcription by regulating nuclear factor kappaB and Ets activation. J Biol Chem. 1999;274(45):31930–40.
32. Delgado M, Ganea D. Vasoactive intestinal peptide prevents activated microglia-induced neurodegeneration under inflammatory conditions: potential therapeutic role in brain trauma. FASEB J. 2003;7:1922–4.
33. Delgado M, Munoz-Elias EJ, Kan Y, Gozes I, Fridkin M, Brenneman DE, et al. Vasoactive intestinal peptide and pituitary adenylate cyclase-activating polypeptide inhibit tumor necrosis factor alpha transcriptional activation by regulating nuclear factor-kB and cAMP response element-binding protein/c-Jun. J Biol Chem. 1998;273:31427–36.
34. Delgado M, Munoz-Elias EJ, Gomariz RP, Ganea D. Vasoactive intestinal peptide and pituitary adenylate cyclase-activating polypeptide enhance IL-10 production by murine macrophages: in vitro and in vivo studies. J Immunol. 1999;162:1707–16.

35. Delgado M, Pozo D, Martinez C, Leceta J, Calvo JR, Ganea D, et al. Vasoactive intestinal peptide and pituitary adenylate cyclase-activating polypeptide inhibit endotoxin-induced TNF-alpha production by macrophages: in vitro and in vivo studies. J Immunol. 1999;162:2358–67.
36. Delgado M, Munoz-Elias EJ, Gomariz RP, Ganea D. VIP and PACAP inhibit IL-12 production in LPS-stimulated macrophages. Subsequent effect on IFNgamma synthesis by T cells. J Neuroimmunol. 1999;96:167–81.
37. Delgado M, Leceta J, Ganea D. Vasoactive intestinal peptide and pituitary adenylate cyclase-activating polypeptide inhibit the production of inflammatory mediators by activated microglia. J Leukoc Biol. 2003;73:155–64.
38. Kim WK, Kan Y, Ganea D, Hart RP, Gozes I, Jonakait GM. Vasoactive intestinal peptide and pituitary adenylyl cyclase-activating polypeptide inhibit tumor necrosis factor-alpha production in injured spinal cord and in activated microglia via a cAMP-dependent pathway. J Neurosci. 2000;20:3622–30.
39. Martínez C, Delgado M, Pozo D, Leceta J, Calvo JR, Ganea D, et al. Vasoactive intestinal peptide and pituitary adenylate cyclase-activating polypeptide modulate endotoxin-induced IL-6 production by murine peritoneal macrophages. J Leukoc Biol. 1998;63:591–601.
40. Delgado M, Munoz-Elias EJ, Gomariz RP, Ganea D. Vasoactive intestinal peptide and pituitary adenylate cyclase-activating polypeptide prevent inducible nitric oxide synthase transcription in macrophages by inhibiting NF-kappa B and IFN regulatory factor 1 activation. J Immunol. 1999;162:4685–96.
41. Gonzalez-Rey E, Delgado M. Vasoactive intestinal peptide inhibits cyclooxygenase-2 expression in activated macrophages, microglia, and dendritic cells. Brain Behav Immun. 2008;22:35–41.
42. Delgado M, Ganea D. Inhibition of endotoxin-induced macrophage chemokine production by vasoactive intestinal peptide and pituitary adenylate cyclase-activating polypeptide in vitro and in vivo. J Immunol. 2001;167:966–75.
43. Delgado M, Ganea D. Vasoactive intestinal peptide inhibits IL-8 production in human monocytes. Biochem Biophys Res Commun. 2003;301:825–32.
44. Delgado M, Jonakait GM, Ganea D. Vasoactive intestinal peptide and pituitary adenylate cyclase-activating polypeptide inhibit chemokine production in activated microglia. Glia. 2002;39:148–61.
45. Tang Y, Lv B, Wang H, Xiao X, Zuo X. PACAP inhibit the release and cytokine activity of HMGB1 and improve the survival during lethal endotoxemia. Int Immunopharmacol. 2008;8:1646–51.
46. Delgado M, Varela N, Gonzalez-Rey E. Vasoactive intestinal peptide protects against beta-amyloid-induced neurodegeneration by inhibiting microglia activation at multiple levels. Glia. 2008;56:1091–103.
47. Mao SS, Hua R, Zhao XP, Qin X, Sun ZQ, Zhang Y, et al. Exogenous administration of PACAP alleviates traumatic brain injury in rats through a mechanism involving the TLR4/MyD88/NF-kB pathway. J Neurotrauma. 2012;29:1941–59.
48. Reglodi D, Renaud J, Tamas A, Tizabi Y, Socías B, Del-Bel E, et al. Novel tactics for neuroprotection in Parkinson's disease: role of antibiotics, polyphenols and neuropeptides. Prog Neurobiol. 2015; pii: S0301-0082(15)00128-8.
49. Azuma YT, Hagi K, Shintani N, Kuwamura M, Nakajima H, Hashimoto H, et al. PACAP provides colonic protection against dextran sodium sulphate induced colitis. J Cell Physiol. 2008;216:111–9.
50. Botz B, Bölcskei K, Kereskai L, Kovács M, Németh T, Szigeti K, et al. Differential regulatory role of pituitary adenylate cyclase-activating polypeptide in the serum-transfer arthritis model. Arthritis Rheumatol. 2014;66:2739–50.
51. Nemetz N, Abad C, Lawson G, Nobuta H, Chhith S, Duong L, et al. Induction of colitis and rapid development of colorectal tumors in mice deficient in the neuropeptide PACAP. Int J Cancer. 2008;122:1803–9.

52. Delgado M. Inhibition of interferon (IFN) gamma-induced Jak-STAT1 activation in microglia by vasoactive intestinal peptide: inhibitory effect on CD40, IFN-induced protein-10, and inducible nitric-oxide synthase expression. J Biol Chem. 2003;278:27620–9.
53. Delgado M, Gonzalez-Rey E, Ganea D. VIP/PACAP preferentially attract Th2 effectors through differential regulation of chemokine production by dendritic cells. FASEB J. 2004;18:1453–5.
54. Delgado M, Reduta A, Sharma V, Ganea D. VIP/PACAP oppositely affects immature and mature dendritic cell expression of CD80/CD86 and the stimulatory activity for CD4(+) T cells. J Leukoc Biol. 2004;75:1122–30.
55. Delgado M. Vasoactive intestinal peptide and pituitary adenylate cyclase-activating polypeptide inhibit the MEKK1/MEK4/JNK signaling pathway in endotoxin-activated microglia. Biochem Biophys Res Commun. 2002;293:771–6.
56. Delgado M, Ganea D. Inhibition of IFN-gamma-induced janus kinase-1-STAT1 activation in macrophages by vasoactive intestinal peptide and pituitary adenylate cyclase-activating polypeptide. J Immunol. 2000;165:3051–7.
57. Delgado M, Ganea D. Vasoactive intestinal peptide and pituitary adenylate cyclase activating polypeptide inhibit the MEKK1/MEK4/JNK signaling pathway in LPS-stimulated macrophages. J Neuroimmunol. 2000;110:97–105.
58. Delgado M, Ganea D. Vasoactive intestinal peptide and pituitary adenylate cyclase-activating polypeptide inhibit nuclear factor-kappa B-dependent gene activation at multiple levels in the human monocytic cell line THP-1. J Biol Chem. 2001;276:369–80.
59. Chorny A, Gonzalez-Rey E, Varela N, Robledo G, Delgado M. Signaling mechanisms of vasoactive intestinal peptide in inflammatory conditions. Regul Pept. 2006;137:67–74.
60. Delgado M, Leceta J, Abad C, Martinez C, Ganea D, Gomariz RP. Shedding of membrane-bound CD14 from lipopolysaccharide-stimulated macrophages by vasoactive intestinal peptide and pituitary adenylate cyclase activating polypeptide. J Neuroimmunol. 1999;99:61–71.
61. Delgado M, Sun W, Leceta J, Ganea D. VIP and PACAP differentially regulate the costimulatory activity of resting and activated macrophages through the modulation of B.71 and B.72 expression. J Immunol. 1999;163:4213–23.
62. Ganea D, Rodriguez R, Delgado M. Vasoactive intestinal peptide and pituitary adenylate cyclase-activating polypeptide: players in innate and adaptive immunity. Cell Mol Biol. 2003;49:127–42.
63. Martinez C, Delgado M, Gomariz RP, Ganea D. Vasoactive intestinal peptide and pituitary adenylate cyclase-activating polypeptide-38 inhibit IL-10 production in murine T lymphocytes. J Immunol. 1996;156:4128–36.
64. Wang HY, Jiang XM, Ganea D. The neuropeptides VIP and PACAP inhibit IL-2 transcription by decreasing c-Jun and increasing JunB expression in T cells. J Neuroimmunol. 2000;104:68–78.
65. Delgado M, Leceta J, Ganea D. Vasoactive intestinal peptide and pituitary adenylate cyclase-activating polypeptide promote in vivo generation of memory Th2 cells. FASEB J. 2002;16:1844–6.
66. Tan YV, Abad C, Lopez R, Dong H, Liu S, Lee A, et al. Pituitary adenylyl cyclase-activating polypeptide is an intrinsic regulator of Treg abundance and protects against experimental autoimmune encephalomyelitis. Proc Natl Acad Sci U S A. 2009;106:2012–7.
67. Voice JK, Grinninger C, Kong Y, Bangale Y, Paul S, Goetzl EJ. Roles of vasoactive intestinal peptide (VIP) in the expression of different immune phenotypes by wild-type mice and T cell-targeted type II VIP receptor transgenic mice. J Immunol. 2003;170:308–14.
68. Voice J, Donnelly S, Dorsam G, Dolganov G, Paul S, Goetzl EJ. c-Maf and JunB mediation of Th2 differentiation induced by the type 2G protein-coupled receptor (VPAC2) for vasoactive intestinal peptide. J Immunol. 2004;172:7289–96.
69. Delgado M, Ganea D. Vasoactive intestinal peptide and pituitary adenylate cyclase-activating polypeptide inhibit T cell-mediated cytotoxicity by inhibiting Fas ligand expression. J Immunol. 2001;165:114–23.

70. Delgado M, Ganea D. VIP and PACAP inhibit Fas ligand-mediated bystander lysis by CD4(+) T cells. J Neuroimmunol. 2001;112:78–88.
71. Delgado M, Ganea D. Vasoactive intestinal peptide and pituitary adenylate cyclase-activating polypeptide inhibit expression of Fas ligand in activated T lymphocytes by regulating c-Myc, NF-kappa B, NF-AT, and early growth factors 2/3. J Immunol. 2001;166:1028–40.
72. Sharma V, Delgado M, Ganea D. Granzyme B, a new player in activation-induced cell death, is down-regulated by vasoactive intestinal peptide in Th2 but not Th1 effectors. J Immunol. 2006;176:97–110.
73. Jiang X, Jing H, Ganea D. VIP and PACAP down-regulate CXCL10 (IP-10) and up-regulate CCL22 (MDC) in spleen cells. J Neuroimmunol. 2002;133:81–94.
74. Harrington LE, Hatton RD, Mangan PR, Turner H, Murphy TL, Murphy KM, et al. Interleukin 17-producing CD4+ effector T cells develop via a lineage distinct from the T helper type 1 and 2 lineages. Nat Immunol. 2005;6:1123–32.
75. Delgado M, Chorny A, Gonzalez-Rey E, Ganea D. Vasoactive intestinal peptide generates CD4+CD25+ regulatory T cells in vivo. J Leukoc Biol. 2005;78:1327–38.
76. Heimesaat MM, Dunay IR, Schulze S, Fischer A, Grundmann U, Alutis M, et al. Pituitary adenylate cyclase-activating polypeptide ameliorates experimental acute ileitis and extra-intestinal sequelae. PLoS One. 2014;9:e108389.
77. Tan YV, Abad C, Wang Y, Lopez R, Waschek JA. Pituitary adenylate cyclase activating peptide deficient mice exhibit impaired thymic and extrathymic regulatory T cell proliferation during EAE. PLoS One. 2013;8:e61200.
78. Pozo D, Anderson P, Gonzalez-Rey E. Induction of alloantigen-specific human T regulatory cells by vasoactive intestinal peptide. J Immunol. 2009;183:4346–59.
79. Prasse A, Zissel G, Lutzen N, Schupp J, Schmiedlin R, Gonzalez-Rey E, et al. Inhaled vasoactive intestinal peptide exerts immunoregulatory effects in sarcoidosis. Am J Respir Crit Care Med. 2010;182:540–8.
80. Anderson P, Gonzalez-Rey E. Vasoactive intestinal peptide induces cell cycle arrest and regulatory functions in human T cells at multiple levels. Mol Cell Biol. 2010;30:2537–51.
81. Maldonado RA, von Andrian UH. How tolerogenic dendritic cells induce regulatory T cells. Adv Immunol. 2010;108:111–65.
82. Chorny A, Gonzalez-Rey E, Fernandez-Martin A, Pozo D, Ganea D, Delgado M. Vasoactive intestinal peptide induces regulatory dendritic cells with therapeutic effects on autoimmune disorders. Proc Natl Acad Sci U S A. 2005;102:13562–7.
83. Delgado M, Gonzalez-Rey E, Ganea D. The neuropeptide vasoactive intestinal peptide generates tolerogenic dendritic cells. J Immunol. 2005;175:7311–24.
84. Gonzalez-Rey E, Chorny A, Fernandez-Martin A, Ganea D, Delgado M. Vasoactive intestinal peptide generates human tolerogenic dendritic cells that induce CD4 and CD8 regulatory T cells. Blood. 2006;107:3632–8.
85. Abad C, Martinez C, Leceta J, Gomariz RP, Delgado M. Pituitary adenylate cyclase-activating polypeptide inhibits collagen-induced arthritis: an experimental immunomodulatory therapy. J Immunol. 2001;167:3182–9.
86. Abad C, Martinez C, Juarranz MG, Arranz A, Leceta J, Delgado M, et al. Therapeutic effects of vasoactive intestinal peptide in the trinitrobenzene sulfonic acid mice model of Crohn's disease. Gastroenterology. 2003;124:961–71.
87. Delgado M, Martinez C, Pozo D, Calvo JR, Leceta J, Ganea D, et al. Vasoactive intestinal peptide (VIP) and pituitary adenylate cyclase-activation polypeptide (PACAP) protect mice from lethal endotoxemia through the inhibition of TNF-alpha and IL-6. J Immunol. 1999;162:1200–5.
88. Elekes K, Sandor K, Moricz A, Kereskai L, Kemeny A, Szoke E, et al. Pituitary adenylate cyclase-activating polypeptide plays an anti-inflammatory role in endotoxin-induced airway inflammation: in vivo study with gene-deleted mice. Peptides. 2011;32:1439–46.
89. Gonzalez-Rey E, Fernandez-Martin A, Chorny A, Martin J, Pozo D, Ganea D, et al. Therapeutic effect of vasoactive intestinal peptide on experimental autoimmune encephalomyelitis: down-regulation of inflammatory and autoimmune responses. Am J Pathol. 2006;168:1179–88.

90. Ji H, Zhang Y, Shen XD, Gao F, Huang CY, Abad C, et al. Neuropeptide PACAP in mouse liver ischemia and reperfusion injury: immunomodulation by the cAMP-PKA pathway. Hepatology. 2013;57:1225–37.
91. Dickinson T, Mitchell R, Robberecht P, Fleetwood-Walker SM. The role of VIP/PACAP receptor subtypes in spinal somatosensory processing in rats with an experimental peripheral mononeuropathy. Neuropharmacology. 1999;38:167–80.
92. Favrais G, Couvineau A, Laburthe M, Gressens P, Lelievre V. Involvement of VIP and PACAP in neonatal brain lesions generated by a combined excitotoxic/inflammatory challenge. Peptides. 2007;28:1727–37.
93. Reglodi D, Kiss P, Szabadfi K, Atlasz T, Gabriel R, Horvath G, et al. PACAP is an endogenous protective factor-insights from PACAP-deficient mice. J Mol Neurosci. 2012;48:482–92.
94. Gonzalez-Rey E, Delgado-Maroto V, Souza Moreira L, Delgado M. Neuropeptides as therapeutic approach to autoimmune diseases. Curr Pharm Des. 2010;16:3158–72.

Part XIII
PACAP in Stress and Mood Disorders

Chapter 41
PACAPergic Synaptic Signaling and Circuitry Mediating Mammalian Responses to Psychogenic and Systemic Stressors

Sunny Z. Jiang and Lee E. Eiden

Abstract This short review is focused on the central and peripheral functional neuroanatomy of PACAPergic circuitry mediating stress responses, which we define here as "circuits mediating catecholamine and corticosterone elevation in response to systemic and psychogenic stressors," and the underlying cell signaling pathways, insofar as they are known, through which PACAP signaling in stress occurs. We will focus on the underlying neurobiology of PACAP signaling in stress derived from studies in rodents, and in human and rodent cells in culture, and consider their relevance to PACAPergic neurotransmission in human anxiety, depression, and posttraumatic stress disorder (PTSD).

Keywords PACAP-38 • PAC1 • GPCR • Cyclic AMP • ERK • NCS/Rapgef2 • Stress circuitry • Depressive behavior • Anxiety

Introduction

A comprehensive picture of stress neurotransmission remains to be effectively built up from facts about chemical neuroanatomy, neuropharmacology, electrophysiology, cell signaling, and biochemistry; genetic effects on these observable properties of cells and circuits; and the penetrance of neuronal properties to behaviors, which can then be linked to stress-related diseases. Such a comprehensive understanding

S.Z. Jiang
Section on Molecular Neuroscience, National Institute of Mental Health, Bethesda, MD, USA

L.E. Eiden (✉)
Section on Molecular Neuroscience, National Institute of Mental Health, Bethesda, MD, USA

Section on Molecular Neuroscience, LCMR, NIMH-IRP, Building 49, Room 5A-38, 9000 Rockville Pike, Bethesda, MD 20892, USA
e-mail: eidenl@mail.nih.gov

D. Reglodi, A. Tamas (eds.), *Pituitary Adenylate Cyclase Activating Polypeptide — PACAP*, Current Topics in Neurotoxicity 11,
DOI 10.1007/978-3-319-35135-3_41

is likely to result in identification of druggable targets for depression, PTSD, and anxiety disorders. This essentially translational enterprise is a highly integrative one, in which correct premises can be followed to highly impactful conclusions, but incorrect ones can be amplified to create unhelpful or even false conceptualizations resonating across multiple areas of research. Here, we illustrate how current research on PACAPergic neurons and their function in transduction of responses to stressors both centrally and peripherally is culminating in testable hypotheses about how engagement of agonists and antagonists with central and peripheral PACAP receptor targets might affect behavioral outcomes of chronic stress.

The original conceptualization of stress responding, by the founders of the field including Cannon and Selye, focused not on behavioral states resulting from chronic or traumatic/acute stress, but on physiological responses to ostensibly stressful stimuli by the major "stress organs," the adrenal cortex and medulla, and the sympathetic nervous system, and their secretory products, corticosterone/cortisol (CORT) and catecholamines (CAs) [1, 2]. Detailed study of effects of *stressors* (stimuli or conditions leading to increased CORT and CA secretion) led to the seminal discoveries of the endocrine and neuronal chemical anatomy of the hypothalamo-pituitary-adrenal (HPA), and hypothalamo-sympathoadrenal (HSA), comprising the adrenomedullary and sympathoneuronal (AHS & SNS) axes (see Fig. 41.1). Identification of these output circuits led in turn to discovery of central and peripheral neurotransmitters, pathways, and circuits influencing stressor effects on CORT and CA secretion [3]. Modern conceptualizations of stress responding in mammals, particularly when mobilized for clinical understanding of depression, PTSD, and anxiety disorders, have generally focused on how pathobehavioral outcomes of chronic stress exposure develop from a repertoire of essentially normal homeostatic responses to stressors. This has involved the invocation of fear learning, and unlearning; the role of cortical function in modulating the salience of aversive stimuli; and the cumulative effects of environmental training, learning, or catastrophic experiencing that can cause chronic overstimulation of homeostatic mechanisms appropriate to survival [4]. This is called by some allostatic load or allostatic wear and tear [5], and it is considered to erode resilience, or flexible adaptation to experiencing stress (or simply coping with life's demands). This latter-day formulation of stress responding is focused on the affective state of the individual, and the circuits supporting states of anxiety, fearfulness, or aggression. Hyperactivation, or dysfunctional plasticity, within these circuits is deemed, in aggregate, to generate stress-associated behavioral pathologies. These include self-medication with addictive drugs, delusional interactions with others, and even more manifestly dysfunctional outcomes, such as catatonia, other types of breakdown of daily living performance, suicidality, and homicidality. The physiological response to stressors is of somewhat less importance, from this perspective, than the altered function of what could be called the "poststressed brain," and stimuli that can become embedded as triggers for dysfunctional behavior [6].

It is our aim here to provide a brief narrative recapitulating how basic research focused on the neuropeptide PACAP has added detail to the landscape of stressor responding, centrally and peripherally, with some new facts about its chemical

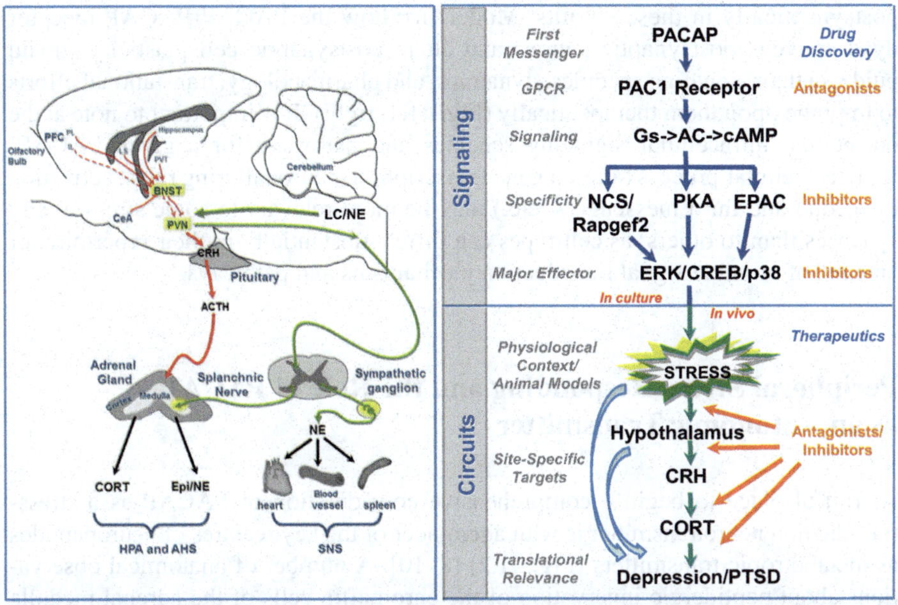

Fig. 41.1 PACAP regulation of central and peripheral stress circuits. *Left panel.* Stress circuits are defined operationally as neuronal and neuroendocrine cell ensembles that mediate adaptation to psychological or systemic perturbogens outside of the normal physiological range. Psychogenic stressors, in laboratory animals, can include social defeat, anxiety generated by threat, and immobilization. Systemic stressors can include hypoglycemia, inflammation, cold, heat and noise. In mammals, stress responses are mediated by the hypothalamic-pituitary-adrenal (HPA) axis, the adrenomedullary hormonal system (AHS) and the sympathetic nervous system (SNS). The chief effector molecules are glucocorticoids (CORT, or cortisol in humans and corticosterone in rodents), epinephrine (Epi) and norepinephrine (NE). PACAPergic synapses operating within these circuits are highlighted in *yellow.* Peripherally, PACAP release at the adrenomedullary synapse regulates catecholamine and neuropeptide synthesis and release from chromaffin cells in response to systemic and psychogenic stressors. Centrally, PACAP release onto corticotropin-releasing hormone (CRH)-containing neurons of the hypothalamus regulates CRH biosynthesis in response to psychogenic stressors. *Right panel.* PACAP mediated cAMP signaling pathways have been identified in neuronal and endocrine cell culture: cAMP elevation caused by PACAP/PAC1 activation, conveys to three distinct cAMP sensors—the neuritogenic cAMP sensor (NCS/Rapgef2), protein kinase A (PKA), and Epac, and which mediate downstream signaling to ERK, CREB, and p38, respectively. Investigation of PACAPergic neuronal circuits in responses to physiological and psychogenic stress in vivo is aimed, in part, at facilitating development of innovative pharmacological strategies for treatment of stress-associated disorders. Adapted from [7].

neuroanatomy. At the systems and circuit levels this has, somewhat unexpectedly, provoked reexamination of the basis for our understanding of interoceptive mechanisms of stress conditioning (vida supra) leading to behavioral pathology. The current picture of PACAP signaling within classical stress response circuits is summarized in Fig. 41.1 (Fig. 41.1, left, adapted from [7]). At the cellular level, recent developments in PACAP signaling, primarily through the PAC1 receptor, have provided some fresh insights into how PACAP, and potentially other neuropeptides, function postsynaptically in these circuits. Models for how the PACAP-PACAP receptor dyad conveys postsynaptic actions and alters postsynaptic cell plasticity should guide, in their signaling specifics, dynamics, and pharmacology, translational efforts to impinge upon them therapeutically (Fig. 41.1, right). It is important to note at the outset that intracellular signaling schemes and pathways for a given PACAP-initiated cellular process (altered gene transcription, neuronal firing rates, activation of second and third messengers, etc.) may be more relevant to some stress-circuit synapses than to others, as cell types can differ substantially in their repertoire, or "menu" of possible signal transduction mechanisms and pathways.

Peripheral Stress Responding and the Role of PACAP as an Autonomic Transmitter

A fruitful place to begin a comprehensive consideration of PACAP as a stress-transduction neurotransmitter is with a reminder of the key features of neuropeptides as metabotropic transmitters (Fig. 41.2) [8–10]. A number of anatomical observations about peptidergic innervation of the chromaffin cells of the adrenal medulla led up to the consideration of PACAP as the major neurotransmitter transducing stress responses across the adrenomedullary synapse (reviewed in [10]). Briefly, PACAP infusion causes epinephrine release from the adrenal medulla in vivo and ex vivo, and from chromaffin cells in culture [11–13]; PACAP is found within cholinergic terminals innervating chromaffin cells [14]; catecholamine release

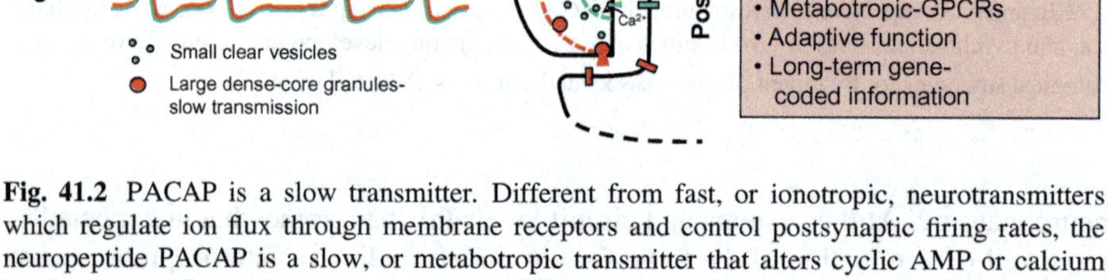

Fig. 41.2 PACAP is a slow transmitter. Different from fast, or ionotropic, neurotransmitters which regulate ion flux through membrane receptors and control postsynaptic firing rates, the neuropeptide PACAP is a slow, or metabotropic transmitter that alters cyclic AMP or calcium levels through G-protein coupled receptors (GPCRs), controlling gene transcription and other cellular functions. While fast transmission is triggered by low-frequency firing and encodes immediate electrical information, slow transmission is triggered by high-frequency ("stressed") neuronal firing and encodes synaptic experience over longer periods. This figure is adapted from Smith and Eiden 2012 and incorporates concepts promulgated by Hokfelt et al. 2003, and properties of slow and fast transmission, and small and large secretory vesicles, summarized by Morgan and Burgoyne 1997

during systemic stress (insulin shock) is greatly blunted in PACAP-deficient mice [14]; and electrical stimulation of the splanchnic nerve ex vivo causes catecholamine release mimicked by PACAP administration and blocked by PACAP(6–38) [10, 12, 14–17]. There is also a prolific literature on PACAPergic transmission across sympathetic ganglia in the peripheral nervous system, indicating the expression of PACAP in preganglionic innervation of the superior cervical ganglion (SCG); the existence of PAC1 receptor expression in postganglionic sympathetic neurons in the same tissue; and effects of exogenous PACAP on catecholamine and neuropeptide release and synthesis in cultured postganglionic principal cells of the SCG [18–22]. The actual role of PACAP as a sympathetic preganglionic transmitter in stress transduction in vivo remains, surprisingly, largely undetermined. Indirect evidence suggests that PACAP probably acts similarly at sympathoadrenal and sympathetic ganglionic synapses [18, 21, 23, 24]. Nevertheless, there is considerable plasticity of neuropeptide expression throughout the sympathetic nervous system, and across species [25], so that further detailed mapping of the sympathetic preganglionic terminals that release PACAP in adult mammals, and the role of PACAP in modulating or stimulating the release of norepinephrine from postganglionic sympathetic terminals, especially during stress responses that are biased toward the sympathetic, versus the sympathoadrenal systems [26], is warranted. These investigations are of particular interest in light of recent provocative evidence for an apparent role of peripheral PACAP release in human (female) PTSD [27]. A defined function for PACAPergic neurotransmission in vegetative (parasympathetic) responding to stress likewise remains unexplored in vivo, albeit Hannibal et al. have fairly clearly established that PACAP is a preganglionic transmitter at both sympathetic and parasympathetic ganglia [28].

PACAP is a sympathoadrenal transmitter, and PAC1 receptors are present on chromaffin cells and cell line congeners such as PC12 cells. This has allowed detailed exploration of how PACAP is likely to signal across the adrenomedullary synapse to cause both secretion and compensatory increase in the expression of genes encoding secretory prohormones, and catecholamine biosynthetic enzymes (stimulus-secretion-synthesis coupling). It has been noted earlier that stress-induced catecholamine secretion from the adrenal medulla in vivo is PACAP dependent [10, 14, 15]. In addition, the induction of catecholamine biosynthetic enzyme activity, and mRNAs encoding catecholamine biosynthetic enzymes, and other secretory proteins including progalanin and proNPY are PACAP dependent (i.e., abrogated during the stress response in PACAP-deficient mice), or elevated upon exposure of chromaffin cells to PACAP in cellula and ex vivo [29–34]. It has been demonstrated that the adrenal medulla in fact releases into the bloodstream its entire complement of catecholamines during several hours of continuous splanchnic stimulation [24], which would clearly lead to exhaustion of catecholamine content in the absence of compensatorily increased biosynthesis during acute prolonged stress over a period of hours, as well as stress prolonged episodically over a period of several days. Thus, the ability of PACAP neurotransmission to support both catecholamine (and neuropeptide) secretion, and coupled enhancement of biosynthesis, is key to its role in sustained secretion from the adrenal medulla during prolonged stress.

The mechanisms of CA release by PACAP have been studied by many laboratories. Smith and colleagues aver, with strong evidence adduced in mouse adrenal slices, that PACAP causes calcium influx and catecholamine release by a partially depolarizing mechanism in which voltage-gated calcium channels (VGCCs) are activated by movement of the membrane potential from −70 to around −50 mV, near the activation threshold for T- and other calcium channels. Calcium influx through these channels then causes further movement of the membrane potential to that for activation of classical (L, P, N, Q) VGCCs that provide sufficient calcium influx to trigger exocytosis [16, 35] (see also Fig. 41.3). Mustafa et al. have shown that the hop isoform of the PAC1 receptor can mediate catecholamine secretion, and calcium influx, equally blocked by combinations of VGCC inhibitors, consistent with the Smith model [36, 37]. Thus, PACAP-induced secretion at the adrenomedullary synapse appears to involve, via the PAC1hop receptor, a processive decrease in membrane potential ultimately triggering the opening of voltage-dependent calcium channels (around −45 mV), calcium influx, and exocytotic secretion. May and colleagues

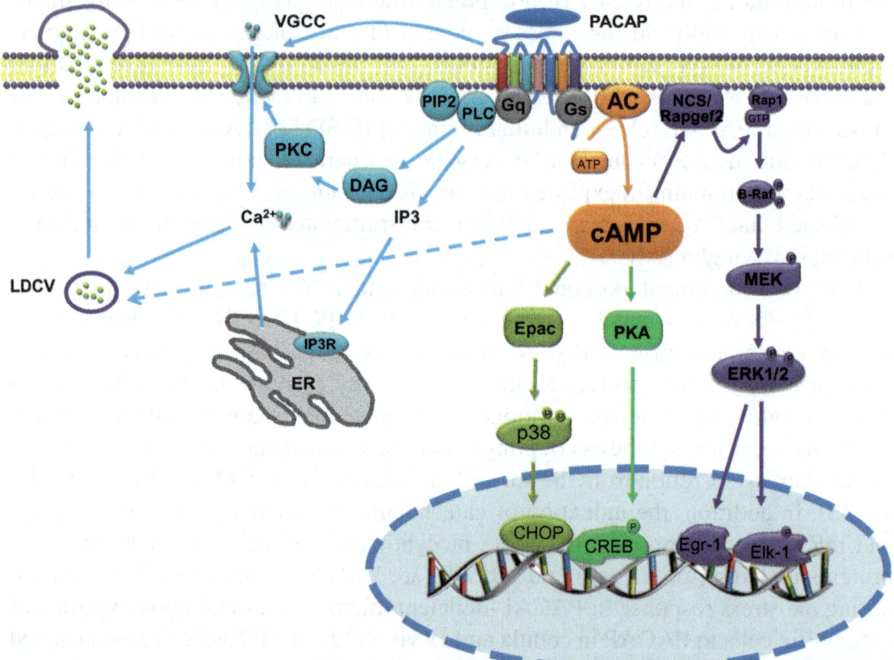

Fig. 41.3 Signaling pathways leading to PACAP-stimulated catecholamine release and PACAP mediated gene regulation in chromaffin cells. PACAP-induced exocytotic catecholamine release from LDCVs (large dense-core vesicles) is evoked by elevated cytosolic calcium due to PACAP-induced extracellular calcium influx via VGCC (voltage-gated calcium channels) and IP3R-induced calcium release from ER stores, while PACAP-mediated gene regulation can occur via multiple pathways: those responding to cAMP elevation include the cAMP sensors NCS/Rapgef2 (activating the MAP kinase ERK), PKA (activating CREB) and Epac2 (activating the MAP kinase p38)

have posited that PACAP-dependent release of catecholamines from *sympathetic* neurons occurs via frank depolarization of sympathetic neurons [21, 22]. As the latter experiments have been carried out in developing sympathetic neurons in culture, rather than in mature postganglion neurons, it is not yet clear if this mechanism also pertains to mature sympathetic neurons responding to PACAP activation of PAC1 in vivo during stress. This is of particular importance in understanding the neurochemical anatomy of peripheral stress circuits, because clear differences exist between sympathoadrenal and sympathetic stimulation by various systemic stressors. Cold stress, for example, likely activates descending autonomic pathways to the intermediolateral column that are specific for *sympathetic* activation (NE release predominates over Epi release), while other stressors (e.g., immobilization), more highly correlated with HPA axis activation, are specific for *sympathoadrenal* activation (Epi release predominating over NE release) [26].

A reasonable summary of how PACAP causes neurotransmitter release (at least, of catecholamine release from chromaffin cells, Fig. 41.3) was advanced by O'Farrell and Marley in 1996: "The present study suggests catecholamine secretion induced by PACAP requires calcium entry through several different types of voltage-sensitive calcium channels." Their further comments are both trenchant and prescient: "This raises the important question of the molecular mechanism by which a G protein-coupled receptor agonist can depolarize chromaffin cells to activate these voltage-sensitive calcium channels" (VSCCs) [38]. Surprisingly, 20 years on, we still do not know the exact molecular mechanism by which PAC1 receptor occupancy leads to chromaffin cell VSCC activation and catecholamine release [39], much less how PACAP affects cell depolarization and calcium entry in other PACAP-receptive cells within stress-responsive circuits in vivo [22]. The current unresolved state of PACAP signaling leading to neurotransmitter release, and calcium-dependent signaling in general, is likely due in equal parts to lack of precise knowledge about the receptor and channel types populating the cells in which experiments have been conducted; the specific signaling pathways to calcium entry available in a given cell type; and a lack of clean pharmacological tools for blocking and activating the plethora of ion channels and signaling molecules that can cause intracellular calcium mobilization, and plasma membrane calcium channel opening, upon PACAP receptor stimulation in a given cell type.

PACAP-induced neurotrophic effects on differentiation and survival of autonomic neurons, as well as the synthesis and release of neuropeptide transmitters and catecholamine biosynthetic enzymes, have been carefully studied in neuronal cultures of the rat SCG [40], using PC12 cell lines [41, 42], and in various in vivo paradigms [23]. May and colleagues have posited that PACAP signals largely through an Akt-dependent pathway to enhance the survival of developing sympathetic neurons [40]. In the neuroendocrine NS-1 cell line, and chromaffin cells in primary culture, a tripartite signaling via PAC1 activation that is wholly cAMP dependent has been identified that leads to neuritogenesis, growth arrest, and survival upon serum withdrawal via three distinct cAMP sensors: PKA, Epac2, and NCS/Rapgef2 (see Fig. 41.3). Delineating the cell- and synapse-specific pathways through which PACAP signals for various short- and long-term outcomes is likely to be important in understanding how PACAP signaling converges onto PACAP-dependent behaviors in stress-related disorders [43]. Therefore, how PACAPergic cellular biochemistry investigated in cells and cell lines in culture apply to specific postsynaptic actions of PACAP in the adult central nervous system in vivo will be an important focus of future research on PACAPergic signaling in stress [44].

The conceptualization for PACAP/PAC1 signaling in neuronal and endocrine cells summarized in Figs. 41.3 and 41.4 is a departure from the classical picture for PACAP signaling drawn from a "canonical" view of GPCR signaling in general, and is based on cell culture experiments performed in PAC1-expressing cells and cell lines (see, e.g., [43]). This new signaling paradigm highlights the role of NCS/Rapgef2 in connecting cAMP elevation via Gs-coupled GPCRs such as the PAC1 receptor to ERK activation in neuronal and endocrine cells. Drawn from experiments conducted with cultured chromaffin cells, pheochromocytoma-derived cell lines, and primary central hypothalamic, hippocampal and cortical neurons, it emphasizes the uniqueness of the NCS/Rapgef2-dependent pathway in cAMP-dependent ERK activation that is consequential for changes in gene transcription in neuronal and

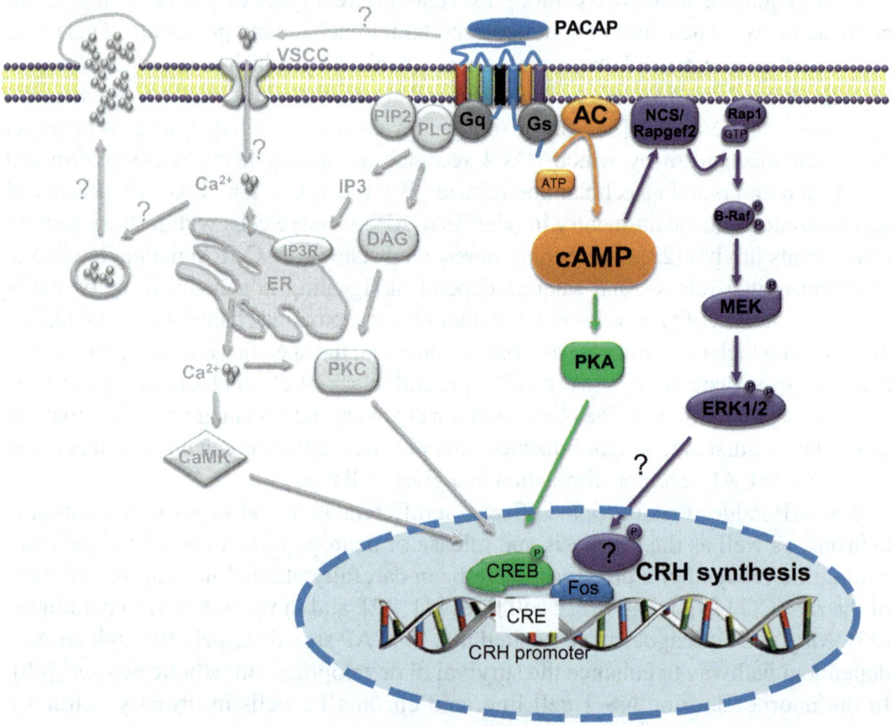

Fig. 41.4 PACAP mediated cAMP- and calcium-dependent CRH biosynthesis in the central CRH neuron. Mechanisms of PACAP regulated CRH release in response to psychogenic stressors are largely unknown, and in fact emerging evidence suggests that PACAPergic innervation may play only a minor role in CRH release in response to stress in vivo (see text). CRH gene transcription stimulated by psychogenic stress is strongly PACAP-dependent, and PACAP may signal to the CRH gene through multiple cAMP-dependent pathways that converge at PKA/CREB- and ERK-dependent responsive elements on the CRH gene promoter

endocrine cells (see, e.g., [42, 45]). PACAP is also reported to signal, via PAC1, to ERK convergently via PKC, PKA, and beta-arrestin, and through both Gq and Gs coupling [46, 47]. This signaling may be consequential for either cytoplasmic [48] or nuclear [45, 49, 50] signaling by ERK. Both Rapgef2/ERK- and PKA/CREB-dependent gene regulation (for example, of the Stc1 and BDNF genes, respectively) seem to be initiated by PACAP in CNS-derived cultured neurons [45, 51]. Which modes of PACAP signaling, through both cAMP and calcium, and through PKA, Epac, NCS/Rapgef2, Akt and other pathways, are operative in neurons responding to PACAP at synapses activated and altered by stress in vivo? This question carries import both for PACAP neurotransmission as a paradigm for neuropeptide signaling in stress, and for translational focusing of PACAP receptors as targets for stress-related affective illnesses.

Central Stress Responding, and the Role of PACAP as a CNS Neurotransmitter

PACAP release from the splanchnic nerve, and ACTH release from the pituitary, constitute the final common pathways for CORT secretion from the adrenal cortex, and CA secretion from the adrenal medulla, in response to both systemic stressors such as LPS and hypoglycemia, and psychogenic stressors such as immobilization and social defeat. However, at the level of CNS modulation of both the HPA axis and behaviors elicited by both acute and chronic stress, clear differences between the effects of systemic and psychogenic stressors emerge. The relative dependence on PACAPergic CNS circuitry has become a key factor in clearly distinguishing these pathways. For example, insulin-induced hypoglycemia has a PACAP-dependent effect on catecholamine secretion, but CORT elevation following insulin hypoglycemia is unaffected by PACAP (i.e., is similar in wild-type and PACAP-deficient mice) [14]. Similarly, HPA axis activation after acute (0.5–1 h) or prolonged (1–5 h) systemic stressors is unaffected in PACAP-deficient mice, while CORT elevation after prolonged or chronic psychogenic stress is markedly PACAP dependent [14, 34, 52, 53].

Reported effects of PACAP deficiency on acute HPA axis activation by psychogenic stressors such as restraint (CORT elevation within minutes of onset of psychogenic stress) originally suggested PACAPergic control, at the level of the paraventricular nucleus (PVN), of both CRH secretion and CRH biosynthesis [33, 34, 52]. This view may now require modification. We have reexamined previous results indicating a strong PACAPergic component to both acute (0–1 h) and prolonged (1–6 h) restraint stress [34, 52], and found that when CORT is measured in blood sampled from the tail rather than the trunk (following cervical separation), a dependence of CORT elevation on PACAP emerges only after prolonged (>1 h) restraint [54]. PACAP-dependent effects on serum CORT elevation after acute (<1 h) restraint stress may be mediated peripherally [55, 56] and are evident only in mice that have undergone stress for a sufficient period to allow PACAP-dependent increase in

CORT biosynthesis within the adrenal cortex, and its release upon splanchnic disinhibition during cervical separation. In fact, central PACAPergic mechanisms appear to drive mainly biosynthesis of CRH, via induction of CRH mRNA, rather than CRH secretion in response to stress. This PACAPergic modulation of CRH neurons of the PVN becomes apparent, as attenuation of CORT elevation, only following *prolonged* (>1 h) or *chronic* (1–2 h daily for several days) psychogenic stress, while the *acute* (0–1 h) release of CRH, the proximate trigger for ACTH secretion from the pituitary and CORT elevation in serum, is largely PACAP independent [54, 101]. Exposure of cultured hypothalamic neurons and hypothalamic CRH gene-transcribing cell lines to PACAP suggests that PACAP affects CRH gene transcription via cAMP elevation [52]. Cyclic AMP activation of ERK through NCS/Rapgef2 may mediate enhanced transcription of CRH mRNA by PACAP at the level of the hypothalamus [45], although PKA-dependent mechanisms are clearly documented for CRH gene transcription and may also play a role [57, 58]. Figures 41.3 and 41.4 summarize the differential effects of PACAP stimulation in peripheral chromaffin cells (stimulus-secretion-synthesis coupling; Fig. 41.3) and central CRH neurons (stimulus-transcription coupling; Fig. 41.4).

For the reasons outlined earlier, the distinction between the PACAP dependence of psychogenic, but not systemic, stress may be more properly assigned as a distinction between stressors causing acute CRH release from the median eminence, and stressors causing more prolonged CRH release, the PACAP dependence of the latter emerging only when (PACAP-dependent) CRH biosynthesis fails to keep pace with (PACAP independent) CRH secretion from the median eminence. Since few systemic stressors remain purely nonpsychogenic after a first application, it will be difficult to fully parse this issue without close experimental examination, perhaps using hypoglycemia as a systemic stressor unlikely to provoke aversive enteroceptive cues from which the perception of psychogenic stress is presumably communicated to amygdalar and cortical CNS components of the stress response.

There is a growing body of evidence for involvement of PACAP signaling in the behavioral sequelae of psychogenic stress emerging from both fundamental, preclinical, and clinical studies of stress-related behavioral dysfunction. Thus, despite potential difficulties of experimental design, pinning down the multiple synaptic loci at which PACAPergic signaling in the CNS mediates separate aspects of stress responding is of critical importance. Abrogation of PACAPergic neurotransmission (i.e., PACAP knock-out) attenuates or abolishes depressive effects of both chronic social defeat (including decreased exploration, social interaction, and mobility in the forced swim test) [53], and chronic restraint stress (including decreased feeding and weight loss), and the depressogenic effects of the restraint stress are also significantly reduced in PAC1-deficient mice [54]. Neuropharmacological data obtained in the rat generally reinforce the notion that PACAP, acting as an anxiogenic neuropeptide at the locus of the bed nucleus of the stria terminalis (BNST), as a stimulator of HPA axis activation at the locus of the PVN, or both, is likely to mediate the depressogenic effects of chronic psychogenic stress. At least one research group has reported antidepressive and anxiogenic, rather than depressive and anxiolytic, effects of PACAP deficiency in mice [59, 60]. These results might be explicable in part due to the use of unstable genetic backgrounds [61], or due to genuine differences in strain dependence of the relative roles of PACAP at various brain loci. Identifying appropriate animal models to parse the likely effects of PACAP agonists, and antagonists, on behavioral pathology associated with chronic stress thus remains an ongoing challenge in this field.

The translational implications of the linkage between PACAP signaling and the depressogenic, and anxiogenic, effects of chronic stress are becoming increasingly obvious. Since PACAP deficiency apparently allows a full acute stress response, while blunting only a chronic one, patients are far more likely to tolerate full target engagement by PAC1 antagonists than by CRHR1 antagonists [62–64], without compromising HPA axis activation required for responding to systemic stressors such as sepsis, hypoglycemia, and acute psychogenic stressors such as actual external threat. In addition, tracing the neuronal circuitry for PACAP inputs to the PVN responsible for HPA axis modulation during chronic psychogenic stress, currently believed to arise either in the BNST, other areas of the hypothalamus, or both (see [43]), will likely provide additional anatomical insights relevant to PACAP's dual roles in stress transduction at the level of the hypothalamus, and in control of fear and anxiety at the level of the extended amygdala (vide infra). These in turn, depending on the PACAP receptors involved, may provide additional avenues for translating PACAPergic neuropharmacology into potential treatments for stress-associated affective illnesses in humans.

As mentioned earlier, PACAP has important actions at suprahypothalamic levels, and these have been worked out primarily in the rat. The BNST is an important extrahypothalamic site of PACAP action [65–71]. PACAP is highly expressed in BNST [72], where its expression, and that of the PAC1 receptor, is up-regulated by chronic stress [65]. PACAP is considered to be anxiogenic when released within, or perfused into the BNST [65, 66, 71], which is functionally connected with the PVN and acts as a relay center for several limbic and neocortical pathways to modulate PVN activity [73]. Neuronal activity within the BNST has been implicated in mediating anxiety-like behaviors in several mammalian species [74–78]. Upregulation of mRNA encoding PACAP and PAC1, in both BNST and PVN, is induced by chronic stress [65], reinforcing the hypothesis that PACAPergic neurotransmission within both brain nuclei is important in the psychogenic stress response.

The BNST has reciprocal connections with the centromedial amygdala and receives projections from hippocampus and medial prefrontal cortex (mPFC), limbic regions important for fear memory formation [79]. The CeA has a high level of expression of both PACAP and CRH [72, 80], and PACAP-innervated neurons in CeA are activated by stress [81]. PACAP infusion into the CeA is reported to increase anxiety-like behaviors (increased passive avoidance) [82], and PACAP(6–38) infusion in BLA blocks consolidation of contextual fear conditioning [83]. The mPFC, which plays an important role in fear consolidation and extinction, provides inhibitory control of the PVN response to stress through indirect projections that are relayed via the BNST. Interestingly, an increase in PAC1 mRNA was reported in amygdala and medial prefrontal cortex during consolidation of fear in male rats [27]. PACAP-deficient mice exhibit an increase in FosB expression, after chronic stress, in both prefrontal cortex and BNST, that is greater than neuronal activation seen in wild-type mice similarly stressed [53]. Recently, the parabrachial nucleus of the brain stem has been suggested as a source of PACAPergic neurons projecting to the extended amygdala and regulating pain perception, and pain-related behaviors [80]. How the features of PACAP signaling to neurons of the extended amygdala differ from PAC1-dependent PACAP signaling to chromaffin cells of the adrenal medulla (Fig. 41.3) and CRH neurons of the PVN (Fig. 41.4) is as yet unknown, and underscores the need for additional information, at the cellular level, about how extrahypothalamic PACAPergic signaling might act postsynaptically to regulate stress responses.

A neurochemical–neuroanatomical model in which PACAP causes both anxiety and chronic stress-driven depressogenic effects via control of CORT release has been considered, based on the Gold-Chrousos hypothesis of a causal role for CORT elevation in some types of clinical depression [84–86], and clearly documented "feedback-like" effects of CORT treatment in vivo on various features of PACAPergic neuronal signaling in the brain [70]. So has an anatomical model in which multiple loci in extended amygdala and hypothalamus could have additional effects on stress-associated behaviors beyond direct (hypothalamic) actions on the "common final pathway" of the HPA axis [79]. A further complication is the observation that PACAP appears to be more critical in modulation of contextual (hippocampal dependent), than cue (amygdala dependent) dependent fear conditioning [87, 88]. Whether this observation can be integrated into a working model for the relationship between fear, anxiety, and potentially PACAP-dependent human psychiatric disorders, including anxiety, depression, and PTSD [89–91] remains to be seen. Nevertheless, studies in this area hold the promise of further melding the neuroanatomical and neurochemical features of PACAPergic neurons, and the circuits in which they are embedded, into a fuller understanding of their roles in the various behavioral and endocrine responses that determine how acute and chronic stressors are homeostatically or allostatically experienced.

Pharmacology of PACAP Receptors: Current Status of Tools

A lack of comprehensive in vivo pharmacology, and insufficient deployment of the pharmacology already existing, for PAC1, VPAC1, and VPAC2 receptor activation and blockade contribute greatly to the lack of translationally focused progress in this area. Specific peptide and peptide-based agonists for PAC1 (maxadilan), VPAC1 (VIP-GRF hybrid and otherwise modified peptides), and VPAC2 (the cyclic VIP analogs Ro25-1392 and Ro24-1553) exist [92], but have unfavorable pharmacological properties for use in vivo. The PAC1 antagonist PACAP(6–38), while highly specific for PAC1 versus VPAC1 receptors [93], is however also a potent VPAC2 and CART receptor antagonist [93–95], limiting its usefulness for perfusion and other in vivo and in cellular investigations of PACAP receptor-specific signaling. While maxadilan derivatives have been invaluable as highly specific PAC1 antagonist agents [96, 97], their decidedly undrug-like properties, like those of the various peptide and peptoid PAC/VPAC agonists, have limited their usefulness for probing PACAP receptor specificity in vivo. Thus, small-molecule inhibitors of the PAC1, as well as VPAC1 and VPAC2 receptors, are urgently sought for these therapeutic targets both centrally and peripherally [98].

Conclusion: Unlocking the Translational Potential of PACAP Signaling in Stress

The signaling mechanisms underlying PACAP modulation of CRH, catecholaminergic, and other stress-transducing neuroendocrine cells, and the molecules that carry them out, need to be fully understood as a basis for determining how PACAP agonists might be developed as treatments for depression, PTSD, or anxiety. The synaptic locations of PACAPergic stress-transducing neurons--both peripherally, and in the brain—also require more precise elucidation, in particular those PACAPergic neurons influencing mainly CRH neurons in the extended amygdala, or those affecting CRH target neurons in PVN. Regiospecific knock-out of PACAP and PACAP receptors in mouse CNS; development of more specific and drug-like receptor agonists and antagonists; better behavioral and neurochemical integration of data obtained in mice, rats, primates, and in the clinic; and more comprehensive analysis of gender as well as species differences in PACAPergic stress responding, are all likely to play a role in focusing our expanding knowledge of PACAP's role in stress onto therapeutic targets. More complete integration of behavioral and neurochemical information about PACAPergic signaling in stress also seems called for, including full "reverse translation" to rodent models of the pronounced female specificity that has been reported in humans; more explicit stratification of clinical outcomes with respect to stressor intensity, frequency, and type across gender; and full phenocopying of findings obtained in PACAP-deficient mice, in PAC1-, VPAC1-, and VPAC2-deficient mice, similarly to the parsing of the roles of PACAP and VIP and their receptors in circadian regulation using multiple-knockout phenocopying [99, 100]. These findings, when they come, will guide both a better understanding of the neurobiology of stress responding, and the link between stress responding and human psychopathology that may perhaps be broken by PACAP receptor-directed therapeutics.

References

1. Cannon WB. Organization for physiological homeostasis. Physiol Rev. 1929;9:399–431.
2. Selye H. The evolution of the stress concept. Am Sci. 1973;61(6):692–9.
3. Pacak K, Palkovits M, Yadid G, Kvetnansky R, Kopin IJ, Goldstein DS. Heterogeneous neurochemical responses to different stressors: a test of Selye's doctrine of nonspecificity. Am J Physiol. 1998;275(4 Pt 2):R1247–55.

4. McEwen BS. Central effects of stress hormones in health and disease: understanding the protective and damaging effects of stress and stress mediators. Eur J Pharmacol. 2008; 583(2-3):174–85.
5. McEwen BS. Protection and damage from acute and chronic stress: allostasis and allostatic overload and relevance to the pathophysiology of psychiatric disorders. Ann N Y Acad Sci. 2004;1032:1–7.
6. Armario A, Escorihuela RM, Nadal R. Long-term neuroendocrine and behavioural effects of a single exposure to stress in adult animals. Neurosci Biobehav Rev. 2008;32:1121–35.
7. Stroth N, Holighaus Y, Ait-Ali D, Eiden LE. PACAP: a master regulator of neuroendocrine stress circuits and the cellular stress response. Ann N Y Acad Sci. 2011;1220:49–59.
8. Hokfelt T, Bartfai T, Bloom F. Neuropeptides:opportunities for drug discovery. Lancet Neurol. 2003;2:463–72.
9. Morgan A, Burgoyne RD. Common mechanisms for regulated exocytosis in the chromaffin cell and the synapse. Semin Cell Dev Biol. 1997;8:141–9.
10. Smith CB, Eiden LE. Is PACAP the major neurotransmitter for stress transduction at the adrenomedullary synapse? J Mol Neurosci. 2012;48:403–12.
11. Lamouche S, Martineau D, Yamaguchi N. Modulation of adrenal catecholamine release by PACAP in vivo. Regul Integr Comp Physiol. 1999;45:R162–70.
12. Watanabe T, Masuo Y, Matsumoto H, Suzuki N, Ohtaki T, Masuda Y, et al. Pituitary adenylate cyclase activating polypeptide provokes cultured rat chromaffin cells to secrete adrenaline. Biochem Biophys Res Commun. 1992;182:403–11.
13. Fukushima Y, Hikichi H, Mizukami K, Nagayama T, Yoshida M, Suzuki-Kusaba M, et al. Role of endogenous PACAP in catecholamine secretion from the rat adrenal gland. Am J Physiol Regul Integr Comp Physiol. 2001;281:R1562–7.
14. Hamelink C, Tjurmina O, Damadzic R, Young WS, Weihe E, Lee H-W, et al. Pituitary adenylate cyclase activating polypeptide is a sympathoadrenal neurotransmitter involved in catecholamine regulation and glucohomeostasis. Proc Natl Acad Sci U S A. 2002;99:461–6.
15. Stroth N, Kuri BA, Mustafa T, Chan SA, Smith CB, Eiden LE. PACAP controls adrenomedullary catecholamine secretion and expression of catecholamine biosynthetic enzymes at high splanchnic nerve firing rates characteristic of stress transduction in male mice. Endocrinology. 2013;154:330–9.
16. Kuri BA, Chan SA, Smith CB. PACAP regulates immediate catecholamine release from adrenal chromaffin cells in an activity-dependent manner through a protein kinase C-dependent pathway. J Neurochem. 2009;110:1214–25.
17. Edwards AV, Jones CT. Adrenal responses to the peptide PACAP in conscious functionally hypophysectomized calves. Am J Physiol. 1994;266:E870–6.
18. Braas KM, May V. Pituitary adenylate cyclase-activating polypeptides, PACAP-38 and PACAP-27, regulation of sympathetic neuron catecholamine, and neuropeptide Y expression through activation of type I PACAP/VIP receptor isoforms. Ann N Y Acad Sci. 1996;805:204–16.
19. Brandenburg CA, May V, Braas KM. Identification of endogenous sympathetic neuron pituitary adenylate cyclase-activating polypeptide (PACAP): depolarization regulates production and secretion through induction of multiple neuropeptide transcripts. J Neurosci. 1997;17:4045–55.
20. Beaudet MM, Braas KM, May V. Pituitary adenylate cyclase activating polypeptide (PACAP) expression in sympathetic preganglionic projection neurons to the superior cervical ganglion. J Neurobiol. 1998;36:325–36.
21. Braas KM, May V. Pituitary adenylate cyclase-activating polypeptides directly stimulate sympathetic neuron neuropeptide Y release through PAC(1) receptor isoform activation of specific intracellular signaling pathways. J Biol Chem. 1999;274(39):27702–10.
22. Beaudet MM, Parsons RL, Braas KM, May V. Mechanisms mediating pituitary adenylate cyclase-activating polypeptide depolarization of rat sympathetic neurons. J Neurosci. 2000;20:7353–61.
23. Ip NY, Perlman RL, Zigmond RE. Acute transsynaptic regulation of tyrosine 3-monooxygenase activity in the rat superior cervical ganglion: evidence for both cholinergic and noncholinergic mechanisms. Proc Natl Acad Sci U S A. 1983;80:2081–5.

24. Wakade AR. Non-cholinergic transmitter(s) maintains secretion of catecholamines from rat adrenal medulla for several hours of continuous stimulation of splanchnic neurons. J Neurochem. 1988;50:1302–8.
25. Benarroch EE. Neuropeptides in the sympathetic system: presence, plasticity, modulation and implications. Ann Neurol. 1994;36:6–13.
26. Goldstein DS, Kopin IJ. Adrenomedullary, adrenocortical, and sympathoneural responses to stressors: a meta-analysis. Endocr Regul. 2008;42:111–9.
27. Ressler KJ, Mercer KB, Bradley B, Jovanovic T, Mahan A, Kerley K, et al. Post-traumatic stress disorder is associated with PACAP and the PAC1 receptor. Nature. 2011;470:492–7.
28. Sundler F, Ekblad E, Hannibal J, Moller K, Zhang Y-Z, Mulder H, et al. Pituitary adenylate cyclase-activating peptide in sensory and autonomic ganglia: localization and regulation. Ann N Y Acad Sci. 1996;805:410–28.
29. Przywara DA, Xi G, Angelilli L, Wakade TD, Wakade AR. A noncholinergic transmitter, pituitary adenylate cyclase activating polypeptide, utilizes a novel mechanism to evoke catecholamine secretion in rat adrenal chromaffin cells. J Biol Chem. 1996;271:10545–50.
30. Hamelink C, Lee H-W, Grimaldi M, Eiden LE. Coincident elevation of cyclic AMP and calcium influx by PACAP-27 synergistically regulates VIP gene transcription through a novel PKA-independent signaling pathway. J Neurosci. 2002;22:5310–20.
31. Samal B, Gerdin MJ, Huddleston D, Hsu CM, Elkahloun AG, Stroth N, et al. Meta-analysis of microarray-derived data from PACAP-deficient adrenal gland in vivo and PACAP-treated chromaffin cells identifies distinct classes of PACAP-regulated genes. Peptides. 2007;28:1871–82.
32. Eiden LE, Samal B, Gerdin MJ, Mustafa T, Stroth N. Discovery of PACAP-related genes through microarray analyses in cell culture and in vivo. Ann N Y Acad Sci. 2008; 1144:6–20.
33. Stroth N, Eiden LE. Stress hormone synthesis in mouse hypothalamus and adrenal gland triggered by restraint is dependent on pituitary adenylate cyclase-activating polypeptide signaling. Neuroscience. 2010;165:1025–30.
34. Tsukiyama N, Saida Y, Kakuda M, Shintani N, Hayata A, Morita Y, et al. PACAP centrally mediates emotional stress-induced corticosterone responses in mice. Stress. 2011;14: 368–75.
35. Kuri BA, Khan SA, Chan SA, Prabhakar NR, Smith CB. Increased secretory capacity of mouse adrenal chromaffin cells by chronic intermittent hypoxia: involvement of protein kinase C. J Physiol. 2007;584(Pt 1):313–9.
36. Mustafa T, Grimaldi M, Eiden LE. The hop cassette of the PAC1 receptor confers coupling to Ca2+ elevation required for pituitary adenylate cyclase-activating polypeptide-evoked neurosecretion. J Biol Chem. 2007;282:8079–91.
37. Mustafa T, Walsh J, Grimaldi M, Eiden LE. PAC1hop receptor activation facilitates catecholamine secretion selectively through 2-APB-sensitive Ca(2+) channels in PC12 cells. Cell Signal. 2010;22(10):1420–6.
38. O'Farrell M, Marley PD. Multiple calcium channels are required for pituitary adenylate cyclase-activating polypeptide-induced catecholamine secretion from bovine cultured adrenal chromaffin cells. Naunyn Schmiedebergs Arch Pharmacol. 1997;356:536–42.
39. Osipenko ON, Barrie AP, Allen JM, Gurney AM. Pituitary adenylyl cyclase-activating peptide activates multiple intracellular signaling pathways to regulate ion channels in PC12 cells. J Biol Chem. 2000;275:16626–31.
40. May V, Lutz E, MacKenzie C, Schutz KC, Dozark K, Braas KM. Pituitary adenylate cyclase-activating polypeptide (PACAP)/PAC1HOP1 receptor activation coordinates multiple neurotrophic signaling pathways: Akt activation through phosphatidylinositol 3-kinase gamma and vesicle endocytosis for neuronal survival. J Biol Chem. 2010;285:9749–61.
41. Lazarovici P, Jiang H, Fink Jr D. The 38-amino-acid form of pituitary adenylate cyclase-activating polypeptide induces neurite outgrowth in PC12 cells that is dependent on protein kinase C and extracellular signal-regulated kinase but not on protein kinase A, nerve growth factor receptor tyrosine kinase, p21ras G protein, and pp60c-src cytoplasmic tyrosine kinase. Mol Pharmacol. 1998;54:547–58.

42. Emery AC, Eiden MV, Eiden LE. Separate cyclic AMP sensors for neuritogenesis, growth arrest, and survival of neuroendocrine cells. J Biol Chem. 2014;289:10126–39.
43. Hammack SE, May V. Pituitary adenylate cyclase activating polypeptide in stress-related disorders: data convergence from animal and human studies. Biol Psychiatry. 2015;78:167–77.
44. Amir-Zilberstein L, Blechman J, Sztainberg Y, Norton WH, Reuveny A, Borodovsky N, et al. Homeodomain protein otp and activity-dependent splicing modulate neuronal adaptation to stress. Neuron. 2012;73:279–91.
45. Emery A, Eiden MV, Mustafa T, Eiden LE. GPCR-Gs signaling to ERK is controlled by the cAMP-sensing guanine nucleotide exchange factor NCS/Rapgef2 in neuronal and endocrine cells. Sci Signal. 2013;6:ra51.
46. May V, Clason TA, Buttolph TR, Girard BM, Parsons RL. Calcium influx, but not intracellular calcium release, supports PACAP-mediated ERK activation in HEK PAC1 receptor cells. J Mol Neurosci. 2014;54:342–50.
47. May V, Buttolph TR, Girard BM, Clason TA, Parsons RL. PACAP-induced ERK activation in HEK cells expressing PAC1 receptors involves both receptor internalization and PKC signaling. Am J Physiol Cell Physiol. 2014;306:C1068–79.
48. Shenoy SK, Drake MT, Nelson CD, Houtz DA, Xiao K, Madabushi S, et al. beta-arrestin-dependent, G protein-independent ERK1/2 activation by the beta2 adrenergic receptor. J Biol Chem. 2006;281:1261–73.
49. Marshall CJ. Specificity of receptor tyrosine kinase signaling: transient versus sustained extracellular signal-regulated kinase activation. Cell. 1995;80:179–85.
50. Robinson MJ, Stippec SA, Goldsmith E, White MA, Cobb MH. A constitutively active and nuclear form of the MAP kinase ERK2 is sufficient for neurite outgrowth and cell transformation. Curr Biol. 1998;8:1141–50.
51. Holighaus Y, Weihe E, Eiden LE. STC1 induction by PACAP is mediated through cAMP and ERK1/2 but not PKA in cultured cortical neurons. J Mol Neurosci. 2012;46:75–87.
52. Stroth N, Liu Y, Aguilera G, Eiden LE. Pituitary adenylate cyclase-activating polypeptide (PACAP) controls stimulus-transcription coupling in the hypothalamic-pituitary-adrenal axis to mediate sustained hormone secretion during stress. J Neuroendocrinol. 2011;23:944–55.
53. Lehmann ML, Mustafa T, Eiden AM, Herkenham M, Eiden LE. PACAP-deficient mice show attenuated corticosterone secretion and fail to develop depressive behavior during chronic social defeat stress. Psychoneuroendocrinology. 2013;38:702–15.
54. Mustafa T, Jiang SZ, Eiden AM, Weihe E, Thistlethwaite I, Eiden LE. Impact of PACAP and PAC1 receptor deficiency on the neurochemical and behavioral effects of acute and chronic restraint stress in male C57BL/6 mice. Stress. 2015;18:408–18.
55. Mravec B, Bodnar I, Uhereczky G, Kvetnansky R, Palkovits M. Effect of lesions of A5 or A7 noradrenergic cell group or surgical transection of brainstem catecholamine pathways on plasma catecholamine levels in rats injected subcutaneously by formalin. Gen Physiol Biophys. 2012;31:247–54.
56. Popper CW, Chiueh CC, Kopin IJ. Plasma catecholamine concentrations in unanesthetized rats during sleep, wakefulness, immobilization and after decapitation. J Pharmacol Exp Ther. 1977;202:144–8.
57. Liu Y, Aguilera G. Cyclic AMP inducible early repressor mediates the termination of corticotropin releasing hormone transcription in hypothalamic neurons. Cell Mol Neurobiol. 2009;29:1275–81.
58. Aguilera G, Liu Y. The molecular physiology of CRH neurons. Front Neuroendocrinol. 2012;33:67–84.
59. Gaszner B, Kormos V, Kozicz T, Hashimoto H, Reglodi D, Helyes Z. The behavioral phenotype of pituitary adenylate-cyclase activating polypeptide-deficient mice in anxiety and depression tests is accompanied by blunted c-Fos expression in the bed nucleus of the stria terminalis, central projecting Edinger-Westphal nucleus, ventral lateral septum, and dorsal raphe nucleus. Neuroscience. 2012;202:283–99.
60. Hashimoto H, Hashimoto R, Shintani N, Tanaka K, Yamamoto A, Hatanaka M, et al. Depression-like behavior in the forced swimming test in PACAP-deficient mice: amelioration by the atypical antipsychotic risperidone. J Neurochem. 2009;110:595–602.

61. Hattori S, Takao K, Tanda K, Toyama K, Shintani N, Baba A, et al. Comprehensive behavioral analysis of pituitary adenylate cyclase-activating polypeptide (PACAP) knockout mice. Front Behav Neurosci. 2012;6:58.
62. Wong ML, Webster EL, Spokes H, Phu P, Ehrhart-Bornstein M, Bornstein S, et al. Chronic administration of the non-peptide CRH type 1 receptor antagonist antalarmin does not blunt hypothalamic-pituitary-adrenal axis responses to acute immobilization stress. Life Sci. 1999;65:PL53–8.
63. Habib KE, Weld KP, Rice KC, Pushkas J, Champoux M, Listwak S, et al. Oral administration of a corticotropin-releasing hormone receptor antagonist significantly attenuates behavioral, neuroendocrine, and autonomic responses to stress in primates. Proc Natl Acad Sci U S A. 2000;97:6079–84.
64. Nemeroff CB, Vale WW. The neurobiology of depression: inroads to treatment and new drug discovery. J Clin Psychiatry. 2005;66 Suppl 7:5–13.
65. Hammack SE, Cheung J, Rhodes KM, Schutz KC, Falls WA, Braas KM, et al. Chronic stress increases pituitary adenylate cyclase-activating peptide (PACAP) and brain-derived neurotrophic factor (BDNF) mRNA expression in the bed nucleus of the stria terminalis (BNST): roles for PACAP in anxiety-like behavior. Psychoneuroendocrinology. 2009;34:833–43.
66. Hammack SE, Roman CW, Lezak KR, Kocho-Shellenberg M, Grimmig B, Falls WA, et al. Roles for pituitary adenylate cyclase-activating peptide (PACAP) expression and signaling in the bed nucleus of the stria terminalis (BNST) in mediating the behavioral consequences of chronic stress. J Mol Neurosci. 2010;42:327–40.
67. Dore R, Iemolo A, Smith KL, Wang X, Cottone P, Sabino V. CRF mediates the anxiogenic and anti-rewarding, but not the anorectic effects of PACAP. Neuropsychopharmacology. 2013;38:2160–9.
68. Kocho-Schellenberg M, Lezak KR, Harris OM, Roelke E, Gick N, Choi I, et al. PACAP in the BNST produces anorexia and weight loss in male and female rats. Neuropsychopharmacology. 2014;39:1614–23.
69. Lezak KR, Roelke E, Harris OM, Choi I, Edwards S, Gick N, et al. Pituitary adenylate cyclase-activating polypeptide (PACAP) in the bed nucleus of the stria terminalis (BNST) increases corticosterone in male and female rats. Psychoneuroendocrinology. 2014;45:11–20.
70. Lezak KR, Roman CW, Braas KM, Schutz KC, Falls WA, Schulkin J, et al. Regulation of bed nucleus of the stria terminalis PACAP expression by stress and corticosterone. J Mol Neurosci. 2014;45:11–20.
71. Roman CW, Lezak KR, Hartsock MJ, Falls WA, Braas KM, Howard AB, et al. PAC1 receptor antagonism in the bed nucleus of the stria terminalis (BNST) attenuates the endocrine and behavioral consequences of chronic stress. Psychoneuroendocrinology. 2014;47:151–65.
72. Piggins HD, Stamp JA, Burns J, Rusak B, Semba K. Distribution of pituitary adenylate cyclase activating polypeptide (PACAP) immunoreactivity in the hypothalamus and extended amygdala of the rat. J Comp Neurol. 1996;376:278–94.
73. Herman JP, Ostrander MM, Mueller NK, Figueiredo H. Limbic system mechanisms of stress regulation: hypothalamo-pituitary-adrenocortical axis. Prog Neuropsychopharmacol Biol Psychiatry. 2005;29:1201–13.
74. Walker DL, Davis M. Double dissociation between the involvement of the bed nucleus of the stria terminalis and the central nucleus of the amygdala in startle increases produced by conditioned versus unconditioned fear. J Neurosci. 1997;17:9375–83.
75. Hammack SE, Richey KJ, Watkins LR, Maier SF. Chemical lesion of the bed nucleus of the stria terminalis blocks the behavioral consequences of uncontrollable stress. Behav Neurosci. 2004;118:443–8.
76. Kalin NH, Shelton SE, Fox AS, Oakes TR, Davidson RJ. Brain regions associated with the expression and contextual regulation of anxiety in primates. Biol Psychiatry. 2005;58:796–804.
77. Waddell J, Morris RW, Bouton ME. Effects of bed nucleus of the stria terminalis lesions on conditioned anxiety: aversive conditioning with long-duration conditional stimuli and reinstatement of extinguished fear. Behav Neurosci. 2006;120(2):324–36.

78. Somerville LH, Whalen PJ, Kelley WM. Human bed nucleus of the stria terminalis indexes hypervigilant threat monitoring. Biol Psychiatry. 2010;68:416–24.
79. Radley JJ, Sawchenko PE. A common substrate for prefrontal and hippocampal inhibition of the neuroendocrine stress response. J Neurosci. 2011;31:9683–95.
80. Missig G, Roman CW, Vizzard MA, Braas KM, Hammack SE, May V. Parabrachial nucleus (PBn) pituitary adenylate cyclase activating polypeptide (PACAP) signaling in the amygdala: implication for the sensory and behavioral effects of pain. Neuropharmacology. 2014;86:38–48.
81. Kozicz T, Arimura A. Dopamine- and cyclic AMP-regulated phosphoprotein-immunoreactive neurons activated by acute stress are innervated by fiber terminals immunopositive for pituitary adenylate cyclase-activating polypeptide in the extended amygdala in the rat. Regul Pept. 2002;109:63–70.
82. Legradi G, Das M, Giunta B, Hirani K, Mitchell EA, Diamond DM. Microinfusion of pituitary adenylate cyclase-activating polypeptide into the central nucleus of amygdala of the rat produces a shift from an active to passive mode of coping in the shock-probe fear/defensive burying test. Neural Plast. 2007;2007:79102.
83. Schmidt SD, Myskiw JC, Furini CR, Schmidt BE, Cavalcante LE, Izquierdo I. PACAP modulates the consolidation and extinction of the contextual fear conditioning through NMDA receptors. Neurobiol Learn Mem. 2015;118C:120–4.
84. Gold PW, Chrousos GP. Clinical studies with corticotropin releasing factor: implications for the diagnosis and pathophysiology of depression, Cushing's disease, and adrenal insufficiency. Psychoneuroendocrinology. 1985;10:401–19.
85. Chrousos GP, Gold PW. The concepts of stress and stress system disorders. Overview of physical and behavioral homeostasis. JAMA. 1992;267(9):1244–52.
86. Gold PW, Chrousos GP. Organization of the stress system and its dysregulation in melancholic and atypical depression: high vs low CRH/NE states. Mol Psychiatry. 2002;7:254–75.
87. Otto C, Kovalchuk Y, Wolfer DP, Gass P, Martin M, Zuschratter W, et al. Impairment of mossy fiber long-term potentiation and associative learning in pituitary adenylate cyclase activating polypeptide type I receptor-deficient mice. J Neurosci. 2001;21:5520–7.
88. Nicot A, Otto T, Brabet P, Dicicco-Bloom EM. Altered social behavior in pituitary adenylate cyclase-activating polypeptide type I receptor-deficient mice. J Neurosci. 2004;24:8786–95.
89. Andero R, Ressler KJ. Fear extinction and BDNF: translating animal models of PTSD to the clinic. Genes Brain Behav. 2012;11:503–12.
90. Dias BG, Ressler KJ. PACAP and the PAC1 receptor in post-traumatic stress disorder. Neuropsychopharmacology. 2013;38:245–6.
91. Stevens JS, Almli LM, Fani N, Gutman DA, Bradley B, Norrholm SD, et al. PACAP receptor gene polymorphism impacts fear responses in the amygdala and hippocampus. Proc Natl Acad Sci U S A. 2014;111:3158–63.
92. Robberecht P, Vertongen P, Langer I, Perret J. Development of selective ligans for PAC1, VPAC1 and VPAC2 receptors. In: Vaudry H, Arimura A, editors. Pituitary adenylate cyclase-activating polypeptide. Norwell, MA: Kluwer Academic Publishers; 2003. p. 49–67.
93. Gourlet P, Vandermeers A, Vandermeers-Piret MC, Rathe J, De Neef P, Robberecht P. Fragments of pituitary adenylate cyclase activating polypeptide discriminate between type I and II recombinant receptors. Eur J Pharmacol. 1995;287:7–11.
94. Lin Y, Hall RA, Kuhar MJ. CART peptide stimulation of G protein-mediated signaling in differentiated PC12 cells: identification of PACAP 6-38 as a CART receptor antagonist. Neuropeptides. 2011;45:351–8.
95. Burgos JR, Iresjo BM, Smedh U. Pituitary adenylate cyclase-activating polypeptide 6-38 blocks cocaine- and amphetamine-regulated transcript peptide-induced hypophagia in rats. PLoS One. 2013;8(8):e72347.
96. Moro O, Lerner EA. Maxadilan, the vasodilator from sand flies, is a specific pituitary adenylate cyclase activating peptide type I receptor agonist. J Biol Chem. 1997;272:966–70.
97. Uchida D, Tatsuno I, Tanaka T, Hirai A, Saito Y, Moro O, et al. Maxadilan is a specific agonist and its deleted peptide (M65) is a specific antagonist for PACAP type 1 receptor. Ann N Y Acad Sci. 1998;865:253–8.

98. Akerman S, Goadsby PJ. Neuronal PAC1 receptors mediate delayed activation and sensitization of trigeminocervical neurons: relevance to migraine. Sci Transl Med. 2015;7:308ra157.
99. Harmar AJ, Marston HM, Shen S, Spratt C, West KM, Sheward WJ, et al. The VPAC2 receptor is essential for circadian function in the mouse suprachiasmatic nuclei. Cell. 2002;109:497–508.
100. Aton SJ, Colwell CS, Harmar AJ, Waschek J, Herzog ED. Vasoactive intestinal polypeptide mediates circadian rhythmicity and synchrony in mammalian clock neurons. Nat Neurosci. 2005;8:476–83.
101. Jiang SZ, Eiden LE. (in press). PACAPergic synaptic signaling and circuitry mediating mammalian responses to psychogenic and systemic stressors. In: Reglodi D, Tamas A, editors. Pituitary adenylate cyclase activating polypeptide – PACAP. New York: Springer Nature.

Chapter 42
Using PACAP Heterozygous Mice as Models of the Three Hit Theory of Depression

Jozsef Farkas, Laszlo A. Kovacs, Tamas Gaszner, and Balazs Gaszner

Abstract Depression and mood disorders belong to the most common reasons for disability. Despite several decades of stress research many questions remained to be answered. One reason for slow progress in this field may be explained by the lack of equivocally accepted, from all points of view reliable animal models. Chronic unpredictable stress (CVMS) exposure of rodents as models for environmental challenges is a commonly used depression model. Early life stress has been also well studied. Maternal deprivation of pups induces epigenetic alterations shaping stress response in later life. In the last two decades numerous genetically modified mouse strains have been generated to assess the effects of gene mutations on mood disorders. All these approaches have some limitations. More recently the three hit theory for human depression has become more widely accepted.

The aim of our work was to establish a mouse model for depression according to the three hit theory. Genetically predisposed (first hit, the lack of one or both normal PACAP genes) mice were exposed to maternal deprivation (second hit). CVMS (third hit) in adult mice with maternal separation history with mutated PACAP allele(s) carry all risk factors.

Our results revealed that about half of the PACAP KO mice did not survive the triple combination of stress exposure. PACAP heterozygous mice with all risk factors showed both physical and endocrinological consequences of stress exposure. Functional morphological results supported that corticotropin releasing factor (CRF) expressing neurons of the oval division of bed nucleus of the stria terminalis (BSTov) increased their FosB neuronal activity. In contrast, urocortin1 (Ucn1) expressing neurons of the centrally projecting Edinger–Westphal nucleus (cpEW) and serotonin (5HT) neurons of the dorsal raphe nucleus (DR) showed blunted neuronal activity in CVMS exposed PACAP HZ mice with maternal deprivation history.

In summary, the three hit theory of depression seems to be an applicable model in PACAP heterozygous mice. Increased BSTov-CRF in addition to blunted cpEW-

J. Farkas • L.A. Kovacs • T. Gaszner • B. Gaszner (✉)
Department of Anatomy, MTA-PTE "Lendület" PACAP Research Team, Medical School, Pecs University, Szigeti u. 12, Pecs 7624, Hungary
e-mail: balazs.b.gaszner@aok.pte.hu

D. Reglodi, A. Tamas (eds.), *Pituitary Adenylate Cyclase Activating Polypeptide — PACAP*, Current Topics in Neurotoxicity 11,
DOI 10.1007/978-3-319-35135-3_42

Ucn1 and 5HT-DR activity proves that at least three players of stress adaptation systems may be affected in an inverse manner.

Keywords Chronic stress • Maternal deprivation • Corticotropin releasing factor • Urocortin1 • Serotonin • FosB

Mood disorders, including major depression, are the most common causes of chronic disability [1] in the communities of Western type. Despite several decades of scientific research many questions on the neurobiology of depression remain to be answered. For instance, diagnostic criteria of major depression include pairs of symptoms resembling in part opposing phenomena supporting the same diagnosis (i.e., weight gain vs. loss, hypersomnia vs. insomnia) (DSM 5 [2]). This suggests that the great variance of the clinical symptomatology of major depression is explainable by diverse anomalies in multiple stress related adaptation systems. Nevertheless, there is only a single widely used pharmacological therapeutic strategy available targeting the monoaminergic systems. Not surprisingly, these pharmacons do not provide the required effectivity which is exemplified by the high proportion of patients who do not respond to the therapy at all, or later, they relapse [3].

There is no doubt that new therapeutical approaches should be discovered providing new possible pharmacological targets. Studies on several neuropeptide systems have proven that physiology of stress (mal)adaptation is not restricted to the corticotropin releasing factor (CRF) controlled hypothalamic–pituitary–adrenal (HPA) axis (for review see ref. [4]); however, it is clear, that not in all, but in most of the cases it is involved [5, 6]. These data further support that major depression is a multifactorial condition. Indeed, neuroscience research applies a great variety of animal models to replicate the symptoms or biological phenomena found in human disease.

Stress Models

Environmental stress applied in laboratory animals has been studied for several decades. A great variety of acute and chronic stress paradigms have been developed to which uncountable modifications were applied to increase the effectivity. The high number of various protocols used suggests that none of them is equivocally accepted. One frequently used stressor is the chronic variable mild stress (CVMS) protocol (for review see ref. [7]). Most of the laboratories report that CVMS exposed mice or rats show the physical, endocrinological, and behavioral consequences of chronic stress exposure.

Body, thymus, and adrenal weight data belong to the physical parameters (for reviews see refs. [8, 9]). Due to increased HPA-axis activity the catabolic effect of glucocorticoids is reflected by the weight loss. As adrenals become hypertrophic in chronic stress, the body weight corrected adrenal tissue weight also increases.

Prolonged high glucocorticoid levels cause decreased thymus size. Endocrinologically, the overactivity of the HPA axis may be tested by plasma corticosterone measurements. Behavioral assessment of depression level is based mostly on forced swim (FST, [10]), tail suspension (TST [11]), or sucrose preference tests (SPT [12]). Anxiety levels may be compared using marble burying (MBT [13]), or light–dark box test (LDT [14]). Animals showing increased immobility time in FST and TST, accompanied with reduced consumption of sweetened water are postulated to show depression-like behavior. Increased anxiety is characterized in animals by high number of marbles buried in MBT and increased time spent in the dark chamber of LDT equipment.

Epigenetic Factors

One of the widely used models for depression applies maternal deprivation of mouse or rat pups in the first two weeks of postnatal developmental period (for review see refs. [15, 16]). This protocol has deep and long lasting impact on stress adaptation ability which influences a great variety of players involved in stress adaptation [17] including CRF [18], Ucn1 [19], and serotoninergic systems [20, 21] More recent studies uncovered that this type of stress exposure shapes gene expression profiles in several systems (for review see ref. [17]), without affecting the genomic sequence proper. Now it is no dispute that the long term gene expression regulation depends at least in part on enzymatic modifications (i.e., acetylation and methylation) of histones. In the last decade the studies of such epigenetic changes have proven that early life stress and significant life events may be associated with mood disorders, even with transgenerational effects (for reviews see refs. [22, 23]).

Genetic Models

The observation that relatives of patients suffering on major depression are more often affected than others, is not new [24]. A great number of gene mutations/polymorphisms are thought to be associated with psychiatric disorders including major depression [25]. Therefore, genetically modified animals are used often as depression models (for review see ref. [4]). Their behavioral phenotype is often evaluated by FST, TST SPT and sometimes they show depression-like phenotype without stress exposure. Besides many others, pituitary adenylate cyclase-activating polypeptide (PACAP) mutant mice (KO) were shown to possess depression-like phenotype [26, 27]. Interestingly, PACAP and PAC1 receptor gene polymorphisms were shown to occur more frequently in patients suffering on mood disorders [28, 29].

The above listed approaches to use as models are not without limitations. For instance, it is well known that experimental animals show individual differences in stress vulnerability [30]. Some strains of mice [31] and rats [32] were shown to differ in their stress reactivity. The reliability of behavioral tests was also debated, and alternative interpretations may be considered [33, 34]. Another important limitation of genetic models is that developmental compensations may occur. For instance, the deletion of CRF gene causes the upregulation of the CRF peptide family member urocortin 1 (Ucn1) [35], and in line with this, artificial overexpression of CRF decreases Ucn1 mRNA expression [36]. Third, a great number of genes and their polymorphisms have been implicated in the pathogenesis of mood disorders (for review see ref. [37]). The inheritance of most of the susceptibility increasing alleles does not abolish the gene expression completely, but it leads to a gene product with slightly different effectivity. Based on this, the translational value of studies applying null mice with two mutated alleles, consequently lacking the gene product may also be debated. As some depression-related alleles increase the susceptibility in heterozygous (HZ) subjects also [25], the use of HZ animals with one normal and one mutated allele may provide a model with greater translational relevance.

The Three Hit Theory of Depression

The three hit theory of major depression postulates (for reviews see refs. [38–40]) that the symptoms are precipitated by (a) genetic predisposition, (b) epigenetic changes, and (c) environmental stress. Genetic predisposition causes an increased vulnerability in numerous stress adaptation systems. For instance, glucocorticoid receptor [41], serotonin transporter gene [42], CRF receptor [43] genes may be altered, leading to slightly altered adaptation ability. A significant life event changes the epigenome, leading to further anomalies in gene expression and adaptation ability. Subjects with history of genetic predisposition and altered epigenome may develop depression upon chronic stress exposure. Based on this, the triple combination of genetic, epigenetic, and environmental stress models in rodents may be a useful tool to set up and validate an animal model.

The Setup of Our Three Hit Model for Depression in Mice

Seventy female PACAP HZ mice were mated with 70 PACAP HZ males. Thirty-seven females delivered pups within a time period of 36 h. These litters were cross fostered, and 9–10 pups were randomly redistributed to the dams. According to the Mendelian rules we anticipated that the proportion of genotypes in the offspring will be 25 % wild type (WT), 25 % KO, and 50 % HZ.

To cause epigenetic alterations litters after the first postnatal day (PD) were exposed to the maternal deprivation paradigm [19, 44]. One third of the litters was separated from their dams for 180 min from the PD1 to PD14 (i.e., maternal deprivation group, MD180). The second third of mouse litters was briefly (15 min) separated (MS15) from the dam for 2 weeks, which represents natural circumstances. The dam leaves pups for shorter periods to search for food [45]. This short separa-

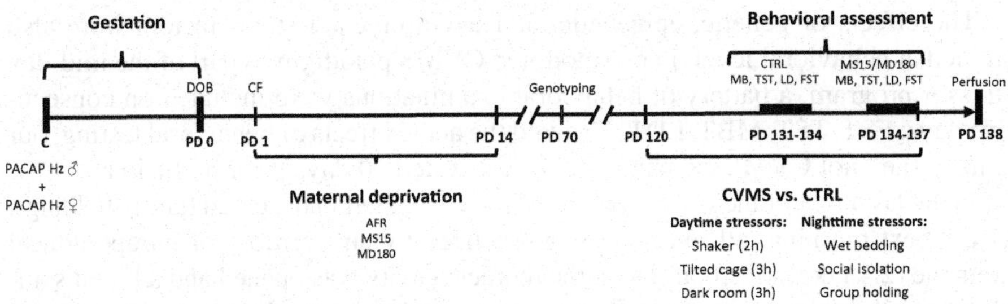

Fig. 42.1 Experimental setup of our three hit model. PACAP heterozygous pairs of mice were mated (*C*—day of conception). The day of birth (*DOB*) was defined as postnatal day (*PD*) zero. On PD 1 litters were cross fostered (*CF*). Between PD1 and PD14 15 min maternal separation (*MS15*) or 180 min maternal deprivation (*MD180*) was applied on the selected litters. The third group was left undisturbed and handled according to the routine animal facility protocol: animal facility rearing (*AFR*). On PD70 tail clipping was performed for genotyping. On PD120 the male offspring in AFR, MS15, and MD180 groups were equally assigned to chronic variable mild stress (*CVMS*) and control (*CTRL*) groups (see also Table 42.1). The CVMS consisted of daytime and nighttime stressors. Daytime stressors: shaker (cage was placed on a laboratory orbital shaker set to 60 rpm for 2 h), tilted cage (the cage was placed obliquely in 45° angle for 3 h), dark room (the cages were placed in a completely dark room for 3 h). Nighttime stressors: wet bedding (animals were kept overnight in cages where the nesting material was soaked with tap water), social isolation (subjects were kept overnight in separate cages), group holding (no stressor was applied). The CTRL groups were subjected to behavior tests between PD131 and PD134 (one test per day): marble burying (*MB*), tail suspension test (*TST*), light–dark test (*LD*), and forced swim test (*FST*). The same tests were conducted in CVMS groups between PD134 and PD137 and were considered as daytime stressors. On PD138 animals were euthanized and transcardially perfused

Table 42.1 Summary of experimental groups in our three hit model

Animal facility rearing						15 min maternal separation						180 min maternal deprivation					
Control			CVMS			Control			CVMS			Control			CVMS		
WT	HZ	KO	WT	HZ	KO	WT	HZ	KO	WT	HZ	KO	WT	HZ	KO	WT	HZ	KO

WT PACAP wild type, *HZ* PACAP heterozygous type, *KO* PACAP KO, *CMVS* chronic variable mild stress

tion is considered to have beneficial effect by reducing HPA axis sensitivity [44]. The last third of litters was left undisturbed except the normal cage cleaning according to the protocol (animal facility rearing group, AFR).

Litters were weaned on PD 28 according to the regular protocol. Only male offspring was used in the following part of the study. The genotype of mice was determined by PCR on tail tissue samples collected on the tenth postnatal weeks. One half of adult offspring in all main groups was exposed to chronic variable mild stress (CVMS). Controls were left undisturbed except that they were weighed once a week at the time of the regular cage cleaning procedure and that they were subjected to a battery of behavioral tests 10 days before perfusion (see details later). The CVMS paradigm consists of a mid-day stressor (Fig. 42.1) applied between 10 am and 14 pm. After this, mice were returned to their original home cages. At 6 pm the overnight stressor ("wet bedding" or "social isolation") was started (Fig. 42.1) except for "group holding" nights, in which case mice were left undisturbed.

The effects of genetic, epigenetic, and environmental stress factors were also studied at behavioral level. In our modified CVMS paradigm as part of the mid-day stressor program, a battery of behavioral examinations were included on consecutive days: FST, TST, MBT, LDT. To avoid the acute effects of behavioral testing, our control (i.e., not CVMS exposed mice) were tested 10 days prior perfusion.

In the last night of the CVMS paradigm mice were left undisturbed (group holding). In the next morning, all mice were anesthetized by an overdose of intraperitoneal urethane (2.4 g/kg) injection. Then, the thoracic cavity was opened and a blood sample was collected for corticosterone measurement. Subsequently, mice were perfused by phosphate buffered saline followed by 4 % paraformaldehyde solution. Brains were dissected and post-fixed. Adrenal and thymus gland weights were measured.

Survival of Mice in the Three Hit Model

Thirty-seven litters, with 362 pups in total were included into the experiment. Thirteen litters were exposed to MD180; 12 litters were briefly separated (MS15); 12 litters were left undisturbed (AFR group). Ten weeks old adult mail offspring ($n = 170$ in total) from all litters were genotyped and 56 WT, 89 HZ, and 25 KO animals were identified. Figure 42.2 shows the number and proportion of mice

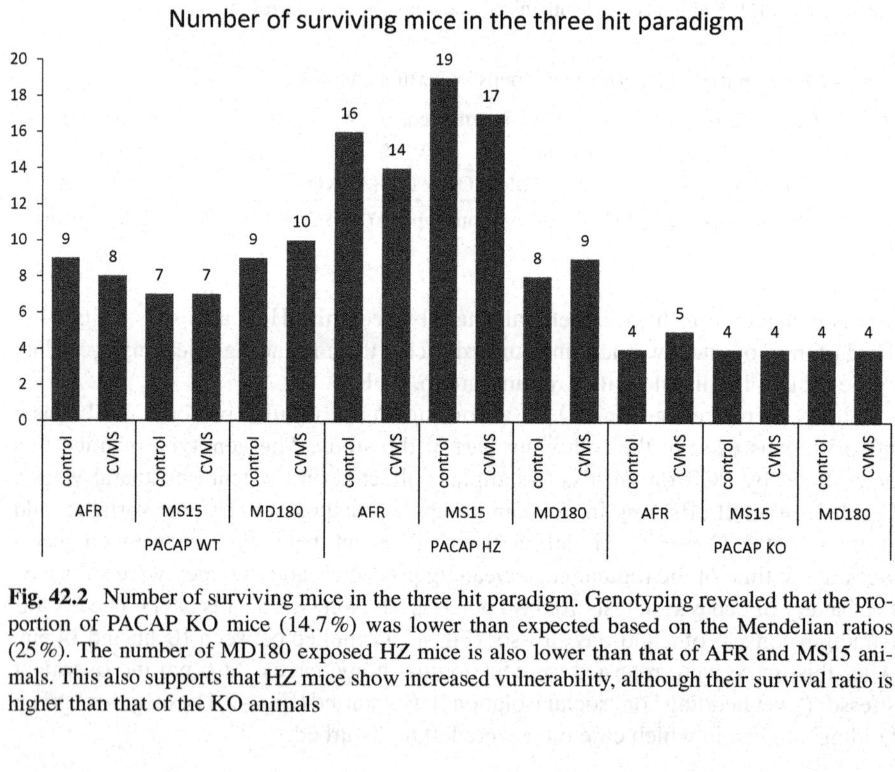

Fig. 42.2 Number of surviving mice in the three hit paradigm. Genotyping revealed that the proportion of PACAP KO mice (14.7 %) was lower than expected based on the Mendelian ratios (25 %). The number of MD180 exposed HZ mice is also lower than that of AFR and MS15 animals. This also supports that HZ mice show increased vulnerability, although their survival ratio is higher than that of the KO animals

which survived the experiment. Survival data show that in contrast to the expected Mendelian ratios the proportion of KO mice was only 14.7 %, while the proportion of WT and HZ mice was 32.95 and 52.35 % respectively.

This unexpected genotype ratio may be explained by our observation that numerous pups did not survive the MD period. The comparison of number of pups lost in total, compared to the difference between expected and registered genotype ratios revealed that the lost pups were mostly KO mice. Later, mice of all genotypes and maternal care groups were assigned to control and CVMS subgroups according to Table 42.1.

The reduced adaptation ability of PACAP KO mice was shown in several studies (for review see ref. [46]). Our [26] and others results [47–50] also support that KO mice do not have normal stress adaptation response. In this model, the coincidence of three factors exhausted the adaptation ability of KO mice, and half of them died.

HZ mice carry one mutated allele and express about 70 % less PACAP in the brain [27] which may increase their vulnerability. The association between gene polymorphisms and human depression is well known (for review see ref. [37]). The significant effect of some genetic alterations was proven even in HZ subjects [25]. Therefore, the translational value of findings in HZ mice in the three hit model may also be high.

The validity of the model is supported by physical, endocrinological, and behavioral data. The comparison of body and adrenal gland weight data support that the CVMS exposure has a significant effect. Maternal care obviously influenced the effect of CVMS, as mice with MD180 history had the greatest adrenal gland weight, when HZ groups were compared [51], suggesting their increased HPA-axis activity. Blood corticosterone values further supported that the HPA-axis activity was elevated [51]. Mice carrying all three risk factors showed increased immobility time in FST paradigm [52].

The three hit model of depression in PACAP HZ and KO mice affects extrahypothalamic CRF, Ucn1, and serotoninergic systems.

Increased CRF neuronal activity in the oval nucleus of the bed nucleus of the stria terminalis (BSTov) was observed in HZ mice with MD180 history upon CVMS expression, which was concomitant with elevated CRF immunosignal in the BSTov (Fig. 42.3b [51]).

The CRF neuropeptide family member Ucn1 expressed in the centrally-projecting Edinger–Westphal nucleus (cpEW) was also studied. Here we found that Ucn1 neurons show increased FosB activity marker expression upon CVMS. However, mice with MD180 history failed to react with FosB expression (Fig 42.3d [51, 52]).

Serotoninergic neurons in the DR were also studied. CVMS exposure effectively increased the FosB expression in the DR. Interestingly, MD180 seems to cause a constantly elevated FosB expression as seen in mice without CVMS exposure. Interestingly, this elevated basal FosB expression in mice with MD180 history is reversed by CVMS exposure in all genotypes examined (Fig. 42.3f [52]).

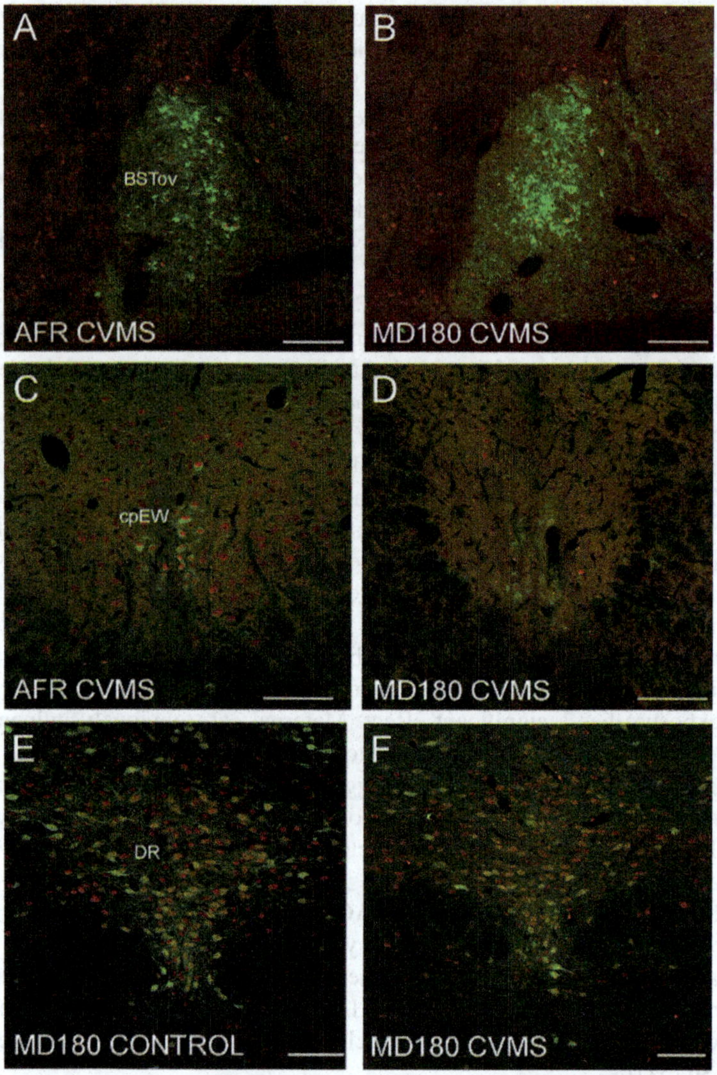

Fig. 42.3 Representative photomicrographs of the oval nucleus of the bed nucleus of the stria terminalis (BSTov) (**a** and **b**), the dorsal raphe nucleus (DR) (**c** and **d**) and in the central projecting Edinger–Westphal nucleus (cpEW) (**e** and **f**) in PACAP heterozygous mice with animal facility rearing (AFR) and with maternal deprivation (MD180) history. Chronic variable mild stress (CVMS) exposure resulted in an increase of corticotropin releasing factor immunoreactivity (*green*) and elevated FosB expression (*red*) in the BSTov of PACAP heterozygous mice with MD180 history (**b**) compared to respective AFR stressed mice (**a**). CVMS caused a clear FosB expression (*red*) in urocortin1 neurons (*green*) of the cpEW in AFR mice (**c**). Mice with MD180 history failed to develop an activation of urocortin1 neurons in the cpEW (**d**). FosB immunoreactivity (red nuclear signal) was remarkable in mice with MD180 history without CVMS exposure in the DR (**e**). CVMS exposure blunted both serotonin (*green*) and FosB immunosignal in the DR (**f**). Scale bars denote 100 μm (**a**–**f**)

Summary

The three hit theory of depression is one of the accepted models for the human mood disorder. The aim of our research program was to prove this hypothesis using PACAP mutant mice subjected to MD 180 and CVMS. Using several physical, endocrinological, and behavioral tools the validity of the model has been supported. Functional-morphological results proved that BSTov-CRF, cpEW-Ucn1, and DR-5HT systems are affected. An inverse relationship is characteristic between increased CRF activity in the BSTov, and blunted cpEW-Ucn1 and DR-serotonin response. Taking the reduced adaptation ability of PACAP KO mice in consideration, PACAP HZ mice seem to be a more reliable tool to study major depression according to the three hit theory.

Acknowledgements This work was supported by the Hungarian Grant OTKA PD 100706 by the "Lendület" research program of the Hungarian Academy of Sciences, Hungarian Brain Research Program—Grant No. KTIA_13_NAP-A-III/5 and PTE AOK KA Research Grant. This work is dedicated to the 650th anniversary of the University of Pecs.

References

1. WHO. 2012. http://www.who.int/mental_health/management/depression/wfmh_paper_depression_wmhd_2012.pdf. Accessed 22 Oct 2015.
2. American Psychiatric Association (2013) DSM-5: diagnostic and statistical manual of mental disorders. Amarican Psychiatric Publishing, ISBN 978-0-89042-555-8.
3. Catena-Dell'Osso M, Fagiolini A, Marazziti D, Baroni S, Bellantuono C. Non-monoaminergic targets for the development of antidepressants: focus on neuropeptides. Mini-Rev Med Chem. 2013;13:2–10.
4. Kormos V, Gaszner B. Role of neuropeptides in anxiety, stress, and depression: from animals to humans. Neuropeptides. 2013;47:401–19.
5. Hartline KM, Owens MJ, Nemeroff CB. Postmortem and cerebrospinal fluid studies of corticotropin-releasing factor in humans. Ann N Y Acad Sci. 1996;780:96–105.
6. Nemeroff CB, Widerlov E, Bissette G, Walleus H, Karlsson I, Eklund K, et al. Elevated concentrations of CSF corticotropin-releasing factor-like immunoreactivity in depressed patients. Science. 1984;226:1342–4.
7. Willner P. Chronic mild stress (CMS) revisited: consistency and behavioural-neurobiological concordance in the effects of CMS. Neuropsychobiology. 2005;52:90–110.
8. Harris RB. Chronic and acute effects of stress on energy balance: are there appropriate animal models? Am J Physiol Regul Integr Comp Physiol. 2015;308:250–65.
9. Makara GB, Mergl Z, Zelena D. The role of vasopressin in hypothalamo-pituitary-adrenal axis activation during stress: an assessment of the evidence. Ann N Y Acad Sci. 2004;1018:151–61.
10. Porsolt RD, Le Pichon M, Jalfre M. Depression: a new animal model sensitive to antidepressant treatments. Nature. 1977;266:730–732.
11. Nielsen DM, Carey GJ, Gold LH. Antidepressant-like activity of corticotropin-releasing factor type-1 receptor antagonists in mice. Eur J Pharmacol. 2004;499(1-2):135–46.
12. Jürgenson M, Aonurm-Helm A, Zharkovsky A. Partial reduction in neural cell adhesion molecule (NCAM) in heterozygous mice induces depression-related behavior without cognitive impairment. Brain Res. 2012;1447:106–18.

13. Njung'e K, Handley SL. Evaluation of marble-burying behavior as a model of anxiety. Pharmacol Biochem Behav. 1991;38:63–7.
14. Ihne JL, Fitzgerald PJ, Hefner KR, Holmes A. Pharmacological modulation of stress-induced behavioral changes in the light/dark exploration test in male C57BL/6J mice. Neuropharmacology. 2012;62(1):464–73.
15. de Kloet ER, Sibug RM, Helmerhorst FM, Schmidt MV. Stress, genes and the mechanism of programming the brain for later life. Neurosci Biobehav Rev. 2005;29:271–81.
16. Marco EM, Llorente R, López-Gallardo M, Mela V, Llorente-Berzal Á, Prada C, et al. The maternal deprivation animal model revisited. Neurosci Biobehav Rev. 2015;51:151–63.
17. Murgatroyd CA, Peña CJ, Podda G, Nestler EJ, Nephew BC. Early life social stress induced changes in depression and anxiety associated neural pathways which are correlated with impaired maternal care. Neuropeptides. 2015;52:103–11.
18. Desbonnet L, Garrett L, Daly E, McDermott KW, Dinan TG. Sexually dimorphic effects of maternal separation stress on corticotrophin-releasing factor and vasopressin systems in the adult rat brain. Int J Dev Neurosci. 2008;26:259–68.
19. Gaszner B, Jensen KO, Farkas J, Reglodi D, Csernus V, Roubos EW, et al. Effects of maternal separation on dynamics of urocortin 1 and brain-derived neurotrophic factor in the rat non-preganglionic Edinger-Westphal nucleus. Int J Dev Neurosci. 2009;5:439–51.
20. van der Doelen RH, Deschamps W, D'Annibale C, Peeters D, Wevers RA, Zelena D, et al. Early life adversity and serotonin transporter gene variation interact at the level of the adrenal gland to affect the adult hypothalamo-pituitary-adrenal axis. Transl Psychiatry. 2014;4:409.
21. Ohta K, Miki T, Warita K, Suzuki S, Kusaka T, Yakura T, et al. Prolonged maternal separation disturbs the serotonergic system during early brain development. Int J Dev Neurosci. 2014;33:15–21.
22. Bale TL. Epigenetic and transgenerational reprogramming of brain development. Nat Rev Neurosci. 2015;16(6):332–44.
23. Constantinof A, Moisiadis VG, Matthews SG. Programming of stress pathways: a transgenerational perspective. J Steroid Biochem Mol Biol. 2015;15:30110–2.
24. Baker M, Dorzab J, Winokur G, Cadoret R. Depressive disease. Evidence favoring polygenic inheritance based on an analysis of ancestral cases. Arch Gen Psychiatry. 1972;27:320–7.
25. Caspi A, Sugden K, Moffitt TE, Taylor A, Craig IW, Harrington H, et al. Influence of life stress on depression: moderation by a polymorphism in the 5-HTT gene. Science. 2003;301:386–9.
26. Gaszner B, Kormos V, Kozicz T, Hashimoto H, Reglodi D, Helyes Z. The behavioral phenotype of pituitary adenylate-cyclase activating polypeptide-deficient mice in anxiety and depression tests is accompanied by blunted c-Fos expression in the bed nucleus of the stria terminalis, central projecting Edinger-Westphal nucleus, ventral lateral septum, and dorsal raphe nucleus. Neuroscience. 2012;202:283–99.
27. Hashimoto H, Shintani N, Tanaka K, Mori W, Hirose M, Matsuda T, et al. Altered psychomotor behaviors in mice lacking pituitary adenylate cyclase-activating polypeptide (PACAP). Proc Natl Acad Sci U S A. 2001;98:13355–60.
28. Almli LM, Mercer KB, Kerley K, Feng H, Bradley B, Conneely KN, et al. ADCYAP1R1 genotype associates with post-traumatic stress symptoms in highly traumatized African–American females. Am J Med Genet B Neuropsychiatr Genet. 2013;162B:262–72.
29. Ressler KJ, Mercer KB, Bradley B, Jovanovic T, Mahan A, Kerley K, et al. Post-traumatic stress disorder is associated with PACAP and the PAC1 receptor. Nature. 2011;470:492–7.
30. Christensen T, Bisgaard CF, Wiborg O. Biomarkers of anhedonic-like behavior, antidepressant drug refraction, and stress resilience in a rat model of depression. Neuroscience. 2011;196:66–79.
31. Ibarguen-Vargas Y, Surget A, Touma C, Palme R, Belzung C. Multifaceted strain-specific effects in a mouse model of depression and of antidepressant reversal. Psychoneuroendocrinology. 2008;33:1357–68.
32. Wigger A, Sánchez MM, Mathys KC, Ebner K, Frank E, Liu D, et al. Alterations in central neuropeptide expression, release, and receptor binding in rats bred for high anxiety: critical role of vasopressin. Neuropsychopharmacology. 2004;29(1):1–14.

33. Geyer MA, Markou A. Animal models of psychiatric disorders. In: BloomFE, Kupfer DJ (eds) Psychopharmacology: the fourth generation of progress. New York, NY: Raven; 1995. p. 787–98.
34. Schmidt MV, Sterlemann V, Wagner K, Niederleitner B, Ganea K, Liebl C, et al. Postnatal glucocorticoid excess due to pituitary glucocorticoid receptor deficiency: differential short- and long-term consequences. Endocrinology. 2009;150:2709–16.
35. Weninger SC, Peters LL, Majzoub JA. Urocortin expression in the Edinger-Westphal nucleus is up-regulated by stress and corticotropin-releasing hormone deficiency. Endocrinology. 2000;141:256–63.
36. Kozicz T, Korosi A, Korsman C, Tilburg-Ouwens D, Groenink L, Veening J, et al. Urocortin expression in the Edinger–Westphal nucleus is down-regulated in transgenic mice over-expressing neuronal corticotropin-releasing factor. Neuroscience. 2004;123:589–94.
37. Flint J, Kendler KS. The genetics of major depression. Neuron. 2014;81(3):484–503.
38. Daskalakis NP, Bagot RC, Parker KJ, Vinkers CH, de Kloet ER. The three-hit concept of vulnerability and resilience: toward understanding adaptation to early-life adversity outcome. Psychoneuroendocrinology. 2013;38(9):1858–73.
39. de Kloet ER, Derijk RH, Meijer OC. TherapyiInsight: is there an imbalanced response of mineralocorticoid and glucocorticoid receptors in depression? Nat Clin Pract Endocrinol Metab. 2007;3(2):168–79.
40. Mill J, Petronis A. Molecular studies of major depressive disorder: the epigenetic perspective. Mol Psychiatry. 2007;12(9):799–814.
41. Claes S. Glucocorticoid receptor polymorphisms in major depression. Ann N Y Acad Sci. 2009;1179:216–28.
42. Sharpley CF, Palanisamy SK, Glyde NS, Dillingham PW, Agnew LL. An update on the interaction between the serotonin transporter promoter variant (5-HTTLPR), stress and depression, plus an exploration of non-confirming findings. Behav Brain Res. 2014;273:89–105.
43. Starr LR, Hammen C, Conway CC, Raposa E, Brennan PA. Sensitizing effect of early adversity on depressive reactions to later proximal stress: moderation by polymorphisms in serotonin transporter and corticotropin releasing hormone receptor genes in a 20-year longitudinal study. Dev Psychopathol. 2014;26:1241–54.
44. Plotsky PM, Thrivikraman KV, Nemeroff CB, Caldji C, Sharma S, Meaney MJ. Long-term consequences of neonatal rearing on central corticotropin-releasing factor systems in adult male rat offspring. Neuropsychopharmacology. 2005;30:2192–204.
45. Jans J, Woodside BC. Nest temperature: effects on maternal behavior, pup development, and interactions with handling. Dev Psychobiol. 1990;23:519–34.
46. Reglodi D, Kiss P, Szabadfi K, Atlasz T, Gabriel R, Horvath G, et al. PACAP is an endogenous protective factor-insights from PACAP-deficient mice. J Mol Neurosci. 2012;48:482–92.
47. Lehmann ML, Mustafa T, Eiden AM, Herkenham M, Eiden LE. PACAP-deficient mice show attenuated corticosterone secretion and fail to develop depressive behavior during chronic social defeat stress. Psychoneuroendocrinology. 2013;38:702–15.
48. Mustafa T, Jiang SZ, Eiden AM, Weihe E, Thistlethwaite I, Eiden LE. Impact of PACAP and PAC1 receptor deficiency on the neurochemical and behavioral effects of acute and chronic restraint stress in male C57BL/6 mice. Stress. 2015;8:1–11.
49. Stroth N, Eiden LE. Stress hormone synthesis in mouse hypothalamus and adrenal gland triggered by restraint is dependent on pituitary adenylate cyclase-activating polypeptide signaling. Neuroscience. 2010;165:1025–30.
50. Tsukiyama N, Saida Y, Kakuda M, Shintani N, Hayata A, Morita Y, et al. PACAP centrally mediates emotional stress-induced corticosterone responses in mice. Stress. 2011;14:368–75.
51. Gaszner B, Farkas J, Bodnar G, Kovacs L, Gaszner T. Gaspar L, et al. Application of the three hit theory in PACAP heterozygote mice: Maternal separation and chronic stress influence BNST CRF and cpEW Ucn1 in an inverse manner. 43nd Annual meeting for neuroscience, San Diegio, 2013; Abstract 382.23.
52. Gaszner B, Kovacs L, Gaszner T, Gaspar L, Reglodi D, Lorincz K, et al. PACAP transgenic mice in the three hit model of depression: the involvement of BNST – CRF, cpEW – urocortin1 and DR – serotonin 44th Annual meeting for neuroscience, Washington, DC, 2014; Abstract 7

Chapter 43
PACAP and Depression

Albert Pinhasov, Izhak Michaelevski, Igor Koman, and Elimelech Nesher

Abstract Depression represents an important global public health concern. It is a complex phenomenon characterized by multifarious etiology and many clinical subtypes. Recent studies show that neuropeptides acting as neurotransmitters and neuromodulators may be involved in the etiology of depression. Pituitary adenylate cyclase-activating polypeptide (PACAP) is a neuropeptide linked to a wide range of biological activities. PACAP regulates monoamine metabolism, modulates HPA axis activity, and plays a critical role in the regulation of distinct neurotransmission pathways, the deterioration of which may lead to behavioral abnormalities, including depression. This mini-review outlines the current state of knowledge regarding the involvement of the PACAP signaling system in depressive disorders.

Keywords PACAP • MDD • Depression • Antidepressant • Stress • Neurotransmission • Neurotrophin • Neuropeptide • HPA axis • BDNF

Depression represents an important global public health concern, evidenced by its status as the most prevalent lifetime psychiatric disorder [1–3]. It is a heterogeneous phenomenon with many subtypes. Because of its complex etiology, depression may occur at any age; it may be chronic, recurrent, and progressive [4, 5]. According to the diagnostic and statistical manual (DSM-V) of mental disorders, the essential diagnostic criteria for major depressive disorders (MDD) include a dysphoric mood or loss of interest or pleasure in all usual activities in combination with at least four of the following seven symptoms: (1) appetite and body weight disturbance; (2) insomnia or hypersomnia; (3) psychomotor retardation or agitation; (4) reduced energy and fatigue; (5) personal underestimation and feelings of guilt; (6) failure in decision making and problems with memory; and (7) thoughts of death or suicide or actual suicide attempts [6].

A. Pinhasov (✉) • I. Koman • E. Nesher
Department of Molecular Biology, Ariel University, Ariel 4070000, Israel
e-mail: albertpi@ariel.ac.il

I. Michaelevski
Department of Biochemistry and Molecular Biology, Tel-Aviv University, Tel-Aviv, Israel

D. Reglodi, A. Tamas (eds.), *Pituitary Adenylate Cyclase Activating Polypeptide — PACAP*, Current Topics in Neurotoxicity 11,
DOI 10.1007/978-3-319-35135-3_43

MDD had been shown to contain a genetic background [7, 8], including changes in allelic variations in the neurotransmission mediating genes [9, 10], thereby linking depression to the proper functioning of hypothalamic–pituitary–adrenal (HPA) axis, neurotrophins, and neuropeptides [11–13].

Neuropeptides acting mostly as neuromodulators are implicated in the fine-tuning of neurophysiology of behavior [14–18]. In this chapter we will focus on the role of pituitary adenylate cyclase-activating polypeptide (PACAP) in depression since it is responsible for numerous physiological processes related to the disease [19]. From an evolutionary standpoint, PACAP is the most conserved member of the neuropeptide family that also includes glucagon, secretin, vasoactive intestinal peptide (VIP), histidine–isoleucine, growth hormone-releasing hormone, and helodermin [20–22]. There are two known isoforms of PACAP: one consisting of 38 amino acids (i.e., the predominant form) and another of 27 amino acids, both being found in the structures related to depression, such as hypothalamus, the cerebral cortex, hippocampus, and striatum [23–25].

Since PACAP exhibits both neuroprotective [26, 27] and neurotrophic [28–30] effects, it is involved in the pathophysiology of various behavioral abnormalities, including MDD and bipolar disorder [31–41]. The pleiotropic activity of PACAP as a neurohormone, neurotransmitter, and neuromodulator [21, 42–49] is mediated by three high affinity G-protein coupled seven-transmembrane family receptors: PAC1, VPAC1, and VPAC2, all of which are widely distributed throughout the brain [22, 50–58], especially in regions engaged in emotional regulation [59, 60]. Through activation of these receptors, PACAP facilitates membrane depolarization, increases action potential frequency, and enhances neurotransmitter synthesis and release [61–63]. Interaction between the PACAP-ergic system and monoamine neurotransmission, as well as its involvement in adaptive response to stress [9, 64] and modulation of the HPA axis [65], further marks the PACAP system as critically important for the development of MDD.

PACAP and Neuroanatomy of Depression

Multiple brain regions are known to be implicated in depression pathophysiology [66, 67]. Deterioration in specific neuronal networks may recruit particular combinations of depression symptomatology [66]. Hypometabolic changes observed in cingulate and orbitofrontal cortices, the medial limbic system, and hippocampus, confirmed by deep brain stimulation and functional imaging using positron emission tomography (PET), were considered to contribute to the clinical expression of depression [68–71]. Reduced hippocampal volume as well as alteration of its pyramidal neurons in the CA3 region was reported in depression [72–75]. In addition, findings made using an animal model of depression [76–78] demonstrated a strong link between depressive-like behavior and modification of synaptic protein expression affiliated with changes in synaptic plasticity in the hippocampus [79], resulting in variable neurotransmitter activity.

PACAP and its receptors are widely distributed throughout different brain regions [23, 24, 58, 80–82]: In particular, they are found in the dorsal raphe nucleus enriched with serotonergic neurons [24]; the locus coeruleus enriched with noradrenergic neurons; the substantia nigra [23], the region of dopamine synthesis; and the nucleus accumbens, which is strongly linked to motivation and rewards, as well as in the hypothalamus, hippocampus [58], striatum [58], and prefrontal cortex [58].

PACAP, Monoamines, and Depression

Our focus, from among competing theories describing the etiology of depression, is on the monoamine hypothesis, in which serotonergic pathways are considered a major affected circuit. This theory incorporates the following mechanisms: (1) reduction of monoamines to presynaptic levels provoking depressive behavior [83]; (2) dietary limitation of tryptophan (a serotonin precursor) and inhibition of serotonin synthesis leading to depressed mood [84–86]; (3) reduction in monoamine metabolites concentrated in the cerebrospinal fluid (CSF) of MDD patients [87]; and (4) augmentation of serotonin clearance from the synaptic cleft observed in MDD patients accompanied with a regional 5-hydroxytryptamine (5HT) transport enhancement [88]. Indeed, pharmacological and anatomical entanglement of monoaminergic neurotransmission provides a neuronal network basis of such cognitive and executive functions as mood, pleasure, motivation, and reward. Deterioration in this network may form a precursor to the onset and progression of MDD [89].

The impact of monoamine versatility, especially serotonin, on behavior is mediated by fourteen different types of receptors, some of which were found to be involved in depression. Abnormalities in the somatodendritic autoreceptor, 5HT1A, subtype of the 5-HT receptor, that binds the endogenous neurotransmitter serotonin, were proven to be important for determination of the effect of antidepressants on MDD patients [90, 91]. A short-term effect of antidepressants is associated with 5HT1A, and provoked reduction of serotonin release in response to increasing extracellular levels of serotonin [92, 93]. Long-term use of antidepressants leads to increase of serotonin in the synaptic cleft, that, in turn, downregulates 5HT1A. The resulting elevated serotonin release can then cause antidepressive effects [94–96]. It was also shown that desensitization of the 5-HT1A by selective serotonin reuptake inhibitors (SSRIs) may activate the dopaminergic (DA-ergic) system associated with reward and motivation typically lacking in MDD patients [97]. The 5HT2A receptor of glutamatergic, GABA-ergic, and DA-ergic neurons, widely distributed in the cortex through presynaptic inhibitory action, increased in MDD patients, and becomes desensitized under chronic antidepressant treatment [97, 98]. Other factors that influence serotonin neurotransmission are dopamine (D) and adrenergic (α) receptors, with the latter known to be hypersensitized in patients with depression. Antidepressant treatment downregulates β-adrenergic receptors and desensitizes inhibitory serotonin receptors [99–101]. Although activation of D2 and α1 receptors positively correlates with serotonin neurotransmission, presynaptic α2 adrenoreceptors are thought to reduce serotonin signaling [89, 102–104].

The role of PACAP and its receptors in monoamine metabolism has been confirmed in a series of experiments [16, 33, 40, 105–108]. Hence, PACAP knockout (KO) mice showed decreased levels of serotonin metabolite in the form of 5-hydroxyindole acetic acid (5-HIAA) in the cortex and striatum, regions associated with affective disorders [33]. Moreover, blocking 5HT2 in PACAP-deficient mice attenuated their depressive-like behavior [40]. Selective 5HT7 receptor blockade effectively suppressed abnormal behavior in PACAP-deficient mice and alleviated immobility in the forced swim test [41]. In addition, antidepressants acting through modulation of monoamines activity increased PACAP and its receptor levels in vitro and in vivo [16].

PACAP reduces the dopaminergic cell loss in a 6-hydroxydopamine-induced lesion of the substantia nigra in young male rats [106]. In mesencephalic cultures, PACAP increases the number of tyrosine hydroxylase (TH) immunoreactive neurons, and enhances dopamine uptake [107]. PACAP expressed in adrenal medulla stimulates activation of phospholipase C and protein kinase C-dependent pathways to facilitate calcium influx via T-type Ca^{2+}-channels, resulting in catecholamine release [105]. In addition, PACAP regulates the gene expression of catecholamine synthesizing enzymes and enhances the enzymatic activity of tyrosine hydroxylase by promoting its phosphorylation [109]. Prolonged PACAP exposure produces a strong and persistent catecholamine release from PC12 cells [110] as well as from isolated adrenal chromaffin cells [111]. This data unambiguously validates PACAP as an important factor of monoamine neurotransmission regulation.

Hypothalamic–Pituitary–Adrenal Axis and Depression

The fundamental role of the HPA axis in depression has also received considerable scientific attention. Depressed patients are characterized by both increased free cortisol urine concentration [112] and failure to reduce cortisol and adenocorticotropic hormone (ACTH) upon dexamethasone administration [113] due to elevated corticotropin-releasing hormone (CRH) secretion [114]. Hypersecretion of hypothalamic CRH leads, among other things, to activation of the sympathetic system [115] and locus coeruleus [116] as well as inhibition of the thyroid hormone axes [117, 118].

In terms of hormonal levels, CRH facilitates glucocorticoid (GC) production in adrenal glands. GC act through activation of low affinity glucocorticoid receptors (GR) involved in negative feedback on excessive elevation of GC level and play a central role in mood regulation [119]. Partial impairment of GR function, GR expression reduction, and GR coding gene polymorphisms are associated with depressive-like behavior and MDD [120]. In the long-term, GCs lead to depressive disorders and hamper neurogenesis in the medial temporal lobe, and this may explain exaggerative loss of hippocampal neurons found in the brains of deceased MDD patients [121, 122]. Failure of GR to reduce GC levels exposed to chronic stress may further exaggerate HPA axis activity negatively affecting the hippocampus [123]. In both MDD patients and animal models of depression, upregulation of GR leads to increase of the GR target gene: Serum and glucocorticoid-inducible kinase 1 (SGK1) which is implicated in hippocampal neurogenesis inhibition [124] and considered to be critical for the efficacy of antidepressant therapy [125, 126]. Furthermore, it has been suggested that known antidepressants exert their activity in part by reducing brain GR levels, thus normalizing the HPA axis function [127].

PACAP is able to modulate HPA axis activity due to its high expression in the hypothalamus and pituitary and chromaffin cells of adrenal glands [65, 128–130]. In the adrenal medulla, PACAP enhances catecholamine release [128, 131, 132], proven by its disruption in PACAP KO mice [130, 131]. The same mice model (PACAP KO mice) is used to demonstrate that CRH mRNA was not upregulated in the paraventricular nucleus of the hypothalamus (PVN) [130], subsequently leading to GC reduction. PACAP enhanced CRH activation resulted from direct modulating of pituitary ACTH release [133]. Moreover, PACAP-ergic innervation of CRH-positive neurons [134] stimulates CRH production and secretion [135]. PACAP was also found to regulate central autonomous pathways resulting in sympathetic activation and subsequent GC release [136, 137]. In wild type mice CRH mRNA in PVN expression was elevated, while in PACAP KO mice, where this phenomenon was not observed, corticosterone release was blunted [130]. In addition, in PACAP KO mice, exposed to social defeat in a chronic stress paradigm, expression of the FosB gene, one of the most prominent biomarkers of depression, was reduced in PVN, preventing HPA axis activation [38]. Interestingly, PVN responsiveness to stressor stimuli is regulated independently not only by PACAP-ergic, but by noradrenergic brain circuits as well [38, 138, 139]. Finally, as long as PACAP exhibits an integrating effect on the functioning of the HPA axis, as a master regulator of the behavioral and neuroendocrine adaptive response to stress [135, 140], employing these pathways, it may be suggested, to control both somatic and psychogenic functioning of the HPA axis.

PACAP, BDNF, and Depression

Among known neurotrophins—proteins that are able to support neuronal survival and differentiation—brain-derived neurotrophic factor (BDNF) plays an important role in the pathophysiology of depression [141, 142]. Many studies demonstrated that BDNF blood level decreased among MDD patients [143–148]. Chronic stress downregulates expression of both BDNF and BDNF-regulated genes in the dentate gyrus of the hippocampus [149]. In addition, BDNF receptor TrkB was found to be reduced in the post-mortem brains of MDD patients [150]. BDNF blood levels in human patients are also induced in acute antidepressant treatment [143, 151]. Moreover, BDNF expression is upregulated in response to antidepressants, electroconvulsive therapy, and transcranial magnetic stimulation [143, 152, 153]. Among other neurotrophic factors, expression of Neurotrophin-3 (NT3) and glial cell-derived neurotrophic factor (GDNF) were found to be decreased in the blood of MDD patients [154]. Such trophic factors as fibroblast growth factor-2 (FGF-2) and vascular endothelial growth factor (VEGF) were suggested to exhibit endogenous antidepressant effects [155, 156].

It has been previously shown by different research groups that PACAP is activated, at least partially, through the release of BDNF [16, 27, 157]. Expressed in germinative areas of the developing brain, PACAP [49] modulates levels of BDNF in mixed [158] and enriched rat cortical neurons [16]. PACAP and BDNF levels are both increased in response to glutamate neurotoxicity in primary cortical neuronal culture [27]. Reduction of PACAP activity by its potent antagonist, PACAP 6-38, resulted in reduced BDNF expression in the hippocampus [27]. The same phenomenon was observed in PAC1 receptor KO mice [159]. In melanotrope cells activation of the VPAC1 receptor was shown to regulate BDNF exon IV expression [160]. Moreover, a PACAP-dependent release of the receptor for the activated C kinase 1 (Rack1) from the N-methyl-D-aspartate receptor (NMDA) complex was shown to translocate to the nucleus, activating BDNF expression [161]. PACAP was suggested to transactivate the TrkB receptor through Src kinase activation bypassing neurotrophin pathways [162]. PACAP may also initiate mitogen-activated protein kinase (MAPK) and PI3K-AKT pathways exhibiting a BDNF synergistic effect in trophic signaling during neurodevelopment, survival, and neuroplasticity [163–165]. Thus, conclusive evidence indicates that a complex synergistic interaction of PACAP and BDNF may play a critical role in regulation of distinct neurotransmission pathways, the deterioration of which may lead to behavioral abnormalities including depression.

Conclusion and Future Directions

Although direct indicators pointing to a link between depression and PACAP exist, the nature of their interaction still needs to be elucidated. Considerable evidence presented in this mini-review shows PACAP-mediated pathways are implicated in neuroanatomical, biochemical, and molecular changes observed in depression. Considering that proper PACAP-ergic signaling is essential for the avoidance of behavioral abnormalities, detailed analysis of the PACAP signaling system, employing both animal models of depressive disorders and patient studies, should illuminate understanding of its role in depression. Finally, progress in pharmacological and molecular approaches for the mediation of PACAP activity would be helpful for clarifying PACAP's physiological role, as well as for the development of new treatments.

References

1. Penninx BW, Milaneschi Y, Lamers F, Vogelzangs N. Understanding the somatic consequences of depression: biological mechanisms and the role of depression symptom profile. BMC Med. 2013;11:129.
2. Kessler RC, Chiu WT, Demler O, Merikangas KR, Walters EE. Prevalence, severity, and comorbidity of 12-month DSM-IV disorders in the National Comorbidity Survey Replication. Arch Gen Psychiatry. 2005;62:617–27.

3. Centers for Disease C, Prevention. Current depression among adults—United States, 2006 and 2008. MMWR Morb Mortal Wkly Rep. 2010;59:1229–35.
4. McKnight PE, Kashdan TB. The importance of functional impairment to mental health outcomes: a case for reassessing our goals in depression treatment research. Clin Psychol Rev. 2009;29:243–59.
5. Kessler RC. The costs of depression. Psychiatr Clin North Am. 2012;35:1–14.
6. Association AP. Diagnostic and statistical manual of mental disorders, (DSM-5®). Arlington, VA: American Psychiatric Publishing; 2013.
7. Sullivan PF, Neale MC, Kendler KS. Genetic epidemiology of major depression: review and meta-analysis. Am J Psychiatry. 2000;157:1552–62.
8. Soleimani L, Lapidus KA, Iosifescu DV. Diagnosis and treatment of major depressive disorder. Neurol Clin. 2011;29:177–93. ix.
9. Charney DS, Manji HK. Life stress, genes, and depression: multiple pathways lead to increased risk and new opportunities for intervention. Sci STKE. 2004;2004(225):5.
10. Dunlop BW, Nemeroff CB. The role of dopamine in the pathophysiology of depression. Arch Gen Psychiatry. 2007;64:327–37.
11. Ogilvie AD, Battersby S, Bubb VJ, Fink G, Harmar AJ, Goodwim GM, et al. Polymorphism in serotonin transporter gene associated with susceptibility to major depression. Lancet. 1996;347(9003):731–3.
12. Bosker FJ, Hartman CA, Nolte IM, Prins BP, Terpstra P, Posthuma D, et al. Poor replication of candidate genes for major depressive disorder using genome-wide association data. Mol Psychiatry. 2011;16:516–32.
13. Papiol S, Arias B, Gastó C, Gutiérrez B, Catalán R, Fañanás L. Genetic variability at HPA axis in major depression and clinical response to antidepressant treatment. J Affect Disord. 2007;104:83–90.
14. Hokfelt T, Bartfai T, Bloom F. Neuropeptides: opportunities for drug discovery. Lancet Neurol. 2003;2:463–72.
15. Said SI. Influence of neuropeptides on airway smooth muscle. Am Rev Respir Dis. 1987; 136:S52–8.
16. Reichenstein M, Rehavi M, Pinhasov A. Involvement of pituitary adenylate cyclase activating polypeptide (PACAP) and its receptors in the mechanism of antidepressant action. J Mol Neurosci. 2008;36:330–8.
17. Chaki S, Okubo T, Sekiguchi Y. Non-monoamine-based approach for the treatment of depression and anxiety disorders. Recent Pat CNS Drug Discov. 2006;1:1–27.
18. Holmes A, Heilig M, Rupniak NM, Steckler T, Griebel G. Neuropeptide systems as novel therapeutic targets for depression and anxiety disorders. Trends Pharmacol Sci. 2003; 24:580–8.
19. Pinhasov A, Nesher E, Gross M, Turgeman G, Kreinin A, Yadid G. 2011. The Role of the PACAP Signaling System in Depression. Curr Pharm Des. 17(10):990–1001.
20. Sherwood NM, Krueckl SL, McRory JE. The origin and function of the pituitary adenylate cyclase-activating polypeptide (PACAP)/glucagon superfamily. Endocr Rev. 2000;21:619–70.
21. Vaudry D, Falluel-Morel A, Bourgault S, Basille M, Burel D, Wurtz O, et al. Pituitary adenylate cyclase-activating polypeptide and its receptors: 20 years after the discovery. Pharmacol Rev. 2009;61:283–357.
22. Laburthe M, Couvineau A, Gaudin P, Maoret JJ, Rouyer-Fessard C, Nicole P. Receptors for VIP, PACAP, secretin, GRF, glucagon, GLP-1, and other members of their new family of G protein-linked receptors: structure-function relationship with special reference to the human VIP-1 receptor. Ann N Y Acad Sci. 1996;805:94–109.
23. Ghatei MA, Takahashi K, Suzuki Y, Gardiner J, Jones PM, Bloom SR. Distribution, molecular characterization of pituitary adenylate cyclase-activating polypeptide and its precursor encoding messenger RNA in human and rat tissues. J Endocrinol. 1993;136:159–66.
24. Masuo Y, Suzuki N, Matsumoto H, Tokito F, Matsumoto Y, Tsuda M, et al. Regional distribution of pituitary adenylate cyclase activating polypeptide (PACAP) in the rat central nervous system as determined by sandwich-enzyme immunoassay. Brain Res. 1993;602:57–63.

25. Koves K, Arimura A, Gorcs TG, Somogyvari-Vigh A. Comparative distribution of immunoreactive pituitary adenylate cyclase activating polypeptide and vasoactive intestinal polypeptide in rat forebrain. Neuroendocrinology. 1991;54:159–69.
26. Chen Y, Samal B, Hamelink CR, Xiang CC, Chen Y, Chen M, et al. Neuroprotection by endogenous and exogenous PACAP following stroke. Regul Pept. 2006;137:4–19.
27. Shintani N, Suetake S, Hashimoto H, Koga K, Kasai A, Kawaguchi C, et al. Neuroprotective action of endogenous PACAP in cultured rat cortical neurons. Regul Pept. 2005;126:123–8.
28. Vaudry D, Gonzalez BJ, Basille M, Fournier A, Vaudry H. Neurotrophic activity of pituitary adenylate cyclase-activating polypeptide on rat cerebellar cortex during development. Proc Natl Acad Sci U S A. 1999;96:9415–20.
29. Ravni A, Bourgault S, Lebon A, Chan P, Galas L, Fournier A, et al. The neurotrophic effects of PACAP in PC12 cells: control by multiple transduction pathways. J Neurochem. 2006;98:321–9.
30. Vaudry D, Gonzalez B, Basille M, Pamantung T, Fournier A, Vaudry H. PACAP acts as a neurotrophic factor during histogenesis of the rat cerebellar cortex. Ann N Y Acad Sci. 2000;921:293–9.
31. Hammack SE, Cheung J, Rhodes KM, Schutz KC, Falls WA, Braas KM, et al. Chronic stress increases pituitary adenylate cyclase-activating peptide (PACAP) and brain-derived neurotrophic factor (BDNF) mRNA expression in the bed nucleus of the stria terminalis (BNST): roles for PACAP in anxiety-like behavior. Psychoneuroendocrinology. 2009;34:833–43.
32. Hammack SE, Roman CW, Lezak KR, Kocho-Shellenberg M, Grimmig B, Falls WA, et al. Roles for pituitary adenylate cyclase-activating peptide (PACAP) expression and signaling in the bed nucleus of the stria terminalis (BNST) in mediating the behavioral consequences of chronic stress. J Mol Neurosci. 2010;42:327–40.
33. Hashimoto H, Shintani N, Tanaka K, Mori W, Hirose M, Matsuda T, et al. Altered psychomotor behaviors in mice lacking pituitary adenylate cyclase-activating polypeptide (PACAP). Proc Natl Acad Sci U S A. 2001;98:13355–60.
34. Ishihama T, Ago Y, Shintani N, Hashimoto H, Baba A, Takuma K, et al. Environmental factors during early developmental period influence psychobehavioral abnormalities in adult PACAP-deficient mice. Behav Brain Res. 2010;209:274–80.
35. Tanaka K, Shintani N, Hashimoto H, Kawagishi N, Ago Y, Matsuda T, et al. Psychostimulant-induced attenuation of hyperactivity and prepulse inhibition deficits in Adcyap1-deficient mice. J Neurosci. 2006;26:5091–7.
36. Hattori S, Takao K, Tanda K, Toyama K, Shintani N, Baba A, et al. Comprehensive behavioral analysis of pituitary adenylate cyclase-activating polypeptide (PACAP) knockout mice. Front Behav Neurosci. 2012;6:58.
37. Acquaah-Mensah GK, Taylor RC, Bhave SV. PACAP interactions in the mouse brain: Implications for behavioral and other disorders. Gene. 2012;491:224–31.
38. Lehmann ML, Mustafa T, Eiden AM, Herkenham M, Eiden LE. PACAP-deficient mice show attenuated corticosterone secretion and fail to develop depressive behavior during chronic social defeat stress. Psychoneuroendocrinology. 2013;38:702–15.
39. Brandish PE, Su M, Holder DJ, Hodor P, Szumiloski J, Kleinhanz RR, et al. Regulation of gene expression by lithium and depletion of inositol in slices of adult rat cortex. Neuron. 2005;45:861–72.
40. Hashimoto H, Hashimoto R, Shintani N, Tanaka K, Yamamoto A, Hatanaka M, et al. Depression-like behavior in the forced swimming test in PACAP-deficient mice: amelioration by the atypical antipsychotic risperidone. J Neurochem. 2009;110:595–602.
41. Tajiri M, Hayata-Takano A, Seiriki K, Ogata K, Hazama K, Shintani N, et al. Serotonin 5-HT(7) receptor blockade reverses behavioral abnormalities in PACAP-deficient mice and receptor activation promotes neurite extension in primary embryonic hippocampal neurons: therapeutic implications for psychiatric disorders. J Mol Neurosci. 2012;48:473–81.
42. Botia B, Jolivel V, Burel D, Le Joncour V, Roy V, Naassila M, et al. Neuroprotective effects of PACAP against ethanol-induced toxicity in the developing rat cerebellum. Neurotox Res. 2011;19:423–34.

43. Morio H, Tatsuno I, Hirai A, Tamura Y, Saito Y. Pituitary adenylate cyclase-activating polypeptide protects rat-cultured cortical neurons from glutamate-induced cytotoxicity. Brain Res. 1996;741:82–8.
44. Onoue S, Endo K, Ohshima K, Yajima T, Kashimoto K. The neuropeptide PACAP attenuates beta-amyloid (1-42)-induced toxicity in PC12 cells. Peptides. 2002;23:1471–8.
45. Takei N, Torres E, Yuhara A, Jongsma H, Otto C, Korhonen L, et al. Pituitary adenylate cyclase-activating polypeptide promotes the survival of basal forebrain cholinergic neurons in vitro and in vivo: comparison with effects of nerve growth factor. Eur J Neurosci. 2000;12:2273–80.
46. Lindholm D, Skoglosa Y, Takei N. Developmental regulation of pituitary adenylate cyclase activating polypeptide (PACAP) and its receptor 1 in rat brain: function of PACAP as a neurotrophic factor. Ann N Y Acad Sci. 1998;865:189–96.
47. Atlasz T, Szabadfi K, Kiss P, Tamas A, Toth G, Reglodi D, et al. Evaluation of the protective effects of PACAP with cell-specific markers in ischemia-induced retinal degeneration. Brain Res Bull. 2010;81:497–504.
48. Deguil J, Chavant F, Lafay-Chebassier C, Perault-Pochat MC, Fauconneau B, Pain S. Neuroprotective effect of PACAP on translational control alteration and cognitive decline in MPTP parkinsonian mice. Neurotox Res. 2010;17:142–55.
49. Vaudry D, Gonzalez BJ, Basille M, Yon L, Fournier A, Vaudry H. Pituitary adenylate cyclase-activating polypeptide and its receptors: from structure to functions. Pharmacol Rev. 2000;52:269–324.
50. Harmar AJ, Arimura A, Gozes I, Journot L, Laburthe M, Pisegna JR, et al. International Union of Pharmacology. XVIII. Nomenclature of receptors for vasoactive intestinal peptide and pituitary adenylate cyclase-activating polypeptide. Pharmacol Rev. 1998;50:265–70.
51. Ohtaki T, Masuda Y, Ishibashi Y, Kitada C, Arimura A, Fujino M. Purification and characterization of the receptor for pituitary adenylate cyclase-activating polypeptide. J Biol Chem. 1993;268:26650–7.
52. Ohtaki T, Watanabe T, Ishibashi Y, Kitada C, Tsuda M, Gottschall PE, et al. Molecular identification of receptor for pituitary adenylate cyclase activating polypeptide. Biochem Biophys Res Commun. 1990;171:838–44.
53. McCulloch DA, Lutz EM, Johnson MS, Robertson DN, MacKenzie CJ, Holland PJ, et al. ADP-ribosylation factor-dependent phospholipase D activation by VPAC receptors and a PAC(1) receptor splice variant. Mol Pharmacol. 2001;59:1523–32.
54. Dickson L, Finlayson K. VPAC and PAC receptors: from ligands to function. Pharmacol Ther. 2009;121:294–316.
55. Hashimoto H, Nogi H, Mori K, Ohishi H, Shigemoto R, Yamamoto K, et al. Distribution of the mRNA for a pituitary adenylate cyclase-activating polypeptide receptor in the rat brain: an in situ hybridization study. J Comp Neurol. 1996;371:567–77.
56. Caberlotto L, Jimenez P, Overstreet DH, Hurd YL, Mathe AA, Fuxe K. Alterations in neuropeptide Y levels and Y1 binding sites in the flinders sensitive line rats, a genetic animal model of depression. Neurosci Lett. 1999;265:191–4.
57. Otto C, Zuschratter W, Gass P, Schutz G. Presynaptic localization of the PACAP-type I-receptor in hippocampal and cerebellar mossy fibres. Mol Brain Res. 1999;66:163–74.
58. Shneider Y, Shtrauss Y, Yadid G, Pinhasov A. Differential expression of PACAP receptors in postnatal rat brain. Neuropeptides. 2010;44:509–14.
59. Otto C, Martin M, Wolfer DP, Lipp HP, Maldonado R, Schutz G. Altered emotional behavior in PACAP-type-I-receptor-deficient mice. Mol Brain Res. 2001;92:78–84.
60. Nicot A, Otto T, Brabet P, Dicicco-Bloom EM. Altered social behavior in pituitary adenylate cyclase-activating polypeptide type I receptor-deficient mice. J Neurosci. 2004;24:8786–95.
61. Harmar AJ, Fahrenkrug J, Gozes I, Laburthe M, May V, Pisegna JR, et al. Pharmacology and functions of receptors for vasoactive intestinal peptide and pituitary adenylate cyclase-activating polypeptide: IUPHAR review 1. Br J Pharmacol. 2012;166:4–17.
62. Blechman J, Levkowitz G. Alternative splicing of the pituitary adenylate cyclase-activating polypeptide receptor PAC1: mechanisms of fine tuning of brain activity. Front Endocrinol. 2013;4:55.

63. Spengler D, Waeber C, Pantaloni C, Holsboer F, Bockaert J, Seeburg PH, et al. Differential signal transduction by five splice variants of the PACAP receptor. Nature. 1993;365:170–5.
64. Duman RS, Malberg J, Nakagawa S, D'Sa C. Neuronal plasticity and survival in mood disorders. Biol Psychiatry. 2000;48:732–9.
65. Nussdorfer GG, Malendowicz LK. Role of VIP, PACAP, and related peptides in the regulation of the hypothalamo-pituitary-adrenal axis. Peptides. 1998;19:1443–67.
66. Drevets WC, Price JL, Furey ML. Brain structural and functional abnormalities in mood disorders: implications for neurocircuitry models of depression. Brain Struct Funct. 2008;213:93–118.
67. Koenigs M, Grafman J. The functional neuroanatomy of depression: distinct roles for ventromedial and dorsolateral prefrontal cortex. Behav Brain Res. 2009;201:239–43.
68. Davidson RJ, Pizzagalli D, Nitschke JB, Putnam K. Depression: perspectives from affective neuroscience. Annu Rev Psychol. 2002;53:545–74.
69. Drevets WC. Functional neuroimaging studies of depression: the anatomy of melancholia. Annu Rev Med. 1998;49:341–61.
70. Drevets WC. Orbitofrontal cortex function and structure in depression. Ann N Y Acad Sci. 2007;1121:499–527.
71. Kennedy SH, Evans KR, Kruger S, Mayberg HS, Meyer JH, McCann S, et al. Changes in regional brain glucose metabolism measured with positron emission tomography after paroxetine treatment of major depression. Am J Psychiatry. 2001;158:899–905.
72. Koolschijn PC, van Haren NE, Lensvelt-Mulders GJ, Hulshoff Pol HE, Kahn RS. Brain volume abnormalities in major depressive disorder: a meta-analysis of magnetic resonance imaging studies. Hum Brain Mapp. 2009;30:3719–35.
73. McKinnon MC, Yucel K, Nazarov A, MacQueen GM. A meta-analysis examining clinical predictors of hippocampal volume in patients with major depressive disorder. J Psychiatry Neurosci. 2009;34:41–54.
74. Sapolsky RM. Depression, antidepressants, and the shrinking hippocampus. Proc Natl Acad Sci U S A. 2001;98:12320–2.
75. Campbell S, Macqueen G. The role of the hippocampus in the pathophysiology of major depression. J Psychiatry Neurosci. 2004;29:417–26.
76. Feder Y, Nesher E, Ogran A, Kreinin A, Malatynska E, Yadid G, et al. Selective breeding for dominant and submissive behavior in Sabra mice. J Affect Disord. 2010;126:214–22.
77. Moussaieff A, Gross M, Nesher E, Tikhonov T, Yadid G, Pinhasov A. Incensole acetate reduces depressive-like behavior and modulates hippocampal BDNF and CRF expression of submissive animals. J Psychopharmacol. 2012;26:1584–93.
78. Nesher E, Gross M, Lisson S, Tikhonov T, Yadid G, Pinhasov A. Differential responses to distinct psychotropic agents of selectively bred dominant and submissive animals. Behav Brain Res. 2013;236:225–35.
79. Gross M, Sheinin A, Nesher E, Tikhonov T, Baranes D, Pinhasov A, et al. Early onset of cognitive impairment is associated with altered synaptic plasticity and enhanced hippocampal GluA1 expression in a mouse model of depression. Neurobiol Aging. 2015;36:1938–52.
80. Watanabe J, Nakamachi T, Matsuno R, Hayashi D, Nakamura M, Kikuyama S, et al. Localization, characterization and function of pituitary adenylate cyclase-activating polypeptide during brain development. Peptides. 2007;28:1713–9.
81. Tabarin A, Chen D, Hakanson R, Sundler F. Pituitary adenylate cyclase-activating peptide in the adrenal gland of mammals: distribution, characterization and responses to drugs. Neuroendocrinology. 1994;59:113–9.
82. Sundler F, Ekblad E, Absood A, Hakanson R, Koves K, Arimura A. Pituitary adenylate cyclase activating peptide: a novel vasoactive intestinal peptide-like neuropeptide in the gut. Neuroscience. 1992;46:439–54.
83. Hasler G. Pathophysiology of depression: do we have any solid evidence of interest to clinicians? World Psychiatry. 2010;9:155–61.
84. Bell C, Abrams J, Nutt D. Tryptophan depletion and its implications for psychiatry. Br J Psychiatry. 2001;178:399–405.

85. Shopsin B, Friedman E, Gershon S. Parachlorophenylalanine reversal of tranylcypromine effects in depressed patients. Arch Gen Psychiatry. 1976;33:811–9.
86. Shopsin B, Gershon S, Goldstein M, Friedman E, Wilk S. Use of synthesis inhibitors in defining a role for biogenic amines during imipramine treatment in depressed patients. Psychopharmacol Commun. 1975;1:239–49.
87. Asberg M, Bertilsson L, Martensson B, Scalia-Tomba GP, Thoren P, Traskman-Bendz L. CSF monoamine metabolites in melancholia. Acta Psychiatr Scand. 1984;69:201–19.
88. Owens MJ, Nemeroff CB. Role of serotonin in the pathophysiology of depression: focus on the serotonin transporter. Clin Chem. 1994;40:288–95.
89. Hamon M, Blier P. Monoamine neurocircuitry in depression and strategies for new treatments. Progr Neuropsychopharmacol Biol Psychiatry. 2013;45:54–63.
90. Lemonde S, Turecki G, Bakish D, Du L, Hrdina PD, Bown CD, et al. Impaired repression at a 5-hydroxytryptamine 1A receptor gene polymorphism associated with major depression and suicide. J Neurosci. 2003;23:8788–99.
91. Neff CD, Abkevich V, Packer JCL, Chen Y, Potter J, Riley R, et al. Evidence for HTR1A and LHPP as interacting genetic risk factors in major depression. Mol Psychiatry. 2009;14:621–30.
92. Artigas F, Adell A, Celada P. Pindolol augmentation of antidepressant response. Curr Drug Targets. 2006;7:139–47.
93. Romero L, Bel N, Artigas F, De Montigny C, Blier P. Effect of pindolol on the function of pre- and postsynaptic 5-HT(1A) receptors: In vivo microdialysis and electrophysiological studies in the rat brain. Neuropsychopharmacology. 1996;15:349–60.
94. Blier P, de Montigny C. Current advances and trends in the treatment of depression. Trends Pharmacol Sci. 1994;15:220–6.
95. Hervás I, Vilaró MT, Romero L, Scorza MC, Mengod G, Artigas F. Desensitization of 5-HT(1A) autoreceptors by a low chronic fluoxetine dose effect of the concurrent administration of WAY-100635. Neuropsychopharmacology. 2001;24:11–20.
96. El Mansari M, Sánchez C, Chouvet G, Renaud B, Haddjeri N. Effects of acute and long-term administration of escitalopram and citalopram on serotonin neurotransmission: an in vivo electrophysiological study in rat brain. Neuropsychopharmacology. 2005;30:1269–77.
97. Artigas F. Serotonin receptors involved in antidepressant effects. Pharmacol Ther. 2013;137:119–31.
98. Carr GV, Lucki I. The role of serotonin receptor subtypes in treating depression: a review of animal studies. Psychopharmacology (Berl). 2011;213:265–87.
99. Prins J, Olivier B, Korte SM. Triple reuptake inhibitors for treating subtypes of major depressive disorder: the monoamine hypothesis revisited. Expert Opin Investig Drugs. 2011;20:1107–30.
100. Catapano LA, Manji HK. G Protein-coupled receptors in major psychiatric disorders. Biochim Biophys Acta. 1768;2007:976–93.
101. Pandey GN, Sudershan P, Davis JM. Beta adrenergic receptor function in depression and the effect of antidepressant drugs. Acta Pharmacol Toxicol. 1985;56 Suppl 1:66–79.
102. Aman TK, Shen RY, Haj-Dahmane S. D2-like dopamine receptors depolarize dorsal raphe serotonin neurons through the activation of nonselective cationic conductance. J Pharmacol Exp Ther. 2007;320:376–85.
103. O'Leary OF, Cryan JF. The behavioral genetics of serotonin: relevance to anxiety and depression. Handb Behav Neurosci. 2010;21:749–89.
104. Adell A, Bortolozzi A, Díaz-Mataix L, Santana N, Celada P, Artigas F. Serotonin interaction with other transmitter systems. Handb Behav Neurosci. 2010;21:259–76.
105. Kuri BA, Chan SA, Smith CB. PACAP regulates immediate catecholamine release from adrenal chromaffin cells in an activity-dependent manner through a protein kinase C-dependent pathway. J Neurochem. 2009;110:1214–25.
106. Reglodi D, Tamas A, Lengvari I, Toth G, Szalontay L, Lubics A. Comparative study of the effects of PACAP in young, aging, and castrated males in a rat model of Parkinson's disease. Ann N Y Acad Sci. 2006;1070:518–24.

107. Takei N, Skoglosa Y, Lindholm D. Neurotrophic and neuroprotective effects of pituitary adenylate cyclase-activating polypeptide (PACAP) on mesencephalic dopaminergic neurons. J Neurosci Res. 1998;54:698–706.
108. Guillot TS, Richardson JR, Wang MZ, Li YJ, Taylor TN, Ciliax BJ, et al. PACAP38 increases vesicular monoamine transporter 2 (VMAT2) expression and attenuates methamphetamine toxicity. Neuropeptides. 2008;42:423–34.
109. Isobe K, Yukimasa N, Nakai T, Takuwa Y. Pituitary adenylate cyclase-activating polypeptide induces gene expression of the catecholamine synthesizing enzymes, tyrosine hydroxylase, and dopamine beta-hydroxylase in cultured porcine adrenal medullary chromaffin cells. Ann N Y Acad Sci. 1996;805:464–9.
110. Taupenot L, Mahata M, Mahata SK, O'Connor DT. Time-dependent effects of the neuropeptide PACAP on catecholamine secretion : stimulation and desensitization. Hypertension. 1999;34:1152–62.
111. Tornoe K, Hannibal J, Jensen TB, Georg B, Rickelt LF, Andreasen MB, et al. PACAP-(1-38) as neurotransmitter in the porcine adrenal glands. Am J Physiol Endocrinol Metab. 2000;279:E1413–25.
112. Otte C, Marmar CR, Pipkin SS, Moos R, Browner WS, Whooley MA. Depression and 24-hour urinary cortisol in medical outpatients with coronary heart disease: The Heart and Soul Study. Biol Psychiatry. 2004;56:241–7.
113. Ferrier IN, Pascual J, Charlton BG, Wright C, Leake A, Griffiths HW, et al. Cortisol, ACTH, and dexamethasone concentrations in a psychogeriatric population. Biol Psychiatry. 1988;23:252–60.
114. Barden N. Implication of the hypothalamic-pituitary-adrenal axis in the physiopathology of depression. J Psychiatry Neurosci. 2004;29:185–93.
115. Gold PW, Wong ML, Goldstein DS, Gold HK, Ronsaville DS, Esler M, et al. Cardiac implications of increased arterial entry and reversible 24-h central and peripheral norepinephrine levels in melancholia. Proc Natl Acad Sci U S A. 2005;102:8303–8.
116. Aston-Jones G, Cohen JD. An integrative theory of locus coeruleus-norepinephrine function: adaptive gain and optimal performance. Annu Rev Neurosci. 2005;28:403–50.
117. Gold PW, Goodwin FK, Chrousos GP. Clinical and biochemical manifestations of depression. Relation to the neurobiology of stress (2). New Engl J Med. 1988;319:413–20.
118. Chrousos GP, Gold PW. The concepts of stress and stress system disorders. Overview of physical and behavioral homeostasis. JAMA. 1992;267:1244–52.
119. Pariante CM. The glucocorticoid receptor: part of the solution or part of the problem? J Psychopharmacol. 2006;20(4 Suppl):79–84.
120. Claes S. Glucocorticoid receptor polymorphisms in major depression. Ann N Y Acad Sci. 2009;1179:216–28.
121. Saaltink DJ, Vreugdenhil E. Stress, glucocorticoid receptors, and adult neurogenesis: a balance between excitation and inhibition? Cell Mol Life Sci. 2014;71:2499–515.
122. Herbert J. Cortisol and depression: three questions for psychiatry. Psychol Med. 2013;43:449–69.
123. Mongeau R, Hamon M, Lanfumey L. How can stress alter emotional balance through its interaction with the serotonergic system? The handbook of stress: neuropsychological effects on the brain. New York, NY: Wiley; 2011. p. 480–504.
124. Anacker C, Cattaneo A, Musaelyan K, Zunszain PA, Horowitz M, Molteni R, et al. Role for the kinase SGK1 in stress, depression, and glucocorticoid effects on hippocampal neurogenesis. Proc Natl Acad Sci U S A. 2013;110:8708–13.
125. Santarelli L, Saxe M, Gross C, Surget A, Battaglia F, Dulawa S, et al. Requirement of hippocampal neurogenesis for the behavioral effects of antidepressants. Science. 2003;301:805–9.
126. Paizanis E, Kelaï S, Renoir T, Hamon M, Lanfumey L. Life-long hippocampal neurogenesis: Environmental, pharmacological and neurochemical modulations. Neurochem Res. 2007;32:1762–71.
127. Weber CC, Eckert GP, Muller WE. Effects of antidepressants on the brain/plasma distribution of corticosterone. Neuropsychopharmacology. 2006;31:2443–8.

128. Stroth N, Liu Y, Aguilera G, Eiden LE. Pituitary adenylate cyclase-activating polypeptide controls stimulus-transcription coupling in the hypothalamic-pituitary-adrenal axis to mediate sustained hormone secretion during stress. J Neuroendocrinol. 2011;23:944–55.
129. Stroth N, Holighaus Y, Ait-Ali D, Eiden LE. PACAP: a master regulator of neuroendocrine stress circuits and the cellular stress response. Ann N Y Acad Sci. 2011;1220:49–59.
130. Stroth N, Eiden LE. Stress hormone synthesis in mouse hypothalamus and adrenal gland triggered by restraint is dependent on pituitary adenylate cyclase-activating polypeptide signaling. Neuroscience. 2010;165:1025–30.
131. Stroth N, Kuri BA, Mustafa T, Chan SA, Smith CB, Eiden LE. PACAP controls adrenomedullary catecholamine secretion and expression of catecholamine biosynthetic enzymes at high splanchnic nerve firing rates characteristic of stress transduction in male mice. Endocrinology. 2013;154:330–9.
132. Tönshoff C, Hemmick L, Evinger MJ. Pituitary adenylate cyclase activating polypeptide (PACAP) regulates expression of catecholamine biosynthetic enzyme genes in bovine adrenal chromaffin cells. J Mol Neurosci. 1997;9:127–40.
133. Miyata A, Arimura A, Dahl RR, Minamino N, Uehara A, Jiang L, et al. Isolation of a novel 38 residue-hypothalamic polypeptide which stimulates adenylate cyclase in pituitary cells. Biochem Biophys Res Commun. 1989;164:567–74.
134. Legradi G, Hannibal J, Lechan RM. Pituitary adenylate cyclase-activating polypeptide-nerve terminals densely innervate corticotropin-releasing hormone-neurons in the hypothalamic paraventricular nucleus of the rat. Neurosci Lett. 1998;246:145–8.
135. Agarwal A, Halvorson LM, Legradi G. Pituitary adenylate cyclase-activating polypeptide (PACAP) mimics neuroendocrine and behavioral manifestations of stress: Evidence for PKA-mediated expression of the corticotropin-releasing hormone (CRH) gene. Mol Brain Res. 2005;138:45–57.
136. Farnham MM, Inglott MA, Pilowsky PM. Intrathecal PACAP-38 causes increases in sympathetic nerve activity and heart rate but not blood pressure in the spontaneously hypertensive rat. Am J Physiol Heart Circ Physiol. 2011;300:H214–22.
137. Tanida M, Shintani N, Morita Y, Tsukiyama N, Hatanaka M, Hashimoto H, et al. Regulation of autonomic nerve activities by central pituitary adenylate cyclase-activating polypeptide. Regul Pept. 2010;161:73–80.
138. Khan AM, Kaminski KL, Sanchez-Watts G, Ponzio TA, Kuzmiski JB, Bains JS, et al. MAP kinases couple hindbrain-derived catecholamine signals to hypothalamic adrenocortical control mechanisms during glycemia-related challenges. J Neurosci. 2011;31:18479–91.
139. Khan AM, Ponzio TA, Sanchez-Watts G, Stanley BG, Hatton GI, Watts AG. Catecholaminergic control of mitogen-activated protein kinase signaling in paraventricular neuroendocrine neurons in vivo and in vitro: a proposed role during glycemic challenges. J Neurosci. 2007;27:7344–60.
140. Norrholm SD, Das M, Legradi G. Behavioral effects of local microinfusion of pituitary adenylate cyclase activating polypeptide (PACAP) into the paraventricular nucleus of the hypothalamus (PVN). Regul Pept. 2005;128:33–41.
141. Yu H, Chen ZY. The role of BDNF in depression on the basis of its location in the neural circuitry. Acta Pharmacol Sin. 2011;32:3–11.
142. Numakawa T, Richards M, Nakajima S, Adachi N, Furuta M, Odaka H, et al. The role of brain-derived neurotrophic factor (BDNF) in comorbid depression: possible linkage with steroid hormones, cytokines, and nutrition. Front Psychiatry. 2014;5:136.
143. Kreinin A, Lisson S, Nesher E, Schneider J, Bergman J, Farhat K, et al. Blood BDNF level is gender specific in severe depression. PLoS One. 2015;10, e0127643.
144. Shimizu E, Hashimoto K, Okamura N, Koike K, Komatsu N, Kumakiri C, et al. Alterations of serum levels of brain-derived neurotrophic factor (BDNF) in depressed patients with or without antidepressants. Biol Psychiatry. 2003;54:70–5.
145. Molendijk ML, Bus BA, Spinhoven P, Penninx BW, Kenis G, Prickaerts J, et al. Serum levels of brain-derived neurotrophic factor in major depressive disorder: state-trait issues, clinical features and pharmacological treatment. Mol Psychiatry. 2011;16:1088–95.

146. Karlovic D, Serretti A, Jevtovic S, Vrkic N, Seric V, Peles AM. Diagnostic accuracy of serum brain derived neurotrophic factor concentration in antidepressant naive patients with first major depression episode. J Psychiatr Res. 2013;47:162–7.
147. Ladea M, Bran M. Brain derived neurotrophic factor (BDNF) levels in depressed women treated with open-label escitalopram. Psychiatr Danub. 2013;25:128–32.
148. Yoshida T, Ishikawa M, Niitsu T, Nakazato M, Watanabe H, Shiraishi T, et al. Decreased serum levels of mature brain-derived neurotrophic factor (BDNF), but not its precursor proBDNF, in patients with major depressive disorder. PLoS One. 2012;7, e42676.
149. Smith MA, Makino S, Kvetnansky R, Post RM. Stress and glucocorticoids affect the expression of brain-derived neurotrophic factor and neurotrophin-3 mRNAs in the hippocampus. J Neurosci. 1995;15:1768–77.
150. Ernst C, Deleva V, Deng X, Sequeira A, Pomarenski A, Klempan T, et al. Alternative splicing, methylation state, and expression profile of tropomyosin-related kinase B in the frontal cortex of suicide completers. Arch Gen Psychiatry. 2009;66:22–32.
151. Rantamäki T, Hendolin P, Kankaanpää A, Mijatovic J, Piepponen P, Domenici E, et al. Pharmacologically diverse antidepressants rapidly activate brain-derived neurotrophic factor receptor TrkB and induce phospholipase-Cγ signaling pathways in mouse brain. Neuropsychopharmacology. 2007;32:2152–62.
152. Altar CA, Whitehead RE, Chen R, Wortwein G, Madsen TM. Effects of electroconvulsive seizures and antidepressant drugs on brain-derived neurotrophic factor protein in rat brain. Biol Psychiatry. 2003;54:703–9.
153. Chen B, Dowlatshahi D, MacQueen GM, Wang JF, Young LT. Increased hippocampal BDNF immunoreactivity in subjects treated with antidepressant medication. Biol Psychiatry. 2001;50:260–5.
154. Otsuki K, Uchida S, Watanuki T, Wakabayashi Y, Fujimoto M, Matsubara T, et al. Altered expression of neurotrophic factors in patients with major depression. J Psychiatr Res. 2008;42:1145–53.
155. Turner CA, Watson SJ, Akil H. Fibroblast growth factor-2: an endogenous antidepressant and anxiolytic molecule? Biol Psychiatry. 2012;72:254–5.
156. Udo H, Hamasu K, Furuse M, Sugiyama H. VEGF-induced antidepressant effects involve modulation of norepinephrine and serotonin systems. Behav Brain Res. 2014;275:107–13.
157. Frechilla D, Garcia-Osta A, Palacios S, Cenarruzabeitia E, Del Rio J. BDNF mediates the neuroprotective effect of PACAP-38 on rat cortical neurons. Neuroreport. 2001;12:919–23.
158. Pellegri G, Magistretti PJ, Martin JL. VIP and PACAP potentiate the action of glutamate on BDNF expression in mouse cortical neurones. Eur J Neurosci. 1998;10:272–80.
159. Zink M, Otto C, Zorner B, Zacher C, Schutz G, Henn FA, et al. Reduced expression of brain-derived neurotrophic factor in mice deficient for pituitary adenylate cyclase activating polypeptide type-I-receptor. Neurosci Lett. 2004;360:106–8.
160. Kidane AH, Roubos EW, Jenks BG. Pituitary adenylate cyclase-activating polypeptide regulates brain-derived neurotrophic factor exon IV expression through the VPAC1 receptor in the amphibian melanotrope cell. Endocrinology. 2008;149:4177–82.
161. Yaka R, He DY, Phamluong K, Ron D. Pituitary adenylate cyclase-activating polypeptide (PACAP(1-38)) enhances N-methyl-D-aspartate receptor function and brain-derived neurotrophic factor expression via RACK1. J Biol Chem. 2003;278:9630–8.
162. Andero R, Ressler KJ. Fear extinction and BDNF: translating animal models of PTSD to the clinic. Genes Brain Behav. 2012;11:503–12.
163. Pugh PC, Margiotta JF. PACAP support of neuronal survival requires MAPK- and activity-generated signals. Mol Cell Neurosci. 2006;31:586–95.
164. Villalba M, Bockaert J, Journot L. Pituitary adenylate cyclase-activating polypeptide (PACAP-38) protects cerebellar granule neurons from apoptosis by activating the mitogen-activated protein kinase (MAP kinase) pathway. J Neurosci. 1997;17:83–90.
165. May V, Lutz E, MacKenzie C, Schutz KC, Dozark K, Braas KM. Pituitary adenylate cyclase-activating polypeptide (PACAP)/PAC1HOP1 receptor activation coordinates multiple neurotrophic signaling pathways: Akt activation through phosphatidylinositol 3-kinase gamma and vesicle endocytosis for neuronal survival. J Biol Chem. 2010;285:9749–61.

Chapter 44
Implications of PACAP Signaling in Psychiatric Disorders

Hitoshi Hashimoto, Norihito Shintani, Yukio Ago, Atsuko Hayata-Takano, Takanobu Nakazawa, Ryota Hashimoto, Shinsuke Matsuzaki, Taiichi Katayama, Masaya Tohyama, Toshio Matsuda, and Akemichi Baba

Abstract Pituitary adenylate cyclase-activating polypeptide (PACAP) has been postulated to be involved in the development of psychiatric disorders. PACAP-deficient mice show behavioral and neurophysiological abnormalities including novelty-induced hyperlocomotion, emotional lability, depression-like behaviors, and memory impairments that are ameliorated with atypical antipsychotic drugs

H. Hashimoto (✉)
Laboratory of Molecular Neuropharmacology, Graduate School of Pharmaceutical Sciences, Osaka University, 1-6 Yamadaoka, Suita, Osaka 565-0871, Japan

Molecular Research Center for Children's Mental Development, United Graduate School of Child Development, Osaka University, Kanazawa University, Hamamatsu University, Hamamatsu University School of Medicine, Chiba University and University of Fukui, 2-2 Yamadaoka, Suita, Osaka 565-0871, Japan

iPS Cell-based Research Project on Brain Neuropharmacology and Toxicology, Graduate School of Pharmaceutical Sciences, Osaka University, 1-6 Yamadaoka, Suita, Osaka 565-0871, Japan
e-mail: hasimoto@phs.osaka-u.ac.jp

N. Shintani
Laboratory of Molecular Neuropharmacology, Graduate School of Pharmaceutical Sciences, Osaka University, 1-6 Yamadaoka, Suita, Osaka 565-0871, Japan

Y. Ago
Laboratory of Medicinal Pharmacology, Graduate School of Pharmaceutical Sciences, Osaka University, 1-6 Yamadaoka, Suita, Osaka 565-0871, Japan

A. Hayata-Takano
Laboratory of Molecular Neuropharmacology, Graduate School of Pharmaceutical Sciences, Osaka University, 1-6 Yamadaoka, Suita, Osaka 565-0871, Japan

Molecular Research Center for Children's Mental Development, United Graduate School of Child Development, Osaka University, Kanazawa University, Hamamatsu University School of Medicine, Chiba University and University of Fukui, 2-2 Yamadaoka, Suita, Osaka 565-0871, Japan

Kanazawa University, Suita, Osaka 565-0871, Japan

D. Reglodi, A. Tamas (eds.), *Pituitary Adenylate Cyclase Activating Polypeptide — PACAP*, Current Topics in Neurotoxicity 11,
DOI 10.1007/978-3-319-35135-3_44

(serotonin–dopamine antagonists). Targeted deletion of the PACAP receptor PAC1 leads to several of these abnormalities. Genetic studies in humans suggest that single nucleotide polymorphisms and copy number variations of the PACAP and/or PACAP receptor genes may be a genetic risk factor for psychiatric disorders including schizophrenia. Although these findings in patients need further replication and

---

Hamamatsu University School of Medicine, Chiba University, Suita, Osaka 565-0871, Japan

University of Fukui, Suita, Osaka 565-0871, Japan

T. Nakazawa
iPS Cell-based Research Project on Brain Neuropharmacology and Toxicology, Graduate School of Pharmaceutical Sciences, Osaka University,
1-6 Yamadaoka, Suita, Osaka 565-0871, Japan

R. Hashimoto
Molecular Research Center for Children's Mental Development, United Graduate School of Child Development, Osaka University, 2-2 Yamadaoka, Suita, Osaka 565-0871, Japan

Kanazawa University, Suita, Osaka 565-0871, Japan

Hamamatsu University School of Medicine, Chiba University, Suita, Osaka 565-0871, Japan

University of Fukui, Suita, Osaka 565-0871, Japan

Department of Psychiatry, Graduate School of Medicine, Osaka University,
2-2 Yamadaoka, Suita, Osaka 565-0871, Japan

S. Matsuzaki
Molecular Research Center for Children's Mental Development, United Graduate School of Child Development, Osaka University, 2-2 Yamadaoka, Suita, Osaka 565-0871, Japan

Kanazawa University, Suita, Osaka 565-0871, Japan

Hamamatsu University School of Medicine, Chiba University, Suita, Osaka 565-0871, Japan

University of Fukui, Suita, Osaka 565-0871, Japan

Department of Child Development and Molecular Brain Science, United Graduate School of Child Development, Osaka University, 2-2 Yamadaoka, Suita, Osaka 565-0871, Japan

Department of Anatomy and Neuroscience, Graduate School of Medicine, Osaka University,
2-2 Yamadaoka, Suita, Osaka 565-0871, Japan

T. Katayama
Kanazawa University, Suita, Osaka 565-0871, Japan

Hamamatsu University School of Medicine, Chiba University, Suita, Osaka 565-0871, Japan

University of Fukui, Suita, Osaka 565-0871, Japan

Department of Child Development and Molecular Brain Science, United Graduate School of Child Development, Osaka University, 2-2 Yamadaoka, Suita, Osaka 565-0871, Japan

M. Tohyama
Division of Molecular Brain Science, Research Institute of Traditional Asian Medicine,
Kinki University, 337-2 Ohno-higashi, Osaka-Sayama, Osaka 589-8511, Japan

Osaka Prefectural Hospital Organization,
3-1-56, Bandai-higashi, Sumiyoshi-ku, Osaka 558-8558, Japan

the exact mechanisms for PACAP-dependent psychomotor regulation remain to be determined, several potential mechanisms have been proposed. Thus, the convergent evidence from animal models and humans suggest that PACAP signaling can be an intrinsic regulator of psychoneurological functions and an attractive therapeutic target for the disorders.

Keywords Pituitary adenylate cyclase-activating polypeptide (PACAP) • Vasoactive intestinal peptide (VIP) • PAC1 receptor • VPAC2 receptor • Animal model • Behavior • Endophenotype • Psychiatric disorder • Schizophrenia • Post-traumatic stress disorder (PTSD) • Antipsychotic drug • Therapeutic target

Abbreviations

CNV	Copy number variation
DBZ	DISC1-binding zinc finger protein
DISC1	Disrupted-In-Schizophrenia 1
PAC1 receptor	PACAP receptor
PACAP	Pituitary adenylate cyclase-activating polypeptide
PTSD	Post-traumatic stress disorder
SNP	Single nucleotide polymorphism
VIP	Vasoactive intestinal peptide
VPAC2 receptor	VIP receptor 2

T. Matsuda
Molecular Research Center for Children's Mental Development, United Graduate School of Child Development, Osaka University, 2-2 Yamadaoka, Suita, Osaka 565-0871, Japan

Kanazawa University, Suita, Osaka 565-0871, Japan

Hamamatsu University School of Medicine, Chiba University, Suita, Osaka 565-0871, Japan

University of Fukui, Suita, Osaka 565-0871, Japan

Laboratory of Medicinal Pharmacology, Graduate School of Pharmaceutical Sciences, Osaka University, 1-6 Yamadaoka, Suita, Osaka 565-0871, Japan

A. Baba
Laboratory of Molecular Neuropharmacology, Graduate School of Pharmaceutical Sciences, Osaka University, 1-6 Yamadaoka, Suita, Osaka 565-0871, Japan

Faculty of Pharmaceutical Sciences, Hyogo University of Health Science, 1-3-6 Minatojima, Chuo-ku, Kobe, Hyogo 650-8530, Japan

Introduction

Pituitary adenylate cyclase-activating polypeptide (PACAP) [1] is widely distributed in the central nervous system and can act as a neurohormone, neurotransmitter, or trophic factor [2]. PACAP exerts multiple activities via three subtypes of G protein-coupled receptors, one PACAP-specific (PAC1) receptor and two receptors that are shared with vasoactive intestinal peptide (VIP) (VPAC1 and VPAC2). Accumulating evidence implicates PACAP signaling as an important regulator of psychiatric functions and suggests that genetic variation of PACAP and VPAC2 may be a risk factor for psychiatric disorders including schizophrenia and stress-related disorders. Such evidence has come from studies in both animal models and human clinical research, as well as through their bidirectional translation.

In this chapter, we briefly review our current knowledge in terms of the psychiatric implication of PACAP. Although we do not cover all contributions to the field, for detailed information, excellent reviews are available in this book and others [3–9].

Animal Studies

PACAP-deficient mice show behavioral and neurophysiological abnormalities that can potentially be correlated with psychiatric disorders and their related endophenotypes (Fig. 44.1). PACAP-deficient mice show hyperlocomotion and abnormally high-frequent jumping, increased novelty-seeking behavior, decreased anxiety-like behavior, and impaired social interaction [10–14]. The mutants have been shown to exhibit exaggerated depression-like behavior in the forced-swimming test [14, 15] and impaired prepulse inhibition of acoustic startle response [16]. However, these mice also demonstrated normal levels of depression-like behavior [17] and prepulse inhibition [13]. These discrepancies may reflect differences in behavioral protocols (behavioral tests each conducted in naïve mice vs. behavioral test batteries) and/or mouse genetic background. In addition, PACAP mutant mice show deficits in working memory [13, 18] and long-term potentiation in the dentate gyrus in vivo [19].

Mutant mice with complete (ubiquitous) or forebrain-specific deletion of the PACAP receptor PAC1 show a deficit in contextual fear conditioning, a hippocampal-dependent associative learning task, and impaired long-term potentiation at mossy fiber-CA3 synapses [20]. Hyperlocomotion and markedly reduced anxiety-like behavior have also been described in PAC1 receptor-deficient mice, but not in forebrain-specific PAC1 receptor mutant mice [21]. Furthermore, PAC1 receptor-deficient mice display markedly abnormal social behaviors [22].

The diversified behavioral and neurophysiological abnormalities in PACAP and PAC1 receptor mutant mice correlate with psychiatric disorders and their related endophenotypes and can be grouped into psychotic agitation (hyperlocomotion and abnormally high-frequent jumping), emotional (affective) components (anxiety-like behavior, depression-like behavior, and impaired social interaction), and cognitive deficits (memory deficits and impaired prepulse inhibition), although impaired social interaction may also be grouped as cognitive deficits.

Interestingly, some of the abnormalities in PACAP-deficient mice are clearly ameliorated by antipsychotic drugs, particularly with risperidone, an atypical antipsychotic drug and serotonin dopamine antagonist [10, 15, 16]. In addition, psychomotor and memory deficits in PACAP-deficient mice are ameliorated by MGS0028, a selective agonist of the metabotropic glutamate 2/3 receptor [23], and by SB-269970, a selective antagonist of the serotonin 7 receptor [18] (Fig. 44.1). Both the metabotropic glutamate 2/3 receptor and serotonin 7 receptor attract much attention owing to their therapeutic potential for psychiatric diseases. Thus, these results suggest that PACAP-deficient mice can be used as an animal model to develop new drugs for reversing neurobehavioral deficits.

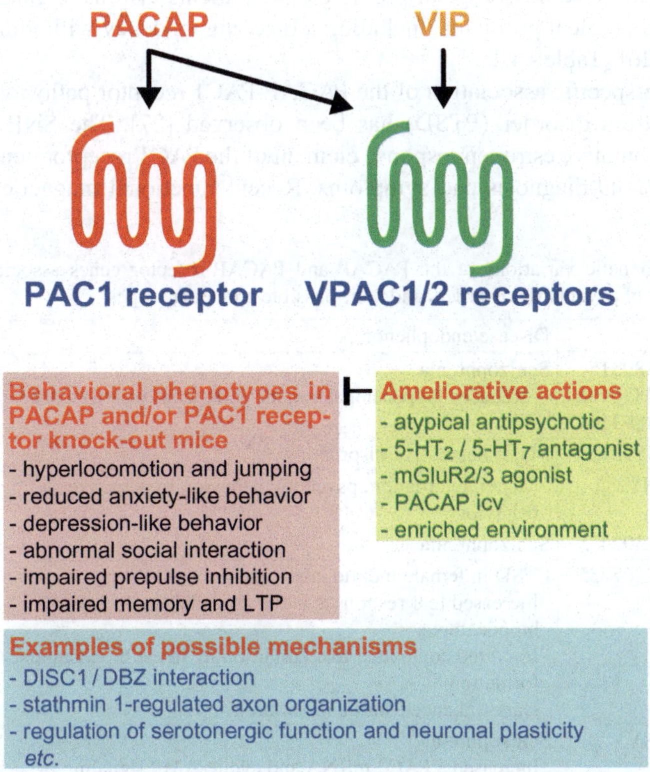

Fig. 44.1 Behavioral phenotypes in PACAP- and PAC1 receptor-mutant mice, treatments ameliorative against the behavioral phenotypes of PACAP-knockout mice, and potential mechanisms of PACAP-dependent behavioral and neurophysiological regulation. *LTP* long-term potentiation, *5-HT* serotonin, *mGluR* metabotropic glutamate receptor, *icv* intracerebroventricular

Genetic Studies in Humans

Table 44.1 shows some, but not all, genetic variations of the genes for PACAP and the PACAP receptors, PAC1 and VPAC2, which have been suggested to be associated with an increased risk of psychiatric disorders and their postulated endophenotypes.

The single nucleotide polymorphisms (SNPs) in the PACAP gene (rs1893154 and rs2856966) and PAC1 receptor gene (rs2302475) were shown to be associated with schizophrenia [24]. Homozygous carriers of the G allele of the SNP rs1893154 of the PACAP gene had smaller bilateral hippocampal volumes and lower visual associative memory performance compared with A-carriers [24]. rs1893154 was also associated with major depressive disorder [25] (Table 44.1).

A copy number variation (CNV) of the PACAP gene owing to a partial trisomy 18p and monosomy 20p has been identified in two related patients. These patients have three copies of the PACAP gene and elevated PACAP concentrations in plasma and suffer from mental retardation, psychotic behavior, and hyperactive behavior. In addition to these neurological problems, the patients also have gastrointestinal and endocrinological problems, including a bleeding tendency with mild thrombocytopenia [26] (Table 44.1).

A female-specific association of the PACAP-PAC1 receptor pathway with post-traumatic stress disorder (PTSD) has been observed [27]. The SNP rs2267735 located in a putative estrogen response element of the PAC1 receptor gene is associated with PTSD diagnosis and symptoms. Recent functional magnetic resonance

Table 44.1 Genetic variations of the PACAP and PACAP receptor genes associated with an increased risk of psychiatric disorders and their possible endophenotypes[a]

Gene		Disease/endophenotype	Reference
PACAP	rs1893154, rs2856966 rs1893154	Schizophrenia Reduced hippocampal volume and memory performance	[24]
	rs1893154	Major depressive disorder	[25]
	CNV	Mental retardation, psychotic behavior, hyperactive behavior[b]	[26]
PAC1	rs2302475	Schizophrenia	[24]
	rs2267735	PTSD in female individuals Increased fear responses in the amygdala and hippocampus Impaired contextual fear conditioning in the hippocampal formation Dark-enhanced startle in children	[27] [28] [29] [30]
VPAC2	CNV	Schizophrenia Increased VPAC2 mRNA and cyclic-AMP signaling in cultured lymphocytes	[32]

CNV copy number variation, *PTSD* post-traumatic stress disorder, *VPAC2* VIP receptor 2 (VIPR2)

[a]Please note that there are independent studies that do not replicate these associations

[b]Patients also experienced multiple neurological, gastrointestinal, and endocrinological problems, including a bleeding tendency with mild thrombocytopenia

imaging studies showed that rs2267735 was associated with increased reactivity of the amygdala and hippocampus to threat stimuli and decreased functional connectivity between these brain regions in a traumatized cohort of women [28], as well as with decreased hippocampal activity during contextual fear conditioning in healthy females [29]. rs2267735 has also been shown to be associated with dark-enhanced startle in children [30] (Table 44.1). In addition to genetic variants, blood levels of PACAP have been shown to associate with PTSD in females, proposing that PACAP levels and PAC1 receptor gene SNPs may serve as biomarkers of PTSD [27].

The VPAC2 receptor, also known as VIPR2, is a common G-protein coupled receptor for PACAP and VIP [2]. Recently, the VPAC2 receptor has been shown to localize to primary cilia both in neurons and glial cells, malfunction of which is known to participate in neurological deficits [31]. Rare CNVs (copy number gains) at chromosome 7q36.3 were found to significantly associate with schizophrenia, which results in increased VPAC2 mRNA and cyclic AMP signaling in cultured lymphocytes from patients [32] (Table 44.1).

It should be noted that there are independent studies that do not replicate these associations [37, 38]. These clinical findings need further replication and investigation to clarify their functional importance.

Mechanistic Perspectives and Outlook

The exact mechanisms for PACAP-dependent psychomotor regulation remain to be determined, but several potential mechanisms have been proposed.

Abnormalities in PACAP-deficient mice show clear responses to environmental conditions during their juvenile stage with increased aggressiveness by isolation rearing and amelioration of hyperactivity, jumping behavior, depression-like behavior, and decreased social interaction by enriched environmental rearing [12]. Pharmacological studies have shown that PACAP heterozygous mutant mice show an exaggerated head-twitch response and prepulse inhibition disruption induced by the serotonin 2 agonist and hallucinogenic drug (±)-2,5-dimethoxy-4-iodoamphetamine (DOI) [33] and PACAP-deficient mice show diminished hypothermic response to serotonin 1A agonists [16]. Furthermore, the forced swim test-induced c-Fos expression was markedly blunted in several stress-related brain areas including the dorsal raphe nucleus and the bed nucleus of the stria terminalis in PACAP-deficient mice [14]. These results further support the validity of PACAP-deficient mice as an animal model to investigate mechanisms underlying gene-environment interactions and aberrant regulation of serotonergic function and neuronal plasticity (Fig. 44.1).

Disrupted-In-Schizophrenia 1 (DISC1) is a risk gene candidate for major psychiatric disorders. PACAP markedly decreases the association between DISC1 and DISC1-binding zinc finger protein (DBZ) leading to enhanced neurite outgrowth [34]. Increased expression of stathmin 1 induces abnormal arborization of axons in primary cultured hippocampal neurons. In PACAP-deficient mice, stathmin 1 is upregulated in the dentate gyrus at both the mRNA and protein levels [35]. In an independent study, stathmin 1 has been shown to be involved in the neurotrophic effect of PACAP in PC12 cells [36]. These results propose possible molecular mechanisms for altered PACAP signaling leading to abolished neuronal circuits and behavioral and neurophysiological abnormalities [9].

Although the results described above and in other literature provide diversified and dispersed features for the etiological mechanisms, taken collectively, the convergent evidence from animal models and clinical studies suggest that PACAP signaling can be an intrinsic regulator of psychoneurological functions and may propose an attractive therapeutic target for psychiatric disorders (Fig. 44.1).

Acknowledgments This work was supported in part by the Japan Society for the Promotion of Science Grants-in-Aid for Scientific Research, KAKENHI, the Funding Program for Next Generation World-Leading Researchers, Grant No. LS081, and the Program for Advancing Strategic International Networks to Accelerate the Circulation of Talented Researchers, Grant No. S2603.

References

1. Miyata A, Arimura A, Dahl RR, Minamino N, Uehara A, Jiang L, et al. Isolation of a novel 38 residue-hypothalamic polypeptide which stimulates adenylate cyclase in pituitary cells. Biochem Biophys Res Commun. 1989;164:567–74.
2. Vaudry D, Falluel-Morel A, Bourgault S, Basille M, Burel D, Wurtz O, et al. Pituitary adenylate cyclase-activating polypeptide and its receptors: 20 years after the discovery. Pharmacol Rev. 2009;61:283–357.
3. Eiden LE. Neuropeptide-catecholamine interactions in stress. Adv Pharmacol. 2013;68:399–404.
4. Hammack SE, May V. Pituitary adenylate cyclase activating polypeptide in stress-related disorders: data convergence from animal and human studies. Biol Psychiatry. 2015;78:167–77.
5. Harmar AJ, Fahrenkrug J, Gozes I, Laburthe M, May V, Pisegna JR, et al. Pharmacology and functions of receptors for vasoactive intestinal peptide and pituitary adenylate cyclase-activating polypeptide: IUPHAR review 1. Br J Pharmacol. 2012;166:4–17.
6. Hashimoto H, Shintani N, Tanida M, Hayata A, Hashimoto R, Baba A. PACAP is implicated in the stress axes. Curr Pharm Des. 2011;17:985–9.
7. Matsuda K, Azuma M, Maruyama K, Shioda S. Neuroendocrine control of feeding behavior and psychomotor activity by pituitary adenylate cyclase-activating polypeptide (PACAP) in vertebrates. Obes Res Clin Pract. 2013;7:e1–7.
8. Pinhasov A, Nesher E, Gross M, Turgeman G, Kreinin A, Yadid G. The role of the PACAP signaling system in depression. Curr Pharm Des. 2011;17:990–1001.
9. Tohyama M, Miyata S, Hattori T, Shimizu S, Matsuzaki S. Molecular basis of major psychiatric diseases such as schizophrenia and depression. Anat Sci Int. 2015;90(3):137–43.
10. Hashimoto H, Shintani N, Tanaka K, Mori W, Hirose M, Matsuda T, et al. Altered psychomotor behaviors in mice lacking pituitary adenylate cyclase-activating polypeptide (PACAP). Proc Natl Acad Sci U S A. 2001;98:13355–60.
11. Colwell CS, Michel S, Itri J, Rodriguez W, Tam J, Lelievre V, et al. Selective deficits in the circadian light response in mice lacking PACAP. Am J Physiol Regul Integr Comp Physiol. 2004;287:1194–201.
12. Ishihama T, Ago Y, Shintani N, Hashimoto H, Baba A, Takuma K, et al. Environmental factors during early developmental period influence psychobehavioral abnormalities in adult PACAP-deficient mice. Behav Brain Res. 2010;209:274–80.

13. Hattori S, Takao K, Tanda K, Toyama K, Shintani N, Baba A, et al. Comprehensive behavioral analysis of pituitary adenylate cyclase-activating polypeptide (PACAP) knockout mice. Front Behav Neurosci. 2012;6:58.
14. Gaszner B, Kormos V, Kozicz T, Hashimoto H, Reglodi D, Helyes Z. The behavioral phenotype of pituitary adenylate-cyclase activating polypeptide-deficient mice in anxiety and depression tests is accompanied by blunted c-Fos expression in the bed nucleus of the stria terminalis, central projecting Edinger-Westphal nucleus, ventral lateral septum, and dorsal raphe nucleus. Neuroscience. 2012;202:283–99.
15. Hashimoto H, Hashimoto R, Shintani N, Tanaka K, Yamamoto A, Hatanaka M, et al. Depression-like behavior in the forced swimming test in PACAP-deficient mice: amelioration by the atypical antipsychotic risperidone. J Neurochem. 2009;110:595–602.
16. Tanaka K, Shintani N, Hashimoto H, Kawagishi N, Ago Y, Matsuda T, et al. Psychostimulant-induced attenuation of hyperactivity and prepulse inhibition deficits in Adcyap1-deficient mice. J Neurosci. 2006;26:5091–7.
17. Lehmann ML, Mustafa T, Eiden AM, Herkenham M, Eiden LE. PACAP-deficient mice show attenuated corticosterone secretion and fail to develop depressive behavior during chronic social defeat stress. Psychoneuroendocrinology. 2013;38:702–15.
18. Tajiri M, Hayata-Takano A, Seiriki K, Ogata K, Hazama K, Shintani N, et al. Serotonin 5-HT(7) receptor blockade reverses behavioral abnormalities in PACAP-deficient mice and receptor activation promotes neurite extension in primary embryonic hippocampal neurons: therapeutic implications for psychiatric disorders. J Mol Neurosci. 2012;48:473–81.
19. Matsuyama S, Matsumoto A, Hashimoto H, Shintani N, Baba A. Impaired long-term potentiation in vivo in the dentate gyrus of pituitary adenylate cyclase-activating polypeptide (PACAP) or PACAP type 1 receptor-mutant mice. Neuroreport. 2003;14:2095–8.
20. Otto C, Kovalchuk Y, Wolfer DP, Gass P, Martin M, Zuschratter W, et al. Impairment of mossy fiber long-term potentiation and associative learning in pituitary adenylate cyclase activating polypeptide type I receptor-deficient mice. J Neurosci. 2001;21:5520–7.
21. Otto C, Martin M, Wolfer DP, Lipp HP, Maldonado R, Schutz G. Altered emotional behavior in PACAP-type-I-receptor-deficient mice. Brain Res Mol Brain Res. 2001;92:78–84.
22. Nicot A, Otto T, Brabet P, Dicicco-Bloom EM. Altered social behavior in pituitary adenylate cyclase-activating polypeptide type I receptor-deficient mice. J Neurosci. 2004;24:8786–95.
23. Ago Y, Hiramatsu N, Ishihama T, Hazama K, Hayata-Takano A, Shibasaki Y, et al. The selective metabotropic glutamate 2/3 receptor agonist MGS0028 reverses psychomotor abnormalities and recognition memory deficits in mice lacking the pituitary adenylate cyclase-activating polypeptide. Behav Pharmacol. 2013;24:74–7.
24. Hashimoto R, Hashimoto H, Shintani N, Chiba S, Hattori S, Okada T, et al. Pituitary adenylate cyclase-activating polypeptide is associated with schizophrenia. Mol Psychiatry. 2007;12:1026–32.
25. Hashimoto R, Hashimoto H, Shintani N, Ohi K, Hori H, Saitoh O, et al. Possible association between the pituitary adenylate cyclase-activating polypeptide (PACAP) gene and major depressive disorder. Neurosci Lett. 2010;468:300–2.
26. Freson K, Hashimoto H, Thys C, Wittevrongel C, Danloy S, Morita Y, et al. The pituitary adenylate cyclase-activating polypeptide is a physiological inhibitor of platelet activation. J Clin Invest. 2004;113:905–12.
27. Ressler KJ, Mercer KB, Bradley B, Jovanovic T, Mahan A, Kerley K, et al. Post-traumatic stress disorder is associated with PACAP and the PAC1 receptor. Nature. 2011;470:492–7.
28. Stevens JS, Almli LM, Fani N, Gutman DA, Bradley B, Norrholm SD, et al. PACAP receptor gene polymorphism impacts fear responses in the amygdala and hippocampus. Proc Natl Acad Sci U S A. 2014;111:3158–63.
29. Pohlack ST, Nees F, Ruttorf M, Cacciaglia R, Winkelmann T, Schad LR, et al. Neural mechanism of a sex-specific risk variant for posttraumatic stress disorder in the type I receptor of the pituitary adenylate cyclase activating polypeptide. Biol Psychiatry. 2015;78(12):840–7.

30. Jovanovic T, Norrholm SD, Davis J, Mercer KB, Almli L, Nelson A, et al. PAC1 receptor (ADCYAP1R1) genotype is associated with dark-enhanced startle in children. Mol Psychiatry. 2013;18:742–3.
31. Soetedjo L, Glover DA, Jin H. Targeting of vasoactive intestinal peptide receptor 2, VPAC2, a secretin family G-protein coupled receptor, to primary cilia. Biol Open. 2013;2:686–94.
32. Vacic V, McCarthy S, Malhotra D, Murray F, Chou HH, Peoples A, et al. Duplications of the neuropeptide receptor gene VIPR2 confer significant risk for schizophrenia. Nature. 2011;471:499–503.
33. Hazama K, Hayata-Takano A, Uetsuki K, Kasai A, Encho N, Shintani N, et al. Increased behavioral and neuronal responses to a hallucinogenic drug in PACAP heterozygous mutant mice. PLoS One. 2014;9:e89153.
34. Hattori T, Baba K, Matsuzaki S, Honda A, Miyoshi K, Inoue K, et al. A novel DISC1-interacting partner DISC1-Binding Zinc-finger protein: implication in the modulation of DISC1-dependent neurite outgrowth. Mol Psychiatry. 2007;12:398–407.
35. Yamada K, Matsuzaki S, Hattori T, Kuwahara R, Taniguchi M, Hashimoto H, et al. Increased stathmin1 expression in the dentate gyrus of mice causes abnormal axonal arborizations. PLoS One. 2010;5:e8596.
36. Dejda A, Chan P, Seaborn T, Coquet L, Jouenne T, Fournier A, et al. Involvement of stathmin 1 in the neurotrophic effects of PACAP in PC12 cells. J Neurochem. 2010;114:1498–510.
37. Koga M, Ishiguro H, Horiuchi Y, Inada T, Ujike H, Itokawa M, et al. Replication study of association between ADCYAP1 gene polymorphisms and schizophrenia. Psychiatr Genet. 2010;20:123–5.
38. Chang SC, Xie P, Anton RF, De Vivo I, Farrer LA, Kranzler HR, et al. No association between ADCYAP1R1 and post-traumatic stress disorder in two independent samples. Mol Psychiatry. 2012;17:239–41.

Chapter 45
Mechanisms of PACAP in PTSD and Stress-Related Disorders in Humans

Lauren A.M. Lebois and Kerry J. Ressler

Abstract Pituitary adenylate cyclase-activating polypeptide (PACAP) is a pleiotropic neuropeptide, a neuronal signaling molecule that affects many distinct phenotypic traits. Despite the diversity in PACAP's functioning, by examining PACAP in the typical stress response, we can identify when PACAP levels and signaling become dysregulated. These findings can serve as clues to help identify the mechanisms of psychiatric and clinical disorders. Furthermore, accumulating evidence suggests that PACAP itself may represent a novel treatment target for a variety of disorders. In this chapter, we provide a brief overview of PACAP in the stress response, and review evidence that PACAP levels, signaling, genetic and epigenetic variations may be important mechanisms underlying human illnesses. Posttraumatic stress disorder and other stress-related medical conditions including migraines, Alzheimer's disease, multiple sclerosis, sudden infant death syndrome, and asthma are discussed.

Keywords Alzheimer's disease • Asthma • Migraine • Multiple sclerosis • PACAP • PAC1 • PTSD • Stress • Sudden infant death syndrome

Overview of PACAP and PAC1

Pituitary adenylate cyclase-activating polypeptide (PACAP) is part of the secretin glucagon vasoactive intestinal polypeptide family (VIP) [1]. The two major biologically active forms of PACAP are PACAP-38 and PACAP-27, which are 38 and 27 amino acids long, respectively [1]. PACAP is preferentially released at higher nerve firing rates, which often occur during the experience of a stressor [2, 3]. PACAP is encoded by the adenylate cyclase activating peptide, *ADCYAP1*, gene in humans [4] or the highly conserved rodent equivalent, *Adcyap1*.

Pituitary adenylate cyclase-activating polypeptide type I receptor (PAC1) is a G protein coupled receptor (GPCR), a class of transmembrane receptors that mediate

L.A.M. Lebois • K.J. Ressler (✉)
McLean Hospital, Harvard Medical School,
115 Mill Street, Mailstop 339, Belmont, MA 02478, USA
e-mail: kressler@partners.org

D. Reglodi, A. Tamas (eds.), *Pituitary Adenylate Cyclase Activating Polypeptide — PACAP*, Current Topics in Neurotoxicity 11,
DOI 10.1007/978-3-319-35135-3_45

cell signaling processes [4]. PAC1 receptors are selective for PACAP [4] (though see evidence that they may be less selective [5]). The PAC1 receptor is encoded by the *ADCYAP1R1* gene or the highly conserved rodent equivalent, *Adcyap1r1* [6].

The General Function of the PACAP Ligand and PACAP-PAC1 Signaling

PACAP and PACAP-PAC1 signaling are involved in diverse biological actions. Overall, PACAP is a neurotrophin, facilitating the development, function, and survival of neurons [3, 7–9]. PACAP's cellular functions also include: (a) synaptic transmission, (b) neuroendocrine signaling, and (c) cytoprotection [8]. PACAP signals both presynaptically and postsynaptically through the PAC1 receptor, and acts as both a neuromodulator, sensitizing cells to incoming signals, and a neurotransmitter, directly conveying signals [10]. PACAP-PAC1 signaling can both facilitate transmission at low doses and depress transmission at higher doses (e.g., in hippocampal CA1 synapses and possibly other brain regions) [11]. Because of this dose-dependent synaptic transmission, PACAP can both facilitate and inhibit neuroendocrine release through a cascade effect [10]. Finally, PACAP improves the likelihood that cells will survive when affronted by stressors such as environmental toxins, hypoxia, and excitotoxins, for example, by reducing oxidative stress [3, 7, 9]. More broadly PACAP-PAC1 signaling is central in development (e.g., high expression during embryogenesis), reproductive functions, thermoregulation, feeding, urinary reflexes, metabolism, and perhaps even the regulation of social interaction [4, 7].

PACAP-PAC1 signaling subserves different functions in the central vs. peripheral nervous system. Within the central nervous system, the PAC1 receptor is highly expressed in the lateral septum, amygdala (central and basolateral), bed nucleus of stria terminalis (BNST), paraventricular nucleus of the hypothalamus, hippocampus, locus coeruleus, and periaqueductal gray, among other regions [1, 4]. PACAP-PAC1 signaling in these neural structures and their associated networks plays a role in learning, memory, behavior, circadian rhythm, and the stress response [4]. In the periphery, PACAP-PAC1 signaling is integral to immune function, the inflammatory response, insulin secretion, and catecholamine release [4].

Assessing PACAP and PAC1 Genetic Variation

PACAP and PAC1 can be measured in a number of ways. Here we briefly overview three ways PACAP can be assessed as they pertain to findings discussed in our chapter. First, genetic variation can be measured, for example, by examining common or rare single nucleotide polymorphisms in a gene. Certain polymorphisms may be "risk" factors for specific disorders. Second, epigenetic variation can be assessed, for example, by measuring the methylation status of a gene. Methylation status is thought to modulate gene expression dynamically. Third, variation in gene expression can be measured at two different levels (a) directly through mRNA levels or (b) one step removed by detecting protein production, that is, the amount or levels of the PACAP protein or PAC1 receptor in the blood, cerebrospinal fluid, specific tissue (e.g., in the amygdala), or on individual cell bodies.

Brief Overview of PACAP/PAC1 in the Stress Response

PACAP-PAC1 signaling plays a modulatory role in the function of the autonomic nervous system, and stress systems (e.g., Hypothalamic-Pituitary-Adrenal axis, sympathoadrenomedullary system) [12]. In general, increased PACAP expression is associated with an up-regulated stress response (and vice versa), and decreased PACAP is associated with a down-regulated stress response [3, 8, 12, 13]. In the extreme, dysregulation in PACAP levels and signaling are associated with the development of various psychiatric disorders [8]. First we briefly review PACAP-PAC1 signaling in the healthy stress response, and then we address its mechanisms in a dysregulated stress response associated with Posttraumatic Stress Disorder and other medical conditions (also see Eiden Chap. 41 of this handbook for more details on the PACAPergic circuits underlying responses to stress).

Autonomic Nervous System

PACAP is present in most areas of the brainstem, especially autonomic sensorimotor areas (e.g., locus coeruleus) [1]. Within the autonomic nervous system, injections of PACAP into the cerebrospinal fluid were associated with increased sympathetic nerve activity, blood pressure, and heart rate, and suppressed parasympathetic nerve activity in the mouse [12, 14]. This suggests that part of PACAP's modulatory role is to maintain homeostasis by keeping a balance between sympathetic and parasympathetic nerve activity [12, 14].

Stress Systems

As mentioned earlier, PACAP is present in many limbic brain regions involved in emotional and stress responding (e.g., amygdala, BNST, hypothalamus) [12]. The Hypothalamic-Pituitary-Adrenal (HPA) axis is a cascade of stress hormone release and feedback that occurs between the hypothalamus, pituitary, and adrenal glands once an event or stimulus is consciously or unconsciously categorized as stressful [15]. There is evidence to suggest that PACAP has modulatory control over the HPA axis [4, 8]. PACAP is expressed in neuroanatomical regions associated with HPA axis potentiation, activity, and control, for example, the hypothalamus, and amygdala [1]. Also, PACAP stimulates *CRH* gene expression [1, 12], promoting the release of corticotropin-releasing hormone, a neurotransmitter in the HPA axis cascade [16].

The sympathoadrenomedullary system is also activated during the stress response. During chronic stress, high frequency sympathetic nerve firing in the adrenal medulla secretes catecholamines recurrently [2, 16]. PACAP may be the primary neurotransmitter regulating the adrenomedullary synapse central to this process, facilitating the ability to release catecholamines in a sustained manner [2, 8, 12].

PACAP's influence on these systems, however, may be context-dependent based on the type of stressor an organism is experiencing [8]. For example, PACAP may only modulate the HPA axis in response to psychogenic stressors (also termed "processive stressors"), that is, those mediated by cortical and limbic structures in the brain and requiring some top-down processing (e.g., in humans, one's reputation is threatened; in rodents, a well-lit environment) [8, 15]. PACAP receptors densely populate the prefrontal cortex and hippocampus, areas related to inhibitory control, top-down regulation of HPA axis stress circuitry, and memory function [10, 15]. This may be part of the reason PACAP modulates HPA axis activity only for top-down psychogenic stressors. In contrast, PACAP may only modulate the sympathoadrenomedullary system in response to systemic stressors [3, 8]. Systemic stressors pose a direct threat to survival, and are relayed directly to the hypothalamus most likely through the brainstem (e.g., in humans and rodents, a predatory attack) [8, 15]. PACAP's context-dependent stress system modulation may provide insight regarding potential mechanisms of dysregulation in psychiatric and medical conditions, and what systems to target for treatment in each context.

Overall, PACAP/PAC1 expression and signaling are necessary for an appropriate acute or chronic stress response, helping to preserve cellular function under stress. Abnormalities in PACAP/PAC1 expression, signaling, or feedback homeostasis can dysregulate the stress response systems, however, and have been associated with the development of psychiatric disorders [4]. Dysfunction in the PACAP system can occur for a myriad of reasons. For example, chronic stress could alter dendritic structure in areas where PACAP is heavily expressed (e.g., BNST) or risk alleles could lead to hyper- or hypo-production of PACAP in response to stressors. Here we focus on variation in PACAP protein blood levels, genetic and epigenetic variation associated with posttraumatic stress disorder (PTSD) and other stress-related conditions.

PACAP and the Dysfunctional Stress Response

Chronic activation of the stress response is a potent disruptor of cognition. It accumulates wear and tear on the body and brain (e.g., allostatic load), and is linked to increased risk for developing psychiatric disorders like anxiety, depression, and schizophrenia [17–19]. The development of these maladaptive cognitive and behavioral patterns appears to be highly PACAP-dependent [8]. An excess of PACAP, for example, is associated with PTSD; in contrast, a PACAP deficiency is associated with schizophrenia [6, 20]. As illustrated in Fig. 45.1, here we review the association between PACAP and PTSD, and a number of other medical conditions, including migraines, Alzheimer's disease, multiple sclerosis, sudden infant death syndrome, and asthma.

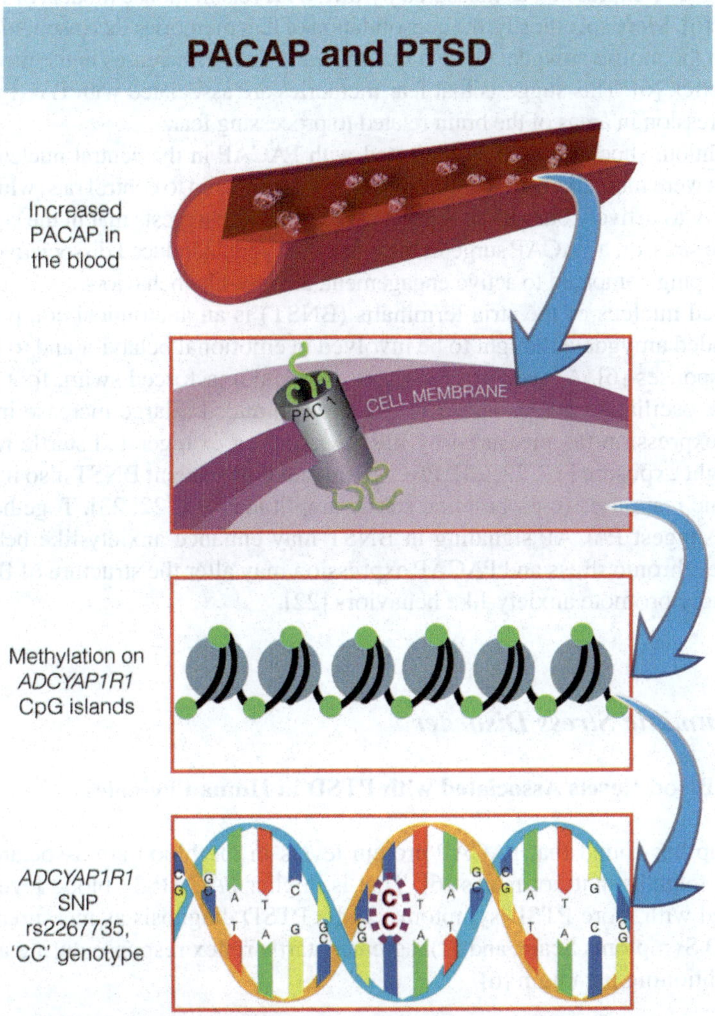

Fig. 45.1 Schematic Overview of PACAP, PAC1, and PTSD. Depiction of the three identified PACAP–PAC1 mechanisms related to posttraumatic stress disorder, including increased PACAP protein levels in the blood, methylated CpG islands on the PAC1 receptor gene (*ADCYAP1R1*), and the rs2267735 SNP CC genotype on *ADCYAP1R1*. The *small circles* in the methylation panel indicate methylation in the PAC1 receptor gene methylation panel. *SNP* single nucleotide polymorphism. Note the schematic depictions are not drawn to scale

Rodent Model Evidence for PACAP-PAC1 Signaling and Stress-Related Symptoms

Evidence from rodent models supports the connection between PACAP, PACAP-PAC1 signaling, activity in the amygdala and BNST, and PTSD-like symptomatology. Our group has found, for example, that there is a correlation between levels of fear learning and *Adcyap1r1* expression as measured by mRNA levels in mouse models of fear conditioning [6]. More specifically, the consolidation of fear memories increased *Adcyap1r1* mRNA in the mouse amygdala with a trend toward similar increases in the medial prefrontal cortex [6]. This suggests that fear memories are associated with PAC1 receptor gene expression in areas of the brain related to processing fear.

In addition, shock-stressed rats injected with PACAP in the central nucleus of the amygdala were more likely to withdraw and freeze compared to control rats, which were more likely to actively bury a shock probe [12, 21]. This suggests that in the context of an intense stressor, a PACAP surge in the central amygdala induced dissociative-like or passive coping compared to active engagement, anxiety-like behaviors.

The bed nucleus of the stria terminalis (BNST) is an anatomical component of the extended amygdala thought to be involved in emotional behavior and to mediate stress responses [6]. A week of chronic stress including forced swim, foot shocks, restraints, oscillation stress, and pedestal stress induced a large increase in BNST PACAP expression (as measured by mRNA), and an exaggerated startle response during light exposure [12, 22, 23]. PACAP injected into rodent BNST also increased anxiogenic responses (e.g., baseline startle amplitude) [12, 22, 23]. Together these findings suggest PACAP signaling in BNST may enhance anxiety-like behaviors. Overtime, chronic stress and PACAP expression may alter the structure of BNST to increasingly promote anxiety-like behaviors [22].

Posttraumatic Stress Disorder

PACAP Blood Levels Associated with PTSD in Human Females

Our group has found that PACAP protein levels in the blood are associated with PTSD in females, but not males [6]. That is, higher PACAP-38 blood levels were associated with more PTSD symptoms, and a PTSD diagnosis as measured by (1) the PTSD Symptoms Scale, and (2) a greater startle reflex response during a classic fear conditioning paradigm [6].

PAC1 Genetic Variation Associated with PTSD in Human Females

We have also found that genetic variation is associated with PTSD symptoms on a diverse set of measures. Although PACAP gene (*ADCYAP1*) variation was not associated with PTSD symptomatology, PAC1 receptor gene (*ADCYAP1R1*) variation was associated with PTSD in females (but not males) [6]. We found that an *ADCYAP1R1* single nucleotide polymorphism, rs2267735, was associated with increased likelihood of PTSD diagnosis and symptoms, spanning the intrusive, hyperarousal, and avoidance/numbing symptom clusters. Specifically, across several studies, the rs2267735 'CC' genotype is associated with higher levels of PTSD hyperarousal symptoms compared to 'G' carries as measured by the (1) PTSD Symptoms Scale, (2) increased amygdala and hippocampal reactivity to threat, (3) decreased functional connectivity between the amygdala and hippocampus, and (4) an increased startle response (regular and dark enhanced) in a classic fear conditioning paradigm [6, 24–26]. Notably, the positive relationship between *ADCYAP1R1*, other PTSD "risk" genotypes (e.g., *FKBP5*, *CRHR1*, *DBH*, *DRD2*, *NPY*, *NTRK2*) and PTSD symptoms was removed if an early exposure intervention was administered in the immediate aftermath of a traumatic experience [27]. These findings suggest that genetic risk for PTSD in the aftermath of trauma may be ameliorated with environmental or therapeutic interventions, preventing the development of trauma-related sequelae.

This association between *ADCYAP1R1* genotype and PTSD symptom severity (as measured by the PTSD Symptoms Scale and Clinician-Administered PTSD Scale) has been replicated, but only when participant trauma load was high or traumas included childhood maltreatment [24, 28]. The association between *ADCYAP1R1* genotype and PTSD symptom severity was not found in those with a lower trauma load [24, 28–30]. Other work with an index trauma of child loss has found *ADCYAP1R1* risk polymorphisms predict PTSD numbing symptoms, but not overall PTSD symptom severity [30]. Together this evidence suggests that differences in types of trauma exposure and trauma load may interact with the relationship between PACAP-PAC1 signaling and PTSD.

The association between *ADCYAP1R1* genotype and PTSD symptom severity as measured by startle response has also been replicated in adult females and both male and female children [31]. The association between PACAP and PTSD symptomatology in both male and female preadolescent children is especially intriguing given adult sex differences in PTSD. We return to this point later on in this chapter.

PAC1 Epigenetic Methylation Status Associated with PTSD in Humans

Similar to the genetic variation results, variation in DNA methylation of the PACAP gene (*ADCYAP1*) was not associated with PTSD symptomatology, but PAC1 receptor gene (*ADCYAP1R1*) methylation variation was associated with PTSD symptoms and diagnosis in males and females [6]. Specifically, methylation on *ADCYAP1R1* CpG islands was associated with more PTSD symptoms and a PTSD diagnosis [6, 25].

Notably, the aforementioned results appear to be relatively specific to PTSD. That is, PACAP protein blood levels and PAC1 receptor genetic and epigenetic variation did not predict other Axis I psychiatric disorders (e.g., depression, schizophrenia) or neurodegenerative diseases (e.g., Alzheimer's disease) within the same cohort [6, 28].

PACAP and PTSD Sex Differences in Adults

Although women are more likely to experience fewer traumatic events over their lifetime, they are nearly two times more likely to receive a PTSD diagnosis compared to men [25, 32]. It is possible that genetic variation contributes to these sex differences. In particular, dynamic PAC1 receptor regulation by the gonadal hormone estrogen may be a central figure in PTSD-related sex differences. There is evidence, for example, that estrogen induces expression of PACAP and PAC1 receptor genes in rats [6]. Specifically, estradiol increased *Adcyap1* and *Adcyap1r1* mRNA levels in the BNST [6].

As mentioned earlier, *ADCYAP1R1* polymorphisms are associated with greater startle responses in a fear conditioning paradigm for both male and female human children, but in adults, the association between *ADCYAP1R1* polymorphisms and fear conditioning is only seen in females, not males [31]. Given greater estrogen levels in adult females compared to preadolescent females (or more similar estrogen levels in female and male children vs. adults) this is further evidence supporting the dynamic relationship between estrogen, fear, trauma, and PACAP-PAC1 signaling. Ongoing work is examining estrogen's stress-dependent mechanistic role in the regulation of *ADCYAP1R1* expression. These data suggest that the rs2267735 polymorphism may exert its effect in females by differentially modulating the ability of estrogen receptors to bind to, and thus properly modulate, an estrogen response element within the *ADCYAP1R1* gene.

PACAP and Other Stress-Related Medical Conditions

PACAP dysregulation and genetic variation have also been linked to many other conditions including migraines, multiple sclerosis, Alzheimer's disease, sudden infant death syndrome, and asthma. PACAP's involvement in the ability to mount an adaptive stress response and maintain homeostasis may be the common mechanism linking PACAP and these conditions in which there is some stress-related dysregulation. Migraines, for example, can be triggered by stress, and are associated with increased PACAP levels. In contrast, Multiple sclerosis and Alzheimer's disease are associated with massive neuroinflammation, and deficits in PACAP levels. Sudden infant death syndrome and asthma are associated with childhood stress, and linked to genetic variation in the PACAP and PAC1 receptor genes. Below we address each of these in turn.

Migraines Associated with Increased PACAP Levels

Migraines are a neurological disorder characterized by moderate to severe headaches, which can be associated with nausea, light or sound sensitivity [33]. In a sample of individuals who experienced migraines, "migraineurs," PACAP levels were increased in the plasma during migraine attacks compared to levels when no migraine occurred [34]. Interestingly, migraineurs' migraine-related PACAP levels were comparable to control participants' non-migraine PACAP levels [33, 34]. The injection of PACAP also induced headaches and migraines in both healthy controls and migraineurs [35]. Migraineurs were more likely to develop a migraine after PACAP injection compared to healthy controls, however [35]. Although the exact mechanism by which PACAP facilitates migraines is unclear, at the very least PACAP is part of a signaling pathway in migraines, and is a possible target in migraine therapy [33, 36]. This topic is addressed further in Edvinsson Chap. 35 of this handbook.

Alzheimer's Disease Associated with PACAP Deficits

Alzheimer's disease (AD) is a neurodegenerative disease associated with neuroinflammation and severe deficits in cognitive functioning [37]. Recent rodent and postmortem human research has linked PACAP deficits to AD pathology. Rodent AD models and postmortem human AD patient brains have shown that reduced PACAP protein levels are associated with greater Alzheimer's pathology in the form of β-amyloid and tau protein levels [9]. Human postmortem brains with late-onset AD compared to cognitively healthy controls had reduced expression of the PACAP gene, *ADCYAP1*, throughout the brain, and more specifically, in middle temporal gyrus, superior frontal gyrus, and primary visual cortex (as measured by RNA levels) [38]. The levels of the PACAP protein in these aforementioned regions in addition to entorhinal cortex and cerebrospinal fluid were also reduced [38]. The PACAP deficit in entorhinal cortex and superior frontal gyrus, regions associated with memory and executive functioning [39, 40], was correlated with higher AD neuropathology (e.g., β-amyloid plaque density and neurofibrillary tangles) [38]. Notably, it is the entorhinal-hippocampal afferents that are the earliest to degenerate in AD [41]. In addition, reductions in cerebrospinal fluid PACAP protein levels were correlated with lower scores on a battery of cognitive functioning tasks (Mattis Dementia Rating Scale–Revised scores) [38]. Taken together, these findings demonstrate PACAP dysregulation is associated with AD neuropathology and lower cognitive functioning.

PACAP protein deficits have also been replicated in brains of patients with Mild Cognitive Impairment that also exhibited AD pathology [42]. Interestingly, PAC1 receptor levels were increased in the superior frontal gyrus of these Mild Cognitive Impairment brains compared to control and full blown AD brains, suggesting a compensatory mechanism in which the presence of PAC1 receptors are increased to counteract reduced PACAP levels or pathology interfering with PACAP functioning (or both) [42]. In addition, cortical neuronal cultures prepared from wild-type mouse brains incubated with PACAP and β-amyloid peptides were protected against β-amyloid toxicity (i.e., AD pathology) [9]. Together these findings suggest that PACAP's neurotrophic functioning is protective against AD pathology.

The mechanism for PACAP's protective effects is unclear, however. Some research suggests that PACAP stimulates alpha-secretase activity, which is in a cascade of activity that can reduce AD pathology (e.g., formation of β-amyloid peptides, and plaque) [43]. It may be that certain risk alleles are associated with lower PACAP levels and those individuals are at greater risk for developing Alzheimer's pathology or AD overwhelms the protective PACAP functioning or any number of causational combinations. More research is needed to address how PACAP deficits contribute to AD pathology, and how PACAP could be targeted as a treatment mechanism for this disease, particularly given its neurotrophic function and localization in circuitry central to AD (e.g., entorhinal cortex) [9, 38].

Multiple Sclerosis Associated with Reduced PACAP Levels

Multiple sclerosis (MS) is characterized by chronic neuroinflammatory and neurodegenerative pathology [44]. PACAP's documented immunomodulatory and neurotrophic functions suggest that it could be a therapeutic target for both the inflammatory and degenerative aspects of MS [44, 45]. Indeed, recent evidence demonstrated reduced PACAP-38 levels in the cerebrospinal fluid of individuals with MS compared to healthy controls [46].

Sudden Infant Death Syndrome Associated with Reduced PACAP and PACAP Genetic Variation

Sudden infant death syndrome (SIDS) is the unexpected death of an infant without an identifiable explanation [47]. Despite its unidentifiable cause, SIDS risk factors have been recognized, including infant vulnerability related to: (a) in utero stressors (e.g., maternal smoking), (b) developmental stage, and (c) environmental conditions (e.g., increased heat) [47]. Growing evidence suggests PACAP's role in the development of healthy breathing and an adaptive stress response may in part underlie these vulnerabilities [48]. For example, mice lacking PACAP were more likely to die suddenly within two weeks after birth, mirroring SIDS in humans [48, 49]. Genetic tests in humans reveal an association between PACAP polymorphisms and SIDS. Specifically, a single nucleotide polymorphism in the coding region of PACAP was associated with SIDS [49]. This association was present for African Americans, but not for Caucasians [49], which may be part of the reason why there is a higher incidence of SIDS in African American infants [50]. Together this evidence suggests PACAP dysfunction may be an underlying cause of SIDS, and "risk" alleles may increase the likelihood of SIDS.

Asthma Associated with PAC1 Genetic Variation

Asthma is chronic inflammation of the lungs associated with difficulty breathing [51]. Psychosocial stress is linked to the occurrence of asthma in children and adults [52, 53]. Recent research demonstrates variation in the PAC1 receptor gene,

ADCYAP1R1, is associated with asthma. That is, the single nucleotide polymorphism rs2267735 'CC' genotype is associated with increased odds of developing asthma compared to 'G' carriers [52]. Recent evidence also shows that high childhood stress and a different single nucleotide polymorphism within the *ADCYAP1R1* gene are associated with reduced bronchodilator response, an objective measure of lung function, in children with asthma [54]. Together, these data suggest an important link between childhood stress and asthma may be mediated by genetic differences in PAC1 signaling.

General Conclusions

PACAP is a protein necessary for healthy development and an adaptive stress response. Dysregulation in PACAP/PAC1 levels or their signaling is associated with stress-related pathology in psychiatric and medical conditions. This dysregulation can occur at many levels—from existing genetic variation to epigenetic changes associated with life experiences, each in turn causing cascades of physiological alterations in the functioning of stress systems and neural networks. With continued characterization of PACAP/PAC1 mechanisms in these disorders we hope to better understand how to identify pathology, enable early intervention, and develop novel treatment strategies.

Acknowledgements Thanks to Richard McDonough for the graphic design of Fig. 45.1. *Declaration of conflicting interests*: The authors declare that they have no conflicts of interest with respect to their authorship or the publication of this chapter. *Funding*: Funding for this work was provided by the National Institutes of Mental Health (MH096764 to K.J.R.), and the Howard Hughes Medical Institute.

References

1. Kormos V, Gaszner B. Role of neuropeptides in anxiety, stress, and depression: from animals to humans. Neuropeptides. 2013;47:401–19.
2. Smith CB, Eiden LE. Is PACAP the major neurotransmitter for stress transduction at the adrenomedullary synapse? J Mol Neurosci. 2012;48:403–12.
3. Stroth N, Holighaus Y, Ait-Ali D, Eiden LE. PACAP: a master regulator of neuroendocrine stress circuits and the cellular stress response. Ann N Y Acad Sci. 2011;1220:49–59.
4. Shen S, Gehlert DR, Collier DA. PACAP and PAC1 receptor in brain development and behavior. Neuropeptides. 2013;47:421–30.
5. Dautzenberg FM, Mevenkamp G, Wille S, Hauger RL. N-Terminal splice variants of the type I PACAP receptor: isolation, characterization and ligand binding/selectivity determinants. J Neuroendocrinol. 1999;11:941–9.
6. Ressler KJ, Mercer KB, Bradley B, Jovanovic T, Mahan A, Kerley K, et al. Post-traumatic stress disorder is associated with PACAP and the PAC1 receptor. Nature. 2011;470:492–7.
7. Mester L, Kovacs K, Racz B, Solti I, Atlasz T, Szabadfi K, et al. Pituitary adenylate cyclase-activating polypeptide is protective against oxidative stress in human retinal pigment epithelial cells. J Mol Neurosci. 2011;43:35–43.

8. Mustafa T. Pituitary adenylate cyclase-activating polypeptide (PACAP): a master regulator in central and peripheral stress responses. Adv Pharmacol. 2013;68:445–57.
9. Han P, Tang Z, Yin J, Maalouf M, Beach TG, Reiman EM, et al. Pituitary adenylate cyclase-activating polypeptide protects against β-amyloid toxicity. Neurobiol Aging. 2014;35:2064–71.
10. Vaudry D, Falluel-Morel A, Bourgault S, Basille M, Burel D, Wurtz O, et al. Pituitary adenylate cyclase-activating polypeptide and its receptors: 20 years after the discovery. Pharmacol Rev. 2009;61:283–357.
11. Roberto M, Scuri R, Brunelli M. Differential effects of PACAP-38 on synaptic responses in rat hippocampal CA1 region. Learn Mem. 2001;8:265–71.
12. Hashimoto H, Shintani N, Tanida M, Hayata A, Hashimoto R, Baba A. PACAP is implicated in the stress axes. Curr Pharm Des. 2011;17:985–9.
13. Lehmann ML, Mustafa T, Eiden AM, Herkenham M, Eiden LE. PACAP-deficient mice show attenuated corticosterone secretion and fail to develop depressive behavior during chronic social defeat stress. Psychoneuroendocrinology. 2013;38:702–15.
14. Tanida M, Shintani N, Morita Y, Tsukiyama N, Hatanaka M, Hashimoto H, et al. Regulation of autonomic nerve activities by central pituitary adenylate cyclase-activating polypeptide. Regul Pept. 2010;161:73–80.
15. Herman JP, Figueiredo H, Mueller NK, Ulrich-Lai Y, Ostrander MM, Choi DC, et al. Central mechanisms of stress integration: hierarchical circuitry controlling hypothalamo-pituitary-adrenocortical responsiveness. Front Neuroendocrinol. 2003;24:151–80.
16. Eiden LE. Neuropeptide-catecholamine interactions in stress. Adv Pharmacol. 2013;68: 399–404.
17. Ganzel BL, Morris PA, Wethington E. Allostasis and the human brain: integrating models of stress from the social and life sciences. Psychol Rev. 2010;117:134–74.
18. Juster RP, McEwen BS, Lupien SJ. Allostatic load biomarkers of chronic stress and impact on health and cognition. Neurosci Biobehav Rev. 2010;35:2–16.
19. Rodrigues SM, LeDoux JE, Sapolsky RM. The influence of stress hormones on fear circuitry. Annu Rev Neurosci. 2009;32:289–313.
20. Matsuzaki S, Tohyama M. Regulation of pituitary adenylyl cyclase-activating polypeptide (PACAP, ADCYAP1: adenylyl cyclase-activating polypeptide 1) in the treatment of schizophrenia. Expert Opin Ther Targets. 2008;12:1097–108.
21. Legradi G, Das M, Giunta B, Hirani K, Mitchell EA, Diamond DM. Microinfusion of pituitary adenylate cyclase-activating polypeptide into the central nucleus of amygdala of the rat produces a shift from an active to passive mode of coping in the shock-probe fear/defensive burying test. Neural Plast. 2007;2007:79102.
22. Hammack S, Cheung J, Rhodes KM, Schutz KC, Falls WA, Braas KM, et al. Chronic stress increases pituitary adenylate cyclase-activating peptide (PACAP) and brain-derived neurotrophic factor (BDNF) mRNA expression in the bed nucleus of the stria terminalis (BNST): roles for PACAP in anxiety-like behavior. Psychoneuroendocrinology. 2009;34:833–43.
23. Hammack S, Roman CW, Lezak KR, Kocho-Shellenberg M, Grimmig B, Falls WA, et al. Roles for pituitary adenylate cyclase-activating peptide (PACAP) expression and signaling in the bed nucleus of the stria terminalis (BNST) in mediating the behavioral consequences of chronic stress. J Mol Neurosci. 2010;42:327–40.
24. Almli LM, Mercer KB, Kerley K, Feng H, Bradley B, Conneely KN, et al. ADCYAP1R1 genotype associates with post-traumatic stress symptoms in highly traumatized African-American females. Am J Med Genet. 2013;162:262–72.
25. Dias BG, Ressler KJ. PACAP and the PAC1 receptor in post-traumatic stress disorder. Neuropsychopharmacology. 2013;38:245–6.
26. Stevens JS, Almli LM, Fani N, Gutman DA, Bradley B, Norrholm SD, et al. PACAP receptor gene polymorphism impacts fear responses in the amygdala and hippocampus. Proc Natl Acad Sci U S A. 2013;111:3158–63.
27. Rothbaum BO, Kearns MC, Reiser E, Davis JS, Kerley KA, Rothbaum AO, et al. Early intervention following trauma may mitigate genetic risk for PTSD in civilians. J Clin Psychiatry. 2014;1380–7.

28. Uddin M, Chang SC, Zhang C, Ressler K, Mercer KB, Galea S, et al. Adcyap1r1 genotype, posttraumatic stress disorder, and depression among women exposed to childhood maltreatment. Depress Anxiety. 2013;30:251–8.
29. Chang SC, Xie P, Anton RF, De Vivo I, Farrer LA, Kranzler HR, et al. No association between ADCYAP1R1 and post-traumatic stress disorder in two independent samples. Mol Psychiatry. 2012;17:239–41.
30. Wang L, Cao C, Wang R, Qing Y, Zhang J, Zhang XY. PAC1 receptor (ADCYAP1R1) genotype is associated with PTSD's emotional numbing symptoms in Chinese earthquake survivors. J Affect Disord. 2013;150:156–9.
31. Jovanovic T, Norrholm SD, Davis J, Mercer KB, Almli L, Nelson A, et al. PAC1 receptor (ADCYAP1R1) genotype is associated with dark-enhanced startle in children. Mol Psychiatry. 2012;18:742–3.
32. Tolin DF, Foa EB. Sex differences in trauma and posttraumatic stress disorder: a quantitative review of 25 years of research. Psychol Bull. 2006;132:959–92.
33. Kaiser EA, Russo AF. CGRP and migraine: could PACAP play a role too? Neuropeptides. 2013;47:451–61.
34. Tuka B, Helyes Z, Markovics A, Bagoly T, Szolcsanyi J, Szabo N, et al. Alterations in PACAP-38-like immunoreactivity in the plasma during ictal and interictal periods of migraine patients. Cephalalgia. 2013;33:1085–95.
35. Schytz HW, Birk S, Wienecke T, Kruuse C, Olesen J, Ashina M. PACAP38 induces migraine-like attacks in patients with migraine without aura. Brain. 2009;132:16–25.
36. Schytz HW, Olesen J, Ashina M. The PACAP receptor: a novel target for migraine treatment. Neurotherapeutics. 2010;7:191–6.
37. Perry RJ, Hodges JR. Attention and executive deficits in Alzheimer's disease: a critical review. Brain. 1999;122:383–404.
38. Han P, Liang W, Baxter LC, Yin J, Tang Z, Beach TG, et al. Pituitary adenylate cyclase-activating polypeptide is reduced in Alzheimer disease. Neurology. 2014;82:1724–8.
39. Hafting T, Fyhn M, Molden S, Moser MB, Moser EI. Microstructure of a spatial map in the entorhinal cortex. Nature. 2005;436:801–6.
40. du Boisgueheneuc F, Levy R, Volle E, Seassau M, Duffau H, Kinkingnehun S, et al. Functions of the left superior frontal gyrus in humans: a lesion study. Brain. 2006;129:3315–28.
41. Braak H, Braak E. Neuropathological stageing of Alzheimer-related changes. Acta Neuropathol. 1991;82:239–59.
42. Han P, Caselli RJ, Baxter L, Serrano G, Yin J, Beach TG, et al. Association of pituitary adenylate cyclase-activating polypeptide with cognitive decline in mild cognitive impairment due to Alzheimer disease. JAMA Neurol. 2015;72:333–9.
43. Kojro E, Postina R, Buro C, Meiringer C, Gehrig-Burger K, Fahrenholz F. The neuropeptide PACAP promotes the alpha-secretase pathway for processing the Alzheimer amyloid precursor protein. FASEB J. 2006;20:512–4.
44. Tan Y, Waschek JA. Targeting VIP and PACAP receptor signalling: new therapeutic strategies in multiple sclerosis. ASN Neuro. 2011;3.
45. Abad C, Waschek JA. Immunomodulatory roles of VIP and PACAP in models of multiple sclerosis. Curr Pharm Des. 2011;17:1025–35.
46. Baranowska-Bik A, Kochanowski J, Uchman D, Wolinska-Witort E, Kalisz M, Martynska L, et al. Vasoactive intestinal peptide (VIP) and pituitary adenylate cyclase activating polypeptide (PACAP) in humans with multiple sclerosis. J Neuroimmunol. 2013;263:159–61.
47. Fleming PJ, Blair PS, Pease A. Sudden unexpected death in infancy: aetiology, pathophysiology, epidemiology and prevention in 2015. Arch Dis Child. 2015;100(10):984–8.
48. Wilson RJA, Cumming KJ. Pituitary adenylate cyclase-activating polypeptide is vital for neonatal survival and the neuronal control of breathing. Respir Physiol Neurobiol. 2008;164:168–78.
49. Cummings KJ, Klotz C, Liu W-Q, Weese-Mayer DE, Marazita ML, Cooper ME, et al. Sudden infant death syndrome (SIDS) in African Americans: polymorphisms in the gene encoding the stress peptide pituitary adenylate cyclase-activating polypeptide (PACAP). Acta Paediatr. 2009;98:482–9.

50. Black L, David RJ, Brouillette RT, Hunt CE. Effects of birth weight and ethnicity on incidence of sudden infant death syndrome. J Pediatr. 1986;108:209–14.
51. What is asthma? 2014. http://www.nhlbi.nih.gov/health/health-topics/topics/asthma. Accessed 2015.
52. Chen W, Boutaoui N, Brehm JM, Han YY, Schmitz C, Cressley A, et al. ADCYAP1R1 and asthma in Puerto Rican children. Am J Respir Crit Care Med. 2013;187:584–8.
53. Yii AC, Koh MS. A review of psychological dysfunction in asthma: affective, behavioral and cognitive factors. J Asthma. 2013;50:915–21.
54. Brehm J, Ramratnam SK, Tse SM, Croteau-Chonka DC, Pino-Yanes M, Rosas-Salazar C, et al. Stress and bronchodilator response in children with asthma. Am J Respir Crit Care Med. 2015;192:47–56.

Chapter 46
PACAP, VIP, and ADNP: Autism and Schizophrenia

Illana Gozes

Abstract Thirty years ago we cloned the gene for vasoactive intestinal peptide (VIP), coding for a major brain neuropeptide. We have also established the first VIP transgenic mice as well as then novel VIP agonists and antagonists leading to the discoveries of VIP functions in brain development, learning and memory as well as social interactions. Fifteen years ago we discovered activity-dependent neuroprotective protein (ADNP), as a VIP regulated protein, and showed that it is essential for brain formation/function. Follow-up studies identified pituitary adenylate cyclase activating polypeptide (PACAP) also as an ADNP regulating neuropeptide. We have further identified ADNP as a member of the chromatin remodeling complex, SWI/SNF also associated with alternative splicing of tau and prediction of tauopathy. In neurons, ADNP is found in the nucleus as well as in the neuronal processed. We have identified cytoplasmic ADNP interactions with the autophagy regulating microtubule-associated protein 1 light chain 3 (LC3) and with microtubule end binding (EB) proteins. The ADNP-EB-binding Ser-Ile-Pro (SIP) domain is shared with the ADNP snippet drug candidate, NAPV*SIP*Q termed NAP (davunetide). Recently, multiple mutations in ADNP were found in children defined within the autism spectrum and exhibiting cognitive dysfunctions. Better understanding of VIP-ADNP interactions should shade light on brain development and function toward better management of the autism spectrum disorders (ASDs). Our further findings indicated dysregulation in the autophagy pathway in schizophrenia, coupled to ADNP dysregulation and amelioration by NAP (davunetide treatment).

Keywords VIP • PACAP • ADNP • NAP • Autism • Autism spectrum disorder (ASD) • Microtubules • Autophagy • Chromatin remodeling complex • Alzheimer's disease • Tau • Schizophrenia

---

I. Gozes (✉)
Department of Human Molecular Genetics and Biochemistry,
Sackler Medical School, Adams Super Center for Brain Studies & Sagol
School of Neuroscience, Tel Aviv University, Tel Aviv 69978, Israel
e-mail: igozes@post.tau.ac.il

D. Reglodi, A. Tamas (eds.), *Pituitary Adenylate Cyclase Activating Polypeptide — PACAP*, Current Topics in Neurotoxicity 11,
DOI 10.1007/978-3-319-35135-3_46

VIP

Vasoactive intestinal peptide (VIP) was first discovered as a potent vasodilator [1] and then identified as a major brain neuropeptide [2, 3]. We have concentrated on VIP in the brain from a molecular and a functional point of view, in terms of learning and memory as well as social interactions [2]. VIP is important for brain development and key to circadian rhythmicity. As a developmental regulator it is also associated with cancer propagation, and regulates other key proteins, such as activity-dependent neuroprotective protein (ADNP). Our major contributions to the VIP field include:

1. The first cloning of the human and the mouse VIP genes [4, 5].
2. The identification of VIP expression during late embryonic development, peaking at the postnatal time of synapse formation, directly regulating synapse formation, and decreasing with aging [6, 7].
3. The establishment of the first VIP transgenic mice showing sexual and learning deficits [8].
4. The design and synthesis of a potent VIP-neurotensin hybrid antagonist, inhibiting sexual interactions [9], learning and memory [10] circadian [11] and social activity at the behavioral level [12] and brain/synaptic development at the functional level [13], as well as producing cell death, at the cellular level [14], also inhibiting cancer growth [15].
5. The development of potent VIP lipophilic analogs with increased skin and brain bioavailability and effects on learning and memory and sexual behavior [16–18].
6. The discovery of activity-dependent neurotrophic factor (ADNF) and activity-dependent neuroprotective protein (ADNP) as delineated below.

Based on our extensive studies as formerly reviewed, we have hypothesized that VIP is involved with autism. Indeed, and citing almost verbatim, recent studies encompassing whole genome sequencing identified a de novo mutation in VIP (VIP p.Tyr73*) [19]. The clinical description indicated average intelligence by standardized testing and a clinical diagnosis at 3.2 years of age of Asperger syndrome. A younger male sibling also carried the mutation. He presented with expressive language delay, above-average receptive vocabulary, and a history of "clumsiness." The mutation was not found in either parent, which might reflect gonadal mosaicism in one of them. The mother presented hypothyroidism and allergies, probably unrelated [19]. However, it should be kept in mind that studies have indicated an influence of the pregnant mother genotype on the development of the embryo, especially in association with VIP [20], as well as the implications of environmental effects on brain development. In this respect, VIP was also linked to the inflammatory response [21].

VIP activates two receptor molecules, VPAC1 and VPAC2 [22, 23]. VPAC1 has been associated with cancer and VPAC2 has been associated with neurodevelopment and diurnal rhythms. Recent clinical studies indicated that microduplications of VPAC2 confer significant risk for schizophrenia and ASD [24–27]. Lymphocytes from patients with these mutations exhibited higher VPAC2 gene expression and VIP responsiveness (cAMP induction). In a recent study, after application of the VPAC2 receptor agonist Ro 25–1553 to C57BL/6 mice from postnatal day 1 (P1) to P14, Western blot analyses on P21 revealed significant reductions of synaptophysin and postsynaptic density protein 95 (PSD-95) in the prefrontal cortex, but not in the hippocampus in the treated mice. The same postnatally restricted treatment resulted in a disruption in prepulse inhibition of the acoustic startle measured in adult mice. No effects were observed in open-field locomotor activity, sociability in the three-chamber social interaction test, or fear conditioning or extinction. These results together with the clinical results further implicate the VIP system in mental disorders, including schizophrenia and ASD [24].

The Discovery of ADNP

Our original VIP studies showed that VIP expression peaks at the time of synapse formation [7], suggesting a direct involvement. At the same time, Douglas E. Brenneman and Lee Eiden at the National Institutes of Health, USA, showed that VIP protects nerve cells against electrical blockade [28]. The Brenneman laboratory further showed that for neuroprotection VIP required glial cells, specifically astrocytes [29], and this is when we teamed together [14] to discover ADNF [30–33] and ADNP [34–36], as glial proteins regulated by VIP to promote neuronal survival, enhance synapse formation and protect synapses [37, 38] and allow for synaptic plasticity [39].

We have also shown the involvement of VPAC2 in the regulation of ADNP expression in astrocytes [40], further connecting the VIP and the ADNP systems.

I have recently reviewed ADNP [41], from our original cloning in mouse [34], with similarities to the previously identified ADNF [30, 42, 43] and pointed to the high conservation with human ADNP (hADNP) [36], and other vertebrates [44]. The protein structure contains nine zinc fingers, a proline-rich region, a nuclear bipartite localization signal, and a homeobox domain profile, suggesting a transcription factor function, while nuclear secretion and cellular secretion signals confirm ADNP cytoplasmic as well as extracellular functions [36, 45, 46]. Comparative analysis identified ADNP2 (33 % identity and 46 % similarity), indicating that these genes belong to a protein family with nine zinc finger motifs followed by a homeobox domain [36, 47, 48]. The hADNP gene structure spans ~40 kb and includes five exons and four introns with alternative splicing of an untranslated second exon (chromosome 20q12-13.2, a region associated with aggressive tumor growth). Human ADNP (hADNP) is mutated in cancer [41] and autism [49], as discussed below.

The PACAP Connection

Before delving into further characterization of ADNP, it is interesting to note that not only VIP regulates ADNP expression, the VIP-like, pituitary adenylate cyclase activating polypeptide (PACAP) also regulates ADNP synthesis. We have identified regulation of ADNP by PACAP in astrocytes [40], through the VPAC2 receptor. Furthermore, the 38 amino acid long, PACAP induced ADNP mRNA expression in a bimodal fashion at subpicomolar and nanomolar concentrations with greater response at subpicomolar level, in mouse neuron–glia cocultures. The response was attenuated by a PAC1 receptor antagonist (mediated by IP3/PLC signaling) and by VPAC1 receptor antagonist at nanomolar concentration (mediated by PKA signaling) [50]. Further immunohistochemical analyses identified co-localization of PAC1 and VPAC1 with ADNP immunoreactivity in neurons and glial cells in the mouse brain [51]. Thus, homeostasis of VIP and PACAP is associated with normative ADNP expression and PACAP signals through the two VIP receptors, VPAC1 and VPAC2 as well as the PACAP specific receptor, PAC1 [22, 23].

Importantly, PAC1-deficient mice exhibited a faster decrease in social investigation after repeated exposure to social cues compared with PAC1-intact mice. PAC1-deficient females exhibited delayed affiliative behavior when housed with novel males, and PAC1-deficient males displayed excessive sexual mounting toward both females and males as well as reduced aggression toward intruder males. Together, these results suggest an involvement of PACAP signaling in social behavior and autism, with a potential association with ADNP regulation [52].

At the level of the peptide, altered psychomotor behaviors were discovered in mice lacking PACAP, which were accompanied by depression-like behavior in the forced swimming test and amelioration by the atypical antipsychotic risperidone [53]. The deficient mice also showed deficits in prepulse inhibition [54]. However, these findings were challenged by later investigations, suggesting a complex impact [55]. Blunted c-Fos expression in selected brain areas, such as the bed nucleus of the stria terminalis, were also observed in the PACAP-deficient mice [56]. In PACAP heterozygous mutant mice, increased behavioral and neuronal responses to the hallucinogenic drug, the 5-HT2 receptor agonist, (+/−)-2,5-dimethoxy-4-iodoamphetamine, were observed [57]. Together, the results suggest an involvement of PACAP in emotional behavior associated with autism and schizophrenia and further results suggest also neuroprotective impacts on cognitive behavior [58]. It is still a matter of future studies to evaluate the impact of autism-related ADNP mutations on PACAP expression and activity.

ADNP Protein Interactions

Our recent studies have identified cytoplasmic ADNP interactions with the microtubule end binding (EB) proteins [39]. The interaction with EBs was discovered to be through the SIP domain shared with the ADNP snippet drug candidate, NAPVSIPQ termed NAP (davunetide) as well as SALLRSIPA (ADNF-9) [39, 43]. Using further biochemical and cell biology techniques, we showed multiple crucial interactions for ADNP, including heterochromatin protein 1 alpha [59], BRG1, BAF250a, and BAF170, all components of the SWItch/Sucrose NonFermentable (SWI/SNF) chromatin remodeling complex [60], the RNA splicing machinery, specifically Brahma (Brm), and the polypyrimidine tract-binding protein (PTB)-associated splicing factor (PSF)-binding [61], protein translation, binding and regulating the eukaryotic translation initiation factor 4E (eIF4E) [62], as well as the microtubule associated protein 1 light chain 3 (LC3) regulating and participating in autophagy [63]. Interestingly, we found autophagy to be deregulated in schizophrenia (reduction in expression in key proteins like the autophagy regulator beclin 1 [63, 64] and others found it to be deregulated in autism, associated with the beclin 1 regulator, Ambra 1 [65].

It should be noted that eIF4E has been linked to autism [66–69], and ADNP differentially regulates it in males and females, which may be associated, in part, with the higher prevalence of autism in males [62]. Furthermore, our HP1alpha–ADNP interaction results were confirmed and extended [70], and interestingly, HP1alpha (CBX5) was found to be de novo mutated in autism [71] as well as in schizophrenia [72].

Sex Influence

Our results have indicated sex-dependent VIP expression in the brain, with estrogen regulating VIP in the female hypothalamus, but orchidectomy, not affecting VIP expression in the male hippocampus [73]. RNA expression analysis revealed sexual dimorphism in hypothalamic ADNP mRNA expression as well as fluctuations within the estrus cycle. ADNP-like immunoreactivity, in the arcuate nucleus of the hypothalamus, exhibited fluctuations during the estrus cycle (proestrus—the most immunoreactive; estrus—the least). Male arcuate nucleus ADNP-like immunoreactivity was significantly lower than that of the female estrus. These results suggest that ADNP plays a part in the estrus cycle [74]. We have extended these findings to the hippocampus, where ADNP expression was higher in males (mice and men) [62]. Furthermore, the hippocampal transcript for apolipoprotein E (the major risk gene for AD) was doubled in female mice compared with males, and further doubled in the ADNP(+/–) females, contrasting a decrease in ADNP(+/–) males. Behaviorally, ADNP(+/–) male mice exhibited deficiencies in object recognition and social memory compared with ADNP(+/+) mice, while ADNP(+/–) females were partially spared. Contrasting males, which preferred novel over familiar mice, ADNP(+/+) females showed no preference to novel mice and ADNP(+/–) females did not prefer mice over object. Coupled to the description above regarding eIF4E, with ADNP regulating eIF4E in males, we introduced a novel concept of hippocampal gene-regulated sexual dimorphism with ADNP playing a central role [62]. As stated above, ADNP is associated with the regulation of autophagy [63, 64]. Interestingly, a much stronger reduction of Ambra1 protein in the cortex of Ambra1 (+/–) female mice compared to males, which was coupled to an autistic behavior only in the Ambra1 (+/–) females [65].

The male predominance of autism, and more generally, autism spectrum disorder is well known, it is suggested that this is an interplay between genes and environment [75], we maintain that among the influential genes in this respect are VIP and ADNP are differentially expressed in the song bird brain [76]. Zebra finches (Taeniopygia guttata) are highly social songbirds that have a sexually dichotomous anatomical brain structure, with males demonstrating a developed song system, we recently discovered sexually dichotomous and age related differences in ADNP mRNA expression in three different brain regions, cerebellum, cerebrum, and brain stem. Higher levels of ADNP mRNA were specifically found in young male compared to the female cerebrum, while aging caused a significant two- and threefold decrease in the female and male cerebrum, respectively. Furthermore, unique sex-dependent ADNP mRNA distribution patterns, affected by aging were discovered [76].

ADNP Mutations in Autism Spectrum Disorder

An original paper showed that deletion of ADNP and adjacent genes may cause mental retardation [77]. Further original studies have identified de novo p.Lys408Valfs*31 [78] and p.Tyr719* ADNP mutation [79] followed by >20 additional cases, with hot spot recurrent mutations at p.Tyr719* (cDNA 2157) preceding the homoeobox profile in ADNP and cDNA 2491 deletion of—TTAA or 2496, TAAA resulting in a frameshift and stop mutation after the homeobox domain. Helsmoortel et al. [49] identified a total of ten patients with mutations in ADNP, including the two patients identified in both earlier studies and highlighted the cognitive and characteristic facial features, in the children (see also [80, 81], Coe et al. [82], De Rubeis et al. [83]) and more [84]).

Interestingly, the recurrent de novo, protein truncating p.Tyr719* ADNP mutation exhibits features such as visual system impairments. This provides a potential role for ADNP in the development of the visual system given its high prenatal expression. Furthermore, mood disorder is also associated with this mutation [80]. These findings are in line with our original observations of ADNP regulating PAX6 expression (a gene critical for eye and cortical development) [85], as well as our studies associating deregulation of ADNP expression in schizophrenia [63, 64, 86].

Further studies are required to decipher the role of each of the ADNP mutations and try to solve the mystery of de novo mutations affecting cognition.

NAP

NAP (NAPVSIPQ) is a short 8 amino acid peptide derived from ADNP that has shown neuroprotection against ADNP deficiency. It was tested in vitro and in vivo [87]. Given the complex ADNP interaction and the importance of the ADNP regulation during development, it may be that NAP can replace, in part, ADNP deficiency; however, this needs to be further tested.

A few past examples—in general:

1. NAP enhances ADNP LC3 [63] and ADNP EB association [39].
2. NAP compensates for ADNP deficiency (cognitive and microtubule deficiency in and ADNP-deficient mouse model [87].

In particular in developing systems (a few selected examples):

1. NAP enhances neural tube closure [88].
2. NAP protects against fetal alcohol syndrome in vivo [89].
3. NAP accelerated postnatal development in the face of genetic deficiencies (ApoE deficiency) [34] and ischemia [90].
4. NAP analogs protected cognition in a mouse model of Down's syndrome [91].

Interestingly, the ADNP regulating VIP analog stearyl-Nle17-VIP (SNV) [92] also accelerate postnatal development [93], perhaps through ADNP? This opens the door for future studies.

Regarding schizophrenia, NAP protected against cognitive impairment in a microtubule-deficient schizophrenia model, the stable tubule only polypeptide (STOP)/microtubule associated protein 6 (Map6) haploinsufficient mice [64, 94]. The mice showed hippocampal ADNP deficiency coupled with beclin 1 deficiency, and NAP treatment increased beclin 1 expression, while providing cognitive protection [64]. Clinical studies showed significant NAP (davunetide) protection against deficits in functional activities of daily living in the patients [95] as well as protection of choline levels in the patients' brains as measured by magnetic resonance spectroscopy [96]. These results are supported by additional preclinical tests [97], suggesting that further clinical evaluations are warranted.

Acknowledgements Supported by the AMN Foundation, the Lily and Avraham Gildor Chair for the Investigation of Growth Factors and the Elton Laboratory for Molecular Neuroendocrinology.

References

1. Said SI, Mutt V. Polypeptide with broad biological activity: isolation from small intestine. Science. 1970;169:1217–8.
2. Gozes I. VIP, from gene to behavior and back: summarizing my 25 years of research. J Mol Neurosci. 2008;36:115–24.
3. Gozes I, Brenneman DE. VIP: molecular biology and neurobiological function. Mol Neurobiol. 1989;3:201–36.
4. Bodner M, Fridkin M, Gozes I. Coding sequences for vasoactive intestinal peptide and PHM-27 peptide are located on two adjacent exons in the human genome. Proc Natl Acad Sci U S A. 1985;82:3548–51.
5. Giladi E, Shani Y, Gozes I. The complete structure of the rat VIP gene. Mol Brain Res. 1990;7:261–7.
6. Gozes I, Schachter P, Shani Y, Giladi E. Vasoactive intestinal peptide gene expression from embryos to aging rats. Neuroendocrinology. 1988;47:27–31.
7. Gozes I, Shani Y, Rostene WH. Developmental expression of the VIP-gene in brain and intestine. Brain Res. 1987;388:137–48.

8. Gozes I, Glowa J, Brenneman DE, McCune SK, Lee E, Westphal H. Learning and sexual deficiencies in transgenic mice carrying a chimeric vasoactive intestinal peptide gene. J Mol Neurosci. 1993;4:185–93.
9. Gozes I, Meltzer E, Rubinrout S, Brenneman DE, Fridkin M. Vasoactive intestinal peptide potentiates sexual behavior: inhibition by novel antagonist. Endocrinology. 1989;125:2945–9.
10. Glowa JR, Panlilio LV, Brenneman DE, Gozes I, Fridkin M, Hill JM. Learning impairment following intracerebral administration of the HIV envelope protein gp120 or a VIP antagonist. Brain Res. 1992;570:49–53.
11. Gozes I, Lilling G, Glazer R, Ticher A, Ashkenazi IE, Davidson A, et al. Superactive lipophilic peptides discriminate multiple vasoactive intestinal peptide receptors. J Pharmacol Exp Ther. 1995;273:161–7.
12. Hill JM, Hauser JM, Sheppard LM, Abebe D, Spivak-Pohis I, Kushnir M, et al. Blockage of VIP during mouse embryogenesis modifies adult behavior and results in permanent changes in brain chemistry. J Mol Neurosci. 2007;31:183–200.
13. Gressens P, Hill JM, Gozes I, Fridkin M, Brenneman DE. Growth factor function of vasoactive intestinal peptide in whole cultured mouse embryos. Nature. 1993;362:155–8.
14. Gozes I, McCune SK, Jacobson L, Warren D, Moody TW, Fridkin M, et al. An antagonist to vasoactive intestinal peptide affects cellular functions in the central nervous system. J Pharmacol Exp Ther. 1991;257:959–66.
15. Moody TW, Zia F, Draoui M, Brenneman DE, Fridkin M, Davidson A, et al. A vasoactive intestinal peptide antagonist inhibits non-small cell lung cancer growth. Proc Natl Acad Sci U S A. 1993;90:4345–9.
16. Gozes I, Bardea A, Reshef A, Zamostiano R, Zhukovsky S, Rubinraut S, et al. Neuroprotective strategy for Alzheimer disease: intranasal administration of a fatty neuropeptide. Proc Natl Acad Sci U S A. 1996;93:427–32.
17. Gozes I, Fridkin M. A fatty neuropeptide. Potential drug for noninvasive impotence treatment in a rat model. J Clin Invest. 1992;90:810–4.
18. Gozes I, Perl O, Giladi E, Davidson A, Ashur-Fabian O, Rubinraut S, et al. Mapping the active site in vasoactive intestinal peptide to a core of four amino acids: neuroprotective drug design. Proc Natl Acad Sci U S A. 1999;96:4143–8.
19. Jiang YH, Yuen RK, Jin X, Wang M, Chen N, Wu X, et al. Detection of clinically relevant genetic variants in autism spectrum disorder by whole-genome sequencing. Am J Hum Genet. 2013;93:249–63.
20. Giladi E, Hill JM, Dresner E, Stack CM, Gozes I. Vasoactive intestinal peptide (VIP) regulates activity-dependent neuroprotective protein (ADNP) expression in vivo. J Mol Neurosci. 2007;33:278–83.
21. Delgado M, Munoz-Elias EJ, Kan Y, Gozes I, Fridkin M, Brenneman DE, et al. Vasoactive intestinal peptide and pituitary adenylate cyclase-activating polypeptide inhibit tumor necrosis factor alpha transcriptional activation by regulating nuclear factor-kB and cAMP response element-binding protein/c-Jun. J Biol Chem. 1998;273:31427–36.
22. Harmar AJ, Arimura A, Gozes I, Journot L, Laburthe M, Pisegna JR, et al. International Union of Pharmacology. XVIII. Nomenclature of receptors for vasoactive intestinal peptide and pituitary adenylate cyclase-activating polypeptide. Pharmacol Rev. 1998;50:265–70.
23. Harmar AJ, Fahrenkrug J, Gozes I, Laburthe M, May V, Pisegna JR, et al. Pharmacology and functions of receptors for vasoactive intestinal peptide and pituitary adenylate cyclase-activating polypeptide: IUPHAR review 1. Br J Pharmacol. 2012;166:4–17.
24. Ago Y, Condro MC, Tan YV, Ghiani CA, Colwell CS, Cushman JD, et al. Reductions in synaptic proteins and selective alteration of prepulse inhibition in male C57BL/6 mice after postnatal administration of a VIP receptor (VIPR2) agonist. Psychopharmacology (Berl). 2015;232:2181–9.
25. Beri S, Bonaglia MC, Giorda R. Low-copy repeats at the human VIPR2 gene predispose to recurrent and nonrecurrent rearrangements. Eur J Hum Genet. 2013;21:757–61.
26. Nieratschker V, Meyer-Lindenberg A, Witt SH. Genome-wide investigation of rare structural variants identifies VIPR2 as a new candidate gene for schizophrenia. Expert Rev Neurother. 2011;11:937–41.

27. Vacic V, McCarthy S, Malhotra D, Murray F, Chou HH, Peoples A, et al. Duplications of the neuropeptide receptor gene VIPR2 confer significant risk for schizophrenia. Nature. 2011;471:499–503.
28. Brenneman DE, Eiden LE. Vasoactive intestinal peptide and electrical activity influence neuronal survival. Proc Natl Acad Sci U S A. 1986;83:1159–62.
29. Brenneman DE, Nicol T, Warren D, Bowers LM. Vasoactive intestinal peptide: a neurotrophic releasing agent and an astroglial mitogen. J Neurosci Res. 1990;25:386–94.
30. Brenneman DE, Gozes I. A femtomolar-acting neuroprotective peptide. J Clin Invest. 1996;97:2299–307.
31. Brenneman DE, Hauser J, Phillips TM, Davidson A, Bassan M, Gozes I. Vasoactive intestinal peptide. Link between electrical activity and glia-mediated neurotrophism. Ann N Y Acad Sci. 1999;897:17–26.
32. Gozes I, Brenneman DE. Activity-dependent neurotrophic factor (ADNF). An extracellular neuroprotective chaperonin? J Mol Neurosci. 1996;7:235–44.
33. Gozes I, Davidson A, Gozes Y, Mascolo R, Barth R, Warren D, et al. Antiserum to activity-dependent neurotrophic factor produces neuronal cell death in CNS cultures: immunological and biological specificity. Dev Brain Res. 1997;99:167–75.
34. Bassan M, Zamostiano R, Davidson A, Pinhasov A, Giladi E, Perl O, et al. Complete sequence of a novel protein containing a femtomolar-activity-dependent neuroprotective peptide. J Neurochem. 1999;72:1283–93.
35. Steingart RA, Gozes I. Recombinant activity-dependent neuroprotective protein protects cells against oxidative stress. Mol Cell Endocrinol. 2006;252:148–53.
36. Zamostiano R, Pinhasov A, Gelber E, Steingart RA, Seroussi E, Giladi E, et al. Cloning and characterization of the human activity-dependent neuroprotective protein. J Biol Chem. 2011;276:708–14.
37. Blondel O, Collin C, McCarran WJ, Zhu S, Zamostiano R, Gozes I, et al. A glia-derived signal regulating neuronal differentiation. J Neurosci. 2000;20:8012–20.
38. Idan-Feldman A, Schirer Y, Polyzoidou E, Touloumi O, Lagoudaki R, Grigoriadis NC, et al. Davunetide (NAP) as a preventative treatment for central nervous system complications in a diabetes rat model. Neurobiol Dis. 2011;44:327–39.
39. Oz S, Kapitansky O, Ivashco-Pachima Y, Malishkevich A, Giladi E, Skalka N, et al. The NAP motif of activity-dependent neuroprotective protein (ADNP) regulates dendritic spines through microtubule end binding proteins. Mol Psychiatry. 2014;19:1115–24.
40. Zusev M, Gozes I. Differential regulation of activity-dependent neuroprotective protein in rat astrocytes by VIP and PACAP. Regul Pept. 2004;123:33–41.
41. Gozes I. The cytoskeleton as a drug target for neuroprotection: the case of the autism-mutated ADNP. Biol Chem. 2016;397(3):177–84.
42. Brenneman DE, Hauser J, Neale E, Rubinraut S, Fridkin M, Davidson A, et al. Activity-dependent neurotrophic factor: structure-activity relationships of femtomolar-acting peptides. J Pharmacol Exp Ther. 1998;285:619–27.
43. Holtser-Cochav M, Divinski I, Gozes I. Tubulin is the target binding site for NAP-related peptides: ADNF-9, D-NAP, and D-SAL. J Mol Neurosci. 2006;28:303–7.
44. Gozes I, Yeheskel A, Pasmanik-Chor M. Activity-dependent neuroprotective protein (ADNP): a case study for highly conserved chordata-specific genes shaping the brain and mutated in cancer. J Alzheimers Dis. 2015;45:57–73.
45. Furman S, Steingart RA, Mandel S, Hauser JM, Brenneman DE, Gozes I. Subcellular localization and secretion of activity-dependent neuroprotective protein in astrocytes. Neuron Glia Biol. 2004;1:193–9.
46. Mandel S, Spivak-Pohis I, Gozes I. ADNP differential nucleus/cytoplasm localization in neurons suggests multiple roles in neuronal differentiation and maintenance. J Mol Neurosci. 2008;35:127–41.
47. Dresner E, Malishkevich A, Arviv C, Leibman Barak S, Alon S, et al. Novel evolutionary-conserved role for the activity-dependent neuroprotective protein (ADNP) family that is important for erythropoiesis. J Biol Chem. 2012;287:40173–85.

48. Kushnir M, Dresner E, Mandel S, Gozes I. Silencing of the ADNP-family member, ADNP2, results in changes in cellular viability under oxidative stress. J Neurochem. 2008;105: 537–45.
49. Helsmoortel C, Vulto-van Silfhout AT, Coe BP, Vandeweyer G, Rooms L, van den Ende J, et al. A SWI/SNF-related autism syndrome caused by de novo mutations in ADNP. Nat Genet. 2014;46:380–4.
50. Nakamachi T, Li M, Shioda S, Arimura A. Signaling involved in pituitary adenylate cyclase-activating polypeptide-stimulated ADNP expression. Peptides. 2006;27:1859–64.
51. Nakamachi T, Ohtaki H, Yofu S, Dohi K, Watanabe J, Hayashi D, et al. Pituitary adenylate cyclase-activating polypeptide (PACAP) type 1 receptor (PAC1R) co-localizes with activity-dependent neuroprotective protein (ADNP) in the mouse brains. Regul Pept. 2008;145:88–95.
52. Nicot A, Otto T, Brabet P, DiCicco-Bloom EM. Altered social behavior in pituitary adenylate cyclase-activating polypeptide type I receptor-deficient mice. J Neurosci. 2004;24:8786–95.
53. Hashimoto H, Hashimoto R, Shintani N, Tanaka K, Yamamoto A, Hatanaka M, et al. Depression-like behavior in the forced swimming test in PACAP-deficient mice: amelioration by the atypical antipsychotic risperidone. J Neurochem. 2009;110:595–602.
54. Tanaka K, Shintani N, Hashimoto H, Kawagishi N, Ago Y, Matsuda T, et al. Psychostimulant-induced attenuation of hyperactivity and prepulse inhibition deficits in Adcyap1-deficient mice. J Neurosci. 2006;26:5091–7.
55. Hattori S, Takao K, Tanda K, Toyama K, Shintani N, Baba A, et al. Comprehensive behavioral analysis of pituitary adenylate cyclase-activating polypeptide (PACAP) knockout mice. Front Behav Neurosci. 2012;6:58.
56. Gaszner B, Kormos V, Kozicz T, Hashimoto H, Reglodi D, Helyes Z. The behavioral phenotype of pituitary adenylate-cyclase activating polypeptide-deficient mice in anxiety and depression tests is accompanied by blunted c-Fos expression in the bed nucleus of the stria terminalis, central projecting Edinger-Westphal nucleus, ventral lateral septum, and dorsal raphe nucleus. Neuroscience. 2012;202:283–99.
57. Hazama K, Hayata-Takano A, Uetsuki K, Kasai A, Encho N, Shintani N, et al. Increased behavioral and neuronal responses to a hallucinogenic drug in PACAP heterozygous mutant mice. PLoS One. 2014;9, e89153.
58. Deguil J, Chavant F, Lafay-Chebassier C, Perault-Pochat MC, Fauconneau B, Pain S. Neuroprotective effect of PACAP on translational control alteration and cognitive decline in MPTP parkinsonian mice. Neurotox Res. 2010;17:142–55.
59. Mandel S, Rechavi G, Gozes I. Activity-dependent neuroprotective protein (ADNP) differentially interacts with chromatin to regulate genes essential for embryogenesis. Dev Biol. 2007;303:814–24.
60. Mandel S, Gozes I. Activity-dependent neuroprotective protein constitutes a novel element in the SWI/SNF chromatin remodeling complex. J Biol Chem. 2007;282:34448–56.
61. Schirer Y, Malishkevich A, Ophir Y, Lewis J, Giladi E, Gozes I. Novel marker for the onset of frontotemporal dementia: early increase in activity-dependent neuroprotective protein (ADNP) in the face of Tau mutation. PLoS One. 2014;9, e87383.
62. Malishkevich A, Amram N, Hacohen-Kleiman G, Magen I, Giladi E, Gozes I. Activity-dependent neuroprotective protein (ADNP) exhibits striking sexual dichotomy impacting on autistic and Alzheimer's pathologies. Transl Psychiatry. 2015;5, e501.
63. Merenlender-Wagner A, Malishkevich A, Shemer Z, Udawela M, Gibbons A, Scarr E, et al. Autophagy has a key role in the pathophysiology of schizophrenia. Mol Psychiatry. 2015;20:126–32.
64. Merenlender-Wagner A, Shemer Z, Touloumi O, Lagoudaki R, Giladi E, Andrieux A, et al. New horizons in schizophrenia treatment: autophagy protection is coupled with behavioral improvements in a mouse model of schizophrenia. Autophagy. 2014;10:2324–32.
65. Dere E, Dahm L, Lu D, Hammerschmidt K, Ju A, Tantra M, et al. Heterozygous ambra1 deficiency in mice: a genetic trait with autism-like behavior restricted to the female gender. Front Behav Neurosci. 2014;8:181.

66. Gkogkas CG, Khoutorsky A, Cao R, Jafarnejad SM, Prager-Khoutorsky M, Giannakas N, et al. Pharmacogenetic inhibition of eIF4E-dependent Mmp9 mRNA translation reverses fragile X syndrome-like phenotypes. Cell Rep. 2014;9:1742–55.
67. Gkogkas CG, Khoutorsky A, Ran I, Rampakakis E, Nevarko T, Weatherill DB, et al. Autism-related deficits via dysregulated eIF4E-dependent translational control. Nature. 2013;493:371–7.
68. Neves-Pereira M, Muller B, Massie D, Williams JH, O'Brien PC, Hughes A, et al. Deregulation of EIF4E: a novel mechanism for autism. J Med Gen. 2009;46:759–65.
69. Waltes R, Gfesser J, Haslinger D, Schneider-Momm K, Biscaldi M, Voran A, et al. Common EIF4E variants modulate risk for autism spectrum disorders in the high-functioning range. J Neural Transm. 2014;121:1107–16.
70. Mosch K, Franz H, Soeroes S, Singh PB, Fischle W. HP1 recruits activity-dependent neuroprotective protein to H3K9me3 marked pericentromeric heterochromatin for silencing of major satellite repeats. PLoS One. 2011;6, e15894.
71. Iossifov I, Ronemus M, Levy D, Wang Z, Hakker I, Rosenbaum J, et al. De novo gene disruptions in children on the autistic spectrum. Neuron. 2012;74:285–99.
72. Gulsuner S, Walsh T, Watts AC, Lee MK, Thornton AM, Casadei S, et al. Spatial and temporal mapping of de novo mutations in schizophrenia to a fetal prefrontal cortical network. Cell. 2013;154:518–29.
73. Gozes I, Werner H, Fawzi M, Abdelatty A, Shani Y, Fridkin M, et al. Estrogen regulation of vasoactive intestinal peptide mRNA in rat hypothalamus. J Mol Neurosci. 1989;1:55–61.
74. Furman S, Hill JM, Vulih I, Zaltzman R, Hauser JM, Brenneman DE, et al. Sexual dimorphism of activity-dependent neuroprotective protein in the mouse arcuate nucleus. Neurosci Lett. 2005;373:73–8.
75. Schaafsma SM, Pfaff DW. Etiologies underlying sex differences in Autism spectrum disorders. Front Neuroendocrinol. 2014;35:255–71.
76. Kleiman GH, Barnea A, Gozes I. ADNP: a major autism mutated gene is differentially distributed (age and gender) in the songbird brain. Peptides. 2015;72:75–9.
77. Borozdin W, Graham Jr JM, Bohm D, Bamshad MJ, Spranger S, Burke L, et al. Multigene deletions on chromosome 20q13.13-q13.2 including SALL4 result in an expanded phenotype of Okihiro syndrome plus developmental delay. Hum Mutat. 2007;28:830.
78. O'Roak BJ, Vives L, Girirajan S, Karakoc E, Krumm N, Coe BP, et al. Sporadic autism exomes reveal a highly interconnected protein network of de novo mutations. Nature. 2012;485:246–50.
79. O'Roak BJ, Vives L, Fu W, Egertson JD, Stanaway IB, Phelps IG, et al. Multiplex targeted sequencing identifies recurrently mutated genes in autism spectrum disorders. Science. 2012;338:1619–22.
80. Pescosolido MF, Schwede M, Johnson Harrison A, Schmidt M, Gamsiz ED, et al. Expansion of the clinical phenotype associated with mutations in activity-dependent neuroprotective protein. J Med Gen. 2014;51:587–9.
81. Vandeweyer G, Helsmoortel C, Van Dijck A, Vulto-van Silfhout AT, Coe BP, Bernier R, et al. The transcriptional regulator ADNP links the BAF (SWI/SNF) complexes with autism. Am J Med Gen C Sem Med Gen. 2014;166C:315–26.
82. Coe BP, Witherspoon K, Rosenfeld JA, van Bon BW, Vulto-van Silfhout AT, Bosco P, et al. Refining analyses of copy number variation identifies specific genes associated with developmental delay. Nat Genet. 2004;46:1063–71.
83. De Rubeis S, He X, Goldberg AP, Poultney CS, Samocha K, Ercument Cicek A, et al. Synaptic, transcriptional and chromatin genes disrupted in autism. Nature. 2014;515:209–15.
84. Deciphering Developmental Disorders S. Large-scale discovery of novel genetic causes of developmental disorders. Nature. 2015;519:223–8.
85. Pinhasov A, Mandel S, Torchinsky A, Giladi E, Pittel Z, Goldsweig AM, et al. Activity-dependent neuroprotective protein: a novel gene essential for brain formation. Dev Brain Res. 2003;144:83–90.

86. Dresner E, Agam G, Gozes I. Activity-dependent neuroprotective protein (ADNP) expression level is correlated with the expression of the sister protein ADNP2: deregulation in schizophrenia. Eur Neuropsychopharmacol. 2011;21:355–61.
87. Vulih-Shultzman I, Pinhasov A, Mandel S, Grigoriadis N, Touloumi O, Pittel Z, et al. Activity-dependent neuroprotective protein snippet NAP reduces tau hyperphosphorylation and enhances learning in a novel transgenic mouse model. J Pharmacol Exp Ther. 2007;323: 438–49.
88. Chen SY, Charness ME, Wilkemeyer MF, Sulik KK. Peptide-mediated protection from ethanol-induced neural tube defects. Dev Neurosci. 2005;27:13–9.
89. Spong CY, Abebe DT, Gozes I, Brenneman DE, Hill JM. Prevention of fetal demise and growth restriction in a mouse model of fetal alcohol syndrome. J Pharmacol Exp Ther. 2001;297:774–9.
90. Rotstein M, Bassan H, Kariv N, Speiser Z, Harel S, Gozes I. NAP enhances neurodevelopment of newborn apolipoprotein E-deficient mice subjected to hypoxia. J Pharmacol Exp Ther. 2006;319:332–9.
91. Incerti M, Toso L, Vink J, Roberson R, Nold C, Abebe D, et al. Prevention of learning deficit in a Down syndrome model. Obstet Gynecol. 2011;117:354–61.
92. Sigalov E, Fridkin M, Brenneman DE, Gozes I. VIP-Related protection against iodoacetate toxicity in pheochromocytoma (PC12) cells: a model for ischemic/hypoxic injury. J Mol Neurosci. 2000;15:147–54.
93. Gozes I, Bachar M, Bardea A, Davidson A, Rubinraut S, Fridkin M, et al. Protection against developmental retardation in apolipoprotein E-deficient mice by a fatty neuropeptide: implications for early treatment of Alzheimer's disease. J Neurobiol. 1997;33:329–42.
94. Merenlender-Wagner A, Pikman R, Giladi E, Andrieux A, Gozes I. NAP (davunetide) enhances cognitive behavior in the STOP heterozygous mouse—a microtubule-deficient model of schizophrenia. Peptides. 2010;31:1368–73.
95. Javitt DC, Buchanan RW, Keefe RS, Kern R, McMahon RP, Green MF, et al. Effect of the neuroprotective peptide davunetide (AL-108) on cognition and functional capacity in schizophrenia. Schizophr Res. 2012;136:25–31.
96. Jarskog LF, Dong Z, Kangarlu A, Colibazzi T, Girgis RR, Kegeles LS, et al. Effects of davunetide on N-acetylaspartate and choline in dorsolateral prefrontal cortex in patients with schizophrenia. Neuropsychopharmacology. 2013;38:1245–52.
97. Vaisburd S, Shemer Z, Yeheskel A2 Giladi E, Gozes I. Risperidone and NAP protect cognition and normalize gene expression in a schizophrenia mouse model. Sci Rep. 2015 Nov 10;5: 16300.

Part XIV
Clinical Relations of PACAP

Chapter 47
PACAP and Cancer

Terry W. Moody and Robert T. Jensen

Abstract The ADCYAP1 gene, which is expressed in certain cancer cells, encodes for pituitary adenylate cyclase activating polypeptide (PACAP). The translated protein products for the ADCYAP1 gene include prepro PACAP (176 amino acids) as well as the biologically active PACAP-27 or PACAP-38. When secreted from cancer cells, PACAP-27 or PACAP-38 binds with high affinity to VPAC1, VPAC2, and PAC1. The activated PAC1 elevates cAMP and causes phosphatidylinositol turnover. PAC1 regulates the phosphorylation of numerous proteins in lung cancer including protein kinase C, paxillin, focal adhesion kinase, Src, the epidermal growth factor receptor, and extracellular signal-regulated kinase (ERK). The phosphorylated ERK enters the nucleus and increases the expression of the nuclear oncogenes c-fos and c-jun. The c-fos and c-jun form heterodimers altering the expression of growth factor genes. PACAP-27 or PACAP-38 stimulates the clonal growth of cancer cells. A PAC1 antagonist is PACAP(6–38), which inhibits the ability of PACAP-27 or PACAP-38 to increase protein phosphorylation, nuclear oncogene expression, and the proliferation of cancer cells.

Keywords PACAP • Lung cancer • Proliferation • PAC1 • PACAP(6–38) • Signal transduction • Tyrosine phosphorylation • EGFR • ERK • Transactivation

Abbreviations

AP	Activator protein
BB	Bombesin
DDC	Dopa decarboxylase

T.W. Moody (✉)
Department of Health and Human Services, National Cancer Institute,
Center for Cancer Research, Office of the Director, 9609 Medical Center Drive,
Room 2W-130, Bethesda, MD 20892, USA
e-mail: moodyt@mail.nih.gov

R.T. Jensen
Digestive Diseases Branch, National Institute of Diabetes, Digestive and Kidney Disease,
Bethesda, MD 20892, USA

D. Reglodi, A. Tamas (eds.), *Pituitary Adenylate Cyclase Activating Polypeptide — PACAP*, Current Topics in Neurotoxicity 11,
DOI 10.1007/978-3-319-35135-3_47

EC	Extracellular
EGFR	Epidermal growth factor receptor
ERK	Extracellular signal-regulated kinase
FAK	Focal adhesion kinase
GPCR	G Protein coupled receptor
IC	Intracellular
MEK	Mitogen/extracellular signal-regulated kinase
MMP	Matrix metalloprotease
NSE	Neuron specific enolase
PACAP	Pituitary adenylate cyclase activating polypeptide
PI	Phosphatidyl inositol
PK	Protein kinase
PL	Phospholipase
PRP	PACAP related peptide
PYK	Proline-rich tyrosine kinase
SCLC	Small cell lung cancer
SV	Splice variant
TGF	Transforming growth factor
TKI	Tyrosine kinase inhibitor
TM	Transmembrane
VEGF	Vascular endothelial growth factor
VIP	Vasoactive intestinal peptide

Introduction

Pituitary adenylate cyclase-activating polypeptide (PACAP)-27 has sequence homology (67 %) with vasoactive intestinal peptide (VIP) [1]. PACAP-27, PACAP related peptide (PRP), and PACAP-38 are derived from a 176-amino acid precursor protein preproPACAP [2]. PACAP-38 contains PACAP-27 plus an addition 11 amino acids at the C-terminal [3, 4]. PACAP-38 and PACAP-27 were isolated from ovine hypothalamus and elevate cAMP in rat pituitary cells in culture. The PACAP-38 amino acid sequence is highly conserved and is identical in mammals [5]. In cellular extracts approximately an order of magnitude more PACAP-38 is detected than PRP or PACAP-27. PACAP-38 and PACAP-27 have approximately an order of magnitude more biological activity than does PRP [5].

PACAP-27 and PACAP-38 bind with high affinity to G-protein coupled receptors (GPCR; VPAC1, VPAC2, and PAC1) which are members of the class II or class B secretin-like receptors [6]. The activated VPAC1, VPAC2, or PAC1 interact with a stimulatory guanine nucleotide binding protein (Gs) increasing adenylylcyclase activity resulting in elevated cellular cAMP [7]. The increased cAMP activates protein kinase (PK) A causing phosphorylation of various proteins such as CREB leading to altered gene expression [8]. In addition, PAC1 interacts with Gq causing phospholipase (PL) C activation [9]. The PLC metabolizes phosphatidyl inositol

(PI) 4,5 bisphosphate to inositol-1,4,5-trisphosphate and diacylglycerol which increase cytosolic calcium and PKC activity, respectively. PACAP-27 addition to cancer cells increases the phosphorylation of numerous proteins in lung cancer including PKC, paxillin, focal adhesion kinase (FAK), Src, the epidermal growth factor receptor (EGFR), and extracellular signal-regulated kinase (ERK) [10]. The phosphorylated ERK enters the nucleus and increases the expression of the nuclear oncogenes c-fos and c-jun [11, 12]. The c-fos and c-jun form heterodimers altering the expression of growth factor genes. PACAP-27 or PACAP-38 stimulates the growth of brain cancer [13], colon cancer [14], neuroendocrine tumors [15, 16], pancreatic cancer [17] and SCLC [18]. A PAC1 antagonist, PACAP(6–38), inhibits the growth of lung, breast and prostate cancer cells [19–22]. PAC1 has been detected in a wide variety of tumors including brain, breast, colon, lung, neuroendocrine, pancreatic, pituitary, and prostate cancer [23–25]. In this chapter, the role of PACAP in cancer is presented.

Lung Cancer

PACAP-38 has been detected in tumors of the colon, lung, and prostate [8, 26, 27]. PACAP binds with high affinity to small cell lung cancer (SCLC) cell lines [18]. SCLC is a neuroendocrine tumor which kills approximately 25,000 United States (US) citizens annually. Patients with SCLC are traditionally treated with chemotherapy and/or radiation therapy. While the therapy is initially effective, relapse occurs and the median survival time is less than 1 year [28]. SCLC contains high levels of dopa decarboxylase (DDC), neuron specific enolase (NSE), and bombesin (BB)-like peptides [29]. Specific binding of ^{125}I-PACAP-27 to NCI-H345 or NCI-N417 SCLC cells was inhibited with high affinity by PACAP-27 or PACAP-38 and the IC_{50} values were 5 and 10 nM respectively [18]. In contrast, PACAP(6–38), PACAPhybrid and VIP bind to PAC1 with IC_{50} values of 1500, 750 and 500 nM, respectively (Table 47.1). The results indicate that PACAP-27 binds to PAC1 with approximately 2-orders of magnitude greater affinity than does VIP. PCR products for VIP, PACAP, VPAC1, and PAC1 were present in six of eight lung cancer cell lines tested, however, only two out of eight cell lines had VPAC2 [30]. PACAP-38 immunoreactivity is higher in the human lung cancer than normal lung biopsy specimens [27].

A general problem is that the VPAC1 density (100,000/cell) is much greater than that of PAC1 (14,000/cell) in lung cancer and PACAP binds with high affinity to PAC1, VPAC1 and VPAC2. The 61-amino acid maxadilan is selective for PAC1; however, it has little sequence homology to PACAP [31]. Recently 46 PACAP analogs were synthesized and evaluated for selectivity for PAC1 relative to VPAC1 and VPAC2 [32]. Fifteen of the 46 analogs tested had selectivity for PAC1 relative to VPAC1. (Iac^{1}, Ala^{16,17}, DLys^{38})PACAP-38 bound with higher affinity to PAC1 (IC_{50}=0.3 nM) relative to VPAC1 (IC_{50}=23 nM) or VPAC2 (IC_{50}=3 nM). (Iac^{1}, Ala^{16,17}, DLys^{38})PACAP-38 [(IAAD)PACAP-38] was an agonist which was more potent at elevating cAMP using cells containing PAC1 (EC_{50}=0.05 nM) relative to

Table 47.1 Binding of PACAP analogs

Peptide	IC_{50}, nM
PACAP-27	5 ± 1
PACAP-38	10 ± 1
PACAP(6–38)	1500 ± 170
PACAPhybrid	750 ± 90
VIP	500 ± 60
PACAP(16–38)	>10,000

The mean IC_{50} ± S.D. of three determinations to inhibit specific ^{125}I-PACAP-27 binding to SCLC cell line NCI-N417 is indicated. The peptide structures are shown below. Sequence homologies relative to PACAP-27 are in italics

PACAP-27: *His-Ser-Asp-Gly-Ile-Phe-Thr-Asp-Ser-Tyr-Ser--Arg-Tyr-Arg-Lys-Gln-Met-Ala-Val-Lys-Lys-Tyr-Leu-Ala-Ala-Val-Leu*-NH_2

PACAP-38: *His-Ser-Asp-Gly-Ile-Phe-Thr-Asp-Ser-Tyr-Ser--Arg-Tyr-Arg-Lys-Gln-Met-Ala-Val-Lys-Lys-Tyr-Leu-Ala-Ala-Val-Leu*-Gly-Lys-Arg-Tyr-Lys-Gln-Arg-Val-Lys-Asn-Lys-NH_2

PACAP-hybrid: Lys-Pro-Arg-Arg-Pro-Tyr-Thr-*Asp-Ser-Tyr-Ser-Arg-Tyr-Thr-Arg-Lys-Gln-Met-Ala-Val-Lys-Lys-Tyr-Leu-Ala-Ala-Val-Leu*-Gly-Lys-Arg-Tyr-Lys-Gln-Arg-Val-Lys-Asn-Lys-NH_2

VIP: *His-Ser-Asp*-Ala-Val-*Phe-Thr-Asp*-Asn-*Tyr*-Thr-*Arg*-Leu-*Arg-Lys-Gln-Met-Ala-Val-Lys-Lys*-Ala-*Leu*-Asn-Ser-Ile-*Leu*-Asn-NH_2

VPAC1 (EC_{50} = 1.5 nM) or VPAC2 (EC_{50} = 2.5 nM). The results indicate that (IAAD) PACAP-38 prefers PAC1 relative to VPAC1 by 50- to 80-fold and prefers PAC1 relative to VPAC2 by 10- to 50-fold. Previously it was found that PACAP-38 is metabolized by dipeptidyl peptidase IV [33]. Acetyl-[Ala^{15}, Ala^{20}]PACAP-38 propylamide and acetyl-PACAP-27-propylamide had improved biological activity and metabolic stability relative to PACAP-27 or PACAP-38. The results indicate that PACAP analogs are available which prefer PAC1 relative to VPAC1 or VPAC2.

PACAP-27 or PACAP-38 has a secondary structure in that it has a β-turn at amino acids 9–12 followed by an α-helix at amino acids 12–14, 15–20, and 22–24 [34, 35]. Also, an α-helix forms at amino acids 28–38 of PACAP-38. Site-directed mutagenesis studies indicate that several amino acids in the N-terminal of PAC1 such as Tyr^{138} and Glu^{137} are essential for high affinity PACAP binding [35]. Mutation of Val^{19}, Tyr^{22}, Val^{26}, or Leu^{27} in PACAP impaired high affinity binding to PAC1. Modeling studies suggest that the C-terminal of PACAP binds to the N-terminal of PAC1, whereas the N-terminal of PACAP binds to the PAC1 extracellular (EC) loops and transmembrane (TM) domains [36].

PAC1 is a GPCR with 467 amino acids (null; [37]) which has seven TM domains. Splice variants (SV) have been identified and PAC1 SV-1 or the hip receptor has an additional 28-amino acid insert in intracellular loop (IC) 3 [38]. PAC1 SV-2 or the hop receptor (HOP1) contains 495 amino acids and has a different 28-amino acid

Table 47.2 PACAP elevates cAMP in lung cancer cells

Addition	cAMP, fmol
None	7.0 ± 1.0***
PACAP-27, 1 nM	40.1 ± 6.0**
PACAP-38, 1 nM	37.8 ± 5.6**
PACAP-27 + 1 μM PACAP(6–38)	14.5 ± 2.5*,***
PACAP-27 + 1 μM PACAPhybrid	10.5 ± 2.7***
PACAP(16–38), 1 nM	6.8 ± 0.9***

The ability of peptides to alter cAMP was investigated in SCLC cell line NCI-H345. The mean value ± S.D. of three determinations each performed in duplicate is shown; *$p < 0.05$; **$p < 0.01$ relative to control; ***$p < 0.01$ relative to PACAP-27

insert in IC3. PAC1 SV-3 or the hip-hop receptor contains 523 amino acids and has both 28-amino acid inserts in IC3. All PAC1SVs interact with Gs and stimulate adenylyl cyclase. PAC1 SV-2 interacted strongly with Gq causing phosphatidylinositol (PI) turnover whereas PAC1 null, PAC1 SV-1 or PAC1 SV-3 have a weaker response [9]. C-Fos expression was strongly increased when PACAP-27 was added to cells containing PAC1 SV-2 relative to PAC1 null, PAC1 SV-1, or PAC1 SV-3 [39]. The Ca^{2+} response was maximal when PACAP-27 was added to cells containing PAC1 SV-2 relative to PAC1null, PAC1 SV-1 or PAC1 SV-3 [39] . The results indicate that PAC1 null, PAC1 SV-1, PAC1 SV-2, or PAC1 SV-3 interact with Gs, however, PAC1 SV-2 interacts better than PAC1 null, PAC1 SV-1, or PAC1 SV-3 with Gq.

When PAC1 activates Gs it stimulates adenylyl cyclase increasing the cAMP 5.5-fold (Table 47.2). Both PACAP-27 and PACAP-38 (1 nM) increased significantly the cAMP 5.5-fold after addition to lung cancer cells, however, PACAP(16–38) is inactive (Table 47.2). The increase in cAMP caused by PACAP-27 was antagonized significantly by 1 μM PACAPhybrid or PACAP(6–38) [20]. The results indicate that nM concentrations of PACAP-27 are required to elevate cAMP and that PACAP(6–38) or PACAPhybrid are PAC1 antagonists in cancer cells.

PACAP-27 or PACAP-38 (100 nM) increased the cytosolic Ca^{2+} in lung cancer cells. The increase in cytosolic Ca^{2+} caused by PACAP-27 was antagonized by 1 μM PACAPhybrid or PACAP(6–38) [20]. Figure 47.1 shows a dose–response curve for (Iac^1, $Ala^{16,17}$, $DLys^{38}$)PACAP-38 to elevate cytosolic Ca^{2+}. (Iac^1, $Ala^{16,17}$, $DLys^{38}$) PACAP-38 had little effect on NCI-H838 cytosolic Ca^{2+} at a 0.1 or 1 nM dose, however, it moderately and strongly increased cytosolic Ca^{2+} at a 10 or 100 nM concentration, respectively. The results indicate that 10 nM doses of PACAP-27, PACAP-38, and (Iac^1, $Ala^{16,17}$, $DLys^{38}$)PACAP-38 are needed to elevate the cytosolic Ca^{2+} in lung cancer cells and whereas lower doses are needed to elevate cAMP [32].

PACAP has additional signal transduction mechanisms. Addition of 10 nM PACAP-27 to non-SCLC (NSCLC) cell line NCI-H1299 cells increased ERK tyrosine phosphorylation [12]. The increase in ERK tyrosine phosphorylation caused by PACAP was reversed by PACAP(6–38) or PD98059, a MEK inhibitor. Phosphorylated ERK can enter the nucleus and alter gene expression. PACAP-27 or

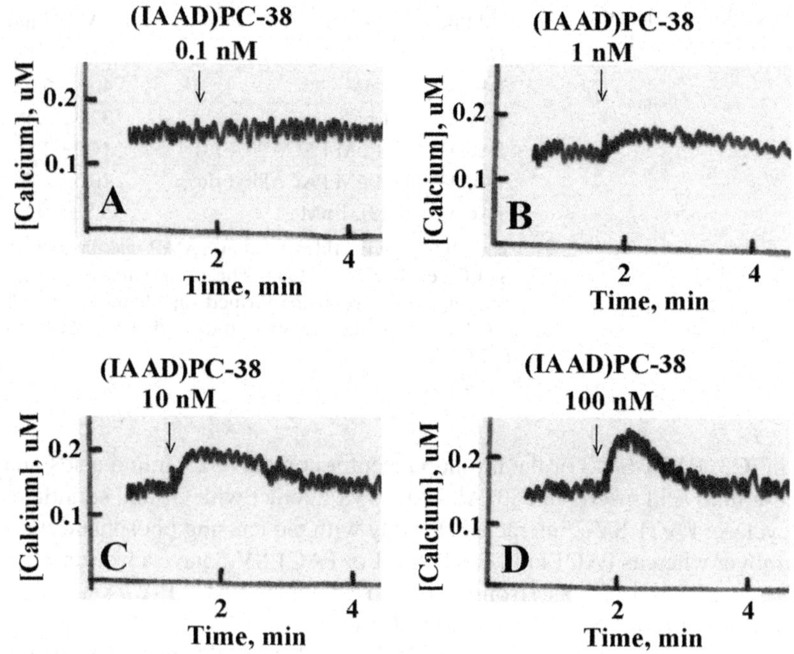

Fig. 47.1 Cytosolic Ca^{2+}. The ability of 0.1 (**a**), 1 (**b**), 10 (**c**), or 100 nM (**d**) (Iac^{1}, $Ala^{16,17}$, $DLys^{38}$) PACAP-38 [(IAAD)PC-38] to elevate cytosolic Ca^{2+} in Fura-2AM loaded NCI-H838 cells was investigated as a function of time. This experiment is representative of three others

PACAP-38 increased c-fos mRNA in NCI-N417 cells after 4 h [11]. The increase in c-fos expression caused by PACAP was reversed by PACAP(6–38) or H7, a PKC inhibitor. C-fos and c-jun can form heterodimers and increase expression of genes with AP-1 sites. PACAP-27 (100 nM) addition to NCI-H1299 cells increased expression of vascular endothelial growth factor (VEGF) after 8 h [12]. The increase in VEGF mRNA caused by PACAP-27 was inhibited by PACAP(6–38) or PD98059. VEGF is an angiogenic factor which increases the vascularization of tumors.

PACAP-27 (10 nM) stimulated the clonal growth of lung cancer cells significantly to 178 % (Table 47.3). The increase in clonal growth stimulated by PACAP-27 was inhibited significantly by 1 μM PACAP(6–38). PACAP(6–38) reduced basal colony number to 63 %. Injection of 0.4 mg/kg of PACAP(6–38) into nude mice reduced NCI-H838 xenograft growth significantly to 67 %. The results suggest that PAC1 regulates the growth of lung cancer cells [19].

PACAP addition to NSCLC cells causes phosphorylation of numerous proteins. PACAP-27 addition to lung cancer cells causes tyrosine phosphorylation of focal adhesion kinase (FAK) and paxillin [40]. Tyrosine phosphorylation of FAK and paxillin are associated with reorganization of the actin cytoskeleton and assembly of focal adhesion plaques [41]. The ability of PACAP to increase FAK phosphorylation was reversed by PP2 (Src inhibitor), U73122 (PLC inhibitor), or PACAP(6–38) [40]. PACAP-27 addition to lung cancer cells caused tyrosine phosphorylation of

Table 47.3 Effect of ligands on lung cancer proliferation

Addition	Relative % proliferation	
	In vitro	In vivo
None	100±13***	100±9
PACAP-27	178±21*	n.d.
PACAP(6–38)	63±4*,***	67±5*
PACAP-27+PACAP(6–38)	113±14***	n.d.
Gefitinib	51±9***	n.d.
PACAP(6–38)+Gefitinib	18±3***	n.d.

The mean in vitro value±S.D. of three determinations each repeated in triplicate is shown; ** $p<0.01$; * $p<0.05$ relative to control; *** $p<0.01$ relative to 10 nM PACAP-27. NSCLC NCI-H838 cells were treated with 1 μM PACAP(6–38) and/or 1 μg/ml gefitinib. The mean value±S.D. of six mice bearing NCI-H838 xenografts is indicated; * $p<0.05$ relative to control using 0.4 mg/kg PACAP(6–38) subcutaneously; *n.d.* not determined

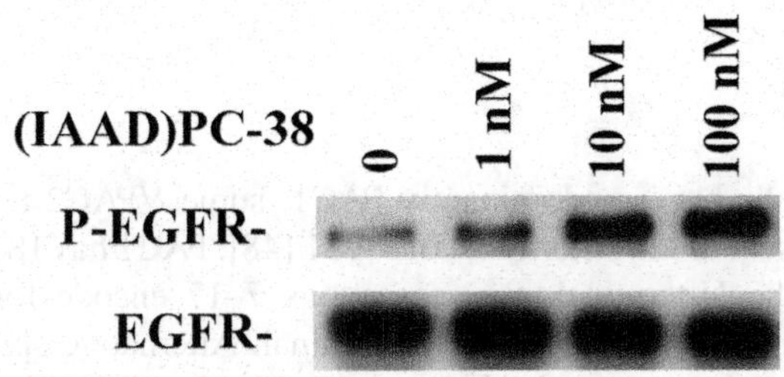

Fig. 47.2 EGFR transactivation. (*Top*) The ability of 1, 10, or 100 nM (IAAD)PC-38 to cause EGFR tyrosine phosphorylation was investigated 2 min after the addition of ligand to NCI-H838 cells. (*Bottom*) (IAAD)PC38 had no effect on total EGFR. This experiment is representative of two others

proline-rich tyrosine kinase (PYK)2 [42]. PYK-2 interacts with paxillin which coordinates Rho family GTPases regulating the actin skeleton [43]. The ability of PACAP to increase PYK-2 phosphorylation was reversed by PP2, PACAP(6–38), U73122 or BAPTA, a Ca^{2+} chelator [42]. Finally PACAP-27 addition to lung cancer cells caused tyrosine phosphorylation of the EGFR. The ability of PACAP to cause EGFR transactivation was inhibited by PP2, PACAP(6–38) or gefitinib, an EGFR tyrosine kinase inhibitor (TKI) [44].

(IAAD)PACAP-38 addition to NSCLC cells increased EGFR tyrosine phosphorylation in a concentration-dependent manner (Fig. 47.2). Addition of 1, 10, or 100 nM (IAAD)PACAP-38 to NCI-H838 cells weakly, moderately and strongly, respectively increased EGFR tyrosine phosphorylation (Fig. 47.2, top) whereas total EGFR was unaffected (Fig. 47.2, bottom). The rapid transactivation of the EGFR caused by PACAP addition to lung cancer cells is matrix metalloprotease (MMP) dependent [10, 44]. Addition of the MMP inhibitor GM6001 to lung cancer cells inhibited the PAC1 regulation of EGFR transactivation [44]. Also, PACAP increased secretion of the EGFR ligand transforming growth factor (TGF) α from lung cancer cells that was inhibited significantly by GM6001. MMP may metabolize inactive proTGFα in lung cancer cells to biologically active TGFα. The TGFα binds with high affinity to the EGFR causing phosphorylated homodimers to form in lung cancer cells [45]. The results suggest that PACAP may stimulate lung cancer growth in an EGFR-dependent manner.

Gefitinib or erlotinib are EGFR TKI that are used to treat NSCLC patients which have certain EGFR mutations [46, 47]. Approximately 13 % of the lung cancer patients who respond to EGFR TKI have activating mutations. A goal is to increase the sensitivity of gefitinib in patients with wild type EGFR. Table 47.3 shows that gefitinib or PACAP(6–38) moderately inhibits the growth of NCI-H838 cells. Addition of both PACAP(6–38) and gefitinib strongly inhibits lung cancer growth suggesting that PACAP(6–38) is synergistic with gefitinib at inhibiting lung cancer growth in vitro. It remains to be determined if PACAP(6–38) is synergistic with gefitinib at inhibiting lung cancer growth in vivo.

Brain Tumors

Neuroblastoma cells have predominantly PAC1, some VPAC2 and little VPAC1. PAC1SVs are present in the N-terminal of PAC1 [48]. PAC1 has 18 exons and exons 1–6 encode for the N-terminal, whereas exons 7–17 encode for the seven TM domains and exon 18 encodes for the C-terminal. Alternative splicing of exon 5, exons 5, 6 or exons 4–6 result in deletions of 7, 21 (short) or 57 amino acids (very short) in the N-terminal of PAC1 [49]. Addition of PACAP-38 to neuroblastoma cell lines SH-SY5Y, SK-N-Be or NB100 increased the cAMP with EC_{50} values ranging from 0.05 to 1 nM [48]. The PAC1SVs which predominated in SH-SY5Y cells were the δ5,6null; δ5,6hop; and δ4,5,6null. Cells which expressed PAC1null, PAC1δ5null, PAC1δ5,6null but not PAC1δ4,5,6null bound ^{125}I-PACAP-27 with high affinity [50]. These results indicate that the very short PAC1 is missing essential amino acids from the N-terminal of PAC1 that are critical to interact with the C-terminal of PACAP.

The ability of PAC1 SV to elevate cAMP and cause PI turnover was investigated. PACAP-38 was more potent at elevating cAMP in cells containing PAC1null or PAC1hop than PAC1hip. Similarly PACAP-38 was more potent at elevating cAMP in PAC1δ5null or PAC1δ5hop than PAC1δ5hip [51]. Also PACAP-38 was more potent at elevating cAMP in cells containing PAC1δ5,6null or PAC1δ5,6hop than PAC1δ5,6hip. The results indicate that the PAC1SV1 does not stimulate adenylylcyclase as well as PAC1null or PAC1SV2 regardless of whether the PAC1δ5 or PAC1δ5,6 is present. Previously it was found that PAC1hop was better than PAC1null or PAC1hip at causing PI turnover [9]. PACAP-38 was more potent at causing PI turnover in cells containing PAC1δ5hop than PAC1δ5null whereas PAC1δ5hip was ineffective. Also, PACAP-38 was more potent at causing PI turnover in cells containing PAC1δ5,6hop than PAC1δ5,6null whereas PAC1δ5,6hip was ineffective. The results indicate that PAC1hop strongly activates PLCβ, PAC1null moderately activates PLCβ whereas PAC1hip weakly activates PLCβ regardless of whether PAC1δ5 or PAC1δ5,6 is present. The results indicate that PAC1 IC3 regulates adenylylcyclase and/or PLC activity.

The signal transduction mechanisms of PAC1 SV were investigated in NG108-15 cells (mouse neuroblastoma × rat glioma hybrid). PACAP-38, 1 nM, strongly increased cAMP in cells transfected with PAC1null, PAC1hip, and PAC1hop. PACAP-38, 100 nM, strongly increased cytosolic Ca^{2+} in cells transfected with PAC1hop and PAC1null but not PAC1hip [52]. All PAC1SVs treated with PACAP-38 increased VIP gene expression and this expression was a function of cAMP/PKA activation [53]. All PAC1SVs treated with PACAP-38 increased stanniocalcin (STC1) expression which was dependent on ERK but not PKA activation [52]. The results indicate that the mechanism by which PAC1 alters gene expression is a function of the gene product.

PACAP has a biphasic dose–response curve on neuroblastoma cells with stimulation of proliferation at low nM concentrations and inhibition of proliferation at high μM concentrations [54]. Stimulation of proliferation was mediated by the MAP kinase pathway whereas inhibition of proliferation was mediated by PKA. PAC1 and PACAP mRNA were detected in numerous neuroblastoma tumors and cell lines [55]. PACAP inhibited cell growth and induced morphological differentiation of neuroblastoma cells in a cAMP dependent manner [56, 57]. It remains to be determined if the ability of PACAP to alter cancer growth varies as a function of PAC1SV status.

PAC1 is present in numerous brain tumors including astrocytomas, ependymomas, glioblastomas, and neuroblastomas [55, 58]. PAC1 mRNA is present in U87, U138, and U373 human glioblastoma cell lines [59]. Specific ^{125}I-PACAP-27 binding was inhibited with high affinity by PACAP-27 but not VIP (IC_{50} values of 10 and 1500 nM, respectively). Addition of 30 nM PACAP-27 to U87 cells elevated the cAMP and cytosolic Ca^{2+}. Addition of 10 nM PACAP-27 stimulated the clonal growth of human U87, U118, or U373 cells [59]. Also PACAP stimulated the proliferation of rat C6 glioma cells [60]. PACAP bound with high affinity to 70 % of the astroglial neoplasms, ependymomas, and oligodendroglioma membranes examined [58] and elevated cAMP. When PACAP-27 or PACAP-38 is added to human glioblastoma T98G cells, which have VPAC2, proliferation is reduced [61]. VPAC2 regulates increases in cAMP but not cytosolic Ca^{2+}.

Disruption of a single copy of the PACAP gene increased medulloblastoma incidence in ptc1 mutant mice 2.5-fold [62]. Ptc1 is a sonic hedgehog receptor gene which is overexpressed in humans and mice with medulloblastoma. PACAP reduced gli1 gene expression and proliferation of primary medulloblastoma tumor spheres in a PKA-dependent manner [63]. The results indicate that PAC1 is present in many brain cancers.

Neuroendocrine Tumors

PAC1 is present on many neuroendocrine tumors (NETs) and regulates prominent growth effects [15, 16, 64–66]. In BON cells, PAC1 is present and its activation increases cAMP, cytosolic Ca^{2+}, ERK tyrosine phosphorylation, EGFR tyrosine phosphorylation, release of serotonin, and growth [64–66]. Gene expression profiling indicated that BON cells have numerous neuroendocrine markers including NSE, DDC, substance P, somatostatin 2 receptors, and muscarinic acetylcholine receptor [67]. PACAP plays a critical role in hypergastrinemia resulting in the proliferation of enterochromaffin cells [15, 16]. PAC1 is important in regulating the proliferation of NET cells especially those secreting insulin [68].

PACAP causes PC12 cells, a rat pheochromocytoma, to decrease proliferation, increase survival, and induce neurite outgrowth [69, 70]. After activation, PAC1 is translocated into caveolae and stimulates neuritogenesis through a cAMP-Rap-1 dependent pathway [71, 72]. PACAP causes PC12 cellular differentiation through the ERK/MAPK cascade [73]. PACAP addition to PC12 adrenal pheochromocytoma cells increases the cAMP [74] and causes catecholamine secretion [75]. PAC1hop facilitated catecholamine secretion through 2-ABP-sensitive Ca^{2+} channels in PC12 cells [76].

PACAP addition to PC12 cells increases TrkA tyrosine phosphorylation [77, 78]. When PACAP was added to PC12 cells, Akt was phosphorylated in a TrkA-dependent manner. The TrkA and Akt phosphorylation caused by PACAP was reversed by the TKI K252a. Also, TrkA transactivation caused by PACAP was inhibited by PP1, an inhibitor of Src. Neuronal differentiation was stimulated by PACAP which activated Galpha/cAMP/exchange protein (Epac) in a cAMP-dependent manner utilizing Rit guanine nucleotide exchange factor [79]. The PACAP-mediated Rit activation involves the Src family kinase-dependent TrkA receptor transactivation. Addition of the Src inhibitor PP1 inhibited the ability of PACAP to cause Rit activation in a TrkA-dependent manner. The increase in TrkA and Akt tyrosine phosphorylation occurred 10 min after addition of PACAP-38 to PC12 cells. Further PACAP promotes survival and neuritogenesis in PC12 cells through the NF-kB pathway [80]. The results indicate that PAC1 regulation of Trk-A phosphorylation is essential for Epac to activate Rit GTPase leading to neuronal differentiation.

Prostate Cancer

PACAP and PAC1 immunoreactivity are present in prostate tissue [81]. In prostate cancer tissues anti-PAC1 antibody stained the apical portion of the prostate cancer cells. By RT-PCR, PAC1null mRNA was more abundant than PAC1SV1 or PAC1SV2. Addition of PACAP-27 of PACAP-38 to prostate cancer biopsy specimens increased adenylyl cyclase activity [82]. PACAP-27, PACAP-38, and PACAP(6–38) inhibited specific binding of ^{125}I-PACAP-27 to PC-3 cells with IC_{50} values of 15, 10, and 300 nM, respectively [21]. PACAP-27 (1 nM) addition to PC-3 cells increased the cAMP threefold which was antagonized significantly by 1 μM PACAP(6–38). Addition of 10 nM PACAP-27 significantly increased PC-3 colony number which was antagonized by 1 μM PACAP(6–38). Injection of 0.4 mg/kg of PACAP(6–38) into nude mice significantly reduced xenograft proliferation by 50 %. Treatment of LNCaP cells with PACAP-27 short-term stimulated proliferation whereas chronic treatment resulted in elevated cAMP resulting in proliferation arrest and neuroendocrine differentiation [83]. Because PACAP binds with high affinity to membranes derived from human benign hyperplastic prostate, VPAC and PAC1 may be expressed early in the carcinogenic process [84].

Pituitary Tumors

PACAP regulates hormone secretion and cell proliferation in pituitary tumor cells. In human pituitary adenoma HP75 cells, PACAP-27 or PACAP-38 inhibits TGF-β induced apoptosis [85]. In human pituitary adenomas, PACAP administration increased ACTH release from three corticotrope tumors as well as GH release from a somatotrope tumor [86]. In mouse pituitary corticotroph AtT20 cells, nM concentrations of PACAP-27 or PACAP-38 increased cAMP, POMC gene transcription and ACTH release [87, 88]. PACAP addition to rat lactotrope 235–1 cells increases PRL release to 260 % and increases proliferation to 162 % [89]. PACAP stimulates LH and FSH release from gonadotrophinomas and potentiated the effects of TRH [90]. PACAP-27 or PACAP-38 addition to lactotrope/somatotrope rat GH3 cells increased PRL and GH release [91]. PACAP increases PRL mRNA is a cAMP/PKA/ERK dependent mechanism [92]. PACAP addition to somatolactotrope BH5C1 cells increase PRL gene expression in a cAMP/PKA/ERK/Rap1 dependent manner [93].

Breast Cancer

A 19.9 kDa prepro-PACAP was detected in human breast cancer biopsy specimens [26]. Also, a smaller molecular weight product was identified suggesting the prepro-PACAP can be processed by breast cancer cells. Also PAC1 null and PAC1 SV3 mRNA were present in breast cancer biopsy specimens [8]. Antibodies to PAC1 stained breast cancer biopsy specimens and recognized a 60-kDa protein [22]. PACAP-27, PACAP-38, and PACAP (6–38) inhibited specific ^{125}I-PACAP-27 binding to T47D cells with IC_{50} values of 8, 17, and 750 nM respectively. Addition of 1 nM PACAP-27 to T47D cells increased the cAMP tenfold and the increase caused by PACAP-27 was significantly inhibited by 1 μM PACAP(6–38). PACAP(6–38) (0.4 mg/kg) inhibited significantly the growth of T47D xenografts in nude mice by 39 % [22].

A ^{99m}Tc-labeled VIP analog named TP-3645 has been developed to image tumors in breast cancer patients [94]. Subsequently a ^{64}Cu-labeled VIP analog was developed named TP-3982 to image breast cancer patients [95]. It remains to be determined if a radiolabeled PACAP analog will be developed for early detection of cancer.

Pancreatic Cancer

The effects of PACAP were investigated on rat pancreatic carcinoma AR4-2J cells. Nanomolar concentration stimulated the growth and increased ornithine decarboxylase activity [17]. The effects of PACAP-27 on proliferation and ODC activity were reversed by the somatostatin analog SMS 201–995 or pertussis toxin. PACAP-27 addition to AR4-2J cells increased proliferation and PLD activity that was inhibited by SMS201-995 or wortmannin, a PI-3-K inhibitor [96]. The growth effects of PACAP on AR4-2J cells are preceded by increased expression of c-fos as well as c-jun [97]. The c-fos and c-jun activate the heterodimeric transcription factor AP-1. The effects of PACAP on c-fos/c-jun/AP-1 are reversed by PKC and PKA inhibitors as well as PACAP(6–38).

Colon Cancer

PACAP-38 immunoreactivity was detected in colon cancer biopsy specimens [27]. PAC1 was present in HCT8 human colonic tumor cell line HCT8 [14]. Addition of PACAP-38 to NCT8 cells increased the cAMP and cytosolic Ca^{2+} leading to increased proliferation. The effects of PACAP on HCT8 cells were antagonized by PACAP(6–38). The effects on PACAP on HCT8 cells on proliferation was impaired if the cells overexpressed PAC1SV1 [98]. Addition of PACAP-38 to the transfected cells increased cAMP but not cytosolic Ca^{2+} or proliferation. The results suggest that PACAP must cause PI turnover to increase colon cancer proliferation.

Other Cancers

In cervical cancer multiple cytosine guanine dinucleotides are found in the promoter region of the PACAP gene and these CpG sites are methylated early in the carcinogenic process [99]. It remains to be determined if methylation of the PACAP gene can be used for early detection of cervical cancer. Overexpression of the PACAP gene in cervical cancer cell lines reduced expression of secretory clusterin, an anti-apoptotic protein [100].

In human choriocarcinoma cell line JAR, PACAP facilitated the ability of hydrogen peroxide (H_2O_2) or $CoCl_2$ to decrease survival [101]. Treatment of JAR cells with H_2O_2 plus PACAP decreased phosphorylation of Akt, p-38-MAPK and ERK. PACAP-38 or PACAP(6–38), however, it had little effect on the ability of methotrexate to cause apoptosis or necrosis of JAR cells [102]. The results suggest that PACAP may increase oxidative stress caused by H_2O_2.

In multiple myeloma proliferation of plasma cells in the bone marrow overproduce immunoglobulin light chains leading to secretion of interleukin 6 (IL-6) by bone marrow stromal cells (BMSC). PACAP suppressed light chain immunoglobulin myeloma release of IL-6 by BMSC [103]. PACAP infusion into the blood of a multiple myeloma patient increased the intravenous level of PACAP to 0.2 nM resulting in a reduction of immunoglobulin light chains in the urine. There were few side effects associated with the PACAP infusion and the half-life was approximately 5 min [104]. PACAP inhibited p38 MAPK and translocation of NFkappaB in renal proximal tubule cells [105]. PACAP-38 prevents renal injury in myeloma models in experimental animals. PACAP preserves renal function in a number of models of renal injury [106].

Summary

PACAP is a neuropeptide growth factor present in numerous cancers. PACAP stimulates the proliferation of both SCLC, a neuroendocrine tumor, and NSCLC, an epithelial tumor. PACAP-27 or PACAP-38, which can be derived from the precursor preproPACAP, binds with high affinity to PAC1, VPAC1, and VPAC2. Also, VIP, which is produced by SCLC and NSCLC cells, binds with high affinity to VPAC1 and VPAC2 but not PAC1. It is important to determine which ligand and which receptor regulates cancer proliferation.

For the first time, selective peptide agonists are now available for PAC1. (Iac^{1}, $Ala^{16,17}$, $DLys^{38}$)PACAP-38 is a synthetic agonist which prefers PAC1 relative to VPAC1 by 50- to 80-fold and prefers PAC1 relative to VPAC2 by 10- to 50-fold [32]. Maxidilan is a 61-amino acid peptide from the sand fly is selective for PAC1, but has little sequence homology to PACAP and is not readily available [31]. (Lys^{17}, Arg^{18}, Leu^{27})$VIP^{1\text{-}7}GRF^{8\text{-}27}$ is a selective VPAC1 agonist and RO25-1553 is a selective VPAC2 agonist [105]. The selective peptide agonists can be used to determine which receptor regulates cancer cellular proliferation. For SCLC and NSCLC, PAC1 and VPAC1 but not VPAC2 agonists stimulated proliferation [107]. PACAP(6–38) is a peptide antagonist for PAC1. While it appears to be specific for PAC1 relative to VPAC1 or VPAC2, the IC_{50} value is in the μM range. Thus PACAP(6–38) will dissociate rapidly from PAC1. High affinity non-peptide antagonists for PAC1 are needed.

The signal transduction mechanisms for PACAP in cancer are complex. Figure 47.3 shows that PACAP addition to NSCLC cells increases cytosolic Ca^{2+} and activates PKC as well as PKA with seconds after addition to NSCLC cells. As a result Src is activated which leads to activation of MMP as well as PYK2, FAK, and paxillin phosphorylation leading to increased cellular motility, migration, and secretion. MMP metabolizes membrane-bound precursors into EGFR ligands such as TGFα, amphiregulin, and heparin binding-EGF [108]. When TGFα binds to NSCLC cells, proteins such as PI3K and the EGFR are tyrosine phosphorylated. PI3K activation leads to Akt and mTor phosphorylation leading to increased cellular survival and/or differentiation. The EGFR activates the Ras, Raf, MEK, ERK pathway, leading to increased cellular proliferation. When ERK or Creb are activated in the

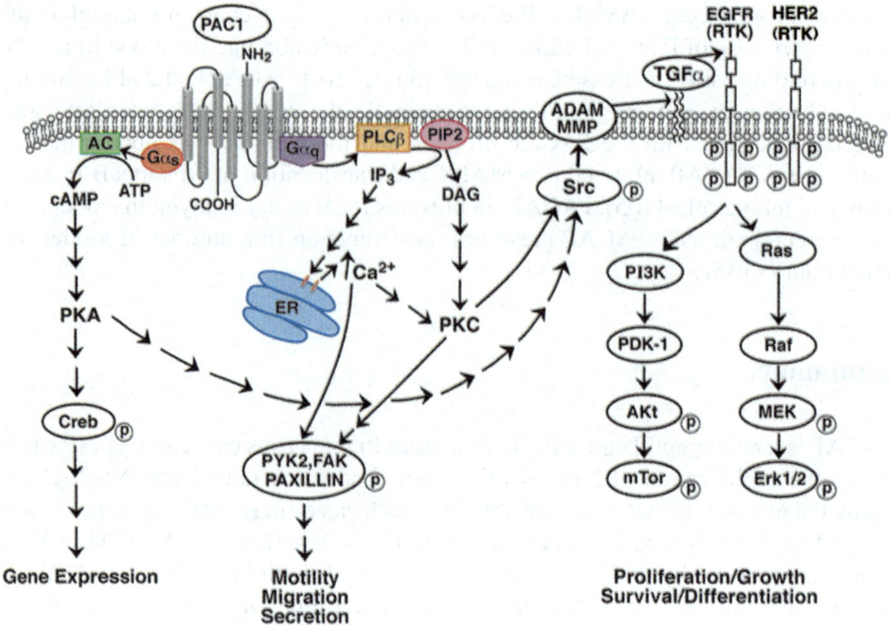

Fig. 47.3 PAC1 signal transduction. Activated PAC1 can interact with Gαs stimulating adenylyl cyclase leading to elevated cAMP, PKA activation, Creb phosphorylation and altered gene expression. Activated PAC1 can interact with Gαq stimulating PLC causing PI turnover. The DAG activates PKC leading to Src phosphorylation whereas the IP3 causes Ca^{2+} release from the ER. Src can tyrosine phosphorylate FAK, paxillin, and PYK2, leading to altered cellular motility and migration. Src can activate MMP metabolizing TGFα from inactive precursors. The TGFα activates the EGFR leading to formation of homodimers with itself or heterodimers with HER2. This results in the tyrosine phosphorylation of the EGFR and protein substrates such as PI3K. The PI3K can activate PDK-1 leading to Akt and mTor phosphorylation increasing cellular survival. Also the EGFR can activate Ras and Raf leading to MEK and ERK tyrosine phosphorylation increasing cellular proliferation

nucleus, gene expression is altered increasing proliferation. In NSCLC, PAC1 may regulate growth in an EGFR-dependent manner.

In NSCLC, nM concentrations of PACAP increase colony number whereas higher concentrations have little effect [109]. The high concentration may chronically increase cAMP leading to cellular differentiation. Undifferentiated cancer cells growth rapidly, whereas proliferation is impaired when cells are differentiated. The role of PAC1 SVs in cancer cellular proliferation remains to be determined. In neuroblastoma PAC1 SVs exist in the N-terminal and well as IC loop3.

PACAP(6–38) is a PAC1 antagonist which inhibits NSCLC proliferation in vitro and in vivo. Unfortunately PACAP(6–38) is cytostatic in that when PACAP(6–38) administration was discontinued into nude mice, NSCLC tumors rapidly regrew. In contrast, the TKI gefitinib is cytotoxic for NSCLC cells. PACAP(6–38) potentiated the cytotoxicity of gefitinib in vitro. It remains to be determined if GPCR antagonists increase the potency of TKI in NSCLC patients.

Acknowledgements This research is supported by the intramural program on NCI and NIDDK of the NIH.

References

1. Arimura A. Pituitary adenylate cyclase activating polypeptide (PACAP): discovery and current status of research. Regul Pept. 1992;37:287–303.
2. Sherwood NM, Krueckl SL, McRory JE. The origin and function of the pituitary adenylate cyclase activating polypeptide (PACAP)/glucagon superfamily. Endocr Rev. 2000;21:619–70.
3. Miyata A, Arimura A, Dahl RR, Minamino N, Uehara A, Jiang L, et al. Isolation of a novel 38 residue-hypothalamic polypeptide which stimulates adenylate cyclase in pituitary cells. Biochem Biophys Res Commun. 1989;164:567–74.
4. Miyata A, Jiang L, Dahl RD, Kitada C, Kubo K, Fujino M, et al. Isolation of a neuropeptide corresponding to the N-terminal 27 residues of the pituitary adenylate cyclase activating polypeptide with 38 residues (PACAP38). Biochem Biophys Res Commun. 1990;170:643–8.
5. Fahrenkrug J. VIP and PACAP. Results Probl Cell Differ. 2010;50:221–34.
6. Harmar AJ, Fahrenkrug J, Gozes I, Laburthe M, May V, Pisegna JR, et al. Pharmacology and functions of receptors for vasoactive intestinal peptide and pituitary adenylate cyclase –activating polypeptide. IUPHAR review 1. Br J Pharmacol. 2012;166:4–17.
7. Moody TW, Jensen RT. VIP and PACAP as autocrine growth factors in breast and lung cancer. In: Kastin AJ, editor. Handbook of biologically active peptides. Amsterdam: Elsevier; 2006. p. 493–8.
8. Moody TW, Chan D, Fahrenkrug J, Jensen RT. Neuropeptides as autocrine growth factors in cancer cells. Curr Pharm Des. 2003;9:495–509.
9. Pisegna J, Wank J. Cloning and characterization of the signal transduction of four splice variants of the human pituitary adenylate cyclase activation polypeptide receptor: evidence for dual coupling to adenylate cyclase and phospholipase C. J Biol Chem. 1996;271:17267–74.
10. Moody TW, Nuche-Berenguer B, Nakamura T, Jensen RT. EGFR transactivation by peptide G Protein-Coupled receptors in cancer. Curr Drug Targets. 2016;17(5):520–8.
11. Draoui M, Hida T, Jakowlew S, Birrer M, Zia F, Moody TW. PACAP stimulates c-fos mRNAs in small cell lung cancer cells. Life Sci. 1996;59:307–13.
12. Moody TW, Leyton J, Casibang M, Pisegna J, Jensen RT. PACAP-27 tyrosine phosphorylates mitogen activated protein kinase and increases VEGF mRNAs in human lung cancer cells. Regul Pept. 2002;109:135–40.
13. Nakamachi T, Sugiyama K, Watanabe J, Imai N, Kagami N, Hori M, et al. Comparison of expression and proliferative effect of pituitary adenylate cyclase-activating polypeptide (PACAP) and its receptors on human astrocytoma cell lines. J Mol Neurosci. 2014;54:388–94.
14. Le SV, Yamaguchi DJ, McArdle CA, Tachiki K, Pisegna JR, Germano P. PAC1 and PACAP expression, signaling and effect on the growth of HCT8 human colonic tumor cells. Regul Pept. 2002;109:115–25.
15. Lauffer JM, Modlin IMH, Inoue T, Kidd M, Zhang T, Schmid SW, et al. Pituitary adenylate cyclase-activating polypeptide modulates gastric enterochromaffin-like proliferation in rats. Gastroenterology. 1999;116:623–35.
16. Lauffer JM, Tang LH, Zhang T, Hinoue T, Rahbar S, Odo M, et al. PACAP mediates the neural proliferative pathway of Mastomys enterochromaffin-like cell transformation. Regul Pept. 2001;102:157–64.
17. Buscail L, Cambillau C, Seva C, Scemama JL, De Neef P, Robberecht P, et al. Stimulation of rat pancreatic tumoral AR4-2J cell proliferation by pituitary adenylate cyclase-activating peptide. Gastroenterology. 1992;103:1002–8.
18. Moody TW, Zia F, Makheja A. Pituitary adenylate cyclase activating polypeptide receptors are present on small cell lung cancer cells. Peptides. 1993;14:241–6.

19. Zia F, Fagarasan M, Bitar K, Coy DH, Pisegna JR, Wank SA, et al. Pituitary adenylate cyclase activating peptide receptors regulate the growth of non-small cell lung cancer cells. Cancer Res. 1995;55:4886–91.
20. Pisegna JR, Leyton J, Coelho T, Hida T, Jakowlew S, Birrer M, et al. PACAP hybrid: a new PACAP receptor antagonist. Life Sci. 1997;61:631–9.
21. Leyton J, Coelho T, Coy DH, Jakowlew S, Birrer MJ, Moody TW. PACAP(6-38) inhibits the growth of prostate cancer cells. Cancer Lett. 1998;125:131–9.
22. Leyton J, Gozes Y, Pisegna J, Coy D, Purdom S, Casibang M, et al. PACAP(6-38) is a PACAP receptor antagonist for breast cancer cells. Breast Cancer Res Treat. 1999;56:177–86.
23. Reubi JC. In vitro evaluation of VIP/PACAP receptors in healthy and diseased human tissues. Clinical implications. Ann N Y Acad Sci. 2000;921:1–25.
24. Reubi JC, Laderach U, Waser B, Gebbers JO, Robberecht P, Laissue JA. Vasoactive intestinal peptide/pituitary adenylate cyclase-activating peptide receptor subtypes in human tumors and their tissues of origin. Cancer Res. 2000;60:3105–12.
25. Schultz S, Rocken C, Mawrin C, Weise W, Holit V, Schulz S. Immunocytochemical identification of VPAC1, VPAC2 and PAC1 receptors in normal and neoplastic human tissues with subtype-specific antibodies. Clin Cancer Res. 2004;10:8235–42.
26. Garcia-Fernandez MA, Bodega G, Ruiz-Villaespesa A, Cortes J, Prieto JC, Carmena MJ. PACAP expression and distribution in human breast cancer and healthy tissue. Cancer Lett. 2004;205:189–95.
27. Szanto Z, Sarszegi Z, Reglodi D, Nemeth J, Szabadfi K, Kiss P, et al. PACAP immunoreactivity in human malignant tumor samples and cardiac diseases. J Mol Neurosci. 2012;48:667–73.
28. Kaufman J, Horn L, Carbone D. Molecular biology of lung cancer. In: DeVita Jr V, Lawrence TS, Rosenberg SA, editors. Cancer: principles and practice of oncology. Williams & Wilkins Philadelphia, PA: Lippincott; 2011. p. 789–98.
29. Carney DN, Gazdar AF, Bepler G, Guccion JG, Marangos PJ, Moody T, et al. Establishment and identification of small cell lung cancer cell lines having classic and variant features. Cancer Res. 1985;45:2913–23.
30. Moody TW, Walters J, Casibang M, Zia F, Gozes Y. VPAC1 receptors and lung cancer. Ann N Y Acad Sci. 2000;921:26–32.
31. Lerner EA, Iuga AO, Reddy VB. Maxadilan, a PAC1 receptor agonist from sand flies. Peptides. 2007;28:1651–4.
32. Ramos-Alvarez I, Mantey SA, Nakamura T, Nuche-Berenguer B, Moreno P, Moody TW, et al. A structure-function study of PACAP using conformationally restricted analogs: identification of PAC1 receptor-selective PACAP agonists. Peptides. 2015;66:26–42.
33. Bourgault S, Vaudry D, Botia B, Couvineau A, Laburthe M, Vaudry H, et al. Novel stable PACAP analogs with potent activity towards the PAC1 receptor. Peptides. 2008;29:919–32.
34. Wray B, Kakoschke C, Nokihara K, Naruse S. Solution structure of pituitary adenylate cyclase activating polypeptide by nuclear magnetic resonance spectroscopy. Biochemistry (Mosc). 1993;32:5832–41.
35. Sun C, Song D, Davis-Taber RA, Barrett LW, Scott VE, Richardson PL, et al. Solution structure and mutational analysis of pituitary adenylate cyclase-activating polypeptide binding to the extracellular domain of PAC1-R. Proc Natl Acad Sci U S A. 2007;104:7875–80.
36. Furness SG, Wootten D, Christopoulos A, Sexton PM. Consequences of splice variation on secretin family G protein coupled receptor function. Br J Pharmacol. 2012;166:98–109.
37. Pisegna JR, Wank SA. Molecular cloning and functional expression of the pituitary adenylate cyclase-activating polypeptide type 1 receptor. Proc Natl Acad Sci U S A. 1993;90:6345–9.
38. Spengler D, Waeber C, Pantaloni C, Holsboer F, Bochaert J, Seeberg PH, et al. Differential signal transduction by five slice variants of the PACAP receptor. Nature. 1993;365:170–5.
39. Pisegna JR, Moody TW, Wank SA. Differential signalling and immediate-early gene activation by four splice variants of the human pituitary adenylate cyclase activating polypeptide receptor. In: Arimura A, editor. VIP, PACAP and related regulatory peptides. Ann N Y Acad Sci. 1996;805:54–64.

40. Moody TW, Leyton J, Jensen RT. Pituitary adenylate cyclase-activating polypeptide causes increased tyrosine phosphorylation of focal adhesion kinase and paxillin. J Mol Neurosci. 2012;46:68–74.
41. Hanks SK, Calalb MD, Harper MC, Patel SK. Focal adhesion protein-tyrosine phosphorylation in response to cell attachment to fibronectin. Proc Natl Acad Sci U S A. 1992;89:8487–91.
42. Moody TW, diFlorio A, Jensen RT. PYK-2 is tyrosine phosphorylated after activation of pituitary adenylate cyclase activating polypeptide receptors lung cancer cells. J Mol Neurosci. 2012;48:660–6.
43. Bellis SL, Perotta JA, Curtis MS, Turner CE. Adhesion of fibroblasts to fibronectin stimulates both serine and tyrosine phosphorylation of paxillin. Biochem J. 1997;325:375–81.
44. Moody TW, Osefo N, Nuche-Berenguer B, Ridnour L, Wink D, Jensen RT. Pituitary adenylate cyclase activating polypeptide causes tyrosine phosphorylation of the EGF receptor in lung cancer cells. J Pharmacol Exp Ther. 2012;341:873–81.
45. Lemmon MA, Schlessinger J. Cell signaling by receptor tyrosine kinases. Cell. 2010;141:1117–34.
46. Lynch TJ, Bell DW, Sordella R, Gurubhagavatula S, Okimoto RA, Brannigan BW, et al. Activating mutations in the epidermal growth factor receptor underlying responsiveness of non-small-cell lung cancer to gefitinib. N Engl J Med. 2004;350:2129–39.
47. Paez JG, Janne PA, Lee JC, Tracey S, Greulich H, Gabriel S, et al. EGFR mutations in lung cancer: correlation with clinical response to gefitinib therapy. Science. 2004;304:1497–500.
48. Lutz EM, Ronaldson E, Shaw P, Johnson MS, Holland PJ, Mitchell R. Characterization of novel splice variants of the PAC1 receptor in human neuroblastoma cells: consequences for signaling by VIP and PACAP. Mol Cell Neurosci. 2006;31:193–209.
49. Blechman J, Levkowitz G. Alternative splicing of the pituitary adenylate cyclase-activating polypeptide receptor PAC1: mechanisms of fine tuning of brain activity. Front Endocrinol. 2013;4:55.
50. Ushiyama M, Ikeda R, Sugawara H, Yoshida M, Mori K, Kangawa K, et al. Differential intracellular signaling through PAC1 isoforms as a result of alternative splicing in the first extracellular domain and the third intracellular loop. Mol Pharmacol. 2007;72:103–11.
51. Ushiyama M, Ikeda R, Yoshida M, Mori K, Kangawa K, Sugarwara H, et al. Alternative splicing of the pituitary adenylate cyclase-activating polypeptide (PACAP) receptor contributes to the function of PACAP-27. J Mol Neurosci. 2010;42:341–8.
52. Holighaus Y, Mustafa T, Eiden LE. PAC1hop, null and hip receptors mediate differential signaling through cyclic AMP and calcium leading to splice variant specific gene induction in neural crest cells. Peptides. 2011;32:1647–55.
53. Falktoft B, Georg B, Fahrenkrug J. Signaling pathways in PACAP regulation of VIP gene expression in human neuroblastoma cells. Neuropeptides. 2009;43:53–61.
54. Lelievre V, Pineau N, Du J, Wen CH, Nguyen T, Janet T, et al. Differential effects of peptide histidine isoleucine (PHI) and related peptides on stimulation and suppression of neuroblastoma cell proliferation. J Biol Chem. 1998;273:19685–90.
55. Vertongen P, Devaick C, Sariban E, De Laet MH, Martelli H, Paraf F, et al. Pituitary adenylate cyclase activating peptide and its receptors are expressed in human neuroblastomas. J Cell Physiol. 1996;167:36–46.
56. Hoshino M, Li M, Zheng LQ, Suzuki M, Mochizuki T, Yanaihara N. Pituitary adenylate cyclase activation peptide and vasoactive intestinal polypeptide: differentiation effects on human neuroblastoma NB-OK-1 cells. Neurosci Lett. 1993;159:35–8.
57. Monaghan TK, Pou C, MacKenzie CJ, Plevin R, Lutz EM. Neurotropic actions of PACAP-38 and LIF on Human neuroblastoma SH-SY5Y cells. J Mol Neurosci. 2008;36:45–56.
58. Robberecht P, Woussen-Colle MC, Vertongen P, De Neef P, Hou X, Salmon I, et al. Expression of pituitary adenylate cyclase activating polypeptide (PACAP) receptors in human glial cell tumors. Peptides. 1994;15:661–5.
59. Sharma A, Walters J, Gozes Y, Fridkin M, Brenneman D, Gozes I, et al. A vasoactive intestinal peptide antagonist inhibits the growth of glioblastoma cells. J Mol Neurosci. 2001;17:331–9.

60. Dufes C, Alleaume C, Montoni A, Olivier JC, Muller JM. Effects of the vasoactive intestinal peptide (VIP) and related peptide on glioblastoma cell growth in vitro. J Mol Neurosci. 2003;21:91–102.
61. Vertongen P, Camby I, Darro F, Kiss R, Robberecht P. VIP and pituitary adenylate cyclase activating polypeptide (PACAP) have an antiproliferative effect on the T98G human glioblastoma cell line through interaction with VIP2 receptor. Neuropeptides. 1996;30:491–6.
62. Lelievre V, Seksenyan A, Nobuta H, Yong HW, Chhith S, Niewiadomski P, et al. Disruption of the PACAP gene promotes medulloblastoma in ptc1 mutant mice. Dev Biol. 2008;313:359–70.
63. Cohen JR, Resnick DZ, Niewiadomski P, Dong H, Liau LM, Waschek JA. Pituitary adenylyl cyclase activating polypeptide inhibits gli1 gene expression and proliferation in primary medulloblastoma derived tumorsphere cultures. BMC Cancer. 2010;10:676.
64. Lieu SN, Oh DS, Pisegna JR, Germano PM. Neuroendocrine tumors express PAC1 receptors. Ann N Y Acad Sci. 2006;1070:399–404.
65. Germano PM, Lieu SN, Xue J, Cooke HJ, Christiofi GL, Lu Y, et al. PACAP induces signaling and stimulation of 5-hydroxytrypamine release and growth in neuroendocrine tumor cells. J Mol Neurosci. 2009;39:391–401.
66. DiFlorio A, Sancho V, Moreno P, Delle Fave G, Jensen RT. Gastrointestinal hormones stimulate growth of foregut neuroendocrine tumors by transactivating the EGF receptor. Biochim Biophys Acta. 1833;2013:573–82.
67. Siddique ZL, Drozdov I, Floch J, Gustafsson BI, Stunes K, Pfragner R, et al. KRJ-I and BON cell lines: defining an appropriate enterochromaffin cell neuroendocrine tumor model. Neuroendocrinology. 2009;89:458–70.
68. Andersson M, Sillard R, Rokaeus A. Demonstration of [125]VIP binding sites and effects of VIP on cAMP-formation in rat insulinoma (RINm5F and RIN14B) cells. Regul Pept. 1992;40:41–9.
69. Deutsch PJ, Sun Y. The 38 amino acid pituitary adenylate cyclase activating polyeptide stimulates dual signaling cascades in PC12 cells and promotes neurite outgrowth. J Biol Chem. 1992;267:5108–13.
70. Vaudry D, Falluel-Morel A, Bourgault S, Basille M, Burel D, Wurtz O, et al. Pituitary adenylate cyclase-activating polypeptide and its receptors: 20 years after the discovery. Pharmacol Rev. 2009;61:283–357.
71. Zhang W, Duan W, Cheung NS, Huang Z, Shao K, Li QT. Pituitary adenylate cyclase activating polypeptide induces translocation of its G-protein coupled receptor into caveolin-enriched membrane microdomains, leading to enhanced cyclc AMP generation and neurite outgrowth in PC12 cells. J Neurochem. 2007;103:1157–67.
72. Ravni A, Bourgault S, Lebon A, Chan P, Galas L, Fournier A, et al. The neurotrophic effects of PACAP in PC12 cells: control by multiple transduction pathways. J Neurochem. 2006;98:321–9.
73. Vaudry D, Stork PJ, Lazarovici P, Eiden LE. Signaling pathways for PC12 cell differentiation: making the right connections. Science. 2002;296:1648–9.
74. Watanabe T, Ohtaki T, Kitada C, Tsuda M, Fujino M. Adrenal pheochromocytoma PC12 cells respond to pituitary adenylate cyclase-activating polypeptide. Biochem Biophys Res Commun. 1990;173:252–8.
75. Taupenot L, Mahata M, Mahata SK, O'Connor DT. Time-dependent effects of the neuropeptide PACAP on catecholamine secretion: stimulation and desensitization. Hypertension. 1999;34:1152–62.
76. Mustafa T, Walsh J, Grimaldi M, Eiden LE. PAC1hop receptor activation facilitates catecholamine secretion selectively through 2-APB-sensitive Ca(2+) channels in PC12 cells. Cell Signal. 2010;22:1420–6.
77. Lee FS, Rajagopal R, Kim AH, Chang PC, Chao MV. Activation of Trk neurotrophin receptor signaling by pituitary adenylate cyclase-activating polypeptides. J Biol Chem. 2002;277:9096–102.

78. Rajagopal R, Chen ZY, Lee FS, Chao MV. Transactivation of Trk neurotrophin receptors by G-protein-coupled receptor ligands occurs on intracellular membranes. J Neurosci. 2004;24:6650–8.
79. Shi GX, Jin L, Andres DA. Src-dependent TrkA transactivation is required for pituitary adenylate cyclase-activating polypeptide 38-mediated Rit activation and neuronal differentiation. Mol Biol Cell. 2010;21:1597–608.
80. Manecka DL, Mahmood SF, Grumolato L, Lihrmann I, Anouar Y. Pituitary adenylate cyclase-activating polypeptide (PACAP) promotes both survival and neuritogenesis in PC12 cells through activation of nuclear factor KB (NF-kB) pathway: involvement of extracelluar signal regulated kinase (ERK), calcium and c-REL. J Biol Chem. 2013;288:14936–48.
81. Mammi C, Frajese GV, Vespasiani G, Mariani S, Gnessi L, Farini D, et al. PAC1-R null isoform expression in human prostate cancer tissue. Prostate. 2006;66:514–21.
82. Garcia-Fernandez MO, Solano RM, Carmena MJ, Busto R, Bodega G, Ruiz-Villaespesa A, et al. Expression of functional PACAP/VIP receptors in human prostate cancer and healthy tissue. Peptides. 2003;24:893–902.
83. Farini D, Puglianiello A, Mammi C, Siracusa G, Moretti C. Dual effect of pituitary adenylate cyclase activating polypeptide on prostate tumor LNCaP cells: Short and long-term exposure affect proliferation and neuroendocrine differentiation. Endocrinology. 2003;144:1632–43.
84. Solano RM, Carmena MJ, Carrero I, Cavallaro S, Roman F, Hueso C, et al. Characterization of vasoactive intestinal peptide/pituitary adenylate cyclase-activating peptide receptors in human benign hyperplastic prostate. Endocrinology. 1996;137:2815–22.
85. Oka H, Jin L, Kulig E, Scheithauer BW, Lloyd RV. Pituitary adenylate cyclase- activating polypeptide inhibits transforming growth factor-beta 1 induced apoptosis in a human pituitary adenoma cell line. Am J Pathol. 1999;155:1893–900.
86. Desai BJ, Monson JP, Holdstock JG, Aylwin SJ, Geddes JF, Wood DF, et al. Effects of pituitary adenylate cyclase-activating polypeptide on hormone secretion by human pituitary adenomas in vitro. J Clin Endocrinol Metab. 1994;79:1771–7.
87. Koch B, Lutz-Bucher B. Vasopressin, unlike phorbol ester, fails to synergistically interact with pituitary adenylate cyclase activating polypeptide (PACAP) in stimulating cyclic AMP formation and ACTH secretion in cultured anterior pituitary cells. Mol Cell Endocrinol. 1993;92:175–81.
88. Boutillier AL, Monnier D, Koch B, Loeffler JP. Pituitary adenyl cyclase-activating polypeptide: a hypophysiotropic factor that stimulates proopiomelanocortin gene transcription and proopiomelanocortin-derived peptide secretion in corticotropic cells. Neuroendocrinology. 1994;60:493–502.
89. Hammond PJ, Smith DM, Akinsanya KO, Mufti WA, Wynick D, Bloom SR. Signalling pathways mediating secretory and mitogenic responses to galanin and pituitary adenylate cyclase-activating polypeptide in the 235-1 clonal rat lactotroph cell line. J Neuroendocrinol. 1996;8:457–64.
90. Peterson B, Buchfelder M, Fahlbusch R, Adams EF. Pituitary adenylate cyclase-activating polypeptide directly stimulates LH and FSH secretion by human pituitary gonadotrophinomas. Exp Clin Endocrinol Diabetes. 1996;104:250–5.
91. Propato-Mussafiri R, Kanse SM, Ghatei MA, Bloom SR. Pituitary adenylate cyclase activating-polypeptide releases 7B2, adrenocorticotrophin, growth hormone and prolactin from the mouse and rat clonal pituitary cell lines AtT-20 and GH3. J Endocrinol. 1992;132:107–13.
92. Yonehara T, Kanasaki H, Yamamoto H, Fukunaga K, Miyazaki K, Miyamoto E. Involvement of mitogen-activated protein kinase in cyclic adenosine 3′,5′monophosphate induced hormone gene expression in rat pituitary GH(3) cells. Endocrinology. 2001;142:2811–9.
93. Romano D, Magalon K, Ciampini A, Talet C, Enjalbert A, Gerard C. Differential involvement of the Ras and Rap1 small GTPases in vasoactive intestinal and pituitary adenylyl cyclase activating polypeptide control of the prolactin gene. J Biol Chem. 2003;278: 51386–94.

94. Pallela VR, Thakur ML, Chakder S, Rattan S. 99mTc-labeled vasoactive intestinal peptide receptor agonist: functional studies. J Nucl Med. 1999;40:352–60.
95. Zhang K, Aruva MR, Shanthly N, Cardi CA, Patel CA, Rattan S, et al. Vasoactive intestinal peptide (VIP) and pituitary adenylate cyclase activating peptide (PACAP) receptor specific peptide analogues for PET imaging of breast cancer: In vitro/in vivo evaluation. Regul Pept. 2007;144:91–100.
96. Morisset J, Douziech N, Rydzewska G, Buscail L, Rivard N. Cell signaling pathway involved in PACAP-induced AR4-2J cell proliferation. Cell Signal. 1995;7:193–205.
97. Schafer H, Zheng J, Gundlach F, Gunther R, Schmidt WE. PACAP stimulates transcription of c-fos and c-jun and activates the AP-1 transcription factor in rat pancreatic carcinoma cells. Biochem Biophys Res Commun. 1996;221:111–6.
98. Germano PM, Le SV, Oh DS, Fan R, Lieu S, Siu A, et al. Differential coupling of the PAC1 SV1 splice variant on human colonic tumors to the activation of intracellular cAMP but not intracellular Ca2+ does not activate tumor proliferation. J Mol Neurosci. 2004;22:83–92.
99. Jung S, Yi L, Jeong D, Kim J, An S, Oh TJ, et al. The role of ADCYAP1, adenylate cyclase activating polypeptide 1, as a methylation biomarker for the early detection of cervical cancer. Oncol Rep. 2011;25:245–52.
100. Lee JH, Lee JY, Tho SB, Choi SJ, Lee DG, An S, et al. PACAP inhibits tumor growth and interferes with clusterin in cervical carcinomas. FEBS Lett. 2014;588:4730–9.
101. Boronkai A, Brubel R, Racz B, Tamas A, Kiss P, Horvath G, et al. Effects of pituitary adenylate cyclase activating polypeptide on the survival and signal transduction pathways in human choriocarcinoma cells. Ann N Y Acad Sci. 2009;1163:353–7.
102. Brubel R, Bornokai A, Reglodi D, Racz B, Nemeth J, Kiss P, et al. Changes in the expression of pituitary adenylate cyclase activating polypeptide in the human placenta during pregnancy and its effects on the survival of JAR choriocarcinoma cells. J Mol Neurosci. 2010;42: 450–8.
103. Li M, Cortez S, Nakamachi T, Batuman V, Arimura A. Pituitary adenylate cyclase-activating polypeptide is a potent inhibitor of the growth of light chain-secreting human multiple myeloma cells. Cancer Res. 2006;66:8796–803.
104. Li M, Maderdrut JL, Lertora JJ, Batuman V. Intravenous infusion of pituitary adenylate cyclase-activating polypeptide (PACAP) in a patient with multiple myeloma and myeloma kidney: a case study. Peptides. 2007;28:1891–5.
105. Arimura A, Li M, Batuman V. Treatment of renal failure associated with multiple myeloma and other diseases by PACAP-38. Ann N Y Acad Sci. 2006;1070:1–4.
106. Arimura A, Li M, Batuman V. Potential protective action of pituitary adenylate cyclase-activating polypeptide. Blood. 2006;107:681–8.
107. Gourlet P, Vertongen P, Vandermeer P, Vandermeer-Pirot MC, Rathe J, DeNeef P. The long-lasting vasoactive intestinal polypeptide agonist RO 25-1553 is highly selective of the VIP2 receptor subclass. Peptides. 1997;18:403–8.
108. Bhola NE, Grandis JR. Crosstalk between G-protein-coupled receptors and epidermal growth factor receptor in cancer. Front Biosci. 2008;13:1857–65.
109. Moody TW, Gozes I. Vasoactive intestinal peptide receptors: a molecular target in breast and lung cancer. Curr Pharm Design. 2007;13:1099–104.

Chapter 48
PACAP as a Potential Biomarker: Alterations of PACAP Levels in Human Physiological and Pathological Conditions

Dora Reglodi, Zsuzsanna Helyes, Jozsef Nemeth, Reka A. Vass, and Andrea Tamas

Abstract Pituitary adenylate cyclase activating polypeptide (PACAP) is a neuropeptide with wide distribution and diverse effects. Its presence has not only been revealed in the nervous system and peripheral organs, but also in the blood and other biological fluids. Although PACAP is rapidly cleaved in the blood by peptidases, increasing number of evidence has been published in the last 10 years that show alterations of PACAP levels in different pathological conditions. The present review summarizes results obtained in the human blood and other fluids under normal, physiological and pathological conditions with clinical relevance, such as pregnancy and delivery, as well as in several diseases. Most clinical data have been obtained so far in neurological diseases, such as Alzheimer's disease, migraine, traumatic brain injury, multiple sclerosis, and cerebral hemorrhage, but some other diseases have also been shown to be accompanied by alterations of PACAP levels. These results indicate that the altered levels may have diagnostic and/or prognostic values in several diseases.

Keywords Serum • Plasma • RIA • ELISA • Biomarker • Neuropeptide

---

D. Reglodi (✉) • R.A. Vass • A. Tamas
Department of Anatomy, MTA-PTE "Lendulet" PACAP Research Team, University of Pecs, Szigeti u 12, Pecs 7624, Hungary
e-mail: reglodi@aok.pte.hu

Z. Helyes
Department of Pharmacology and Pharmacotherapy, University of Pecs, Pecs, Hungary

Janos Szentagothai Research Centre, MTA-PTE NAP B Chronic Pain Research Group, University of Pecs, Pecs, Hungary

J. Nemeth
Department of Pharmacology and Pharmacotherapeutics, University of Debrecen, Debrecen, Hungary

D. Reglodi, A. Tamas (eds.), *Pituitary Adenylate Cyclase Activating Polypeptide — PACAP*, Current Topics in Neurotoxicity 11,
DOI 10.1007/978-3-319-35135-3_48

Introduction

The pitfalls of neuropeptide measurements in biological samples are well known, and it is suggested to use neuropeptide immunoreactivity to describe what we measure with immunoassay techniques. This situation is further complicated in biological fluids: the stable tissue levels of neuropeptides change once the peptides enter the circulation and are broken down by peptidases. In addition, different pathological conditions and medications influence the levels and breakdown of peptides. In spite of difficulties and uncertainties, various peptides have been implicated as potential biomarkers in human diseases and their measurement in biological fluids have also helped to elucidate their involvement in several disease pathomechanisms [1].

Soon after the isolation of pituitary adenylate cyclase activating polypeptide (PACAP) by Miyata et al. [2], certain effects of PACAP in humans and its localization in different human tissues were described [3, 4]. The source of PACAP in the blood is not known with certainty. The most logical sources of circulating PACAP are tissues known to contain the highest level of PACAP, such as brain and endocrine glands. It has also been described that PACAP levels are 1.5 times higher in the mouse serum than in the plasma [5]. The authors suggested that platelets significantly contribute to circulating PACAP [5]. However, we do not know, for example in the case of the brain, how different barriers influence the passage of the peptide into the blood (see Chap. 25 by Banks). Another important issue to take into consideration is the short plasma elimination half-life of the peptide (5–10 min), similarly to other members of the VIP–secretin–glucagon–GHRH peptide family [6].

PACAP levels have been measured in human blood plasma in several studies. In healthy volunteers, plasma PACAP levels showed no difference between radial artery, internal or external jugular vein or cubital vein [7] (Table 48.1). In addition to the baseline levels, it was shown that continuous infusion of PACAP caused rapid increase of plasma PACAP levels, and a rapid decrease after cessation of the infusion in a male patient, while inhalation of the peptide did not lead to increased plasma levels [6, 9]. Subsequent reports also demonstrated that intravenous infusion in volunteers led to increased plasma levels [13]. PACAP is cleaved by dipeptidyl-peptidase IV (DPP IV), which also cleaves several other peptides. The half-life of infused PACAP is estimated to be 3.5 min when infused to healthy young volunteers for 20 min [8], while others suggested a longer half-life of about 5–10 min [6] (Table 48.1).

Increasing evidence shows that plasma PACAP levels and PACAP in several other body fluids change under different physiological and pathological conditions and these levels may have diagnostic/prognostic values in certain diseases. The present review summarizes data obtained in human blood and other fluids under normal and pathological conditions (summarized in Tables 48.1, 48.2, and 48.3).

Table 48.1 Changes in PACAP levels in human physiological conditions

Conditions	Patient number	Sample/ method	Level ranges	Change of PACAP	Reference
Baseline levels	17	Plasma, RIA	0.54, 0.58, 0.62, and 0.46 pmol/l in radial artery, internal, external jugular, and cubital veins, respectively	No difference between different arterial and venous sources	[7]
PACAP infusion (4 pmol/kg/min, 60 min)	1	Plasma, ELISA	Baseline: 0.01 nmol/l Infusion plateau: 0.197 nmol/l	↑ During continuous infusion, rapid ↓ thereafter, half-life 5–10 min	[6]
PACAP infusion (10 pmol/kg/min, 2 h)	12	Plasma, RIA	Baseline: below 2 pmol/l, median peak at 20 min: 37 pmol/l	↑, Half-life of 3.5 min	[8]
PACAP inhalation	12	Plasma, RIA	36 pg/ml	No change	[9]
Pregnancy	12 (vs. 18 controls)	Plasma, RIA	Between 310 and 460 fmol/ml	↑ in second–third trimesters	[10]
Pregnancy	8 (vs. 13 controls)	Plasma, ELISA	136 vs. 55 pg/ml	↓ Between 12 and 28 weeks	[11]
During delivery	10	Plasma, RIA	150 fmol/ml	70 % ↓	[10]
Postmenopausal women	12	Plasma, ELISA	105 pg/ml	No change	[11]
Newborn babies	10	Plasma, RIA	400 fmol/ml in plasma, in umbilical vessels: 190–250 fmol/ml	Peripheral blood: similar levels to adults Higher in umbilical artery than vein	[10]
Fetal cord blood	29 (vs. 5 postmenopausal controls)	Plasma, RIA	31 ng/ml (vs. 39 ng/ml)	No sex difference, levels similar to adults, slightly higher after cesarian delivery (compared to vaginal)	[12]

Table 48.2 Changes in PACAP levels in human neurological diseases

Condition or disease	Patient number	Sample/method	Level ranges	Change of PACAP	Reference
Spontaneous basal ganglia hemorrhage	150 hemorrhage patients (vs. 150 controls)	Plasma, ELISA	268 vs. 72 pg/ml	↑	[14]
Aneurysmal subarachnoid hemorrhage	118 (vs. 118 controls)	Plasma, ELISA	296 vs. 77 pg/ml	↑, Correlation with clinical symptoms	[15]
Alzheimer's disease	9 (vs. 7 controls)	CSF, ELISA	1.8 vs. 2.1 ng/ml	↓	[16, 17]
Mild cognitive impairment preceding Alzheimer's disease	9 vs. 10 controls (16 more severe cases)	CSF, ELISA	1.8 vs. 2.1 ng/ml (vs. 1.6 in more severe cases)	↓	[18]
Parkinson's disease with dementia	8 (vs. 7 controls)	CSF, ELISA	2.1 ng/ml in both	No change (vs. controls)	[16]
Frontotemporal lobar degeneration	7 (vs. 7 controls)	CSF, ELISA	2 vs. 2.1 ng/ml	No change (vs. controls)	[16]
Children with mental retardation (autistic spectrum disorder, retardation without autism, and cerebral palsy)	69, 60, and 63, respectively (vs. 54 controls)	Neonatal blood, immuno-affinity chromatography	26, 27, and 31 pg/ml, respectively, vs. 25 pg/ml in controls	No alteration	[19]
Severe mental retardation with partial trisomy 18p	2	Plasma, ELISA	10 pg vs. 2 pg/0.01 μl (in controls)	Marked ↑	[5]
Borderline IQ	2	Plasma, ELISA	5–6 pg vs. 2 pg/0.01 μl (in controls)	Moderate ↑	[5]
Migraine	87 (vs. 40 controls)	Plasma, RIA	interictal: 24.6 ictal: 27.3 vs. 26.5 fmol/ml	↓ in interictal period, ↑ in ictal period	[20]
Migraine	22	Plasma, RIA	10.2 vs. 4.3 pmol/l (after infusion with or without attack)	↑ After infusion. more ↑ after PACAP infusion in patients with delayed migraine attacks	[13]

Migraine with triptan treatment	15	Plasma, RIA	36, 26, 21 pmol/l in migraine, after triptan treatment and interictally, respectively	↑ Migraine, no change in healthy volunteers, ↓ after sumatriptan treatment	[21]
Sumatriptan treatment in healthy subjects	16	Plasma, RIA	0.46–0.62 pmol/l	No change in radial artery and different venous sources	[7]
Migraine and tension-type headache	133 migraine, 106 tension-type headache, 50 controls	Plasma, ELISA	(32, 39 respectively, vs. controls: 42 pg/ml)	↓ in interictal period in migraineurs, lower than tension-type headache patients or controls	[18]
Multiple sclerosis	20 (vs. 27 controls)	Plasma, CSF, RIA	Plasma: 4.2 pg/ml both groups; CSF: 6.7 vs. 4.1 (controls) pg/ml	↓ in CSF no change in plasma	[22]
Post-traumatic stress disorder	64	Plasma, RIA	Low: < 20 pM High: > 20 pM	↑ in females only	[23]
Traumatic brain injury	38 vs. 14 controls	Plasma, CSF, RIA	Plasma: 30-51 vs. 14 fmol/ml in controls; CSF: 25-39 vs. 16 fmol/ml in controls	↑ in plasma and CSF, correlation with mortality in CSF	[24]

Table 48.3 Changes in PACAP levels in human non-neurological diseases

Condition or disease	Patient number	Sample/ method	Level ranges	Change of PACAP	Reference
Chronic hepatitis B patients under antiviral therapy	25 treated, 22 healthy controls	Plasma, RIA	Controls: 65, hepatitis: 22 after treatment: 49 pg/ml	↓ in hepatitis patients, ↑ after antiviral therapy	[25]
Nephrotic syndrome in children	28 vs. 10 controls	Plasma, urine, Western blot	Results given as % of controls, absolute values not available	Plasma PACAP ↓, urinary levels ↑ in nephrotic states	[26]
Nephrectomy	3	Plasma, Western blot	As above	↑ After nephrectomy	[26]
Premature ovarian failure	5 (vs. 13 controls)	Plasma, ELISA	58 (vs. 136) pg/ml	↓	[11]
Idiopathic hypogonadotropic hypogonadism	9 (vs. 13 controls)	Plasma, ELISA	92 (vs. 136) pg/ml	↓	[11]
Superovulation treatment	132	Follicular, fluid RIA	143 fmol/ml (28–690 fmol/ ml), cutoff value: 290 fmol/ml	Correlation with the number of retrieved oocytes: lower levels associated with higher number of oocytes	[27]

PACAP Levels Under Physiological Conditions

PACAP levels under normal circumstances are summarized in Table 48.1.

Early studies measured plasma PACAP levels in rats. A pioneer paper showed that the hypohyseal portal blood contained significantly higher PACAP immunoreactivity than the peripheral blood, with higher hypophyseal levels in females than in males ([28]; 50–100 pmol/l vs. 23–24 pmol/l). This study proved that PACAP was released into the hypophyseal portal vessel blood and, therefore, can be regarded as a hypothalamic pituitary regulatory factor, with PACAP38 being the major form [28].

Pregnancy and Delivery

During pregnancy, PACAP levels show a moderate but significant elevation during the second and third trimesters compared to earlier gestational periods or nonpregnant women [10]. From the same volunteer population, during delivery, a significant, about 70 %, decrease was observed. Levels in the plasma of mothers returned to baseline 3 days after delivery. Plasma levels were also measured in the newborn babies of these volunteer mothers. It was found that the levels of PACAP were in the same range as in adults in the peripheral blood, but not in the blood obtained from the umbilical arteries, where PACAP level was lower [10]. The levels were higher in the umbilical arteries than in the umbilical vein. These findings indicate that during pregnancy PACAP is synthesized by the placenta or other maternal tissues in higher amounts, possibly necessary for the fetal and/or maternal tissue development. The sudden drop during delivery might indicate a role in the uteroplacental circulation and/or uterine contraction. The higher level in the umbilical artery (carrying venous blood from the fetus) than in the vein (carrying blood from the placenta to the fetus) indicates a fetal synthesis of the peptide [10].

Another study also investigated the level of PACAP in fetal cord blood and newborn babies [12]. The authors found no difference between male and female babies or between newborns and postmenopausal women between 64 and 83 years. It was also found that there was a slightly elevated level after caesarian delivery compared to vaginal birth, but there was no difference between PACAP levels in babies whose mothers progressed through labor or those who did not [12]. PACAP levels were unrelated to gestational age or Apgar score. The level of PACAP showed a slight increasing tendency with increasing birth weight. Based on the function of PACAP in stress adaptation and stress responses, it could be hypothesized that PACAP levels would differ in perinatal stress, but these data did not support this hypothesis.

In contrast to the above-mentioned results, Kanasaki and colleagues showed that plasma PACAP levels were lower in pregnant women compared to nonpregnant women [11]. PACAP levels were measured parallel with levels of luteinizing hormones, but no correlation was found between variations in PACAP and gonadotropin levels [11].

Related to pregnancy and childcare, other fluids also have to be mentioned: breast milk, amniotic fluid, and follicular fluid. Follicular fluid contained relatively stable levels of PACAP, with PACAP levels correlating with the number of retrieved oocytes during hyperstimulation treatment, the details of which are discussed below [27, 29]. In the amniotic fluid, the presence of the unchanged PACAP molecule could not be detected by mass spectrometry, but RIA measurements confirmed PACAP-like immunoreactivity (LI) [29]. These results show that an altered or bound form of the peptide was present in the amniotic fluid, the significance of which still awaits further investigation. PACAP levels were markedly higher in the breast milk than in the plasma [30]. This topic is reviewed in a separate chapter (Chap. 49) of this book by Tamas et al.

PACAP Levels Under Pathological Conditions

Neurological Diseases

Changes in PACAP levels in neurological diseases are summarized in Table 48.2.

Alzheimer's Disease

In postmortem human brains, Han and coworkers demonstrated a marked reduction in PACAP levels in several brain areas of Alzheimer's disease patients [16, 17]. It was found that PACAP levels were reduced in the middle temporal gyrus, superior frontal gyrus, entorhinal cortex and primary visual cortex both at the mRNA and protein levels. Furthermore, the authors found that PACAP reduction was associated with pathological hallmarks of Alzheimer's disease: PACAP levels were reduced with higher amyloid plaque scores in the entorhinal cortex and superior temporal gyrus, but not in the primary visual cortex, a region spared in most cases of the disease. Levels were lower in advanced Braak stages than in moderate stages. Lower PACAP levels were measured in postmortem cisternal cerebrospinal fluid (CSF) of the same patients parallel with the reduced brain tissue levels [16]. Additionally, PACAP concentrations in patients with Alzheimer's disease strongly correlated with the dementia rating scores. The inverse correlation between PACAP and the pathological hallmarks of the disease suggests that PACAP is not only reduced in Alzheimer's disease, but also represents the severity of this pathology [16, 17]. These data suggest that the strongly reduced neurotrophic effect of PACAP may be an important contributing factor to Alzheimer pathology. It supports the observations made in mouse models of Alzheimer's disease, where one of the down-regulated genes important for neuronal survival was PACAP [31] and PACAP levels were strongly reduced in the brains of mutant mice resembling Alzheimer pathology [17].

In a subsequent study, the same research group showed that PACAP decline was associated with mild cognitive impairment preceding Alzheimer's disease in the above described brain areas [18]. It was found that PACAP levels were lower not only in the superior frontal gyrus and middle temporal gyrus but also in the CSF [18], although to a lesser extent than in fully developed Alzheimer's disease. Values were inversely correlated with the dementia rating score. These data suggest a progressive decline of PACAP levels in Alzheimer-related dementias. The expression level of the PAC1 receptor was also evaluated and it was shown to be upregulated in mild cognitive impairment, but not in fully developed Alzheimer's disease in the superior frontal gyrus [18]. This implies a transient, possibly compensatory upregulation of PAC1 receptor in mild impairment, which is lost at more progressed states [18]. The authors suggest that CSF PACAP could be a good candidate of disease progression marker given the close correlation with disease scores and the specificity, since no reduction was observed in other types of dementia (see below).

Other Dementias

An early study compared blood levels of several neuropeptides, including that of PACAP, in neonatal blood of children with autism spectrum disorder, mental retardation without autism and in children with cerebral palsy [19]. There was no difference in PACAP levels between the groups, in contrast to VIP, which showed higher levels in children with autism [19]. Han and coworkers measured CSF PACAP levels not only in Alzheimer's disease, but also in other types of dementia, such as Parkinson's disease with dementia and frontotemporal lobar degeneration. In contrast to the significant reduction observed in Alzheimer's disease, no reduction was observed in the CSF of other dementia patients postmortem [16]. In a family study, where members showed mental retardation, some displaying partial trisomy 18p with three copies of the PACAP gene [5]. It was found that two family members with severe mental retardation and accompanying other neurological problems had significantly higher level of PACAP, and two members with borderline IQs had moderately increased PACAP levels [5].

Post-traumatic Stress Disorder

Regarding the potential use of PACAP as a biomarker in post-traumatic stress disorder, a pioneer study was published in 2011 [23]. The authors found, for the first time, that there was a sex-specific association of PACAP blood levels with fear physiology and stress symptoms in females. Post-traumatic stress disorder symptoms significantly correlated with PACAP38 levels in females, but not in males, and the disease diagnosis was associated with PACAP levels in females, with higher PACAP38 found in the stress-disorder cohort. PACAP38 levels predicted differential response on the symptoms necessary to fulfill the criteria in females, like avoidance, hyperarousal, and reexperiencing [23]. They also found association with gene polymorphism of PACAP and PAC1 receptor, reinforced in a number of subsequent studies (further discussed in Chap. 45 of this book, by Ressler et al.).

Acute Spontaneous Basal Ganglia Hemorrhage

A recent study showed that plasma PACAP concentrations were significantly higher in patients after acute spontaneous basal ganglia hemorrhage than in healthy control subjects [14]. This study examined a high number of patients: 150 patients with stroke and 150 age- and gender-matched control individuals. In addition to elevated PACAP levels, a positive association was also shown between PACAP concentrations and neurological score, as well as with hematoma volume. Patients, who died within the first week after admission, had higher PACAP levels. A cutoff value was identified that predicted 1-week mortality with high sensitivity and specificity, and the predictive value was found to be similar to the neurological scores. Similarly, in patients with 6-month survival, elevated PACAP levels were found and it also predicted mortality with high sensitivity and specificity. Studying the overall survival, plasma PACAP concentration proved to be an independent predictor for overall survival similarly to scores and hematoma volumes. Overall survival times were significantly shorter in patients with high PACAP levels than those with low concentrations [14]. The authors suggested that PACAP could be a good prognostic predictor in hemorrhage patients and hypothesized that the elevated blood PACAP levels may reflect an overproduction of PACAP as a pathological response to the loss of neural tissue in the CNS and it might be associated with the neuroprotective effects of the neuropeptide. Another recent study has confirmed these findings in aneurysmal subarachnoid hemorrhage [15]. These authors examined 118 hemorrhage patients and compared them to 118 controls. Hemorrhage patients had significantly elevated PACAP levels on admission and plasma PACAP levels were independently associated with clinical severity scores. The neuropeptide was identified as an independent predictor of survival [15], suggesting that PACAP might be a potential prognostic biomarker of subarachnoid hemorrhage.

Traumatic Brain Injury

Our group demonstrated that PACAP could be reliably measured from the human CSF (25–80 fmol/ml) [24]. We measured PACAP levels in the plasma and CSF of patients who suffered severe traumatic brain injury. Patients were enrolled with a Glasgow Coma Scale ≤ 8 on admission. We detected higher concentrations of PACAP in the CSF after traumatic brain injury and both plasma and CSF levels increased within the first 24–48 h after the injury. Thereafter, levels stayed high and did not increase further. In surviving patients, plasma and CSF levels were parallel, possibly indicating a damage in the blood–brain barrier. However, in patients, who died within the first week, plasma levels were almost double compared to levels of the CSF. There was no correlation with other clinical parameters [24]. These data show a relationship between first week mortality and plasma PACAP levels and support the possible hypothesis that PACAP levels are upregulated by the peri-injury areas, similarly to earlier immunohistochemical data from postmortem brains of traumatic brain injury patients [32]. However, similarly feasible explanation can be the relationship between the degree of blood–brain barrier injury and the elevated plasma PACAP levels.

Multiple Sclerosis

The role of PACAP in neuro-inflammatory processes was proposed on the basis of experimental data. Administration of PACAP suppressed experimental autoimmune encephalomyelitis, a model of multiple sclerosis [33]. Studies in mice supported these observations by showing that PACAP knockout mice exhibited exacerbated encephalomyelitis, thus, the endogenous production of the peptide protects against multiple sclerosis, probably by modulating regulatory T cells [34, 35]. A clinical study investigated PACAP and VIP levels in the plasma and CSF of multiple sclerosis patients [22]. It was found that PACAP levels decreased in the CSF, while plasma PACAP levels did not change. These data suggest that PACAP may influence the course of multiple sclerosis.

Migraine

The role of PACAP in migraine is evidenced both by animal and human studies. Due to the strong vasodilatory effect of PACAP and its sensitization function in the trigeminovascular system, PACAP is suggested to play an essential role in the induction of migraine [36]. Infusion of PACAP caused headache in healthy patients and migraine-like attacks in migraine patients [13, 37]. Animal studies supported these observations: PACAP38 was present in the trigeminal ganglion and caudal trigeminus nucleus, and plasma PACAP levels increased after electrical stimulation of the trigeminal ganglion [38] or after dural stimulation in cats [21]. PACAP knockout animals showed reduced responses to nitroglycerol [39].

As far as the plasma levels are concerned, Tuka et al. [20] found that plasma PACAP38 levels were significantly lower in the plasma of migraineurs during the interictal period than in healthy subjects, but elevated levels were shown during the ictal phase relative to the attack-free periods. In 21 migraineurs, samples were measured from the same subjects during attack and interictally. It was found that plasma PACAP38 levels were significantly higher during the ictal period. A negative correlation was observed between the interictal PACAP38 level and the disease duration [20]. Similarly to these data significantly lower plasma PACAP38-LI was detected in the inter-bout period of cluster headache patients than in healthy controls. Plasma PACAP38-LI was remarkably elevated during the attack compared to the attack-free period of patients (Tuka et al. unpublished observations). These results show that PACAP does not only play a role in the pathological mechanisms of different types of primary headaches, but its plasma levels are associated with the disease and show correlation with the migraine phases [40, 41]. Another recent study has further confirmed the significance of PACAP in the pathogenesis of migraine, rather than in tension-type headaches [18]. The authors confirmed significantly lower interictal levels of PACAP in migraineurs than in healthy controls or in tension-type headache patients. Unlike migraineurs, there was no difference in PACAP levels between samples from tension-type headache patients and healthy controls [18].

Another study also confirmed the elevation in PACAP levels during headache and lower levels between attacks [21]. In healthy volunteers, an anti-migraine medication (sumatriptan) did not lead to any change in circulating PACAP and VIP levels without trigeminovascular stimulation [7]. However, in migraine patients, sumatriptan treatment resulted in decreases in PACAP levels 1 h postinjection [21]. A recent study not only confirmed that PACAP38 infusion could trigger migraine attacks, but also found more increase in plasma PACAP38 levels in patients who reported delayed migraine attacks than in those patients who did not have headaches after PACAP infusion [13]. This observation indicates that migraine attacks were preceded by elevated plasma levels of PACAP38, similarly to the observation by others. The PACAP infusion-induced plasma elevations, in addition, indicate a de novo synthesis or release of the peptide. This human study raised interesting points on the source of elevated plasma PACAP levels: PACAP38 infusion to volunteers induced elevated plasma PACAP levels only in those patients who reported delayed migraine attacks. They found that after 20 min infusion, plasma levels in these patients were still elevated after 40 min, which is beyond the 3.5 min half-life of the peptide. They suggest that the source of the peptide under normal conditions may represent neuronal release from PACAP-ergic nerve terminals. This can be triggered in migraine patients, which can result in further elevation [13].

Other Diseases

Changes in PACAP levels in non-neurological diseases are summarized in Table 48.3.

Chronic Hepatitis

PACAP levels were investigated in patients with chronic hepatitis under oral antiviral therapy (lamivudine) with RIA method [25]. The authors found no correlation between plasma PACAP38 levels and histological finding (liver biopsy), serum alanine amino transferase (ALT) level or serum hepatitis B virus -DNA level. However, they found that PACAP38 levels were significantly higher in healthy controls than in hepatitis B patients at baseline. This decrease was diminished after 52 weeks of antiviral therapy: PACAP levels significantly increased in treated patients. The authors suggested that the lower PACAP levels observed in hepatitis patients indicated the altered T1/T2 immune responses and related cytokine expression and 1 year of antiviral therapy could induce a marked amelioration also indicated by increased PACAP levels.

Nephrotic Syndrome

A recent study [26] has investigated PACAP levels in the plasma and urine in children with nephrotic syndrome. It was found that during the nephrotic state, PACAP was lost in the urine, compared with non-nephrotic state, when PACAP could not be reliably shown in the urine with Western blot. Parallel with the urinary loss, plasma PACAP levels decreased during these phases. Together with PACAP, the loss of its binding protein, ceruloplasmin, was also lost. These data indicate urinary loss of PACAP and its binding protein, thus resulting in a PACAP-deficient state. This was also accompanied by increased megakaryopoiesis and increased platelet aggregation [26]. Addition of PACAP inhibited the differentiation of hematopoetic stem cells into megakaryocytes during the nephrotic state, while addition of the plasma of congenital nephrotic patients during the nephrotic state directly promoted megakaryopoiesis when stem cells of healthy donors were used [26]. In a few patients, plasma PACAP levels were measured after bilateral nephrectomy, when gradually increasing levels were found together with decreased platelet count (with a strong inverse correlation in congenital nephrotic patients).

Gynecological Diseases

Lower levels of PACAP were found in patients with premature ovarian failure and in patients with idiopathic hypogonadotropic hypogonadism [11]. PACAP was also investigated in the follicular fluid of women undergoing superovulation treatment, where the presence of PACAP was confirmed with mass spectrometry analysis [27, 29]. Superovulation treatment is a usual procedure in controlled ovarian hyperstimulation during the in vitro fertilization procedure. After treatment, the number of harvested oocytes was recorded. All corresponding PACAP concentrations were below 290 fmol/ml in cases when the number of harvested oocytes exceeded 14 per patient, while in all cases above this cutoff value, the number of oocytes was below 14. Thus, lower levels of PACAP might correlate with a markedly higher number of ova retrieved and may predict a higher chance for ovarian hyperstimulation, a potentially life-threatening condition from excessive ovarian stimulation [27].

Concluding Remarks: Future Perspectives and Limitations

As summarized in the present review, increasing number of evidence suggests the potential clinical use of PACAP as a diagnostic and prognostic biomarker. Numerous data have been published in the last few years, showing that it is a rapidly developing, hot and promising topic. However, several limitations arise at the moment, which are briefly outlined below.

Presently we do not know where the PACAP measured in biological fluids, especially in the plasma, is exactly derived from. As already mentioned in the introductory part of this review, the most probable sources are the organs where PACAP occurs at high concentrations, like the nervous systems and the endocrine glands. In both physiological and pathological situations, PACAP levels in tissues can rapidly and drastically change and therefore, the levels measured under those conditions may represent PACAP from other sources as well. PACAP is upregulated in the central nervous system and peripheral tissues upon several injuries (rev. [42]). Serum PACAP levels have been shown to be increased after sensory nerve stimulation in rats (0.38–0.8 fmol/ml), or superior sagittal sinus stimulation in cats (112–287 pmol/l) indicating that levels can indeed rapidly change in response to physiological or pathological stimuli [21, 43]. In the skin of psoriatic patients, PACAP levels showed a 20-fold elevation [44]. Significantly higher tissue levels were also shown in cardiac tissue obtained from patients with ischemic heart disease [45], while lower levels/downregulation were found in other diseases: in lung and colon cancer samples [45]. In another recent study, no alterations in PACAP-LI between healthy and tumoral samples from the urinary bladder and testis were detected, but lower PACAP38-LI was found in kidney tumors compared with healthy tissue samples, and higher PACAP27-LI in prostatic cancer compared to samples from benign prostatic hyperplasia [46]. These data indicate that PACAP levels of different tissue samples are altered under pathological conditions suggesting a potential role of PACAP on tumor growth [46]. Reduction of PACAP-containing nerve fibers was described in a few conditions, such as in women with polycystic ovaries [47]. To what extent these human tissue changes are reflected in alterations of serum PACAP levels is not known at the moment, but animal data suggest that acute local changes can cause changes of circulating PACAP as well.

Several recent studies have shown alterations in human plasma and other fluids in relation to certain diseases as detailed above. However, the question whether PACAP alterations are consequences or contributing factors of the disease still remains open [17]. One possibility is that the triggering cause of a disease leads to PACAP decline resulting in a higher vulnerability of the cells, as the neuroprotective effect of PACAP is reduced. However, it is equally possible that the PACAP deficit—for an unknown reason—predisposes neurons and other cells to certain pathological changes [17].

Several studies suggest that future biomarkers will be used in panels including several markers of different sensitivity and specificity [48]. Whether PACAP will indeed be a biomarker is a question to be answered by future studies, but based on our current knowledge it seems that PACAP is a good candidate for certain diseases and can be included in a panel of biomarkers after mapping the conditions where PACAP alterations are present or absent.

One main limitation of the presently available data pool is that we cannot compare results from different laboratories using different methods [1]. Even the same method, but from different manufacturers, can yield non-comparable results. Different PACAP antisera developed for immunohistochemistry, RIA or ELISA are most of the times not highly specific, are directed against different parts of the peptide or the prepro-peptide, in which case other cleavage products might also be bound by the antibody. Some authors interpret their results as PACAP-immunoreactivity even though the antiserum was developed against prepro-PACAP, in other cases the fact that the antiserum was produced against prepro-PACAP is not even mentioned in the paper, it only turns out by studying the manufacturer's description. Due to the cleavage by DPPIV, inactive PACAP3-38 and PACAP5-38 are yielded. The different antibodies used for detection of PACAP may also recognize these inactive forms, so we do not know at the moment whether PACAP-immunoreactivity measured in the blood plasma/serum always indicates the active peptide. As pointed out by Edvinsson et al. [1], this mixture of peptide fragments also belong to the physiological picture and, therefore, should not be neglected. If we use the term "PACAP- or PACAP-LI", it can cover all these forms. Revealing changes in PACAP-like immunoreactivity can yield important results as for the biological and pathological significance of the peptide. In addition, PACAP binds to plasma proteins, and the best documented carrier candidate among them is ceruloplasmin [49]. It has not been clearly demonstrated whether the measured PACAP reflects total, free, and/or protein-bound PACAP.

Several further technical limitations limit the comparison of results between groups. One of them is the way how human samples are obtained and stored. As PACAP is quickly degraded by DPPIV, peptidase inhibitors are required to use. However, this is not always indicated in the papers. Some data indicate that PACAP27 is much more resistant to DPPIV degradation than PACAP38 and therefore has a much longer half-life [50]. Thus, the ratio between the two peptides may also change with time, which is not taken into consideration by any clinical report. The different anticoagulants used can lead to different values, as shown by Freson et al. [5]. The time passed between sample collection and measurement is also critical. Some authors have described PACAP levels in postmortem CSF samples [16], which obviously cannot be compared to fresh samples taken from patients. Another issue that needs to be taken into consideration is the range within which PACAP levels change. In some studies very small changes are significant and authors draw final conclusions, while in other studies much bigger alterations are not significant and therefore are interpreted as no change. In some studies the sample sizes are very high, with more than a hundred involved patients, while in other studies, only a few patients are included in one group and even case studies are available with a single patient's values.

One further limitation is the lack of a fast and sensitive analytical method that is able to measure PACAP from small sample volumes. In spite of all these factors that limit the comparison of data, the alterations within one study can yield valuable results showing changes of PACAP levels, which can not only lead to biomarker development but can also reveal important functions of PACAP in physiological and pathological processes. In order to obtain a solid proof of PACAP as a biomarker, multicenter, controlled studies will be necessary in the future, but the growing body of evidence for the diagnostic and prognostic value of the peptide summarized in the current review justify these efforts.

Acknowledgements This study was supported by the following grants: OTKA K104984, 115874, 119759, Arimura Foundation, TAMOP 4.2.4.A/2-11-1-2012-0001 "National Excellence Program", New National Excellence Program (UNKP), Janos Bolyai Research Scholarship of Hungarian Academy of Sciences, MTA-PTE "Lendulet" Program, PTE AOK KA Research Grant, and the National Brain Research Programmes A and B (KTIA_NAP_13-2014-0022 and KTIA_13_NAP-A-III/5, Research site ID number: 888819, Hungary). This work is dedicated to the 650th anniversary of the University of Pecs.

References

1. Edvinsson L, Ekman R, Goadsby PJ. Measurement of vasoactive neuropeptides in biological materials: problems and pitfalls from 30 years of experience and novel future approaches. Cephalalgia. 2010;30:761–6.
2. Miyata A, Arimura A, Dahl RR, Minamino N, Uehara A, Jiang L, et al. Isolation of a novel 38 residue-hypothalamic polypeptide which stimulates adenylate cyclase in pituitary cells. Biochem Biophys Res Commun. 1989;164:567–74.

3. Palkovits M, Somogyvari-Vigh A, Arimura A. Concentrations of pituitary adenylate cyclase activating polypeptide (PACAP) in human brain nuclei. Brain Res. 1995;699:116–20.
4. Warren JB, Cockcroft JR, Larkin SW, Kajekar R, Macrae A, Ghatei MA, et al. Pituitary adenylate cyclase activating polypeptide is a potent vasodilator in humans. J Cardiovasc Pharmacol. 1992;20:83–7.
5. Freson K, Hashimoto H, Thys C, Wittevrongel C, Danloy S, Morita Y, et al. The pituitary adenylate cyclase-activating polypeptide is a physiological inhibitor of platelet activation. J Clin Invest. 2004;113:905–12.
6. Li M, Maderdrut JL, Lertora JJ, Batuman V. Intravenous infusion of pituitary adenylate cyclase-activating polypeptide (PACAP) in a patient with multiple myeloma and myeloma kidney: a case study. Peptides. 2007;28:1891–5.
7. Hansen JM, Fahrenkrug J, Petersen J, Wienecke T, Olsen KS, Ashina M. Vasoactive intestinal peptide (VIP) and pituitary adenylate cyclase-activating polypeptide (PACAP) in the circulation after sumatriptan. Scand J Pain. 2013;4:211–6.
8. Birk S, Sitarz JT, Petersen KA, Oturai PS, Kruuse C, Fahrenkrug J, et al. The effect of intravenous PACAP38 on cerebral hemodynamics in healthy volunteers. Regul Pept. 2007;140:185–91.
9. Doberer D, Gschwandtner M, Mosgoeller W, Bieglmayer C, Heinzl H, Petkov V. Pulmonary and systemic effects of inhaled PACAP38 in healthy male subjects. Eur J Clin Invest. 2007;37:665–72.
10. Reglodi D, Gyarmati J, Ertl T, Borzsei R, Bodis J, Tamas A, et al. Alterations of pituitary adenylate cyclase-activating polypeptide-like immunoreactivity in the human plasma during pregnancy and after birth. J Endocrinol Invest. 2010;33:443–5.
11. Kanasaki H, Purwana IN, Oride A, Mijiddorj T, Sukhbaatar U, Miyazaki K. Circulating kisspeptin and pituitary adenylate cyclase-activating polypeptide (PACAP) do not correlate with gonadotropin serum levels. Gynecol Endocrinol. 2013;29:583–7.
12. Winters SJ, King JC, Brees CK, Moore Jr JP. Pituitary adenylate cyclase-activating polypeptide (PACAP) in fetal cord blood. Early Hum Dev. 2014;90:451–3.
13. Amin FM, Hougaard A, Schytz HW, Asghar MS, Lundholm E, Parvaiz AI, et al. Investigation of the pathophysiological mechanisms of migraine attacks induced by pituitary adenylate cyclase-activating polypeptide-38. Brain. 2014;137:779–94.
14. Ma BQ, Zhang M, Ba L. Plasma pituitary adenylate cyclase-activating polypeptide concentrations and mortality after acute spontaneous basal ganglia hemorrhage. Clin Chim Acta. 2015;439:102–6.
15. Jiang L, Wang WH, Dong XQ, Yu WH, Du Q, Yang DB, et al. The change of plasma pituitary adenylate cyclase-activating polypeptide levels after aneurysmal subarachnoid hemorrhage. Acta Neurol Scand. 2016;197:35–42.
16. Han P, Liang W, Baxter LC, Yin J, Tang Z, Beach TG, et al. Pituitary adenylate cyclase-activating polypeptide is reduced in Alzheimer disease. Neurology. 2014;82:1724–8.
17. Han P, Tang Z, Yin J, Maalouf M, Beach TG, Reiman EM, et al. Pituitary adenylate cyclase-activating polypeptide protects against β-amyloid toxicity. Neurobiol Aging. 2014;35:2064–71.
18. Han P, Caselli RJ, Baxter L, Serrano G, Yin J, Beach TG, et al. Association of pituitary adenylate cyclase-activating polypeptide with cognitive decline in mild cognitive impairment due to Alzheimer disease. JAMA Neurol. 2015;72:333–9.
19. Nelson KB, Grether JK, Croen LA, Dambrosia JM, Dickens BF, Jelliffe LL, et al. Neuropeptides and neurotrophins in neonatal blood of children with autism or mental retardation. Ann Neurol. 2001;49:597–606.
20. Tuka B, Helyes Z, Markovics A, Bagoly T, Szolcsanyi J, et al. Alterations in PACAP38-like immunoreactivity in the plasma during ictal and interictal periods of migraine patients. Cephalalgia. 2013;33:1085–95.
21. Zagami AS, Edvinsson L, Goadsby PJ. Pituitary adenylate cyclase activating polypeptide and migraine. Ann Clin Transl Neurol. 2014;1:1036–40.
22. Baranowska-Bik A, Kochanowski J, Uchman D, Wolinska-Witort E, Kalisz M, Martynska L, et al. Vasoactive intestinal peptide (VIP) and pituitary adenylate cyclase activating polypeptide (PACAP) in humans with multiple sclerosis. J Neuroimmunol. 2013;263:159–61.

23. Ressler KJ, Mercer KB, Bradley B, Jovanovic T, Mahan A, Kerley K, et al. Post-traumatic stress disorder is associated with PACAP and the PAC1 receptor. Nature. 2011;470:492–7.
24. Bukovics P, Czeiter E, Amrein K, Kovacs N, Pal J, Tamas A, et al. Changes of PACAP level in cerebrospinal fluid and plasma of patients with severe traumatic brain injury. Peptides. 2014;60:18–22.
25. Elefsiniotis IS, Ketikoglou I, Kafiri G, Pantazis KD, Moulakakis A, Mavrogiannis C. Plasma pituitary adenylate cyclase activating polypeptide (PACAP) levels in chronic hepatitis B patients under lamivudine treatment. Eur J Gastroenterol Hepatol. 2003;15:1209–16.
26. Eneman B, Freson K, van den Heuvel L, van Hoyweghen E, Collard L, Vande Walle J, et al. Pituitary adenylate cyclase-activating polypeptide deficiency associated with increased platelet count and aggregability in nephrotic syndrome. J Thromb Haemost. 2015;13:755–67.
27. Koppan M, Varnagy A, Reglodi D, Brubel R, Nemeth J, Tamas A, et al. Correlation between oocyte number and follicular fluid concentration of pituitary adenylate cyclase-activating polypeptide (PACAP) in women after superovulation treatment. J Mol Neurosci. 2012;48: 617–22.
28. Dow RC, Bennie J, Fink G. Pituitary adenylate cyclase-activating peptide-38 (PACAP)-38 is released into hypophysial portal blood in the normal male and female rat. J Endocrinol. 1994;142:R1–4.
29. Brubel R, Reglodi D, Jambor E, Koppan M, Varnagy A, Biro Z, et al. Investigation of pituitary adenylate cyclase activating polypeptide in human gynecological and other biological fluids by using MALDI TOF mass spectrometry. J Mass Spectrom. 2011;46:189–94.
30. Borzsei R, Mark L, Tamas A, Bagoly T, Bay C, Csanaky K, et al. Presence of pituitary adenylate cyclase activating polypeptide-38 in human plasma and milk. Eur J Endocrinol. 2009;160:561–5.
31. Wu ZL, Ciallella JR, Flood DG, O'Kane TM, Bozyczko-Coyne D, Savage MJ. Comparative analysis of cortical gene expression in mouse models of Alzheimer's disease. Neurobiol Aging. 2006;27:377–86.
32. van Landeghem FK, Weiss T, Oehmichen M, von Deimling A. Cellular localization of pituitary adenylate cyclase-activating peptide (PACAP) following traumatic brain injury in humans. Acta Neuropathol. 2007;113:683–93.
33. Kato H, Ito A, Kawanokuchi J, Jin S, Mizuno T, Ojika K, et al. Pituitary adenylate cyclase-activating polypeptide (PACAP) ameliorates experimental autoimmune encephalomyelitis by suppressing the functions of antigen presenting cells. Mult Scler. 2004;10:651–9.
34. Tan YV, Abad C, Wang Y, Lopez R, Waschek JA. VPAC2 (vasoactive intestinal peptide receptor type 2) receptor deficient mice develop exacerbated experimental autoimmune encephalomyelitis with increased Th1/Th17 and reduced Th2/Treg responses. Brain Behav Immun. 2015;44:167–75.
35. Tan YV, Waschek JA. Targeting VIP and PACAP receptor signalling: new therapeutic strategies in multiple sclerosis. ASN Neuro. 2011;3(4):e00065.
36. Vecsei L, Tuka B, Tajti J. Role of PACAP in migraine headaches. Brain. 2014;137:650–1.
37. Schytz HW, Birk S, Wienecke T, Kruuse C, Olesen J, Ashina M. PACAP38 induces migraine-like attacks in patients with migraine without aura. Brain. 2009;132:16–25.
38. Tuka B, Helyes Z, Markovics A, Bagoly T, Nemeth J, Mark L, et al. Peripheral and central alterations of pituitary adenylate cyclase activating polypeptide-like immunoreactivity in the rat in response to activation of the trigeminovascular system. Peptides. 2012;33:307–16.
39. Markovics A, Kormos V, Gaszner B, Lashgarara A, Szoke E, Sandor K, et al. Pituitary adenylate cyclase-activating polypeptide plays a key role in nitroglycerol-induced trigeminovascular activation in mice. Neurobiol Dis. 2012;45:633–44.
40. Tajti J, Tuka B, Botz B, Helyes Z, Vecsei L. Role of pituitary adenylate cyclase-activating polypeptide in nociception and migraine. CNS Neurol Disord Drug Targets. 2015;14:540–53.
41. Tajti J, Szok D, Majlath Z, Tuka B, Csati A, Vecsei L. Migraine and neuropeptides. Neuropeptides. 2015;52:19–30.
42. Somogyvari-Vigh A, Reglodi D. Pituitary adenylate cyclase activating polypeptide: a potential neuroprotective peptide. Curr Pharm Des. 2004;10:2861–89. Review.

43. Nemeth J, Reglodi D, Pozsgai G, Szabo A, Elekes K, Pinter E, et al. Effect of pituitary adenylate cyclase activating polypeptide-38 on sensory neuropeptide release and neurogenic inflammation in rats and mice. Neuroscience. 2006;143:223–30.
44. Steinhoff M, McGregor GP, Radleff-Schlimme A, Steinhoff A, Jarry H, Schmidt WE. Identification of pituitary adenylate cyclase activating polypeptide (PACAP) and PACAP type 1 receptor in human skin: expression of PACAP38 is increased in patients with psoriasis. Regul Pept. 1999;80:49–55.
45. Szanto Z, Sarszegi Z, Reglodi D, Nemeth J, Szabadfi K, Kiss P, et al. PACAP immunoreactivity in human malignant tumor samples and cardiac diseases. J Mol Neurosci. 2012;48:667–73.
46. Tamas A, Javorhazy A, Reglodi D, Sarlos DP, Banyai D, Semjen D, et al. Examination of PACAP-like immunoreactivity in urogenital tumor samples. J Mol Neurosci. 2015;2015: PMID:26454744.
47. Wojtkiewicz J, Jana B, Kozłowska A, Crayton R, Majewski M, Zalecki M, et al. Innervation pattern of polycystic ovaries in the women. J Chem Neuroanat. 2014;6162:147–52.
48. Chan MK, Gottschalk MG, Haenisch F, Tomasik J, Ruland T, Rahmoune H, et al. Applications of blood-based protein biomarker strategies in the study of psychiatric disorders. Prog Neurobiol. 2014;122:45–72.
49. Tams JW, Johnsen AH, Fahrenkrug J. Identification of pituitary adenylate cyclase-activating polypeptide1-38-binding factor in human plasma, as ceruloplasmin. Biochem J. 1999;341: 271–6.
50. Bourgault S, Vaudry D, Botia B, Couvineau A, Laburthe M, Vaudry H, et al. Novel stable PACAP analogs with potent activity towards the PAC1 receptor. Peptides. 2008;29:919–32.

Chapter 49
Examination of PACAP During Lactation

Andrea Tamas, Reka A. Vass, Zsuzsanna Helyes, Katalin Csanaky, Zalan Szanto, Jozsef Nemeth, and Dora Reglodi

Abstract Breast milk is a complex biological fluid that contains nutritional components and non-nutritive bioactive factors promoting survival and healthy development of the newborn. PACAP has important functions, as a bioactive factor, in reproductive and developmental processes. Shortly after its discovery, PACAP and its receptors were identified in normal and cancerous mammary gland samples. The present review summarizes data obtained in breast milk during different periods of lactation by radioimmunoassay. Our group showed, for the first time, that PACAP is present in the human milk at levels 5- to 20-fold higher than in the respective plasma samples. PACAP-like immunoreactivity (LI) is higher in colostrum compared to transitional and mature human milk samples. PACAP level seems stable until the 10th month of lactation and there after, a significant increase can be observed between samples obtained in the interval 11th–17th months of breastfeeding. The presence of PACAP can also be confirmed in milk and plasma samples from the most commonly used ruminant domestic animals (cow, sheep and goat), pasteurized cow milk and infant formula samples. Similarly to the human results, the concentration of PACAP in the milk whey is almost ten times higher than in the plasma of the respective animals, while pasteurized cow milk and infant formula samples contain PACAP-LI levels comparable to human milk samples. The exact function of PACAP in the milk is not known at the moment. We hypothesize that PACAP (1) is essential for the growth and development of the

A. Tamas (✉) • R.A. Vass • K. Csanaky • D. Reglodi
Department of Anatomy, MTA-PTE "Lendulet" PACAP Research Team, University of Pecs, Pecs 7624, Hungary
e-mail: andreatamassz@gmail.com

Z. Helyes
Department of Pharmacology and Pharmacotherapy, Janos Szentagothai Research Center, Pecs, Hungary

Z. Szanto
Department of Surgery, University of Pecs, Pecs, Hungary

J. Nemeth
Department of Pharmacology and Pharmacotherapeutics, University of Debrecen, Debrecen, Hungary

D. Reglodi, A. Tamas (eds.), *Pituitary Adenylate Cyclase Activating Polypeptide — PACAP*, Current Topics in Neurotoxicity 11,
DOI 10.1007/978-3-319-35135-3_49

newborn, (2) may be required for the development of the immune system and immunological microenvironment of the gastrointestinal tract, and (3) could be important in the growth and function of the mammary gland.

Keywords PACAP-like immunoreactivity • Milk • Lactation • Mammary gland

Based on the recommendation of the World Health Organization the exclusive breastfeeding is the normative standard for infant feeding for the first 6 months of life [1, 2]. Human milk is a complex biological fluid [3, 4], it contains nutritional components and non-nutritive bioactive factors that promote survival and healthy development of newborn [5]. The nutritional components of human milk vary depending on maternal diet and the period of lactation. The mean macronutrient composition of the mature milk is approximately 0.9–1.2 g/dl for protein, 3.2–3.6 g/dl for fat, and 6.7–7.8 g/dl for lactose. Many micronutrients are also present in the milk including vitamins A, B1, B2, B6, B12, D and iodine [5]. The definition of the bioactive components of the nutrition is "the elements that affect biological processes or substrates and hence have an impact on body function or condition and ultimately health" [6]. The bioactive factors in the human milk originate from different sources: (1) secreted by the mammary epithelium, (2) produced by cells carried within the milk, (3) drawn from maternal serum and transported across the mammary epithelium [5]. Several bioactive factors have been described in the milk, including different hormones (adrenal, gonadal, gastrointestinal, pituitary, hypothalamic), growth factors, prostaglandins, immunoglobulins, cytokines, chemokines, and different neuropeptides [3, 5]. Before the isolation of pituitary adenylate cyclase activating polypeptide (PACAP) Werner et al. [7] described the presence of vasoactive intestinal peptide (VIP), the neuropeptide structurally the closest to PACAP, in milk samples.

Shortly after the discovery of PACAP [8, 9], Skakkebaek et al. [10] investigated the occurrence and distribution of PACAP immunoreactivity in the mammary gland of lactating and non-lactating rats by radioimmunoassay and immunohistochemistry. They showed PACAP-immunopositive nerve fibers associated with blood vessels and smooth muscle surrounding the lactiferous duct of the nipple. PACAP-immunoreactive fibers were present in the subepidermal connective tissue of the nipple, in the mammary parenchyma and around secretory alveoli. Although there was no significant change in PACAP-immunohistochemical level during pregnancy and lactation, the concentration of PACAP38-immunoreactivity was elevated in the extract of mammary gland during lactation [10]. Double immunostaining demonstrated that PACAP-immunoreactive fibers were co-localized with VIP and calcitonin gene-related peptide (CGRP) originated from the neurons of sensory ganglia indicating the role of PACAP in the transport of suckling stimuli centrally [10]. The presence of both PACAP mRNA and PACAP immunoreactivity was also demonstrated in human normal mammary gland samples and breast carcinoma [11]. Normal, peritumoral and tumoral mammary gland samples expressed both preproPACAP mRNA and protein, the levels of which increased from normal to tumoral breast tissue. Immunohistochemistry showed PACAP-immunoreactivity both in normal and tumoral tissue in the alveolar epithelial cells, but not in the connective tissue. In tumoral samples duct-like structures of some invasive tumors expressed very strong PACAP-immunoreactivity supporting the important role of this peptide not only in physiological conditions but also in tumorigenesis [11–13].

All three PACAP receptors (PAC1, VPAC1, and VPAC2) were identified in normal and cancerous human mammary glands [13–18]. Both mRNA and protein of VPAC1 and VPAC2 receptors, as well as different isoforms (null, hip/hop) of PAC1 receptors were identified in normal, peritumoral, and tumoral breast tissue samples. In normal tissue immunoreactivity to all three PACAP receptors were located in the ductal and glandular epithelial cells. On the other hand, tumoral tissue expressed stronger immunopositivity compared to normal tissue samples. There were no immunopositive signals in the connective tissue stroma of the mammary gland [14]. Zhang et al. [19] demonstrated that the analogs of VPAC1, VPAC2, and PAC1 receptors are potent, have biological activity and it is suitable of further evaluation for accurate PET imaging of benign and malignant lesions of human breast cancers.

In a set of earlier studies, our research group examined the changes of PACAP38-like immunoreactivity (LI) in the serum and milk samples of different species during pregnancy and lactation. First we investigated the concentration of PACAP38 in human plasma of healthy male and female volunteers, pregnant women and lactating women having 1- to 6-month-old babies using radioimmunoassay (RIA) analysis [20, 21]. We found relatively small interindividual differences among healthy volunteers (both sexes, age between 20 and 40 years), there were no significant differences between PACAP38 level of females with different age or hormone cycle. However, in the second and third trimester of pregnancy and during lactation a significant elevation could be observed compared to the earlier gestational period and nonpregnant healthy volunteers [20, 21]. We detected 5- to 20-fold higher concentration of PACAP38 in the milk whey compared to the respective plasma samples [20]. Similarly to our results numerous studies measured higher concentration of bioactive factors in the milk than in the plasma. Several hormones such as estrogen, gonadotropin releasing hormone (GnRH), thyrotropin releasing hormone, VIP, somatostatin, bombesin, neurotensin, oxytocin, prolactin, and different growth factors such as insulin-like, epidermal, transforming growth factors are present in higher concentrations in the milk than in the plasma [3].

It is well known that the composition of milk is changing during lactation based on the requirements of the newborns [5, 22]. Therefore, we next investigated the differences in PACAP38-LI in human milk samples from different periods of lactation. We collected colostrum samples at the beginning of lactation (1–3 days), transitional milk at fourth day of lactation, and mature milk samples every month during the whole period of lactation (1–17 months). Similarly to earlier results showing higher levels of different bioactive factors and nutrients in the colostrum we found significantly higher PACAP38-LI in the colostrum samples compared to transitional and mature milk samples [23]. Higher levels of hormones, growth factors, oligosaccharides, cytokines, and immune regulating factors in the colostrum have important role in the development of the immune system and protective effects in allergic and inflammatory disorders [24–26]. The first 6 months of lactation have special importance for the development of the newborn, therefore, the WHO recommends exclusive breastfeeding in the first half year of life. We did not find significant alterations in the level of PACAP38-LI during the first 6 months of lactation. We measured a stable PACAP38-LI level until the tenth month and after a significant increase was observed in the samples from 11th to 17th months of lactation [23]. The composition of the milk also depends on the time that passes between the successive sucklings. At the beginning of the suckling the foremilk has lower fat content and later the hindmilk contains more fat and lactose. Several bioactive factors have different concentration in the foremilk compared to the hindmilk. Endothelin-1, ghrelin and cholesterol have higher level, and triglyceride, leptin, retinol have lower level in the foremilk compared to hindmilk [27, 28]. However, we did not find significant difference in PACAP38-LI between foremilk and hindmilk samples similarly to atrial natriuretic peptide and granulocyte colony-stimulating factor [29–31].

Investigation of bioactive factors in the milk of domestic animals has important nutritional value and agricultural significance [32]. Numerous studies investigated different growth factors, hormones in the milk of pig, cow, sheep and goats [33–38]. Therefore, we aimed to compare the PACAP38 level of milk and plasma samples from the most commonly used ruminant domestic animals: cow, sheep and goat. Similarly to our earlier human results the concentration of PACAP in the milk whey was almost ten times higher than in the plasma of the respective animals, and the PACAP38-LI did not change during the 3-month-period of lactation [39]. Fresh cow milk had similar PACAP content as human samples [31]. We also examined PACAP38-LI in the homogenates of sheep udder biopsies with RIA. Localization of PAC1 receptor was also investigated in the lactating and non-lactating sheep mammary gland samples with immunohistochemistry compared to non-lactating human mammary gland samples. In the non-lactating human and sheep samples, a very weak PAC1 receptor immunopositivity was detected in the glandular epithelial cells [23]. In contrast, significantly increased PAC1 receptor expression was detected in lactating sheep mammary gland epithelial cells compared to non-lactating samples, while the surrounding connective tissue remained unstained in all samples [23, 39]. RIA examination of the homogenates of lactating mammary gland samples from sheep showed significantly higher PACAP38-LI 7 days after delivery compared with samples from postpartum 30th days [39].

When mother's milk is unavailable cow-milk based infant formulas are necessary to use. Therefore, in the next step we aimed to measure the PACAP38-LI in pasteurized cow milk and two basic infant formulas, the hypoantigenic and the non-hypoantigenic infant formula samples by RIA. The infant formula samples contain minerals, vitamins, lactose, fat and some proteins for example parathyroid hormone-like peptide [40], parathyroid hormone-related protein [41], but most of the proteins are hydrolyzed during the preparation process. There are many bioactive factors, which cannot be detected in the infant formula, for example insulin-like growth factors, GnRH, antibodies, and enzymes [3, 41]. In our earlier experiment we showed that the pasteurized cow milk and infant formula samples contain PACAP38-LI at

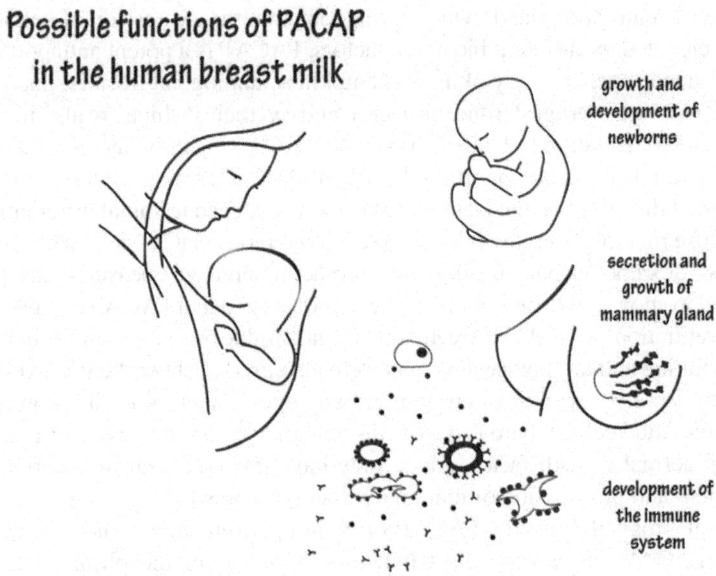

Fig. 49.1 Possible functions of PACAP in the breast milk during lactation

levels comparable to human milk samples. This observation means that because of the low molecular weight PACAP might withstand hydrolysis and the manufacturing processes. Interestingly, we detected higher PACAP level in the hypoantigenic formulas compared to non-hypo antigenic formula samples. These results suggest that PACAP38 could have a carrier not yet known in the milk (such as ceruloplamin) and during the extensive hydrolysis it might be released from the carrier molecule [31].

The exact function of PACAP in the milk is not known at the moment. We hypothesize that PACAP (1) is essential for the growth and development of the newborn; (2) may be required for the development of the immune system and immunological microenvironment of the gastrointestinal tract; (3) could be important in the growth and function of mammary gland (Fig. 49.1).

It is well known that PACAP plays a very important role in the development of the nervous system and other internal organs. PACAP is a potent neurotrophic factor, it plays an important role in neurogenesis, myelinization, neuronal differentiation and migration. A separate chapter of this book summarizes the function of PACAP in neuronal development (Chap. 6. Watanabe et al.). The source of PACAP required for development of the nervous system of the newborn could be partially originating from the milk.

PACAP has well known immunomodulatory effects (Chap. 40. Delgado et al.). We suggest that as other bioactive components of milk, PACAP has a role in protecting newborns against infections and stimulating the development of the immune system and immunological microenvironment of the gastrointestinal tract.

The third possible function of PACAP in the human breast is the regulation of secretion and growth of the mammary gland. Apoptosis plays an important role in the involution of mammary gland, while during lactation the expression of antiapoptotic factors is elevated by different bioactive factors. PACAP is a potent antiapoptotic factor, and it has influence on cytokines, chemokines, angiogenic factors, and different hormones (estrogen, progesterone, prolactin, and oxytocin) which are also involved in the regulation of lactation [42, 43]. Intravenous PACAP injection increases the prolactin level in human plasma supporting the role of PACAP during lactation [44]. Earlier we examined the effect on the lactogenic hormone-induced terminal differentiation of HC11 mouse mammary epithelial cells. We treated mouse HC11 cells with lactogenic hormones (dexamethasone, insulin, and prolactin) and we measured the β-casein expression to show cell differentiation. We demonstrated that PACAP had no effect on cell differentiation, as PACAP treatment did not influence the β-casein expression. Mouse cytokine and angiogenesis arrays were also used to show the PACAP-induced changes in secreted cytokines, growth and angiogenic factors in differentiated and non-differentiated cells, where PACAP was able to decrease the levels of amphiregulin and epidermal growth factor, which may have physiological implications in the development and progression of mammary gland diseases [45].

The oral bioavailability of PACAP originating from milk is not known at the moment. PACAP has a very short half-life (minutes) in the plasma; exogenous PACAP38 is rapidly degraded in the presence of dipeptidyl peptidase IV [46, 47]. In contrast to the plasma samples PACAP was stable in milk samples, because we could detect PACAP38-LI from milk samples stored in −20 °C without protease inhibitor treatment several days or weeks after the collection. It is known that the mammary gland produces different protease inhibitors which are responsible for the stability of different bioactive proteins and peptides in the milk. In neonates the proteolytic enzymes (including dipeptidyl peptidase) have lower activity and the intestinal epithelium has higher permeability for the macromolecules, and therefore, we suggest that PACAP could be utilized during breastfeeding.

Taken together, the experimental data suggest that PACAP could have an important role during lactation based on its antiapoptotic, immunomodulatory, and neurotrophic effects, but further examinations are necessary to describe the exact functions.

Acknowledgments This study was supported by the following grants: OTKA K104984, 119759, Arimura Foundation, TAMOP 4.2.4.A/2-11-1-2012-0001 "National Excellence Program", New National Excellence Program (UNKP), Janos Bolyai Research Scholarship of Hungarian Academy of Sciences, MTA-PTE "Lendulet" Program, PTE AOK KA Research Grant. This work is dedicated to the 650th anniversary of the University of Pecs. The authors also thank all the volunteers.

References

1. The American Academy of Pediatrics. Section on Breastfeeding. Breastfeeding and the use of human milk. Pediatrics. 2012;129:827–41.
2. WHO. Global strategy for infant and young child feeding. Geneva: WHO; 2013.
3. Grosvenor CE, Picciano MF, Baumrucker CR. Hormones and growth factors in milk. Endocr Rev. 1993;14:710–28.
4. Jenness R. The composition of human milk. Semin Perinatol. 1979;3:225–39.

5. Ballard O, Morrow AL. Human milk composition: nutrients and bioactive factors. Pediatr Clin North Am. 2013;60:49–74. Review.
6. Schrezenmeir J, Korhonen H, Williams C, Gill HS, Shah N. Foreword. Br J Nutr. 2000;84(S1):1.
7. Werner H, Koch Y, Fridkin M, Fahrenkrug J, Gozes I. High levels of vasoactive intestinal peptide in human milk. Biochem Biophys Res Commun. 1985;133:228–32.
8. Miyata A, Arimura A, Dahl RR, Minamino N, Uehara A, Jiang L, et al. Isolation of a novel 38 residue-hypothalamic polypeptide which stimulates adenylate cyclase in pituitary cells. Biochem Biophys Res Commun. 1989;164:567–74.
9. Vaudry D, Falluel-Morel A, Bourgault S, Basille M, Burel D, Wurtz O, et al. Pituitary adenylate cyclase-activating polypeptide and its receptors: 20 years after the discovery. Pharmacol Rev. 2009;61:283–357.
10. Skakkebaek M, Hannibal J, Fahrenkrug J. Pituitary adenylate cyclase activating polypeptide (PACAP) in the rat mammary gland. Cell Tissue Res. 1999;298:153–9.
11. Garcia-Fernandez MO, Bodega G, Ruiz-Villaespesa A, Cortes J, Prieto JC, Carmena MJ. PACAP expression and distribution in human breast cancer and healthy tissue. Cancer Lett. 2004;205:189–95.
12. Leyton J, Gozes Y, Pisegna J, Coy D, Purdom S, Casibang M, et al. PACAP(6-38) is a PACAP receptor antagonist for breast cancer cells. Breast Cancer Res Treat. 1999;56:177–86.
13. Moody TW, Gozes I. Vasoactive intestinal peptide receptors: a molecular target in breast and lung cancer. Curr Pharm Des. 2007;13:1099–104. Review.
14. Garcia-Fernandez MO, Collado B, Bodega G, Cortes J, Ruiz Villaespesa A, Carmena MJ, et al. Pituitary adenylate cyclase activating peptide/vasoactive intestinal peptide receptors in human normal mammary gland and breast cancer tissue. Gynecol Endocrinol. 2005;20:327–33.
15. Moody TW, Leyton J, Gozes I, Lang L, Eckelman WC. VIP and breast cancer. Ann N Y Acad Sci. 1998;865:290–6.
16. Reubi JC. In vitro evaluation of VIP/PACAP receptors in healthy and diseased human tissues. Clinical implications. Ann N Y Acad Sci. 2000;921:1–25.
17. Reubi JC, Laderach U, Waser B, Gebbers JO, Robberecht P, Laissue JA. Vasoactive intestinal peptide/pituitary adenylate cyclase-activating peptide receptor subtypes in human tumors and their tissues of origin. Cancer Res. 2000;60:3105–12.
18. Zia H, Hida T, Jakowlew S, Birrer M, Gozes Y, Reubi JC, et al. Breast cancer growth is inhibited by vasoactive intestinal peptide (VIP) hybrid, a synthetic VIP receptor antagonist. Cancer Res. 1996;56:3486–9.
19. Zhang K, Aruva MR, Shanthly N, Cardi CA, Patel CA, Rattan S, et al. Vasoactive intestinal peptide (VIP) and pituitary adenylate cyclase activating peptide (PACAP) receptor specific peptide analogues for PET imaging of breast cancer: In vitro/in vivo evaluation. Regul Pept. 2007;144:91–100.
20. Borzsei R, Mark L, Tamas A, Bagoly T, Bay C, Csanaky K, et al. Presence of pituitary adenylate cyclase activating polypeptide-38 in human plasma and milk. Eur J Endocrinol. 2009;160:561–5.
21. Reglodi D, Gyarmati J, Ertl T, Borzsei R, Bodis J, Tamas A, et al. Alterations of pituitary adenylate cyclase-activating polypeptide-like immunoreactivity in the human plasma during pregnancy and after birth. J Endocrinol Invest. 2010;33:443–5.
22. Aydin S, Aydin S, Ozkan Y, Kumru S. Ghrelin is present in human colostrum, transitional and mature milk. Peptides. 2006;27:878–82.
23. Csanaky K, Banki E, Szabadfi K, Reglodi D, Tarcai I, Czegledi L, et al. Changes in PACAP immunoreactivity in human milk and presence of PAC1 receptor in mammary gland during lactation. J Mol Neurosci. 2012;48:631–7.
24. Kelly GS. Bovine colostrums: a review of clinical uses. Altern Med Rev. 2003;8:378–94. Review.
25. Kelly D, Coutts AG. Early nutrition and the development of immune function in the neonate. Proc Nutr Soc. 2000;59:177–85. Review.
26. Yoshioka Y, Kudo S, Nishimura H, Yajima T, Kishihara K, Saito K, et al. Oral administration of bovine colostrum stimulates intestinal intraepithelial lymphocytes to polarize Th1-type in mice. Int Immunopharmacol. 2005;5:581–90.

27. Karatas Z, Durmus Aydogdu S, Dinleyici EC, Colak O, Dogruel N. Breastmilk ghrelin, leptin, and fat levels changing foremilk to hindmilk: is that important for self-control of feeding? Eur J Pediatr. 2011;170:1273–80.
28. Ribeiro KD, Dimenstein R. Foremilk and hindmilk retinol levels. Rev Panam Salud Publica. 2004;16:19–22.
29. Calhoun DA, Lunoe M, Du Y, Christensen RD. Granulocyte colony-stimulating factor is present in human milk and its receptor is present in human fetal intestine. Pediatrics. 2000;105, e7.
30. Ken-Dror S, Weintraub Z, Yechiely H, Kahana L. Atrial natriuretic peptide and endothelin concentrations in human milk during postpartum lactation. Acta Paediatr. 1997;86:793–5.
31. Csanaky K, Reglodi D, Banki E, Tarcai I, Mark L, Helyes Z, et al. Examination of PACAP38-like immunoreactivity in different milk and infant formula samples. Acta Physiol Hung. 2013;100:28–36.
32. Odle J, Zijlstra RT, Donovan SM. Intestinal effects of milkborne growth factors in neonates of agricultural importance. J Anim Sci. 1996;74:2509–22.
33. Algers A, Madej A, Rojanasthien S, Uvnas-Moberg K. Quantitative relationships between suckling-induced teat stimulation and the release of prolactin, gastrin, somatostatin, insulin, glucagon and vasoactive intestinal polypeptide in sows. Vet Res Commun. 1991;15:395–407.
34. Dehnhard M, Claus R, Munz O, Weiler U. Course of epidermal growth factor (EGF) and insulin-like growth factor I (IGF-I) in mammary secretions of the goat during end-pregnancy ad early lactation. J Vet Med A Physiol Pathol Clin Med. 2000;47:533–40.
35. Drackova M, Hadra L, Janstova B, Navratilova P, Pridalova H, Vorlova L. Analysis of goat milk by infrared spectroscopy. Acta Vet Brno. 2008;77:415–22.
36. Foschino R, Invernizzi A, Barucco R, Stradiotto K. Microbial composition, including the incidence of pathogens, of goat milk from the bergamo region of Italy during a lactation year. J Dairy Res. 2002;69:213–25.
37. Lara-Villoslada F, Olivares M, Jimenez J, Boza J, Xaus J. Goat milk is less immunogenic than cow milk in a murine model of atopy. J Pediatr Gastroenterol Nutr. 2004;39:354–60.
38. Whitley NC, Walker EL, Harley SA, Keisler DH, Jackson DJ. Correlation between blood and milk serum leptin in goats and growth of their offspring. J Anim Sci. 2005;83:1854–9.
39. Czegledi L, Tamas A, Borzsei R, Bagoly T, Kiss P, Horvath G, et al. Presence of pituitary adenylate cyclase-activating polypeptide (PACAP) in the plasma and milk of ruminant animals. Gen Comp Endocrinol. 2011;172:115–9.
40. Budayr AA, Halloran BP, King JC, Diep D, Nissenson RA, Strewler GJ. High levels of a parathyroid hormone-like protein in milk. Proc Natl Acad Sci U S A. 1989;86:7183–5.
41. Ducroc R, Rubio S, Garzon B, Brunel-Riveau B, Couraud JY. Immunoreactive substance P and calcitonin-gene-related peptide (CGRP) in rat milk and in human milk and infants formulas. Am J Clin Nutr. 1995;62:554–8.
42. Khaled WT, Read EK, Nicholson SE, Baxter FO, Brennan AJ, Came PJ, et al. The IL-4/IL-13/Stat6 signalling pathway promotes luminal mammary epithelial cell development. Development. 2007;134:2739–50.
43. Watson CJ, Oliver CH, Khaled WT. Cytokine signalling in mammary gland development. J Reprod Immunol. 2011;88:124–9.
44. Chiodera P, Volpi R, Capretti L, Caffarri G, Magotti MG, Coiro V. Effects of intravenously infused pituitary adenylate cyclase-activating polypeptide on adenohypophyseal hormone secretion in normal men. Neuroendocrinology. 1996;64:242–6.
45. Csanaky K, Doppler W, Tamas A, Kovacs K, Toth G, Reglodi D. Influence of terminal differentiation and PACAP on the cytokine, chemokine, and growth factor secretion of mammary epithelial cells. J Mol Neurosci. 2014;52:28–36.
46. Bourgault S, Vaudry D, Botia B, Couvineau A, Laburthe M, Vaudry H, et al. Novel stable PACAP analogs with potent activity towards the PAC1 receptor. Peptides. 2008;29:919–32.
47. Zhu L, Tamvakopoulos C, Xie D, Dragovic J, Shen X, Fenyk-Melody JE, et al. The role of dipeptidyl peptidase IV in the cleavage of glucagon family peptides: in vivo metabolism of pituitary adenylate cyclase activating polypeptide-(1–38). J Biol Chem. 2003;278:22418–23.

CPSIA information can be obtained
at www.ICGtesting.com
Printed in the USA
LVOW05*2127151216
517432LV00002B/49/P

9 783319 351339